Formulas from Geometry

Formulas for Area (A), Perimeter (P), Circumference (C), and Volume (V):

Square

$A = s^2$

$P = 4s$

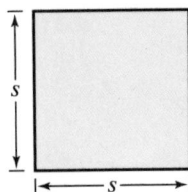

Rectangle

$A = lw$

$P = 2l + 2w$

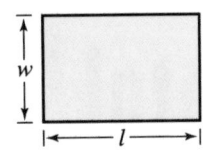

Circle

$A = \pi r^2$

$C = 2\pi r$

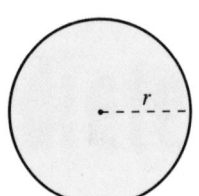

Triangle

$A = \frac{1}{2}bh$

$P = a + b + c$

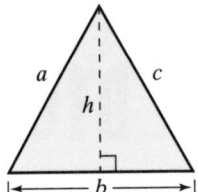

Trapezoid

$A = \frac{1}{2}h(b_1 + b_2)$

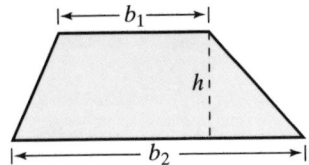

Parallelogram

$A = bh$

$P = 2a + 2b$

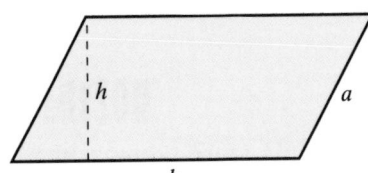

Pythagorean Theorem

$a^2 + b^2 = c^2$

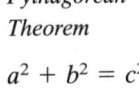

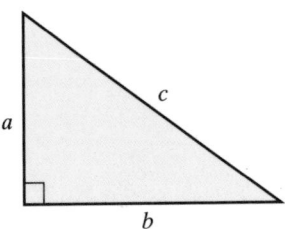

Cube

$V = s^3$

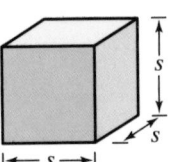

Rectangular Solid

$V = lwh$

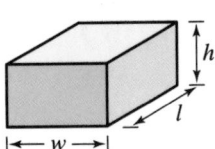

Circular Cylinder

$V = \pi r^2 h$

Sphere

$V = \frac{4}{3}\pi r^3$

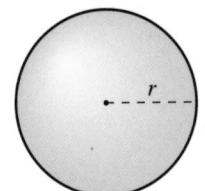

Publisher: Jack Shira
Managing Editor: Cathy Cantin
Development Manager: Maureen Ross
Development Editor: Laura Wheel
Assistant Editor: Rosalind Martin
Supervising Editor: Karen Carter
Senior Project Editor: Patty Bergin
Editorial Assistant: Meghan Lydon
Production Technology Supervisor: Gary Crespo
Executive Marketing Manager: Michael Busnach
Marketing Manager: Ben Rivera
Marketing Associate: Alexandra Shaw
Cover Design: Harold Burch Design/NYC

Photo credits: See page A26

Printed in the U.S.A.

Library of Congress Catalog Control Number: 2001133297

ISBN: 0-618-21878-5

123456789-VH-06 05 04 03 02

Contents

CONTENTS

Appendices   A1

Answers

Indices

A Word from the Authors

Welcome to *Intermediate Algebra: Graphs and Functions*, Third Edition. In this revision we enhanced the presentation of the material to provide a more accessible text for students. Our focus was to include a comprehensive study thread throughout the text specifically designed to help students achieve success.

Over the years we have taken care to write a text for the student. We paid careful attention to the presentation, using precise mathematical language and clear writing, to create an effective learning tool. We believe that every student can learn mathematics and we are committed to providing a text that makes the mathematics within it accessible to all students. In the Third Edition, we have revised and improved upon many text features designed for this purpose. Our approach now includes presenting solutions in selected examples from multiple perspectives— algebraic, graphical, and numerical. The side-by-side format allows students to see that a problem can be solved in more than one way and to compare the accuracy of the solution methods. In addition, we have retained and improved many of the text features. The *Technology*, *Exploration*, and *Study Tip* features have been expanded. *Mid-Chapter Quizzes* and *Chapter Tests* found in each chapter offer students a self-assessment tool midway through the chapter and at the conclusion of the chapter. *Cumulative Tests* found in every chapter (except Chapter 1) following the *Chapter Test*, have been expanded to give students a more comprehensive assessment tool for reviewing material from preceding chapters. The exercise sets have been reorganized and are now categorized by exercise type— skill, applied, conceptual, and review. Students also have access to several media resources that accompany this text—videotapes, tutorial software (packaged with the text), and a free text website—which provide additional support.

During the past 30 years of teaching and writing, we have learned many things about how students learn mathematics successfully. We found that students are most successful when they know what they are expected to learn and why it is important to learn. With that in mind, we have restructured the Third Edition to include a thematic study thread throughout the entire text. Each chapter begins with a study guide, which includes a comprehensive overview of the chapter concepts (*The Big Picture*) and a list of *Key Terms* that are integral to learning *The Big Picture* concepts. The study guide helps students to get organized and prepare for the chapter. We have also included a set of learning objectives in every section that outlines what students are expected to learn (*What you should learn*), followed by an interesting real-life application used to illustrate why it is important to learn the concepts in that section (*Why you should learn it*). The chapter summary has been completely revised in the Third Edition. Organized by section, the chapter summary recaps the main concepts of the chapter providing students

PREFACE

a convenient study tool. Finally, the *Review Exercises*, which provide additional study and support at the end of the chapter, have been reorganized and are now correlated by section and tied directly to the *What you should learn* objectives in those sections.

In response to intermediate algebra instructors, changes in the coverage of topics have been made in the Third Edition. To cut back on the amount of Elementary Algebra review at the beginning of the text, Section 1.1 *Operations with Real Numbers* and Section 1.2 *Properties of Real Numbers* of the Second Edition have been moved from Chapter 1 to an appendix (Appendix A *Real Numbers*) in the Third Edition. Also, several additional sections have been removed from the text and placed in appendices that are available on the website that accompanies the text. Appendix B *Mathematical Modeling* contains Section 3.2 *Modeling Data with Linear Functions*, Section 6.6 *Modeling Data with Quadratic Functions*, and Section 9.6 *Modeling Data* from the Second Edition; Appendix C *Linear Programming* contains the Linear Programming material from Section 4.6 *Systems of Inequalities and Linear Programming* from the Section Edition; and Appendix D *Counting Principles and Probability* contains Section 10.5 *Counting Principles* and Section 10.6 *Probability* from the Second Edition. The appendices that appeared in the Second Edition (Appendix A *Introduction to Logic* and Appendix B *Graphing Utilities*) have been revised and moved to the website as well. We have also included new appendix material on the website for students and instructors. The Third Edition now has a geometry appendix (Appendix F *Further Concepts in Geometry*) and a statistics appendix (Appendix G *Further Concepts in Statistics*).

From the first time we began writing in the early 1970s, we have always viewed part of our authoring role as that of providing instructors with flexible teaching programs. The optional features within the text and the additional instructor support materials allow instructors with different pedagogical approaches to design their course to meet both their instructional needs and the needs of their students.

We hope you will enjoy the Third Edition.

Ron Larson

Ron Larson

Robert P. Hostetler

Robert P. Hostetler

Carolyn F. Neptune

Carolyn F. Neptune

Acknowledgments

We would like to thank our colleagues who have helped us develop this project throughout this and previous editions. Their encouragement, criticisms, and suggestions have been invaluable to us.

Reviewers

Mary K. Alter, University of Maryland—College Park; William Archibald, Lehigh Carbon Community College; Mary Jean Brod, University of Montana; Clark Brown, Mohave Community College; Martin Brown, Jefferson Community College; John W. Burns, Mt. San Antonio College; Bradd Clark, University of Southwestern Louisiana; D. J. Clark, Portland Community College; Linda Crabtree, Metropolitan Community College; Patricia Dalton, Montgomery College; Paul A. Dirks, Miami-Dade Community College; Lionel Geller, Dawson College; William Grimes, Central Missouri State University; Ingrid Holzner, University of Wisconsin—Milwaukee; Dr. Alan Jian, Solano Community College; Rosalyn Jones, Albany State College; Donald Kinney, University of Minnesota; Barbara Kistler, Lehigh Carbon Community College; Debra Landre, San Joaquin Delta College; Antonio M. Lopez, Jr., Loyola University; David Lunsford, Grossmont College; Giles Wilson Maloof, Boise State University; Myrna F. Manley, El Camino Community College; James I. McCullough, Arapahoe Community College; Katherine McLain, Cosumners River College; Karen S. Norwood, North Carolina State University; Louise Olshan, County College of Morris; Eric Preibisius, Cuyamaca College; William Radulovich, Florida Community College; Jack Rotman, Lansing Community College; Nora I. Schukei, University of South Carolina at Beaufort; Minnie Shuler, Gulf Coast Community College; Julia Simms, Southern Illinois University—Edwardsville; Judith D. Smalling, St. Petersburg Junior College; Kay Stroope, Phillips County Community College; James Tarvin, Grossmont College; Gwen Terwilliger, University of Toledo; Frances Ventola, Brookdale Community College; Jo Fitzsimmons Warner, Eastern Michigan University

In addition, we would like to thank the staff at Larson Texts, Inc. and Meridian Creative Group who assisted with the proofreading of the manuscript, preparing and proofreading the art package, and checking and typesetting the supplements.

A special note of appreciation goes to all the instructors and students who have used previous editions of the text.

On a personal level, we would like to thank our families, especially Deanna Gilbert Larson, Eloise Hostetler, and Harold Neptune, for their love, patience, and support. Also, a special thanks goes to R. Scott O'Neil.

If you have suggestions for improving the text, please feel free to write to us. Over the past 20 years, we have received many useful comments from both instructors and students.

Ron Larson
Robert P. Hostetler
Carolyn F. Neptune

Highlights of Features

Introduction to Graphs and Functions

2

THE BIG PICTURE

In this chapter you will learn how to:

■ plot points on a rectangular coordinate system.

■ sketch and use graphs of real-life data.

■ use the Distance and Midpoint Formulas.

■ sketch graphs of equations and find the *x*- and *y*-intercepts.

■ sketch graphs of lines using slopes and *y*-intercepts.

■ evaluate functions and find their domains.

■ sketch graphs of functions.

■ identify and sketch transformations of graphs of functions.

Key Terms

As you encounter each new vocabulary term in this chapter, add the term and its definition to your notebook glossary.

Additional text-specific resources are available to help you do well in this course. See page xvi for details.

79

New Chapter Openers Include:

The Big Picture
The Big Picture is an objective-based overview of the main concepts of the chapter.

Key Terms
The Key Terms is a list of all of the mathematical vocabulary that is essential to learning the concepts outlined in The Big Picture.

2.1 Describing Data Graphically

What you should learn:

• How to plot points on a rectangular coordinate system

• How to determine whether ordered pairs are solutions of equations

• How to use the Distance Formula to find the distance between two points

• How to use the Midpoint Formula to find the midpoints of line segments

Why you should learn it:

A rectangular coordinate system can be used to represent relationships between two variables. For example, Exercise 98 on page 94 shows the relationship between the various speeds of a car and the corresponding fuel efficiencies.

The Rectangular Coordinate System

Just as you can represent real numbers by points on the real number line, you can represent **ordered pairs** of real numbers by points in a plane. This plane is called a **rectangular coordinate system** or the **Cartesian plane,** after the French mathematician René Descartes.

A rectangular coordinate system is formed by two real lines intersecting at right angles, as shown in Figure 2.1. The horizontal number line is usually called the **x-axis,** and the vertical number line is usually called the **y-axis.** (The plural of axis is *axes.*) The point of intersection of the two axes is called the **origin,** and the axes separate the plane into four regions called **quadrants.**

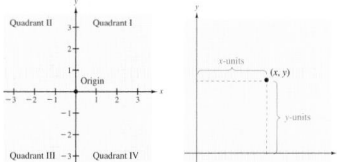

Figure 2.1   Figure 2.2

Each point in the plane corresponds to an **ordered pair** (x, y) of real numbers x and y, called the **coordinates** of the point. The first number (or **x-coordinate**) tells how far to the left or right the point is from the vertical axis, and the second number (or **y-coordinate**) tells how far up or down the point is from the horizontal axis, as shown in Figure 2.2.

A positive x-coordinate implies that the point lies to the *right* of the vertical axis; a negative x-coordinate implies that the point lies to the *left* of the vertical axis; and an x-coordinate of zero implies that the point lies *on* the vertical axis. Similar statements can be made about y-coordinates. A positive y-coordinate implies that the point lies *above* the horizontal axis; a negative y-coordinate implies that the point lies *below* the horizontal axis; and a y-coordinate of zero implies that the point lies *on* the horizontal axis.

NOTE   The signs of the coordinates tell you which quadrant the point lies in. For instance, if x and y are positive, the point (x, y) lies in Quadrant I.

Locating a given point in a plane is called **plotting** the point. Example 1 shows how this is done.

New Section Openers Include:

"What you should learn"
A set of objectives outlining the main concepts of the section, "What you should learn" helps keep students focused on the Big Picture.

"Why you should learn it"
"Why you should learn it" emphasizes the relevance of the section's content in the real world through a real-life application of reference to other branches of mathematics.

FEATURES

Section 5.2   Rational Exponents and Radicals   **319**

You already know from Section 2.4 that the domain of the square root function $f(x) = \sqrt{x}$ is the set of all nonnegative real numbers. The **domain** of the radical function $f(x) = \sqrt[n]{x}$ is the set of all real numbers such that x has a principal nth root.

> **Domain of a Radical Function**
> Let n be an integer that is greater than or equal to 2.
> 1. If n is odd, the domain of $f(x) = \sqrt[n]{x}$ is the set of all real numbers.
> 2. If n is even, the domain of $f(x) = \sqrt[n]{x}$ is the set of all nonnegative real numbers.

Example 10 ■ Finding the Domain of a Radical Function

Describe the domain of each function.

(a) $f(x) = \sqrt[3]{x}$ (b) $f(x) = \sqrt{x^2}$ (c) $f(x) = \sqrt{x^3}$

Solution

(a) The domain of $f(x) = \sqrt[3]{x}$ is the set of all real numbers because for any real number x, the expression $\sqrt[3]{x}$ is a real number.

(b) The domain of $f(x) = \sqrt{x^2}$ is the set of all real numbers because for any real number x, the expression x^2 is a nonnegative real number.

(c) The domain of $f(x) = \sqrt{x^3}$ is the set of all nonnegative real numbers. For instance, 1 is in the domain, but -1 is not because $\sqrt{(-1)^3} = \sqrt{-1}$ is not a real number.

In general, the domain of a radical function where the index n is even includes all real values for which the expression under the radicand is greater than or equal to zero.

Example 11 ■ Finding the Domain of a Radical Function

Find the domain of $f(x) = \sqrt{2x - 5}$.

Algebraic Solution

The domain of f consists of all x such that $2x - 5 \geq 0$. Using the methods described in Section 3.4, you can solve this inequality as follows.

$$2x - 5 \geq 0 \qquad \text{Write original inequality.}$$
$$2x \geq 5 \qquad \text{Add 5 to each side.}$$
$$x \geq \tfrac{5}{2} \qquad \text{Divide each side by 2.}$$

So, the domain is the set of all real numbers x such that $x \geq \tfrac{5}{2}$.

Graphical Solution

Use a graphing utility to graph $y = \sqrt{2x - 5}$, as shown in Figure 5.2. From the graph you can see that this function is defined only for x-values for which $x \geq \tfrac{5}{2}$.

Figure 5.2

New Algebraic, Graphical, and Numerical Approach

Solutions to selected examples are presented from multiple approaches—algebraic/graphical, graphical/numerical, and algebraic/numerical. The side-by-side, multiple approach format shows students how different solution methods can be used to arrive at the same answer.

Real-Life Applications

A wide variety of real-life applications, many using current, real data are integrated throughout examples and exercises. The icon 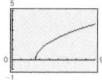 indicates an example that involves a real-life application.

82   Chapter 2   Introduction to Graphs and Functions

As a consumer today, you are presented almost daily with vast amounts of data given in various forms. Data is given in *numerical* form using lists and tables and in *graphical* form using scatter plots, line graphs, circle graphs, and bar graphs. Graphical forms are more visual and make wide use of Descartes's rectangular coordinate system to show the relationship between two variables. Today, Descartes's ideas are commonly used in virtually every scientific and business-related field.

In the next example, data is represented graphically by points plotted on a rectangular coordinate system. This type of graph is called a **scatter plot**.

> **Historical Note**
> Descartes introduced his analytical geometry in 1637. Most of the terminology we use comes from the Cartesian-coordinate system (*linear, quadratic*, etc.). Prior to this time there was no distinct tie between algebra and geometry.

Example 3 ■ Sketching a Scatter Plot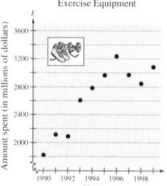

From 1990 through 1999, the amount E (in millions of dollars) spent on exercise equipment in the United States is shown in the table, where t represents the year. Sketch a scatter plot of the data. *(Source: National Sporting Goods Association)*

t	1990	1991	1992	1993	1994	1995	1996	1997	1998	1999
E	1824	2106	2078	2602	2781	2960	3232	2968	2850	3078

Solution

To sketch a scatter plot, begin by choosing which variable will be plotted on the horizontal axis and which will be plotted on the vertical axis. For this data, it seems natural to plot the years on the horizontal axis (which means that the amount must be plotted on the vertical axis). Next, use the data in the table to form ordered pairs. For instance, the first three ordered pairs are

(1990, 1824), (1991, 2106), and (1992, 2078).

All 10 points are shown in Figure 2.5. Note that the break in the x-axis indicates that the numbers between 0 and 1990 have been omitted. The break in the y-axis indicates that the numbers between 0 and 1800 have been omitted.

> **NOTE** In Example 3, you could have let $t = 0$ represent the year 1990. In that case, the horizontal axis would not have been broken, and the tick marks would have been labeled 0 through 9.

Exercise Equipment

Figure 2.5

90   Chapter 2   Introduction to Graphs and Functions

EXPLORATION

Plot the points $A(-1, -3)$ and $B(5, 2)$ and sketch the line segment from A to B. How could you verify that point $C(2, -0.5)$ is the midpoint of the segment? Why is it not sufficient to show that the distances from A to C and from C to B are equal?

Figure 2.16

The Midpoint Formula

The **midpoint** of a line segment that joins two points is the point that divides the segment into two equal parts. To find the midpoint of the line segment that joins two points in a coordinate plane, you can simply find the average values of the respective coordinates of the two endpoints using the **Midpoint Formula.**

The Midpoint Formula

The midpoint of the line segment joining the points (x_1, y_1) and (x_2, y_2) is given by the Midpoint Formula

$$\text{Midpoint} = \left(\frac{x_1 + x_2}{2}, \frac{y_1 + y_2}{2}\right).$$

Example 11 ■ Finding the Midpoint of a Line Segment

Find the midpoint of the line segment joining the points $(-5, -3)$ and $(9, 3)$.

Solution

Let $(x_1, y_1) = (-5, -3)$ and $(x_2, y_2) = (9, 3)$.

$$\text{Midpoint} = \left(\frac{x_1 + x_2}{2}, \frac{y_1 + y_2}{2}\right)$$ Midpoint Formula
$$= \left(\frac{-5 + 9}{2}, \frac{-3 + 3}{2}\right)$$ Substitute for x_1, y_1, x_2, and y_2.
$$= (2, 0)$$ Simplify.

The line segment and the midpoint are shown in Figure 2.16.

Collaborate!

Extending the Concept

Three or more points are **collinear** if they all lie on the same line. Use the steps below to determine if the set of points

$$\{A(3, 1), B(5, 4), C(9, 10)\}$$

and the set of points

$$\{A(2, 2), B(4, 3), C(5, 4)\}$$

are collinear.

a. For each set of points, use the Distance Formula to find the distances from A to B, from B to C, and from A to C. What relationship exists among these distances for each set of points?

b. Plot each set of points on a rectangular coordinate system. Do all the points of either set appear to lie on the same line?

c. Compare your conclusions from part (a) with the conclusions you made from the graphs in part (b). Make a general statement about how to use the Distance Formula to determine collinearity.

New Collaborate!

Ideal for small group work or class discussions, *Collaborate!* appears at the end of many sections to give students an opportunity to think, talk, and write about mathematics.

New Looking Further

At the end of each section exercise set, *Looking Further* presents meaningful problems that expand upon mathematical concepts presented in the section. These multi-part explorations and applications enhance the development of critical thinking and problem solving skills.

Section 4.1   Systems of Linear Equations in Two Variables   **247**

111. *Driving Distance*   Two people share the driving for a trip of 300 miles. One person drives three times as far as the other. Find the distance that each person drives.

112. *Air Speed*   An airplane flying into a headwind travels the 3000-mile flying distance between two cities in 6 hours and 15 minutes. On the return flight, the distance is traveled in 5 hours. Find the speed of the plane in still air and the speed of the wind, assuming that both remain constant throughout the round trip.

Explaining Concepts

113. If you graph a system of equations and the graphs do not intersect, what can you conclude about the solution of the system?

114. You graph a system of equations on a graphing utility and see only one line. Describe how the system could have (a) a unique solution, (b) infinitely many solutions, or (c) no solution.

115. Is it possible for a consistent system of linear equations to have exactly two solutions? Explain.

116. How can you recognize that a system of linear equations has no solution? Give an example.

Ongoing Review

In Exercises 117–120, sketch the graph of the equation.

117. $y = \frac{1}{3}x - 2$ **118.** $y = -2(x - 3)$

119. $y + 6 = -4(x + 1)$ **120.** $y - 5 = \frac{3}{4}(x - 10)$

In Exercises 121 and 122, solve the equation.

121. $5x + 2(x - 5) = 4$

122. $9y - 5(4 + 3y) = 2$

In Exercises 123 and 124, find an equation of the line passing through the two points.

123. $(3, 7), (-1, -5)$ **124.** $(0, 3), (-8, 11)$

125. *Cost*   The annual operating cost C, of a truck is modeled by $C = 0.45m + 6200$, where m is the number of miles traveled by the truck each year. What number of miles will yield an annual operating cost that is less than \$15,000?

126. *Wages*   You must choose between two plans of payment when working for a company. One plan pays \$2500 per month. The second pays \$1500 per month plus a commission of 4% of your gross sales. Write an inequality whose solution is such that the second option gives the greater monthly wage. Solve the inequality.

Looking Further

Vietnam Veterans Memorial "The Wall" in Washington, D.C., was designed by Maya Ying Lin when she was a student at Yale University. This monument has two vertical, triangular sections of black granite with a common side (see figure). The top of each section is level with the ground. The bottoms of the two sections can be approximately modeled by the equations $2x + 50y = -505$ and $2x - 50y = 505$ when the x-axis is superimposed on the top of the wall. Each unit in the coordinate system represents 1 foot.

How deep is the memorial at the point where the two sections meet? How long is each section?

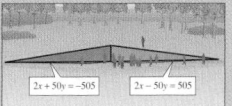

$2x + 50y = -505$ $2x - 50y = 505$

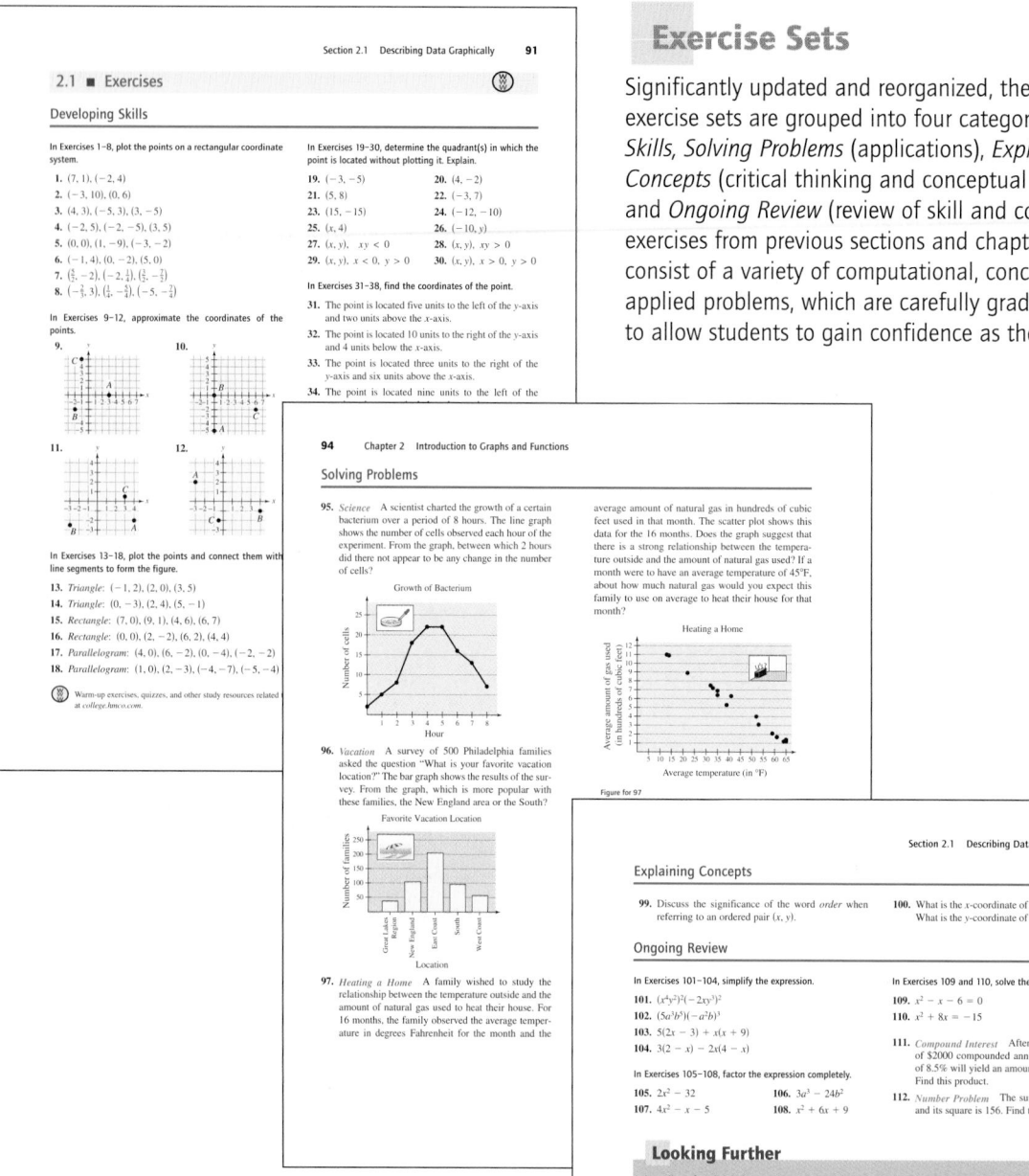

Exercise Sets

Significantly updated and reorganized, the section exercise sets are grouped into four categories: *Developing Skills, Solving Problems* (applications), *Explaining Concepts* (critical thinking and conceptual exercises), and *Ongoing Review* (review of skill and conceptual exercises from previous sections and chapters). Exercises consist of a variety of computational, conceptual, and applied problems, which are carefully graded in difficulty to allow students to gain confidence as they progress.

152 Chapter 2 Introduction to Graphs and Functions

Chapter Summary: Key Concepts

2.1 ■ Guidelines for verifying solutions
To verify that an ordered pair (x, y) is a solution of an equation with variables x and y, use the steps below.

1. Substitute the values of x and y into the equation.
2. Simplify each side of the equation.
3. If each side simplifies to the same number, the ordered pair is a solution. If the two sides yield different numbers, the ordered pair is not a solution.

2.1 ■ The Distance Formula
The distance d between two points (x_1, y_1) and (x_2, y_2) is

$$d = \sqrt{(x_2 - x_1)^2 + (y_2 - y_1)^2}.$$

2.1 ■ The Midpoint Formula
The midpoint of the line segment joining the points (x_1, y_1) and (x_2, y_2) is given by the Midpoint Formula

$$\text{Midpoint} = \left(\frac{x_1 + x_2}{2}, \frac{y_1 + y_2}{2} \right).$$

2.2 ■ The point-plotting method of sketching a graph

1. If possible, rewrite the equation by isolating one of the variables.
2. Make a table of values showing several solution points.
3. Plot these points on a rectangular coordinate system.
4. Connect the points with a smooth curve or line.

2.2 ■ Using a graphical check of a solution
The solution of an equation involving one variable x can be checked graphically with the steps below.

1. Write the equation so that all nonzero terms are on one side and zero is on the other side.
2. Sketch the graph of $y = $ (nonzero terms).
3. The solution of the one-variable equation is the x-intercept of the graph of the two-variable equation.

2.3 ■ Slope and the equations of lines

1. Slope of a line through (x_1, y_1) and (x_2, y_2):

$$m = \frac{y_2 - y_1}{x_2 - x_1}$$

2. Slope-intercept form of the equation of a line:

$$y = mx + b$$

3. Parallel lines (equal slopes): $m_1 = m_2$
4. Perpendicular lines (negative reciprocal slopes):

$$m_1 = -\frac{1}{m_2}$$

2.4 ■ Characteristics of a function

1. Each element in the domain A must be matched with an element in the range, which is contained in the set B.
2. Some elements in the set B may not be matched with any element in the domain A.
3. Two or more elements of the domain may be matched with the same element in the range.
4. No element of the domain is matched with two different elements in the range.

2.5 ■ Vertical Line Test for functions
A set of points on a rectangular coordinate system is the graph of y as a function of x if and only if no vertical line intersects the graph at more than one point.

2.6 ■ Vertical and horizontal shifts
Let c be a positive real number. Vertical and horizontal shifts of the graph of the function $y = f(x)$ are represented as follows.

1. Vertical shift c units upward: $h(x) = f(x) + c$
2. Vertical shift c units downward: $h(x) = f(x) - c$
3. Horizontal shift c units to the right: $h(x) = f(x - c)$
4. Horizontal shift c units to the left: $h(x) = f(x + c)$

2.6 ■ Reflections in the coordinate axes
Reflections of the graph of $y = f(x)$ are represented as follows:

1. Reflection in the x-axis: $h(x) = -f(x)$
2. Reflection in the y-axis: $h(x) = f(-x)$

New Chapter Summary: Key Concepts

The Chapter Summary is a concise, section-by-section summary of the main concepts of the chapter, providing a convenient study tool for students.

Review Exercises **153**

Review Exercises

Reviewing Skills

2.1 ■ *How to plot points on a rectangular coordinate system*

In Exercises 1 and 2, plot the points on a rectangular coordinate system.

1. $(0, -3), (\frac{5}{2}, 5), (-2, -4)$
2. $(1, -\frac{3}{2}), (-2, 2), (5, 10)$

In Exercises 3 and 4, plot the points and connect them with line segments to form the figure.

3. *Right triangle:* $(1, 1), (12, 9), (4, 20)$
4. *Parallelogram:* $(0, 0), (7, 1), (8, 4), (1, 3)$

In Exercises 5–8, determine the quadrant(s) in which the point is located.

5. $(2, -6)$ **6.** $(-4.8, -2)$
7. $(4, y)$ **8.** $(x, y), xy > 0$

■ *How to determine whether ordered pairs are solutions of equations*

In Exercises 9 and 10, determine whether the ordered pairs are solution points of the given equation.

9. $y - 4 - \frac{1}{4}x$
   (a) $(4, 2)$ (b) $(-1, 5)$
   (c) $(-4, 0)$ (d) $(8, 0)$
10. $3x - 2y + 18 = 0$
   (a) $(3, 10)$ (b) $(0, 9)$
   (c) $(-4, 3)$ (d) $(-8, 0)$

■ *How to use the Distance Formula to find the distance between two points*
■ *How to use the Midpoint Formula to find the midpoints of line segments*

In Exercises 11–14, (a) plot the points on a rectangular coordinate system, (b) find the distance between the points, and (c) find the coordinates of the midpoint.

11. $(-2, -3), (4, 5)$ **12.** $(-5, 1), (-1, 7)$
13. $(2, 0), (14, 5)$ **14.** $(0, 7), (6, -8)$

2.2 ■ *How to sketch graphs of equations using the point-plotting method*

In Exercises 15–18, sketch the graph of the equation.

15. $y = 5 - \frac{3}{2}x$ **16.** $y = x^2 + 4$
17. $y = |x| + 4$ **18.** $y = \sqrt{x + 4}$

■ *How to find and use x- and y-intercepts as aids to sketching graphs*

In Exercises 19–26, find the x- and y-intercepts of the graph of the equation. Sketch the graph of the equation and label the x- and y-intercepts.

19. $y = 6 - \frac{1}{3}x$ **20.** $y = \frac{3}{4}x - 2$
21. $3y - 2x - 3 = 0$ **22.** $3x + 4y + 12 = 0$
23. $x = |y - 3|$ **24.** $y = 7 - |x|$
25. $y = 1 - x^2$ **26.** $y = (x + 3)^2$

■ *How to use graphical checking to check solutions of equations involving one variable*

In Exercises 27–30, solve the equation. Use a graphing utility to verify your solution.

27. $4x - 9 = 7$ **28.** $2(x - 3) = 10$
29. $x^2 - 3x = 18$ **30.** $x^2 + 2x = -1$

2.3 ■ *How to determine the slope of a line given two points on the line*

In Exercises 31–36, find the slope of the line through the points.

31. $(-1, 1), (6, 3)$ **32.** $(-2, 5), (3, -8)$
33. $(-1, 3), (4, 3)$ **34.** $(7, 2), (7, 8)$
35. $(0, 6), (8, 0)$ **36.** $(0, 0), (\frac{5}{2}, 6)$

In Exercises 37–42, a point on a line and the slope of the line are given. Find two additional points on the line. (There are many correct answers.)

37. $(2, -4)$ **38.** $(-4, \frac{1}{2})$
   $m = -3$ $m = 2$

Review Exercises

Found at the end of each chapter, Review Exercises have been reorganized and grouped into two categories: *Reviewing Skills* and *Problem Solving*. Exercises are not only correlated by section, but also by "What you should learn" objectives within each section, allowing students easily to identify sections and concepts for study and review.

Additional Features

Carefully crafted learning tools designed to create a rich learning environment can be found throughout the text. These learning tools include Explorations, Technology Tips, Study Tips, Historical Notes, Mid-Chapter Quizzes, Chapter Tests, Cumulative Tests, and an extensive art program.

FEATURES

Supplements

Resources

Website (college.hmco.com/mathematics)
Additional text-specific resources are available for students and instructors at the website that accompanies this text.

BlackBoard Course Cartridge

WebCT e-pack

For the Student

Student Solutions Guide *by Carolyn F. Neptune (Johnson County Community College)*

Graphing Calculator Keystroke Guide *by Benjamin Levy and Laurel Technical Services*

Instructional Videotapes/DVDs *by Dana Mosely*
Comprehensive section-by-section coverage, with detailed explanations of important concepts

HM^3 Tutorial CD (Windows)

***SMARTHINKING™.com* Live, On-Line Tutoring**

For the Instructor

Instructor's Annotated Edition

Instructor's Solution Guide *by Carolyn F. Neptune (Johnson County Community College)*

HM Testing (Windows, Macintosh)
Computerized test generator with algorithmically generated test items

Test Item File
Print component of the HM Testing software, containing an example of each algorithm

HMClassPrep™ Instructor's CD-ROM

For more information on these and other resources available, go to our website at *college.hmco.com.*

An Introduction to Graphing Utilities

Graphing utilities such as graphing calculators and computers with graphing software are very valuable tools for visualizing mathematical principles, verifying solutions to equations, exploring mathematical ideas, and developing mathematical models. Although graphing utilities are extremely helpful in learning mathematics, their use does not mean that learning algebra is any less important. In fact, the combination of knowledge of mathematics and the use of graphing utilities allows you to explore mathematics more easily and to a greater depth. If you are using a graphing utility in this course, it is up to you to learn its capabilities and to practice using this tool to enhance your mathematical learning.

In this text there are many opportunities to use a graphing utility, some of which are described below.

Some Uses of a Graphing Utility

A graphing utility can be used to

- check or validate answers to problems obtained using algebraic methods.
- discover and explore algebraic properties, rules, and concepts.
- graph functions and approximate solutions to equations involving functions.
- efficiently perform complicated mathematical procedures such as those found in many real-life applications.
- find mathematical models for sets of data.

In this introduction, the features of graphing utilities are discussed from a generic perspective. To learn how to use the features of a specific graphing utility, consult your user's manual. Additionally, keystroke guides are available for most graphing utilities, and your college library may have a videotape on how to use your graphing utility.

The Equation Editor

Many graphing utilities are designed to act as "function graphers." In this course, you will study functions and their graphs in detail. As you will learn in Chapter 2, a function can be thought of as a rule that describes the relationship between two variables. These rules are frequently written in terms of x and y. For example, the equation $y = 3x + 5$ represents y as a function of x.

Many graphing utilities have an equation editor that requires an equation to be written in "y =" form in order to be entered, as shown in Figure 1. (You should note that your equation editor screen may not look like the screen shown in Figure 1.) To determine exactly how to enter an equation into your graphing utility, consult your user's manual.

Figure 1

The Table Feature

Most graphing utilities are capable of displaying a table of values with x-values and one or more corresponding y-values. These tables can be used to check solutions of an equation and to generate ordered pairs to assist in graphing an equation.

To use the *table* feature, enter an equation into the equation editor in "$y =$" form. The table may have a setup screen, which allows you to select the starting x-value and the table step or x-increment. See page 85 for detailed steps on how to create a table with a graphing utility.

You may have the option of building your own table using the *ask* mode. In the *ask* mode, you enter a value for x and the graphing utility displays the y-value.

With the equation

$$y = \frac{3x}{x + 2}$$

in the equation editor, set the table to *ask* mode. In this mode you do not need to set the starting x-value or the table step, because you are entering any value you choose for x. You may enter any real value for x—integers, fractions, decimals, irrational numbers, and so forth. If you enter $x = 1 + \sqrt{3}$, the graphing utility may rewrite the number as a decimal approximation, as shown in Figure 2. You can continue to build your own table by entering additional x-values in order to generate y-values.

If you have several equations in the equation editor, the table may generate y-values for each equation.

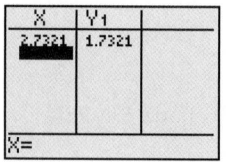

Figure 2

Creating a Viewing Window

A **viewing window** for a graph is a rectangular portion of the coordinate plane. A viewing window is determined by the following six values.

Xmin = the smallest value of x

Xmax = the largest value of x

Xscl = the number of units per tick mark on the x-axis

Ymin = the smallest value of y

Ymax = the largest value of y

Yscl = the number of units per tick mark on the y-axis

When you enter these six values into a graphing utility, you are setting the viewing window. See page 99 for a more detailed description on creating a viewing window.

Zoom and Trace Features

When you graph an equation, you can move from point to point along its graph using the *trace* feature. As you trace the graph, the coordinates of each point are displayed, as shown in Figure 3. The *trace* feature combined with the *zoom* feature allows you to obtain better and better approximations of desired points on

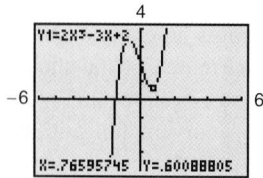

Figure 3

a graph. For instance, you can use the *zoom* feature of a graphing utility to approximate the *x*-intercept(s) of a graph [the point(s) where the graph crosses the *x*-axis]. Suppose you want to approximate the *x*-intercept(s) of the graph of $y = 2x^3 - 3x + 2$.

Begin by graphing the equation, as shown in Figure 4(a). From the viewing window shown, the graph appears to have only one *x*-intercept. This intercept lies between -2 and -1. By zooming in on the intercept, you can improve the approximation, as shown in Figure 4(b). To three decimal places, the intercept is $x \approx -1.476$.

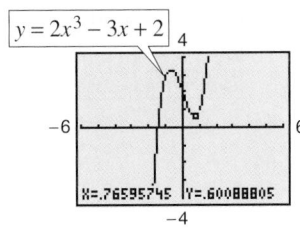
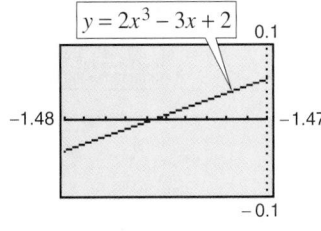

(a)   (b)

Figure 4

Here are some suggestions for using the *zoom* feature.

1. With each successive zoom-in, adjust the *x*-scale so that the viewing window shows at least one tick mark on each side of the *x*-intercept.

2. The error in your approximation will be less than the distance between two scale marks.

3. The *trace* feature can usually be used to add one more decimal place of accuracy without changing the viewing window.

Figure 5(a) shows the graph of $y = x^2 - 5x + 3$. Figures 5(b) and 5(c) show "zoom-in views" of the two *x*-intercepts. From these views, you can approximate the *x*-intercepts to be $x \approx 0.697$ and $x \approx 4.303$.

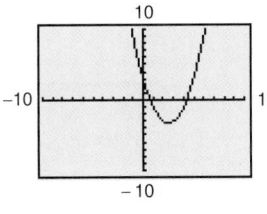
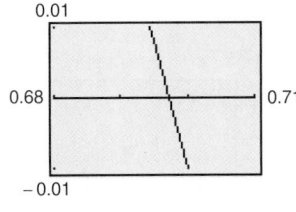
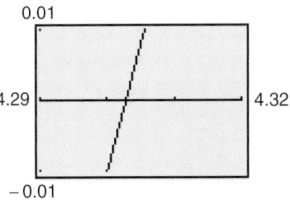

(a)   (b)   (c)

Figure 5

Zero or Root Feature

Using the zero or root feature, you can find the real zeros of functions of the various types studied in this text—polynomial, exponential, and logarithmic functions. To find the zeros of a function such as $f(x) = \frac{3}{4}x - 2$, first enter the function as $y_1 = \frac{3}{4}x - 2$. Then use the *zero* or *root* feature, which may require entering lower and upper bound estimates of the root, as shown in Figure 6.

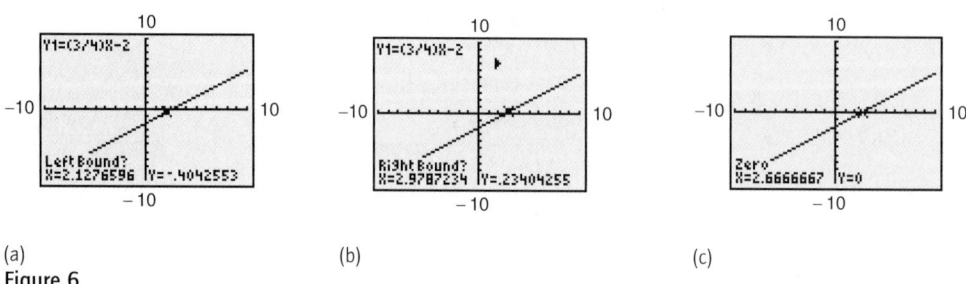

(a) (b) (c)

Figure 6

In Figure 6(c), you can see that the zero is $x = 2.6666667 \approx 2\frac{2}{3}$.

Intersect Feature

To find the points of intersection of two graphs, you can use the *intersect* feature. For instance, to find the points of intersection of the graphs of $y_1 = -x + 2$ and $y_2 = x + 4$, enter these two functions and use the *intersect* feature, as shown in Figure 12.

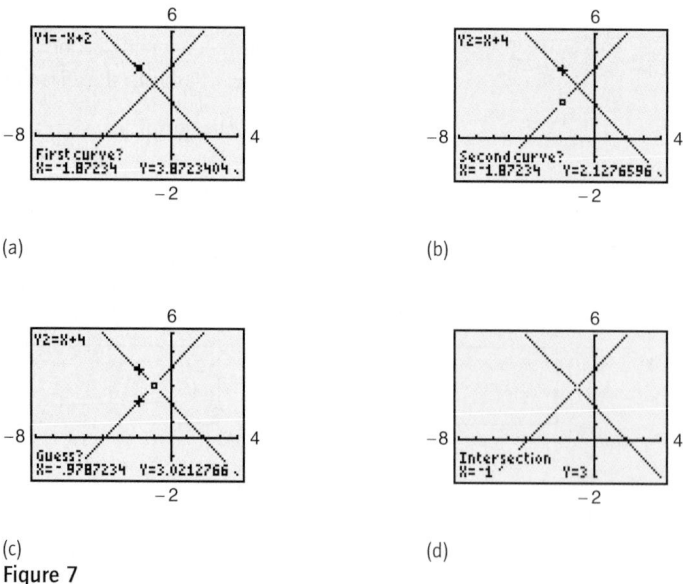

(a) (b)

(c) (d)

Figure 7

From Figure 7(d), you can see that the point of intersection is $(-1, 3)$.

Concepts of Elementary Algebra

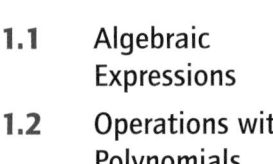

1.1 Algebraic Expressions

1.2 Operations with Polynomials

1.3 Factoring Polynomials

1.4 Factoring Trinomials

1.5 Solving Linear Equations

1.6 Solving Equations by Factoring

THE BIG PICTURE

In this chapter you will learn how to:

- simplify and evaluate algebraic expressions.

- add, subtract, and multiply polynomials.

- solve linear and literal equations.

- factor expressions completely and solve polynomial equations by factoring.

- use expressions or equations to solve application problems.

Key Terms

As you encounter each new vocabulary term in this chapter, add the term and its definition to your notebook glossary.

Additional text-specific resources are available to help you do well in this course.
See page xvi for details.

1.1 Algebraic Expressions

What you should learn:

- How to identify the terms and coefficients of algebraic expressions
- How to identify the properties of algebra
- How to apply the properties of exponents to simplify algebraic expressions
- How to simplify algebraic expressions by combining like terms and removing symbols of grouping
- How to evaluate algebraic expressions

Why you should learn it:

Algebraic expressions can help you construct tables of values. For instance, in Example 14 on page 10, you can determine the hourly wages of miners using an expression and a table of values.

Algebraic Expressions

A basic characteristic of algebra is the use of letters (or combinations of letters) to represent numbers. The letters used to represent the numbers are **variables,** and combinations of letters and numbers are **algebraic expressions.** Here are a few examples.

$$3x, \quad x + 2, \quad \frac{x}{x^2 + 1}, \quad 2x - 3y$$

Definition of Algebraic Expression

A collection of letters (called **variables**) and real numbers (called **constants**) combined using the operations of addition, subtraction, multiplication, and division is called an **algebraic expression.**

The **terms** of an algebraic expression are those parts that are separated by *addition.* For example, the algebraic expression $x^2 - 3x + 6$ has three terms: x^2, $-3x$, and 6. Note that $-3x$ is a term, rather than $3x$, because

$$x^2 - 3x + 6 = x^2 + (-3x) + 6. \qquad \text{Think of subtraction as a form of addition.}$$

The terms x^2 and $-3x$ are called the **variable terms** of the expression, and 6 is called the **constant term** of the expression. The numerical factor of a variable term is called the **coefficient** of the variable term. For instance, the coefficient of the variable term $-3x$ is -3, and the coefficient of the variable term x^2 is 1. (The constant term of an expression is also considered to be a coefficient.)

NOTE It is important to understand the difference between a *term* and a *factor.* Terms are separated by addition whereas factors are separated by multiplication. For instance, the expression $4x(x + 2)$ has three factors: 4, x, and $(x + 2)$.

Example 1 ■ Identifying Terms and Coefficients

Identify the terms and coefficients in each algebraic expression.

(a) $5x - \dfrac{1}{3}$ (b) $4y + 6x - 9$ (c) $x^2y - \dfrac{1}{x} + 3y$

Solution

	Terms	Coefficients
(a)	$5x, \ -\dfrac{1}{3}$	$5, \ -\dfrac{1}{3}$
(b)	$4y, 6x, -9$	$4, 6, -9$
(c)	$x^2y, \ -\dfrac{1}{x}, 3y$	$1, -1, 3$

Properties of Algebra

The properties of real numbers (see Appendix A) can be used to rewrite algebraic expressions. The following list is similar to those given in Appendix A, except that the examples involve algebraic expressions. In other words, the properties are true for variables and algebraic expressions as well as for real numbers.

Historical Note

The French mathematician François Viète (1540–1603) was the first to use letters to represent numbers. He used vowels to represent unknown quantities and consonants to represent known quantities.

Properties of Algebra

Let a, b, and c be real numbers, variables, or algebraic expressions.

Property   *Example*

Commutative Property of Addition

$$a + b = b + a$$
$$5x + x^2 = x^2 + 5x$$

Commutative Property of Multiplication

$$ab = ba$$
$$(3 + x)x^3 = x^3(3 + x)$$

Associative Property of Addition

$$(a + b) + c = a + (b + c)$$
$$(-x + 6) + 3x^2 = -x + (6 + 3x^2)$$

Associative Property of Multiplication

$$(ab)c = a(bc)$$
$$(5x \cdot 4y)(6) = (5x)(4y \cdot 6)$$

Distributive Property

$$a(b + c) = ab + ac$$
$$(a + b)c = ac + bc$$
$$2x(4 + 3x) = 2x \cdot 4 + 2x \cdot 3x$$
$$(y + 6)y = y \cdot y + 6 \cdot y$$

Additive Identity Property

$$a + 0 = 0 + a = a$$
$$4y^2 + 0 = 0 + 4y^2 = 4y^2$$

Multiplicative Identity Property

$$a \cdot 1 = 1 \cdot a = a$$
$$(-5x^3)(1) = (1)(-5x^3) = -5x^3$$

Additive Inverse Property

$$a + (-a) = 0$$
$$4x^2 + (-4x^2) = 0$$

Multiplicative Inverse Property

$$a \cdot \frac{1}{a} = 1, a \neq 0$$
$$(x^2 + 1)\left(\frac{1}{x^2 + 1}\right) = 1$$

NOTE Because subtraction is defined as "adding the opposite," the Distributive Property is also true for subtraction. For instance, the "subtraction form" of $a(b + c) = ab + ac$ is

$$a(b - c) = a[b + (-c)]$$
$$= ab + a(-c)$$
$$= ab - ac.$$

NOTE The operations of subtraction and division are neither commutative nor associative. The examples

$$7 - 3 \neq 3 - 7 \text{ and } 20 \div 4 \neq 4 \div 20$$

show that subtraction and division are not commutative. Similarly

$$5 - (3 - 2) \neq (5 - 3) - 2 \text{ and } 16 \div (4 \div 2) \neq (16 \div 4) \div 2$$

demonstrate that subtraction and division are not associative.

In addition to these properties, the properties of equality, zero, and negation given in Appendix A are also valid for algebraic expressions. The next example illustrates the use of a variety of these properties.

Example 2 ■ Identifying the Properties of Algebra

Identify the property of algebra illustrated in each statement.

(a) $(5x^2)3 = 3(5x^2)$

(b) $(3x^2 + x) - (3x^2 + x) = 0$

(c) $3x + 3y^2 = 3(x + y^2)$

(d) $(5 + x^2) + 4x^2 = 5 + (x^2 + 4x^2)$

(e) $5x \cdot \dfrac{1}{5x} = 1, x \neq 0$

(f) $(y - 6)3 + (y - 6)y = (y - 6)(3 + y)$

Solution

(a) This statement illustrates the Commutative Property of Multiplication. In other words, you obtain the same result whether you multiply $5x^2$ by 3, or 3 by $5x^2$.

(b) This statement illustrates the Additive Inverse Property. In terms of subtraction, this property simply states that when any expression is subtracted from itself the result is zero.

(c) This statement illustrates the Distributive Property. In other words, multiplication is distributed over addition.

(d) This statement illustrates the Associative Property of Addition. In other words, to form the sum

$$5 + x^2 + 4x^2$$

it does not matter whether 5 and x^2 are added first or x^2 and $4x^2$ are added first.

(e) This statement illustrates the Multiplicative Inverse Property. Note that it is important that x be a nonzero number. If x were zero, the reciprocal of x would be undefined.

(f) This statement illustrates the Distributive Property in reverse order.

$$ab + ac = a(b + c) \qquad \text{Distributive Property}$$
$$(y - 6)3 + (y - 6)y = (y - 6)(3 + y)$$

Note in this case that $a = y - 6$, $b = 3$, and $c = y$.

EXPLORATION

Discovering Properties of Exponents   Write each of the following as a single power of 2. Explain how you obtained your answer. Then generalize your procedure by completing the statement *"When you multiply exponential expressions that have the same base, you. . . ."*

a. $2^2 \cdot 2^3$ b. $2^4 \cdot 2^1$ c. $2^5 \cdot 2^2$ d. $2^3 \cdot 2^4$ e. $2^1 \cdot 2^5$

Historical Note

Originally, Arabian mathematicians used their words for colors to represent quantities (*cosa, censa, cubo*). These words were eventually abbreviated to *co, ce, cu*. René Descartes (1596–1650) simplified this even further by introducing the symbols x, x^2, and x^3.

Properties of Exponents

Just as multiplication by a positive integer can be described as repeated addition, *repeated multiplication* can be written in what is called **exponential form** (see Appendix A). Let n be a positive integer and let a be a real number. Then the product of n factors of a is given by

$$a^n = \underbrace{a \cdot a \cdot a \cdots a}_{n \text{ factors}}.$$

NOTE   In the exponential form a^n, a is the *base* and n is the *exponent*.

When multiplying two exponential expressions that have the *same base*, you add exponents. To see why this is true, consider the product $a^3 \cdot a^2$. Because the first expression represents $a \cdot a \cdot a$ and the second represents $a \cdot a$, the product of the two expressions represents $a \cdot a \cdot a \cdot a \cdot a$, as follows.

$$a^3 \cdot a^2 = \underbrace{(a \cdot a \cdot a)}_{3 \text{ factors}} \cdot \underbrace{(a \cdot a)}_{2 \text{ factors}} = \underbrace{(a \cdot a \cdot a \cdot a \cdot a)}_{5 \text{ factors}} = a^{3+2} = a^5$$

NOTE   The first and second properties can be extended to products involving three or more factors. For example,

$$a^m \cdot a^n \cdot a^k = a^{m+n+k}$$

and

$$(abc)^m = a^m b^m c^m.$$

Properties of Exponents

Let m and n be positive integers, and let a and b represent real numbers, variables, or algebraic expressions.

Property	*Example*
1. $a^m \cdot a^n = a^{m+n}$	$x^5(x^4) = x^{5+4} = x^9$
2. $(ab)^m = a^m \cdot b^m$	$(2x)^3 = 2^3(x^3) = 8x^3$
3. $(a^m)^n = a^{mn}$	$(x^2)^3 = x^{2 \cdot 3} = x^6$
4. $\dfrac{a^m}{a^n} = a^{m-n},\ m > n,\ a \neq 0$	$\dfrac{x^6}{x^2} = x^{6-2} = x^4,\ x \neq 0$
5. $\left(\dfrac{a}{b}\right)^m = \dfrac{a^m}{b^m},\ b \neq 0$	$\left(\dfrac{x}{2}\right)^3 = \dfrac{x^3}{2^3} = \dfrac{x^3}{8}$

Example 3   ■   Illustrating the Properties of Exponents

(a) To multiply exponential expressions that have the *same base*, add exponents.

$$x^2 \cdot x^4 = \underbrace{x \cdot x}_{2 \text{ factors}} \cdot \underbrace{x \cdot x \cdot x \cdot x}_{4 \text{ factors}} = \underbrace{x \cdot x \cdot x \cdot x \cdot x \cdot x}_{6 \text{ factors}} = x^{2+4} = x^6$$

(b) To raise the product of two factors to the *same power*, raise each factor to the power and multiply the results.

$$(3x)^3 = \underbrace{3x \cdot 3x \cdot 3x}_{3 \text{ factors}} = \underbrace{3 \cdot 3 \cdot 3}_{3 \text{ factors}} \cdot \underbrace{x \cdot x \cdot x}_{3 \text{ factors}} = 3^3 \cdot x^3 = 27x^3$$

(c) To raise an exponential expression to a power, multiply exponents.

$$(x^3)^2 = \underbrace{(x \cdot x \cdot x)}_{3 \text{ factors}} \cdot \underbrace{(x \cdot x \cdot x)}_{3 \text{ factors}} = \underbrace{(x \cdot x \cdot x \cdot x \cdot x \cdot x)}_{6 \text{ factors}} = x^{3 \cdot 2} = x^6$$

Example 4 ■ Illustrating the Properties of Exponents

(a) To divide exponential expressions that have the *same base*, subtract exponents.

$$\frac{x^5}{x^2} = \frac{\overbrace{x \cdot x \cdot x \cdot x \cdot x}^{5 \text{ factors}}}{\underbrace{x \cdot x}_{2 \text{ factors}}} = x^{5-2} = x^3$$

(b) To raise the quotient of two expressions to the *same power*, raise each expression to the power and divide the results.

$$\left(\frac{x}{3}\right)^3 = \frac{x}{3} \cdot \frac{x}{3} \cdot \frac{x}{3} = \frac{\overbrace{x \cdot x \cdot x}^{3 \text{ factors}}}{\underbrace{3 \cdot 3 \cdot 3}_{3 \text{ factors}}} = \frac{x^3}{3^3} = \frac{x^3}{27}$$

Example 5 ■ Applying Properties of Exponents

Use the properties of exponents to simplify each expression.

(a) $(x^2y^4)(3x)$ (b) $-2(y^2)^3$ (c) $(-2y^2)^3$ (d) $(3x^2)(-5x)^3$

Solution

(a) $(x^2y^4)(3x) = 3(x^2 \cdot x)(y^4) = 3(x^{2+1})(y^4) = 3x^3y^4$

(b) $-2(y^2)^3 = (-2)(y^{2 \cdot 3}) = -2y^6$

(c) $(-2y^2)^3 = (-2)^3(y^2)^3 = -8(y^{2 \cdot 3}) = -8y^6$

(d) $(3x^2)(-5x)^3 = 3(-5)^3(x^2 \cdot x^3) = 3(-125)(x^{2+3}) = -375x^5$

Example 6 ■ Applying Properties of Exponents

Use the properties of exponents to simplify each expression.

(a) $\dfrac{14a^5b^3}{7a^2b^2}$ (b) $\left(\dfrac{x^2}{2y}\right)^3$ (c) $\dfrac{x^n y^{3n}}{x^2 y^4}$ (d) $\dfrac{(2a^2b^3)^2}{a^3b^2}$

Solution

(a) $\dfrac{14a^5b^3}{7a^2b^2} = 2(a^{5-2})(b^{3-2}) = 2a^3b$

(b) $\left(\dfrac{x^2}{2y}\right)^3 = \dfrac{(x^2)^3}{(2y)^3} = \dfrac{x^{2 \cdot 3}}{2^3y^3} = \dfrac{x^6}{8y^3}$

(c) $\dfrac{x^n y^{3n}}{x^2 y^4} = x^{n-2} y^{3n-4}$

(d) $\dfrac{(2a^2b^3)^2}{a^3b^2} = \dfrac{2^2(a^{2 \cdot 2})(b^{3 \cdot 2})}{a^3b^2} = \dfrac{4a^4b^6}{a^3b^2} = 4(a^{4-3})(b^{6-2}) = 4ab^4$

Simplifying Algebraic Expressions

One common use of the basic properties of algebra is to rewrite an algebraic expression in a simpler form. To **simplify** an algebraic expression generally means to *remove symbols of grouping* such as parentheses or brackets and *combine like terms.*

Two or more terms of an algebraic expression can be combined only if they are *like terms.* In an algebraic expression, two terms are said to be **like terms** if they are both constant terms or if they have the same variable factor(s). For example, the terms $4x$ and $-2x$ are like terms because they have the same variable factor, x. Similarly, $2x^2y$, $-x^2y$, and $\frac{1}{2}(x^2y)$ are like terms because they have the same variable factor, x^2y. Note that $4x^2y$ and $-x^2y^2$ are not like terms because their variable factors are different.

To combine like terms in an algebraic expression, simply add their respective coefficients and attach the common variable factor. This is actually an application of the Distributive Property, as shown in Example 7.

Study Tip

As you gain experience with the properties of algebra, you may want to combine some of the steps in your work. For instance, you might feel comfortable listing only the following steps to solve Example 7(c).

$$5x + 3y - 4x$$
$$= (5x - 4x) + 3y \quad \text{Group like terms.}$$
$$= x + 3y \quad \text{Combine like terms.}$$

Example 7 ■ Combining Like Terms

Simplify each expression by combining like terms.

(a) $2x + 3x - 4$ (b) $-3 + 5 + 2y - 7y$ (c) $5x + 3y - 4x$

Solution

(a) $2x + 3x - 4 = (2 + 3)x - 4$ Distributive Property
$$= 5x - 4 \quad \text{Simplest form}$$

(b) $-3 + 5 + 2y - 7y = (-3 + 5) + (2 - 7)y$ Distributive Property
$$= 2 - 5y \quad \text{Simplest form}$$

(c) $5x + 3y - 4x = 5x - 4x + 3y$ Commutative Property
$$= (5x - 4x) + 3y \quad \text{Associative Property}$$
$$= (5 - 4)x + 3y \quad \text{Distributive Property}$$
$$= x + 3y \quad \text{Simplest form}$$

Example 8 ■ Combining Like Terms

(a) $7x + 7y - 4x - y = (7x - 4x) + (7y - y)$ Group like terms.
$$= 3x + 6y \quad \text{Combine like terms.}$$

(b) $2x^2 + 3x - 5x^2 - x = (2x^2 - 5x^2) + (3x - x)$ Group like terms.
$$= -3x^2 + 2x \quad \text{Combine like terms.}$$

(c) $3xy^2 - 4x^2y^2 + 2xy^2 + (xy)^2$
$$= (3xy^2 + 2xy^2) + (-4x^2y^2 + x^2y^2)$$
$$= 5xy^2 - 3x^2y^2$$

A set of parentheses preceded by a *minus* sign can be removed by changing the sign of each term inside the parentheses. For instance,

$$8x - (5x - 4) = 8x - 5x + 4 = 3x + 4.$$

A set of parentheses preceded by a *plus* sign can be removed without changing the signs of the terms inside the parentheses. For instance,

$$8x + (5x - 4) = 8x + 5x - 4 = 13x - 4.$$

Example 9 ■ Removing Symbols of Grouping

Simplify $3(x - 5) - (2x - 7)$.

Solution

$$
\begin{aligned}
3(x - 5) - (2x - 7) &= 3x - 15 - 2x + 7 & \text{Distributive Property} \\
&= (3x - 2x) + (-15 + 7) & \text{Group like terms.} \\
&= x - 8 & \text{Combine like terms.}
\end{aligned}
$$

Example 10 ■ Geometry: Perimeter and Area of a Region

Write and simplify an expression for the perimeter and for the area of each region shown in Figure 1.1.

Solution

(a) Perimeter of a rectangle $= 2 \cdot \text{length} + 2 \cdot \text{width}$

$$
\begin{aligned}
&= 2(3x - 5) + 2x & \text{Substitute.} \\
&= 6x - 10 + 2x & \text{Distributive Property} \\
&= (6x + 2x) - 10 & \text{Group like terms.} \\
&= 8x - 10 & \text{Combine like terms.}
\end{aligned}
$$

Area of a rectangle $= \text{length} \cdot \text{width}$

$$
\begin{aligned}
&= (3x - 5)x & \text{Substitute.} \\
&= 3x^2 - 5x & \text{Distributive Property}
\end{aligned}
$$

(b) Perimeter of a triangle $= \text{sum of the three sides}$

$$
\begin{aligned}
&= 2x + (2x + 4) + (x + 5) & \text{Substitute.} \\
&= (2x + 2x + x) + (4 + 5) & \text{Group like terms.} \\
&= 5x + 9 & \text{Combine like terms.}
\end{aligned}
$$

Area of a triangle $= \frac{1}{2} \cdot \text{base} \cdot \text{height}$

$$
\begin{aligned}
&= \tfrac{1}{2}(x + 5)(2x) & \text{Substitute.} \\
&= \tfrac{1}{2}(2x)(x + 5) & \text{Commutative Property of Multiplication} \\
&= x(x + 5) & \text{Multiply.} \\
&= x^2 + 5x & \text{Distributive Property}
\end{aligned}
$$

When removing symbols of grouping, combine like terms within the innermost symbols of grouping first, as shown in the next example.

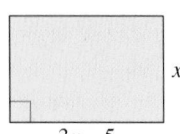

(a)

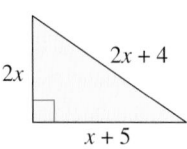

(b)

Figure 1.1

Example 11 ■ Removing Symbols of Grouping

Simplify each expression by combining like terms.

(a) $5x - 2x[3 + 2(x - 7)]$ (b) $-3x(5x^4) + (2x)^5$

Solution

(a)
$$\begin{aligned}
5x - 2x[3 + 2(x - 7)] &= 5x - 2x[3 + 2x - 14] \\
&= 5x - 2x[2x - 11] \\
&= 5x - 4x^2 + 22x \\
&= -4x^2 + 27x
\end{aligned}$$

Remove parentheses.

Combine like terms in brackets.

Remove brackets.

Combine like terms.

(b)
$$\begin{aligned}
-3x(5x^4) + (2x)^5 &= -15x^5 + (2^5)(x^5) \\
&= -15x^5 + 32x^5 \\
&= 17x^5
\end{aligned}$$

Evaluating Algebraic Expressions

To **evaluate** an algebraic expression, substitute numerical values for each of the variables in the expression. Here are some examples.

Expression	*Value of Variable*	*Substitute*	*Value of Expression*
$3x + 2$	$x = 2$	$3(2) + 2$	$6 + 2 = 8$
$4x^2 + 2x - 1$	$x = -1$	$4(-1)^2 + 2(-1) - 1$	$4 - 2 - 1 = 1$
$2x(x + 4)$	$x = -2$	$2(-2)(-2 + 4)$	$2(-2)(2) = -8$

Example 12 ■ Evaluating Algebraic Expressions

Evaluate each algebraic expression when $x = -2$ and $y = 5$.

(a) $2y - 3x$ (b) $5 + x^2$ (c) $5 - x^2$

Solution

(a) When $x = -2$ and $y = 5$, the expression $2y - 3x$ has a value of
$$\begin{aligned}
2y - 3x &= 2(5) - 3(-2) \\
&= 10 + 6 \\
&= 16.
\end{aligned}$$

Substitute for x and y.

Simplify.

Simplify.

(b) When $x = -2$, the expression $5 + x^2$ has a value of
$$\begin{aligned}
5 + x^2 &= 5 + (-2)^2 \\
&= 5 + 4 \\
&= 9.
\end{aligned}$$

Substitute for x.

Simplify.

Simplify.

(c) When $x = -2$, the expression $5 - x^2$ has a value of
$$\begin{aligned}
5 - x^2 &= 5 - (-2)^2 \\
&= 5 - 4 \\
&= 1.
\end{aligned}$$

Substitute for x.

Simplify.

Simplify.

Example 13 ■ Evaluating Algebraic Expressions

Evaluate each algebraic expression when $x = 2$ and $y = -1$.

(a) $x^2 - 2xy + y^2$ (b) $|y - x|$ (c) $\dfrac{2xy}{5x + y}$

Solution

(a) When $x = 2$ and $y = -1$, the expression $x^2 - 2xy + y^2$ has a value of

$$x^2 - 2xy + y^2 = 2^2 - 2(2)(-1) + (-1)^2 \qquad \text{Substitute for } x \text{ and } y.$$
$$= 4 + 4 + 1 = 9. \qquad \text{Simplify.}$$

(b) When $x = 2$ and $y = -1$, the expression $|y - x|$ has a value of

$$|y - x| = |-1 - 2| \qquad \text{Substitute for } x \text{ and } y.$$
$$= |-3| = 3. \qquad \text{Simplify.}$$

(c) When $x = 2$ and $y = -1$, the expression $2xy/(5x + y)$ has a value of

$$\frac{2xy}{5x + y} = \frac{2(2)(-1)}{5(2) + (-1)} \qquad \text{Substitute for } x \text{ and } y.$$
$$= \frac{-4}{10 - 1} = -\frac{4}{9}. \qquad \text{Simplify.}$$

Example 14 ■ Using a Mathematical Model

From 1992 to 1999, the average hourly wage for miners in the United States can be modeled by the expression

$$0.0290t^2 + 0.076t + 14.18, \quad 2 \le t \le 9$$

where $t = 2$ represents 1992. Create a table that shows the average hourly wages for these years.   *(Source: U.S. Bureau of Labor Statistics)*

Solution

To create a table of values that shows the average hourly wages for the years 1992 to 1999, evaluate the expression

$$0.0290t^2 + 0.076t + 14.18$$

for each integer value of t from $t = 2$ to $t = 9$.

Year	1992	1993	1994	1995
t	2	3	4	5
Hourly wage	$14.45	$14.67	$14.95	$15.29

Year	1996	1997	1998	1999
t	6	7	8	9
Hourly wage	$15.68	$16.13	$16.64	$17.21

1.1 ■ Exercises

Developing Skills

In Exercises 1–8, identify the terms of the algebraic expression.

1. $10x + 5$

2. $-16t^2 + 48$

3. $-3y^2 + 2y - 8$

4. $25z^3 - 4.8z^2$

5. $4x^2 - 3y^2 - 5x + 2y$

6. $14u^2 + 25uv - 3v^2$

7. $x^2 - 2.5x - \dfrac{1}{x}$

8. $\dfrac{3}{t^2} - \dfrac{4}{t} + 6$

In Exercises 9–12, identify the coefficient of the term.

9. $5y^3$

10. $4x^6$

11. $-\frac{3}{4}t^2$

12. $-8.4x$

In Exercises 13–22, identify the property of algebra that is illustrated by the statement.

13. $4 - 3x = -3x + 4$

14. $(10 + x) - y = 10 + (x - y)$

15. $-5(2x) = (-5 \cdot 2)x$

16. $(x - 2)(3) = 3(x - 2)$

17. $(x + 5) \cdot \dfrac{1}{x + 5} = 1, \quad x \neq -5$

18. $(x^2 + 1) - (x^2 + 1) = 0$

19. $5(y^3 + 3) = 5y^3 + 5 \cdot 3$

20. $10x^3y + 0 = 10x^3y$

21. $(16t^4) \cdot 1 = 16t^4$

22. $-32(u^2 - 3u) = -32u^2 + 96u$

In Exercises 23–28, use the property to rewrite the expression.

23. (a) Distributive Property

$5(x + 6) =$

(b) Commutative Property of Multiplication

$5(x + 6) =$

24. (a) Distributive Property

$6x + 6 =$

(b) Commutative Property of Addition

$6x + 6 =$

25. (a) Commutative Property of Multiplication

$6(xy) =$

(b) Associative Property of Multiplication

$6(xy) =$

26. (a) Additive Identity Property

$3ab + 0 =$

(b) Commutative Property of Addition

$3ab + 0 =$

27. (a) Additive Inverse Property

$4t^2 + (-4t^2) =$

(b) Commutative Property of Addition

$4t^2 + (-4t^2) =$

28. (a) Associative Property of Addition

$(3 + 6) + (-9) =$

(b) Additive Inverse Property

$9 + (-9) =$

In Exercises 29–38, use the definition of exponents to write the expression as a repeated multiplication.

29. $x^3 \cdot x^4$

30. $z^2 \cdot z^5$

31. $(2y)^3$

32. $(4t)^4$

33. $(-2x)^3$

34. $(-5y)^4$

35. $\left(\dfrac{y}{5}\right)^4$

36. $\left(\dfrac{z}{2}\right)^3$

37. $\left(\dfrac{6}{x}\right)^4$

38. $\left(\dfrac{3}{t}\right)^5$

In Exercises 39–44, write the expression using exponential notation.

39. $(5x)(5x)(5x)(5x)$

40. $(2y)(2y)(2y)(2y)(2y)$

41. $(y \cdot y \cdot y)(y \cdot y \cdot y \cdot y)$

42. $(x \cdot x \cdot x)(y \cdot y \cdot y)$

43. $(-z)(-z)(-z)(-z)(-z)(-z)(-z)$

44. $(-9t)(-9t)(-9t)(-9t)(-9t)(-9t)$

In Exercises 45–90, use the properties of exponents to simplify the expression.

45. $x^5 \cdot x^2$

46. $y^3 \cdot y^4$

47. $(a^2)^4$

48. $(x^5)^3$

49. $\dfrac{x^7}{x^3}$

50. $\dfrac{z^8}{z^5}$

51. $\left(\dfrac{a^2}{b}\right)^2$

52. $\left(\dfrac{x^3}{y^2}\right)^3$

53. $3^3 y^4 \cdot y^2$

54. $6^2 x^3 \cdot x^5$

55. $(-4x)^2$

56. $(-4x)^3$

57. $(-5z^2)^3$

58. $(-5z^3)^2$

59. $(a^4 b^4)^4$

60. $(y^2 z^5)^3$

61. $(x^3)(-x)$

62. $(-z^2)(z^3)$

63. $(2xy)(3x^2 y^3)$

64. $(-5a^2 b^3)(2ab^4)$

65. $\dfrac{3^7 x^5}{3^3 x^3}$

66. $\dfrac{2^4 y^5}{2^2 y^3}$

67. $\dfrac{(2xy)^5}{6(xy)^3}$

68. $\dfrac{4^3(ab)^6}{4(ab)^2}$

69. $(5y)^2(-y^4)$

70. $(3y)^3(2y^2)$

71. $-5z^4(-5z)^4$

72. $(-6n)(-3n^2)$

73. $(-2a)^2(-2a)^2$

74. $(-2a^2)(-8a)$

75. $\dfrac{(2x)^4 y^2}{2x^3 y}$

76. $\dfrac{3a^5 b^7}{a^3(3b)^2}$

77. $\dfrac{6(a^3 b)^3}{(3ab)^2}$

78. $\dfrac{(-3c^5 d^3)^2}{(-2cd)^3}$

79. $\dfrac{-x^2 y^3}{(-xy)^2}$

80. $\dfrac{(-x^2)^2 z^8}{-x^3 z^7}$

81. $-\left(\dfrac{2x^4}{5y}\right)^2$

82. $-\left(\dfrac{3a^3}{2b^5}\right)^3$

83. $\dfrac{x^{n+1}}{x^n}$

84. $\dfrac{a^{m+3}}{a^3}$

85. $(x^n)^4$

86. $(a^3)^k$

87. $x^n \cdot x^3$

88. $y^m \cdot y^2$

89. $\dfrac{r^n s^m}{rs^3}$

90. $\left(\dfrac{x^{2n} y^m}{x^4 y^3}\right)^2$

In Exercises 91–102, simplify the expression by combining like terms.

91. $3x + 4x$

92. $-2x^2 + 4x^2$

93. $9y - 5y + 4y$

94. $8y + 7y - y$

95. $3x - 2y + 5x + 20y$

96. $-2a + \frac{1}{3}b - 7a - b$

97. $8z^2 + \frac{3}{2}z - \frac{5}{2}z^2 + 10$

98. $-5y^3 + 3y - 6y^2 + 8y^3 + y - 4$

99. $2uv + 5u^2 v^2 - uv - (uv)^2$

100. $3m^2 n^2 - 4mn - n(5m) + 2(mn)^2$

101. $5(ab)^2 + 2ab - 4ab$

102. $3xy - xy + 8$

In Exercises 103–120, simplify the algebraic expression.

103. $10(x - 3) + 2x - 5$

104. $3(x^2 + 1) + x^2 - 6$

105. $(4x + 1) - (2x + 2)$

106. $(7y^2 + 5) - (8y^2 + 4)$

107. $-(3z^2 - 2z + 4) + (z^2 - z - 2)$

108. $(t^2 + 10t - 3) - (5t - 8)$

109. $-3(y^2 + 3y - 1) + 2(y - 5)$

110. $5(a + 6) - 4(a^2 - 2a - 1)$

111. $4[5 - 3(x^2 + 10)]$

112. $2[5x^2 - (x^3 + 5)]$

113. $2[3(b - 5) - (b^2 + b + 3)]$

114. $-[4(t + 1) - (t^2 - 2t - 5)]$

115. $y^2(y + 1) + y(y^2 + 1)$

116. $z^2(z^4 - z^2) + 4z^3(z + 1)$

117. $x(xy^2 + y) - 2xy(xy + 1)$

118. $2ab(b^2 - 3) - ab(b^2 + 2)$

119. $-2a(3a^2)^3 + \dfrac{9a^8}{3a}$

120. $5y^3 + \dfrac{4y^5}{2y^2} - (7y)y^2$

In Exercises 121–136, evaluate the algebraic expression for the specified values of the variable(s). If not possible, state the reason.

Expression	Values	
121. $5 - 3x$	(a) $x = 5$	(b) $x = \frac{2}{3}$
122. $\frac{3}{2}x - 2$	(a) $x = 6$	(b) $x = -3$
123. $10 - |x|$	(a) $x = 3$	(b) $x = -3$
124. $-|x| + 6$	(a) $x = -4$	(b) $x = 5$
125. $3x^2 - x + 7$	(a) $x = -1$	(b) $x = \frac{1}{3}$
126. $2x^2 + 5x - 3$	(a) $x = -3$	(b) $x = \frac{1}{2}$

Expression	Values
127. $\dfrac{x}{x^2 + 1}$	(a) $x = 0$ (b) $x = 3$
128. $5 - \dfrac{3}{x}$	(a) $x = 0$ (b) $x = -6$
129. $3x + 2y$	(a) $x = 1,\ y = 5$
	(b) $x = -6,\ y = -9$
130. $4x - y$	(a) $x = 2,\ y = 0$
	(b) $x = -2,\ y = -5$
131. $x^2 + 3xy - y^2$	(a) $x = -1,\ y = 2$
	(b) $x = -6,\ y = -3$
132. $x^2 - xy + y^2$	(a) $x = 2,\ y = -1$
	(b) $x = -3,\ y = -2$

Expression	Values
133. $|y - x|$	(a) $x = 2,\ y = 5$
	(b) $x = -2,\ y = -2$
134. $|x - y|$	(a) $x = 0,\ y = 10$
	(b) $x = 4,\ y = 4$
135. rt	(a) $r = 40,\ t = 5\frac{1}{4}$
	(b) $r = 35,\ t = 4$
136. Prt	(a) $P = \$5000,\ r = 0.085,$ $t = 10$
	(b) $P = \$750,\ r = 0.07,$ $t = 3$

Solving Problems

Geometry   In Exercises 137–140, (a) write and simplify an expression for the perimeter of the region and (b) write and simplify an expression for the area of the region.

137.

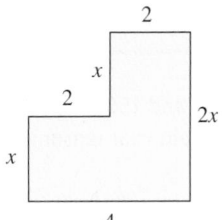

138.

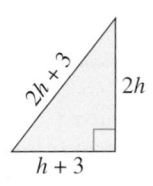

139.

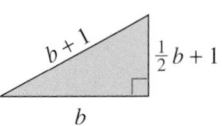

140.

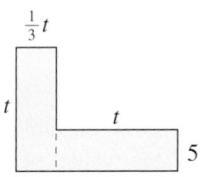

Geometry   In Exercises 141–144, write an expression for the area of the figure. Then evaluate the expression for the given value of the variable.

141. $b = 15$

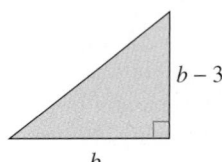

142. $x = 3$

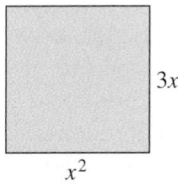

143. $h = 12$

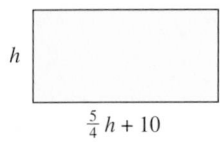

144. $y = 20$

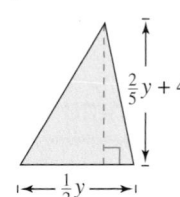

Using a Model   In Exercises 145 and 146, use the model, which approximates the annual sales (in millions of dollars) of recreational vehicles in the United States from 1992 through 1999 (see figure).

$$\text{Sales} = 51.65t^2 + 82.8t + 4142,\ 2 \le t \le 9$$

In this formula, $t = 2$ represents 1992. *(Source: National Sporting Goods Association)*

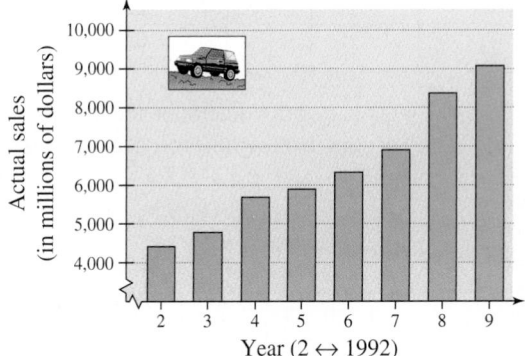

145. Graphically approximate the sales of recreational vehicles in 1996. Then use the model to confirm your estimate algebraically.

146. Graphically approximate the sales of recreational vehicles in 1998. Then use the model to confirm your estimate algebraically.

147. *Geometry*   The area of the trapezoid with parallel bases of lengths b_1 and b_2, and height h, is $\frac{1}{2}(b_1 + b_2)h$. Use the Distributive Property to show that the area can also be expressed as $b_1h + \frac{1}{2}(b_2 - b_1)h$.

148. *Geometry*   Use both formulas given in Exercise 147 to find the area of a trapezoid with $b_1 = 7$, $b_2 = 12$, and $h = 3$.

149. *Area of a Rectangle*   The figure shows two adjoining rectangles. Demonstrate the Distributive Property by filling in the blanks to express the total area of the two rectangles in two ways.

$$\boxed{}\ (\ \boxed{}\ +\ \boxed{}\ ) = \boxed{}\ +\ \boxed{}$$

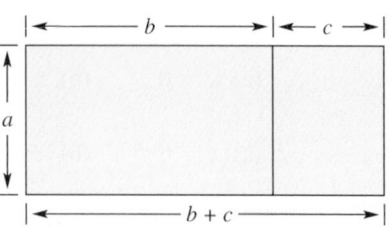

Figure for 149

150. *Area of a Rectangle*   The figure shows two adjoining rectangles. Demonstrate the "subtraction version" of the Distributive Property by filling in the blanks to express the area of the left rectangle in two ways.

$$\boxed{}\ (\ \boxed{}\ -\ \boxed{}\ ) = \boxed{}\ -\ \boxed{}$$

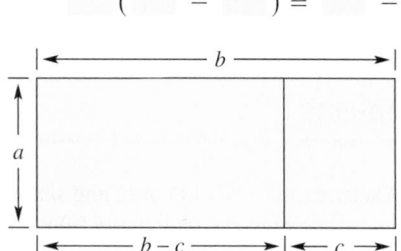

Explaining Concepts

151. Explain the difference between constants and variables in an algebraic expression.

152. Write, from memory, the properties of exponents.

153. Explain the difference between $(2x)^3$ and $2x^3$.

True or False?   In Exercises 154 and 155, decide whether the statement is true or false. Explain your reasoning.

154. $(3a)b = 3(ab)$

155. $2x(5 - y) = 2x(5) + 2x(-y)$

Looking Further

(a) Complete the table by evaluating $2x - 5$.

x	-1	0	1	2	3	4
$2x - 5$						

(b) From the table in part (a), determine the increase in the value of the expression for each one-unit increase in x.

(c) Complete the table by using the *table* feature of a graphing utility to evaluate $3x + 2$.

x	-1	0	1	2	3	4
$3x + 2$						

(d) From the table in part (c), determine the increase in the value of the expression for each one-unit increase in x.

(e) Use the results in parts (a) through (d) to make a conjecture about the increase in the algebraic expression $7x + 4$ for each one-unit increase in x. Use the *table* feature of a graphing utility to confirm your result.

(f) Use the results in parts (a) through (d) to make a conjecture about the increase in the algebraic expression $-3x + 1$ for each one-unit increase in x. Use the *table* feature of a graphing utility to confirm your result.

(g) In general, what does the coefficient of the x-term represent in the expressions above?

1.2 Operations with Polynomials

What you should learn:

- How to identify the leading coefficients and degrees of polynomials
- How to add and subtract polynomials
- How to multiply polynomials
- How to find special products
- How to use operations with polynomials in application problems

Why you should learn it:

Operations with polynomials enable you to model various aspects of the physical world, such as the position of a free-falling object, as shown in Exercises 163–168 on page 27.

Basic Definitions

An algebraic expression containing only terms of the form ax^k, where a is any real number and k is a nonnegative integer, is called a **polynomial in one variable** or simply a **polynomial.** Here are some examples of polynomials in one variable.

$$3x - 8, \quad x^4 + 3x^3 - x^2 - 8x + 1, \quad x^3 + 5, \quad \text{and} \quad 9x^5$$

In the term ax^k, a is called the **coefficient,** and k the **degree,** of the term. Note that the degree of the term ax is 1, and the degree of a constant term is 0. Because a polynomial is an algebraic *sum*, the coefficients take on the signs between the terms. For instance,

$$x^3 - 4x^2 + 3 = (1)x^3 + (-4)x^2 + (0)x + 3$$

has coefficients $1, -4, 0,$ and 3. Polynomials are usually written in order of descending powers of the variable. This is referred to as **standard form.** For example, the standard form of $3x^2 - 5 - x^3 + 2x$ is

$$-x^3 + 3x^2 + 2x - 5. \qquad \text{Standard form}$$

The **degree of a polynomial** is defined as the degree of the term with the highest power, and the coefficient of this term is called the **leading coefficient** of the polynomial. For instance, the polynomial

$$-3x^4 + 4x^2 + x + 7$$

is of fourth degree and its leading coefficient is -3.

Definition of Polynomial in *x*

Let $a_n, \ldots, a_2, a_1, a_0$ be real numbers and let n be a *nonnegative integer*. A **polynomial in *x*** is an expression of the form

$$a_n x^n + a_{n-1} x^{n-1} + \cdots + a_2 x^2 + a_1 x + a_0$$

where $a_n \neq 0$. The polynomial is of **degree *n*,** and the number a_n is called the **leading coefficient.** The number a_0 is called the **constant term.**

The following are *not* polynomials, for the reasons stated.

- The expression

    $$2x^{-1} + 5$$

    is not a polynomial because the exponent in $2x^{-1}$ is negative.

- The expression

    $$x^3 + 3x^{1/2}$$

    is not a polynomial because the exponent in $3x^{1/2}$ is not an integer.

Example 1 ■ Identifying Leading Coefficients and Degrees

Write the polynomial in standard form and identify the degree and leading coefficient of the polynomial.

(a) $5x^2 - 2x^7 + 4 - 2x$ (b) $16 - 8x^3$ (c) $5 + x^4 - 6x^3$

Solution

Polynomial	Standard Form	Degree	Leading Coefficient
(a) $5x^2 - 2x^7 + 4 - 2x$	$-2x^7 + 5x^2 - 2x + 4$	7	-2
(b) $16 - 8x^3$	$-8x^3 + 16$	3	-8
(c) $5 + x^4 - 6x^3$	$x^4 - 6x^3 + 5$	4	1

A polynomial with only one term is a **monomial.** Polynomials with two *unlike* terms are **binomials,** and those with three *unlike* terms are **trinomials.** Here are some examples.

   Monomial: $5x^3$ *Binomial:* $-4x + 3$ *Trinomial:* $2x^2 + 3x - 7$

NOTE   The prefix *mono* means one, the prefix *bi* means two, and the prefix *tri* means three.

Example 2 ■ Evaluating a Polynomial

Find the value of $x^3 - 5x^2 + 6x - 3$ when $x = 4$.

Solution

When $x = 4$, the value of $x^3 - 5x^2 + 6x - 3$ is

$$x^3 - 5x^2 + 6x - 3 = 4^3 - 5(4)^2 + 6(4) - 3 \qquad \text{Substitute 4 for } x.$$
$$= 64 - 80 + 24 - 3 \qquad \text{Evaluate terms.}$$
$$= 5 \qquad \text{Simplify.}$$

Adding and Subtracting Polynomials

To add two polynomials, simply combine like terms. This can be done in either a horizontal or a vertical format, as shown in Examples 3 and 4.

Example 3 ■ Adding Polynomials Horizontally

Use a horizontal format to add $2x^3 + x^2 - 5$ and $x^2 + x + 6$.

Solution

$$(2x^3 + x^2 - 5) + (x^2 + x + 6) \qquad \text{Write original polynomials.}$$
$$= (2x^3) + (x^2 + x^2) + (x) + (-5 + 6) \qquad \text{Group like terms.}$$
$$= 2x^3 + 2x^2 + x + 1 \qquad \text{Combine like terms.}$$

To use a vertical format to add polynomials, align the terms of the polynomials by their degrees, as shown in the following example.

Example 4 ■ Using a Vertical Format to Add Polynomials

Use a vertical format to add $(5x^3 + 2x^2 - x + 7)$, $(3x^2 - 4x + 7)$, and $(-x^3 + 4x^2 - 8)$.

Solution

$$
\begin{array}{r}
5x^3 + 2x^2 - x + 7 \\
3x^2 - 4x + 7 \\
-x^3 + 4x^2 \phantom{{}- 4x} - 8 \\
\hline
4x^3 + 9x^2 - 5x + 6
\end{array}
$$

To subtract one polynomial from another, *add the opposite*. You can do this by changing the sign of each term of the polynomial that is being subtracted and then adding the resulting like terms.

Example 5 ■ Subtracting Polynomials Horizontally

Use a horizontal format to subtract $x^3 + 2x^2 - x - 4$ from $3x^3 - 5x^2 + 3$.

Solution

$(3x^3 - 5x^2 + 3) - (x^3 + 2x^2 - x - 4)$	Write original polynomials.
$= 3x^3 - 5x^2 + 3 - x^3 - 2x^2 + x + 4$	Add the opposite.
$= (3x^3 - x^3) + (-5x^2 - 2x^2) + (x) + (3 + 4)$	Group like terms.
$= 2x^3 - 7x^2 + x + 7$	Combine like terms.

Be especially careful to get the correct signs when you are subtracting one polynomial from another. One of the most common mistakes in algebra is to forget to change signs correctly when subtracting one expression from another. Here is an example.

Wrong sign

$$(x^2 - 2x + 3) - (x^2 + 2x - 2) \neq x^2 - 2x + 3 - x^2 + 2x - 2$$

Common error

Wrong sign

Study Tip

The common error illustrated to the right is forgetting to change two of the signs in the polynomial that is being subtracted. When subtracting polynomials, remember to add the *opposite* of every term of the subtracted polynomial.

Example 6 ■ Using a Vertical Format to Subtract Polynomials

Use a vertical format to subtract $3x^4 - 2x^3 + 3x - 4$ from $4x^4 - 2x^3 + 5x^2 - x + 8$.

Solution

$$
\begin{array}{r}
(4x^4 - 2x^3 + 5x^2 - x + 8) \\
-(3x^4 - 2x^3 \phantom{{}+ 5x^2} + 3x - 4) \\
\hline
\end{array}
\quad\Longrightarrow\quad
\begin{array}{r}
4x^4 - 2x^3 + 5x^2 - x + 8 \\
-3x^4 + 2x^3 \phantom{{}+ 5x^2} - 3x + 4 \\
\hline
x^4 \phantom{{}- 2x^3} + 5x^2 - 4x + 12
\end{array}
$$

Multiplying Polynomials

The simplest type of polynomial multiplication involves a monomial multiplier. The product is obtained by direct application of the Distributive Property. For instance, to multiply the monomial $3x$ by the polynomial $(2x^2 - 5x + 3)$, multiply *each* term of the polynomial by $3x$.

$$(3x)(2x^2 - 5x + 3) = (3x)(2x^2) - (3x)(5x) + (3x)(3)$$
$$= 6x^3 - 15x^2 + 9x$$

Example 7 ■ Finding Products with Monomial Multipliers

Multiply the polynomial by the monomial.

(a) $(2x - 7)(-3x)$ (b) $4x^2(-2x^3 + 3x + 1)$

Solution

(a) $(2x - 7)(-3x) = 2x(-3x) - 7(-3x)$ Distributive Property

$= -6x^2 + 21x$ Properties of exponents

(b) $4x^2(-2x^3 + 3x + 1) = 4x^2(-2x^3) + 4x^2(3x) + 4x^2(1)$ Distributive Property

$= -8x^5 + 12x^3 + 4x^2$ Properties of exponents

To multiply two binomials, you can use both (left and right) forms of the Distributive Property. For example, if you treat the binomial $(2x + 7)$ as a single quantity, you can multiply $(3x - 2)$ by $(2x + 7)$ as follows.

$$(3x - 2)(2x + 7) = 3x(2x + 7) - (2)(2x + 7)$$
$$= (3x)(2x) + (3x)(7) - (2)(2x) - (2)(7)$$
$$= 6x^2 + 21x - 4x - 14$$

Product of First terms	Product of Outer terms	Product of Inner terms	Product of Last terms

$$= 6x^2 + 17x - 14$$

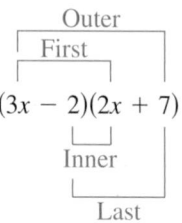

FOIL Diagram

The four products in the boxes above suggest that you can put the product of two binomials in the FOIL form in just one step. This is called the **FOIL Method.** Note that the words *first, outer, inner,* and *last* refer to the positions of the terms in the original product.

Example 8 ■ Multiplying Binomials (Distributive Property)

Use the Distributive Property to multiply $x + 2$ by $x - 3$.

Solution

$$(x + 2)(x - 3) = x(x - 3) + 2(x - 3)$$ Distributive Property
$$= x^2 - 3x + 2x - 6$$ Distributive Property
$$= x^2 - x - 6$$ Combine like terms.

Example 9 ■ Multiplying Binomials (FOIL Method)

Use the FOIL method to multiply the binomials.

(a) $(x - 3)(x - 9)$ (b) $(3x + 4)(2x + 1)$

Solution

$$
\begin{array}{cccc}
\text{F} & \text{O} & \text{I} & \text{L}
\end{array}
$$

(a) $(x - 3)(x - 9) = x^2 - 9x - 3x + 27 = x^2 - 12x + 27$

$$
\begin{array}{cccc}
\text{F} & \text{O} & \text{I} & \text{L}
\end{array}
$$

(b) $(3x + 4)(2x + 1) = 6x^2 + 3x + 8x + 4 = 6x^2 + 11x + 4$

To multiply two polynomials that have three or more terms, you can use the same basic principle that you use when multiplying monomials and binomials. That is, *each term of one polynomial must be multiplied by each term of the other polynomial.* This can be done using either a horizontal or a vertical format.

Example 10 ■ Multiplying Polynomials (Horizontal Format)

$$
\begin{aligned}
(4x^2 &- 3x - 1)(2x - 5) \\
&= 4x^2(2x - 5) - 3x(2x - 5) - 1(2x - 5) && \text{Distributive Property} \\
&= 8x^3 - 20x^2 - 6x^2 + 15x - 2x + 5 && \text{Distributive Property} \\
&= 8x^3 - 26x^2 + 13x + 5 && \text{Combine like terms.}
\end{aligned}
$$

When multiplying two polynomials, it is best to write each in standard form before using either the horizontal or vertical format. This is illustrated in the next example.

Example 11 ■ Multiplying Polynomials (Vertical Format)

Write the polynomials in standard form and use a vertical format to multiply.

$$(4x^2 + x - 2)(5 + 3x - x^2)$$

Solution

With a vertical format, line up like terms in the same vertical columns, much as you align digits in whole-number multiplication.

$$
\begin{array}{r}
4x^2 + x - 2 \\
\times \quad -x^2 + 3x + 5 \\
\hline
20x^2 + 5x - 10 \\
12x^3 + 3x^2 - 6x \\
-4x^4 - x^3 + 2x^2 \\
\hline
-4x^4 + 11x^3 + 25x^2 - x - 10
\end{array}
$$

Standard form
Standard form
$5(4x^2 + x - 2)$
$3x(4x^2 + x - 2)$
$-x^2(4x^2 + x - 2)$

EXPLORATION

Use the FOIL Method to find the product of

$$(x + a)(x - a)$$

where a is a constant.
What do you notice about the number of terms in your product? What degree are the terms in your product?

Polynomials are often written with exponents. As shown in the next example, the properties of algebra are used to simplify these expressions.

Example 12 ■ Multiplying Polynomials

Expand $(x - 4)^3$.

Solution

$$
\begin{aligned}
(x - 4)^3 &= (x - 4)(x - 4)(x - 4) &&\text{Write each factor.}\\
&= [(x - 4)(x - 4)](x - 4) &&\text{Associative Property of Multiplication}\\
&= (x^2 - 4x - 4x + 16)(x - 4) &&\text{Find } (x - 4)(x - 4).\\
&= (x^2 - 8x + 16)(x - 4) &&\text{Combine like terms.}\\
&= x^2(x - 4) - 8x(x - 4) + 16(x - 4) &&\text{Distributive Property}\\
&= x^3 - 4x^2 - 8x^2 + 32x + 16x - 64 &&\text{Distributive Property}\\
&= x^3 - 12x^2 + 48x - 64 &&\text{Combine like terms.}
\end{aligned}
$$

Example 13 ■ An Area Model for Multiplying Polynomials

Show that $(x + 2)(2x + 1) = 2x^2 + 5x + 2$.

Solution

An appropriate area model to demonstrate the multiplication of two binomials would be $A = lw$, the area formula for a rectangle. Think of a rectangle whose sides are $x + 2$ and $2x + 1$. The area of this rectangle is

$$(x + 2)(2x + 1). \qquad \text{Area = (width)(length)}$$

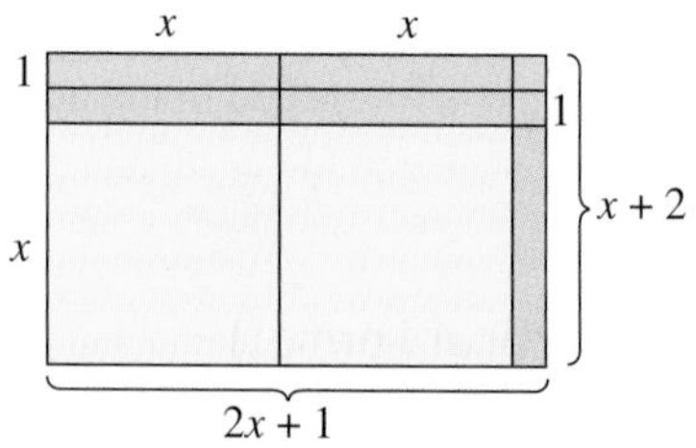

Another way to find the area is to add the areas of the rectangular parts, as shown in Figure 1.2. There are two squares whose sides are x, five rectangles whose sides are x and 1, and two squares whose sides are 1. The total area of these nine rectangles is

$$2x^2 + 5x + 2. \qquad \text{Area = sum of rectangular areas}$$

Because each method must produce the same area, you can conclude that

$$(x + 2)(2x + 1) = 2x^2 + 5x + 2.$$

Figure 1.2

Special Products

Some binomial products have special forms that occur frequently in algebra. For instance, the product $(x + 3)(x - 3)$ is called the **product of the sum and difference of two terms.** With such products, the two middle terms cancel, as follows.

$$
\begin{aligned}
(x + 3)(x - 3) &= x^2 - 3x + 3x - 9 &&\text{Sum and difference of two terms}\\
&= x^2 - 9 &&\text{Product has no middle term.}
\end{aligned}
$$

Another common type of product is the **square of a binomial.** With this type of product, the middle term is always twice the product of the terms in the binomial.

$$(2x + 5)^2 = (2x + 5)(2x + 5)$$ Square of a binomial

$$= 4x^2 + 10x + 10x + 25$$ Outer and inner terms are equal.

$$= 4x^2 + 20x + 25$$ Middle term is twice the product of the terms in the binomial.

Special Products

Let u and v be real numbers, variables, or algebraic expressions. Then the following formulas are true.

Sum and Difference of Two Terms   *Example*

$(u + v)(u - v) = u^2 - v^2$ $(3x - 4)(3x + 4) = (3x)^2 - 4^2$

$$= 9x^2 - 16$$

Square of a Binomial   *Example*

$(u + v)^2 = u^2 + 2uv + v^2$ $(4x + 9)^2 = (4x)^2 + 2(4x)(9) + 9^2$

$$= 16x^2 + 72x + 81$$

$(u - v)^2 = u^2 - 2uv + v^2$ $(x - 6)^2 = x^2 - 2(x)(6) + 6^2$

$$= x^2 - 12x + 36$$

The square of a binomial can also be demonstrated geometrically. Consider a square, each of whose sides are of length $a + b$. (See Figure 1.3). The total area includes one square of area a^2, two rectangles of area ab each, and one square of area b^2. So, the total area is $a^2 + 2ab + b^2$.

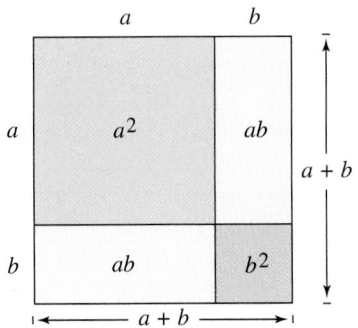

Figure 1.3

Example 14 ■ Finding Special Products

Multiply the polynomials.

(a) $(3x - 2)(3x + 2)$

(b) $(2x - 7)^2$

(c) $[(a - 2) + b]^2$

Solution

(a) $(3x - 2)(3x + 2) = (3x)^2 - 2^2$ Special product

$$= 9x^2 - 4$$ Simplify.

(b) $(2x - 7)^2 = (2x)^2 - 2(2x)(7) + 7^2$ Special product

$$= 4x^2 - 28x + 49$$ Simplify.

(c) $[(a - 2) + b]^2 = (a - 2)^2 + 2(a - 2)b + b^2$ Special product

$$= a^2 - 4a + 4 + 2ab - 4b + b^2$$ Simplify.

Applications

There are many applications that require the evaluation of polynomials. One commonly used second-degree polynomial is called a **position polynomial.** This polynomial has the form

$$-16t^2 + v_0 t + s_0 \qquad \text{Position polynomial}$$

where t is the time, measured in seconds. The value of this polynomial gives the height (in feet) of a free-falling object above the ground, assuming no air resistance. The coefficient of t, v_0, is called the **initial velocity** of the object, and the constant term, s_0, is called the **initial height** of the object. If the initial velocity is positive, the object was projected upward (at $t = 0$), and if the initial velocity is negative, the object was projected downward.

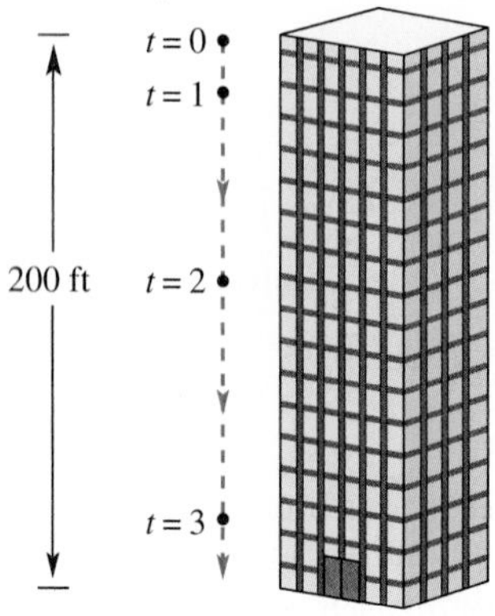

Example 15 ■ Finding the Height of a Free-Falling Object

An object is thrown downward from the top of a 200-foot building. The initial velocity is -10 feet per second. Use the position polynomial

$$-16t^2 - 10t + 200$$

to find the height of the object when $t = 1$, $t = 2$, and $t = 3$ (see Figure 1.4).

Solution

When $t = 1$, the height of the object is

$$\begin{aligned}
\text{Height} &= -16(1)^2 - 10(1) + 200 \\
&= -16 - 10 + 200 \\
&= 174 \text{ feet.}
\end{aligned}$$

When $t = 2$, the height of the object is

$$\begin{aligned}
\text{Height} &= -16(2)^2 - 10(2) + 200 \\
&= -64 - 20 + 200 \\
&= 116 \text{ feet.}
\end{aligned}$$

When $t = 3$, the height of the object is

$$\begin{aligned}
\text{Height} &= -16(3)^2 - 10(3) + 200 \\
&= -144 - 30 + 200 \\
&= 26 \text{ feet.}
\end{aligned}$$

The figure in the left margin shows a building with time markers $t = 0$, $t = 1$, $t = 2$, $t = 3$ and a height of 200 ft.

Figure 1.4

NOTE In Example 15, the initial velocity is -10 feet per second. The value is negative because the object was thrown downward. If it had been thrown upward, the initial velocity would have been positive. If it had been dropped, the initial velocity would have been zero.

Use your calculator to determine the height of the object in Example 15 when $t = 3.2368$. What can you conclude?

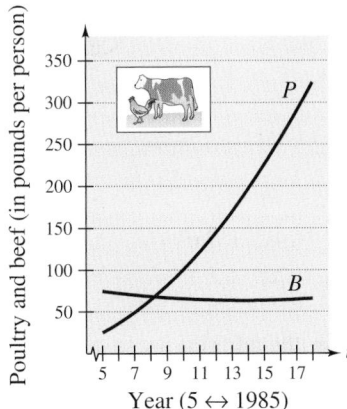

Poultry and beef (in pounds per person)

Year (5 ↔ 1985)

Figure 1.5

Example 16 ■ Using Polynomial Models

The numbers of pounds of poultry P and of beef B consumed per person in the United States from 1985 to 1998 (see Figure 1.5) can be modeled by

$$P = -0.084t^2 + 3.43t + 30.5, \; 5 \le t \le 18 \qquad \text{Poultry (pounds per person)}$$
$$B = 0.143t^2 - 3.95t + 90.4, \; 5 \le t \le 18 \qquad \text{Beef (pounds per person)}$$

where $t = 5$ represents 1985. Find a model that represents the total amount T of poultry *and* beef consumed from 1985 to 1998. Estimate the total amount T consumed in 1996.   *(Source: U.S. Department of Agriculture)*

Solution

The sum of the two polynomial models would be

$$P + B = (-0.084t^2 + 3.43t + 30.5) + (0.143t^2 - 3.95t + 90.4)$$
$$= 0.059t^2 - 0.52t + 120.9.$$

The model for the total consumption of poultry and beef is

$$T = P + B$$
$$= 0.059t^2 - 0.52t + 120.9 \qquad \text{Total (pounds per person)}$$

Using this model, and substituting $t = 16$, you can estimate the 1996 consumption to be

$$T = 0.059(16)^2 - 0.52(16) + 120.9.$$
$$= 127.7 \text{ pounds per person.}$$

Example 17 ■ Geometry: Finding the Area of a Shaded Region

Find an expression for the area of the shaded portion of the figure.

$2x + 5$

$x + 3$

$x - 3$

$x + 1$

Solution

First find the area of the large rectangle A_1 and the area of the small rectangle A_2.

$$A_1 = (2x + 5)(x + 1) \quad \text{and} \quad A_2 = (x + 3)(x - 3)$$

Then to find the area A of the shaded portion, subtract A_2 from A_1.

$$A = A_1 - A_2 \qquad \text{Write formula.}$$
$$= (2x + 5)(x + 1) - (x + 3)(x - 3) \qquad \text{Substitute.}$$
$$= 2x^2 + 7x + 5 - (x^2 - 9) \qquad \text{Use FOIL Method.}$$
$$= 2x^2 + 7x + 5 - x^2 + 9 \qquad \text{Distributive Property.}$$
$$= x^2 + 7x + 14 \qquad \text{Combine like terms.}$$

1.2 ■ Exercises

Developing Skills

In Exercises 1–12, write the polynomial in standard form, and find its degree and leading coefficient.

1. $10x - 4$
2. $3x^2 + 8$
3. $5 - 3y^4$
4. $-3x^3 - 2x^2 - 3$
5. $8z - 16z^2$
6. $35t - 16t^2$
7. $6t + 4t^5 - t^2 + 3$
8. $10 + 3x^2 - 15x^5 - 7x$
9. $x - 5 + x^3 - 5x^2$
10. $16 + z^2 - 8z - 4z^3$
11. x
12. -4

In Exercises 13–18, determine whether the polynomial is a monomial, binomial, or trinomial.

13. $12 - 5y^2$
14. t^3
15. $x^3 + 2x^2 - 4$
16. $2u^7 - 9u^3$
17. $1.3x^2$
18. $2 + x^4 - 4z^2$

In Exercises 19–26, give an example of a polynomial in one variable satisfying the conditions. (*Note:* There is more than one correct answer.)

19. A monomial of degree 3
20. A trinomial of degree 3
21. A trinomial of degree 4 and leading coefficient -2
22. A binomial of degree 2 and leading coefficient 8
23. A monomial of degree 1 and leading coefficient 7
24. A binomial of degree 5 and leading coefficient -3
25. A monomial of degree 0
26. A monomial of degree 2 and leading coefficient 9

In Exercises 27–30, find the values of the polynomial at the given values of the variable.

27. $x^3 - 12x$ (a) $x = -2$ (b) $x = 0$
28. $\frac{1}{4}x^4 - 2x^2$ (a) $x = 2$ (b) $x = -2$
29. $x^4 - 4x^3 + 16x - 16$ (a) $x = -1$ (b) $x = \frac{5}{2}$
30. $3t^4 + 4t^3$ (a) $t = 1$ (b) $t = -\frac{2}{3}$

In Exercises 31–34, perform the addition using a horizontal format.

31. $(2x^2 - 3) + (5x^2 + 6)$

32. $(3x^3 - 2x + 8) + (3x - 5)$
33. $(x^2 - 3x + 8) + (2x^2 - 4x) + 3x^2$
34. $(5y + 6) + (4y^2 - 6y - 3)$

In Exercises 35–38, perform the addition using a vertical format.

35. $(5x^2 - 3x + 4) + (-3x^2 - 4)$
36. $(4x^3 - 2x^2 + 8x) + (4x^2 + x - 6)$
37. $(2b - 3) + (b^2 - 2b) + (7 - b^2)$
38. $(v^2 + v - 3) + (4v + 1) + (2v^2 - 3v)$

In Exercises 39–42, perform the subtraction using a horizontal format.

39. $(3x^2 - 2x + 1) - (2x^2 + x - 1)$
40. $(5y^4 - 2) - (3y^4 + 2)$
41. $(10x^3 + 15) - (6x^3 - x + 11)$
42. $(y^2 + 3y^4) - (y^4 - y^2)$

In Exercises 43–46, perform the subtraction using a vertical format.

43. $(x^2 - x + 3) - (x - 2)$
44. $(3z^2 + z) - (z^3 + 2z^2 + z)$
45. $(-2x^3 - 15x + 25) - (2x^3 - 13x + 12)$
46. $(0.2t^4 - 5t^2) - (-t^4 + 0.3t^2 - 1.4)$

In Exercises 47–64, perform the operations.

47. $(3x^2 + 8) + (7 - 5x^2)$
48. $(20s - 12s^2 - 32) + (15s^2 + 6s)$
49. $(4x^2 + 5x - 6) - (2x^2 - 4x + 5)$
50. $(13x^3 - 9x^2 + 4x - 5) - (5x^3 + 7x + 3)$
51. $(10x^2 - 11) - (-7x^3 - 12x^2 - 15)$
52. $(15y^4 - 18y - 18) - (-11y^4 - 8y - 8)$
53. $5s - [6s - (30s + 8)]$
54. $3x^2 - 2[3x + (9 - x^2)]$
55. $(8x^3 - 4x^2 + 3x) - [(x^3 - 4x^2 + 5) + (x - 5)]$
56. $(5y^2 - 2y) - [(y^2 + y) - (3y^2 - 6y + 2)]$
57. $[5(2x^3 + 1) - (3x^3 - 12x^2 + 4x + 2)] +$
$3(x^2 + 2x - 1)$

58. $2[(y^2 + 3y - 9) + 3(4y + 4)] -$
$\qquad\qquad 5(y^2 + 2y - 3)$

59. $2(t^2 + 12) - 5(t^2 + 5) + 6(t^2 + 5)$

60. $-10(v + 2) + 8(v - 1) - 3(v - 9)$

61. $(2z^2 + z - 11) + 3(z^2 + 4z + 5) -$
$\qquad\qquad 2(2z^2 - 5z + 10)$

62. $7(3t^4 + 2t^2 - t) - (5t^4 + 9t^2 - 4t) +$
$\qquad\qquad 3(8t^2 + 5t)$

63. $2(5x^3 - 13x) + (4x^3 - 9x^2 + 3x) -$
$\qquad\qquad 3(x^3 - 2x^2 + 6x - 5)$

64. $5(t^3 + 2t^2 - t - 8) - (3t^3 - t^2 - 4t + 2) +$
$\qquad\qquad 4(2t^2 + 3t - 1) - (t^3 + 1)$

In Exercises 65–68, use a calculator to perform the operations.

65. $8.04x^2 - 9.37x^2 + 5.62x^2$

66. $-11.98y^3 + 4.63y^3 - 6.79y^3$

67. $(4.098a^2 + 6.349a) - (11.246a^2 - 9.342a)$

68. $(27.433k^2 - 19.018k) + (-14.61k^2 + 3.814k)$

In Exercises 69–96, perform the multiplication and simplify.

69. $(-2a^2)(-8a)$

70. $(-6n)(3n^2)$

71. $2y(5 - y)$

72. $5z(2z - 7)$

73. $4x^3(2x^2 - 3x + 5)$

74. $3y^2(-3y^2 + 7y - 3)$

75. $-2x^2(5 + 3x^2 - 7x^3)$

76. $-3a^2(11a - 3)$

77. $(x + 7)(x - 4)$

78. $(y - 2)(y + 3)$

79. $(5 - x)(3 + x)$

80. $(2 - y)(4 - y)$

81. $(2t - 1)(t + 8)$

82. $(3z + 5)(2z - 7)$

83. $(3a^4 + 5)(2a^4 - 7)$

84. $(8b^5 - 1)(3b^5 - 2)$

85. $(2x + y)(3x + 2y)$

86. $(2x - y)(3x - 2y)$

87. $(5x^2 - 3y)(x^2 + y)$

88. $(a^3 + 2b^5)(4a^3 + 3b^5)$

89. $\left(4y - \frac{1}{3}\right)(12y + 9)$

90. $\left(5t - \frac{3}{4}\right)(2t - 16)$

91. $-3x(-5x)(5x + 2)$

92. $4t(-3t)(t^2 - 1)$

93. $5a(a + 2) - 3a(2a - 3)$

94. $4x(2x - 1) + 9x(x - 3)$

95. $(2t - 1)(t + 1) + 3(2t - 5)$

96. $5(8y + 3) + (2y - 1)(y - 7)$

In Exercises 97–100, perform the multiplication using a horizontal format.

97. $(x^3 - 3x + 2)(x - 2)$

98. $(t + 3)(t^2 - 5t + 1)$

99. $(u + 5)(2u^2 + 3u - 4)$

100. $(x - 1)(x^2 - 4x + 6)$

In Exercises 101–104, perform the multiplication using a vertical format.

101. $(7x^2 - 14x + 9)(x + 3)$

102. $(4x^4 - 6x^2 + 9)(2x^2 + 3)$

103. $(-x^2 + 2x - 1)(2x + 1)$

104. $(2s^2 - 5s + 6)(3s - 4)$

In Exercises 105–138, perform the multiplication.

105. $(x - 4)(x + 4)$

106. $(y + 7)(y - 7)$

107. $(a - 6c)(a + 6c)$

108. $(8n - m)(8n + m)$

109. $(2t + 9)(2t - 9)$

110. $(5z - 1)(5z + 1)$

111. $\left(2x - \frac{1}{4}\right)\left(2x + \frac{1}{4}\right)$

112. $\left(\frac{2}{3}x + 7\right)\left(\frac{2}{3}x - 7\right)$

113. $(0.2t + 0.5)(0.2t - 0.5)$

114. $(4a - 0.1b)(4a + 0.1b)$

115. $(x^3 + 4)(x^3 - 4)$

116. $(a^5 - 3)(a^5 + 3)$

117. $(x + 5)^2$

118. $(x + 9)^2$

119. $(5x - 2)^2$

120. $(3x - 8)^2$

121. $(2a + 3b)^2$

122. $(4x - 5y)^2$

123. $(2x^4 - 3)^2$

124. $(y^7 + 4z)^2$

125. $(x^2 + 4)(x^2 - 2x - 4)$

126. $(2x^2 - 3)(2x^2 - 2x + 3)$

127. $(t^2 + 5t - 1)(2t^2 - 5)$

128. $(2z^2 - 3z - 7)(3z + 4)$

129. $(a + 5)^3$

130. $(y - 2)^3$

131. $(2x - 3)^3$

132. $(3y + 4)^3$

133. $(a^2 + 9a - 5)(a^2 - a + 3)$

134. $(t^2 - 2t - 7)(2t^2 + 8t - 3)$

135. $[(x + 2) - y]^2$

136. $[(x - 4) + y]^2$

137. $[2z + (y + 1)]^2$

138. $[u - (v - 3)]^2$

In Exercises 139–142, perform the operations and simplify.

139. $(x + 3)(x - 3) - (x^2 + 8x - 2)$

140. $(k - 8)(k + 8) - (k^2 - k + 3)$

141. $(t + 3)^2 - (t - 3)^2$

142. $(a + 6)^2 + (a - 6)^2$

Solving Problems

Geometry In Exercises 143–146, find an expression for the perimeter or circumference of the figure.

143.

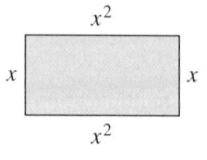

144.

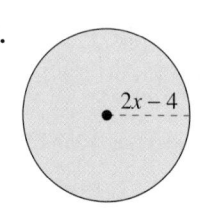

145.

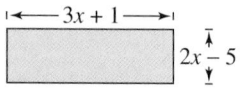

146.

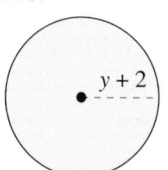

Geometry In Exercises 147–152, find an expression for the area of the shaded portion of the figure.

147.

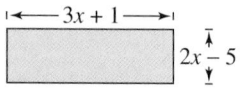

148.

149.

150.

151.

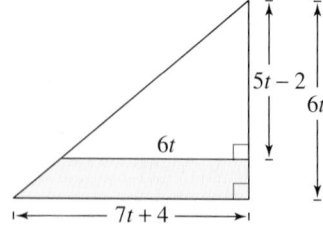

152.

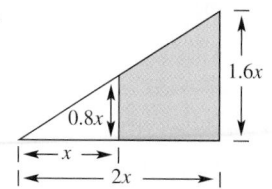

Geometric Modeling In Exercises 153–156, (a) perform the multiplication algebraically and (b) use a geometric area model to verify your solution to part (a).

153. $x(x + 3)$ **154.** $2y(y + 1)$

155. $(t + 3)(t + 2)$ **156.** $(2z + 5)(z + 1)$

Geometric Modeling In Exercises 157 and 158, use the area model to write two different expressions for the total area. Then equate the two expressions and name the algebraic property that is illustrated.

157.

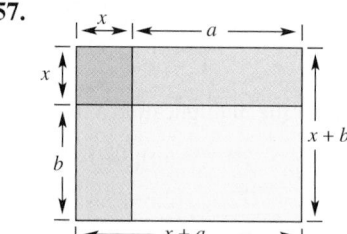

158.

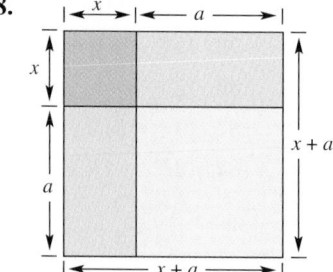

159. *Geometry* The length of a rectangle is $1\frac{1}{2}$ times its width w. Find expressions for (a) the perimeter and (b) the area of the rectangle.

160. *Geometry* The base of a triangle is $3x$ and its height is $x + 5$. Find an expression for the area A of the triangle.

161. *Personal Finance* After 2 years, an investment of $1000 compounded annually at an interest rate of r will yield an amount $1000(1 + r)^2$. Find this product.

162. *Personal Finance*   After 2 years, an investment of $1000 compounded annually at an interest rate of 9.5% will yield an amount $1000(1 + 0.095)^2$. Find this product.

Free-Falling Object   In Exercises 163–166, use the position polynomial to determine whether the free-falling object was dropped, thrown upward, or thrown downward. Also determine the height of the object at time $t = 0$.

163. $-16t^2 + 100$ **164.** $-16t^2 + 50t$

165. $-16t^2 - 24t + 50$ **166.** $-16t^2 + 32t + 300$

167. *Free-Falling Object*   An object is thrown upward from the top of a 200-foot building (see figure). The initial velocity is 40 feet per second. Use the position polynomial $-16t^2 + 40t + 200$ to find the height of the object when $t = 1$, $t = 2$, and $t = 3$.

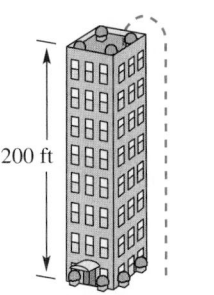 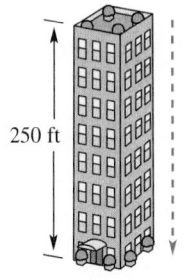

Figure for 167   Figure for 168

168. *Free-Falling Object*   An object is thrown downward from the top of a 250-foot building (see figure). The initial velocity is -25 feet per second. Use the position polynomial $-16t^2 - 25t + 250$ to find the height of the object when $t = 1$, $t = 2$, and $t = 3$.

Explaining Concepts

169. Explain why $x^2 - 3\sqrt{x}$ is not a polynomial.

170. Explain why $(x + y)^2$ is not equal to $x^2 + y^2$.

171. Is every trinomial a second-degree polynomial? Explain.

172. Can two third-degree polynomials be added to produce a second-degree polynomial? If so, give an example.

173. Perform the multiplications.

(a) $(x - 1)(x + 1)$ (b) $(x - 1)(x^2 + x + 1)$

(c) $(x - 1)(x^3 + x^2 + x + 1)$

From the pattern formed by these products, can you predict the result of $(x - 1)(x^4 + x^3 + x^2 + x + 1)$?

Looking Further

The per capita consumption (average consumption per person) of whole milk W and lowfat milk L in the United States between 1990 and 1999 can be approximated by these two polynomial models.

$W = 0.024t^2 - 0.46t + 10.2$ $0 \le t \le 8$ Whole milk

$L = -0.016t^2 + 0.35t + 4.9$ $0 \le t \le 8$ Low-fat milk

In these models, W and L represent the average consumption per person in gallons and t represents the year, with $t = 0$ corresponding to 1990. *(Source: U.S. Department of Agriculture)*

(a) Find a polynomial model that represents the per capita consumption of milk (of both types) during this time period. Use this model to find the per capita consumption of milk in 1993 and in 1997.

(b) During the given period, the per capita consumption of whole milk was decreasing and the per capita consumption of lowfat milk was increasing (see figure). Was the combined per capita consumption of milk increasing or decreasing?

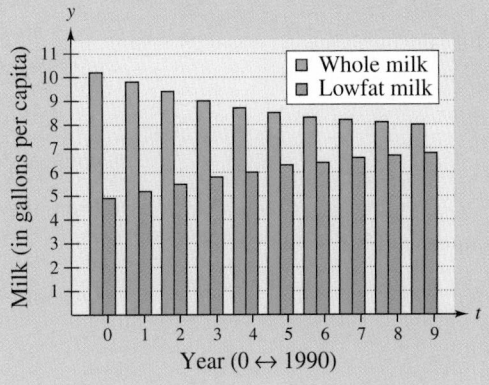

1.3 Factoring Polynomials

What you should learn:

- How to factor polynomials with common factors
- How to factor polynomials by grouping terms
- How to factor the difference of two squares
- How to factor the sum or difference of two cubes
- How to factor polynomials completely

Why you should learn it:

In some cases, factoring a polynomial enables you to determine unknown quantities. For example, in Exercise 113 on page 36, you will factor the expression for the area of a rectangle to determine the length of the rectangle.

Factoring Polynomials with Common Factors

Now we will switch from the process of multiplying polynomials to the *reverse* process—**factoring polynomials.** This section and the next section deal only with polynomials that have integer coefficients. Remember that in Section 1.1 you used the Distributive Property to *multiply* and *remove* parentheses, as follows.

$$3x(4 - 5x) = 12x - 15x^2 \qquad \text{Distributive Property}$$

In this and the next section, you will use the Distributive Property in the reverse direction to *factor* and *create* parentheses.

$$12x - 15x^2 = 3x(4 - 5x) \qquad \text{Distributive Property}$$

Factoring an expression (by the Distributive Property) changes a *sum of terms* into a *product of factors*. Later you will see that this is an important strategy for solving equations and for simplifying algebraic expressions.

To be efficient in factoring, you need to understand the concept of the **greatest common factor** (or **GCF**). Recall from arithmetic that every integer can be factored into a product of prime numbers. The *greatest common factor* of two or more integers is the greatest integer that divides evenly into each integer. To find the greatest common factor of two integers or two expressions, begin by writing each as a product of prime factors. The greatest common factor is the product of the *common* prime factors. For instance, from the factorizations

$$18 = 2 \cdot 3 \cdot 3 \qquad 42 = 2 \cdot 3 \cdot 7$$

you can see that the common prime factors of 18 and 42 are 2 and 3. So, it follows that the greatest common factor is $2 \cdot 3$ or 6.

Example 1 ■ Finding the Greatest Common Factor

Find the greatest common factors of (a) 36, $8x$, $64y$ and (b) $6x^5$, $30x^4$, $12x^3$.

Solution

(a) From the factorizations

$$36 = 2 \cdot 2 \cdot 3 \cdot 3 = 4(9)$$
$$8x = 2 \cdot 2 \cdot 2 \cdot x = 4(2x)$$
$$64y = 2 \cdot 2 \cdot 2 \cdot 2 \cdot 2 \cdot 2 \cdot y = 4(16y)$$

you can conclude that the greatest common factor is 4.

(b) From the factorizations

$$6x^5 = 2 \cdot 3 \cdot x \cdot x \cdot x \cdot x \cdot x = (6x^3)(x^2)$$
$$30x^4 = 2 \cdot 3 \cdot 5 \cdot x \cdot x \cdot x \cdot x = (6x^3)(5x)$$
$$12x^3 = 2 \cdot 2 \cdot 3 \cdot x \cdot x \cdot x = (6x^3)(2)$$

you can conclude that the greatest common factor is $6x^3$.

Consider the three terms given in Example 1(b) as terms of the polynomial

$$6x^5 + 30x^4 + 12x^3.$$

The greatest common factor, $6x^3$, of these terms is called the **greatest common monomial factor** of the polynomial. When you use the Distributive Property to remove this factor from each term of the polynomial, you are **factoring out** the greatest common monomial factor.

$$6x^5 + 30x^4 + 12x^3 = 6x^3(x^2) + 6x^3(5x) + 6x^3(2) \qquad \text{Factor each term.}$$
$$= 6x^3(x^2 + 5x + 2) \qquad \text{Factor out common monomial factor.}$$

If a polynomial in x (with integer coefficients) has a greatest common monomial factor of the form ax^n, the statements below must be true.

1. The coefficient a of the greatest common monomial factor must be the greatest integer that *divides* each of the coefficients in the polynomial.

2. The variable factor x^n of the greatest common monomial factor has the *lowest* power of x of all terms of the polynomial.

Example 2 ■ Factoring out a Greatest Common Monomial Factor

Factor out the greatest common monomial factor from $24x^3 - 32x^2$.

Solution

For the terms $24x^3$ and $32x^2$, 8 is the greatest integer of 24 and 32 and x^2 is the highest-powered variable factor common to x^3 and x^2. So, the greatest common monomial factor of $24x^3$ and $32x^2$ is $8x^2$. You can factor the given polynomial as follows.

$$24x^3 - 32x^2 = (8x^2)(3x) - (8x^2)(4) = 8x^2(3x - 4)$$

The greatest common monomial factor of a polynomial is usually considered to have a positive coefficient. However, sometimes it is convenient to factor a negative number out of a polynomial, as shown in the next example.

Example 3 ■ A Negative Common Monomial Factor

Factor the polynomial $-3x^2 + 12x - 18$ in two ways.

(a) Factor out a 3.   (b) Factor out a -3.

Study Tip

Whenever you are factoring a polynomial, remember that you can check your results by multiplying. That is, if you multiply the factors, you should obtain the original polynomial.

Solution

(a) By factoring out the common monomial factor of 3, you obtain

$$-3x^2 + 12x - 18 = 3(-x^2) + 3(4x) + 3(-6) \qquad \text{Factor each term.}$$
$$= 3(-x^2 + 4x - 6). \qquad \text{Factored form}$$

(b) By factoring out the common monomial factor of -3, you obtain

$$-3x^2 + 12x - 18 = -3(x^2) + (-3)(-4x) + (-3)(6) \qquad \text{Factor each term.}$$
$$= -3(x^2 - 4x + 6). \qquad \text{Factored form}$$

Factoring by Grouping

Some polynomials have common factors that are not simple monomials. For instance, the polynomial $x^2(2x - 3) + 4(2x - 3)$ has the common *binomial* factor $(2x - 3)$. Factoring out this common factor produces

$$x^2(2x - 3) + 4(2x - 3) = (2x - 3)(x^2 + 4).$$

This type of factoring is part of a more general procedure called **factoring by grouping.**

Example 4 ■ A Common Binomial Factor

Factor the polynomial $5x^2(6x - 5) - 2(6x - 5)$.

Solution

Each of the terms of this polynomial has a binomial factor of $(6x - 5)$. Factoring this binomial out of each term produces

$$5x^2(6x - 5) - 2(6x - 5) = (6x - 5)(5x^2 - 2).$$

In Example 4, the given polynomial was already grouped, and so it was easy to determine the common binomial factor. In practice, you will have to do the grouping as well as the factoring. To see how this works, consider the expression $x^3 - 3x^2 - 5x + 15$ and try to *factor* it. Note first that there is no common monomial factor to take out of all four terms. But suppose you *group* the first two terms together and the last two terms together. Then you have

$$\begin{aligned} x^3 - 3x^2 - 5x + 15 &= (x^3 - 3x^2) - (5x - 15) && \text{Group terms.}\\ &= x^2(x - 3) - 5(x - 3) && \text{Factor out common monomial factor in each group.}\\ &= (x - 3)(x^2 - 5). && \text{Factor out common binomial factor.} \end{aligned}$$

Example 5 ■ Factoring by Grouping

Factor the polynomial $x^3 - 5x^2 + x - 5$.

Solution

$$\begin{aligned} x^3 - 5x^2 + x - 5 &= (x^3 - 5x^2) + (x - 5) && \text{Group terms.}\\ &= x^2(x - 5) + 1(x - 5) && \text{Factor out common monomial factor in each group.}\\ &= (x - 5)(x^2 + 1) && \text{Factored form} \end{aligned}$$

Note that in Example 5 the polynomial is factored by grouping the first and second terms and the third and fourth terms. You could just as easily have grouped the first and third terms and the second and fourth terms, as follows.

$$\begin{aligned} x^3 - 5x^2 + x - 5 &= (x^3 + x) - (5x^2 + 5)\\ &= x(x^2 + 1) - 5(x^2 + 1)\\ &= (x^2 + 1)(x - 5) \end{aligned}$$

Factoring the Difference of Two Squares

Some polynomials have special forms that you should learn to recognize so that they can be factored easily. Here are some examples of forms that you should be able to recognize by the time you have completed this section.

$$x^2 - 9 = (x + 3)(x - 3)$$ Difference of two squares

$$x^3 + 8 = (x + 2)(x^2 - 2x + 4)$$ Sum of two cubes

$$x^3 - 1 = (x - 1)(x^2 + x + 1)$$ Difference of two cubes

One of the easiest special polynomial forms to recognize and to factor is the form $u^2 - v^2$, called a **difference of two squares.** This form arises from the special product $(u + v)(u - v)$ in Section 1.2.

Difference of Two Squares

Let u and v be real numbers, variables, or algebraic expressions. Then the expression $u^2 - v^2$ can be factored using the following pattern.

$$u^2 - v^2 = (u + v)(u - v)$$

Difference   Opposite signs

To recognize perfect squares, look for coefficients that are squares of integers and for variables raised to *even* powers. Here are some examples.

Original Polynomial		*Think: Difference of Squares*		*Factored Form*
$x^2 - 4$	⇨	$x^2 - 2^2$	⇨	$(x + 2)(x - 2)$
$4x^2 - 25$	⇨	$(2x)^2 - 5^2$	⇨	$(2x + 5)(2x - 5)$
$25 - 49x^2$	⇨	$5^2 - (7x)^2$	⇨	$(5 + 7x)(5 - 7x)$
$x^{10} - 16y^2$	⇨	$(x^5)^2 - (4y)^2$	⇨	$(x^5 + 4y)(x^5 - 4y)$

Example 6 ■ Factoring the Difference of Two Squares

Factor (a) $x^2 - 64$ and (b) $4y^2 - 49$.

Solution

(a) Because x^2 and 64 are both perfect squares, you can recognize this polynomial as the difference of two squares. So, the polynomial factors as follows.

$$x^2 - 64 = x^2 - 8^2$$ Write as difference of two squares.

$$= (x + 8)(x - 8)$$ Factored form

(b) Because both $4y^2$ and 49 are perfect squares, you can recognize the difference of two squares. So, the polynomial factors as follows.

$$4y^2 - 49 = (2y)^2 - 7^2$$ Write as difference of two squares.

$$= (2y + 7)(2y - 7)$$ Factored form

Example 7 ■ Factoring the Difference of Two Squares

Factor the polynomial $49x^2 - 81y^2$.

Solution

Because $49x^2$ and $81y^2$ are both perfect squares, you can recognize this polynomial as the difference of two squares. So, the polynomial factors as follows.

$$49x^2 - 81y^2 = (7x)^2 - (9y)^2 \qquad \text{Write as difference of two squares.}$$
$$= (7x + 9y)(7x - 9y) \qquad \text{Factored form}$$

Remember that the rule $u^2 - v^2 = (u + v)(u - v)$ applies to polynomials or expressions in which u and v are themselves expressions. The next example illustrates this possibility.

Example 8 ■ Factoring the Difference of Two Squares

Factor the difference of two squares.

(a) $(x + 2)^2 - 9$ (b) $(x - 5)^2 - 16$

Solution

(a) $(x + 2)^2 - 9 = (x + 2)^2 - 3^2$ Write as difference of two squares.

$= [(x + 2) + 3][(x + 2) - 3]$ Factored form

$= (x + 5)(x - 1)$ Simplify.

(b) $(x - 5)^2 - 16 = (x - 5)^2 - 4^2$ Write as difference of two squares.

$= [(x - 5) + 4][(x - 5) - 4]$ Factored form

$= (x - 1)(x - 9)$ Simplify.

To check this result, write the original polynomial in standard form. Then multiply the factored form to see that you obtain the same standard form.

Factoring the Sum or Difference of Two Cubes

The last type of special factoring discussed in this section is factoring of the sum or difference of two cubes. The patterns for these two special forms are summarized below. In these patterns, pay particular attention to the signs of the terms.

Sum or Difference of Two Cubes

Let u and v be real numbers, variables, or algebraic expressions. Then the expressions $u^3 + v^3$ and $u^3 - v^3$ can be factored as follows.

Like signs   Like signs

$$u^3 + v^3 = (u + v)(u^2 - uv + v^2) \qquad u^3 - v^3 = (u - v)(u^2 + uv + v^2)$$

Unlike signs   Unlike signs

Study Tip

Remember that you can check a factoring result by multiplying. For instance, you can check the first two results in Example 9 as follows.

(a)
$$
\begin{array}{r}
x^2 + 5x + 25 \\
x - 5 \\
\hline
-5x^2 - 25x - 125 \\
x^3 + 5x^2 + 25x \\
\hline
x^3 \qquad\qquad - 125
\end{array}
$$

(b)
$$
\begin{array}{r}
4y^2 - 2y + 1 \\
2y + 1 \\
\hline
4y^2 - 2y + 1 \\
8y^3 - 4y^2 + 2y \\
\hline
8y^3 \qquad\qquad + 1
\end{array}
$$

EXPLORATION

Find a formula for completely factoring $u^6 - v^6$ using the formulas from this section. Use your formula to factor completely $x^6 - 1$ and $x^6 - 64$.

Example 9 ■ Factoring Sums and Differences of Cubes

Factor each polynomial.

(a) $x^3 - 125$ (b) $8y^3 + 1$ (c) $27x^3 - 64y^3$

Solution

(a) This polynomial is the difference of two cubes, because x^3 is the cube of x and 125 is the cube of 5. So, you can factor the polynomial as follows.

$$
\begin{aligned}
x^3 - 125 &= x^3 - 5^3 && \text{Write as difference of two cubes.} \\
&= (x - 5)(x^2 + 5x + 5^2) && \text{Factored form} \\
&= (x - 5)(x^2 + 5x + 25) && \text{Simplify.}
\end{aligned}
$$

(b) This polynomial is the sum of two cubes, because $8y^3$ is the cube of $2y$ and 1 is its own cube. So, you can factor the polynomial as follows.

$$
\begin{aligned}
8y^3 + 1 &= (2y)^3 + 1^3 && \text{Write as sum of two cubes.} \\
&= (2y + 1)[(2y)^2 - (2y)(1) + 1^2] && \text{Factored form} \\
&= (2y + 1)(4y^2 - 2y + 1) && \text{Simplify.}
\end{aligned}
$$

(c) Because $27x^3$ is the cube of $3x$ and $64y^3$ is the cube of $4y$, this polynomial is the difference of two cubes. So, you can factor the polynomial as follows.

$$
\begin{aligned}
27x^3 - 64y^3 &= (3x)^3 - (4y)^3 && \text{Write as difference of two cubes.} \\
&= (3x - 4y)[(3x)^2 + (3x)(4y) + (4y)^2] && \text{Factored form} \\
&= (3x - 4y)(9x^2 + 12xy + 16y^2) && \text{Simplify.}
\end{aligned}
$$

Factoring Completely

Sometimes the difference of two squares can be hidden by the presence of a common monomial factor. Remember that with *all* factoring techniques, you should first factor out any common monomial factors.

Example 10 ■ Factoring out a Common Monomial Factor First

Factor the polynomial $125x^2 - 80$ completely.

Solution

The polynomial $125x^2 - 80$ has a common monomial factor of 5. After factoring out this factor, the remaining polynomial is the difference of two squares.

$$
\begin{aligned}
125x^2 - 80 &= 5(25x^2 - 16) && \text{Factor out common monomial factor.} \\
&= 5[(5x)^2 - 4^2] && \text{Write as difference of two squares.} \\
&= 5(5x + 4)(5x - 4) && \text{Factored form}
\end{aligned}
$$

The polynomial in Example 10 is said to be **completely factored** because none of its factors can be factored further using integer coefficients.

Example 11 ■ Factoring out a Common Monomial Factor First

Factor the polynomial $3x^4 + 81x$ completely.

Solution

$$3x^4 + 81x = 3x(x^3 + 27) \qquad \text{Factor out common monomial factor.}$$
$$= 3x(x^3 + 3^3) \qquad \text{Write as sum of two cubes.}$$
$$= 3x(x + 3)(x^2 - 3x + 9) \qquad \text{Factored form}$$

To factor a polynomial completely, always check to see whether the factors obtained might themselves be factorable. For instance, after factoring the polynomial $x^4 - 16$ once as the difference of two squares

$$x^4 - 16 = (x^2)^2 - 4^2 = (x^2 + 4)(x^2 - 4)$$

you can see that the second factor is itself the difference of two squares. So, to factor the polynomial *completely*, you must continue factoring, as follows.

$$x^4 - 16 = (x^2 + 4)(x^2 - 4) = (x^2 + 4)(x + 2)(x - 2)$$

Example 12 ■ Factoring Completely

Factor (a) $x^4 - y^4$ and (b) $81m^4 - 1$ completely.

Solution

(a) Recognizing $x^4 - y^4$ as a difference of two squares, you can write

$$x^4 - y^4 = (x^2)^2 - (y^2)^2 = (x^2 + y^2)(x^2 - y^2).$$

Note that the second factor $(x^2 - y^2)$ is itself a difference of two squares and you therefore obtain

$$x^4 - y^4 = (x^2 + y^2)(x^2 - y^2) = (x^2 + y^2)(x + y)(x - y).$$

(b) Recognizing $81m^4 - 1$ as a difference of two squares, you can write

$$81m^4 - 1 = (9m^2)^2 - (1)^2 = (9m^2 + 1)(9m^2 - 1).$$

Note that the second factor $(9m^2 - 1)$ is itself a difference of two squares and you therefore obtain

$$81m^4 - 1 = (9m^2 + 1)(9m^2 - 1) = (9m^2 + 1)(3m + 1)(3m - 1).$$

> **NOTE**   Note in Example 12 that the *sum of two squares* does not factor further. A second-degree polynomial that is the sum of two squares, such as $x^2 + y^2$ or $9m^2 + 1$, is *not factorable*, which means that it is not factorable *using integer coefficients*.

Example 13 ■ Using Factoring to Find the Length of a Rectangle

The area of a rectangle of width $5x$ is given by $15x^2 + 40x$, as shown in Figure 1.6. Factor this expression to determine the length of the rectangle.

Solution

Factoring out the common monomial factor of $5x$ from the area polynomial $15x^2 + 40x$ results in the polynomial $3x + 8$. So, the length of the rectangle is $3x + 8$. To check this result, multiply $5x(3x + 8)$ to see that you obtain the area polynomial $15x^2 + 40x$.

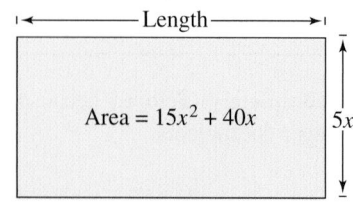

Figure 1.6

> **Collaborate!**
>
> **Factoring Higher-Degree Polynomials**
>
> Have each person in your group create a cubic polynomial by multiplying three polynomials of the form
>
> $$(x - a)(x + a)(x + b).$$
>
> Then write the result in standard form on a piece of paper and give it to another person in the group to factor. When each person has factored his or her polynomial, compare the results with the original factors used to create the polynomials.

1.3 ■ Exercises

Developing Skills

In Exercises 1–12, find the greatest common factor of the expressions.

1. 48, 90

2. 36, 150, 100

3. $3x^2$, $12x$

4. $27x^4$, $18x^3$

5. $30z^2$, $12z^3$

6. $45y$, $150y^3$

7. $28ab^2$, $14a^2b^3$, $42a^2b^5$

8. $16x^2y$, $84xy^2$, $36x^2y^2$

9. $42(x + 8)^2$, $63(x + 8)^3$

10. $66(3 - y)$, $44(3 - y)^2$

11. $4x(1 - z)^2$, $x^2(1 - z)^3$

12. $2(x + 5)$, $8(x + 5)$

In Exercises 13–34, factor out the greatest common monomial factor. (Some of the polynomials have no common monomial factor other than 1 or -1.)

13. $8z - 8$

14. $5x + 5$

15. $24x^2 - 18$

16. $14z^3 + 21$

17. $2x^2 + x$

18. $-a^3 - 4a$

19. $21u^2 - 14u$

20. $36y^4 + 24y^2$

21. $11u^2 + 9$

22. $16x^2 - 3y^3$

23. $3x^2y^2 - 15y$

24. $4uv + 6u^2v^2$

25. $28x^2 + 16x - 8$

26. $9 - 27y - 15y^2$

27. $14x^4 + 21x^3 + 9x^2$

28. $25z^6 - 15z^4 + 35z^3$

29. $17x^5y^3 - xy^2 + 34y^2$

30. $8y^3 + 6y^3z - 2y^2z^2$

31. $9x^3y + 6xy^2$

32. $24y^4z^2 + 12y^2z^2$

33. $3x^3y^3 - 2x^2y^2 + 5xy$

34. $18a^2b^4 + 24a^2b^2 - 12a^2b$

In Exercises 35–42, factor a negative real number from the polynomial and then write the polynomial factor in standard form.

35. $10 - 5x$

36. $32 - 4x$

37. $7 - 14x$

38. $15 - 5x$

39. $8 + 4x - 2x^2$

40. $12x - 6x^2 - 18$

41. $2t - 15 - 4t^2$

42. $16 + 32s^2 - 5s^4$

In Exercises 43–46, factor the expression.

43. $\frac{3}{2}x + \frac{5}{4} = \frac{1}{4}\big(\quad\big)$

44. $\frac{1}{3}x - \frac{5}{6} = \frac{1}{6}\big(\quad\big)$

45. $\frac{5}{6}x + \frac{5}{9}y = \frac{5}{18}\big(\quad\big)$

46. $\frac{7}{12}u - \frac{21}{8}v = \frac{7}{24}\big(\quad\big)$

In Exercises 47–52, factor the polynomial by factoring out the common binomial factor.

47. $2y(y - 3) + 5(y - 3)$

48. $7t(s + 9) - 6(s + 9)$

49. $5t(t^2 + 1) - 4(t^2 + 1)$

50. $3a(a^2 - 3) + 10(a^2 - 3)$

51. $a(a + 6) + (2a - 5)(a + 6)$

52. $(5x + y)(x - y) - 5x(x - y)$

In Exercises 53–64, factor the polynomial by grouping.

53. $y^3 - 6y^2 + 2y - 12$

54. $4x^3 - 2x^2 + 6x - 3$

55. $14z^3 + 21z^2 - 6z - 9$

56. $7t^3 + 5t^2 - 35t - 25$

57. $x^3 + 2x^2 + x + 2$

58. $t^3 - 11t^2 + t - 11$

59. $a^3 - 4a^2 + 2a - 8$

60. $3s^3 + 6s^2 + 5s + 10$

61. $z^4 + 3z^3 - 2z - 6$

62. $4u^4 - 2u^3 - 6u + 3$

63. $cd + 3c - 3d - 9$

64. $u^2 + uv - 4u - 4v$

In Exercises 65–82, factor the difference of two squares.

65. $x^2 - 64$

66. $y^2 - 144$

67. $100 - 9y^2$

68. $625 - 49x^2$

69. $121 - y^2$

70. $81 - z^2$

71. $16y^2 - 9z^2$

72. $9z^2 - 25w^2$

73. $x^2 - 4y^2$

74. $81a^2 - b^6$

75. $a^8 - 36$

76. $y^{10} - 64$

77. $a^2b^2 - 16$

78. $u^2v^2 - 25$

79. $(a + 4)^2 - 49$

80. $(x - 3)^2 - 4$

81. $81 - (z + 5)^2$

82. $100 - (y - 3)^2$

In Exercises 83–94, factor the sum or difference of two cubes.

83. $x^3 - 8$

84. $t^3 - 27$

85. $y^3 + 125$

86. $z^3 + 216$

87. $8t^3 - 27$

88. $64a^3 - 1$

89. $27s^3 + 64$

90. $125x^3 + 343$

91. $8x^3 - y^3$

92. $a^3 - 216b^3$

93. $y^3 + 64z^3$

94. $z^3 + 125w^3$

In Exercises 95–106, factor the polynomial completely.

95. $50x^2 - 8$

96. $3z^2 - 192$

97. $x^3 - 144x$

98. $a^3 - 16a$

99. $b^4 - 16$

100. $a^4 - 625$

101. $y^4 - 81x^4$

102. $u^4 - 256v^4$

103. $2x^3 - 54$

104. $5y^3 - 625$

105. $3a^3 + 192$

106. $7b^3 + 56$

In Exercises 107 and 108, factor the expression. (Assume $n > 0$.)

107. $4x^{2n} - 25$

108. $81 - 16y^{4n}$

In Exercises 109 and 110, show all the different groupings that can be used to factor the polynomial completely. Carry out the various factorizations to show that they yield the same result.

109. $3x^3 + 4x^2 - 3x - 4$

110. $6x^3 - 8x^2 + 9x - 12$

Solving Problems

111. *Personal Finance* The total amount of money from a principal of P invested at $r\%$ simple interest for t years is given by $P + Prt$. Factor this expression.

112. *Business* The revenue from selling x units of a product at a price of p dollars per unit is given by xp. For a particular commodity, the revenue is

$$R = 800x - 0.25x^2.$$

Factor the expression for the revenue and determine an expression for the price in terms of x.

113. *Geometry* The area of a rectangle of width w is given by $32w - w^2$. Factor this expression to determine the length of the rectangle. Draw a diagram to illustrate your answer.

114. *Geometry* The area of a rectangle of length l is given by $45l - l^2$. Factor this expression to deter-

mine the width of the rectangle. Draw a diagram to illustrate your answer.

115. *Geometry* A rectangle has area and width as indicated in the figure. Factor the expression for area to find the length of the rectangle.

$A = w^2 + 8w$ w

?

$A = 27t^2 - 18t$?

$9t$

Figure for 115   Figure for 116

116. *Geometry* A rectangle has area and length as indicated in the figure. Find the width of the rectangle.

117. *Geometry*   The area of a rectangle of width $2x^2$ is given by $10x^3 + 4x^2$. Factor this expression to determine the length of the rectangle. Draw a diagram to illustrate your answer.

118. *Geometry*   The area of a rectangle of length $y + 5$ is given by $3y^2 + 15y$. Factor this expression to determine the width of the rectangle. Draw a diagram to illustrate your answer.

119. *Geometry*   The surface area of a right circular cylinder is $S = 2\pi r^2 + 2\pi rh$ (see figure). Factor the expression for the surface area.

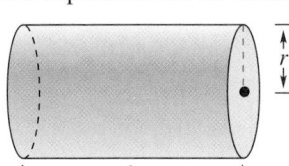

120. *Manufacturing*   A washer on the drive train of a car has an inside radius of r centimeters and an outside radius of R centimeters (see figure). Find the area of one of the flat surfaces of the washer and express the area in factored form.

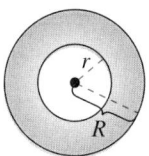

121. *Chemistry*   The rate of change of a certain chemical reaction is $kQx - kx^2$, where Q is the amount of the original substance, x is the amount of substance formed, and k is a constant of proportionality. Factor the expression for this rate of change.

Explaining Concepts

122. Explain the relationship between using the Distributive Property to multiply polynomials and using the Distributive Property to factor algebraic expressions.

123. Explain what is meant by saying that a polynomial is in factored form.

124. Explain how the word *factor* can be used as a noun and as a verb.

In Exercises 125 and 126, explain how the product could be obtained mentally using the sample as a model.

$$48 \cdot 52 = (50 - 2)(50 + 2) = 50^2 - 2^2 = 2496$$

125. $79 \cdot 81 = 6399$ **126.** $18 \cdot 22 = 396$

Looking Further

The cube shown in the figure is formed by four solids: I, II, III, and IV.

(a) Explain how you could determine each expression for volume.

	Volume
Entire cube	a^3
Solid I	$(a - b)a^2$
Solid II	$(a - b)ab$
Solid III	$(a - b)b^2$
Solid IV	b^3

(b) Add the volumes of solids I, II, and III. Factor the result to show that the total volume can be expressed as $(a - b)(a^2 + ab + b^2)$.

(c) Explain why the total volume of solids I, II, and III can also be expressed as $a^3 - b^3$. Then explain how the figure can be used as a geometric model for the *difference of two cubes* factoring pattern.

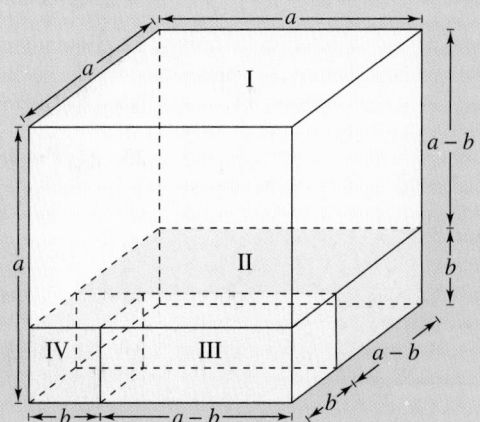

Mid-Chapter Quiz

Take this quiz as you would take a quiz in class. After you are done, check your work against the answers given in the back of the book.

In Exercises 1–3, identify the property of algebra that is illustrated by the statement.

1. $5(x + 3) = 5 \cdot x + 5 \cdot 3$

2. $6y \cdot 1 = 6y$

3. $(z + 3) - (z + 3) = 0$

4. Write the expression $(5x)(5x)(5x)(5x)$ using exponential notation.

In Exercises 5–18, perform the operations and/or simplify the expression.

5. $4x^2 \cdot x^3$

6. $(-2x)^4$

7. $\left(\dfrac{y^2}{3}\right)^3$

8. $\dfrac{18x^2y^3}{12xy}$

9. $4x^2 - 3xy + 5xy - 5x^2$

10. $(4x^2 - x + 7) - (x^2 - 3x - 1)$

11. $3(x - 5) + 4x$

12. $3[x - 2x(x^2 + 1)]$

13. $(6r + 5s)(6r - 5s)$

14. $(2x^2 - 4)(x + 3)$

15. $(2x - 3y)^2$

16. $(x + 1)(x^2 - x + 1)$

17. $2z(z + 5) - 7(z + 5)$

18. $(v - 3)^2 - (v + 3)^2$

19. Evaluate the expression $10x^2 - |5x|$ when (a) $x = 5$ and (b) $x = -1$.

20. Find an expression for the perimeter of the rectangle shown at the left.

21. Write the polynomial $4 + 7x^2 - 16x^4$ in standard form and identify its degree and leading coefficient.

22. Find an expression for the area of the shaded region shown at the left.

In Exercises 23–30, factor the expression completely.

23. $24x^3 + 28x$

24. $3y(y - 5) + (y - 5)$

25. $x^3 - 9x^2 + 5x - 45$

26. $64a^3 + b^3$

27. $-16x^3 + 54$

28. $81x^2 - 16y^2$

29. $x^3 - 9xy^2$

30. $t^3 + 2t^2 - t - 2$

31. An object is thrown upward from the top of a 350-foot building. The initial velocity is 45 feet per second. Use the position polynomial

$$-16t^2 + 45t + 350$$

to find the height of the object when $t = 1$, $t = 2$, and $t = 3$.

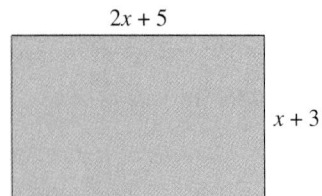

2x + 5

x + 3

Figure for 20

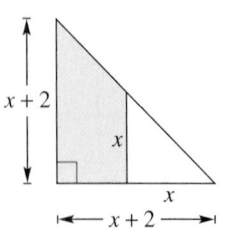

x + 2

x

x

x + 2

Figure for 22

1.4 Factoring Trinomials

What you should learn:

- How to factor trinomials of the form $x^2 + bx + c$
- How to factor trinomials of the form $ax^2 + bx + c$
- How to factor trinomials by grouping
- How to factor perfect square trinomials
- How to select the best factoring technique using the guidelines for factoring polynomials

Why you should learn it:

The techniques for factoring trinomials will help you in solving quadratic equations in Section 1.6.

Factoring Trinomials of the Form $x^2 + bx + c$

Try covering the factored forms in the left-hand column below. Can you determine the factored forms from the trinomial forms?

Factored Form	**F**	**O**	**I**	**L**	*Trinomial Form*

$$(x - 1)(x + 4) = x^2 + 4x - x - 4 = x^2 + 3x - 4$$
$$(x - 3)(x - 2) = x^2 - 2x - 3x + 6 = x^2 - 5x + 6$$
$$(3x + 5)(x + 1) = 3x^2 + 3x + 5x + 5 = 3x^2 + 8x + 5$$

Your goal here is to factor trinomials of the form $x^2 + bx + c$. To begin, consider the factorization

$$x^2 + bx + c = (x + m)(x + n).$$

By multiplying the right-hand side, you obtain the following result.

$$(x + m)(x + n) = x^2 + nx + mx + mn$$
$$= x^2 + (m + n)x + mn$$

Sum of terms ↓   Product of terms ↓

$$= x^2 + \boxed{b}\, x + \boxed{c}$$

So, to *factor* a trinomial $x^2 + bx + c$ into a product of two binomials, you must find two factors of c with a sum of b.

Study Tip

Use a list to help you find the two numbers with the required product and sum. For Example 1(a):

Factors of -8	*Sum*
$1, -8$	-7
$-1, 8$	7
$2, -4$	-2
$-2, 4$	2

Because -2 is the required sum, the correct factorization is

$$x^2 - 2x - 8 = (x - 4)(x + 2).$$

Example 1 ■ Factoring Trinomials

Factor the trinomials (a) $x^2 - 2x - 8$ and (b) $x^2 - 5x + 6$.

Solution

(a) You need to find two factors whose product is -8 and whose sum is -2.

The product of -4 and 2 is -8.

$$x^2 - 2x - 8 = (x - 4)(x + 2)$$

The sum of -4 and 2 is -2.

(b) You need to find two factors whose product is 6 and whose sum is -5.

The product of -3 and -2 is 6

$$x^2 - 5x + 6 = (x - 3)(x - 2)$$

The sum of -3 and -2 is -5.

Note that when the constant term of the trinomial is positive, its factors must have *like* signs; otherwise, its factors have *unlike* signs.

Factors of -24	*Sum of Factors*
$(1)(-24)$	$1 - 24 = -23$
$(-1)(24)$	$-1 + 24 = 23$
$(2)(-12)$	$2 - 12 = -10$
$(-2)(12)$	$-2 + 12 = 10$
$(3)(-8)$	$3 - 8 = -5$
$(-3)(8)$	$-3 + 8 = 5$
$(4)(-6)$	$4 - 6 = -2$
$(-4)(6)$	$-4 + 6 = 2$

When factoring a trinomial of the form $x^2 + bx + c$, if you have trouble finding two factors of c with a sum of b, it may be helpful to list all of the distinct pairs of factors and then choose the appropriate pair from the list. For instance, consider the trinomial

$$x^2 - 5x - 24.$$

For this trinomial, $c = -24$ and $b = -5$. So, you need to find two factors of -24 with a sum of -5, as shown at the left. With experience, you will be able to narrow this list down *mentally* to only two or three possibilities whose sums can then be tested to determine the correct factorization, which is $x^2 - 5x - 24 = (x + 3)(x - 8)$.

Study Tip

With *any* factoring problem, remember that you can check your result by multiplying. For instance, in Example 2, you can check the result by multiplying $(x - 9)$ by $(x + 2)$ to see that you obtain $x^2 - 7x - 18$.

Remember that not all trinomials are factorable using integers. For instance, $x^2 - 2x - 4$ is not factorable using integers because there is no pair of factors of -4 whose sum is -2.

Example 2 ■ Factoring a Trinomial

Factor the trinomial $x^2 - 7x - 18$.

Solution

To factor this trinomial, you need to find two factors whose product is -18 and whose sum is -7.

The product of 2 and -9 is -18.

$$x^2 - 7x - 18 = (x + 2)(x - 9)$$

The sum of 2 and -9 is -7.

Applications of algebra sometimes involve trinomials that have a common monomial factor. To factor such trinomials completely, first factor out the common monomial factor. Then try to factor the resulting trinomial by the methods given in this section.

Example 3 ■ Factoring Completely

Factor the trinomials (a) $4x^3 - 8x^2 - 60x$ and (b) $5x^2y + 20xy^2 + 15y^3$ completely.

Solution

(a) This trinomial has a common monomial factor of $4x$. So, you should start the factoring process by factoring $4x$ out of each term.

$$4x^3 - 8x^2 - 60x = 4x(x^2 - 2x - 15)$$

Factor out common monomial factor.

$$= 4x(x +)(x +)$$

Think: You need two factors of -15 with a sum of -2.

$$= 4x(x + 3)(x - 5)$$

$(3)(-5) = -15$, $3 - 5 = -2$

(b) This trinomial has a common monomial factor of $5y$. So, you should start the factoring process by factoring $5y$ out of each term.

$$5x^2y + 20xy^2 + 15y^3 = 5y(x^2 + 4xy + 3y^2)$$

Factor out common monomial factor.

$$= 5y(x +)(x +)$$

Think: You need two factors of 3 with a sum of 4.

$$= 5y(x + y)(x + 3y)$$

$(1)(3) = 3$, $1 + 3 = 4$

Factoring Trinomials of the Form $ax^2 + bx + c$

To factor a trinomial whose leading coefficient is not 1, use the following pattern.

Factors of a

$$ax^2 + bx + c = (\quad x + \quad)(\quad x + \quad)$$

Factors of c

The goal is to find a combination of factors of a and c such that the outer and inner products add up to the middle term bx. For instance, in the trinomial $6x^2 + 17x + 5$, $a = 6$, $c = 5$, and $b = 17$. After some experimentation, you can determine that the factorization is

$$6x^2 + 17x + 5 = (2x + 5)(3x + 1).$$

Example 4 ■ Factoring a Trinomial of the Form $ax^2 + bx + c$

Factor the trinomial $6x^2 + 5x - 4$.

Solution

First, observe that $6x^2 + 5x - 4$ has no common monomial factor. For this trinomial, you have $ax^2 + bx + c = 6x^2 + 5x - 4$, which implies that $a = 6$, $c = -4$, and $b = 5$. The possible factors of the leading coefficient 6 are $(1)(6)$ and $(2)(3)$, and the possible factors of -4 are $(-1)(4)$, $(1)(-4)$, and $(2)(-2)$. By trying the *many* different combinations of these factors, you obtain the following list.

$$(x + 1)(6x - 4) = 6x^2 + 2x - 4$$
$$(x - 1)(6x + 4) = 6x^2 - 2x - 4$$
$$(x + 4)(6x - 1) = 6x^2 + 23x - 4$$
$$(x - 4)(6x + 1) = 6x^2 - 23x - 4$$
$$(x + 2)(6x - 2) = 6x^2 + 10x - 4$$
$$(x - 2)(6x + 2) = 6x^2 - 10x - 4$$
$$(2x + 1)(3x - 4) = 6x^2 - 5x - 4$$
$$(2x - 1)(3x + 4) = 6x^2 + 5x - 4 \quad \Longleftarrow \quad \text{Correct factorization}$$
$$(2x + 4)(3x - 1) = 6x^2 + 10x - 4$$
$$(2x - 4)(3x + 1) = 6x^2 - 10x - 4$$
$$(2x + 2)(3x - 2) = 6x^2 + 2x - 4$$
$$(2x - 2)(3x + 2) = 6x^2 - 2x - 4$$

So, you can conclude that the correct factorization is

$$6x^2 + 5x - 4 = (2x - 1)(3x + 4).$$

Check this result by multiplying $(2x - 1)$ by $(3x + 4)$.

Study Tip

If the original trinomial has no common monomial factor, its binomial factors cannot have common monomial factors. So, in Example 4, you do not have to test factors, such as $(6x - 4)$, that have a common factor of 2. Which of the other factors in Example 4 did not need to be tested?

To help shorten the list of *possible* factorizations of a trinomial of the form $ax^2 + bx + c$, use the guidelines presented below.

Guidelines for Limiting Possible Trinomial Factorizations

1. If the trinomial has a *common monomial factor*, you should factor out the monomial factor before trying to find binomial factors. For instance, the trinomial $12x^2 + 10x - 8$ has a common factor of 2. By removing this common factor, you obtain $12x^2 + 10x - 8 = 2(6x^2 + 5x - 4)$.

2. Do not switch the signs of the factors of c unless the middle term is correct except in sign. In Example 4, after determining that $(x + 4)(6x - 1)$ is not the correct factorization, for instance, it is unnecessary to test $(x - 4)(6x + 1)$.

3. Do not use binomial factors that have a common monomial factor. Such a factor cannot be correct, because the trinomial has no common monomial factor. (Any common monomial factor should already have been factored out, in accordance with Guideline 1.) For instance, in Example 4, it is unnecessary to test $(x + 1)(6x - 4) = 6x^2 + 2x - 4$ because the factor $(6x - 4)$ has a common factor of 2.

Using these guidelines, you could shorten the list given in Example 4 to the following three possible factorizations.

$$(x + 4)(6x - 1) = 6x^2 + 23x - 4$$
$$(2x + 1)(3x - 4) = 6x^2 - 5x - 4$$
$$(2x - 1)(3x + 4) = 6x^2 + 5x - 4 \quad \Longleftarrow \quad \text{Correct factorization}$$

Do you see why you can cut the list from 12 possible factorizations to only three?

Example 5 ■ Factoring a Trinomial of the Form $ax^2 + bx + c$

Factor the trinomial $2x^2 - x - 21$.

Solution

First observe that $2x^2 - x - 21$ has no common monomial factor. For this trinomial, $a = 2$, which factors as $(1)(2)$, and $c = -21$, which factors as $(1)(-21)$, $(-1)(21)$, $(3)(-7)$, or $(-7)(3)$.

$$(2x + 1)(x - 21) = 2x^2 - 41x - 21$$
$$(2x + 21)(x - 1) = 2x^2 + 19x - 21$$
$$(2x + 3)(x - 7) = 2x^2 - 11x - 21$$
$$(2x + 7)(x - 3) = 2x^2 + x - 21 \qquad\qquad \text{Middle term has incorrect sign.}$$
$$(2x - 7)(x + 3) = 2x^2 - x - 21 \quad \Longleftarrow \quad \text{Correct factorization}$$

So, the correct factorization of $2x^2 - x - 21$ is $(2x - 7)(x + 3)$. Check this result by multiplying $(2x - 7)$ by $(x + 3)$.

NOTE   Remember that if the middle term is correct except in sign, you need only change the signs of the factors of c, as in Example 5.

Example 6 ■ Factoring Trinomials

Factor the trinomials (a) $3x^2 + 11x + 10$ and (b) $5x^2 - 9xy + 4y^2$.

Solution

(a) First, observe that $3x^2 + 11x + 10$ has no common monomial factor. For this trinomial, $a = 3$, which factors as $(1)(3)$, and $c = 10$, which factors as $(1)(10)$ or $(2)(5)$. You can test the possible factors as follows.

$$(x + 10)(3x + 1) = 3x^2 + 31x + 10$$
$$(x + 1)(3x + 10) = 3x^2 + 13x + 10$$
$$(x + 5)(3x + 2) = 3x^2 + 17x + 10$$
$$(x + 2)(3x + 5) = 3x^2 + 11x + 10 \qquad \Longleftarrow \qquad \text{Correct factorization}$$

So, the correct factorization is

$$3x^2 + 11x + 10 = (x + 2)(3x + 5).$$

(b) First observe that $5x^2 - 9xy + 4y^2$ has no common monomial factor. For this trinomial, $a = 5$, which factors as $(1)(5)$, and $c = 4$, which factors as $(1)(4)$ or $(2)(2)$. You can test the possible factors as follows.

$$(x - 4y)(5x - y) = 5x^2 - 21xy + 4y^2$$
$$(x - y)(5x - 4y) = 5x^2 - 9xy + 4y^2 \qquad \Longleftarrow \qquad \text{Correct factorization}$$

So, the correct factorization is

$$5x^2 - 9xy + 4y^2 = (x - y)(5x - 4y).$$

Remember that if a trinomial has a common monomial factor, the common monomial factor should be removed first. This is illustrated in the next two examples.

Example 7 ■ Factoring Completely

Factor $8x^2y - 60xy + 28y$ completely.

Solution

Begin by factoring out the common monomial factor $4y$.

$$8x^2y - 60xy + 28y = 4y(2x^2 - 15x + 7)$$

Now, for the new trinomial $2x^2 - 15x + 7$, $a = 2$ and $c = 7$. The possible factorizations of this trinomial are as follows.

$$(2x - 7)(x - 1) = 2x^2 - 9x + 7$$
$$(2x - 1)(x - 7) = 2x^2 - 15x + 7 \qquad \Longleftarrow \qquad \text{Correct factorization}$$

So, the complete factorization of the original trinomial is

$$8x^2y - 60xy + 28y = 4y(2x^2 - 15x + 7)$$
$$= 4y(2x - 1)(x - 7).$$

Example 8 ■ Factoring Completely

Factor $4x^4 + 2x^3y - 6x^2y^2$ completely.

Solution

Begin by factoring out the common monomial factor $2x^2$.

$$4x^4 + 2x^3y - 6x^2y^2 = 2x^2(2x^2 + xy - 3y^2)$$

Now, for the new trinomial $2x^2 + xy - 3y^2$, $a = 2$, and $c = -3$. The possible factorizations of this trinomial are as follows.

$$(2x - 3y)(x + y) = 2x^2 - xy - 3y^2$$
$$(2x - y)(x + 3y) = 2x^2 + 5xy - 3y^2$$
$$(2x + 3y)(x - y) = 2x^2 + xy - 3y^2 \qquad \Longleftarrow \quad \text{Correct factorization}$$

So, the complete factorization of the original trinomial is

$$4x^4 + 2x^3y - 6x^2y^2 = 2x^2(2x^2 + xy - 3y^2)$$
$$= 2x^2(2x + 3y)(x - y).$$

Example 9 ■ A Trinomial with a Negative Leading Coefficient

Factor the trinomial $-3x^2 + 16x + 35$.

Solution

This trinomial has a negative leading coefficient, so you should begin by factoring (-1) out of the trinomial.

$$-3x^2 + 16x + 35 = (-1)(3x^2 - 16x - 35)$$

Now, for the new trinomial $3x^2 - 16x - 35$, you have $a = 3$ and $c = -35$. The possible factorizations of this trinomial are as follows.

$$(3x - 1)(x + 35) = 3x^2 + 104x - 35$$
$$(3x - 35)(x + 1) = 3x^2 - 32x - 35$$
$$(3x - 5)(x + 7) = 3x^2 + 16x - 35 \qquad \text{Middle term has incorrect sign.}$$
$$(3x + 5)(x - 7) = 3x^2 - 16x - 35 \qquad \Longleftarrow \quad \text{Correct factorization}$$

So, the correct factorization is

$$-3x^2 + 16x + 35 = (-1)(3x + 5)(x - 7).$$

Alternative forms of this factorization include

$$(3x + 5)(-x + 7) \text{ and } (-3x - 5)(x - 7).$$

NOTE When factoring a trinomial with a negative leading coefficient, first factor -1 out of the trinomial, as demonstrated in Example 9.

Not all trinomials are factorable using only integers. For instance, to factor

$$x^2 + 3x + 5$$

you need factors of 5 that add up to 3. This is not possible, because the only integer factors of 5 are 1 and 5, and their sum is not 3. Such a trinomial is *not factorable* over the integers. Polynomials that cannot be factored using integer coefficients are called **prime** with respect to the integers. Some other examples of prime polynomials are $2x^2 - 3x + 2$, $4x^2 + 9$, and $2x^2 - xy + 7y^2$. Watch for other trinomials that are not factorable in the exercises for this section.

Factoring Trinomials by Grouping (Optional)

In this section, you have seen that factoring a trinomial can involve quite a bit of trial and error. An alternative technique that some people like is to use *factoring by grouping*. For instance, suppose you rewrite the trinomial $2x^2 + x - 15$ as

$$2x^2 + x - 15 = 2x^2 + 6x - 5x - 15.$$

Then, by grouping the first and second terms and the third and fourth terms, you can factor the polynomial as follows.

$$
\begin{aligned}
2x^2 + x - 15 &= 2x^2 + 6x - 5x - 15 && \text{Rewrite middle term.} \\
&= (2x^2 + 6x) - (5x + 15) && \text{Group terms.} \\
&= 2x(x + 3) - 5(x + 3) && \text{Factor out common monomial factor in each group.} \\
&= (x + 3)(2x - 5) && \text{Distributive Property}
\end{aligned}
$$

The key to this method of factoring is knowing how to rewrite the middle term. In general, *to factor a trinomial $ax^2 + bx + c$ by grouping, choose factors of the product ac that add up to b and use these factors to rewrite the middle term.* This technique is illustrated in Example 10.

Study Tip

You should put a polynomial in standard form before trying to factor by grouping. Then group and remove a common monomial factor from the first two terms and the last two terms. Finally, if possible, factor out the common binomial factor.

Example 10 ■ Factoring a Trinomial by Grouping

Use factoring by grouping to factor the trinomials completely.

(a) $2x^2 + 5x - 3$ (b) $6y^2 + 5y - 4$

Solution

(a) In the trinomial $2x^2 + 5x - 3$, $a = 2$ and $c = -3$, which implies that the product ac is -6. Now, because -6 factors as $(6)(-1)$, and $6 - 1 = 5 = b$, you can rewrite the middle term as $5x = 6x - x$. This produces the following result.

$$
\begin{aligned}
2x^2 + 5x - 3 &= 2x^2 + 6x - x - 3 && \text{Rewrite middle term.} \\
&= (2x^2 + 6x) - (x + 3) && \text{Group terms.} \\
&= 2x(x + 3) - (x + 3) && \text{Factor out common monomial factor in first group.} \\
&= (x + 3)(2x - 1) && \text{Distributive Property}
\end{aligned}
$$

So, the trinomial factors as $2x^2 + 5x - 3 = (x + 3)(2x - 1)$.

(b) In the trinomial $6y^2 + 5y - 4$, $a = 6$ and $c = -4$, which implies that the product of ac is -24. Now, because -24 factors as $(8)(-3)$, and $8 - 3 = 5 = b$, you can rewrite the middle term as $5y = 8y - 3y$. This produces the following result.

$$
\begin{aligned}
6y^2 + 5y - 4 &= 6y^2 + 8y - 3y - 4 && \text{Rewrite middle term.} \\
&= (6y^2 + 8y) - (3y + 4) && \text{Group terms.} \\
&= 2y(3y + 4) - (3y + 4) && \text{Factor out common monomial factor in first group.} \\
&= (3y + 4)(2y - 1) && \text{Distributive Property}
\end{aligned}
$$

So, the trinomial factors as $6y^2 + 5y - 4 = (3y + 4)(2y - 1)$.

Factoring Perfect Square Trinomials

A **perfect square trinomial** is the square of a binomial. For instance,

$$x^2 + 6x + 9 = (x + 3)(x + 3) = (x + 3)^2$$

is the square of the binomial $(x + 3)$, and

$$4x^2 - 20x + 25 = (2x - 5)(2x - 5) = (2x - 5)^2$$

is the square of the binomial $(2x - 5)$. Perfect square trinomials come in two forms: one in which the middle term is positive, and the other in which the middle term is negative.

Perfect Square Trinomials

Let u and v represent real numbers, variables, or algebraic expressions. Then the perfect square trinomials below can be factored as indicated.

$$u^2 + 2uv + v^2 = (u + v)^2 \qquad\qquad u^2 - 2uv + v^2 = (u - v)^2$$

Same sign   Same sign

To recognize a perfect square trinomial, remember that the first and last terms must be perfect squares and positive, and that the middle term must be twice the product of u and v. (The middle term can be positive or negative.)

Example 11 ■ Factoring Perfect Square Trinomials

Rewrite each trinomial as a perfect square trinomial. Then factor the trinomial.

(a) $x^2 - 4x + 4$ (b) $16y^2 + 24y + 9$ (c) $9x^2 - 30xy + 25y^2$

Solution

(a) $x^2 - 4x + 4 = x^2 - 2(x)(2) + 2^2 = (x - 2)^2$

(b) $16y^2 + 24y + 9 = (4y)^2 + 2(4y)(3) + 3^2 = (4y + 3)^2$

(c) $9x^2 - 30xy + 25y^2 = (3x)^2 - 2(3x)(5y) + (5y)^2 = (3x - 5y)^2$

Example 12 ■ Factoring out a Common Monomial Factor First

Factor out any common monomial factors from each trinomial. Then factor completely.

(a) $3x^2 - 30x + 75$ (b) $16y^3 + 80y^2 + 100y$

Solution

(a) $3x^2 - 30x + 75 = 3(x^2 - 10x + 25)$ Factor out common monomial factor.

$= 3(x - 5)^2$ Factor as perfect square trinomial.

(b) $16y^3 + 80y^2 + 100y = 4y(4y^2 + 20y + 25)$ Factor out common monomial factor.

$= 4y(2y + 5)^2$ Factor as perfect square trinomial.

Summary of Factoring

Although the basic factoring techniques have been discussed one at a time, from this point on you must decide which technique to apply to any given problem situation. The guidelines below should assist you in this selection process.

Guidelines for Factoring Polynomials

1. Factor out any common factors.

2. Factor according to one of the special polynomial forms: difference of squares, sum or difference of cubes, or perfect square trinomial.

3. Factor trinomials, $ax^2 + bx + c$, using the methods for $a = 1$ or $a \neq 1$.

4. Factor by grouping—for polynomials with four terms.

5. Check to see if the factors themselves can be factored further.

6. Check the results by multiplying the factors.

Example 13 ■ Factoring Polynomials

Factor each polynomial completely.

(a) $3x^2 - 108$ (b) $4x^3 - 32x^2 + 64x$ (c) $6x^3 + 27x^2 - 15x$

(d) $x^3 - 3x^2 - 4x + 12$ (e) $x^2 + 6x + 9 - y^2$

Solution

(a) $3x^2 - 108 = 3(x^2 - 36)$ — Factor out common factor.

$= 3(x + 6)(x - 6)$ — Difference of two squares

(b) $4x^3 - 32x^2 + 64x = 4x(x^2 - 8x + 16)$ — Factor out common factor.

$= 4x(x - 4)^2$ — Perfect square trinomial

(c) $6x^3 + 27x^2 - 15x = 3x(2x^2 + 9x - 5)$ — Factor out common factor.

$= 3x(2x - 1)(x + 5)$ — Factor trinomial.

(d) $x^3 - 3x^2 - 4x + 12 = (x^3 - 3x^2) - (4x - 12)$ — Group terms.

$= x^2(x - 3) - 4(x - 3)$ — Factor out common factors.

$= (x - 3)(x^2 - 4)$ — Distributive Property

$= (x - 3)(x + 2)(x - 2)$ — Difference of two squares

(e) $x^2 + 6x + 9 - y^2 = (x^2 + 6x + 9) - y^2$ — Group terms.

$= (x + 3)^2 - y^2$ — Perfect square trinomial

$= [(x + 3) + y][(x + 3) - y]$ — Difference of two squares

$= (x + 3 + y)(x + 3 - y)$ — Simplify.

Notice the grouping in the first step of Example 13(e). The first three terms form a perfect square trinomial, which enables you to further factor the polynomial as the difference of two squares.

1.4 ■ Exercises

Developing Skills

In Exercises 1–6, fill in the missing factor.

1. $x^2 + 5x + 4 = (x + 4)()$

2. $a^2 + 2a - 8 = (a + 4)()$

3. $y^2 - y - 20 = (y + 4)()$

4. $y^2 + 6y + 8 = (y + 4)()$

5. $z^2 - 6z + 8 = (z - 4)()$

6. $z^2 + 2z - 24 = (z - 4)()$

In Exercises 7–22, factor the trinomial.

7. $x^2 + 4x + 3$ **8.** $x^2 - 10x + 24$

9. $y^2 + 7y - 30$ **10.** $m^2 - 3m - 10$

11. $t^2 - 4t - 21$ **12.** $x^2 + 4x - 12$

13. $x^2 - 12x + 20$ **14.** $y^2 - 8y - 33$

15. $t^2 - 17t + 60$ **16.** $z^2 - 13z + 36$

17. $x^2 - 20x + 96$ **18.** $y^2 + 22y + 96$

19. $u^2 + 5uv + 6v^2$ **20.** $x^2 - 7xy + 12y^2$

21. $x^2 - 2xy - 35y^2$ **22.** $a^2 - 21ab + 110b^2$

In Exercises 23–28, fill in the missing factor.

23. $5x^2 + 18x + 9 = (x + 3)()$

24. $5x^2 + 19x + 12 = (x + 3)()$

25. $5a^2 + 12a - 9 = (a + 3)()$

26. $5c^2 + 11c - 12 = (c + 3)()$

27. $2y^2 - 3y - 27 = (y + 3)()$

28. $3y^2 - y - 30 = (y + 3)()$

In Exercises 29–48, factor the trinomial, if possible. Some of the expressions are not factorable using integer coefficients.

29. $3x^2 + 4x + 1$ **30.** $5x^2 + 7x + 2$

31. $8t^2 - 6t - 5$ **32.** $2x^2 - 6x + 2$

33. $6b^2 + 19b - 7$ **34.** $12a^2 + 2a - 4$

35. $2a^2 - 13a + 20$ **36.** $9t^2 - 51t + 30$

37. $2x^2 + 5xy + 3y^2$ **38.** $5a^2 - 8ab + 3b^2$

39. $11y^2 - 43yz - 4z^2$ **40.** $6u^2 - 7uv - 5v^2$

41. $6a^2 + ab - 2b^2$ **42.** $24x^2 - 14xy - 3y^2$

43. $20x^2 + x - 12$ **44.** $10x^2 + 9xy - 9y^2$

45. $2u^2 + 9uv - 35v^2$ **46.** $5r^2 - 4rs - 9s^2$

47. $15x^2 + 3xy - 8y^2$ **48.** $9u^2 - 5uv + 6v^2$

In Exercises 49–52, factor the trinomial by first factoring (-1) from each term.

49. $-2x^2 - x + 6$ **50.** $-6x^2 + 5x + 6$

51. $1 - 11x - 60x^2$ **52.** $2 + 5x - 12x^2$

In Exercises 53–58, factor the trinomial by grouping.

53. $3x^2 + 10x + 8$ **54.** $2x^2 + 9x + 9$

55. $6x^2 + x - 2$ **56.** $6x^2 - x - 15$

57. $15x^2 - 11x + 2$ **58.** $12x^2 - 28x + 15$

In Exercises 59–70, factor the perfect square trinomial.

59. $x^2 + 4x + 4$ **60.** $z^2 + 6z + 9$

61. $a^2 - 12a + 36$ **62.** $y^2 - 14y + 49$

63. $25y^2 - 10y + 1$ **64.** $4z^2 + 28z + 49$

65. $9b^2 + 12b + 4$ **66.** $4x^2 - 4x + 1$

67. $4x^2 - 4xy + y^2$ **68.** $m^2 + 6mn + 9n^2$

69. $u^2 + 8uv + 16v^2$ **70.** $4y^2 + 20yz + 25z^2$

In Exercises 71–108, factor the algebraic expression completely.

71. $3x^5 - 12x^3$ **72.** $20y^2 - 45$

73. $10t^3 + 2t^2 - 36t$ **74.** $3x^2 - 12x - 36$

75. $5a^2 - 25a + 30$ **76.** $4t^2 + 8t - 32$

77. $6u^2 + 3u - 63$ **78.** $10x^2 - 22x - 24$

79. $9y^2 - 66y + 121$ **80.** $16z^2 - 56z + 49$

81. $4x(3x - 2) + (3x - 2)^2$

82. $(2x + 5)^2 - x(2x + 5)$

83. $x^2(x + 3) + 5x(x + 3)$

84. $6y^3(2y - 5) - 3y^2(2y - 5)$

85. $36 - (z + 3)^2$ **86.** $(x + 4)^2 - 25$

87. $(2y - 1)^2 - 9$ **88.** $81 - (3t + 2)^2$

89. $54x^3 - 2$ **90.** $3t^3 - 24$

91. $v^3 + 3v^2 + 5v$ **92.** $2a^3 - 8a^2 + 10a$

93. $2x^3y - 2x^2y^2 - 84xy^3$

94. $8m^3n + 20m^2n^2 - 48mn^3$

95. $5x^2y - 20y^3$

96. $27a^3 - 3ab^2$

97. $x^3 + 2x^2 - 16x - 32$

98. $x^3 - 7x^2 - 4x + 28$

99. $x^3 - 6x^2 - 9x + 54$

100. $x^3 + 10x^2 - 16x - 160$

101. $x^2 - 10x + 25 - y^2$

102. $9y^2 + 12y + 4 - z^2$

103. $a^2 - 2ab + b^2 - 16$

104. $x^2 + 14xy + 49y^2 - 4a^2$

105. $x^8 - 1$

106. $3y^3 - 192$

107. $b^4 - 216b$

108. $x^4 - 16y^4$

In Exercises 109–114, find two real numbers b such that the algebraic expression is a perfect square trinomial.

109. $x^2 + bx + 81$

110. $x^2 + bx + 16$

111. $9y^2 + by + 1$

112. $36z^2 + bz + 1$

113. $4x^2 + bx + 9$

114. $16x^2 + bxy + 25y^2$

In Exercises 115–120, find a real number c such that the algebraic expression is a perfect square trinomial.

115. $x^2 + 8x + c$

116. $x^2 + 12x + c$

117. $y^2 - 6y + c$

118. $z^2 - 20z + c$

119. $16a^2 + 40a + c$

120. $9t^2 - 12t + c$

In Exercises 121 and 122, fill in the missing number.

121. $x^2 + 12x + 50 = (x + 6)^2 + $ ▢

122. $x^2 + 10x + 22 = (x + 5)^2 + $ ▢

In Exercises 123–128, find all integers b such that the trinomial can be factored.

123. $x^2 + bx + 18$

124. $x^2 + bx + 14$

125. $x^2 + bx - 21$

126. $x^2 + bx - 7$

127. $5x^2 + bx + 8$

128. $3x^2 + bx - 10$

In Exercises 129–134, find two integers c such that the trinomial can be factored. (There are many correct answers.)

129. $x^2 + 6x + c$

130. $x^2 + 9x + c$

131. $x^2 - 3x + c$

132. $x^2 - 12x + c$

133. $t^2 - 4t + c$

134. $s^2 + s + c$

Solving Problems

Geometry In Exercises 135–138, write, in factored form, an expression for the shaded portion of the figure.

135.

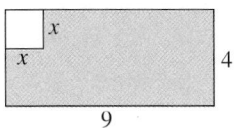

136.

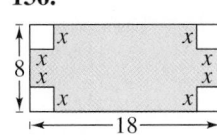

137.

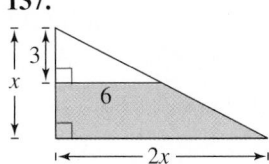

138.

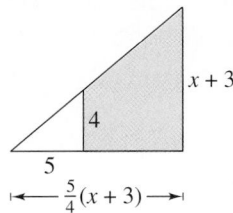

139. *Geometry* A rectangle has area and width as indicated in the figure. Factor to find the length.

$2x - 5$

$A = 2x^2 + x - 15$?

140. *Geometry* A circle has area as indicated in the figure. Factor to find the radius.

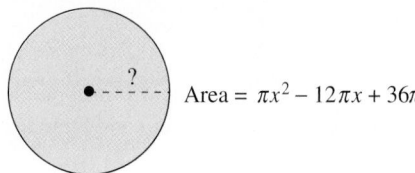

Area $= \pi x^2 - 12\pi x + 36\pi$

In Exercises 141–144, match the geometric factoring model with the correct factoring formula. [The models are labeled (a), (b), (c), and (d).]

141. $a^2 - b^2 = (a + b)(a - b)$

142. $a^2 + 2ab + b^2 = (a + b)^2$

143. $a^2 + 2a + 1 = (a + 1)^2$

144. $ab + a + b + 1 = (a + 1)(b + 1)$

(a)

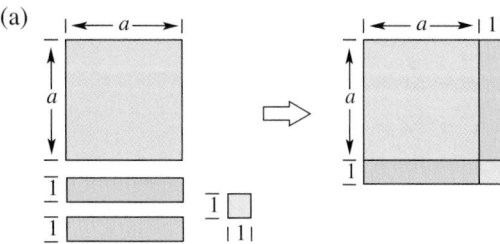

(b)

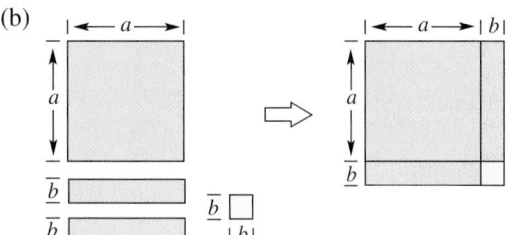

(c)

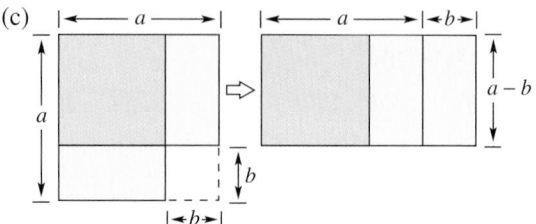

(d)

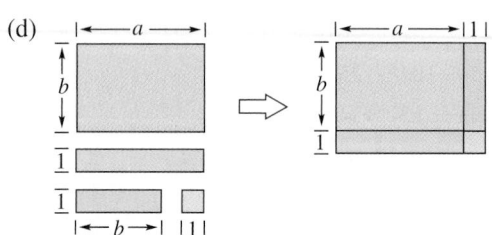

In Exercises 145 and 146, factor the trinomial and represent the result with a geometric factoring model.

145. $x^2 + 4x + 3$ **146.** $x^2 + 5x + 4$

Explaining Concepts

147. Identify the error.

$$9x^2 - 9x - 54 = (3x + 6)(3x - 9)$$
$$= 3(x + 2)(x - 3)$$

148. Is $x(x + 2) - 2(x + 2)$ completely factored? Explain.

In Exercises 149 and 150, explain how the same result could be obtained mentally using the sample below as a model.

$$29^2 = (30 - 1)^2 = 30^2 - 2(30)(1) + 1^2$$
$$= 900 - 60 + 1$$
$$= 841$$

149. $52^2 = 2704$

150. $39^2 = 1521$

Looking Further

(a) Factor the expression

$$4x^3 - 36x^2 + 80x$$

completely.

(b) An open box is to be made from a rectangular piece of material 10 inches long and 8 inches wide by cutting equal squares of side x from each corner and turning up the sides. Show that the volume of the box can be described by the factored expression in part (a).

1.5 Solving Linear Equations

What you should learn:

- How to identify types of equations and check solutions of equations
- How to solve linear equations
- How to solve linear equations containing parentheses
- How to solve linear equations containing fractions or decimals
- How to solve literal equations

Why you should learn it:

Linear equations are used in many real-life applications. For instance, Exercise 120 on page 62 shows how a linear equation can model the number of pieces of First-Class Mail handled by the U.S. Postal Service.

Equations and Solutions of Equations

An **equation** is a statement that equates two mathematical expressions. Some examples are

$$x = 4, \quad 4x + 5 = 17, \quad 2x - 8 = 2(x - 4), \quad \text{and} \quad x^2 - 16 = 0.$$

Solving an equation involving x means finding all values of x for which the equation is true. Such values are **solutions** and are said to **satisfy** the equation. For instance, 3 is a solution of $4x + 5 = 17$ because $4(3) + 5 = 17$ is a true statement.

The **solution set** of an equation is the set of all solutions of the equation. Sometimes, an equation will have the set of all real numbers as its solution set. Such an equation is an **identity.** For instance, the equation

$$2x - 8 = 2(x - 4) \qquad \text{Identity}$$

is an identity because the equation is true for all real values of x. Try values such as 0, 1, -2, and 5 in this equation to see that each one is a solution.

An equation whose solution set is not the entire set of real numbers is called a **conditional equation.** For instance, the equation

$$x^2 - 16 = 0 \qquad \text{Conditional equation}$$

is a conditional equation because it has only two solutions, 4 and -4. Example 1 shows how to **check** whether a given value is a solution.

Example 1 ■ Checking a Solution of an Equation

Determine whether (a) $x = -3$ or (b) $x = -2$ is a solution of $-3x - 5 = 4x + 16$.

Solution

(a)
$$
\begin{aligned}
-3x - 5 &= 4x + 16 && \text{Write original equation.}\\
-3(-3) - 5 &\overset{?}{=} 4(-3) + 16 && \text{Substitute } -3 \text{ for } x.\\
9 - 5 &\overset{?}{=} -12 + 16 && \text{Simplify.}\\
4 &= 4 && \text{Solution checks. } \checkmark
\end{aligned}
$$

Because each side turns out to be the same number, you can conclude that $x = -3$ *is* a solution of the original equation.

(b)
$$
\begin{aligned}
-3x - 5 &= 4x + 16 && \text{Write original equation.}\\
-3(-2) - 5 &\overset{?}{=} 4(-2) + 16 && \text{Substitute } -2 \text{ for } x.\\
6 - 5 &\overset{?}{=} -8 + 16 && \text{Simplify.}\\
1 &\neq 8 && \text{Solution } does\ not \text{ check. } \times
\end{aligned}
$$

Because the numbers on each side are different, you can conclude that $x = -2$ *is not* a solution of the original equation.

Study Tip

When checking a solution, you should write a question mark over the equal sign to indicate that you are uncertain whether the "equation" is true.

It is helpful to think of an equation as having two sides that are "in balance." Consequently, when you try to solve an equation, you must be careful to maintain that balance by performing the same operation(s) on each side.

Two equations that have the same set of solutions are **equivalent equations.** For instance, the equations $x = 3$ and $x - 3 = 0$ are equivalent because both equations have only one solution—the number 3. When any one of the four techniques listed below is applied to an equation, the resulting equation is *equivalent* to the original equation.

Historical Note

In the late 1800s, a movement was begun to identify a complete set of axioms for each branch of mathematics from which all other propositions could be deduced. In 1931, Kurt Gödel (1906–1978), a faculty member at the University of Vienna, showed that this goal was unattainable. He proved that a complete set of axioms could never be identified for a branch of mathematics such that all of its propositions could be proven or disproven on the basis of those axioms. Although this closed one avenue of research, Gödel also pointed out new directions for the future.

Forming Equivalent Equations: Properties of Equality

An equation can be transformed into an *equivalent equation* using one or more of the procedures listed below.

	Original Equation	Equivalent Equation
1. *Simplify Each Side:* Remove symbols of grouping, combine like terms, or simplify fractions on one or both sides of the equation.	$3x - x = 8$	$2x = 8$
2. *Apply the Addition Property of Equality:* Add (or subtract) the same quantity to (or from) *each* side of the equation.	$x - 3 = 5$	$x = 8$
3. *Apply the Multiplication Property of Equality:* Multiply (or divide) *each* side of the equation by the same nonzero quantity.	$3x = 9$	$x = 3$
4. *Interchange Sides:* Interchange the two sides of the equation.	$7 = x$	$x = 7$

When solving an equation, you can use any of the four techniques for forming equivalent equations to eliminate terms or factors in the equation. For example, to solve the equation

$$x + 4 = 2$$

you need to remove the term $+4$ from the left side. This is accomplished by subtracting 4 from each side.

$x + 4 = 2$	Write original equation.
$x + 4 - 4 = 2 - 4$	Subtract 4 from each side.
$x + 0 = -2$	Combine like terms.
$x = -2$	Simplify.

Although this solution involved subtracting 4 from each side, you could just as easily have added -4 to each side. Both techniques are legitimate—the one you decide to use is a matter of personal preference. Remember the goal when you *solve an equation* is to get the *variable* on one side of the equal sign and everything else on the other side.

Solving Linear Equations

The most common type of equation in one variable is a **linear equation.**

> **Definition of Linear Equation**
>
> A **linear equation** in one variable x is an equation that can be written in the standard form
>
> $$ax + b = 0$$
>
> where a and b are real numbers with $a \neq 0$.

A linear equation in one variable is also called a **first-degree equation** because its variable has an implied exponent of 1. Some examples of linear equations in the standard form $ax + b = 0$ are $3x + 2 = 0$ and $5x - 4 = 0$.

Remember that to *solve* an equation in x means to find the values of x that satisfy the equation. For a linear equation in the standard form $ax + b = 0$, the goal is to **isolate** x by rewriting the standard equation in the form

$$x = (\text{a number}).$$

Beginning with the original equation, you write a sequence of equivalent equations, each having the same solution as the original equation.

Example 2 ■ Solving a Linear Equation

Solve the equation. Then check the solution.

$$4x - 12 = 0$$

Solution

$4x - 12 = 0$	Write original equation.
$4x - 12 + 12 = 0 + 12$	Add 12 to each side.
$4x = 12$	Combine like terms.
$\dfrac{4x}{4} = \dfrac{12}{4}$	Divide each side by 4.
$x = 3$	Simplify.

It appears that the solution is $x = 3$. You can check this as follows.

Check

$4x - 12 = 0$	Write original equation.
$4(3) - 12 \overset{?}{=} 0$	Substitute 3 for x.
$12 - 12 \overset{?}{=} 0$	Simplify.
$0 = 0$	Solution checks. ✓

NOTE   Be sure you see that solving an equation such as the one in Example 2 has two basic steps. The first step is to *find* the solution(s). The second step is to *check* that each solution you find actually satisfies the original equation.

Technology

A graphing utility can be used to check solutions of equations. Consult the user's guide for your calculator or graphing utility for the proper keystrokes.

Solve the equation $3(x - 2) + 5 = -10$. Check the solution using a graphing utility.

For more information on how to use a graphing utility, refer to the *Graphing Technology Guide* and *Graphing Calculator Videotape* that accompany this text.

NOTE In Example 3(b), the equation was solved by isolating the variable on the left side. You could just as easily have isolated the variable on the right side, as follows.

$$x + 2 = 2x - 6$$
$$-x + x + 2 = -x + 2x - 6$$
$$2 = x - 6$$
$$2 + 6 = x - 6 + 6$$
$$8 = x$$

Example 3 ■ Solving Linear Equations

Solve the equation, then check the solution.

(a) $2x - 3 = -5$ (b) $x + 2 = 2x - 6$

Solution

(a)

$2x - 3 = -5$	Write original equation.
$2x - 3 + 3 = -5 + 3$	Add 3 to each side.
$2x = -2$	Combine like terms.
$\dfrac{2x}{2} = \dfrac{-2}{2}$	Divide each side by 2.
$x = -1$	Simplify.

It appears that the solution is $x = -1$.

(b)

$x + 2 = 2x - 6$	Write original equation.
$-2x + x + 2 = -2x + 2x - 6$	Add $-2x$ to each side.
$-x + 2 = -6$	Combine like terms.
$-x + 2 - 2 = -6 - 2$	Subtract 2 from each side.
$-x = -8$	Combine like terms.
$(-1)(-x) = (-1)(-8)$	Multiply each side by -1.
$x = 8$	Simplify.

It appears that the solution is $x = 8$.

Check

(a)

$2x - 3 = -5$	Write original equation.
$2(-1) - 3 \stackrel{?}{=} -5$	Substitute 3 for x.
$-2 - 3 \stackrel{?}{=} -5$	Simplify.
$-5 = -5$	Solution checks. ✓

(b)

$x + 2 = 2x - 6$	Write original equation.
$8 + 2 \stackrel{?}{=} 2(8) - 6$	Substitute 8 for x.
$10 \stackrel{?}{=} 16 - 6$	Simplify.
$10 = 10$	Solution checks. ✓

As you gain experience in solving linear equations, you will probably find that you can perform some of the solution steps in your head. For instance, you might solve the equation given in Example 3(a) by performing two of the steps mentally, and writing only three steps, as follows.

$2x - 3 = -5$	Write original equation.
$2x = -2$	Add 3 to each side.
$x = -1$	Divide each side by 2.

Remember, however, that you should not skip the final step—checking your solution. You may find your calculator useful for checking solutions.

Study Tip

Avoid the temptation to first divide an equation by x. You may obtain an incorrect solution, as in the example below.

$$7x = -4x \qquad \text{Write original equation.}$$

$$\frac{7x}{x} = -\frac{4x}{x} \qquad \text{Divide each side by } x.$$

$$7 = -4 \qquad \text{False statement}$$

The false statement indicates that there is no solution. However, when the equation is solved correctly, the solution is $x = 0$.

$$7x = -4x$$

$$7x + 4x = -4x + 4x$$

$$11x = 0$$

$$\frac{11x}{11} = \frac{0}{11}$$

$$x = 0$$

NOTE   An equation that has no solution is said to have an **empty solution set,** which is denoted by $\{\ \}$ or $\emptyset$.

You now know that 8 is a solution of the equation in Example 3(b), but at this point you might be asking, "How can I be sure that the equation does not have other solutions?" There are actually three different situations that can be encountered when solving linear equations in one variable. The first situation occurs when the linear equation has *exactly one* solution. You can show this with the steps below.

$$ax + b = 0 \qquad \text{Write original equation, with } a \neq 0.$$

$$ax + b - b = 0 - b \qquad \text{Subtract } b \text{ from each side.}$$

$$ax = -b \qquad \text{Combine like terms.}$$

$$\frac{ax}{a} = \frac{-b}{a} \qquad \text{Divide each side by } a.$$

$$x = \frac{-b}{a} \qquad \text{Simplify.}$$

It is clear that the last equation has only one solution, $x = -b/a$. The other two situations are the possibilities for the equation to have either *no solution* or *infinitely many solutions*. These two special cases are demonstrated in Example 4.

Example 4   ■   Solving Linear Equations: Special Cases

Solve (a) $2x - 4 = 2(x - 3)$ and (b) $3x + 2(x - 10) = 5(x - 4)$.

Solution

(a) $2x - 4 = 2(x - 3) \qquad \text{Write original equation.}$

$\quad\ \ 2x - 4 = 2x - 6 \qquad \text{Distributive Property}$

$\qquad\quad -4 = -6 \qquad \text{Subtract } 2x \text{ from each side.}$

Because the last equation is a false statement, you can conclude that the original equation has no solution.

(b) $3x + 2(x - 10) = 5(x - 4) \qquad \text{Write original equation.}$

$\quad\ \ 3x + 2x - 20 = 5x - 20 \qquad \text{Distributive Property.}$

$\qquad\quad 5x - 20 = 5x - 20 \qquad \text{Simplify.}$

$\qquad\qquad\ \ -20 = -20 \qquad \text{Subtract } 5x \text{ from each side.}$

Because the last equation is true for any value of x, the equation is an identity and you can conclude that the equation has infinitely many solutions.

Solving Linear Equations Containing Parentheses

Linear equations often contain parentheses or other symbols of grouping. Here is an example.

$$6(y - 1) + 4y = 3(7y + 1)$$

In most cases, it helps to remove symbols of grouping as a first step to solving an equation. This is illustrated in Example 5.

Example 5 ■ Solving a Linear Equation Containing Parentheses

Solve the equation $6(y - 1) + 4y = 3(7y + 1)$.

Solution

$$6(y - 1) + 4y = 3(7y + 1) \qquad \text{Write original equation.}$$

$$6y - 6 + 4y = 21y + 3 \qquad \text{Distributive Property.}$$

$$10y - 6 = 21y + 3 \qquad \text{Combine like terms.}$$

$$10y - 21y - 6 = 21y - 21y + 3 \qquad \text{Subtract } 21y \text{ from each side.}$$

$$-11y - 6 = 3 \qquad \text{Combine like terms.}$$

$$-11y - 6 + 6 = 3 + 6 \qquad \text{Add 6 to each side.}$$

$$-11y = 9 \qquad \text{Combine like terms.}$$

$$\frac{-11y}{-11} = \frac{9}{-11} \qquad \text{Divide each side by } -11.$$

$$y = -\frac{9}{11} \qquad \text{Simplify.}$$

The solution is $y = -\frac{9}{11}$. Check this in the original equation.

Solving Linear Equations Containing Fractions or Decimals

If a linear equation contains fractions, we suggest that you first *clear the equation of fractions* by multiplying both sides of the equation by the least common denominator (LCD) of the fractions.

Example 6 ■ Solving a Linear Equation That Contains Fractions

Solve the equation $\dfrac{x}{18} + \dfrac{3x}{4} = 2$.

Solution

$$\frac{x}{18} + \frac{3x}{4} = 2 \qquad \text{Write original equation.}$$

$$36\left(\frac{x}{18} + \frac{3x}{4}\right) = 36(2) \qquad \text{Multiply each side by LCD of 36.}$$

$$36 \cdot \frac{x}{18} + 36 \cdot \frac{3x}{4} = 36(2) \qquad \text{Distributive Property}$$

$$2x + 27x = 72 \qquad \text{Simplify.}$$

$$29x = 72 \qquad \text{Combine like terms.}$$

$$\frac{29x}{29} = \frac{72}{29} \qquad \text{Divide each side by 29.}$$

$$x = \frac{72}{29} \qquad \text{Simplify.}$$

The solution is $x = \frac{72}{29}$. Check this in the original equation.

NOTE A different approach to Example 7 would be to begin by multiplying each side of the equation by 100, thus clearing the equation of decimals and producing

$$12x + 9(5000 - x) = 51{,}300.$$

Try solving this equation to see that you obtain the same solution.

Study Tip

There is no general agreement as to the "best" or "simplest" way to write solutions of literal equations. In Example 8(b), the solution

$$t = \frac{8 - 5s}{-2}$$

was rewritten as

$$t = \frac{5s - 8}{2}$$

because the latter form seems to be simpler. This, however, is partly a matter of personal preference. For instance, you might prefer to keep the solution in the first form *or* you might prefer to write the solution as $t = \frac{5}{2}s - 4$. The actual form that you use is not particularly important. It is important, however, to realize that many forms can be used *and* that you are able to convert from one form to another.

The next example shows how to solve a linear equation involving decimals. The procedure is basically the same, but the arithmetic can be messier.

Example 7 ■ Solving a Linear Equation Involving Decimals

Solve the equation $0.12x + 0.09(5000 - x) = 513$

Solution

$0.12x + 0.09(5000 - x) = 513$	Write original equation.
$0.12x + 450 - 0.09x = 513$	Distributive Property
$(0.12x - 0.09x) + 450 = 513$	Group like terms.
$0.03x + 450 - 450 = 513 - 450$	Subtract 450 from each side.
$0.03x = 63$	Combine like terms.
$\dfrac{0.03x}{0.03} = \dfrac{63}{0.03}$	Divide each side by 0.03.
$x = 2100$	Simplify.

The solution is 2100. Check this in the original equation.

Solving Literal Equations

A **literal equation** is an equation that has more than one variable. For instance, $5x + 2y = 7$ is a literal equation because it has two variables, x and y. The word *literal* comes from the Latin word for "letter." To **solve a literal equation** for one of its variables means to write an equivalent equation in which the "solved variable" is isolated on one side of the equation.

Example 8 ■ Solving a Literal Equation

Solve the equation $5s - 2t = 8$ (a) for s and (b) for t.

Solution

(a) You can solve the equation for s as follows.

$5s - 2t = 8$	Write original equation.
$5s = 8 + 2t$	Add $2t$ to each side.
$s = \dfrac{8 + 2t}{5}$	Divide each side by 5.

(b) You can solve the equation for t as follows.

$5s - 2t = 8$	Write original equation.
$-2t = 8 - 5s$	Subtract $5s$ from each side.
$t = \dfrac{8 - 5s}{-2}$	Divide each side by -2.
$t = \dfrac{5s - 8}{2}$	Simplify.

Example 9 ■ Solving a Literal Equation

Solve the literal equation $3y + 2x - 4 = 5y - 3x + 2$ for y.

Solution

$$3y + 2x - 4 = 5y - 3x + 2 \qquad \text{Write original equation.}$$
$$-2y + 2x - 4 = -3x + 2 \qquad \text{Subtract } 5y \text{ from each side.}$$
$$-2y - 4 = -5x + 2 \qquad \text{Subtract } 2x \text{ from each side.}$$
$$-2y = -5x + 6 \qquad \text{Add 4 to each side.}$$
$$y = \frac{-5x + 6}{-2} \qquad \text{Divide each side by } -2.$$
$$y = \frac{5x - 6}{2} \qquad \text{Simplify.}$$

Example 10 ■ Solving a Literal Equation

Solve the literal equation $A = P + Prt$ for P.

Solution

Notice how the Distributive Property is used to factor P out of the terms on the right side of the equation.

$$A = P + Prt \qquad \text{Write original equation.}$$
$$A = P(1 + rt) \qquad \text{Distributive Property}$$
$$\frac{A}{1 + rt} = P \qquad \text{Divide each side by } 1 + rt.$$

NOTE In Example 10, notice that it is not necessary to isolate the variable on the left side of the equation. Sometimes it is more convenient to isolate the variable on the right side.

Example 11 ■ Solving a Literal Equation

The formula for the perimeter of a rectangle is given by

$$P = 2l + 2w.$$

Solve the equation for l.

Solution

$$P = 2l + 2w \qquad \text{Write original equation.}$$
$$P - 2w = 2l \qquad \text{Subtract } 2w \text{ from each side.}$$
$$\frac{P - 2w}{2} = l \qquad \text{Divide each side by 2.}$$

NOTE In Example 11, the solution can also be written in the equivalent form $l = \frac{1}{2}P - w$.

1.5 ■ Exercises

Developing Skills

In Exercises 1–6, determine whether the values of the variable are solutions of the equation.

Equation	Values	
1. $3x - 7 = 2$	(a) $x = 0$	(b) $x = 3$
2. $5x + 9 = 4$	(a) $x = -1$	(b) $x = 2$
3. $3 - 2x = 21$	(a) $x = -3$	(b) $x = -9$
4. $10x - 3 = 7x$	(a) $x = 0$	(b) $x = -1$
5. $3x + 3 = 2(x - 4)$	(a) $x = -11$	(b) $x = 5$
6. $7x - 1 = 5(x - 5)$	(a) $x = 2$	(b) $x = -2$

In Exercises 7–10, identify the equation as a conditional equation, an identity, or an equation with no solution.

7. $3(x - 1) = 3x$

8. $2x + 8 = 6x$

9. $5(x + 3) = 2x + 3(x + 5)$

10. $4(x + 2) = 3x - 1$

In Exercises 11–14, determine whether the equation is linear. If not, state why.

11. $3x + 4 = 10$

12. $x^2 + 3 = 8$

13. $\dfrac{4}{x} - 3 = 5x$

14. $3(x - 2) = 4x$

In Exercises 15–18, justify each step of the solution.

15.
$$3x + 15 = 0 \qquad \text{Original equation}$$
$$3x + 15 - 15 = 0 - 15$$
$$3x = -15$$
$$\frac{3x}{3} = \frac{-15}{3}$$
$$x = -5$$

16.
$$7x - 21 = 0 \qquad \text{Original equation}$$
$$7x - 21 + 21 = 0 + 21$$
$$7x = 21$$
$$\frac{7x}{7} = \frac{21}{7}$$
$$x = 3$$

17.
$$-2x + 5 = 12 \qquad \text{Original equation}$$
$$-2x + 5 - 5 = 12 - 5$$
$$-2x = 7$$
$$\frac{-2x}{-2} = \frac{7}{-2}$$
$$x = -\frac{7}{2}$$

18.
$$25 - 3x = 10 \qquad \text{Original equation}$$
$$25 - 3x - 25 = 10 - 25$$
$$-3x = -15$$
$$\frac{-3x}{-3} = \frac{-15}{-3}$$
$$x = 5$$

In Exercises 19–76, solve the equation and check the result. (If not possible, state the reason.)

19. $y + 7 = 0$

20. $x - 10 = 0$

21. $3x = 12$

22. $8z = -16$

23. $23x - 4 = 42$

24. $15x - 18 = 27$

25. $12y + 7 = 31$

26. $25y - 29 = -4$

27. $7 - 8x = 13x$

28. $2s - 16 = 34s$

29. $15t = 0$

30. $-93x = 0$

31. $6a + 2 = 6a$

32. $-8t + 7 = -8t$

33. $4x = -12x$

34. $-y = 7y$

35. $4x - 7 = x + 11$

36. $5x + 3 = 2x - 21$

37. $2 - 3x = 10 + x$

38. $4 + 2x = 7 - 4x$

39. $8(x - 8) = 24$

40. $6(x + 2) = 30$

41. $5 - (2y - 4) = 15$

42. $26 - (3x - 10) = 6$

43. $8x - 3(x - 2) = 12$

44. $28 = 2(3x + 5) - 9$

45. $2(x + 3) = 7(x + 3)$

46. $-25(x - 100) = 16(x - 100)$

47. $-3(x + 2) = 2(2x + 4)$

48. $8(x - 3) = -(x - 12)$

49. $3(x + 5) - 2x = 3 - (4x - 2)$

50. $4x + 5(2x - 4) = 8 - 10(x + 1)$

51. $5(t + 2) - (3t + 4) = 0$

52. $12 = 6(y + 1) - 8(1 - y)$

53. $\dfrac{u}{5} = 10$

54. $-\dfrac{z}{2} = 7$

55. $t - \dfrac{2}{5} = \dfrac{3}{2}$

56. $z + \dfrac{1}{15} = -\dfrac{3}{10}$

57. $\dfrac{t}{14} + \dfrac{2}{7} = \dfrac{2}{7}$

58. $\dfrac{11x}{6} + \dfrac{1}{3} = 0$

59. $\dfrac{1}{9}x + \dfrac{1}{3} = \dfrac{11}{8}x$

60. $\dfrac{2}{5}y - \dfrac{6}{7} = \dfrac{1}{4}$

61. $\dfrac{11x}{6} + \dfrac{1}{3} = 2x$

62. $3z - \dfrac{3}{7} = \dfrac{5z}{6}$

63. $\dfrac{t}{5} - \dfrac{t}{2} = 1$

64. $\dfrac{t}{6} + \dfrac{t}{8} = 1$

65. $\dfrac{8x}{5} - \dfrac{x}{4} = -3$

66. $\dfrac{1}{3}x + 1 = \dfrac{1}{12}x - 4$

67. $\dfrac{4u}{3} = \dfrac{5u}{4} + 6$

68. $\dfrac{-3x}{4} - 4 = \dfrac{x}{6}$

69. $\dfrac{8 - 3x}{4} - 4 = \dfrac{x}{6}$

70. $\dfrac{25 - 4u}{3} = \dfrac{5u + 12}{4} + 6$

71. $\dfrac{2}{3}(2x - 4) = \dfrac{1}{2}(x + 3) - 4$

72. $\dfrac{3}{4}(6 - x) = \dfrac{1}{3}(4x + 5) + 2$

73. $0.3x + 1.5 = 8.4$

74. $16.3 - 0.2x = 7.1$

75. $1.2(x - 3) = 10.8$

76. $6.5(1 - 2x) = 13$

In Exercises 77–84, solve the equation (a) for x and (b) for y.

77. $2x - 3y = 6$

78. $-5x + 3y = 15$

79. $7x + 4 = 10y - 7$

80. $8y + 9 = 12x - 5$

81. $12(x - 2) + 7(y + 1) = 25$

82. $32 = 6(x + 9) - 3(y - 4)$

83. $\dfrac{x}{2} + \dfrac{y}{5} = 1$

84. $\dfrac{3x}{4} + \dfrac{y}{6} = 5$

In Exercises 85–90, verify that the solution can be written in the rewritten form.

	Solution	Rewritten Form
85.	$\dfrac{y - 5}{-1}$	$5 - y$
86.	$u + 3uv$	$u(1 + 3v)$
87.	$\dfrac{h + 4\pi}{2}$	$\dfrac{h}{2} + 2\pi$
88.	$\dfrac{2t - 5}{-2}$	$\dfrac{5}{2} - t$
89.	$\dfrac{3x - 7}{6}$	$\dfrac{1}{2}x - \dfrac{7}{6}$
90.	$\dfrac{-7(3 - x)}{2}$	$\dfrac{7}{2}(x - 3)$

In Exercises 91–112, solve for the indicated variable.

91. Solve for R

     Ohm's Law: $E = IR$

92. Solve for l

     Area of a rectangle: $A = lw$

93. Solve for h

     Volume of a rectangular prism: $V = lwh$

94. Solve for t

     Simple interest: $I = Prt$

95. Solve for r

     Circumference of a circle: $C = 2\pi r$

96. Solve for h

     Volume of a right circular cylinder: $V = \pi r^2 h$

97. Solve for a

     Perimeter of a triangle: $P = a + b + c$

98. Solve for C

     Sum of the angles of a triangle: $A + B + C = 180$

99. Solve for w

     Area of a rectangle: $A = lw$

100. Solve for r

     Investment at simple interest: $A = P + Prt$

101. Solve for h

     Area of a triangle: $A = \dfrac{1}{2}bh$

102. Solve for b

     Area of a trapezoid: $A = \dfrac{1}{2}(a + b)h$

103. Solve for n

     Arithmetic progression: $S = \dfrac{n}{2}(a_1 + a_n)$

104. Solve for P

     Investment at compound interest: $A = P\left(1 + \dfrac{r}{n}\right)^{nt}$

105. Solve for C

     Markup: $S = C + rC$

106. Solve for L

Discount: $S = L - rL$

107. Solve for P

Investment at simple interest: $A = P + Prt$

108. Solve for h

Area of a trapezoid: $A = \frac{1}{2}hb_1 + \frac{1}{2}hb_2$

109. Solve for n

Arithmetic progression: $L = a + (n - 1)d$

110. Solve for r

Geometric progression: $S = \dfrac{a_1}{1 - r}$

111. Solve for m_2

Newton's Law of Universal Gravitation: $F = \dfrac{km_1m_2}{r^2}$

112. Solve for x_1

Arithmetic mean of two numbers: $\bar{x} = \dfrac{x_1 + x_2}{2}$

Solving Problems

113. *Investigation*   The length of a rectangle is t times its width (see figure). So, the perimeter P is given by $P = 2w + 2(tw)$, where w is the width of the rectangle. The perimeter of the rectangle is 1200 meters.

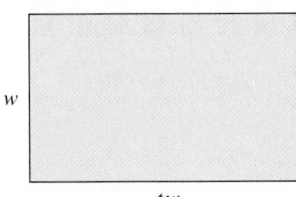

(a) Complete the table of widths, lengths, and areas of the rectangle by solving the original equation.

t	1	1.5	2	3	4	5
Width						
Length						
Area						

(b) Use the table to write a short paragraph describing the relationship among the width, length, and area of a rectangle that has a *fixed* perimeter.

114. *Geometry*   A rectangular plot of land with length 12 feet and width 10 feet is to be made into a garden with a walkway around it. Let x represent the width of the walkway (see figure).

(a) Find an expression for the perimeter P of the garden.

(b) Solve for x when the perimeter is fixed at 20 feet.

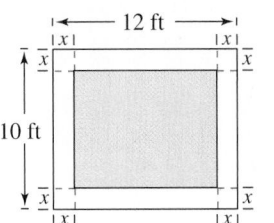

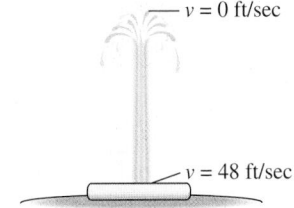

Figure for 114   Figure for 115

115. *Physics*   Consider the water fountain shown in the figure. The initial velocity of the stream of water is 48 feet per second. The velocity v of the water at any time t (in seconds) is then given by $v = 48 - 32t$. Find the time for a drop of water to travel from the base to the maximum height of the fountain. (*Hint:* The maximum height is reached when $v = 0$.)

116. *Physics*   The velocity v of an object projected vertically upward with an initial velocity of 64 feet per second is given by $v = 64 - 32t$, where t is the time in seconds. When does the object reach its maximum height? Explain.

117. *Work*   Two people can complete a task in t hours, where t must satisfy the equation

$$\frac{t}{10} + \frac{t}{15} = 1.$$

Find the required time t.

118. *Work*   Two people can complete a task in t hours, where t must satisfy the equation

$$\frac{t}{21} + \frac{t}{28} = 1.$$

Find the required time t.

119. *Using a Model* The average salary y (in dollars) of elementary school principals from 1990 to 1999 can be modeled by

$$y = 2069.5t + 48{,}435, \ 0 \le t \le 9$$

where t represents the year, with $t = 0$ corresponding to 1990 (see figure). Graphically determine the year during which the salary reached $50,500. Then algebraically determine the year, rounding your answer to the nearest year. *(Source: Educational Research Service)*

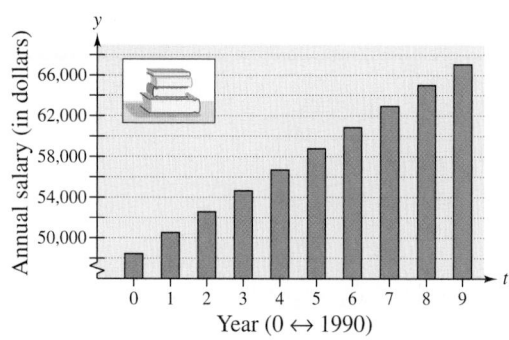

120. *Using a Model* The number of pieces of First-Class Mail y (in millions) handled by the U.S. Postal Service from 1990 to 1999 can be modeled by

$$y = 1417.5t + 89{,}366, \ 0 \le t \le 9$$

where t represents the year, with $t = 0$ corresponding to 1990 (see figure). Graphically determine the year during which the number of pieces of First-Class Mail reached 100,700 million pieces. Then algebraically determine the year, rounding your answer to the nearest year. *(Source: U.S. Postal Service)*

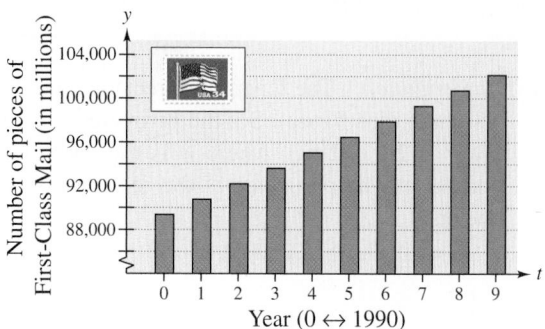

Explaining Concepts

121. Explain the difference between (a) an expression and an equation and (b) a conditional equation and an identity.

122. What is meant by equivalent equations? Give an example of two equivalent equations.

Looking Further

Use the *two-part* model below, which approximates the number of cable television subscribers y (in thousands) in the United States between 1985 and 2000.

$$y = 3511.4t + 15{,}681, \quad 5 \le t \le 10$$
$$y = 1795.6t + 31{,}618, \quad 11 \le t \le 20$$

In this model, t represents the year, with $t = 5$ corresponding to 1985. *(Source: Warren Communications News, Inc.)*

(a) Make a table that shows the number of subscribers from 1985 through 2000.

(b) Draw a bar graph of the data in the table.

(c) Which year had 43,772 (thousand) subscribers?

(d) Which year had 58,552 (thousand) subscribers?

(e) In parts (c) and (d), did you use the model, the table, or the bar graph? Could you have used any of the three? Explain.

1.6 Solving Equations by Factoring

Quadratic Equations and the Zero-Factor Property

In Section 1.5, you studied techniques for solving first-degree polynomial equations (linear equations). In this section, you will study a technique for solving **second-degree** polynomial equations (*quadratic* equations) and *higher-degree* polynomial equations.

Definition of Quadratic Equation

A **quadratic equation in x** is an equation that can be written in the general form

$$ax^2 + bx + c = 0 \qquad \text{Quadratic equation}$$

where a, b, and c are real numbers with $a \neq 0$.

In Section 1.4, you reviewed techniques for factoring trinomials of the form

$$ax^2 + bx + c. \qquad \text{Trinomial}$$

These skills can be combined with the **Zero-Factor Property** to *solve* quadratic equations.

Zero-Factor Property

Let u and v represent real numbers, variables, or algebraic expressions. If u and v are factors such that

$$uv = 0$$

then $u = 0$ or $v = 0$. This property applies to three or more factors as well.

NOTE　The Zero-Factor Property is just another way of saying that the only way the product of two (or more) real numbers can be zero is if one (or more) of the real numbers is zero.

The Zero-Factor Property is an important property that is used to solve equations in algebra. For instance, to solve the equation

$$(x - 2)(x + 3) = 0 \qquad \text{Original equation}$$

you can use the Zero-Factor Property to conclude that either $(x - 2)$ or $(x + 3)$ must be zero. Setting the first factor equal to zero implies that $x = 2$ is a solution. That is,

$$x - 2 = 0 \quad \Longrightarrow \quad x = 2. \qquad \text{First solution}$$

Similarly, setting the second factor equal to zero implies that $x = -3$ is a solution. That is,

$$x + 3 = 0 \quad \Longrightarrow \quad x = -3. \qquad \text{Second solution}$$

So, the equation $(x - 2)(x + 3) = 0$ has exactly two solutions: $x = 2$ and $x = -3$. You can check these solutions by substituting them into the original equation.

Check

$$(x - 2)(x + 3) = 0 \qquad \text{Write original equation.}$$
$$(2 - 2)(2 + 3) \stackrel{?}{=} 0 \qquad \text{Substitute 2 for } x.$$
$$(0)(5) = 0 \qquad \text{First solution checks. } \checkmark$$
$$(-3 - 2)(-3 + 3) \stackrel{?}{=} 0 \qquad \text{Substitute } -3 \text{ for } x.$$
$$(-5)(0) = 0 \qquad \text{Second solution checks. } \checkmark$$

Solving Quadratic Equations by Factoring

Factoring and the Zero-Factor Property allow you to solve a quadratic equation by converting it into two *linear* equations (which you already know how to solve). This is a common strategy of algebra—to break down a problem into simpler parts, each solvable by previously learned methods.

In each of the examples that follow, note how factoring skills are combined with the Zero-Factor Property to solve equations.

Example 1 ■ Using Factoring to Solve an Equation

Solve the quadratic equation $x^2 - x - 12 = 0$.

Solution

Begin by checking to see that the right side of the equation is zero. Next, factor the left side of the equation. Finally, apply the Zero-Factor Property to find the solutions.

$$x^2 - x - 12 = 0 \qquad \text{Write original equation.}$$
$$(x + 3)(x - 4) = 0 \qquad \text{Factor left side of equation.}$$
$$x + 3 = 0 \implies x = -3 \qquad \text{Set 1st factor equal to 0 and solve for } x.$$
$$x - 4 = 0 \implies x = 4 \qquad \text{Set 2nd factor equal to 0 and solve for } x.$$

So, the given equation has two solutions: $x = -3$ and $x = 4$.

Check First Solution

$$x^2 - x - 12 = 0 \qquad \text{Write original equation.}$$
$$(-3)^2 - (-3) - 12 \stackrel{?}{=} 0 \qquad \text{Substitute } -3 \text{ for } x.$$
$$9 + 3 - 12 \stackrel{?}{=} 0 \qquad \text{Simplify.}$$
$$0 = 0 \qquad \text{Solution checks. } \checkmark$$

Check Second Solution

$$x^2 - x - 12 = 0 \qquad \text{Write original equation.}$$
$$(4)^2 - (4) - 12 \stackrel{?}{=} 0 \qquad \text{Substitute 4 for } x.$$
$$16 - 4 - 12 \stackrel{?}{=} 0 \qquad \text{Simplify.}$$
$$0 = 0 \qquad \text{Solution checks. } \checkmark$$

In order for the Zero-Factor Property to be used, a quadratic equation *must* be written in *general form*. That is, the quadratic must be on the left side of the equation and zero must be the only term on the right side of the equation. For instance, to write the equation $x^2 - 3x = 18$ in general form, you must subtract 18 from both sides of the equation, as follows.

$$x^2 - 3x = 18 \qquad \text{Write original equation.}$$
$$x^2 - 3x - 18 = 18 - 18 \qquad \text{Subtract 18 from each side.}$$
$$x^2 - 3x - 18 = 0 \qquad \text{General form}$$

To solve this equation, factor the left side as $(x + 3)(x - 6)$ and then form the linear equations $x + 3 = 0$ and $x - 6 = 0$. The solutions of these two linear equations are $x = -3$ and $x = 6$, respectively. The general strategy for solving a quadratic equation by factoring is summarized in the guidelines below.

Guidelines for Solving Quadratic Equations

1. Write the quadratic equation in general form.

2. Factor the left side of the equation.

3. Set each factor with a variable equal to zero.

4. Solve each linear equation.

5. Check each solution in the original equation.

Example 2 ■ Solving a Quadratic Equation by Factoring

Solve the quadratic equation

$$3x^2 + 5x = 12.$$

Solution

$$3x^2 + 5x = 12 \qquad \text{Write original equation.}$$
$$3x^2 + 5x - 12 = 0 \qquad \text{Write in general form.}$$
$$(3x - 4)(x + 3) = 0 \qquad \text{Factor left side of equation.}$$
$$3x - 4 = 0 \implies x = \tfrac{4}{3} \qquad \text{Set 1st factor equal to 0 and solve for } x.$$
$$x + 3 = 0 \implies x = -3 \qquad \text{Set 2nd factor equal to 0 and solve for } x.$$

So, the solutions are $x = \tfrac{4}{3}$ and $x = -3$. The solutions can be checked as follows.

Check First Solution

$$3x^2 + 5x = 12$$
$$3\left(\tfrac{4}{3}\right)^2 + 5\left(\tfrac{4}{3}\right) \stackrel{?}{=} 12$$
$$\tfrac{16}{3} + \tfrac{20}{3} \stackrel{?}{=} 12$$
$$\tfrac{36}{3} = 12 \ \checkmark$$

Check Second Solution

$$3x^2 + 5x = 12$$
$$3(-3)^2 + 5(-3) \stackrel{?}{=} 12$$
$$27 - 15 \stackrel{?}{=} 12$$
$$12 = 12 \ \checkmark$$

After converting a quadratic equation to general form, you should check to see whether the left side of the equation has a common *numerical* factor. If it does, you can divide each side of the equation by this factor without "losing" any of the solutions. For instance, each of the equations

$$5x^2 + 30x + 40 = 0 \qquad \text{Write original equation.}$$
$$5(x^2 + 6x + 8) = 0 \qquad \text{Factor out common monomial factor.}$$
$$x^2 + 6x + 8 = 0 \qquad \text{Divide each side by 5.}$$

has the same solutions. Be sure, however, that you *do not divide each side of an equation by a variable factor*. This will usually result in "losing" one or more of the solutions.

Example 3 ■ Equations with Common Monomial Factors

Solve each equation.

(a) $2x^2 - 2x - 24 = 0$

(b) $2x^2 = -10x$

Solution

(a) $\quad 2x^2 - 2x - 24 = 0 \qquad$ Write original equation.

$\quad\quad 2(x^2 - x - 12) = 0 \qquad$ Factor out common monomial factor.

$\quad\quad 2(x - 4)(x + 3) = 0 \qquad$ Factor trinomial.

$\quad\quad\quad\quad\quad\quad 2 \neq 0 \qquad$ 1st factor does not contain a variable.

$\quad\quad\quad\quad x - 4 = 0 \implies x = 4 \qquad$ Set 2nd factor equal to 0 and solve for x.

$\quad\quad\quad\quad x + 3 = 0 \implies x = -3 \qquad$ Set 3rd factor equal to 0 and solve for x.

The solutions are $x = 4$ and $x = -3$. Check these in the original equation. Notice that you do not obtain a solution by setting a *constant* factor equal to zero.

(b) $\quad\quad\quad 2x^2 = -10x \qquad$ Write original equation.

$\quad\quad 2x^2 + 10x = 0 \qquad$ Add $10x$ to each side.

$\quad\quad 2x(x + 5) = 0 \qquad$ Factor out common monomial factor.

$\quad\quad\quad\quad 2x = 0 \implies x = 0 \qquad$ Set 1st factor equal to 0 and solve for x.

$\quad\quad\quad x + 5 = 0 \implies x = -5 \qquad$ Set 2nd factor equal to 0 and solve for x.

The solutions are $x = 0$ and $x = -5$. Check these in the original equation. Notice that you obtain a solution of zero by setting a factor of the form kx equal to zero.

NOTE In Example 3(a), notice that you could divide each side of the equation by the *constant* factor 2 without losing a solution. In Example 3(b), however, dividing each side of the equation by the variable factor $2x$ would result in losing the solution $x = 0$.

In Examples 1, 2, and 3, the original equations each involved a second-degree (quadratic) polynomial and each had *two different* solutions. You will sometimes encounter second-degree polynomial equations that have only one (repeated) solution. This occurs when the left side of the equation is a perfect square trinomial, as shown in Example 4.

Example 4 ■ A Quadratic Equation with a Repeated Solution

Solve the quadratic equation

$$x^2 - 6x + 11 = 2.$$

Solution

$x^2 - 6x + 11 = 2$	Write original equation.
$x^2 - 6x + 9 = 0$	Write in general form.
$(x - 3)(x - 3) = 0$	Factor left side of equation.
$x - 3 = 0$ or $x - 3 = 0$	Set factors equal to 0.
$x = 3$	Repeated solution is 3.

Note that even though the left side of this equation has two factors, the two factors are the same. So, you can conclude that the only solution of the equation is $x = 3$. (This solution is called a **repeated solution.**) Check this solution in the original equation, as shown below.

Check

$x^2 - 6x + 11 = 2$	Write original equation.
$(3)^2 - 6(3) + 11 \overset{?}{=} 2$	Substitute 3 for x.
$9 - 18 + 11 \overset{?}{=} 2$	Simplify.
$2 = 2$	Solution checks. ✓

Technology

Some graphing utilities have a built-in "equation solver." Consult the user's guide for your calculator or graphing utility for the proper keystrokes. Use the *solve* feature of a graphing utility to solve each equation.

(a) $x^2 + 4x + 3 = 0$

(b) $3x^2 - x - 2 = 0$

(c) $2x^2 + 11x - 6 = 0$

Be sure you see that the Zero-Factor Property can be applied only to a product that is equal to *zero*. For instance, the following solution is incorrect.

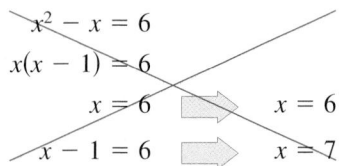

$$x^2 - x = 6$$
$$x(x - 1) = 6$$
$$x = 6 \qquad x = 6$$
$$x - 1 = 6 \qquad x = 7$$

Instead, you must first write the equation in general form and then factor the left side, as follows.

$$x^2 - x = 6$$
$$x^2 - x - 6 = 0$$
$$(x - 3)(x + 2) = 0$$
$$x - 3 = 0 \implies x = 3$$
$$x + 2 = 0 \implies x = -2$$

Try checking that $x = 6$ and $x = 7$ *are not* solutions of $x^2 - x = 6$, and that $x = 3$ and $x = -2$ *are* solutions.

Example 5 ■ Solving a Quadratic Equation by Factoring

Solve the quadratic equation

$$(x + 2)(x + 4) = 3.$$

Solution

$(x + 2)(x + 4) = 3$	Write original equation.
$x^2 + 6x + 8 = 3$	Multiply factors.
$x^2 + 6x + 5 = 0$	Write in general form.
$(x + 1)(x + 5) = 0$	Factor.
$x + 1 = 0 \implies x = -1$	Set 1st factor equal to 0.
$x + 5 = 0 \implies x = -5$	Set 2nd factor equal to 0.

So, the equation has two solutions: $x = -1$ and $x = -5$. Check these solutions in the original equation, as shown below.

Check

$(x + 2)(x + 4) = 3$	Write original equation.
$(-1 + 2)(-1 + 4) \overset{?}{=} 3$	Substitute -1 for x.
$(1)(3) = 3$	First solution checks. ✓
$(-5 + 2)(-5 + 4) \overset{?}{=} 3$	Substitute -5 for x.
$(-3)(-1) = 3$	Second solution checks. ✓

Some quadratic equations do not have solutions that are real numbers. For instance, there is no real number that satisfies the equation $x^2 = -4$. The reason that this equation has no real solution is that there is no real number that can be multiplied by itself to produce -4. This type of equation will be discussed further in Chapter 5.

You might remember from an earlier course in algebra that a polynomial equation can have *at most* as many solutions as its degree. For instance, a second-degree equation can have zero, one, or two real solutions, but it cannot have three or more solutions.

Example 6 ■ Solving Linear and Quadratic Equations

Describe the technique you would use to solve each equation.

(a) $7x + 1 = 2x - 3$

(b) $7x + 1 = 2x^2 - 3$

Solution

(a) This is a linear equation, so you would solve it by isolating x. After doing this, you would find that the solution is $x = -\frac{4}{5}$.

(b) This is a quadratic equation. To solve it, you would write the equation in general form, factor the quadratic, and apply the Zero-Factor Property to conclude that the solutions are $x = 4$ and $x = -\frac{1}{2}$.

Solving Higher-Degree Equations by Factoring

The Zero-Factor Property can be used to solve polynomial equations of degree three or higher. To do this, use the same strategy you used with quadratic equations. That is, write the equation in general form with the polynomial on the left and zero on the right. Then factor the left side of the equation. Finally, set each factor equal to zero to obtain the solutions.

Example 7 ■ Solving a Polynomial Equation with Three Factors

Solve $3x^3 = 15x^2 + 18x$.

Solution

$3x^3 = 15x^2 + 18x$	Write original equation.
$3x^3 - 15x^2 - 18x = 0$	Write in general form.
$3x(x^2 - 5x - 6) = 0$	Factor out common monomial factor.
$3x(x - 6)(x + 1) = 0$	Factor.
$3x = 0 \implies x = 0$	Set 1st factor equal to 0.
$x - 6 = 0 \implies x = 6$	Set 2nd factor equal to 0.
$x + 1 = 0 \implies x = -1$	Set 3rd factor equal to 0.

There are three solutions: $x = 0$, $x = 6$, and $x = -1$. Check these in the original equation.

Notice that the equation in Example 7 is a third-degree equation and has three solutions. Observe that there are four solutions to the fourth-degree equation in Example 8.

Example 8 ■ Solving a Polynomial Equation with Four Factors

Solve $x^4 + x^3 - 4x^2 - 4x = 0$.

Solution

$x^4 + x^3 - 4x^2 - 4x = 0$	Write original equation.
$x(x^3 + x^2 - 4x - 4) = 0$	Factor out common monomial factor.
$x[(x^3 + x^2) - (4x + 4)] = 0$	Group terms.
$x[x^2(x + 1) - 4(x + 1)] = 0$	Factor by grouping.
$x[(x + 1)(x^2 - 4)] = 0$	Distributive Property
$x(x + 1)(x + 2)(x - 2) = 0$	Factor.
$x = 0 \implies x = 0$	Set 1st factor equal to 0.
$x + 1 = 0 \implies x = -1$	Set 2nd factor equal to 0.
$x + 2 = 0 \implies x = -2$	Set 3rd factor equal to 0.
$x - 2 = 0 \implies x = 2$	Set 4th factor equal to 0.

There are four solutions: $x = 0$, $x = -1$, $x = -2$, and $x = 2$. Check these in the original equation.

Applications

Example 9 ■ Free-Falling Object

The height h of a rock that is dropped into a well that is 64 feet deep above the water level is given by the position equation $h = -16t^2 + 64$, where the height is measured in feet and the time t is measured in seconds. (See Figure 1.7.) How long will it take the rock to hit the water at the bottom of the well?

Solution

In Figure 1.7, note that the water level of the well corresponds to a height of 0 feet. So, substitute a height of 0 for h in the equation and solve for t.

$$0 = -16t^2 + 64 \qquad \text{Substitute 0 for } h.$$
$$16t^2 - 64 = 0 \qquad \text{Write in general form.}$$
$$16(t^2 - 4) = 0 \qquad \text{Factor out common factor.}$$
$$16(t + 2)(t - 2) = 0 \qquad \text{Difference of two squares}$$
$$t = -2 \quad \text{or} \quad t = 2 \qquad \text{Solutions using Zero-Factor Property}$$

Because a time of -2 seconds does not make sense in this problem, choose the positive solution $t = 2$, and conclude that the rock hits the water 2 seconds after it is dropped. Check this solution in the original statement of the problem.

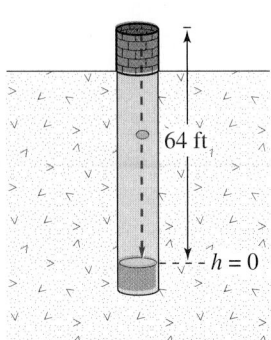

Figure 1.7

Example 10 ■ Geometry

A triangular roadside sign has a base that is to be 4 feet less than twice its height, as shown in Figure 1.8. A local zoning ordinance restricts the area of signs to a maximum of 24 square feet. Find the base and height of the largest sign that meets the zoning ordinance.

Solution

The area of a triangle is given by $A = \frac{1}{2}bh$, where h is the height measured in feet and b is the base measured in feet. Substitute $2h - 4$ for b and 24 for A in the equation and solve for h.

$$\frac{1}{2}bh = A \qquad \text{Equation for the area of a triangle}$$
$$\frac{1}{2}(2h - 4)(h) = 24 \qquad \text{Substitute } 2h - 4 \text{ for } b \text{ and 24 for } A.$$
$$h^2 - 2h = 24 \qquad \text{Distributive Property}$$
$$h^2 - 2h - 24 = 0 \qquad \text{Write in general form.}$$
$$(h + 4)(h - 6) = 0 \qquad \text{Factor.}$$
$$h = -4 \quad \text{or} \quad h = 6 \qquad \text{Solutions using Zero-Factor Property}$$

Because the height must be positive, discard -4 as a solution and choose $h = 6$. So, the height of the sign is 6 feet, and the base is

$$b = 2h - 4 \qquad \text{Equation for base}$$
$$= 2(6) - 4 \qquad \text{Substitute 6 for } h.$$
$$= 8 \text{ feet.} \qquad \text{Simplify.}$$

Check this solution in the original statement of the problem.

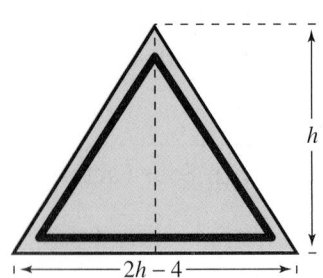

Figure 1.8

Collaborate!

Extending the Concept

Determine how you could solve each equation for the indicated variable.

Solve for a in $(a + b)^2 - 7(a + b) + 12 = 0$

Solve for m in $(m - n)^2 - 4(m - n) - 32 = 0$

Solve for k in $3(w + k)^2 + 19(w + k) - 14 = 0$

Then solve each equation below for the indicated variable.

Solve for x in $x^2 + 2xy + y^2 + 6x + 6y + 8 = 0$

Solve for c in $c^2 - 2cd + d^2 + 8c - 8d - 20 = 0$

Solve for t in $2r^2 + 4rt + 2t^2 - 15r - 15t + 18 = 0$

1.6 ■ Exercises

Developing Skills

In Exercises 1–10, use the Zero-Factor Property to solve the equation.

1. $2x(x - 8) = 0$ **2.** $z(z + 6) = 0$

3. $(y - 3)(y + 10) = 0$ **4.** $(s - 16)(s + 15) = 0$

5. $25(a + 4)(a - 2) = 0$

6. $17(t - 3)(t + 8) = 0$

7. $4x(2x - 3)(2x + 25) = 0$

8. $\frac{1}{5}x(x - 2)(3x + 4) = 0$

9. $(x - 3)(2x + 1)(x + 4) = 0$

10. $(y - 39)(2y + 7)(y + 12) = 0$

In Exercises 11–48, solve the polynomial equation by factoring.

11. $x^2 - 3x - 10 = 0$ **12.** $y^2 + 12y + 27 = 0$

13. $y^2 + 20 = 9y$ **14.** $x^2 + x = 12$

15. $3x^2 + 9x = 0$ **16.** $5y - y^2 = 0$

17. $x^2 - 25 = 0$ **18.** $x^2 - 81 = 0$

19. $3x^2 - 300 = 0$ **20.** $4b^3 - 36b = 0$

21. $m^2 - 8m = -16$

22. $a^2 + 4a + 4 = 0$

23. $4z^2 + 9 = 12z$

24. $2x^2 + 24x + 72 = 0$

25. $7 + 13x - 2x^2 = 0$

26. $11 + 32y - 3y^2 = 0$

27. $x(x - 3) = 10$

28. $s(s + 4) = 96$

29. $y(y + 6) = 72$

30. $x(x - 2) = 15$

31. $x(x + 2) - 10(x + 2) = 0$

32. $x(x - 15) + 3(x - 15) = 0$

33. $(t - 2)^2 - 16 = 0$

34. $(x + 4)^2 - 49 = 0$

35. $6t^3 - t^2 - t = 0$

36. $3u^3 - 5u^2 - 2u = 0$

37. $x^3 - 19x^2 + 84x = 0$

38. $x^3 + 18x^2 + 45x = 0$

39. $x^2(x - 25) - 16(x - 25) = 0$

40. $y^2(y + 250) - (y + 250) = 0$

41. $z^2(z + 2) - 4(z + 2) = 0$

42. $16(3 - u) - u^2(3 - u) = 0$

43. $c^3 - 3c^2 - 9c + 27 = 0$

44. $v^3 + 4v^2 - 4v - 16 = 0$

45. $a^3 + 2a^2 - 9a - 18 = 0$

46. $x^3 - 2x^2 - 4x + 8 = 0$

47. $x^4 - 5x^3 - 9x^2 + 45x = 0$

48. $2x^4 + 6x^3 - 50x^2 - 150x = 0$

Solving Problems

49. *Investigation*   Solve the equation

$$3(x + 4)^2 + (x + 4) - 2 = 0$$

in the following two ways.

  (a) Let $u = x + 4$, and solve the resulting equation for u. Then find the corresponding values of x that are solutions of the original equation.

  (b) Expand and collect like terms in the original equation, and solve the resulting equation for x.

  (c) Which method is easier? Explain.

Investigation   In Exercises 50 and 51, solve the equation using either method (or both methods) described in Exercise 49.

50. $3(x + 6)^2 - 10(x + 6) - 8 = 0$

51. $8(x + 2)^2 - 18(x + 2) + 9 = 0$

52. Let a and b be real numbers such that $a \neq 0$. Find the solutions of $ax^2 + bx = 0$.

53. Let a be a nonzero real number. Find the solutions of $ax^2 - ax = 0$.

Think About It   In Exercises 54–57, find a quadratic equation with the indicated solutions.

54. $x = -3, x = 5$ **55.** $x = 1, x = 6$

56. $x = 7, \ x = 4$ **57.** $x = -2, \ x = -10$

58. *Number Problem*   The sum of a positive number and its square is 240. Find the number.

59. *Number Problem*   The sum of a positive number and its square is 72. Find the number.

60. *Free-Falling Object*   An object is dropped from a weather balloon 6400 feet above the ground (see figure). Find the time t for the object to reach the ground by solving the equation $-16t^2 + 6400 = 0$.

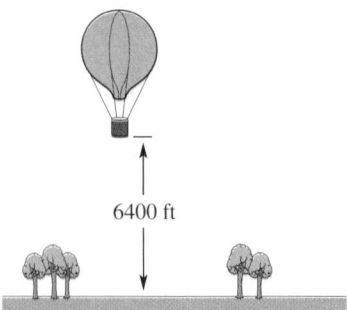

6400 ft

61. *Free-Falling Object*   An object is thrown upward from the Royal Gorge Bridge in Colorado, 1053 feet above the Arkansas River, with an initial velocity of 48 feet per second. Find the time t for the object to reach the ground by solving the equation $-16t^2 + 48t + 1053 = 0$.

62. *Geometry*   The rectangular floor of a storage shed has an area of 330 square feet. The length of the floor is 7 feet more than its width (see figure). Find the dimensions of the floor.

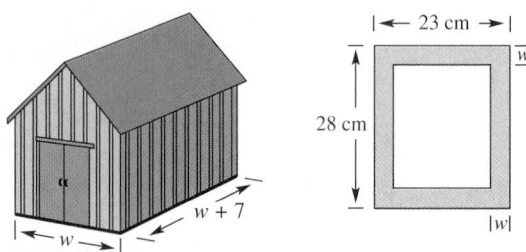

Figure for 62   Figure for 63

63. *Geometry*   The outside dimensions of a picture frame are 28 centimeters and 23 centimeters (see figure). The area of the exposed part of the picture is 414 square centimeters. Find the width w of the frame.

64. *Geometry*   The triangular cross section of a machined part must have an area of 48 square inches (see figure). Find the base and height of the triangle if the height is $1\frac{1}{2}$ times the base.

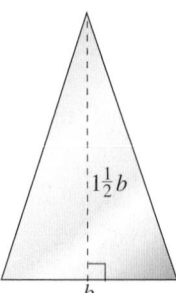

$1\frac{1}{2}b$

b

65. *Geometry*   The height of a triangle is 4 inches less than its base. Find the base and height of the triangle if its area is 70 square inches.

66. *Geometry*   An open box with a square base is to be constructed from 405 square centimeters of material (see figure). What should the dimensions of the base be if the height of the box is to be 3 centimeters? (*Hint:* The surface area is given by $S = x^2 + 4xh$.)

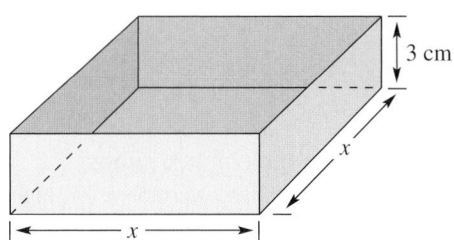

Figure for 66

67. *Geometry*   An open box with a square base is to be constructed from 880 square inches of material (see figure). What should the dimensions of the base be if the height of the box is to be 6 inches? (*Hint:* The surface area is given by $S = x^2 + 4xh$.)

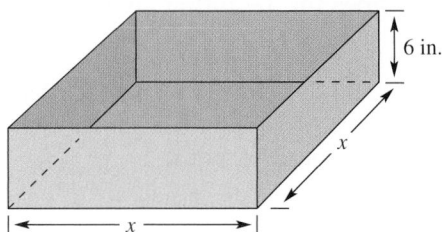

68. *Business*   The revenue R from the sale of x units of a product is given by $R = 90x - x^2$. The cost of producing x units of the product is $C = 200 + 60x$. How many units of the product must be produced and sold in order to break even $(R = C)$?

69. *Business*   The revenue R from the sale of x units of a product is given by $R = x^2 - 35x$. The cost of producing x units of the product is $C = 150 + 12x$. How many units of the product must be produced and sold in order to break even $(R = C)$?

70. *Numerical Reasoning*   Consider the product $P = (x + 5)(x - 4)$.

(a) Complete the table.

x	3	4	5	6	7	8
P						

(b) Use the table to determine how P changes for each one-unit increase in x.

(c) Use your answer to part (b) to solve the equation $(x + 5)(x - 4) = 70$.

Explaining Concepts

71. *True or False?*   If $(2x - 5)(x + 4) = 1$, then $2x - 5 = 1$ or $x + 4 = 1$. Explain.

72. Is it possible for a quadratic equation to have only one solution? Explain.

73. What is the maximum number of solutions of an nth-degree polynomial equation? Give an example of a third-degree equation that has only one real-number solution.

Looking Further

An open box is to be made from a piece of material 5 meters long and 4 meters wide. The box is made by cutting squares of dimension x from the corners and turning up the sides (see figure). The volume V of a rectangular solid is the product of its length, width, and height.

(a) Show that the volume is given by
$$V = (5 - 2x)(4 - 2x)x.$$

(b) Determine the values of x for which $V = 0$.

(c) Use the *table* feature of a graphing utility to complete the table.

x	0.25	0.50	0.75	1.00	1.25	1.50	1.75
V							

(d) Use the table to determine x if $V = 3$. Verify the result algebraically.

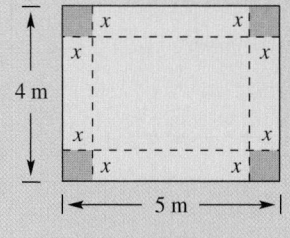
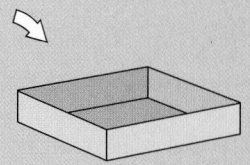

Chapter Summary: Key Concepts

1.1 ■ Properties of algebra

Let a, b, and c represent real numbers, variables, or algebraic expressions.

Commutative Property of Addition: $a + b = b + a$

Commutative Property of Multiplication: $ab = ba$

Associative Property of Addition:

$(a + b) + c = a + (b + c)$

Associative Property of Multiplication: $(ab)c = a(bc)$

Distributive Property:

$a(b \pm c) = ab \pm ac$

$(a \pm b)c = ac \pm bc$

Additive Identity Property: $a + 0 = 0 + a = a$

Multiplicative Identity Property: $a \cdot 1 = 1 \cdot a = a$

Additive Inverse Property: $a + (-a) = 0$

Multiplicative Inverse Property: $a \cdot \dfrac{1}{a} = 1, \quad a \neq 0$

1.1 ■ Properties of exponents

Let m and n be positive integers, and let a and b be real numbers, variables, or algebraic expressions.

1. $a^m \cdot a^n = a^{m+n}$ 2. $(ab)^m = a^m \cdot b^m$

3. $(a^m)^n = a^{mn}$

4. $\dfrac{a^m}{a^n} = a^{m-n}, m > n, a \neq 0$

5. $\left(\dfrac{a}{b}\right)^m = \dfrac{a^m}{b^m}, b \neq 0$

1.2 ■ Special products

Let u and v be real numbers, variables, or algebraic expressions. Then the formulas below are true.

1. Sum and Difference of Two Terms:

$(u + v)(u - v) = u^2 - v^2$

2. Square of a Binomial: $(u \pm v)^2 = u^2 \pm 2uv + v^2$

1.3 ■ Difference of two squares

Let u and v be real numbers, variables, or algebraic expressions. Then the expression $u^2 - v^2$ can be factored as follows: $u^2 - v^2 = (u + v)(u - v)$.

1.3 ■ Sum or difference of two cubes

Let u and v be real numbers, variables, or algebraic expressions. Then the expressions $u^3 \pm v^3$ can be factored as follows: $u^3 \pm v^3 = (u \pm v)(u^2 \mp uv + v^2)$.

1.4 ■ Perfect square trinomials

Let u and v be real numbers, variables, or algebraic expressions. Then the expressions $u^2 \pm 2uv + v^2$ can be factored as follows: $u^2 \pm 2uv + v^2 = (u \pm v)^2$.

1.4 ■ Guidelines for factoring polynomials

1. Factor out any common factors.

2. Factor according to one of the special polynomial forms: difference of squares, sum or difference of cubes, or perfect square trinomial.

3. Factor trinomials using the methods for $a = 1$ or $a \neq 1$.

4. Factor by grouping—for polynomials with four terms.

5. Check to see if the factors themselves can be factored further.

1.5 ■ Forming equivalent equations: properties of equality

An equation can be transformed into an equivalent equation using one or more of the procedures listed below.

1. Simplify each side.

2. Apply the Addition Property of Equality.

3. Apply the Multiplication Property of Equality.

4. Interchange sides.

1.6 ■ Zero-Factor Property

Let u and v be real numbers, variables, or algebraic expressions. If u and v are factors such that $uv = 0$, then $u = 0$ or $v = 0$. This property also applies to three or more factors.

1.6 ■ Guidelines for solving quadratic equations

1. Write the quadratic equation in general form.

2. Factor the left side of the equation.

3. Set each factor with a variable equal to zero.

4. Solve each linear equation.

5. Check each solution in the original equation.

Review Exercises

Reviewing Skills

1.1 ■ *How to identify the terms and coefficients of algebraic expressions*

In Exercises 1–4, identify the terms and coefficients of the algebraic expression.

1. $14y^2 - 9$

2. $-7x + 13$

3. $15t^3 - 2t^2 + 19t$

4. $22z^2 + z - 15$

■ *How to identify the properties of algebra*

In Exercises 5–8, identify the property of algebra that is illustrated in each statement.

5. $5 + (4 - y) = (5 + 4) - y$

6. $(u - v)(2) = 2(u - v)$

7. $(x + y) + 0 = x + y$

8. $ab \cdot \dfrac{1}{ab} = 1$

■ *How to apply the properties of exponents to simplify algebraic expressions*

In Exercises 9–20, use the properties of exponents to simplify the expression.

9. $x^2 \cdot x^3 \cdot x$

10. $y \cdot y^5 \cdot y^4$

11. $y^3(-2y^2)$

12. $2x^2(5x^4)$

13. $(-2a^2)^3(8a)$

14. $(7x^4)^2(x^3)^3$

15. $(xy)(-3x^2y^3)$

16. $3uv(-2uv^2)^2$

17. $-(u^2v)^2(-4u^3v)$

18. $(12x^2y)(3x^2y^4)$

19. $\dfrac{120u^5v^3}{15u^3v}$

20. $-\dfrac{(-2x^2y^3)^2}{-3xy^2}$

■ *How to simplify algebraic expressions by combining like terms and removing symbols of grouping*

In Exercises 21–32, simplify the algebraic expression.

21. $5(2x - 4) - 7x$

22. $-4(3x + 7y) + 5y$

23. $y(3y - 10) + y^2$

24. $x(3x + 4y) - 7xy$

25. $30x - (10x + 80)$

26. $(5x - 7) + (3x - 2)$

27. $3x - (y - 2x)$

28. $(9x + 5y) - (4x + y)$

29. $-2(11x - y) + 7(2x + 3y)$

30. $8(4x + 5y) - 3(12x - y)$

31. $3[b + 5(b - a)]$

32. $-2t[8 - (6 - t)]$

■ *How to evaluate algebraic expressions*

In Exercises 33–36, evaluate the algebraic expression for the specified values of the variable(s).

33. $2x^2 + x - 6$ (a) $x = -4$ (b) $x = \frac{1}{2}$

34. $y^2 - 5y + 1$ (a) $y = -2$ (b) $y = \frac{2}{3}$

35. $4x - xy$ (a) $x = 1$ (b) $x = -3$
                         $y = 5$ $y = 4$

36. $3x^2 + 2y^2 - 9$ (a) $x = 6$ (b) $x = 10$
                              $y = -1$ $y = 7$

1.2 ■ *How to identify the leading coefficients and degrees of polynomials*

In Exercises 37 and 38, write the polynomial in standard form, and find its degree and leading coefficient.

37. $x + 12x^6 - 4x^2 + 15$ **38.** $11t^4 - t^2 + 7t^3 + 21$

■ *How to add and subtract polynomials*

■ *How to multiply polynomials*

In Exercises 39–54, perform the operations and simplify.

39. $(5x + 3x^2) + (x - 4x^2)$

40. $(3t^2 - 5t) + (t^2 + 5t - 4)$

41. $(5x^2 - 2x + 7) - (3x - 10)$

42. $(12x^4 + 7x^2 - 13) - (8x^4 + 11)$

43. $(-x^3 - 3x) - 4(2x^3 - 3x + 1)$

44. $(7z^2 + 6z) - 3(5z^2 + 2z)$

45. $3y^2 - [2y - 3(y^2 + 5)]$

46. $(16a^3 + 5a) - 5[a + (2a^3 - 1)]$

47. $(-2x)^3(x + 4)$ **48.** $3y(-4y)(y - 2)$

49. $(2z + 3)(3z - 5)$ **50.** $(6t + 1)(t - 11)$

51. $(5x + 3)(3x - 4)$ **52.** $(3y^2 + 2)(4y^2 - 5)$

53. $(2x^2 - 3x + 2)(2x + 3)$

54. $(5s^2 + 4s - 3)(4s - 5)$

■ *How to find special products*

In Exercises 55–58, find the product by using the special product formulas.

55. $(4x - 7)^2$ **56.** $(8 - 3x)^2$

57. $(5u - 8)(5u + 8)$ **58.** $(7a + 4)(7a - 4)$

1.3 ■ *How to factor polynomials with common factors*

In Exercises 59–62, factor the expression by factoring out a common factor.

59. $6x^2 + 15x^3$ **60.** $8y - 12y^4$

61. $28(x + 5) - 70(x + 5)^2$

62. $(u - 9v)(u - v) + v(u - 9v)$

■ *How to factor polynomials by grouping terms*

In Exercises 63 and 64, factor the expression by grouping terms.

63. $y^3 + 5y^2 + 7y + 35$ **64.** $v^3 - 2v^2 - 4v + 8$

■ *How to factor the difference of two squares*

In Exercises 65–70, factor the expression by factoring the difference of squares.

65. $x^2 - 16$ **66.** $4y^2 - 1$

67. $9a^2 - 100$ **68.** $25x^2 - 16y^2$

69. $(y - 3)^2 - 16$ **70.** $4 - (x + 2)^2$

■ *How to factor the sum or difference of two cubes*

In Exercises 71–74, factor the expression by factoring the sum or difference of two cubes.

71. $u^3 - 1$ **72.** $8t^3 - 27$

73. $27x^3 + 64$ **74.** $y^3 + z^3$

■ *How to factor polynomials completely*

In Exercises 75–78, factor the expression completely.

75. $x^4 + 7x^3 - 9x^2 - 63x$

76. $y^4 + 2y^3 - 16y^2 - 32y$

77. $x^2 + 18x + 81 - 4y^2$ **78.** $x^2 - 6x + 9 - y^2$

1.4 ■ *How to factor trinomials of the form $x^2 + bx + c$*

In Exercises 79 and 80, factor the trinomial.

79. $x^2 - 11x + 24$ **80.** $a^2 + 7a - 18$

■ *How to factor trinomials of the form $ax^2 + bx + c$*

In Exercises 81 and 82, factor the trinomial.

81. $3x^2 + 23x - 8$ **82.** $10x^2 - 9x - 7$

■ *How to factor trinomials by grouping*

In Exercises 83 and 84, factor the trinomial by grouping.

83. $2x^2 - x - 15$ **84.** $6x^2 + x - 12$

■ *How to factor perfect square trinomials*

In Exercises 85 and 86, factor the perfect square trinomial.

85. $a^2 + 4ab + 4b^2$ **86.** $4u^2 - 28u + 49$

■ *How to select the best factoring technique using the guidelines for factoring polynomials*

In Exercises 87–92, factor the polynomial completely.

87. $18y^3 - 32y$

88. $250a^3 - 2b^3$

89. $6h^3 - 23h^2 - 13h$

90. $35y^3 - 10y^2 - 25y$

91. $x^4 - 4x^3 + x^2 - 4x$

92. $3u - 2u^2 + 6 - u^3$

1.5 ■ *How to identify types of equations and check solutions of equations*

In Exercises 93 and 94, identify the equation as a conditional equation, an identity, or an equation with no solution.

93. $7x = 7(x + 5)$

94. $3(x + 8) = 2x$

In Exercises 95 and 96, determine whether the values of the variable are solutions of the equation.

95. $10x - 3 = 17$ (a) $x = 2$ (b) $x = -1$

96. $-2x + 9 = 11$ (a) $x = 0$ (b) $x = -1$

■ *How to solve linear equations*

In Exercises 97–102, solve the linear equation.

97. $5x - 2 = 13$ **98.** $12x + 15 = 3$

99. $8x + 4 = 6x - 10$ **100.** $3x + 10 = 6x - 8$

101. $8 - 5t = 20 + t$ **102.** $3y + 14 = y + 20$

■ *How to solve linear equations containing parentheses*

In Exercises 103–108, solve the linear equation.

103. $4y - 6(y - 5) = 2$ **104.** $7x + 2(7 - x) = 8$

105. $3(x + 1) = 5(x - 9)$

106. $8(x - 2) = 3(x - 2)$

107. $2[(x + 7) - 9] = 5(x - 4)$

108. $6[x - (5x - 7)] = 4 - 5x$

■ *How to solve linear equations containing fractions or decimals*

In Exercises 109–112, solve the linear equation.

109. $\frac{4}{5}x - \frac{1}{10} = \frac{3}{2}$ **110.** $\frac{1}{4}s + \frac{3}{8} = \frac{5}{2}$

111. $4.1(2x + 1.5) = 16.8$ **112.** $1.2(x - 3) = 10.8$

■ *How to solve literal equations*

In Exercises 113–116, solve the equation for the specified variable.

113. $V = \pi r^2 h$; solve for h.

114. $S = 2\pi r^2 + 2\pi rh$; solve for h.

115. $-3x + 8 = 5y - 11$; solve for y.

116. $5(x + 3) - 2(3y - 10) = 0$; solve for x.

1.6 ■ *How to use the Zero-Factor Property to solve equations*

In Exercises 117–120, solve the equation.

117. $10x(x - 3) = 0$ **118.** $3x(4x + 7) = 0$

119. $(x - 4)(x + 5) = 0$

120. $(u - 8)(u + 10) = 0$

■ *How to solve quadratic equations by factoring*

In Exercises 121–130, solve the quadratic equation by factoring.

121. $x^2 - 5x + 6 = 0$ **122.** $3s^2 - 2s - 8 = 0$

123. $7 + 13x - 2x^2 = 0$

124. $11 + 32y - 3y^2 = 0$

125. $z(5 - z) + 36 = 0$ **126.** $y(y + 8) + 7 = 0$

127. $v^2 - 100 = 0$ **128.** $x^2 - 121 = 0$

129. $3y^2 - 48 = 0$ **130.** $(x + 3)^2 - 25 = 0$

■ *How to solve higher-degree equations by factoring*

In Exercises 131–134, solve the equation by factoring.

131. $2y^3 - 2y^2 - 24y = 0$ **132.** $t^3 + 3t^2 - 18t = 0$

133. $b^3 - 6b^2 - b + 6 = 0$

134. $x^3 + x^2 - 9x - 9 = 0$

Solving Problems

Geometry **In Exercises 135 and 136, write expressions for the perimeter and area of the region, and then simplify.**

135.

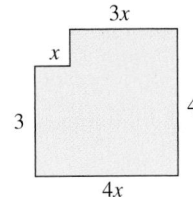

136.

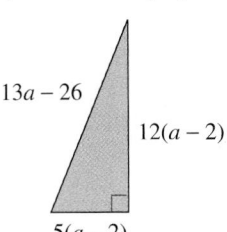

137. *Geometry* The length of a rectangle is five times its width w. Find algebraic expressions that represent (a) the perimeter and (b) the area of the rectangle.

138. *Personal Finance* After 2 years, an investment of $750 compounded annually at an interest rate of r will yield an amount $750(1 + r)^2$. Find this product.

139. *Geometry* The area of a rectangle of length l is given by $16l - l^2$. Factor this expression to deter-

mine the width of the rectangle. Draw a diagram to illustrate your answer.

140. *Geometry* Write, in factored form, an expression for the area of the shaded portion of the figure.

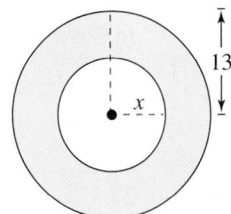

141. *Geometry* The width of a rectangle is three-fourths of its length. Find the dimensions of the rectangle if its area is 432 square inches.

142. *Free-Falling Object* An object is thrown upward from a height of 48 feet with an initial velocity of 32 feet per second. Find the time t for the object to reach the ground by solving the equation

$$-16t^2 + 32t + 48 = 0.$$

1 Chapter Test

Take this test as you would take a test in class. After you are done, check your work against the answers given in the back of the book.

In Exercises 1–10, simplify the expression.

1. $(3x^2y)(-xy)^2$

2. $(5x^2y^3)^2(-2xy)^3$

3. $\dfrac{12x^3y^5}{4x^2y}$

4. $3x^2 - 2x - 5x^2 + 7x - 1$

5. $-5x + 4x^2 - 6x + 3 + 2x^2$

6. $(x^2 - 7x + 4) + (9x^2 + x + 1)$

7. $(16 - y^2) - (16 + 2y + y^2)$

8. $-2(2x^4 - 5) + 4x(x^3 + 2x - 1)$

9. $4t - [3t - (10t + 7)]$

10. $2y\left(\dfrac{y}{4}\right)^2$

In Exercises 11–14, multiply the polynomials and simplify.

11. $(2x - 3y)(x + 5y)$

12. $(2s - 3)(3s^2 - 4s + 7)$

13. $(4x - 3)^2$

14. $(6 - 4y)(6 + 4y)$

In Exercises 15–24, factor completely.

15. $18y^2 - 12y$

16. $5x^3 - 10x^2 - 6x + 12$

17. $9u^2 - 6u + 1$

18. $6x^2 - 26x - 20$

19. $b^2 - 2b - 48$

20. $49x^2 - 36$

21. $(a + 2)^2 - 9b^2$

22. $2y^2 + 15y + 18$

23. $2x^3 + 26x^2 + 24x$

24. $2x^3 - 128y^3$

In Exercises 25–30, solve the equation.

25. $6x - 5 = 19$

26. $15 - 7(1 - x) = 3(x + 8)$

27. $\dfrac{2x}{3} = \dfrac{x}{2} + 4$

28. $3y^2 - 5y = 12$

29. $(x + 5)(x - 2) = 60$

30. $2x^3 + 10x^2 + 8x = 0$

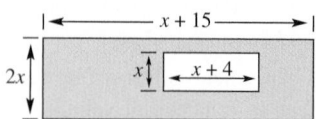

Figure for 31

31. Find the area of the shaded region shown at the left.

32. Solve the literal equation $5a + 2b - 10 = 8a + 7$ for b.

Introduction to Graphs and Functions

2

- **2.1** Describing Data Graphically
- **2.2** Graphs of Equations
- **2.3** Slope: An Aid to Graphing Lines
- **2.4** Relations, Functions, and Function Notation
- **2.5** Graphs of Functions
- **2.6** Transformations of Functions

THE BIG PICTURE

In this chapter you will learn how to:

- plot points on a rectangular coordinate system.

- sketch and use graphs of real-life data.

- use the Distance and Midpoint Formulas.

- sketch graphs of equations and find the *x*- and *y*-intercepts.

- sketch graphs of lines using slopes and *y*-intercepts.

- evaluate functions and find their domains.

- sketch graphs of functions.

- identify and sketch transformations of graphs of functions.

Key Terms

As you encounter each new vocabulary term in this chapter, add the term and its definition to your notebook glossary.

ordered pairs (p. 80)
rectangular coordinate system (p. 80)
Cartesian plane (p. 80)
x-axis (p. 80)
y-axis (p. 80)
origin (p. 80)
quadrants (p. 80)
coordinates (p. 80)
scatter plot (p. 82)
solution point (p. 84)
Pythagorean Theorem (p. 87)
Distance Formula (p. 88)
Midpoint Formula (p. 90)
collinear (p. 90)
graph (p. 96)
linear equation (p. 96)
intercepts (p. 100)

x-intercept (p. 100)
y-intercept (p. 100)
slope (p. 109)
slope-intercept form (p. 113)
constant rate of change (p. 116)
average rate of change (p. 116)
relation (p. 123)
domain of the relation (p. 123)
range of the relation (p. 123)
function (p. 124)
domain of the function (p. 124)
range of the function (p. 124)
independent variable (p. 126)
dependent variable (p. 126)
piecewise-defined function (p. 128)
linear function (p. 135)
constant function (p. 137)

Additional text-specific resources are available to help you do well in this course.
See page xvi for details.

2.1 Describing Data Graphically

What you should learn:

- How to plot points on a rectangular coordinate system
- How to determine whether ordered pairs are solutions of equations
- How to use the Distance Formula to find the distance between two points
- How to use the Midpoint Formula to find the midpoints of line segments

Why you should learn it:

A rectangular coordinate system can be used to represent relationships between two variables. For example, Exercise 98 on page 94 shows the relationship between the various speeds of a car and the corresponding fuel efficiencies.

The Rectangular Coordinate System

Just as you can represent real numbers by points on the real number line, you can represent **ordered pairs** of real numbers by points in a plane. This plane is called a **rectangular coordinate system** or the **Cartesian plane,** after the French mathematician René Descartes.

A rectangular coordinate system is formed by two real lines intersecting at right angles, as shown in Figure 2.1. The horizontal number line is usually called the **x-axis,** and the vertical number line is usually called the **y-axis.** (The plural of axis is *axes*.) The point of intersection of the two axes is called the **origin,** and the axes separate the plane into four regions called **quadrants.**

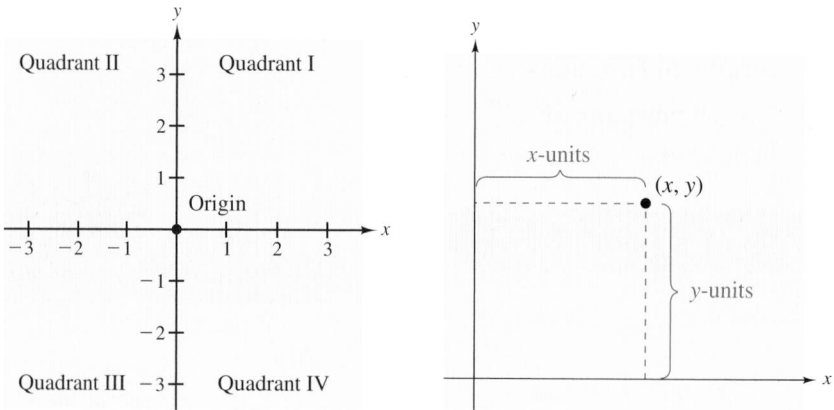

Figure 2.1   Figure 2.2

Each point in the plane corresponds to an **ordered pair** (x, y) of real numbers x and y, called the **coordinates** of the point. The first number (or **x-coordinate**) tells how far to the left or right the point is from the vertical axis, and the second number (or **y-coordinate**) tells how far up or down the point is from the horizontal axis, as shown in Figure 2.2.

A positive x-coordinate implies that the point lies to the *right* of the vertical axis; a negative x-coordinate implies that the point lies to the *left* of the vertical axis; and an x-coordinate of zero implies that the point lies *on* the vertical axis. Similar statements can be made about y-coordinates. A positive y-coordinate implies that the point lies *above* the horizontal axis; a negative y-coordinate implies that the point lies *below* the horizontal axis; and a y-coordinate of zero implies that the point lies *on* the horizontal axis.

NOTE   The signs of the coordinates tell you which quadrant the point lies in. For instance, if x and y are positive, the point (x, y) lies in Quadrant I.

Locating a given point in a plane is called **plotting** the point. Example 1 shows how this is done.

Example 1 ■ Plotting Points on a Rectangular Coordinate System

Plot the points

$$(-2, 1), (4, 0), (3, -1), (4, 3), (0, 0), \text{ and } (-1, -3)$$

on a rectangular coordinate system.

Solution

The point $(-2, 1)$ is two units to the *left* of the vertical axis and one unit *above* the horizontal axis.

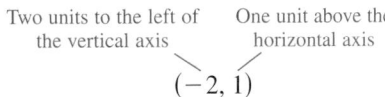

Similarly, the point $(4, 0)$ is four units to the *right* of the vertical axis and *on* the horizontal axis. (It is on the horizontal axis because its y-coordinate is zero.)

The other four points can be plotted in a similar way, as shown in Figure 2.3.

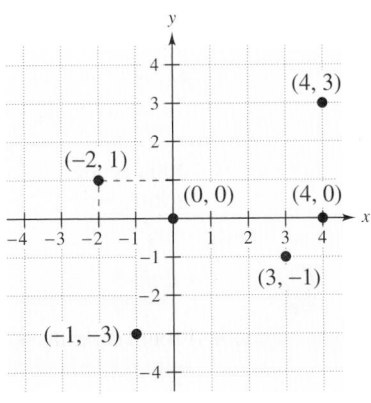

Figure 2.3

In Example 1, you were given the coordinates of several points and were asked to plot the points on a rectangular coordinate system. Example 2 looks at the reverse problem. That is, you are given points on a rectangular coordinate system and are asked to determine their coordinates.

Example 2 ■ Finding Coordinates of Points

Determine the coordinates of each of the points shown in Figure 2.4.

Solution

Point A lies two units to the *right* of the vertical axis and one unit *below* the horizontal axis. So, point A must be given by the ordered pair $(2, -1)$. The coordinates of the other five points can be determined in a similar way; the results are summarized as follows.

Point	Position	Coordinates
A	Two units *right*, one unit *down*	$(2, -1)$
B	One unit *left*, five units *up*	$(-1, 5)$
C	Zero units *right* (or *left*), two units *up*	$(0, 2)$
D	Three units *left*, two units *down*	$(-3, -2)$
E	Two units *right*, four units *up*	$(2, 4)$
F	One unit *left*, two units *up*	$(-1, 2)$

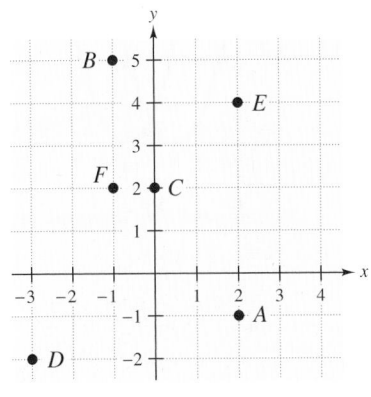

Figure 2.4

In Example 2, note that point A $(2, -1)$ and point F $(-1, 2)$ are different points. The order in which the numbers appear in an ordered pair is important. Notice that because point C lies on the y-axis, it has an x-coordinate of 0.

As a consumer today, you are presented almost daily with vast amounts of data given in various forms. Data is given in *numerical* form using lists and tables and in *graphical* form using scatter plots, line graphs, circle graphs, and bar graphs. Graphical forms are more visual and make wide use of Descartes's rectangular coordinate system to show the relationship between two variables. Today, Descartes's ideas are commonly used in virtually every scientific and business-related field.

In the next example, data is represented graphically by points plotted on a rectangular coordinate system. This type of graph is called a **scatter plot.**

Historical Note

Descartes introduced his analytical geometry in 1637. Most of the terminology we use comes from the Cartesian-coordinate system (*linear, quadratic,* etc.). Prior to this time there was no distinct tie between algebra and geometry.

Example 3 ■ Sketching a Scatter Plot

From 1990 through 1999, the amount E (in millions of dollars) spent on exercise equipment in the United States is shown in the table, where t represents the year. Sketch a scatter plot of the data. *(Source: National Sporting Goods Association)*

t	1990	1991	1992	1993	1994	1995	1996	1997	1998	1999
E	1824	2106	2078	2602	2781	2960	3232	2968	2850	3078

Solution

To sketch a scatter plot, begin by choosing which variable will be plotted on the horizontal axis and which will be plotted on the vertical axis. For this data, it seems natural to plot the years on the horizontal axis (which means that the amount must be plotted on the vertical axis). Next, use the data in the table to form ordered pairs. For instance, the first three ordered pairs are

(1990, 1824), (1991, 2106), and (1992, 2078).

All 10 points are shown in Figure 2.5. Note that the break in the x-axis indicates that the numbers between 0 and 1990 have been omitted. The break in the y-axis indicates that the numbers between 0 and 1800 have been omitted.

NOTE   In Example 3, you could have let $t = 0$ represent the year 1990. In that case, the horizontal axis would not have been broken, and the tick marks would have been labeled 0 through 9.

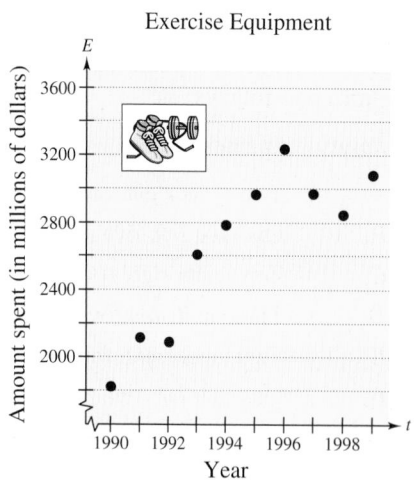

Figure 2.5

In Example 3, the data for the amount spent on exercise equipment was represented graphically by a *scatter plot*. This is only one of the many ways to represent data graphically. The same data could be represented by a **bar graph** or by a **line graph,** as shown in Figures 2.6 and 2.7.

NOTE   When data is presented graphically without a table, the exact values are not evident. However, you can use the graph to estimate the values. For instance, using the bar graph in Figure 2.6, you could estimate that the amount spent on exercise equipment in 1996 was approximately $3225 million.

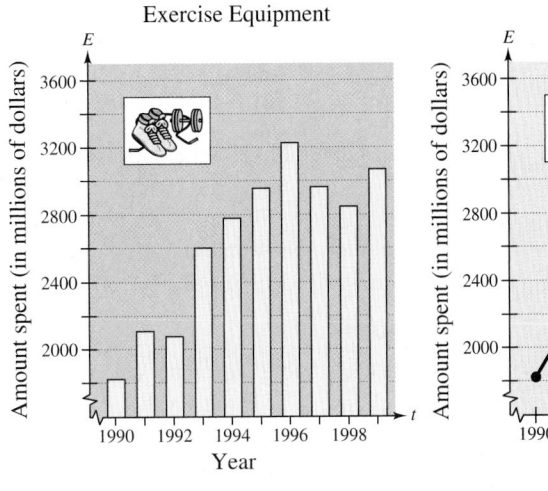

Figure 2.6   *Bar graph*   Figure 2.7   *Line graph*

Study Tip

Use a bar graph when the data fall into distinct categories and you want to compare totals. Use a line graph when you want to show the relationship between consecutive amounts of data over time.

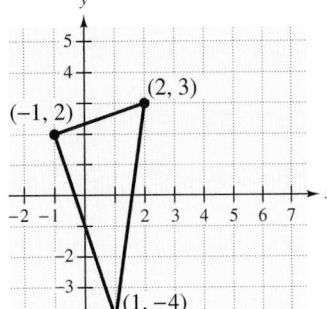

(a)

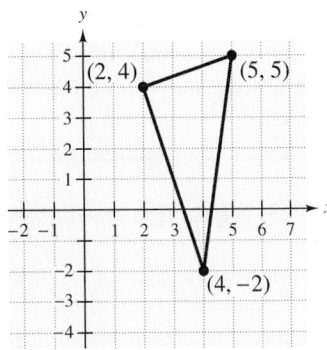

(b)
Figure 2.8

Example 4 ■ Translating Points in a Plane

The triangle shown in Figure 2.8(a) has vertices at the points

$$(-1, 2), (1, -4) \text{ and } (2, 3).$$

Shift the triangle three units to the right and two units up and find the vertices of the shifted triangle.

Solution

To shift the vertices three units to the right, add 3 to each of the x-coordinates. To shift the vertices two units up, add 2 to each of the y-coordinates.

Original Point	*Translated Point*
$(-1, 2)$	$(-1 + 3, 2 + 2) = (2, 4)$
$(1, -4)$	$(1 + 3, -4 + 2) = (4, -2)$
$(2, 3)$	$(2 + 3, 3 + 2) = (5, 5)$

The shifted triangle and its vertices are shown in Figure 2.8(b).

Ordered Pairs as Solutions

In Example 3, the relationship between the year and the amount spent on exercise equipment was given by a **table of values.** In mathematics, the relationship between the variables x and y is often given by an equation. From the equation, you can construct your own table of values.

Example 5 ■ Constructing a Table of Values

Construct a table of values for

$$y = 3x + 2.$$

Then plot the solution points on a rectangular coordinate system. Choose x-values of $-3, -2, -1, 0, 1, 2,$ and 3.

Solution

For each x-value, you must calculate the corresponding y-value. For example, if you choose $x = 1$, the y-value is

$$y = 3(1) + 2$$
$$= 5.$$

The ordered pair $(x, y) = (1, 5)$ is a **solution point** (or **solution**) of the equation.

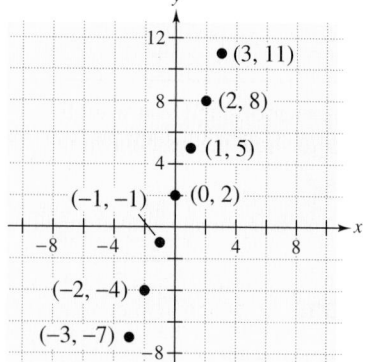

Figure 2.9

Choose x	Calculate y from $y = 3x + 2$	Solution point
$x = -3$	$y = 3(-3) + 2 = -7$	$(-3, -7)$
$x = -2$	$y = 3(-2) + 2 = -4$	$(-2, -4)$
$x = -1$	$y = 3(-1) + 2 = -1$	$(-1, -1)$
$x = 0$	$y = 3(0) + 2 = 2$	$(0, 2)$
$x = 1$	$y = 3(1) + 2 = 5$	$(1, 5)$
$x = 2$	$y = 3(2) + 2 = 8$	$(2, 8)$
$x = 3$	$y = 3(3) + 2 = 11$	$(3, 11)$

Once you have constructed a table of values, you can get a visual idea of the relationship between the variables x and y by plotting the solution points on a rectangular coordinate system, as shown in Figure 2.9.

When making a table of values for an equation, it is helpful first to solve the equation for y. Here is an example.

$$5x + 3y = 4 \qquad \text{Write original equation.}$$
$$3y = -5x + 4 \qquad \text{Subtract } 5x \text{ from each side.}$$
$$y = -\frac{5}{3}x + \frac{4}{3} \qquad \text{Divide each side by 3.}$$

Technology

1. Y =

2. TblSet

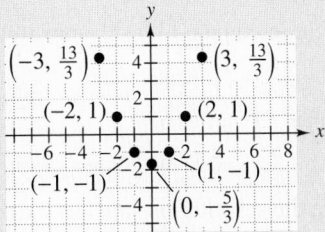

3. Table

Creating a Table with a Graphing Utility

Some graphing utilities have built-in programs that can create tables. Suppose, for instance, that you want to create a table of values for the equation $2x^2 - 3y = 5$. To begin, solve the equation for y.

$$2x^2 - 3y = 5 \qquad \text{Write original equation.}$$
$$-3y = -2x^2 + 5 \qquad \text{Subtract } 2x^2 \text{ from each side.}$$
$$y = \frac{2}{3}x^2 - \frac{5}{3} \qquad \text{Divide each side by } -3.$$

Now, using the equation $y = \frac{2}{3}x^2 - \frac{5}{3}$, you can use the steps below.

1. Enter the equation into the graphing utility.

2. Enter the beginning x-value you want displayed in the table, and the value of the increment for the x-values in the table. For instance, using a beginning x-value of -3 and an increment of 1, the table has x-values of $-3, -2, -1$, and so on.

3. Obtain the table of values. You can use the cursor keys to view x- and y-values that are not shown on the default screen.

You can confirm the values shown on the graphing utility's screen by substituting the appropriate x-values, as shown in the table below. (Note that the graphing utility lists decimal approximations of fractional values. For instance, $-\frac{5}{3}$ is listed as -1.667.)

Choose x	Calculate y from $y = \frac{2}{3}x^2 - \frac{5}{3}$	Solution point
$x = -3$	$y = \frac{2}{3}(-3)^2 - \frac{5}{3} = \frac{13}{3}$	$\left(-3, \frac{13}{3}\right)$
$x = -2$	$y = \frac{2}{3}(-2)^2 - \frac{5}{3} = 1$	$(-2, 1)$
$x = -1$	$y = \frac{2}{3}(-1)^2 - \frac{5}{3} = -1$	$(-1, -1)$
$x = 0$	$y = \frac{2}{3}(0)^2 - \frac{5}{3} = -\frac{5}{3}$	$\left(0, -\frac{5}{3}\right)$
$x = 1$	$y = \frac{2}{3}(1)^2 - \frac{5}{3} = -1$	$(1, -1)$
$x = 2$	$y = \frac{2}{3}(2)^2 - \frac{5}{3} = 1$	$(2, 1)$
$x = 3$	$y = \frac{2}{3}(3)^2 - \frac{5}{3} = \frac{13}{3}$	$\left(3, \frac{13}{3}\right)$

After creating the table, you can plot the points, as shown below.

In the next example, you are given several ordered pairs and are asked to determine whether they are solutions of the original equation. To do this, you need to substitute the values of x and y into the equation. If the substitution produces a true equation, the ordered pair (x, y) is a solution and is said to **satisfy** the equation.

Guidelines for Verifying Solutions

To verify that an ordered pair (x, y) is a solution of an equation with variables x and y, use the steps below.
1. Substitute the values of x and y into the equation.
2. Simplify each side of the equation.
3. If each side simplifies to the same number, the ordered pair is a solution. If the two sides yield different numbers, the ordered pair is not a solution.

Example 6 ■ Verifying Solutions of an Equation

Which of the ordered pairs are solutions of $x^2 - 2y = 6$?

(a) $(2, 1)$

(b) $(0, -3)$

(c) $\left(1, -\frac{5}{2}\right)$

Solution

(a) For the ordered pair $(2, 1)$, substitute $x = 2$ and $y = 1$ into the equation.

$$x^2 - 2y = 6 \qquad \text{Write original equation.}$$
$$(2)^2 - 2(1) \overset{?}{=} 6 \qquad \text{Substitute 2 for } x \text{ and 1 for } y.$$
$$2 \neq 6 \qquad \text{Is not a solution } ✗$$

Because the substitution does not satisfy the original equation, you can conclude that the ordered pair $(2, 1)$ *is not* a solution of the original equation.

(b) For the ordered pair $(0, -3)$, substitute $x = 0$ and $y = -3$ into the equation.

$$x^2 - 2y = 6 \qquad \text{Write original equation.}$$
$$(0)^2 - 2(-3) \overset{?}{=} 6 \qquad \text{Substitute 0 for } x \text{ and } -3 \text{ for } y.$$
$$6 = 6 \qquad \text{Is a solution } ✓$$

Because the substitution satisfies the original equation, you can conclude that the ordered pair $(0, -3)$ *is* a solution of the original equation.

(c) For the ordered pair $\left(1, -\frac{5}{2}\right)$, substitute $x = 1$ and $y = -\frac{5}{2}$ into the equation.

$$x^2 - 2y = 6 \qquad \text{Write original equation.}$$
$$(1)^2 - 2\left(-\frac{5}{2}\right) \overset{?}{=} 6 \qquad \text{Substitute 1 for } x \text{ and } -\frac{5}{2} \text{ for } y.$$
$$6 = 6 \qquad \text{Is a solution } ✓$$

Because the substitution satisfies the original equation, you can conclude that the ordered pair $\left(1, -\frac{5}{2}\right)$ *is* a solution of the original equation.

The Distance Formula

Recall that the distance d between two points a and b on the real number line is simply

$$d = |b - a|.$$

The same "absolute value rule" is used to find the distance between two points that lie on the same *vertical* or *horizontal* line, as shown in Example 7.

Example 7 ■ Finding Horizontal and Vertical Distances

(a) Find the distance between the points $(2, -2)$ and $(2, 4)$.

(b) Find the distance between the points $(-3, -2)$ and $(2, -2)$.

Solution

(a) Because the x-coordinates are equal, you can visualize a vertical line through the points $(2, -2)$ and $(2, 4)$, as shown in Figure 2.10. The distance between these two points is the absolute value of the difference of their y-coordinates.

$$\text{Vertical distance} = |4 - (-2)| \qquad \text{Subtract } y\text{-coordinates.}$$
$$= |6| \qquad \text{Simplify.}$$
$$= 6 \qquad \text{Evaluate absolute value.}$$

(b) Because the y-coordinates are equal, you can visualize a horizontal line through the points $(-3, -2)$ and $(2, -2)$, as shown in Figure 2.10. The distance between these two points is the absolute value of the difference of their x-coordinates.

$$\text{Horizontal distance} = |2 - (-3)| \qquad \text{Subtract } x\text{-coordinates.}$$
$$= |5| \qquad \text{Simplify.}$$
$$= 5 \qquad \text{Evaluate absolute value.}$$

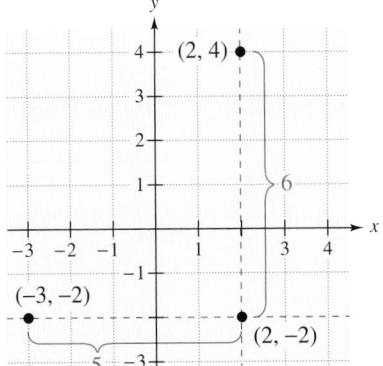

Figure 2.10

NOTE In Figure 2.10, note that the horizontal distance between the points $(-3, -2)$ and $(2, -2)$ is the absolute value of the difference of the x-coordinates, and the vertical distance between the points $(2, -2)$ and $(2, 4)$ is the absolute value of the difference of the y-coordinates.

The technique applied in Example 7 can be used to develop a general formula for finding the distance between two points in the plane. This general formula will work for any two points, even if they do not lie on the same vertical or horizontal line. To develop the formula, you use the **Pythagorean Theorem,** which states that for a right triangle, the hypotenuse c and sides a and b are related by the formula

$$a^2 + b^2 = c^2 \qquad \text{Pythagorean Theorem}$$

as shown in Figure 2.11. (The converse is also true. That is, if $a^2 + b^2 = c^2$, the triangle is a right triangle.)

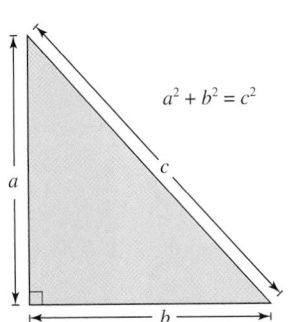

Figure 2.11 *Pythagorean Theorem*

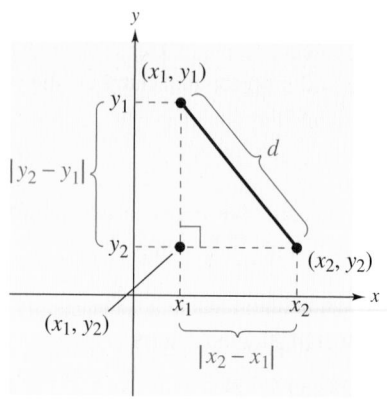

Figure 2.12   *Distance between two points*

NOTE   For the special case in which the two points lie on the same vertical or horizontal line, the Distance Formula still works.

To develop a general formula for the distance between two points, let (x_1, y_1) and (x_2, y_2) represent two points in the plane (that do not lie on the same horizontal or vertical line). With these two points, a right triangle can be formed, as shown in Figure 2.12. Note that the third vertex of the triangle is (x_1, y_2). Because (x_1, y_1) and (x_1, y_2) lie on the same vertical line, the length of the vertical side of the triangle is $|y_2 - y_1|$. Similarly, the length of the horizontal side is $|x_2 - x_1|$. By the Pythagorean Theorem, the square of the distance between (x_1, y_1) and (x_2, y_2) is

$$d^2 = |x_2 - x_1|^2 + |y_2 - y_1|^2.$$

Because the distance d must be positive, choose the positive square root and write

$$d = \sqrt{|x_2 - x_1|^2 + |y_2 - y_1|^2}.$$

Finally, replacing $|x_2 - x_1|^2$ and $|y_2 - y_1|^2$ by the equivalent expressions $(x_2 - x_1)^2$ and $(y_2 - y_1)^2$ gives the **Distance Formula.**

The Distance Formula

The distance d between two points (x_1, y_1) and (x_2, y_2) is

$$d = \sqrt{(x_2 - x_1)^2 + (y_2 - y_1)^2}.$$

Example 8   ■   Finding the Distance Between Two Points

Find the distance between the points $(-2, 1)$ and $(3, 4)$.

Algebraic Solution

Let $(x_1, y_1) = (-2, 1)$ and $(x_2, y_2) = (3, 4)$. Then apply the Distance Formula as follows.

$$d = \sqrt{(x_2 - x_1)^2 + (y_2 - y_1)^2} \quad \text{Distance Formula}$$

$$= \sqrt{[3 - (-2)]^2 + (4 - 1)^2} \quad \begin{matrix}\text{Substitute for } x_1, y_1,\\ x_2, \text{ and } y_2.\end{matrix}$$

$$= \sqrt{(5)^2 + (3)^2} \quad \text{Simplify.}$$

$$= \sqrt{34} \approx 5.83 \quad \text{Use a calculator.}$$

So, the distance between the points is about 5.83 units.

You can use the Pythagorean Theorem to check that the distance is correct.

$$d^2 \overset{?}{=} 3^2 + 5^2 \quad \begin{matrix}\text{Pythagorean}\\ \text{Theorem}\end{matrix}$$

$$\left(\sqrt{34}\right)^2 \overset{?}{=} 3^2 + 5^2 \quad \begin{matrix}\text{Substitute } \sqrt{34}\\ \text{for } d.\end{matrix}$$

$$34 = 34 \quad \text{Distance checks. } \checkmark$$

Graphical Solution

Use centimeter graph paper to plot the points $(-2, 1)$ and $(3, 4)$. Carefully sketch the line segment from one point to the other. Then use a centimeter ruler to measure the length of the segment.

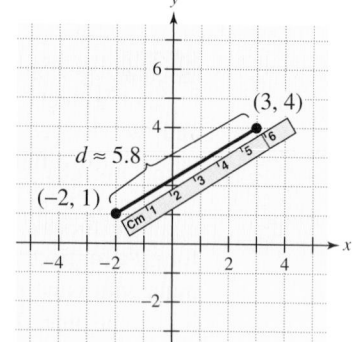

Figure 2.13

The line segment measures about 5.8 centimeters, as shown in Figure 2.13. So, the distance between the points is about 5.8 units.

The Distance Formula has many applications in mathematics. For instance, the next example shows how you can use the Distance Formula and the converse of the Pythagorean Theorem to decide whether three points form the vertices of a right triangle.

Example 9 ■ Verifying a Right Triangle

Show that the points $(1, 2)$, $(3, 1)$, and $(4, 3)$ are vertices of a right triangle.

Solution

The three points are plotted in Figure 2.14. Using the Distance Formula, you can find the lengths of the three sides of the triangle.

$$d_1 = \sqrt{(3 - 1)^2 + (1 - 2)^2} = \sqrt{4 + 1} = \sqrt{5}$$
$$d_2 = \sqrt{(4 - 3)^2 + (3 - 1)^2} = \sqrt{1 + 4} = \sqrt{5}$$
$$d_3 = \sqrt{(4 - 1)^2 + (3 - 2)^2} = \sqrt{9 + 1} = \sqrt{10}$$

Because $d_1{}^2 + d_2{}^2 = 5 + 5 = 10 = d_3{}^2$, you can conclude from the converse of the Pythagorean Theorem that the triangle is a right triangle.

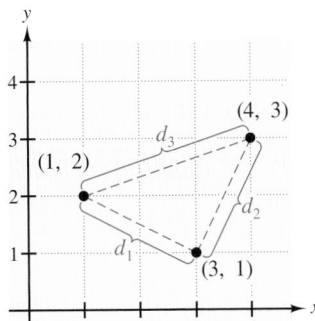

Figure 2.14

Example 10 ■ Finding the Length of a Pass

A football quarterback throws a pass from the five-yard line, 20 yards from the sideline. The pass is caught by a wide receiver on the 45-yard line, 50 yards from the same sideline, as shown in Figure 2.15. How long is the pass?

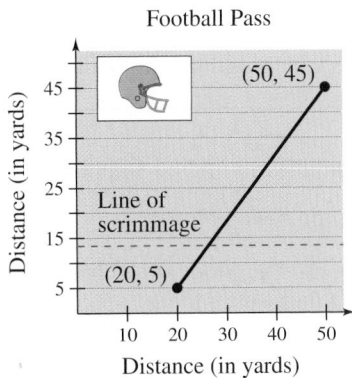

Figure 2.15

Solution

You can find the length of the pass by finding the distance between points $(20, 5)$ and $(50, 45)$.

$$d = \sqrt{(50 - 20)^2 + (45 - 5)^2} \qquad \text{Distance Formula}$$
$$= \sqrt{900 + 1600}$$
$$= 50 \qquad\qquad\qquad\qquad \text{Simplify.}$$

So, the pass is 50 yards long.

EXPLORATION

Plot the points $A(-1, -3)$ and $B(5, 2)$ and sketch the line segment from A to B. How could you verify that point $C(2, -0.5)$ is the midpoint of the segment? Why is it not sufficient to show that the distances from A to C and from C to B are equal?

The Midpoint Formula

The **midpoint** of a line segment that joins two points is the point that divides the segment into two equal parts. To find the midpoint of the line segment that joins two points in a coordinate plane, you can simply find the average values of the respective coordinates of the two endpoints using the **Midpoint Formula.**

The Midpoint Formula

The midpoint of the line segment joining the points (x_1, y_1) and (x_2, y_2) is given by the Midpoint Formula

$$\text{Midpoint} = \left(\frac{x_1 + x_2}{2}, \frac{y_1 + y_2}{2}\right).$$

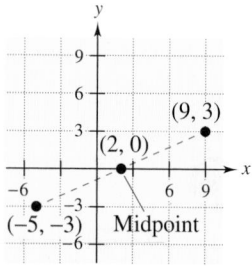

Figure 2.16

Example 11 ■ Finding the Midpoint of a Line Segment

Find the midpoint of the line segment joining the points $(-5, -3)$ and $(9, 3)$.

Solution

Let $(x_1, y_1) = (-5, -3)$ and $(x_2, y_2) = (9, 3)$.

$$\text{Midpoint} = \left(\frac{x_1 + x_2}{2}, \frac{y_1 + y_2}{2}\right). \qquad \text{Midpoint Formula}$$
$$= \left(\frac{-5 + 9}{2}, \frac{-3 + 3}{2}\right) \qquad \text{Substitute for } x_1, y_1, x_2, \text{ and } y_2.$$
$$= (2, 0) \qquad \text{Simplify.}$$

The line segment and the midpoint are shown in Figure 2.16.

Collaborate!

Extending the Concept

Three or more points are **collinear** if they all lie on the same line. Use the steps below to determine if the set of points

$$\{A(3, 1), B(5, 4), C(9, 10)\}$$

and the set of points

$$\{A(2, 2), B(4, 3), C(5, 4)\}$$

are collinear.

a. For each set of points, use the Distance Formula to find the distances from A to B, from B to C, and from A to C. What relationship exists among these distances for each set of points?

b. Plot each set of points on a rectangular coordinate system. Do all the points of either set appear to lie on the same line?

c. Compare your conclusions from part (a) with the conclusions you made from the graphs in part (b). Make a general statement about how to use the Distance Formula to determine collinearity.

2.1 ■ Exercises

Developing Skills

In Exercises 1–8, plot the points on a rectangular coordinate system.

1. $(7, 1), (-2, 4)$

2. $(-3, 10), (0, 6)$

3. $(4, 3), (-5, 3), (3, -5)$

4. $(-2, 5), (-2, -5), (3, 5)$

5. $(0, 0), (1, -9), (-3, -2)$

6. $(-1, 4), (0, -2), (5, 0)$

7. $\left(\frac{5}{2}, -2\right), \left(-2, \frac{1}{4}\right), \left(\frac{3}{2}, -\frac{7}{2}\right)$

8. $\left(-\frac{2}{3}, 3\right), \left(\frac{1}{4}, -\frac{5}{4}\right), \left(-5, -\frac{7}{4}\right)$

In Exercises 9–12, approximate the coordinates of the points.

9.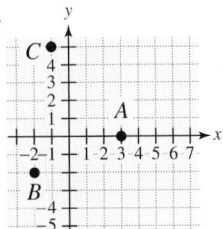

10.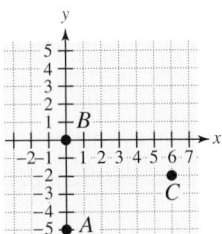

11.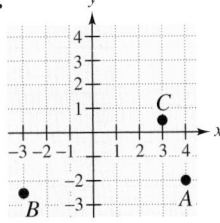

12.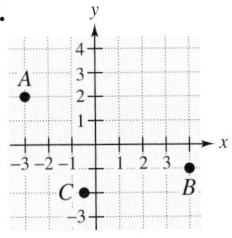

In Exercises 13–18, plot the points and connect them with line segments to form the figure.

13. *Triangle:* $(-1, 2), (2, 0), (3, 5)$

14. *Triangle:* $(0, -3), (2, 4), (5, -1)$

15. *Rectangle:* $(7, 0), (9, 1), (4, 6), (6, 7)$

16. *Rectangle:* $(0, 0), (2, -2), (6, 2), (4, 4)$

17. *Parallelogram:* $(4, 0), (6, -2), (0, -4), (-2, -2)$

18. *Parallelogram:* $(1, 0), (2, -3), (-4, -7), (-5, -4)$

In Exercises 19–30, determine the quadrant(s) in which the point is located without plotting it. Explain.

19. $(-3, -5)$

20. $(4, -2)$

21. $(5, 8)$

22. $(-3, 7)$

23. $(15, -15)$

24. $(-12, -10)$

25. $(x, 4)$

26. $(-10, y)$

27. $(x, y), \quad xy < 0$

28. $(x, y), \ xy > 0$

29. $(x, y), \ x < 0, \ y > 0$

30. $(x, y), \ x > 0, \ y > 0$

In Exercises 31–38, find the coordinates of the point.

31. The point is located five units to the left of the y-axis and two units above the x-axis.

32. The point is located 10 units to the right of the y-axis and 4 units below the x-axis.

33. The point is located three units to the right of the y-axis and six units above the x-axis.

34. The point is located nine units to the left of the y-axis and seven units below the x-axis.

35. The point is on the positive x-axis 10 units from the origin.

36. The point is on the negative y-axis five units from the origin.

37. The point is on the positive y-axis 12 units from the origin.

38. The point is on the negative x-axis one unit from the origin.

In Exercises 39–42, sketch a scatter plot of the points whose coordinates are shown in the table.

39. *Exam Scores*   The table shows the times x in hours invested in studying for five different algebra exams and the resulting exam scores y.

x	5	2	3	6.5	4
y	81	71	88	92	86

 Warm-up exercises, quizzes, and other study resources related to this section are available at *college.hmco.com*.

40. *Temperature*   The table shows the temperature y in degrees Fahrenheit at different times of the day x.

x	1:00	2:00	3:00	4:00	5:00	6:00
y	62°	65°	64°	64°	61°	58°

41. *Fuel Efficiency*   The table shows various speeds x of a car in miles per hour and the corresponding approximate fuel efficiencies y in miles per gallon.

x	50	55	60	65	70
y	28	26.4	24.8	23.4	22

42. *Average Temperature*   The table shows the average temperature y (degrees Fahrenheit) for Duluth, Minnesota, for each month of a year, with $x = 1$ representing January.   *(Source: NOAA)*

x	1	2	3	4	5	6
y	7.0	12.3	24.4	38.6	50.8	59.8

x	7	8	9	10	11	12
y	66.1	63.7	54.2	43.7	28.4	12.8

In Exercises 43–48, determine whether the ordered pairs are solutions of the equation.

43. $y = 3x + 8$

   (a) $(3, 17)$ (b) $(-1, 10)$

   (c) $(0, 0)$ (d) $(-2, 2)$

44. $5x - 2y + 50 = 0$

   (a) $(-10, 0)$ (b) $(-5, 5)$

   (c) $(0, 25)$ (d) $(20, -2)$

45. $y = \frac{7}{8}x$

   (a) $\left(\frac{8}{7}, 1\right)$ (b) $\left(4, \frac{7}{2}\right)$

   (c) $(0, 0)$ (d) $(-16, 14)$

46. $y = \frac{5}{8}x - 2$

   (a) $(0, 0)$ (b) $(2, 2)$

   (c) $(-4, -7)$ (d) $(32, 49)$

47. $4y - 2x + 1 = 0$

   (a) $(0, 0)$ (b) $\left(\frac{1}{2}, 0\right)$

   (c) $\left(-3, -\frac{7}{4}\right)$ (d) $\left(1, -\frac{3}{4}\right)$

48. $y = 10x - 7$

   (a) $(2, 10)$ (b) $(-2, -27)$

   (c) $(5, 43)$ (d) $(1, 5)$

In Exercises 49–52, complete the table of values. Then plot the solution points on a rectangular coordinate system.

49.

x	-2	0	2	4	6
$y = 5x - 1$					

50.

x	-2	0	2	4	6
$y = 3x + 2$					

51.

x	-4	$\frac{2}{5}$	4	8	12
$y = -\frac{5}{2}x + 4$					

52.

x	-6	-3	0	$\frac{3}{4}$	10
$y = \frac{4}{3}x - \frac{1}{3}$					

In Exercises 53 and 54, use the *table* feature of a graphing utility to complete the table of values.

53.

x	-2	0	2	4	6
$y = 4x^2 + x - 2$					

54.

x	-2	0	2	4	6
$y = \frac{4}{3}x - \frac{1}{3}$					

55. *Investigation*   Plot the points $(2, 1)$, $(-3, 5)$, and $(7, -3)$ on a rectangular coordinate system. Then change the sign of the x-coordinate of each point and plot the three new points on the same rectangular coordinate system. What conjecture can you make about the location of a point when the sign of its x-coordinate is changed?

The symbol ⊞ indicates an exercise in which you are instructed to use a calculator or graphing utility.
The solutions of other exercises may also be facilitated by the use of appropriate technology.

56. *Investigation*   Plot the points $(2, 1)$, $(-3, 5)$, and $(7, -3)$ on a rectangular coordinate system. Then change the sign of the y-coordinate of each point and plot the three new points on the same rectangular coordinate system. What conjecture can you make about the location of a point when the sign of its y-coordinate is changed?

Shifting a Graph   In Exercises 57–60, the figure is shifted to a new location in the plane. Find the coordinates of the vertices of the figure in its new location.

57.

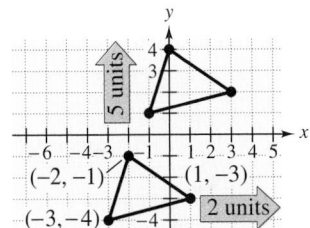

58.

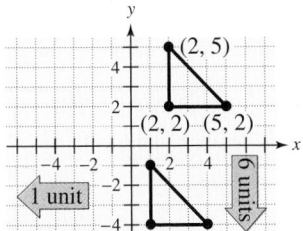

59.

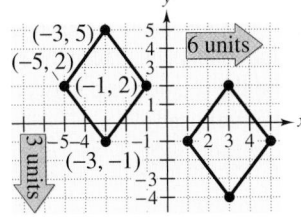

60.

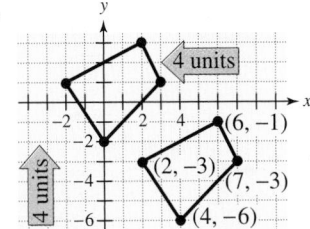

In Exercises 61–64, plot the points and find the distance between them. State whether the points lie on a horizontal or a vertical line.

61. $(3, -2)$, $(3, 5)$

62. $(-2, 8)$, $(-2, 1)$

63. $(3, 2)$, $(10, 2)$

64. $(-120, -2)$, $(130, -2)$

In Exercises 65–78, find the distance between the points.

65. $(1, 3)$, $(5, 6)$

66. $(3, 10)$, $(15, 5)$

67. $(0, 0)$, $(12, -9)$

68. $(-5, 0)$, $(3, 15)$

69. $(-9, -9)$, $(-4, 3)$

70. $(4, -5)$, $(-20, 2)$

71. $(-2, -3)$, $(4, 2)$

72. $(-5, 4)$, $(10, -3)$

73. $(1, 3)$, $(3, -2)$

74. $(-3, 11)$, $(6, 7)$

75. $\left(\frac{1}{3}, -1\right)$, $\left(\frac{10}{3}, -5\right)$

76. $\left(-5, \frac{4}{5}\right)$, $\left(3, \frac{34}{5}\right)$

77. $\left(\frac{1}{2}, 1\right)$, $\left(\frac{3}{2}, 2\right)$

78. $\left(5, \frac{2}{3}\right)$, $\left(0, \frac{8}{3}\right)$

Geometry   In Exercises 79 and 80, the coordinates of two of the vertices of a right triangle are given. (a) Find the length of the horizontal side, the length of the vertical side, and the length of the hypotenuse of the triangle. (b) Use your answers from part (a) to support the Pythagorean Theorem, $a^2 + b^2 = c^2$.

79.

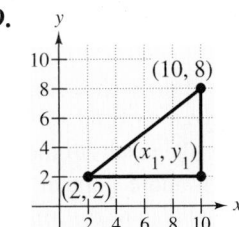

80.

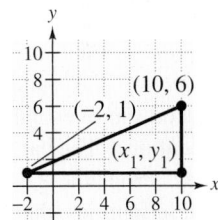

Geometry   In Exercises 81–84, determine whether the points are vertices of a right triangle.

81. $(2, 3)$, $(2, 6)$, $(6, 3)$

82. $(2, 4)$, $(1, 1)$, $(7, -1)$

83. $(8, 3)$, $(5, 2)$, $(1, 9)$

84. $(2, 4)$, $(-1, 6)$, $(-3, 1)$

Geometry   In Exercises 85 and 86, find the perimeter of the triangle with the given vertices.

85. $(-2, 0)$, $(0, 5)$, $(1, 0)$

86. $(-5, -2)$, $(-1, 4)$, $(3, -1)$

In Exercises 87–94, find the midpoint of the line segment joining the points, and then plot the points and the midpoint.

87. $(-2, 0)$, $(4, 8)$

88. $(-3, -2)$, $(7, 2)$

89. $(1, 9)$, $(5, -3)$

90. $(6, -5)$, $(-2, -1)$

91. $(1, 6)$, $(6, 3)$

92. $(2, 7)$, $(9, -1)$

93. $(5, 10)$, $(-2, -2)$

94. $(-4, 3)$, $(3, -4)$

Solving Problems

95. *Science* A scientist charted the growth of a certain bacterium over a period of 8 hours. The line graph shows the number of cells observed each hour of the experiment. From the graph, between which 2 hours did there not appear to be any change in the number of cells?

Growth of Bacterium

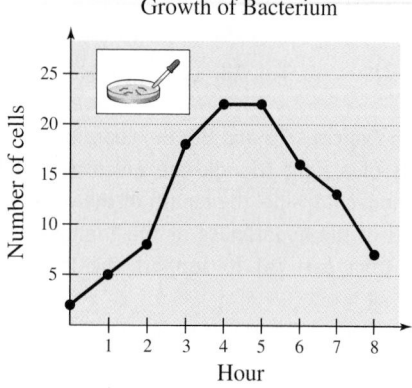

96. *Vacation* A survey of 500 Philadelphia families asked the question "What is your favorite vacation location?" The bar graph shows the results of the survey. From the graph, which is more popular with these families, the New England area or the South?

Favorite Vacation Location

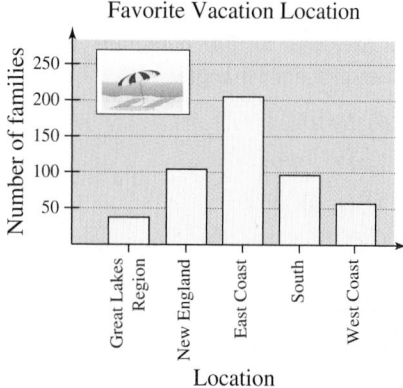

97. *Heating a Home* A family wished to study the relationship between the temperature outside and the amount of natural gas used to heat their house. For 16 months, the family observed the average temperature in degrees Fahrenheit for the month and the

average amount of natural gas in hundreds of cubic feet used in that month. The scatter plot shows this data for the 16 months. Does the graph suggest that there is a strong relationship between the temperature outside and the amount of natural gas used? If a month were to have an average temperature of 45°F, about how much natural gas would you expect this family to use on average to heat their house for that month?

Heating a Home

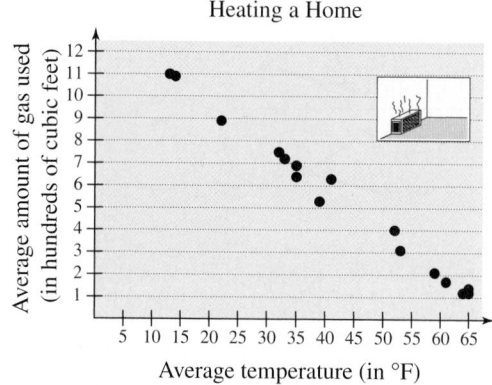

Figure for 97

98. *Fuel Efficiency* The scatter plot shows the speed of a car in kilometers per hour and the amount of fuel, in liters used per 100 kilometers traveled, that the car needs to maintain that speed. From the graph, how would you describe the relationship between the speed of the car and the fuel used?

Fuel Efficiency

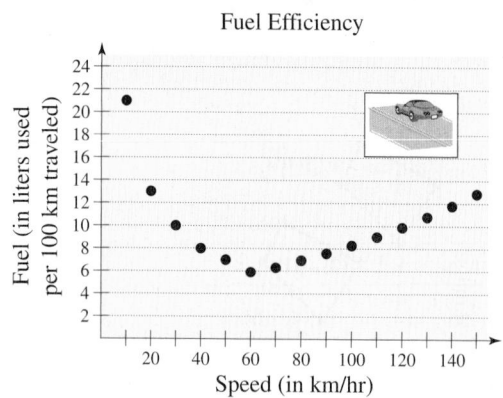

Explaining Concepts

99. Discuss the significance of the word *order* when referring to an ordered pair (x, y).

100. What is the *x*-coordinate of any point on the *y*-axis? What is the *y*-coordinate of any point on the *x*-axis?

Ongoing Review

In Exercises 101–104, simplify the expression.

101. $(x^4y^2)^2(-2xy^3)^2$

102. $(5a^3b^5)(-a^2b)^3$

103. $5(2x - 3) + x(x + 9)$

104. $3(2 - x) - 2x(4 - x)$

In Exercises 105–108, factor the expression completely.

105. $2x^2 - 32$

106. $3a^3 - 24b^2$

107. $4x^2 - x - 5$

108. $x^2 + 6x + 9$

In Exercises 109 and 110, solve the equation.

109. $x^2 - x - 6 = 0$

110. $x^2 + 8x = -15$

111. *Compound Interest*   After 2 years, an investment of $2000 compounded annually at an interest rate of 8.5% will yield an amount of $2000(1 + 0.085)^2$. Find this product.

112. *Number Problem*   The sum of a positive number and its square is 156. Find the number.

Looking Further

A median of a triangle is a segment whose endpoints are a vertex of the triangle and the midpoint of the opposite side. For instance, in $\triangle ABC$ shown below, D is the midpoint of side $\overline{BC}$. So, $\overline{AD}$ is a median of the triangle.

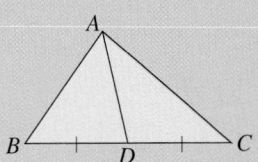

Every triangle has three medians. A theorem in geometry states that the three medians of any triangle will intersect at a point that is two-thirds of the distance from each vertex to the midpoint of the opposite side.

Complete parts (a)–(d) to show that the theorem is true for the triangle below.

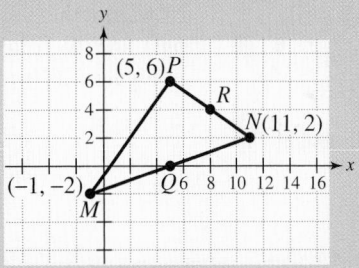

(a) Find the coordinates of Q, the midpoint of $\overline{MN}$.

(b) Find the length of the median $\overline{PQ}$.

(c) Find the coordinates of the point at which the three medians of the triangle would intersect. Label the point T.

(d) Find the coordinates of R, the midpoint of $\overline{PN}$. Show that the length of $\overline{MT}$ is $\frac{2}{3}$ the length of $\overline{MR}$.

2.2 Graphs of Equations

What you should learn:

- How to sketch graphs of equations using the point-plotting method
- How to find and use x- and y-intercepts as aids to sketching graphs
- How to use graphical checking to check solutions of equations involving one variable
- How to use graphs of equations in real-life problems

Why you should learn it:

The graph of an equation can help you see relationships between real-life quantities. For instance, in Exercise 78 on page 107 a graph can be used to relate the change in membership in the Girl Scouts of America to time.

The Graph of an Equation

In Section 2.1, you saw that the solutions of an equation in x and y can be represented by points on a rectangular coordinate system. The set of *all* solution points of an equation is called its **graph.** In this section, you will study a basic technique for sketching the graph of an equation—the **point-plotting method.**

Example 1 ■ Sketching the Graph of an Equation

Sketch the graph of $3x - y = 2$.

Solution

To begin, solve the equation for y to obtain $y = 3x - 2$. Next, create a table of values. The choice of x-values to use in the table is somewhat arbitrary. However, the more x-values you choose, the easier it will be to recognize a pattern.

x	-2	-1	0	1	2	3
$y = 3x - 2$	-8	-5	-2	1	4	7
Solution point	$(-2, -8)$	$(-1, -5)$	$(0, -2)$	$(1, 1)$	$(2, 4)$	$(3, 7)$

Now, plot the points, as shown in Figure 2.17(a). It appears that all six points lie on a line, so complete the sketch by drawing a line through the points, as shown in Figure 2.17(b).

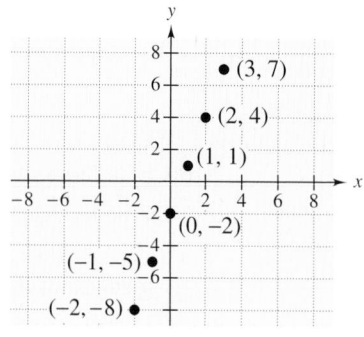

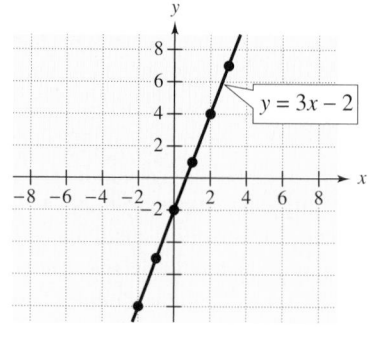

(a) (b)

Figure 2.17

The equation in Example 1 is an example of a **linear equation** in two variables—it is of first degree in both variables and its graph is a line. By drawing a line (curve) through the plotted points, we are implying that every point on this line (curve) is a solution point of the given equation and, conversely, that every solution point is on the line (curve).

The Point-Plotting Method of Sketching a Graph

1. If possible, rewrite the equation by isolating one of the variables.

2. Make a table of values showing several solution points.

3. Plot these points on a rectangular coordinate system.

4. Connect the points with a smooth curve or line.

Example 2 ■ Sketching the Graph of a Nonlinear Equation

Sketch the graph of

$$-x^2 + 2x + y = 0.$$

Solution

Begin by solving the equation for y to obtain $y = x^2 - 2x$. Next, create a table of values.

Study Tip

It is possible to have an equation with only one variable—for example, $y = 3$. When you make a table of values for the graph of $y = 3$, you will notice that y is 3 regardless of the value of x. This produces a horizontal line, as shown below.

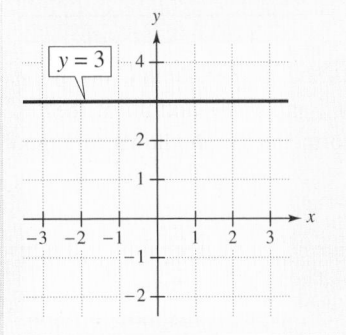

x	-2	-1	0	1	2	3	4
$y = x^2 - 2x$	8	3	0	-1	0	3	8
Solution point	$(-2, 8)$	$(-1, 3)$	$(0, 0)$	$(1, -1)$	$(2, 0)$	$(3, 3)$	$(4, 8)$

Now plot the seven solution points, as shown in Figure 2.18(a). Finally, connect the points with a smooth curve, as shown in Figure 2.18(b).

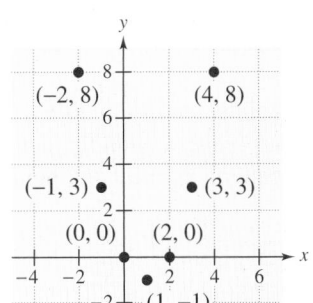

(a)

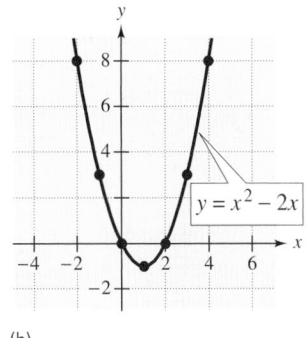

(b)

Figure 2.18

The graph of the equation given in Example 2 is called a **parabola.** You will study this type of graph in detail in Section 6.5.

NOTE   Example 2 shows three common ways to represent the relationship between two variables. The equation $y = x^2 - 2x$ is the *analytical* or *algebraic* representation, the table of values is the *numerical* representation, and the graph in Figure 2.18(b) is the *graphical* representation. You will see and use analytical, numerical, and graphical representations throughout this course.

Example 3 looks at the graph of an equation that involves an absolute value. Remember that the absolute value of a number is its distance from zero on the real number line. For instance,

$$|-5| = 5, \quad |2| = 2, \quad \text{and} \quad |0| = 0.$$

EXPLORATION

Use a graphing utility to graph each equation, and then answer the questions.

i. $y = 3x + 2$

ii. $y = 4 - x$

iii. $y = x^2 + 3x$

iv. $y = x^2 - 5$

v. $y = |x - 4|$

vi. $y = |x + 1|$

a. Which of the graphs are lines?
b. Which of the graphs are U-shaped?
c. Which of the graphs are V-shaped?
d. Describe the graph of the equation $y = x^2 + 7$ before you graph it. Use a graphing utility to confirm your answer.

Example 3 ■ The Graph of an Absolute Value Equation

Sketch the graph of $y = |x - 2|$.

Solution

This equation is already written in a form with y isolated on the left. So begin by creating a table of values. Be sure that you understand how the absolute value is evaluated. For instance, when $x = -2$, the value of y is

$$y = |-2 - 2| \qquad \text{Substitute } -2 \text{ for } x.$$
$$= |-4| \qquad \text{Simplify.}$$
$$= 4 \qquad \text{Simplify.}$$

and when $x = 3$, the value of y is

$$y = |3 - 2| \qquad \text{Substitute } 3 \text{ for } x.$$
$$= |1| \qquad \text{Simplify.}$$
$$= 1. \qquad \text{Simplify.}$$

x	-2	-1	0	1	2	3	4	5
$y = |x - 2|$	4	3	2	1	0	1	2	3
Solution point	$(-2, 4)$	$(-1, 3)$	$(0, 2)$	$(1, 1)$	$(2, 0)$	$(3, 1)$	$(4, 2)$	$(5, 3)$

Next, plot the points, as shown in Figure 2.19(a). It appears that the points lie in a "V-shaped" pattern, with the point $(2, 0)$ lying at the bottom of the "V." Following this pattern, connect the points to form the graph shown in Figure 2.19(b).

Technology

You can use a graphing utility to verify the graph in Example 3. Enter the equation into the graphing utility. Use a standard setting to display the graph as shown below.

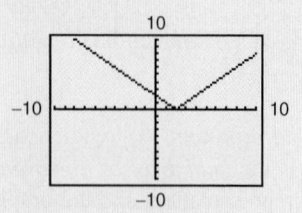

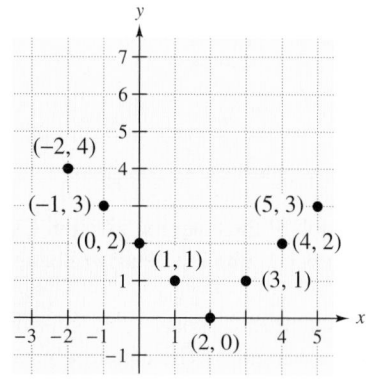

(a)

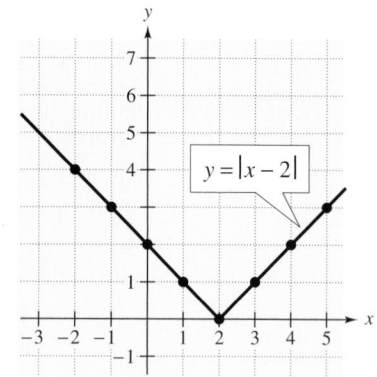

(b)

Figure 2.19

Technology

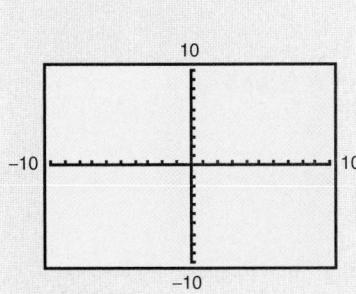

Setting the viewing window

A **viewing window** for a graph is a rectangular portion of the coordinate plane. A viewing window is determined by six values: the minimum x-value, the maximum x-value, the x-scale, the minimum y-value, the maximum y-value, and the y-scale, as shown at the left. When you enter these six values into a graphing utility, you are setting the **range** or **window.** Some graphing utilities have a standard viewing window, as shown at center left.

By choosing different viewing windows for a graph, it is possible to obtain very different impressions of the graph's shape. For instance, below are four different viewing windows for the graph of

$$y = -x^2 + 12x - 6.$$

Of these, the view in part (d) is the most complete.

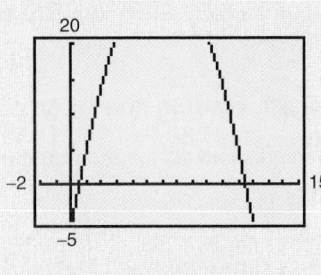

(a)

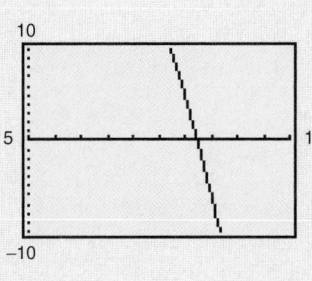

(b)

Standard viewing window

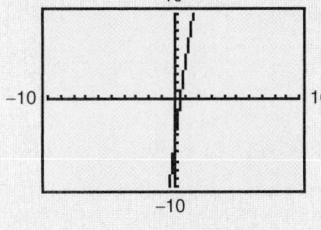

(c)

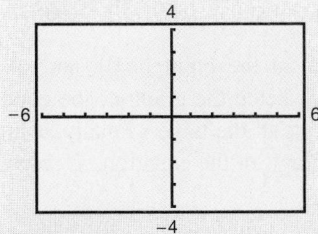

Square setting

(d)

On most graphing utilities, the display screen is two-thirds as high as it is wide. On such screens, you can obtain a graph with a true geometric perspective by using a **square setting**—one in which

$$\frac{Y_{max} - Y_{min}}{X_{max} - X_{min}} = \frac{2}{3}.$$

One such setting is shown at the left. Notice that the x and y tick marks are equally spaced on a square setting, but not on a standard one.

Intercepts: An Aid to Sketching Graphs

Two types of solution points that are especially useful are those having zero as the y-coordinate and those having zero as the x-coordinate. Such points are called **intercepts** because they are the points at which the graph intersects, respectively, the x- and y-axes.

Study Tip

When creating a table of values for a graph, choose a span of x-values that lie at least one unit to the left and right of the intercepts of the graph. This helps to give a more complete view of the graph.

Definitions of Intercepts

The point $(a, 0)$ is called an **x-intercept** of the graph of an equation if it is a solution point of the equation. To find the x-intercepts, let $y = 0$ and solve the equation for x.

The point $(0, b)$ is called a **y-intercept** of the graph of an equation if it is a solution point of the equation. To find the y-intercepts, let $x = 0$ and solve the equation for y.

Example 4 ■ Finding the Intercepts of a Graph

Find the intercepts and sketch the graph of

$$y = 2x - 3.$$

Solution

Find the x-intercept by letting $y = 0$ and solving for x.

$y = 2x - 3$	Write original equation.
$0 = 2x - 3$	Substitute 0 for y.
$3 = 2x$	Add 3 to each side.
$\frac{3}{2} = x$	Solve for x.

Find the y-intercept by letting $x = 0$ and solving for y.

$y = 2x - 3$	Write original equation.
$y = 2(0) - 3$	Substitute 0 for x.
$y = -3$	Solve for y.

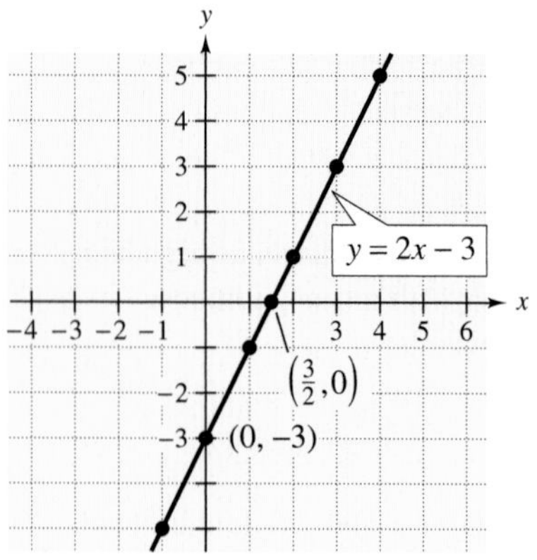

So, the graph has one x-intercept, which occurs at the point $\left(\frac{3}{2}, 0\right)$, and one y-intercept, which occurs at the point $(0, -3)$. To sketch the graph of the equation, create a table of values. (Include the intercepts in the table.) Finally, using the solution points given in the table, sketch the graph of the equation, as shown in Figure 2.20.

Figure 2.20

x	-1	0	1	$\frac{3}{2}$	2	3	4
$y = 2x - 3$	-5	-3	-1	0	1	3	5
Solution point	$(-1, -5)$	$(0, -3)$	$(1, -1)$	$\left(\frac{3}{2}, 0\right)$	$(2, 1)$	$(3, 3)$	$(4, 5)$

It is possible for a graph to have no intercepts or several intercepts. For instance, consider the three graphs in Figure 2.21.

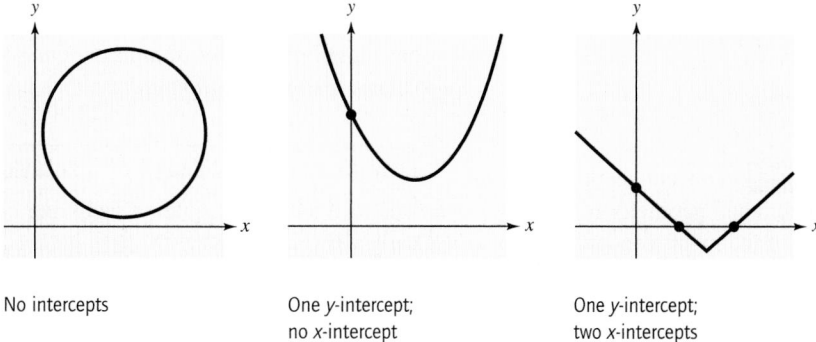

No intercepts

One *y*-intercept;
no *x*-intercept

One *y*-intercept;
two *x*-intercepts

Figure 2.21

Example 5 ■ A Graph That Has Two *x*-Intercepts

Find the intercepts and sketch the graph of

$$y = x^2 - 5x + 4.$$

Solution

Find the *x*-intercepts by letting $y = 0$ and solving for *x*.

$y = x^2 - 5x + 4$	Write original equation.
$0 = x^2 - 5x + 4$	Substitute 0 for *y*.
$0 = (x - 4)(x - 1)$	Factor.
$x - 4 = 0 \Longrightarrow x = 4$	Set 1st factor equal to 0 and solve for *x*.
$x - 1 = 0 \Longrightarrow x = 1$	Set 2nd factor equal to 0 and solve for *x*.

Find the *y*-intercept by letting $x = 0$ and solving for *y*.

$y = x^2 - 5x + 4$	Write original equation.
$y = 0^2 - 5(0) + 4$	Substitute 0 for *x*.
$y = 4$	Solve for *y*.

So, the graph has two *x*-intercepts, the points $(4, 0)$ and $(1, 0)$, and one *y*-intercept, the point $(0, 4)$. To sketch the graph of the equation, create a table of values. (Include the intercepts in the table.)

x	-1	0	1	2	3	4	5
$y = x^2 - 5x + 4$	10	4	0	-2	-2	0	4
Solution point	$(-1, 10)$	$(0, 4)$	$(1, 0)$	$(2, -2)$	$(3, -2)$	$(4, 0)$	$(5, 4)$

Next, plot the points given in the table, as shown in Figure 2.22(a). Then connect them with a smooth curve, as shown in Figure 2.22(b).

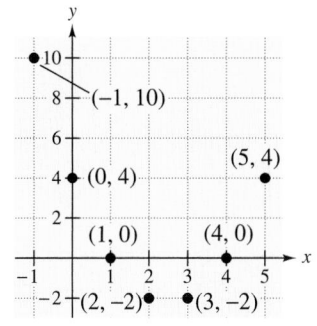

(a)

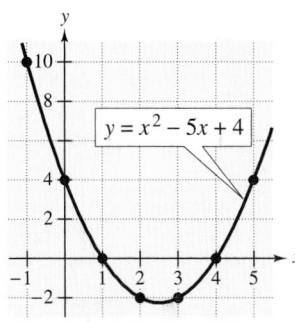

(b)

Figure 2.22

The Connection Between Solutions and x-Intercepts

In Chapter 1, we emphasized the importance of checking the solutions of equations algebraically. *You should continue to do this.* However, there is also a way to check a solution *graphically*.

NOTE This connection between algebra and geometry represents one of the most wonderful discoveries ever made in mathematics. Before René Descartes introduced the coordinate plane in 1637, mathematicians had no easy way of "seeing" a solution of an algebraic equation.

Using a Graphical Check of a Solution

The solution of an equation involving one variable x can be checked graphically with the steps below.

1. Write the equation so that all nonzero terms are on one side and zero is on the other side.

2. Sketch the graph of $y =$ (nonzero terms).

3. The solution of the one-variable equation is the x-intercept of the graph of the two-variable equation.

Example 6 ■ Using a Graphical Check of a Solution

Solve the equation $3x + 1 = -8$. Check your solution graphically and algebraically.

Solution

$$3x + 1 = -8 \qquad \text{Write original equation.}$$
$$3x = -9 \qquad \text{Subtract 1 from each side.}$$
$$x = -3 \qquad \text{Divide each side by 3.}$$

Graphical Check

Rewrite the equation so that all nonzero terms are on the left side.

$$3x + 1 = -8 \qquad \text{Write original equation.}$$
$$3x + 9 = 0 \qquad \text{Add 8 to each side.}$$

Now, use a graphing utility to graph the equation $y = 3x + 9$, as shown in Figure 2.23. Notice that the x-intercept is $(-3, 0)$ (where $3x + 9 = 0$), which checks with the algebraic solution.

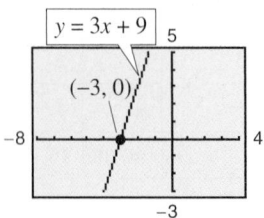

$y = 3x + 9$

$(-3, 0)$

Figure 2.23

Algebraic Check

Check the solution, $x = -3$, by substituting -3 for x in the original equation.

$$3x + 1 = -8 \qquad \text{Write original equation.}$$
$$3(-3) + 1 \overset{?}{=} -8 \qquad \text{Substitute } -3 \text{ for } x.$$
$$-9 + 1 \overset{?}{=} -8 \qquad \text{Simplify.}$$
$$-8 = -8 \qquad \text{Solution checks. } \checkmark$$

This close connection between x-intercepts and solutions is crucial to the study of algebra, and you can take advantage of this connection in two basic ways. You can use your algebraic "equation-solving skills" to find the x-intercepts of a graph and your "graphing skills" to approximate the solutions of an equation.

Example 7 ■ Finding Solutions and Using a Graphical Check of Solutions

Solve the equation $x^2 + 3x = 10$. Check your solution graphically and algebraically.

Solution

$x^2 + 3x = 10$	Write original equation.
$x^2 + 3x - 10 = 0$	Subtract 10 from each side.
$(x + 5)(x - 2) = 0$	Factor.
$x + 5 = 0 \implies x = -5$	Set 1st factor equal to 0 and solve for x.
$x - 2 = 0 \implies x = 2$	Set 2nd factor equal to 0 and solve for x.

Graphical Check

Rewrite the equation so that all nonzero terms are on the left side.

$x^2 + 3x = 10$	Write original equation.
$x^2 + 3x - 10 = 0$	Subtract 10 from each side.

Now use a graphing utility to graph the equation $y = x^2 + 3x - 10$, as shown in Figure 2.24. Notice that the x-intercepts are $(-5, 0)$ and $(2, 0)$ (where $x^2 + 3x - 10 = 0$), which checks with the algebraic solution.

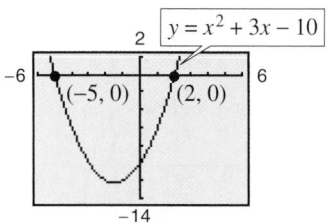

$y = x^2 + 3x - 10$

Figure 2.24

Algebraic Check

Check the first solution, $x = -5$, by substituting -5 for x in the original equation.

$x^2 + 3x = 10$	Write original equation.
$(-5)^2 + 3(-5) \stackrel{?}{=} 10$	Substitute -5 for x.
$10 = 10$	Solution checks. ✓

Check the second solution, $x = 2$, in a similar manner.

$x^2 + 3x = 10$	Write original equation.
$(2)^2 + 3(2) \stackrel{?}{=} 10$	Substitute 2 for x.
$10 = 10$	Solution checks. ✓

Note that the graph of $y = x^2 + 3x - 10$ in Figure 2.24 has x-*intercepts* of $(-5, 0)$ and $(2, 0)$ because the equation $x^2 + 3x - 10 = 0$ has *solutions* $x = -5$ and $x = 2$.

EXPLORATION

Use a graphing utility to graph the equation $3x^2 + x - 2 = 0$. Use the *zoom* and *trace* features of the graphing utility to find graphically the solutions to the equation. Solve the equation algebraically to verify your solutions. Discuss the accuracy of finding solutions to an equation graphically.

Real-Life Application of Graphs

There are many ways to approach a problem. Three common approaches are as follows.

A Numerical Approach: Construct and use a table.

A Graphical Approach: Draw and use a graph.

An Analytic Approach: Use the rules of algebra.

It is strongly recommended that you develop the habit of using at least two approaches with every problem. This helps build your intuition and helps you check that your answer is reasonable. Because newspapers and news magazines frequently use graphs to show real-life relationships between variables, Example 8 shows how a *graphical* approach can help you visualize the concept. The algebraic or analytic approach is then used to verify your conclusions.

Example 8 ■ Recommended Weight

The median recommended weight y (in pounds) for men of medium frame who are 25 to 59 years old can be approximated by the mathematical model

$$y = 0.073x^2 - 7.0x + 289, \quad 62 \le x \le 76$$

where x is the man's height in inches. *(Source: Metropolitan Life Insurance Company)*

(a) Construct a table of values that shows the median recommended weights for men with heights of 62, 64, 66, 68, 70, 72, 74, and 76 inches.

(b) Use the table of values to graph the model. Then use the graph to estimate *graphically* the median recommended weight for a man whose height is 71 inches.

(c) Use the model to confirm *algebraically* the estimate you found in part (b).

Solution

(a) You can use a calculator to complete the table, as shown below.

x	62	64	66	68	70	72	74	76
y	135.6	140.0	145.0	150.6	156.7	163.4	170.8	178.7

(b) The table of values can be used to graph the equation as shown in Figure 2.25. From the graph, you can estimate that a height of 71 inches corresponds to a weight of about 160 pounds.

(c) To confirm algebraically the estimate found in part (b), you can substitute 71 for x in the model.

$$y = 0.073x^2 - 7.0x + 289 \qquad \text{Write original model.}$$
$$= 0.073(71)^2 - 7.0(71) + 289 \qquad \text{Substitute 71 for } x.$$
$$\approx 160 \qquad \text{Use a calculator.}$$

So, the graphical estimate of 160 pounds is fairly good.

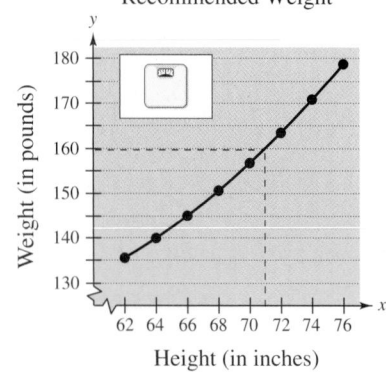

Recommended Weight

Figure 2.25

2.2 ■ Exercises

Developing Skills

In Exercises 1–6, match the equation with its graph. [The graphs are labeled (a), (b), (c), (d), (e), and (f).]

(a)

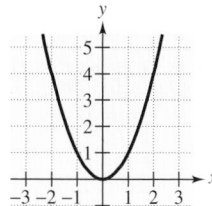

(b)

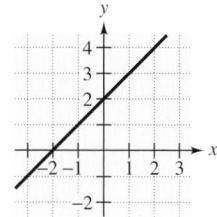

(c)

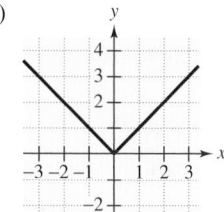

(d)

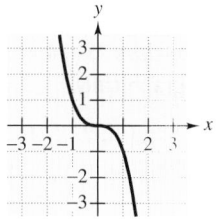

(e)

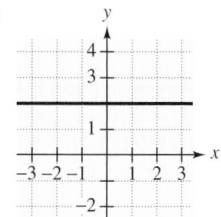

(f)

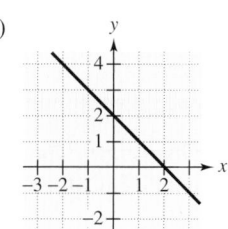

1. $y = 2$

2. $y = 2 + x$

3. $y = 2 - x$

4. $y = x^2$

5. $y = -x^3$

6. $y = |x|$

In Exercises 7–10, complete the table and use the results to sketch the graph of the equation.

7. $2x + y = 3$

x	-4			2	4
y			7	3	

8. $2x - 3y = 6$

x	-3	0		3	
y			$-\frac{2}{3}$		5

9. $y = 4 - x^2$

x		-1		2	
y	0		4		-5

10. $y = \frac{1}{2}x^3 - 4$

x	-1		1		3
y		-4		0	

In Exercises 11–20, sketch the graph of the equation.

11. $y = 3x$

12. $y = \frac{1}{3}x$

13. $y = 2x - 3$

14. $y = -x + 2$

15. $y = x^2 - 1$

16. $y = -x^2$

17. $y = x^2 - 6x + 9$

18. $y = 6 - x - x^2$

19. $y = |x| - 1$

20. $y = |x - 1|$

In Exercises 21–24, graphically estimate the x- and y-intercepts of the graph. Then check your results algebraically.

21. $y = x^2 + 3$

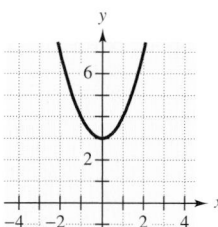

22. $x = 3$

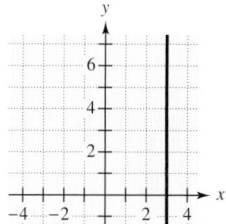

23. $y = |x - 2|$

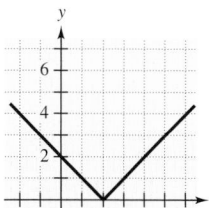

24. $y = (x - 3)(x - 4)$

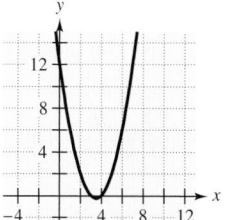

In Exercises 25–36, find the x- and y-intercepts (if any) of the graph of the equation.

25. $x + 2y = 10$

26. $3x - 2y + 12 = 0$

27. $15x - 18y + 20 = 0$

28. $-4x + 9y = 6$

29. $y = \frac{3}{4}x + 15$

30. $y = 12 - \frac{2}{5}x$

31. $y = (x - 4)(2x + 7)$ **32.** $y = (x + 5)(x - 5)$

33. $y = x^2 + x - 42$ **34.** $y = x^2 - 10x + 16$

35. $y = x^3 - 16x$ **36.** $y = 2x^3 - 7x^2 - 15x$

In Exercises 37–46, sketch the graph of the equation and show the coordinates of three solution points (including x- and y-intercepts).

37. $y = 3 - x$ **38.** $y = x - 3$

39. $y = 4$ **40.** $x = -6$

41. $4x + y = 3$ **42.** $y - 2x = -4$

43. $y = x^2 - 4$ **44.** $y = 1 - x^2$

45. $y = |x + 2|$ **46.** $y = |x| + 2$

In Exercises 47–56, use a graphing utility to graph the equation. Estimate the y-intercept.

47. $y = 2x - 6$ **48.** $y = x - 2$

49. $y = x^2 - 3$ **50.** $y = 6 - x^2$

51. $y = 1 - x^3$ **52.** $y = x^4 - 4$

53. $y = \sqrt{x + 4}$ **54.** $y = 2\sqrt{2x + 1}$

55. $y = |x| - 6$ **56.** $y = |x - 6|$

In Exercises 57–60, use a graphing utility to graph the equation and find any x-intercepts of the graph. Verify algebraically that any x-intercepts are solutions of the polynomial equation when $y = 0$.

57. $y = \frac{1}{2}x - 2$ **58.** $y = -3x + 6$

59. $y = x^2 - 6x$ **60.** $y = x^2 - 11x + 28$

In Exercises 61–70, use a graphing utility to solve the equation graphically. Verify the solution algebraically.

61. $7 - 2(x - 1) = 0$ **62.** $2x - 1 = 3(x + 1)$

63. $4 - x^2 = 0$ **64.** $x^2 + 2x = 0$

65. $x^2 - 2x + 1 = 0$ **66.** $1 - (x - 2)^2 = 0$

67. $2x^2 + 5x - 12 = 0$ **68.** $(x - 2)^2 - 9 = 0$

69. $x^3 - 4x = 0$ **70.** $2 + x - 2x^2 - x^3 = 0$

In Exercises 71–74, use a graphing utility to solve the equation graphically.

71. $5 - 0.2x^2 = 0$ **72.** $0.3x^2 - 4 = 0$

73. $0.2x^2 - 0.4x - 5 = 0$ **74.** $8 + 1.5x - 0.4x^2 = 0$

Solving Problems

75. *Physics*   The force F (in pounds) to stretch a spring x inches from its natural length is given by

$$F = \tfrac{4}{3}x, \quad 0 \le x \le 12.$$

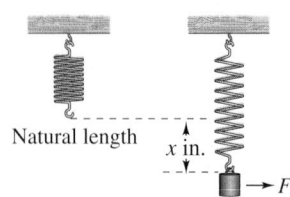

Natural length x in. $\longrightarrow F$

(a) Use the model to complete the table.

x	0	3	6	9	12
F					

(b) Sketch a graph of the model.

(c) Use the graph in part (b) to determine how the length of the spring changes each time the force is doubled.

76. *Geometry*   The area A (in square feet) of a square with sides of length s (in feet) is given by

$$A = s^2, \quad s > 0.$$

(a) Use the equation to complete the table.

s	1	2	4	8	16
A					

(b) Sketch a graph of the equation.

(c) Use the graph in part (b) to determine how the area of the square changes each time the side length is doubled.

77. *Sports*   The average salaries of professional basketball players in the United States have been increasing. The averages of the salaries S (in thousands of dollars) for the years 1994 through 1998 are shown in the table.

Year	1994	1995	1996	1997	1998
S	1700	1900	2000	2200	2600

A model for this data is $S = 210t + 820$, where t is the time in years, with $t = 4$ corresponding to 1994. *(Source: National Basketball Association)*

(a) Use a graphing utility to plot the data and graph the model in the same viewing window.

(b) How well does the model represent the data? Explain your reasoning.

(c) Use the model to predict the average salary of professional basketball players in 2005.

(d) Explain why the model may not be accurate in the future.

78. *Scouting* The numbers of girls N (in thousands) who were members of the Girl Scouts of America in the years 1995 through 1999 are shown in the table.

Year	1995	1996	1997	1998	1999
N	2534	2584	2671	2708	2749

A model for this data is $N = -4.86t^2 + 123.4t + 2033$, where t is the time in years, with $t = 5$ corresponding to 1995. *(Source: Girl Scouts of the United States of America)*

(a) Use a graphing utility to plot the data and graph the model in the same viewing window.

(b) Use the model to predict the membership of the Girl Scouts of America in 2006.

(c) Explain why the model may not be accurate.

79. *Manufacturing* A manufacturing plant purchases a new molding machine for $225,000. The depreciated value y after t years is given by

$$y = 225{,}000 - 20{,}000t, \quad 0 \le t \le 8.$$

Sketch the graph of this model.

80. *Physics* The velocity of a ball thrown upward from ground level is given by

$$v = -32t + 80, \quad 0 \le t \le 5$$

where t is time in seconds and v is velocity in feet per second. Sketch the graph of this model.

81. *Geometry* A rectangle of length l and width w has a perimeter of 12 meters.

(a) Show that the width of the rectangle is $w = 6 - l$ and its area is $A = l(6 - l)$.

(b) Sketch the graph of the equation for the area.

(c) Use the graph in part (b) to estimate the area of the rectangle if the length is 4 meters.

82. *Geometry* A farmer wants to fence in a rectangular area of width w and length l using 36 feet of fence. The farmer plans to use the side of a barn as one side of the rectangle (see figure).

(a) Show that the length of the rectangle is $l = 36 - 2w$ and its area is $A = w(36 - 2w)$.

(b) Sketch the graph of the equation for the area.

(c) Use the graph in part (b) to estimate the area of the rectangular enclosure if the width is 9 feet.

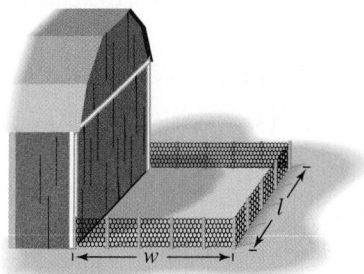

Explaining Concepts

In Exercises 83–86, explain how to use a graph to verify that $y_1 = y_2$. Then identify the rule of algebra that is illustrated.

83. $y_1 = \frac{1}{2}(x - 4)$
$y_2 = \frac{1}{2}x - 2$

84. $y_1 = 3x - x^2$
$y_2 = -x^2 + 3x$

85. $y_1 = x + (2x - 1)$
$y_2 = (x + 2x) - 1$

86. $y_1 = 2x + 0$
$y_2 = 2x$

In Exercises 87–90, explain how the x-intercepts of the graph correspond to the solutions of the polynomial equation when $y = 0$.

87. $y = x^2 - 9$

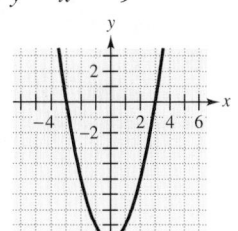

88. $y = x^2 - 4x + 4$

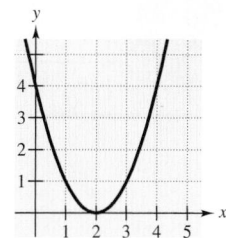

89. $y = x^2 - 2x - 3$

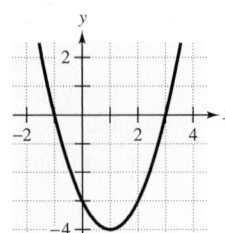

90. $y = x^3 - 3x^2 - x + 3$

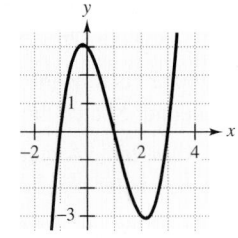

In Exercises 91 and 92, graph the equations on the same set of coordinate axes. What conclusions can you make by comparing the graphs?

91. (a) $y = x^2$
   (b) $y = (x - 2)^2$
   (c) $y = (x + 2)^2$
   (d) $y = (x + 4)^2$

92. (a) $y = x^2$
   (b) $y = x^2 - 2$
   (c) $y = x^2 - 4$
   (d) $y = x^2 + 4$

Ongoing Review

In Exercises 93–96, plot the points on a rectangular coordinate system.

93. $(5, 1), (-2, 3)$

94. $(1, 9), (-4, -5)$

95. $\left(\frac{1}{2}, \frac{9}{2}\right), \left(\frac{5}{3}, 0\right)$

96. $\left(\frac{2}{3}, \frac{1}{5}\right), \left(-2, \frac{11}{2}\right)$

In Exercises 97 and 98, solve the equation.

97. $\frac{3}{4}x + 8 = 0$

98. $20 - \frac{1}{4}x = 0$

In Exercises 99–102, find the missing coordinate of the solution point.

99. $y = \frac{9}{7}x - 1, \quad \left(14, \quad \right)$

100. $y = 8 + \frac{3}{4}x, \quad \left(4, \quad \right)$

101. $y = 3.8 - 1.2x, \quad \left(\quad , -3\right)$

102. $y = 11 + 4.6x, \quad \left(\quad , 8.8\right)$

103. *Construction* A house is 30 feet wide and the ridge of the roof is 7 feet above the tops of the walls (see figure). Find the length of the rafters if they overhang the edges of the walls by 2 feet.

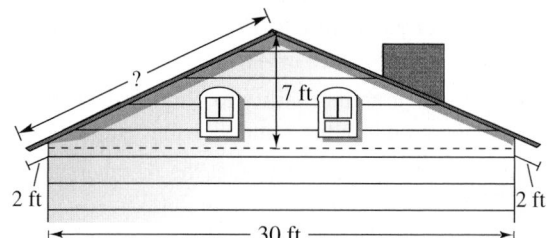

Looking Further

Misleading Graphs Graphs can help us visualize relationships between two variables, but they can also be misused to imply results that are not correct. In each pair of graphs below, both graphs represent the *same* data.

(a) Which graph is misleading and why?

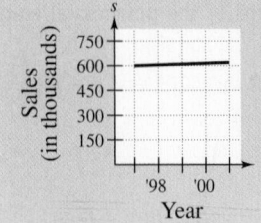

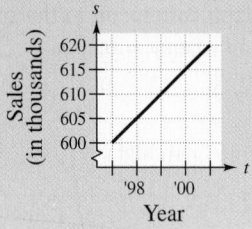

(b) Which graph is misleading and why?

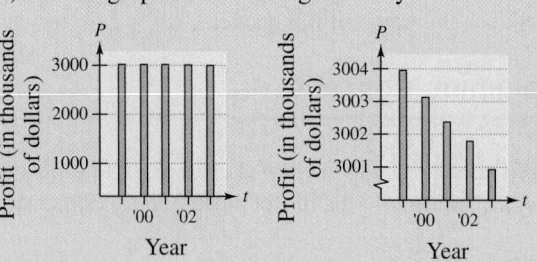

(c) Look for an example of a misleading graph in a newspaper or magazine. Explain how the graph could represent the data more accurately.

2.3 Slope: An Aid to Graphing Lines

The Slope of a Line

The **slope** of a nonvertical line represents the number of units the line rises or falls vertically for each unit of horizontal change from left to right. For example, the line in Figure 2.26 rises two units for each unit of horizontal change from left to right, and we say that this line has a slope of $m = 2$. Next consider the two points (x_1, y_1) and (x_2, y_2) on the line shown in Figure 2.27. As you move from left to right along this line, a change of $(y_2 - y_1)$ units in the vertical direction corresponds to a change of $(x_2 - x_1)$ units in the horizontal direction. That is,

$$y_2 - y_1 = \text{the change in } y$$

and

$$x_2 - x_1 = \text{the change in } x.$$

The slope of the line is given by the ratio of these two changes.

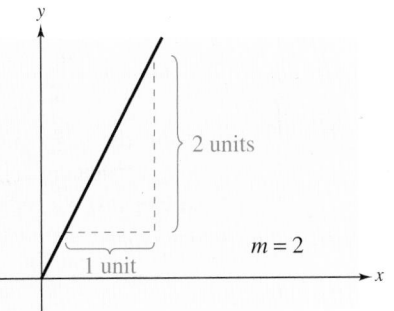

Figure 2.26

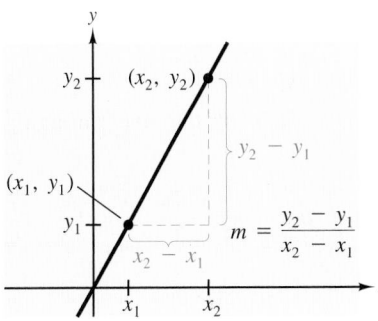

Figure 2.27

Definition of the Slope of a Line

The **slope** m of the nonvertical line passing through the points (x_1, y_1) and (x_2, y_2) is

$$m = \frac{y_2 - y_1}{x_2 - x_1} = \frac{\text{change in } y}{\text{change in } x} = \frac{\text{rise}}{\text{run}}$$

where $x_1 \neq x_2$ (see Figure 2.27).

When the formula for slope is used, the *order of subtraction* is important. Given two points on a line, you are free to label either of them (x_1, y_1) and the other (x_2, y_2). However, once this is done, you must form the numerator and denominator using the same order of subtraction.

$$m = \frac{y_2 - y_1}{x_2 - x_1} \qquad m = \frac{y_1 - y_2}{x_1 - x_2} \qquad m = \frac{y_2 - y_1}{x_1 - x_2}$$

Correct   Correct   Incorrect

Example 1 ■ Finding the Slope of a Line Through Two Points

Find the slope of the line passing through each pair of points.

(a) $(3, 4)$ and $(1, -2)$ (b) $(-2, 4)$ and $(3, 4)$ (c) $(2, -3)$ and $(0, 1)$ (d) $(3, 1)$ and $(3, 3)$

Algebraic Solution

(a) Let $(x_1, y_1) = (3, 4)$ and $(x_2, y_2) = (1, -2)$.

$$m = \frac{y_2 - y_1}{x_2 - x_1}$$ ⟵ Difference in y-values
                                        ⟵ Difference in x-values

$$= \frac{-2 - 4}{1 - 3}$$

$$= \frac{-6}{-2} = 3$$

(b) The slope of the line through $(-2, 4)$ and $(3, 4)$ is

$$m = \frac{4 - 4}{3 - (-2)}$$

$$= \frac{0}{5} = 0.$$

(c) The slope of the line through $(2, -3)$ and $(0, 1)$ is

$$m = \frac{1 - (-3)}{0 - 2}$$

$$= \frac{4}{-2} = -2.$$

(d) The slope of the line through $(3, 1)$ and $(3, 3)$ is undefined. Applying the formula for slope, you have

$$\frac{3 - 1}{3 - 3} = \frac{2}{0}.$$ Undefined

Because division by zero is not defined, the slope of a vertical line is not defined.

Graphical Solution

(a) Using the graph in Figure 2.28(a), you can see that the vertical change from $(3, 4)$ to $(1, -2)$ is -6 and the horizontal change is -2. So, the slope is

$$\frac{\text{rise}}{\text{run}} = \frac{-6}{-2} = 3.$$

(b) Using the graph in Figure 2.28(b), you can see that the vertical change from $(-2, 4)$ to $(3, 4)$ is 0 and the horizontal change is 5. So, the slope is

$$\frac{\text{rise}}{\text{run}} = \frac{0}{5} = 0.$$

(c) Using the graph in Figure 2.28(c), you can see that the vertical change from $(2, -3)$ to $(0, 1)$ is 4 and the horizontal change is -2. So, the slope is

$$\frac{\text{rise}}{\text{run}} = \frac{4}{-2} = -2.$$

(d) Using the graph in Figure 2.28(d), you can see that the vertical change from $(3, 1)$ to $(3, 3)$ is 2 and the horizontal change is 0. So, the slope is

$$\frac{\text{rise}}{\text{run}} = \frac{2}{0}$$

which is undefined.

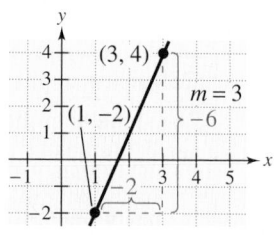

(a)

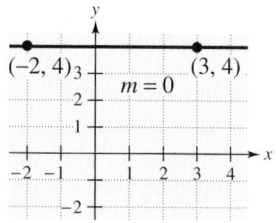

(b)

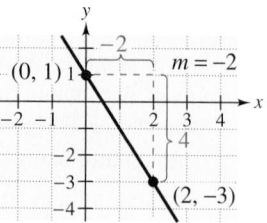

(c)

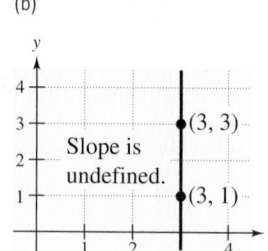

(d)

Figure 2.28

From the slopes of the lines shown in Example 1, you can make the following generalizations about the slope of a line.

1. A line with positive slope ($m > 0$) *rises* from left to right.

2. A line with negative slope ($m < 0$) *falls* from left to right.

3. A line with zero slope ($m = 0$) is *horizontal*.

4. A line with undefined slope is *vertical*.

Example 2 ■ Using Slope to Describe Lines

Describe the line through each pair of points.

(a) $(2, -1), (2, 3)$ (b) $(-2, 4), (3, 1)$ (c) $(1, 3), (4, 3)$ (d) $(-1, 1), (2, 5)$

Solution

(a) Let $(x_1, y_1) = (2, -1)$ and $(x_2, y_2) = (2, 3)$.

$$m = \frac{3 - (-1)}{2 - 2} = \frac{4}{0} \qquad \text{Undefined slope [see Figure 2.29(a)]}$$

Because the slope is undefined, the line is vertical.

(b) Let $(x_1, y_1) = (-2, 4)$ and $(x_2, y_2) = (3, 1)$.

$$m = \frac{1 - 4}{3 - (-2)} = -\frac{3}{5} < 0 \qquad \text{Negative slope [see Figure 2.29(b)]}$$

Because the slope is negative, the line falls from left to right.

(c) Let $(x_1, y_1) = (1, 3)$ and $(x_2, y_2) = (4, 3)$.

$$m = \frac{3 - 3}{4 - 1} = \frac{0}{3} = 0 \qquad \text{Zero slope [see Figure 2.29(c)]}$$

Because the slope is zero, the line is horizontal.

(d) Let $(x_1, y_1) = (-1, 1)$ and $(x_2, y_2) = (2, 5)$.

$$m = \frac{5 - 1}{2 - (-1)} = \frac{4}{3} > 0 \qquad \text{Positive slope [see Figure 2.29(d)]}$$

Because the slope is positive, the line rises from left to right.

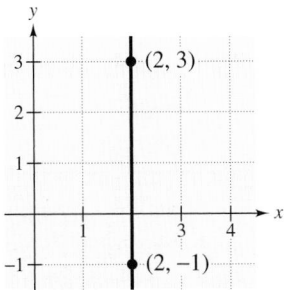

(a) Vertical line: undefined slope

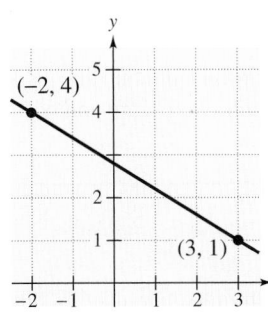

(b) Line falls: negative slope

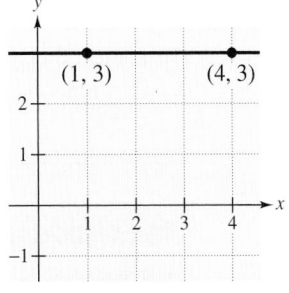

(c) Horizontal line: zero slope

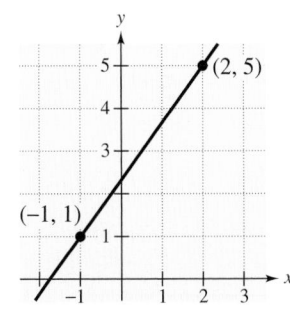

(d) Line rises: positive slope

Figure 2.29

Any two points on a nonvertical line can be used to calculate its slope. This is demonstrated in the next example.

Example 3 ■ Finding the Slope of a Line

Sketch the graph of the line given by $2x + 3y = 6$. Then find the slope of the line. (Choose two different pairs of points on the line and show that the same slope is obtained from either pair.)

Solution

Begin by solving the given equation for y.

$$2x + 3y = 6 \qquad \text{Write original equation.}$$
$$3y = -2x + 6 \qquad \text{Subtract } 2x \text{ from each side.}$$
$$\frac{3y}{3} = \frac{-2x + 6}{3} \qquad \text{Divide each side by 3.}$$
$$y = -\frac{2}{3}x + 2 \qquad \text{Simplify.}$$

Then construct a table of values, as shown below.

x	-3	0	3	6
$y = -\frac{2}{3}x + 2$	4	2	0	-2
Solution point	$(-3, 4)$	$(0, 2)$	$(3, 0)$	$(6, -2)$

From the solution points shown in the table, sketch the graph of the line, as shown in Figure 2.30. To calculate the slope of the line using two different sets of points, first use the points $(-3, 4)$ and $(0, 2)$, as shown in Figure 2.30(a), and obtain a slope of

$$m = \frac{2 - 4}{0 - (-3)} = -\frac{2}{3}.$$

Next, use the points $(3, 0)$ and $(6, -2)$, as shown in Figure 2.30(b), and obtain a slope of

$$m = \frac{-2 - 0}{6 - 3} = -\frac{2}{3}.$$

Try some other pairs of points on the line to see that you obtain a slope of $m = -\frac{2}{3}$ regardless of which two points you use.

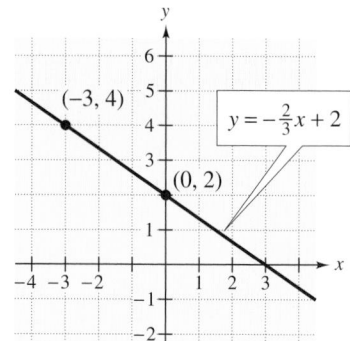

(a)

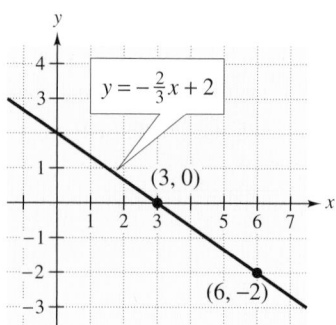

(b)

Figure 2.30

Technology

The setting you use for the viewing window on a graphing utility may affect the appearance of a line's slope. When you are using a graphing utility, you cannot judge whether a slope is steep or shallow *unless* you use a square setting.

EXPLORATION

a. Use a graphing utility to graph the equations below in the same viewing window. How are the equations similar? How are the graphs similar? What does this imply?

$$y_1 = 2x + 4$$
$$y_2 = x + 4$$
$$y_3 = -2x + 4$$
$$y_4 = -x + 4$$

b. Use a graphing utility to graph the equations below in the same viewing window. How are the equations similar? How are the graphs similar? What does this imply?

$$y_1 = 2x + 4$$
$$y_2 = 2x + 2$$
$$y_3 = 2x - 1$$
$$y_4 = 2x - 3$$

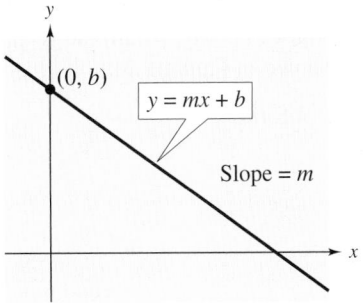

Figure 2.31

Slope as a Graphing Aid

You have seen that, before creating a table of values for an equation, you should first solve the equation for y. When you do this for a linear equation, you obtain some very useful information. Consider the results of Example 3. When the equation

$$2x + 3y = 6$$

is solved for y, you see that

$$y = -\frac{2}{3}x + 2.$$

Observe that the coefficient of x is the slope of the graph of this equation (see Example 3). Moreover, the constant term, 2, gives the y-intercept of the graph.

$$y = -\frac{2}{3}x + 2$$

Slope ⟋ y-intercept $(0, 2)$

This form is called the **slope-intercept form** of the equation of the line.

> **Slope-Intercept Form of the Equation of a Line**
>
> The graph of the equation
>
> $$y = mx + b$$
>
> is a line with a slope of m and a y-intercept of $(0, b)$. (See Figure 2.31.)

NOTE When you substitute zero for x in the slope-intercept form of the equation of a line, $y = mx + b$, you obtain $y = m(0) + b$ or $y = b$. So, $(0, b)$ is the point where the line crosses the y-axis.

Example 4 ■ Slope and y-Intercept of a Line

Find the slope and y-intercept of the graph of the equation

$$4x - 5y = 15.$$

Solution

Begin by writing the equation in slope-intercept form, as follows.

$4x - 5y = 15$	Write original equation.
$-5y = -4x + 15$	Subtract $4x$ from each side.
$y = \dfrac{-4x + 15}{-5}$	Divide each side by -5.
$y = \dfrac{4}{5}x - 3$	Slope-intercept form

From the slope-intercept form, you can see that $m = \frac{4}{5}$ and $b = -3$. So, the slope of the graph of the equation is $\frac{4}{5}$ and the y-intercept is $(0, -3)$.

So far, you have been plotting several points in order to sketch the equation of a line. However, now that you can recognize equations of lines, you don't have to plot as many points—two points are enough. (You might remember from geometry that *two points are all that are necessary to determine a line.*)

Example 5 ■ Using the Slope and *y*-Intercept to Sketch a Line

Use the slope and *y*-intercept to sketch the graph of each equation.

(a) $y = \dfrac{5}{2}x - 3$ (b) $6x + 2y = 10$

Solution

(a) The equation is already in slope-intercept form, $y = mx + b$.

$$y = \frac{5}{2}x - 3 \qquad \text{Slope-intercept form}$$

So, the slope of the line is $m = \frac{5}{2}$ and the *y*-intercept is $(0, b) = (0, -3)$. Now you can sketch the graph of the line as follows. First, plot the *y*-intercept. Then, using a slope of $\frac{5}{2}$

$$m = \frac{5}{2}$$

$$= \frac{\text{change in } y}{\text{change in } x}$$

locate a second point on the line by moving two units to the right and five units up (or five units up and two units to the right). Finally, obtain the graph by drawing a line through the two points, as shown in Figure 2.32.

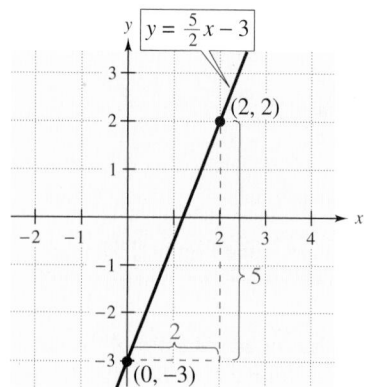

Figure 2.32

(b) Begin by writing the equation in slope-intercept form.

$$6x + 2y = 10 \qquad \text{Write original equation.}$$

$$2y = -6x + 10 \qquad \text{Subtract } 6x \text{ from each side.}$$

$$y = \frac{-6x + 10}{2} \qquad \text{Divide each side by 2.}$$

$$y = -3x + 5 \qquad \text{Slope-intercept form}$$

So, the slope of the line is $m = -3$ and the *y*-intercept is $(0, b) = (0, 5)$. Now you can sketch the graph of the line as follows. First, plot the *y*-intercept. Then, using a slope of -3

$$m = \frac{-3}{1}$$

$$= \frac{\text{change in } y}{\text{change in } x}$$

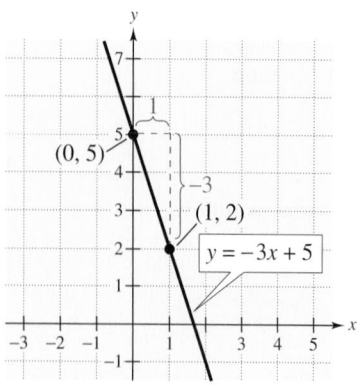

Figure 2.33

locate a second point on the line by moving one unit to the right and three units down (or three units down and one unit to the right). Finally, obtain the graph by drawing a line through the two points, as shown in Figure 2.33.

Parallel and Perpendicular Lines

You know from geometry that two lines in a plane are *parallel* if they do not intersect. What this means in terms of their slopes is suggested in Example 6.

Example 6 ■ Lines That Have the Same Slope

On the same set of coordinate axes, sketch the lines given by

$$y = 2x \quad \text{and} \quad y = 2x - 3.$$

Solution

For the line given by $y = 2x$, the slope is $m = 2$ and the y-intercept is $(0, 0)$. For the line given by $y = 2x - 3$, the slope is also $m = 2$ and the y-intercept is $(0, -3)$. The graphs of these two lines are shown in Figure 2.34.

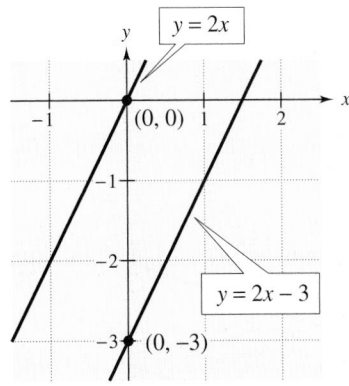

Figure 2.34

In Example 6, notice that the two lines have the same slope *and* appear to be parallel. The following rule states that this is always the case.

> **Parallel Lines**
>
> Two distinct nonvertical lines are parallel if and only if they have the same slope.

NOTE   The phrase "if and only if" in this rule is used in mathematics as a way to write two statements in one. The first statement says that *if two distinct nonvertical lines have the same slope, they must be parallel.* The second statement says that *if two distinct nonvertical lines are parallel, they must have the same slope.*

Another rule from geometry is that two lines in a plane are *perpendicular* if and only if they intersect at right angles. In terms of their slopes, this means that two nonvertical lines are perpendicular if their slopes are negative reciprocals of each other. For instance, the negative reciprocal of 5 is $-\frac{1}{5}$, so the lines

$$y = 5x + 2 \quad \text{and} \quad y = -\frac{1}{5}x - 4$$

are perpendicular to each other, as shown in Figure 2.35.

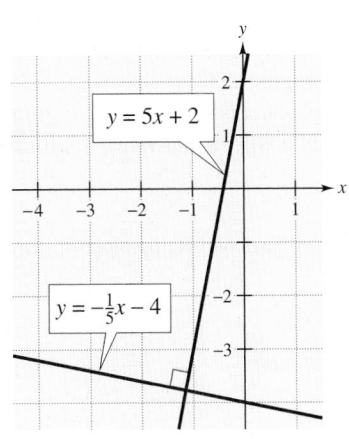

Figure 2.35

> **Perpendicular Lines**
>
> Consider two nonvertical lines whose slopes are m_1 and m_2. The two lines are perpendicular if and only if their slopes are *negative reciprocals* of each other. That is
>
> $$m_1 = -\frac{1}{m_2} \quad \text{or, equivalently,} \quad m_1 \cdot m_2 = -1.$$

Example 7 ■ Parallel or Perpendicular?

Are the pairs of lines parallel, perpendicular, or neither?

(a) $y = -2x + 4$, $y = \frac{1}{2}x + 1$ (b) $y = \frac{1}{3}x + 2$, $y = \frac{1}{3}x - 3$

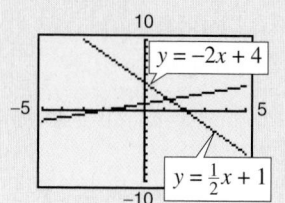
Solution

(a) The first line has a slope of $m_1 = -2$ and the second line has a slope of $m_2 = \frac{1}{2}$. Because these slopes are negative reciprocals of each other, the two lines must be perpendicular, as shown in Figure 2.36.

(b) Each of these two lines has a slope of $m = \frac{1}{3}$. So, the two lines must be parallel, as shown in Figure 2.37.

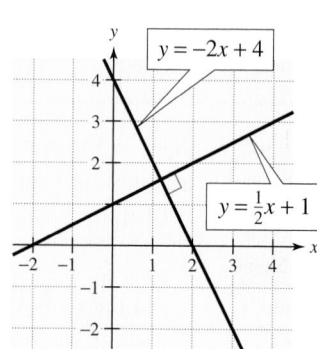

Figure 2.36

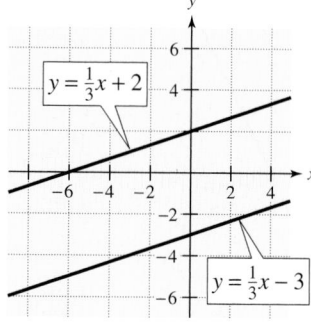

Figure 2.37

Slope as a Rate of Change

In real-life problems, slope is often used to describe a **constant rate of change** or an **average rate of change.** In such cases, units of measure are assigned, such as miles per hour or dollars per year. For instance, Figure 2.38(a) shows the rate of change in the total cost of producing x units of a product is $25 per unit. Figure 2.38(b) shows the rate of change in the uphill slope of a wheelchair ramp is 1 inch per foot.

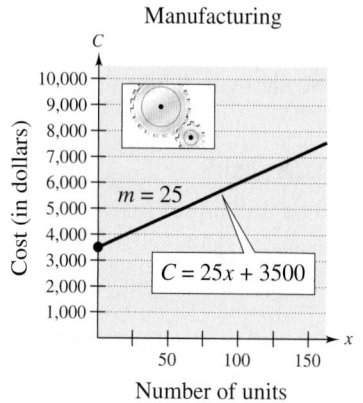

(a)

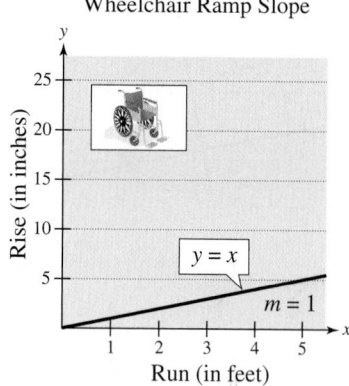

(b)

Figure 2.38

Women Engineers

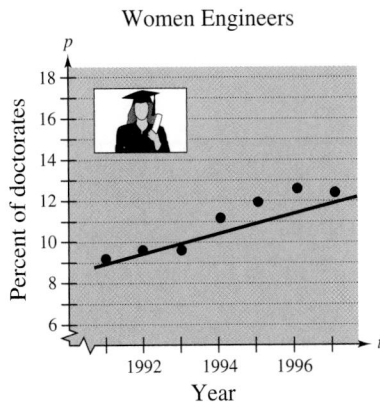

Figure 2.39

Example 8   ■   Slope as a Rate of Change

In 1991, 9.2% of the doctorates in engineering were awarded to women. By 1997, 12.3% were being earned by women. Find the average rate of change in the percent of engineering doctorates earned by women from 1991 to 1997. *(Source: U.S. National Center for Education Statistics)*

Solution

Let p represent the percent of engineering doctorates earned by women and let t represent the year. The two given data points are represented by (t_1, p_1) and (t_2, p_2).

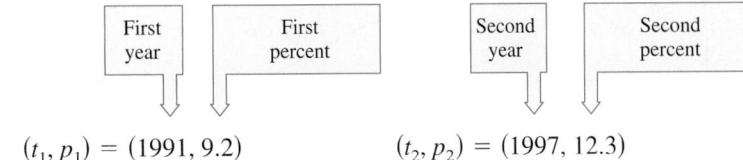

$$(t_1, p_1) = (1991, 9.2) \qquad (t_2, p_2) = (1997, 12.3)$$

Now, use the formula for slope to find the average rate of change.

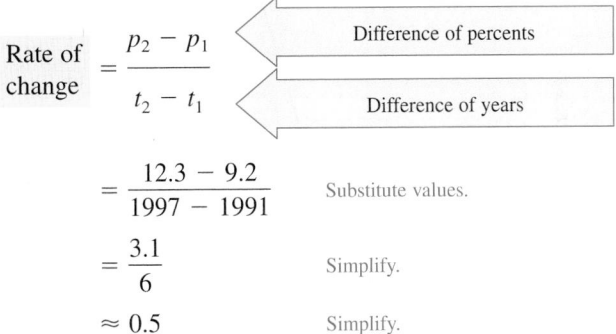

$$\text{Rate of change} = \frac{p_2 - p_1}{t_2 - t_1}$$

$$= \frac{12.3 - 9.2}{1997 - 1991} \qquad \text{Substitute values.}$$

$$= \frac{3.1}{6} \qquad \text{Simplify.}$$

$$\approx 0.5 \qquad \text{Simplify.}$$

From 1991 through 1997, the *average rate of change* in the percent of engineering doctorates awarded to women was about $\frac{1}{2}\%$ per year. (The exact changes in percent varied from one year to the next, as shown in the scatter plot in Figure 2.39.)

NOTE   Make sure you understand that the answer in Example 8 is $\frac{1}{2}\%$ or 0.005, not 50% or 0.5, because you were finding the *average rate of change in the percent.*

Collaborate!

Collinear Points

You can use slope to determine if three points are collinear. Consider any three points A, B, and C. If the slope of the line through points A and B is the same as the slope of the line through points B and C, the three points are collinear. Determine if the points are collinear.

a. $(3, -1), (0, -3), (12, 5)$

b. $\left(2, -\frac{1}{2}\right), (-4, 4), (6, -3)$

2.3 ■ Exercises

Developing Skills

In Exercises 1–6, estimate the slope of the line from its graph.

1.
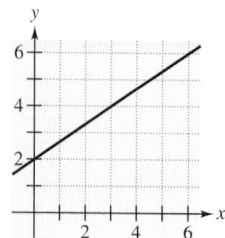

2.
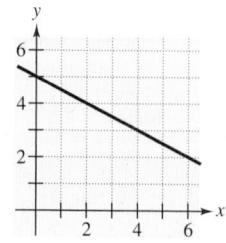

3.
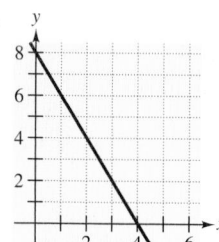

4.
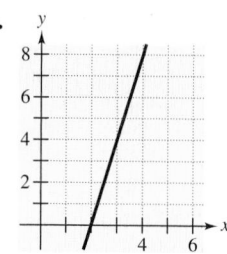

5.
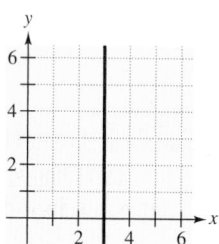

6.
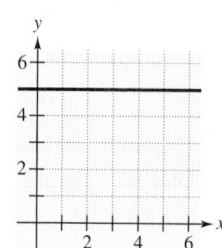

In Exercises 7–26, plot the points and find the slope (if possible) of the line passing through them. Then describe the line as rising, falling, horizontal, or vertical.

7. $(0, 12), (8, 0)$

8. $(0, -4), (6, 0)$

9. $(-2, -3), (6, 1)$

10. $(3, 6), (5, -2)$

11. $(-7, 6), (5, -3)$

12. $(2, -3), (-12, 15)$

13. $(-5, 4), (-9, -6)$

14. $(10, -8), (-5, 4)$

15. $(2, -5), (7, -5)$

16. $(-2, 1), (-4, -3)$

17. $(-3, -5), (3, 5)$

18. $(-5, -3), (-5, 4)$

19. $\left(\frac{5}{6}, \frac{2}{5}\right), \left(-\frac{1}{3}, \frac{1}{5}\right)$

20. $\left(-\frac{3}{2}, -\frac{1}{2}\right), \left(\frac{5}{8}, \frac{1}{2}\right)$

21. $\left(\frac{3}{4}, 2\right), \left(5, -\frac{5}{2}\right)$

22. $\left(-1, \frac{4}{9}\right), \left(\frac{2}{7}, 3\right)$

23. $(4.2, -1), (-4.2, 6)$

24. $(3.4, 0), (3.4, 1)$

25. $(0, 4.5), (3, 4.5)$

26. $(2.5, -2), (4.75, 5.25)$

In Exercises 27 and 28, solve for x so that the line through the points has the given slope.

27. $(4, 5), (x, 7); m = -\frac{2}{3}$ **28.** $(x, -2), (5, 0); m = \frac{3}{4}$

In Exercises 29 and 30, solve for y so that the line through the points has the given slope.

29. $(-3, y), (9, 3); m = \frac{3}{2}$

30. $(-3, 20), (2, y); m = -6$

In Exercises 31–34, sketch the graph of a line through the point $(3, 2)$ having the given slope.

31. $m = 3$

32. $m = \frac{3}{2}$

33. $m = -\frac{1}{3}$

34. $m = 0$

In Exercises 35–38, sketch the graph of a line through the point $(0, 1)$ having the given slope.

35. m is undefined.

36. $m = -1$

37. $m = -\frac{4}{3}$

38. $m = \frac{2}{3}$

In Exercises 39–46, a point on a line and the slope of the line are given. Find two additional points on the line. (There are many correct answers.)

39. $(5, 2)$
$m = 0$

40. $(-4, 3)$
m is undefined.

41. $(3, -4)$
$m = 3$

42. $(-1, -5)$
$m = 2$

43. $(0, 3)$
$m = -1$

44. $(-2, 6)$
$m = -3$

45. $(-5, 0)$
$m = \frac{4}{3}$

46. $(-1, 1)$
$m = -\frac{3}{4}$

In Exercises 47–54, write the equation of the line in slope-intercept form. Then state the slope and the y-intercept of the line.

47. $6x - 3y = 9$

48. $2x + 4y = 16$

49. $4y - x = -4$

50. $3x - 2y = -10$

51. $2x + 5y - 3 = 0$

52. $8x - 6y + 1 = 0$

53. $x = 2y - 4$

54. $x = -\frac{3}{2}y + \frac{2}{3}$

In Exercises 55–66, write the equation of the line in slope-intercept form, and then use the slope and y-intercept to sketch the line. Use a graphing utility to confirm your sketch.

55. $3x - y - 2 = 0$

56. $x - y - 5 = 0$

57. $x + y = 0$

58. $x - y = 0$

59. $3x + 2y - 2 = 0$

60. $x - 2y - 2 = 0$

61. $x - 4y + 2 = 0$

62. $8x + 6y - 3 = 0$

63. $y - 2 = 0$

64. $y + 4 = 0$

65. $x - 0.2y - 1 = 0$

66. $0.5x + 0.6y - 3 = 0$

In Exercises 67–74, determine whether the lines are parallel, perpendicular, or neither.

67. $y = \frac{1}{2}x - 2$

$y = \frac{1}{2}x + 3$

68. $y = 3x - 2$

$y = 3x + 1$

69. $y = \frac{3}{4}x - 3$

$y = -\frac{4}{3}x + 1$

70. $y = -\frac{2}{3}x - 5$

$y = \frac{3}{2}x + 1$

71. $y = -\frac{2}{5}x + 1$

$y = -\frac{5}{2}x - 7$

72. $y = 4x + 3$

$y = -4x + 9$

73. $x + 2y - 3 = 0$

$-2x - 4y + 1 = 0$

74. $3x - 4y - 1 = 0$

$4x + 3y + 2 = 0$

In Exercises 75–78, determine whether the lines L_1 and L_2 passing through the given pairs of points are parallel, perpendicular, or neither.

75. L_1: $(1, 3), (2, 1)$

L_2: $(0, 0), (4, 2)$

76. L_1: $(-3, -3), (1, 7)$

L_2: $(0, 4), (5, -2)$

77. L_1: $(-2, 0), (4, 4)$

L_2: $(1, -2), (4, 0)$

78. L_1: $(-5, 3), (3, 0)$

L_2: $(1, 2), \left(3, \frac{22}{3}\right)$

In Exercises 79–86, write the equation of the line in slope-intercept form, and then find the slopes of lines that are (a) parallel and (b) perpendicular to the given line.

79. $x + y = 9$

80. $5x + y = 14$

81. $14x - 7y = 11$

82. $-8x + 2y = -5$

83. $5x - 2y = -26$

84. $12x + 9y = 2$

85. $18x + 12y + 10 = 0$

86. $-25x + 10y - 1 = 0$

Solving Problems

In Exercises 87–90, match the description with its graph. Determine the slope and interpret its meaning in the context of the problem. [The graphs are labeled (a), (b), (c), and (d).]

(a)
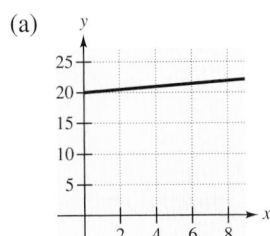

(b)
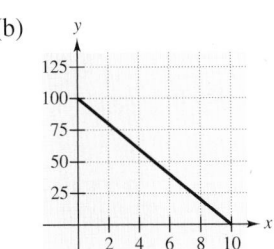

(c)
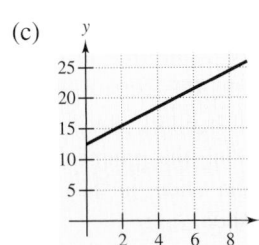

(d)
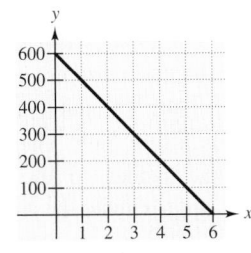

87. A person is paying $10 per week to a friend to repay a $100 loan.

88. An employee is paid $12.50 per hour plus $1.50 for each unit produced per hour.

89. A sales representative receives $20 per day for food plus $0.25 for each mile traveled.

90. A word processor that was purchased for $600 depreciates $100 per year.

91. *Temperature*   The graph shows the temperature in degrees Fahrenheit for various measurements on the Celsius scale. Use the graph to approximate the rate of change in the temperature measured on the Fahrenheit scale in relation to the temperature measured on the Celsius scale.

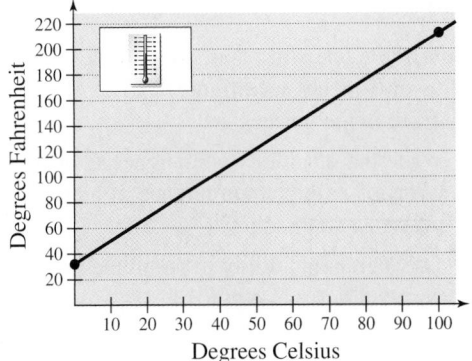

92. *Business*   The graph shows the depreciated value (in dollars) of a molding machine during an eight-year period. Use the graph to approximate the rate of change in the value of the equipment during the 8 years it was used.

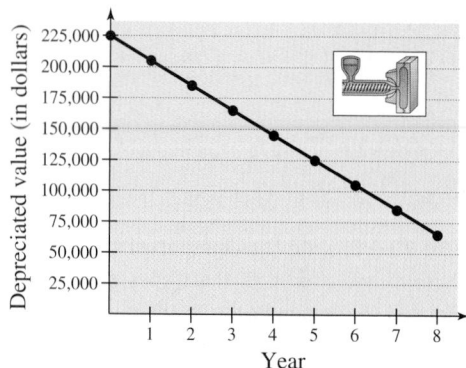

93. *Geometry*   The length and width of a rectangular flower garden are 40 feet and 30 feet (see figure). A walkway of width x surrounds the garden.

   (a) Write the outside perimeter P of the walkway in terms of x.

   (b) Use a graphing utility to graph the model for the perimeter in part (a).

   (c) Determine the slope of the graph in part (b). For each additional one-foot increase in the width of the walkway, determine the increase in its outside perimeter.

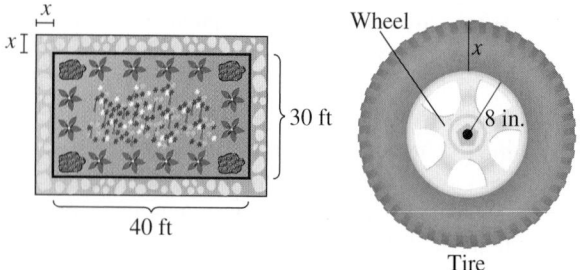

Figure for 93   Figure for 94

94. *Geometry*   The rim of an automobile wheel has a radius of 8 inches (see figure). A tire of height x is mounted on the wheel.

   (a) Write the outside circumference C of the tire in terms of x (the circumference of a circle is $2\pi r$, where r is the radius).

   (b) Use a graphing utility to graph the model for the circumference in part (a).

   (c) Determine the slope of the graph in part (b). For each additional one-inch increase in the height of the tire, determine the change in its outside circumference.

95. *Road Signs*   When driving down a mountain road, you notice warning signs indicating a "12% grade." This means that the slope of the road is $-\frac{12}{100}$. Over a stretch of road, your elevation drops by 2000 feet. What is the horizontal change in your position?

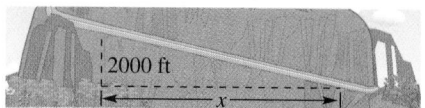

96. *Road Signs*   On a stretch of road that climbs up a steep hill, there is a road sign that reads "9% grade." This means that the slope of the road is $\frac{9}{100}$. Over a certain length of your trip on this road, your elevation increases by 200 feet. What is the horizontal change in your position?

97. *Leaning Tower of Pisa*   When it was built, the Leaning Tower of Pisa in Italy was 180 feet tall. Since then, one side of the base has sunk 1 foot into the ground, causing the top of the tower to lean 16 feet off center (see figure). Approximate the slope of the side of the tower.

The tower leans because it was built on a layer of unstable soil—a mixture of clay, sand, and water.

98. *Slide*   The ladder of a straight slide in a playground is 8 feet high. The distance along the ground from the ladder to the foot of the slide is 12 feet. Approximate the slope of the slide.

99. *Construction*   The slope, or pitch, of a roof is such that it rises (or falls) 3 feet for every 4 feet of horizontal distance. Determine the maximum height in the attic of the house if the house is 30 feet wide (see figure).

100. *Construction*   Determine the maximum height in the attic of a house as in Exercise 99 if the house is 50 feet wide and the pitch of the roof is such that it rises (or falls) 4 feet for every 5 feet of horizontal distance.

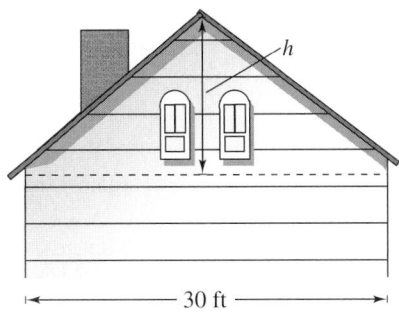

Figure for 99

Explaining Concepts

101. In your own words, give interpretations of a negative slope, a zero slope, and a positive slope.

102. Can any pair of points on a line be used to calculate the slope of the line? Explain.

103. Is it possible for two lines with positive slopes to be perpendicular to each other? Explain.

104. In the form $y = mx + b$, what does m represent? What does b represent?

Ongoing Review

In Exercises 105 and 106, evaluate the algebraic expression for the indicated values.

105. $\dfrac{a - b}{c - d}$, $a = 5$, $b = 2$, $c = 3$, and $d = -1$

106. $\dfrac{a - b}{c - d}$, $a = 7$, $b = -2$, $c = -3$, and $d = 15$

In Exercises 107 and 108, solve for *y* in terms of *x*.

107. $y - 6 = 4[x - (-10)]$

108. $y - (-8) = \frac{1}{2}(x - 16)$

109. *Geometry*   Write an algebraic expression for the perimeter of the figure.

110. *Geometry*   Write an algebraic expression for the area of the figure.

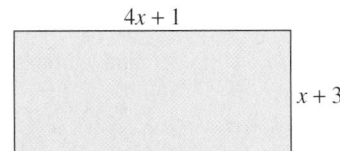

Figure for 109 and 110

Looking Further

Spending Your Pocket Money   Consider the ordered pair (x, y), where y represents your pocket money and x is the time in days, with $x = 0$ corresponding to payday. On payday, you have $200. Eight days later, you have $46.

(a) Plot the two points modeling the given information and sketch a line segment connecting the points.

(b) Determine the slope of the line segment in part (a).

(c) At what average rate did you spend your pocket money over the eight-day period?

(d) On the following payday, you again have $200 of pocket money. Three days later, you have $95. Plot these two points and sketch a line segment connecting the points.

(e) Determine the slope of the line segment in part (d).

(f) At what average rate did you spend your pocket money over this three-day period?

(g) Explain how slope is related to an average rate of change in this example.

Mid-Chapter Quiz

Take this quiz as you would take a quiz in class. After you are done, check your work against the answers given in the back of the book.

In Exercises 1 and 2, (a) plot the points on a rectangular coordinate system, (b) find the distance between the points, (c) find the coordinates of the midpoint, and (d) find the slope of the line through the points.

1. $(-1, 5), (3, 2)$

2. $(-3, -2), (2, 10)$

3. Find the coordinates of the point that lies 10 units to the right of the y-axis and three units below the x-axis.

4. Determine whether the ordered pairs are solution points of the equation $4x - 3y = 10$.

(a) $(2, 1)$ (b) $(1, -2)$ (c) $(2.5, 0)$ (d) $\left(2, -\frac{2}{3}\right)$

In Exercises 5 and 6, use the point-plotting method to sketch the graph of the equation. Show the coordinates of three solution points (including x- and y-intercepts).

5. $y = 6x - x^2$

6. $y = |x| - 3$

7. Use a graphing utility to graph the equation $y = 2x^3 - 6x^2 + 4x$. Graphically estimate the x- and y-intercepts of the graph. Verify algebraically.

8. Sketch the line passing through $(0, -6)$ with a slope of $m = \frac{4}{3}$.

In Exercises 9 and 10, write the equation in slope-intercept form. State the slope and y-intercept and then use them to sketch the graph of the line.

9. $3x + y - 6 = 0$

10. $8x - 6y = 30$

In Exercises 11–13, determine whether the lines are parallel, perpendicular, or neither.

11. $y = 3x + 2$

$y = -\frac{1}{3}x - 4$

12. $y = 2x - 3$

$y = -2x - 3$

13. $y = 4x + 3$

$y = \frac{1}{2}(8x + 5)$

14. Write the equation $5x + 3y - 9 = 0$ in slope-intercept form. Find the slope of lines that are (a) parallel and (b) perpendicular to this line.

15. A company purchases a new printing press for $85,000. For tax purposes, the printing press will be depreciated over a 10-year period. At the end of 10 years, the salvage value of the printing press is expected to be $4000. Find the average rate of change in the value of the printing press over the 10-year period.

16. As part of a science experiment, a student records the outside temperature every hour from 8:00 A.M. to 4:00 P.M. The line graph shows the data collected. From the graph, what was the highest temperature during the eight-hour period? At what time did it occur? What was the lowest temperature during the eight-hour period? At what time did it occur?

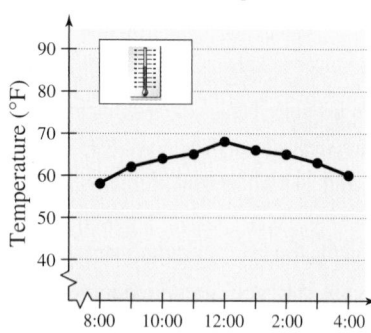

Outside Temperature

Figure for 16

2.4 Relations, Functions, and Function Notation

Relations

Many everyday occurrences involve two quantities that are paired or matched with each other by some rule of correspondence. The mathematical term for such a correspondence is a **relation.**

> **Definition of a Relation**
>
> A **relation** is any set of ordered pairs. The set of first components in the ordered pairs is the **domain of the relation,** and the set of second components is the **range of the relation.**

Example 1 ■ Analyzing a Relation

Find the domain and range of each relation.

(a) $\{(0, 1), (1, 3), (2, 5), (3, 5), (0, 3)\}$

(b) $\{(-2, 3), (-1, 3), (0, 3), (4, 3), (10, 3)\}$

Solution

(a) The domain is the set of all first components of the relation, and the range is the set of all second components.

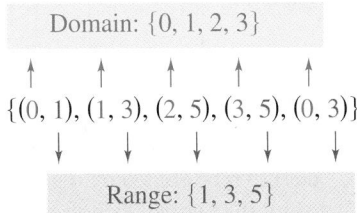

$$\text{Domain: } \{0, 1, 2, 3\}$$
$$\{(0, 1), (1, 3), (2, 5), (3, 5), (0, 3)\}$$
$$\text{Range: } \{1, 3, 5\}$$

A graphical representation is shown in Figure 2.40.

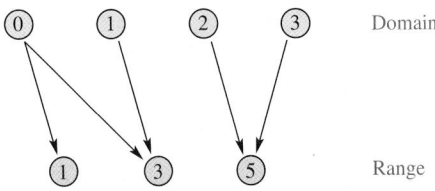

Figure 2.40

NOTE It is not necessary to list repeating components of the domain and range of a relation.

(b) The domain is the set of all first components of the relation, and the range is the set of all second components.

Domain: $\{-2, -1, 0, 4, 10\}$

Range: $\{3\}$

Historical Note

Although the concept of function dates back to the Babylonians as early as 2000 B.C., the concept was fine-tuned in the mid-1600s. Descartes is often credited as the first to use the term *function*.

Functions

In modeling real-life situations, you will work with a special type of relation called a function. A **function** is a relation in which no two ordered pairs have the same first component and different second components. For instance, (2, 3) and (2, 4) could not be ordered pairs of a function.

> ### Definition of a Function
>
> A **function** f from a set A to a set B is a rule of correspondence that assigns to each element x in the set A exactly one element y in the set B. The set A is called the **domain** (or set of inputs) **of the function** f, and the set B contains the **range** (or set of outputs) **of the function.**

The rule of correspondence for a function establishes a set of "input-output" ordered pairs of the form (x, y), where x is an input and y is the corresponding output. In some cases, the rule may generate only a finite set of ordered pairs, whereas in other cases the rule may generate an infinite set of ordered pairs.

Functions are commonly represented in four ways.

1. *Verbally* by a sentence that describes how the input variable is related to the output variable

   Example: Each positive integer less than 7 is paired with its square.

2. *Numerically* by a table or a list of ordered pairs that matches input values with output values

   Example:

Input x	1	2	3	4	5
Output y	11	10	8	5	1

3. *Graphically* by points on a graph in a coordinate plane in which the input values are represented by the horizontal axis and the output values are represented by the vertical axis

   Example:

   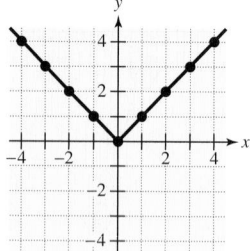

4. *Algebraically* by an equation in two variables

   Example: $y = 3x + 4$

A function has certain characteristics that distinguish it from a relation. Use the list below to determine whether a relation is a function.

Characteristics of a Function

1. Each element in the domain A must be matched with an element in the range, which is contained in the set B.
2. Some elements in the set B may not be matched with any element in the domain A.
3. Two or more elements of the domain may be matched with the same element in the range.
4. No element of the domain is matched with two different elements in the range.

Example 2 ■ Testing for Functions

Decide whether the description represents a function.

(a) The input value x is the number of representatives from a state and the output value y is the number of senators.

(b) $\{(1, 1), (2, 4), (3, 9), (4, 16), (5, 25), (6, 36)\}$

(c)

Input, x	-1	3	4	3
Output, y	7	2	0	4

(d) Let $A = \{a, b, c\}$ and let $B = \{1, 2, 3, 4, 5\}$.

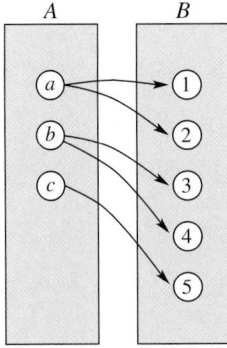

Solution

(a) This verbal description *does* represent a function. Regardless of the value of x, the value of y is always 2. Such functions are called constant functions.

(b) This set of ordered pairs *does* represent a function. No input value is matched with two output values.

(c) This table *does not* represent a function. The input value 3 is matched with two different output values, 2 and 4.

(d) This diagram *does not* represent a function. The element a in set A is matched with two elements, 1 and 2, in set B. This is also true of b.

Representing functions by sets of ordered pairs is a common practice in the study of *discrete mathematics*, which deals mainly with finite sets of data or with finite subsets of the set of real numbers. In algebra, however, it is more common to represent functions by equations or formulas involving two variables. For instance, the equation

$$y = x^2 \qquad \text{Squaring function}$$

represents the variable y as a function of the variable x. The variable x is the **independent variable** and the variable y is the **dependent variable.** In this context, the domain of the function is the set of all *allowable* real values for the independent variable x, and the range of the function is the *resulting* set of all values taken on by the dependent variable y.

Example 3   ■   Testing for Functions Represented by Equations

Which of the equations represents y as a function of x?

(a) $y = x^2 + 1$ (b) $x - y^2 = 2$ (c) $-2x + 3y = 4$

Solution

(a) For the equation

$$y = x^2 + 1$$

there corresponds just one value of y for each value of x. For instance, when $x = 1$, the value of y is

$$y = 1^2 + 1 = 2$$

So, y *is* a function of x.

(b) By writing the equation $x - y^2 = 2$ in the form

$$y^2 = x - 2$$

you can see that some values of x correspond to *two* values of y. For instance, when $x = 3$

$$y^2 = 3 - 2$$
$$y^2 = 1$$
$$y = 1 \quad \text{or} \quad y = -1.$$

So, the solution points $(3, 1)$ and $(3, -1)$ show that y *is not* a function of x.

(c) By writing the equation $-2x + 3y = 4$ in the form

$$y = \frac{2}{3}x + \frac{4}{3}$$

you can see that there corresponds just one value of y for each value of x. For instance, when $x = 2$, the value of y is $\frac{4}{3} + \frac{4}{3} = \frac{8}{3}$. So, y *is* a function of x.

NOTE   An equation that defines y as a function of x may or may not also define x as a function of y. For instance, the equation in part (a) of Example 3 does not define x as a function of y, but the equations in parts (b) and (c) do.

Function Notation

When an equation is used to represent a function, it is convenient to name the function so that it can be easily referenced. For example, the function $y = x^2 + 1$ in Example 3(a) can be given the name "f" and written in **function notation** as

$$f(x) = x^2 + 1.$$

NOTE Often y and $f(x)$ are used to represent the same quantity. That is, each represents the value of a function for a given value of x.

Function Notation

In the notation $f(x)$:

f is the **name** of the function,

x is the **domain** (or input) value, and

$f(x)$ is a **range** (or output) value y for a given x.

The symbol $f(x)$ is read as *the value of f at x* or simply *f of x.*

The process of finding the value of $f(x)$ for a given value of x is called **evaluating a function.** This is accomplished by substituting a given x-value (input) into the equation to obtain the value of $f(x)$ (output). Here is an example.

Function	*x-Value*	*Function Value*
$f(x) = 3 - 4x$	$x = -1$	$f(-1) = 3 - 4(-1)$
		$= 3 + 4$
		$= 7$

Although f is often used as a convenient function name and x as the independent variable, you can use other letters. For instance, the equations

$$f(x) = 2x^2 + 5, \quad f(t) = 2t^2 + 5, \quad \text{and} \quad g(s) = 2s^2 + 5$$

all define the same function. In fact, the letters used are simply "placeholders" and this same function is well described by the form

$$f(\quad) = 2(\quad)^2 + 5$$

where the parentheses are used in place of a letter. To evaluate $f(-2)$, simply place -2 in each set of parentheses, as follows.

$$f(-2) = 2(-2)^2 + 5$$
$$= 8 + 5$$
$$= 13$$

When evaluating a function, you are not restricted to substituting only numerical values into the parentheses. For instance, the value of $f(3x)$ is

$$f(3x) = 2(3x)^2 + 5$$
$$= 2(9x^2) + 5$$
$$= 18x^2 + 5.$$

Example 4 ■ Evaluating a Function

Let $g(x) = 3x - x^2$. Find each value of the function.

(a) $g(1)$ (b) $g(-5)$ (c) $g(t)$ (d) $g(x + 1)$ (e) $g(x) + g(1)$

Solution

(a) Replacing x with 1 produces $g(1) = 3(1) - (1)^2 = 3 - 1 = 2$.

(b) Replacing x with -5 produces

$$g(-5) = 3(-5) - (-5)^2 = -15 - 25 = -40.$$

(c) Replacing x with t produces $g(t) = 3(t) - (t)^2 = 3t - t^2$.

(d) Replacing x with $x + 1$ produces

$$
\begin{aligned}
g(x + 1) &= 3(x + 1) - (x + 1)^2 \\
&= 3x + 3 - (x^2 + 2x + 1) \\
&= 3x + 3 - x^2 - 2x - 1 \\
&= -x^2 + x + 2.
\end{aligned}
$$

(e) Using the result of part (a) for $g(1)$, you have

$$g(x) + g(1) = (3x - x^2) + 2 = -x^2 + 3x + 2.$$

> ## Study Tip
>
> Note that $g(x + 1) \neq g(x) + g(1)$. In general, $g(a + b)$ is not equal to $g(a) + g(b)$.

Sometimes a function is defined by more than one equation, each of which is given a portion of the domain. Such a function is called a **piecewise-defined function.** To evaluate a piecewise-defined function f for a given value of x, first determine the portion of the domain in which the x-value lies and then use the corresponding equation to evaluate f. This is illustrated in Example 5.

Example 5 ■ A Function Defined by Two Equations

Let $f(t) = \begin{cases} t^2 + 1, & t < 0 \\ t - 2, & t \geq 0 \end{cases}$. Find each value of the function.

(a) $f(-4)$ (b) $f(0)$ (c) $f(10) + f(-2)$

Solution

(a) Because $t = -4 < 0$, use $f(t) = t^2 + 1$ to obtain

$$f(-4) = (-4)^2 + 1 = 17.$$

(b) Because $t = 0 \geq 0$, use $f(t) = t - 2$ to obtain

$$f(0) = 0 - 2 = -2.$$

(c) Because $t = 10 \geq 0$, use $f(t) = t - 2$ to obtain

$$f(10) = 10 - 2 = 8.$$

Because $t = -2 < 0$, use $f(t) = t^2 + 1$ to obtain

$$f(-2) = (-2)^2 + 1 = 5.$$

So, $f(10) + f(-2) = 8 + 5 = 13.$

Finding the Domain of a Function

The domain of a function may be explicitly described along with the function, or it may be *implied* by the expression used to define the function. The **implied domain** is the set of all real numbers (inputs) that yield real number values for the function. For instance, the function given by

NOTE   Note from the discussion at the right that the two most common mathematical reasons for restricting the domain of a function are to (a) avoid division by zero and (b) avoid taking the square root of a negative number.

$$f(x) = \frac{1}{x^2 - 9} \qquad \text{Domain: all } x \neq \pm 3$$

has an implied domain that consists of all real values of x other than $x = \pm 3$. These two values are excluded from the domain because division by zero is undefined. Another common type of implied domain is that used to avoid even roots of negative numbers. For instance, the function given by

$$f(x) = \sqrt{x} \qquad \text{Domain: all } x \geq 0$$

is defined only for $x \geq 0$. So, its implied domain is the set of all real numbers $x \geq 0$. More will be said about the domains of square root functions in Chapter 5.

Example 6 ■ Finding the Domain of a Function

Find the domain of each function.

(a) $f: \{(-3, 0), (-1, 2), (0, 4), (2, 4), (4, -1)\}$ (b) $f(x) = \dfrac{x - 1}{x + 4}$

(c) Area of a circle: $A = \pi r^2$ (d) $f(x) = \sqrt{x - 3}$

(e) $f(x) = 5x + 7$

Solution

(a) The domain of f consists of all first coordinates in the set of ordered pairs. So, the domain is

$$\text{Domain} = \{-3, -1, 0, 2, 4\}.$$

(b) Excluding x-values that yield zero in the denominator, the domain of f is the set of all real numbers such that $x \neq -4$.

(c) For the area of a circle, you must choose positive values for the radius r. So, the domain is the set of all real numbers r such that $r > 0$.

(d) This function is defined only for x-values that do not result in taking the square root of a negative number. By simple inspection you can see that any x-value less than 3 will result in the square root of a negative number. So, the domain consists of all real numbers x such that $x \geq 3$.

(e) The domain of the function $f(x) = 5x + 7$ is the set of all real numbers.

Note in Example 6(c) that the domain of a function can be implied by a physical context. For instance, from the equation $A = \pi r^2$, you would have no reason to restrict r to positive values. However, because you know this function represents the area of a circle, you can conclude that the radius must be positive.

2.4 ■ Exercises

Developing Skills

In Exercises 1–4, give the domain and range of the relation. Then draw a graphical representation of the relation.

1. $\{(-2, 0), (0, 1), (1, 4), (0, -1)\}$

2. $\{(3, 10), (4, 5), (6, -2), (8, 3)\}$

3. $\{(0, 0), (4, -3), (2, 8), (5, 5), (6, 5)\}$

4. $\{(-3, 6), (-3, 2), (-3, 5)\}$

In Exercises 5–10, write a set of ordered pairs that represents the rule of correspondence.

5. In a given week, a salesperson travels a distance d in t hours at an average speed of 50 miles per hour. The travel times for each day are 3 hours, 2 hours, 8 hours, 6 hours, and $\frac{1}{2}$ hour.

6. A high school teacher has five different classes full of students. It takes the teacher t minutes to grade all of the homework papers for a class with n students in it at an average rate of 3 minutes per student. The total numbers of students in each of the classes are 20, 24, 32, 27, and 25.

7. The cubes of all positive integers less than 8

8. The squares of all negative integers greater than -10

9. The winners of the World Series from 1997 through 2001

10. The men inaugurated president of the United States in 1973, 1977, 1981, 1985, 1989, 1993, 1997, and 2001

In Exercises 11–18, determine if the relation is a function.

11.
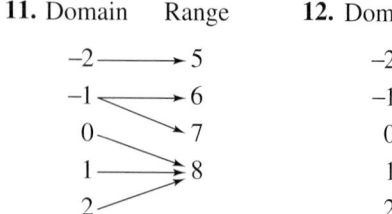

12.

13.
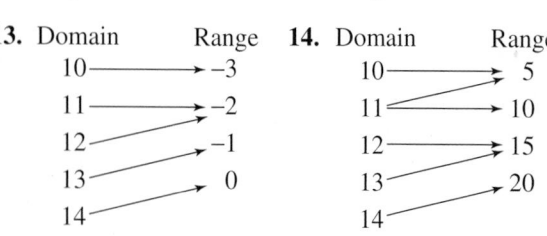

14.

15.

Domain	Range
CBS	60 Minutes / Survivor / Dan Rather
ABC	Spin City / NYPD Blue / Peter Jennings

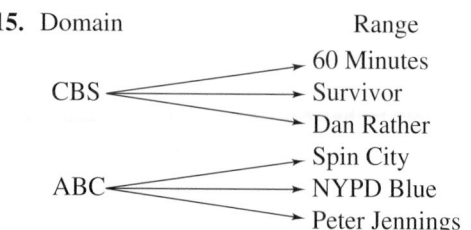

16.

Domain (Year)	Range (Married women in the labor force, in millions)
1994	32.9
1995	33.4
1996	33.6
1997	33.8
1998	34.4
1999	

(Source: U.S. Bureau of Labor Statistics)

17.

Input value	4	7	9	7	4
Output value	2	4	6	8	10

18.

Input value	0	2	4	6	8
Output value	5	5	5	5	5

In Exercises 19 and 20, determine which sets of ordered pairs represent functions from A to B.

19. $A = \{0, 1, 2, 3\}$
$B = \{-2, -1, 0, 1, 2\}$
(a) $\{(0, 1), (1, -2), (2, 0), (3, 2)\}$
(b) $\{(0, -1), (2, 2), (1, -2), (3, 0), (1, 1)\}$
(c) $\{(0, 0), (1, 0), (2, 0), (3, 0)\}$
(d) $\{(0, 2), (3, 0), (1, 1)\}$

20. $A = \{1, 2, 3\}$
$B = \{9, 10, 11, 12\}$
(a) $\{(1, 10), (3, 11), (3, 12), (2, 12)\}$
(b) $\{(1, 10), (2, 11), (3, 12)\}$
(c) $\{(1, 10), (1, 9), (3, 11), (2, 12)\}$
(d) $\{(3, 9), (2, 9), (1, 12)\}$

In Exercises 21–24, show that both ordered pairs are solutions of the equation and explain why this implies that y is not a function of x.

21. $x^2 + y^2 = 25$, $(0, 5)$, $(0, -5)$

22. $x^2 + 4y^2 = 16$, $(0, 2)$, $(0, -2)$

23. $|y| = x + 2$, $(1, 3)$, $(1, -3)$

24. $|y - 2| = x$, $(2, 4)$, $(2, 0)$

Interpreting a Graph   In Exercises 25 and 26, use the graph, which shows the numbers of high school and college students in the United States. *(Source: U.S. National Center for Education Statistics)*

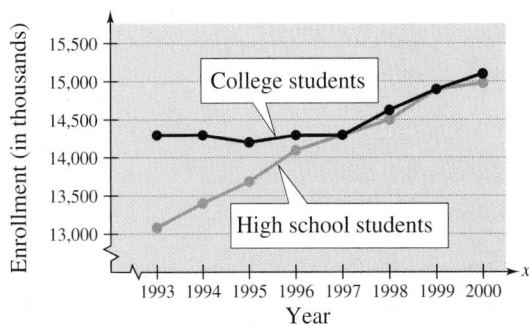

Number of Students in U.S.

25. Is the high school enrollment a function of the year? Is the college enrollment a function of the year? Explain.

26. Let $f(x)$ represent the number of high school students in year x. Approximate $f(1996)$.

In Exercises 27 and 28, fill in the blanks and simplify.

27. $f(x) = 3x + 5$

   (a) $f(2) = 3(\quad) + 5$

   (b) $f(-2) = 3(\quad) + 5$

   (c) $f(k) = 3(\quad) + 5$

   (d) $f(k + 1) = 3(\quad) + 5$

28. $f(x) = \sqrt{x + 8}$

   (a) $f(1) = \sqrt{(\quad) + 8}$

   (b) $f(-4) = \sqrt{(\quad) + 8}$

   (c) $f(h) = \sqrt{(\quad) + 8}$

   (d) $f(h - 8) = \sqrt{(\quad) + 8}$

In Exercises 29–40, evaluate the function as indicated, and simplify.

29. $f(x) = 12x - 7$

   (a) $f(3)$ (b) $f\left(\frac{3}{2}\right)$

   (c) $f(a) + f(1)$ (d) $f(a + 1)$

30. $f(x) = 3 - 7x$

   (a) $f(-1)$ (b) $f\left(\frac{1}{2}\right)$

   (c) $f(t) + f(-2)$ (d) $f(w)$

31. $f(x) = \sqrt{x + 5}$

   (a) $f(-1)$ (b) $f(4)$

   (c) $f\left(\frac{16}{3}\right)$ (d) $f(5z)$

32. $g(x) = 8 - \sqrt{x + 4}$

   (a) $g(0)$ (b) $g(-2)$

   (c) $g\left(-\frac{11}{9}\right)$ (d) $g(2t)$

33. $f(x) = \dfrac{3x}{x - 5}$

   (a) $f(0)$ (b) $f\left(\frac{5}{3}\right)$

   (c) $f(2) - f(-1)$ (d) $f(x + 4)$

34. $g(x) = \dfrac{x + 2}{x + 1}$

   (a) $g(0)$ (b) $g\left(-\frac{1}{3}\right)$

   (c) $g(3) + g(-5)$ (d) $g(x - 2)$

35. $f(x) = \begin{cases} x + 8, & x < 0 \\ 10 - 2x, & x \geq 0 \end{cases}$

   (a) $f(4)$ (b) $f(-10)$

   (c) $f(0)$ (d) $f(6) - f(-2)$

36. $g(x) = \begin{cases} 3x + 5, & x < 2 \\ -x + 10, & x \geq 2 \end{cases}$

   (a) $g(2)$ (b) $g(0)$

   (c) $g(-5)$ (d) $g(-1) + g(5)$

37. $f(x) = \begin{cases} 12, & x < -2 \\ 5x - 4, & -2 \leq x < 2 \\ 10x, & x \geq 2 \end{cases}$

   (a) $f(-2)$ (b) $f(2)$

   (c) $f(1) + f(3)$ (d) $f(-5) + f(-4)$

38. $g(x) = \begin{cases} -6x + 1, & x < 0 \\ 0, & 0 \leq x \leq 3 \\ 12x - 5, & x > 3 \end{cases}$

   (a) $g(0)$ (b) $g(3)$

   (c) $g(-3) + g(5)$ (d) $g(2) - g(4)$

39. $f(x) = 2x^2 + 5$

(a) $f(x + 2) - f(2)$ (b) $\dfrac{f(x - 3) - f(3)}{x}$

40. $f(x) = x^2 + 4$

(a) $f(x + 1) - f(1)$ (b) $\dfrac{f(x - 5) - f(x)}{5}$

In Exercises 41–46, find the domain and range of the function.

41. $f: \{(0, 0), (2, 1), (4, 8), (6, 27)\}$

42. $f: \left\{\left(-3, -\frac{17}{2}\right), \left(-1, -\frac{5}{2}\right), (4, 2), (10, 11)\right\}$

43. $g: \{(-5, 9), (-2, 3), (0, -2), (3, 3), (6, 9)\}$

44. $g: \left\{(-1, 26), (0, 30), (1, 35), (2, 29), \left(3, \frac{61}{2}\right)\right\}$

45. Circumference of a circle: $C = 2\pi r$

46. Area of a square of side s: $A = s^2$

In Exercises 47–64, find the domain of the function.

47. $h(x) = 4x - 3$

48. $f(x) = 18 - 5x$

49. $f(x) = \dfrac{2x}{x - 3}$

50. $g(x) = \dfrac{x + 5}{x + 4}$

51. $f(t) = \dfrac{t + 3}{t(t + 2)}$

52. $g(s) = \dfrac{s - 2}{(s - 6)(s - 10)}$

53. $g(x) = \dfrac{5x}{x^2 - 3x + 2}$

54. $g(x) = \dfrac{-3x + 1}{x^2 - x - 12}$

55. $h(x) = \dfrac{9}{x^2 - 1}$

56. $f(x) = \dfrac{4}{x^2 - 4}$

57. $f(x) = \sqrt{x - 2}$

58. $f(x) = \sqrt{x - 5}$

59. $f(x) = \sqrt{3 - x}$

60. $g(x) = \sqrt{8 - x}$

61. $g(t) = |t - 4|$

62. $f(x) = |x + 3|$

63. $f(t) = t^2 + 4t - 1$

64. $g(t) = 7t^3 + 10$

Solving Problems

65. *Geometry*   Express the perimeter P of a square as a function of the length x of a side.

66. *Geometry*   Express the surface area S of a cube as a function of the length x of one of its edges.

67. *Geometry*   An open box is to be made from a square piece of material 24 inches on a side by cutting equal squares from the corners and turning up the sides (see figure). Write the volume V of the box as a function of x.

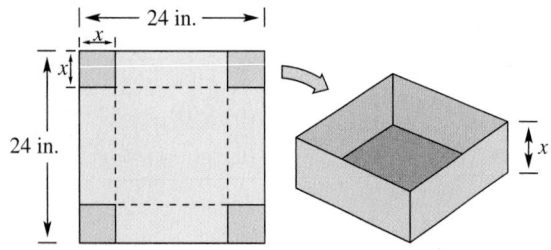

68. *Geometry*   A box with a lid is to be made from a rectangular piece of material 60 inches by 100 inches by cutting equal squares from the corners and the middles of the longer sides (see figure). Write the volume V of the box as a function of x.

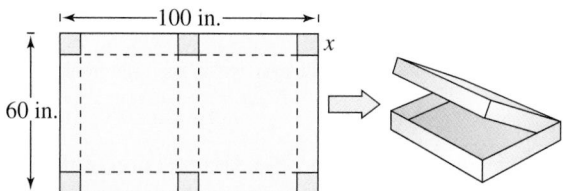

Figure for 68

69. *Geometry*   Strips of width x are cut from the four sides of a square that is 32 inches on a side (see figure). Write the area A of the remaining square as a function of x.

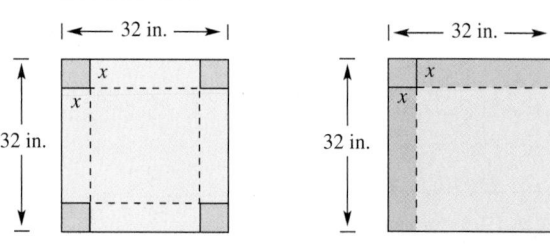

Figure for 69   Figure for 70

70. *Geometry*   Strips of width x are cut from two adjacent sides of a square that is 32 inches on a side (see figure). Write the area A of the remaining square as a function of x.

71. *Distance*   A plane is flying at a speed of 230 miles per hour. Express the distance d traveled by the plane as a function of the time t in hours.

72. *Business*   The inventor of a new game believes that the variable cost of producing the game is $1.95 per unit and the fixed costs are $8000. Write the total cost C as a function of x, the number of games produced.

73. *Business*   The marketing department of a business has determined that the profit for selling x units of a product is approximated by the model $P(x) = 50\sqrt{x} - 0.5x - 500$. Find (a) $P(1600)$ and (b) $P(2500)$.

74. *Physics*   A solid rectangular beam has a height of 6 inches and a width of 4 inches. The safe load S of the beam with the load at the center is a function of its length L and is approximated by the model

$$S(L) = \frac{128{,}160}{L}$$

where S is measured in pounds and L is measured in feet. Find (a) $S(12)$ and (b) $S(16)$.

Explaining Concepts

75. Explain the difference between a relation and a function.

76. Is it true that every relation is a function? Is it true that every function is a relation? Explain.

In Exercises 77 and 78, determine whether the statements use the word *function* in ways that are *mathematically correct*.

77. (a) The sales tax on a purchased item is a function of the selling price.

(b) Your score on the next algebra exam is a function of the number of hours you study the night before the exam.

78. (a) The amount in your savings account is a function of your salary.

(b) The speed at which a free-falling baseball strikes the ground is a function of the height from which it was dropped.

Ongoing Review

In Exercises 79–82, simplify the expression.

79. $5(x + t) + 8 - (3x + 1)$

80. $4(y + h) - 3 - (8y - 5)$

81. $7(x + 4)^2 - 8$ **82.** $-2(x - 6)^2 + 13$

In Exercises 83–86, evaluate the expression at each value of the variable. (If not possible, state the reason.)

83. $8 - 5x^2$ (a) $x = 1$ (b) $x = -2$

84. $2 + (x - 1)^2$ (a) $x = 0$ (b) $x = -8$

85. $\dfrac{x + 5}{x - 12}$ (a) $x = -5$ (b) $x = 12$

86. $5x - \dfrac{7}{2x}$ (a) $x = -7$ (b) $x = 0$

87. *Geometry*   The rectangular floor of a storage shed has an area of 308 square feet. The length of the floor is 8 feet more than its width. Find the dimensions of the floor.

Looking Further

Wages   A wage earner is paid $12.00 per hour for regular time and time-and-a-half for overtime. The weekly wage function is

$$W(h) = \begin{cases} 12h, & 0 < h \le 40 \\ 18(h - 40) + 480, & h > 40 \end{cases}$$

where h represents the number of hours worked in a week.

Evaluate the weekly wage function for each value of h.

(a) $h = 30$ (b) $h = 40$ (c) $h = 45$ (d) $h = 50$

Could you use values of h for which $h < 0$ in this model? Why or why not? If the company increased the regular work week to 45 hours, what would the new weekly wage function be?

2.5 Graphs of Functions

What you should learn:

- How to sketch graphs of functions on rectangular coordinate systems using the point-plotting method
- How to sketch the graphs of linear functions
- How to use the Vertical Line Test to determine if graphs represent functions
- How to find domains and ranges graphically
- How to sketch graphs of piecewise-defined functions

Why you should learn it:

Graphs of functions can help you visualize relationships between variables in real-life situations. For instance, in Exercise 62 on page 142 the graph of a function visually represents the total assets of Individual Retirement Accounts in the United States.

The Graph of a Function

Consider a function f whose domain and range are the set of real numbers. The **graph** of f is the set of ordered pairs $(x, f(x))$, where x is in the domain of f.

$$x = x\text{-coordinate of the ordered pair}$$
$$f(x) = y\text{-coordinate of the ordered pair}$$

Figure 2.41 shows a typical graph of such a function.

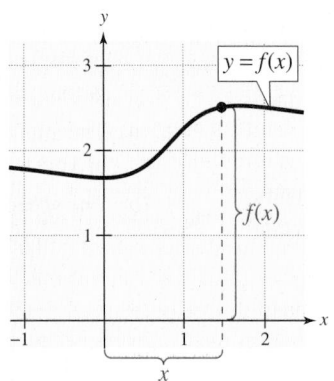

Figure 2.41

Example 1 ■ Sketching the Graph of a Function

Sketch the graph of $f(x) = x^2 - 4x + 3$.

Solution

One way to sketch the graph is to begin by making a table of values.

x	-1	0	1	2	3	4	5
$f(x)$	8	3	0	-1	0	3	8

Next, plot the seven points shown in the table. Finally, connect the points with a smooth curve, as shown in Figure 2.42.

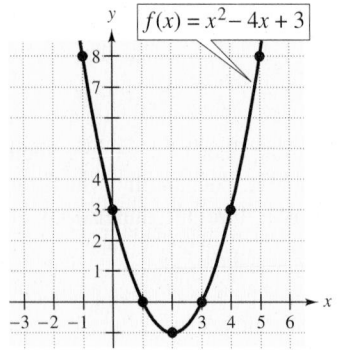

Figure 2.42

A graphing utility can be a big help in sketching the graph of a function. The difficulty, however, is to find a viewing window that provides a good view of the overall characteristics of the function. Use a graphing utility to find a good viewing window for each of the functions. How are the three graphs similar?

a. $f(x) = -2x^2 + 4x + 12$ b. $f(x) = x^2 - 5x$ c. $f(x) = x^2 - 24$

Graphs of Linear Functions

There are three basic strategies for sketching the graph of a function.

1. Use the point-plotting method.

2. Use a graphing utility.

3. Use an algebraic approach.

The first two strategies are addressed on the previous page: the point-plotting method is illustrated in Example 1, and the use of a graphing utility is discussed in the Exploration feature. In the third strategy—*an algebraic approach*—you write the function in a form that allows you to recognize its graph. You might recognize the graph to be a line, a parabola, a V-shaped graph, or some other shape. For instance, in Section 2.3 you learned that the graph of an equation of the form

$$y = mx + b$$

is a line with a slope of m and a y-intercept of $(0, b)$. This type of equation is called a **linear function.**

Definition of a Linear Function

A **linear function** of x is one that can be written in the form

$$f(x) = mx + b$$

where $m \neq 0$. The graph of a linear function is a line with a slope of m and a y-intercept of $(0, b)$.

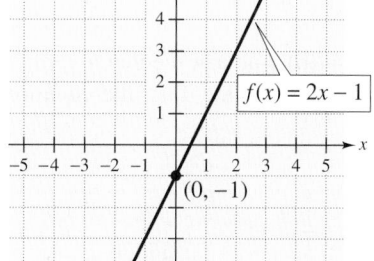

(a)

Example 2 ■ Graphing Linear Functions

Sketch the graph of each function.

(a) $f(x) = 2x - 1$ (b) $f(x) = -\frac{1}{3}x + 2$

Solution

(a) This is a linear function.

Slope is 2. y-intercept is $(0, -1)$.

$$f(x) = 2x - 1$$

Its graph is a line with a slope of 2 and a y-intercept of $(0, -1)$, as shown in Figure 2.43(a).

(b) This is a linear function.

Slope is $-\frac{1}{3}$. y-intercept is $(0, 2)$.

$$f(x) = -\frac{1}{3}x + 2$$

Its graph is a line with a slope of $-\frac{1}{3}$ and a y-intercept of $(0, 2)$, as shown in Figure 2.43(b).

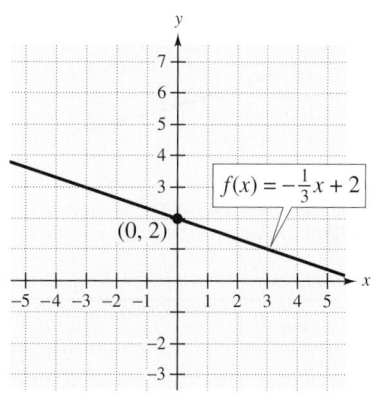

(b)

Figure 2.43

The Vertical Line Test

By the definition of a function, at most one y-value corresponds to a given x-value. This implies that any vertical line can intersect the graph of a function at most once.

> **Vertical Line Test for Functions**
>
> A set of points on a rectangular coordinate system is the graph of y as a function of x if and only if no vertical line intersects the graph at more than one point.

Example 3 ■ Using the Vertical Line Test

Decide whether each equation represents y as a function of x.

(a) $y = |x| - 2$ (b) $x = y^2 - 1$

Solution

(a) From the graph of the equation in Figure 2.44(a), you can see that every vertical line intersects the graph at most once. So, by the Vertical Line Test, the equation *does* represent y as a function of x.

(b) From the graph of the equation in Figure 2.44(b), you can see that a vertical line intersects the graph twice. So, by the Vertical Line Test, the equation *does not* represent y as a function of x.

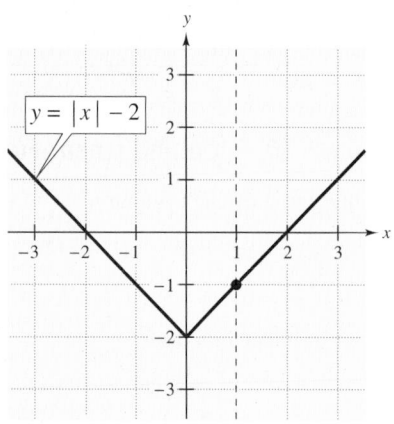

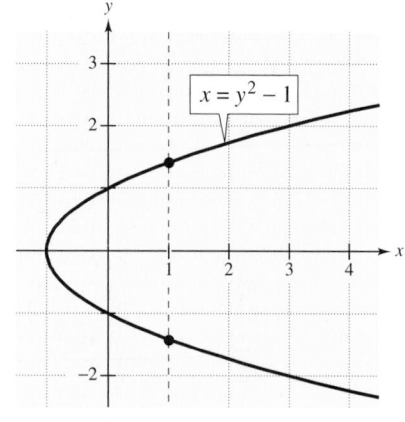

(a) Graph of a function of x;
   vertical line intersects once.

(b) Not a graph of a function of x;
   vertical line intersects twice.

Figure 2.44

Technology

The Vertical Line Test can be illustrated with a graphing utility. Once the equation is entered and graphed, you can obtain a vertical line from the DRAW menu of your graphing utility. You can decide whether an equation represents y as a function of x by moving the vertical line to the left and to the right with the arrow keys.

Example 4 ■ Using the Vertical Line Test

Which of the equations represent y as a function of x? Explain your reasoning.

(a) $x^2 = y - 5$

(b) $|y| = x - 5$

(c) $y = 2$

Solution

(a) Solving the equation for y, you obtain

$$y = x^2 + 5.$$

Because you are able to isolate y, it appears that the equation *does* represent y as a function of x. You can confirm this conclusion by the Vertical Line Test. Notice in Figure 2.45(a) that no vertical line intersects the graph more than once.

(b) The graph of this equation is shown in Figure 2.45(b). Note that it is possible for a vertical line to intersect the graph at more than one point. So, this equation *does not* represent y as a function of x. Remember that if an equation represents y as a function of x, then no two different solution points of the equation can have the same x-value. For this equation, you can see that the points $(6, 1)$ and $(6, -1)$ are both solution points of the equation. Because these two points have the same x-value, it follows that the equation

$$|y| = x - 5$$

does not represent y as a function of x.

(c) At first glance, you might be tempted to say that this equation does not represent y as a function of x, because x does not appear in the equation. In Figure 2.45(c), you can see that the graph passes the Vertical Line Test. So, the equation $y = 2$ *does* represent y as a function of x. This type of function is called a **constant function.**

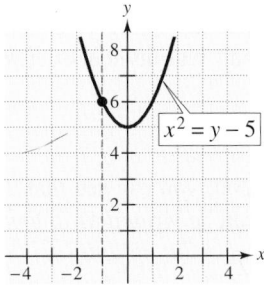

(a)

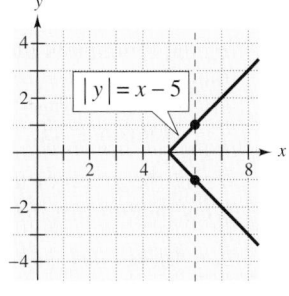

(b)

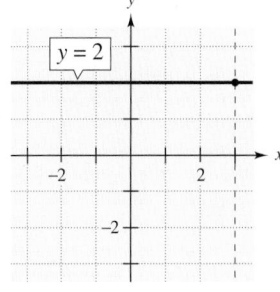

(c)

Figure 2.45

Finding Domains and Ranges Graphically

Recall from Section 2.4 that the *domain* of a function of x is the set of all x-values for which the function is defined. Unless specified differently, the domain of a linear function is the set of all real numbers.

If you want to restrict the domain of a function, you can write a restriction to the right of the equation. For instance, the domain of the function

$$f(x) = 4x + 5, \qquad x \geq 0$$

is the set of all nonnegative real numbers. As x varies over the domain of a function, the values of $f(x)$ form the *range* of the function. The graph of a function can help you describe its domain and its range.

Example 5 ■ Finding a Function's Domain and Range

Describe the domain and range of $f(x) = |x - 3|$.

Solution

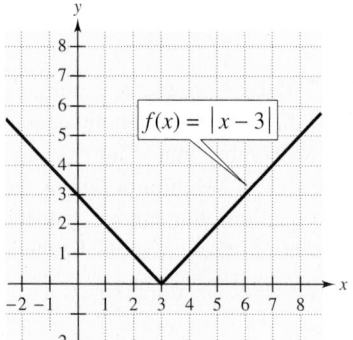

Figure 2.46

The graph of this function is V-shaped, as shown in Figure 2.46. The domain of the function is the set of all real numbers. In other words, you can substitute *any* real number for x and obtain a corresponding y-value.

Domain $= \{x: x$ is a real number.$\}$

From the graph of the function, it appears that the y-values are never negative. This is confirmed by the fact that the absolute value of a number can never be negative. The range of this function is the set of all nonnegative real numbers.

Range $= \{y: y \geq 0\}$

Example 6 ■ Finding a Function's Domain and Range

Describe the domain and range of $f(x) = \sqrt{4 - x^2}$.

Solution

The graph of this function is a half circle, as shown in Figure 2.47. The domain of the function is the set of all real numbers between -2 and 2, including -2 and 2.

Domain $= \{x: -2 \leq x \leq 2\}$

From the graph of the function, you can see that the y-values vary from 0 to 2. This implies that the range can be written as follows.

Range $= \{y: 0 \leq y \leq 2\}$

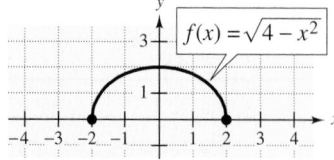

Figure 2.47

Graphs of Piecewise-Defined Functions

Most graphing utilities can graph a piecewise-defined function. For instance, to graph the function in Example 7(a) you can enter the expression below.

$$Y_1 = (2x + 1)(x \leq 0) +$$
$$(x - 3)(x > 0)$$

Be sure to place parentheses around each piece of the function and around each domain.

In Example 5 in Section 2.4, you saw that a *piecewise-defined* function is defined by two or more equations. To sketch the graph of a piecewise-defined function, you must consider each of the equations used to define the function.

Example 7 ■ Sketching Piecewise-Defined Functions

Sketch the graph of each function.

(a) $f(x) = \begin{cases} 2x + 1, & x \leq 0 \\ x - 3, & x > 0 \end{cases}$

(b) $f(x) = \begin{cases} 2, & x \leq -2 \\ -x, & -2 < x < 1 \\ -1, & x \geq 1 \end{cases}$

Solution

(a) The graph is made up of parts of two lines, as shown in Figure 2.48(a). To the left of 0 (including 0), the graph is represented by the line

$$y = 2x + 1. \qquad \text{Left portion of graph}$$

To the right of 0 (not including 0), the graph is represented by the line

$$y = x - 3. \qquad \text{Right portion of graph}$$

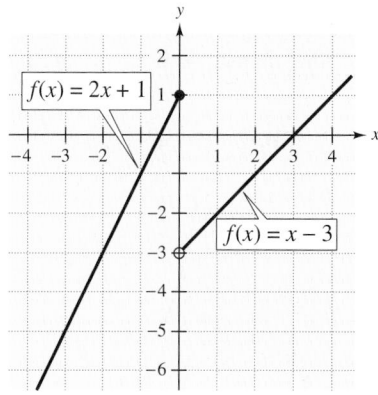

(a)

(b) The graph is made up of parts of three lines, as shown in Figure 2.48(b). To the left of -2 (including -2), the graph is represented by the horizontal line

$$y = 2. \qquad \text{Left portion of graph}$$

Between -2 and 1 (not including -2 and 1), the graph is represented by the line

$$y = -x. \qquad \text{Center portion of graph}$$

Finally, to the right of 1 (including 1), the graph is represented by the horizontal line

$$y = -1. \qquad \text{Right portion of graph}$$

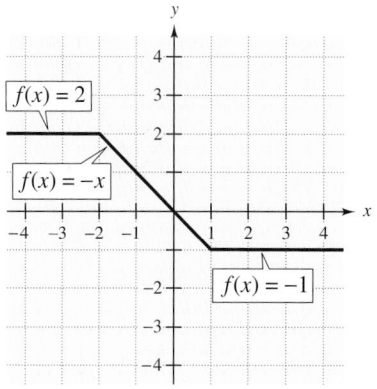

(b)
Figure 2.48

Collaborate!

Exploring with Technology

One person in your group should enter a linear function with *integer coefficients* into a graphing utility. The person should then graph the function using a viewing window that shows all intercepts of the function. From looking at only the graph of the function, the other members of the group should try to determine the equation of the linear function.

Take turns graphing functions and asking the other group members to find the equation that was graphed.

2.5 ■ Exercises

Developing Skills

In Exercises 1–6, use the slope and y-intercept to sketch the graph of the linear function.

1. $f(x) = 2x - 6$ **2.** $g(x) = -3x + 6$

3. $f(x) = -x - 4$ **4.** $g(x) = 5x + 1$

5. $g(x) = -\frac{3}{4}x + 1$ **6.** $f(x) = 0.8x + 2$

In Exercises 7–10, use function notation to write y as a function of x. Then use the slope and y-intercept to sketch the graph of the linear function.

7. $3x - y + 10 = 0$ **8.** $x + 2y - 8 = 0$

9. $0.2x + 0.8y - 4.5 = 0$ **10.** $\frac{2}{3}x - \frac{3}{4}y = 0$

In Exercises 11–18, use the Vertical Line Test to determine whether y is a function of x.

11. $y = \frac{1}{3}x^3$

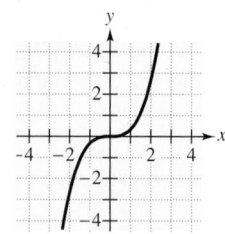

12. $y = x^2 - 2x$

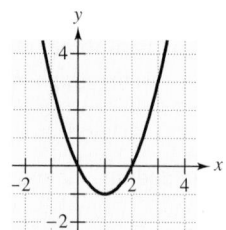

13. $y^2 = x$

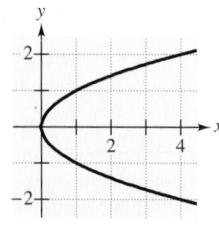

14. $y = |x|$

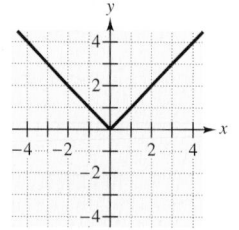

15. $x^2 + y^2 = 16$

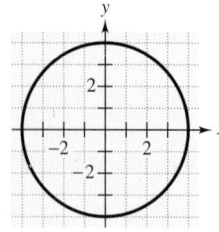

16. $x - 2y^2 = 0$

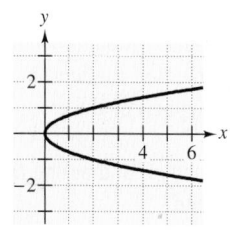

17. $y = (x + 2)^2$

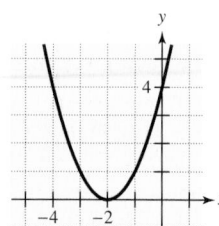

18. $|y| = x$

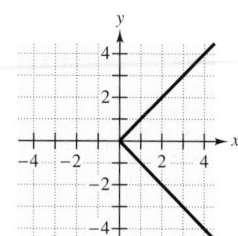

In Exercises 19–24, find the domain and range of the function.

19. $f(x) = 6 - 2x$

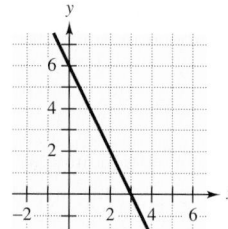

20. $h(x) = x^2 + 2x$

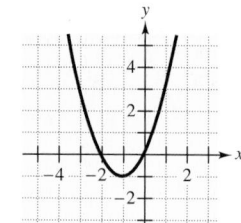

21. $f(x) = \sqrt{x^2 - 4}$

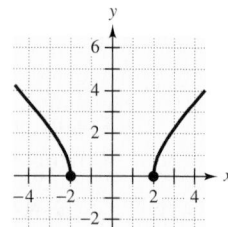

22. $g(x) = \frac{3}{2}|x - 2|$

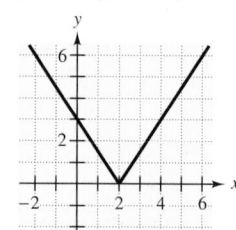

23. $g(x) = \sqrt{9 - x^2}$

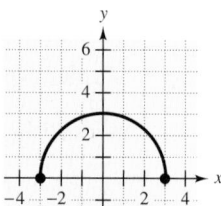

24. $h(x) = 2\sqrt{x + 3}$

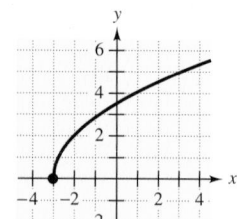

In Exercises 25–32, use a graphing utility to graph the function and find its domain and range.

25. $g(x) = 1 - x^2$

26. $f(x) = x^2 - 6x + 10$

27. $f(x) = |x + 1|$

28. $g(x) = \frac{1}{2}|x| + 3$

29. $f(x) = \sqrt{x - 2}$

30. $h(x) = \sqrt{x + 5} - 2$

31. $h(t) = \sqrt{4 - t^2}$

32. $f(x) = \sqrt{16 - x^2} + 1$

In Exercises 33–36, match the function with its graph. [The graphs are labeled (a), (b), (c), and (d).]

(a)

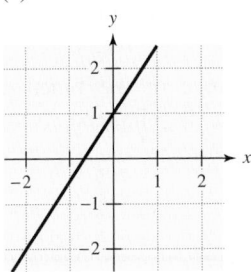

(b)

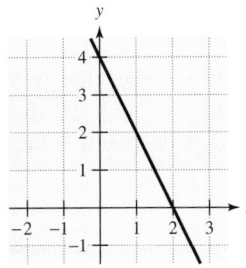

(c)

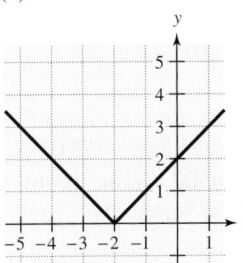

(d)

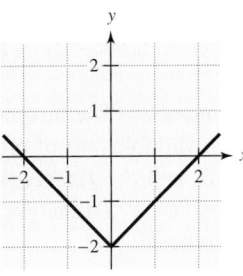

33. $f(x) = 4 - 2x$

34. $h(x) = |x + 2|$

35. $h(x) = |x| - 2$

36. $f(x) = \frac{3}{2}x + 1$

In Exercises 37–58, sketch the graph of the function and use the graph to determine its domain and range.

37. $g(x) = 4$

38. $f(x) = -2$

39. $f(x) = 2x - 7$

40. $f(x) = 3 - 2x$

41. $g(x) = \frac{1}{2}x^2$

42. $h(x) = \frac{1}{4}x^2 - 1$

43. $f(x) = -(x - 1)^2$

44. $g(x) = (x + 2)^2 + 3$

45. $K(s) = |s - 4| + 1$

46. $Q(t) = 1 - |t + 1|$

47. $f(t) = \sqrt{t - 2}$

48. $h(x) = \sqrt{4 - x}$

49. $g(s) = \frac{1}{2}s^3$

50. $f(x) = x^3 - 4$

51. $f(x) = 6 - 3x, \qquad 0 \le x \le 2$

52. $f(x) = \frac{1}{3}x - 2, \qquad 6 \le x \le 12$

53. $h(x) = x^3, \qquad -2 \le x \le 2$

54. $h(x) = x(6 - x), \qquad 0 \le x \le 6$

55. $f(t) = \begin{cases} 4 + t, & t < 0 \\ 4 - t, & t \ge 0 \end{cases}$

56. $f(x) = \begin{cases} -x, & x < 0 \\ 2x + 1, & x \ge 0 \end{cases}$

57. $g(x) = \begin{cases} 5x + 1, & x \le -2 \\ x, & -2 < x < 3 \\ -2x + 3, & x \ge 3 \end{cases}$

58. $h(x) = \begin{cases} 7, & x < 0 \\ 10x - 6, & 0 \le x < 6 \\ 8 - x, & x \ge 6 \end{cases}$

Solving Problems

59. *Geometry*   The perimeter of a rectangle is 200 meters.

(a) Show that the area of the rectangle is given by $A = x(100 - x)$, where x is its length.

(b) Use a graphing utility to graph the area function in part (a).

(c) Approximate the value of x that yields the largest value of A. Interpret the results.

60. *Geometry*   The perimeter of a rectangle is 280 yards.

(a) Show that the area of the rectangle is given by $A = 140x - x^2$, where x is its length.

(b) Use a graphing utility to graph the area function in part (a).

(c) Approximate the value of x that yields the largest value of A.

61. *Business*   The profit P when x units of a product are sold is given by $P(x) = 0.47x - 100$ for x in the interval $0 \le x \le 1000$.

(a) Use a graphing utility to graph the profit function over the specified domain.

(b) Approximately how many units must be sold for the company to break even $(P = 0)$?

(c) Approximately how many units must be sold for the company to make a profit of $300?

62. *Personal Savings* The total assets of Individual Retirement Accounts (IRAs) in billions of dollars in the United States from 1992 to 1999 can be modeled by

$$A = 25.40t^2 - 57.7t + 903$$

where t is the year, with $t = 2$ corresponding to 1992. *(Source: Investment Company Institute)*

(a) What is the domain of the model?

(b) Use a graphing utility to graph the model.

(c) Use the model to approximate the total assets of IRAs in 1997.

Explaining Concepts

63. Does the graph in Exercise 11 represent x as a function of y? Explain.

64. Does the graph in Exercise 12 represent x as a function of y? Explain.

65. In your own words, explain how to use the Vertical Line Test.

66. Explain the change in the range of the function $f(x) = x^2 - 1$ if the domain is changed from $-1 \leq x \leq 3$ to $-3 \leq x \leq 0$.

Ongoing Review

In Exercises 67–70, solve the equation.

67. $x^2 - 13x + 40 = 0$

68. $y^2 + 5y - 14 = 0$

69. $3a^3 - 12a = 0$

70. $5x^3 + 40x^2 = 0$

In Exercises 71–74, evaluate the function at each value.

71. $f(x) = 4x^2$ (a) $f(-3)$ (b) $f(0)$

72. $g(x) = 7x + 10$ (a) $g(-2)$ (b) $g(x - 2) + g(1)$

73. $f(x) = \dfrac{x + 2}{x - 5}$ (a) $f(8)$ (b) $f(t + 1)$

74. $g(x) = \sqrt{19 - 2x}$ (a) $g(5)$ (b) $g(a - 2)$

75. *Business* The inventor of a new game believes that the variable cost of producing the game is $2.05 per unit and the fixed costs are $9500. Write the total cost C as a function of x, the number of games produced.

Looking Further

Aircraft Sales The table shows the net sales N (in billions of dollars) of completed aircraft and parts in the United States for the years 1993 through 1998. *(Source: U.S. Census Bureau)*

Year	1993	1994	1995	1996	1997	1998
N	109.9	104.3	102.8	103.1	114.9	120.7

A model for this data is given by

$$N = 1.968t^2 - 19.19t + 149.5$$

where t is the time in years, with $t = 3$ corresponding to 1993.

(a) What is the domain of the function?

(b) Use a graphing utility to make a scatter plot of the data and graph the model in the same viewing window.

(c) For which year does the model most accurately estimate the actual data? For which year is it least accurate?

2.6 Transformations of Functions

Graphs of Basic Functions

To become good at sketching the graphs of functions, it helps to be familiar with the graphs of some basic functions. The functions shown in Figure 2.49, and variations of them, occur frequently in applications.

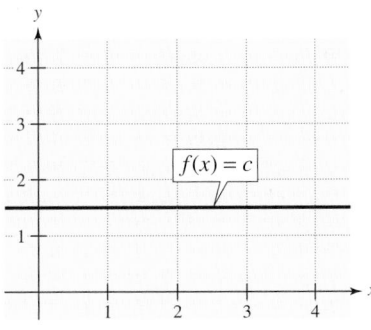

(a) Constant function

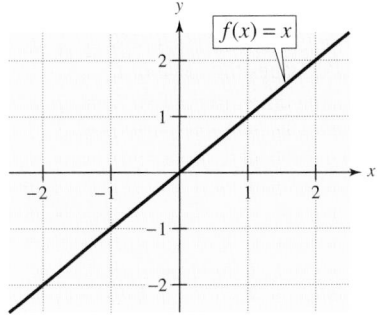

(b) Identity function

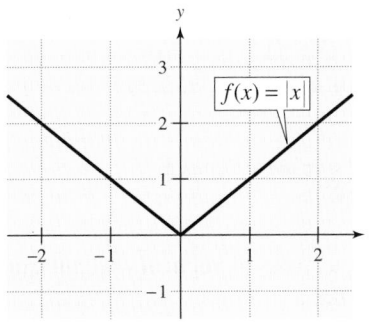

(c) Absolute value function

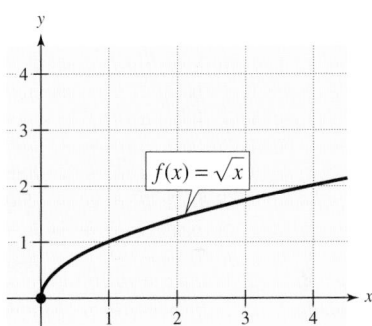

(d) Square root function

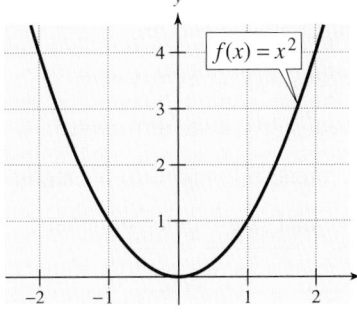

(e) Squaring function

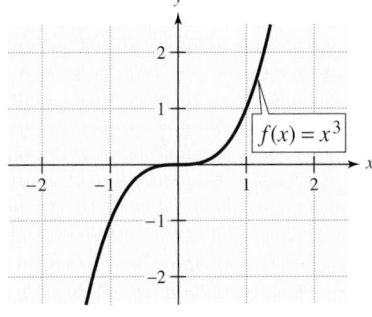

(f) Cubing function

Figure 2.49

Be sure you can recognize each of the graphs in Figure 2.49. If you have a graphing utility, try using it to confirm the shape of each graph.

Transformations of Graphs of Functions

EXPLORATION

Use a graphing utility to display the graphs of $y = x^2 + c$ where c is equal to -2, 0, 2, and 4. What effect does the value of c have on the graph of the function?

Use a graphing utility to display the graphs of $y = (x + c)^2$ where c is equal to -3, -1, 0, 1, and 3. What effect does the value of c have on the graph of the function?

Many functions have graphs that are simple transformations of the basic graphs shown in Figure 2.49. For example, you can obtain the graph of $h(x) = x^2 + 1$ by shifting the graph of $f(x) = x^2$ *upward* one unit, as shown in Figure 2.50(a). In function notation, h and f are related as follows.

$$h(x) = x^2 + 1 = f(x) + 1 \qquad \text{Upward shift of 1}$$

Similarly, you can obtain the graph of $g(x) = (x - 1)^2$ by shifting the graph of $f(x) = x^2$ to the *right* one unit, as shown in Figure 2.50(b). In this case, the functions g and f have the following relationship.

$$g(x) = (x - 1)^2 = f(x - 1) \qquad \text{Right shift of 1}$$

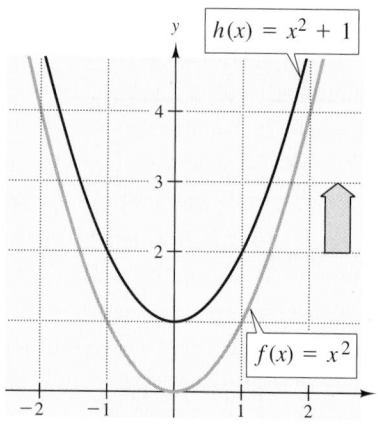

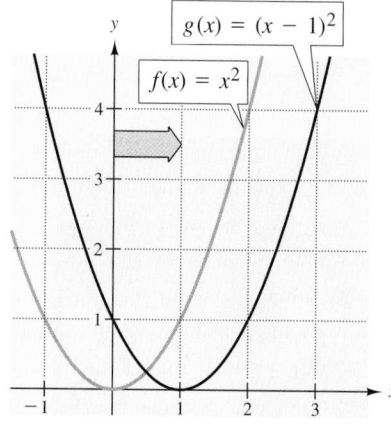

(a) Vertical shift: one unit upward

(b) Horizontal shift: one unit right

Figure 2.50

The various types of **vertical and horizontal shifts** of the graphs of functions are summarized as follows.

Vertical and Horizontal Shifts

Let c be a positive real number. **Vertical and horizontal shifts** of the graph of the function $y = f(x)$ are represented as follows.

1. Vertical shift c units **upward:** $\qquad h(x) = f(x) + c$

2. Vertical shift c units **downward:** $\qquad h(x) = f(x) - c$

3. Horizontal shift c units to the **right:** $\qquad h(x) = f(x - c)$

4. Horizontal shift c units to the **left:** $\qquad h(x) = f(x + c)$

Note that for a vertical transformation the addition of a positive number c yields a shift upward (in the positive direction) and the subtraction of a positive number c yields a shift downward (in the negative direction). For a horizontal transformation, the addition of a positive number c yields a shift to the left (in the negative direction) and the subtraction of a positive number c yields a shift to the right (in the positive direction).

Example 1 ■ Shifts of the Graphs of Functions

Use the graph of $f(x) = x^2$ to sketch the graph of each function.

(a) $g(x) = x^2 - 2$

(b) $h(x) = (x + 3)^2$

Solution

(a) Relative to the graph of $f(x) = x^2$, the graph of

$$g(x) = x^2 - 2$$

represents a shift of two units *downward*, as shown in Figure 2.51(a).

(b) Relative to the graph of $f(x) = x^2$, the graph of

$$h(x) = (x + 3)^2$$

represents a shift of three units to the *left*, as shown in Figure 2.51(b).

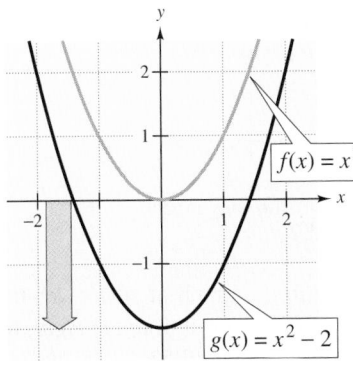
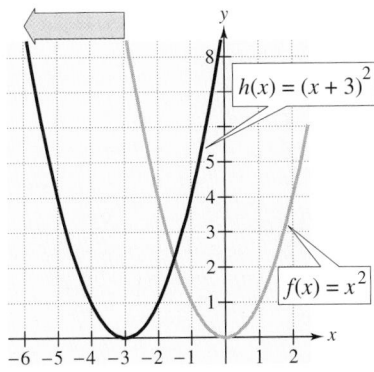

(a) Vertical shift: two units downward　　(b) Horizontal shift: three units left

Figure 2.51

NOTE　In Example 1(a), you should note that $g(x) = f(x) - 2$ and that in Example 1(b), $h(x) = f(x + 3)$.

EXPLORATION

Combining Vertical and Horizontal Shifts　The graph of each of the following can be formed by shifting the graph of one of the six basic functions *twice—once* vertically and once horizontally. Describe the graph of each function and use your description to sketch the graph. If you have a graphing utility, use it to confirm your result.

a. $f(x) = (x - 3)^2 - 1$　　　　　b. $f(x) = |x + 1| - 2$

c. $f(x) = (x - 1)^3 + 2$　　　　　d. $f(x) = \sqrt{x + 4} + 1$

e. $f(x) = (x + 2)^2 + 2$　　　　　f. $f(x) = |x - 1| - 4$

Some graphs can be obtained from *combinations* of vertical and horizontal shifts, as shown in part (b) of the next example.

Example 2 ■ Shifts of the Graphs of Functions

Use the graph of $f(x) = x^3$ to sketch the graph of each function.

(a) $g(x) = x^3 + 2$ (b) $h(x) = (x - 1)^3 + 2$

Solution

(a) Relative to the graph of $f(x) = x^3$, the graph of $g(x) = x^3 + 2$ represents a shift of two units *upward*, as shown in Figure 2.52.

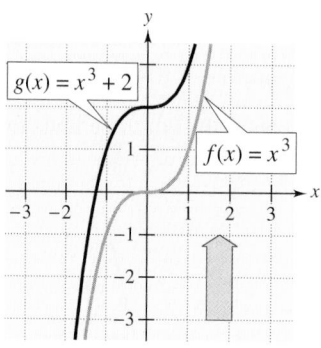

Vertical shift: two units upward
Figure 2.52

(b) Relative to the graph of $f(x) = x^3$, the graph of $h(x) = (x - 1)^3 + 2$ represents a shift of one unit to the *right*, followed by a shift of two units *upward*, as shown in Figure 2.53.

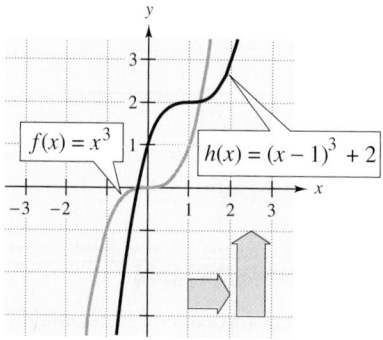

Horizontal shift: one unit right
Vertical shift: two units upward
Figure 2.53

NOTE   When you are combining horizontal and vertical shifts, it does not matter which you do first. For instance, in the graph shown in Figure 2.53, you would obtain the same result with either of the following: (a) first shift the graph of $f(x) = x^3$ to the right one unit, then shift this graph upward two units; (b) first shift the graph of $f(x) = x^3$ upward two units, then shift this graph to the right one unit.

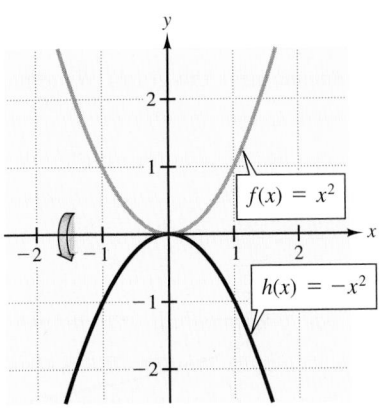

Figure 2.54   *Reflection*

The second basic type of transformation is called a **reflection.** For instance, if you imagine that the *x*-axis represents a mirror, then the graph of

$$h(x) = -x^2$$

is the mirror image (or reflection) of the graph of

$$f(x) = x^2$$

as shown in Figure 2.54.

Reflections in the Coordinate Axes

Reflections of the graph of $y = f(x)$ are represented as follows.

1. Reflection in the *x*-axis: $h(x) = -f(x)$

2. Reflection in the *y*-axis: $h(x) = f(-x)$

Example 3 ■ Reflections of the Graphs of Functions

Use the graph of $f(x) = \sqrt{x}$ to sketch the graph of each function.

(a) $g(x) = -\sqrt{x}$ (b) $h(x) = \sqrt{-x}$

Solution

(a) Relative to the graph of $f(x) = \sqrt{x}$, the graph of

$$g(x) = -\sqrt{x} = -f(x)$$

represents a *reflection in the x-axis*, as shown in Figure 2.55(a).

(b) Relative to the graph of $f(x) = \sqrt{x}$, the graph of

$$h(x) = \sqrt{-x} = f(-x)$$

represents a *reflection in the y-axis*, as shown in Figure 2.55(b).

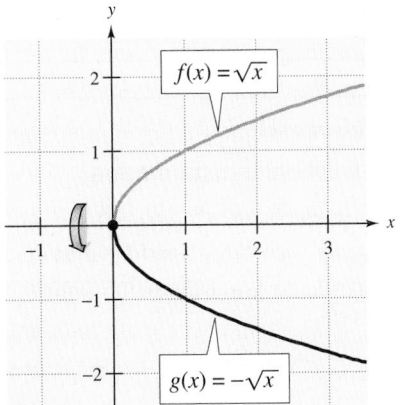

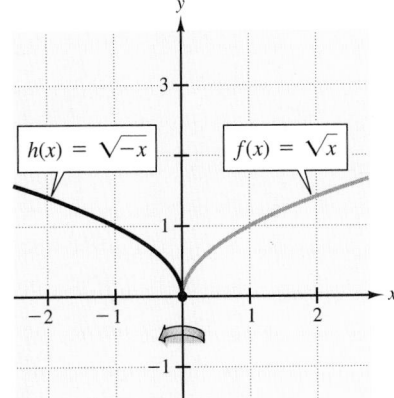

(a) Reflection in *x*-axis

(b) Reflection in *y*-axis

Figure 2.55

Example 4 ■ Graphical Reasoning

Identify the basic function, and any transformation shown in the graph. Write the equation for the graphed function.

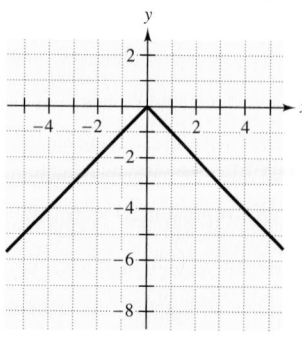

(a)

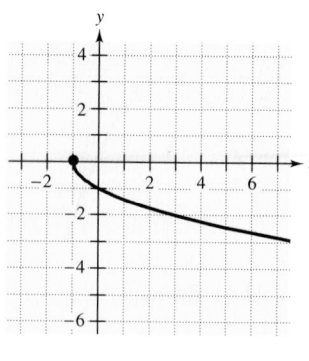

(b)   (c)

Figure 2.56

Solution

(a) In Figure 2.56(a), the basic function is $f(x) = |x|$. The transformation is a reflection in the x-axis, so the equation of the graphed function is

$$g(x) = -|x|.$$

(b) In Figure 2.56(b), the basic function is $f(x) = x^3$. The transformation is a horizontal shift four units to the left and a vertical shift one unit upward. The equation of the graphed function is

$$g(x) = (x + 4)^3 + 1.$$

(c) In Figure 2.56(c), the basic function is $f(x) = \sqrt{x}$. The transformation is a reflection in the x-axis and a horizontal shift one unit to the left. The equation of the graphed function is

$$g(x) = -\sqrt{x + 1}.$$

Collaborate!

Constructing Transformations

Use a graphing utility to graph $f(x) = x^2 + 2$. Decide how to alter this function to produce each of the transformation descriptions. Graph each transformation in the same viewing window with f; confirm that the transformation moved f as described.

a. The graph of f shifted to the left three units.

b. The graph of f shifted downward five units.

c. The graph of f shifted upward one unit.

d. The graph of f shifted to the right two units.

2.6 ■ Exercises

Developing Skills

In Exercises 1–12, sketch the graphs of the three functions *by hand* on the same rectangular coordinate system. Verify your result with a graphing utility.

1. $f(x) = x$
   $g(x) = x - 3$
   $h(x) = 3x$

2. $f(x) = \frac{1}{2}x$
   $g(x) = \frac{1}{2}x + 1$
   $h(x) = \frac{1}{2}(x - 1)$

3. $f(x) = x^2$
   $g(x) = x^2 + 3$
   $h(x) = (x - 3)^2$

4. $f(x) = x^2$
   $g(x) = x^2 - 2$
   $h(x) = (x - 2)^2 + 1$

5. $f(x) = -x^2$
   $g(x) = -x^2 + 2$
   $h(x) = -(x - 1)^2$

6. $f(x) = (x + 1)^2$
   $g(x) = (x + 1)^2 + 2$
   $h(x) = -(x + 1)^2 + 4$

7. $f(x) = x^2$
   $g(x) = \left(\frac{1}{2}x\right)^2$
   $h(x) = (2x)^2$

8. $f(x) = x^2$
   $g(x) = \left(\frac{1}{4}x\right)^2 + 3$
   $h(x) = -\left(\frac{1}{4}x\right)^2$

9. $f(x) = |x|$
   $g(x) = |x| - 2$
   $h(x) = |x - 2|$

10. $f(x) = |x|$
   $g(x) = \frac{1}{2}|x|$
   $h(x) = 1 - |x|$

11. $f(x) = \sqrt{x}$
   $g(x) = \sqrt{x} + 2$
   $h(x) = \sqrt{x - 2}$

12. $f(x) = \sqrt{x}$
   $g(x) = \sqrt{x + 3}$
   $h(x) = \sqrt{x - 2} + 1$

In Exercises 13–26, identify the basic function, and any transformation shown in the graph. Write the equation for the graphed function.

13.

14.
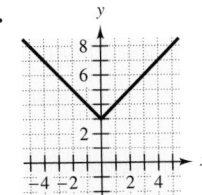

15.
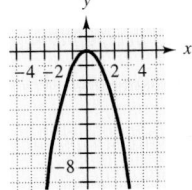

16.
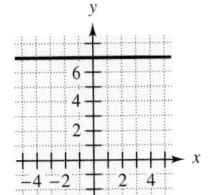

17.
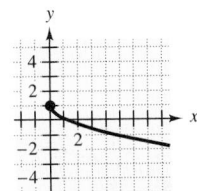

18.
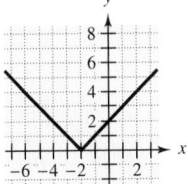

19.
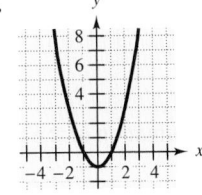

20.
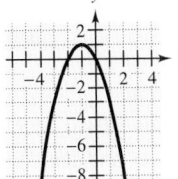

21.
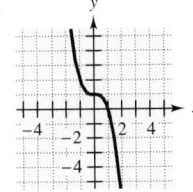

22.
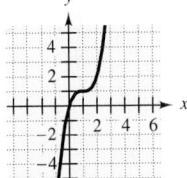

23.
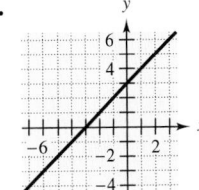

24.
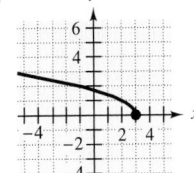

25.
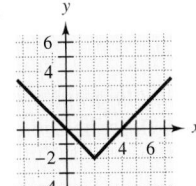

26.
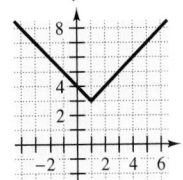

In Exercises 27–32, use the graph of $f(x) = \sqrt{x}$ to write a function that represents the graph.

27.
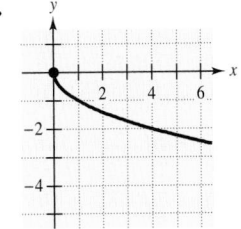

28.
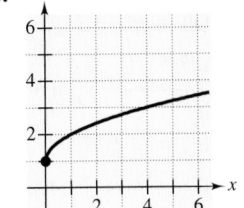

29.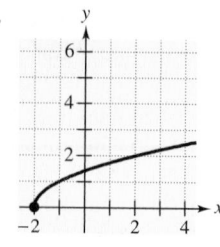

30.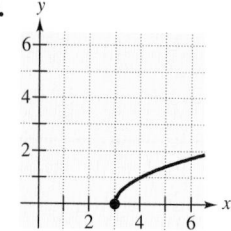

31.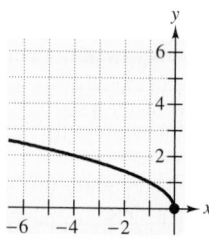

32.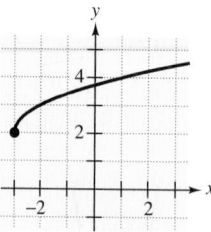

In Exercises 33–40, identify the transformation of the graph of $f(x) = x^3$ and sketch the graph of h.

33. $h(x) = x^3 + 3$

34. $h(x) = x^3 - 5$

35. $h(x) = (x - 3)^3$

36. $h(x) = (x + 2)^3$

37. $h(x) = (-x)^3$

38. $h(x) = -x^3$

39. $h(x) = 2 - (x - 1)^3$

40. $h(x) = (x + 2)^3 - 3$

In Exercises 41–46, identify the transformation of the graph of $f(x) = |x|$ and sketch the graph of h.

41. $h(x) = |x - 5|$

42. $h(x) = |x + 3|$

43. $h(x) = |x| - 5$

44. $h(x) = |-x|$

45. $h(x) = -|x|$

46. $h(x) = 5 - |x|$

In Exercises 47–50, use a graphing utility to graph the two functions in the same viewing window. Describe the graph of g relative to the graph of f.

47. $f(x) = x^3 - 3x^2$

$g(x) = f(x + 2)$

48. $f(x) = x^3 - 3x^2 + 2$

$g(x) = f(x - 1)$

49. $f(x) = x^3 - 3x^2$

$g(x) = f(-x)$

50. $f(x) = x^3 - 3x^2 + 2$

$g(x) = -f(x)$

In Exercises 51 and 52, use the graph of $f(x) = x^3 - 3x^2$ to write an equation for the function g shown in the graph.

51.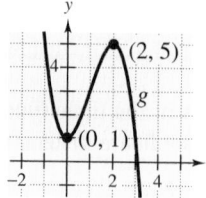

52.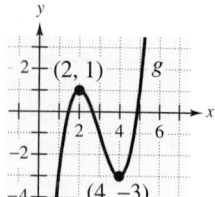

53. Use the graph of f to sketch the graphs.

(a) $y = f(x) + 2$

(b) $y = -f(x)$

(c) $y = f(x - 2)$

(d) $y = f(x + 3)$

(e) $y = f(x) - 1$

(f) $y = f(-x)$

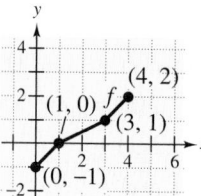

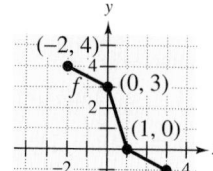

Figure for 53   Figure for 54

54. Use the graph of f to sketch the graphs.

(a) $y = f(x) - 1$

(b) $y = f(x + 1)$

(c) $y = f(x - 1)$

(d) $y = -f(x - 2)$

(e) $y = f(-x)$

(f) $y = f(x) + 2$

55. *Investigation* Use a graphing utility to graph each function. Describe any similarities and differences you observe among the graphs.

(a) $y = x$

(b) $y = x^2$

(c) $y = x^3$

(d) $y = x^4$

(e) $y = x^5$

(f) $y = x^6$

56. *Conjecture* Use the results from Exercise 55 to make a conjecture about the graphs of $y = x^7$ and $y = x^8$. Use a graphing utility to verify your conjecture.

57. Use the results from Exercise 55 to sketch the graph of

$$y = (x - 3)^5$$

by hand. Use a graphing utility to verify your graph.

58. Use the results from Exercise 55 to sketch the graph of

$$y = (x + 1)^6$$

by hand. Use a graphing utility to verify your graph.

Solving Problems

59. *Population*   For the years 1950 through 2000, the civilian population P (in thousands) of the United States can be modeled by

$$P_1(t) = 2.91t^2 + 2374.0t + 153,084$$

$$0 \le t \le 50$$

where $t = 0$ represents 1950. *(Source: U.S. Census Bureau)*

(a) Use a graphing utility to graph the function over the appropriate domain.

(b) In the transformation of the population function

$$P_2(t) = 2.91(t + 30)^2 + 2374.0(t + 30) + 153,084,$$

$t = 0$ corresponds to what calendar year? Explain.

(c) Use a graphing utility to graph P_2 over the appropriate domain.

Explaining Concepts

In Exercises 60–63, use the results from Exercises 55–58 to make a conjecture about the shape of the graph of the function. Use a graphing utility to verify your conjecture.

60. $f(x) = x^2(x - 6)^2$

61. $f(x) = x^3(x - 6)^2$

62. $f(x) = x^2(x - 6)^3$

63. $f(x) = x^3(x - 6)^3$

64. Explain the four types of shifts of the graph of a function.

Ongoing Review

In Exercises 65–74, simplify the expression.

65. $4(x + 9) - 2$

66. $-3(y - 4) + 7$

67. $2(x + 6) - 5(x + 6)$

68. $(t - 2) + 3(t - 2)$

69. $x(x + 7) - 4(x + 7)$

70. $2x(x - 5) + 3(x - 5)$

71. $5(z - 2) + (z - 2)^2$

72. $4(3 - u) - 2(3 - u)^2$

73. $8(x + 4) - (x + 4)^2$

74. $3(t - 3)^3 + 5(t - 3)^2$

Geometry   In Exercises 75 and 76, find an expression for the area of the shaded portion of the figure.

75.

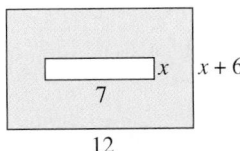

76.

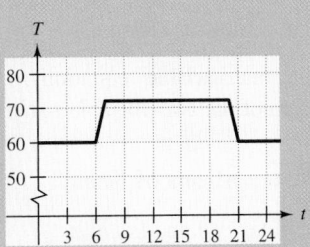

Looking Further

Graphical Reasoning   An electronically controlled thermostat in a home is programmed to lower the temperature automatically during the night. The temperature in the house T, in degrees Fahrenheit, is given in terms of t, the time in hours on a 24-hour clock.

(a) Explain why T is a function of t.

(b) Find $T(4)$ and $T(15)$.

(c) Suppose the thermostat were reprogrammed to produce a temperature H where $H(t) = T(t - 1)$. Explain how this would change the temperature in the house.

(d) Suppose the thermostat were reprogrammed to produce a temperature H where $H(t) = T(t) - 1$. Explain how this would change the temperature in the house.

Chapter Summary: Key Concepts

2.1 ■ Guidelines for verifying solutions

To verify that an ordered pair (x, y) is a solution of an equation with variables x and y, use the steps below.

1. Substitute the values of x and y into the equation.

2. Simplify each side of the equation.

3. If each side simplifies to the same number, the ordered pair is a solution. If the two sides yield different numbers, the ordered pair is not a solution.

2.1 ■ The Distance Formula

The distance d between two points (x_1, y_1) and (x_2, y_2) is

$$d = \sqrt{(x_2 - x_1)^2 + (y_2 - y_1)^2}.$$

2.1 ■ The Midpoint Formula

The midpoint of the line segment joining the points (x_1, y_1) and (x_2, y_2) is given by the Midpoint Formula

$$\text{Midpoint} = \left(\frac{x_1 + x_2}{2}, \frac{y_1 + y_2}{2} \right).$$

2.2 ■ The point-plotting method of sketching a graph

1. If possible, rewrite the equation by isolating one of the variables.

2. Make a table of values showing several solution points.

3. Plot these points on a rectangular coordinate system.

4. Connect the points with a smooth curve or line.

2.2 ■ Using a graphical check of a solution

The solution of an equation involving one variable x can be checked graphically with the steps below.

1. Write the equation so that all nonzero terms are on one side and zero is on the other side.

2. Sketch the graph of $y = $ (nonzero terms).

3. The solution of the one-variable equation is the x-intercept of the graph of the two-variable equation.

2.3 ■ Slope and the equations of lines

1. Slope of a line through (x_1, y_1) and (x_2, y_2):

$$m = \frac{y_2 - y_1}{x_2 - x_1}$$

2. Slope-intercept form of the equation of a line:

$$y = mx + b$$

3. Parallel lines (equal slopes): $m_1 = m_2$

4. Perpendicular lines (negative reciprocal slopes):

$$m_1 = -\frac{1}{m_2}$$

2.4 ■ Characteristics of a function

1. Each element in the domain A must be matched with an element in the range, which is contained in the set B.

2. Some elements in the set B may not be matched with any element in the domain A.

3. Two or more elements of the domain may be matched with the same element in the range.

4. No element of the domain is matched with two different elements in the range.

2.5 ■ Vertical Line Test for functions

A set of points on a rectangular coordinate system is the graph of y as a function of x if and only if no vertical line intersects the graph at more than one point.

2.6 ■ Vertical and horizontal shifts

Let c be a positive real number. Vertical and horizontal shifts of the graph of the function $y = f(x)$ are represented as follows.

1. Vertical shift c units upward: $h(x) = f(x) + c$

2. Vertical shift c units downward: $h(x) = f(x) - c$

3. Horizontal shift c units to the right: $h(x) = f(x - c)$

4. Horizontal shift c units to the left: $h(x) = f(x + c)$

2.6 ■ Reflections in the coordinate axes

Reflections of the graph of $y = f(x)$ are represented as follows:

1. Reflection in the x-axis: $h(x) = -f(x)$

2. Reflection in the y-axis: $h(x) = f(-x)$

Review Exercises

Reviewing Skills

2.1 ■ *How to plot points on a rectangular coordinate system*

In Exercises 1 and 2, plot the points on a rectangular coordinate system.

1. $(0, -3), \left(\frac{5}{2}, 5\right), (-2, -4)$

2. $\left(1, -\frac{3}{2}\right), (-2, 2), (5, 10)$

In Exercises 3 and 4, plot the points and connect them with line segments to form the figure.

3. *Right triangle:* $(1, 1), (12, 9), (4, 20)$

4. *Parallelogram:* $(0, 0), (7, 1), (8, 4), (1, 3)$

In Exercises 5–8, determine the quadrant(s) in which the point is located.

5. $(2, -6)$

6. $(-4.8, -2)$

7. $(4, y)$

8. $(x, y), \ xy > 0$

■ *How to determine whether ordered pairs are solutions of equations*

In Exercises 9 and 10, determine whether the ordered pairs are solution points of the given equation.

9. $y = 4 - \frac{1}{2}x$

   (a) $(4, 2)$ (b) $(-1, 5)$

   (c) $(-4, 0)$ (d) $(8, 0)$

10. $3x - 2y + 18 = 0$

   (a) $(3, 10)$ (b) $(0, 9)$

   (c) $(-4, 3)$ (d) $(-8, 0)$

■ *How to use the Distance Formula to find the distance between two points*

■ *How to use the Midpoint Formula to find the midpoints of line segments*

In Exercises 11–14, (a) plot the points on a rectangular coordinate system, (b) find the distance between the points, and (c) find the coordinates of the midpoint.

11. $(-2, -3), (4, 5)$

12. $(-5, 1), (-1, 7)$

13. $(2, 0), (14, 5)$

14. $(0, 7), (6, -8)$

2.2 ■ *How to sketch graphs of equations using the point-plotting method*

In Exercises 15–18, sketch the graph of the equation.

15. $y = 5 - \frac{3}{2}x$

16. $y = x^2 + 4$

17. $y = |x| + 4$

18. $y = \sqrt{x + 4}$

■ *How to find and use x- and y-intercepts as aids to sketching graphs*

In Exercises 19–26, find the *x*- and *y*-intercepts of the graph of the equation. Sketch the graph of the equation and label the *x*- and *y*-intercepts.

19. $y = 6 - \frac{1}{3}x$

20. $y = \frac{3}{4}x - 2$

21. $3y - 2x - 3 = 0$

22. $3x + 4y + 12 = 0$

23. $x = |y - 3|$

24. $y = 7 - |x|$

25. $y = 1 - x^2$

26. $y = (x + 3)^2$

■ *How to use graphical checking to check solutions of equations involving one variable*

In Exercises 27–30, solve the equation. Use a graphing utility to verify your solution.

27. $4x - 9 = 7$

28. $2(x - 3) = 10$

29. $x^2 - 3x = 18$

30. $x^2 + 2x = -1$

2.3 ■ *How to determine the slope of a line given two points on the line*

In Exercises 31–36, find the slope of the line through the points.

31. $(-1, 1), (6, 3)$

32. $(-2, 5), (3, -8)$

33. $(-1, 3), (4, 3)$

34. $(7, 2), (7, 8)$

35. $(0, 6), (8, 0)$

36. $(0, 0), \left(\frac{7}{2}, 6\right)$

In Exercises 37–42, a point on a line and the slope of the line are given. Find two additional points on the line. (There are many correct answers.)

37. $(2, -4)$

   $m = -3$

38. $\left(-4, \frac{1}{2}\right)$

   $m = 2$

39. $(3, 1)$

$m = \frac{5}{4}$

40. $\left(-3, -\frac{3}{2}\right)$

$m = -\frac{1}{3}$

41. $(3, 7)$

m is undefined.

42. $(7, -2)$

$m = 0$

■ *How to write linear equations in slope-intercept form and graph the equations*

In Exercises 43–46, write the equation of the line in slope-intercept form and sketch the line.

43. $5x - 2y - 4 = 0$

44. $x - 3y - 6 = 0$

45. $x + 2y - 2 = 0$

46. $y - 6 = 0$

■ *How to use slopes to determine whether lines are parallel, perpendicular, or neither*

In Exercises 47–52, are the lines parallel, perpendicular, or neither?

47. $y = \frac{3}{2}x + 1$

$y = \frac{2}{3}x - 1$

48. $y = 2x - 5$

$y = 2x + 3$

49. $y = \frac{3}{2}x - 2$

$y = -\frac{2}{3}x + 1$

50. $y = -0.3x - 2$

$y = 0.3x + 1$

51. $5x + 20y = 30$

$-12x + 3y = -10$

52. $24x - 3y + 5 = 0$

$-16x + 2y = 0$

2.4 ■ *How to identify the domains and ranges of relations*

In Exercises 53 and 54, give the domain and range of the relation. Then draw a graphical representation of the relation.

53. $\{(1, 6), (2, 10), (3, 8), (3, 7)\}$

54. $\{(-2, 5), (0, -1), (1, 3), (3, 5)\}$

■ *How to determine if relations are functions by inspection*

In Exercises 55–64, determine if the relation is a function.

55. Domain   Range

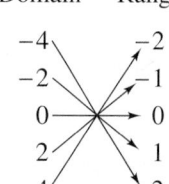

56. Domain   Range

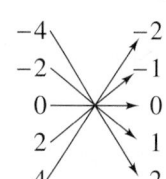

57. Domain   Range

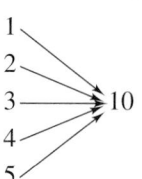

58. Domain   Range

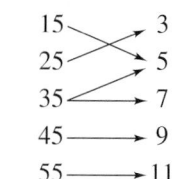

59. $\{(-10, 2), (-5, 6), (0, 8), (5, 9)\}$

60. $\{(-4, 0), (-3, 1), (0, -2), (4, 1)\}$

61. $\{(3, 18), (5, 26), (-2, 18), (3, 10)\}$

62. $\{(1, 1), (2, 4), (3, 9), (0, 0)\}$

63.

Input value	0	1	2	3	4
Output value	4	3	2	1	0

64.

Input value	3	2	1	2	3
Output value	−4	−3	−2	−1	0

■ *How to use function notation and evaluate functions*

In Exercises 65–72, evaluate the function as indicated, and simplify.

65. $f(x) = 4 - \frac{5}{2}x$

(a) $f(-10)$

(b) $f\left(\frac{2}{5}\right)$

(c) $f(t) + f(-4)$

(d) $f(x + h)$

66. $h(x) = x(x - 8)$

(a) $h(8)$

(b) $h(10)$

(c) $h(-3)$

(d) $h(t + 4)$

67. $f(t) = \sqrt{5 - t}$

(a) $f(-4)$

(b) $f(5)$

(c) $f(3)$

(d) $f(5z)$

68. $g(x) = 2\sqrt{x + 3}$

(a) $g(1)$

(b) $g(-1)$

(c) $g(10)$

(d) $g(2z)$

69. $g(x) = \dfrac{|x + 4|}{4}$

(a) $g(0)$

(b) $g(-8)$

(c) $g(2) - g(-5)$

(d) $g(x - 2)$

70. $f(x) = \dfrac{x}{|x - 3|}$

(a) $f(0)$

(b) $f(-2)$

(c) $f(2) - f(-4)$

(d) $f(x + 3)$

71. $f(x) = \begin{cases} -3x, & x \le 0 \\ 1 - x^2, & x > 0 \end{cases}$

   (a) $f(2)$ (b) $f\left(-\frac{2}{3}\right)$

   (c) $f(1)$ (d) $f(4) - f(3)$

72. $h(x) = \begin{cases} x^3, & x \le 1 \\ (x - 1)^2 + 1, & x > 1 \end{cases}$

   (a) $h(2)$ (b) $h\left(-\frac{1}{2}\right)$

   (c) $h(0)$ (d) $h(4) - h(3)$

■ *How to identify the domains and ranges of functions.*

In Exercises 73–76, find the domain and range of the function.

73. $f: \{(1, 2), (3, 8), (5, 7), (7, 2)\}$

74. $f: \left\{(0, 10), (-3, -5), (5, 12), \left(-1, \frac{1}{2}\right)\right\}$

75. Perimeter of a square of side s: $P = 4s$

76. Volume of a sphere of radius r: $V = \frac{4}{3}\pi r^3$

In Exercises 77–80, find the domain of the function.

77. $h(x) = 4x^2 - 7$ **78.** $g(s) = \dfrac{s + 1}{(s - 1)(s + 5)}$

79. $f(x) = \sqrt{x - 2}$ **80.** $f(x) = 6x + 10$

2.5 ■ *How to sketch graphs of functions on rectangular coordinate systems using the point-plotting method*

In Exercises 81–88, sketch the graph of the function by plotting points. Use a graphing utility to confirm your result.

81. $g(x) = \frac{1}{8}x^2$

82. $y = \frac{1}{2}x^2$

83. $y = 3(x - 2)^2$

84. $h(x) = 9 - 2x^2$

85. $y = 8 - 2|x|$

86. $f(x) = 2|x + 1|$

87. $g(x) = \frac{1}{4}x^3, \quad -2 \le x \le 2$

88. $h(x) = x(4 - x), \quad 0 \le x \le 4$

■ *How to sketch the graphs of linear functions*

In Exercises 89–92, use the slope and y-intercept to sketch the graph of the linear function.

89. $f(x) = -\frac{3}{2}x + 5$ **90.** $g(x) = \frac{4}{5}x - 1$

91. $f(x) = -0.4x - 3$ **92.** $h(x) = 1.2x + 4$

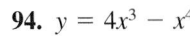 ■ *How to use the Vertical Line Test to determine if graphs represent functions*

In Exercises 93–98, use the Vertical Line Test to determine if the graph represents y as a function of x.

93. $9y^2 = 4x^3$ **94.** $y = 4x^3 - x^4$

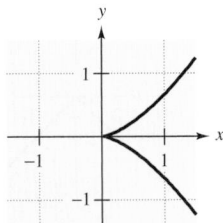   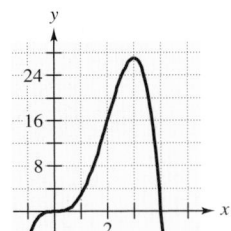

95. $y = x^2(x - 3)$ **96.** $x^3 + y^3 - 6xy = 0$

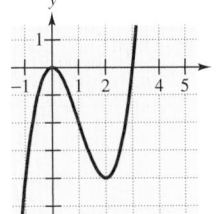   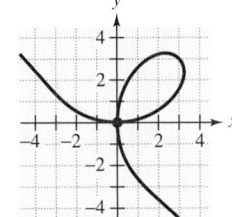

97. $y = x^4 - 1$ **98.** $\dfrac{x^2}{4} + \dfrac{y^2}{9} = 1$

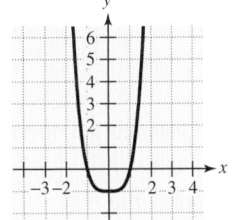   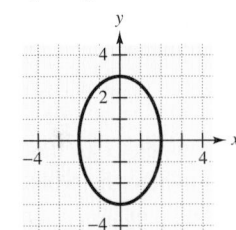

■ *How to find domains and ranges graphically*

In Exercises 99–102, sketch the graph of the function and use the graph to determine its domain and range.

99. $f(x) = x^2 + 8x + 13$ **100.** $g(x) = \sqrt{8 - x}$

101. $h(x) = |x - 3|$ **102.** $f(x) = 2x^3$

■ *How to sketch graphs of piecewise-defined functions*

In Exercises 103 and 104, sketch the graph of the piecewise-defined function.

103. $f(x) = \begin{cases} x + 5, & x < 1 \\ 3x, & x \ge 1 \end{cases}$

104. $f(x) = \begin{cases} 2x, & x \le 0 \\ x^2 + 1, & x > 0 \end{cases}$

2.6 ■ *How to identify the graphs of basic functions*

In Exercises 105–110, identify the basic function, and any transformation shown in the graph. Write the equation for the graphed function.

105.

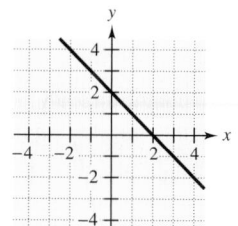

106.

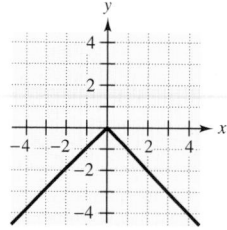

107.

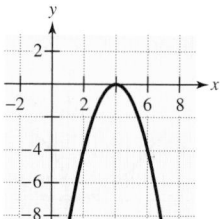

108.

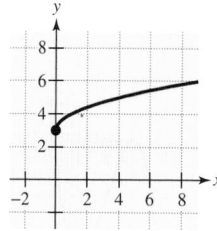

109.

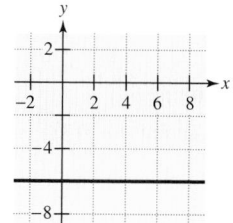

110.

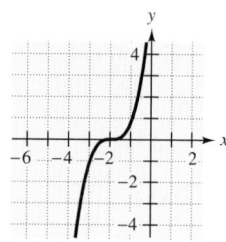

■ *How to use vertical and horizontal shifts and reflections to sketch graphs of functions*

In Exercises 111–114, identify the transformation of the graph of $f(x) = x^2$ and sketch the graph of h.

111. $h(x) = -x^2$

112. $h(x) = x^2 + 2$

113. $h(x) = (x - 1)^2$

114. $h(x) = 1 - x^2$

Solving Problems

115. *Physics*   The height y (in feet) of a projectile is given by

$$y = -\frac{1}{16}x^2 + 5x$$

where x is the horizontal distance (in feet) from where the projectile was launched.

(a) Sketch the path of the projectile.

(b) How far from the launch point does the projectile strike the ground?

116. *Physics*   The velocity of a ball thrown upward from ground level is given by $v = -32t + 80$, where t is the time in seconds and v is the velocity in feet per second.

(a) Find the velocity when $t = 2$.

(b) Find the time when the ball reaches its maximum height. (*Hint:* Find the time when $v = 0$.)

(c) Find the velocity when $t = 3$.

(d) Sketch a graph of the model and compare your answers to (a), (b), and (c) with your graph.

117. *Wages*   A wage earner is paid $14.00 per hour for regular time and time-and-a-half for overtime. The weekly wage function is

$$W(h) = \begin{cases} 14h, & 0 < h \le 40 \\ 21(h - 40) + 560, & h > 40 \end{cases}$$

where h represents the number of hours worked in a week.

(a) Use a graphing utility to graph the model.

(b) Evaluate $W(35)$, $W(40)$, $W(45)$, and $W(50)$.

(c) Could you use values of h for which $h < 0$ in this model? Why or why not?

118. *Geometry*   A wire 100 inches long is to be cut into four pieces to form a rectangle whose shortest side has a length of x. Express the area A of the rectangle as a function of x.

119. *Personal Income*   Your salary was $28,500 in 1998 and $33,900 in 2001. What is the average rate of change of your salary each year?

120. *Business*   A business purchases a piece of equipment for $875. After 5 years the equipment will be outdated and have no value. What is the average rate of change of the value of the equipment?

2 Chapter Test

Take this test as you would take a test in class. After you are done, check your work against the answers given in the back of the book.

1. Determine the quadrant in which the point (x, y) lies if $x > 0$ and $y < 0$.

2. Is $(-2, 1)$ a solution to the equation $x^2 - y = 3$?

3. Find (a) the distance between the points $(0, 9)$ and $(3, 1)$ and (b) the coordinates of the midpoint of the line segment joining the points.

4. Find the x- and y-intercepts of the graph of the equation $y = -3(x + 1)$.

5. Use the point-plotting method to sketch the graph of the equation $y = x^2 - x - 6$.

6. Find the slope (if possible) of the line passing through each pair of points.

(a) $(-4, 7), (2, 3)$ (b) $(3, -2), (3, 6)$

7. Write the equation $2x - 4y = 12$ in slope-intercept form and sketch the graph of the line.

8. The slope of a line is $\frac{7}{5}$.

(a) What is the slope of a line parallel to the line?

(b) What is the slope of a line perpendicular to the line?

9. Use the Vertical Line Test to determine whether the equation $y^2(4 - x) = x^3$, shown at the left, represents y as a function of x.

10. Evaluate (if possible) the function $g(x) = x/(x - 3)$ for the indicated values of the independent variable.

(a) $g(2)$ (b) $g\left(\frac{7}{2}\right)$ (c) $g(3)$ (d) $g(x + 2)$

11. Let $f(x) = \begin{cases} 3x - 1, & x < 5 \\ x^2 + 4, & x \geq 5 \end{cases}$ and find the indicated values.

(a) $f(10)$ (b) $f(-8)$ (c) $f(5)$ (d) $f(0)$

12. Find the domain of each function.

(a) $h(t) = \sqrt{t + 9}$ (b) $f(x) = \dfrac{x + 1}{x - 4}$ (c) $g(r) = 3r + 5$

13. Sketch the graph of the function $f(x) = x^2 + 3$.

14. Use the graph of $f(x) = |x|$ to write an equation for each graph. Use a graphing utility to verify your result.

(a)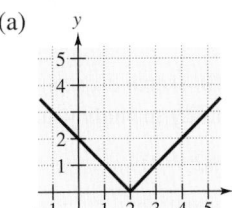

(b)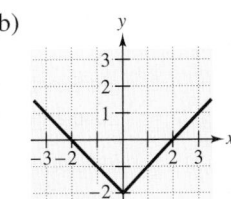

(c)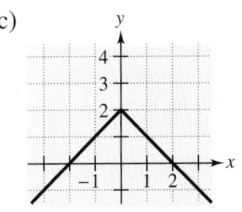

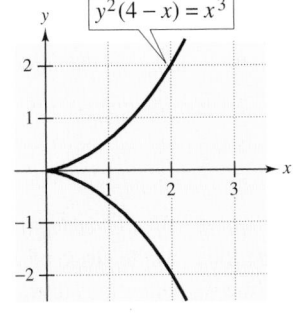

$y^2(4 - x) = x^3$

Figure for 9

Cumulative Test: Chapters 1–2

Take this test as you would take a test in class. After you are done, check your work against the answers given in the back of the book.

1. Evaluate the expression $\dfrac{a^2 - 2ab}{a + b}$ for $a = -4$ and $b = 7$.

In Exercises 2–4, perform the operations and simplify.

2. (a) $(2a^2b)^3(-ab^2)^2$ (b) $3(x^2y)^2(-2xy^3)^3$

3. (a) $t(3t - 1) - 2t(t + 4)$ (b) $3x(x^2 - 2) - x(x^2 + 5)$

4. (a) $(3x + 7)(x^2 - 2x + 5)$ (b) $[2 + (x - y)]^2$

In Exercises 5 and 6, solve the equation.

5. (a) $12 - 5(3 - x) = x + 3$ (b) $1 - \dfrac{x + 2}{4} = \dfrac{7}{8}$

6. (a) $y^2 - 64 = 0$ (b) $2t^2 - 5t - 3 = 0$

7. Solve for y in the equation $2x - 3y + 9 = 0$.

8. Given the line $y = -4x + 7$, find (a) the slope of the line, (b) the slope of a line parallel to the given line, and (c) the slope of a line perpendicular to the given line.

9. Name the property of algebra that justifies the statement.

   (a) $8x \cdot \dfrac{1}{8x} = 1$ (b) $5 + (-3 + x) = (5 - 3) + x$

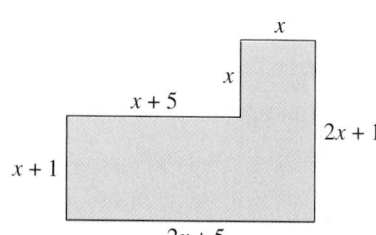

Figure for 10

10. Find and simplify the expression for the perimeter of the figure shown at the left.

11. Factor $y^3 - 3y^2 - 9y + 27$. 12. Factor $3x^2 - 8x - 35$.

13. For the points $(-3, 8)$ and $(1, 5)$, (a) plot the points, (b) find the distance between the points, (c) find the coordinates of the midpoint of the line segment joining the points, and (d) find the slope of the line passing through the points.

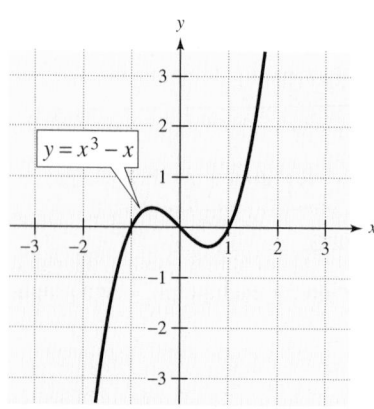

Figure for 14

14. Use the Vertical Line Test and the graph at the left to determine whether the equation $y = x^3 - x$ represents y as a function of x.

15. Find the domain of the function $f(x) = \sqrt{x - 2}$.

16. Given $f(x) = x^2 - 3x$, find (a) $f(4)$ and (b) $f(c + 3)$.

In Exercises 17–20, sketch a graph of the function and label the x- and y-intercepts. Use a graphing utility to verify your result.

17. $4x + 3y - 12 = 0$ 18. $f(x) = -(x - 2)^2$

19. $f(x) = |x + 3| + 2$ 20. $y = x^3 - 1$

Linear Functions, Equations, and Inequalities

3

3.1 Writing Equations of Lines

3.2 Applications of Linear Equations

3.3 Business and Scientific Problems

3.4 Linear Inequalities in One Variable

3.5 Absolute Value Equations and Inequalities

THE BIG PICTURE

In this chapter you will learn how to:

- write the equation of a line given a description of its graph.

- translate verbal statements into mathematical models to solve applied problems.

- solve business, rate, and mixture problems.

- solve linear inequalities in one variable and sketch the graphs of their solution sets.

- solve absolute value equations and inequalities.

Key Terms

As you encounter each new vocabulary term in this chapter, add the term and its definition to your notebook glossary.

point-slope form (p. 162)
general form (p. 165)
linear extrapolation (p. 166)
linear interpolation (p. 166)
consecutive integers (p. 174)
mathematical modeling (p. 174)
angle (p. 176)
initial side (p. 176)
terminal side (p. 176)
vertex (p. 176)
right angles (p. 176)
straight angles (p. 176)
complementary angles (p. 176)
supplementary angles (p. 176)
rate (p. 178)
ratio (p. 178)
proportion (p. 178)
extremes (p. 178)
means (p. 178)

cross multiplying (p. 178)
formulas (p. 187)
rate problems (p. 189)
rate of work (p. 190)
mixture problems (p. 191)
algebraic inequalities (p. 200)
solution set (p. 200)
graph of an inequality (p. 200)
bounded intervals (p. 200)
endpoints (p. 200)
unbounded intervals (p. 201)
equivalent inequalities (p. 202)
compound inequality (p. 205)
double inequality (p. 205)
conjunctive (p. 205)
disjunctive (p. 205)
intersection (p. 206)
union (p. 206)

Additional text-specific resources are available to help you do well in this course. See page xvi for details.

3.1 Writing Equations of Lines

Slope-Intercept Equation of a Line

There are two basic types of problems in coordinate geometry.

1. Given an algebraic equation, sketch its graph.

2. Given a description of a graph, write its equation.

The first type of problem can be thought of as moving from algebra to geometry, whereas the second type can be thought of as moving the other way—from geometry to algebra. So far in the text, you have been working primarily with the first type of problem. In this section, you will look at the second type.

In Section 2.3, you used the slope-intercept form of the equation of a line to sketch a line. Here you will write the equation of a line in slope-intercept form using information from the graph of the line.

Example 1 ■ Writing an Equation of a Line

Write the slope-intercept form of the equation of the line that has a slope of $-\frac{3}{5}$ and a y-intercept of $(0, 6)$. (See Figure 3.1).

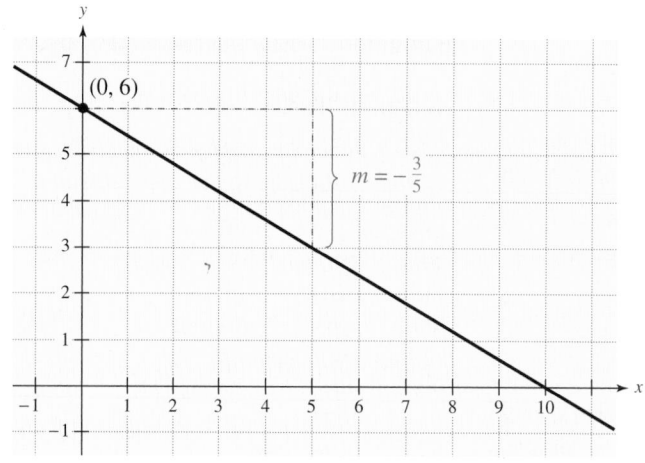

Figure 3.1

Solution

Begin by writing the slope-intercept form of the equation of a line, then substitute $-\frac{3}{5}$ for m and 6 for b.

$$y = mx + b \qquad \text{Slope-intercept form}$$

$$y = -\frac{3}{5}x + 6 \qquad \text{Substitute for } m \text{ and } b.$$

So, the equation of the line in slope-intercept form is $y = -\frac{3}{5}x + 6$.

Horizontal and Vertical Lines

Recall from Section 2.3 that a horizontal line has a slope of zero. From the slope-intercept form of the equation of a line, you can see that a horizontal line has an equation of the form

$$y = (0)x + b \qquad \text{or} \qquad y = b. \qquad \text{Horizontal line}$$

This is consistent with the fact that each point on a horizontal line through $(0, b)$ has a y-coordinate of b, as shown in Figure 3.2. Similarly, each point on a vertical line through $(a, 0)$ has an x-coordinate of a, as shown in Figure 3.3. Because you know that a vertical line has an undefined slope, you know that it has an equation of the form

$$x = a. \qquad \text{Vertical line}$$

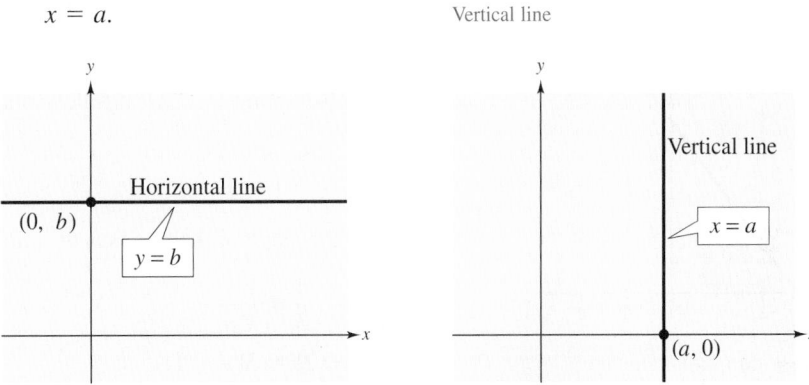

Figure 3.2   Figure 3.3

Example 2 ■ Writing Equations of Horizontal and Vertical Lines

Write an equation for each line.

(a) Vertical line through $(-2, 4)$

(b) Horizontal line through $(7, 9)$

(c) Line passing through $(-2, 3)$ and $(3, 3)$

(d) Line passing through $(-1, 2)$ and $(-1, 3)$

Solution

(a) Because the line is vertical and passes through the point $(-2, 4)$, you know that every point on the line has an x-coordinate of -2. So, the equation is $x = -2$.

(b) Because the line is horizontal and passes through the point $(7, 9)$, you know that every point on the line has a y-coordinate of 9. So, the equation of the line is $y = 9$.

(c) The line through $(-2, 3)$ and $(3, 3)$ is horizontal, and every point on the line has a y-coordinate of 3. So, its equation is $y = 3$.

(d) The line through $(-1, 2)$ and $(-1, 3)$ is vertical, and every point on the line has an x-coordinate of -1. So, its equation is $x = -1$.

Point-Slope Equation of a Line

If you know the slope of a line *and* you also know the coordinates of any point on the line, you can find an equation for the line. Before giving a general formula for doing this, let's look at an example.

Example 3 ■ Writing an Equation of a Line

A line has a slope of $\frac{4}{3}$ and passes through the point $(-2, 1)$. Find an equation of this line.

Solution

Begin by sketching the line, as shown in Figure 3.4. You know that the slope of a line is the same through any two points on the line. So, to find an equation of the line, let (x, y) represent *any* point on the line. Using the representative point (x, y) and the given point $(-2, 1)$, it follows that the slope of the line is

$$m = \frac{y - 1}{x - (-2)}.$$

⟵ Difference in *y*-values

⟵ Difference in *x*-values

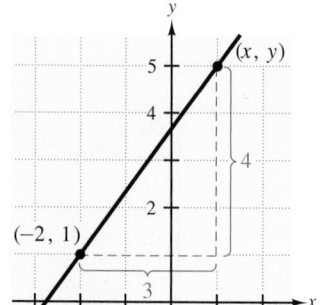

Figure 3.4

Because the slope of the line is $m = \frac{4}{3}$, this equation can be rewritten as follows.

$$\frac{4}{3} = \frac{y - 1}{x + 2}$$ Slope formula

$$4(x + 2) = 3(y - 1)$$ Cross multiply.

$$4x + 8 = 3y - 3$$ Distributive Property

$$4x = 3y - 11$$ Subtract 8 from each side.

$$4x - 3y = -11$$ Subtract 3y from each side.

An equation of the line is $4x - 3y = -11$.

The procedure in Example 3 can be used to derive a *formula* for the equation of a line, given its slope and a point on the line. In Figure 3.5, let (x_1, y_1) be a given point on the line whose slope is m. If (x, y) is any *other* point on the line, it follows from Section 2.3 that the slope of the line is

$$\frac{y - y_1}{x - x_1} = m.$$

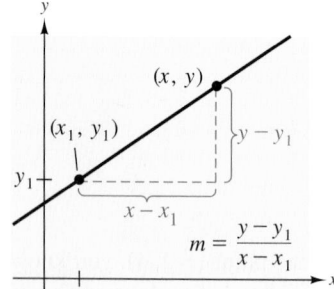

Figure 3.5

This equation in variables x and y can be rewritten in the form

$$y - y_1 = m(x - x_1)$$

which is called the **point-slope form** of the equation of a line.

> **Point-Slope Form of the Equation of a Line**
>
> The **point-slope form** of the equation of the line that passes through the point (x_1, y_1) and has a slope of m is
>
> $$y - y_1 = m(x - x_1).$$

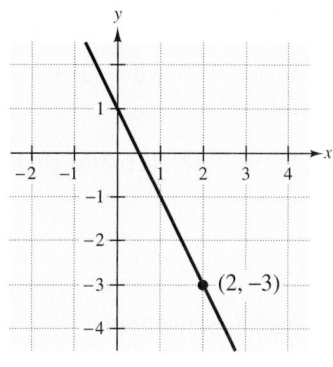

(a)

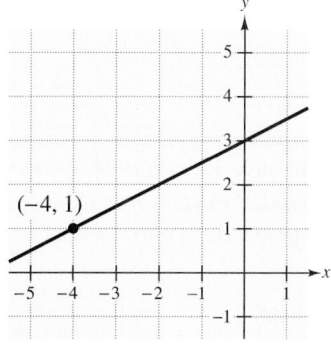

(b)

Figure 3.6

Example 4 ■ The Point-Slope Form of the Equation of a Line

Find an equation of the line that passes through the point and has the specified slope.

(a) $(2, -3)$, $m = -2$ (b) $(-4, 1)$, $m = \frac{1}{2}$

Solution

(a) Use the point-slope form with $(x_1, y_1) = (2, -3)$ and $m = -2$.

$$y - y_1 = m(x - x_1) \qquad \text{Point-slope form}$$
$$y - (-3) = -2(x - 2) \qquad \text{Substitute } y_1 = -3, x_1 = 2, \text{ and } m = -2.$$
$$y + 3 = -2x + 4 \qquad \text{Simplify.}$$
$$y = -2x + 1 \qquad \text{Subtract 3 from each side.}$$

The graph of this line is shown in Figure 3.6(a).

(b) Use the point-slope form with $(x_1, y_1) = (-4, 1)$ and $m = \frac{1}{2}$.

$$y - y_1 = m(x - x_1) \qquad \text{Point-slope form}$$
$$y - 1 = \frac{1}{2}[x - (-4)] \qquad \text{Substitute } y_1 = 1, x_1 = -4, \text{ and } m = \frac{1}{2}.$$
$$y - 1 = \frac{1}{2}(x + 4) \qquad \text{Simplify.}$$
$$y - 1 = \frac{1}{2}x + 2 \qquad \text{Simplify.}$$
$$y = \frac{1}{2}x + 3 \qquad \text{Add 1 to each side.}$$

The graph of this line is shown in Figure 3.6(b).

In Example 4(a), notice that the final equation is written in slope-intercept form.

$$y = mx + b \qquad \text{Slope-intercept form}$$

You can use this form to check your work. First, observe that the slope is $m = -2$. Then substitute the coordinates of the given point $(2, -3)$ to see that the equation is satisfied.

The point-slope form can be used to find the equation of a line passing through two points (x_1, y_1) and (x_2, y_2). First, use the formula for the slope of a line passing through two points.

$$m = \frac{y_2 - y_1}{x_2 - x_1}$$

Then, once you know the slope, use the point-slope form to obtain the equation

$$y - y_1 = \frac{y_2 - y_1}{x_2 - x_1}(x - x_1). \qquad \text{Two-point form}$$

This is sometimes called the **two-point form** of the equation of a line.

Technology

A program for several models of graphing utilities that uses the two-point form to find an equation of a line can be found in Appendix H at our website *college.hmco.com*. The program outputs the slope and y-intercept of the line that passes through the two points.

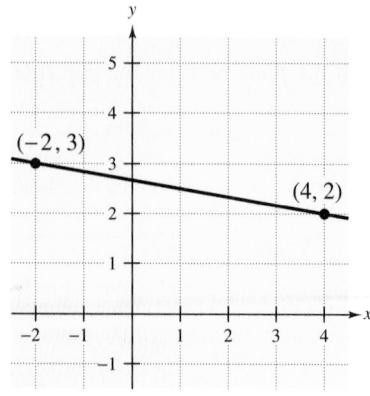

Figure 3.7

NOTE   In Example 5, it does not matter which of the two points is labeled (x_1, y_1) and which is labeled (x_2, y_2). Try switching these labels to

$$(x_1, y_1) = (-2, 3)$$

and

$$(x_2, y_2) = (4, 2)$$

and reworking the problem to see that you obtain the same equation.

Example 5   ■   An Equation of a Line Passing Through Two Points

Find an equation of the line that passes through the points $(4, 2)$ and $(-2, 3)$.

Solution

Let $(x_1, y_1) = (4, 2)$ and $(x_2, y_2) = (-2, 3)$. Then apply the formula for the slope of a line passing through two points, as follows.

$$m = \frac{y_2 - y_1}{x_2 - x_1} = \frac{3 - 2}{-2 - 4} = \frac{1}{-6} = -\frac{1}{6}$$

Now, using the point-slope form, you can find the equation of the line.

$y - y_1 = m(x - x_1)$	Point-slope form
$y - 2 = -\frac{1}{6}(x - 4)$	Substitute $y_1 = 2, x_1 = 4$, and $m = -\frac{1}{6}$.
$y - 2 = -\frac{1}{6}x + \frac{2}{3}$	Simplify.
$y = -\frac{1}{6}x + \frac{8}{3}$	Add 2 to each side.

The graph of this line is shown in Figure 3.7.

In Section 2.3, you learned that parallel lines have the same slope and perpendicular lines have slopes that are negative reciprocals of each other. You can use these facts to find an equation of a line parallel or perpendicular to a given line.

Example 6   ■   Parallel and Perpendicular Lines

Find equations of the lines that pass through the point $(3, -2)$ and are (a) parallel and (b) perpendicular to the line $x - 4y = 6$, as shown in Figure 3.8.

Solution

Write the equation in slope-intercept form to determine the slope of the line.

$x - 4y = 6$	Write original equation.
$-4y = -x + 6$	Subtract x from each side.
$y = \frac{1}{4}x - \frac{3}{2}$	Slope-intercept form

So, the line has a slope of $\frac{1}{4}$.

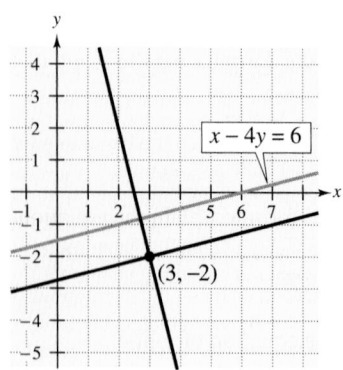

Figure 3.8

(a) Any line parallel to the given line must also have a slope of $\frac{1}{4}$. The line through $(3, -2)$ with slope $m = \frac{1}{4}$ has the equation below.

$y - y_1 = m(x - x_1)$	Point-slope form
$y - (-2) = \frac{1}{4}(x - 3)$	Substitute $y_1 = -2, x_1 = 3$, and $m = \frac{1}{4}$.
$y + 2 = \frac{1}{4}x - \frac{3}{4}$	Simplify.
$y = \frac{1}{4}x - \frac{11}{4}$	Equation of parallel line

(b) Any line perpendicular to the given line must have a slope of -4. The line through $(3, -2)$ with slope $m = -4$ has the equation below.

$y - y_1 = m(x - x_1)$	Point-slope form
$y - (-2) = -4(x - 3)$	Substitute $y_1 = -2, x_1 = 3$, and $m = -4$.
$y + 2 = -4x + 12$	Simplify.
$y = -4x + 10$	Equation of perpendicular line

Summary of Equations of Lines

The equation of a vertical line cannot be written in slope-intercept form because the slope of a vertical line is undefined. However, *every* line has an equation that can be written in the **general form**

$$ax + by + c = 0 \qquad \text{General form}$$

where a and b are not *both* zero.

Summary of Equations of Lines

1. Slope of line through (x_1, y_1) and (x_2, y_2): $\qquad m = \dfrac{y_2 - y_1}{x_2 - x_1}$

2. General form of equation of line: $\qquad ax + by + c = 0$

3. Equation of vertical line: $\qquad x = a$

4. Equation of horizontal line: $\qquad y = b$

5. Slope-intercept form of equation of line: $\qquad y = mx + b$

6. Point-slope form of equation of line: $\qquad y - y_1 = m(x - x_1)$

7. Parallel lines (equal slopes): $\qquad m_1 = m_2$

8. Perpendicular lines
   (negative reciprocal slopes): $\qquad m_2 = -\dfrac{1}{m_1}$

Applications

Linear equations can be used as models to represent real-life data. The next example shows you how to use real-life data to write a linear model.

Example 7 ■ Writing a Linear Model

In 1990, 31 women held elected offices in the United States Congress. By 2002, the number had increased to 73. Write a linear model for the number n of women in Congress between 1990 and 2002. (Let $t = 0$ represent 1990.) *(Source: Center for American Women and Politics)*

Solution

One way to create such a model is to interpret the given information as two points on a line. Using $n = 31$ in 1990 $(t = 0)$, you can determine that one point on the line is $(0, 31)$. Using $n = 73$ in 2002 $(t = 12)$, you can determine that a second point on the line is $(12, 73)$. (See Figure 3.9.) So, the slope of the line is

$$m = \frac{73 - 31}{12 - 0} = \frac{42}{12} = \frac{7}{2}.$$

Because the n-intercept of the line is $(0, 31)$, you can conclude that an equation of the line is

$$n = mt + b \qquad \text{Slope-intercept form—use } t \text{ and } n.$$
$$n = \tfrac{7}{2}t + 31. \qquad \text{Substitute } m = \tfrac{7}{2} \text{ and } b = 31.$$

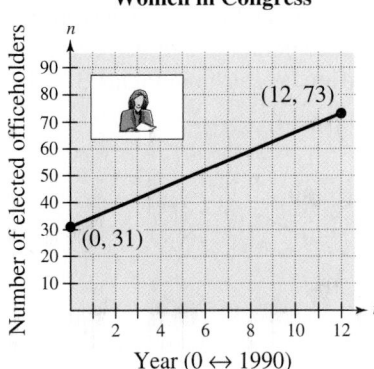

Women in Congress

Number of elected officeholders

Year (0 ↔ 1990)

Figure 3.9

Linear equations are used frequently as mathematical models in business. The next example gives you one idea of how useful such a model can be.

Example 8 ■ Total Sales

The total sales of a new computer software company were $500,000 for the second year and $1,000,000 for the fourth year. Using only this information, what would you estimate the total sales to be during the fifth year?

Solution

To solve this problem, use a *linear model*, with y representing the total sales (in thousands of dollars) and t representing the year. That is, in Figure 3.10, let $(2, 500)$ and $(4, 1000)$ be two points on the line representing the total sales for the company. The slope of the line passing through these points is

$$m = \frac{1000 - 500}{4 - 2}$$

$$= \frac{500}{2}$$

$$= 250.$$

Now, using the point-slope form, the equation of the line is

$$
\begin{aligned}
y - y_1 &= m(t - t_1) && \text{Point-slope form} \\
y - 500 &= 250(t - 2) && \text{Substitute } y_1 = 500, t_1 = 2, \text{ and } m = 250. \\
y - 500 &= 250t - 500 && \text{Distributive Property} \\
y &= 250t. && \text{Linear model for sales}
\end{aligned}
$$

Finally, estimate the total sales during the fifth year ($t = 5$) to be

$$
\begin{aligned}
y &= 250(5) \\
&= \$1250 \text{ thousand} \\
&= \$1{,}250{,}000.
\end{aligned}
$$

The estimation method illustrated in Example 8 is called **linear extrapolation.** Note in Figure 3.11(a) that for linear extrapolation, the estimated point lies to the right of the given points. When the estimated point lies *between* two given points, the procedure is called **linear interpolation,** as shown in Figure 3.11(b).

Total sales (in thousands of dollars) vs. Year, with points $(2, 500)$, $(4, 1000)$, and $(5, 1250)$.

Figure 3.10

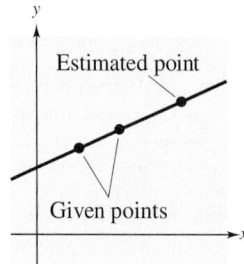

(a) Linear extrapolation

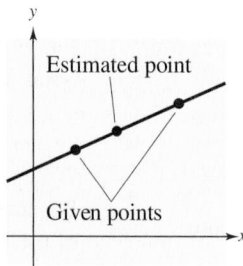

(b) Linear interpolation

Figure 3.11

In the linear equation $y = mx + b$, you know that m represents the slope of the line. In applications, the slope of a line can often be interpreted as the rate of change of y with respect to x. Rates of change should always be described with appropriate units of measure.

Example 9 ■ Height of a Mountain Climber

A mountain climber is climbing up a 500-foot cliff. By 1 P.M., the mountain climber has climbed 115 feet up the cliff. By 4 P.M., the climber has reached a height of 280 feet, as shown in Figure 3.12. Find the average rate of change of the climber and use this rate of change to find a linear model that relates the height of the climber to the time. Use the model to estimate the time when the climber will reach the top of the cliff.

Solution

Let y represent the height of the climber and let t represent the time. Then the two points that represent the climber's two positions are $(t_1, y_1) = (1, 115)$ and $(t_2, y_2) = (4, 280)$. So, the average rate of change of the climber is

$$\text{Average rate of change} = \frac{y_2 - y_1}{t_2 - t_1} = \frac{280 - 115}{4 - 1} = 55 \text{ feet per hour.}$$

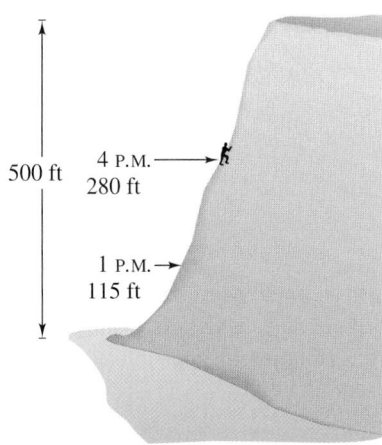

500 ft

4 P.M. → *k*
280 ft

1 P.M. →
115 ft

Figure 3.12

So, an equation that relates the height of the climber to the time is

$$y - y_1 = m(t - t_1) \qquad \text{Point-slope form}$$
$$y - 115 = 55(t - 1) \qquad \text{Substitute } y_1 = 115, \, t_1 = 1, \text{ and } m = 55.$$
$$y = 55t + 60. \qquad \text{Linear model}$$

To find the time when the climber reaches the top of the cliff, let $y = 500$ and solve for t to obtain

$$500 = 55t + 60$$
$$t = 8.$$

So, continuing at the same rate, the climber will reach the top of the cliff at 8 P.M.

Collaborate!

Problem Solving

Your manager asks you to make sense of the set of data shown below, in which y represents the number of daily newspapers in the United States and x represents the year from 1993 through 1999, with $x = 3$ corresponding to 1993. You think that a mathematical model will help you understand the trend in the data and may be useful in predicting what could happen in the future. Plot the data. Do you think that a linear model would represent the data well? If so, find the equation of the best-fitting line. Interpret the meaning of the slope in the context of the data. Use your model to predict the number of daily newspapers in 2004 (assuming that the trend continues). *(Source: Editor & Publisher Co.)*

x	3	4	5	6	7	8	9
y	1556	1548	1533	1520	1509	1489	1483

3.1 ■ Exercises

Developing Skills

In Exercises 1–6, determine the slope and the *y*-intercept of the line.

1. $y = \frac{2}{3}x - 2$

2. $y = -5x + 12$

3. $3x - 2y = 0$

4. $y = -2(6x - 1)$

5. $5x - 2y + 24 = 0$

6. $3x + 4y - 16 = 0$

In Exercises 7–10, write the slope-intercept form of the equation of the line that has the specified *y*-intercept and slope.

7. $(0, 0), m = -\frac{1}{2}$

8. $(0, 0), m = -2$

9. $(0, -4), m = 3$

10. $(0, 9), m = -\frac{1}{3}$

In Exercises 11–14, write an equation of the line.

11. Vertical line through $(10, -6)$

12. Vertical line through $(-8, 7)$

13. Horizontal line through $(2, 3)$

14. Horizontal line through $(-5, -9)$

In Exercises 15–20, match the equation with its graph. [The graphs are labeled (a), (b), (c), (d), (e), and (f).]

15. $y = \frac{2}{3}x + 2$

16. $y = \frac{2}{3}x - 2$

17. $y = -\frac{3}{2}x + 2$

18. $y = -3x + 2$

19. $y = 5$

20. $x = 1$

(a)
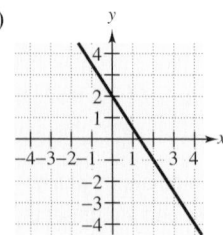

(b)
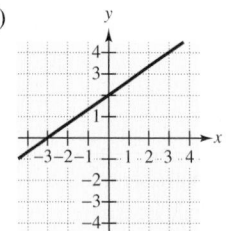

(c)
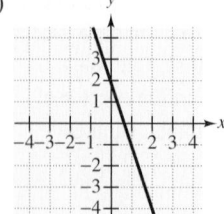

(d)
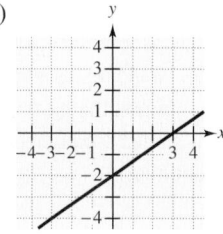

(e)
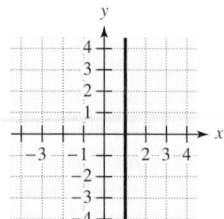

(f)
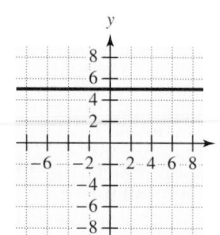

In Exercises 21–32, write an equation of the line that passes through the given point and has the specified slope. When possible, write the equation in slope-intercept form. Sketch the line.

21. $(5, 6), m = 2$

22. $(3, 7), m = 4$

23. $(-8, 1), m = \frac{3}{4}$

24. $(6, -9), m = \frac{5}{2}$

25. $(5, -3), m = \frac{2}{3}$

26. $(-3, -7), m = \frac{5}{4}$

27. $(-8, 5), m = 0$

28. $(4, -1), m = 0$

29. $(2, -1),$

    m is undefined.

30. $(2, -6),$

    m is undefined.

31. $\left(\frac{3}{4}, \frac{5}{2}\right), m = \frac{4}{3}$

32. $\left(-\frac{3}{2}, \frac{1}{2}\right), m = -3$

In Exercises 33–50, write an equation of the line that passes through the two points. When possible, write the equation in slope-intercept form.

33. $(0, 0), (2, 3)$

34. $(0, 0), (3, -5)$

35. $(7, 3), (5, -5)$

36. $(-9, 1), (-6, 10)$

37. $(-5, 2), (5, -2)$

38. $(-3, -2), (6, -8)$

39. $(-2, 12), (6, 12)$

40. $(4, -3), (-1, -3)$

41. $(-2, 3), (5, 0)$

42. $(5, 4), (-3, 5)$

43. $(1, -2), (1, 8)$

44. $(-7, 4), (-7, 6)$

45. $(-5, 0.6), (3, -3.4)$

46. $(7.5, 2), (7.5, 9)$

47. $\left(\frac{3}{2}, 3\right), \left(\frac{9}{2}, -4\right)$

48. $\left(4, \frac{7}{3}\right), \left(-1, \frac{2}{3}\right)$

49. $\left(\frac{3}{5}, 9\right), \left(-4, \frac{1}{2}\right)$

50. $(2, -8), \left(6, \frac{8}{3}\right)$

In Exercises 51–54, write an equation of the line passing through the two points. Use function notation to write *y* as a function of *x*. Use a graphing utility to graph the linear function.

51. $(-2, 2), (4, 5)$

52. $(0, 10), (5, 0)$

53. $(-2, 3), (4, 3)$ **54.** $(-6, -3), (4, 3)$

In Exercises 55–62, write equations of the lines through the point (a) parallel and (b) perpendicular to the given line.

55. $(2, 1)$
$-9x + 3y = 6$

56. $(-8, 3)$
$4x - y = 8$

57. $(-6, 2)$
$3x + 2y = 4$

58. $(-3, 4)$
$x + 6y = 12$

59. $(1, -7)$
$4x - 3y = 9$

60. $(-5, 4)$
$5x + 4y = 24$

61. $(-1, 2)$
$y = -5$

62. $(3, -4)$
$x = 10$

In Exercises 63–74, write an equation of the line.

63. Line with slope $m = \frac{1}{5}$ and y-intercept $(0, 3)$

64. Line with slope $m = -\frac{2}{3}$ and y-intercept $(0, -2)$

65. Line passing through $(7, -4)$ with slope $m = -2$

66. Line passing through $(1, 5)$ with slope $m = 3$

67. Line passing through $(-2, 3)$ and perpendicular to $7x + 3y = 8$

68. Line passing through $(7, -4)$ and perpendicular to $-3x + 5y = 9$

69. Line passing through $(5, -6)$ and parallel to $x - 4y = 10$

70. Line passing through $(3, 4)$ and parallel to $-6x - 12y = 11$

71. Horizontal line passing through $(-5, 1)$

72. Horizontal line passing through $(10, -3)$

73. Vertical line passing through $(9, 2)$

74. Vertical line passing through $(-3, 7)$

Solving Problems

75. *Business*   A small business purchases a computer system for $12,500. After 5 years, its depreciated value will be $1000.

(a) Assuming straight-line depreciation, write a linear model giving the value V of the computer system in terms of the time t.

(b) Use the model in part (a) to find the value of the computer system after 3 years.

76. *Business*   A small business purchases a photocopier for $7400. After 4 years, its depreciated value will be $1500.

(a) Assuming straight-line depreciation, write a linear model giving the value V of the copier in terms of the time t.

(b) Use the model in part (a) to find the value of the copier after 2 years.

77. *Business*   A real estate office handles an apartment complex with 50 units. When the rent is $450 per month, all 50 units are occupied. However, when the rent is $525 per month, the average number of occupied units drops to 45. Assume that the relationship between the monthly rent p and the demand x is linear.

(a) Write a linear model giving the demand x in terms of the rent p.

(b) Use a graphing utility to graph the model in part (a).

(c) *Linear Extrapolation*   Use the model in part (a) to predict the number of units occupied if the rent is raised to $570.

(d) *Linear Interpolation*   Use the model in part (a) to estimate the number of units occupied if the rent is $480.

78. *Business*   When soft drinks sold for $0.80 per can at football games, approximately 6000 cans were sold. When the price was raised to $1.00 per can, the demand dropped to 4000. Assume that the relationship between the price p and the demand x is linear.

(a) Write a linear model giving the demand x in terms of the price p.

(b) Use a graphing utility to graph the model in part (a).

(c) *Linear Extrapolation*   Use the model in part (a) to predict the number of cans of soft drink sold if the price is raised to $1.10.

(d) *Linear Interpolation*   Use the model in part (a) to estimate the number of cans of soft drink sold if the price is $0.90.

The symbol ▦ indicates an exercise in which you are instructed to use a calculator or graphing utility.

79. *College Enrollment* A small college had an enrollment of 1500 students in 1990. During the next 10 years, the enrollment increased by approximately 60 students per year.

(a) Write a linear model giving the enrollment N in terms of the year t. (Let $t = 0$ represent 1990.)

(b) Use a graphing utility to graph the model in part (a).

(c) *Linear Extrapolation* Use the model in part (a) to predict the enrollment in the year 2010.

(d) *Linear Interpolation* Use the model in part (a) to estimate the enrollment in 1995.

80. *Sales* A small business had total sales of $250,000 in the year 1990. During the next 10 years, the sales increased by $25,000 per year.

(a) Write a linear model giving the total sales S in terms of the year t. (Let $t = 0$ represent 1990.)

(b) Use a graphing utility to graph the model in part (a).

(c) *Linear Extrapolation* Use the model in part (a) to predict the total sales in the year 2015.

(d) *Linear Interpolation* Use the model in part (a) to estimate the total sales in 1998.

Explaining Concepts

81. Can any pair of points on a line be used to determine an equation of the line? Explain.

82. Write, from memory, the point-slope form, the slope-intercept form, and the general form of an equation of a line.

Ongoing Review

In Exercises 83 and 84, sketch the lines through the given point with the indicated slopes, on the same set of coordinate axes.

Point   Slopes

83. $(1, 4)$ (a) 1   (b) -2 (c) $\frac{3}{2}$ (d) 0

84. $(-1, 3)$ (a) 3   (b) $-\frac{5}{9}$ (c) $\frac{2}{5}$ (d) undefined

85. *Business* The inventor of a new game believes that the cost of producing the game is $5.75 per unit plus a fixed cost of $12,000. If x is the number of games produced, express the total cost C as a function of x.

86. *Geometry* The length of a rectangle is $1\frac{1}{2}$ times its width. Express the perimeter P of the rectangle as a function of the rectangle's width w.

Looking Further

The intercepts of a line are $(a, 0)$ and $(0, b)$, $a \neq 0$, $b \neq 0$.

(a) Find the slope of the line.

(b) Write the equation of the line in slope-intercept form.

(c) Using the result in part (b), clear the equation of fractions and show that the equation can be written in the form $bx + ay = ab$.

(d) Use the result in part (c) to show that the equation of the line can be written in the form

$$\frac{x}{a} + \frac{y}{b} = 1.$$

Use the result in part (d) to find an equation of the line with the indicated intercepts.

(i) *x*-intercept: $(3, 0)$ (ii) *x*-intercept: $(-6, 0)$

    y-intercept: $(0, 2)$ *y*-intercept: $(0, 2)$

(iii) *x*-intercept: $\left(-\frac{5}{6}, 0\right)$ (iv) *x*-intercept: $\left(-\frac{8}{3}, 0\right)$

     y-intercept: $\left(0, -\frac{7}{3}\right)$ *y*-intercept: $(0, -4)$

3.2 Applications of Linear Equations

Translating Phrases

In this section, you will study ways to *construct* algebraic expressions. When you translate a verbal sentence or phrase into an algebraic expression, watch for key words and phrases that indicate the four different operations of arithmetic.

Translating Key Words and Phrases

Key Words and Phrases	Verbal Description	Algebraic Expression
Addition:		
Sum, plus, greater than, increased by, more than,	The sum of 5 and x	$5 + x$
exceeds, total of	Seven more than y	$y + 7$
Subtraction:		
Difference, minus, less than, decreased by,	b is subtracted from 4.	$4 - b$
subtracted from, reduced by, the remainder	Three less than z	$z - 3$
Multiplication:		
Product, multiplied by, twice, times, percent of	Twice x	$2x$
Division:		
Quotient, divided by, ratio, per	The ratio of x and 8	$\dfrac{x}{8}$

Example 1 ■ Translating Verbal Phrases

(a) *Verbal Description:* Seven more than three times x

  Algebraic Expression: $3x + 7$

(b) *Verbal Description:* Four less than the product of 6 and n

  Algebraic Expression: $6n - 4$

(c) *Verbal Description:* Four times the sum of y and 9

  Algebraic Expression: $4(y + 9)$

(d) *Verbal Description:* Three times the ratio of x and 7

  Algebraic Expression: $3\left(\dfrac{x}{7}\right)$

In Example 1, the verbal description specified the name of the variable. In most real-life situations, however, the variables are not specified and it is your task to assign variables to the *appropriate* quantities.

Example 2 ■ Translating Verbal Phrases

(a) *Verbal Description:* The sum of 7 and a number

Label: The number $= x$

Algebraic Expression: $7 + x$

(b) *Verbal Description:* Four decreased by the product of 2 and a number

Label: The number $= x$

Algebraic Expression: $4 - 2x$

(c) *Verbal Description:* One more than the product of 2 and a number, all divided by 3

Label: The number $= x$

Algebraic Expression: $\dfrac{2x + 1}{3}$

(d) *Verbal Description:* Seven less than twice the sum of a number and 5

Label: The number $= x$

Algebraic Expression: $2(x + 5) - 7$

A good way to learn algebra is to do it both forward and backward. For instance, the next example translates algebraic expressions into verbal form. Keep in mind that other key words can be used to describe the operations in each expression.

Example 3 ■ Translating Expressions into Verbal Phrases

Without using a variable, write a verbal description for each expression.

(a) $5x - 10$ (b) $\dfrac{3 + x}{4}$

(c) $4(x - 11)$ (d) $(6 + x) - 2$

Solution

(a) Ten less than the product of 5 and a number

(b) The sum of 3 and some number, all divided by 4

(c) Four times the difference of a number and 11

(d) Two less than the sum of 6 and a number

Using Verbal Models

Study Tip

Most real-life problems do not contain verbal expressions that clearly identify the arithmetic operations involved in the problem. You need to rely on past experience and the physical nature of the problem in order to identify the operations hidden in the problem statement.

When verbal phrases are translated into algebraic expressions, products are often overlooked. Watch for hidden products in the next example.

Example 4 ■ Using Verbal Models

(a) A cash register contains x quarters. Write an expression for this amount of money in dollars.

(b) A cash register contains n nickels and d dimes. Write an expression for this amount of money in cents.

(c) A person riding a bicycle travels at a constant rate of 12 miles per hour. Write an expression showing how far the person can ride in t hours.

Solution

(a) *Verbal Model:* | Value of coin | · | Number of coins |

   Labels: Value of coin = 0.25   (dollars per quarter)
   Number of coins = x (quarters)

   Expression: $0.25x$ (dollars)

(b) *Verbal Model:* | Value of nickel | · | Number of nickels | + | Value of dime | · | Number of dimes |

   Labels: Value of nickel = 5   (cents per nickel)
   Number of nickels = n (nickels)
   Value of dime = 10   (cents per dime)
   Number of dimes = d (dimes)

   Expression: $5n + 10d$ (cents)

(c) For this problem, use the formula (distance) = (rate)(time).

   Verbal Model: | Rate | · | Time |

   Labels: Rate = 12   (miles per hour)
   Time = t (hours)

   Expression: $12t$ (miles)

NOTE Using unit analysis, you can see that the expression in Example 4(c) has *miles* as its unit of measure.

$$12 \frac{\text{miles}}{\text{hour}} \cdot t \text{ hours}$$

In Example 4(b), the final expression $5n + 10d$ is measured in cents. This makes "sense" as described below.

$$\frac{5 \text{ cents}}{\text{nickel}} \cdot n \text{ nickels} + \frac{10 \text{ cents}}{\text{dime}} \cdot d \text{ dimes}$$

Note that the nickels and dimes "divide out," leaving cents as the unit of measure for each term. This technique is called *unit analysis*, and it can be very helpful in determining the final unit of measure.

Example 5 ■ Using a Verbal Model

The width of a rectangle is w inches. The length of the rectangle is 7 inches more than twice its width. Write expressions for the perimeter and area.

Solution

First draw a rectangle as shown in Figure 3.13. Next, use a verbal model to solve the problem.

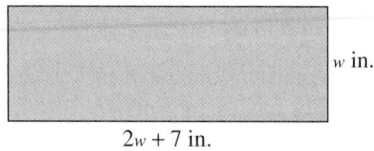

w in.

$2w + 7$ in.

Figure 3.13

For the perimeter, use the formula (perimeter) = 2(length) + 2(width).

Verbal Model: $2 \cdot$ Length $+ 2 \cdot$ Width

Labels: Length $= 2w + 7$ (inches)

          Width $= w$ (inches)

Expression: $2(2w + 7) + 2w = 4w + 14 + 2w = 6w + 14$ (inches)

For the area, use the formula (area) = (length)(width).

Verbal Model: Length $\cdot$ Width

Labels: Length $= 2w + 7$ (inches)

          Width $= w$ (inches)

Expression: $(2w + 7)(w) = 2w^2 + 7w$ (square inches)

Mathematical Modeling

So far in this section, you have translated *verbal phrases* into *algebraic expressions*. When algebra is used to solve real-life problems, you usually have to carry the process one step further—by translating *verbal sentences* into *algebraic equations*. The process of translating phrases or sentences into algebraic expressions or equations is called **mathematical modeling.**

A good approach is to use a *verbal model* by using the given verbal description of the problem. Then, after assigning labels to the unknown quantities in the verbal model, you can form a *mathematical model* or *algebraic equation*.

In mathematics it is useful to know how to represent certain types of integers algebraically.

Two integers are called **consecutive integers** if they differ by 1. So, for any integer n, the next consecutive integer is $n + 1$.

> **Labels for Consecutive Integers**
>
> Let n represent an integer.
>
> 1. $\{n, n + 1, n + 2, \ldots\}$ denotes a set of *consecutive* integers.
>
> 2. If n is an even integer, then $\{n, n + 2, n + 4, \ldots\}$ denotes a set of *consecutive even integers*.
>
> 3. If n is an odd integer, then $\{n, n + 2, n + 4, \ldots\}$ denotes a set of *consecutive odd integers*.

Example 6 ■ Constructing a Verbal Model

Construct a verbal model. Write and solve an equation that represents the problem.

(a) Find two consecutive integers such that the sum of the first integer and three times the second is 87.

(b) Find two consecutive odd integers such that the difference of five times the first integer and two times the second is 23.

Solution

(a) *Verbal Model:*   First integer $+ \, 3 \cdot$ Second integer $= 87$

     Labels: First integer $= n$

                Second integer $= n + 1$

Equation: $n + 3(n + 1) = 87$		Algebraic model
$n + 3n + 3 = 87$		Distributive Property
$4n + 3 = 87$		Combine like terms.
$4n = 84$		Subtract 3 from each side.
$n = 21$		Divide each side by 4.

So, the first integer is 21, and the second integer is $21 + 1 = 22$. You can check this by substituting 21 and 22 as the two consecutive integers in the original problem.

Study Tip

It is helpful to break a verbal sentence into parts separated by the word "is." In application problems, "is" often represents an equal sign. To write a verbal model, first identify where the word "is" appears in the sentence.

(b) *Verbal Model:* $5 \cdot$ First integer $- \, 2 \cdot$ Second integer $= 23$

     Labels: First integer $= n$

                Second integer $= n + 2$

Equation: $5n - 2(n + 2) = 23$		Algebraic model
$5n - 2n - 4 = 23$		Distributive Property
$3n - 4 = 23$		Combine like terms.
$3n = 27$		Add 4 to each side.
$n = 9$		Divide each side by 3.

So, the first integer is 9, and the second integer is $9 + 2 = 11$. You can check this by substituting 9 and 11 as the two consecutive odd integers in the original problem.

Example 7 ■ A Percent Application

A real estate agency receives a commission of $8092.50 for the sale of a $124,500 house. What percent commission is this?

Solution

Verbal Model: $$\text{Commission} = \frac{\text{Percent}}{\text{(decimal form)}} \cdot \frac{\text{Sale}}{\text{price}}$$

Labels: Commission = 8092.50   (dollars)
             Percent = p (decimal form)
             Sale price = 124,500   (dollars)

Equation: $8092.50 = p(124{,}500)$

$$\frac{8092.50}{124{,}500} = p$$

$$0.065 = p$$

The real estate agency receives a commission of 6.5%. Use your calculator to check this solution in the original statement of the problem.

Recall from geometry that an **angle** is determined by rotating a ray (half-line) about its endpoint. The starting position of the ray is the **initial side** of the angle, and the position after rotation is the **terminal side,** as shown in Figure 3.14. The endpoint of the ray is the **vertex** of the angle. Angles are labeled with Greek letters α (alpha), β (beta), and θ (theta), as well as uppercase letters A, B, and C.

The most common unit of angle measure is the degree, denoted by the symbol °. A measure of one degree (1°) is equivalent to a rotation of $\frac{1}{360}$ of a complete revolution about the vertex. Figure 3.15 shows that angles measuring 90° are called **right angles** and angles measuring 180° are called **straight angles.**

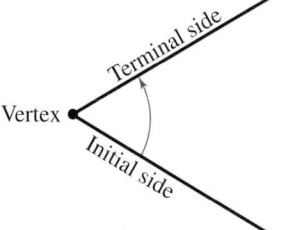

Figure 3.14   *Angle*

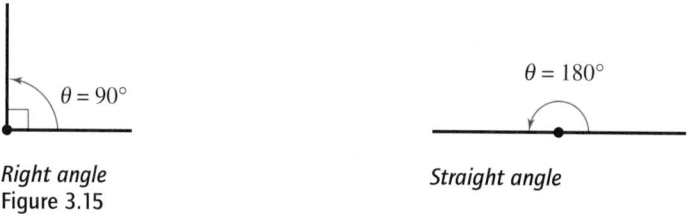

Right angle
Figure 3.15

$\theta = 90°$ $\theta = 180°$

Straight angle

Two angles are **complementary** (complements of each other) if their sum is 90°. Two angles are **supplementary** (supplements of each other) if their sum is 180°. (See Figure 3.16.)

Complementary angles: $\alpha + \beta = 90°$
Figure 3.16

Supplementary angles: $\alpha + \beta = 180°$

Example 8 ■ A Geometry Application

In a pair of complementary angles, one angle measures $36°$ more than the other angle. Find the measure of each angle.

Solution

First draw a diagram of the angles, as shown in Figure 3.17. Next, use a verbal model to solve the problem.

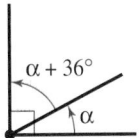

Figure 3.17

Verbal Model:	Measure of smaller angle	$+$	Measure of larger angle	$= 90°$

Labels: Measure of smaller angle $= \alpha$ (degrees)

Measure of larger angle $= \alpha + 36$ (degrees)

Equation:	$\alpha + \alpha + 36 = 90$	Algebraic model
	$2\alpha + 36 = 90$	Combine like terms.
	$2\alpha = 54$	Subtract 36 from each side.
	$\alpha = 27$	Divide each side by 2.

So, the smaller angle is $27°$ and the larger angle is $27° + 36° = 63°$. Check this in the original statement of the problem.

The next example uses the fact that the sum of the interior angles of a triangle is $180°$.

Example 9 ■ A Geometry Application

One angle of a triangle measures $12°$ more than the smallest angle. The third angle is twice as large as the smallest angle. Find the measure of each angle.

Solution

First draw a diagram of the triangle, as shown in Figure 3.18. Next, use a verbal model to solve the problem.

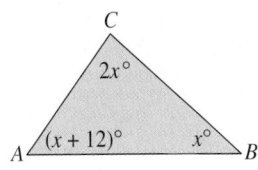

Figure 3.18

Verbal Model:	Measure of Angle A	$+$	Measure of Angle B	$+$	Measure of Angle C	$= 180°$

Labels: Measure of Angle $A = x + 12$ (degrees)

Measure of Angle $B = x$ (degrees)

Measure of Angle $C = 2x$ (degrees)

Equation:	$x + 12 + x + 2x = 180$	Algebraic model
	$4x + 12 = 180$	Combine like terms.
	$4x = 168$	Subtract 12 from each side.
	$x = 42$	Divide each side by 4.

So, angle A measures $42° + 12° = 54°$, angle B measures $42°$, and angle C measures $2 \cdot 42° = 84°$. Check that the sum of the three measures is $180°$.

Proportions

In real-life applications, the quotient a/b is called a **rate** if a and b have different units, and is called a **ratio** if a and b have the same unit. Note the *order* implied by a ratio. The ratio of a to b means a/b, whereas the ratio of b to a means b/a. Situations describing rates and ratios are shown below.

1. You have driven 110 miles in 2 hours. Your average rate for the trip can be expressed as

$$\text{Rate} = \frac{110 \text{ miles}}{2 \text{ hours}}$$

$$= 55 \text{ miles per hour.}$$

2. You have driven 110 miles in 2 hours and your friend has driven 100 miles in the same length of time. The ratio of the distance you have traveled to the distance your friend has traveled is

$$\text{Ratio} = \frac{110 \text{ miles}}{100 \text{ miles}}$$

$$= \frac{11}{10}.$$

NOTE When comparing two *measurements* by means of a ratio, be sure to use the same unit of measurement in both the numerator and the denominator.

A **proportion** is a statement that equates two ratios. For example, if the ratio of a to b is the same as the ratio of c to d, you can write the proportion as

$$\frac{a}{b} = \frac{c}{d}.$$

In typical problems, you know three of the values and need to find the fourth. The quantities a and d are called the **extremes** of the proportion, and the quantities b and c are called the **means** of the proportion. In a proportion, the product of the extremes is equal to the product of the means. Rewriting a proportion in the form $ad = bc$ is called **cross multiplying.**

Example 10 ■ Solving a Proportion

The ratio of 8 to x is the same as the ratio of 5 to 2. What is x?

Solution

$$\frac{8}{x} = \frac{5}{2} \qquad \text{Set up proportion.}$$

$$16 = 5x \qquad \text{Cross multiply.}$$

$$\frac{16}{5} = x \qquad \text{Divide each side by 5.}$$

So, $x = \frac{16}{5}$. Check this in the original statement of the problem.

Proportions are often used in geometric applications involving similar triangles. Similar triangles have the same shape, but they may differ in size. The corresponding sides of similar triangles are proportional.

Example 11 ■ Solving a Proportion in Geometry

The triangles shown in Figure 3.19 are similar triangles. Use this fact to find the lengths of the unknown sides.

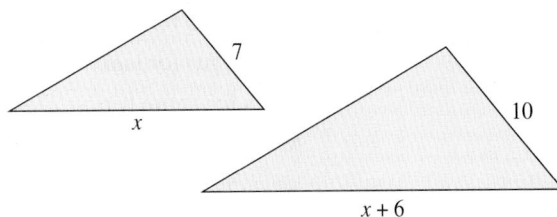

Figure 3.19

Solution

NOTE The proportion in Example 11 could also have been written as

$$\frac{10}{7} = \frac{x + 6}{x}.$$

After cross multiplying, you obtain the equation $10x = 7(x + 6)$.

$$\frac{10}{x + 6} = \frac{7}{x} \qquad \text{Set up proportion.}$$

$$10x = 7(x + 6) \qquad \text{Cross multiply.}$$

$$10x = 7x + 42 \qquad \text{Distributive Property}$$

$$3x = 42 \qquad \text{Subtract } 7x \text{ from each side.}$$

$$x = 14 \qquad \text{Divide each side by 3.}$$

So, the length of the unknown side of the smaller triangle is 14 units and the length of the unknown side of the larger triangle is

$$x + 6 = 14 + 6$$
$$= 20 \text{ units.}$$

Check this in the original statement of the problem.

Example 12 ■ An Application of Proportion

At a club meeting in July, 20 people attended the meeting and drank 184 ounces of lemonade. In August, 45 people are expected to attend the meeting. How much lemonade is needed at the August meeting?

Solution

$$\frac{20}{184} = \frac{45}{x} \qquad \text{Set up proportion.}$$

$$20x = 8280 \qquad \text{Cross multiply.}$$

$$x = 414 \qquad \text{Divide each side by 20.}$$

So, 414 ounces of lemonade is needed at the August meeting. Check this in the original statement of the problem.

Example 13 ■ An Application of Proportion

Study Tip

You can write a proportion in several ways. Just be sure to put like quantities in similar positions on each side of the proportion.

You are driving from Arizona to New York, a trip of 2750 miles. You begin the trip with a full tank of gas and after traveling 424 miles, you refill the tank for $22.00. How much should you plan to spend on gasoline for the entire trip?

Solution

Verbal Model: $\dfrac{\text{Dollars for trip}}{\text{Dollars for tank}} = \dfrac{\text{Miles for trip}}{\text{Miles for tank}}$

Labels: Cost of gas for entire trip $= x$ (dollars)

Cost of gas for tank $= 22$ (dollars)

Miles for entire trip $= 2750$ (miles)

Miles for tank $= 424$ (miles)

Proportion: $\dfrac{x}{22} = \dfrac{2750}{424}$

$$x = 22\left(\dfrac{2750}{424}\right)$$

$$x \approx 142.69$$

You should plan to spend approximately $142.69 for gasoline on the trip. Check this in the original statement of the problem.

The list below summarizes a strategy for modeling and solving real-life problems.

> **Strategy for Solving Word Problems**
>
> 1. Ask yourself what you need to know to solve the problem. Then *write a verbal model* that will give you what you need to know.
>
> 2. *Assign labels* to each part of the verbal model—numbers to the known quantities and letters (or expressions) to the variable quantities.
>
> 3. Use the labels to *write an algebraic model* based on the verbal model.
>
> 4. *Solve* the resulting algebraic equation.
>
> 5. *Answer* the original question and *check* that your answer satisfies the original problem as stated.

In previous mathematics courses, you studied several other problem-solving strategies, such as *drawing a diagram, making a table, looking for a pattern,* and *solving a simpler problem.* Each of these strategies can also help you to solve problems in algebra.

3.2 ■ Exercises

Developing Skills

In Exercises 1–14, translate the statement into an algebraic expression.

1. The sum of 8 and a number n

2. Six less than a number n

3. Fifteen decreased by three times a number n

4. Six less than four times a number n

5. One-third of a number n

6. Seven-fifths of a number n

7. Thirty percent of the list price L

8. Forty percent of the cost C

9. The quotient of a number x and 6

10. The ratio of y to 3

11. The sum of 3 and four times a number x, all divided by 8

12. The product of a number y and 10, decreased by 35

13. The absolute value of the difference between a number n and 5

14. The absolute value of the quotient of y and 4

In Exercises 15–22, write a verbal description of the algebraic expression without using the variable.

15. $3x + 2$

16. $4x - 5$

17. $8(x - 5)$

18. $3(x + 2)$

19. $\dfrac{y}{8}$

20. $\dfrac{4x}{5}$

21. $\dfrac{x + 10}{3}$

22. $25 + \dfrac{x}{6}$

In Exercises 23–36, write an algebraic expression that represents the specified quantity in the verbal statement and simplify if possible.

23. The amount of money (in dollars) represented by n quarters

24. The amount of money (in cents) represented by m dimes and n quarters

25. The distance traveled in t hours at an average speed of 55 miles per hour

26. The distance traveled in 5 hours at an average speed of r miles per hour

27. The time required to travel 100 miles at an average speed of r miles per hour

28. The average rate of speed for a journey of 360 miles in t hours

29. The amount of antifreeze in a cooling system containing y gallons of coolant that is 45% antifreeze

30. The amount of water in q quarts of food product that is 65% water

31. The amount of wage tax due for a taxable income of I dollars that is taxed at a rate of 1.25%

32. The amount of sales tax on a purchase valued at L dollars if the tax rate is 6%

33. The sale price of a coat that has a list price of L dollars if it is a "20% off" sale

34. The total cost for a family to stay one night at a campground if the charge is $18 for the parents plus $3 for each of the n children

35. The total hourly wage for an employee when the base pay is $8.25 per hour and an additional $0.60 is paid for each of q units produced per hour

36. The total hourly wage for an employee when the base pay is $11.65 per hour and an additional $0.80 is paid for each of q units produced per hour

In Exercises 37–48, solve the proportion.

37. $\dfrac{x}{6} = \dfrac{2}{3}$

38. $\dfrac{y}{36} = \dfrac{6}{7}$

39. $\dfrac{5}{4} = \dfrac{t}{6}$

40. $\dfrac{7}{8} = \dfrac{x}{2}$

41. $\dfrac{y + 5}{6} = \dfrac{y - 2}{4}$

42. $\dfrac{x - 1}{5} = \dfrac{x + 9}{4}$

43. $\dfrac{z - 3}{3} = \dfrac{z + 8}{12}$

44. $\dfrac{x + 5}{2} = \dfrac{x - 6}{3}$

45. $\dfrac{2x - 3}{7} = \dfrac{x + 4}{2}$

46. $\dfrac{x - 6}{9} = \dfrac{3x + 1}{10}$

47. $\dfrac{4x - 2}{3} = \dfrac{2x + 1}{4}$

48. $\dfrac{5x + 3}{5} = \dfrac{3 - x}{2}$

Solving Problems

Geometry In Exercises 49–52, write an expression for the perimeter of the region. Then simplify the expression.

49.

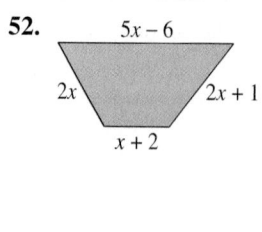

w

$2w$

50.

$2x$ $2x - 2$

$x + 10$

51.

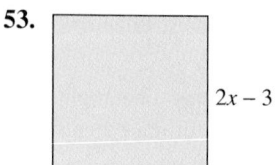

3

x

3

$2x$

x

6

52.

$5x - 6$

$2x$ $2x + 1$

$x + 2$

Geometry In Exercises 53–56, write an expression for the area of the region.

53.

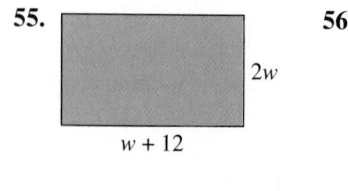

$2x - 3$

$2x - 3$

54.

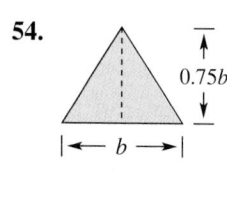

$0.75b$

$\mid\!\!\leftarrow b \rightarrow\!\!\mid$

55.

$2w$

$w + 12$

56.

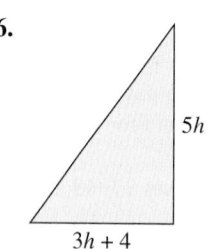

$5h$

$3h + 4$

57. *Advertising Banner* An advertising banner has a width of w and a length of $6w$, where w is measured in meters (see figure). Find an algebraic expression that represents the area of the banner. What is the unit of measure for the area?

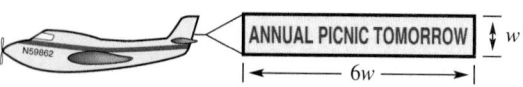

ANNUAL PICNIC TOMORROW w

$\mid\!\!\leftarrow 6w \rightarrow\!\!\mid$

58. *Billiard Table* The top of a billiard table has a length of l and a width of $l - 4$, where l is measured in feet (see figure). Find an algebraic expression that represents the area of the top of the billiard table. What is the unit of measure for the area?

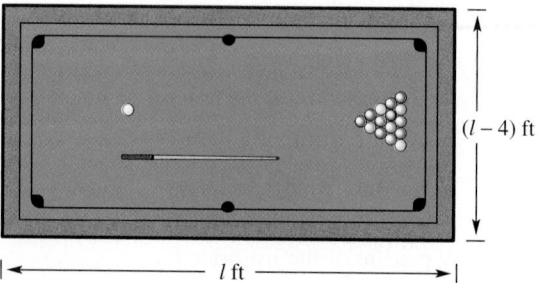

$(l - 4)$ ft

l ft

59. *Geometry* The width of a rectangle is w feet. The length of the rectangle is 2 feet more than three times its width. Write expressions for the perimeter and area of the rectangle.

60. *Geometry* The width of a rectangle is w inches. The length of the rectangle is 8 inches less than five times its width. Write expressions for the perimeter and area of the rectangle.

Number Problems In Exercises 61–74, solve the number problem.

61. The sum of three consecutive integers is 60. Find the integers.

62. The sum of three consecutive integers is 81. Find the integers.

63. The sum of three consecutive even integers is 138. Find the integers.

64. The sum of three consecutive odd integers is 237. Find the integers.

65. Eight times the sum of a number and 6 is 128. What is the number?

66. Six times 4 less than a number is 138. What is the number?

67. When the sum of a number and 18 is divided by 5, the quotient is 12. Find the number.

68. When the sum of a number and 11 is divided by 7, the quotient is 15. Find the number.

69. Find a number such that the sum of that number and 30 is 82.

70. Find a number such that the sum of three times the number and 26 is 38.

71. Find a number such that six times the difference of the number and 12 is 300.

72. Find a number such that four times the difference of the number and 111 is 364.

73. The difference of five times an odd integer and three times the next consecutive odd integer is 24. Find the integers.

74. The sum of three times an even integer and eight times the next consecutive even integer is 170. Find the integers.

75. *Geometry*   The "Slow Moving Vehicle" sign has the shape of a hexagon surrounding an equilateral triangle. The triangle has a perimeter of 129 centimeters. Find the length of each side of the triangle.

76. *Geometry*   The length of a rectangle is three times its width. The perimeter of the rectangle is 64 inches. Find the dimensions of the rectangle.

77. *Labor*   The bill for the repair of an automobile is $380. Included in this bill is a charge of $275 for parts. If the remainder of the bill is for labor at a rate of $35 per hour, how many hours were spent in repairing the car?

78. *Labor*   You have a job on an assembly line for which you are paid $10 per hour plus $0.75 per unit produced. Find the number of units produced in an eight-hour day if your earnings are $146.

79. *Employment*   Because of slumping sales, a small company laid off 25 of its 160 employees. What percent of the work force was laid off?

80. *Real Estate*   A real estate agency receives a commission of $9100 for the sale of a $130,000 house. What percent commission is this?

81. *Defective Parts*   A quality control engineer reported that 1.5% of a sample of parts were defective. Find the size of the sample if the engineer detected three defective parts.

82. *Inflation*   The price of a new van is approximately 115% of what it was 3 years ago. What was the approximate price 3 years ago if the current price is $25,750?

83. *Geometry*   In a pair of supplementary angles, one angle measures 24° more than twice the other angle. Find the measure of each angle.

84. *Geometry*   In a pair of complementary angles, one angle measures 18° less than three times the other angle. Find the measure of each angle.

85. *Geometry*   Use the figure to find the measure of each angle.

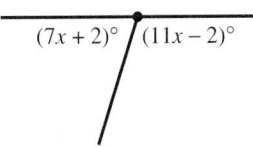

86. *Geometry*   Use the figure to find the measure of each angle.

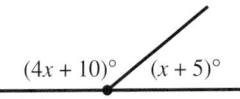

87. *Geometry*   Use the figure to find the measure of each angle.

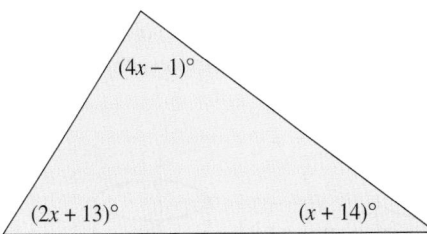

88. *Geometry*   Use the figure to find the measure of each angle.

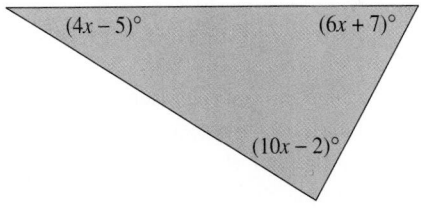

Geometry   In Exercises 89 and 90, solve for the length x of the side of the triangle by using the fact that corresponding sides of similar triangles are proportional.

89.

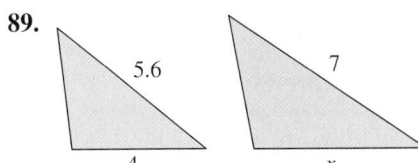

90.

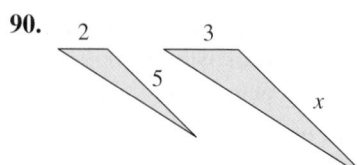

91. *Population* In the 1980 census, the population of Honolulu, Hawaii was 365,000. The population grew 3.3% during the decade of the eighties and grew 5.0% during the decade of the nineties. *(Source: U.S. Census Bureau)*

(a) Use the given information to approximate the populations of the city at the time of the 1990 census and at the time of the 2000 census.

(b) Use the results in part (a) to approximate the percent increase in the population between the 1980 census and the 2000 census. Explain why it is not 8.3%.

92. *Physics* The compression ratio of a cylinder is the ratio of its expanded volume to its compressed volume (see figure). The expanded volume of one cylinder of a small diesel engine is 425 cubic centimeters, and its compressed volume is 20 cubic centimeters. Find the compression ratio of this cylinder.

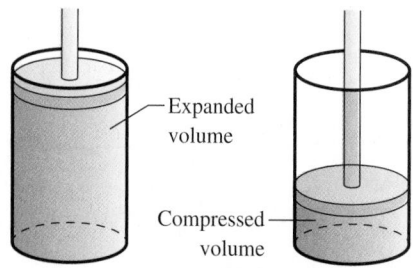

93. *Investment* The ratio of the price of a stock to its earnings is called the price-earnings ratio. Find the price-earnings ratio of stock that sells for $56.25 per share and earns $6.25 per share.

94. *Tax* You have $12.50 of state tax withheld from your paycheck per week when your gross pay is $625. Find the ratio of tax to gross pay. (Write this as a percent.)

95. *Mechanics* The gear ratio of two gears is the ratio of the number of teeth in one gear to the number of teeth in the other gear (see figure). If two gears in a gear box have 60 teeth and 40 teeth, find the gear ratio.

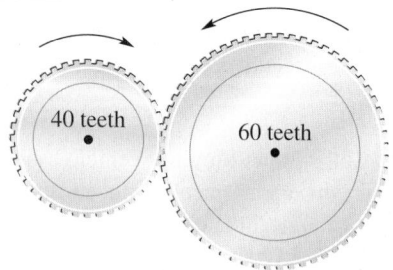

96. *Tax* The taxes on property with an assessed value of $75,000 are $1125. Find the taxes on property with an assessed value of $120,000.

97. *Politics* In a public opinion poll, 870 people from a sample of 1500 indicated that they would vote for a specific candidate. Assuming this poll to be a correct indicator of the electorate, how many votes can the candidate expect to receive from a total of 80,000 votes cast?

98. *Fuel* The gasoline-to-oil ratio of a two-cycle engine is 40 to 1. Determine the amount of gasoline required to produce a mixture that contains $\frac{1}{2}$ pint of oil.

99. *Baking* Three cups of flour are required to make one batch of cookies. How many cups are required to make $3\frac{1}{2}$ batches?

100. *Map Scale* One-quarter inch represents 60 miles on a map. Approximate the distance between two cities that are $1\frac{1}{3}$ inches apart on the map.

Explaining Concepts

101. Is it true that $\frac{1}{2}\% = 50\%$? Explain.

102. In your own words, describe the meaning of *mathematical modeling*. Give an example.

103. Define the term *ratio*. Give an example of a ratio.

104. During a year of financial difficulties, your company reduces your salary by 7%. What percent increase in this reduced salary is required to raise your salary to the amount it was prior to the reduction? Why isn't the percent increase the same as the percent of the reduction?

Ongoing Review

In Exercises 105–108, use the properties of algebra to simplify the expression.

105. $12 - 4(x - 1)$ **106.** $3(a + 5) - 2(a - 7)$

107. $9(3y + 2) - 4(2y - 1)$

108. $-3(3x - 5) + 6(x + 2)$

In Exercises 109–116, solve the equation.

109. $3x + 7 = 13$ **110.** $2x - 8 = 2$

111. $\dfrac{x}{4} - 1 = 2$ **112.** $6 - \dfrac{x}{3} = 5$

113. $5(x - 2) = 3(x + 2)$

114. $6(x + 1) = 2 - 4(x + 4)$

115. *Distance*   The length of a relay race is 2.5 miles. The last change of runners occurs at the 1.8-mile marker. How far does the last person run?

116. *Agriculture*   During the months of January, February, and March, a farmer bought $34\frac{1}{3}$ tons, $18\frac{1}{5}$ tons, and $25\frac{5}{6}$ tons of soybeans, respectively. Find the total amount of soybeans purchased during the first quarter of the year.

Looking Further

You can answer each of the following questions by writing and solving a proportion using the fact that corresponding sides of similar triangles are proportional.

Geometry   Solve for the length x of the side of the triangle (see figure).

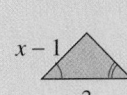 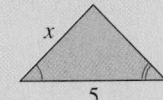

Geometry   Solve for the length x of the side of the triangle (see figure).

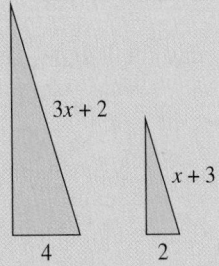

Tree Height   A man who is 6 feet tall walks directly toward the tip of a shadow of a tree. When the man is 75 feet from the tree, he notices his own shadow beyond the shadow of the tree. Find the height of the tree, if the length of the shadow of the tree beyond this point is 11 feet (see figure).

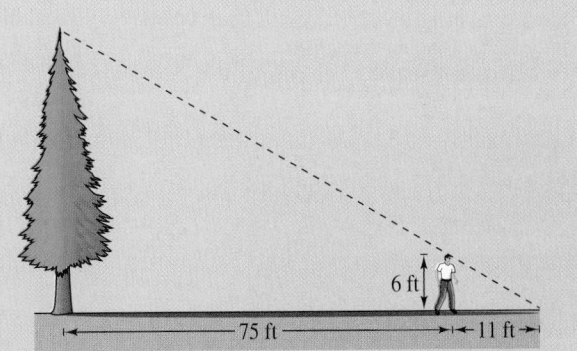

Shadow Length   Find the length of the shadow of a man who is 6 feet tall and is standing 15 feet from a streetlight that is 20 feet high (see figure).

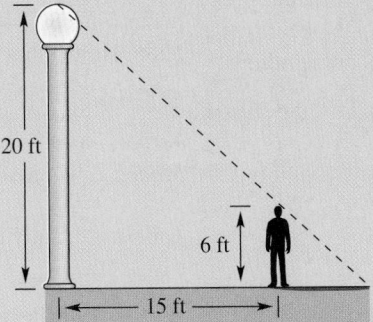

3.3 Business and Scientific Problems

What you should learn:

- How to use mathematical models to solve business-related problems
- How to use formulas to solve application problems
- How to use mathematical models to solve rate problems
- How to use mathematical models to solve mixture problems

Why you should learn it:

Mathematical models can be used to solve a wide variety of real-life problems. For instance, you can find the annual premium of an insurance policy using mathematical modeling. See Exercise 17 on page 194.

Business Problems

Many business problems can be represented by mathematical models involving the sum of a fixed term and a variable term.

Example 1 ■ Finding the Hours of Labor

An auto repair bill of $338 lists $170 for parts and the rest for labor. The labor charge is $28 per hour. How many hours did it take to repair the auto?

Solution

Verbal Model:

| Total bill | = | Charge for parts | + | Charge for labor |

Labels:

Total bill = 338	(dollars)
Charge for parts = 170	(dollars)
Hours of labor = x	(hours)
Hourly rate for labor = 28	(dollars per hour)
Charge for labor = $28x$	(dollars)

Equation: $338 = 170 + 28x$

$6 = x$

So, it took 6 hours to repair the auto.

NOTE The variable term in a mathematical model is often a *hidden product* in which one of the factors is a percent or some other type of rate.

Example 2 ■ Reimbursed Expenses

A company reimburses its sales representatives $150 per day for lodging and meals, plus $0.34 per mile driven. The daily cost for a five-day business trip was $286. How many miles were driven per day?

Solution

Verbal Model:

| Daily cost | = | Cost for lodging and meals | + | Cost per mile |

Labels:

Daily cost = 286	(dollars)
Cost for lodging and meals = 150	(dollars)
Mileage rate = 0.34	(dollar per mile)
Miles driven = x	(miles)
Cost per mile = $0.34x$	(dollars)

Equation: $286 = 150 + 0.34x$

$400 = x$

So, 400 miles were driven per day on the five-day business trip.

Study Tip

Remember to check your solution of an application problem in the original statement of the problem.

Formulas

Many common types of geometric, scientific, and investment problems use ready-made equations called **formulas.** Knowing formulas such as those in the lists below will help you translate and solve a wide variety of real-life problems involving perimeter, area, volume, temperature, interest, and distance.

Common Formulas for Area, Perimeter, and Volume

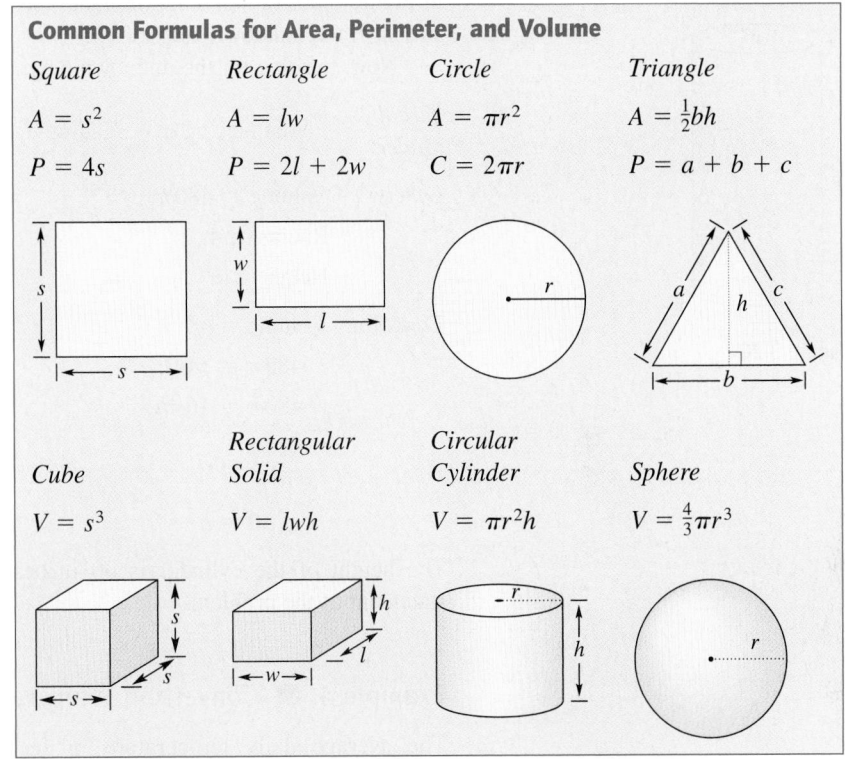

Square	Rectangle	Circle	Triangle
$A = s^2$	$A = lw$	$A = \pi r^2$	$A = \frac{1}{2}bh$
$P = 4s$	$P = 2l + 2w$	$C = 2\pi r$	$P = a + b + c$

Cube	Rectangular Solid	Circular Cylinder	Sphere
$V = s^3$	$V = lwh$	$V = \pi r^2 h$	$V = \frac{4}{3}\pi r^3$

Miscellaneous Common Formulas

Temperature: F = degrees Fahrenheit, C = degrees Celsius

$$F = \frac{9}{5}C + 32$$

Simple Interest: I = interest, P = principal, r = interest rate, t = time

$$I = Prt$$

Distance: d = distance traveled, r = rate, t = time

$$d = rt$$

When working with applied problems, you often need to rewrite one of the common formulas. For instance, the formula for the perimeter of a rectangle, $P = 2l + 2w$, can be rewritten or solved for w as $w = \frac{1}{2}(P - 2l)$.

Example 3 ■ Using a Geometric Formula

The volume of a circular cylinder is 480π cubic inches and the radius of the cylinder is 4 inches. What is the height of the cylinder?

Solution

In a problem such as this, it is helpful to begin by drawing a diagram, as shown in Figure 3.20. In this diagram, label the radius of the cylinder as $r = 4$ inches, and label the unknown height as h.

   Now, to solve for the unknown height, use the steps below.

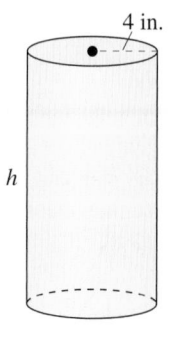

4 in.

h

Figure 3.20

Verbal Model: Volume $= \pi \cdot$ Radius $^2 \cdot$ Height

Labels: Volume $= 480\pi$ (cubic inches)

Radius $= 4$ (inches)

Height $= h$ (inches)

Equation: Volume $= \pi r^2 h$

$$480\pi = \pi(4)^2 \cdot h$$

$$480\pi = 16\pi h$$

$$\frac{480\pi}{16\pi} = h$$

$$30 = h$$

The height of the cylinder is 30 inches. Check this solution in the original statement of the problem.

Example 4 ■ Converting Temperature

The average daily temperature in Jacksonville, Florida is 68°F. What is Jacksonville's average daily temperature in degrees Celsius?

Solution

Verbal Model: $\dfrac{\text{Fahrenheit}}{\text{temperature}} = \dfrac{9}{5} \cdot \dfrac{\text{Celsius}}{\text{temperature}} + 32$

Labels: Fahrenheit temperature $= 68$ (degrees Fahrenheit)

Celsius temperature $= C$ (degrees Celsius)

Equation: $68 = \dfrac{9}{5}C + 32$

$$36 = \frac{9}{5}C$$

$$20 = C$$

The average daily temperature in Jacksonville is 20°C. Check this solution in the original statement of the problem.

Example 5 ■ Simple Interest

A deposit of $8000 earned $300 in interest in 6 months. What was the annual interest rate for this account?

Solution

Verbal Model: | Interest | = | Principal | · | Rate | · | Time |

Labels: Interest = 300   (dollars)

Principal = 8000   (dollars)

Time = $\frac{1}{2}$ (year)

Annual interest rate = r (percent in decimal form)

Equation: $300 = 8000(r)\left(\dfrac{1}{2}\right)$

$\dfrac{2(300)}{8000} = r$

$0.075 = r$

The annual interest rate is $r = 0.075$ (or 7.5%). Check this in the original statement of the problem.

Technology

You can use a graphing utility to solve simple interest problems by using the program found in Appendix H at our website *college.hmco.com.* Use the program and the Guess, Check, and Revise method to find P when $I = \$3330$, $r = 6\%$, and $t = 3$ years.

Rate Problems

Time-dependent problems such as those involving distance can be classified as **rate problems.** They fit the verbal model ·

| Distance | = | Rate | · | Time | .

For instance, if you travel at a constant (or average) rate of 55 miles per hour for 45 minutes, the total distance you travel is given by

$$\left(55\ \frac{\text{miles}}{\text{hour}}\right)\left(\frac{45}{60}\ \text{hour}\right) = 41.25\ \text{miles.}$$

As with all problems involving applications, be sure to check that the units in the verbal model make sense. For instance, in this problem the rate is given in *miles per hour*. So, in order for the solution to be given in *miles*, you must convert the time (from minutes) to *hours*. In the model, you can think of the 2 "hours" as divided out, as follows.

$$\left(55\ \frac{\text{miles}}{\text{hour}}\right)\left(\frac{45}{60}\ \text{hour}\right) = 41.25\ \text{miles}$$

> **NOTE**   To convert minutes to hours, use the fact that there are 60 minutes in each hour. So, 45 minutes is equivalent to $\frac{45}{60}$ of 1 hour. In general, x minutes is $x/60$ of 1 hour.

Example 6 ■ Distance-Rate Problem

Students are traveling in two cars to a football game 150 miles away. The first car leaves on time and travels at an average speed of 48 miles per hour. The second car starts $\frac{1}{2}$ hour later and travels at an average speed of 58 miles per hour. At these speeds, how long will it take the second car to catch up to the first car?

Solution

Verbal Model: | Distance of first car | = | Distance of second car |

Labels: Time for first car $= t$ (hours)

Distance of first car $= 48t$ (miles)

Time for second car $= t - \frac{1}{2}$ (hours)

Distance of second car $= 58\left(t - \frac{1}{2}\right)$ (miles)

Equation:
$$48t = 58\left(t - \frac{1}{2}\right)$$
$$48t = 58t - 29$$
$$29 = 10t$$
$$2.9 = t$$

> **NOTE** In work problems, the **rate of work** is the reciprocal of the time needed to do the entire job. For instance, if it takes 5 hours to complete a job, the per-hour work rate is
>
> $$\frac{1}{5} \text{ job per hour.}$$

After the first car travels for 2.9 hours, the second car catches up to it. So, it takes the second car $t - 0.5 = 2.9 - 0.5 = 2.4$ hours to catch up to the first car.

Example 7 ■ Work-Rate Problem

Consider two machines in a paper manufacturing plant. Machine 1 can complete one job (2000 pounds of paper) in 5 hours. Machine 2 is newer and can complete one job in 2 hours. How long will it take the two machines working together to complete one job?

Solution

Verbal Model: | Work done | = | Portion done by machine 1 | + | Portion done by machine 2 |

Labels: Work done by both machines $= 1$ (job)

Time for each machine $= t$ (hours)

Rate for machine 1 $= \frac{1}{5}$ (job per hour)

Rate for machine 2 $= \frac{1}{2}$ (job per hour)

Equation:
$$1 = \tfrac{1}{5}t + \tfrac{1}{2}t$$
$$10(1) = 10\left(\tfrac{1}{5}t + \tfrac{1}{2}t\right) \qquad \text{Multiply each side by LCD of 10.}$$
$$10 = 2t + 5t$$
$$10 = 7t$$
$$\tfrac{10}{7} = t$$

> ## Study Tip
>
> In Example 7, the "2000 pounds" of paper is unnecessary information. The 2000 pounds is represented as "one job." This type of unnecessary information in an applied problem is sometimes called a *red herring*. The 150 miles given in Example 6 is also a red herring.

It will take $\frac{10}{7}$ hours (or about 1.43 hours) for both machines to complete the job.

Mixture Problems

Many real-life problems involve combinations of two or more quantities that make up new or different quantities. Such problems are called **mixture problems.** A mixture problem is usually composed of the sum of two or more "hidden products" that involve rate factors. Here is the generic form of the verbal model for a mixture problem.

$$\boxed{\begin{array}{c}\text{First}\\\text{rate}\end{array}} \cdot \boxed{\text{Amount}} + \boxed{\begin{array}{c}\text{Second}\\\text{rate}\end{array}} \cdot \boxed{\text{Amount}} = \boxed{\begin{array}{c}\text{Final}\\\text{rate}\end{array}} \cdot \boxed{\begin{array}{c}\text{Final}\\\text{amount}\end{array}}$$

Example 8 ■ A Mixture Problem

A nursery wants to mix two types of lawn seed; one type sells for $10 per pound and the other type sells for $15 per pound. To obtain 20 pounds of a mixture at $12 per pound, how many pounds of each type of seed are needed?

Solution

Verbal Model:

$$\boxed{\begin{array}{c}\text{Total cost}\\\text{of \$10 seed}\end{array}} + \boxed{\begin{array}{c}\text{Total cost}\\\text{of \$15 seed}\end{array}} = \boxed{\begin{array}{c}\text{Total cost}\\\text{of \$12 seed}\end{array}}$$

Labels:

Cost of $10 seed = 10	(dollars per pound)
Pounds of $10 seed = x	(pounds)
Cost of $15 seed = 15	(dollars per pound)
Pounds of $15 seed = $20 - x$	(pounds)
Cost of $12 seed = 12	(dollars per pound)
Pounds of $12 seed = 20	(pounds)

Equation:

$$10x + 15(20 - x) = 12(20)$$
$$10x + 300 - 15x = 240$$
$$300 - 5x = 240$$
$$-5x = -60$$
$$x = 12$$

The mixture should contain 12 pounds of the $10 seed and $20 - x = 20 - 12 = 8$ pounds of the $15 seed.

Remember that when you have found a solution, you should always go back to the original statement of the problem and check to see that the solution makes sense. For instance, you can check the result of Example 8 as follows.

$$\overbrace{\left(\begin{array}{c}\$10 \text{ per}\\\text{pound}\end{array}\right)\left(\begin{array}{c}12\\\text{pounds}\end{array}\right)}^{\$10 \text{ seed}} + \overbrace{\left(\begin{array}{c}\$15 \text{ per}\\\text{pound}\end{array}\right)\left(\begin{array}{c}8\\\text{pounds}\end{array}\right)}^{\$15 \text{ seed}} = \overbrace{\left(\begin{array}{c}\$12 \text{ per}\\\text{pound}\end{array}\right)\left(\begin{array}{c}20\\\text{pounds}\end{array}\right)}^{\$12 \text{ seed}}$$

$$\$120 + \$120 = \$240$$

Study Tip

When you set up a verbal model, be sure to check that you are working with the *same type of units* in each part of the model. In Example 8, for instance, note that each of the three parts of the verbal model measures cost. (If two parts measured cost and the other part measured pounds, you would know that the model was incorrect.)

Example 9 ■ A Solution Mixture Problem

A chemist needs to strengthen a 20% alcohol solution with a 50% solution to obtain a 30% solution. How much of the 50% solution should be added to 240 milliliters of the 20% solution (see Figure 3.21)?

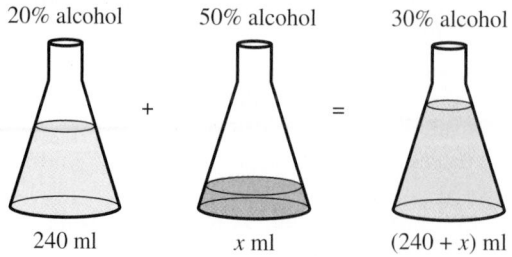

| 20% alcohol | 50% alcohol | 30% alcohol |

240 ml x ml $(240 + x)$ ml

Figure 3.21

Solution

Verbal Model:

$$\boxed{\text{Amount of 20\% alcohol solution}} + \boxed{\text{Amount of 50\% alcohol solution}} = \boxed{\text{Amount of final alcohol solution}}$$

Labels: 20% solution: Percent alcohol = 0.20   (decimal form)
                        Amount of alcohol solution = 240   (milliliters)
            50% solution: Percent alcohol = 0.50   (decimal form)
                        Amount of alcohol solution = x (milliliters)
            Final solution: Percent alcohol = 0.30   (decimal form)
                        Amount of alcohol solution = $240 + x$ (milliliters)

Equation: $0.20(240) + 0.50(x) = 0.30(240 + x)$

$$48 + 0.50x = 72 + 0.30x$$

$$0.20x = 24$$

$$x = \frac{24}{0.20}$$

$$= 120 \text{ milliliters}$$

So, the chemist should add 120 milliliters of the 50% solution to the 20% solution. Check this solution in the original problem.

In Example 9, the original solution was strengthened by adding a 50% alcohol solution. To dilute the solution, you could add a solution containing 0% alcohol. How much solution containing 0% alcohol would you need to add to the original solution to dilute it to a 15% solution? You can determine the answer, 80 milliliters, by solving the equation as follows.

$$0.20(240) + 0.0(x) = 0.15(240 + x)$$

$$48 + 0 = 36 + 0.15x$$

$$12 = 0.15x$$

$$80 = x$$

Mixture problems can also involve a "mix" of investments, as shown in the next example.

Example 10 ■ Investment Mixture

You invested a total of $10,000 at $4\frac{1}{2}\%$ and $5\frac{1}{2}\%$ simple interest. During 1 year, the two accounts earned $508.75. How much did you invest in each?

Solution

Verbal Model:

$$\boxed{\text{Interest earned from } 4\frac{1}{2}\%} + \boxed{\text{Interest earned from } 5\frac{1}{2}\%} = \boxed{\begin{array}{c}\text{Total interest}\\\text{earned}\end{array}}$$

Labels:

Amount invested at $4\frac{1}{2}\% = x$	(dollars)
Amount invested at $5\frac{1}{2}\% = 10{,}000 - x$	(dollars)
Interest earned from $4\frac{1}{2}\% = (x)(0.045)(1)$	(dollars)
Interest earned from $5\frac{1}{2}\% = (10{,}000 - x)(0.055)(1)$	(dollars)
Total interest earned $= 508.75$	(dollars)

Equation:

$$0.045x + 0.055(10{,}000 - x) = 508.75$$
$$0.045x + 550 - 0.055x = 508.75$$
$$550 - 0.01x = 508.75$$
$$-0.01x = -41.25$$
$$x = 4125$$

So, you invested $4125 at $4\frac{1}{2}\%$ and $10{,}000 - x = 10{,}000 - 4125 = \$5875$ at $5\frac{1}{2}\%$. Check this in the original statement of the problem.

Collaborate!

Communicating Mathematically

Use the information provided in the statement below to write a mathematical formula for the 10-second pulse count.

"The target heart rate is the heartbeat rate a person should have during aerobic exercise to get the full benefit of the exercise for cardiovascular conditioning. . . . Using the American College of Sports Medicine method to calculate one's target heart rate, an individual should subtract his or her age from 220, then multiply by the desired intensity level (as a percent—sedentary persons may want to use 60% and highly fit individuals may want to use 85% to 95%) of the workout. Then divide the answer by 6 for a 10-second pulse count. (The 10-second pulse count is useful for checking whether the target heart rate is being achieved during the workout. One can easily check one's pulse—at the wrist or side of the neck—by counting the number of beats in 10 seconds.)" *(Source: Aerobic Fitness Association of America)*

Use the formula you have written to find your own 10-second pulse count.

3.3 ■ Exercises

Developing Skills

In Exercises 1–4, determine the unknown distance, rate, or time.

Distance, d	Rate, r	Time, t
1.	650 mi/hr	$3\frac{1}{2}$ hr
2. 1000 km	110 km/hr	

Distance, d	Rate, r	Time, t
3. 250 ft	32 ft/sec	
4. 1000 ft		$\frac{3}{2}$ sec

Solving Problems

5. *Consumer Awareness*   A department store is offering a discount of 20% on a sewing machine with a list price of $279.95. A mail-order catalog has the same machine for $228.95 plus $4.32 for shipping. Which is the better buy?

6. *Consumer Awareness*   A hardware store is offering a discount of 15% on a 4000-watt generator with a list price of $699.99. A mail-order company has the same generator for $549.95 plus $14.32 for shipping. Which is the better buy?

7. *Labor*   An auto repair bill of $216.37 lists $136.37 for parts and the rest for labor. The labor rate is $32 per hour. How many hours did it take to repair the auto?

8. *Labor*   An appliance repair store charges $35 for the first half hour of a service call. For each additional half hour of labor there is a charge of $18. Find the length of a service call for which the charge is $89.

9. *Long Distance Phone Calls*   The weekday rate for a telephone call is $0.75 for the first minute plus $0.55 for each additional minute. Determine the length of a call that costs $5.15. What would have been the cost of the call if it had been made during the weekend when there is a 60% discount?

10. *Overtime*   Last week you earned $740. If you are paid $14.50 per hour for the first 40 hours and $20 for each hour over 40, how many hours of overtime did you work?

11. *Tip Rate*   A customer left a total of $10 for a meal that costs $8.45. Determine the tip rate.

12. *Tip Rate*   A customer left a total of $40 for a meal that costs $34.73. Determine the tip rate.

13. *Commission Rate*   Determine the commission rate for an employee who earned $450 in commissions for sales of $5000.

14. *Commission Rate*   Determine the commission rate for an employee who earned $1014 in commissions for sales of $8450.

15. *Weekly Pay*   The weekly salary of an employee is $250 plus a 6% commission on the employee's total sales. Find the weekly pay for a week in which the sales are $5500.

16. *Weekly Pay*   The weekly salary of an employee is $325 plus a 9% commission on the employee's total sales. Find the weekly pay for a week in which the sales are $4050.

17. *Insurance*   The annual insurance premium for a policyholder is $862. Find the amount of a 20% surcharge the policyholder had to pay because of an accident. Find the total amount of the annual premium with the surcharge.

18. *Amount Financed*   A customer bought a lawn tractor that cost $4450 plus 6% sales tax. Find the amount of the sales tax and the total bill. Find the amount financed if a down payment of $1000 was made.

19. *Temperature*   The average daily temperature in Sacramento, California is 60.8°F. What is Sacramento's average daily temperature in degrees Celsius? *(Source: U.S. National Oceanic and Atmospheric Administration)*

20. *Temperature*   The average daily temperature in Hartford, Connecticut is 9.9°C. What is Hartford's average daily temperature in degrees Fahrenheit? *(Source: U.S. National Oceanic and Atmospheric Administration)*

21. *Volume*   Find the volume of a circular cylinder with a radius of 3.5 centimeters and a height of 12 centimeters.

22. *Volume*   Find the volume of a sphere with a radius of 6 meters.

23. *Average Speed*   An Olympic runner completes a 5000-meter race in 13 minutes and 20 seconds. What is the average speed of the runner?

24. *Travel Time*   Determine the time for the space shuttle to travel a distance of 5000 miles in orbit when its average speed is 17,000 miles per hour (see figure).

25. *Speed of Light*   Determine the time for light to travel from the sun to Earth. The distance between the sun and Earth is 93,000,000 miles and the speed of light is 186,282.369 miles per second (see figure).

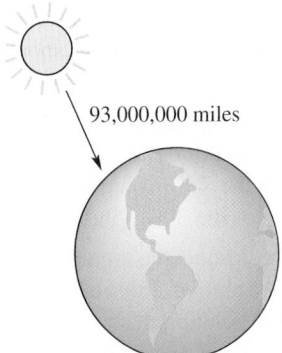

93,000,000 miles

26. *Speed of Sound*   A window is broken 500 feet away from you. You hear the sound of the breaking glass 0.448 second after it happened. Determine the speed of sound.

27. *Distance*   Two planes leave an airport at approximately the same time and fly in opposite directions. How far apart are the planes after $1\frac{1}{3}$ hours if their speeds are 480 miles per hour and 600 miles per hour?

28. *Distance*   Two trains leave the same city at the same time and travel in opposite directions. How far apart are the trains after $2\frac{1}{2}$ hours if their speeds are 120 miles per hour and 150 miles per hour?

29. *Distance*   Two people walk from point A to point B. The first person walks at a rate of 3 miles per hour. The second person leaves an hour after the first person, and walks at a rate of 4 miles per hour. How long will it take the second person to catch up to the first person?

30. *Distance*   One jogger begins his run on a path and jogs at a rate of 6 miles per hour. A second jogger starts at the same point of the path as the first jogger, but jogs at a rate of 6.5 miles per hour. The second jogger begins 15 minutes after the first jogger. How long will it take the second jogger to catch up to the first jogger?

31. *Distance*   A bus leaves a city at 12:00 P.M. and travels 50 miles per hour. A car leaves the same city at 12:30 P.M. and travels the same route as the bus, but travels at a rate of 60 miles per hour. What time will the car catch up to the bus?

32. *Distance*   A boat leaves a harbor and sails east at a speed of 18 miles per hour. A second boat leaves the same harbor and sails east at a speed of 28 miles per hour. The first boat leaves the harbor at 2:00 P.M. and the second boat leaves at 3:00 P.M. What time will the second boat catch up to the first boat?

33. *Travel Time*   On the first part of a 317-mile trip, a sales representative averaged 58 miles per hour. The sales representative averaged only 52 miles per hour on the last part of the trip because of an increased volume of traffic (see figure). Find the amount of driving time at each speed if the total time was 5 hours and 45 minutes.

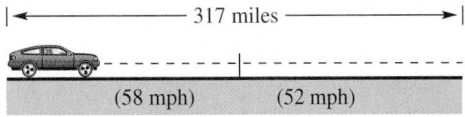

|◄──────── 317 miles ────────►|

(58 mph)   (52 mph)

34. *Travel Time*   On the first part of a 280-mile trip, a sales representative averaged 63 miles per hour. The sales representative averaged only 54 miles per hour on the last part of the trip because of an increased volume of traffic.

(a) Express the total time for the trip as a function of the distance x traveled at an average speed of 63 miles per hour.

(b) Use a graphing utility to graph the time function. What is the domain of the function?

(c) Approximate the number of miles traveled at 63 miles per hour if the total time was 4 hours and 45 minutes.

35. *Work-Rate Problem* You can complete a typing project in 5 hours, and a friend estimates that it would take him 8 hours. What fractional part of the project can be completed by each typist in 1 hour? If you both work on the project, in how many hours can it be completed?

36. *Work-Rate Problem* It takes 30 minutes for a pump to remove an amount of water from a basement. A larger pump can remove the same amount of water in half the time. If both pumps were operating, how long would it take to remove the water?

37. *Work-Rate Problem* You can mow a lawn in 3 hours, and your friend can mow it in 4 hours. How long will it take both of you to mow the lawn working together?

38. *Work-Rate Problem* You can paint a room in 6 hours, and your friend can paint it in 8 hours. How long will it take both of you to paint the room?

39. *Work-Rate Problem* It takes you and your sister 3 hours to clean the entire house when you work together. If you can do the job alone in 4 hours, how long would it take your sister to do it by herself?

40. *Work-Rate Problem* It takes 90 minutes to remove an amount of water from a basement when two pumps of different sizes are operating. It takes 2 hours to remove the same amount of water when only the larger pump is running. How long would it take to remove the same amount if only the smaller pump were running?

41. *Number of Stamps* You have a set of 70 stamps with a total value of $20.96. If the set includes 20¢ stamps and 32¢ stamps, find the number of each type.

42. *Coin Problem* A person has 50 coins in dimes and quarters with a combined value of $10.25. Determine the number of coins of each type.

43. *Ticket Sales* Ticket sales for a play total $2200. There are three times as many adult tickets sold as child tickets, and the prices of the tickets for adults and children are $6 and $4. Find the number of child tickets sold.

44. *Nut Mixture* A grocer mixes two kinds of nuts that cost $3.88 per pound and $4.88 per pound, to make 100 pounds of a mixture that costs $4.13 per pound. How many pounds of each kind of nut are in the mixture?

45. *Simple Interest* You invested $6000 at $4\frac{1}{2}$% and $5\frac{1}{2}$% simple interest. During 1 year, the two accounts earned $305. How much did you invest in each account?

46. *Simple Interest* You invested $12,000 at $10\frac{1}{2}$% and 13% simple interest. During 1 year, the two accounts earned $1447.50. How much did you invest in each account?

47. *Simple Interest* An inheritance of $40,000 is divided into two investments earning 8% and 10% simple interest. (There is more risk in the 10% fund.) Your objective is to obtain a total annual interest income of $3500 from the investments.

    (a) Find the interest accumulated after 3 years on $40,000 for each interest rate.

    (b) What is the smallest amount you can invest at 10% in order to meet your objective?

48. *Simple Interest* An amount of $4000 is divided into two investments earning $6\frac{1}{2}$% and 9% simple interest. (There is more risk in the 9% fund.) Your objective is to obtain a total annual interest income of $300 from the investments.

    (a) Find the interest accumulated after 3 years on $4000 for each interest rate.

    (b) What is the smallest amount you can invest at 9% in order to meet your objective?

49. *Mixture Problem* A chemist needs to strengthen a 25% solution with a 60% solution to obtain a 30% solution. How much of the 60% solution should be added to 300 milliliters of the 25% solution?

50. *Mixture Problem* A chemist needs to strengthen a 30% solution with a 75% solution to obtain a 40% solution. How much of the 75% solution should be added to 280 milliliters of the 30% solution?

51. *Mixture Problem* Determine the number of gallons of a 20% alcohol solution and the number of gallons of a 60% alcohol solution that are required to make 100 gallons of a 40% alcohol solution.

52. *Mixture Problem* Determine the number of liters of a 50% alcohol solution and the number of liters of a 75% alcohol solution that are required to make 10 liters of a 60% alcohol solution.

53. *Mixture Problem* Determine the number of quarts of a 15% concentrated solution and the number of quarts of a 60% concentrated solution that are required to make 24 quarts of a 45% concentrated solution.

54. *Mixture Problem* Determine the number of gallons of a 60% concentrated solution and the number of gallons of an 80% concentrated solution that are required to make 55 gallons of a 75% concentrated solution.

55. *Mixture Problem*   The cooling system on a truck contains 5 gallons of coolant that is 40% antifreeze. How much must be withdrawn and replaced with 100% antifreeze to bring the coolant in the system to 50% antifreeze?

56. *Mixture Problem*   You mix gasoline and oil to obtain $2\frac{1}{2}$ gallons of mixture for an engine. The mixture is 40 parts gasoline and one part two-cycle oil. How much gasoline must be added to bring the mixture to 50 parts gasoline and 1 part oil?

57. *Politics*   An opinion poll was conducted to determine the public's preference of three political candidates. The poll was administered to 1320 people. Candidate B was preferred by twice the number of people as candidate A. Candidate C was preferred by three times as many people as A. Determine the number of people in the sample that preferred candidate A.

58. *Politics*   Fourteen hundred people were surveyed in an opinion poll. Political candidates A and B received approximately the same preference, but candidate C was preferred by twice the number of people as either A or B. Determine the number of people in the sample that preferred candidate C.

59. *Height of a Picture Frame*   A rectangular picture frame has a perimeter of 3 feet. The width of the frame is 0.62 times its height. Find the height of the frame.

60. *Area of a Square*   The figure shows three squares. The perimeter of square I is 12 inches and the area of square II is 36 square inches. Find the area of square III.

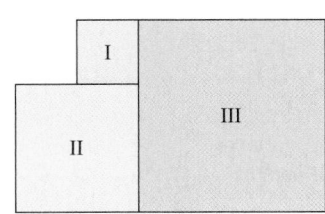

61. *Average Wage*   The average hourly wage for construction workers in the United States between 1990 and 1999 can be approximated by the linear model

$$y = 13.46 + 0.371t, \ 0 \le t \le 9$$

where *y* represents the hourly wage (in dollars) and *t* represents the year, with $t = 0$ corresponding to 1990 (see figure). *(Source: U. S. Bureau of Labor Statistics)*

(a) During which year was the average hourly wage $14.38?

(b) What was the average annual hourly raise for construction workers during this 10-year period? Explain how you determined your answer.

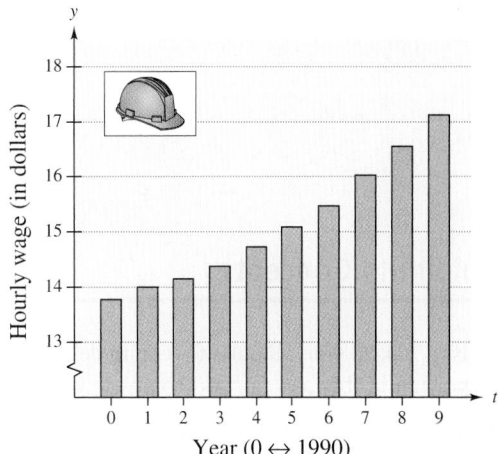

Figure for 61

62. *Average Wage*   The average hourly wage for workers in retail trade in the United States between 1990 and 1999 can be approximated by the linear model

$$y = 6.60 + 0.255t, \ 0 \le t \le 9$$

where *y* represents the hourly wage (in dollars) and *t* represents the year, with $t = 0$ corresponding to 1990 (see figure). *(Source: U.S. Bureau of Labor Statistics)*

(a) During which year was the average hourly wage $7.49?

(b) What was the average annual hourly raise for retail workers during this 10-year period? Explain how you determined your answer.

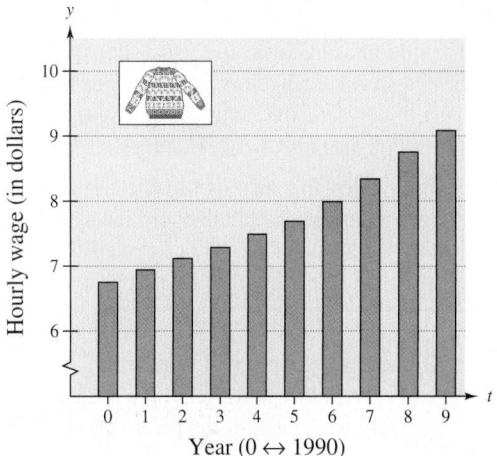

63. *Comparing Wage Increases*   Use the information given in Exercises 61 and 62 to determine which of the two groups' average wages was increasing at a greater annual rate during the 10-year period from 1990 to 1999. Explain.

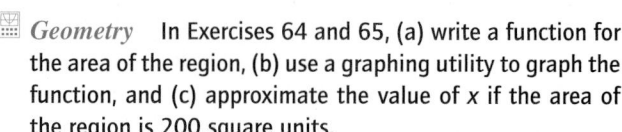 *Geometry*   In Exercises 64 and 65, (a) write a function for the area of the region, (b) use a graphing utility to graph the function, and (c) approximate the value of *x* if the area of the region is 200 square units.

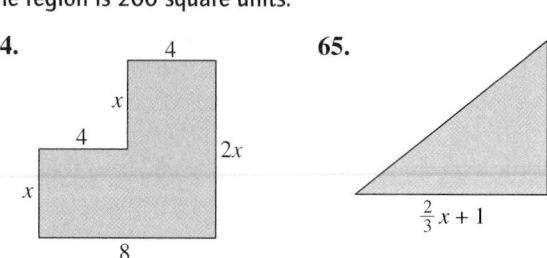

64.

65.

Explaining Concepts

66. If it takes you *t* hours to complete a task, what portion of the task can you complete in 1 hour?

67. If the sides of a square are doubled, does the perimeter double? Explain.

68. If the sides of a square are doubled, does the area double? Explain.

Ongoing Review

In Exercises 69–76, solve the equation.

69. $32 - 20x = 0$

70. $15 + 3x = 2$

71. $4(x + 7) - 1 = 0$

72. $8 - 2(x + 3) = 5$

73. $\dfrac{2x}{5} - \dfrac{x}{2} = 4$

74. $\dfrac{x}{8} + \dfrac{3x}{4} = 6$

75. $\dfrac{5x}{9} + 3 = \dfrac{x}{2} - 1$

76. $\dfrac{2x}{3} - 4 = \dfrac{2x}{5} + 6$

77. *Fuel Usage*   A tractor uses 5 gallons of diesel fuel to plow for 105 minutes. Assuming conditions remain the same, determine the number of gallons of fuel used in 6 hours.

78. *Physics*   A force of 32 pounds stretches a spring 6 inches. Determine the number of pounds of force required to stretch the spring 9 inches.

Looking Further

Feed Mixture   A rancher must purchase 500 bushels of a feed mixture for cattle and is considering oats and corn, which cost $2.70 and $4.00 per bushel, respectively. Complete the table at the right, where *x* is the number of bushels of oats in the mixture.

(a) How does the increase in the number of bushels of oats affect the number of bushels of corn in the mixture?

(b) How does the increase in the number of bushels of oats affect the price per bushel of the mixture?

(c) If there were an equal number of bushels of oats and corn in the mixture, how would the price of the mixture be related to the price of each component?

Oats x	Corn $500 - x$	Price/bushel of the mixture
0		
100		
200		
300		
400		
500		

Mid-Chapter Quiz

Take this quiz as you would take a quiz in class. After you are done, check your work against the answers given in the back of the book.

In Exercises 1–4, write an equation of the line passing through the point with the indicated slope.

1. $\left(0, -\frac{3}{2}\right)$

$m = 2$

2. $(4, 7)$

$m = \frac{1}{2}$

3. $\left(\frac{5}{2}, 6\right)$

$m = -\frac{3}{4}$

4. $(-3.5, -1.8)$

$m = 3$

In Exercises 5–8, write an equation of the line passing through the points.

5. $(2, 1), (4, 5)$

6. $(0, 0.8), (3, -2.3)$

7. $(3, -1), (10, -1)$

8. $\left(4, \frac{5}{3}\right), (4, 8)$

9. Write an equation of the line that passes through the point $(3, 5)$ and is parallel to the line given by $2x - 3y = 1$.

10. Write an equation of the line that passes through the point $(-1, 4)$ and is perpendicular to the line given by $6x + 8y = -9$.

11. A company produces a product for which the variable cost is $5.60 and the fixed costs are $24,000. The product sells for $9.20. Write the profit P as a linear function of x.

12. Translate the statement below into an algebraic expression.

"The product of a number n and 5 is decreased by 8."

13. The length of a rectangle is l feet. The width is 1 foot less than 0.6 times the length. Write algebraic expressions for the perimeter and area of the rectangle shown at the left. Simplify the expressions.

14. Write an algebraic expression for the sum of three consecutive even integers, the first of which is n.

15. A quality control engineer for a manufacturer finds one defective unit in a sample of 300. At this rate, what is the expected number of defective units in a shipment of 600,000?

16. Fifty gallons of a 30% acid solution is obtained by combining solutions that are 25% acid and 50% acid. How much of each solution is required?

17. You can paint a room in 3 hours, and your friend can paint it in 4 hours. How long will it take both of you to paint the room?

18. Two cars start at a given point and travel in the same direction at average speeds of 40 miles per hour and 55 miles per hour. How much time must elapse before the cars are 5 miles apart?

$0.6l - 1$

l

Figure for 13

3.4 Linear Inequalities in One Variable

What you should learn:

- How to sketch the graphs of inequalities
- How to solve linear inequalities in one variable
- How to solve compound inequalities
- How to solve application problems involving inequalities

Why you should learn it:

Linear inequalities can be used to model and solve real-life problems. For example, writing an inequality in Exercise 97 on page 212 allows you to determine a payment plan that gives a greater hourly wage.

Historical Note

The first recorded use of the inequality symbols < and > is in *Artis Analyticae Praxis*, published in 1631 and written by the English mathematician Thomas Harriot (1560–1621).

Intervals on the Real Line

In this section you will study **algebraic inequalities,** which are inequalities that contain one or more variable terms. Some examples are

$$x \leq 4, \qquad x \geq -3, \qquad x + 2 < 7, \qquad \text{and} \qquad 4x - 6 < 3x + 8.$$

As with an equation, you **solve** an inequality in the variable x by finding all values of x for which the inequality is true. Such values are called **solutions** and are said to **satisfy** the inequality. The set of all solutions of an inequality is the **solution set** of the inequality. The **graph** of an inequality is obtained by plotting its solution set on the real number line. Often, these graphs are intervals—either bounded or unbounded.

Bounded Intervals on the Real Number Line

Let a and b be real numbers such that $a < b$. The intervals on the real number line listed below are called **bounded intervals.** The numbers a and b are the **endpoints** of each interval.

Notation	Interval Type	Inequality	Graph
$[a, b]$	Closed	$a \leq x \leq b$	
(a, b)	Open	$a < x < b$	
$[a, b)$	—	$a \leq x < b$	
$(a, b]$	—	$a < x \leq b$	

NOTE In the list above, note that a closed interval contains both of its endpoints and an open interval does not contain either of its endpoints.

The **length** of the interval $[a, b]$ is the distance between its endpoints: $b - a$. For instance, the length of the interval $[3, 8]$ is

$$b - a = 8 - 3$$
$$= 5.$$

The lengths of $[a, b]$, (a, b), $[a, b)$, and $(a, b]$ are the same. The reason that these four types of intervals are called bounded is that each has a finite length. An interval that *does not* have a finite length is **unbounded** (or **infinite**).

Unbounded Intervals on the Real Number Line

Let a and b be real numbers. The intervals on the real number line listed below are called **unbounded intervals.**

Notation	Interval Type	Inequality	Graph
$[a, \infty)$	—	$x \geq a$	
(a, ∞)	Open	$x > a$	
$(-\infty, b]$	—	$x \leq b$	
$(-\infty, b)$	Open	$x < b$	
$(-\infty, \infty)$	Entire real line		

> **NOTE**   The symbols ∞ (**positive infinity**) and $-\infty$ (**negative infinity**) do not represent real numbers. They are simply convenient symbols used to describe the unboundedness of an interval such as $(1, \infty)$.

Example 1 ■ Graphs of Inequalities

Sketch the graph of each inequality.

(a) $-3 < x \leq 1$ (b) $0 < x < 2$ (c) $-3 < x$ (d) $x \leq 2$

Solution

(a) The graph of $-3 < x \leq 1$ is a bounded interval.

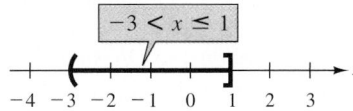

(b) The graph of $0 < x < 2$ is a bounded interval.

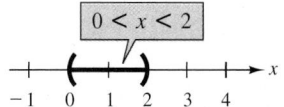

(c) The graph of $-3 < x$ is an unbounded interval.

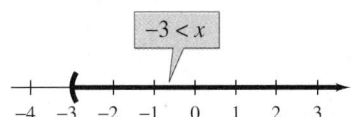

(d) The graph of $x \leq 2$ is an unbounded interval.

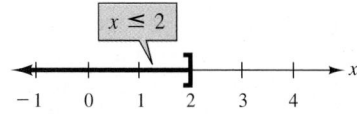

> **NOTE**   In Example 1(c), the inequality $-3 < x$ can also be written as $x > -3$. In other words, saying "-3 is less than x" is the same thing as saying "x is greater than -3."

Solving a Linear Inequality

An inequality in one variable is **linear** if it can be written in one of the forms shown below.

$$ax + b \leq 0, \qquad ax + b < 0, \qquad ax + b \geq 0, \qquad ax + b > 0$$

Solving a linear inequality is much like solving a linear equation. To isolate the variable, you make use of **properties of inequalities.** These properties are similar to the properties of equality, but there are two important exceptions. When both sides of an inequality are multiplied or divided by a negative number, the direction of the inequality symbol must be reversed. Here is an example.

$-2 < 5$	Write original inequality.
$(-3)(-2) > (-3)(5)$	Multiply each side by -3 and reverse the inequality symbol.
$6 > -15$	Simplify.

Two inequalities that have the same solution set are **equivalent inequalities.** The list below describes operations that can be used to create equivalent inequalities.

Properties of Inequalities

1. *Addition and Subtraction Properties*

   Adding the same quantity to, or subtracting the same quantity from, each side of an inequality produces an equivalent inequality.

   If $a < b$, then $a + c < b + c$.

   If $a < b$, then $a - c < b - c$.

2. *Multiplication and Division Properties: Positive Quantities*

   Multiplying or dividing each side of an inequality by a *positive* quantity produces an equivalent inequality.

   If $a < b$ and c is positive, then $ac < bc$.

   If $a < b$ and c is positive, then $\dfrac{a}{c} < \dfrac{b}{c}$.

3. *Multiplication and Division Properties: Negative Quantities*

   Multiplying or dividing each side of an inequality by a *negative* quantity produces an equivalent inequality in which the inequality symbol is reversed.

   If $a < b$ and c is negative, then $ac > bc$.   Reverse inequality.

   If $a < b$ and c is negative, then $\dfrac{a}{c} > \dfrac{b}{c}$.   Reverse inequality.

4. *Transitive Property*

   Consider three quantities of which the first is less than the second, and the second is less than the third. It follows that the first quantity must be less than the third quantity.

   If $a < b$ and $b < c$, then $a < c$.

NOTE   These properties remain true if the symbols $<$ and $>$ are replaced by $\leq$ and $\geq$. Moreover, a, b, and c can represent real numbers, variables, or expressions. Note that you cannot multiply or divide each side of an inequality by zero.

As you study the following examples, pay special attention to the steps in which the inequality symbol is reversed. Remember that when you multiply or divide an inequality by a negative number, you must reverse the inequality symbol.

Example 2　■　Solving a Linear Inequality

Solve the inequality $x + 6 < 9$.

Algebraic Solution

$$x + 6 < 9 \qquad \text{Write original inequality.}$$
$$x + 6 - 6 < 9 - 6 \qquad \text{Subtract 6 from each side.}$$
$$x < 3 \qquad \text{Simplify.}$$

The solution set consists of all real numbers that are less than 3. The interval notation for the solution set is $(-\infty, 3)$. The graph is shown in Figure 3.22.

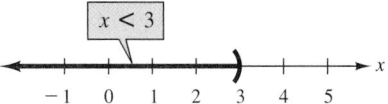

Figure 3.22

Graphical Solution

Begin by sketching the graphs of $y = x + 6$ and $y = 9$ on the same set of coordinate axes, as shown in Figure 3.23. In the figure, notice that the graphs intersect at $x = 3$. The graph of $y = x + 6$ lies below the line $y = 9$ for all values of x less than 3. So, you can conclude that $x + 6 < 9$ for all x such that $x < 3$.

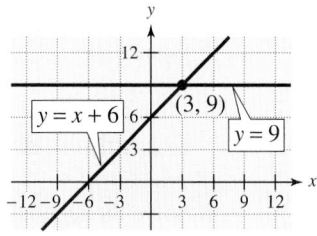

Figure 3.23

Study Tip

Checking the solution set of an inequality is not as simple as checking the solution set of an equation. (There are usually too many x-values to substitute back into the original inequality.) You can, however, get an indication of the validity of a solution set by substituting a few convenient values of x. For instance, in Example 2, try checking that $x = 0$ satisfies the original inequality, whereas $x = 4$ does not.

Example 3　■　Solving a Linear Inequality

Solve the inequality $8 - 3x \leq 20$.

Solution

$$8 - 3x \leq 20 \qquad \text{Write original inequality.}$$
$$8 - 8 - 3x \leq 20 - 8 \qquad \text{Subtract 8 from each side.}$$
$$-3x \leq 12 \qquad \text{Simplify.}$$
$$x \geq -4 \qquad \text{Divide each side by } -3 \text{ and reverse the inequality symbol.}$$

The solution set consists of all real numbers that are greater than or equal to -4. The interval notation for the solution set is $[-4, \infty)$. The graph is shown in Figure 3.24.

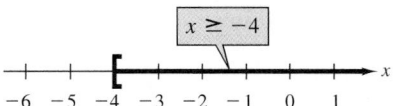

Figure 3.24

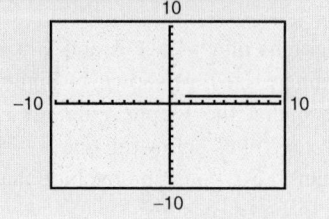
Example 4 ■ Solving a Linear Inequality

Solve the inequality $5x + 2 < 9x - 4$.

Solution

$5x + 2 < 9x - 4$	Write original inequality.
$-4x + 2 < -4$	Subtract $9x$ from each side.
$-4x < -6$	Subtract 2 from each side.
$x > \dfrac{3}{2}$	Divide each side by -4 and reverse the inequality symbol.

The solution set consists of all real numbers that are greater than $\frac{3}{2}$. The interval notation for the solution set is $\left(\frac{3}{2}, \infty\right)$. The graph of this inequality is an unbounded interval, as shown in Figure 3.25.

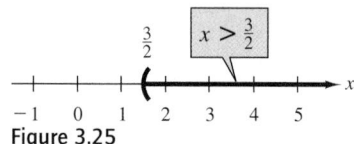

Figure 3.25

Example 5 ■ Solving a Linear Inequality

Solve the inequality $\dfrac{2x}{3} + 12 < \dfrac{x}{6} + 18$.

Solution

$\dfrac{2x}{3} + 12 < \dfrac{x}{6} + 18$	Write original inequality.
$6 \cdot \left(\dfrac{2x}{3} + 12\right) < 6 \cdot \left(\dfrac{x}{6} + 18\right)$	Multiply each side by LCD of 6.
$6 \cdot \dfrac{2x}{3} + 6 \cdot 12 < 6 \cdot \dfrac{x}{6} + 6 \cdot 18$	Distributive Property
$4x + 72 < x + 108$	Simplify.
$3x + 72 < 108$	Subtract x from each side.
$3x < 36$	Subtract 72 from each side.
$x < 12$	Divide each side by 3.

The solution set consists of all real numbers that are less than 12. The interval notation for the solution set is $(-\infty, 12)$. The graph is shown in Figure 3.26.

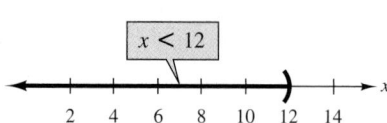

Figure 3.26

Study Tip

When one or more of the terms in a linear inequality involve constant denominators, it helps to multiply each side of the inequality by the least common denominator. This clears the inequality of fractions, as shown in Example 5.

Solving a Compound Inequality

Two inequalities joined by the word *and* or the word *or* constitute a **compound inequality.** When two inequalities are joined by the word *and*, the solution set consists of all real numbers that satisfy both inequalities. The solution set for the compound inequality $-4 \leq 5x - 2$ *and* $5x - 2 < 7$ can be written more simply as the **double inequality**

$$-4 \leq 5x - 2 < 7.$$

A compound inequality formed by the word *and* is called **conjunctive** and is the only kind that has the potential to form a double inequality. A compound inequality joined by the word *or* is called **disjunctive** and cannot be re-formed into a double inequality.

NOTE   When performing an operation on a double inequality, you must apply the operation to *all* three parts of the inequality, as shown in Example 6.

Example 6 ■ Solving a Double Inequality

Write the inequalities $-7 \leq 5x - 2$ and $5x - 2 < 8$ as a double inequality. Then solve the double inequality to find the set of all real numbers that satisfy *both* inequalities.

Solution

$$-7 \leq 5x - 2 < 8 \qquad \text{Write original inequality.}$$

$$-7 + 2 \leq 5x - 2 + 2 < 8 + 2 \qquad \text{Add 2 to all three parts.}$$

$$-5 \leq 5x < 10 \qquad \text{Simplify.}$$

$$\frac{-5}{5} \leq \frac{5x}{5} < \frac{10}{5} \qquad \text{Divide all three parts by 5.}$$

$$-1 \leq x < 2 \qquad \text{Simplify.}$$

The solution set consists of all real numbers that are greater than or equal to -1 and less than 2. The interval notation for the solution set is $[-1, 2)$. The graph is shown in Figure 3.27.

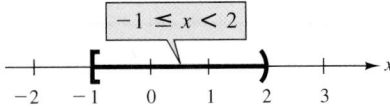

Figure 3.27

The double inequality in Example 6 could have been solved in two parts as follows.

$$-7 \leq 5x - 2 \qquad \text{and} \qquad 5x - 2 < 8$$
$$-5 \leq 5x \qquad\qquad\qquad 5x < 10$$
$$-1 \leq x \qquad\qquad\qquad x < 2$$

The solution set consists of all real numbers that satisfy *both* inequalities. In other words, the solution set is the set of all values of x for which $-1 \leq x < 2$.

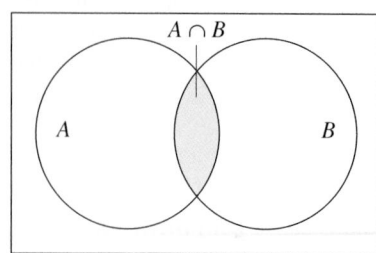

Intersection of two sets

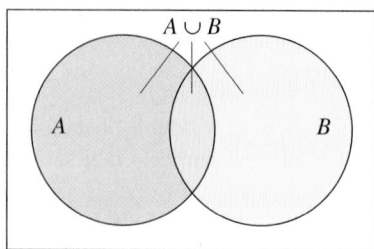

Union of two sets
Figure 3.28

Compound inequalities can be written using *symbols*. For compound inequalities, the word *and* is represented by the symbol ∩, which is read as **intersection.** The word *or* is represented by the symbol ∪, which is read as **union.** A graphical representation is shown in Figure 3.28. If *A* and *B* are sets, then x is in $A \cap B$ if it is in both *A and B*, and x is in $A \cup B$ if it is in *A or B* or possibly both.

Example 7 ■ Writing a Solution Set Using Union

A solution set is shown on the number line in Figure 3.29.

(a) Write the solution set as a compound inequality.

(b) Write the solution set using the union symbol.

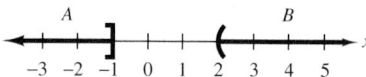

Figure 3.29

Solution

(a) As a compound inequality, the solution set can be written as $x \leq -1$ *or* $x > 2$.

(b) Using the union symbol, you can write the entire solution set as $A \cup B$.

Example 8 ■ Writing a Solution Set Using Intersection

A solution set is shown on the number line in Figure 3.30.

(a) Write the solution set as a compound inequality.

(b) Write the solution set using the intersection symbol.

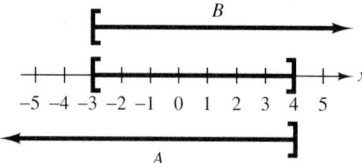

Figure 3.30

Solution

(a) As a compound inequality, the solution set can be written as $x \leq 4$ *and* $x \geq -3$.

(b) Using the intersection symbol, you can write the solution set as $A \cap B$.

Example 9 ■ Solving a Conjunctive Inequality

Solve the compound inequality

$$3x + 5 \geq 8 \text{ and } 2x - 3 < 5.$$

Solution

$$
\begin{array}{ll}
3x + 5 \geq 8 & \text{and} \quad 2x - 3 < 5 \\
3x \geq 3 & \qquad 2x < 8 \\
\dfrac{3x}{3} \geq \dfrac{3}{3} & \qquad \dfrac{2x}{2} < \dfrac{8}{2} \\
x \geq 1 & \qquad x < 4
\end{array}
$$

The solution set is $x \geq 1$ and $x < 4$. This solution set can be written as $1 \leq x < 4$. The graph of the solution set is shown in Figure 3.31.

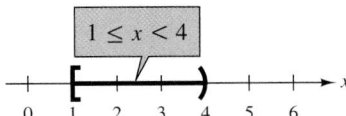

Figure 3.31

Example 10 ■ Solving a Disjunctive Inequality

Solve the compound inequality

$$10 - 6x \leq 2 \quad \text{or} \quad -3x + 6 \geq 7.$$

Solution

$$
\begin{array}{ll}
10 - 6x \leq 2 & \text{or} \qquad -3x + 6 \geq 7 \\
10 - 10 - 6x \leq 2 - 10 & \quad -3x + 6 - 6 \geq 7 - 6 \\
-6x \leq -8 & \qquad -3x \geq 1 \\
\dfrac{-6x}{-6} \geq \dfrac{-8}{-6} & \qquad \dfrac{-3x}{-3} \leq \dfrac{1}{-3} \\
x \geq \dfrac{4}{3} & \qquad x \leq -\dfrac{1}{3}
\end{array}
$$

The solution set is $x \leq -\frac{1}{3}$ or $x \geq \frac{4}{3}$. The graph of the solution set is shown in Figure 3.32.

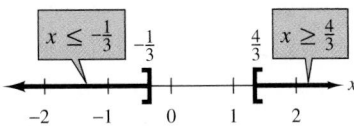

Figure 3.32

Applications

Linear inequalities in real-life problems arise from statements that involve phrases such as "at least," "no more than," "minimum value," and so on.

Example 11 ■ Translating Verbal Statements

Verbal Statement	*Inequality*
(a) x is at most 3.	$x \leq 3$
(b) x is no more than 3.	$x \leq 3$
(c) x is at least 3.	$x \geq 3$
(d) x is more than 3.	$x > 3$
(e) x is less than 3.	$x < 3$
(f) x is a minimum of 3.	$x \geq 3$
(g) x is no less than 3.	$x \geq 3$
(h) x is at least 2, but less than 7.	$2 \leq x < 7$
(i) x is greater than 2, but no more than 7.	$2 < x \leq 7$

> **NOTE**　In Example 11(a), note that "at most" means "less than or equal to." Similarly, in Example 11(c), note that "at least" means "greater than or equal to."

To solve real-life problems involving inequalities, you can use the same "verbal-model approach" you use with equations.

Example 12 ■ Finding the Maximum Width of a Package

An overnight delivery service will not accept any package whose combined length and girth (perimeter of a cross section) exceeds 132 inches. Suppose that you are sending a rectangular package that has square cross sections. If the length of the package is 68 inches, what is the maximum width of the sides of its square cross sections?

Solution

Begin by making a sketch. In Figure 3.33, notice that the length of the package is 68 inches, and each side is x inches wide.

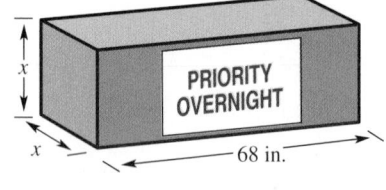

Figure 3.33

Verbal Model:　Length　+　Girth　$\leq$　132 inches

Labels:	Width of a side $= x$	(inches)
	Length $= 68$	(inches)
	Girth $= 4x$	(inches)

Inequality:　$68 + 4x \leq 132$

$$4x \leq 64$$
$$x \leq 16$$

The width of each side of the package must be less than or equal to 16 inches.

Example 13 ■ Comparing Costs

A subcompact car can be rented from Company A for $240 per week with no extra charge for mileage. A similar car can be rented from Company B for $100 per week, plus an additional 25 cents for each mile driven. How many miles must you drive in a week so that the rental fee for Company B is more than that for Company A?

Algebraic Solution

Verbal Model:

Weekly cost for Company B	>	Weekly cost for Company A

Labels:

Number of miles driven in 1 week $= m$ (miles)

Weekly cost for Company A $= 240$ (dollars)

Weekly cost for Company B $= 100 + 0.25m$ (dollars)

Inequality: $100 + 0.25m > 240$

$$0.25m > 140$$

$$m > 560$$

The car from Company B is more expensive if you drive more than 560 miles in a week.

Numerical Solution

You can use a table to determine when the rental fee for Company B is more than that for Company A by evaluating the weekly cost of renting a car from each rental company.

Miles driven	520	530	540
Company A	$240.00	$240.00	$240.00
Company B	$230.00	$232.50	$235.00

Miles driven	550	560	570
Company A	$240.00	$240.00	$240.00
Company B	$237.50	$240.00	$242.50

From the table you can see that once you drive more than 560 miles in a week, you can assume the car from Company B is more expensive. You can improve this estimate by making a table similar to the one below.

Miles driven	559	560	561
Company A	$240.00	$240.00	$240.00
Company B	$239.75	$240.00	$240.25

The table above confirms that once you drive more than 560 miles in a week, the car from Company B is more expensive.

Collaborate!

Communicating Mathematically

Suppose your group owns a small business and must choose between two carriers for long distance telephone service. Create realistic data for cost of the first minute of a call and cost per additional minute for each carrier, and decide what question(s) would be most helpful to ask when making such a choice. Solve the problem your group has created. Write a short memo to your company's business manager outlining the situation, explaining your mathematical solution and summarizing your recommendations.

3.4 ■ Exercises

Developing Skills

In Exercises 1–4, determine whether the given values of x satisfy the inequality.

Inequality	Values

1. $7x - 10 > 0$ (a) $x = 3$ (b) $x = -2$

(c) $x = \frac{5}{2}$ (d) $x = \frac{1}{2}$

2. $3x + 2 < \dfrac{7x}{5}$ (a) $x = 0$ (b) $x = 4$

(c) $x = -4$ (d) $x = -1$

3. $0 < \dfrac{x + 5}{6} < 2$ (a) $x = 10$ (b) $x = 4$

(c) $x = 0$ (d) $x = -6$

4. $-2 < \dfrac{3 - x}{2} \leq 2$ (a) $x = 0$ (b) $x = 3$

(c) $x = 9$ (d) $x = -12$

In Exercises 5–8, match the inequality with its graph. [The graphs are labeled (a), (b), (c), and (d).]

(a)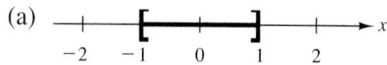

(b)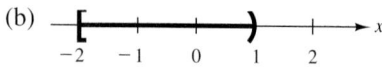

(c)

(d)

5. $-1 < x \leq 2$ **6.** $-1 \leq x \leq 1$

7. $-1 < x < 2$ **8.** $-2 \leq x < 1$

In Exercises 9–12, match the inequality with its graph. [The graphs are labeled (a), (b), (c), and (d).]

(a)

(b)

(c)

(d)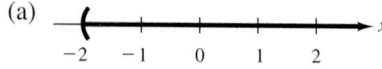

9. $x > 2$ **10.** $x > -2$

11. $x < -2$ **12.** $x < 2$

In Exercises 13–16, match the inequality with its graph. [The graphs are labeled (a), (b), (c), and (d).]

(a)

(b)

(c)

(d)

13. $x \leq -1$ or $x \geq 1$ **14.** $x < -1$ or $x > 1$

15. $x \leq -2$ or $x > 2$ **16.** $x < -2$ or $x \geq 2$

In Exercises 17–30, sketch the graph of the inequality. Write the interval notation for the inequality.

17. $x \leq 2$ **18.** $x > -6$

19. $x > 3.5$ **20.** $x \leq -2.5$

21. $-5 < x \leq 3$ **22.** $4 > x \geq 1$

23. $\frac{3}{2} \geq x > 0$ **24.** $-7 < x \leq -3$

25. $-\frac{15}{4} < x < -\frac{5}{2}$ **26.** $-\pi > x > -5$

27. $x < -5$ or $x > -1$ **28.** $x \leq -4$ or $x > 0$

29. $x \leq 3$ or $x > 7$ **30.** $x \leq -1$ or $x \geq 1$

In Exercises 31–70, solve the inequality and sketch the solution on the real number line.

31. $x + 7 \leq 9$ **32.** $z - 4 > 0$

33. $4x < 22$ **34.** $2x > 5$

35. $-9x \geq 36$ **36.** $-6x \leq 24$

37. $-\frac{3}{4}x < -6$ **38.** $-\frac{1}{5}x > -2$

39. $2x - 5 > 9$ **40.** $3x + 4 \leq 22$

41. $5 - x \leq -2$ **42.** $1 - y \geq -5$

43. $5 - 3x < 7$ **44.** $12 - 5x > 5$

45. $3x - 11 > -x + 7$

46. $21x - 11 \leq 6x + 19$

47. $5 - 3(2x + 6) < 9x + 7$

48. $11 - 5x \geq 16 + 2(8 - 3x)$

49. $16 < 4(y + 2) - 5(2 - y)$

50. $4[z - 2(z + 1)] < 2 - 7z$

51. $-3(y + 10) \geq 4(y + 10)$

52. $2(4 - z) \geq 8(1 + z)$

53. $\dfrac{x}{6} - \dfrac{x}{4} \leq 1$

54. $\dfrac{x + 3}{6} + \dfrac{x}{8} \geq 1$

55. $0 < 2x - 5 < 9$

56. $-4 \leq 2 - 3(x + 2) < 11$

57. $-3 < \dfrac{2x - 3}{2} < 3$ **58.** $0 \leq \dfrac{x - 5}{2} < 4$

59. $1 > \dfrac{x - 4}{-3} > -2$ **60.** $-\dfrac{2}{3} < \dfrac{x - 4}{-6} \leq \dfrac{1}{3}$

61. $\dfrac{2}{5} < x + 1 < \dfrac{4}{5}$ **62.** $-1 < -\dfrac{x}{6} < 1$

63. $2x - 4 \leq 4$ and $2x + 8 > 6$

64. $8 - 3x > 5$ and $x - 5 \geq -10$

65. $7 + 4x < -5 + x$ and $2x + 10 \leq -2$

66. $9 - x \leq 3 + 2x$ and $3x - 7 \leq -22$

67. $6 - \dfrac{x}{2} \geq 1$ or $\dfrac{5}{4}x - 6 \geq 4$

68. $\dfrac{x}{3} - 2 \geq 1$ or $5 + \dfrac{3}{4}x \leq -4$

69. $7x + 11 < 3 + 4x$ or $\dfrac{5}{2}x - 1 \geq 9 - \dfrac{3}{2}x$

70. $3x + 10 \leq -x - 6$ or $\dfrac{x}{2} + 5 < \dfrac{5}{2}x - 4$

In Exercises 71–76, rewrite the statement using inequality notation.

71. x is nonnegative.

72. y is more than -2.

73. z is at least 2.

74. m is at least 4.

75. n is no more than 16.

76. x is at least 450 but no more than 500.

In Exercises 77–80, write a verbal description of the inequality.

77. $x \geq \frac{5}{2}$ **78.** $t < 4$

79. $0 < z \leq \pi$ **80.** $-4 \leq t \leq 4$

81. Four times a number n must be at least 12 and no more than 30. What interval contains this number?

82. Five times a number n must be at least 15 and no more than 45. What interval contains this number?

Solving Problems

83. *Budget*   A student group has $4500 budgeted for a field trip. The cost of transportation for the trip is $1900. To stay within the budget, all other costs C must be no more than what amount?

84. *Budget*   You have budgeted $1800 per month for your total expenses. The cost of rent per month is $600 and the cost of food is $350. To stay within your budget, all other costs C must be no more than what amount?

85. *Temperature*   The average temperature in Miami is greater than the average temperature in Washington, D.C., and the average temperature in Washington, D.C. is greater than the average temperature in New York. How does the average temperature in Miami compare with the average temperature in New York?

86. *Elevation*   The elevation (above sea level) of San Francisco is less than the elevation of Dallas, and the elevation of Dallas is less than the elevation of Denver. How does the elevation of San Francisco compare with the elevation of Denver?

87. *Operating Costs*   A utility company has a fleet of vans. The annual operating cost per van is

$$C = 0.28m + 2900$$

where m is the number of miles traveled by a van in a year. What is the maximum number of miles that will yield an annual operating cost that is less than $10,000?

88. *Operating Costs*   A fuel company has a fleet of trucks. The annual operating cost per truck is

$$C = 0.58m + 7800$$

where m is the number of miles traveled by a truck in a year. What is the maximum number of miles that will yield an annual operating cost that is less than $25,000?

Profit In Exercises 89 and 90, the revenue *R* for selling *x* units and the cost *C* of producing *x* units of a product are given. In order to obtain a profit, the revenue must be greater than the cost. For what values of *x* will this product produce a profit? Use a graphing utility to graph the inequality to confirm the result.

89. $R = 89.95x$

    $C = 61x + 875$

90. $R = 105.45x$

    $C = 78x + 25,850$

91. *Long Distance Phone Calls* The cost of a long distance telephone call is $0.96 for the first minute and $0.75 for each additional minute. If the total cost of the call cannot exceed $5, find the interval of time that is available for the call. Use a graphing utility to graph the inequality to confirm the result.

92. *Long Distance Phone Calls* The cost of a long distance telephone call is $1.45 for the first minute and $0.95 for each additional minute. If the total cost of the call cannot exceed $15, find the interval of time that is available for the call. Use a graphing utility to graph the inequality to confirm the result.

93. *Distance* You live 5 miles from school and your friend lives 3 miles from you. The distance *d* that your friend lives from school is in what interval?

94. *Distance* You live 10 miles from work and work is 5 miles from school. The distance *d* that you live from school is in what interval?

95. *Geometry* The perimeter of the rectangle in the figure must be at least 36 centimeters and not more than 64 centimeters. Find the interval for *x*.

Figure for 95   Figure for 96

96. *Geometry* The perimeter of the rectangle in the figure must be at least 100 meters and not more than 120 meters. Find the interval for *x*.

97. *Wages* You must select one of two plans for payment when working for a company. The first plan pays a straight $12.50 per hour. The second plan pays $8.00 per hour plus $0.75 per unit produced per hour. Write an inequality yielding the number of units that must be produced per hour so that the second plan gives the greater hourly wage. Solve the inequality.

98. *Wages* You must select one of two plans for payment when working for a company. The first plan pays a straight $3000 per month. The second plan pays $1000 per month plus a commission of 4% of your gross sales. Write an inequality yielding the gross sales per month so that the second plan gives the greater monthly wage. Solve the inequality.

99. *Medicine* The equation below models the number of nurses in the United States from 1990 to 1998 (see figure).

$$y = 1799 + 58.3t, \quad 0 \le t \le 8$$

In this model, *y* represents the number of nurses in thousands and *t* represents the year, with $t = 0$ corresponding to 1990. *(Source: U.S. Department of Health and Human Services)*

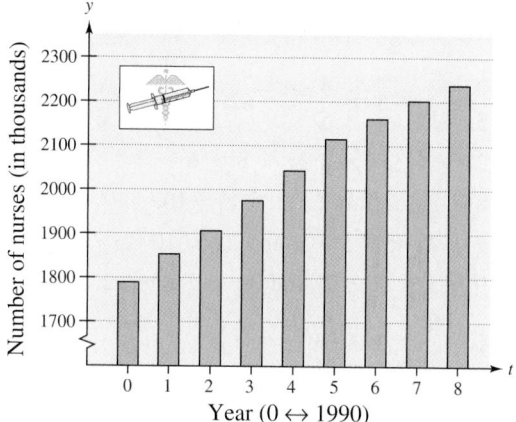

(a) During which years from 1990 to 1998 was the number of nurses greater than 2 million?

(b) During which years from 1990 to 1998 was the number of nurses less than 1,900,000?

100. *Scientific Research* The equation below models the amount of money (in billions of dollars) the United States government spent on mathematics and computer sciences research from 1991 to 2000 (see figure).

$$y = 0.822 + 0.1292t, \quad 1 \le t \le 10$$

In this model, *y* represents the amount of money in billions of dollars and *t* represents the year, with $t = 1$ corresponding to 1991. *(Source: U.S. National Science Foundation)*

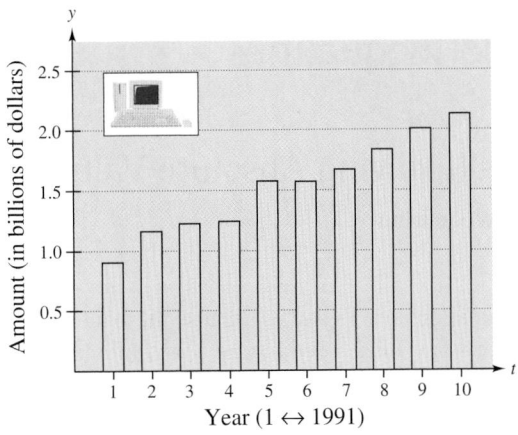

Figure for 100

(a) During which years from 1991 to 2000 was the amount of money spent on mathematics and computer sciences research greater than $1.8 billion?

(b) During which years from 1991 to 2000 was the amount of money spent on mathematics and computer sciences research less than $1.6 billion?

Explaining Concepts

101. Describe the differences between properties of equality and properties of inequality.

102. Give an example of an inequality with an unbounded solution set.

103. If $t < 8$, then $-t$ must be in what interval?

104. If $-3 < x \le 10$, then $-x$ must be in what interval?

Ongoing Review

In Exercises 105–108, place the correct inequality symbol ($<$ or $>$) between the two real numbers.

105. $\frac{7}{8}$ ___ 4

106. $-\frac{11}{3}$ ___ 5

107. -10 ___ $\frac{1}{10}$

108. 7 ___ $\frac{25}{3}$

In Exercises 109–112, solve the equation.

109. $-3x - 4 = 8$

110. $12 + 5t = -3$

111. $12 - 5(2 - x) = 6$

112. $7 + 3(x - 1) = -2$

Geometry   In Exercises 113 and 114, find the area of the trapezoid. The area of a trapezoid with parallel bases b_1 and b_2 and height h is $A = \frac{1}{2}(b_1 + b_2)h$.

113.

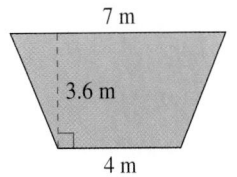

7 m

3.6 m

4 m

114.

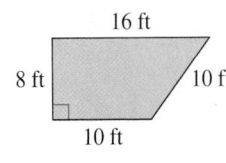

16 ft

8 ft   10 ft

10 ft

Looking Further

Hearing Frequencies   The range of a human's voice frequency h (in cycles per second) is about $85 \le h \le 1100$. The range of a human's hearing frequency H is about $20 \le H \le 20{,}000$.

(a) The relationship between a human's voice frequency h and a bat's voice frequency b is

$$h = 85 + \frac{203}{22{,}000}(b - 10{,}000).$$

Find the range of a bat's voice frequency.

(b) The relationship between a human's hearing frequency H and a bat's hearing frequency B is

$$H = 20 + \frac{999}{5950}(B - 1000).$$

Find the range of a bat's hearing frequency.

(c) Use your answer in part (a) to determine whether a human could hear a bat's voice.

(d) Use your answer in part (b) to determine whether a bat could hear a human's voice.

3.5 Absolute Value Equations and Inequalities

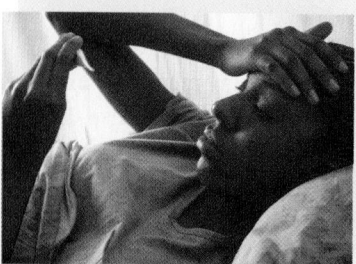

Solving Equations Involving Absolute Value

Consider the **absolute value equation**

$$|x| = 3.$$

The only solutions of this equation are -3 and 3, because these are the only two real numbers whose distance from 0 is 3. (See Figure 3.34.) In other words, the absolute value equation $|x| = 3$ has exactly two solutions: $x = -3$ and $x = 3$.

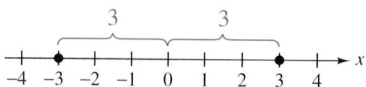

Figure 3.34

Solving an Absolute Value Equation

Let x be a variable or an algebraic expression and let a be a real number such that $a \geq 0$. The solutions of the equation $|x| = a$ are given by $x = -a$ and $x = a$. That is,

$$|x| = a \implies x = -a \quad \text{or} \quad x = a.$$

Example 1 ■ Solving Absolute Value Equations

Solve each absolute value equation.

(a) $|x| = 10$

(b) $|x| = 0$

(c) $|y| = -1$

Solution

(a) This equation is equivalent to the two linear equations

$$x = -10 \quad \text{or} \quad x = 10. \quad \text{Equivalent linear equations}$$

So, the absolute value equation has two solutions: $x = -10$ and $x = 10$.

(b) This equation is equivalent to the two linear equations

$$x = -0 \quad \text{or} \quad x = 0. \quad \text{Equivalent linear equations}$$

Because both equations are equivalent, you can conclude that the absolute value equation has only one solution: $x = 0$.

(c) This absolute value equation has *no solution* because it is not possible for the absolute value of a real number to be negative.

Example 2 ■ Solving an Absolute Value Equation

Solve $|3x + 4| = 10$.

Solution

$|3x + 4| = 10$		Write original equation.
$3x + 4 = -10$ or $3x + 4 = 10$		Equivalent equations
$3x = -14$ $3x = 6$		Subtract 4 from each side.
$x = -\dfrac{14}{3}$ $x = 2$		Divide each side by 3.

The solutions are $x = -\frac{14}{3}$ and $x = 2$. Check these in the original equation.

EXPLORATION

Solve each absolute value equation using the technique shown in Example 2.

a. $|5x - 6| = -2$

b. $|2x + 1| = -3$

Substitute the resulting solution(s) into the original equation. Do the results check? Why or why not? What can you conclude about absolute value equations of the form $|ax + b| = c, c < 0$?

NOTE   When you are solving absolute value equations, remember that it is possible that they have no solutions. For instance, the equation

$$|3x + 4| = -10$$

has no solution because the absolute value of a real number cannot be negative. Do not make the mistake of trying to solve such an equation by writing the "equivalent" linear equations $3x + 4 = -10$ and $3x + 4 = 10$. These equations have solutions, but they are both extraneous.

The equations in the next example are not given in the **standard form**

$$|ax + b| = c, \qquad c \geq 0.$$

Notice that the first step in solving such equations is to write them in standard form.

Example 3 ■ Absolute Value Equations in Nonstandard Form

Solve (a) $|2x - 1| + 3 = 8$ and (b) $-2|4x + 5| = -10$

Solution

(a) $|2x - 1| + 3 = 8$		Write original equation.
$|2x - 1| = 5$		Write in standard form.
$2x - 1 = -5$ or $2x - 1 = 5$		Equivalent equations
$2x = -4$ $2x = 6$		Add 1 to each side.
$x = -2$ $x = 3$		Divide each side by 2.

The solutions are $x = -2$ and $x = 3$. Check these in the original equation.

NOTE   Notice that Example 3(b) has two solutions even though c is negative. Remember that there is no solution to an absolute value equation when the equation is in standard form and c is a negative number.

(b) $-2|4x + 5| = -10$		Write original equation.
$|4x + 5| = 5$		Write in standard form.
$4x + 5 = -5$ or $4x + 5 = 5$		Equivalent equations
$4x = -10$ $4x = 0$		Subtract 5 from each side.
$x = -\frac{5}{2}$ $x = 0$		Divide each side by 4.

The solutions are $x = -\frac{5}{2}$ and $x = 0$. Check these in the original equation.

If two algebraic expressions are equal in absolute value, they must either be *equal* to each other or be the *opposites* of each other. So, you can solve equations of the form $|ax + b| = |cx + d|$ by forming the two linear equations

Expressions opposite

Expressions equal

$$ax + b = -(cx + d) \quad \text{and} \quad ax + b = cx + d.$$

Example 4 ■ Solving an Equation Involving Two Absolute Values

Solve $|3x - 4| = |7x - 16|$.

Algebraic Solution

$$|3x - 4| = |7x - 16|$$

$$3x - 4 = -(7x - 16) \quad \text{or} \quad 3x - 4 = 7x - 16$$

$$3x - 4 = -7x + 16 \qquad\qquad 3x = 7x - 12$$

$$10x = 20 \qquad\qquad -4x = -12$$

$$x = 2 \qquad\qquad x = 3$$

The solutions are $x = 2$ and $x = 3$. Check these in the original equation.

Graphical Solution

Use a graphing utility to graph the equations

$$y_1 = |3x - 4| \quad \text{and} \quad y_2 = |7x - 16|$$

in the same viewing window. From Figure 3.35, the graphs appear to have two points of intersection. Use the *intersect* feature of the graphing utility to find the points of intersection to be $(2, 2)$ and $(3, 5)$. So, the solutions are $x = 2$ and $x = 3$.

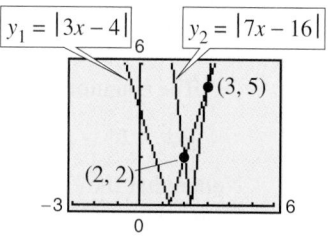

Figure 3.35

Example 5 ■ Solving an Equation Involving Two Absolute Values

Solve $|x + 5| = |x + 11|$.

Solution

$$x + 5 = -(x + 11) \quad \text{or} \quad x + 5 = x + 11$$

$$x + 5 = -x - 11 \qquad\qquad x + 5 = x + 11$$

$$x = -x - 16 \qquad\qquad x = x + 6$$

$$2x = -16 \qquad\qquad 0x = 6$$

$$x = -8 \qquad\qquad 0 = 6$$

The only solution is $x = -8$ because the second linear equation results in the false statement $0 = 6$. Check this solution in the original equation.

NOTE When solving equations of the form

$$|ax + b| = |cx + d|$$

it is possible that one of the resulting equations will not have a solution. Note this occurrence in Example 5.

Solving Inequalities Involving Absolute Value

To see how to solve inequalities involving absolute value, consider the comparisons below.

$$|x| = 2$$
$$x = -2 \text{ or } x = 2$$

$$|x| < 2$$
$$-2 < x < 2$$

$$|x| > 2$$
$$x < -2 \text{ or } x > 2$$

These comparisons suggest the following rules for solving inequalities involving absolute value.

EXPLORATION

Consider the inequalities below. Describe all numbers that are solutions of each inequality. Then describe the graph of each inequality. Explain your reasoning.

a. $|x| \geq 3$

b. $|x| \leq 3$

Solving an Absolute Value Inequality

Let x be a variable or an algebraic expression and let a be a real number such that $a > 0$.

1. The solutions of $|x| < a$ are all values of x that lie between $-a$ and a. That is,

$$|x| < a \quad \text{if and only if} \quad -a < x < a.$$

2. The solutions of $|x| > a$ are all values of x that are *less than* $-a$ *or greater than* a. That is,

$$|x| > a \quad \text{if and only if} \quad x < -a \text{ or } x > a.$$

These rules are also valid if $<$ is replaced by $\leq$ and $>$ is replaced by $\geq$.

Example 6 ■ Solving an Absolute Value Inequality

Solve $|x - 5| < 2$.

Solution

$|x - 5| < 2$	Write original inequality.
$-2 < x - 5 < 2$	Equivalent double inequality
$-2 + 5 < x - 5 + 5 < 2 + 5$	Add 5 to all three parts.
$3 < x < 7$	Simplify.

The solution set consists of all real numbers that are greater than 3 and less than 7. The interval notation for this solution set is $(3, 7)$. The graph is shown in Figure 3.36.

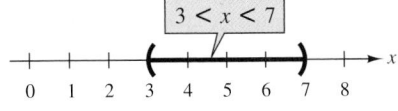

Figure 3.36

To verify the solution of an absolute value inequality, you need to check values in the solution set and outside of the solution set. For instance, in Example 6 you can check that $x = 4$ is in the solution set and that $x = 2$ and $x = 8$ are not in the solution set.

Study Tip

In Example 6, note that absolute value inequalities of the form $|x| < a$ (or $|x| \leq a$) can be solved with a double inequality. Inequalities of the form $|x| > a$ (or $|x| \geq a$) cannot be solved with a double inequality. Instead, you must solve two separate inequalities, as demonstrated in Example 7. You should always check to make sure the solution set makes sense.

Example 7 ■ Solving an Absolute Value Inequality

Solve $|3x - 4| \geq 5$.

Solution

$|3x - 4| \geq 5$		Write original inequality.
$3x - 4 \leq -5$ or $\quad$ $3x - 4 \geq 5$		Equivalent inequalities
$3x - 4 + 4 \leq -5 + 4 \quad 3x - 4 + 4 \geq 5 + 4$		Add 4 to each side.
$3x \leq -1 \qquad\qquad\qquad 3x \geq 9$		Simplify.
$\dfrac{3x}{3} \leq \dfrac{-1}{3} \qquad\qquad \dfrac{3x}{3} \geq \dfrac{9}{3}$		Divide each side by 3.
$x \leq -\dfrac{1}{3} \qquad\qquad\quad x \geq 3$		Simplify.

The solution set consists of all real numbers that are less than or equal to $-\frac{1}{3}$ or greater than or equal to 3. The interval notation for this solution set is $\left(-\infty, -\frac{1}{3}\right] \cup [3, \infty)$. The graph is shown in Figure 3.37.

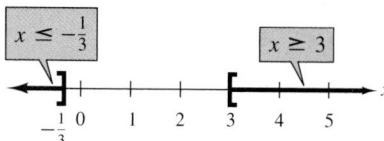

Figure 3.37

NOTE The symbol $\cup$ is called a **union** symbol, and it is used to denote the combining of two sets.

Example 8 ■ Solving an Absolute Value Inequality

Solve $\left|2 - \dfrac{x}{3}\right| \leq 0.01$.

Solution

$\left|2 - \dfrac{x}{3}\right| \leq 0.01$	Write original inequality.
$-0.01 \leq 2 - \dfrac{x}{3} \leq 0.01$	Equivalent double inequality
$-2.01 \leq -\dfrac{x}{3} \leq -1.99$	Subtract 2 from all three parts.
$6.03 \geq x \geq 5.97$	Multiply all three parts by -3 and reverse both inequality symbols.
$5.97 \leq x \leq 6.03$	Solution set in standard form

The solution set consists of all real numbers that are greater than or equal to 5.97 and less than or equal to 6.03. The interval notation for this solution set is $[5.97, 6.03]$. The graph is shown in Figure 3.38.

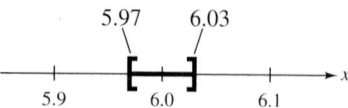

Figure 3.38

Example 9 ■ Solving an Absolute Value Inequality

Solve $|2x + 1| < 3$.

Algebraic Solution

$$|2x + 1| < 3 \qquad \text{Write original inequality.}$$
$$-3 < 2x + 1 < 3 \qquad \text{Equivalent double inequality}$$
$$-4 < 2x < 2 \qquad \text{Subtract 1 from all three parts.}$$
$$-2 < x < 1 \qquad \text{Simplify.}$$

The solution set consists of all real numbers that are greater than -2 and less than 1. The interval notation for this solution set is $(-2, 1)$.

Graphical Solution

Begin by sketching the graphs of

$$y = |2x + 1| \text{ and } y = 3$$

on the same set of coordinate axes, as shown in Figure 3.39. In the figure, notice that for all values of x between -2 and 1, the graph of $y = |2x + 1|$ lies below the line given by $y = 3$. So, you can conclude that $|2x + 1| < 3$ for all x such that $-2 < x < 1$.

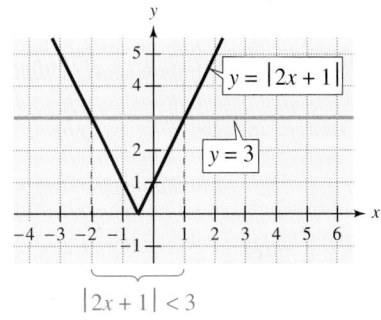

Figure 3.39

The pit organs of a rattlesnake can detect temperature changes as small as 0.005°F. By moving its head back and forth, a rattlesnake can detect warm prey, even in the dark.

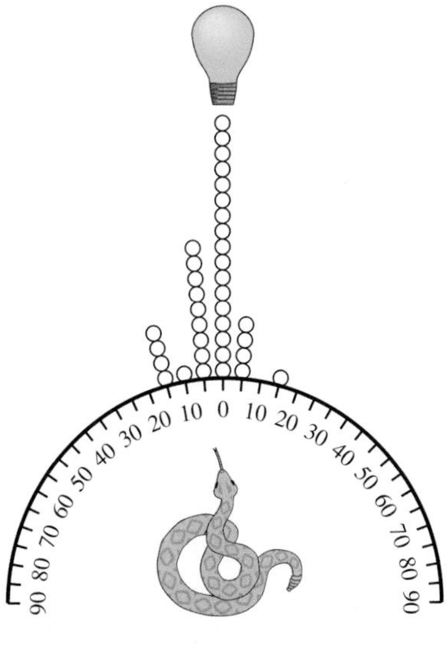

Example 10 ■ Creating a Model

To test the accuracy of a rattlesnake's "pit-organ sensory system," a biologist blindfolded a rattlesnake and presented the snake with a warm "target." Of 36 strikes, the snake was on target 17 times. In fact, the snake was within 5 degrees of the target for 30 of the strikes. Let A represent the number of degrees by which the snake is off target. Then $A = 0$ represents a strike that is aimed directly at the target. Positive values of A represent strikes to the right of the target and negative values of A represent strikes to the left of the target. Use the diagram shown in Figure 3.40 to write an absolute value inequality that describes the interval in which the 36 strikes occurred.

Solution

From the diagram, you can see that the snake was never off by more than 15 degrees in either direction. As a compound inequality, this can be represented by

$$-15 \le A \le 15.$$

As an absolute value inequality, the interval in which the strikes occurred can be represented by

$$|A| \le 15.$$

Figure 3.40

3.5 ■ Exercises

Developing Skills

In Exercises 1–4, determine whether the value is a solution of the equation.

Equation	Value
1. $\lvert 4x + 5 \rvert = 10$	$x = -3$
2. $\lvert 2x - 16 \rvert = 10$	$x = 3$
3. $\lvert 6 - 2w \rvert = 2$	$w = 4$
4. $\lvert \frac{1}{2}t + 4 \rvert = 8$	$t = 6$

In Exercises 5–8, transform the absolute value equation into two linear equations.

5. $\lvert x - 10 \rvert = 17$ **6.** $\lvert 7 - 2t \rvert = 5$

7. $\lvert 4x + 1 \rvert = \frac{1}{2}$ **8.** $\lvert 22k + 6 \rvert = 9$

In Exercises 9–50, solve the equation. (Some equations have no solution.)

9. $\lvert t \rvert = 45$ **10.** $\lvert s \rvert = 16$

11. $\lvert h \rvert = 0$ **12.** $\lvert x \rvert = -82$

13. $\lvert x - 16 \rvert = 5$ **14.** $\lvert z - 100 \rvert = 100$

15. $\lvert 2s + 3 \rvert = 25$ **16.** $\lvert 7a + 6 \rvert = 8$

17. $\lvert 32 - 3y \rvert = 16$ **18.** $\lvert 3 - 5x \rvert = 13$

19. $\lvert 3x + 4 \rvert = -16$ **20.** $\lvert 20 - 5t \rvert = 50$

21. $\lvert 2x + 13 \rvert = 4$ **22.** $\lvert 9x - 2 \rvert = -7$

23. $\lvert 4 - 3x \rvert = 0$ **24.** $\lvert 3x - 2 \rvert = -5$

25. $\lvert \frac{2}{3}x + 4 \rvert = 9$ **26.** $\lvert \frac{3}{2} - \frac{4}{5}x \rvert = 1$

27. $\lvert 0.32x - 2 \rvert = 4$

28. $\lvert 3.2 - 1.054x \rvert = 2$

29. $\lvert 5x - 3 \rvert + 8 = 22$

30. $\lvert 4x - 2 \rvert + 1 = 11$

31. $\lvert 7x + 9 \rvert - 5 = 16$

32. $\lvert 13x - 1 \rvert - 9 = 3$

33. $\lvert 2x + 10 \rvert - 7 = -2$

34. $\lvert 15x - 6 \rvert - 4 = 6$

35. $4\lvert x + 5 \rvert = 9$ **36.** $2\lvert x - 1 \rvert = 6$

37. $8\lvert x + 5 \rvert = 20$ **38.** $3\lvert 2x - 3 \rvert = 15$

39. $5\lvert 8x + 11 \rvert = -10$ **40.** $9\lvert x - 7 \rvert = -9$

41. $3\lvert 4x + 3 \rvert + 2 = 15$

42. $5\lvert x - 10 \rvert - 6 = 20$

43. $7\lvert 5x + 9 \rvert - 12 = -5$

44. $2\lvert 8x - 1 \rvert - 7 = -3$

45. $\lvert x + 8 \rvert = \lvert 2x + 1 \rvert$

46. $\lvert 10 - 3x \rvert = \lvert x + 7 \rvert$

47. $\lvert 45 - 4x \rvert = \lvert 32 - 3x \rvert$

48. $\lvert 5x + 4 \rvert = \lvert 3x + 25 \rvert$

49. $\lvert x + 2 \rvert = \lvert 3x - 1 \rvert$

50. $\lvert x - 2 \rvert = \lvert 2x - 15 \rvert$

In Exercises 51 and 52, write a single equation that is equivalent to the two equations.

51. $2x + 3 = 5, \ 2x + 3 = -5$

52. $4x - 6 = 7, \ 4x - 6 = -7$

In Exercises 53–60, determine whether the x-values are solutions of the inequality.

	Inequality	Values	
53.	$\lvert x \rvert < 3$	(a) $x = 2$	(b) $x = -4$
54.	$\lvert x \rvert \leq 5$	(a) $x = -7$	(b) $x = -4$
55.	$\lvert x \rvert \geq 3$	(a) $x = 2$	(b) $x = -4$
56.	$\lvert x \rvert > 5$	(a) $x = -7$	(b) $x = -4$
57.	$\lvert x - 7 \rvert < 3$	(a) $x = 9$	(b) $x = -4$
58.	$\lvert x - 3 \rvert \leq 5$	(a) $x = 16$	(b) $x = 3$
59.	$\lvert x - 7 \rvert \geq 3$	(a) $x = 9$	(b) $x = -4$
60.	$\lvert x - 3 \rvert > 5$	(a) $x = 16$	(b) $x = 3$

In Exercises 61–64, transform the absolute value inequality into a double inequality or two separate inequalities.

61. $\lvert y + 5 \rvert < 3$ **62.** $\lvert 6x + 7 \rvert \leq 5$

63. $\lvert 7 - 2h \rvert \geq 9$ **64.** $\lvert 8 - x \rvert > 25$

In Exercises 65–68, sketch a graph that shows the real numbers that satisfy the statement.

65. All real numbers greater than -2 *and* less than 5

66. All real numbers greater than or equal to 3 *and* less than 10

67. All real numbers less than or equal to 4 *or* greater than 7

68. All real numbers less than -6 *or* greater than or equal to 6

In Exercises 69–104, solve the inequality and sketch the solution on the real number line.

69. $|y| < 4$

70. $|x| < 6$

71. $|2x| < 14$

72. $|4z| \leq 9$

73. $\left|\dfrac{y}{3}\right| \leq 3$

74. $\left|\dfrac{t}{2}\right| < 4$

75. $|y - 2| \leq 4$

76. $|x - 3| \leq 6$

77. $|y + 5| < 8$

78. $|x + 7| < 9$

79. $|4 - y| > 6$

80. $|2 - x| > 10$

81. $\left|y + \frac{1}{2}\right| \geq 4$

82. $\left|x + \frac{1}{3}\right| \geq 6$

83. $|6t + 15| \geq 30$

84. $|3t + 1| > 5$

85. $\left|\dfrac{z}{10} - 3\right| > 8$

86. $\left|\dfrac{x}{8} + 1\right| < 12$

87. $|0.2x - 3| < 4$

88. $|1.5t - 8| \leq 16$

89. $2|x + 4| > 8$

90. $3|2x - 5| \leq 12$

91. $5|9 - 4x| < 25$

92. $6|x - 14| > 48$

93. $\dfrac{|x + 2|}{10} \leq 8$

94. $\dfrac{|y - 16|}{4} < 30$

95. $\dfrac{|s - 3|}{5} > 4$

96. $\dfrac{|a + 6|}{2} \geq 16$

97. $|12x + 5| - 2 \geq 7$

98. $|9x - 3| + 4 > 11$

99. $|16x - 9| + 6 < 7$

100. $|4x - 10| - 8 \leq -3$

101. $2|x + 9| - 1 > 15$

102. $3|2x - 1| + 4 < 10$

103. $5|4x + 18| - 7 \leq -2$

104. $9|6x - 2| + 1 \geq 28$

In Exercises 105–110, use a graphing utility to solve the inequality. Verify your solution algebraically.

105. $|3x + 2| < 4$

106. $|2x - 1| \leq 3$

107. $|x - 5| + 3 \leq 5$

108. $|a + 1| - 4 < 0$

109. $|2x + 3| > 9$

110. $|7r - 3| > 11$

In Exercises 111–114, match the inequality with its graph. [The graphs are labeled (a), (b), (c), and (d).]

(a)

(b)

(c)

(d)

111. $|x - 4| \leq 4$

112. $|x - 4| < 1$

113. $\frac{1}{2}|x - 4| > 4$

114. $|2(x - 4)| \geq 4$

In Exercises 115–120, write an absolute value inequality that represents the interval.

115.

116.

117.

118.

119.

120.

Solving Problems

121. *Temperature*   The operating temperature of an electronic device must satisfy the inequality

$$|t - 72| < 10$$

where t is given in degrees Fahrenheit. Sketch the graph of the solution of the inequality.

122. *Time Study*   A time study was conducted to determine the length of time required to perform a particular task in a manufacturing process. The times required by approximately two-thirds of the workers in the study satisfied the inequality

$$\left|\dfrac{t - 15.6}{1.9}\right| < 1$$

where t is the time in minutes. Sketch the graph of the solution of the inequality.

123. *Body Temperature*   Physicians consider an adult's body temperature to be normal if it is between $97.6°F$ and $99.6°F$. Write an absolute value inequality that describes this normal temperature range.

124. *Accuracy of Measurements*   In woodshop class, you must cut several pieces of wood to within $\frac{3}{16}$ inch of the teacher's specifications. Let $(s - x)$ rep-resent the difference between the specification s and the measured length x of a cut piece.

   (a) Write an absolute value inequality that describes the values of x that are within the specifications.

   (b) The length of one piece of wood is specified to be $s = 5\frac{1}{8}$ inches. Describe the acceptable lengths for this piece.

Explaining Concepts

125. Give a graphical description of the absolute value of a real number.

126. Give an example of an absolute value equation that has only one solution.

127. In your own words, explain how to solve an absolute value equation. Illustrate your explanation with an example.

128. When you buy a 16-ounce bag of chips, you probably don't expect to get *precisely* 16 ounces. Suppose that the actual weight w (in ounces) of a "16-ounce" bag of chips is given by $|w - 16| \le \frac{1}{2}$. If you buy four 16-ounce bags, what is the greatest amount you can expect to get? What is the smallest amount? Explain.

Ongoing Review

In Exercises 129–134, evaluate the expression.

129. $|-8|$

130. $|-5.4|$

131. $-|26|$

132. $-|-13|$

133. $|35 - 43|$

134. $-|29 - 10.4|$

In Exercises 135–138, solve the inequality.

135. $-3 \le 5x + 9 \le 7$ **136.** $-10 < 6 - x < 4$

137. $2x + 1 > 5$ or $5x - 2 < -2$

138. $12 - 7x < 5$ or $16 - 2x > 19$

139. *Business*   A printing company had annual sales of $696.5 million in 1999 and $1308.7 million in 2001. Use the Midpoint Formula to estimate the sales in 2000.

140. *Business*   A clothing retailer had annual sales of $1118.7 million in 1998 and $1371.4 million in 2002. Use the Midpoint Formula to estimate the sales in 2000.

Looking Further

The absolute value inequality

$$|x - 3| < 2$$

represents the set of all real numbers x whose distance from 3 is less than two units. Give a similar description of $|x - 4| < 1$.

The absolute value inequality

$$|y - 1| > 3$$

represents the set of all real numbers x whose distance from 1 is more than three units. Give a similar description of $|y + 2| > 4$.

Write an absolute value inequality that represents each verbal statement. Then solve the inequality and sketch the graph of the solution.

(a) The set of all real numbers x whose distance from 0 is less than 3

(b) The set of all real numbers x whose distance from 0 is more than 2

(c) The set of all real numbers x whose distance from 5 is more than 6

(d) The set of all real numbers x whose distance from 16 is less than 5

Chapter Summary: Key Concepts

3.1 ■ Summary of equations of lines

1. Slope of line through (x_1, y_1) and (x_2, y_2):

$$m = \frac{y_2 - y_1}{x_2 - x_1}$$

2. General form of equation of line: $ax + by + c = 0$

3. Equation of vertical line: $x = a$

4. Equation of horizontal line: $y = b$

5. Slope-intercept form of equation of line:

$$y = mx + b$$

6. Point-slope form of equation of line:

$$y - y_1 = m(x - x_1)$$

7. Parallel lines (equal slopes): $m_1 = m_2$

8. Perpendicular lines (negative reciprocal slopes):

$$m_2 = -\frac{1}{m_1}$$

3.2 ■ Translating key words and phrases
See summary box on page 171.

3.2 ■ Labels for consecutive integers
Let n represent an integer.

1. $\{n, n + 1, n + 2, \ldots\}$ denotes a set of consecutive integers.

2. If n is an even integer, then $\{n, n + 2, n + 4, \ldots\}$ denotes a set of consecutive even integers.

3. If n is an odd integer, then $\{n, n + 2, n + 4, \ldots\}$ denotes a set of consecutive odd integers.

3.2 ■ Strategy for solving word problems

1. Ask yourself what you need to know to solve the problem. Then write a verbal model that will give you what you need to know.

2. Assign labels to each part of the verbal model—numbers to the known quantities and letters (or expressions) to the variable quantities.

3. Use the labels to write an algebraic model based on the verbal model.

4. Solve the resulting algebraic equation.

5. Answer the original question and check that your answer satisfies the original problem as stated.

3.3 ■ Common formulas
See the summary boxes on page 187.

3.4 ■ Properties of inequalities

1. Addition and subtraction properties

If $a < b$, then $a + c < b + c$.

If $a < b$, then $a - c < b - c$.

2. Multiplication and division properties: positive quantities

If $a < b$ and c is positive, then $ac < bc$.

If $a < b$ and c is positive, then $\dfrac{a}{c} < \dfrac{b}{c}$.

3. Multiplication and division properties: negative quantities

If $a < b$ and c is negative, then $ac > bc$.

If $a < b$ and c is negative, then $\dfrac{a}{c} > \dfrac{b}{c}$.

4. Transitive property

If $a < b$ and $b < c$, then $a < c$.

3.5 ■ Solving an absolute value equation
Let x be a variable or an algebraic expression and let a be a real number such that $a \geq 0$. The solutions of the equation $|x| = a$ are given by $x = -a$ and $x = a$. That is,

$$|x| = a \Rightarrow x = -a \text{ or } x = a.$$

3.5 ■ Solving an absolute value inequality
Let x be a variable or an algebraic expression and let a be a real number such that $a > 0$.

1. The solutions of $|x| < a$ are all values of x that lie between $-a$ and a. That is,

$$|x| < a \text{ if and only if } -a < x < a.$$

2. The solutions of $|x| > a$ are all values of x that are less than $-a$ or greater than a. That is,

$$|x| > a \text{ if and only if } x < -a \text{ or } x > a.$$

Review Exercises

Reviewing Skills

3.1 ■ *How to write equations of lines in slope-intercept form*

In Exercises 1–4, write the equation in slope-intercept form, then state the slope of the line and the *y*-intercept of the line.

1. $2x + 6y = 18$ **2.** $9x - 3y = 0$

3. $7x - 9y + 27 = 0$ **4.** $x + 7y - 49 = 0$

In Exercises 5 and 6, write the slope-intercept form of the equation of the line with the specified slope and *y*-intercept.

5. $m = \frac{3}{2}, (0, -7)$

6. $m = -8, \left(0, \frac{2}{5}\right)$

■ *How to write equations of horizontal and vertical lines*

In Exercises 7 and 8, write equations of the lines passing through the point that are (a) horizontal and (b) vertical.

7. $(3, -4)$ **8.** $(9, 16)$

■ *How to write equations of lines using point-slope form*

In Exercises 9–16, write an equation of the line passing through the point with the specified slope.

9. $(1, -4)$ **10.** $(-5, -5)$

$m = 2$ $m = 3$

11. $(-1, 4)$ **12.** $(5, -2)$

$m = -4$ $m = -2$

13. $\left(\frac{5}{2}, 4\right)$ **14.** $\left(-2, -\frac{4}{3}\right)$

$m = -\frac{2}{3}$ $m = \frac{3}{2}$

15. $(7, 8)$ **16.** $(-6, 5)$

m is undefined. $m = 0$

In Exercises 17–22, write an equation of the line passing through the two points.

17. $(-6, 0), (0, -3)$ **18.** $(-2, -3), (4, 6)$

19. $(-10, 2), (4, -7)$ **20.** $(0, 10), (6, 10)$

21. $\left(\frac{5}{2}, 0\right), \left(\frac{5}{2}, 5\right)$ **22.** $\left(\frac{4}{3}, \frac{1}{6}\right), \left(4, \frac{7}{6}\right)$

In Exercises 23–26, find equations of the lines passing through the point that are (a) parallel and (b) perpendicular to the line.

23. $(-1, 5)$ **24.** $\left(\frac{3}{8}, 3\right)$

$2x + 4y = 1$ $4x - 3y = 12$

25. $\left(\frac{3}{5}, -\frac{4}{5}\right)$ **26.** $(12, 1)$

$3x + y = 2$ $5x = 3$

3.2 ■ *How to translate verbal phrases into algebraic expressions and vice versa*

In Exercises 27–32, translate the phrase into a mathematical expression. (Let *n* represent the arbitrary real number.)

27. Two hundred decreased by three times a number

28. One hundred increased by the product of 15 and a number

29. The sum of the square of a number and 49

30. The sum of a number and the square of the number

31. The absolute value of the sum of a number and 10

32. The absolute value of the quotient of a number and 5

In Exercises 33–36, write a verbal description of the algebraic expression without using the variable.

33. $2y + 7$ **34.** $5u - 3$

35. $\dfrac{x - 5}{4}$ **36.** $-3(a - 10)$

■ *How to write algebraic expressions that represent specified quantities using verbal models*

In Exercises 37–42, write an algebraic expression that represents the quantity given by the verbal statement.

37. The amount of income tax on a taxable income of *I* dollars when the tax rate is 18%

38. The distance traveled in 8 hours at an average speed of *r* miles per hour

39. The area of a rectangle whose length is *l* inches and whose width is 5 inches less than the length

40. The sum of three consecutive odd integers, the first of which is *n*

41. The cost of 30 acres of land if the price per acre is p dollars

42. The time it takes to copy z pages if the copy rate is 8 pages per minute

■ *How to translate verbal sentences into algebraic equations*

In Exercises 43–46, write an algebraic equation that represents the verbal description.

43. When the sum of a number and 9 is divided by 2, the quotient is 15.

44. Find a number such that four times the difference of the number and 11 is 112.

45. A salesperson gets paid a certain commission for each sale, as well as a weekly wage of $350. For one particular week, the salesperson made 32 sales and earned a total of $590.

46. A car rental costs $20 per day plus a rate of $0.35 for each mile driven. One customer's bill for a day was $76.

3.4 ■ *How to sketch the graphs of inequalities*

In Exercises 47–54, write an inequality for the given statement. Then graph the inequality.

47. z is no more than 10.

48. x is nonnegative.

49. y is at least 7 but less than 14.

50. x is more than -2 and less than 5.

51. The volume V is less than 27 cubic feet.

52. The height h is at least 48 inches.

53. x is no more than 3 or x is greater than 10.

54. q is negative or q is more than 4.

■ *How to solve linear inequalities in one variable*

In Exercises 55–58, solve the inequality and sketch the solution on the real number line.

55. $7x + 3 > 2x + 18$　　**56.** $-11x - 6 \geq 38$

57. $\frac{1}{3} - \frac{1}{2}y < 12$

58. $3(2 - y) \geq 2(1 + y)$

■ *How to solve compound inequalities*

In Exercises 59–66, solve the compound inequality and sketch the solution on the real number line.

59. $-4 < \dfrac{x}{5} \leq 4$　　**60.** $-13 \leq 3 - 4x < 13$

61. $5 > \dfrac{x + 1}{-3} > 0$　　**62.** $12 \geq \dfrac{x - 3}{2} > 1$

63. $8 - 2x < -3$ or $5x + 1 > 11$

64. $9x + 2 \geq 15$ or $3x - 4 < 5$

65. $3(x + 2) \leq 0$ or $5(x - 4) > 20$

66. $6(2x + 7) > 12$ or $9(6x - 1) \leq -18$

3.5 ■ *How to solve absolute value equations*

In Exercises 67–74, solve the equation.

67. $|x - 2| - 2 = 4$　　**68.** $|2x + 3| = 7$

69. $4|2x + 3| = 20$　　**70.** $5|6x - 7| = 45$

71. $7|x - 12| + 8 = 22$　　**72.** $2|9x + 10| - 3 = 15$

73. $|3x - 4| = |x + 2|$　　**74.** $|5x + 6| = |2x - 1|$

■ *How to solve inequalities involving absolute value*

In Exercises 75–80, solve the inequality.

75. $|2x - 7| < 15$　　**76.** $|5x - 1| < 9$

77. $|x - 4| > 3$　　**78.** $|t + 3| > 2$

79. $|b + 2| - 6 \geq 1$　　**80.** $\left|\dfrac{t}{3}\right| + 4 < 5$

Solving Problems

81. *Apartment Rent*　The rent for a two-bedroom apartment was $625 in 1997 and $800 in 2000.

(a) Assuming the rent for a two-bedroom apartment follows a linear growth pattern, write a linear model giving the rent R of the apartment in terms or the year t, where $t = 7$ represents 1997.

(b) Use a graphing utility to graph the model in part (a).

(c) *Linear Extrapolation* Use the model in part (a) to predict the rent in 2005.

(d) *Linear Interpolation* Use the model in part (a) to estimate the rent in 1999.

82. *Annual Salary* Suppose that your salary was $28,500 in 1998 and $31,800 in 2001.

  (a) Assuming your salary follows a linear growth pattern, write a linear model giving your salary S in terms of the year t, where $t = 8$ represents 1998.

  (b) Use a graphing utility to graph the model in part (a).

  (c) *Linear Extrapolation* Use the model in part (a) to predict your salary in 2005.

  (d) *Linear Interpolation* Use the model in part (a) to estimate your salary in 2000.

83. *Tax* The tax on a property with an assessed value of $80,000 is $1350. Find the tax on a property with an assessed value of $110,000.

84. *Cooking* One and a half cups of milk is required to make one batch of pudding. How much is required to make $2\frac{1}{2}$ batches?

85. *Map Scale* One-third inch represents 50 miles on a map. Approximate the distance between two cities that are $3\frac{1}{4}$ inches apart on the map.

86. *Fuel Mixture* The gasoline-to-oil ratio is 50 to 1 for a lawn mower engine. Determine the amount of gasoline required if $\frac{1}{2}$ pint of oil is used in producing the mixture.

Similar Triangles In Exercises 87 and 88, solve for the length x of the side of a triangle by using the fact that corresponding sides of similar triangles are proportional. (*Note:* Assume that the triangles are similar.)

87.

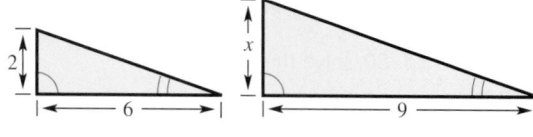

88.

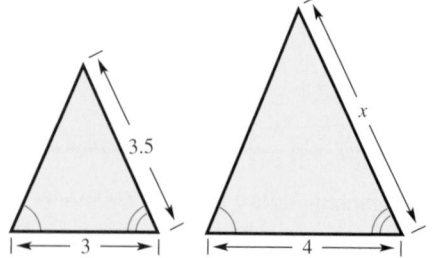

89. *Height of a Building* You want to determine the height of a building. To do this, you measure the building's shadow and find that it is 20 feet long. You

are 6 feet tall and your shadow is $1\frac{1}{2}$ feet long. How tall is the building?

90. *Height of a Flagpole* You want to determine the height of a flagpole. To do this, you measure the flagpole's shadow and find that it is 30 feet long. You also measure a 5-foot lamppost's shadow and find that it is 3 feet long (see figure). How tall is the flagpole?

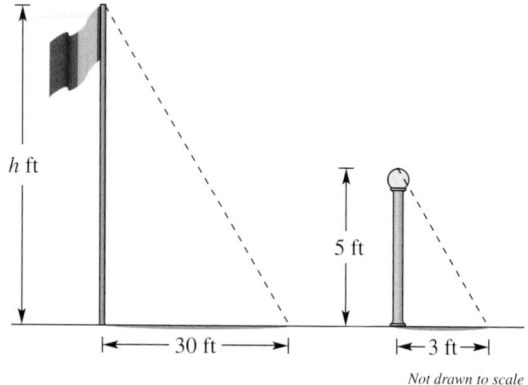

Not drawn to scale

91. *Expenses* Use the information in the figure to approximate the percent increase in the expenses (in thousands of dollars) for a local television broadcasting company from 1980 to 2000.

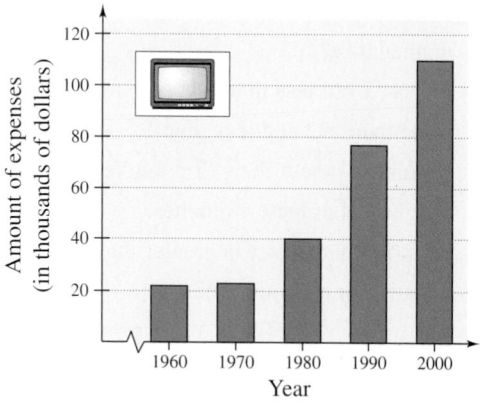

92. *Revenue* The revenues for a corporation in millions of dollars for the years 2000 and 2001 were $4521.4 and $4679.0, respectively. Determine the percent increase in revenue from 2000 to 2001.

93. *Price Increase* The manufacturer's suggested retail price for a certain truck model is $25,750. Estimate the price of a comparably equipped truck for the next model year if it is projected that truck prices will increase by $5\frac{1}{2}\%$.

94. *Retail Price*   A camera that costs a retailer $259.95 is marked up by 35%. Find the price to the consumer.

95. *Geometry*   Use the figure to find the measure of each angle.

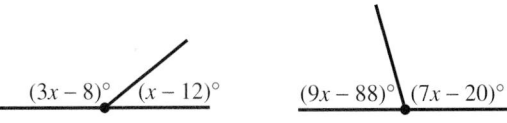

$(3x - 8)°$ $(x - 12)°$ $(9x - 88)°$ $(7x - 20)°$

Figure for 95   Figure for 96

96. *Geometry*   Use the figure to find the measure of each angle.

97. *Mixture Problem*   Determine the number of liters of a 30% saline solution and the number of liters of a 60% saline solution that are required to make 10 liters of a 50% saline solution.

98. *Mixture Problem*   Determine the number of gallons of a 25% alcohol solution and the number of gallons of a 50% alcohol solution that are required to make 8 gallons of a 40% alcohol solution.

99. *Travel Time*   Determine the time for a bus to travel 330 miles if its average speed is 52 miles per hour.

100. *Average Speed*   An Olympic cross-country skier completed the 15-kilometer event in 41 minutes and 20 seconds. What was the average speed of the skier?

101. *Distance*   Determine the distance an Air Force jet can travel in $2\frac{1}{3}$ hours if its average speed is 1200 miles per hour.

102. *Average Speed*   For 2 hours of a 400-mile trip, your average speed was 40 miles per hour. Determine the average speed that must be maintained for the remainder of the trip if you want the average speed for the entire trip to be 50 miles per hour.

103. *Work-Rate Problem*   Find the time for two people working together to complete half a task if it takes them 8 hours and 10 hours to complete the entire task working individually.

104. *Work-Rate Problem*   Find the time for two people working together to complete a task if it takes them 4.5 hours and 6 hours working individually.

105. *Simple Interest*   Find the total simple interest you will earn on a $1000 corporate bond that matures in 4 years and has an 8.5% interest rate.

106. *Simple Interest*   Find the annual simple interest rate on a certificate of deposit that pays $37.50 per year in interest on a principal of $500.

107. *Simple Interest*   Find the principal required to have an annual interest income of $20,000 if the annual simple interest rate on the principal is 9.5%.

108. *Simple Interest*   A corporation borrows 3.25 million dollars for 2 years to modernize one of its manufacturing facilities. If it pays an annual simple interest rate of 12%, what will be the total principal and interest that must be repaid?

109. *Simple Interest*   An inheritance of $50,000 is divided between two investments earning 8.5% and 10% simple interest. (The 10% investment has greater risk.) Determine the minimum amount that can be invested at 10% if you want an annual interest income of $4700 from the investment.

110. *Simple Interest*   You invest $1000 in a certificate of deposit that has an annual simple interest rate of 7%. After 6 months the interest is computed and added to the principal. During the second 6 months, the interest is computed using the original investment plus the interest earned during the first 6 months. What is the total interest earned during the first year of the investment?

111. *Geometry*   The area of the rectangle is 48 square inches. Find the dimensions of the rectangle.

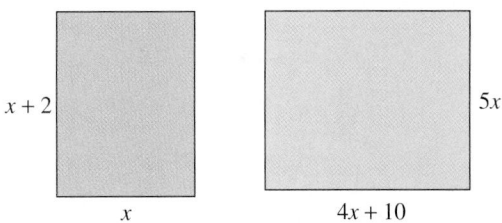

$x + 2$ $5x$

x $4x + 10$

Figure for 111   Figure for 112

112. *Geometry*   The perimeter of the rectangle is 110 feet. Find x.

113. *Distance*   You live 8 miles from the shopping mall. Your friend live 3 miles from you. Write an inequality that represents the distance d your friend lives from the shopping mall.

114. *Long Distance Phone Calls*   The cost of an international phone call is $0.99 for the first minute and $0.49 for each additional minute. If the total cost of the call is not to exceed $7.50, find the interval of time that is available for the call.

3 Chapter Test

Take this test as you would take a test in class. After you are done, check your work against the answers given in the back of the book.

1. Write an equation of the line passing through $(25, -15)$ and $(75, 10)$.

2. Write an equation of the line passing through $(10, 2)$ that is perpendicular to the line $5x + 3y - 9 = 0$.

3. After 4 years, a $26,000 car has a depreciated value of $10,000. Write a linear model giving the value V of the car in terms of t, the number of years. Use the model to find the time t when the value of the car is $16,000.

4. One afternoon a meteorologist observed that the temperature rose at an average rate of 3 degrees per hour between noon and 5 P.M. The temperature at noon was $72°$ F. Write a mathematical model that gives the afternoon temperature T in terms of the hour t, with $t = 0$ corresponding to noon.

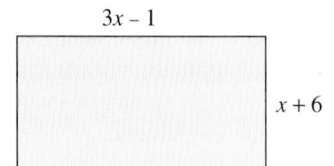

$3x - 1$

$x + 6$

Figure for 5

5. Find the length and width of the rectangle shown at the left if the perimeter is 114 inches.

6. When the product of a number n and 5 is decreased by 8, the result is 27. Find the number.

7. The sum of two consecutive even integers is 54. Find the integers.

8. The tax on a property with an assessed value of $90,000 is $1200. Estimate the tax on a property with an assessed value of $110,000.

9. The bill (including parts and labor) for the repair of a home appliance was $165. The cost for parts was $85. How many hours were spent repairing the appliance if the cost of labor was $16 per half hour?

10. Two solutions (10% concentration and 40% concentration) are mixed to create 100 liters of a 30% solution. Determine the numbers of liters of the 10% solution and the 40% solution that are required.

11. Two cars start at a given time and travel in the same direction at average speeds of 40 miles per hour and 55 miles per hour. How much time will elapse before the two cars are 10 miles apart?

12. Find the principal required to earn $300 in simple interest in 2 years if the annual interest rate is 7.5%.

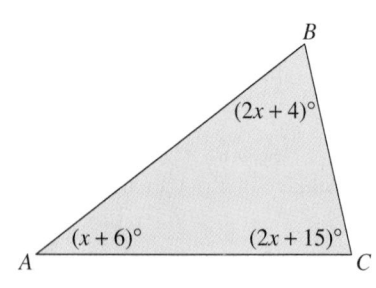

B

$(2x + 4)°$

$(x + 6)°$ $(2x + 15)°$

A C

Figure for 13

13. Use the figure at the left to find the measure of each angle.

14. Solve $|x + 6| - 3 = 8$.   15. Solve $|4x - 1| = |2x + 7|$.

In Exercises 16–20, solve the inequality and sketch the solution on the real number line.

16. $1 + 2x > 7 - x$ 17. $0 \le \dfrac{1 - x}{4} < 2$

18. $5(x + 7) < -4$ or $2(3x - 8) + 6 \ge 10$

19. $|x - 3| \le 2$ 20. $|3x + 13| > 4$

Cumulative Test: Chapters 1–3

Take this test as you would take a test in class. After you are done, check your work against the answers given in the back of the book.

In Exercises 1–8, simplify the expression.

1. $5^2 x^3 \cdot 5x^4$

2. $(-7z^3)^4(2z^3)$

3. $\dfrac{(2a^2b^7)^2}{-5(ab^4)^3}$

4. $-\left(\dfrac{3x^6}{2y^2}\right)^3$

5. $2(x + 5) - 3 - (2x - 3)$

6. $4 - 2[3 + 4(x + 1)]$

7. $(2x - 1)(3x + 4)$

8. $(5x + 1)(5x - 1)$

In Exercises 9–12, factor the expression.

9. $x^3 + 5x^2 - 8x - 40$

10. $4x^2 - 28xy + 49y^2$

11. $8x^3 + 27$

12. $12x^2 - 75$

In Exercises 13–20, solve the equation.

13. $5x - 7 = 2(x + 10)$

14. $3(6x - 1) + 4 = 17 - (x - 6)$

15. $3x + 6 = x + 4$

16. $4x - 2 = 2(x - 8) + 1$

17. $3x^2 + x - 24 = 0$

18. $7x^3 - 63x = 0$

19. $|2x + 6| = 16$

20. $|3x - 5| = |6x - 1|$

In Exercises 21 and 22, find the *x*- and *y*-intercepts (if any) of the graph of the equation.

21. $5x + 4y - 20 = 0$

22. $y = (x + 3)(x - 7)$

In Exercises 23 and 24, graph the equation.

23. $4x + 3y - 12 = 0$

24. $y = 1 - (x - 2)^2$

25. Evaluate the function $f(x) = 3 - 2x$ for (a) $f(5)$ and (b) $f(x + 3) - f(3)$.

26. Use the Vertical Line Test to determine whether the equation $x + 3y^2 = 9$ represents y as a function of x. (See figure at left.)

27. Find the domain and range of the function $f(x) = \sqrt{x - 4} + 2$ shown in the figure at the left.

28. Use the graph of $f(x) = \sqrt{x}$ to write an equation for the graph shown at the left.

29. Determine the slope of a line perpendicular to the line $3x + 7y = 20$.

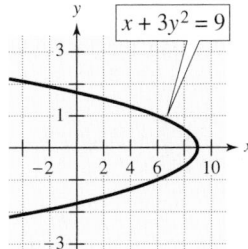

$x + 3y^2 = 9$

Figure for 26

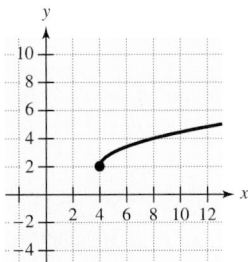

Figure for 27

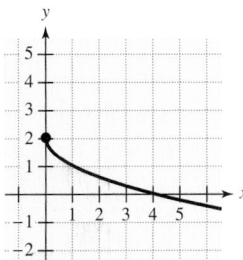

Figure for 28

30. Write an equation of the line that passes through $(3, -6)$ and is parallel to the line $2x + 8y = 5$.

31. Write an equation of the line passing through the points $(-2, 7)$ and $(5, 5)$.

In Exercises 32–34, solve the inequality and sketch the solution.

32. $7 - 3x > 4 - x$ **33.** $-1 < \dfrac{3x + 4}{5} < 5$ **34.** $|x - 2| \geq 3$

35. The revenue R from the sale of x units of a product is given by $R = x^2 - 36x$. The cost of producing x units of the product is $C = 232 + 18x$. How many units of the product must be produced and sold in order to break even $(R = C)$?

36. The inventor of a new game believes that the variable cost for producing the game is \$5.75 per unit and the fixed costs are \$12,000. Let x represent the number of games produced. Express the total cost C as a function of x.

37. Find the time for two people working together to complete half a task if it takes them 9 hours and 12 hours to complete the entire task working individually.

38. The triangles shown below are similar. Solve for x by using the fact that corresponding sides of similar triangles are proportional.

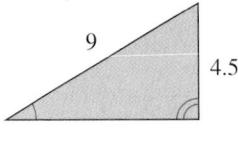

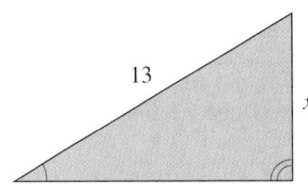

39. A combined total of \$24,000 is invested in two bonds that pay 7.5% and 9% simple interest. The total annual interest is \$1935. How much is invested in each bond?

40. An electronic device is to be operated in a room with relative humidity h defined by

$$|h - 50| \leq 30.$$

What are the minimum and maximum relative humidities for the operation of the device?

Systems of Linear Equations and Inequalities

4

THE BIG PICTURE

In this chapter you will learn how to:

- solve systems of linear equations by substitution, by elimination, by Gaussian elimination, by using matrices, and graphically.

- evaluate the determinant of a matrix.

- sketch graphs of solutions of linear inequalities and systems of linear inequalities.

- model real-life situations using systems of linear equations and inequalities.

Key Terms

As you encounter each new vocabulary term in this chapter, add the term and its definition to your notebook glossary.

system of equations (p. 232)
solution of a system of equations (p. 232)
consistent system (p. 234)
dependent system (p. 234)
inconsistent system (p. 234)
method of substitution (p. 236)
method of elimination (p. 237)
ordered triple (p. 248)
row-echelon form (p. 250)
equivalent systems (p. 251)
row operations (p. 251)
Gaussian elimination (p. 252)
matrix (p. 262)
order (of a matrix) (p. 262)
entry (of a matrix) (p. 262)
square matrix (p. 262)

augmented matrix (p. 262)
coefficient matrix (p. 262)
elementary row operations (p. 264)
row-equivalent (p. 264)
determinant (p. 274)
minor (of an entry) (p. 275)
Cramer's Rule (p. 277)
linear inequality (in two variables) (p. 286)
solution of a linear inequality (p. 286)
graph of an inequality (p. 287)
half-planes (p. 287)
systems of linear inequalities (p. 290)
solution of a system of linear inequalities (p. 290)
vertex (p. 290)

Additional text-specific resources are available to help you do well in this course. See page xvi for details.

4.1 Systems of Linear Equations in Two Variables

What you should learn:

- How to determine if ordered pairs are solutions of systems of linear equations
- How to solve systems of linear equations graphically
- How to solve systems of linear equations using the method of substitution
- How to solve systems of linear equations using the method of elimination
- How to use systems of equations to model and solve real-life problems

Why you should learn it:

The method of substitution is one method of solving a system of linear equations. For instance, in Exercise 99 on page 246, a system of linear equations can be used to find the prices per gallon of two different grades of gasoline.

Systems of Equations

Up to this point in the text, most problems have involved a single equation with one or two variables. Many problems in business and science involve a **system of equations** that consists of two or more equations, each involving two or more variables. A **solution** of a system of equations in two variables x and y is an ordered pair (x, y) of real numbers that satisfies *each* equation in the system. When you find the set of all solutions of the system of equations, you are **solving the system of equations.**

Example 1 ■ Checking Solutions of a System of Equations

Check whether each ordered pair is a solution of the system of equations.

(a) $(3, 3)$　　(b) $(4, 2)$

$$\begin{cases} x + y = 6 & \text{Equation 1} \\ 2x - 5y = -2 & \text{Equation 2} \end{cases}$$

Solution

(a) To determine whether the ordered pair $(3, 3)$ is a solution of the given system of equations, you must substitute $x = 3$ and $y = 3$ into *each* of the given equations. Substituting into Equation 1 produces

$$x + y = 6 \qquad \text{Equation 1}$$
$$3 + 3 = 6. \checkmark \qquad \text{Substitute 3 for } x \text{ and 3 for } y.$$

Similarly, substituting into Equation 2 produces

$$2x - 5y = -2 \qquad \text{Equation 2}$$
$$2(3) - 5(3) = -9 \neq -2. ✗ \qquad \text{Substitute 3 for } x \text{ and 3 for } y.$$

Because the ordered pair $(3, 3)$ fails to check in *both* equations, you can conclude that it *is not* a solution of the system of equations.

(b) Substituting the coordinates of the ordered pair $(4, 2)$ into Equation 1 produces

$$x + y = 6 \qquad \text{Equation 1}$$
$$4 + 2 = 6. \checkmark \qquad \text{Substitute 4 for } x \text{ and 2 for } y.$$

Similarly, substituting into Equation 2 produces

$$2x - 5y = -2 \qquad \text{Equation 2}$$
$$2(4) - 5(2) = -2. \checkmark \qquad \text{Substitute 4 for } x \text{ and 2 for } y.$$

Because the ordered pair $(4, 2)$ checks in both equations, you can conclude that it *is* a solution of the system of equations.

Solving Systems of Equations by Graphing

You can gain insight about the location and number of solutions of a system of equations by sketching the graph of each equation in the same coordinate plane. The solutions of the system correspond to the **points of intersection** of the graphs.

Example 2 ■ The Graphical Method of Solving a System

Use the graphical method to solve the system of equations.

$$\begin{cases} 2x + 3y = 7 & \text{Equation 1} \\ 2x - 5y = -1 & \text{Equation 2} \end{cases}$$

Solution

Because both equations in the given system are linear, you know that they have graphs that are straight lines. To sketch these lines, write each equation in slope-intercept form, as follows.

$$y = -\tfrac{2}{3}x + \tfrac{7}{3} \qquad \text{Equation 1}$$

$$y = \tfrac{2}{5}x + \tfrac{1}{5} \qquad \text{Equation 2}$$

The lines corresponding to these two equations are shown in Figure 4.1. From this figure, it appears that the two lines intersect in a single point, and that the coordinates of the point are approximately $(2, 1)$. A check will show that this point is a solution of each of the original equations.

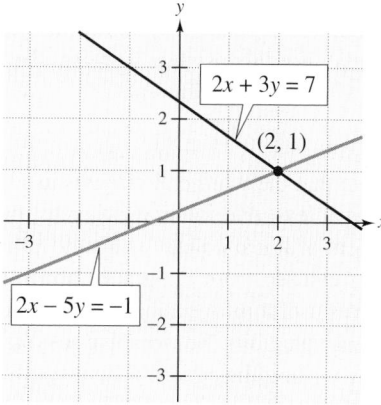

Figure 4.1

A system of linear equations can have exactly one solution, infinitely many solutions, or no solution. To see why this is true, consider the graphical interpretations of three systems of two linear equations shown below.

Graphs			
Graphical Interpretation	The two lines intersect.	The two lines coincide (are identical).	The two lines are parallel.
Intersection	Single point of intersection	Infinitely many points of intersection	No point of intersection
Slopes of Lines	Slopes are not equal.	Slopes are equal.	Slopes are equal.
Number of Solutions	Exactly one solution	Infinitely many solutions	No solution
Type of System	**Consistent system**	**Dependent (consistent) system**	**Inconsistent system**

Study Tip

Note that for dependent systems, the slopes of the lines and the y-intercepts are equal. For inconsistent systems, the slopes are equal, but the y-intercepts of the two lines are different.

Note that the word *consistent* is used to mean that the system of linear equations has at least one solution, whereas the word *inconsistent* is used to mean that the system of linear equations has no solution.

You can see from the graphs above that a comparison of the slopes of two lines gives useful information about the number of solutions of the corresponding system of equations. So, to solve a system of equations graphically, it helps to begin by writing the equations in slope-intercept form.

EXPLORATION

Use a graphing utility to graph the system of equations.

$$\begin{cases} 3x - 2y = -4 \\ 9x + 4y = 13 \end{cases}$$

Use the *intersect* feature or the *zoom* and *trace* features to approximate the point of intersection of the two graphs. Is the solution an approximate solution or an exact solution? Is the graphical method a good method to use when you want to find an exact solution to a system of equations? Is the graphical method a good method to use when you want to verify an exact solution to a system of equations? Explain.

Solving Systems of Equations by Substitution

One way to solve a system of two equations in two variables algebraically is to convert the system to *one* equation in *one* variable by an appropriate substitution. This procedure is illustrated in Example 3.

Example 3 ■ The Method of Substitution: One-Solution Case

Solve the system of equations.

$$\begin{cases} -x + y = 3 & \text{Equation 1} \\ 3x + y = -1 & \text{Equation 2} \end{cases}$$

Solution

Begin by solving for y in Equation 1.

$$y = x + 3 \qquad \text{Solve for } y \text{ in Equation 1.}$$

Next, substitute this expression for y into Equation 2.

$$3x + y = -1 \qquad \text{Equation 2}$$
$$3x + (x + 3) = -1 \qquad \text{Substitute } x + 3 \text{ for } y.$$
$$4x + 3 = -1 \qquad \text{Combine like terms.}$$
$$4x = -4 \qquad \text{Subtract 3 from each side.}$$
$$x = -1 \qquad \text{Divide each side by 4.}$$

At this point, you know that the x-coordinate of the solution is -1. To find the y-coordinate, back-substitute the x-value into the revised Equation 1.

$$y = x + 3 \qquad \text{Revised Equation 1}$$
$$y = -1 + 3 \qquad \text{Substitute } -1 \text{ for } x.$$
$$y = 2 \qquad \text{Simplify.}$$

So, the solution is $(-1, 2)$. Check to see that it satisfies both original equations. You can also check the solution graphically, as shown in Figure 4.2.

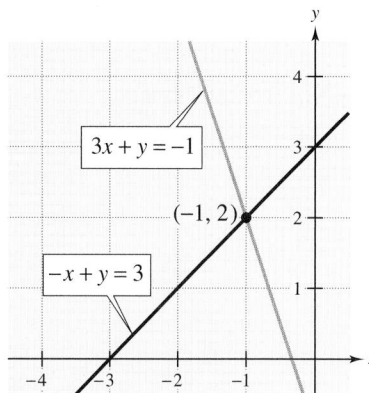

$3x + y = -1$

$(-1, 2)$

$-x + y = 3$

Figure 4.2

NOTE The term **back-substitute** implies that you work backwards. After finding a value for one of the variables, substitute that value back into one of the equations in the original (or revised) system to find the value of the other variable.

When using the method of substitution, it does not matter which variable you choose to solve for first. Whether you solve for y first or x first, you will obtain the same solution. When making your choice, you should choose the variable and equation that is easier to work with. For instance, in the system

$$\begin{cases} 3x - 2y = 1 & \text{Equation 1} \\ x + 4y = 3 & \text{Equation 2} \end{cases}$$

it is easier to begin by solving for x in the second equation. But in the system

$$\begin{cases} 2x + y = 5 & \text{Equation 1} \\ 3x - 2y = 11 & \text{Equation 2} \end{cases}$$

it is easier to begin by solving for y in the first equation.

The steps for using the **method of substitution** to solve a system of two equations involving two variables are summarized as follows.

The Method of Substitution

To solve a system of two equations in two variables, use the steps below.

1. Solve one of the equations for one variable in terms of the other variable.

2. Substitute the expression found in Step 1 into the other equation to obtain an equation in one variable.

3. Solve the equation obtained in Step 2.

4. Back-substitute the solution from Step 3 into the expression obtained in Step 1 to find the value of the other variable.

5. Check the solution to ensure that it satisfies *both* of the original equations.

Example 4 ■ A System with No Solution

Solve the system of equations.

$$\begin{cases} 2x - 2y = 0 & \text{Equation 1} \\ x - y = 1 & \text{Equation 2} \end{cases}$$

Algebraic Solution

Begin by solving for y in Equation 2.

$$y = x - 1$$

Next, substitute this expression for y into Equation 1.

$2x - 2y = 0$	Equation 1
$2x - 2(x - 1) = 0$	Substitute $x - 1$ for y.
$2x - 2x + 2 = 0$	Distributive Property
$2 = 0$	False statement

Because the substitution process produced a false statement $(2 = 0)$, you can conclude that the original system of equations is inconsistent and has no solution.

Graphical Solution

Solve each equation for y.

$$y_1 = x \text{ and } y_2 = x - 1$$

Notice that the lines have the same slope, so they are parallel. You can use a graphing utility to verify this by graphing both equations as shown in Figure 4.3. Then try using the *intersect* feature to find a point of intersection. Because the graphing utility cannot find a point of intersection, you will get an error. Use the *zoom* feature and then try using the *intersect* feature again. Because repeated zooming yields the same error message, you can conclude that the lines are parallel and, so, the system has no solution.

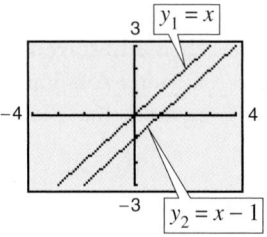

Figure 4.3

The Method of Elimination

A third method of solving a system of linear equations is called the **method of elimination.** The key step in the method of elimination is to obtain, for one of the variables, coefficients that differ only in sign, so that by *adding* the two equations this variable will be eliminated. This system contains such coefficients for x.

$$
\begin{array}{ll}
2x + 7y = 16 & \text{Equation 1} \\
\underline{-2x - 3y = -8} & \text{Equation 2} \\
4y = 8 & \text{Add equations.}
\end{array}
$$

Note that by adding the two equations, you eliminate the variable x and obtain a single equation in y. Solving this equation for y produces

$$
\begin{aligned}
4y &= 8 \\
y &= 2
\end{aligned}
$$

which can be back-substituted into one of the original equations to solve for x, as follows.

$$
\begin{array}{ll}
2x + 7(2) = 16 & \text{Substitute 2 for } y. \\
2x = 2 & \text{Subtract 14 from each side.} \\
x = 1 & \text{Divide each side by 1.}
\end{array}
$$

So, the ordered pair $(1, 2)$ is a solution of the system. The method of elimination is summarized as follows.

The Method of Elimination

To use the **method of elimination** to solve a system of two linear equations in x and y, use the steps below.

1. Write each equation in the form $Ax + By = C$.

2. Obtain coefficients for x (or y) that differ only in sign by multiplying all terms of one or both equations by suitable constants.

3. Add the equations to eliminate one variable and solve the resulting equation.

4. Back-substitute the value obtained in Step 3 into either of the original equations and solve for the other variable.

5. Check your solution in *both* of the original equations.

To obtain coefficients (for one of the variables) that differ only in sign, you often need to multiply one or both of the equations by a suitable constant. For instance, in the system

$$
\begin{cases}
2x + 3y = 10 & \text{Equation 1} \\
4x - y = 5 & \text{Equation 2}
\end{cases}
$$

you can multiply each side of Equation 1 by -2 (which will produce x-coefficients that differ only in sign) or you can multiply each side of Equation 2 by 3 (which will produce y-coefficients that differ only in sign). This is further demonstrated in the next example.

Example 5 ■ The Method of Elimination

Solve each linear system.

(a) $\begin{cases} 4x - 5y = 13 \\ 3x - y = 7 \end{cases}$

(b) $\begin{cases} 5x + 3y = 8 \\ 2x - 4y = 11 \end{cases}$

Solution

(a) For this system, you can obtain coefficients of y that differ only in sign by multiplying Equation 2 by -5.

$$\begin{cases} 4x - 5y = 13 \\ 3x - y = 7 \end{cases} \Rightarrow \begin{array}{r} 4x - 5y = 13 \\ -15x + 5y = -35 \\ \hline -11x = -22 \end{array} \quad \begin{array}{l} \text{Equation 1} \\ \text{Multiply Equation 2 by } -5. \\ \text{Add equations.} \end{array}$$

So, you can see that $x = 2$. By back-substituting this value of x into Equation 2, you can solve for y, as follows.

$3x - y = 7$	Equation 2
$3(2) - y = 7$	Substitute 2 for x.
$6 - y = 7$	Simplify.
$-y = 1$	Subtract 6 from each side.
$y = -1$	Multiply each side by -1.

So, the solution is $(2, -1)$. Check to see that this solution satisfies both original equations. The solution is verified graphically in Figure 4.4.

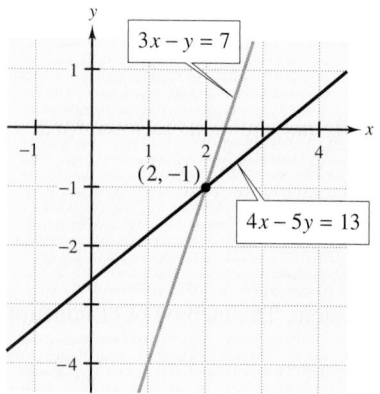

Figure 4.4

(b) You can obtain coefficients of y that differ only in sign by multiplying Equation 1 by 4 and Equation 2 by 3, in order to eliminate the y-term in both equations.

$$\begin{cases} 5x + 3y = 8 \\ 2x - 4y = 11 \end{cases} \Rightarrow \begin{array}{r} 20x + 12y = 32 \\ 6x - 12y = 33 \\ \hline 26x = 65 \end{array} \quad \begin{array}{l} \text{Multiply Equation 1 by 4.} \\ \text{Multiply Equation 2 by 3.} \\ \text{Add equations.} \end{array}$$

From this equation, you can determine that $x = \frac{5}{2}$. By back-substituting this value of x into Equation 2, you can solve for y, as follows.

$2x - 4y = 11$	Equation 2
$2\left(\frac{5}{2}\right) - 4y = 11$	Substitute $\frac{5}{2}$ for x.
$5 - 4y = 11$	Simplify.
$-4y = 6$	Subtract 5 from each side.
$y = -\frac{3}{2}$	Divide each side by -4.

So, the solution is $\left(\frac{5}{2}, -\frac{3}{2}\right)$. Check this solution in both of the original equations. The solution is verified graphically in Figure 4.5.

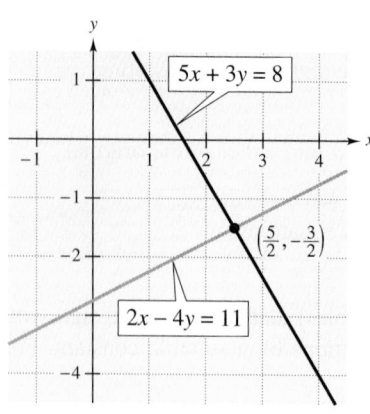

Figure 4.5

Example 6 ■ The Method of Elimination: No-Solution Case

Solve the system of linear equations.

$$\begin{cases} 3x + 9y = 8 & \text{Equation 1} \\ 2x + 6y = 7 & \text{Equation 2} \end{cases}$$

Solution

To obtain coefficients of x that differ only in sign, multiply Equation 1 by 2 and Equation 2 by -3.

$$\begin{cases} 3x + 9y = 8 \\ 2x + 6y = 7 \end{cases} \implies \begin{aligned} 6x + 18y &= 16 & \text{Multiply Equation 1 by 2.} \\ -6x - 18y &= -21 & \text{Multiply Equation 2 by } -3. \\ \hline 0 &= -5 & \text{False statement} \end{aligned}$$

Because the method of elimination produced the false statement $0 = -5$, you can conclude that the system is inconsistent and has no solution. The lines corresponding to the two equations of this system are shown in Figure 4.6. Note that the two lines are parallel and so have no point of intersection.

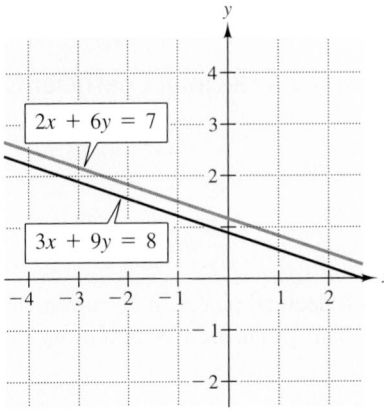

Figure 4.6

Example 7 shows how the method of elimination works with a system that has infinitely many solutions.

Example 7 ■ A System with Infinitely Many Solutions

Solve the system of linear equations.

$$\begin{cases} -2x + 6y = 3 & \text{Equation 1} \\ -12y = -6 - 4x & \text{Equation 2} \end{cases}$$

Algebraic Solution

To begin, write Equation 2 in standard form. Then, to obtain coefficients of x that differ only in sign, multiply Equation 1 by 2.

$$\begin{cases} -2x + 6y = 3 \\ 4x - 12y = -6 \end{cases} \implies \begin{aligned} -4x + 12y &= 6 \\ 4x - 12y &= -6 \\ \hline 0 &= 0 \end{aligned}$$

Because $0 = 0$ is a true statement, you can conclude that the system has infinitely many solutions. The solution set consists of all points (x, y) lying on the line $-2x + 6y = 3$.

Graphical Solution

Solve each equation for y.

$$y_1 = \tfrac{1}{3}x + \tfrac{1}{2} \text{ and } y_2 = \tfrac{1}{3}x + \tfrac{1}{2}$$

These two lines are identical, as shown in Figure 4.7. So, you can conclude that the system has infinitely many solutions. The solution set consists of all points (x, y) lying on the line $-2x + 6y = 3$.

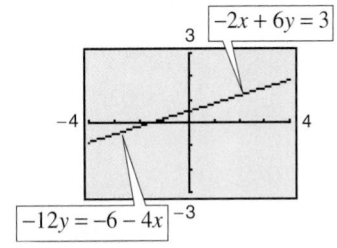

Figure 4.7

The next example shows how the method of elimination works with a system of linear equations having decimal coefficients.

Example 8 ■ Solving a Linear System Having Decimal Coefficients

Solve the system of linear equations.

$$\begin{cases} 0.02x - 0.05y = -0.38 & \text{Equation 1} \\ 0.03x + 0.04y = 1.04 & \text{Equation 2} \end{cases}$$

Solution

Because the coefficients in this system have two decimal places, it is convenient to begin by multiplying each equation by 100. This produces a system in which the coefficients are all integers.

$$\begin{cases} 2x - 5y = -38 & \text{Revised Equation 1} \\ 3x + 4y = 104 & \text{Revised Equation 2} \end{cases}$$

Now, to obtain coefficients of x that differ only in sign, multiply Equation 1 by 3 and Equation 2 by -2.

$$\begin{cases} 2x - 5y = -38 \\ 3x + 4y = 104 \end{cases}$$

$$\begin{array}{rl} 6x - 15y = -114 & \text{Multiply Equation 1 by 3.} \\ -6x - 8y = -208 & \text{Multiply Equation 2 by } -2. \\ \hline -23y = -322 & \text{Add equations.} \end{array}$$

So, the value of y is $\dfrac{-322}{-23} = 14$.

By back-substituting this value into revised Equation 2, you can solve for x as follows.

$$\begin{array}{ll} 3x + 4y = 104 & \text{Revised Equation 2} \\ 3x + 4(14) = 104 & \text{Substitute 14 for } y. \\ 3x + 56 = 104 & \text{Simplify.} \\ 3x = 48 & \text{Subtract 56 from each side.} \\ x = 16 & \text{Divide each side by 3.} \end{array}$$

So, the solution is $(16, 14)$. You can check this solution in both of the original equations in the system, as follows.

Check

Substitute into 1st Equation

$$0.02(16) - 0.05(14) \stackrel{?}{=} -0.38 \qquad \text{Substitute 16 for } x \text{ and 14 for } y \text{ in Equation 1.}$$
$$0.32 - 0.70 = -0.38 \ \checkmark \qquad \text{Solution checks.}$$

Substitute into 2nd Equation

$$0.03(16) + 0.04(14) \stackrel{?}{=} 1.04 \qquad \text{Substitute 16 for } x \text{ and 14 for } y \text{ in Equation 2.}$$
$$0.48 + 0.56 = 1.04 \ \checkmark \qquad \text{Solution checks.}$$

Figure 4.8 shows the solution to the system of linear equations in Example 8 using the *intersect* feature of a graphing utility.

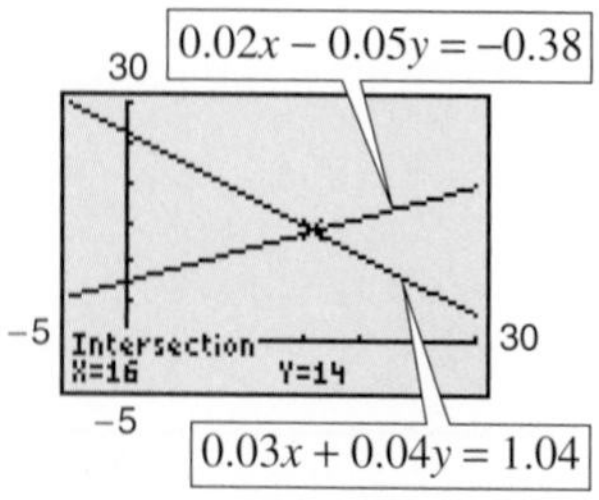

Figure 4.8

Applications

To determine whether a real-life problem can be solved using a system of linear equations, consider these questions:

(1) Does the problem involve more than one unknown quantity?

(2) Are there two (or more) equations or conditions to be satisfied?

If one or both of these conditions occur, the appropriate mathematical model for the problem may be a system of linear equations.

Example 9 ■ A Mixture Problem

A company with two stores buys six large delivery vans and five small delivery vans. The first store receives four of the large vans and two of the small vans for a total cost of $160,000. The second store receives two of the large vans and three of the small vans for a total cost of $128,000. What is the cost of each type of van?

Solution

The two unknowns in this problem are the costs of the two types of vans.

Verbal Model:

$$4 \left(\begin{array}{c} \text{Cost of} \\ \text{large van} \end{array} \right) + 2 \left(\begin{array}{c} \text{Cost of} \\ \text{small van} \end{array} \right) = \$160,000$$

$$2 \left(\begin{array}{c} \text{Cost of} \\ \text{large van} \end{array} \right) + 3 \left(\begin{array}{c} \text{Cost of} \\ \text{small van} \end{array} \right) = \$128,000$$

Labels: $x = $ Cost of large van   (dollars)

$y = $ Cost of small van   (dollars)

System of Equations: $\begin{cases} 4x + 2y = 160,000 & \text{Equation 1} \\ 2x + 3y = 128,000 & \text{Equation 2} \end{cases}$

To solve this system of linear equations, you can use the method of elimination. To obtain coefficients of x that differ only in sign, multiply Equation 2 by -2.

$$\begin{cases} 4x + 2y = 160,000 \\ 2x + 3y = 128,000 \end{cases} \implies \begin{array}{r} 4x + 2y = 160,000 \\ -4x - 6y = -256,000 \\ \hline -4y = -96,000 \end{array}$$

So, the cost of each small van is $y = \$24,000$. By back-substituting this value into Equation 1, you can find the cost of each large van, as follows.

$$\begin{array}{ll} 4x + 2y = 160,000 & \text{Equation 1} \\ 4x + 2(24,000) = 160,000 & \text{Substitute 24,000 for } y. \\ 4x + 48,000 = 160,000 & \text{Simplify.} \\ 4x = 112,000 & \text{Subtract 48,000 from each side.} \\ x = 28,000 & \text{Divide each side by 4.} \end{array}$$

So, the cost of each large van is $x = \$28,000$, and the cost of each small van is $y = \$24,000$. Check this solution in the original problem.

Study Tip

When solving application problems, make sure your answers make sense. For instance, a negative result for the x- or y-value in Example 9 would not make sense.

Example 10 ■ Investment

A total of $12,000 was invested in two funds paying 6% and 8% simple interest. If the interest for 1 year is $880, how much of the $12,000 was invested in each fund?

Solution

Verbal Model:

| Amount of money invested in 6% account | + | Amount of money invested in 8% account | = | Total amount invested |

| Interest from 6% account | + | Interest from 8% account | = | Total interest |

Labels: x = Amount invested at 6%   (dollars)

y = Amount invested at 8%   (dollars)

System of Equations:
$$\begin{cases} x + y = 12,000 & \text{Equation 1} \\ 0.06x + 0.08y = 880 & \text{Equation 2} \end{cases}$$

To solve this system, solve Equation 1 for y to obtain

$$y = 12,000 - x.$$

Next, substitute this expression into Equation 2.

$0.06x + 0.08y = 880$	Equation 2
$0.06x + 0.08(12,000 - x) = 880$	Substitute $12,000 - x$ for y.
$0.06x + 960 - 0.08x = 880$	Distributive Property
$-0.02x + 960 = 880$	Simplify.
$-0.02x = -80$	Subtract 960 from each side.
$x = 4000$	Divide each side by -0.02.

Back-substitute this value for x into revised Equation 1.

$y = 12,000 - x$	Revised Equation 1
$y = 12,000 - 4000$	Substitute 4000 for x.
$y = 8000$	Simplify.

So, you can conclude that $4000 was invested in the 6% fund and $8000 was invested in the 8% fund.

NOTE Notice that mathematical modeling, presented in Chapter 3, is used to solve the application problems in Examples 9, 10, and 11.

The total cost C of producing x units of a product usually has two components—the initial cost and the cost per unit. When enough units have been sold so that the total revenue R equals the total cost C, the sales have reached the **break-even point.** You can find this break-even point by setting C equal to R and solving for x. In other words, the break-even point corresponds to the point of intersection of the cost and revenue graphs.

Example 11 ■ Break-Even Analysis

A small business invests \$14,000 in equipment to produce a product. Each unit of the product costs \$0.80 to produce and is sold for \$1.50. How many items must be sold before the business breaks even?

Solution

Verbal Model:

$$\boxed{\text{Total cost}} = \boxed{\text{Cost per unit}} \cdot \boxed{\text{Number of units}} + \boxed{\text{Initial cost}}$$

$$\boxed{\text{Total revenue}} = \boxed{\text{Price per unit}} \cdot \boxed{\text{Number of units}}$$

Labels:
Total Cost $= C$ (dollars)
Cost per unit $= 0.80$ (dollars per unit)
Number of units $= x$ (units)
Initial cost $= 14,000$ (dollars)
Total revenue $= R$ (dollars)
Price per unit $= 1.50$ (dollars per unit)

System:
$$\begin{cases} C = 0.80x + 14,000 & \text{Equation 1} \\ R = 1.50x & \text{Equation 2} \end{cases}$$

Because the break-even point occurs when $R = C$, you have

$$1.50x = 0.80x + 14,000 \qquad \text{\small R = C}$$
$$0.7x = 14,000 \qquad \text{\small Subtract 0.80x from each side.}$$
$$x = 20,000. \qquad \text{\small Divide each side by 0.7.}$$

So, the business must sell 20,000 items before it breaks even. The graphs of the two equations are shown in Figure 4.9.

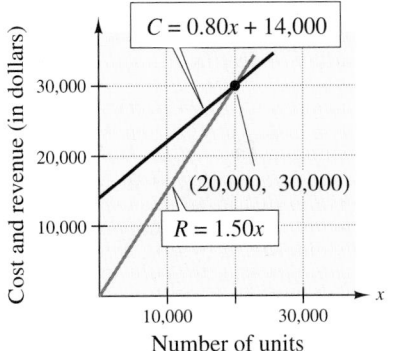

Figure 4.9

    Profit P (or loss) for the business can be determined by the equation $P = R - C$. Note in Figure 4.9 that sales less than the break-even point correspond to a loss for the business, whereas sales greater than the break-even point correspond to a profit for the business.

Collaborate!

Writing Systems of Equations

Suppose you are tutoring another student and want to create several systems of equations that the student can use for practice. To begin, you want to find some equations that have relatively simple solutions. For instance, ask each person in your group to try to write a system of equations that has $(3, -2)$ as a solution.

    Now, compare your systems. Discuss how you obtained your systems. Did everyone in the group write the same system? How many systems of equations have $(3, -2)$ as a solution? Explain the reasoning behind your answer.

4.1 ■ Exercises

Developing Skills

In Exercises 1–4, determine whether each ordered pair is a solution of the system of linear equations.

1. $\begin{cases} x + 2y = 9 \\ -2x + 3y = 10 \end{cases}$ (a) $(1, 4)$

                            (b) $(3, -1)$

2. $\begin{cases} 5x - 4y = 34 \\ x - 2y = 8 \end{cases}$ (a) $(0, 3)$

                            (b) $(6, -1)$

3. $\begin{cases} -2x + 7y = 46 \\ y = -3x \end{cases}$ (a) $(-3, 2)$

                            (b) $(-2, 6)$

4. $\begin{cases} -5x - 2y = 23 \\ 4y = -x - 19 \end{cases}$ (a) $(-3, -4)$

                            (b) $(3, 7)$

In Exercises 5–10, use the graphs of the equations to determine whether the system has any solutions. Find any solutions that exist.

5. $\begin{cases} x + y = 4 \\ x + y = -1 \end{cases}$

6. $\begin{cases} -x + y = 5 \\ x + 2y = 4 \end{cases}$

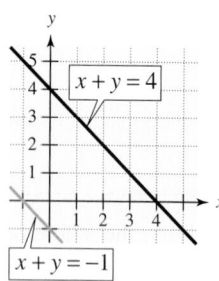

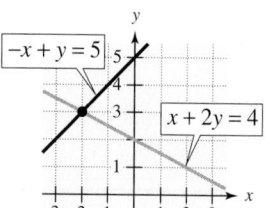

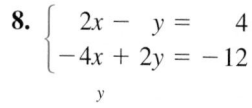

7. $\begin{cases} 5x - 3y = 4 \\ 2x + 3y = 3 \end{cases}$

8. $\begin{cases} 2x - y = 4 \\ -4x + 2y = -12 \end{cases}$

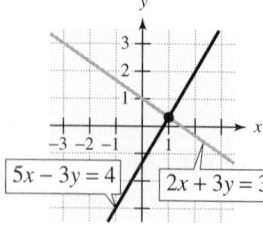

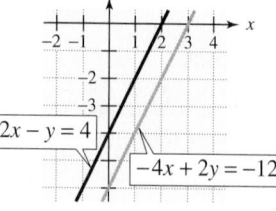

9. $\begin{cases} -x + 2y = 5 \\ 2x - 4y = -10 \end{cases}$

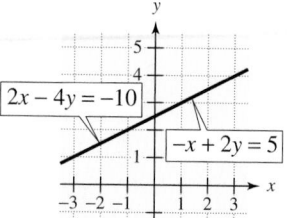

10. $\begin{cases} x = 5 \\ y = -3 \end{cases}$

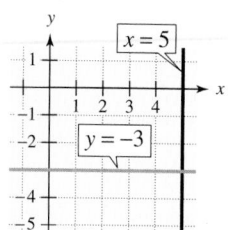

In Exercises 11–22, sketch the graphs of the equations and approximate any solutions of the system of linear equations.

11. $\begin{cases} -2x + y = 1 \\ x - 3y = 2 \end{cases}$

12. $\begin{cases} 5x - 6y = -30 \\ 5x + 4y = 20 \end{cases}$

13. $\begin{cases} 2x - 5y = 20 \\ 4x - 5y = 40 \end{cases}$

14. $\begin{cases} 5x + 3y = 24 \\ x - 2y = 10 \end{cases}$

15. $\begin{cases} x - y = -3 \\ 2x - y = 6 \end{cases}$

16. $\begin{cases} x = 4 \\ y = 3 \end{cases}$

17. $\begin{cases} x - 2y = 0 \\ 3x + 2y = 8 \end{cases}$

18. $\begin{cases} x - y = 0 \\ 5x - 2y = 6 \end{cases}$

19. $\begin{cases} -x + 2y = 1 \\ x - y = 2 \end{cases}$

20. $\begin{cases} x + y = 0 \\ 3x - 2y = 10 \end{cases}$

21. $\begin{cases} x - y = 1 \\ -3x + 3y = 8 \end{cases}$

22. $\begin{cases} x - 3y = 5 \\ -2x + 6y = -10 \end{cases}$

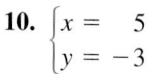 In Exercises 23–28, use a graphing utility to graph the equations and approximate any solutions of the system of linear equations.

23. $\begin{cases} -5x + 3y = 15 \\ x + y = 1 \end{cases}$

24. $\begin{cases} 9x + 4y = 8 \\ 7x - 4y = 8 \end{cases}$

25. $\begin{cases} 5x + 4y = 35 \\ -x + 3y = 12 \end{cases}$

26. $\begin{cases} 5x - 4y = 0 \\ -3x + 8y = 14 \end{cases}$

27. $\begin{cases} 4x - y = 3 \\ 6x + 2y = 1 \end{cases}$

28. $\begin{cases} x - 6y = 2 \\ 2x + 3y = 9 \end{cases}$

The symbol 🖩 indicates an exercise in which you are instructed to use a calculator or graphing utility.

 Warm-up exercises, quizzes, and other study resources related to this section are available at *college.hmco.com.*

In Exercises 29–32, decide whether the system is consistent or inconsistent.

29. $\begin{cases} 4x - 5y = 3 \\ -8x + 10y = -6 \end{cases}$　**30.** $\begin{cases} 4x - 5y = 3 \\ -8x + 10y = 14 \end{cases}$

31. $\begin{cases} -2x + 5y = 3 \\ 5x + 2y = 8 \end{cases}$　**32.** $\begin{cases} x + 10y = 12 \\ -2x + 5y = 2 \end{cases}$

In Exercises 33–36, use a graphing utility to graph the equations in the system. Use the graphs to determine whether the system is consistent or inconsistent. If the system is consistent, determine the number of solutions.

33. $\begin{cases} \frac{1}{3}x - \frac{1}{2}y = 1 \\ -2x + 3y = 6 \end{cases}$　**34.** $\begin{cases} x + y = 5 \\ x - y = 5 \end{cases}$

35. $\begin{cases} -2x + 3y = 6 \\ x - y = -1 \end{cases}$　**36.** $\begin{cases} 2x - 4y = 9 \\ x - 2y = 4.5 \end{cases}$

In Exercises 37–40, use a graphing utility to graph the equations in the system. The graphs appear to be parallel. Yet, from the slope-intercept forms of the lines, you find that the slopes are not equal and the graphs intersect. Find the point of intersection of the two lines.

37. $\begin{cases} x - 100y = -200 \\ 3x - 275y = 198 \end{cases}$　**38.** $\begin{cases} 35x - 33y = 0 \\ 12x - 11y = 92 \end{cases}$

39. $\begin{cases} 3x - 25y = 50 \\ 9x - 100y = 50 \end{cases}$　**40.** $\begin{cases} x + 40y = 80 \\ 2x + 150y = 195 \end{cases}$

In Exercises 41–54, solve the system by the method of substitution.

41. $\begin{cases} x = 4 \\ x - 2y = -2 \end{cases}$　**42.** $\begin{cases} y = 2 \\ x - 6y = -6 \end{cases}$

43. $\begin{cases} x + y = 3 \\ 2x - y = 0 \end{cases}$　**44.** $\begin{cases} -x + y = 5 \\ x - 4y = 0 \end{cases}$

45. $\begin{cases} x + y = 2 \\ x - 4y = 12 \end{cases}$　**46.** $\begin{cases} x - 2y = -1 \\ x - 5y = 2 \end{cases}$

47. $\begin{cases} x + 6y = 19 \\ x - 7y = -7 \end{cases}$　**48.** $\begin{cases} x - 5y = -6 \\ 4x - 3y = 10 \end{cases}$

49. $\begin{cases} 2x + 5y = 29 \\ 5x + 2y = 13 \end{cases}$

50. $\begin{cases} -13x + 16y = 10 \\ 5x + 16y = -26 \end{cases}$

51. $\begin{cases} 4x - 14y = -15 \\ 18x - 12y = 9 \end{cases}$

52. $\begin{cases} 5x - 24y = -12 \\ 17x - 24y = 36 \end{cases}$

53. $\begin{cases} 7x + 8y = 24 \\ x - 8y = 8 \end{cases}$　**54.** $\begin{cases} x - 3y = -2 \\ 5x + 3y = 17 \end{cases}$

In Exercises 55–74, solve the system by the method of elimination.

55. $\begin{cases} 3x - 2y = 5 \\ x + 2y = 7 \end{cases}$　**56.** $\begin{cases} -x + 2y = 9 \\ x + 3y = 16 \end{cases}$

57. $\begin{cases} 4x + y = -3 \\ -4x + 3y = 23 \end{cases}$　**58.** $\begin{cases} -3x + 5y = -23 \\ 2x - 5y = 22 \end{cases}$

59. $\begin{cases} x - 3y = 2 \\ 3x - 7y = 4 \end{cases}$　**60.** $\begin{cases} 7r - s = -25 \\ 2r + 5s = 14 \end{cases}$

61. $\begin{cases} 2u + 3v = 8 \\ 3u + 4v = 13 \end{cases}$　**62.** $\begin{cases} 4x - 3y = 25 \\ -3x + 8y = 10 \end{cases}$

63. $\begin{cases} 12x - 5y = 2 \\ -24x + 10y = 6 \end{cases}$　**64.** $\begin{cases} -2x + 3y = 9 \\ 6x - 9y = -27 \end{cases}$

65. $\begin{cases} \frac{2}{3}r - s = 0 \\ 10r + 4s = 19 \end{cases}$　**66.** $\begin{cases} x - y = -\frac{1}{2} \\ 4x - 48y = -35 \end{cases}$

67. $\begin{cases} 0.7u - v = -0.4 \\ 0.3u - 0.8v = 0.2 \end{cases}$

68. $\begin{cases} 0.15x - 0.35y = -0.5 \\ -0.12x + 0.25y = 0.1 \end{cases}$

69. $\begin{cases} 5x + 7y = 25 \\ x + 1.4y = 5 \end{cases}$　**70.** $\begin{cases} 12b - 13m = 2 \\ -6b + 6.5m = -2 \end{cases}$

71. $\begin{cases} 2x = 25 \\ 4x - 10y = 0.52 \end{cases}$　**72.** $\begin{cases} 6x - 6y = 25 \\ 3y = 11 \end{cases}$

73. $\begin{cases} \frac{3}{2}x - y = 4 \\ -x + \frac{2}{3}y = -1 \end{cases}$　**74.** $\begin{cases} 12x - 3y = 6 \\ 4x - y = 2 \end{cases}$

In Exercises 75–88, use a graphing utility to graph and solve the system. Verify your solution algebraically.

75. $\begin{cases} 3x + y = 3 \\ 2x - y = 7 \end{cases}$　**76.** $\begin{cases} -x + 2y = 2 \\ 3x + y = 15 \end{cases}$

77. $\begin{cases} y = 5x - 3 \\ y = -2x + 11 \end{cases}$　**78.** $\begin{cases} 4x + y = -2 \\ -6x + y = 18 \end{cases}$

79. $\begin{cases} 2x - y = 20 \\ -x + y = -5 \end{cases}$　**80.** $\begin{cases} 3x - 2y = -20 \\ 5x + 6y = 32 \end{cases}$

81. $\begin{cases} 3y = 2x + 21 \\ x = 50 - 4y \end{cases}$ **82.** $\begin{cases} x - 4y = 5 \\ 5x + 4y = 7 \end{cases}$

83. $\begin{cases} 2x + 3y = 0 \\ 3x + 5y = -1000 \end{cases}$ **84.** $\begin{cases} 3x + 2y = 5 \\ y = 2x + 13 \end{cases}$

85. $\begin{cases} x + 2y = 4 \\ \frac{1}{2}x + \frac{1}{3}y = 1 \end{cases}$ **86.** $\begin{cases} \frac{3}{2}x + 2y = 12 \\ \frac{1}{4}x + y = 4 \end{cases}$

87. $\begin{cases} 3x + 4y = 2 \\ 0.6x + 0.8y = 1.6 \end{cases}$ **88.** $\begin{cases} 0.4u + 1v = 800 \\ 0.7u + 2v = 1850 \end{cases}$

In Exercises 89 and 90, determine the value of k such that the system of linear equations is inconsistent.

89. $\begin{cases} 5x - 10y = 40 \\ -2x + ky = 30 \end{cases}$ **90.** $\begin{cases} 12x - 18y = 5 \\ -18x + ky = 10 \end{cases}$

In Exercises 91–94, find a system of linear equations that has the given solution. (There are many correct answers.)

91. $(5, 2)$ **92.** $(-8, 12)$

93. $\left(3, -\frac{3}{2}\right)$ **94.** $\left(-1, \frac{4}{5}\right)$

Solving Problems

Geometry In Exercises 95 and 96, find the dimensions of the rectangle meeting the specified conditions.

	Perimeter	Condition
95.	220 meters	Length is 120% of the width.
96.	280 feet	Width is 75% of the length.

97. *Geometry* Two angles are complementary. The measure of the first angle is 10 degrees less than three times the measure of the second angle. Find the measure of each angle.

98. *Geometry* Two angles are supplementary. The measure of the first angle is 4 degrees more than seven times the measure of the second angle. Find the measure of each angle.

99. *Gasoline Mixture* Twelve gallons of regular unleaded gasoline plus 8 gallons of premium unleaded gasoline cost $23.08. The price of premium unleaded is 11 cents more per gallon than the price of regular unleaded. Find the price per gallon for each grade of gasoline.

100. *Seed Mixture* Ten pounds of mixed birdseed sell for $6.97 per pound. The mixture is obtained from two kinds of birdseed, with one variety priced at $5.65 per pound and the other at $8.95 per pound. How many pounds of each variety of birdseed are used in the mixture?

101. *Hay Mixture* How many tons of hay at $125 per ton must be purchased with hay at $75 per ton to have 100 tons of hay with a value of $90 per ton?

102. *Alcohol Mixture* How many liters of a 40% alcohol solution must be mixed with a 65% solution to obtain 20 liters of a 50% solution?

103. *Ticket Sales* Eight hundred tickets were sold for a theater production and the receipts for the performance were $8600. The tickets for adults and students sold for $12.50 and $7.50, respectively. How many of each kind of ticket were sold?

104. *Ticket Sales* A fundraising dinner was held on two consecutive nights. On the first night, 100 adult tickets and 175 children's tickets were sold, for a total of $937.50. On the second night, 200 adult tickets and 316 children's tickets were sold, for a total of $1790.00. Find the price of each type of ticket.

105. *Simple Interest* A total of $20,000 is invested in two bonds that pay 8% and 9.5% simple interest. The annual interest is $1675. How much is invested in each bond?

106. *Simple Interest* A total of $4500 is invested in two funds paying 4% and 5% simple interest. The annual interest is $210. How much is invested in each fund?

107. *Break-Even Analysis* A small business invests $8000 in equipment to produce a product. Each unit of the product costs $1.20 to produce and is sold for $2.00. How many items must be sold before the business breaks even?

108. *Break-Even Analysis* You are planning to open a small business. You need an initial investment of $85,400. You estimate your weekly costs to be about $7400. If your projected weekly revenue is $8100, how many weeks will it take to break even?

109. *Rope Length* You must cut a rope that is 160 inches long into two pieces so that one piece is four times as long as the other. Find the length of each piece.

110. *Rope Length* You must cut a rope that is 25 feet long into two pieces so that one piece is 15 feet longer than the other piece. Find the length of each piece.

111. *Driving Distance* Two people share the driving for a trip of 300 miles. One person drives three times as far as the other. Find the distance that each person drives.

112. *Air Speed* An airplane flying into a headwind travels the 3000-mile flying distance between two cities in 6 hours and 15 minutes. On the return flight, the distance is traveled in 5 hours. Find the speed of the plane in still air and the speed of the wind, assuming that both remain constant throughout the round trip.

Explaining Concepts

113. If you graph a system of equations and the graphs do not intersect, what can you conclude about the solution of the system?

114. You graph a system of equations on a graphing utility and see only one line. Describe how the system could have (a) a unique solution, (b) infinitely many solutions, or (c) no solution.

115. Is it possible for a consistent system of linear equations to have exactly two solutions? Explain.

116. How can you recognize that a system of linear equations has no solution? Give an example.

Ongoing Review

In Exercises 117–120, sketch the graph of the equation.

117. $y = \frac{3}{5}x - 2$

118. $y = -2(x - 3)$

119. $y + 6 = -4(x + 1)$

120. $y - 5 = \frac{3}{4}(x - 10)$

In Exercises 121 and 122, solve the equation.

121. $5x + 2(x - 5) = 4$

122. $9y - 5(4 + 3y) = 2$

In Exercises 123 and 124, find an equation of the line passing through the two points.

123. $(3, 7), (-1, -5)$

124. $(0, 3), (-8, 11)$

125. *Cost* The annual operating cost C, of a truck is modeled by $C = 0.45m + 6200$, where m is the number of miles traveled by the truck each year. What number of miles will yield an annual operating cost that is less than $15,000?

126. *Wages* You must choose between two plans of payment when working for a company. One plan pays $2500 per month. The second pays $1500 per month plus a commission of 4% of your gross sales. Write an inequality whose solution is such that the second option gives the greater monthly wage. Solve the inequality.

Looking Further

Vietnam Veterans Memorial "The Wall" in Washington, D.C., was designed by Maya Ying Lin when she was a student at Yale University. This monument has two vertical, triangular sections of black granite with a common side (see figure). The top of each section is level with the ground. The bottoms of the two sections can be approximately modeled by the equations $2x + 50y = -505$ and $2x - 50y = 505$ when the x-axis is superimposed on the top of the wall. Each unit in the coordinate system represents 1 foot.

How deep is the memorial at the point where the two sections meet? How long is each section?

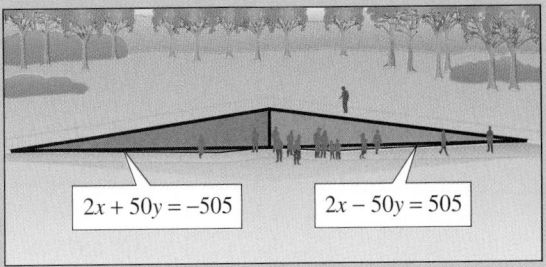

$2x + 50y = -505$ $2x - 50y = 505$

4.2 Systems of Linear Equations in Three Variables

What you should learn:

- How to determine if ordered triples are solutions of systems of linear equations
- How to solve systems of linear equations using row-echelon form with back-substitution
- How to solve systems of linear equations using the method of Gaussian elimination
- How to solve application problems using elimination with back-substitution

Why you should learn it:

Systems of linear equations in three variables can be used to model and solve real-life problems. For instance, in Exercise 53 on page 260, a system of linear equations can be used to analyze a chemical mixture for a pesticide.

Systems of Linear Equations in Three Variables

Each of the systems of linear equations discussed in Section 4.1 had only two variables. In real-life applications, many systems have more than two variables. Here are two examples.

$$\begin{cases} 2x + 5y + z = 0 \\ 3x + 2y - 4z = 0 \\ y + z = 0 \end{cases}$$

$$\begin{cases} 3x - 2y + z = 9 \\ x + y - 2z = -8 \\ -x - y + 3z = 12 \end{cases}$$

A **solution** of such a system is an **ordered triple,** (x, y, z), of real numbers that satisfies each equation in the system.

Example 1 ■ Checking Solutions

Check whether each of the ordered triples is a solution of the system of equations.

(a) $(4, -2, 2)$ (b) $(-2, -1, 1)$

$$\begin{cases} 2x + 5y + z = 0 & \text{Equation 1} \\ 3x + 2y - 4z = 0 & \text{Equation 2} \\ y + z = 0 & \text{Equation 3} \end{cases}$$

Solution

(a) To check the ordered triple $(4, -2, 2)$, substitute 4 for x, -2 for y, and 2 for z in each of the original equations.

     Equation 1: $2(4) + 5(-2) + (2) = 8 - 10 + 2 = 0$ ✓

     Equation 2: $3(4) + 2(-2) - 4(2) = 12 - 4 - 8 = 0$ ✓

     Equation 3: $(-2) + (2) = -2 + 2 = 0$ ✓

     Because $(4, -2, 2)$ is a solution of *each* equation, it *is* a solution of the system.

(b) To check the ordered triple $(-2, -1, 1)$, substitute -2 for x, -1 for y, and 1 for z in each of the original equations.

     Equation 1: $2(-2) + 5(-1) + (1) = -4 - 5 + 1 = -8$ ✗

     Equation 2: $3(-2) + 2(-1) - 4(1) = -6 - 2 - 4 = -12$ ✗

     Equation 3: $(-1) + (1) = -1 + 1 = 0$ ✓

     Because $(-2, -1, 1)$ is not a solution of *each* equation, it *is not* a solution of the system.

One way to solve a system of linear equations in three variables is to add multiples of the equations to each other to eliminate one of the variables. Your goal is to reduce the system to one that has only two equations and two variables. This procedure is demonstrated in Example 2.

Example 2 ■ Solving a System of Three Linear Equations

Solve the linear system.

$$\begin{cases} 3x - 2y + z = 9 & \text{Equation 1} \\ x + y - 2z = -8 & \text{Equation 2} \\ -x - 2y + 3z = 13 & \text{Equation 3} \end{cases}$$

Solution

There are many ways to solve this system. One way is to try to eliminate the variable x. Begin by multiplying Equation 2 by -3 and add the result to Equation 1.

$$\begin{array}{ll} 3x - 2y + z = 9 & \text{Equation 1} \\ \underline{-3x - 3y + 6z = 24} & \text{Multiply Equation 2 by } -3. \\ -5y + 7z = 33 & \text{New Equation 1} \end{array}$$

Then add Equation 2 and Equation 3.

$$\begin{array}{ll} x + y - 2z = -8 & \text{Equation 2} \\ \underline{-x - 2y + 3z = 13} & \text{Equation 3} \\ -y + z = 5 & \text{New Equation 2} \end{array}$$

You now have a new system of linear equations that has only two equations and two variables. You can solve this system using the techniques in Section 4.1.

$$\begin{cases} -5y + 7z = 33 \\ -y + z = 5 \end{cases} \implies \begin{array}{ll} -5y + 7z = 33 & \text{New Equation 1} \\ \underline{5y - 5z = -25} & \text{Multiply new Equation 2 by } -5. \\ 2z = 8 & \text{Add equations.} \end{array}$$

So, you can see that $z = 4$. By back-substituting this value of z into new Equation 1, you can solve for y.

$$-5y + 7(4) = 33 \implies y = -1$$

Now, knowing that $y = -1$ and $z = 4$, you can back-substitute these values into any of the original three equations to solve for x.

$$3x - 2(-1) + (4) = 9 \implies x = 1$$

So, $(1, -1, 4)$ is the solution of the system.

In Example 2, you can check the solution by substituting $x = 1$, $y = -1$, and $z = 4$ into each original equation as follows.

$$\begin{array}{ll} \text{Equation 1:} & 3(1) - 2(-1) + (4) = 3 + 2 + 4 = 9 \checkmark \\ \text{Equation 2:} & (1) + (-1) - 2(4) = 1 - 1 - 8 = -8 \checkmark \\ \text{Equation 3:} & -(1) - 2(-1) + 3(4) = -1 + 2 + 12 = 13 \checkmark \end{array}$$

Row-Echelon Form

Consider the systems of linear equations shown below.

$$\begin{cases} x - 2y + 2z = 9 \\ -x + 3y = -4 \\ 2x - 5y + z = 10 \end{cases} \qquad \begin{cases} x - 2y + 2z = 9 \\ y + 2z = 5 \\ z = 3 \end{cases}$$

Which one of these two systems do you think is easier to solve? After comparing the two systems, it should be clear that it is easier to solve the system on the right. The system on the right is said to be in **row-echelon form,** which means that it has a "stair-step" pattern with leading coefficients of 1.

Back-substitution can be used to solve a system in row-echelon form, as shown in Example 3.

Example 3 ■ Solving a System in Row-Echelon Form

Solve the system of linear equations.

$$\begin{cases} x - 2y + 2z = 9 & \text{Equation 1} \\ y + 2z = 5 & \text{Equation 2} \\ z = 3 & \text{Equation 3} \end{cases}$$

Solution

From Equation 3, you know the value of z. To solve for y, substitute $z = 3$ into Equation 2.

$$y + 2z = 5 \qquad \text{Equation 2}$$
$$y + 2(3) = 5 \qquad \text{Substitute 3 for } z.$$
$$y + 6 = 5 \qquad \text{Simplify.}$$
$$y = -1 \qquad \text{Subtract 6 from each side.}$$

Finally, substitute $y = -1$ and $z = 3$ into Equation 1.

$$x - 2y + 2z = 9 \qquad \text{Equation 1}$$
$$x - 2(-1) + 2(3) = 9 \qquad \text{Substitute } -1 \text{ for } y \text{ and 3 for } z.$$
$$x + 8 = 9 \qquad \text{Simplify.}$$
$$x = 1 \qquad \text{Subtract 8 from each side.}$$

Study Tip

When checking a solution, remember that the solution must satisfy each equation in the original system.

The solution is $x = 1$, $y = -1$, and $z = 3$, which can also be written as the ordered triple $(1, -1, 3)$. You can check this in the original system of equations, as follows.

Check

Equation 1: $\quad x - 2y + 2z = 9$
$$(1) - 2(-1) + 2(3) = 9 \checkmark$$

Equation 2: $\quad\quad y + 2z = 5$
$$(-1) + 2(3) = 5 \checkmark$$

Equation 3: $z = 3$
$$3 = 3 \checkmark$$

The Method of Gaussian Elimination

Two systems of equations are **equivalent systems** if they have the same solution set. To solve a system that is not in row-echelon form, you must first convert it to an *equivalent* system that is in row-echelon form. To see how this is done, let's take another look at the method of elimination, as applied to a system of two linear equations.

Example 4 ■ The Method of Elimination

Solve the system of linear equations.

$$\begin{cases} 3x - 2y = -1 \\ x - y = 0 \end{cases}$$

Equation 1

Equation 2

Solution

$$\begin{cases} x - y = 0 \\ 3x - 2y = -1 \end{cases}$$

You can interchange two equations in the system.

$$\begin{aligned} -3x + 3y &= 0 \\ \underline{3x - 2y} &= \underline{-1} \\ y &= -1 \end{aligned}$$

Multiply the first equation by -3.

You can add the multiple of the first equation to the second equation to obtain a new second equation.

$$\begin{cases} x - y = 0 \\ y = -1 \end{cases}$$

New system in row-echelon form

Now, using back-substitution, you can determine that the solution is $y = -1$ and $x = -1$, which can be written as the ordered pair $(-1, -1)$. Check the solution in the original system of equations, as follows.

Check

Equation 1: $\qquad 3x - 2y = -1$

$$3(-1) - 2(-1) = -1 \checkmark$$

Equation 2: $\qquad\qquad x - y = 0$

$$(-1) - (-1) = 0 \checkmark$$

Operations That Produce Equivalent Systems

Each of the **row operations** on a system of linear equations below produces an *equivalent* system of linear equations.

1. Interchange two equations.

2. Multiply one of the equations by a nonzero constant.

3. Add a multiple of one of the equations to another equation to replace the latter equation.

As shown in Example 4, rewriting a system of linear equations in row-echelon form usually involves a *chain* of equivalent systems, each of which is obtained by using one of the three basic row operations. This process is called **Gaussian elimination,** after the German mathematician Carl Friedrich Gauss (1777–1855). This method of Gaussian elimination easily adapts to computer use for solving systems of linear equations with dozens of variables.

Example 5 ■ Using Gaussian Elimination to Solve a System

Solve the system of linear equations.

$$\begin{cases} x - 2y + 2z = 9 & \text{Equation 1} \\ -x + 3y = -4 & \text{Equation 2} \\ 2x - 5y + z = 10 & \text{Equation 3} \end{cases}$$

Solution

Because the leading coefficient of the first equation is 1, you can begin by saving the x in the upper left position and eliminating the other x terms from the first column as follows.

$$\begin{cases} x - 2y + 2z = 9 \\ y + 2z = 5 \\ 2x - 5y + z = 10 \end{cases}$$

> Adding the first equation to the second equation produces a new second equation.

$$\begin{cases} x - 2y + 2z = 9 \\ y + 2z = 5 \\ -y - 3z = -8 \end{cases}$$

> Adding -2 times the first equation to the third equation produces a new third equation.

Now that all but the first x have been eliminated from the first column, go to work on the second column. (You need to eliminate y from the third equation.)

$$\begin{cases} x - 2y + 2z = 9 \\ y + 2z = 5 \\ -z = -3 \end{cases}$$

> Adding the second equation to the third equation produces a new third equation.

Finally, you need a coefficient of 1 for z in the third equation.

$$\begin{cases} x - 2y + 2z = 9 \\ y + 2z = 5 \\ z = 3 \end{cases}$$

> Multiplying the third equation by -1 produces a new third equation.

This is the same system that was solved in Example 3, and, as in that example, you can conclude by back-substitution that the solution is $x = 1$, $y = -1$, and $z = 3$. The solution can be written as the ordered triple $(1, -1, 3)$. You can check the solution by substituting $x = 1$, $y = -1$, and $z = 3$ into each equation of the original system as follows.

Check

Equation 1: $(1) - 2(-1) + 2(3) = 9$ ✓

Equation 2: $-(1) + 3(-1) = -4$ ✓

Equation 3: $2(1) - 5(-1) + (3) = 10$ ✓

Example 6 ■ Using Gaussian Elimination to Solve a System

Solve the system of linear equations.

$$\begin{cases} 2x + 4y + z = 6 & \text{Equation 1} \\ 3x + y + 3z = 17 & \text{Equation 2} \\ -5x - 8y + z = -10 & \text{Equation 3} \end{cases}$$

Solution

$$\begin{cases} x + 2y + \frac{1}{2}z = 3 \\ 3x + y + 3z = 17 \\ -5x - 8y + z = -10 \end{cases}$$

Multiplying the first equation by $\frac{1}{2}$ produces a new first equation.

$$\begin{cases} x + 2y + \frac{1}{2}z = 3 \\ -5y + \frac{3}{2}z = 8 \\ -5x - 8y + z = -10 \end{cases}$$

Adding -3 times the first equation to the second equation produces a new second equation.

$$\begin{cases} x + 2y + \frac{1}{2}z = 3 \\ -5y + \frac{3}{2}z = 8 \\ 2y + \frac{7}{2}z = 5 \end{cases}$$

Adding 5 times the first equation to the third equation produces a new third equation.

$$\begin{cases} x + 2y + \frac{1}{2}z = 3 \\ y - \frac{3}{10}z = -\frac{8}{5} \\ 2y + \frac{7}{2}z = 5 \end{cases}$$

Multiplying the second equation by $-\frac{1}{5}$ produces a new second equation.

$$\begin{cases} x + 2y + \frac{1}{2}z = 3 \\ y - \frac{3}{10}z = -\frac{8}{5} \\ \frac{41}{10}z = \frac{41}{5} \end{cases}$$

Adding -2 times the second equation to the third equation produces a new third equation.

$$\begin{cases} x + 2y + \frac{1}{2}z = 3 \\ y - \frac{3}{10}z = -\frac{8}{5} \\ z = 2 \end{cases}$$

Multiplying the third equation by $\frac{10}{41}$ produces a new third equation.

Now that the system of equations is in row-echelon form, you can see that $z = 2$. By back-substituting into the second equation, you can determine the value of y.

$$y - \frac{3}{10}(2) = -\frac{8}{5} \implies y = -1$$

By back-substituting $y = -1$ and $z = 2$ into the first equation, you can solve for x.

$$x + 2(-1) + \frac{1}{2}(2) = 3 \implies x = 4$$

So, the solution is $x = 4$, $y = -1$, and $z = 2$, which can be written as the ordered triple $(4, -1, 2)$. You can check this solution as follows.

Check

Equation 1: $2(4) + 4(-1) + (2) = 6$ ✓

Equation 2: $3(4) + (-1) + 3(2) = 17$ ✓

Equation 3: $-5(4) - 8(-1) + (2) = -10$ ✓

The next example involves an inconsistent system—one that has no solution. The key to recognizing an inconsistent system is that, at some stage in the elimination process, you obtain a false statement such as $0 = 6$. Watch for such statements as you do the exercises for this section.

Example 7 ■ An Inconsistent System

Solve the system of linear equations.

$$\begin{cases} x - 3y + z = 1 \\ 2x - y - 2z = 2 \\ x + 2y - 3z = -1 \end{cases}$$

Equation 1
Equation 2
Equation 3

Solution

$$\begin{cases} x - 3y + z = 1 \\ 5y - 4z = 0 \\ x + 2y - 3z = -1 \end{cases}$$

Adding -2 times the first equation to the second equation produces a new second equation.

$$\begin{cases} x - 3y + z = 1 \\ 5y - 4z = 0 \\ 5y - 4z = -2 \end{cases}$$

Adding -1 times the first equation to the third equation produces a new third equation.

$$\begin{cases} x - 3y + z = 1 \\ 5y - 4z = 0 \\ 0 = -2 \end{cases}$$

Adding -1 times the second equation to the third equation produces a new third equation.

Because the third "equation" is a false statement, you can conclude that this system is inconsistent and so has no solution. Moreover, because this system is equivalent to the original system, you can conclude that the original system also has no solution.

As with a system of linear equations in two variables, the solution(s) of a system of linear equations in more than two variables must fall into one of three categories.

> **The Number of Solutions of a Linear System**
>
> For a system of linear equations, exactly one of the statements below is true.
>
> 1. There is exactly one solution.
>
> 2. There are infinitely many solutions.
>
> 3. There is no solution.

The graph of a system of three linear equations in three variables consists of *three planes.* When these planes intersect in a single point, the system has exactly one solution. [See Figure 4.10(a).] When the three planes intersect in a line or a plane, the system has infinitely many solutions. [See Figures 4.10(b) and (c).] When the three planes have no point in common, the system has no solution. [See Figures 4.10(d) and (e).]

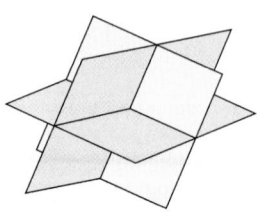

(a) Solution: one point

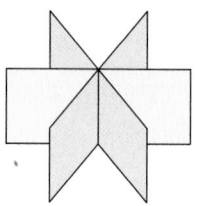

(b) Solution: one line

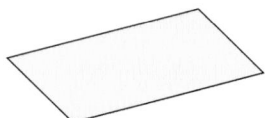

(c) Solution: one plane

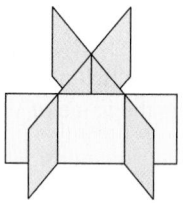

(d) Solution: none

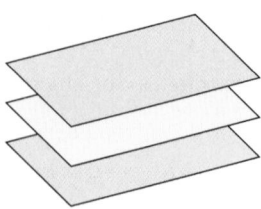

(e) Solution: none
Figure 4.10

Example 8 ■ A System with Infinitely Many Solutions

Solve the system of linear equations.

$$\begin{cases} x + y - 3z = -1 & \text{Equation 1} \\ y - z = 0 & \text{Equation 2} \\ -x + 2y = 1 & \text{Equation 3} \end{cases}$$

Solution

Begin by rewriting the system in row-echelon form.

$$\begin{cases} x + y - 3z = -1 \\ y - z = 0 \\ 3y - 3z = 0 \end{cases}$$

> Adding the first equation to the third equation produces a new third equation.

$$\begin{cases} x + y - 3z = -1 \\ y - z = 0 \\ 0 = 0 \end{cases}$$

> Adding -3 times the second equation to the third equation produces a new third equation.

This means that Equation 3 depends on Equations 1 and 2 in the sense that it gives us no additional information about the variables. So, the original system is equivalent to the system

$$\begin{cases} x + y - 3z = -1 \\ y - z = 0. \end{cases}$$

In this last equation, solve for y in terms of z to obtain $y = z$. Then, back-substituting for y in the previous equation produces $x = 2z - 1$. Finally, letting $z = a$, where a is any real number, you can see that the solutions to the given system are all of the form

$$x = 2a - 1, \qquad y = a, \qquad \text{and} \qquad z = a.$$

So, every ordered triple of the form

$$(2a - 1, a, a), \qquad a \text{ is a real number}$$

is a solution of the system.

In Example 8, there are other ways to write the same infinite set of solutions. For instance, letting $x = b$, the solutions could have been written as

$$\left(b, \tfrac{1}{2}(b + 1), \tfrac{1}{2}(b + 1)\right), \qquad b \text{ is a real number}.$$

To convince yourself that this description produces the same set of solutions, consider the comparison shown below.

Substitution	*Solution*	
$a = 0$	$(2(0) - 1, 0, 0) = (-1, 0, 0)$	Same solution
$b = -1$	$\left(-1, \tfrac{1}{2}(-1 + 1), \tfrac{1}{2}(-1 + 1)\right) = (-1, 0, 0)$	
$a = 1$	$(2(1) - 1, 1, 1) = (1, 1, 1)$	Same solution
$b = 1$	$\left(1, \tfrac{1}{2}(1 + 1), \tfrac{1}{2}(1 + 1)\right) = (1, 1, 1)$	

Study Tip

When comparing descriptions of an infinite solution set, keep in mind that there is more than one way to describe the set.

Applications

Example 9 ■ Moving Object

The height at time t of an object that is moving in a (vertical) line with constant acceleration a is given by the **position equation**

$$s = \tfrac{1}{2}at^2 + v_0t + s_0.$$

The height s is measured in feet, the acceleration a is measured in feet per second squared, the time t is measured in seconds, v_0 is the initial velocity (at time $t = 0$), and s_0 is the initial height. Find the values of a, v_0, and s_0 if $s = 164$ feet at 1 second, $s = 180$ feet at 2 seconds, and $s = 164$ feet at 3 seconds.

Solution

By substituting the three values of t and s into the position equation, you obtain three linear equations in a, v_0, and s_0.

When $t = 1$, $s = 164$: $\quad \dfrac{1}{2}a(1)^2 + v_0(1) + s_0 = 164$

When $t = 2$, $s = 180$: $\quad \dfrac{1}{2}a(2)^2 + v_0(2) + s_0 = 180$

When $t = 3$, $s = 164$: $\quad \dfrac{1}{2}a(3)^2 + v_0(3) + s_0 = 164$

By multiplying Equation 1 and Equation 3 by 2, this system can be rewritten as

$$\begin{cases} a + 2v_0 + 2s_0 = 328 & \text{Equation 1} \\ 2a + 2v_0 + s_0 = 180 & \text{Equation 2} \\ 9a + 6v_0 + 2s_0 = 328 & \text{Equation 3} \end{cases}$$

and you can apply elimination to obtain

$$\begin{cases} a + 2v_0 + 2s_0 = 328 & \text{Equation 1} \\ -2v_0 - 3s_0 = -476. & \text{Equation 2} \\ 2s_0 = 232 & \text{Equation 3} \end{cases}$$

From the third equation $s_0 = 116$, so back-substituting into the second equation yields

$$\begin{aligned} -2v_0 - 3s_0 &= -476 & &\text{Equation 2} \\ -2v_0 - 3(116) &= -476 & &\text{Substitute 116 for } s_0. \\ -2v_0 &= -128 & &\text{Add 348 to each side.} \\ v_0 &= 64. & &\text{Divide each side by } -2. \end{aligned}$$

Finally, back-substituting $s_0 = 116$ and $v_0 = 64$ into the first equation yields

$$\begin{aligned} a + 2v_0 + 2s_0 &= 328 & &\text{Equation 1} \\ a + 2(64) + 2(116) &= 328 & &\text{Substitute 116 for } s_0 \text{ and 64 for } v_0. \\ a &= -32. & &\text{Solve for } a. \end{aligned}$$

So, the solution is $a = -32$, $v_0 = 64$, and $s_0 = 116$, which can be written as $(-32, 64, 116)$. The position equation for this object is $s = -16t^2 + 64t + 116$.

Example 10 ■ Data Analysis: Curve-Fitting

Find a quadratic equation $y = ax^2 + bx + c$ whose graph passes through the points $(-1, 3)$, $(1, 1)$, and $(2, 6)$.

Solution

Because the graph of $y = ax^2 + bx + c$ passes through the points $(-1, 3)$, $(1, 1)$, and $(2, 6)$, you can write three linear equations as follows.

$$\text{When } x = -1, y = 3: \quad a(-1)^2 + b(-1) + c = 3$$
$$\text{When } x = 1, y = 1: \quad a(1)^2 + b(1) + c = 1$$
$$\text{When } x = 2, y = 6: \quad a(2)^2 + b(2) + c = 6$$

This produces the system of linear equations below.

$$\begin{cases} a - b + c = 3 & \text{Equation 1} \\ a + b + c = 1 & \text{Equation 2} \\ 4a + 2b + c = 6 & \text{Equation 3} \end{cases}$$

The solution of this system is $a = 2$, $b = -1$, and $c = 0$ (try solving this on your own). So, the quadratic equation is $y = 2x^2 - x$. Recall from Section 2.2 that the graph shown in Figure 4.11, is called a *parabola*.

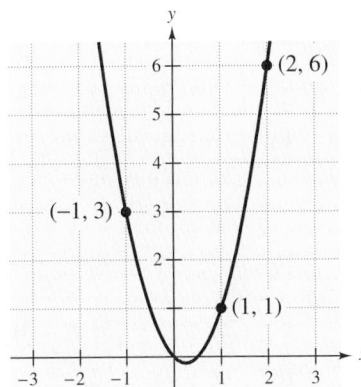

Figure 4.11

Example 11 ■ A Geometry Application

The sum of the measures of two angles of a triangle is twice the measure of the third angle. The measure of the first angle is $18°$ more than the measure of the third angle. Find the measures of the three angles.

Solution

Let x, y, and z represent the measures of the first, second, and third angles, respectively. The sum of the measures of the three angles of a triangle is $180°$. From the given information, you can write the system of equations as follows.

$$\begin{cases} x + y + z = 180 & \text{Equation 1} \\ x + y = 2z & \text{Equation 2} \\ x = z + 18 & \text{Equation 3} \end{cases}$$

By rewriting this system in the standard form you obtain

$$\begin{cases} x + y + z = 180 & \text{Equation 1} \\ x + y - 2z = 0. & \text{Equation 2} \\ x - z = 18 & \text{Equation 3} \end{cases}$$

Using Gaussian elimination to solve this system yields $x = 78$, $y = 42$, and $z = 60$. So, the measures of the three angles are $78°$, $42°$, and $60°$, respectively. You can check these solutions as follows.

Check

$$\text{Equation 1: } 78 + 42 + 60 = 180 \ \checkmark$$
$$\text{Equation 2: } 78 + 42 - 2(60) = 0 \ \checkmark$$
$$\text{Equation 3: } 78 - 60 = 18 \ \checkmark$$

4.2 ■ Exercises

Developing Skills

In Exercises 1 and 2, determine whether each ordered triple is a solution of the system of linear equations.

1. $\begin{cases} x + 3y + 2z = 1 \\ 5x - y + 3z = 16 \\ -3x + 7y + z = -14 \end{cases}$

    (a) $(0, 3, -2)$ (b) $(12, 5, -13)$

    (c) $(1, -2, 3)$ (d) $(-2, 5, -3)$

2. $\begin{cases} 3x - y + 4z = -10 \\ -x + y + 2z = 6 \\ 2x - y + z = -8 \end{cases}$

    (a) $(-2, 4, 0)$ (b) $(0, -3, 10)$

    (c) $(1, -1, 5)$ (d) $(7, 19, -3)$

In Exercises 3–6, use back-substitution to solve the system of linear equations.

3. $\begin{cases} x - 2y + 4z = 4 \\ 3y - z = 2 \\ z = -5 \end{cases}$

4. $\begin{cases} 5x + 4y - z = 0 \\ 10y - 3z = 11 \\ z = 3 \end{cases}$

5. $\begin{cases} x - 2y + 4z = 4 \\ y = 3 \\ y + z = 2 \end{cases}$

6. $\begin{cases} x = 10 \\ 3x + 2y = 2 \\ x + y + 2z = 0 \end{cases}$

In Exercises 7–38, use one or more of the methods illustrated in this section to solve the system of linear equations.

7. $\begin{cases} x + z = 4 \\ y = 2 \\ 4x + z = 7 \end{cases}$

8. $\begin{cases} x = 3 \\ -x + 3y = 3 \\ y + 2z = 4 \end{cases}$

9. $\begin{cases} x + y + z = 6 \\ 2x - y + z = 3 \\ 3x - z = 0 \end{cases}$

10. $\begin{cases} x + y + z = 2 \\ -x + 3y + 2z = 8 \\ 4x + y = 4 \end{cases}$

11. $\begin{cases} x + y + z = -3 \\ 4x + y - 3z = 11 \\ 2x - 3y + 2z = 9 \end{cases}$

12. $\begin{cases} x - y + 2z = -4 \\ 3x + y - 4z = -6 \\ 2x + 3y - 4z = 4 \end{cases}$

13. $\begin{cases} x + 2y + 6z = 5 \\ -x + y - 2z = 3 \\ x - 4y - 2z = 1 \end{cases}$

14. $\begin{cases} x + 6y + 2z = 9 \\ 3x - 2y + 3z = -1 \\ 5x - 5y + 2z = 7 \end{cases}$

15. $\begin{cases} 2x + 2z = 2 \\ 5x + 3y = 4 \\ 3y - 4z = 4 \end{cases}$

16. $\begin{cases} 6y + 4z = -12 \\ 3x + 3y = 9 \\ 2x - 3z = 10 \end{cases}$

17. $\begin{cases} x + y + 8z = 3 \\ 2x + y + 11z = 4 \\ x + 3z = 0 \end{cases}$

18. $\begin{cases} 2x + y + 3z = 1 \\ 2x + 6y + 8z = 3 \\ 6x + 8y + 18z = 5 \end{cases}$

19. $\begin{cases} y + z = 5 \\ 2x + 4z = 4 \\ 2x - 3y = -14 \end{cases}$

20. $\begin{cases} 3x - y - 2z = 5 \\ 2x + y + 3z = 6 \\ 6x - y - 4z = 9 \end{cases}$

21. $\begin{cases} 2x - 4y + z = 0 \\ 3x + 2z = -1 \\ -6x + 3y + 2z = -10 \end{cases}$

22. $\begin{cases} 5x + 2y = -8 \\ z = 5 \\ 3x - y + z = 9 \end{cases}$

23. $\begin{cases} 2x + 6y - 4z = 8 \\ 3x + 10y - 7z = 12 \\ -2x - 6y + 5z = -3 \end{cases}$

24. $\begin{cases} x + 2y - 2z = 4 \\ 2x + 5y - 7z = 5 \\ 3x + 7y - 9z = 10 \end{cases}$

25. $\begin{cases} x - 2y - z = 3 \\ 2x + y - 3z = 1 \\ x + 8y - 3z = -7 \end{cases}$

26. $\begin{cases} x + y + z = 2 \\ 2x + 3y + z = 3 \\ x - y - 2z = -6 \end{cases}$

27. $\begin{cases} x - 2y + 4z = 12 \\ 2x - y + 5z = 18 \\ -x + 3y - 3z = -8 \end{cases}$

28. $\begin{cases} x - y + 3z = 8 \\ 2x - y + 4z = 11 \\ -x + 2y - 4z = -11 \end{cases}$

29. $\begin{cases} 2x - y + 4z = 1 \\ x - y + 3z = 0 \\ -x + 2y - 4z = 2 \end{cases}$

30. $\begin{cases} x + 4y - 2z = 2 \\ -3x + y + z = -2 \\ 5x + 7y - 5z = 6 \end{cases}$

31. $\begin{cases} 2x + z = 1 \\ 5y - 3z = 2 \\ 6x + 20y - 9z = 11 \end{cases}$

32. $\begin{cases} 2x + y - z = 4 \\ y + 3z = 2 \\ 3x + 2y = 4 \end{cases}$

33. $\begin{cases} 2x + 3z = 4 \\ 5x + y + z = 2 \\ 11x + 3y - 3z = 0 \end{cases}$

34. $\begin{cases} 3x + y + z = 2 \\ 4x + 2z = 1 \\ 5x - y + 3z = 0 \end{cases}$

35. $\begin{cases} \frac{1}{3}x + \frac{2}{3}y + 2z = -1 \\ x + 2y + \frac{3}{2}z = \frac{3}{2} \\ \frac{1}{2}x + 2y + \frac{12}{5}z = \frac{1}{10} \end{cases}$

36. $\begin{cases} 6x - 3y + 2z = -1 \\ 3x - \frac{3}{2}y + z = -\frac{1}{2} \\ -2x + y - \frac{2}{3}z = -\frac{1}{3} \end{cases}$

37. $\begin{cases} 0.2x + 1.3y + 0.6z = 0.1 \\ 0.1x + 0.3z = 0.7 \\ 2x + 10y + 8z = 8 \end{cases}$

38. $\begin{cases} 0.3x - 0.1y + 0.2z = 0.35 \\ 2x + y - 2z = -1 \\ 2x + 4y + 3z = 10.5 \end{cases}$

In Exercises 39 and 40, find a system of linear equations in three variables with integer coefficients that has the given point as a solution. (*Note:* There are many correct answers.)

39. $(4, -3, 2)$ **40.** $(5, 7, -10)$

Solving Problems

Curve-Fitting   In Exercises 41–46, find the equation of the parabola $y = ax^2 + bx + c$ that passes through the points.

41. $(0, -4), (1, 1), (2, 10)$ **42.** $(0, 5), (1, 6), (2, 5)$

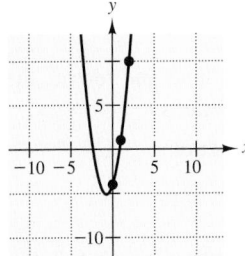   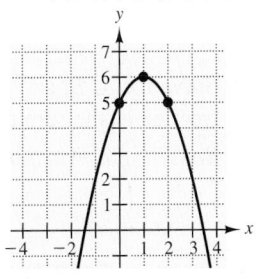

43. $(1, 0), (2, -1), (3, 0)$ **44.** $(1, 2), (2, 1), (3, -4)$

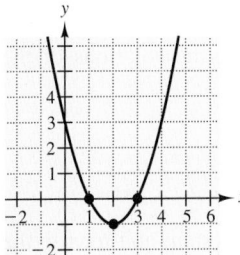   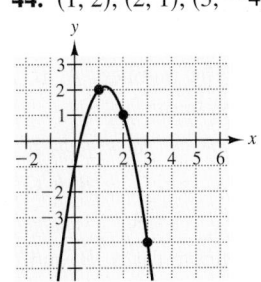

45. $(-1, -3), (1, 1), (2, 0)$
46. $(-1, -1), (1, 1), (2, -4)$

Physics   In Exercises 47–50, find the position equation $s = \frac{1}{2}at^2 + v_0t + s_0$ for an object that has the indicated heights at the specified times.

47. $s = 128$ feet at $t = 1$ second
     $s = 80$ feet at $t = 2$ seconds
     $s = 0$ feet at $t = 3$ seconds

48. $s = 48$ feet at $t = 1$ second
     $s = 64$ feet at $t = 2$ seconds
     $s = 48$ feet at $t = 3$ seconds

49. $s = 32$ feet at $t = 1$ second
     $s = 32$ feet at $t = 2$ seconds
     $s = 0$ feet at $t = 3$ seconds

50. $s = 10$ feet at $t = 0$ seconds
     $s = 54$ feet at $t = 1$ second
     $s = 46$ feet at $t = 3$ seconds

51. *Investment* An inheritance of $16,000 was divided among three investments yielding a total of $940 in interest per year. The interest rates for the three investments were 5%, 6%, and 7%. Find the amount invested at each rate if the amount invested at 6% is $3000 less than the amount invested at 5%.

52. *Investment* You receive a total of $1520 a year in interest from three investments. The interest rates for the three investments are 5%, 7%, and 8%. The 5% investment is half of the 7% investment, and the 7% investment is $1500 less than the 8% investment. What is the amount of each investment?

53. *Chemical Mixture* A mixture of 12 gallons of chemical A, 16 gallons of chemical B, and 26 gallons of chemical C is required to kill a certain destructive crop insect. Commercial spray X contains one, two, and two parts of these chemicals. Spray Y contains only chemical C. Spray Z contains only chemicals A and B in equal amounts. How much of each type of commercial spray is needed to obtain the desired mixture?

54. *Fertilizer Mixture* A mixture of 5 pounds of fertilizer A, 13 pounds of fertilizer B, and 4 pounds of fertilizer C provides the optimal nutrients for a certain kind of plant. Commercial brand X contains equal parts of fertilizer B and fertilizer C. Brand Y contains one part of fertilizer A and two parts of fertilizer B. Brand Z contains two parts of fertilizer A, five parts of fertilizer B, and two parts of fertilizer C. How much of each fertilizer brand is needed to obtain the desired mixture?

55. *Sports* The table shows the percents of each unit of the North High School football team that were chosen for academic honors, as city all-stars, and as county all-stars. Of all the players on the football team, 5 were awarded with academic honors, 13 were named city all-stars, and 4 were named county all-stars. How many members of each unit are there on the football team?

	Defense	Offense	Special teams
Academic honors	0%	10%	20%
City all-stars	10%	20%	50%
County all-stars	10%	0%	20%

56. *Music* The table shows the percents of each section of the North High School orchestra that were chosen to participate in the city orchestra, the county orchestra, and the state orchestra. Thirty members of the city orchestra, 17 members of the county orchestra, and 10 members of the state orchestra are from North High. How many members are in each section of North High's orchestra?

Orchestra	String	Wind	Percussion
City orchestra	40%	30%	50%
County orchestra	20%	25%	25%
State orchestra	10%	15%	25%

57. *Geometry* The sum of the measures of two angles of a triangle is twice the measure of the third angle. The measure of the second angle is 28° less than the measure of the third angle. Find the measures of the three angles.

58. *Geometry* The measure of the third angle of a triangle is 6° more than twice the measure of the first angle. The measure of the second angle is three times the measure of the first angle. Find the measures of the three angles.

59. *Rewriting a Fraction* The fraction

$$\frac{2x^2 - 9x}{(x - 2)^3}$$

can be written as a sum of three fractions as follows.

$$\frac{2x^2 - 9x}{(x - 2)^3} = \frac{A}{x - 2} + \frac{B}{(x - 2)^2} + \frac{C}{(x - 2)^3}$$

The numbers A, B, and C are the solutions of the system

$$\begin{cases} 4A - 2B + C = 0 \\ -4A + B = -9. \\ A = 2 \end{cases}$$

Solve for A, B, and C in the system and rewrite the fraction as the sum of three fractions.

60. *Rewriting a Fraction* The fraction

$$\frac{1}{(x^3 - x)}$$

can be written as a sum of three fractions as follows.

$$\frac{1}{x^3 - x} = \frac{A}{x} + \frac{B}{x + 1} + \frac{C}{x - 1}$$

The numbers A, B, and C are the solutions of the system

$$\begin{cases} A + B + C = 0 \\ -B + C = 0. \\ -A = 1 \end{cases}$$

Solve for A, B, and C in the system and rewrite the fraction as the sum of three fractions.

61. *Investigation* The total numbers of sides and diagonals of regular polygons with three, four, and five sides are 3, 6, and 10, as shown in the figure. Find the function $y = ax^2 + bx + c$ that fits the data. Then check to see if it gives the correct answer (15) for a polygon with six sides.

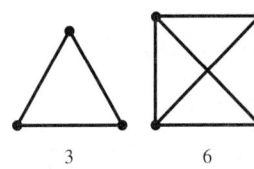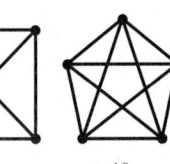

3 6 10 15

Figure for 61

Explaining Concepts

62. Describe a linear system of equations that is in row-echelon form.

63. Are the two systems of linear equations equivalent? Give reasons for your answer.

$$\begin{cases} x + 3y - z = 6 \\ 2x - y + 2z = 1 \\ 3x + 2y - z = 2 \end{cases} \qquad \begin{cases} x + 3y - z = 6 \\ -7y + 4z = 1 \\ -7y - 4z = -16 \end{cases}$$

64. Describe the row operations that are performed on a system of linear equations to produce an equivalent system of equations.

In Exercises 65 and 66, perform the row operation and write the equivalent system of linear equations. Explain what the operation accomplished.

65. Add Equation 1 to Equation 2.

$$\begin{cases} x - 2y + 3z = 5 & \text{Equation 1} \\ -x + 3y - 5z = 4 & \text{Equation 2} \\ 2x \qquad - 3z = 0 & \text{Equation 3} \end{cases}$$

66. Add -2 times Equation 1 to Equation 3.

$$\begin{cases} x - 2y + 3z = 5 & \text{Equation 1} \\ -x + 3y - 5z = 4 & \text{Equation 2} \\ 2x \qquad - 3z = 0 & \text{Equation 3} \end{cases}$$

Ongoing Review

In Exercises 67–70, evaluate the expression.

67. $4(6) - 2(5) + 7(-10)$

68. $8\left(-\frac{1}{2}\right) - 2(-1) + 14(3)$

69. $\dfrac{5(-1) + 3(3) - 10(5)}{-6(7) + 5(-2) - 6(3)}$

70. $\dfrac{2(-6) - 3(-9) + 11(5)}{-7(12) + 6(2) - 8(-3)}$

In Exercises 71–74, simplify the expression.

71. $(5x - 3y + 10z) + 2(2x + y - 5z)$

72. $(8x + 7y - 2z) - 4(2x - 2y + 4z)$

73. $2(3x - 12y + 9z) - 3(2x + 2y + 5z)$

74. $-5(4x + 3y - z) + 3(-x + 5y + 6z)$

75. *Average Speed* A car travels for 2 hours at an average speed of 40 miles per hour. How much longer must the car travel at an average speed of 55 miles per hour so that the average speed for the total trip will be 50 miles per hour?

76. *Ticket Sales* Five hundred tickets were sold for a fundraising dinner. The receipts totaled $3312.50. Adult tickets were $7.50 each and children's tickets were $4.00 each. How many tickets of each type were sold?

Looking Further

Chemistry A chemist needs 10 liters of a 25% acid solution. It is to be mixed from three solutions whose concentrations are 10%, 20%, and 50%. How many liters of each solution will satisfy the following conditions?

(a) Use 2 liters of the 50% solution.

(b) Use as little as possible of the 50% solution.

(c) Use as much as possible of the 50% solution.

4.3 Matrices and Linear Systems

What you should learn:

- How to determine the order of matrices
- How to form coefficient and augmented matrices and form linear systems from the augmented matrices
- How to perform elementary row operations to solve systems of linear equations
- How to use matrices and Gaussian elimination with back-substitution to solve systems of linear equations

Why you should learn it:

Matrices can be used to solve systems of linear equations that model real-life situations. For instance, in Exercise 67 on page 271, a matrix can be used to find a model for the amount of recycled materials in the United States.

NOTE The order of a matrix is always given as *row by column*.

Matrices

In this section, you will study a streamlined technique for solving systems of linear equations. This technique involves the use of a rectangular array of real numbers called a **matrix.** (The plural of matrix is *matrices*.) Here is an example of a matrix.

$$
\begin{array}{cccc}
\text{Column} & \text{Column} & \text{Column} & \text{Column} \\
1 & 2 & 3 & 4
\end{array}
$$

$$
\begin{array}{c}
\text{Row 1} \\
\text{Row 2} \\
\text{Row 3}
\end{array}
\begin{bmatrix}
3 & -2 & 4 & 1 \\
0 & 1 & -1 & 2 \\
2 & 0 & -3 & 0
\end{bmatrix}
$$

This matrix has three rows and four columns, which means that its **order** is 3×4, which is read as "3 by 4." Each number in the matrix is an **entry** of the matrix.

Example 1 ■ Order of Matrices

Determine the order of each matrix.

(a) $\begin{bmatrix} 1 & -2 & 4 \\ 0 & 1 & -2 \end{bmatrix}$ (b) $\begin{bmatrix} 0 & 0 \\ 0 & 0 \end{bmatrix}$ (c) $\begin{bmatrix} 1 & -3 \\ -2 & 0 \\ 4 & -2 \end{bmatrix}$

Solution

(a) This matrix has two rows and three columns, so the order is 2×3.

(b) This matrix has two rows and two columns, so the order is 2×2.

(c) This matrix has three rows and two columns, so the order is 3×2.

A matrix with the same number of rows as columns is called a **square matrix.** For instance, the 2×2 matrix in Example 1(b) is square.

Augmented and Coefficient Matrices

A matrix derived from a system of linear equations (each written in standard form) is the **augmented matrix** of the system. Moreover, the matrix that is derived from the coefficients of the system (but that does not include the constant terms) is the **coefficient matrix** of the system. Here is an example.

$$
\begin{array}{ccc}
\textit{System} & \textit{Coefficient Matrix} & \textit{Augmented Matrix}
\end{array}
$$

$$
\begin{cases}
x - 4y + 3z = 5 \\
-x + 3y - z = -3 \\
2x \quad\quad - 4z = 6
\end{cases}
\quad
\begin{bmatrix}
1 & -4 & 3 \\
-1 & 3 & -1 \\
2 & 0 & -4
\end{bmatrix}
\quad
\begin{bmatrix}
1 & -4 & 3 & \vdots & 5 \\
-1 & 3 & -1 & \vdots & -3 \\
2 & 0 & -4 & \vdots & 6
\end{bmatrix}
$$

Note the use of 0 for the missing y-variable in the third equation, and also note the fourth column of constant terms in the augmented matrix.

When forming either the coefficient matrix or the augmented matrix of a system, you should begin by vertically aligning the variables in the equations.

System	*Align Variables*	*Form Augmented Matrix*

$$\begin{cases} x + 3y = 9 \\ -y + 4z = -2 \\ x - 5z = 0 \end{cases} \qquad \begin{aligned} x + 3y = 9 \\ -y + 4z = -2 \\ x - 5z = 0 \end{aligned} \qquad \begin{bmatrix} 1 & 3 & 0 & \vdots & 9 \\ 0 & -1 & 4 & \vdots & -2 \\ 1 & 0 & -5 & \vdots & 0 \end{bmatrix}$$

Example 2 ■ Forming Coefficient and Augmented Matrices

Form the coefficient matrix and the augmented matrix for each system of linear equations.

(a) $\begin{cases} -x + 5y = 2 \\ 7x - 2y = -6 \end{cases}$ (b) $\begin{cases} 3x + 2y - z = 1 \\ x + 2z = -3 \\ -2x - y = 4 \end{cases}$ (c) $\begin{cases} x = 3y - 1 \\ 2y - 5 = 9x \end{cases}$

Solution

System	*Coefficient Matrix*	*Augmented Matrix*

(a) $\begin{cases} -x + 5y = 2 \\ 7x - 2y = -6 \end{cases} \qquad \begin{bmatrix} -1 & 5 \\ 7 & -2 \end{bmatrix} \qquad \begin{bmatrix} -1 & 5 & \vdots & 2 \\ 7 & -2 & \vdots & -6 \end{bmatrix}$

(b) $\begin{cases} 3x + 2y - z = 1 \\ x + 2z = -3 \\ -2x - y = 4 \end{cases} \qquad \begin{bmatrix} 3 & 2 & -1 \\ 1 & 0 & 2 \\ -2 & -1 & 0 \end{bmatrix} \qquad \begin{bmatrix} 3 & 2 & -1 & \vdots & 1 \\ 1 & 0 & 2 & \vdots & -3 \\ -2 & -1 & 0 & \vdots & 4 \end{bmatrix}$

(c) $\begin{cases} x - 3y = -1 \\ -9x + 2y = 5 \end{cases} \qquad \begin{bmatrix} 1 & -3 \\ -9 & 2 \end{bmatrix} \qquad \begin{bmatrix} 1 & -3 & \vdots & -1 \\ -9 & 2 & \vdots & 5 \end{bmatrix}$

Example 3 ■ Forming Linear Systems from Their Matrices

Write the system of linear equations that is represented by each matrix.

(a) $\begin{bmatrix} 3 & -5 & \vdots & 4 \\ -1 & 2 & \vdots & 0 \end{bmatrix}$ (b) $\begin{bmatrix} 1 & 3 & \vdots & 2 \\ 0 & 1 & \vdots & -3 \end{bmatrix}$ (c) $\begin{bmatrix} 2 & 0 & -8 & \vdots & 1 \\ -1 & 1 & 1 & \vdots & 2 \\ 5 & -1 & 7 & \vdots & 3 \end{bmatrix}$

Solution

(a) $\begin{cases} 3x - 5y = 4 \\ -x + 2y = 0 \end{cases}$ (b) $\begin{cases} x + 3y = 2 \\ y = -3 \end{cases}$ (c) $\begin{cases} 2x - 8z = 1 \\ -x + y + z = 2 \\ 5x - y + 7z = 3 \end{cases}$

NOTE   In Example 3, you could use any two different variables for the equations in parts (a) and (b), and any three different variables for the equations in part (c). For example, the system for part (c) could be written with the variables a, b, and c as follows.

$$\begin{cases} 2a - 8c = 1 \\ -a + b + c = 2 \\ 5a - b + 7c = 3 \end{cases}$$

Elementary Row Operations

In Section 4.2, you studied three operations that can be used on a system of linear equations to produce an equivalent system: (1) interchange two equations, (2) multiply an equation by a nonzero constant, and (3) add a multiple of an equation to another equation. In matrix terminology, these three operations correspond to **elementary row operations.**

Study Tip

Although elementary row operations are simple to perform, they involve a lot of arithmetic. Because it is easy to make a mistake, you should get in the habit of noting the elementary row operations performed in each step so that you can go back and check your work. People use different schemes to denote which elementary row operations have been performed. The scheme that is used in this text is to write an abbreviated version of the row operation to the left of the row that has been changed, as shown in Example 4.

Elementary Row Operations

Any of the **elementary row operations** below performed on an augmented matrix will produce a matrix that is row-equivalent to the original matrix. Two matrices are **row-equivalent** if one can be obtained from the other by a sequence of elementary row operations.

1. Interchange two rows.

2. Multiply a row by a nonzero constant.

3. Add a multiple of a row to another row.

Example 4 ■ Elementary Row Operations

(a) Interchange the first and second rows.

Original Matrix

$$\begin{bmatrix} 0 & 1 & 3 & 4 \\ -1 & 2 & 0 & 3 \\ 2 & -3 & 4 & 1 \end{bmatrix}$$

New Row-Equivalent Matrix

$$\begin{matrix} R_2 \\ R_1 \end{matrix} \begin{bmatrix} -1 & 2 & 0 & 3 \\ 0 & 1 & 3 & 4 \\ 2 & -3 & 4 & 1 \end{bmatrix}$$

(b) Multiply the first row by $\frac{1}{2}$.

Original Matrix

$$\begin{bmatrix} 2 & -4 & 6 & -2 \\ 1 & 3 & -3 & 0 \\ 5 & -2 & 1 & 2 \end{bmatrix}$$

New Row-Equivalent Matrix

$$\frac{1}{2}R_1 \rightarrow \begin{bmatrix} 1 & -2 & 3 & -1 \\ 1 & 3 & -3 & 0 \\ 5 & -2 & 1 & 2 \end{bmatrix}$$

(c) Add -2 times the first row to the third row.

Original Matrix

$$\begin{bmatrix} 1 & 2 & -4 & 3 \\ 0 & 3 & -2 & -1 \\ 2 & 1 & 5 & -2 \end{bmatrix}$$

New Row-Equivalent Matrix

$$-2R_1 + R_3 \rightarrow \begin{bmatrix} 1 & 2 & -4 & 3 \\ 0 & 3 & -2 & -1 \\ 0 & -3 & 13 & -8 \end{bmatrix}$$

(d) Add six times the first row to the second row.

Original Matrix

$$\begin{bmatrix} 1 & 2 & 2 & -4 \\ -6 & -11 & 3 & 18 \\ 0 & 0 & 4 & 7 \end{bmatrix}$$

New Row-Equivalent Matrix

$$6R_1 + R_2 \rightarrow \begin{bmatrix} 1 & 2 & 2 & -4 \\ 0 & 1 & 15 & -6 \\ 0 & 0 & 4 & 7 \end{bmatrix}$$

In Section 4.2, Gaussian elimination was used with back-substitution to solve a system of linear equations. Example 5 demonstrates the matrix version of Gaussian elimination. The two methods are essentially the same. The basic difference is that with matrices you do not need to keep writing the variables.

Technology

Most graphing utilities are capable of performing row operations on matrices. Some graphing utilities have a function that will return the reduced row-echelon form of the matrix. Consult the user's guide of your graphing utility to learn how to perform elementary row operations. Most graphing utilities store the resulting matrix of each step into an answer variable. It is suggested that you store the results of each operation into a matrix variable.

Enter the matrix from Example 5 into your graphing utility and perform the indicated row operations.

Example 5 ■ Solving a System of Linear Equations

Linear System

$$\begin{cases} x - 2y + 2z = 9 \\ -x + 3y = -4 \\ 2x - 5y + z = 10 \end{cases}$$

Associated Augmented Matrix

$$\left[\begin{array}{ccc:c} 1 & -2 & 2 & 9 \\ -1 & 3 & 0 & -4 \\ 2 & -5 & 1 & 10 \end{array}\right]$$

Add the first equation to the second equation.

$$\begin{cases} x - 2y + 2z = 9 \\ y + 2z = 5 \\ 2x - 5y + z = 10 \end{cases}$$

Add the first row to the second row $(R_1 + R_2)$.

$$R_1 + R_2 \rightarrow \left[\begin{array}{ccc:c} 1 & -2 & 2 & 9 \\ 0 & 1 & 2 & 5 \\ 2 & -5 & 1 & 10 \end{array}\right]$$

Add -2 times the first equation to the third equation.

$$\begin{cases} x - 2y + 2z = 9 \\ y + 2z = 5 \\ -y - 3z = -8 \end{cases}$$

Add -2 times the first row to the third row $(-2R_1 + R_3)$.

$$-2R_1 + R_3 \rightarrow \left[\begin{array}{ccc:c} 1 & -2 & 2 & 9 \\ 0 & 1 & 2 & 5 \\ 0 & -1 & -3 & -8 \end{array}\right]$$

Add the second equation to the third equation.

$$\begin{cases} x - 2y + 2z = 9 \\ y + 2z = 5 \\ -z = -3 \end{cases}$$

Add the second row to the third row $(R_2 + R_3)$.

$$R_2 + R_3 \rightarrow \left[\begin{array}{ccc:c} 1 & -2 & 2 & 9 \\ 0 & 1 & 2 & 5 \\ 0 & 0 & -1 & -3 \end{array}\right]$$

Multiply the third equation by -1.

$$\begin{cases} x - 2y + 2z = 9 \\ y + 2z = 5 \\ z = 3 \end{cases}$$

Multiply the third row by -1.

$$-R_3 \rightarrow \left[\begin{array}{ccc:c} 1 & -2 & 2 & 9 \\ 0 & 1 & 2 & 5 \\ 0 & 0 & 1 & 3 \end{array}\right]$$

Now use back-substitution to find that the solution is $x = 1$, $y = -1$, and $z = 3$. The solution can be written as the ordered triple $(1, -1, 3)$.

NOTE The last matrix in Example 5 is in **row-echelon form**. The term *echelon* refers to the stair-step pattern formed by the nonzero elements of the matrix.

Definition of Row-Echelon Form of a Matrix

A matrix in **row-echelon form** has the properties below.

1. All rows consisting entirely of zeros occur at the bottom of the matrix.

2. For each row that does not consist entirely of zeros, the first nonzero entry is 1 (called a **leading 1**).

3. For two successive (nonzero) rows, the leading 1 in the higher row is farther to the left than the leading 1 in the lower row.

Using Matrices to Solve a System of Linear Equations

> **Gaussian Elimination with Back-Substitution**
>
> To use matrices and Gaussian elimination to solve a system of linear equations, use the steps below.
>
> 1. Write the augmented matrix of the system of linear equations.
>
> 2. Use elementary row operations to rewrite the augmented matrix in row-echelon form.
>
> 3. Write the system of linear equations corresponding to the matrix in row-echelon form, and use back-substitution to find the solution.

When you perform Gaussian elimination with back-substitution, you should operate from *left to right by columns*, using elementary row operations to obtain zeros in all entries directly below the leading 1's.

Example 6 ■ Gaussian Elimination with Back-Substitution

Solve the system of linear equations.

$$\begin{cases} 2x - 3y = -2 \\ x + 2y = 13 \end{cases}$$

Solution

$$\begin{bmatrix} 2 & -3 & \vdots & -2 \\ 1 & 2 & \vdots & 13 \end{bmatrix}$$ Augmented matrix for system of linear equations

$$\begin{matrix} R_2 \\ R_1 \end{matrix} \begin{bmatrix} 1 & 2 & \vdots & 13 \\ 2 & -3 & \vdots & -2 \end{bmatrix}$$ First column has leading 1 in upper left corner.

$$-2R_1 + R_2 \rightarrow \begin{bmatrix} 1 & 2 & \vdots & 13 \\ 0 & -7 & \vdots & -28 \end{bmatrix}$$ First column has a zero under its leading 1.

$$-\tfrac{1}{7}R_2 \rightarrow \begin{bmatrix} 1 & 2 & \vdots & 13 \\ 0 & 1 & \vdots & 4 \end{bmatrix}$$ Second column has leading 1 in second row.

The system of linear equations that corresponds to the (row-echelon) matrix is

$$\begin{cases} x + 2y = 13 \\ y = 4 \end{cases}.$$

Using back-substitution, you can find that the solution of the system is $x = 5$ and $y = 4$, which can be written as the ordered pair $(5, 4)$. Check this solution in the original system, as follows.

Check

Equation 1: $2(5) - 3(4) = -2$ ✓

Equation 2: $(5) + 2(4) = 13$ ✓

Example 7 ■ Gaussian Elimination with Back-Substitution

Solve the system of linear equations.

$$\begin{cases} 3x + 3y & = & 9 \\ 2x & - 3z = & 10 \\ 6y + 4z = & -12 \end{cases}$$

Solution

$$\begin{bmatrix} 3 & 3 & 0 & \vdots & 9 \\ 2 & 0 & -3 & \vdots & 10 \\ 0 & 6 & 4 & \vdots & -12 \end{bmatrix}$$ Augmented matrix for system of linear equations

$$\tfrac{1}{3}R_1 \rightarrow \begin{bmatrix} 1 & 1 & 0 & \vdots & 3 \\ 2 & 0 & -3 & \vdots & 10 \\ 0 & 6 & 4 & \vdots & -12 \end{bmatrix}$$ First column has leading 1 in upper left corner.

$$-2R_1 + R_2 \rightarrow \begin{bmatrix} 1 & 1 & 0 & \vdots & 3 \\ 0 & -2 & -3 & \vdots & 4 \\ 0 & 6 & 4 & \vdots & -12 \end{bmatrix}$$ First column has zeros under its leading 1.

$$-\tfrac{1}{2}R_2 \rightarrow \begin{bmatrix} 1 & 1 & 0 & \vdots & 3 \\ 0 & 1 & \tfrac{3}{2} & \vdots & -2 \\ 0 & 6 & 4 & \vdots & -12 \end{bmatrix}$$ Second column has leading 1 in second row.

$$-6R_2 + R_3 \rightarrow \begin{bmatrix} 1 & 1 & 0 & \vdots & 3 \\ 0 & 1 & \tfrac{3}{2} & \vdots & -2 \\ 0 & 0 & -5 & \vdots & 0 \end{bmatrix}$$ Second column has a zero under its leading 1.

$$-\tfrac{1}{5}R_3 \rightarrow \begin{bmatrix} 1 & 1 & 0 & \vdots & 3 \\ 0 & 1 & \tfrac{3}{2} & \vdots & -2 \\ 0 & 0 & 1 & \vdots & 0 \end{bmatrix}$$ Third column has leading 1 in third row.

The system of linear equations that corresponds to the (row-echelon) matrix is

$$\begin{cases} x + y & = & 3 \\ y + \tfrac{3}{2}z & = & -2. \\ z & = & 0 \end{cases}$$

In the third equation, you can see that $z = 0$. By back-substituting into the second equation, you can determine the value of y.

$$y + \tfrac{3}{2}(0) = -2 \quad \Longrightarrow \quad y = -2$$

By back-substituting $y = -2$ into Equation 1, you can solve for x.

$$x + (-2) = 3 \quad \Longrightarrow \quad x = 5$$

So, the solution is $x = 5$, $y = -2$, and $z = 0$, which can be written as the ordered triple $(5, -2, 0)$. Check this in the original system, as follows.

Check

Equation 1: $3(5) + 3(-2) \qquad\qquad = \quad 9$ ✓

Equation 2: $2(5) \qquad\qquad - 3(0) = \quad 10$ ✓

Equation 3: $\qquad\qquad 6(-2) + 4(0) = -12$ ✓

Example 8 ■ A System with No Solution

Solve the system of linear equations.

$$\begin{cases} 6x - 10y = -4 \\ 9x - 15y = 5 \end{cases}$$

Solution

$$\begin{bmatrix} 6 & -10 & \vdots & -4 \\ 9 & -15 & \vdots & 5 \end{bmatrix}$$
Augmented matrix for system of linear equations

$$\tfrac{1}{6}R_1 \rightarrow \begin{bmatrix} 1 & -\tfrac{5}{3} & \vdots & -\tfrac{2}{3} \\ 9 & -15 & \vdots & 5 \end{bmatrix}$$
First column has leading 1 in upper left corner.

$$-9R_1 + R_2 \rightarrow \begin{bmatrix} 1 & -\tfrac{5}{3} & \vdots & -\tfrac{2}{3} \\ 0 & 0 & \vdots & 11 \end{bmatrix}$$
First column has a zero under its leading 1.

The "equation" that corresponds to the second row of the matrix is $0 = 11$. Because this is a false statement, the system of equations has no solution.

Example 9 ■ A System with Infinitely Many Solutions

Solve the system of linear equations.

$$\begin{cases} 12x - 6y = -3 \\ -8x + 4y = 2 \end{cases}$$

Solution

$$\begin{bmatrix} 12 & -6 & \vdots & -3 \\ -8 & 4 & \vdots & 2 \end{bmatrix}$$
Augmented matrix for system of linear equations

$$\tfrac{1}{12}R_1 \rightarrow \begin{bmatrix} 1 & -\tfrac{1}{2} & \vdots & -\tfrac{1}{4} \\ -8 & 4 & \vdots & 2 \end{bmatrix}$$
First column has leading 1 in upper left corner.

$$8R_1 + R_2 \rightarrow \begin{bmatrix} 1 & -\tfrac{1}{2} & \vdots & -\tfrac{1}{4} \\ 0 & 0 & \vdots & 0 \end{bmatrix}$$
First column has a zero under its leading 1.

Because the second row of the matrix is all zeros, you can conclude that the system of equations has an infinite number of solutions, represented by all points (x, y) on the line $x - \tfrac{1}{2}y = -\tfrac{1}{4}$. Because this line can be written as $x = \tfrac{1}{2}y - \tfrac{1}{4}$, you can write the solution set as

$$\left(\tfrac{1}{2}a - \tfrac{1}{4}, a\right), \quad \text{where } a \text{ is any real number.}$$

Collaborate!

Checking the Sensibility of an Answer

Use a graphing utility to graph each system of equations from Example 6, Example 8, and Example 9. Verify the solution given in each example and explain how you may reach the same conclusion by using the graph. Summarize how you may conclude that a system has a unique solution, no solution, or infinitely many solutions when you use Gaussian elimination.

4.3 ■ Exercises

Developing Skills

In Exercises 1–6, determine the order of the matrix.

1. $\begin{bmatrix} 3 & -2 \\ -4 & 0 \\ 2 & -7 \end{bmatrix}$
2. $\begin{bmatrix} 4 & 0 & -5 \\ -1 & 8 & 9 \\ 0 & -3 & 4 \end{bmatrix}$

3. $\begin{bmatrix} 5 & -8 & 32 \\ 7 & 15 & 28 \end{bmatrix}$
4. $\begin{bmatrix} -2 & 5 \\ 0 & -1 \end{bmatrix}$

5. $\begin{bmatrix} 12 & -3 & 1 & 9 \\ 8 & 4 & 0 & 11 \\ 6 & -5 & -4 & 9 \end{bmatrix}$
6. $\begin{bmatrix} 1 & 9 \\ 18 & 14 \\ -6 & 2 \\ 7 & -4 \end{bmatrix}$

In Exercises 7–12, form the augmented matrix for the system of linear equations.

7. $\begin{cases} 4x - 5y = -2 \\ -x + 8y = 10 \end{cases}$
8. $\begin{cases} 8x + 3y = 25 \\ 3x - 9y = 12 \end{cases}$

9. $\begin{cases} x + 10y - 3z = 2 \\ 5x - 3y + 4z = 0 \\ 2x + 4y = 6 \end{cases}$

10. $\begin{cases} 3x + 4y - 10z = 11 \\ 12x - y + 2z = 6 \\ -3x + 4y - 17z = 8 \end{cases}$

11. $\begin{cases} 4x + 13y = -2 \\ 7x - 6y + z = 0 \end{cases}$

12. $\begin{cases} 9x - 3y + z = 13 \\ 12x - 8z = 5 \end{cases}$

In Exercises 13–18, write the system of linear equations represented by the augmented matrix. (Use variables $x, y, z,$ and w.)

13. $\begin{bmatrix} 4 & 3 & \vdots & 8 \\ 1 & -2 & \vdots & 3 \end{bmatrix}$

14. $\begin{bmatrix} 9 & -4 & \vdots & 0 \\ 6 & 1 & \vdots & -4 \end{bmatrix}$

15. $\begin{bmatrix} 1 & 0 & 2 & \vdots & -10 \\ 0 & 3 & -1 & \vdots & 5 \\ 4 & 2 & 0 & \vdots & 3 \end{bmatrix}$

16. $\begin{bmatrix} 7 & 1 & -1 & \vdots & 4 \\ 3 & -8 & 0 & \vdots & 0 \\ 12 & -4 & 1 & \vdots & 1 \end{bmatrix}$

17. $\begin{bmatrix} 15 & -4 & 12 & 3 & \vdots & 6 \\ -1 & -3 & 0 & 10 & \vdots & 14 \\ 9 & 2 & -7 & 0 & \vdots & 5 \end{bmatrix}$

18. $\begin{bmatrix} 5 & 8 & 2 & 0 & \vdots & -1 \\ -2 & 15 & 5 & 1 & \vdots & 9 \\ 1 & 6 & -7 & 0 & \vdots & -3 \end{bmatrix}$

In Exercises 19–22, fill in the blank(s) by using elementary row operations to form a row-equivalent matrix.

19. $\begin{bmatrix} 1 & 4 & 3 \\ 2 & 10 & 5 \end{bmatrix}$
20. $\begin{bmatrix} 3 & 6 & 8 \\ 4 & -3 & 6 \end{bmatrix}$

$\begin{bmatrix} 1 & 4 & 3 \\ 0 & & -1 \end{bmatrix}$
$\begin{bmatrix} 3 & 6 & 8 \\ 1 & -9 & \end{bmatrix}$

21. $\begin{bmatrix} 1 & 1 & 4 & -1 \\ 3 & 8 & 10 & 3 \\ -2 & 1 & 12 & 6 \end{bmatrix}$

$\begin{bmatrix} 1 & 1 & 4 & -1 \\ 0 & 5 & & \\ 0 & 3 & & \end{bmatrix}$

$\begin{bmatrix} 1 & 1 & 4 & -1 \\ 0 & 1 & & \\ 0 & 3 & 20 & 4 \end{bmatrix}$

22. $\begin{bmatrix} 2 & 4 & 8 & 3 \\ 1 & -1 & -3 & 2 \\ 2 & 6 & 4 & 9 \end{bmatrix}$

$\begin{bmatrix} 1 & & & \\ 1 & -1 & -3 & 2 \\ 2 & 6 & 4 & 9 \end{bmatrix}$

$\begin{bmatrix} 1 & 2 & 4 & \frac{3}{2} \\ 0 & & -7 & \frac{1}{2} \\ 0 & 2 & & \end{bmatrix}$

In Exercises 23–28, convert the matrix to row-echelon form. (*Note:* There is more than one correct answer.)

23. $\begin{bmatrix} 1 & 2 & 3 \\ 2 & -1 & -4 \end{bmatrix}$
24. $\begin{bmatrix} 1 & 3 & 6 \\ -4 & -9 & 3 \end{bmatrix}$

25. $\begin{bmatrix} 4 & 6 & 1 \\ -2 & 2 & 5 \end{bmatrix}$
26. $\begin{bmatrix} 3 & 2 & 6 \\ 2 & 3 & -3 \end{bmatrix}$

27. $\begin{bmatrix} 1 & 1 & 0 & 5 \\ -2 & -1 & 2 & -10 \\ 3 & 6 & 7 & 14 \end{bmatrix}$

28. $\begin{bmatrix} 1 & 2 & -1 & 3 \\ 3 & 7 & -5 & 14 \\ -2 & -1 & -3 & 8 \end{bmatrix}$

In Exercises 29–32, use the matrix capabilities of a graphing utility to write the matrix in row-echelon form. (*Note:* There is more than one correct answer.)

29. $\begin{bmatrix} 1 & -1 & -1 & 1 \\ 4 & -4 & 1 & 8 \\ -6 & 8 & 18 & 0 \end{bmatrix}$

30. $\begin{bmatrix} 1 & -3 & 0 & -7 \\ -3 & 10 & 1 & 23 \\ 4 & -10 & 2 & -24 \end{bmatrix}$

31. $\begin{bmatrix} 1 & 1 & -1 & 3 \\ 2 & 1 & 2 & 5 \\ 3 & 2 & 1 & 8 \end{bmatrix}$ 32. $\begin{bmatrix} 1 & -3 & -2 & -8 \\ 1 & 3 & -2 & 17 \\ 1 & 2 & -2 & -5 \end{bmatrix}$

In Exercises 33–36, write the system of linear equations represented by the augmented matrix. Then use back-substitution to find the solution. (Use variables x, y, and z.)

33. $\begin{bmatrix} 1 & -2 & \vdots & 4 \\ 0 & 1 & \vdots & -3 \end{bmatrix}$

34. $\begin{bmatrix} 1 & 5 & \vdots & 0 \\ 0 & 1 & \vdots & -1 \end{bmatrix}$

35. $\begin{bmatrix} 1 & -1 & 2 & \vdots & 4 \\ 0 & 1 & -1 & \vdots & 2 \\ 0 & 0 & 1 & \vdots & -2 \end{bmatrix}$

36. $\begin{bmatrix} 1 & 2 & -2 & \vdots & -1 \\ 0 & 1 & 1 & \vdots & 9 \\ 0 & 0 & 1 & \vdots & -3 \end{bmatrix}$

In Exercises 37–56, use matrices to solve the system of linear equations.

37. $\begin{cases} x + 2y = 7 \\ 3x + y = 11 \end{cases}$ 38. $\begin{cases} 2x + 6y = 16 \\ 2x + 3y = 7 \end{cases}$

39. $\begin{cases} 6x - 4y = 2 \\ 5x + 2y = 7 \end{cases}$ 40. $\begin{cases} 2x - y = 9 \\ 3x + 2y = 10 \end{cases}$

41. $\begin{cases} x - 2y = 5 \\ 3x + y = 1 \end{cases}$ 42. $\begin{cases} 2x + y = 8 \\ 4x - 3y = 6 \end{cases}$

43. $\begin{cases} -x + 2y = 1.5 \\ 2x - 4y = 3 \end{cases}$ 44. $\begin{cases} x - 3y = 5 \\ -2x + 6y = -10 \end{cases}$

45. $\begin{cases} x - 2y - z = 6 \\ y + 4z = 5 \\ 4x + 2y + 3z = 8 \end{cases}$ 46. $\begin{cases} x - 3z = -2 \\ 3x + y - 2z = 5 \\ 2x + 2y + z = 4 \end{cases}$

47. $\begin{cases} x + 2y - z = 1 \\ 2x - y + z = 3 \\ -x + 2y + 3z = 7 \end{cases}$ 48. $\begin{cases} 2x + y + 3z = 1 \\ 4x + 3y + 5z = 1 \\ 6x + 5y + 5z = -3 \end{cases}$

49. $\begin{cases} 3x + 6y - 3z = 6 \\ -2x - 4y - 3z = -1 \\ 3x + 6y - 2z = 10 \end{cases}$

50. $\begin{cases} 3x + 2y + z = 0 \\ -2x + y - z = 2 \\ 2x - y + 2z = -1 \end{cases}$

51. $\begin{cases} x + y - 5z = 3 \\ x - 2z = 1 \\ 2x - y - z = 0 \end{cases}$ 52. $\begin{cases} 2y + z = 3 \\ -4y - 2z = 0 \\ x + y + z = 2 \end{cases}$

53. $\begin{cases} 2x + 4y = 10 \\ 2x + 2y + 3z = 3 \\ -3x + y + 2z = -3 \end{cases}$

54. $\begin{cases} 2x - y + 3z = 24 \\ 2y - z = 14 \\ 7x - 5y = 6 \end{cases}$

55. $\begin{cases} x - 3y + 2z = 8 \\ 2y - z = -4 \\ x + z = 3 \end{cases}$

56. $\begin{cases} 2x + 3z = 3 \\ 4x - 3y + 7z = 5 \\ 8x - 9y + 15z = 9 \end{cases}$

In Exercises 57–62, use the matrix capabilities of a graphing utility to solve the system of equations.

57. $\begin{cases} -2x - 2y - 15z = 0 \\ x + 2y + 2z = 18 \\ 3x + 3y + 22z = 2 \end{cases}$

58. $\begin{cases} 2x + 4y + 5z = 5 \\ x + 3y + 3z = 2 \\ 2x + 4y + 4z = 2 \end{cases}$

59. $\begin{cases} 2x + 4z = 1 \\ x + y + 3z = 0 \\ x + 3y + 5z = 0 \end{cases}$ 60. $\begin{cases} 3x + y - 2z = 2 \\ 6x + 2y - 4z = 1 \\ -3x - y + 2z = 1 \end{cases}$

61. $\begin{cases} x + 3y = 2 \\ 2x + 6y = 4 \\ 2x + 5y + 4z = 3 \end{cases}$ 62. $\begin{cases} 2x + 2y + z = 8 \\ 2x + 3y + z = 7 \\ 6x + 8y + 3z = 22 \end{cases}$

Solving Problems

63. *Simple Interest* A corporation borrowed $1,500,000 to expand its product line. Some of the money was borrowed at 8%, some at 9%, and the remainder at 12%. The annual interest payment to the lenders was $133,000. If the amount borrowed at 8% was four times the amount borrowed at 12%, how much was borrowed at each rate?

64. *Simple Interest* An inheritance of $16,000 was divided among three investments yielding a total of $990 in simple interest per year. The interest rates for the three investments were 5%, 6%, and 7%. Find the amount invested at each rate if the amount invested at 6% is $2000 less than the amount invested at 7%.

Curve-Fitting In Exercises 65 and 66, find the equation of the parabola $y = ax^2 + bx + c$ that passes through the points.

65.

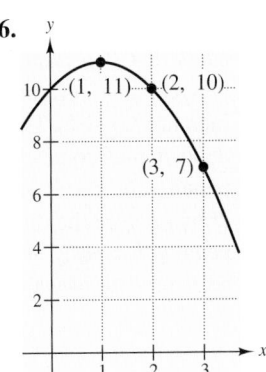

66.

67. *Recycling* The table shows the amount y, in millions of short tons, of recycled materials in the United States for the years 1996 through 1998. (*Source: Franklin Associates, Ltd.*)

Year	1996	1997	1998
y	57.3	59.4	62.2

(a) Find the equation $y = at^2 + bt + c$ that passes through the three points, with $t = 6$ corresponding to 1996.

(b) Use a graphing utility to graph part (a).

(c) Assuming that this trend continues, use the model in part (a) to predict the amount of recycled materials in 2006.

68. *Savings* The table shows the gross savings y (in billions of dollars) in the United States for the years 1997 through 1999. (*Source: U.S. Bureau of Economic Analysis*)

Year	1997	1998	1999
y	1521	1646	1727

(a) Find the equation $y = at^2 + bt + c$ that passes through the three points, with $t = 7$ corresponding to 1997.

(b) Use a graphing utility to graph the model in part (a).

(c) Assuming that this trend continues, use the model in part (a) to predict gross savings in the year 2006.

69. *Mathematical Modeling* A videotape of the path of a ball thrown by a baseball player was analyzed on a television set with a grid covering the screen. The tape was paused three times and the coordinates of the ball were measured each time. The coordinates were approximately $(0, 6)$, $(25, 18.5)$, and $(50, 26)$. (The x-coordinate is the horizontal distance in feet from the player and the y-coordinate is the height in feet of the ball above the ground.)

(a) Find the equation $y = ax^2 + bx + c$ that passes through the three points.

(b) Use a graphing utility to graph the model in part (a). Use the graph to approximate the maximum height of the ball and the point at which it struck the ground.

70. The table gives the number y, of television viewing hours per person per year in the years 1997–1999 in the United States. (*Source: Communications Industry Forecast*)

Year	1997	1998	1999
y	1544	1551	1588

(a) Find the equation $y = at^2 + bt + c$ that passes through the three points, with $t = 7$ corresponding to 1997.

(b) Use a graphing utility to graph part (a).

(c) Use the model in part (a) to predict the number of television viewing hours per person per year in 2005 if the trend continues.

Explaining Concepts

71. Describe the row-echelon form of an augmented matrix that corresponds to a system of linear equations that is inconsistent.

72. Describe the row-echelon form of an augmented matrix that corresponds to a system of linear equations that has an infinite number of solutions.

73. What is meant when it is said that two augmented matrices are *row-equivalent*?

74. Give an example of a matrix in *row-echelon form*.

75. Describe the three elementary row operations that can be performed on an augmented matrix.

76. What is the relationship between the three elementary row operations on an augmented matrix and the row operations on a system of linear equations?

Ongoing Review

In Exercises 77–82, plot the points on a rectangular coordinate system and find the slope (if possible) of the line passing through the points.

77. $(-3, 2), (5, -4)$

78. $(2, 8), (7, -3)$

79. $\left(\frac{5}{2}, \frac{7}{2}\right), \left(\frac{7}{3}, -2\right)$

80. $\left(-\frac{9}{4}, -\frac{1}{4}\right), \left(-3, \frac{9}{2}\right)$

81. $(6, 4), (6, -3)$

82. $(-4, 5), (7, 5)$

In Exercises 83–86, evaluate the expression.

83. $5 - \frac{1}{3}(-6)$

84. $-6 + \frac{3}{8}(10)$

85. $4(-2) - \frac{4}{5}(-3)$

86. $-\frac{1}{4}(3) + \frac{3}{7}(2)$

87. *Membership Drive* Through a membership drive, the membership for a public television station increased by 10%. The current number of members is 8415. How many members did the station have before the membership drive?

88. *Consumer Awareness* A sales representative indicates that if a customer waits another month for a new car that currently costs $23,500, the price will increase by 4%. The customer has a certificate of deposit that comes due in 1 month and will pay a penalty for early withdrawal if the money is withdrawn before the due date. Determine the maximum penalty for early withdrawal that would equal the cost increase of waiting to buy the car.

Looking Further

Investment Portfolio Consider an investment portfolio totaling $500,000 that is to be allocated among the following types of investments: certificates of deposit, municipal bonds, blue-chip stocks, and growth or speculative stocks. How much should be allocated to each type of investment? If there is more than one solution, determine several possible ways to allocate the investments.

(a) The certificates of deposit pay 10% annually, and the municipal bonds pay 8% annually. Over a five-year period, the investor expects the blue-chip stocks to return 12% annually, and expects the growth stocks to return 13% annually. The investor wants a combined annual return of 10% and also wants to have only one-fourth of the portfolio invested in stocks.

(b) The certificates of deposit pay 9% annually, and the municipal bonds pay 5% annually. Over a five-year period, the investor expects the blue-chip stocks to return 12% annually, and expects the growth stocks to return 14% annually. The investor wants a combined annual return of 10% and also wants to have only one-fourth of the portfolio invested in stocks.

Mid-Chapter Quiz

Take this quiz as you would take a quiz in class. After you are done, check your work against the answers given in the back of the book.

In Exercises 1 and 2, solve the system of linear equations graphically.

1. $\begin{cases} x - 2y = 0 \\ 2x + y = 5 \end{cases}$

2. $\begin{cases} 4x + 2y = 8 \\ x + y = 1 \end{cases}$

In Exercises 3 and 4, solve the system by the method of substitution. Use a graphing utility to verify your solution.

3. $\begin{cases} 5x - y = 32 \\ 6x - 9y = 15 \end{cases}$

4. $\begin{cases} 0.2x + 0.7y = 8 \\ -x + 2y = 15 \end{cases}$

In Exercises 5–7, use elimination or Gaussian elimination to solve the linear system.

5. $\begin{cases} x + 10y = 18 \\ 5x + 2y = 42 \end{cases}$

6. $\begin{cases} 3x + 11y = 38 \\ 7x - 5y = -34 \end{cases}$

7. $\begin{cases} x + 4z = 17 \\ -3x + 2y - z = -20 \\ x - 5y + 3z = 19 \end{cases}$

8. Use a matrix to solve the system of linear equations.

$$\begin{cases} x - 3y + z = -3 \\ 3x + 2y - 5z = 18 \\ y + z = -1 \end{cases}$$

In Exercises 9 and 10, set up a system of linear equations that models the problem. (It is not necessary to solve the system.)

9. The linear system of equations has the unique solution $(10, -12)$. (There are many correct answers.)

10. The linear system of equations has the unique solution $(2, -3, 1)$. (There are many correct answers.)

11. Two people share the driving on a 300-mile trip. One person drives three times as far as the other. Find the distance each person drives.

12. Twenty gallons of a 30% brine solution are obtained by mixing a 20% solution with a 50% solution. Let x represent the number of gallons of the 20% solution and let y represent the number of gallons of the 50% solution. How many gallons of each solution are required?

13. Two angles are supplementary. The measure of the second angle is $45°$ less than two times the measure of the first angle. Find the measure of each angle.

14. Find the equation of the parabola $y = ax^2 + bx + c$ that passes through the points $(1, 2)$, $(-1, -4)$, and $(2, 8)$.

4.4 Determinants and Linear Systems

What you should learn:

- How to find determinants of 2×2 matrices and 3×3 matrices
- How to use determinants and Cramer's Rule to solve systems of linear equations
- How to use determinants to find areas of triangles, to test for collinear points, and to find equations of lines

Why you should learn it:

Determinants and Cramer's Rule can be used to find currents of electrical networks, as shown in Exercise 101 on page 284.

The Determinant of a Matrix

Associated with each square matrix is a real number called its **determinant.** The use of determinants arose from special number patterns that occur during the solution of systems of linear equations. For instance, the system

$$\begin{cases} a_1x + b_1y = c_1 \\ a_2x + b_2y = c_2 \end{cases}$$

has a solution given by

$$x = \frac{c_1b_2 - c_2b_1}{a_1b_2 - a_2b_1} \quad \text{and} \quad y = \frac{a_1c_2 - a_2c_1}{a_1b_2 - a_2b_1}$$

provided that $a_1b_2 - a_2b_1 \neq 0$. Note that the denominator of each fraction is the same. The denominator is called the *determinant* of the coefficient matrix of the system.

Coefficient Matrix   *Determinant*

$$A = \begin{bmatrix} a_1 & b_1 \\ a_2 & b_2 \end{bmatrix} \qquad \det(A) = a_1b_2 - a_2b_1$$

The determinant of the matrix A can also be denoted by vertical bars on both sides of the matrix, as indicated in the definition below.

Definition of the Determinant of a 2×2 Matrix

The **determinant** of the matrix

$$A = \begin{bmatrix} a_1 & b_1 \\ a_2 & b_2 \end{bmatrix}$$

is given by

$$\det(A) = |A| = \begin{vmatrix} a_1 & b_1 \\ a_2 & b_2 \end{vmatrix} = a_1b_2 - a_2b_1$$

NOTE Note that $\det(A)$ and $|A|$ are used interchangeably to represent the determinant of A. Although vertical bars are also used to denote the absolute value of a real number, the context will show which use is intended.

A convenient method for remembering the formula for the determinant of a 2×2 matrix is shown in the diagram below.

$$\det(A) = \begin{vmatrix} a_1 & b_1 \\ a_2 & b_2 \end{vmatrix} = a_1b_2 - a_2b_1$$

Note that the determinant is given by the difference of the products of the two diagonals of the matrix.

Historical Note

An ancient Chinese method of solving systems of equations involved the use of bamboo rods to represent coefficients. In 1683, Seki Kowa (1642–1708), the Japanese mathematician, advanced this method by rearranging the rods in a way that resembles the determinant notation of today.

Historical Note

In a letter written in 1693 to Guillaume de L'Hôpital, Gottfried Wilhelm von Leibniz (1646–1716) gave a written notation for determinants in what is considered to be the first formal use of them.

Example 1 ■ The Determinant of a 2 × 2 Matrix

Find the determinant of each matrix.

(a) $A = \begin{bmatrix} 2 & -3 \\ 1 & 4 \end{bmatrix}$ (b) $B = \begin{bmatrix} -1 & 2 \\ 2 & -4 \end{bmatrix}$ (c) $C = \begin{bmatrix} 1 & 3 \\ 2 & 5 \end{bmatrix}$

Solution

(a) $\det(A) = \begin{vmatrix} 2 & -3 \\ 1 & 4 \end{vmatrix} = 2(4) - 1(-3) = 8 + 3 = 11$

(b) $\det(B) = \begin{vmatrix} -1 & 2 \\ 2 & -4 \end{vmatrix} = (-1)(-4) - 2(2) = 4 - 4 = 0$

(c) $\det(C) = \begin{vmatrix} 1 & 3 \\ 2 & 5 \end{vmatrix} = 1(5) - 2(3) = 5 - 6 = -1$

NOTE   Notice in Example 1 that the determinant of a matrix can be positive, zero, or negative.

One way to evaluate the determinant of a 3 × 3 matrix, called **expanding by minors,** allows you to write the determinant of a 3 × 3 matrix in terms of three 2 × 2 determinants. The **minor** of an entry in a 3 × 3 matrix is the determinant of the 2 × 2 matrix that remains after deletion of the row and column in which the entry occurs. Here are three examples.

Technology

A graphing utility with matrix capabilities can be used to evaluate the determinant of a square matrix. Consult the user's guide for your graphing utility to learn how to evaluate a determinant. Use the graphing utility to check the result in Example 1(a). Then try to evaluate the determinant of the 3 × 3 matrix at the right using a graphing utility. Finish the evaluation of the determinant by expanding by minors to check the result.

Determinant	*Entry*	*Minor of Entry*	*Value of Minor*
$\begin{vmatrix} 1 & -1 & 3 \\ 0 & 2 & 5 \\ -2 & 4 & -7 \end{vmatrix}$	1	$\begin{vmatrix} 2 & 5 \\ 4 & -7 \end{vmatrix}$	$2(-7) - 4(5) = -34$
$\begin{vmatrix} 1 & -1 & 3 \\ 0 & 2 & 5 \\ -2 & 4 & -7 \end{vmatrix}$	-1	$\begin{vmatrix} 0 & 5 \\ -2 & -7 \end{vmatrix}$	$0(-7) - (-2)(5) = 10$
$\begin{vmatrix} 1 & -1 & 3 \\ 0 & 2 & 5 \\ -2 & 4 & -7 \end{vmatrix}$	3	$\begin{vmatrix} 0 & 2 \\ -2 & 4 \end{vmatrix}$	$0(4) - (-2)(2) = 4$

Expanding by Minors

$$\det(A) = \begin{vmatrix} a_1 & b_1 & c_1 \\ a_2 & b_2 & c_2 \\ a_3 & b_3 & c_3 \end{vmatrix}$$

$$= a_1(\text{minor of } a_1) - b_1(\text{minor of } b_1) + c_1(\text{minor of } c_1)$$

$$= a_1 \begin{vmatrix} b_2 & c_2 \\ b_3 & c_3 \end{vmatrix} - b_1 \begin{vmatrix} a_2 & c_2 \\ a_3 & c_3 \end{vmatrix} + c_1 \begin{vmatrix} a_2 & b_2 \\ a_3 & b_3 \end{vmatrix}$$

This pattern is called **expanding by minors** along the first row. A similar pattern can be used to expand by minors along any row or column.

$$\begin{bmatrix} + & - & + \\ - & + & - \\ + & - & + \end{bmatrix}$$

Figure 4.12 *Sign pattern for 3 × 3 matrix*

The *signs* of the terms used in expanding by minors follow the alternating pattern shown in Figure 4.12. For instance, the signs used to expand by minors along the second row are $-, +, -$, as follows.

$$\det(A) = \begin{vmatrix} a_1 & b_1 & c_1 \\ a_2 & b_2 & c_2 \\ a_3 & b_3 & c_3 \end{vmatrix}$$

$$= -a_2(\text{minor of } a_2) + b_2(\text{minor of } b_2) - c_2(\text{minor of } c_2)$$

Example 2 ■ Finding the Determinant of a 3 × 3 Matrix

Find the determinant of $A = \begin{bmatrix} -1 & 1 & 2 \\ 0 & 2 & 3 \\ 3 & 4 & 2 \end{bmatrix}$.

Solution

By expanding by minors along the *first column*, you obtain

$$\det(A) = \begin{vmatrix} -1 & 1 & 2 \\ 0 & 2 & 3 \\ 3 & 4 & 2 \end{vmatrix}$$

$$= (-1)\begin{vmatrix} 2 & 3 \\ 4 & 2 \end{vmatrix} - (0)\begin{vmatrix} 1 & 2 \\ 4 & 2 \end{vmatrix} + (3)\begin{vmatrix} 1 & 2 \\ 2 & 3 \end{vmatrix}$$

$$= (-1)(4 - 12) - (0)(2 - 8) + (3)(3 - 4)$$

$$= 8 - 0 - 3$$

$$= 5.$$

Note in the expansion in Example 2 that a zero entry will always yield a zero term when expanding by minors. So, when you are evaluating the determinant of a matrix, you should choose to expand along the row or column that has the most zero entries. This idea is reinforced in Example 3.

Example 3 ■ Finding the Determinant of a 3 × 3 Matrix

Find the determinant of $A = \begin{bmatrix} 1 & 2 & 1 \\ 3 & 0 & 2 \\ 4 & 0 & -1 \end{bmatrix}$.

Solution

By expanding by minors along the *second column*, you obtain

$$\det(A) = \begin{vmatrix} 1 & 2 & 1 \\ 3 & 0 & 2 \\ 4 & 0 & -1 \end{vmatrix}$$

$$= -(2)\begin{vmatrix} 3 & 2 \\ 4 & -1 \end{vmatrix} + (0)\begin{vmatrix} 1 & 1 \\ 4 & -1 \end{vmatrix} - (0)\begin{vmatrix} 1 & 1 \\ 3 & 2 \end{vmatrix}$$

$$= -(2)(-3 - 8) + 0 - 0$$

$$= 22.$$

Cramer's Rule

So far in this chapter, you have studied three methods for solving a system of linear equations: substitution, elimination (with equations), and elimination (with matrices). We now look at one more method, called **Cramer's Rule,** which is named after Gabriel Cramer (1704–1752). This rule uses determinants to write the solution of a system of linear equations.

In Cramer's Rule, the value of a variable is expressed as the quotient of two determinants. The denominator is the determinant of the coefficient matrix of the system. The numerator is the determinant of the matrix formed by using the column of constants as replacements for the coefficients of the variable. In the definition below, note the notation for the different determinants.

NOTE   Cramer's Rule is not as general as the elimination method because Cramer's Rule requires that the coefficient matrix of the system be square *and* that the system have exactly one solution.

Historical Note

Gabriel Cramer invented determinants independently and in 1750 published his method of using them to solve linear systems of equations. He did not, however, use the notation that is used today.

Cramer's Rule

1. For the system of linear equations

$$\begin{cases} a_1x + b_1y = c_1 \\ a_2x + b_2y = c_2 \end{cases}$$

the solution is given by

$$x = \frac{D_x}{D} = \frac{\begin{vmatrix} c_1 & b_1 \\ c_2 & b_2 \end{vmatrix}}{\begin{vmatrix} a_1 & b_1 \\ a_2 & b_2 \end{vmatrix}}, \qquad y = \frac{D_y}{D} = \frac{\begin{vmatrix} a_1 & c_1 \\ a_2 & c_2 \end{vmatrix}}{\begin{vmatrix} a_1 & b_1 \\ a_2 & b_2 \end{vmatrix}}$$

provided that $D \neq 0$.

2. For the system of linear equations

$$\begin{cases} a_1x + b_1y + c_1z = d_1 \\ a_2x + b_2y + c_2z = d_2 \\ a_3x + b_3y + c_3z = d_3 \end{cases}$$

the solution is given by

$$x = \frac{D_x}{D} = \frac{\begin{vmatrix} d_1 & b_1 & c_1 \\ d_2 & b_2 & c_2 \\ d_3 & b_3 & c_3 \end{vmatrix}}{\begin{vmatrix} a_1 & b_1 & c_1 \\ a_2 & b_2 & c_2 \\ a_3 & b_3 & c_3 \end{vmatrix}}, \qquad y = \frac{D_y}{D} = \frac{\begin{vmatrix} a_1 & d_1 & c_1 \\ a_2 & d_2 & c_2 \\ a_3 & d_3 & c_3 \end{vmatrix}}{\begin{vmatrix} a_1 & b_1 & c_1 \\ a_2 & b_2 & c_2 \\ a_3 & b_3 & c_3 \end{vmatrix}},$$

$$z = \frac{D_z}{D} = \frac{\begin{vmatrix} a_1 & b_1 & d_1 \\ a_2 & b_2 & d_2 \\ a_3 & b_3 & d_3 \end{vmatrix}}{\begin{vmatrix} a_1 & b_1 & c_1 \\ a_2 & b_2 & c_2 \\ a_3 & b_3 & c_3 \end{vmatrix}}$$

provided that $D \neq 0$.

Example 4 ■ Using Cramer's Rule for a 2 × 2 System

Use Cramer's Rule to solve the system of linear equations.

$$\begin{cases} 4x - 2y = 10 \\ 3x - 5y = 11 \end{cases}$$

Solution

Begin by finding the determinant of the coefficient matrix.

$$D = \begin{vmatrix} 4 & -2 \\ 3 & -5 \end{vmatrix} = -20 - (-6) = -14$$

NOTE When using Cramer's Rule, remember that the method *does not* apply if the determinant of the coefficient matrix is zero. For instance, the system

$$\begin{cases} 6x - 10y = -4 \\ 9x - 15y = 5 \end{cases}$$

has no solution (see Example 8 in Section 4.3), and the determinant of the coefficient matrix of this system is zero.

$$x = \frac{D_x}{D} = \frac{\begin{vmatrix} 10 & -2 \\ 11 & -5 \end{vmatrix}}{-14} = \frac{(-50) - (-22)}{-14} = \frac{-28}{-14} = 2$$

$$y = \frac{D_y}{D} = \frac{\begin{vmatrix} 4 & 10 \\ 3 & 11 \end{vmatrix}}{-14} = \frac{44 - 30}{-14} = \frac{14}{-14} = -1$$

The solution is $(2, -1)$. Check this in the original system of equations as follows.

Check

Equation 1: $4(2) - 2(-1) = 10$ ✓

Equation 2: $3(2) - 5(-1) = 11$ ✓

Example 5 ■ Using Cramer's Rule for a 3 × 3 System

Use Cramer's Rule to solve the system of linear equations.

$$\begin{cases} -x + 2y - 3z = 1 \\ 2x + z = 0 \\ 3x - 4y + 4z = 2 \end{cases}$$

Solution

The determinant of the coefficient matrix is $D = 10$.

$$x = \frac{D_x}{D} = \frac{\begin{vmatrix} 1 & 2 & -3 \\ 0 & 0 & 1 \\ 2 & -4 & 4 \end{vmatrix}}{10} = \frac{8}{10} = \frac{4}{5}$$

$$y = \frac{D_y}{D} = \frac{\begin{vmatrix} -1 & 1 & -3 \\ 2 & 0 & 1 \\ 3 & 2 & 4 \end{vmatrix}}{10} = \frac{-15}{10} = -\frac{3}{2}$$

$$z = \frac{D_z}{D} = \frac{\begin{vmatrix} -1 & 2 & 1 \\ 2 & 0 & 0 \\ 3 & -4 & 2 \end{vmatrix}}{10} = \frac{-16}{10} = -\frac{8}{5}$$

The solution is $\left(\frac{4}{5}, -\frac{3}{2}, -\frac{8}{5} \right)$. Check this in the original system of equations.

Applications of Determinants

In addition to Cramer's Rule, determinants have many other practical applications. For instance, you can use a determinant to find the area of a triangle whose vertices are given by three points on a rectangular coordinate system.

Area of a Triangle

The area of a triangle with vertices (x_1, y_1), (x_2, y_2), and (x_3, y_3) is

$$\text{Area} = \pm\frac{1}{2}\begin{vmatrix} x_1 & y_1 & 1 \\ x_2 & y_2 & 1 \\ x_3 & y_3 & 1 \end{vmatrix}$$

where the symbol $(\pm)$ indicates that the appropriate sign should be chosen to yield a positive area.

Example 6 ■ Finding the Area of a Triangle

Find the area of the triangle whose vertices are $(2, 0)$, $(1, 3)$, and $(3, 2)$, as shown in Figure 4.13.

Solution

Choose $(x_1, y_1) = (2, 0)$, $(x_2, y_2) = (1, 3)$, and $(x_3, y_3) = (3, 2)$. To find the area of the triangle, evaluate the determinant

$$\begin{vmatrix} x_1 & y_1 & 1 \\ x_2 & y_2 & 1 \\ x_3 & y_3 & 1 \end{vmatrix} = \begin{vmatrix} 2 & 0 & 1 \\ 1 & 3 & 1 \\ 3 & 2 & 1 \end{vmatrix}$$

$$= 2\begin{vmatrix} 3 & 1 \\ 2 & 1 \end{vmatrix} - 0\begin{vmatrix} 1 & 1 \\ 3 & 1 \end{vmatrix} + 1\begin{vmatrix} 1 & 3 \\ 3 & 2 \end{vmatrix}$$

$$= 2(1) - 0 + 1(-7)$$

$$= -5.$$

Using this value, you can conclude that the area of the triangle is

$$\text{Area} = -\frac{1}{2}\begin{vmatrix} 2 & 0 & 1 \\ 1 & 3 & 1 \\ 3 & 2 & 1 \end{vmatrix}$$

$$= -\frac{1}{2}(-5)$$

$$= \frac{5}{2}.$$

NOTE   To see the benefit of the "determinant formula," try finding the area of the triangle in Example 6 using the standard formula:

$$\text{Area} = \tfrac{1}{2}(\text{base})(\text{height}).$$

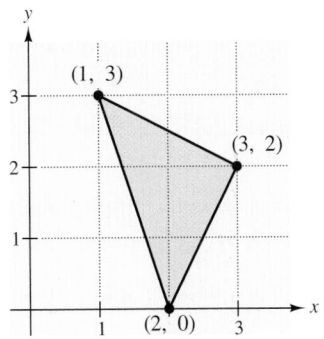

Figure 4.13

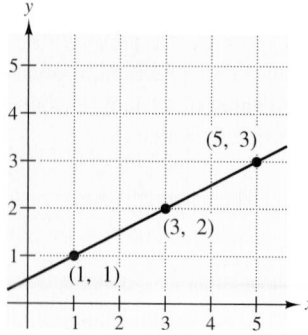

Figure 4.14

Suppose the three points in Example 6 had been on the same line. What would have happened had the area formula been applied to three such points? The answer is that the determinant would have been zero. Consider for instance, the three collinear points $(1, 1)$, $(3, 2)$, and $(5, 3)$, as shown in Figure 4.14. The area of the "triangle" that has these three points as vertices is

$$\frac{1}{2}\begin{vmatrix} 1 & 1 & 1 \\ 3 & 2 & 1 \\ 5 & 3 & 1 \end{vmatrix} = \frac{1}{2}\left(1\begin{vmatrix} 2 & 1 \\ 3 & 1 \end{vmatrix} - 1\begin{vmatrix} 3 & 1 \\ 5 & 1 \end{vmatrix} + 1\begin{vmatrix} 3 & 2 \\ 5 & 3 \end{vmatrix}\right)$$

$$= \frac{1}{2}[-1 - (-2) + (-1)] = 0.$$

This result is generalized as follows.

Test for Collinear Points

Three points (x_1, y_1), (x_2, y_2), and (x_3, y_3) are collinear (lie on the same line) if and only if

$$\begin{vmatrix} x_1 & y_1 & 1 \\ x_2 & y_2 & 1 \\ x_3 & y_3 & 1 \end{vmatrix} = 0.$$

Example 7 ■ Testing for Collinear Points

Determine whether the points $(-2, -2)$, $(1, 1)$, and $(7, 5)$ are collinear. (See Figure 4.15.)

Solution

Letting $(x_1, y_1) = (-2, -2)$, $(x_2, y_2) = (1, 1)$, and $(x_3, y_3) = (7, 5)$, you have

$$\begin{vmatrix} x_1 & y_1 & 1 \\ x_2 & y_2 & 1 \\ x_3 & y_3 & 1 \end{vmatrix} = \begin{vmatrix} -2 & -2 & 1 \\ 1 & 1 & 1 \\ 7 & 5 & 1 \end{vmatrix}$$

$$= -2\begin{vmatrix} 1 & 1 \\ 5 & 1 \end{vmatrix} - (-2)\begin{vmatrix} 1 & 1 \\ 7 & 1 \end{vmatrix} + 1\begin{vmatrix} 1 & 1 \\ 7 & 5 \end{vmatrix}$$

$$= -2(-4) - (-2)(-6) + 1(-2) = -6.$$

Because the value of this determinant *is not* zero, you can conclude that the three points *do not* lie on the same line and are not collinear.

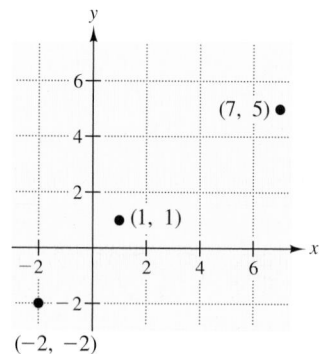

Figure 4.15

As a good review, look at how the slope can be used to verify the result in Example 7. Label the points $A(-2, -2)$, $B(1, 1)$, and $C(7, 5)$. Find the slopes of the lines between points A and B and between points A and C as follows.

$$\text{Slope from } A \text{ to } B = \frac{y_2 - y_1}{x_2 - x_1} = \frac{1 - (-2)}{1 - (-2)} = \frac{3}{3} = 1$$

$$\text{Slope from } A \text{ to } C = \frac{y_2 - y_1}{x_2 - x_1} = \frac{5 - (-2)}{7 - (-2)} = \frac{7}{9}$$

Because the slopes from A to B and from A to C are different, the points are not collinear.

You can also use determinants to find the equation of a line through two points. In this case the first row consists of the variables x and y and the number 1. By expanding by minors along the first row, the resulting 2×2 determinants are the coefficients of the variables x and y and the constant of the linear equation, as shown in Example 8.

Two-Point Form of the Equation of a Line

An equation of the line passing through the distinct points (x_1, y_1) and (x_2, y_2) is given by

$$\begin{vmatrix} x & y & 1 \\ x_1 & y_1 & 1 \\ x_2 & y_2 & 1 \end{vmatrix} = 0.$$

Example 8 ■ Finding an Equation of a Line

Find an equation of the line passing through $(-2, 1)$ and $(3, -2)$.

Solution

$$\begin{vmatrix} x & y & 1 \\ -2 & 1 & 1 \\ 3 & -2 & 1 \end{vmatrix} = 0$$

$$x \begin{vmatrix} 1 & 1 \\ -2 & 1 \end{vmatrix} - y \begin{vmatrix} -2 & 1 \\ 3 & 1 \end{vmatrix} + 1 \begin{vmatrix} -2 & 1 \\ 3 & -2 \end{vmatrix} = 0$$

$$3x + 5y + 1 = 0$$

So, an equation of the line is

$$3x + 5y + 1 = 0.$$

NOTE This method of finding the equation of a line works for all lines, including horizontal and vertical lines. For instance, the equation of the vertical line through $(2, 0)$ and $(2, 2)$ is

$$\begin{vmatrix} x & y & 1 \\ 2 & 0 & 1 \\ 2 & 2 & 1 \end{vmatrix} = 0$$

$$4 - 2x = 0$$

$$x = 2.$$

Collaborate!

Extending the Concept

There is an alternative method for evaluating the determinant of a 3×3 matrix A. (This method works only for 3×3 matrices.) To apply this method, copy the first and second columns of A to form fourth and fifth columns. The determinant of A is then obtained by adding the products of three diagonals and subtracting the products of three diagonals.

$$A = \begin{bmatrix} 0 & 2 & 1 \\ 3 & -1 & 2 \\ 4 & -4 & 1 \end{bmatrix}$$

Subtract these products.

$$|A| = 3 = 0 + 16 - 12 - (-4) - 0 - 6 = 2$$

Add these products.

Work with a partner and try using this technique to find the determinants of the matrices in Examples 2 and 3. Do you think this method is easier than expanding by minors?

4.4 ■ Exercises

Developing Skills

In Exercises 1–12, find the determinant of the matrix.

1. $\begin{bmatrix} 2 & 1 \\ 3 & 4 \end{bmatrix}$
2. $\begin{bmatrix} -3 & 1 \\ 5 & 2 \end{bmatrix}$

3. $\begin{bmatrix} 5 & 2 \\ -6 & 3 \end{bmatrix}$
4. $\begin{bmatrix} 2 & -2 \\ 4 & 3 \end{bmatrix}$

5. $\begin{bmatrix} 5 & -4 \\ -10 & 8 \end{bmatrix}$
6. $\begin{bmatrix} 4 & -3 \\ 0 & 0 \end{bmatrix}$

7. $\begin{bmatrix} 2 & 6 \\ 0 & 3 \end{bmatrix}$
8. $\begin{bmatrix} -2 & 3 \\ 6 & -9 \end{bmatrix}$

9. $\begin{bmatrix} -7 & 6 \\ \frac{1}{2} & 3 \end{bmatrix}$
10. $\begin{bmatrix} \frac{2}{3} & \frac{5}{6} \\ 14 & -2 \end{bmatrix}$

11. $\begin{bmatrix} 0.3 & 0.5 \\ 0.5 & 0.3 \end{bmatrix}$
12. $\begin{bmatrix} -1.2 & 4.5 \\ 0.4 & -0.9 \end{bmatrix}$

In Exercises 13–16, evaluate the determinant of the matrix six different ways by expanding by minors along each row and column.

13. $\begin{bmatrix} 2 & 3 & -1 \\ 6 & 0 & 0 \\ 4 & 1 & 1 \end{bmatrix}$
14. $\begin{bmatrix} 10 & 2 & -4 \\ 8 & 0 & -2 \\ 4 & 0 & 2 \end{bmatrix}$

15. $\begin{bmatrix} 1 & 1 & 2 \\ 3 & 1 & 0 \\ -2 & 0 & 3 \end{bmatrix}$
16. $\begin{bmatrix} 2 & 1 & 3 \\ 1 & 4 & 4 \\ 1 & 0 & 2 \end{bmatrix}$

In Exercises 17–30, evaluate the determinant of the matrix by expanding by minors along the row or column that appears to make the computation easiest.

17. $\begin{bmatrix} 2 & 4 & 6 \\ 0 & 3 & 1 \\ 0 & 0 & -5 \end{bmatrix}$
18. $\begin{bmatrix} 2 & 3 & 1 \\ 0 & 5 & -2 \\ 0 & 0 & -2 \end{bmatrix}$

19. $\begin{bmatrix} -2 & 2 & 3 \\ 1 & -1 & 0 \\ 0 & 1 & 4 \end{bmatrix}$
20. $\begin{bmatrix} 3 & 2 & 2 \\ 2 & 2 & 2 \\ -4 & 4 & 3 \end{bmatrix}$

21. $\begin{bmatrix} 1 & 4 & -2 \\ 3 & 6 & -6 \\ -2 & 1 & 4 \end{bmatrix}$
22. $\begin{bmatrix} 2 & -1 & 0 \\ 4 & 2 & 1 \\ 4 & 2 & 1 \end{bmatrix}$

23. $\begin{bmatrix} -3 & 2 & 1 \\ 4 & 5 & 6 \\ 2 & -3 & 1 \end{bmatrix}$
24. $\begin{bmatrix} -3 & 4 & 2 \\ 6 & 3 & 1 \\ 4 & -7 & -8 \end{bmatrix}$

25. $\begin{bmatrix} 1 & 4 & -2 \\ 3 & 2 & 0 \\ -1 & 4 & 3 \end{bmatrix}$
26. $\begin{bmatrix} 6 & 8 & -7 \\ 0 & 0 & 0 \\ 4 & -6 & 22 \end{bmatrix}$

27. $\begin{bmatrix} 0.1 & 0.2 & 0.3 \\ -0.3 & 0.2 & 0.2 \\ 5 & 4 & 4 \end{bmatrix}$
28. $\begin{bmatrix} -0.4 & 0.4 & 0.3 \\ 0.2 & 0.2 & 0.2 \\ 0.3 & 0.2 & 0.2 \end{bmatrix}$

29. $\begin{bmatrix} x & y & 1 \\ 3 & 1 & 1 \\ -2 & 0 & 1 \end{bmatrix}$
30. $\begin{bmatrix} x & y & 1 \\ -2 & -2 & 1 \\ 1 & 5 & 1 \end{bmatrix}$

In Exercises 31–42, use a graphing utility to evaluate the determinant of the matrix.

31. $\begin{bmatrix} 5 & -3 & 2 \\ 7 & 5 & -7 \\ 0 & 6 & -1 \end{bmatrix}$
32. $\begin{bmatrix} -\frac{1}{2} & -1 & 6 \\ 8 & -\frac{1}{4} & -4 \\ 1 & 2 & 1 \end{bmatrix}$

33. $\begin{bmatrix} 35 & 15 & 70 \\ -8 & 20 & 3 \\ -5 & 6 & 20 \end{bmatrix}$
34. $\begin{bmatrix} 3 & -1 & 2 \\ 1 & -1 & 2 \\ -2 & 3 & 10 \end{bmatrix}$

35. $\begin{bmatrix} 0.4 & 0.3 & 0.3 \\ -0.2 & 0.6 & 0.6 \\ 3 & 1 & 1 \end{bmatrix}$

36. $\begin{bmatrix} 0.3 & -0.2 & 0.5 \\ 0.6 & 0.4 & -0.3 \\ 1.2 & 0 & 0.7 \end{bmatrix}$

37. $\begin{bmatrix} 2 & -3 & 3 \\ \frac{3}{4} & 1 & -\frac{1}{4} \\ 12 & 3 & -\frac{1}{2} \end{bmatrix}$
38. $\begin{bmatrix} 1 & 3 & \frac{2}{3} \\ -\frac{3}{2} & \frac{1}{2} & 5 \\ 5 & 2 & \frac{4}{5} \end{bmatrix}$

39. $\begin{bmatrix} \frac{3}{2} & -\frac{3}{4} & 1 \\ 10 & 8 & 7 \\ 12 & -4 & 12 \end{bmatrix}$
40. $\begin{bmatrix} \frac{1}{2} & \frac{3}{2} & \frac{1}{2} \\ 4 & 8 & 10 \\ -2 & -6 & 12 \end{bmatrix}$

41. $\begin{bmatrix} 0.2 & 0.8 & -0.3 \\ 0.1 & 0.8 & 0.6 \\ -10 & -5 & 1 \end{bmatrix}$

42. $\begin{bmatrix} 250 & -125 & 60 \\ -125 & 200 & -50 \\ 60 & -50 & 150 \end{bmatrix}$

In Exercises 43–58, use Cramer's Rule to solve the system of linear equations. (If not possible, state the reason.)

43. $\begin{cases} x + 2y = 5 \\ -x + y = 1 \end{cases}$ **44.** $\begin{cases} 2x - y = -10 \\ 3x + 2y = -1 \end{cases}$

45. $\begin{cases} 3x + 4y = -2 \\ 5x + 3y = 4 \end{cases}$ **46.** $\begin{cases} 18x + 12y = 13 \\ 30x + 24y = 23 \end{cases}$

47. $\begin{cases} 4x + 8y = 16 \\ 2x - 4y = 5 \end{cases}$ **48.** $\begin{cases} 13x - 6y = 17 \\ 26x - 12y = 8 \end{cases}$

49. $\begin{cases} 2x + 0.8y = 1.1 \\ 1.2x - 2.4y = 2.1 \end{cases}$ **50.** $\begin{cases} -0.4x + 0.8y = 1.6 \\ 0.2x + 0.3y = 2.2 \end{cases}$

51. $\begin{cases} 3u + 6v = 5 \\ 6u + 14v = 11 \end{cases}$ **52.** $\begin{cases} 3x_1 + 2x_2 = 1 \\ 2x_1 + 10x_2 = 6 \end{cases}$

53. $\begin{cases} 4x - y + z = -5 \\ 2x + 2y + 3z = 10 \\ 5x - 2y + 6z = 1 \end{cases}$

54. $\begin{cases} 4x - 2y + 3z = -2 \\ 2x + 2y + 5z = 16 \\ 8x - 5y - 2z = 4 \end{cases}$

55. $\begin{cases} 3x + 4y + 4z = 11 \\ 4x - 4y + 6z = 11 \\ 6x - 6y = 3 \end{cases}$

56. $\begin{cases} 14x_1 - 21x_2 - 7x_3 = 10 \\ -4x_1 + 2x_2 - 2x_3 = 4 \\ 56x_1 - 21x_2 + 7x_3 = 5 \end{cases}$

57. $\begin{cases} 3a + 3b + 4c = 1 \\ 3a + 5b + 9c = 2 \\ 5a + 9b + 17c = 4 \end{cases}$

58. $\begin{cases} 2x + 3y + 5z = 4 \\ 3x + 5y + 9z = 7 \\ 5x + 9y + 17z = 13 \end{cases}$

In Exercises 59–68, solve the system of linear equations using a graphing utility and Cramer's Rule.

59. $\begin{cases} -3x + 10y = 22 \\ 9x - 3y = 0 \end{cases}$ **60.** $\begin{cases} 3x + 7y = 3 \\ 7x + 25y = 11 \end{cases}$

61. $\begin{cases} 4x - y = -2 \\ -2x + y = 3 \end{cases}$ **62.** $\begin{cases} 4x + 8y = 6 \\ x + 26y = 19 \end{cases}$

63. $\begin{cases} x + y - z = 2 \\ 6x + 4y + 3z = 4 \\ 3x + 6z = -3 \end{cases}$

64. $\begin{cases} 3x + 2y - 5z = -10 \\ 6x - z = 8 \\ -y + 3z = -2 \end{cases}$

65. $\begin{cases} 3x + y + z = 6 \\ x - 4y + 2z = -1 \\ x - 3y + z = 0 \end{cases}$

66. $\begin{cases} x - y + 2z = 6 \\ -2x + 3y - z = -7 \\ 3x + 2y + 2z = 5 \end{cases}$

67. $\begin{cases} 3x - 2y + 3z = 8 \\ x + 3y + 6z = -3 \\ x + 2y + 9z = -5 \end{cases}$

68. $\begin{cases} 6x + 4y - 8z = -22 \\ -2x + 2y + 3z = 13 \\ -2x + 2y - z = 5 \end{cases}$

In Exercises 69 and 70, solve the equation.

69. $\begin{vmatrix} 5 - x & 4 \\ 1 & 2 - x \end{vmatrix} = 0$

70. $\begin{vmatrix} 4 - x & -2 \\ 1 & 1 - x \end{vmatrix} = 0$

Solving Problems

Area of a Triangle   In Exercises 71–78, use a determinant to find the area of the triangle with the indicated vertices.

71.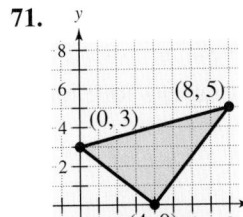

72.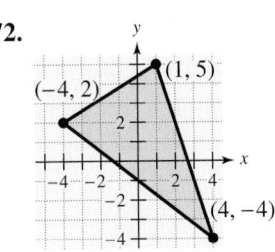

73. $(-2, 1), (3, -1), (1, 6)$

74. $(4, 1), (-3, 0), (2, 6)$

75. $(10, 1), (0, -2), (3, 3)$

76. $(-6, 4), (8, -3), (-2, -5)$

77. $(-3, 4), \left(0, \frac{3}{2}\right), \left(0, \frac{7}{2}\right)$

78. $\left(0, \frac{1}{2}\right), \left(\frac{5}{2}, 0\right), (4, 3)$

Area of a Region   In Exercises 79–82, find the area of the shaded region of the figure.

79.

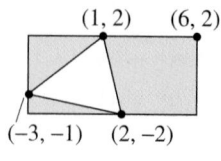

$(1, 2)$ $(6, 2)$

$(-3, -1)$ $(2, -2)$

80.

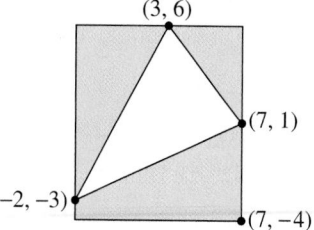

$(3, 6)$

$(7, 1)$

$(-2, -3)$

$(7, -4)$

81.

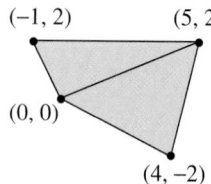

$(-1, 2)$ $(5, 2)$

$(0, 0)$

$(4, -2)$

82.

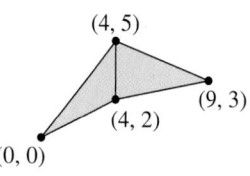

$(4, 5)$

$(9, 3)$

$(4, 2)$

$(0, 0)$

Collinear Points   In Exercises 83–88, determine whether the points are collinear.

83. $(-1, 11), (0, 8), (2, 2)$

84. $(-1, -1), (1, 9), (2, 13)$

85. $(1, 2), (5, 0), (10, -2)$

86. $(-3, 7), (1, 3), (5, -6)$

87. $\left(-2, \frac{1}{3}\right), (2, 1), \left(3, \frac{1}{5}\right)$

88. $\left(0, \frac{1}{2}\right), \left(1, \frac{7}{6}\right), \left(9, \frac{13}{2}\right)$

Equation of a Line   In Exercises 89–96, use a determinant to find the equation of the line through the points.

89. $(0, 0), (5, 3)$

90. $(-4, 3), (2, 1)$

91. $(10, 7), (-2, -7)$

92. $(-6, -4), (-1, 2)$

93. $(2, 7), (10, -1)$

94. $(3, -5), (-3, 12)$

95. $\left(6, \frac{3}{2}\right), \left(-9, -\frac{7}{2}\right)$

96. $\left(-\frac{1}{2}, 3\right), \left(\frac{5}{2}, 1\right)$

Curve-Fitting   In Exercises 97–100, use Cramer's Rule to find the equation of a parabola $y = ax^2 + bx + c$ that passes through the points. Use a graphing utility to plot the points and graph the model.

97. $(0, 1), (1, -3), (-2, 21)$

98. $(2, 3), \left(-1, \frac{9}{2}\right), (-2, 9)$

99. $(1, -1), (-1, -5), \left(\frac{1}{2}, \frac{1}{4}\right)$

100. $(-2, 6), (1, 9), (3, 1)$

101. *Electrical Networks*   When Kirchhoff's Laws are applied to the electrical network in the figure, the currents $I_1, I_2,$ and I_3 are the solution of the system

$$\begin{cases} I_1 - I_2 + I_3 = 0 \\ 3I_1 + 2I_2 \qquad = 7. \\ \qquad 2I_2 + 4I_3 = 8 \end{cases}$$

Find the currents.

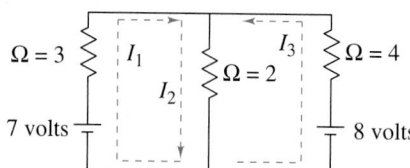

$\Omega = 3$ I_1 I_3 $\Omega = 4$

$\Omega = 2$

I_2

7 volts   8 volts

102. *Electrical Networks*   When Kirchhoff's Laws are applied to the electrical network in the figure, the currents $I_1, I_2,$ and I_3 are the solution of the system

$$\begin{cases} I_1 + I_2 - I_3 = 0 \\ I_1 \qquad + 2I_3 = 12. \\ I_1 - 2I_2 \qquad = -4 \end{cases}$$

Find the currents.

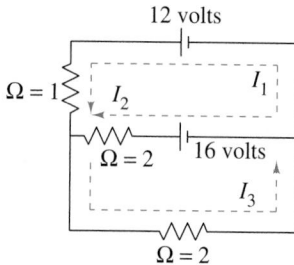

12 volts

$\Omega = 1$ I_2 I_1

$\Omega = 2$ 16 volts

I_3

$\Omega = 2$

103. (a) Use Cramer's Rule to solve the system of linear equations.

$$\begin{cases} kx + (1 - k)y = 1 \\ (1 - k)x + ky = 3 \end{cases}$$

(b) For what value of k can Cramer's Rule not be used?

104. (a) Use Cramer's Rule to solve the system of linear equations.

$$\begin{cases} kx + 3ky = 2 \\ (2 + k)x + ky = 5 \end{cases}$$

(b) For what value of k can Cramer's Rule not be used?

Explaining Concepts

105. Is it possible to find the determinant of a 2×3 matrix? Explain your answer.

106. What conditions must be met in order to use Cramer's Rule to solve a system of linear equations?

107. What is meant by the *minor* of an entry of a square matrix?

108. Explain the difference between a square matrix and its determinant.

Ongoing Review

In Exercises 109 and 110, evaluate the function for the indicated values.

109. $f(x) = \frac{1}{3}x^2$ (a) $f(6)$ (b) $f\left(\frac{3}{4}\right)$

110. $g(x) = \dfrac{x}{x + 10}$ (a) $g(5)$ (b) $g(c - 6)$

In Exercises 111–114, use a graphing utility to graph the function g and identify the transformation of $f(x) = x^5$ represented by g.

111. $g(x) = x^5 - 2$ **112.** $g(x) = (x - 2)^5$

113. $g(x) = -x^5$ **114.** $g(x) = (-x)^5$

In Exercises 115 and 116, write an equation of the line that passes through the two points. Write the equation in slope-intercept form.

115. $(9, -2), (5, 8)$ **116.** $(-4, 1), (-7, -1)$

117. *Number Problem*   The sum of three positive numbers is 33. The second number is three greater than the first, and the third is four times the first. Find the three numbers.

118. *Nut Mixture*   A grocer wishes to mix three kinds of nuts costing \$3.00, \$4.00, and \$6.00 per pound to obtain 50 pounds of a mixture to be priced at \$4.10 per pound. How many pounds of each variety should the grocer use if three-quarters of the mixture is composed of the two least expensive varieties?

Looking Further

Area of a Region   A large region of forest has been infested with gypsy moths. The region is roughly triangular, as shown in the figure. (Dimensions are in miles.)

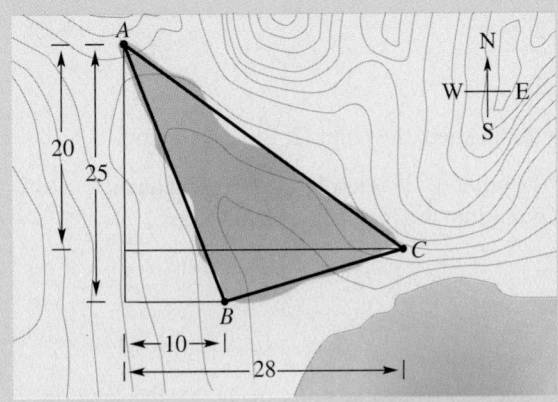

(a) Choose an appropriate point on the figure to use as the origin of a rectangular coordinate system, and determine the coordinates of each vertex of the triangular region. Use a determinant to approximate the number of square miles in this region.

(b) Choose another point on the figure to use as the origin and determine the new coordinates of the vertices. Use a determinant to approximate the number of square miles in the region using the new coordinates.

(c) Do you think the choice of the origin affects the approximation of the area of the figure? Why or why not?

4.5 Linear Inequalities in Two Variables

What you should learn:

- How to verify solutions to linear inequalities in two variables
- How to sketch graphs of linear inequalities in two variables
- How to solve systems of linear inequalities in two variables

Why you should learn it:

Systems of inequalities in two variables can be used to model and solve real-life problems. For instance, in Exercise 82 on page 295, a system of linear inequalities can be used to analyze the compositions of dietary supplements.

EXPLORATION

Sketch the graph of $4x - 3y = 12$. Choose several points to the left and right of the line and evaluate $4x - 3y$ at each point. Which points satisfy the inequality $4x - 3y < 12$? Which points satisfy the inequality $4x - 3y > 12$? What can you conclude about the graph of the solution of a linear inequality?

Linear Inequalities in Two Variables

A **linear inequality** in two variables, x and y, is an inequality that can be written in one of the forms below (where a and b are not both zero).

$$ax + by < c, \qquad ax + by > c, \qquad ax + by \le c, \qquad ax + by \ge c$$

Here are some examples.

$$x - y > -3, \qquad 4x - 3y \le 7, \qquad x < 2, \qquad y \ge -4$$

An ordered pair (x_1, y_1) is a **solution** of a linear inequality in x and y if the inequality is true when x_1 and y_1 are substituted for x and y, respectively. For instance, the ordered pair $(-1, 2)$ is a solution of the inequality $x - y < 1$ because $-1 - 2 < 1$ is a true statement.

Example 1 ■ Verifying Solutions of Linear Inequalities

Determine whether each point is a solution of $2x - 3y \ge -2$.

(a) $(0, 0)$　　(b) $(2, 2)$　　(c) $(0, 1)$

Solution

(a) To determine whether the point $(0, 0)$ is a solution of the inequality, substitute the coordinates of the point into the inequality, as follows.

$$2x - 3y \ge -2 \qquad \text{Write original inequality.}$$
$$2(0) - 3(0) \overset{?}{\ge} -2 \qquad \text{Substitute 0 for } x \text{ and 0 for } y.$$
$$0 \ge -2 \qquad \text{Inequality is satisfied. } \checkmark$$

Because the inequality is satisfied, the point $(0, 0)$ *is* a solution.

(b) To determine whether the point $(2, 2)$ is a solution of the inequality, substitute the coordinates of the point into the inequality, as follows.

$$2x - 3y \ge -2 \qquad \text{Write original inequality.}$$
$$2(2) - 3(2) \overset{?}{\ge} -2 \qquad \text{Substitute 2 for } x \text{ and 2 for } y.$$
$$-2 \ge -2. \qquad \text{Inequality is satisfied. } \checkmark$$

Because the inequality is satisfied, the point $(2, 2)$ *is* a solution.

(c) To determine whether the point $(0, 1)$ is a solution of the inequality, substitute the coordinates of the point into the inequality, as follows.

$$2x - 3y \ge -2 \qquad \text{Write original inequality.}$$
$$2(0) - 3(1) \overset{?}{\ge} -2 \qquad \text{Substitute 0 for } x \text{ and 1 for } y.$$
$$-3 \not\ge -2 \qquad \text{Inequality is not satisfied. } \times$$

Because the inequality is not satisfied, the point $(0, 1)$ *is not* a solution.

The Graph of a Linear Inequality in Two Variables

The **graph** of an inequality is the collection of all solution points of the inequality. To sketch the graph of a linear inequality such as

$$4x - 3y < 12 \qquad \text{Original linear inequality}$$

begin by sketching the graph of the *corresponding linear equation*

$$4x - 3y = 12. \qquad \text{Write corresponding linear equation.}$$

Use *dashed* lines for the inequalities $<$ and $>$ and *solid* lines for the inequalities $\leq$ and $\geq$. The graph of the equation (corresponding to a given linear inequality) separates the plane into two regions, called **half-planes.** In each half-plane, one of the statements below *must* be true.

1. All points in the half-plane are solutions of the inequality.

2. No point in the half-plane is a solution of the inequality.

So, you can determine whether the points in an entire half-plane satisfy the inequality by simply testing *one* point in the region. This graphing procedure is summarized as follows.

Study Tip

When the inequality is strictly less than ($<$) or greater than ($>$), the line of the corresponding equation is dashed because the points on the line *are not* solutions of the inequality. When the inequality is less than or equal to ($\leq$) or greater than or equal to ($\geq$), the line of the corresponding equation is solid because the points on the line *are* solutions of the inequality. The test point used to determine whether the points in a half-plane satisfy the inequality cannot lie on the line of the corresponding equation.

Sketching the Graph of a Linear Inequality in Two Variables

1. Replace the inequality sign by an equal sign, and sketch the graph of the resulting equation. (Use a dashed line for $<$ or $>$ and a solid line for $\leq$ or $\geq$.)

2. Test one point in each of the half-planes formed by the graph in Step 1.

    (a) If the point satisfies the inequality, shade the entire half-plane to denote that every point in the region satisfies the inequality.

    (b) If the point does not satisfy the inequality, then shade the other half-plane.

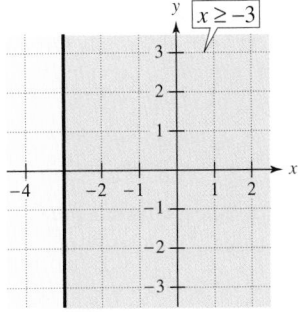

Figure 4.16

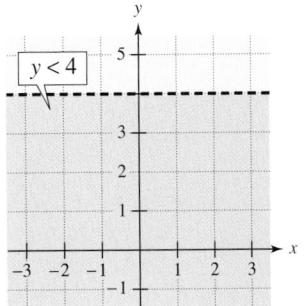

Figure 4.17

Example 2 ■ Sketching the Graph of a Linear Inequality

Sketch the graph of each linear inequality.

(a) $x \geq -3$ (b) $y < 4$

Solution

(a) The graph of the corresponding equation $x = -3$ is a vertical line. The points that satisfy the inequality $x \geq -3$ are those lying to the right of (or on) this line, as shown in Figure 4.16.

(b) The graph of the corresponding equation $y = 4$ is a horizontal line. The points that satisfy the inequality $y < 4$ are those lying below this line, as shown in Figure 4.17.

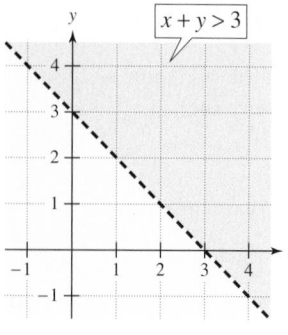

Figure 4.18

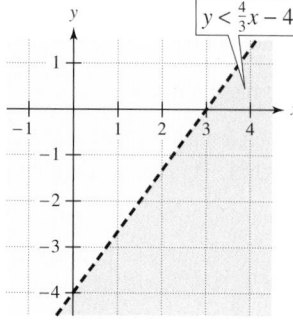

Figure 4.19

Example 3 ■ Sketching the Graph of a Linear Inequality

Sketch the graph of the linear inequality $x + y > 3$.

Solution

The graph of the corresponding equation

$$x + y = 3$$ Write corresponding linear equation.

is a line, as shown in Figure 4.18. Because the origin $(0, 0)$ *does not* satisfy the inequality, the graph consists of the half-plane lying *above* the line. (Try checking a point above the line. Regardless of which point you choose, you will see that it satisfies the inequality.)

For a linear inequality in two variables, you can sometimes simplify the graphing procedure by writing the inequality in *slope-intercept* form. For instance, by writing $x + y > 3$ in the form $y > -x + 3$, you can see that the solution points lie *above* the line $y = -x + 3$, as shown in Figure 4.18. Similarly, by writing the inequality

$$4x - 3y > 12$$ Original linear inequality

in the form

$$y < \frac{4}{3}x - 4$$ Rewrite in slope-intercept form.

you can see that the solutions lie *below* the line $y = \frac{4}{3}x - 4$, as shown in Figure 4.19.

Study Tip

A convenient test point for determining which half-plane contains solutions to the inequality is the origin $(0, 0)$. In Example 3, when you substitute zero for x and zero for y you can easily see that $(0, 0)$ does not satisfy the inequality.

Remember that the origin cannot be used as a test point if it lies on the graph of the corresponding equation.

Technology

Most graphing utilities can graph inequalities in two variables. Consult the user's guide for your graphing utility for specific keystrokes. The graph of

$$y < -\frac{3}{2}x + 2$$

is shown below.

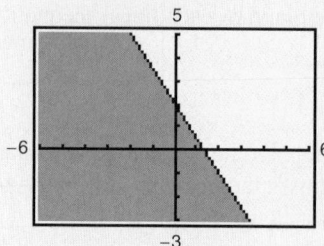

Try using a graphing utility to graph each inequality.

(a) $2x + 3y \geq 3$ (b) $x - 2y \leq 2$

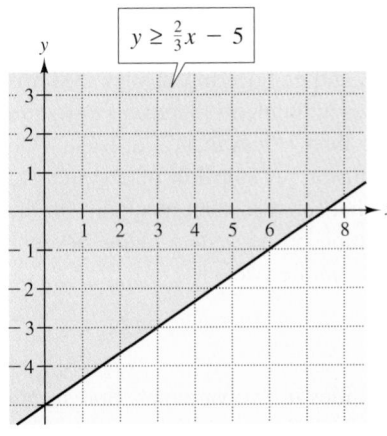

$$y \geq \frac{2}{3}x - 5$$

Figure 4.20

Example 4 ■ Sketching the Graph of a Linear Inequality

Use the slope-intercept form of a linear equation as an aid in sketching the graph of the inequality $2x - 3y \leq 15$.

Solution

To begin, rewrite the inequality in slope-intercept form.

$2x - 3y \leq 15$	Write original inequality.
$-3y \leq -2x + 15$	Subtract $2x$ from each side.
$y \geq \frac{2}{3}x - 5$	Divide each side by -3 and reverse the inequality symbol.

From this form, you can conclude that the solution is the half-plane lying *on* or *above* the line $y = \frac{2}{3}x - 5$. The graph is shown in Figure 4.20.

Example 5 ■ An Application: Working to Meet a Budget

Your budget requires you to earn *at least* $160 per week. You work two part-time jobs. The first job pays $10 per hour and the second job pays $8 per hour. Let x represent the number of hours worked at the first job and let y represent the number of hours worked at the second job.

(a) Write an inequality that represents the number of hours worked at each job in order to meet your budget requirements.

(b) Graph the inequality and identify at least two ordered pairs (x, y) that identify the number of hours you must work at each job in order to meet your budget requirements.

Solution

(a) *Verbal Model:* $10 \cdot \boxed{\text{Number of hours at job 1}} + 8 \cdot \boxed{\text{Number of hours at job 2}} \geq 160$

   Labels: Number of hours at job 1 = x (hours)
   Number of hours at job 2 = y (hours)

   Inequality: $10x + 8y \geq 160$

(b) Rewrite the inequality in slope-intercept form.

$10x + 8y \geq 160$	Write original inequality.
$8y \geq -10x + 160$	Subtract $10x$ from each side.
$y \geq \dfrac{-10x + 160}{8}$	Divide each side by 8.
$y \geq -1.25x + 20.$	Simplify.

Graph the corresponding equation $y = -1.25x + 20$ and shade the half-plane lying above the line, as shown in Figure 4.21. From the graph, you can see that two solutions that will yield the desired weekly earnings of at least $160 are $(8, 10)$ and $(12, 5)$. There are many other solutions.

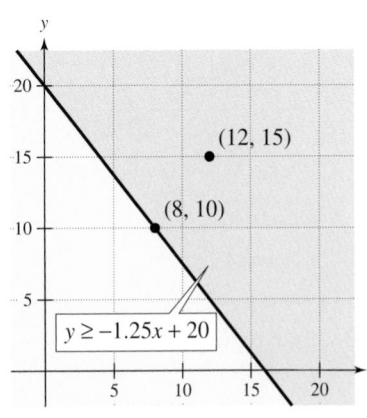

$$y \geq -1.25x + 20$$

Figure 4.21

Systems of Linear Inequalities in Two Variables

Many practical problems in business, science, and engineering involve **systems of linear inequalities.** This type of system arises in problems that have *constraint* statements that contain phrases such as "more than," "less than," "at least," "no more than," "a minimum of," and "a maximum of." A **solution** of a system of linear inequalities in x and y is a point (x, y) that satisfies each inequality in the system. For instance, the point $(2, 4)$ is a solution of the system below because $x = 2$ and $y = 4$ satisfy each of the inequalities in the system.

$$\begin{cases} x + y \le 10 \\ 3x - y \le 2 \end{cases}$$

To sketch the graph of a system of inequalities in two variables, first sketch (on the same coordinate system) the graph of each individual inequality. The **solution set** is the region that is *common* to every graph in the system.

Example 6 ■ Graphing a System of Linear Inequalities

Sketch the graph of the system of linear inequalities.

$$\begin{cases} 2x - y \le 5 \\ x + 2y \ge 2 \end{cases}$$

Solution

Begin by rewriting each inequality in slope-intercept form. Then sketch the graph of each inequality. The graph of $2x - y \le 5$ consists of all points on and above the line $y = 2x - 5$, as shown in Figure 4.22. The graph of $x + 2y \ge 2$ consists of all points on and above the line $y = -\frac{1}{2}x + 1$, as shown in Figure 4.23. The graph of the *system* of linear inequalities consists of the purple wedge-shaped region, which is common to the two half-planes (representing the individual inequalities), shown in Figure 4.24.

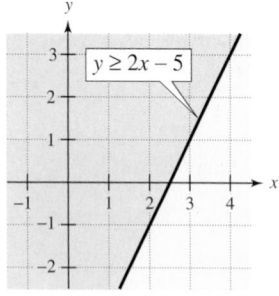

Figure 4.22

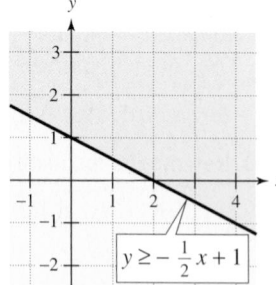

Figure 4.23

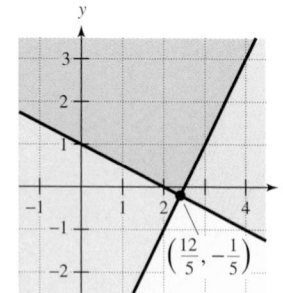

Figure 4.24

In Figure 4.24, note that the two borderlines of the region

$$y = 2x - 5 \quad \text{and} \quad y = -\frac{1}{2}x + 1$$

intersect at the point $\left(\frac{12}{5}, -\frac{1}{5}\right)$. Such a point is called a **vertex** of the region. The region shown in the figure has only one vertex. Some regions, however, have several vertices. When you are sketching the graph of a system of linear inequalities, it is helpful to find and label any vertices of the region.

> **Graphing a System of Linear Inequalities**
>
> 1. Sketch the line that corresponds to each inequality. (Use dashed lines for inequalities with $<$ or $>$ and solid lines for inequalities with $\leq$ or $\geq$.)
>
> 2. Lightly shade the half-plane that is the graph of each linear inequality. (Colored pencils may help you distinguish the different half-planes.)
>
> 3. The graph of the system is the intersection of the half-planes. (If you have used colored pencils, it is the region that is shaded with *every* color.)

Example 7 ■ Graphing a System of Linear Inequalities

Sketch the graph of the system of linear inequalities.

$$\begin{cases} y < 4 \\ y > 1 \end{cases}$$

Solution

The graph of the first inequality is the half-plane *below* the horizontal line

$$y = 4. \qquad \text{Upper boundary}$$

The graph of the second inequality is the half-plane *above* the horizontal line

$$y = 1. \qquad \text{Lower boundary}$$

The graph of the system is the horizontal band that lies *between* the two horizontal lines (where $y < 4$ *and* $y > 1$), as shown in Figure 4.25.

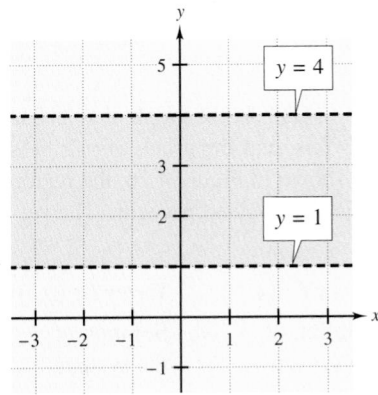

Figure 4.25

Technology

A graphing utility can be used to graph a system of linear inequalities. The graph of

$$\begin{cases} 4y < 2x - 6 \\ x + y \geq 7 \end{cases}$$

is shown below. The shaded region, in which all points satisfy both inequalities, is the solution of the system.

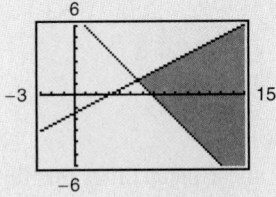

Try using a graphing utility to graph

$$\begin{cases} 3x + y < 1 \\ -2x - 2y < 8 \end{cases}.$$

Example 8 ■ Graphing a System of Linear Inequalities

Sketch the graph of the system of linear inequalities, and label the vertices.

$$\begin{cases} x + y \le 5 \\ 3x + 2y \le 12 \\ x \ge 0 \\ y \ge 0 \end{cases}$$

Solution

Begin by sketching the half-planes represented by the four linear inequalities. The graph of

$$x + y \le 5$$

is the half-plane lying on and below the line $y = -x + 5$, the graph of

$$3x + 2y \le 12$$

is the half-plane lying on and below the line $y = -\frac{3}{2}x + 6$, the graph of $x \ge 0$ is the half-plane lying on and to the right of the y-axis, and the graph of $y \ge 0$ is the half-plane lying on and above the x-axis. As shown in Figure 4.26, the region that is common to all four of these half-planes is a four-sided polygon. The vertices of the region are found as follows.

Vertex A: $(0, 5)$	*Vertex B:* $(2, 3)$	*Vertex C:* $(4, 0)$	*Vertex D:* $(0, 0)$
Solution of the system	Solution of the system	Solution of the system	Solution of the system
$\begin{cases} x + y = 5 \\ x = 0 \end{cases}$	$\begin{cases} x + y = 5 \\ 3x + 2y = 12 \end{cases}$	$\begin{cases} 3x + 2y = 12 \\ y = 0 \end{cases}$	$\begin{cases} x = 0 \\ y = 0 \end{cases}$

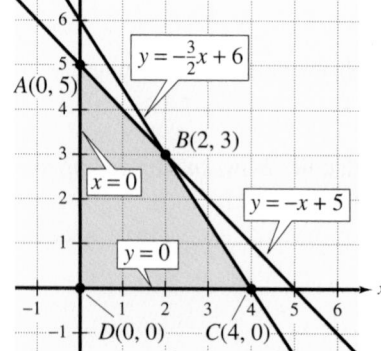

Figure 4.26

Example 9 ■ Finding the Boundaries of a Region

Find a system of inequalities that defines the region shown in Figure 4.27.

Solution

Three of the boundaries of the region are horizontal or vertical—they are easy to find. To find the diagonal boundary line, use the techniques from Section 3.1 to find the equation of the line passing through the points $(4, 4)$ and $(6, 0)$. You can use the formula for slope to find $m = -2$, and then use the point-slope form with point $(6, 0)$ and $m = -2$ to obtain

$$y - 0 = -2(x - 6).$$

So, the equation is

$$y = -2x + 12.$$

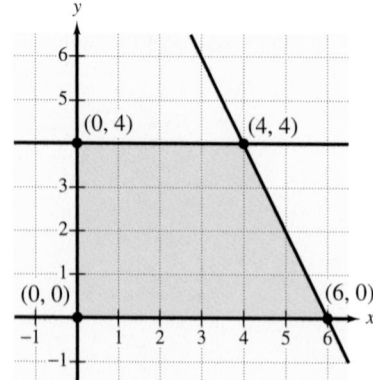

Figure 4.27

The system of linear inequalities that describes the region is as follows.

$$\begin{cases} y \le 4 \\ y \ge 0 \\ x \ge 0 \\ y \le -2x + 12 \end{cases}$$
Region lies on and below line $y = 4$.
Region lies on and above x-axis.
Region lies on and to the right of y-axis.
Region lies on and below line $y = -2x + 12$.

4.5 ■ Exercises

Developing Skills

In Exercises 1–6, match the linear inequality with its graph.
[The graphs are labeled (a), (b), (c), (d), (e), and (f).]

(a)
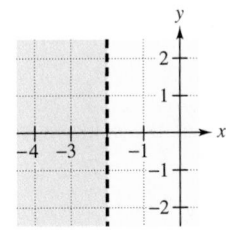

(b)
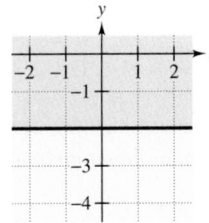

(c)
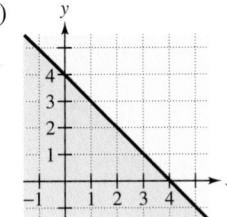

(d)
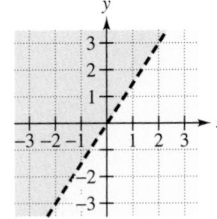

(e)
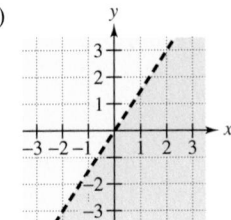

(f)
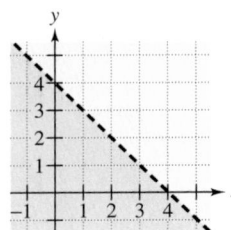

1. $y \geq -2$

2. $x < -2$

3. $3x - 2y < 0$

4. $3x - 2y > 0$

5. $x + y < 4$

6. $x + y \leq 4$

In Exercises 7–14, determine whether the points are solutions of the inequality.

Inequality	Points	
7. $x - 2y < 4$	(a) $(0, 0)$	(b) $(2, -1)$
	(c) $(3, 4)$	(d) $(5, 1)$
8. $x + y < 3$	(a) $(0, 6)$	(b) $(4, 0)$
	(c) $(0, -2)$	(d) $(1, 1)$
9. $3x + y \geq 10$	(a) $(1, 3)$	(b) $(-3, 1)$
	(c) $(3, 1)$	(d) $(2, 15)$
10. $-3x + 5y \geq 6$	(a) $(2, 8)$	(b) $(-10, -3)$
	(c) $(0, 0)$	(d) $(3, 3)$

Inequality	Points	
11. $y > 0.2x - 1$	(a) $(0, 2)$	(b) $(6, 0)$
	(c) $(4, -1)$	(d) $(-2, 7)$
12. $y < -3.5x + 7$	(a) $(1, 5)$	(b) $(5, -1)$
	(c) $(-1, 4)$	(d) $\left(0, \frac{4}{3}\right)$
13. $y \leq 3 - |x|$	(a) $(-1, 4)$	(b) $(2, -2)$
	(c) $(6, 0)$	(d) $(5, -2)$
14. $y \geq |x - 3|$	(a) $(0, 0)$	(b) $(1, 2)$
	(c) $(4, 10)$	(d) $(5, -1)$

In Exercises 15–30, sketch the graph of the solution of the linear inequality.

15. $x \geq 2$ **16.** $y > 2$

17. $y > \frac{1}{2}x$ **18.** $y \leq 2x$

19. $y \leq x + 1$ **20.** $y > 4 - x$

21. $y - 1 > -\frac{1}{2}(x - 2)$ **22.** $y - 2 < -\frac{2}{3}(x - 1)$

23. $\frac{x}{3} + \frac{y}{4} \leq 1$ **24.** $\frac{x}{2} + \frac{y}{6} \geq 1$

25. $x - 2y \geq 6$ **26.** $3x + 5y \leq 15$

27. $3x - 2y \geq 4$ **28.** $x + 3y \leq 5$

29. $0.2x + 0.3y < 2$ **30.** $x - 0.75y > 6$

In Exercises 31–38, use a graphing utility to graph the solution of the linear inequality.

31. $y \geq \frac{3}{4}x - 1$ **32.** $y \leq 9 - 1.5x$

33. $y \leq -\frac{2}{3}x + 6$ **34.** $y \geq \frac{1}{4}x + 3$

35. $x - 2y - 4 \geq 0$ **36.** $2x + 4y - 3 \leq 0$

37. $2x + 3y - 12 \leq 0$ **38.** $x - 3y + 9 \geq 0$

In Exercises 39–44, write an inequality for the shaded region shown in the figure.

39.
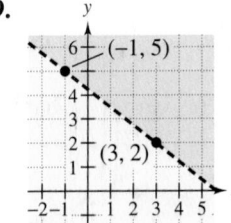

40.
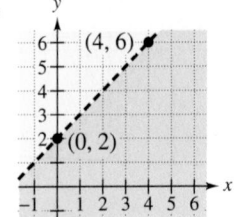

41.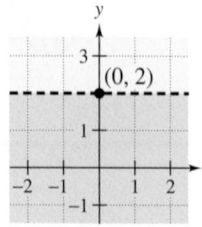

42.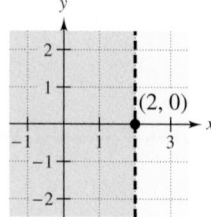

43.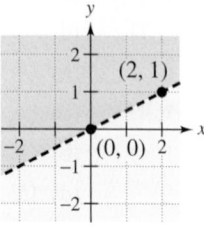

44.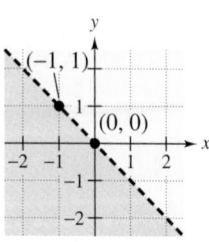

In Exercises 45–50, match the system of linear inequalities with its graph. [The graphs are labeled (a), (b), (c), (d), (e), and (f).]

(a)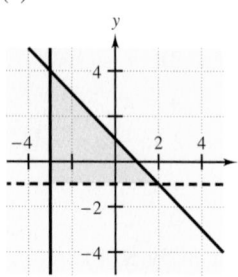

(b)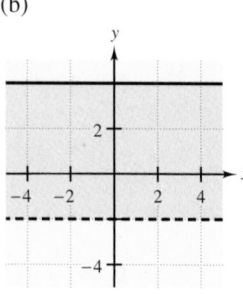

(c)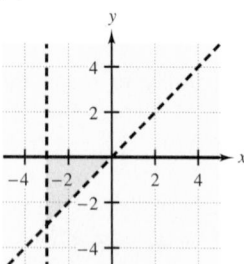

(d)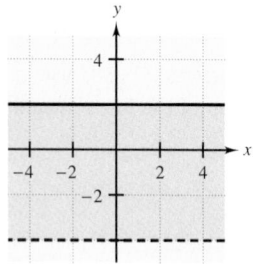

(e)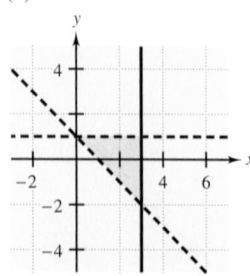

(f)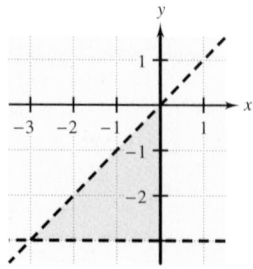

45. $\begin{cases} y > x \\ x > -3 \\ y \le 0 \end{cases}$

46. $\begin{cases} y \le 4 \\ y > -2 \end{cases}$

47. $\begin{cases} y < x \\ y > -3 \\ x \le 0 \end{cases}$

48. $\begin{cases} x \le 3 \\ y < 1 \\ y > -x + 1 \end{cases}$

49. $\begin{cases} y > -1 \\ x \ge -3 \\ y \le -x + 1 \end{cases}$

50. $\begin{cases} y > -4 \\ y \le 2 \end{cases}$

In Exercises 51–64, sketch a graph of the solution of the system of linear inequalities.

51. $\begin{cases} x < 3 \\ x > -2 \end{cases}$

52. $\begin{cases} y > -1 \\ y \le 2 \end{cases}$

53. $\begin{cases} y > -5 \\ x \le 2 \\ y \le x + 2 \end{cases}$

54. $\begin{cases} y \ge -1 \\ x \le 2 \\ y \le x + 2 \end{cases}$

55. $\begin{cases} x + y \le 1 \\ -x + y \le 1 \\ y \ge 0 \end{cases}$

56. $\begin{cases} 3x + 2y < 6 \\ x \ge 0 \\ y \ge 0 \end{cases}$

57. $\begin{cases} x + y \le 5 \\ x \ge 2 \\ y \ge 0 \end{cases}$

58. $\begin{cases} 2x + y \ge 2 \\ x \le 2 \\ y \le 1 \end{cases}$

59. $\begin{cases} -3x + 2y < 6 \\ x - 4y > -2 \\ 2x + y < 3 \end{cases}$

60. $\begin{cases} x - 7y > -36 \\ 5x + 2y > 5 \\ 6x + 5y > 6 \end{cases}$

61. $\begin{cases} 2x + y < 2 \\ 6x + 3y > 2 \end{cases}$

62. $\begin{cases} x - 2y < -6 \\ 5x - 3y > -9 \end{cases}$

63. $\begin{cases} x \ge 1 \\ x - 2y \le 3 \\ 3x + 2y \ge 9 \\ x + y \le 6 \end{cases}$

64. $\begin{cases} x + y \le 4 \\ x + y \ge -1 \\ x - y \ge -2 \\ x - y \le 2 \end{cases}$

In Exercises 65–70, use a graphing utility to graph the solution of the system of linear inequalities.

65. $\begin{cases} 2x - 3y \le 6 \\ y \le 4 \end{cases}$

66. $\begin{cases} 6x + 3y \ge 12 \\ y \le 4 \end{cases}$

67. $\begin{cases} 2x - 2y \le 5 \\ y \le 6 \end{cases}$

68. $\begin{cases} 2x + 3y \ge 12 \\ y \ge 2 \end{cases}$

69. $\begin{cases} 2x + y \le 2 \\ y \ge -4 \end{cases}$

70. $\begin{cases} x - 2y \ge -6 \\ y \le 6 \end{cases}$

In Exercises 71–74, write a system of linear inequalities that describes the shaded region.

71.

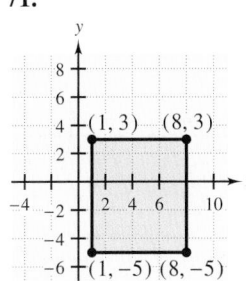

72.

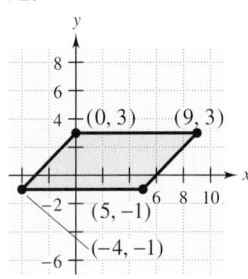

73.

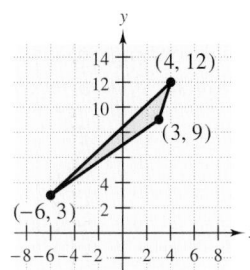

74.

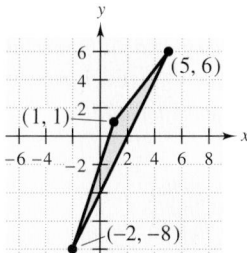

Solving Problems

75. *Geometry*   The perimeter of a rectangle of length x and width y cannot exceed 500 feet. Write a linear inequality for this constraint. Use a graphing utility to graph the inequality.

76. *Geometry*   The perimeter of a rectangle of length x and width y must be 212 feet or more. Write a linear inequality for this constraint. Use a graphing utility to graph the inequality.

77. *Storage Space*   A warehouse for storing chairs and tables has 1000 square feet of floor space. Each chair requires 10 square feet and each table requires 15 square feet. Write a linear inequality for this space constraint if x is the number of chairs and y is the number of tables stored. Sketch the graph of the inequality.

78. *Personal Income*   You have two part-time jobs. One is at a grocery store, which pays $9 per hour, and the other is mowing lawns, which pays $6 per hour. Between the two jobs you want to earn at least $150 per week. Write a linear inequality that shows the different numbers of hours you can work at each job, and sketch the graph of the inequality. From the graph, find several ordered pairs with positive integer coordinates that are solutions of the inequality.

79. *Investment*   A person plans to invest up to $20,000 in two different interest-bearing accounts. Each account is to contain at least $5000. Moreover, one account should have at least twice the amount in the other account. Write a system of linear inequalities describing the various amounts that can be deposited in each account, and sketch the graph of the system.

80. *Cooking*   The time t (in minutes) that it takes to roast a turkey weighing p pounds at 350°F is given by the inequalities below.

For a turkey up to 6 pounds: $t \geq 20p$

For a turkey over 6 pounds: $t \geq 15p + 30$

Sketch the graphs of these inequalities. What are the coordinates for a 12-pound turkey that has been roasting for 3 hours and 40 minutes? Is this turkey fully cooked?

81. *Ticket Sales*   Two types of tickets are to be sold for a concert. One type costs $15 per ticket and the other type costs $25 per ticket. The promoter of the concert must sell at least 15,000 tickets, including at least 8000 of the $15 tickets and at least 4000 of the $25 tickets. Moreover, the gross receipts must total at least $275,000 in order for the concert to be held. Write a system of linear inequalities describing the different numbers of tickets that can be sold. Use a graphing utility to graph the system.

82. *Nutrition*   A dietitian is asked to design a special diet supplement using two different foods. Each ounce of food X contains 20 units of calcium, 15 units of iron, and 10 units of vitamin B. Each ounce of food Y contains 10 units of calcium, 10 units of iron, and 20 units of vitamin B. The minimum daily requirements in the diet are 280 units of calcium, 160 units of iron, and 180 units of vitamin B. Write a system of linear inequalities describing the different amounts of food X and food Y that can be used in the diet. Use a graphing utility to graph the system.

83. *Area of a Region* The figure shows a cross section of a roped-off swimming area at a beach. Write a system of linear inequalities describing the cross section. (Each unit in the coordinate system represents 1 foot.)

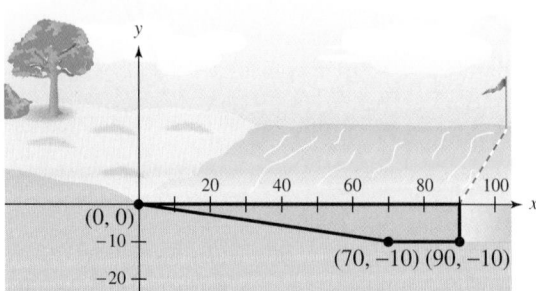

84. *Area of a Region* The figure shows the chorus platform on a stage. Write a system of linear inequalities describing the part of the audience that can see the full chorus. (Each unit in the coordinate system represents 1 meter.)

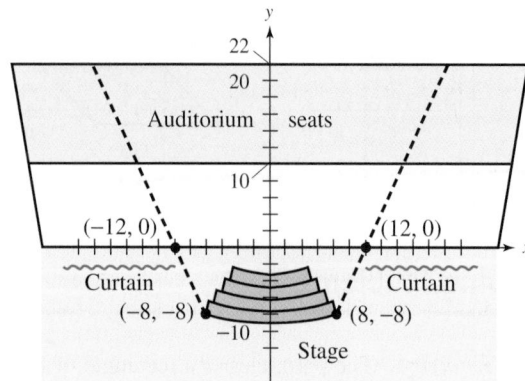

Explaining Concepts

85. Explain the meaning of the term *half-plane*. Give an example of an inequality whose graph is a half-plane.

86. How does the solution set of $x - y > 1$ differ from the solution set of $x - y \geq 1$?

87. After graphing the boundary, explain how you determine which half-plane is the solution set of a linear inequality.

88. Explain the difference between graphing the solution of the inequality $x \leq 3$ on the real number line and on the rectangular coordinate system.

Ongoing Review

In Exercises 89–96, sketch the graph of the equation.

89. $2x + 4y = 8$

90. $3x - 5y = -5$

91. $y = 4(x + 1) - 3$

92. $y = -\frac{3}{4}(x + 4) - 3$

93. $x + 4y = 5$

94. $0.3x - 0.2y = 0.8$

95. $y = -5$

96. $x = 6$

97. *Average Speed* A truck travels for 4 hours at an average speed of 42 miles per hour. How much longer must the truck travel at an average speed of 60 miles per hour so that the average speed for the total trip will be 55 miles per hour?

98. *Speed* An airplane flying into a headwind travels 1800 miles in 3 hours and 36 minutes. On the return flight, the same distance is traveled in 3 hours. Find the speed of the plane in still air and the speed of the wind, assuming that both remain constant throughout the round trip.

Looking Further

Pizza and Soda Pop You and some friends go out for pizza. Together you have $30. You want to order two large pizzas with cheese at $10 each. Each additional topping costs $0.50, and each small soft drink costs $1.10. Write a linear inequality that represents the number of toppings x and drinks y that your group can afford. Sketch the graph of the inequality. What are the coordinates for an order of six soft drinks and two large pizzas with cheese, each with three additional toppings? Is this a solution of the inequality? (Assume there is no sales tax.)

Chapter Summary: Key Concepts

4.1 ■ The method of substitution

1. Solve one of the equations for one variable in terms of the other variable.

2. Substitute the expression found in Step 1 into the other equation to obtain an equation in one variable.

3. Solve the equation obtained on Step 2.

4. Back-substitute the solution from Step 3 into the expression obtained in Step 1 to find the value of the other variable.

5. Check the solution in the original system.

4.1 ■ The method of elimination

1. Write each equation in the form $Ax + By = C$.

2. Obtain coefficients for x (or y) that differ only in sign by multiplying all terms of one or both equations by suitable constants.

3. Add the equations to eliminate one variable and solve the resulting equation.

4. Back-substitute the value obtained in Step 3 into either of the original equations and solve for the other variable.

5. Check your solution in the original system.

4.2 ■ Operations that produce equivalent systems

Each of the row operations on a system of linear equations below produces an equivalent system of linear equations.

1. Interchange two equations.

2. Multiply one of the equations by a nonzero constant.

3. Add a multiple of one of the equations to another equation to replace the latter equation.

4.2 ■ The number of solutions of a linear system

For a system of linear equations, exactly one of the statements below is true.

1. There is exactly one solution.

2. There are infinitely many solutions.

3. There is no solution.

4.3 ■ Elementary row operations

Two matrices are row-equivalent if one can be obtained from the other by a sequence of elementary row operations.

1. Interchange two rows.

2. Multiply a row by a nonzero constant.

3. Add a multiple of a row to another row.

4.3 ■ Gaussian elimination with back-substitution

1. Write the augmented matrix of the system of linear equations.

2. Use elementary row operations to rewrite the augmented matrix in row-echelon form.

3. Write the system of linear equations corresponding to the matrix in row-echelon form, and use back-substitution to find the solution.

4.4 ■ Determinant of a 2 × 2 matrix

$$\det(A) = |A| = \begin{vmatrix} a_1 & b_1 \\ a_2 & b_2 \end{vmatrix} = a_1 b_2 - a_2 b_1$$

4.4 ■ Expanding by minors

The determinant of a 3×3 matrix can be evaluated by expanding by minors.

$$\det(A) = \begin{vmatrix} a_1 & b_1 & c_1 \\ a_2 & b_2 & c_2 \\ a_3 & b_3 & c_3 \end{vmatrix}$$

$$= a_1 \begin{vmatrix} b_2 & c_2 \\ b_3 & c_3 \end{vmatrix} - b_1 \begin{vmatrix} a_2 & c_2 \\ a_3 & c_3 \end{vmatrix} + c_1 \begin{vmatrix} a_2 & b_2 \\ a_3 & b_3 \end{vmatrix}$$

4.5 ■ Sketching the graph of a linear inequality in two variables

1. Replace the inequality sign by an equal sign, and sketch the graph of the resulting equation. (Use a dashed line for $<$ or $>$ and a solid line for $\leq$ or $\geq$.)

2. Test one point in each of the half-planes formed by the graph in Step 1. If the point satisfies the inequality, shade the entire half-plane to denote that every point in the region satisfies the inequality.

4.5 ■ Graphing a system of linear inequalities

1. Sketch the line that corresponds to each inequality.

2. Lightly shade the half-plane that is the graph of each linear inequality.

3. The graph of the system is the intersection of the half-planes.

Review Exercises

Reviewing Skills

4.1 ■ *How to determine if ordered pairs are solutions of systems of linear equations*

In Exercises 1 and 2, determine whether each ordered pair is a solution of the system of linear equations.

1. $\begin{cases} -3x + 2y = -19 \\ 5x - y = 27 \end{cases}$ (a) $(5, -2)$ (b) $(3, 4)$

2. $\begin{cases} 7x + 2y = -1 \\ 3x - 6y = -21 \end{cases}$ (a) $(-2, -4)$ (b) $(-1, 3)$

■ *How to solve systems of linear equations graphically*

In Exercises 3–8, sketch the graphs of the equations and approximate any solutions of the system of linear equations.

3. $\begin{cases} x + y = 2 \\ x - y = 0 \end{cases}$

4. $\begin{cases} 2x = 3y - 3 \\ y = x \end{cases}$

5. $\begin{cases} x - y = 3 \\ -x + y = 1 \end{cases}$

6. $\begin{cases} x + y = -1 \\ 3x + 2y = 0 \end{cases}$

7. $\begin{cases} 2x - y = 0 \\ -x + y = 4 \end{cases}$

8. $\begin{cases} x = y + 3 \\ x = y + 1 \end{cases}$

In Exercises 9–12, use a graphing utility to graph the equations and approximate any solutions of the system of linear equations.

9. $\begin{cases} 5x - 3y = 3 \\ 2x + 2y = 14 \end{cases}$

10. $\begin{cases} 8x + 5y = 1 \\ 3x - 4y = 18 \end{cases}$

11. $\begin{cases} x - 3y = -1 \\ -3x + 2y = -4 \end{cases}$

12. $\begin{cases} 2x + y = 4 \\ -x - y = -1 \end{cases}$

■ *How to solve systems of linear equations using the method of substitution*

In Exercises 13–18, solve the system of linear equations by the method of substitution.

13. $\begin{cases} 2x + 3y = 1 \\ x + 4y = -2 \end{cases}$

14. $\begin{cases} 3x - 7y = 10 \\ -2x + y = -14 \end{cases}$

15. $\begin{cases} -5x + 2y = 4 \\ 10x - 4y = 7 \end{cases}$

16. $\begin{cases} 5x + 2y = 3 \\ 2x + 3y = 10 \end{cases}$

17. $\begin{cases} 3x - 7y = 5 \\ 5x - 9y = -5 \end{cases}$

18. $\begin{cases} 24x - 4y = 20 \\ 6x - y = 5 \end{cases}$

■ *How to solve systems of linear equations using the method of elimination*

In Exercises 19–22, solve the system of equations by the method of elimination.

19. $\begin{cases} x + y = 0 \\ 2x + y = 0 \end{cases}$

20. $\begin{cases} 4x + y = 1 \\ x - y = 4 \end{cases}$

21. $\begin{cases} 0.2x - 0.1y = 0.2 \\ 0.6x + 0.8y = 3.9 \end{cases}$

22. $\begin{cases} 0.2x + 0.3y = 0.14 \\ 0.4x + 0.5y = 0.20 \end{cases}$

4.2 ■ *How to determine if ordered triples are solutions of systems of linear equations*

In Exercises 23 and 24, determine whether each ordered triple is a solution of the system of linear equations.

23. $\begin{cases} 3x - y + 5z = 6 \\ -x + 8z = 16 \\ 10x + 3y - 5z = 2 \end{cases}$ (a) $(8, 1, -3)$ (b) $(0, 4, 2)$

24. $\begin{cases} 4x + 9y - 6z = 15 \\ -2x + y - 4z = 1 \\ 8x - 4y = 28 \end{cases}$ (a) $(3, -1, -2)$ (b) $(5, 0, -2)$

■ *How to solve systems of linear equations using row-echelon form with back-substitution*

In Exercises 25 and 26, solve the system of equations by the method of back-substitution.

25. $\begin{cases} x - 6y + 2z = 13 \\ y - z = -4 \\ z = 3 \end{cases}$

26. $\begin{cases} x + y - 2z = 15 \\ y + 5z = -4 \\ z = -2 \end{cases}$

■ *How to solve systems of linear equations using the method of Gaussian elimination*

In Exercises 27 and 28, solve the system of equations by the method of elimination or Gaussian elimination.

27. $\begin{cases} 2x + 3y + z = 10 \\ 2x - 3y - 3z = 22 \\ 4x - 2y + 3z = -2 \end{cases}$

28. $\begin{cases} -x + y + 2z = 1 \\ 2x + 3y + z = -2 \\ 5x + 4y + 2z = 4 \end{cases}$

4.3 ■ *How to determine the order of matrices*

In Exercises 29–32, determine the order of the matrix.

29. $\begin{bmatrix} 7 & 0 \end{bmatrix}$

30. $\begin{bmatrix} 5 & -3 & 8 & 7 \end{bmatrix}$

31. $\begin{bmatrix} 2 \\ 36 \\ 3 \end{bmatrix}$

32. $\begin{bmatrix} -7 & 6 & 4 \\ 0 & -5 & 1 \end{bmatrix}$

■ *How to form coefficient and augmented matrices and form linear systems from the augmented matrices*

In Exercises 33–36, form (a) the coefficient matrix and (b) the augmented matrix for the system of linear equations.

33. $\begin{cases} 12x - 2y = 6 \\ 3x + 8y = -5 \end{cases}$

34. $\begin{cases} -x + 9y = 10 \\ 15x + 6y = 0 \end{cases}$

35. $\begin{cases} 3x - 3y + z = 1 \\ 10x + 7y = 4 \\ 6x + 14y - 9z = 10 \end{cases}$

36. $\begin{cases} 12x - 15y + z = 1 \\ 7x + 8z = 9 \\ -x + y - 4z = 0 \end{cases}$

In Exercises 37 and 38, write the system of linear equations represented by the augmented matrix.

37. $\begin{bmatrix} 6 & 4 & 1 & \vdots & 0 \\ 12 & 9 & -2 & \vdots & 6 \\ 1 & 8 & -4 & \vdots & -5 \end{bmatrix}$

38. $\begin{bmatrix} 0 & 5 & -4 & \vdots & 4 \\ -2 & 12 & 9 & \vdots & -3 \\ 8 & 12 & 0 & \vdots & 0 \end{bmatrix}$

■ *How to perform elementary row operations to solve systems of linear equations*

In Exercises 39–42, use elementary row operations to solve the system of linear equations.

39. $\begin{cases} x + y = 4 \\ x - y = 2 \end{cases}$

40. $\begin{cases} 2x + y = 3 \\ 4x + 3y = 5 \end{cases}$

41. $\begin{cases} -x + y + 2z = 4 \\ 4x + 2y + z = -2 \\ x + 4y + 2z = 3 \end{cases}$

42. $\begin{cases} 2x + 3y + z = 10 \\ 2x - 3y - 3z = 22 \\ 4x - 2y + 3z = -2 \end{cases}$

■ *How to use matrices and Gaussian elimination with back-substitution to solve systems of linear equations*

In Exercises 43–48, use matrices to solve the system of linear equations.

43. $\begin{cases} 5x + 4y = 2 \\ -x + y = -22 \end{cases}$

44. $\begin{cases} 2x - 5y = 2 \\ 3x - 7y = 1 \end{cases}$

45. $\begin{cases} x + 2y + 6z = 4 \\ -3x + 2y - z = -4 \\ 4x + 2z = 16 \end{cases}$

46. $\begin{cases} -x + 3y - z = -4 \\ 2x + 6z = 14 \\ -3x - y + z = 10 \end{cases}$

47. $\begin{cases} 2x_1 + 3x_2 + 3x_3 = 3 \\ 6x_1 + 6x_2 + 12x_3 = 13 \\ 12x_1 + 9x_2 - x_3 = 2 \end{cases}$

48. $\begin{cases} -x_1 + 2x_2 + 3x_3 = 4 \\ 2x_1 - 4x_2 - x_3 = -13 \\ 3x_1 + 2x_2 - 4x_3 = -1 \end{cases}$

In Exercises 49–52, use the matrix capabilities of a graphing utility to solve the system of equations.

49. $\begin{cases} 0.2x - 0.1y = 0.07 \\ 0.4x - 0.5y = -0.01 \end{cases}$

50. $\begin{cases} 2x + y = 0.3 \\ 3x - y = -1.3 \end{cases}$

51. $\begin{cases} 5x - 3y + 2z = 2 \\ 2x + 2y - 3z = 3 \\ x - 7y + 8z = -4 \end{cases}$

52. $\begin{cases} 3x + 2y + 5z = 4 \\ 4x - 3y - 4z = 1 \\ -8x + 2y + 3z = 0 \end{cases}$

4.4 ■ *How to find determinants of 2 × 2 matrices and 3 × 3 matrices*

In Exercises 53 and 54, find the determinant of the matrix.

53. $\begin{bmatrix} 7 & 10 \\ 10 & 15 \end{bmatrix}$

54. $\begin{bmatrix} -3.4 & 1.2 \\ -5 & 2.5 \end{bmatrix}$

In Exercises 55 and 56, find the determinant of the matrix using any appropriate method.

55. $\begin{bmatrix} 8 & 3 & 2 \\ 1 & -2 & 4 \\ 6 & 0 & 5 \end{bmatrix}$

56. $\begin{bmatrix} 4 & 0 & 10 \\ 0 & 10 & 0 \\ 10 & 0 & 34 \end{bmatrix}$

In Exercises 57 and 58, use the matrix capabilities of a graphing utility to evaluate the determinant.

57. $\begin{bmatrix} 2 & -5 & 0 \\ 4 & 7 & 0 \\ -7 & 25 & 3 \end{bmatrix}$ **58.** $\begin{bmatrix} 8 & 7 & 6 \\ -4 & 0 & 0 \\ 5 & 1 & 4 \end{bmatrix}$

■ *How to use determinants and Cramer's Rule to solve systems of linear equations*

In Exercises 59–64, solve the system of linear equations using Cramer's Rule. (If not possible, state the reason.)

59. $\begin{cases} 7x + 12y = 63 \\ 2x + 3y = 15 \end{cases}$ **60.** $\begin{cases} 12x + 42y = -17 \\ 30x - 18y = 19 \end{cases}$

61. $\begin{cases} 3x - 2y = 16 \\ 12x - 8y = -5 \end{cases}$ **62.** $\begin{cases} 4x + 24y = 20 \\ -3x + 12y = -5 \end{cases}$

63. $\begin{cases} -x + y + 2z = 1 \\ 2x + 3y + z = -2 \\ 5x + 4y + 2z = 4 \end{cases}$

64. $\begin{cases} 2x_1 + x_2 + 2x_3 = 4 \\ 2x_1 + 2x_2 = 5 \\ 2x_1 - x_2 + 6x_3 = 2 \end{cases}$

■ *How to use determinants to find areas of triangles, to test for collinear points, and to find equations of lines*

Area of a Triangle   In Exercises 65–68, use a determinant to find the area of the triangle with the indicated vertices.

65. $(1, 0), (5, 0), (5, 8)$

66. $(-4, 0), (4, 0), (0, 6)$

67. $(1, 2), (4, -5), (3, 2)$

68. $\left(\frac{3}{2}, 1\right), \left(4, -\frac{1}{2}\right), (4, 2)$

Collinear Points   In Exercises 69–72, determine whether the points are collinear.

69. $(-1, -5), (1, -1), (4, 5)$

70. $(-1, 8), (1, 2), (2, 0)$

71. $(-8, 2), (-3, 1), (5, -3)$

72. $(-4, -10), (-1, -3), (2, 4)$

Equation of a Line   In Exercises 73–76, use a determinant to find the equation of the line through the points.

73. $(-4, 0), (4, 4)$

74. $(2, 5), (6, -1)$

75. $\left(-\frac{5}{2}, 3\right), \left(\frac{7}{2}, 1\right)$ **76.** $(-0.8, 0.2), (0.7, 3.2)$

4.5 ■ *How to verify solutions to linear inequalities in two variables*

In Exercises 77 and 78, determine whether the points are solutions of the inequality.

77. $-6x + 2y \le 14$

  (a) $(4, 1)$ (b) $(-3, 2)$

  (c) $(-2, 1)$ (d) $(0, 9)$

78. $3x - 5y > 2$

  (a) $(2, -1)$ (b) $(-3, 4)$

  (c) $(-1, -1)$ (d) $(9, 2)$

■ *How to sketch graphs of linear inequalities in two variables*

In Exercises 79–84, sketch the graph of the solution of the linear inequality in two variables. Use a graphing utility to confirm your result.

79. $x - 2 \ge 0$

80. $y + 3 < 0$

81. $2x + y < 1$

82. $3x - 4y > 2$

83. $x \le 4y - 2$

84. $(y - 3) \ge \frac{2}{3}(x - 5)$

■ *How to solve systems of linear inequalities in two variables*

In Exercises 85–88, sketch the graph of the solution of the system of linear inequalities.

85. $\begin{cases} x + y < 5 \\ x > 2 \\ y \ge 0 \end{cases}$ **86.** $\begin{cases} 2x + y > 2 \\ x < 2 \\ y < 1 \end{cases}$

87. $\begin{cases} x + 2y \le 160 \\ 3x + y \le 180 \\ x \ge 0 \\ y \ge 0 \end{cases}$ **88.** $\begin{cases} 2x + 3y \le 24 \\ 2x + y \le 16 \\ x \ge 0 \\ y \ge 0 \end{cases}$

In Exercises 89 and 90, derive a system of linear inequalities to describe the region.

89. Parallelogram with vertices at $(1, 5), (3, 1), (6, 10),$ and $(8, 6)$

90. Triangle with vertices at $(1, 2), (6, 7),$ and $(8, 1)$

Solving Problems

91. *Break-Even Analysis* A small business invests $25,000 in equipment to produce a product. Each unit of the product costs $3.75 to produce and is sold for $5.25. How many items must be sold before the business breaks even?

92. *Break-Even Analysis* A small business invests $33,000 in equipment to produce a product. Each unit of the product costs $1.70 to produce and is sold for $5.00. How many items must be sold before the business breaks even?

93. *Geometry* The perimeter of a rectangle is 480 meters, and its length is 150% of its width. Find the dimensions of the rectangle.

94. *Geometry* Two angles are complementary. The measure of the first angle is $12°$ more than two times the measure of the second angle. Find the measure of each angle.

95. *Sales* You are the manager of a music store and are going over receipts for the previous week's sales. Six hundred fifty CDs of two different types were sold. One type of CD sold for $9.95, and the other sold for $14.95. The total CD receipts were $7717.50. The cash register that was supposed to record the number of each type of CD sold malfunctioned. Can you recover the information? If so, how many of each type of CD were sold?

96. *Flying Speeds* Two planes leave Pittsburgh and Philadelphia at the same time, each flying to the other city. Because of the wind, one plane flies 25 miles per hour faster than the other. Find the ground speed of each plane if the cities are 275 miles apart and the planes pass one another after 40 minutes.

97. *Geometry* The sum of the measures of two angles of a triangle is twice the measure of the third angle. The measure of the second angle is $10°$ less than the measure of the third angle. Find the measures of the three angles.

98. *Investments* An inheritance of $20,000 is divided among three investments. The interest rates for the three investments are 7%, 9%, and 11%. Find the amount placed in each investment if the second and third are $3000 and $1000 less than the first, respectively.

Curve-Fitting 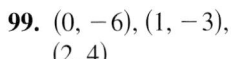 In Exercises 99 and 100, find the equation of the parabola $y = ax^2 + bx + c$ that passes through the points.

99. $(0, -6), (1, -3),$ $(2, 4)$ **100.** $(-5, 0), (1, -6),$ $(2, 14)$

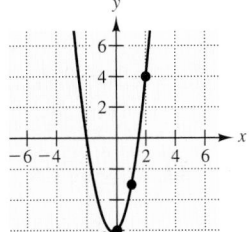 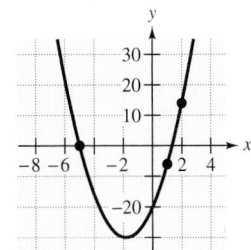

101. *Mathematical Modeling* A child throws a softball over a garage. The locations of the eaves and the peak of the roof are given by $(0, 10)$, $(15, 15)$, and $(30, 10)$.

  (a) Find the equation $y = ax^2 + bx + c$ for the path of the ball if the ball follows a path 1 foot over the eaves and the peak of the roof.

  (b) Use a graphing utility to graph the path of the ball in part (a).

  (c) From the graph, estimate how far from the edge of the garage the child is standing if the ball is at a height of 5 feet when it leaves his hand.

102. *Number Problem* The sum of three positive numbers is 68. The second number is four greater than the first, and the third is twice the first. Find the three numbers.

In Exercises 103 and 104, determine a system of linear inequalities that models the description, and sketch a graph of the solution of the system.

103. *Fruit Distribution* A Pennsylvania fruit grower has up to 1500 bushels of apples that are to be divided between markets in Harrisburg and Philadelphia. These two markets need at least 400 bushels and 600 bushels, respectively.

104. *Inventory Costs* A warehouse operator has up to 24,000 square feet of floor space in which to store two products. Each unit of product I requires 20 square feet of floor space and costs $12 per day to store. Each unit of product II requires 30 square feet of floor space and costs $8 per day to store. The total storage cost per day cannot exceed $12,400.

4 Chapter Test

Take this test as you would take a test in class. After you are done, check your work against the answers given in the back of the book.

In Exercises 1–5, use the indicated method to solve the system.

1. Substitution:

$$\begin{cases} 5x - y = 6 \\ 4x - 3y = -4 \end{cases}$$

2. Elimination:

$$\begin{cases} x + 2y - 3z = 0 \\ 3x + y + z = 5 \\ 3x - y + 5z = 7 \end{cases}$$

3. Matrices:

$$\begin{cases} x + 4y + 3z = 1 \\ 2x + 8y + 11z = 7 \\ x + 6y + 7z = 3 \end{cases}$$

4. Cramer's Rule:

$$\begin{cases} x + 3y + z = -2 \\ 2x + 5y + z = -5 \\ x + 2y + 3z = 6 \end{cases}$$

5. Any method:

$$\begin{cases} x + 2y - 2z = 1 \\ 2x + 5y + z = 9 \\ x + 3y + 4z = 9 \end{cases}$$

6. Evaluate the determinant of the matrix.

$$\begin{bmatrix} 3 & -2 & 0 \\ -1 & 5 & 3 \\ 2 & 7 & 1 \end{bmatrix}$$

7. Find the equation of the parabola $y = ax^2 + bx + c$ that passes through the points $(0, 4)$, $(1, 3)$, and $(2, 6)$.

8. Use a determinant to find the area of the triangle with vertices $(0, 0)$, $(5, 4)$, and $(6, 0)$.

9. Sketch the graph of the inequality $x + 2y \leq 4$.

In Exercises 10 and 11, graph the solution of the system of linear inequalities.

10. $\begin{cases} 3x - y < 4 \\ x > 0 \\ y > 0 \end{cases}$

11. $\begin{cases} x + y < 6 \\ 2x + 3y > 9 \\ x \geq 0 \\ y \geq 0 \end{cases}$

12. Two people share the driving for a 200-mile trip. One person drives four times as far as the other. Write a system of linear equations that models this problem. Find the distance each person drives.

13. The total cost of 8 gallons of regular unleaded gasoline and 12 gallons of premium unleaded gasoline is $27.84. Premium unleaded gasoline costs $0.17 more per gallon than regular unleaded. Find the price per gallon for each grade of gasoline.

14. You have decided to invest some money in certificates of deposit. You have up to $1000 to divide between two different programs, program A and program B. Program A requires a minimum deposit of $300 and program B requires a minimum deposit of $400. Determine a system of linear inequalities that models this situation, and sketch a graph of the solution of the system.

Cumulative Test: Chapters 1–4

Take this test as you would take a test in class. After you are done, check your work against the answers given in the back of the book.

In Exercises 1–4, simplify the expression by using the properties of exponents.

1. $(-3a^5)^2 \cdot (-6a^8)$

2. $\dfrac{(2x^3y)^4}{6x^2y^3}$

3. $-\left(\dfrac{6x^2y}{5z^3}\right)^2$

4. $\dfrac{2a^{2n}b^{m+5}}{a^n b^5}$

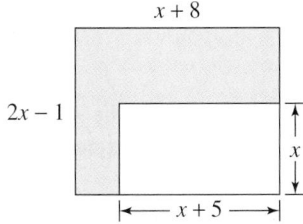

Figure for 5

5. Find expressions for the perimeter and the area of the triangle shown at the left. Then evaluate the expressions for the value of $b = 3$.

6. Find an expression for the area of the shaded portion of the rectangle shown at the left.

Figure for 6

In Exercises 7–10, factor the expression completely.

7. $10a - 15a^4$

8. $2a^2 + ab - 6b^2$

9. $x^3 - 3x^2 - x + 3$

10. $y^3 - 64$

In Exercises 11–16, solve the equation.

11. $x + \dfrac{x}{2} = 4$

12. $5(x - 2) = 3 - 2(x + 3)$

13. $8(x - 1) + 14 = 3(x + 7)$

14. $x^2 + x - 42 = 0$

15. $\dfrac{x - 2}{2} = \dfrac{4x + 1}{14}$

16. $\left|\dfrac{2}{3}x + 2\right| = 10$

In Exercises 17–20, find the domain of the function.

17. $f(x) = 3x + 5$

18. $g(x) = x^2 + 4x + 4$

19. $h(x) = \dfrac{4}{x - 8}$

20. $f(t) = \sqrt{t - 6}$

In Exercises 21–23, identify the transformation of the graph of $f(x) = x^4$. Then use a graphing utility to verify your answer.

21. $g(x) = x^4 - 2$ **22.** $g(x) = (x - 2)^4$ **23.** $g(x) = -x^4$

24. The length and width of the rectangle shown at the left are x centimeters and y centimeters. The perimeter of the rectangle is 500 centimeters.

(a) Write y in terms of x.

(b) Write the area A of the rectangle as a function of x.

Figure for 24

25. Write an equation of the line that passes through the point $(7, -2)$ and is parallel to the line $5x - y = 8$.

26. Write an equation of the vertical line that passes through the point $(-2, 4)$.

27. A small business purchases a typesetting system for \$117,000. After 5 years, its depreciated value will be \$52,000. Assuming straight-line depreciation, write a linear model giving the value V of the typesetting system in terms of the time t. Find the value of the system after 3 years.

28. If you spend 40% of your monthly income on your car payment of \$260, what is your monthly income?

29. Ticket sales for a charity dinner totaled \$1806. There were four times as many adult tickets sold as student tickets, and the tickets for adults and students sold for \$14.50 and \$6.50, respectively. How many of each kind of ticket were sold?

30. Solve the inequality

$$-16 < 6x + 2 \le 5$$

and sketch the graph of the solution set on the real number line.

31. Solve the inequality $|9 - 2x| - 2 < -1$.

In Exercises 32 and 33, use a graphing utility to graph the equations in the same viewing window, and use the graph to approximate any points of intersection. Find the solution of the system algebraically.

32. $\begin{cases} x + 2y = 8 \\ 2x - 4y = -4 \end{cases}$ **33.** $\begin{cases} x + y = 2 \\ 0.2x + y = 6 \end{cases}$

34. Solve the linear system.

$$\begin{cases} 2x + y - 2z = 1 \\ x - z = 1 \\ 3x + 3y + z = 12 \end{cases}$$

35. The perimeter of a rectangle is 68 feet and its width is $\frac{8}{9}$ times its length. Find the dimensions of the rectangle.

36. A small corporation borrowed \$800,000 to expand its product line. Some of the money was borrowed at 8%, some at 9%, and some at 10%. How much was borrowed at each rate if the annual interest was \$67,000 and the amount borrowed at 8% was five times the amount borrowed at 10%?

37. Use a determinant to find an equation of the line through the points $(-2, 5)$ and $(0, 6)$.

38. Determine whether the points $(-4, -2)$, $(-1, 1)$, and $(3, 4)$ are collinear.

39. Sketch a graph of the solution of the system of linear inequalities.

$$\begin{cases} y \ge -2 \\ x < 5 \\ y < x + 4 \end{cases}$$

Radicals and Complex Numbers

5

- **5.1** Integer Exponents and Scientific Notation
- **5.2** Rational Exponents and Radicals
- **5.3** Simplifying and Combining Radicals
- **5.4** Multiplying and Dividing Radicals
- **5.5** Solving Radical Equations
- **5.6** Complex Numbers

THE BIG PICTURE

In this chapter you will learn how to:

- simplify and evaluate expressions involving integer and rational exponents.
- simplify and evaluate expressions involving radicals.
- add, subtract, multiply, and divide radical expressions.
- solve radical equations.
- add, subtract, multiply, and divide complex numbers.

Key Terms

As you encounter each new vocabulary term in this chapter, add the term and its definition to your notebook glossary.

Additional text-specific resources are available to help you do well in this course.
See page xvi for details.

5.1 Integer Exponents and Scientific Notation

What you should learn:

- How to use properties of exponents to simplify algebraic expressions
- How to write very large or very small numbers in scientific notation

Why you should learn it:

Scientific notation can be used to represent very large numbers in real life. For instance, in Exercise 81 on page 312, scientific notation is used to represent the distance from the sun to Earth.

Integer Exponents

So far in the text, all exponents have been positive integers. In this section, the definition of an exponent is extended to include zero and negative integers. If a is a real number such that $a \neq 0$, then a^0 is defined as 1. Moreover, if m is an integer, then a^{-m} is defined as the reciprocal of a^m.

Definitions of Zero Exponents and Negative Exponents

Let a be a real number such that $a \neq 0$, and let m be an integer.

1. $a^0 = 1$ 2. $a^{-m} = \dfrac{1}{a^m}$

These definitions are consistent with the properties of exponents from Section 1.1. For instance, consider the following.

$$x^0 \cdot x^m = x^{0+m} = x^m = 1 \cdot x^m$$

(x^0 is the same as 1.)

EXPLORATION

In Section 1.1, you saw that

$$\frac{a^m}{a^n} = a^{m-n}, \, m > n, a \neq 0.$$

Suppose that this property holds when $m = n$. Then what do the statements below tell you about the value of a^0? Explain your reasoning.

$$\frac{a^6}{a^6} = a^{6-6}$$

$$\frac{a^6}{a^6} = \frac{1(a^6)}{1(a^6)}$$

Using similar reasoning, how could you use the property above to give meaning to the expression a^{-4}?

$\left[\textit{Hint:} \text{ Consider the following.} \right.$

$$\frac{a^2}{a^6} = a^{2-6}$$

$$\left. \frac{a^2}{a^6} = \frac{1(a^2)}{a^4(a^2)} \right]$$

Example 1 ■ Zero Exponents and Negative Exponents

Rewrite each expression without using zero exponents or negative exponents.

(a) 3^0

(b) 0^0

(c) 2^{-1}

(d) 3^{-2}

(e) $\left(\frac{1}{7}\right)^{-2}$

Solution

(a) $3^0 = 1$ Definition of zero exponents

(b) 0^0 is undefined.   Zero cannot have a zero exponent.

(c) $2^{-1} = \dfrac{1}{2^1} = \dfrac{1}{2}$ Definition of negative exponents

(d) $3^{-2} = \dfrac{1}{3^2} = \dfrac{1}{9}$ Definition of negative exponents

(e) $\left(\dfrac{1}{7}\right)^{-2} = \dfrac{1}{\left(\frac{1}{7}\right)^2} = \dfrac{1}{\left(\frac{1}{49}\right)} = 49$ Definition of negative exponents

The properties of exponents listed below are valid for all integer exponents, including integer exponents that are zero or negative. (The first five properties were listed in Section 1.1.)

Properties of Exponents

Let m and n be integers, and let a and b represent real numbers, variables, or algebraic expressions.

Property	*Example*
1. $a^m \cdot a^n = a^{m+n}$	$x^4 \cdot x^3 = x^{4+3} = x^7$
2. $(ab)^m = a^m \cdot b^m$	$(3x)^2 = 3^2 \cdot x^2 = 9x^2$
3. $(a^m)^n = a^{mn}$	$(x^3)^3 = x^{3 \cdot 3} = x^9$
4. $\dfrac{a^m}{a^n} = a^{m-n}, \quad a \neq 0$	$\dfrac{x^3}{x} = x^{3-1} = x^2, \quad x \neq 0$
5. $\left(\dfrac{a}{b}\right)^m = \dfrac{a^m}{b^m}, \quad b \neq 0$	$\left(\dfrac{x}{3}\right)^2 = \dfrac{x^2}{3^2} = \dfrac{x^2}{9}$
6. $\left(\dfrac{a}{b}\right)^{-m} = \left(\dfrac{b}{a}\right)^m, \quad \begin{matrix} a \neq 0 \\ b \neq 0 \end{matrix}$	$\left(\dfrac{x}{3}\right)^{-2} = \left(\dfrac{3}{x}\right)^2 = \dfrac{3^2}{x^2} = \dfrac{9}{x^2}, \quad x \neq 0$
7. $a^{-m} = \dfrac{1}{a^m}, \quad a \neq 0$	$x^{-2} = \dfrac{1}{x^2}, \quad x \neq 0$
8. $a^0 = 1, \quad a \neq 0$	$(x^2 + 1)^0 = 1$

Study Tip

As you become accustomed to working with negative exponents, you will probably not write as many steps as shown in Example 2. For instance, to rewrite a fraction involving exponents, you might use the simplified rule as follows. *To move a factor from the numerator to the denominator or vice versa, change the sign of its exponent.* You can apply this rule to the expression in Example 2(c) by "moving" the factor x^{-2} to the numerator and changing the exponent to 2. That is,

$$\frac{3}{x^{-2}} = 3x^2.$$

This same rule applies to the expression in Example 2(d).

Example 2 ■ Using Properties of Exponents

Rewrite each expression using only positive exponents. (For each expression, assume that $x \neq 0$.)

(a) $2x^{-1}$ (b) $(2x)^{-1}$ (c) $\dfrac{3}{x^{-2}}$ (d) $\dfrac{1}{(3x)^{-2}}$

Solution

(a) $2x^{-1} = 2(x^{-1}) = 2\left(\dfrac{1}{x^1}\right) = \dfrac{2}{x}$ Property 7

(b) $(2x)^{-1} = \dfrac{1}{(2x)^1} = \dfrac{1}{2x}$ Properties 2 and 7

(c) $\dfrac{3}{x^{-2}} = \dfrac{3}{\left(\dfrac{1}{x^2}\right)} = 3\left(\dfrac{x^2}{1}\right) = 3x^2$ Property 7

(d) $\dfrac{1}{(3x)^{-2}} = \dfrac{1}{\left[\dfrac{1}{(3x)^2}\right]} = \dfrac{1}{\left(\dfrac{1}{3^2x^2}\right)} = \dfrac{1}{\left(\dfrac{1}{9x^2}\right)} = \left(\dfrac{9x^2}{1}\right) = 9x^2$ Properties 2 and 7

Example 3 ■ Using Properties of Exponents

Rewrite each expression using only positive exponents. (For each expression, assume that $x \neq 0$ and $y \neq 0$.)

(a) $(-5x^{-3})^2$　　(b) $-\left(\dfrac{7x}{y^2}\right)^{-2}$

Solution

(a) $(-5x^{-3})^2 = (-5)^2(x^{-3})^2$　　　　　　　Property 2

$\qquad\qquad\quad = 25x^{-6}$　　　　　　　　　　Property 3

$\qquad\qquad\quad = \dfrac{25}{x^6}$　　　　　　　　　　Property 7

(b) $-\left(\dfrac{7x}{y^2}\right)^{-2} = -\left(\dfrac{y^2}{7x}\right)^2$　　　　　　Property 6

$\qquad\qquad\quad = -\dfrac{(y^2)^2}{(7x)^2}$　　　　　　Property 5

$\qquad\qquad\quad = -\dfrac{y^4}{49x^2}$　　　　　　Property 3

Example 4 ■ Using Properties of Exponents

Rewrite each expression using only positive exponents. (For each expression, assume that $x \neq 0$ and $y \neq 0$.)

(a) $\left(\dfrac{8x^{-1}y^4}{4x^3y^2}\right)^{-3}$　　(b) $\dfrac{3xy^0}{x^2(5y)^0}$　　(c) $\dfrac{15x^4y^{-3}}{10x^{-1}y^5}$

NOTE Here is another way to simplify the expression in Example 4(a).

$\left(\dfrac{8x^{-1}y^4}{4x^3y^2}\right)^{-3} = \dfrac{8^{-3}x^3y^{-12}}{4^{-3}x^{-9}y^{-6}}$

$\qquad = \dfrac{4^3x^3x^9y^6}{8^3y^{12}}$

$\qquad = \dfrac{64x^{12}y^6}{512y^{12}}$

$\qquad = \dfrac{x^{12}}{8y^6}$

Solution

(a) $\left(\dfrac{8x^{-1}y^4}{4x^3y^2}\right)^{-3} = \left(\dfrac{2y^2}{x^4}\right)^{-3}$　　　　Properties 1, 4, and 7

$\qquad\qquad\quad = \left(\dfrac{x^4}{2y^2}\right)^3$　　　　　Property 6

$\qquad\qquad\quad = \dfrac{(x^4)^3}{(2y^2)^3}$　　　　　Property 5

$\qquad\qquad\quad = \dfrac{x^{12}}{8y^6}$　　　　　　Property 3

(b) $\dfrac{3xy^0}{x^2(5y)^0} = \dfrac{3x(1)}{x^2(1)} = \dfrac{3}{x}$　　　Properties 4, 7, and 8

(c) $\dfrac{15x^4y^{-3}}{10x^{-1}y^5} = \dfrac{3}{2}(x^{4-(-1)})(y^{-3-5})$　　Property 4

$\qquad\qquad\quad = \dfrac{3}{2}x^5y^{-8}$　　　　　Simplify.

$\qquad\qquad\quad = \dfrac{3x^5}{2y^8}$　　　　　　Property 7

Example 5 ■ Using Properties of Exponents

Simplify each expression using only positive exponents. (For each expression, assume that $x \neq 0$ and $y \neq 0$.)

(a) $(5x^{-2}y^{-5})^{-2}(3xy^{-4})$ (b) $\dfrac{4x^{-5}y^3}{(7x^{-2}y)^{-1}}$

Solution

(a) $(5x^{-2}y^{-5})^{-2}(3xy^{-4}) = (5^{-2}x^4y^{10})(3xy^{-4})$ Property 3

$= \dfrac{x^4y^{10}(3x)}{5^2y^4}$ Property 7

$= \dfrac{3x^5y^6}{25}$ Simplify.

(b) $\dfrac{4x^{-5}y^3}{(7x^{-2}y)^{-1}} = \dfrac{4x^{-5}y^3}{7^{-1}x^2y^{-1}}$ Property 3

$= \dfrac{4y^3 \cdot 7y}{x^2x^5}$ Property 7

$= \dfrac{28y^4}{x^7}$ Simplify.

Scientific Notation

Exponents provide an efficient way of writing and computing with very large (or very small) numbers. For instance, a drop of water contains more than 33 billion billion molecules—that is, 33 followed by 18 zeros.

$$33,000,000,000,000,000,000$$

It is convenient to write such numbers in **scientific notation.** This notation has the form $c \times 10^n$, where $1 \leq c < 10$ and n is an integer. So, the number of molecules in a drop of water can be written in scientific notation as

$$3.3 \times 10,000,000,000,000,000,000 = 3.3 \times 10^{19}.$$

The *positive* exponent 19 indicates that the number being written in scientific notation is *large* (10 or more) and that the decimal point has been moved 19 places. A *negative* exponent in scientific notation indicates that the number is *small* (less than 1).

Example 6 ■ Writing Scientific Notation

Write each real number in scientific notation.

(a) 0.0000684   (b) 937,200,000

Solution

(a) $0.0000684 = 6.84 \times 10^{-5}$ (b) $937,200,000.0 = 9.372 \times 10^8$

Five places   Eight places

Example 7 ■ Writing Decimal Notation

Convert each number from scientific notation to decimal notation.

(a) 2.486×10^2 (b) 1.81×10^{-6}

Solution

(a) $2.486 \times 10^2 = 248.6$ (b) $1.81 \times 10^{-6} = 0.00000181$

                                       Two places   Six places

---

Technology

Most scientific calculators and graphing utilities automatically switch to scientific notation when they are showing large (or small) numbers that exceed the display range.

To *enter* numbers in scientific notation, your calculator should have an exponential entry key labeled EE or EXP. Consult the user's guide for your graphing utility for instructions on keystrokes and how numbers in scientific notation are displayed.

Example 8 ■ Using Scientific Notation

Evaluate $\dfrac{(2,400,000,000)(0.0000045)}{(0.00003)(1500)}$.

Solution

Begin by rewriting each number in scientific notation and simplifying.

$$\frac{(2,400,000,000)(0.0000045)}{(0.00003)(1500)} = \frac{(2.4 \times 10^9)(4.5 \times 10^{-6})}{(3 \times 10^{-5})(1.5 \times 10^3)}$$

$$= \frac{(2.4)(4.5)(10^3)}{(4.5)(10^{-2})}$$

$$= (2.4)(10^5)$$

$$= 240,000$$

Example 9 ■ Average Amount Spent on Golf Equipment

The U.S. population in 1998 was 270 million. Use the information shown in Figure 5.1 to find the average amount that an American spent on golf equipment in 1998. *(Source: National Sporting Goods Association)*

Solution

From the bar graph shown in Figure 5.1, you know that the total amount spent on golf equipment in 1998 was about \$3.6 billion. To find the average amount spent on golf equipment in 1998, divide the total amount spent by the number of people.

$$\text{Average amount per person} = \frac{\text{Total amount}}{\text{Number of people}}$$

$$= \frac{3.6 \text{ billion}}{270 \text{ million}}$$

$$= \frac{3.6 \times 10^9}{2.7 \times 10^8}$$

$$= \frac{3.6}{2.7} \times 10^1 \approx 13.33$$

So, the average amount spent per person was approximately \$13.33.

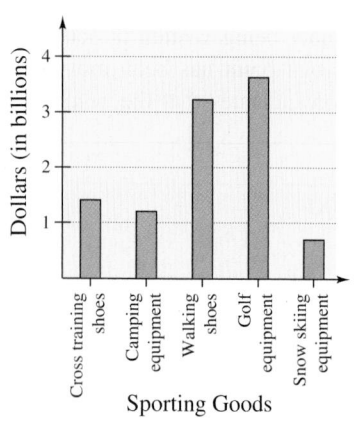

Figure 5.1

5.1 ■ Exercises

Developing Skills

In Exercises 1–12, evaluate the expression.

1. 5^{-2}

2. -20^{-2}

3. $\dfrac{1}{4^{-3}}$

4. $\dfrac{1}{-8^{-2}}$

5. $\left(\dfrac{3}{16}\right)^{0}$

6. $\left(-\dfrac{5}{8}\right)^{-2}$

7. $27 \cdot 3^{-3}$

8. $4^2 \cdot 4^{-3}$

9. $\dfrac{10^3}{10^{-2}}$

10. $\dfrac{10^{-5}}{10^{-6}}$

11. $(4^2 \cdot 4^{-1})^{-2}$

12. $(5^3 \cdot 5^{-4})^{-3}$

In Exercises 13–58, rewrite the expression using only positive exponents, and simplify. (Assume that any variables in the expression are nonzero.)

13. $7x^{-3}$

14. $-6z^{-4}$

15. $y^4 \cdot y^{-2}$

16. $z^5 \cdot z^{-3}$

17. $x^{-2} \cdot x^{-5}$

18. $t^{-1} \cdot t^{-6}$

19. $\dfrac{1}{x^{-6}}$

20. $\dfrac{3}{z^{-8}}$

21. $\dfrac{x^{-3}}{y^{-1}}$

22. $\dfrac{a^{-6}}{a^{-7}}$

23. $\left(\dfrac{x}{10}\right)^{-1}$

24. $\left(\dfrac{4}{z}\right)^{-2}$

25. $\dfrac{8t^{-3}}{6t^{-4}}$

26. $\dfrac{6u^{-2}}{15u^{-1}}$

27. $\dfrac{(4t)^0}{t^{-2}}$

28. $\dfrac{a^{-6}}{(3a^2)^0}$

29. $(2x^2)^{-2}$

30. $(4a^{-2})^{-3}$

31. $(-3x^{-3}y^2)(4x^2y^{-5})$

32. $(3a^4b^{-2})(-a^{-2}b^7)$

33. $-\left(\dfrac{4x^2}{y^3}\right)^{-3}$

34. $\left(\dfrac{5y^3}{z^2}\right)^{-2}$

35. $(3x^2y^{-2})^{-2}$

36. $(-4y^{-3}z)^{-3}$

37. $(4xy^{-3})^{-3}(9x^2y^{-1})$

38. $(3y^{-6}z^2)^{-1}(2y^2z^{-2})$

39. $\dfrac{6^2x^3y^{-3}}{12x^{-2}y}$

40. $\dfrac{2^{-4}y^{-1}z^{-3}}{4^{-2}yz^{-3}}$

41. $\left(\dfrac{3u^2v^{-1}}{3^3u^{-1}v^3}\right)^{-2}$

42. $\left(\dfrac{5^2x^3y^{-3}}{125xy}\right)^{-1}$

43. $\left[\left(\dfrac{2y^2}{x^3}\right)^2\right]^{-2}$

44. $\left[\left(\dfrac{2x^2}{4y}\right)^{-3}\right]^2$

45. $[(x^{-4}y^{-6})^{-1}]^2$

46. $[(2a^3b^{-3})^3]^{-2}$

47. $(4m)^3\left(\dfrac{4}{3m}\right)^{-2}$

48. $\left(\dfrac{3z^2}{z^2}\right)^{-2}(2z)^2$

49. $\left(\dfrac{6x^4}{7y^{-2}}\right)(14x^{-1}y^5)$

50. $(5s^5t^{-5})\left(\dfrac{3s^{-2}}{50t^{-1}}\right)$

51. $(5x^2y^4)^3(xy^{-5})^{-3}$

52. $(s^4t)^{-2}(s^4t)^2$

53. $\dfrac{(2a^{-2}b^4)^3}{(10a^3b)^2}$

54. $\dfrac{(5x^2y^{-5})^{-1}}{2x^{-5}y^4}$

55. $(2x^0y^{-1})(4xy^{-6})$

56. $x^5(3x^0y^4)(7y)^0$

57. $(18x)^0(4xy)^2(3x^{-1})$

58. $(5ab^2)(a^{-3}b^0)(2a^0b)^{-2}$

In Exercises 59–64, write the number in scientific notation.

59. 3,600,000

60. 98,100,000

61. 0.00381

62. 0.0007384

63. *Ocean Area of Earth:* 139,400,000 square miles

64. *Thickness of Soap Bubble:* 0.0000001 meter

In Exercises 65–70, write the number in decimal notation.

65. 6×10^7

66. 5.05×10^{12}

67. 1.359×10^{-7}

68. 8.6×10^{-9}

69. *Number of Air Sacs in Lungs:* 3.5×10^8

70. *Width of Human Hair:* 9.0×10^{-4} meter

In Exercises 71–80, evaluate the expression. Write the result in scientific notation, $c \times 10^n$, with c rounded to two decimal places.

71. $(2 \times 10^9)(3.4 \times 10^{-4})$

72. $(6.5 \times 10^6)(2 \times 10^4)$

73. $\dfrac{3.6 \times 10^9}{9 \times 10^5}$

74. $\dfrac{2.5 \times 10^{-3}}{5 \times 10^2}$

75. $(4,500,000)(2,000,000,000)$

76. $(6,200,000,000)(3,800,000)$

77. $\dfrac{1.357 \times 10^{12}}{(4.2 \times 10^2)(6.87 \times 10^{-3})}$

78. $\dfrac{2.612 \times 10^9}{(6.8 \times 10^3)(2.95 \times 10^{-2})}$

79. $\dfrac{(0.0000565)(2,850,000,000,000)}{0.00465}$

80. $\dfrac{(5,000,000)(0.000037)^2}{(0.005)^4}$

Solving Problems

81. *Light Year*   One light year (the distance light can travel in 1 year) is approximately 9.46×10^{15} meters. Approximate the time for light to travel from the sun to Earth if that distance is approximately 1.49×10^{11} meters.

82. *Distance*   Determine the distance (in meters) to the star Alpha Andromeda if it is 90 light years from Earth. See Exercise 81 for the definition of a light year.

83. *Mass*   The masses of Earth and the sun are approximately 5.975×10^{24} kg and 1.99×10^{30} kg, respectively. The mass of the sun is approximately how many times that of Earth?

84. *Federal Debt*   In July 1999, the estimated population of the United States was 273 million people, and the estimated federal debt was 5606 billion dollars. Use these two numbers to determine the amount each person would have to pay to eliminate the debt.   *(Source: U.S. Census Bureau and U.S. Office of Management and Budget)*

Explaining Concepts

85. Discuss any differences between the expressions $(-2x)^{-4}$ and $-2x^{-4}$.

86. Is the number 32.5×10^5 written in scientific notation? Explain.

Ongoing Review

In Exercises 87–90, find the domain of the function.

87. $f(x) = \dfrac{x}{x + 1}$

88. $f(x) = \dfrac{x + 6}{x - 5}$

89. $g(x) = \sqrt{x - 4}$

90. $g(x) = \sqrt{x - 10}$

91. *Simple Interest*   An inheritance of \$24,000 is invested in two bonds that pay 7.5% and 9% simple interest. The annual interest is \$1935. How much is invested in each bond?

Looking Further

Kepler's Third Law   In 1619, Johannes Kepler, a German astronomer, discovered that the time T (in years) it takes a planet to orbit the sun is related to the planet's mean distance R (in astronomical units) from the sun by the equation

$$\frac{T^2}{R^3} = k.$$

Test Kepler's equation for the nine planets in our solar system, using the table shown at the right. What value do you get for k for each planet? Are the values of k all approximately the same? (Astronomical units relate a

planet's time to orbit the sun and mean distance to Earth's time to orbit the sun and mean distance.)

Planet	Mercury	Venus	Earth	Mars	Jupiter
T	0.241	0.615	1.000	1.881	11.861
R	0.387	0.723	1.000	1.524	5.203

Planet	Saturn	Uranus	Neptune	Pluto
T	29.457	84.008	164.784	248.350
R	9.555	19.191	30.107	39.529

5.2 Rational Exponents and Radicals

What you should learn:

- How to determine the nth roots of numbers and evaluate radical expressions
- How to use the properties of exponents to evaluate or simplify expressions with rational exponents
- How to evaluate radical functions and find the domains of radical functions

Why you should learn it:

Algebraic equations often involve rational exponents. For instance, in Exercise 121 on page 322 you will use an equation involving a rational exponent to find the depreciation rate of a truck.

NOTE "Having the same sign" means that the principal nth root of a is positive if a is positive and negative if a is negative. For example, $\sqrt{4} = 2$ and $\sqrt[3]{-8} = -2$.

Roots and Radicals

A **square root** of a number is one of its two equal factors. For example, 5 is a square root of 25 because 5 is one of the two equal factors of 25. In a similar way, a **cube root** of a number is one of its three equal factors. For example, 5 is a cube root of 125 because 5 is one of the three equal factors of 125.

Definition of nth Root of a Number

Let a and b be real numbers and let $n \geq 2$ be a positive integer. If

$$a = b^n$$

then b is an **nth root of a.** If $n = 2$, the root is a **square root.** If $n = 3$, the root is a **cube root.**

Some numbers have more than one nth root. For example, both 5 and -5 are square roots of 25. The *principal square root* of 25, written as $\sqrt{25}$, is the positive root, 5. The **principal nth root** of a number is defined as follows.

Principal nth Root of a Number

Let a be a real number that has at least one (real number) nth root. The **principal nth root of a** is the nth root that has the same sign as a, and it is denoted by the **radical symbol**

$$\sqrt[n]{a}. \qquad \text{Principal } n\text{th root}$$

The positive integer n is the **index** of the radical, and the number a is the **radicand.** If $n = 2$, omit the index and write $\sqrt{a}$ rather than $\sqrt[2]{a}$.

You need to be aware of the properties of nth roots below. (Remember that for nth roots, n is an integer that is greater than or equal to 2.)

Properties of nth Roots

1. If a is a positive real number and n is *even*, then a has exactly two (real) nth roots, which are denoted by $\sqrt[n]{a}$ and $-\sqrt[n]{a}$.
2. If a is any real number and n is *odd*, then a has only one (real) nth root, which is denoted by $\sqrt[n]{a}$.
3. If a is a negative real number and n is *even*, then a has no (real) nth root.

Integers such as 1, 4, 9, 16, 25, 36, 49, 64, and 81 are called **perfect squares** because they have integer square roots. Similarly, integers such as 1, 8, 27, 64, and 125 are called **perfect cubes** because they have integer cube roots.

Example 1 ■ Evaluating Radical Expressions

Evaluate each radical expression.

(a) $\sqrt{144}$ (b) $-\sqrt{81}$ (c) $-\sqrt[3]{-64}$ (d) $\sqrt[4]{1}$

Solution

(a) $\sqrt{144} = 12$

(b) $-\sqrt{81} = -9$

(c) $-\sqrt[3]{-64} = -(-4) = 4$

(d) $\sqrt[4]{1} = 1$

Example 2 ■ Classifying Perfect *n*th Powers

State whether each number is a perfect square, a perfect cube, both, or neither.

(a) 81   (b) 64   (c) 32   (d) -125

Solution

(a) 81 is a perfect square because $9^2 = 81$. It is not a perfect cube.

(b) 64 is a perfect square because $8^2 = 64$, and it is also a perfect cube because $4^3 = 64$.

(c) 32 is not a perfect square or a perfect cube. (It is a perfect 5th power because $2^5 = 32$.)

(d) -125 is a perfect cube because $(-5)^3 = -125$. It is not a perfect square.

> **NOTE** The square roots of perfect squares are rational numbers, so $\sqrt{25}$, $\sqrt{49}$, and $\sqrt{100}$ are examples of rational numbers. However, square roots such as $\sqrt{5}$, $\sqrt{19}$, and $\sqrt{34}$ are irrational numbers. Similarly, $\sqrt[3]{27}$ and $\sqrt[4]{16}$ are rational numbers, while $\sqrt[3]{6}$ and $\sqrt[4]{21}$ are irrational.

Raising a number to the *n*th power and taking the principal *n*th root of a number can be thought of as *inverse* operations. Here are four examples.

$$\left(\sqrt{4}\right)^2 = 2^2 = 4 \quad \text{and} \quad \sqrt{2^2} = \sqrt{4} = 2$$

$$\left(\sqrt[3]{27}\right)^3 = 3^3 = 27 \quad \text{and} \quad \sqrt[3]{3^3} = \sqrt[3]{27} = 3$$

$$\left(\sqrt[4]{16}\right)^4 = 2^4 = 16 \quad \text{and} \quad \sqrt[4]{2^4} = \sqrt[4]{16} = 2$$

$$\left(\sqrt[5]{-243}\right)^5 = (-3)^5 = -243 \quad \text{and} \quad \sqrt[5]{(-3)^5} = \sqrt[5]{-243} = -3$$

> **Inverse Properties of *n*th Powers and *n*th Roots**
>
> Let a be a real number, and let n be an integer such that $n \geq 2$.
>
> 1. If a has a principal *n*th root, then
> $$\left(\sqrt[n]{a}\right)^n = a.$$
>
> 2. If n is *odd*, then
> $$\sqrt[n]{a^n} = a.$$
>   If n is *even*, then
> $$\sqrt[n]{a^n} = |a|.$$

> **NOTE** In Property 2 at the right, notice the use of absolute value symbols when n is even. Absolute value notation is necessary unless a is known to be nonnegative.

Example 3 ■ Evaluating Radical Expressions

Evaluate each radical expression.

(a) $\sqrt[3]{5^3}$ (b) $\sqrt[3]{(-2)^3}$ (c) $\left(\sqrt{7}\right)^2$ (d) $\sqrt{(-3)^2}$ (e) $\sqrt{-3^2}$

Solution

(a) Because the index of the radical is odd, you can write
$$\sqrt[3]{5^3} = 5.$$

(b) Because the index of the radical is odd, you can write
$$\sqrt[3]{(-2)^3} = -2.$$

(c) Using the inverse property of powers and roots, you can write
$$\left(\sqrt{7}\right)^2 = 7.$$

(d) Because the index of the radical is even, you must include absolute value signs, and write
$$\sqrt{(-3)^2} = |-3| = 3.$$

(e) Because $\sqrt{-3^2} = \sqrt{-9}$ is an even root of a negative number, its value is not a real number.

NOTE In parts (d) and (e) of Example 3, notice that the two expressions inside the radical are different. In $(-3)^2$, the negative sign is part of the base. In -3^2, the negative sign is not part of the base.

Rational Exponents

So far in the text you have worked with algebraic expressions involving only integer exponents. Next you will see that algebraic expressions may also contain **rational exponents.**

> **Definition of Rational Exponents**
>
> Let a be a real number, and let n be an integer such that $n \geq 2$. If the principal nth root of a exists, we define $a^{1/n}$ to be
> $$a^{1/n} = \sqrt[n]{a}.$$
> If m is a positive integer that has no common factor with n, then
> $$a^{m/n} = (a^{1/n})^m = \left(\sqrt[n]{a}\right)^m \quad \text{and} \quad a^{m/n} = (a^m)^{1/n} = \sqrt[n]{a^m}.$$

NOTE The numerator of a rational exponent denotes the power to which the base is raised, and the denominator denotes the *root* to be taken.

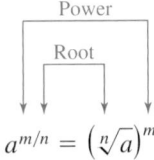

$$a^{m/n} = \left(\sqrt[n]{a}\right)^m$$

It does not matter in which order the two operations are performed, provided the nth root exists. Here is an example.

$$8^{2/3} = \left(\sqrt[3]{8}\right)^2 = 2^2 = 4 \qquad \text{Cube root, then second power}$$
$$8^{2/3} = \sqrt[3]{8^2} = \sqrt[3]{64} = 4 \qquad \text{Second power, then cube root}$$

The properties of exponents listed in Section 5.1 also apply to rational exponents (provided the roots indicated by the denominators exist). The properties are relisted below, with different examples.

EXPLORATION

Use a calculator to evaluate the expressions below.

$$\frac{3.4^{4.6}}{3.4^{3.1}} \text{ and } 3.4^{1.5}$$

How are these two expressions related? Use your calculator to verify some of the other properties of exponents.

Properties of Exponents

Let r and s be rational numbers, and let a and b be real numbers, variables, or algebraic expressions.

Property	*Example*
1. $a^r \cdot a^s = a^{r+s}$	$4^{1/2} \cdot 4^{1/3} = 4^{5/6}$
2. $(ab)^r = a^r \cdot b^r$	$(2x)^{1/2} = 2^{1/2} \cdot x^{1/2}$
3. $(a^r)^s = a^{rs}$	$(x^3)^{1/3} = x$
4. $\dfrac{a^r}{a^s} = a^{r-s}, \quad a \neq 0$	$\dfrac{x^2}{x^{1/2}} = x^{2-(1/2)} = x^{3/2}, \quad x \neq 0$
5. $\left(\dfrac{a}{b}\right)^r = \dfrac{a^r}{b^r}, \quad b \neq 0$	$\left(\dfrac{x}{3}\right)^{1/3} = \dfrac{x^{1/3}}{3^{1/3}}$
6. $\left(\dfrac{a}{b}\right)^{-r} = \left(\dfrac{b}{a}\right)^r, \quad \begin{matrix} a \neq 0 \\ b \neq 0 \end{matrix}$	$\left(\dfrac{x}{25}\right)^{-1/2} = \left(\dfrac{25}{x}\right)^{1/2} = \dfrac{5}{x^{1/2}}, \quad x \neq 0$
7. $a^{-r} = \dfrac{1}{a^r}, \quad a \neq 0$	$9^{-1/2} = \dfrac{1}{9^{1/2}} = \dfrac{1}{3}$
8. $a^0 = 1, \quad a \neq 0$	$(10x^3)^0 = 1$

Example 4 ■ Evaluating Expressions with Rational Exponents

Evaluate each expression.

(a) $8^{4/3}$ (b) $(4^2)^{3/2}$ (c) $25^{-3/2}$

(d) $\left(\dfrac{64}{125}\right)^{2/3}$ (e) $-9^{1/2}$ (f) $(-9)^{1/2}$

Solution

(a) $8^{4/3} = \left(\sqrt[3]{8}\right)^4 = 2^4 = 16$

(b) $(4^2)^{3/2} = \left(\sqrt{4^2}\right)^3 = 4^3 = 64$

(c) $25^{-3/2} = \dfrac{1}{25^{3/2}} = \dfrac{1}{\left(\sqrt{25}\right)^3} = \dfrac{1}{5^3} = \dfrac{1}{125}$

(d) $\left(\dfrac{64}{125}\right)^{2/3} = \dfrac{64^{2/3}}{125^{2/3}} = \dfrac{\left(\sqrt[3]{64}\right)^2}{\left(\sqrt[3]{125}\right)^2} = \dfrac{4^2}{5^2} = \dfrac{16}{25}$

(e) $-9^{1/2} = -\sqrt{9} = -3$

(f) $(-9)^{1/2} = \sqrt{-9}$ is not a real number.

NOTE In parts (e) and (f) of Example 4, be sure that you see the distinction between the expressions $-9^{1/2}$ and $(-9)^{1/2}$.

Example 5 ■ Using Properties of Exponents

Rewrite each expression using rational exponents.

(a) $x\sqrt[4]{x^3}$　　(b) $\dfrac{\sqrt[3]{x^2}}{\sqrt{x^3}}$　　(c) $\sqrt[3]{x^2 y}$

Solution

(a) $x\sqrt[4]{x^3} = x(x^{3/4}) = x^{1+(3/4)} = x^{7/4}$

(b) $\dfrac{\sqrt[3]{x^2}}{\sqrt{x^3}} = \dfrac{x^{2/3}}{x^{3/2}} = x^{(2/3)-(3/2)} = x^{-5/6} = \dfrac{1}{x^{5/6}}$

(c) $\sqrt[3]{x^2 y} = (x^2 y)^{1/3} = (x^2)^{1/3}y^{1/3} = x^{2/3}y^{1/3}$

Example 6 ■ Using Properties of Exponents

Use the properties of exponents to simplify each expression.

(a) $x^{3/5} \cdot x^{2/3}$　　(b) $(x^{3/5})^{2/3}$　　(c) $(6x^{3/5}y^{-3/2})^2$, $y \neq 0$

Solution

(a) $x^{3/5} \cdot x^{2/3} = x^{(3/5)+(2/3)} = x^{(9+10)/15} = x^{19/15}$

(b) $(x^{3/5})^{2/3} = x^{(3/5)(2/3)} = x^{2/5}$

(c) $(6x^{3/5}y^{-3/2})^2 = 6^2 x^{(3/5)2}y^{(-3/2)2} = 36x^{6/5}y^{-3} = \dfrac{36x^{6/5}}{y^3}$

Example 7 ■ Using Properties of Exponents

Use the properties of exponents to simplify each expression.

(a) $\sqrt{\sqrt[3]{x}}$　　(b) $\dfrac{(2x-1)^{4/3}}{\sqrt[3]{2x-1}}$

Solution

(a) $\sqrt{\sqrt[3]{x}} = \sqrt{x^{1/3}} = (x^{1/3})^{1/2} = x^{1/6}$

(b) $\dfrac{(2x-1)^{4/3}}{\sqrt[3]{2x-1}} = \dfrac{(2x-1)^{4/3}}{(2x-1)^{1/3}} = (2x-1)^{(4/3)-(1/3)} = 2x - 1$

Radical Functions

A **radical function** is a function that contains a radical, such as

$$f(x) = \sqrt{x} \qquad \text{or} \qquad g(x) = \sqrt[3]{x}.$$

When evaluating a radical function, note that the radical symbol is a grouping symbol.

Example 8 ■ Evaluating a Radical Function

Evaluate each radical function when $x = 4$.

(a) $f(x) = \sqrt[3]{x - 31}$ (b) $g(x) = \sqrt{16 - 3x}$ (c) $h(x) = \sqrt{(x + 1)^4}$

Solution

(a) $f(x) = \sqrt[3]{x - 31}$ Write original function.

$\quad f(4) = \sqrt[3]{4 - 31}$ Substitute 4 for x.

$\quad\quad\quad = \sqrt[3]{-27}$ Simplify.

$\quad\quad\quad = -3$ Simplify.

(b) $g(x) = \sqrt{16 - 3x}$ Write original function.

$\quad g(4) = \sqrt{16 - 3(4)}$ Substitute 4 for x.

$\quad\quad\quad = \sqrt{16 - 12}$ Multiply.

$\quad\quad\quad = \sqrt{4}$ Simplify.

$\quad\quad\quad = 2$ Simplify.

(c) $h(x) = \sqrt{(x + 1)^4}$ Write original function.

$\quad h(4) = \sqrt{(4 + 1)^4}$ Substitute 4 for x.

$\quad\quad\quad = \sqrt{5^4}$ Simplify.

$\quad\quad\quad = \sqrt{625}$ $5^4 = 625$

$\quad\quad\quad = 25$ Simplify.

Example 9 ■ Finding the Speed of a Ship

The speed s (in knots) of the *Olympias* was found to be related to the power P (in kilowatts) generated by the rowers according to the model

$$s = \sqrt[3]{\frac{100P}{3}}.$$

The volunteer crew was able to generate maximum power of 10.5 kilowatts. What was the ship's greatest speed? *(Source: Scientific American)*

Solution

To find the greatest speed, use the model and substitute 10.5 for P.

$$s = \sqrt[3]{\frac{100P}{3}} \quad\quad \text{Write original model.}$$

$$= \sqrt[3]{\frac{100(10.5)}{3}} \quad\quad \text{Substitute 10.5 for } P.$$

$$= \sqrt[3]{350} \quad\quad \text{Simplify.}$$

$$\approx 7 \quad\quad \text{Use a calculator.}$$

The Olympias *is a reconstruction of a trireme (a Greek galley ship). The ship's triple set of oars was operated by volunteers.*

So, the greatest speed attained by the *Olympias* was about 7 knots (about 8 miles per hour).

You already know from Section 2.4 that the domain of the square root function $f(x) = \sqrt{x}$ is the set of all nonnegative real numbers. The **domain** of the radical function $f(x) = \sqrt[n]{x}$ is the set of all real numbers such that x has a principal nth root.

> **Domain of a Radical Function**
>
> Let n be an integer that is greater than or equal to 2.
>
> 1. If n is odd, the domain of $f(x) = \sqrt[n]{x}$ is the set of all real numbers.
>
> 2. If n is even, the domain of $f(x) = \sqrt[n]{x}$ is the set of all nonnegative real numbers.

Example 10 ■ Finding the Domain of a Radical Function

Describe the domain of each function.

(a) $f(x) = \sqrt[3]{x}$ (b) $f(x) = \sqrt{x^2}$ (c) $f(x) = \sqrt{x^3}$

Solution

(a) The domain of $f(x) = \sqrt[3]{x}$ is the set of all real numbers because for any real number x, the expression $\sqrt[3]{x}$ is a real number.

(b) The domain of $f(x) = \sqrt{x^2}$ is the set of all real numbers because for any real number x, the expression x^2 is a nonnegative real number.

(c) The domain of $f(x) = \sqrt{x^3}$ is the set of all nonnegative real numbers. For instance, 1 is in the domain, but -1 is not because $\sqrt{(-1)^3} = \sqrt{-1}$ is not a real number.

In general, the domain of a radical function where the index n is even includes all real values for which the expression under the radicand is greater than or equal to zero.

Example 11 ■ Finding the Domain of a Radical Function

Find the domain of $f(x) = \sqrt{2x - 5}$.

Algebraic Solution

The domain of f consists of all x such that $2x - 5 \geq 0$. Using the methods described in Section 3.4, you can solve this inequality as follows.

$2x - 5 \geq 0$	Write original inequality.
$2x \geq 5$	Add 5 to each side.
$x \geq \frac{5}{2}$	Divide each side by 2.

So, the domain is the set of all real numbers x such that $x \geq \frac{5}{2}$.

Graphical Solution

Use a graphing utility to graph $y = \sqrt{2x - 5}$, as shown in Figure 5.2. From the graph you can see that this function is defined only for x-values for which $x \geq \frac{5}{2}$.

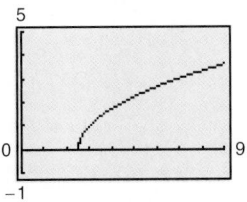

Figure 5.2

Technology

A graphing utility can be used to investigate the graphs of functions of the form $y = x^n$ for $x \geq 0$ and $n > 0$.

After some experimentation, you can see that graphs of functions of this form fall into three categories.

1. If $n = 1$, the graph of $y = x^n$ is a straight line.

2. If $n > 1$, the graph of $y = x^n$ curves upward. (This is called *concave up*.)

3. If $n < 1$, the graph of $y = x^n$ curves downward. (This is called *concave down*.)

Several examples are shown in the calculator screens below. Try confirming this observation by sketching graphs of other functions of the form $y = x^n$.

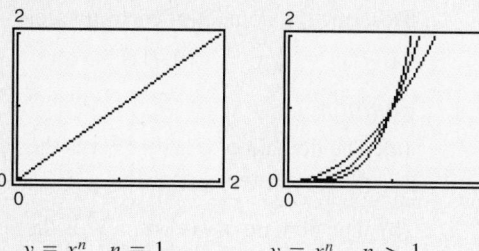

$y = x^n, \quad n = 1$ $y = x^n, \quad n > 1$ $y = x^n, \quad n < 1$

Collaborate!

Exploring with Technology

In your group, discuss the domain and range of each of the functions below. Use a graphing utility to verify your conclusions.

a. $y = x^{3/2}$ b. $y = x^2$ c. $y = x^{1/3}$

d. $y = \left(\sqrt{x}\right)^2$ e. $y = x^{-4/5}$ f. $y = x^5$

Technology

Consider the function $f(x) = x^{2/3}$.

a. What is the domain of the function?

b. Use your graphing utility to graph the following, in order.

$$y_1 = x^{(2 \div 3)}$$
$$y_2 = (x^2)^{1/3} \quad \text{Power, then root}$$
$$y_3 = (x^{1/3})^2 \quad \text{Root, then power}$$

c. Are the graphs all the same? Are their domains all the same?

d. On your graphing utility, which of the forms properly represent the function $f(x) = x^{m/n}$?

$$y_1 = x^{(m \div n)}$$
$$y_2 = (x^m)^{1/n}$$
$$y_3 = (x^{1/n})^m$$

e. Explain how the domains of $f(x) = x^{2/3}$ and $g(x) = x^{-2/3}$ differ.

5.2 ■ Exercises

Developing Skills

In Exercises 1–6, complete the statement.

1. Because $7^2 = 49$,   is a square root of 49.

2. Because $24.5^2 = 600.25$,   is a square root of 600.25.

3. Because $4.2^3 = 74.088$,   is a cube root of 74.088.

4. Because $6^4 = 1296$,   is a fourth root of 1296.

5. Because $45^2 = 2025$, 45 is a   of 2025.

6. Because $12^3 = 1728$, 12 is a   of 1728.

In Exercises 7–24, evaluate without a calculator. (If not possible, state the reason.)

7. $\sqrt{64}$
8. $-\sqrt{100}$
9. $\sqrt{-100}$
10. $\sqrt{144}$
11. $-\sqrt{\frac{4}{9}}$
12. $\sqrt{\frac{9}{16}}$
13. $\sqrt{0.09}$
14. $-\sqrt{0.36}$
15. $\sqrt[3]{125}$
16. $\sqrt[3]{-8}$
17. $\sqrt[3]{1000}$
18. $\sqrt[3]{64}$
19. $\sqrt[3]{-\frac{1}{64}}$
20. $-\sqrt[3]{0.008}$
21. $\sqrt[4]{81}$
22. $\sqrt[5]{32}$
23. $\sqrt[5]{-0.00243}$
24. $-\sqrt[4]{\frac{1}{625}}$

In Exercises 25–30, determine whether the square root is a rational or irrational number.

25. $\sqrt{6}$
26. $\sqrt{144}$
27. $\sqrt{\frac{24}{25}}$
28. $\sqrt{\frac{9}{16}}$
29. $\sqrt{900}$
30. $\sqrt{72}$

In Exercises 31–34, fill in the missing description.

Radical Form	Rational Exponent Form
31. $\sqrt{16} = 4$	
32. $\sqrt[3]{27^2} = 9$	
33.	$125^{1/3} = 5$
34.	$256^{3/4} = 64$

In Exercises 35–42, evaluate without a calculator.

35. $25^{1/2}$
36. $-121^{1/2}$
37. $32^{-2/5}$
38. $81^{-3/4}$
39. $\left(\frac{8}{27}\right)^{2/3}$
40. $\left(\frac{256}{625}\right)^{1/4}$
41. $\left(\frac{121}{9}\right)^{-1/2}$
42. $\left(\frac{27}{1000}\right)^{-4/3}$

In Exercises 43–50, use a calculator to approximate the quantity accurate to four decimal places. (If not possible, state the reason.)

43. $\sqrt{73}$
44. $\sqrt{-532}$
45. $1698^{-3/4}$
46. $962^{2/3}$
47. $\sqrt[4]{342}$
48. $\sqrt[3]{159}$
49. $\sqrt[3]{545^2}$
50. $\sqrt[5]{-35^3}$

In Exercises 51–66, simplify the expression. If necessary, rewrite the expression using rational exponents.

51. $\sqrt{t^2}$
52. $\sqrt[3]{z^3}$
53. $\sqrt[3]{y^9}$
54. $\sqrt[4]{a^8}$
55. $t\sqrt[3]{t^6}$
56. $z\sqrt[4]{z^4}$
57. $x^2\sqrt{x^8}$
58. $y^3\sqrt[5]{y^{15}}$
59. $\dfrac{\sqrt{x}}{\sqrt{x^3}}$
60. $\dfrac{\sqrt[3]{x^2}}{\sqrt[3]{x^4}}$
61. $\sqrt[3]{x^2} \cdot \sqrt[3]{x^7}$
62. $\sqrt[5]{z^3} \cdot \sqrt[5]{z^2}$
63. $\sqrt[4]{x^3y}$
64. $\sqrt[3]{u^4v^2}$
65. $z^2\sqrt{y^5z^4}$
66. $x^2\sqrt[3]{xy^4}$

In Exercises 67–100, simplify the expression.

67. $3^{1/4} \cdot 3^{3/4}$
68. $2^{1/2} \cdot 2^{3/2}$
69. $\dfrac{2^{1/5}}{2^{6/5}}$
70. $\dfrac{5^{-3/4}}{5}$
71. $\left(\frac{2}{3}\right)^{5/3} \cdot \left(\frac{2}{3}\right)^{1/3}$
72. $\left(\frac{8}{9}\right)^{5/4} \cdot \left(\frac{8}{9}\right)^{-1/4}$
73. $\left(6^{2/5}\right)^{10/3}$
74. $\left(4^{1/3}\right)^{9/4}$
75. $x^{2/3} \cdot x^{7/3}$
76. $z^{3/5} \cdot z^{-2/5}$
77. $\left(3x^{-1/3}y^{3/4}\right)^2$
78. $\left(-2u^{3/5}v^{-1/5}\right)^3$
79. $\dfrac{a^{3/4} \cdot a^{1/2}}{a^{5/2}}$
80. $\dfrac{2x^{1/5} \cdot x^{3/10}}{x^{1/2}}$

The symbol   indicates an exercise in which you are instructed to use a calculator or graphing utility.

81. $\dfrac{3a^{-1/2}b^{2/5}}{6a^{5/2}b^{-1/5}}$

82. $\dfrac{18y^{4/3}z^{-1/3}}{24y^{-2/3}z}$

83. $\left(\dfrac{x^{1/4}}{x^{1/6}}\right)^3$

84. $\left(\dfrac{t^{3/5}}{t^{1/4}}\right)^2$

85. $\left(\dfrac{7x^{2/3}y^{3/4}}{14xy^{-1/4}}\right)^3$

86. $\left(\dfrac{3m^{1/6}n^{1/3}}{4n^{-2/3}}\right)^2$

87. $(c^{3/2})^{1/3}$

88. $(k^{-1/3})^{3/2}$

89. $(x^{3/5}y^{-1/2})^{2/3}$

90. $(a^{5/6}b^{1/4})^{3/2}$

91. $\dfrac{x^{4/3}y^{2/3}}{(xy)^{1/3}}$

92. $\dfrac{(2x^3)^{4/3}}{2^{1/3}x^5}$

93. $\sqrt{\sqrt[4]{y}}$

94. $\sqrt[3]{\sqrt{2x}}$

95. $\sqrt{\sqrt[3]{x}}$

96. $\sqrt[5]{\sqrt[4]{3a}}$

97. $\dfrac{(x+y)^{3/4}}{\sqrt[4]{x+y}}$

98. $\dfrac{(a-b)^{1/3}}{\sqrt[3]{a-b}}$

99. $\dfrac{(3u-2v)^{2/3}}{\sqrt{(3u-2v)^3}}$

100. $\dfrac{\sqrt[4]{2x+y}}{(2x+y)^{3/2}}$

In Exercises 101–104, evaluate the function as indicated, and simplify.

101. $g(x) = 3\sqrt{2x}$ (a) $g(0)$ (b) $g(2)$
                            (c) $g(8)$ (d) $g(50)$

102. $f(x) = \sqrt{2x+6}$ (a) $f(4)$ (b) $f(-3)$
                            (c) $f(-1)$ (d) $f(5)$

103. $f(x) = \sqrt[3]{x+1}$ (a) $f(7)$ (b) $f(0)$
                               (c) $f(63)$ (d) $f(-28)$

104. $h(x) = \sqrt[3]{5-x}$ (a) $h(13)$ (b) $h(4)$
                               (c) $h(6)$ (d) $h(-3)$

In Exercises 105–112, determine the domain of the function.

105. $f(x) = 3\sqrt{x}$

106. $h(x) = \sqrt[4]{x}$

107. $g(x) = \dfrac{2}{\sqrt[4]{x}}$

108. $g(x) = \dfrac{10}{\sqrt[3]{x}}$

109. $h(x) = \sqrt{x+10}$

110. $g(x) = \sqrt{6-x}$

111. $f(x) = \sqrt{3x+8}$

112. $g(x) = \sqrt{5x+4}$

In Exercises 113–116, use a graphing utility to graph the function. Find the domain of the function algebraically. Did the graphing utility omit part of the domain? If so, complete the graph by hand.

113. $y = \dfrac{5}{\sqrt[4]{x^3}}$

114. $y = 4\sqrt[3]{x}$

115. $g(x) = 2x^{3/5}$

116. $h(x) = 5x^{2/3}$

In Exercises 117–120, perform the multiplication.

117. $x^{1/2}(2x-3)$

118. $x^{4/3}(3x^2-4x+5)$

119. $y^{-1/3}(y^{1/3}+5y^{4/3})$

120. $(x^{1/2}-3)(x^{1/2}+3)$

Solving Problems

Mathematical Modeling   In Exercises 121 and 122, use the formula for the *declining balances method*

$$r = 1 - \left(\dfrac{S}{C}\right)^{1/n}$$

to find the depreciation rate r. In the formula, n is the useful life of the item (in years), S is the salvage value (in dollars), and C is the original cost (in dollars).

121. A \$75,000 truck is depreciated over an eight-year period, as shown in the graph.

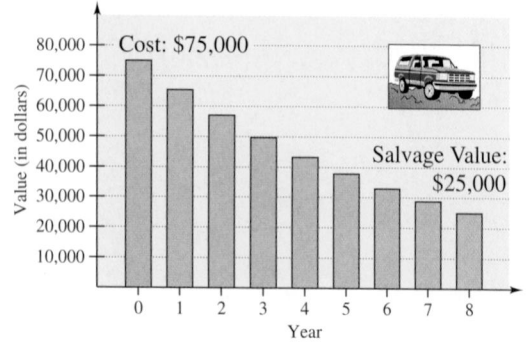

122. A \$125,000 printing press is depreciated over a 10-year period, as shown in the graph.

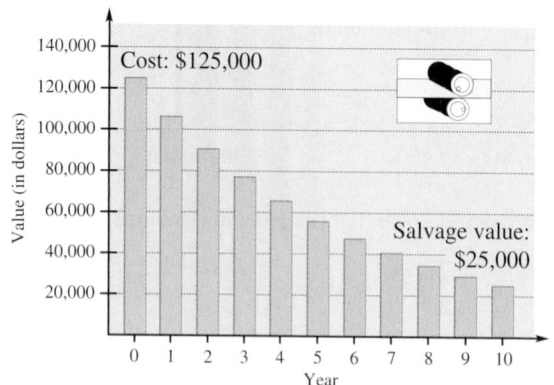

123. *Volume*   The usable space in a particular microwave oven is in the form of a cube. The sales brochure indicates that the interior space of the oven is 2197 cubic inches. Find the inside dimensions of the oven.

124. *Area*   Find the dimensions of a piece of carpet for a classroom with 529 square feet of floor space, assuming the floor is square.

125. *Velocity of a Stream*   A stream of water moving at a rate of v feet per second can carry particles of size $0.03\sqrt{v}$ inches. Find the particle size that can be carried by a stream flowing at a rate of $\frac{3}{4}$ foot per second.

126. *Price Index*   The producer price index for footwear in the United States between 1993 and 1998 can be approximated by the model

$$y = 11.8\sqrt{t + 1.6} + 108.8$$

where t represents the year, with $t = 3$ corresponding to 1993. Use this model to approximate the producer price index for footwear in the United States for the year 1996.   *(Source: U.S. Bureau of Labor Statistics)*

Explaining Concepts

127. In your own words, define an nth root of a number.

128. Is it true that $\sqrt{2} = 1.414$? Explain.

129. Determine the values of x for which $\sqrt{x^2} \ne x$. Explain your answer.

130. Find all possible "last digits" of perfect squares. (For instance, the last digit of 81 is 1 and the last digit of 64 is 4.) Is it possible that 4,322,788,987 is a perfect square?

Ongoing Review

In Exercises 131–134, perform the multiplication and simplify.

131. $(2x + 1)(x - 4)$

132. $(3x - 2)(x + 5)$

133. $(x + 2y)(x - 2y)$

134. $(5a + b)(5a - b)$

In Exercises 135–138, factor the expression completely.

135. $3x^2 - 7x + 4$

136. $4x^2 - 9x + 5$

137. $2x^3 - 16x^2 + 2x - 16$

138. $2x^3 - 8x^2 + 5x - 20$

139. *Average Speed*   A truck driver traveled at an average speed of 54 miles per hour on a 100-mile trip. On the return trip with the truck fully loaded, the average speed was 45 miles per hour. Find the average speed for the round trip.

140. *Quality Control*   A quality control engineer for a certain buyer found two defective units in a sample of 75. At that rate, what is the expected number of defective units in a shipment of 10,000 units?

Looking Further

Investigation   (a) Choose a positive real number $x > 1$. Enter the number into a calculator and find its square root. Then repeatedly take the square root of the result.

$$\sqrt{x}, \ \sqrt{\sqrt{x}}, \ \sqrt{\sqrt{\sqrt{x}}}, \dots$$

What real number does the display appear to be approaching?

(b) Repeat part (a) using a positive real number $x < 1$.

(c) How do the results compare?

5.3 Simplifying and Combining Radicals

What you should learn:

- How to use the Multiplication and Division Properties of Radicals to simplify radical expressions
- How to use rationalization techniques to simplify radical expressions
- How to use the Distributive Property to add and subtract like radicals
- How to simplify radical expressions in application problems

Why you should learn it:

Algebraic equations often involve radicals. For instance, in Exercise 124 on page 331 you will use a radical equation to find the period of a pendulum.

Simplifying Radicals

In this section, you will study ways to simplify and combine radicals. For instance, the expression $\sqrt{12}$ can be simplified as

$$\sqrt{12} = \sqrt{4 \cdot 3}$$
$$= \sqrt{4}\,\sqrt{3}$$
$$= 2\sqrt{3}.$$

This rewriting is based on the rules below for multiplying and dividing radicals.

Multiplication and Division Properties of Radicals

Let u and v be real numbers, variables, or algebraic expressions. If the nth roots of u and v are real, the properties below are true.

1. $\sqrt[n]{u}\,\sqrt[n]{v} = \sqrt[n]{uv}$ Multiplication Property

2. $\dfrac{\sqrt[n]{u}}{\sqrt[n]{v}} = \sqrt[n]{\dfrac{u}{v}}, \qquad v \neq 0$ Division Property

---

You can use these properties of radicals to *simplify* square root expressions by finding the largest perfect square factor and removing it from the radical as follows.

$$\sqrt{48} = \sqrt{16 \cdot 3} = \sqrt{16}\,\sqrt{3} = 4\sqrt{3}$$

This simplification process is called **removing perfect square factors from the radical.**

Example 1 ■ Removing Constant Factors from Radicals

Simplify each radical by removing as many perfect square factors as possible.

(a) $\sqrt{75}$ (b) $\sqrt{72}$ (c) $\sqrt{162}$

Solution

(a) $\sqrt{75} = \sqrt{25 \cdot 3} = \sqrt{25}\,\sqrt{3} = 5\sqrt{3}$ 25 is a perfect square factor of 75.

(b) $\sqrt{72} = \sqrt{36 \cdot 2} = \sqrt{36}\,\sqrt{2} = 6\sqrt{2}$ 36 is a perfect square factor of 72.

(c) $\sqrt{162} = \sqrt{81 \cdot 2} = \sqrt{81}\,\sqrt{2} = 9\sqrt{2}$ 81 is a perfect square factor of 162.

When removing *variable* factors from a square root radical, remember that it is not valid to write $\sqrt{x^2} = x$ *unless* you happen to know that x is nonnegative. Without knowing anything about x, the only way you can simplify $\sqrt{x^2}$ is to include absolute value signs when you remove x from the radical.

$$\sqrt{x^2} = |x|$$ Restricted by absolute value signs

When simplifying the expression $\sqrt{x^3}$, it is not necessary to include absolute value signs because the domain of this expression does not include negative numbers.

$$\sqrt{x^3} = \sqrt{x^2(x)} = x\sqrt{x} \qquad \text{Restricted by domain of radical}$$

Example 2 ■ Removing Variable Factors from Radicals

Simplify each radical expression.

(a) $\sqrt{25x^2}$ (b) $\sqrt{12x^3}, \quad x \geq 0$ (c) $\sqrt{144x^4}$ (d) $\sqrt{72x^3y^2}$

Solution

(a) $\sqrt{25x^2} = \sqrt{5^2x^2} = \sqrt{5^2}\sqrt{x^2} = 5|x|$ $\sqrt{x^2} = |x|$

(b) $\sqrt{12x^3} = \sqrt{2^2x^2(3x)} = 2x\sqrt{3x}$ $\sqrt{2^2}\sqrt{x^2} = 2x, \quad x \geq 0$

(c) $\sqrt{144x^4} = \sqrt{12^2(x^2)^2} = 12x^2$ $\sqrt{12^2}\sqrt{(x^2)^2} = 12|x^2| = 12x^2$

(d) $\sqrt{72x^3y^2} = \sqrt{6^2x^2y^2(2x)}$

$\qquad\qquad = 6x|y|\sqrt{2x}$ $\sqrt{6^2}\sqrt{x^2}\sqrt{y^2} = 6x|y|$

In the same way that perfect squares can be removed from square root radicals, perfect nth powers can be removed from nth root radicals.

Example 3 ■ Removing Constant Factors from Radicals

Simplify each radical expression.

(a) $\sqrt[3]{40}$ (b) $\sqrt[4]{162}$

Solution

(a) $\sqrt[3]{40} = \sqrt[3]{8(5)} = \sqrt[3]{2^3(5)} = 2\sqrt[3]{5}$ $\sqrt[3]{2^3} = 2$

(b) $\sqrt[4]{162} = \sqrt[4]{81(2)} = \sqrt[4]{3^4(2)} = 3\sqrt[4]{2}$ $\sqrt[4]{3^4} = 3$

Example 4 ■ Removing Factors from Radicals

Simplify each radical expression.

(a) $\sqrt[4]{x^5}, \quad x \geq 0$ (b) $\sqrt[3]{54x^3y^5}$ (c) $\sqrt[5]{486x^7}$

Solution

(a) $\sqrt[4]{x^5} = \sqrt[4]{x^4(x)} = x\sqrt[4]{x}$ $\sqrt[4]{x^4} = x, \quad x \geq 0$

(b) $\sqrt[3]{54x^3y^5} = \sqrt[3]{27x^3y^3(2y^2)}$

$\qquad\qquad = \sqrt[3]{3^3x^3y^3(2y^2)}$

$\qquad\qquad = 3xy\sqrt[3]{2y^2}$ $\sqrt[3]{3^3}\sqrt[3]{x^3}\sqrt[3]{y^3} = 3xy$

(c) $\sqrt[5]{486x^7} = \sqrt[5]{243x^5(2x^2)}$

$\qquad\qquad = \sqrt[5]{3^5x^5(2x^2)}$

$\qquad\qquad = 3x\sqrt[5]{2x^2}$ $\sqrt[5]{3^5}\sqrt[5]{x^5} = 3x$

Rationalization Techniques

Removing factors from radicals is only one of two techniques used to simplify radicals. Three conditions must be met in order for a radical expression to be in simplest form. These three conditions are summarized as follows.

> **Simplest Form of Radical Expressions**
>
> A radical expression is said to be in simplest form if all three of the statements below are true.
>
> 1. All possible nth-powered factors have been removed from each radical.
>
> 2. No radical contains a fraction.
>
> 3. No denominator of a fraction contains a radical.

Study Tip

When rationalizing a denominator, remember that for square roots you want a perfect square in the denominator, for cube roots you want a perfect cube, and so on. For instance, to find the radical factor needed to create a perfect square in the denominator of Example 5(c) you can write the prime factorization of 18.

$$18 = 2 \cdot 3 \cdot 3$$
$$= 2 \cdot 3^2$$

From its prime factorization you can see that 3^2 is a square root factor of 18 and you need one more factor of 2 to create a perfect square in the denominator

To meet the last two conditions, you can use a second technique for simplifying radical expressions called **rationalizing the denominator.** This involves multiplying both the numerator and denominator by a factor that creates a perfect nth power in the denominator.

Example 5 ■ Rationalizing the Denominator

Rationalize the denominator in each expression.

(a) $\sqrt{\dfrac{3}{5}}$ (b) $\dfrac{4}{\sqrt[3]{9}}$ (c) $\dfrac{8}{3\sqrt{18}}$

Solution

(a) $\sqrt{\dfrac{3}{5}} = \dfrac{\sqrt{3}}{\sqrt{5}} = \dfrac{\sqrt{3}}{\sqrt{5}} \cdot \dfrac{\sqrt{5}}{\sqrt{5}} = \dfrac{\sqrt{15}}{\sqrt{5^2}} = \dfrac{\sqrt{15}}{5}$ Multiply by $\sqrt{5}/\sqrt{5}$ to create a perfect square in the denominator.

(b) $\dfrac{4}{\sqrt[3]{9}} = \dfrac{4}{\sqrt[3]{9}} \cdot \dfrac{\sqrt[3]{3}}{\sqrt[3]{3}} = \dfrac{4\sqrt[3]{3}}{\sqrt[3]{3^3}} = \dfrac{4\sqrt[3]{3}}{3}$ Multiply by $\sqrt[3]{3}/\sqrt[3]{3}$ to create a perfect cube in the denominator.

(c) $\dfrac{8}{3\sqrt{18}} = \dfrac{8}{3\sqrt{18}} \cdot \dfrac{\sqrt{2}}{\sqrt{2}} = \dfrac{8\sqrt{2}}{3\sqrt{36}} = \dfrac{8\sqrt{2}}{3\sqrt{6^2}} = \dfrac{8\sqrt{2}}{3(6)} = \dfrac{4\sqrt{2}}{9}$

Example 6 ■ Rationalizing the Denominator

Simplify each expression.

(a) $\sqrt{\dfrac{8x}{12y^5}}$ (b) $\sqrt[3]{\dfrac{54x^6y^3}{5z^2}}$

Solution

(a) $\sqrt{\dfrac{8x}{12y^5}} = \sqrt{\dfrac{2x}{3y^5}} = \dfrac{\sqrt{2x}}{\sqrt{3y^5}} \cdot \dfrac{\sqrt{3y}}{\sqrt{3y}} = \dfrac{\sqrt{6xy}}{\sqrt{3^2y^6}} = \dfrac{\sqrt{6xy}}{3|y^3|}$

(b) $\sqrt[3]{\dfrac{54x^6y^3}{5z^2}} = \dfrac{\sqrt[3]{(3^3)(2)(x^6)(y^3)}}{\sqrt[3]{5z^2}} \cdot \dfrac{\sqrt[3]{25z}}{\sqrt[3]{25z}} = \dfrac{3x^2y\sqrt[3]{50z}}{\sqrt[3]{5^3z^3}} = \dfrac{3x^2y\sqrt[3]{50z}}{5z}$

Adding and Subtracting Radicals

Two or more radical expressions are *alike* if they have the same radicand and the same index. For instance, $\sqrt{2}$ and $3\sqrt{2}$ are alike, but $\sqrt{3}$ and $\sqrt[3]{3}$ are not alike. Two radical expressions that are alike can be added or subtracted by adding or subtracting their coefficients.

Example 7 ■ Combining Radicals

Perform the indicated operations and simplify.

(a) $\sqrt{7} + 5\sqrt{7} - 2\sqrt{7}$ (b) $6\sqrt{x} - \sqrt[3]{4} - 5\sqrt{x}$ (c) $3\sqrt[3]{x} + \sqrt[3]{8x}$

Solution

(a) $\sqrt{7} + 5\sqrt{7} - 2\sqrt{7} = (1 + 5 - 2)\sqrt{7} = 4\sqrt{7}$

(b) $6\sqrt{x} - \sqrt[3]{4} - 5\sqrt{x} = 6\sqrt{x} - 5\sqrt{x} - \sqrt[3]{4}$

$$= (6 - 5)\sqrt{x} - \sqrt[3]{4}$$

$$= \sqrt{x} - \sqrt[3]{4}$$

(c) $3\sqrt[3]{x} + \sqrt[3]{8x} = 3\sqrt[3]{x} + 2\sqrt[3]{x} = (3 + 2)\sqrt[3]{x} = 5\sqrt[3]{x}$

> **NOTE**   Notice in Example 7(c) that *before* concluding that two radicals cannot be combined, you should check to see that they are written in simplest form.

Example 8 ■ Simplifying Radical Expressions

(a) $\sqrt{45x} + 3\sqrt{20x} = \sqrt{9 \cdot 5x} + 3\sqrt{4 \cdot 5x} = 3\sqrt{5x} + 6\sqrt{5x} = 9\sqrt{5x}$

(b) $5\sqrt{x^3} - x\sqrt{4x} = 5x\sqrt{x} - 2x\sqrt{x} = 3x\sqrt{x}$

(c) $\sqrt[3]{54y^5} + 4\sqrt[3]{2y^2} = \sqrt[3]{27y^3(2y^2)} + 4\sqrt[3]{2y^2}$

$$= 3y\sqrt[3]{2y^2} + 4\sqrt[3]{2y^2} = (3y + 4)\sqrt[3]{2y^2}$$

In some instances, it may be necessary to rationalize denominators before combining radicals.

Technology

You can use a graphing utility to check simplifications such as those shown in Example 8. For instance, to check the result of Example 8(a), use a graphing utility to graph the equations

$$y_1 = \sqrt{45x} + 3\sqrt{20x}$$

and

$$y_2 = 9\sqrt{5x}.$$

Both equations have the same graph, as shown below.

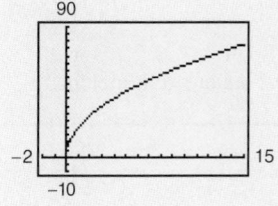

Example 9 ■ Rationalizing Denominators Before Simplifying

Simplify the expression $\sqrt{7} - \dfrac{5}{\sqrt{7}}$.

Solution

$$\sqrt{7} - \frac{5}{\sqrt{7}} = \sqrt{7} - \left(\frac{5}{\sqrt{7}} \cdot \frac{\sqrt{7}}{\sqrt{7}}\right)$$ Multiply by $\sqrt{7}/\sqrt{7}$ to create a perfect square in the denominator.

$$= \sqrt{7} - \frac{5\sqrt{7}}{7}$$ Simplify.

$$= \left(1 - \frac{5}{7}\right)\sqrt{7}$$ Distributive Property

$$= \frac{2}{7}\sqrt{7}$$ Simplify.

Application

A common use of radicals occurs in many geometric applications, such as applications involving right triangles. Recall that a right triangle is one that contains a right (or 90°) angle. The relationship among the three sides of a right triangle is described by the Pythagorean Theorem, which says that if a and b are the lengths of the legs and c is the length of the hypotenuse, then

$$c = \sqrt{a^2 + b^2} \quad \text{and} \quad a = \sqrt{c^2 - b^2} \qquad \text{Pythagorean Theorem: } a^2 + b^2 = c^2$$

For instance, if $a = 6$ and $b = 9$, then

$$c = \sqrt{6^2 + 9^2} = \sqrt{117} = 3\sqrt{13}$$

Radicals are also used in applications involving right circular cones. The lateral surface of a cone consists of all segments that connect the vertex with points on the edge of the base, as shown in Figure 5.3. The lateral surface area S of a right circular cone is given by

$$S = \pi r \sqrt{r^2 + h^2}$$

where r is the radius of the base of the cone and h is the height.

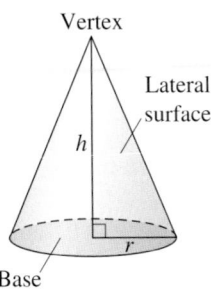

Vertex

Lateral surface

h

r

Base

Figure 5.3

Example 10 ■ An Application of Lateral Surface Area

The radius of a traffic cone is 14 centimeters and the height of the cone is 34 centimeters (see Figure 5.4). What is the lateral surface area of the traffic cone?

Solution

You can use the formula for the lateral surface area of a cone as follows.

$$
\begin{aligned}
S &= \pi r \sqrt{r^2 + h^2} & & \text{Formula for lateral surface area} \\
&= \pi(14)\sqrt{14^2 + 34^2} & & \text{Substitute 14 for } r \text{ and 34 for } h. \\
&= 14\pi\sqrt{1352} & & \text{Simplify.} \\
&= 14\pi\sqrt{676 \cdot 2} & & \text{676 is a perfect square factor of 1352.} \\
&= 14 \cdot 26\pi\sqrt{2} & & \text{Simplify.} \\
&= 364\pi\sqrt{2} \approx 1617.2 & & \text{Simplify.}
\end{aligned}
$$

So, the lateral surface area of the cone is about 1617.2 square centimeters.

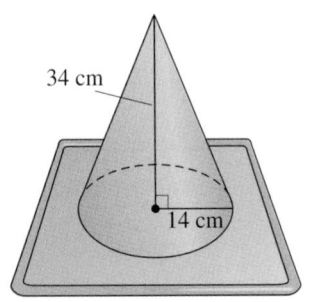

34 cm

14 cm

Figure 5.4

Collaborate!

Extending the Concept

In general, $\sqrt{a + b}$ is *not* equal to $\sqrt{a} + \sqrt{b}$. One convenient way to demonstrate this is to let $a = 9$ and $b = 16$. Then $\sqrt{9 + 16} = \sqrt{25} = 5$, whereas $\sqrt{9} + \sqrt{16} = 3 + 4 = 7$. In your group, try to find another example in which a, b, and $a + b$ are all perfect squares.

5.3 ■ Exercises

Developing Skills

In Exercises 1–18, simplify the radical.

1. $\sqrt{20}$

2. $\sqrt{50}$

3. $\sqrt{27}$

4. $\sqrt{125}$

5. $\sqrt{0.04}$

6. $\sqrt{0.25}$

7. $\sqrt[3]{24}$

8. $\sqrt[3]{54}$

9. $\sqrt[4]{30,000}$

10. $\sqrt[5]{96}$

11. $\sqrt{\frac{15}{4}}$

12. $\sqrt{\frac{5}{36}}$

13. $\sqrt[3]{\frac{35}{64}}$

14. $\sqrt[4]{\frac{5}{16}}$

15. $\sqrt[5]{\frac{15}{243}}$

16. $\sqrt[3]{\frac{1}{1000}}$

17. $\sqrt{\frac{13}{25}}$

18. $\sqrt{\frac{15}{36}}$

In Exercises 19–46, simplify the expression.

19. $\sqrt{a^7}$

20. $\sqrt{b^8}$

21. $\sqrt{9x^5}$

22. $\sqrt{64x^3}$

23. $\sqrt{48y^4}$

24. $\sqrt{32x}$

25. $\sqrt[3]{x^{10}}$

26. $\sqrt[3]{a^{11}}$

27. $\sqrt[3]{8a^7}$

28. $\sqrt[3]{27b^{12}}$

29. $\sqrt[3]{54x^5}$

30. $\sqrt[3]{16y^{17}}$

31. $\sqrt{a^9b^5}$

32. $\sqrt{x^{17}y^{20}}$

33. $\sqrt[3]{x^4y^3}$

34. $\sqrt[3]{a^5b^6}$

35. $\sqrt[4]{128u^4v^7}$

36. $\sqrt[4]{81a^9b^{12}}$

37. $\sqrt[5]{32x^5y^6}$

38. $\sqrt[5]{243u^{18}v^{24}}$

39. $\sqrt[5]{\frac{32x^2}{y^5}}$

40. $\sqrt[3]{\frac{16z^3}{y^6}}$

41. $\sqrt[3]{\frac{54a^4}{b^9}}$

42. $\sqrt[4]{\frac{3u^2}{16v^8}}$

43. $\sqrt{\frac{32a^4}{b^2}}$

44. $\sqrt{\frac{18x^2}{z^6}}$

45. $\sqrt[4]{(3x^2)^4}$

46. $\sqrt[5]{96x^5}$

In Exercises 47–50, use a graphing utility to graph the equations y_1 and y_2 in the same viewing window. Use the graphs to determine if $y_1 = y_2$. If not, explain why.

47. $y_1 = \sqrt{12x^2}$
$y_2 = 2x\sqrt{3}$

48. $y_1 = \sqrt{12x^3}$
$y_2 = 2x\sqrt{3x}$

49. $y_1 = \sqrt[3]{16x^3}$
$y_2 = 2x\sqrt[3]{2}$

50. $y_1 = \sqrt[4]{16x^6}$
$y_2 = 2x\sqrt[4]{x^2}$

In Exercises 51–78, rationalize the denominator and simplify further, if possible.

51. $\sqrt{\frac{1}{3}}$

52. $\sqrt{\frac{1}{5}}$

53. $\frac{12}{\sqrt{3}}$

54. $\frac{5}{\sqrt{10}}$

55. $\sqrt[4]{\frac{5}{4}}$

56. $\sqrt[3]{\frac{9}{25}}$

57. $\frac{6}{\sqrt[3]{32}}$

58. $\frac{10}{\sqrt[5]{16}}$

59. $\frac{1}{\sqrt{y}}$

60. $\frac{1}{\sqrt{2x}}$

61. $\sqrt{\frac{4}{x}}$

62. $\sqrt{\frac{5}{c}}$

63. $\sqrt{\frac{4}{x^3}}$

64. $\frac{5}{\sqrt{8x^5}}$

65. $\sqrt[3]{\frac{2x}{3y}}$

66. $\sqrt[3]{\frac{20x^2}{9y^2}}$

67. $\frac{a^3}{\sqrt[3]{ab^2}}$

68. $\frac{3u^2}{\sqrt[4]{8u^3}}$

69. $\frac{6}{\sqrt{3b^3}}$

70. $\frac{1}{\sqrt{xy}}$

71. $\sqrt{\frac{6a^3}{15b^7}}$

72. $\sqrt{\frac{2x^2}{5y^3}}$

73. $\sqrt[3]{\frac{8u^6}{3v}}$

74. $\sqrt[3]{\frac{16x^8}{25y^5}}$

75. $\sqrt{\frac{12x^3y^4}{4z^7}}$

76. $\sqrt{\frac{16ab^{16}}{3c^3}}$

77. $\sqrt[3]{\frac{7x^{12}y^5}{4z^{10}}}$

78. $\sqrt[3]{\frac{3a^8b^{18}}{50c^5}}$

In Exercises 79 and 80, use a graphing utility to graph the equations in the same viewing window. Use the graphs to verify that the expressions are equivalent. Verify the results algebraically.

79. $y_1 = \sqrt{\frac{3}{x}}$
$y_2 = \frac{\sqrt{3x}}{x}$

80. $y_1 = \frac{4}{\sqrt{2x}}$
$y_2 = \frac{2\sqrt{2x}}{x}$

In Exercises 81–106, combine the radical expressions, if possible.

81. $3\sqrt{2} - \sqrt{2}$

82. $\frac{2}{5}\sqrt{5} - \frac{6}{5}\sqrt{5}$

83. $12\sqrt{8} - 3\sqrt{8}$

84. $4\sqrt{32} + 7\sqrt{32}$

85. $\sqrt{12} + \sqrt{75}$

86. $\sqrt{125} + \sqrt{45}$

87. $5\sqrt{54} + 3\sqrt{24}$

88. $7\sqrt{162} + 4\sqrt{72}$

89. $2\sqrt[3]{54} + 12\sqrt[3]{16}$

90. $4\sqrt[4]{48} - \sqrt[4]{243}$

91. $5\sqrt{9x} - 3\sqrt{x}$

92. $3\sqrt{x+1} + 10\sqrt{x+1}$

93. $\sqrt{25y} + \sqrt{64y}$

94. $\sqrt[3]{4x^2} + \sqrt[3]{32x^2}$

95. $\sqrt{9a^3} + \sqrt{36a^7}$

96. $\sqrt[3]{16t^4} - \sqrt[3]{54t^4}$

97. $10\sqrt[3]{z} - \sqrt[3]{z^4}$

98. $5\sqrt[3]{24u^2} + 2\sqrt[3]{81u^5}$

99. $2y^2\sqrt{y} + 5\sqrt{y^5}$

100. $10a\sqrt{4a^3} - 3\sqrt{9a^5}$

101. $5\sqrt[4]{2x^7} + x\sqrt[4]{2x^3}$

102. $\sqrt[4]{16y^9} + 4y\sqrt[4]{y^5}$

103. $7\sqrt{8x^3y^5} + xy\sqrt{2xy^3}$

104. $12\sqrt{12x^4y^7} - y^2\sqrt{27x^4y^3}$

105. $7\sqrt{80x^3y} - 2x\sqrt{125xy} + 3\sqrt{45x^3y}$

106. $-3y^2\sqrt{48x^2} + 7\sqrt{75x^2y^4} + 8x\sqrt{147y^4}$

In Exercises 107 and 108, use a graphing utility to graph the equations in the same viewing window. Use the graphs to verify that the expressions are equivalent. Verify the results algebraically.

107. $y_1 = 7\sqrt{x^3} - 2x\sqrt{4x}$

$y_2 = 3x\sqrt{x}$

108. $y_1 = \sqrt[3]{8x^4} - \sqrt[3]{x^4}$

$y_2 = x\sqrt[3]{x}$

In Exercises 109–112, perform the addition or subtraction and simplify your answer.

109. $\sqrt{5} - \dfrac{3}{\sqrt{5}}$

110. $\sqrt{10} + \dfrac{5}{\sqrt{10}}$

111. $\dfrac{x}{\sqrt{3x}} + \sqrt{27x}$

112. $\sqrt{20x} - \dfrac{x}{\sqrt{5x}}$

In Exercises 113–116, place the correct symbol (<, >, or =) between the numbers.

113. $\sqrt{7} + \sqrt{18}$ ____ $\sqrt{7+18}$

114. $\sqrt{10} - \sqrt{6}$ ____ $\sqrt{10-6}$

115. 5 ____ $\sqrt{3^2 + 2^2}$

116. 5 ____ $\sqrt{3^2 + 4^2}$

Solving Problems

Geometry In Exercises 117 and 118, find the length of the hypotenuse of the right triangle.

117.

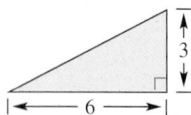

118.

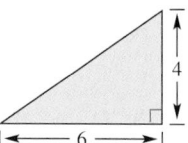

Geometry In Exercises 119 and 120, find the lateral surface area of the cone.

119.

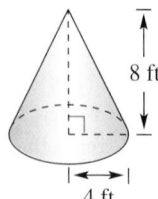

120.

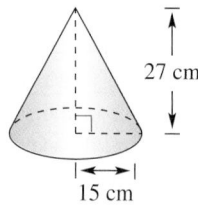

121. *Geometry* The four corners are cut from a four-foot by eight-foot sheet of plywood, as shown in the figure. Find the perimeter of the remaining piece of plywood.

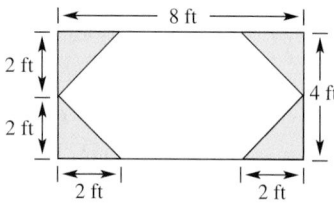

Figure for 121

122. *Geometry* The foundation of a house is 40 feet long and 30 feet wide. The height of the attic is 5 feet (see figure). (a) Use the Pythagorean Theorem to find the length of the hypotenuse of the right triangles formed by the roof line. (b) Use the result of part (a) to determine the total area of the roof.

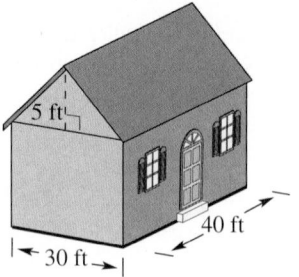

123. *Vibrating String*   The frequency f in cycles per second of a vibrating string is given by

$$f = \frac{1}{100}\sqrt{\frac{400 \times 10^6}{5}}.$$

Use a calculator to approximate this number. (Round the result to two decimal places.)

124. *Period of a Pendulum*   The time T in seconds for a pendulum of length L feet (see figure) to go through one complete cycle (its period) is given by

$$T = 2\pi\sqrt{\frac{L}{32}}.$$

Find the period of a pendulum whose length is 4 feet. (Round the result to two decimal places.)

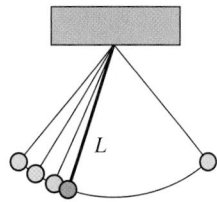

Explaining Concepts

125. Describe the three conditions that characterize a simplified radical expression.

126. Is $\sqrt{2} + \sqrt{18}$ in simplest form? Explain.

127. Explain what it means for two or more radical expressions to be alike.

128. Square the real number $5/\sqrt{3}$ and note that the radical is eliminated from the denominator. Is this equivalent to rationalizing the denominator? Why or why not?

Ongoing Review

In Exercises 129–132, find the distance between each pair of points. Then find the midpoint of the line segment joining the points.

129. $(-4, 2), (1, 12)$ **130.** $(1, -2), (10, 3)$

131. $(3, 6), (-5, -8)$ **132.** $(0, -3), (-6, 9)$

In Exercises 133–136, solve the inequality and graph the solution on the real number line.

133. $10 - 3x \leq 0$

134. $5 - 2x > 5(x + 1)$

135. $|4 - (x - 2)| < 20$

136. $\frac{1}{2}|2x + 3| \geq 5$

137. *Geometry*   Find the dimensions of a square mirror with an area of 1024 square inches.

138. *Number Problem*   Find two consecutive integers such that the sum of two times the first number and three times the second number is 88.

Looking Further

The Square Root Spiral   The square root spiral (see figure) is formed by a sequence of right triangles, each with a side whose length is 1. Let r_n be the length of the hypotenuse of the nth triangle.

(a) Each leg of the first triangle has a length of 1. Use the Pythagorean Theorem to show that $r_1 = \sqrt{2}$.

(b) Find $r_2, r_3, r_4, r_5,$ and r_6.

(c) What can you conclude about r_n?

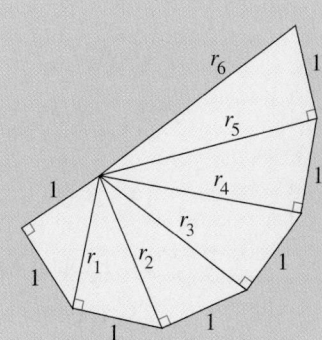

Mid-Chapter Quiz

Take this quiz as you would take a quiz in class. After you are done, check your work against the answers given in the back of the book.

In Exercises 1–4, evaluate the expression.

1. -12^{-2}　　**2.** $\left(\frac{3}{4}\right)^{-3}$　　**3.** $\sqrt{\frac{25}{9}}$　　**4.** $(-64)^{2/3}$

In Exercises 5–8, rewrite the expression using only positive exponents, and simplify. (Assume that any variables in the expression are nonzero.)

5. $(t^3)^{-1/2}(3t^3)$　　**6.** $\dfrac{(10x)^0}{(4x^{-2})^{3/2}}$　　**7.** $\dfrac{10u^{-2}}{15u}$　　**8.** $(3x^2y^{-1})(4x^{-2}y)^{-2}$

9. Write each number in scientific notation.

(a) 13,400,000　　(b) 0.00075

In Exercises 10–13, simplify each expression.

10. (a) $\sqrt{150}$　　(b) $\sqrt[3]{54}$

11. (a) $\sqrt[3]{27x^7}$　　(b) $\sqrt[4]{81x^{11}y^5}$

12. (a) $\sqrt[4]{\dfrac{5}{16}}$　　(b) $\sqrt{\dfrac{24}{49}}$

13. (a) $\sqrt{\dfrac{40u^3}{9z^{10}}}$　　(b) $\sqrt[3]{\dfrac{16a^5}{b^{12}}}$

In Exercises 14 and 15, evaluate the function as indicated, and simplify.

14. $f(x) = \sqrt{x - 12}$　　(a) $f(12)$　　(b) $f(40)$

15. $g(x) = \sqrt{x + 10}$　　(a) $g(26)$　　(b) $g(8)$

16. Determine the domain of each function.

(a) $f(x) = \sqrt{12 - 4x}$　　(b) $g(x) = \sqrt[3]{x + 6}$

In Exercises 17 and 18, rationalize the denominator and simplify further, if possible.

17. (a) $\sqrt{\dfrac{2}{3}}$　　(b) $\dfrac{24}{\sqrt{12}}$　　**18.** (a) $\dfrac{10}{\sqrt{5x}}$　　(b) $\sqrt[3]{\dfrac{3}{2a^2}}$

In Exercises 19–21, combine the radical expressions, if possible.

19. $\sqrt{200y} - 3\sqrt{8y}$　　**20.** $6x\sqrt[3]{5x} + 2\sqrt[3]{40x^4}$　　**21.** $\sqrt{\dfrac{5x}{2}} + \sqrt{10x}$

22. Explain why $\sqrt{5^2 + 12^2} \neq 17$. Determine the correct value of the radical.

23. The four corners are cut from an $8\frac{1}{2}$-inch by 11-inch sheet of paper, as shown in the figure. Find the perimeter of the remaining piece of paper.

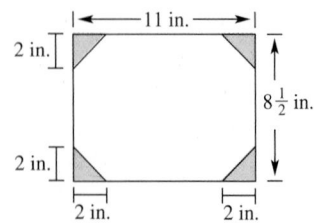

Figure for 23

5.4 Multiplying and Dividing Radicals

Multiplying Radical Expressions

You can multiply radical expressions by using the Distributive Property or the FOIL Method. In both procedures, you also make use of the Multiplication Property of Radicals. Recall from Section 5.3 that the product of two radicals is given by

$$\sqrt[n]{u}\ \sqrt[n]{v} = \sqrt[n]{uv}$$

where u and v are real numbers whose nth roots are also real numbers.

Example 1 ■ Multiplying Radical Expressions

Find each product and simplify.

(a) $\sqrt{6} \cdot \sqrt{3}$

(b) $\sqrt[3]{4} \cdot \sqrt[3]{12}$

Solution

(a) $\sqrt{6} \cdot \sqrt{3} = \sqrt{6 \cdot 3} = \sqrt{18} = \sqrt{9 \cdot 2} = 3\sqrt{2}$

(b) $\sqrt[3]{4} \cdot \sqrt[3]{12} = \sqrt[3]{4 \cdot 12} = \sqrt[3]{48} = \sqrt[3]{8 \cdot 6} = 2\sqrt[3]{6}$

Example 2 ■ Multiplying Radical Expressions

Find each product and simplify.

(a) $\sqrt{3}\,(2 + \sqrt{5})$

(b) $\sqrt{2}\,(4 - \sqrt{8})$

(c) $\sqrt{6}\,(\sqrt{12} - \sqrt{3})$

Solution

(a) $\sqrt{3}(2 + \sqrt{5}) = 2\sqrt{3} + \sqrt{3}\sqrt{5}$ — Distributive Property

$\qquad\qquad\qquad = 2\sqrt{3} + \sqrt{15}$ — Multiplication Property of Radicals

(b) $\sqrt{2}(4 - \sqrt{8}) = 4\sqrt{2} - \sqrt{2}\sqrt{8}$ — Distributive Property

$\qquad\qquad\qquad = 4\sqrt{2} - \sqrt{16}$ — Multiplication Property of Radicals

$\qquad\qquad\qquad = 4\sqrt{2} - 4$ — Simplify.

(c) $\sqrt{6}(\sqrt{12} - \sqrt{3}) = \sqrt{6}\sqrt{12} - \sqrt{6}\sqrt{3}$ — Distributive Property

$\qquad\qquad\qquad = \sqrt{72} - \sqrt{18}$ — Multiplication Property of Radicals

$\qquad\qquad\qquad = 6\sqrt{2} - 3\sqrt{2}$ — Find perfect square factors.

$\qquad\qquad\qquad = 3\sqrt{2}$ — Simplify.

Study Tip

Throughout this chapter, remember that you can use a calculator to check your computations. For instance, in Example 2(c), try using a calculator to evaluate the expressions

$$\sqrt{6}\,(\sqrt{12} - \sqrt{3}) \text{ and } 3\sqrt{2}.$$

For each expression, you should obtain a calculator display of about 4.2426.

In Example 2, the Distributive Property was used to multiply radical expressions. In Example 3, note how the FOIL Method can be used to multiply binomial radical expressions.

Example 3 ■ Multiplying Radical Expressions

Multiply and simplify.

(a) $\left(2\sqrt{7} - 4\right)\left(\sqrt{7} + 1\right)$ (b) $\left(3 - \sqrt{x}\right)\left(1 + \sqrt{x}\right)$

Solution

(a) $\left(2\sqrt{7} - 4\right)\left(\sqrt{7} + 1\right) = 2\left(\sqrt{7}\right)^2 + 2\sqrt{7} - 4\sqrt{7} - 4$ FOIL Method

$\qquad = 2(7) + (2 - 4)\sqrt{7} - 4$ Combine like radicals.

$\qquad = 10 - 2\sqrt{7}$ Simplify.

(b) $\left(3 - \sqrt{x}\right)\left(1 + \sqrt{x}\right) = 3 + 3\sqrt{x} - \sqrt{x} - \left(\sqrt{x}\right)^2$ FOIL Method

$\qquad = 3 + 2\sqrt{x} - x$ Combine like radicals.

Conjugates

The expressions $3 + \sqrt{6}$ and $3 - \sqrt{6}$ are called **conjugates** of each other. Notice that they differ only in the sign between the terms. The product of two conjugates is the difference of two squares, which is given by the special product formula $(a + b)(a - b) = a^2 - b^2$. Here are some other examples.

Expression	Conjugate	Product
$1 - \sqrt{3}$	$1 + \sqrt{3}$	$(1)^2 - \left(\sqrt{3}\right)^2 = 1 - 3 = -2$
$\sqrt{5} + \sqrt{2}$	$\sqrt{5} - \sqrt{2}$	$\left(\sqrt{5}\right)^2 - \left(\sqrt{2}\right)^2 = 5 - 2 = 3$
$\sqrt{10} - 3$	$\sqrt{10} + 3$	$\left(\sqrt{10}\right)^2 - (3)^2 = 10 - 9 = 1$
$\sqrt{x} + 2$	$\sqrt{x} - 2$	$\left(\sqrt{x}\right)^2 - (2)^2 = x - 4, \quad x \geq 0$

Example 4 ■ Multiplying Conjugates

Find the conjugate of the expression and multiply the expression by its conjugate.

(a) $2 - \sqrt{5}$ (b) $\sqrt{3} + \sqrt{x}$

Solution

(a) The conjugate of $2 - \sqrt{5}$ is $2 + \sqrt{5}$.

$\left(2 - \sqrt{5}\right)\left(2 + \sqrt{5}\right) = 2^2 - \left(\sqrt{5}\right)^2$ Special product formula

$\qquad = 4 - 5 = -1$

(b) The conjugate of $\sqrt{3} + \sqrt{x}$ is $\sqrt{3} - \sqrt{x}$

$\left(\sqrt{3} + \sqrt{x}\right)\left(\sqrt{3} - \sqrt{x}\right) = \left(\sqrt{3}\right)^2 - \left(\sqrt{x}\right)^2$ Special product formula

$\qquad = 3 - x, \quad x \geq 0$

Technology

Remember that you can use a graphing utility to check simplifications. For instance, to check the result of Example 4(b), use a graphing utility to graph

$$y_1 = \left(\sqrt{3} + \sqrt{x}\right)\left(\sqrt{3} - \sqrt{x}\right)$$

and

$$y_2 = 3 - x.$$

Both equations have the same graph, as shown below.

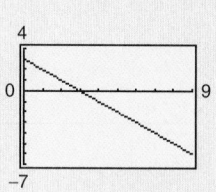

Dividing Radical Expressions

To simplify a *quotient* involving radicals, you rationalize the denominator. For single-term denominators, you can use the rationalizing process described in Section 5.3. To rationalize a denominator involving two terms, multiply both the numerator and denominator by the *conjugate* of the denominator, as demonstrated in Examples 5 and 6.

Example 5 ■ Simplifying Quotients Involving Radicals

Simplify each expression.

(a) $\dfrac{\sqrt{3}}{1 - \sqrt{5}}$ (b) $\dfrac{4}{2 - \sqrt{3}}$ (c) $\dfrac{5\sqrt{2}}{\sqrt{7} + \sqrt{2}}$

Solution

(a)
$$\dfrac{\sqrt{3}}{1 - \sqrt{5}} = \dfrac{\sqrt{3}}{1 - \sqrt{5}} \cdot \dfrac{1 + \sqrt{5}}{1 + \sqrt{5}}$$
Multiply numerator and denominator by conjugate of denominator.

$$= \dfrac{\sqrt{3}\left(1 + \sqrt{5}\right)}{1^2 - \left(\sqrt{5}\right)^2}$$
Special product formula

$$= \dfrac{\sqrt{3} + \sqrt{15}}{1 - 5}$$
Simplify.

$$= \dfrac{\sqrt{3} + \sqrt{15}}{-4}$$
Simplify.

$$= -\dfrac{\sqrt{3} + \sqrt{15}}{4}$$

(b)
$$\dfrac{4}{2 - \sqrt{3}} = \dfrac{4}{2 - \sqrt{3}} \cdot \dfrac{2 + \sqrt{3}}{2 + \sqrt{3}}$$
Multiply numerator and denominator by conjugate of denominator.

$$= \dfrac{4\left(2 + \sqrt{3}\right)}{2^2 - \left(\sqrt{3}\right)^2}$$
Special product formula

$$= \dfrac{8 + 4\sqrt{3}}{4 - 3}$$
Simplify.

$$= 8 + 4\sqrt{3}$$
Simplify.

(c)
$$\dfrac{5\sqrt{2}}{\sqrt{7} + \sqrt{2}} = \dfrac{5\sqrt{2}}{\sqrt{7} + \sqrt{2}} \cdot \dfrac{\sqrt{7} - \sqrt{2}}{\sqrt{7} - \sqrt{2}}$$
Multiply numerator and denominator by conjugate of denominator.

$$= \dfrac{5\left(\sqrt{14} - \sqrt{4}\right)}{\left(\sqrt{7}\right)^2 - \left(\sqrt{2}\right)^2}$$
Special product formula

$$= \dfrac{5\left(\sqrt{14} - 2\right)}{7 - 2}$$
Simplify.

$$= \dfrac{5\left(\sqrt{14} - 2\right)}{5}$$
Divide out common factor.

$$= \sqrt{14} - 2$$
Simplest form

Example 6 ■ Dividing Radical Expressions

Perform each division and simplify.

(a) $6 \div \left(\sqrt{x} - 2\right)$ (b) $\left(5 - \sqrt{3}\right) \div \left(\sqrt{6} + \sqrt{2}\right)$

(c) $1 \div \left(\sqrt{x} - \sqrt{x + 1}\right)$

Solution

(a) $\dfrac{6}{\sqrt{x} - 2} = \dfrac{6}{\sqrt{x} - 2} \cdot \dfrac{\sqrt{x} + 2}{\sqrt{x} + 2}$ Multiply numerator and denominator by conjugate of denominator.

$= \dfrac{6\left(\sqrt{x} + 2\right)}{\left(\sqrt{x}\right)^2 - 2^2}$ Special product formula

$= \dfrac{6\sqrt{x} + 12}{x - 4}$ Simplify.

(b) $\dfrac{5 - \sqrt{3}}{\sqrt{6} + \sqrt{2}} = \dfrac{5 - \sqrt{3}}{\sqrt{6} + \sqrt{2}} \cdot \dfrac{\sqrt{6} - \sqrt{2}}{\sqrt{6} - \sqrt{2}}$ Multiply numerator and denominator by conjugate of denominator.

$= \dfrac{5\sqrt{6} - 5\sqrt{2} - \sqrt{18} + \sqrt{6}}{\left(\sqrt{6}\right)^2 - \left(\sqrt{2}\right)^2}$ FOIL Method and special product formula

$= \dfrac{6\sqrt{6} - 5\sqrt{2} - 3\sqrt{2}}{6 - 2}$ Simplify.

$= \dfrac{6\sqrt{6} - 8\sqrt{2}}{4}$ Simplify.

$= \dfrac{3\sqrt{6} - 4\sqrt{2}}{2}$ Simplify.

(c) $\dfrac{1}{\sqrt{x} - \sqrt{x + 1}} = \dfrac{1}{\sqrt{x} - \sqrt{x + 1}} \cdot \dfrac{\sqrt{x} + \sqrt{x + 1}}{\sqrt{x} + \sqrt{x + 1}}$ Multiply numerator and denominator by conjugate of denominator.

$= \dfrac{\sqrt{x} + \sqrt{x + 1}}{\left(\sqrt{x}\right)^2 - \left(\sqrt{x + 1}\right)^2}$ Special product formula

$= \dfrac{\sqrt{x} + \sqrt{x + 1}}{x - (x + 1)}$ Simplify.

$= \dfrac{\sqrt{x} + \sqrt{x + 1}}{-1}$ Combine like terms.

$= -\sqrt{x} - \sqrt{x + 1}$ Simplify.

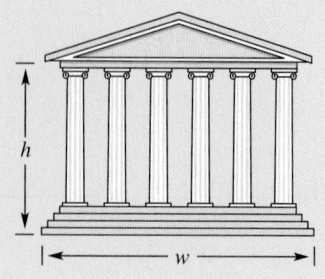

Collaborate!

The Golden Section

The ratio of the width of the Temple of Hephaestus to its height is approximately $2/\left(\sqrt{5} - 1\right)$. This number is called the **golden section**. Early Greeks believed that the most aesthetically pleasing rectangles were those whose sides had this ratio.

a. Rationalize the denominator for this number. Approximate your answer, rounded to two decimal places.

b. Use the Pythagorean Theorem, a straightedge, and a compass to construct a rectangle whose sides have the golden section as their ratio.

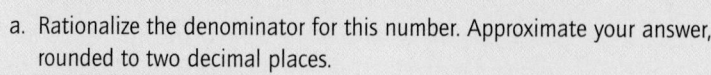

5.4 ■ Exercises

Developing Skills

In Exercises 1–32, multiply and simplify.

1. $\sqrt{2} \cdot \sqrt{8}$ **2.** $\sqrt{6} \cdot \sqrt{18}$

3. $\sqrt{3} \cdot \sqrt{6}$ **4.** $\sqrt{5} \cdot \sqrt{10}$

5. $\sqrt{5}(2 - \sqrt{3})$ **6.** $\sqrt{11}(\sqrt{5} - 3)$

7. $\sqrt{2}(\sqrt{20} + 8)$ **8.** $\sqrt{7}(\sqrt{14} + 3)$

9. $(\sqrt{3} + 2)(\sqrt{3} - 2)$

10. $(\sqrt{15} + 3)(\sqrt{15} - 3)$

11. $(3 - \sqrt{5})(3 + \sqrt{5})$

12. $(\sqrt{11} + 3)(\sqrt{11} - 3)$

13. $(2\sqrt{2} + \sqrt{4})(2\sqrt{2} - \sqrt{4})$

14. $(4\sqrt{3} + \sqrt{2})(4\sqrt{3} - \sqrt{2})$

15. $(\sqrt{20} + 2)^2$ **16.** $(4 - \sqrt{20})^2$

17. $(10 + \sqrt{2x})^2$ **18.** $(5 - \sqrt{3v})^2$

19. $\sqrt{y}(\sqrt{y} + 4)$ **20.** $\sqrt{x}(5 - \sqrt{x})$

21. $(\sqrt{5} + 3)(\sqrt{3} - 5)$ **22.** $(\sqrt{30} + 6)(\sqrt{2} + 6)$

23. $(9\sqrt{x} + 2)(5\sqrt{x} - 3)$ **24.** $(16\sqrt{u} - 3)(\sqrt{u} - 1)$

25. $(\sqrt{x} + \sqrt{y})(\sqrt{x} - \sqrt{y})$

26. $(3\sqrt{u} + \sqrt{3v})(3\sqrt{u} - \sqrt{3v})$

27. $\sqrt[3]{4}(\sqrt[3]{2} - 7)$ **28.** $(\sqrt[3]{9} + 5)(\sqrt[3]{5} - 5)$

29. $(\sqrt[3]{2x} + 5)^2$ **30.** $(\sqrt[3]{y} + 2)(\sqrt[3]{y^2} - 5)$

31. $(\sqrt[3]{2y} + 10)(\sqrt[3]{4y^2} - 10)$

32. $(\sqrt[3]{t} + 1)(\sqrt[3]{t^2} + 4\sqrt[3]{t} - 3)$

In Exercises 33–38, complete the statement.

33. $5x\sqrt{3} + 15\sqrt{3} = 5\sqrt{3}()$

34. $x\sqrt{7} - x^2\sqrt{7} = x\sqrt{7}()$

35. $4\sqrt{12} - 2x\sqrt{27} = 2\sqrt{3}()$

36. $5\sqrt{50} + 10y\sqrt{8} = 5\sqrt{2}()$

37. $6u^2 + \sqrt{18u^3} = 3u()$

38. $12s^3 - \sqrt{32s^4} = 4s^2()$

In Exercises 39–44, simplify the expression.

39. $\dfrac{4 - 8\sqrt{x}}{12}$ **40.** $\dfrac{-3 + 27\sqrt{2y}}{18}$

41. $\dfrac{-2y + \sqrt{12y^3}}{8y}$ **42.** $\dfrac{3x - \sqrt{18x^3}}{3x}$

43. $\dfrac{-4x^2 + \sqrt{28x^3}}{2x}$ **44.** $\dfrac{-t^2 - \sqrt{2t^3}}{3t}$

In Exercises 45–56, find the conjugate of the expression. Then multiply the expression by its conjugate and simplify.

45. $2 + \sqrt{5}$ **46.** $12 - \sqrt{7}$

47. $\sqrt{6} + 10$ **48.** $\sqrt{2} - 9$

49. $\sqrt{11} - \sqrt{3}$ **50.** $\sqrt{10} + \sqrt{7}$

51. $\sqrt{x} - 3$ **52.** $\sqrt{t} + 7$

53. $\sqrt{2u} - \sqrt{3}$ **54.** $\sqrt{5a} + \sqrt{2}$

55. $\sqrt{6x} + \sqrt{y}$ **56.** $\sqrt{a} - \sqrt{7b}$

In Exercises 57–80, rationalize the denominator of the expression and simplify.

57. $\dfrac{6}{\sqrt{11} - 2}$ **58.** $\dfrac{5}{3 - \sqrt{6}}$

59. $\dfrac{7}{5 + \sqrt{3}}$ **60.** $\dfrac{3}{2\sqrt{10} - 5}$

61. $\dfrac{8}{\sqrt{7} + 3}$ **62.** $\dfrac{4}{3\sqrt{5} - 1}$

63. $\dfrac{2}{\sqrt{6} + \sqrt{2}}$ **64.** $\dfrac{10}{\sqrt{9} + \sqrt{5}}$

65. $\dfrac{2}{6 + \sqrt{2}}$ **66.** $\dfrac{44\sqrt{5}}{3\sqrt{5} - 1}$

67. $(\sqrt{7} + 2) \div (\sqrt{7} - 2)$

68. $(5 - 3\sqrt{3}) \div (3 + \sqrt{3})$

69. $\dfrac{3x}{\sqrt{15} - \sqrt{3}}$ **70.** $\dfrac{6(y + 1)}{y^2 + \sqrt{y}}$

71. $\dfrac{2t^2}{\sqrt{5t} - \sqrt{t}}$ **72.** $\dfrac{5x}{\sqrt{x} - \sqrt{2}}$

73. $\dfrac{7z}{\sqrt{5z} - \sqrt{z}}$ **74.** $\dfrac{8a}{\sqrt{3a} + \sqrt{a}}$

75. $\dfrac{2\sqrt{x} - 1}{\sqrt{x} + 2}$ **76.** $\dfrac{3\sqrt{x} - 4}{\sqrt{x} - 5}$

77. $(\sqrt{x} - 5) \div (2\sqrt{x} - 1)$

78. $(2\sqrt{t} + 1) \div (2\sqrt{t} - 1)$

79. $\dfrac{\sqrt{u + v}}{\sqrt{u - v} - \sqrt{u}}$ **80.** $\dfrac{z}{\sqrt{u + z} - \sqrt{u}}$

In Exercises 81–84, evaluate the function.

81. $f(x) = x^2 - 6x + 1$

(a) $f(2 - \sqrt{3})$ (b) $f(3 - 2\sqrt{2})$

82. $g(x) = x^2 + 8x + 11$

(a) $g(-4 + \sqrt{5})$ (b) $g(-4\sqrt{2})$

83. $f(x) = x^2 - 2x - 1$

(a) $f(1 + \sqrt{2})$ (b) $f(\sqrt{4})$

84. $g(x) = x^2 - 4x + 1$

(a) $g(1 + \sqrt{5})$ (b) $g(2 - \sqrt{3})$

In Exercises 85–88, use a graphing utility to graph the equations in the same viewing window. Use the graphs to verify that the expressions are equivalent. Verify your results algebraically.

85. $y_1 = \dfrac{10}{\sqrt{x} + 1}$

$y_2 = \dfrac{10(\sqrt{x} - 1)}{x - 1}$

86. $y_1 = \dfrac{4x}{\sqrt{x} + 4}$

$y_2 = \dfrac{4x(\sqrt{x} - 4)}{x - 16}$

87. $y_1 = \dfrac{2\sqrt{x}}{2 - \sqrt{x}}$

$y_2 = \dfrac{2(2\sqrt{x} + x)}{4 - x}$

88. $y_1 = \dfrac{\sqrt{2x} + 6}{\sqrt{2x} - 2}$

$y_2 = \dfrac{x + 6 + 4\sqrt{2x}}{x - 2}$

Solving Problems

Geometry   In Exercises 89 and 90, find the area of the rectangle.

89.

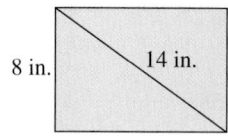

8 in.   14 in.

90.

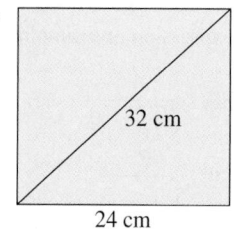

32 cm

24 cm

91. *Area of a Cross Section*   The rectangular cross section of a wooden beam cut from a log of diameter 24 inches (see figure) will have maximum strength if its width w and height h are given by

$$w = 8\sqrt{3}$$

and

$$h = \sqrt{24^2 - (8\sqrt{3})^2}.$$

Find the area of the rectangular cross section and express the area in simplest form.

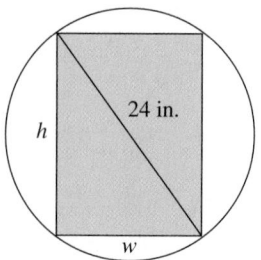

h 24 in. w

92. *Area of a Cross Section*   Suppose that the wooden beam in Exercise 91 is to be cut from the log of diameter 24 inches so that the cross section is square (see figure). Find the dimensions of the square cross section, then find the area of the cross section. Express the area in simplest form.

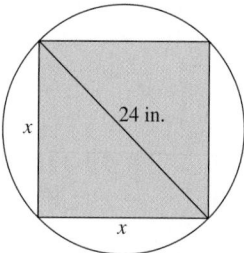

x 24 in. x

93. *Geometry*   The areas of the circles in the figure are 15 square centimeters and 20 square centimeters. Find the ratio of the radius of the small circle to the radius of the large circle.

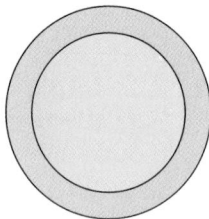

94. *Geometry*   Suppose the circles in Exercise 93 have areas of 18 and 36 square centimeters. Find the ratio of the radius of the small circle to the radius of the large circle.

95. *Force*   The force required to slide a 500-pound steel block across a milling machine is

$$\frac{500k}{\sqrt{k^2 + 1}}$$

where k is the friction constant (see figure). Simplify this expression.

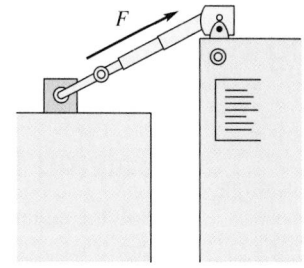

Figure for 95

Explaining Concepts

96. Multiply $\sqrt{3}(1 - \sqrt{6})$. State an algebraic property to justify each step.

97. Describe the differences and similarities of using the FOIL Method with polynomial expressions and with radical expressions.

98. Multiply $3 - \sqrt{2}$ by its conjugate. Explain why the result has no radicals.

Ongoing Review

In Exercises 99–104, perform the indicated operations and simplify.

99. $12 + 4(x - 7)$

100. $15 - 5(2x + 1)$

101. $\left(x + \frac{1}{2}\right)^2$

102. $(5x - 6)^2$

103. $(3x + 8)(3x - 8)$

104. $\left(\frac{1}{3} + 2x\right)(6 - x)$

In Exercises 105–108, identify the transformation of f and sketch a graph of the function g.

105. $f(x) = x^2,\ g(x) = x^2 + 8$

106. $f(x) = x^3,\ g(x) = (x - 5)^3$

107. $f(x) = \sqrt{x},\ g(x) = \sqrt{x - 1} + 1$

108. $f(x) = |x|,\ g(x) = -|x + 4|$

In Exercises 109–112, simplify the expression. (Assume that all variables are positive.)

109. $\sqrt{128x^4y^7}$

110. $\sqrt[3]{54a^5b^7}$

111. $\sqrt{32x^3} + 5x\sqrt{2x}$

112. $\dfrac{9x}{\sqrt{5}}$

113. *Free-Falling Object*   An object is thrown downward from a height of 120 feet with an initial velocity of 56 feet per second. Find the time t for the object to reach the ground by solving the equation

$$-16t^2 - 56t + 120 = 0.$$

114. *Consumer Awareness*   A suit sells for \$375 during a 25% storewide clearance sale. What was the original price of the suit?

115. *Mixture Problem*   Determine the number of gallons of a 30% solution that must be mixed with a 60% solution to obtain 20 gallons of a 40% solution.

Looking Further

Rationalizing Numerators   To simplify a quotient involving radicals, you rationalize the denominator. In the study of calculus, students sometimes rewrite an expression by rationalizing the numerator. For each of the expressions, rationalize the numerator. (*Note:* Your results will not be in simplest radical form.)

(a) $\dfrac{\sqrt{2}}{7}$

(b) $\dfrac{\sqrt{7} + \sqrt{3}}{5}$

(c) $\dfrac{\sqrt{x} + 6}{\sqrt{2}}$

5.5 Solving Radical Equations

Solving Radical Equations

Solving equations involving radicals is somewhat like solving equations that contain fractions—you try to get rid of the radicals and obtain a linear or quadratic equation. Then you solve the equation using the standard procedures. The property presented below plays a key role.

Raising Each Side of an Equation to the *n*th Power

Let u and v be real numbers, variables, or algebraic expressions, and let n be a positive integer. If $u = v$, then it follows that

$$u^n = v^n.$$

This is called **raising each side of an equation to the *n*th power.**

To use this property to solve an equation, first try to isolate one of the radicals on one side of the equation. When you are using this property to solve radical equations, it is critical to check your solutions in the *original* equation.

Example 1 ■ Solving an Equation Having One Radical

Solve $\sqrt{x} - 8 = 0$.

Algebraic Solution

$$\sqrt{x} - 8 = 0 \qquad \text{Write original equation.}$$
$$\sqrt{x} = 8 \qquad \text{Isolate radical.}$$
$$\left(\sqrt{x}\right)^2 = 8^2 \qquad \text{Square each side.}$$
$$x = 64 \qquad \text{Simplify.}$$

Check

$$\sqrt{x} - 8 = 0 \qquad \text{Write original equation.}$$
$$\sqrt{64} - 8 \overset{?}{=} 0 \qquad \text{Substitute 64 for } x.$$
$$8 - 8 = 0 \qquad \text{Solution checks. } \checkmark$$

So, the equation has one solution: $x = 64$.

Graphical Solution

Use a graphing utility to graph

$$y = \sqrt{x} - 8.$$

Use the *zoom* and *trace* features of the graphing utility to approximate the x-intercepts of the graph (the values of x for which y is equal to zero). From the graph in Figure 5.5, it appears that y is equal to zero when $x = 64$. This is the only solution of the equation $\sqrt{x} - 8 = 0$.

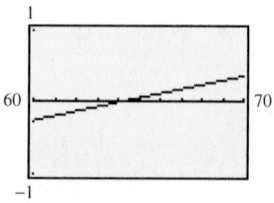

Figure 5.5

The next example demonstrates another reason for checking solutions in the *original* equation. Even with no mistakes in the solution process, it can happen that a "trial solution" does not satisfy the original equation. This type of "solution" is called **extraneous.** An extraneous solution does not satisfy the original equation, and therefore *must not* be listed as an actual solution.

Example 2 ■ Solving an Equation Having One Radical

Solve $\sqrt{3x} + 6 = 0$.

Solution

$$\sqrt{3x} + 6 = 0 \qquad \text{Write original equation.}$$
$$\sqrt{3x} = -6 \qquad \text{Isolate radical.}$$
$$\left(\sqrt{3x}\right)^2 = (-6)^2 \qquad \text{Square each side.}$$
$$3x = 36 \qquad \text{Simplify.}$$
$$x = 12 \qquad \text{Divide each side by 3.}$$

Check

$$\sqrt{3x} + 6 = 0 \qquad \text{Write original equation.}$$
$$\sqrt{3(12)} + 6 \overset{?}{=} 0 \qquad \text{Substitute 12 for } x.$$
$$6 + 6 \neq 0 \qquad \text{Solution does not check. ✗}$$

So, $x = 12$ is an *extraneous* solution, which means that the original equation has no solution. You can also check this graphically, as shown in Figure 5.6. Notice that the graph does not cross the x-axis and so has no x-intercept.

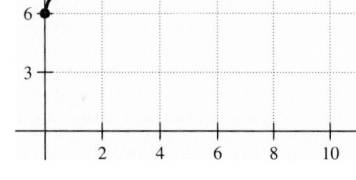

Figure 5.6

Example 3 ■ Solving an Equation Having One Radical

Solve $\sqrt[3]{2x + 1} - 2 = 3$.

Solution

$$\sqrt[3]{2x + 1} - 2 = 3 \qquad \text{Write original equation.}$$
$$\sqrt[3]{2x + 1} = 5 \qquad \text{Isolate radical.}$$
$$\left(\sqrt[3]{2x + 1}\right)^3 = 5^3 \qquad \text{Cube each side.}$$
$$2x + 1 = 125 \qquad \text{Simplify.}$$
$$2x = 124 \qquad \text{Subtract 1 from each side.}$$
$$x = 62 \qquad \text{Divide each side by 2.}$$

Check

$$\sqrt[3]{2x + 1} - 2 = 3 \qquad \text{Write original equation.}$$
$$\sqrt[3]{2(62) + 1} - 2 \overset{?}{=} 3 \qquad \text{Substitute 62 for } x.$$
$$\sqrt[3]{125} - 2 \overset{?}{=} 3 \qquad \text{Simplify.}$$
$$5 - 2 = 3 \qquad \text{Solution checks. ✓}$$

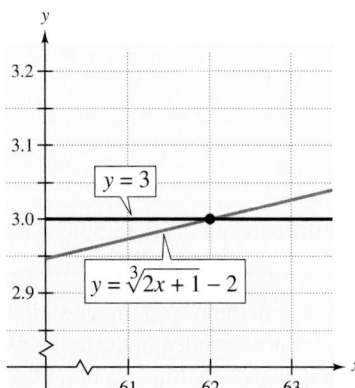

Figure 5.7

So, the equation has one solution: $x = 62$. You can also check the solution graphically by determining the point of intersection of the graphs of $y = \sqrt[3]{2x + 1} - 2$ (left side of equation) and $y = 3$ (right side of equation), as shown in Figure 5.7.

Example 4 ■ Solving an Equation Having Two Radicals

Solve $\sqrt{5x + 3} = \sqrt{x + 11}$.

Algebraic Solution

$$\sqrt{5x + 3} = \sqrt{x + 11} \qquad \text{Write original equation.}$$
$$\left(\sqrt{5x + 3}\right)^2 = \left(\sqrt{x + 11}\right)^2 \qquad \text{Square each side.}$$
$$5x + 3 = x + 11 \qquad \text{Simplify.}$$
$$5x = x + 8 \qquad \text{Subtract 3 from each side.}$$
$$4x = 8 \qquad \text{Subtract } x \text{ from each side.}$$
$$x = 2 \qquad \text{Divide each side by 4.}$$

Check

$$\sqrt{5x + 3} = \sqrt{x + 11} \qquad \text{Write original equation.}$$
$$\sqrt{5(2) + 3} \stackrel{?}{=} \sqrt{2 + 11} \qquad \text{Substitute 2 for } x.$$
$$\sqrt{13} = \sqrt{13} \qquad \text{Solution checks. } \checkmark$$

So, the equation has one solution: $x = 2$.

Graphical Solution

Use a graphing utility to graph $y_1 = \sqrt{5x + 3}$ and $y_2 = \sqrt{x + 11}$ in the same viewing window. Use the *intersect* feature of the graphing utility to approximate the points where the graphs intersect. From Figure 5.8, it appears that the graphs intersect when $x = 2$. This x-coordinate of the intersection point is the solution of the equation $\sqrt{5x + 3} = \sqrt{x + 11}$.

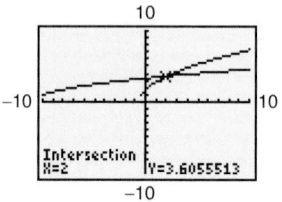

Figure 5.8

Example 5 ■ Solving an Equation Having Two Radicals

Solve $\sqrt[4]{3x} - \sqrt[4]{2x - 5} = 0$.

Solution

$$\sqrt[4]{3x} - \sqrt[4]{2x - 5} = 0 \qquad \text{Write original equation.}$$
$$\sqrt[4]{3x} = \sqrt[4]{2x - 5} \qquad \text{Isolate radicals.}$$
$$\left(\sqrt[4]{3x}\right)^4 = \left(\sqrt[4]{2x - 5}\right)^4 \qquad \text{Raise each side to 4th power.}$$
$$3x = 2x - 5 \qquad \text{Simplify.}$$
$$x = -5 \qquad \text{Subtract } 2x \text{ from each side.}$$

Check

$$\sqrt[4]{3x} - \sqrt[4]{2x - 5} = 0 \qquad \text{Write original equation.}$$
$$\sqrt[4]{3(-5)} - \sqrt[4]{2(-5) - 5} \stackrel{?}{=} 0 \qquad \text{Substitute } -5 \text{ for } x.$$
$$\sqrt[4]{-15} - \sqrt[4]{-15} \neq 0 \qquad \text{Solution does not check. } \times$$

The solution does not check because it yields fourth roots of negative radicands.

Your knowledge of the domain of a function will help you in checking solutions of equations. Notice in Example 5 that -5 is not included in the domains of the radicals in the original equation. So, you can conclude that the equation has no solution.

In the next example you will see that squaring each side of the equation can result in a quadratic equation. Remember that you must check the solutions in the *original* radical equation.

EXPLORATION

Try solving the equation in Example 5 graphically. First isolate the radicals, then use a graphing utility to graph each side of the equation in the same viewing window. Do the graphs intersect? What does an extraneous solution mean graphically?

Example 6 ■ An Equation That Converts to a Quadratic Equation

Solve $\sqrt{x} + 2 = x$.

Solution

$\sqrt{x} + 2 = x$	Write original equation.
$\sqrt{x} = x - 2$	Isolate radical.
$\left(\sqrt{x}\right)^2 = (x - 2)^2$	Square each side.
$x = x^2 - 4x + 4$	Simplify.
$0 = x^2 - 5x + 4$	Write in standard form.
$0 = (x - 4)(x - 1)$	Factor.
$x - 4 = 0 \implies x = 4$	Set 1st factor equal to 0.
$x - 1 = 0 \implies x = 1$	Set 2nd factor equal to 0.

EXPLORATION

Graph the system $y_1 = x$ and $y_2 = \sqrt{x} + c$ on a graphing utility using different values of c. What values of c correspond to one solution? What values of c correspond to two solutions? What values of c correspond to no solution? What effect does the value of c have on the number of solutions to the system?

Check First Solution

$$\sqrt{x} + 2 = x$$
$$\sqrt{4} + 2 \overset{?}{=} 4$$
$$2 + 2 \overset{?}{=} 4$$
$$4 = 4$$

Check Second Solution

$$\sqrt{x} + 2 = x$$
$$\sqrt{1} + 2 \overset{?}{=} 1$$
$$1 + 2 \overset{?}{=} 1$$
$$3 \neq 1$$

From the check you can see that $x = 1$ is an extraneous solution. So, the only solution is $x = 4$.

When an equation contains two radicals, it may not be possible to isolate both. In such cases, you may have to raise each side of the equation to a power at *two* different stages in the solution.

Example 7 ■ Repeatedly Squaring Each Side of an Equation

Solve $\sqrt{3t + 1} = 2 - \sqrt{3t}$.

Solution

$\sqrt{3t + 1} = 2 - \sqrt{3t}$	Write original equation.
$\left(\sqrt{3t + 1}\right)^2 = \left(2 - \sqrt{3t}\right)^2$	Square each side (1st time).
$3t + 1 = 4 - 4\sqrt{3t} + 3t$	Simplify.
$-3 = -4\sqrt{3t}$	Isolate radical.
$(-3)^2 = \left(-4\sqrt{3t}\right)^2$	Square each side (2nd time).
$9 = 16(3t)$	Simplify.
$9 = 48t$	Multiply.
$\dfrac{9}{48} = t$	Divide each side by 48.
$\dfrac{3}{16} = t$	Simplify.

The solution is $t = \frac{3}{16}$. Check this in the original equation.

Example 8 ■ Repeatedly Squaring Each Side of an Equation

Solve $\sqrt{x^2 + 11} - \sqrt{x^2 - 9} = 2$.

Solution

$$\sqrt{x^2 + 11} - \sqrt{x^2 - 9} = 2 \qquad \text{Write original equation.}$$

$$\sqrt{x^2 + 11} = 2 + \sqrt{x^2 - 9} \qquad \text{Isolate one of the radicals.}$$

$$\left(\sqrt{x^2 + 11}\right)^2 = \left(2 + \sqrt{x^2 - 9}\right)^2 \qquad \text{Square each side (1st time).}$$

$$x^2 + 11 = 4 + 4\sqrt{x^2 - 9} + (x^2 - 9)$$

$$16 = 4\sqrt{x^2 - 9} \qquad \text{Simplify.}$$

$$4 = \sqrt{x^2 - 9} \qquad \text{Isolate radical.}$$

$$4^2 = \left(\sqrt{x^2 - 9}\right)^2 \qquad \text{Square each side (2nd time).}$$

$$16 = x^2 - 9 \qquad \text{Simplify.}$$

$$0 = x^2 - 25 \qquad \text{Write in standard form.}$$

$$0 = (x + 5)(x - 5) \qquad \text{Factor.}$$

By setting each factor equal to zero, you can see that this equation has two solutions: $x = 5$ and $x = -5$. Check these in the original equation.

Applications

Example 9 ■ An Application Involving Electricity

The power consumed by an electrical appliance is given by

$$I = \sqrt{\frac{P}{R}}$$

where I is the current measured in amps, R is the resistance measured in ohms, and P is the power measured in watts. Find the power used by an electric heater for which $I = 10$ amps and $R = 16$ ohms.

Solution

$$I = \sqrt{\frac{P}{R}} \qquad \text{Write original equation.}$$

$$10 = \sqrt{\frac{P}{16}} \qquad \text{Substitute 10 for } I \text{ and 16 for } R.$$

$$10^2 = \left(\sqrt{\frac{P}{16}}\right)^2 \qquad \text{Square each side.}$$

$$100 = \frac{P}{16} \qquad \text{Simplify.}$$

$$1600 = P \qquad \text{Multiply each side by 16.}$$

So, the solution is $P = 1600$ watts. Check this in the original equation.

> **NOTE** An alternative way to solve the problem in Example 9 would be first to solve the equation for P.
>
> $$I = \sqrt{\frac{P}{R}}$$
>
> $$I^2 = \left(\sqrt{\frac{P}{R}}\right)^2$$
>
> $$I^2 = \frac{P}{R}$$
>
> $$I^2 R = P$$
>
> At this stage, you can substitute the known values of I and R to obtain
>
> $$P = (10)^2 16 = 1600.$$

Example 10 ■ The Velocity of a Falling Object

The velocity of a free-falling object can be determined from the equation

$$v = \sqrt{2gh}$$

where v is the velocity measured in feet per second, $g = 32$ feet per second per second, and h is the distance (in feet) the object has fallen. Find the height from which a rock has been dropped if it strikes the ground with a velocity of 50 feet per second.

Solution

$v = \sqrt{2gh}$	Write original equation.
$50 = \sqrt{2(32)h}$	Substitute 50 for v and 32 for g.
$(50)^2 = \left(\sqrt{64h}\right)^2$	Square each side.
$2500 = 64h$	Simplify.
$39 \approx h$	Divide each side by 64.

Check

Because the value of h was rounded in the solution, the check will not result in an equality. The expressions on each side of the equal sign will be approximately equal to each other.

$v = \sqrt{2gh}$	Write original equation.
$50 \stackrel{?}{\approx} \sqrt{2(32)(39)}$	Substitute 50 for v, 32 for g, and 39 for h.
$50 \stackrel{?}{\approx} \sqrt{2496}$	Simplify.
$50 \approx 49.96$	Solution checks. ✓

So, the height from which the rock has been dropped is approximately 39 feet.

Collaborate!

Extending the Concept

Without using a stopwatch, you can find the length of time an object has been falling by using the equation from physics

$$t = \sqrt{\frac{h}{384}}$$

where t is the length of time in seconds and h is the height in inches the object has fallen. How far does an object fall in 0.25 second? in 0.10 second?

Use this equation to test how long it takes members of your group to catch a falling ruler. Hold the ruler vertically while another group member holds his or her hands near the lower end of the ruler ready to catch it. Before releasing the ruler, record the mark on the ruler closest to the top of the catcher's hands. Release the ruler. After it has been caught, again note the mark closest to the top of the catcher's hands. (The difference between these two measurements is h.) Which member of your group reacts most quickly?

5.5 ■ Exercises

Developing Skills

In Exercises 1–4, determine whether the values of x are solutions of the radical equation.

Equation	Values of x	
1. $\sqrt{x} - 10 = 0$	(a) $x = -4$	(b) $x = -100$
	(c) $x = \sqrt{10}$	(d) $x = 100$
2. $\sqrt{3x} - 6 = 0$	(a) $x = \frac{2}{3}$	(b) $x = 2$
	(c) $x = 12$	(d) $x = -\frac{1}{3}\sqrt{6}$
3. $\sqrt[3]{x - 4} = 4$	(a) $x = -60$	(b) $x = 68$
	(c) $x = 20$	(d) $x = 0$
4. $\sqrt[4]{2x} + 2 = 6$	(a) $x = 128$	(b) $x = 2$
	(c) $x = -2$	(d) $x = 0$

In Exercises 5–52, solve the equation. (Some of the equations have no solution.)

5. $\sqrt{x} = 20$ **6.** $\sqrt{x} = 5$

7. $\sqrt{y} - 7 = 0$ **8.** $\sqrt{t} - 13 = 0$

9. $\sqrt{u} + 13 = 0$ **10.** $\sqrt{y} + 15 = 0$

11. $\sqrt{a + 100} = 25$ **12.** $\sqrt{b + 12} = 13$

13. $\sqrt{10x} = 30$ **14.** $\sqrt{8x} = 6$

15. $\sqrt{3y + 5} - 3 = 4$ **16.** $\sqrt{5z - 2} + 7 = 10$

17. $5\sqrt{x + 2} = 8$ **18.** $2\sqrt{x + 4} = 7$

19. $\sqrt{x^2 + 5} = x + 3$ **20.** $\sqrt{x^2 - 4} = x - 2$

21. $\sqrt{2x} = x - 4$ **22.** $\sqrt{x} = x - 6$

23. $\sqrt{4y} = 3 - y$ **24.** $\sqrt{3x} = x - 6$

25. $\sqrt{3y + 1} - 4 = 0$ **26.** $\sqrt{3 - 2x} - 2 = 0$

27. $\sqrt{3x + 2} + 5 = 0$ **28.** $\sqrt{1 - x} + 10 = 4$

29. $\sqrt{x + 3} = \sqrt{2x - 1}$ **30.** $\sqrt{3t + 1} = \sqrt{t + 15}$

31. $\sqrt{2x + 5} = \sqrt{x + 7}$

32. $\sqrt{5y - 1} = \sqrt{4y + 1}$

33. $\sqrt{3y - 5} = 3\sqrt{y}$

34. $\sqrt{3x + 6} = 2\sqrt{x}$

35. $\sqrt{10x - 4} - 3\sqrt{x} = 0$

36. $\sqrt{2u + 10} - 2\sqrt{u} = 0$

37. $\sqrt[3]{3x - 4} = \sqrt[3]{x + 10}$

38. $2\sqrt[3]{10 - 3x} = \sqrt[3]{2 - x}$

39. $\sqrt[3]{2x + 15} - \sqrt[3]{x} = 0$

40. $\sqrt[4]{2x} + \sqrt[4]{x + 3} = 0$

41. $\sqrt{8x + 1} = x + 2$

42. $\sqrt{3x + 7} = x + 3$

43. $\sqrt{5x - 4} = 2 - \sqrt{5x}$

44. $\sqrt{1 + 4a} = 2\sqrt{a} - 3$

45. $\sqrt{z + 2} = 1 + \sqrt{z}$

46. $\sqrt{2x + 5} = 7 - \sqrt{2x}$

47. $\sqrt{x} + \sqrt{x + 2} = 2$

48. $\sqrt{x + 5} - \sqrt{x} = 1$

49. $\sqrt{2t + 3} = 3 - \sqrt{2t}$

50. $\sqrt{6x + 7} = 8 - \sqrt{3x}$

51. $\sqrt{x + 3} + \sqrt{x - 2} = 5$

52. $\sqrt{x + 3} - \sqrt{x - 1} = 1$

In Exercises 53–60, use a graphing utility to graph each side of the equation in the same viewing window. Then use the graphs to approximate the solution. Verify your answer algebraically.

53. $\sqrt{2x + 3} = 4x - 3$ **54.** $\sqrt{x} = 2(2 - x)$

55. $\sqrt{x^2 + 1} = 5 - 2x$ **56.** $\sqrt{8 - 3x} = x$

57. $\sqrt{x + 3} = 5 - \sqrt{x}$ **58.** $\sqrt[3]{5x - 8} = 4 - \sqrt[3]{x}$

59. $\sqrt[3]{x + 4} = \sqrt{6 - x}$ **60.** $4\sqrt[3]{x} = 7 - x$

Solving Problems

Length of a Pendulum   In Exercises 61 and 62, use the equation for the time t in seconds for a pendulum of length L feet to go through one complete cycle (its period). The equation is $t = 2\pi\sqrt{L/32}$.

61. How long is the pendulum of a grandfather clock with a period of 1.5 seconds?

62. How long is the pendulum of a mantle clock with a period of 0.75 second?

63. *Surface Area of a Cone*　The lateral surface area of a cone is $S = \pi r \sqrt{r^2 + h^2}$. Solve this equation for h^2.

64. *Airline Passengers*　An airline offers daily flights between Chicago and Denver. The total monthly cost of the flights is $C = \sqrt{0.2x + 1}$, for $x \geq 0$, where C is measured in millions of dollars and x is measured in thousands of passengers. The total cost of the flights for a certain month was 2.5 million dollars. Approximately how many passengers flew that month?

Free-Falling Object　In Exercises 65–68, use the equation for the velocity of a free-falling object $\left(v = \sqrt{2gh}\right)$, as described in Example 10.

65. An object is dropped from a height of 50 feet. Find the velocity of the object when it strikes the ground.

66. An object is dropped from a height of 200 feet. Find the velocity of the object when it strikes the ground.

67. An object that was dropped strikes the ground with a velocity of 60 feet per second. Find the height from which the object was dropped.

68. An object that was dropped strikes the ground with a velocity of 120 feet per second. Find the height from which the object was dropped.

69. *Demand*　The demand equation for a certain product is $p = 50 - \sqrt{0.8(x - 1)}$, where x is the number of units demanded per day and p is the price per unit. Find the demand if the price is $30.02.

70. *Demand*　The demand equation for a certain product is $p = 12 - \sqrt{0.75(x - 12)}$, where x is the number of units demanded per day and p is the price per unit. Find the demand if the price is $6.00.

Explaining Concepts

71. Does raising each side of an equation to the nth power always yield an equivalent equation? Explain.

72. In your own words, describe the steps that can be used to solve a radical equation.

Ongoing Review

In Exercises 73–78, solve the equation.

73. $9x - 4 = 41$

74. $6x + 24 = 2x$

75. $14 + 2(x - 1) = 0$

76. $23 - 3(2x + 3) = 0$

77. $4(x + 7)(x - 11) = 0$

78. $-3(6 - x)(x + 89) = 0$

79. *Loan Repayment*　You borrow $12,000 for 6 months. You agree to pay back the principal and interest in one lump sum. What will be the amount of the payment if the interest rate is 12%?

Looking Further

The equation $x + \sqrt{x} - 6 = 0$ can be solved using the methods discussed in this section. Alternatively, the substitution $a = \sqrt{x}$ can be made, so that the equation transforms to $a^2 + a - 6 = 0$. This transformed equation is a quadratic equation that can be factored and solved as follows.

$$
\begin{array}{ll}
a^2 + a - 6 = 0 & \text{Write original equation.} \\
(a + 3)(a - 2) = 0 & \text{Factor.} \\
a + 3 = 0 \implies a = -3 & \text{Set 1st factor equal to 0.} \\
a - 2 = 0 \implies a = 2 & \text{Set 2nd factor equal to 0.}
\end{array}
$$

Substituting these values for a back into the equation $a = \sqrt{x}$ yields solutions of $x = 9$ and $x = 4$. Check each of these solutions and you will see that $x = 9$ is extraneous.

Try this substitution method on each of the radical equations below.

(a) $x - 7\sqrt{x} + 12 = 0$

(b) $x - 2\sqrt{x} - 15 = 0$

(c) $2x + \sqrt{x} = 3$

5.6 Complex Numbers

What you should learn:

- How to write the square roots of negative numbers in i-form and perform operations on numbers in i-form
- How to determine the equality of two complex numbers
- How to add, subtract, and multiply complex numbers
- How to use complex conjugates to divide complex numbers

Why you should learn it:

Understanding complex numbers can help you in Section 6.3 to identify quadratic equations that do not have real solutions.

The Imaginary Unit i

In Section 5.2, you learned that a negative number has no *real* square root. For instance, $\sqrt{-1}$ is not real because there is no real number x such that $x^2 = -1$. So, as long as you are dealing only with real numbers, the equation

$$x^2 = -1$$

has no solution. To overcome this deficiency, mathematicians have expanded the set of numbers, using the **imaginary unit i,** defined as

$$i = \sqrt{-1}. \qquad \text{Imaginary unit}$$

This number has the property that $i^2 = -1$. So, the imaginary unit i is a solution of the equation $x^2 = -1$.

The Square Root of a Negative Number

Let c be a positive real number. Then the square root of $-c$ is given by

$$\sqrt{-c} = \sqrt{c(-1)} = \sqrt{c}\sqrt{-1} = \sqrt{c}\,i.$$

When writing $\sqrt{-c}$ in the *i*-form, $\sqrt{c}\,i$, note that i is outside the radical.

Example 1 ■ Writing Numbers in i-Form

Write each number in i-form.

(a) $\sqrt{-36}$ (b) $\sqrt{-\dfrac{16}{25}}$

(c) $\sqrt{-5}$ (d) $\sqrt{-54}$

(e) $\dfrac{\sqrt{-48}}{\sqrt{-3}}$ (f) $\dfrac{\sqrt{-18}}{\sqrt{2}}$

Solution

(a) $\sqrt{-36} = \sqrt{36(-1)} = \sqrt{36}\sqrt{-1} = 6i$

(b) $\sqrt{-\dfrac{16}{25}} = \sqrt{\dfrac{16}{25}(-1)} = \sqrt{\dfrac{16}{25}}\sqrt{-1} = \dfrac{4}{5}i$

(c) $\sqrt{-5} = \sqrt{5(-1)} = \sqrt{5}\sqrt{-1} = \sqrt{5}\,i$

(d) $\sqrt{-54} = \sqrt{54(-1)} = \sqrt{54}\sqrt{-1} = \sqrt{9}\sqrt{6}\sqrt{-1} = 3\sqrt{6}\,i$

(e) $\dfrac{\sqrt{-48}}{\sqrt{-3}} = \dfrac{\sqrt{48}\sqrt{-1}}{\sqrt{3}\sqrt{-1}} = \dfrac{\sqrt{48}\,i}{\sqrt{3}\,i} = \sqrt{\dfrac{48}{3}} = \sqrt{16} = 4$

(f) $\dfrac{\sqrt{-18}}{\sqrt{2}} = \dfrac{\sqrt{18}\sqrt{-1}}{\sqrt{2}} = \dfrac{\sqrt{18}\,i}{\sqrt{2}} = \sqrt{\dfrac{18}{2}}\,i = \sqrt{9}\,i = 3i$

To perform operations with square roots of negative numbers, you must *first* write the numbers in *i*-form. Once the numbers are written in *i*-form, you can add, subtract, and multiply as follows.

$$ai + bi = (a + b)i$$ Addition

$$ai - bi = (a - b)i$$ Subtraction

$$(ai)(bi) = ab(i^2) = ab(-1) = -ab$$ Multiplication

Example 2 ■ Adding Square Roots of Negative Numbers

Perform each operation.

(a) $\sqrt{-9} + \sqrt{-49}$ (b) $6\sqrt{-28} - \sqrt{-63}$

Solution

(a) $\sqrt{-9} + \sqrt{-49} = \sqrt{9}\sqrt{-1} + \sqrt{49}\sqrt{-1}$ Property of radicals

$$= 3i + 7i$$ Write in *i*-form.

$$= 10i$$ Simplify.

(b) $6\sqrt{-28} - \sqrt{-63} = 6\sqrt{28}\sqrt{-1} - \sqrt{63}\sqrt{-1}$ Property of radicals

$$= 6 \cdot 2\sqrt{7}i - 3\sqrt{7}i$$ Write in *i*-form.

$$= 12\sqrt{7}i - 3\sqrt{7}i$$ Simplify.

$$= 9\sqrt{7}i$$ Simplify.

In the next example, notice how you can multiply radicals that involve square roots of negative numbers.

Example 3 ■ Multiplying Square Roots of Negative Numbers

Find each product.

(a) $\sqrt{-15}\sqrt{-15}$ (b) $\sqrt{-5}\left(\sqrt{-45} - \sqrt{-4}\right)$

Solution

(a) $\sqrt{-15}\sqrt{-15} = \left(\sqrt{15}i\right)\left(\sqrt{15}i\right)$ Write in *i*-form.

$$= \left(\sqrt{15}\right)^2 i^2$$ Multiply.

$$= 15(-1)$$ $i^2 = -1$

$$= -15$$ Simplify.

(b) $\sqrt{-5}\left(\sqrt{-45} - \sqrt{-4}\right) = \sqrt{5}i\left(3\sqrt{5}i - 2i\right)$ Write in *i*-form.

$$= \left(\sqrt{5}i\right)\left(3\sqrt{5}i\right) - \left(\sqrt{5}i\right)(2i)$$ Distributive Property

$$= 3(5)(-1) - 2\sqrt{5}(-1)$$ Multiply.

$$= -15 + 2\sqrt{5}$$ Simplify.

When multiplying square roots of negative numbers, be sure to write them in *i*-form *before multiplying*. If you do not do this, you can obtain incorrect answers. For instance, in Example 3(a), be sure you see that

$$\sqrt{-15}\sqrt{-15} \neq \sqrt{(-15)(-15)} = \sqrt{225} = 15.$$

EXPLORATION

When performing operations with numbers in *i*-form, you sometimes need to be able to evaluate powers of the imaginary unit *i*. Evaluate the powers of *i* listed below using $i^2 = -1$.

$$i^1, i^2, i^3, i^4, i^5, i^6, i^7, i^8$$

Describe the pattern. Use the pattern to evaluate i^{16} and i^{19}.

Complex Numbers

NOTE Each of the numbers listed below is a complex number.

$5, 2i, 8 + 3i$

A number of the form $a + bi$, where a and b are real numbers, is called a **complex number.** The real number a is called the **real part** of the complex number $a + bi$, and the number bi is called the **imaginary part** of the complex number.

Definition of a Complex Number

If a and b are real numbers, the number $a + bi$ is a **complex number,** and it is said to be written in **standard form.** If $b = 0$, the number $a + bi = a$ is a real number. If $b \neq 0$, the number $a + bi$ is called an **imaginary number.** A number of the form bi, where $b \neq 0$, is called a **pure imaginary number.**

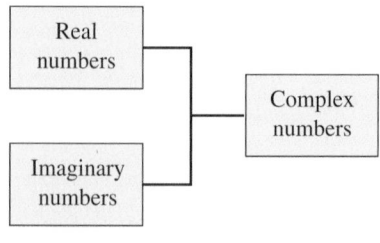

Figure 5.9

A number cannot be both real and imaginary. For instance, the numbers -2, 0, 1, $\frac{1}{2}$, and $\sqrt{2}$ are real numbers (but they are *not* imaginary numbers), and the numbers $-3i$, $2 + 4i$, and $-1 + i$ are imaginary numbers (but they are *not* real numbers). The diagram shown in Figure 5.9 further illustrates the relationships among real, complex, and imaginary numbers.

Two complex numbers $a + bi$ and $c + di$, in standard form, are equal if and only if $a = c$ and $b = d$.

Example 4 ■ Equality of Two Complex Numbers

(a) Are the complex numbers $\sqrt{9} + \sqrt{-48}$ and $3 - 4\sqrt{3}\,i$ equal?

(b) Determine the values of x and y that satisfy the equation

$$3x - \sqrt{-25} = -6 + 3yi.$$

Solution

(a) Begin by writing the first number in standard form.

$$\sqrt{9} + \sqrt{-48} = 3 + \sqrt{48}\sqrt{-1} = 3 + 4\sqrt{3}\,i.$$

From this form, you can see that the two numbers are not equal because they have imaginary parts that differ in sign.

(b) Begin by writing the left side of the equation in standard form.

$$3x - \sqrt{-25} = -6 + 3yi \qquad \text{Write original equation.}$$
$$3x - 5i = -6 + 3yi \qquad \text{Each side in standard form}$$

For these two numbers to be equal, their real parts must be equal to each other and their imaginary parts must be equal to each other.

Real Parts	*Imaginary Parts*
$3x = -6$	$3yi = -5i$
	$3y = -5$
$x = -2$	$y = -\dfrac{5}{3}$

So, $x = -2$ and $y = -\frac{5}{3}$.

Operations with Complex Numbers

To add or subtract two complex numbers, you add (or subtract) the real and imaginary parts separately. This is similar to combining like terms of a polynomial.

$$(a + bi) + (c + di) = (a + c) + (b + d)i \qquad \text{Addition of complex numbers}$$

$$(a + bi) - (c + di) = (a - c) + (b - d)i \qquad \text{Subtraction of complex numbers}$$

Example 5　■　Adding and Subtracting Complex Numbers

Perform each operation and write the result in standard form.

(a) $(3 - i) + (-2 + 4i)$　　(b) $3i + (5 - 3i)$

(c) $4 - (-1 + 5i) + (7 + 2i)$

Solution

(a) $(3 - i) + (-2 + 4i) = (3 - 2) + (-1 + 4)i = 1 + 3i$

(b) $3i + (5 - 3i) = 5 + (3 - 3)i = 5$

(c) $4 - (-1 + 5i) + (7 + 2i) = [4 - (-1) + 7] + (-5 + 2)i = 12 - 3i$

NOTE　In Example 5(b), notice that the sum of two imaginary numbers can be a real number.

The Commutative, Associative, and Distributive Properties of real numbers are also valid for complex numbers.

Example 6　■　Multiplying Complex Numbers

Perform each operation and write the result in standard form.

(a) $(4i)(-7i)$　　　　　(b) $(1 - i)(\sqrt{-9})$

(c) $(2 - i)(4 + 3i)$　　(d) $(3 + 2i)(3 - 2i)$

Solution

(a) $(4i)(-7i) = -28i^2 = -28(-1) = 28$

(b) $(1 - i)(\sqrt{-9}) = (1 - i)(3i)$ 　　　Write in i-form.

$\qquad\qquad = (1)(3i) - (i)(3i)$ 　　Distributive Property

$\qquad\qquad = 3i - 3i^2$ 　　　　　　Simplify.

$\qquad\qquad = 3i - 3(-1)$ 　　　　　$i^2 = -1$

$\qquad\qquad = 3 + 3i$ 　　　　　　Simplify.

(c) $(2 - i)(4 + 3i) = 8 + 6i - 4i - 3i^2$ 　　FOIL Method

$\qquad\qquad = 8 + 6i - 4i - 3(-1)$ 　$i^2 = -1$

$\qquad\qquad = 11 + 2i$ 　　　　　Combine like terms.

(d) $(3 + 2i)(3 - 2i) = 3^2 - (2i)^2$ 　　Special product formula

$\qquad\qquad = 9 - 4i^2$ 　　　　Simplify.

$\qquad\qquad = 9 - 4(-1)$ 　　　$i^2 = -1$

$\qquad\qquad = 13$ 　　　　　　Simplify.

Historical Note

Until very recently it was thought that shapes in nature, such as clouds, coastlines, and mountain ranges, could not be described in mathematical terms. In the 1970s, Benoit Mandelbrot discovered that many of these shapes do have patterns in their irregularity—they are made up of smaller parts that are scaled-down versions of the shapes themselves. Computers using mathematical terms with complex numbers are able to generate the larger images. Mandelbrot coined the term *fractals* for these shapes and for the geometry used to describe them.

Complex Conjugates

In Example 6(d), note that the product of two complex numbers can be a real number. This occurs with pairs of complex numbers of the form $a + bi$ and $a - bi$, called **complex conjugates.** In general, the product of complex conjugates has the form

$$(a + bi)(a - bi) = a^2 - (bi)^2 = a^2 - b^2i^2 = a^2 - b^2(-1) = a^2 + b^2.$$

Here are some examples.

Complex Number	Complex Conjugate	Product
$4 - 5i$	$4 + 5i$	$4^2 + 5^2 = 41$
$3 + 2i$	$3 - 2i$	$3^2 + 2^2 = 13$
$-2 = -2 + 0i$	$-2 = -2 - 0i$	$(-2)^2 + 0^2 = 4$
$i = 0 + i$	$-i = 0 - i$	$0^2 + 1^2 = 1$

Complex conjugates are used to divide one complex number by another. To do this, multiply the numerator and denominator by the *complex conjugate* of the denominator, as shown in Example 7.

Example 7 ■ Division of Complex Numbers

(a) $\dfrac{6 - i}{3i} = \dfrac{6 - i}{3i} \cdot \dfrac{(-3i)}{(-3i)}$ Multiply numerator and denominator by complex conjugate of denominator.

$= \dfrac{-18i + 3i^2}{-9i^2}$ Multiply fractions.

$= \dfrac{-18i + 3(-1)}{-9(-1)}$ $i^2 = -1$

$= \dfrac{-6i - 1}{3}$ Simplify.

$= -\dfrac{1}{3} - 2i$ Write in standard form.

(b) $\dfrac{5}{3 - 2i} = \dfrac{5}{3 - 2i} \cdot \dfrac{3 + 2i}{3 + 2i}$ Multiply numerator and denominator by complex conjugate of denominator.

$= \dfrac{5(3 + 2i)}{(3 - 2i)(3 + 2i)}$ Multiply fractions.

$= \dfrac{5(3 + 2i)}{3^2 + 2^2}$ Product of complex conjugates

$= \dfrac{15 + 10i}{13}$ Simplify.

$= \dfrac{15}{13} + \dfrac{10}{13}i$ Write in standard form.

Example 8 ■ Division of Complex Numbers

Perform each division.

(a) $\dfrac{3 + i}{3 - i}$ (b) $\dfrac{2 + 3i}{4 - 2i}$

Solution

(a) $\dfrac{3 + i}{3 - i} = \dfrac{3 + i}{3 - i} \cdot \dfrac{3 + i}{3 + i}$ Multiply numerator and denominator by complex conjugate of denominator.

$= \dfrac{9 + 6i + i^2}{3^2 + 1^2}$ Multiply fractions.

$= \dfrac{9 + 6i + (-1)}{3^2 + 1^2}$ $i^2 = -1$

$= \dfrac{8 + 6i}{10}$ Simplify.

$= \dfrac{4}{5} + \dfrac{3}{5}i$ Write in standard form.

(b) $\dfrac{2 + 3i}{4 - 2i} = \dfrac{2 + 3i}{4 - 2i} \cdot \dfrac{4 + 2i}{4 + 2i}$ Multiply numerator and denominator by complex conjugate of denominator.

$= \dfrac{8 + 16i + 6i^2}{4^2 + 2^2}$ Multiply fractions.

$= \dfrac{8 + 16i + 6(-1)}{4^2 + 2^2}$ $i^2 = -1$

$= \dfrac{2 + 16i}{20}$ Simplify.

$= \dfrac{1}{10} + \dfrac{4}{5}i$ Write in standard form.

Example 9 ■ Verifying a Complex Solution of an Equation

Show that $x = 2 + i$ is a solution of the equation

$$x^2 - 4x + 5 = 0.$$

Solution

$x^2 - 4x + 5 = 0$ Write original equation.

$(2 + i)^2 - 4(2 + i) + 5 \stackrel{?}{=} 0$ Substitute $2 + i$ for x.

$4 + 4i + i^2 - 8 - 4i + 5 \stackrel{?}{=} 0$ Expand.

$i^2 + 1 \stackrel{?}{=} 0$ Combine like terms.

$(-1) + 1 \stackrel{?}{=} 0$ $i^2 = -1$

$0 = 0$ Solution checks. ✓

So, $x = 2 + i$ is a solution of the original equation.

5.6 ■ Exercises

Developing Skills

In Exercises 1–8, write the number in *i*-form.

1. $\sqrt{-4}$ **2.** $\sqrt{-9}$

3. $\sqrt{-\frac{4}{25}}$ **4.** $-\sqrt{-\frac{36}{121}}$

5. $\sqrt{-8}$ **6.** $\sqrt{-75}$

7. $\sqrt{-7}$ **8.** $\sqrt{-15}$

In Exercises 9–20, perform the operations and write the result in standard form.

9. $\sqrt{-16} + \sqrt{-36}$ **10.** $\sqrt{-50} - \sqrt{-8}$

11. $\sqrt{-8}\sqrt{-2}$ **12.** $\sqrt{-25}\sqrt{-6}$

13. $\sqrt{-18}\sqrt{-3}$ **14.** $\sqrt{-16}\sqrt{-121}$

15. $\sqrt{-3}\left(\sqrt{-3} + \sqrt{-4}\right)$

16. $\sqrt{-12}\left(\sqrt{-3} - \sqrt{-12}\right)$

17. $\sqrt{-5}\left(\sqrt{-16} - \sqrt{-10}\right)$

18. $\sqrt{-24}\left(\sqrt{-9} + \sqrt{-4}\right)$

19. $\left(\sqrt{-16}\right)^2$

20. $\left(\sqrt{-2}\right)^2$

In Exercises 21–28, determine the values of a and b that satisfy the equation.

21. $3 - 4i = a + bi$

22. $-8 + 6i = a + bi$

23. $5 - 4i = (a + 3) + (b - 1)i$

24. $-10 + 12i = 2a + (5b - 3)i$

25. $-4 - \sqrt{-8} = a + bi$

26. $\sqrt{-36} - 3 = a + bi$

27. $(a + 5) + (b - 1)i = 7 - 3i$

28. $(2a + 1) + (2b + 3)i = 5 + 12i$

In Exercises 29–38, perform the operations and write the result in standard form.

29. $(4 - 3i) + (6 + 7i)$

30. $(-10 + 2i) + (4 - 7i)$

31. $(-4 - 7i) + (-10 - 33i)$

32. $(15 + 10i) - (2 + 10i)$

33. $13i - (14 - 7i)$

34. $(-21 - 50i) + (21 - 20i)$

35. $(30 - i) - (18 + 6i) + 3i^2$

36. $(4 + 6i) + (15 + 24i) - (1 - i)$

37. $15i - (3 - 25i) + \sqrt{-81}$

38. $(-1 + i) - \sqrt{2} - \sqrt{-2}$

In Exercises 39–60, perform the operations and write the result in standard form.

39. $(3i)(12i)$ **40.** $(-5i)(4i)$

41. $(-6i)(-i)(6i)$ **42.** $\frac{1}{2}(10i)(12i)(-3i)$

43. $(-3i)^3$ **44.** $(8i)^2$

45. $-5(13 + 2i)$ **46.** $10(8 - 6i)$

47. $4i(-3 - 5i)$ **48.** $-3i(10 - 15i)$

49. $(4 + 3i)(-7 + 4i)$ **50.** $(3 + 5i)(2 + 15i)$

51. $(-7 + 7i)(4 - 2i)$ **52.** $(3 + 5i)(2 - 15i)$

53. $(3 - 4i)^2$ **54.** $(7 + i)^2$

55. $(2 + 5i)^2$ **56.** $(8 - 3i)^2$

57. $\left(-2 + \sqrt{-5}\right)\left(-2 - \sqrt{-5}\right)$

58. $\sqrt{-9}\left(1 + \sqrt{-16}\right)$

59. $(2 + i)^3$

60. $(3 - 2i)^3$

In Exercises 61–70, multiply the number by its complex conjugate.

61. $2 + i$ **62.** $3 + 2i$

63. $-2 - 8i$ **64.** $10 - 3i$

65. $5 - \sqrt{6}i$ **66.** $-4 + \sqrt{2}i$

67. $10i$ **68.** 20

69. $1 + \sqrt{-3}$ **70.** $-3 - \sqrt{-5}$

In Exercises 71–86, perform the operations and write the result in standard form.

71. $\dfrac{4}{1 - i}$ **72.** $\dfrac{20}{3 + i}$

73. $\dfrac{-12}{2 + 7i}$ **74.** $\dfrac{-6}{5 - 2i}$

75. $\dfrac{12i}{7 - 6i}$ **76.** $\dfrac{17i}{5 + 3i}$

77. $\dfrac{20}{2i}$

78. $\dfrac{-8}{3i}$

79. $\dfrac{9 - i}{5i}$

80. $\dfrac{1 + i}{3i}$

81. $\dfrac{4i}{1 - 3i}$

82. $\dfrac{-16i}{3 + 10i}$

83. $\dfrac{-8}{4(i - 5)}$

84. $\dfrac{15}{2(1 - i)}$

85. $\dfrac{2 + 3i}{1 + 2i}$

86. $\dfrac{4 - 5i}{4 + 5i}$

In Exercises 87–90, perform the operations and write the result in standard form.

87. $\dfrac{1}{1 - 2i} + \dfrac{4}{1 + 2i}$

88. $\dfrac{3i}{1 + i} + \dfrac{2}{2 + 3i}$

89. $\dfrac{i}{4 - 3i} - \dfrac{5}{2 + i}$

90. $\dfrac{1 + i}{i} - \dfrac{3}{5 - 2i}$

In Exercises 91–94, decide whether each number is a solution of the equation.

91. $x^2 + 2x + 5 = 0$

    (a) $x = -1 + 2i$ (b) $x = -1 - 2i$

92. $x^2 - 4x + 13 = 0$

    (a) $x = 2 - 3i$ (b) $x = 2 + 3i$

93. $x^3 + 4x^2 + 9x + 36 = 0$

    (a) $x = -4$ (b) $x = -3i$

94. $x^3 - 8x^2 + 25x - 26 = 0$

    (a) $x = 2$ (b) $x = 3 - 2i$

In Exercises 95–98, perform the operations.

95. $(a + bi) + (a - bi)$

96. $(a + bi)(a - bi)$

97. $(a + bi) - (a - bi)$

98. $(a + bi)^2 + (a - bi)^2$

Explaining Concepts

99. Define the imaginary unit i.

100. Explain why the equation $x^2 = -1$ does not have real number solutions.

101. Describe the error.

$$\sqrt{-3}\sqrt{-3} = \sqrt{(-3)(-3)} = \sqrt{9} = 3$$

102. *True or False?* Some numbers are both real and imaginary. Explain.

Ongoing Review

In Exercises 103–106, solve the equation.

103. $x + 21 = 0$

104. $3x + 7 = 0$

105. $3x^2 - 8x - 16 = 0$

106. $2x^2 + 5x - 7 = 0$

In Exercises 107 and 108, translate the phrase into an algebraic equation.

107. The time to travel 360 miles if the average speed is r miles per hour

108. The perimeter of a rectangle of length L and width $\dfrac{L}{3}$

Looking Further

(a) The principal cube root of 125, $\sqrt[3]{125}$, is 5. Evaluate the expression x^3 for each of the values of x.

    (i) $x = \dfrac{-5 + 5\sqrt{3}i}{2}$ (ii) $x = \dfrac{-5 - 5\sqrt{3}i}{2}$

(b) The principal cube root of 27, $\sqrt[3]{27}$, is 3. Evaluate the expression x^3 for each of the values of x.

    (i) $x = \dfrac{-3 + 3\sqrt{3}i}{2}$ (ii) $x = \dfrac{-3 - 3\sqrt{3}i}{2}$

(c) Compare the results in parts (a) and (b). Use the results to find the three cube roots of each number.

    (i) 1   (ii) 8   (iii) 64

    Verify your results algebraically.

Chapter Summary: Key Concepts

5.1 ■ Properties of exponents

Let m and n be integers, and let a and b represent real numbers, variables, or algebraic expressions.

1. $a^m \cdot a^n = a^{m+n}$

2. $(ab)^m = a^m \cdot b^m$

3. $(a^m)^n = a^{mn}$

4. $\dfrac{a^m}{a^n} = a^{m-n}, \ a \neq 0$

5. $\left(\dfrac{a}{b}\right)^m = \dfrac{a^m}{b^m}, \ b \neq 0$

6. $\left(\dfrac{a}{b}\right)^{-m} = \left(\dfrac{b}{a}\right)^m, \ a \neq 0, b \neq 0$

7. $a^{-m} = \dfrac{1}{a^m}, \ a \neq 0$

8. $a^0 = 1, \ a \neq 0$

5.2 ■ Properties of nth roots

1. If a is a positive real number and n is even, then a has exactly two (real) nth roots, which are denoted by $\sqrt[n]{a}$ and $-\sqrt[n]{a}$.

2. If a is any real number and n is odd, then a has only one (real) nth root, which is denoted by $\sqrt[n]{a}$.

3. If a is a negative real number and n is even, then a has no (real) nth root.

5.2 ■ Inverse properties of nth powers and nth roots

Let a be a real number, and let n be an integer such that $n \geq 2$.

1. If a has a principal nth root, then $\left(\sqrt[n]{a}\right)^n = a$.

2. If n is odd, then $\sqrt[n]{a^n} = a$.
   If n is even, then $\sqrt[n]{a^n} = |a|$.

5.2 ■ Properties of exponents

The eight properties of exponents listed above for Section 5.1 also apply to rational exponents.

5.2 ■ Domain of a radical function

Let n be an integer that is greater than or equal to 2.

1. If n is odd, the domain of $f(x) = \sqrt[n]{x}$ is the set of all real numbers.

2. If n is even, the domain of $f(x) = \sqrt[n]{x}$ is the set of all nonnegative real numbers.

5.3 ■ Multiplication and Division Properties of Radicals

Let u and v be real numbers, variables, or algebraic expressions. If the nth roots of u and v are real, the properties below are true.

1. $\sqrt[n]{u}\,\sqrt[n]{v} = \sqrt[n]{uv}$

2. $\dfrac{\sqrt[n]{u}}{\sqrt[n]{v}} = \sqrt[n]{\dfrac{u}{v}}, \ v \neq 0$

5.3 ■ Simplest form of radical expressions

A radical expression is said to be in the simplest form if all three of the statements below are true.

1. All possible nth-powered factors have been removed from each radical.

2. No radical contains a fraction.

3. No denominator of a fraction contains a radical.

5.5 ■ Raising each side of an equation to the nth power

Let u and v be real numbers, variables, or algebraic expressions, and let n be a positive integer. If $u = v$, then it follows that $u^n = v^n$.

5.6 ■ The square root of a negative number

Let c be a positive real number. Then the square root of $-c$ is given by $\sqrt{-c} = \sqrt{c(-1)} = \sqrt{c}\sqrt{-1} = \sqrt{c}\,i$.

5.6 ■ Definition of a complex number

If a and b are real numbers, the number $a + bi$ is a complex number, and it is said to be written in standard form. If $b = 0$, the number $a + bi = a$ is a real number. If $b \neq 0$, the number $a + bi$ is called an imaginary number. A number of the form bi, where $b \neq 0$, is called a pure imaginary number.

Review Exercises

Reviewing Skills

5.1 ■ *How to use properties of exponents to simplify algebraic expressions*

In Exercises 1–4, evaluate the expression.

1. $(2^3 \cdot 3^2)^{-1}$

2. $(2^{-2} \cdot 5^2)^{-2}$

3. $\left(\dfrac{2}{5}\right)^{-3}$

4. $\left(\dfrac{1}{3^{-2}}\right)^2$

In Exercises 5–12, rewrite the expression using only positive exponents.

5. $(x^3y^{-4})^2(x^{-2}y)^{-3}$

6. $(2x^2y^4)^4(3x^{-1}y^2)^2$

7. $-3(a^4b^{-6})^0(-2a^{-3}b^2)^2$

8. $5^2(2u^{-4}v^5)^3(uv^2)^0$

9. $\dfrac{7x^5y^{-2}}{14x^3y^8}$

10. $\dfrac{8y^{-2}z^{-4}}{6y^3z^{-2}}$

11. $\dfrac{15b^6c}{3ab^2c^{-4}}$

12. $\dfrac{4xy^{-5}z^{-1}}{20x^7y^{-2}}$

■ *How to write very large or very small numbers in scientific notation*

In Exercises 13–16, write the number in scientific notation.

13. 1,460,000,000

14. 98,400,000

15. 0.000000641

16. 0.00728

In Exercises 17–20, write the number in decimal notation.

17. 4.09×10^{-12}

18. 8.231×10^{-6}

19. 9.58×10^{10}

20. 3.78×10^{15}

In Exercises 21–24, evaluate the expression. Write the result in scientific notation, $c \times 10^n$.

21. $(6 \times 10^3)^2$

22. $(3 \times 10^{-3})(8 \times 10^7)$

23. $\dfrac{3.5 \times 10^7}{7 \times 10^4}$

24. $\dfrac{1}{(6 \times 10^{-3})^2}$

5.2 ■ *How to determine the nth roots of numbers and evaluate radical expressions*

In Exercises 25–34, evaluate without a calculator. (if not possible, state the reason.)

25. $-\sqrt{16}$

26. $\sqrt{-16}$

27. $\sqrt[3]{-8}$

28. $-\sqrt[3]{-8}$

29. $\sqrt{\dfrac{4}{25}}$

30. $\sqrt{\dfrac{16}{36}}$

31. $\sqrt[3]{0.001}$

32. $\sqrt[3]{-0.000027}$

33. $\sqrt[4]{-625}$

34. $-\sqrt[4]{81}$

■ *How to use the properties of exponents to evaluate or simplify expressions with rational exponents*

In Exercises 35–40, evaluate the expression.

35. $27^{4/3}$

36. $16^{3/4}$

37. $25^{3/2}$

38. $216^{1/3}$

39. $16^{-1/4}$

40. $243^{-2/5}$

In Exercises 41–48, simplify the expression.

41. $x^{3/4} \cdot x^{-1/6}$

42. $x^{-1/2} \cdot 5x^{5/2}$

43. $(4a^0)^{2/3}(4a^{-2})^{1/3}$

44. $(2y^2)^{3/2}(2y^{-4})^{1/2}$

45. $\dfrac{15x^{1/4}y^{3/5}}{5x^{1/2}y}$

46. $\dfrac{48a^2b^{5/2}}{14a^{-3}b^{-1/2}}$

47. $\dfrac{(3x+2)^{2/3}}{\sqrt[3]{3x+2}}$

48. $\dfrac{\sqrt[5]{3x+6}}{(3x+6)^{4/5}}$

In Exercises 49 and 50, evaluate the expression. (Round the result to two decimal places.)

49. $75^{-3/4}$

50. $510^{5/3}$

■ *How to evaluate radical functions and find the domains of radical functions*

In Exercises 51 and 52, evaluate the function as indicated, and simplify.

51. $g(x) = \sqrt{3x - 2}$

    (a) $g(1)$ (b) $g(2)$ (c) $g(9)$ (d) $g(6)$

52. $f(x) = \sqrt[3]{2x - 1}$

    (a) $f(14)$ (b) $f(63)$ (c) $f(0)$ (d) $f(-13)$

In Exercises 53–56, find the domain of the function.

53. $f(x) = \dfrac{1}{\sqrt{2x}}$

54. $g(x) = \dfrac{5}{\sqrt[3]{x}}$

55. $h(x) = \sqrt{2x + 6}$

56. $f(x) = \sqrt{12 - x}$

5.3 ■ *How to use the Multiplication and Division Properties of Radicals to simplify radical expressions*

In Exercises 57–66, simplify the expression.

57. $\sqrt{360}$

58. $\sqrt{\frac{50}{9}}$

59. $\sqrt{50x^4}$

60. $\sqrt{18t^3}$

61. $\sqrt[3]{48a^3b^4}$

62. $\sqrt[3]{54x^8y^{10}}$

63. $\sqrt[4]{81x^{12}y^{15}}$

64. $\sqrt[4]{32u^4v^5}$

65. $\sqrt[5]{32a^{11}b^{16}}$

66. $\sqrt[5]{486x^{18}y^{22}}$

■ *How to use rationalization techniques to simplify radical expressions*

In Exercises 67–78, rationalize the denominator and simplify further, if possible.

67. $\sqrt{\frac{5}{6}}$

68. $\sqrt{\frac{3}{20}}$

69. $\dfrac{3}{\sqrt{12x}}$

70. $\dfrac{4y}{\sqrt{10z}}$

71. $\dfrac{2}{\sqrt[3]{2x}}$

72. $\sqrt[3]{\frac{16t}{s^2}}$

73. $\sqrt{\dfrac{12x^2y^3}{5z}}$

74. $\sqrt{\dfrac{a^5b^3}{2z^3}}$

75. $\dfrac{x^3}{\sqrt[3]{4xy^2}}$

76. $\dfrac{7a}{\sqrt[3]{5a^2b^4}}$

77. $\dfrac{-4x}{\sqrt[4]{8x^3y}}$

78. $\dfrac{15uv}{\sqrt[4]{3u^3v^5}}$

■ *How to use the Distributive Property to add and subtract like radicals*

In Exercises 79–84, perform the operations and simplify.

79. $3\sqrt{40} - 10\sqrt{90}$

80. $9\sqrt{50} - 5\sqrt{8} + \sqrt{48}$

81. $10\sqrt[4]{y+3} - 3\sqrt[4]{y+3}$

82. $\sqrt{25x} + \sqrt{49x} - \sqrt[3]{8x}$

83. $\sqrt{5x} - \dfrac{3}{\sqrt{5x}}$

84. $\dfrac{4}{\sqrt{2x}} + 3\sqrt{2x}$

5.4 ■ *How to use the Distributive Property or the FOIL Method to multiply radical expressions*

In Exercises 85–88, perform the operations and simplify.

85. $\left(\sqrt{5} + 6x\right)^2$

86. $\left(7 - \sqrt{2x}\right)^2$

87. $\left(2\sqrt{x} + 7\right)\left(\sqrt{x} - 5\right)$

88. $\left(4\sqrt{u} - 9\right)\left(3\sqrt{u} + 1\right)$

■ *How to determine the product of conjugates*

In Exercises 89–92, find the conjugate of the expression. Then multiply the expression by its conjugate and simplify.

89. $\left(\sqrt{3} - \sqrt{x}\right)$

90. $\left(\sqrt{2a} + 7\right)$

91. $\left(\sqrt{5u} + \sqrt{v}\right)$

92. $\left(\sqrt{x} + 8\sqrt{2y}\right)$

■ *How to simplify quotients involving radicals by rationalizing the denominators*

In Exercises 93–98, perform the operations and simplify.

93. $\dfrac{15}{\sqrt{x} + 3}$

94. $\dfrac{22}{7 - \sqrt{x}}$

95. $\dfrac{\sqrt{x}}{5 + \sqrt{6x}}$

96. $\dfrac{2\sqrt{y}}{\sqrt{y} - 8}$

97. $\left(\sqrt{x} + 10\right) \div \left(\sqrt{x} - 10\right)$

98. $\left(3\sqrt{s} + 4\right) \div \left(\sqrt{s} + 2\right)$

5.5 ■ *How to solve radical equations*

In Exercises 99–108, solve the equation.

99. $\sqrt{y} = 15$

100. $\sqrt{3x} + 9 = 0$

101. $\sqrt{2(a-7)} = 14$

102. $\sqrt{3(2x+3)} = \sqrt{x+15}$

103. $\sqrt{2(x+5)} = x + 5$

104. $\sqrt{5t} = 1 + \sqrt{5(t-1)}$

105. $\sqrt[3]{5x+2} - \sqrt[3]{7x-8} = 0$

106. $\sqrt[4]{9x-2} - \sqrt[4]{8x} = 0$

107. $\sqrt{1+6x} = 2 - \sqrt{6x}$

108. $\sqrt{2+9b} + 1 = 3\sqrt{b}$

▦ In Exercises 109 and 110, use a graphing utility to approximate the solution of the equation. Verify your solution algebraically.

109. $\sqrt{5(4-3x)} = 10$

110. $\sqrt[3]{5x-7} - 3 = -1$

5.6 ■ *How to write the square roots of negative numbers in i-form and perform operations on numbers in i-form*

In Exercises 111–116, write the number in *i*-form.

111. $\sqrt{-16}$

112. $\sqrt{-25}$

113. $\sqrt{-48}$

114. $\sqrt{-50}$

115. $\sqrt{-0.16}$

116. $\sqrt{-0.81}$

In Exercises 117–124, perform the operations and write the result in standard form.

117. $\sqrt{-25} + \sqrt{-4}$ **118.** $\sqrt{-144} - \sqrt{-49}$

119. $\sqrt{-10}\sqrt{-5}$ **120.** $\sqrt{-40}\sqrt{-8}$

121. $\sqrt{-3}\left(\sqrt{-10} + \sqrt{-15}\right)$

122. $\sqrt{-6}\left(\sqrt{-18} + \sqrt{-20}\right)$

123. $\left(\sqrt{-7}\right)^2$

124. $\left(\sqrt{-19}\right)^2$

■ *How to determine the equality of two complex numbers*

In Exercises 125–128, determine values for a and b that satisfy the equation.

125. $12 - 5i = (a + 2) + (b - 1)i$

126. $-48 + 9i = (a - 5) + (b + 10)i$

127. $\sqrt{-49} + 4 = a + bi$

128. $-3 - \sqrt{-4} = a + bi$

■ *How to add, subtract, and multiply complex numbers*

In Exercises 129–136, perform the operations and write the result in standard form.

129. $(-4 + 5i) - (-12 + 8i)$

130. $5(3 - 8i) + (5 + 12i)$

131. $(-2)(15i)(-3i)$ **132.** $-10i(4 - 7i)$

133. $(4 - 3i)(4 + 3i)$ **134.** $(6 - 5i)^2$

135. $(12 - 5i)(2 + 7i)$ **136.** $(5 + 2i)(9 - 3i)$

■ *How to use complex conjugates to divide complex numbers*

In Exercises 137–140, perform the operations and write the result in standard form.

137. $\dfrac{4}{5i}$ **138.** $\dfrac{-10}{7i}$

139. $\dfrac{5i}{2 + 9i}$ **140.** $\dfrac{2 + i}{1 - 9i}$

Solving Problems

141. *Metal Expansion* When the temperature of an iron steam pipe 200 feet long is increased by 75°C, the length of the pipe will increase by an amount of
$$75(200)(10 \times 10^{-6}).$$
Find this amount of increase in length.

142. *Price Index* The producer price index for mobile homes in the United States between 1993 and 1999 can be approximated by the model
$$y = 14\sqrt{t - 3} + 124.4$$
where t represents the year, with $t = 3$ corresponding to 1993. Use this model to approximate the producer price index for mobile homes in the United States for the year 1998. *(Source: U.S. Bureau of Labor Statistics)*

143. *Geometry* Find the length of the hypotenuse of a right triangle if the base measures 10 inches and the height measures 12 inches.

144. *Perimeter* The four corners are cut from an $8\frac{1}{2}$-inch by 14-inch sheet of paper (see figure). Find the perimeter of the remaining piece of paper.

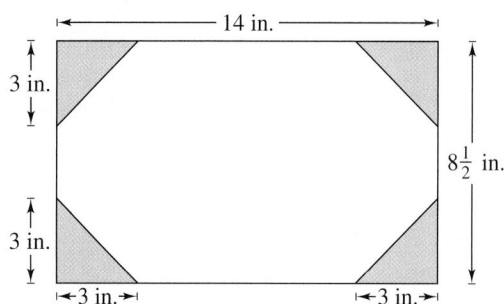

Figure for 144

145. *Length of a Pendulum* The time t in seconds for a pendulum of length L in feet to go through one complete cycle (its period) is
$$t = 2\pi\sqrt{\frac{L}{32}}.$$
How long is the pendulum of a grandfather clock with a period of 1.3 seconds?

146. *Height of a Bridge* The time t in seconds for a free-falling object to fall d feet is given by
$$t = \sqrt{\frac{d}{16}}.$$
A child drops a pebble from a bridge and observes it strike the water after approximately 4 seconds. Estimate the height of the bridge.

5 Chapter Test

Take this test as you would take a test in class. After you are done, check your work against the answers given in the back of the book.

In Exercises 1 and 2, evaluate each expression without using a calculator.

1. (a) $2^{-2} + 2^{-3}$

(b) $\dfrac{6.3 \times 10^{-3}}{2.1 \times 10^{2}}$

2. (a) $27^{-2/3}$

(b) $\left(\dfrac{81}{16}\right)^{1/4}$

3. (a) Write 0.000032 in scientific notation.

(b) Write 3.04×10^{7} in decimal notation.

In Exercises 4–6, simplify each expression.

4. (a) $\dfrac{12s^{5}t^{-2}}{20s^{-2}t^{-1}}$

(b) $(-2x^{5}y^{-2}z^{0})^{-1}$

5. (a) $\left(\dfrac{4x^{1/2}y^{-1/3}}{5x^{1/3}y^{5/6}z}\right)^{2}$

(b) $(5xy^{2})^{1/4}(5xy^{2})^{7/4}$

6. (a) $\sqrt{\dfrac{32x^{5}y}{9x^{2}y^{3}}}$

(b) $\sqrt[3]{24u^{8}v^{14}}$

7. Rationalize the denominator and simplify the result of each expression.

(a) $\dfrac{10}{\sqrt{6} - \sqrt{2}}$

(b) $\dfrac{2}{\sqrt[3]{9y}}$

8. Subtract: $5\sqrt{3x} - 3\sqrt{75x}$.

9. Multiply: $\sqrt{5}(\sqrt{15x} + 3)$.

10. Expand: $(4 - \sqrt{2x})^{2}$.

In Exercises 11–13, solve the equation.

11. $\sqrt{x^{2} - 1} = x - 2$ **12.** $\sqrt{x} - x + 6 = 0$ **13.** $\sqrt{x - 4} = \sqrt{x + 7} - 1$

In Exercises 14–19, perform the operations and simplify.

14. $\left(2 + \sqrt{-9}\right) - \sqrt{-25}$

15. $\sqrt{-16}\left(1 + \sqrt{-4}\right)$

16. $(15 - 3i) + (-8 + 12i)$

17. $(2 - 3i)^{2}$

18. $(3 - 2i)(1 + 5i)$

19. $\dfrac{5 - 2i}{3 + i}$

20. The velocity v (in feet per second) of an object is given by $v = \sqrt{2gh}$, where $g = 32$ feet per second per second and h is the distance (in feet) the object has fallen. Find the height from which a rock has been dropped if it strikes the ground with a velocity of 80 feet per second.

Cumulative Test: Chapters 1–5

Take this test as you would take a test in class. After you are done, check your work against the answers given in the back of the book.

In Exercises 1 and 2, write the statement using inequality notation.

1. y is no more than 45.

2. x is at least 15.

In Exercises 3 and 4, identify the property of algebra illustrated by the statement.

3. $3x + 0 = 3x$

4. $-4(x + 10) = -4 \cdot x + (-4)(10)$

In Exercises 5 and 6, write an algebraic expression that represents the quantity in the verbal statement.

5. The total hourly wage for an employee when the base pay is \$9.35 per hour plus 75 cents for each of q units produced per hour

6. An automobile is traveling at an average speed of 48 miles per hour. Express the distance d traveled by the automobile as a function of time t in hours.

In Exercises 7–10, factor the expression completely.

7. $16x^2 - 121$

8. $9t^2 - 24t + 16$

9. $x(x - 10) - 4(x - 10)$

10. $4x^3 - 12x^2 + 16x$

In Exercises 11–16, solve the equation.

11. $\dfrac{x}{4} - \dfrac{2}{3} = 0$

12. $2x - 3[1 + (4 - x)] = 0$

13. $3x^2 - 13x - 10 = 0$

14. $x(x - 3) = 40$

15. $|2x + 5| = 11$

16. $\sqrt{x - 5} - 6 = 0$

In Exercises 17–19, find the x- and y-intercepts of the graph of the equation.

17. $y = -x + 6$

18. $y = x^2 + 2x - 8$

19. $y = 2|x|$

20. Given the function $f(x) = 3x - x^2$, find and simplify $f(2 + t) - f(2)$.

21. Use the graph of $f(x) = \sqrt{x}$ to write a function g for each of the graphs.

(a)   (b)   (c)

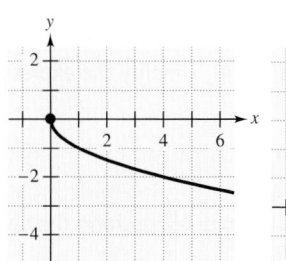
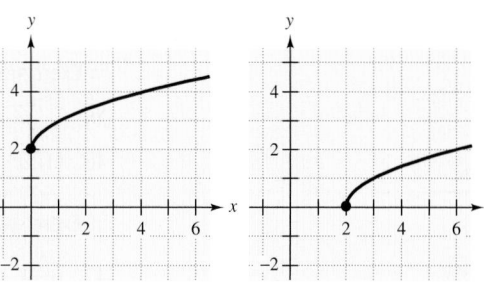

In Exercises 22 and 23, plot the points and find the slope of the line passing through the points.

22. $(-3, 2), (5, -4)$ **23.** $(2, 8), (7, -3)$

In Exercises 24–26, find an equation of the line that passes through the two points.

24. $(-4, -2), (-3, 8)$ **25.** $(1, 5), (-8, 2)$ **26.** $\left(\frac{3}{2}, -1\right), \left(-\frac{1}{3}, 4\right)$

In Exercises 27 and 28, solve for x by using the fact that corresponding sides of similar triangles are proportional.

27. 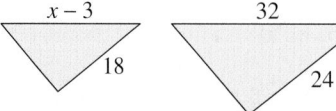   **28.**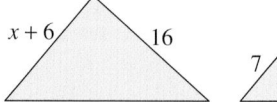

In Exercises 29 and 30, solve the inequality and sketch the solution on the real number line.

29. $3(x - 5) < 4x - 7$ **30.** $|x + 8| - 1 \geq 15$

In Exercises 31 and 32, solve the system of linear equations.

31. $\begin{cases} 4x - y = 0 \\ -3x + 2y = 2 \end{cases}$ **32.** $\begin{cases} x - y = -1 \\ x + 2y - 2z = 3 \\ 3x - y + 2z = 3 \end{cases}$

33. One hundred gallons of a 60% acid solution is obtained by mixing a 75% solution with a 50% solution. How many gallons of each solution must be used to obtain the desired mixture?

34. Find the determinant of the matrix.

$$A = \begin{bmatrix} 4 & 0 & 5 \\ 0 & -7 & 2 \\ 9 & 1 & -1 \end{bmatrix}$$

35. Use a determinant to find the area of the triangle with vertices $(-5, 8)$, $(10, 0)$, and $(3, -4)$.

In Exercises 36 and 37, sketch the graph of the linear inequality.

36. $y \leq 5 - \frac{1}{2}x$ **37.** $3y - x \geq 7$

In Exercises 38–41, simplify the expression.

38. $\sqrt{24x^2y^3}$ **39.** $\sqrt[3]{80a^{15}b^8}$

40. $(12a^{-4}b^6)^{1/2}$ **41.** $(16^{1/3})^{3/4}$

In Exercises 42 and 43, perform the operations and simplify.

42. $(-4 + 11i) - (3 - 5i)$ **43.** $\dfrac{6 + 4i}{1 - 3i}$

Quadratic Functions, Equations, and Inequalities

6

THE BIG PICTURE

In this chapter you will learn how to:

■ solve quadratic equations by factoring, extracting square roots, completing the square, and using the Quadratic Formula.

■ solve equations of quadratic form.

■ write and solve quadratic equations that model real-life problems.

■ sketch the graphs of quadratic functions.

■ solve quadratic inequalities.

Key Terms

As you encounter each new vocabulary term in this chapter, add the term and its definition to your notebook glossary.

repeated solution (p. 364)

extracting square roots (p. 366)

quadratic form (p. 369)

completing the square (p. 376)

Quadratic Formula (p. 384)

discriminant (p. 387)

parabola (p. 401)

vertex of a parabola (p. 401)

axis of a parabola (p. 401)

standard form of a quadratic function (p. 401)

zeros (p. 412)

critical numbers (p. 412)

test intervals (p. 412)

general form of a quadratic inequality (p. 415)

Additional text-specific resources are available to help you do well in this course. See page xvi for details.

6.1 The Factoring and Square Root Methods

Solving Equations by Factoring

In this chapter, you will study methods for solving quadratic equations and equations of quadratic form. To begin, let's review the method of factoring that you studied in Section 1.6.

Remember that the first step in solving a quadratic equation by factoring is to write the equation in general form. Next, factor the left side. Finally, set each factor equal to zero and solve for x.

Example 1 ■ Solving Quadratic Equations by Factoring

Solve each quadratic equation by factoring.

(a) $x^2 + 5x = 24$ (b) $3x^2 = 4 - 11x$ (c) $9x^2 + 12 = 3 + 12x + 5x^2$

Solution

(a)

$x^2 + 5x = 24$	Write original equation.
$x^2 + 5x - 24 = 0$	Write in general form.
$(x + 8)(x - 3) = 0$	Factor.
$x + 8 = 0 \implies x = -8$	Set 1st factor equal to 0.
$x - 3 = 0 \implies x = 3$	Set 2nd factor equal to 0.

The solutions are $x = -8$ and $x = 3$. Check these in the original equation.

(b)

$3x^2 = 4 - 11x$	Write original equation.
$3x^2 + 11x - 4 = 0$	Write in general form.
$(3x - 1)(x + 4) = 0$	Factor.
$3x - 1 = 0 \implies x = \frac{1}{3}$	Set 1st factor equal to 0.
$x + 4 = 0 \implies x = -4$	Set 2nd factor equal to 0.

The solutions are $x = \frac{1}{3}$ and $x = -4$. Check these in the original equation.

(c)

$9x^2 + 12 = 3 + 12x + 5x^2$	Write original equation.
$4x^2 - 12x + 9 = 0$	Write in general form.
$(2x - 3)(2x - 3) = 0$	Factor.
$2x - 3 = 0$	Set factor equal to 0.
$x = \dfrac{3}{2}$	Repeated solution

The only solution is $x = \frac{3}{2}$. Check this in the original equation.

When the two solutions of a quadratic equation are identical, they are called a **double** or **repeated solution.** This occurred in Example 1(c).

EXPLORATION

Use a graphing utility to approximate the solutions of

$$6x^2 - 5x - 4 = 0.$$

Use the *zoom* and *trace* features of the graphing utility to write the approximations with an error of less than 0.01. Then solve the equation algebraically and compare your algebraic solutions with your approximations.

NOTE   Notice in Example 2(b) that you do not set the constant factor of 5 equal to zero. When solving quadratic equations, you only need to set factors containing variables equal to zero.

Example 2 ■ Solving Quadratic Equations by Factoring

Solve each equation by factoring.

(a) $(x + 1)(x - 2) = 18$ (b) $5x^2 = 45$

Solution

(a)

$(x + 1)(x - 2) = 18$	Write original equation.
$x^2 - x - 2 = 18$	Multiply.
$x^2 - x - 20 = 0$	Write in general form.
$(x - 5)(x + 4) = 0$	Factor.
$x - 5 = 0 \implies x = 5$	Set 1st factor equal to 0.
$x + 4 = 0 \implies x = -4$	Set 2nd factor equal to 0.

The solutions are $x = 5$ and $x = -4$. Check these in the original equation.

(b)

$5x^2 = 45$	Write original equation.
$5x^2 - 45 = 0$	Write in general form.
$5(x^2 - 9) = 0$	Common monomial factor
$5(x + 3)(x - 3) = 0$	Factor as difference of squares.
$x + 3 = 0 \implies x = -3$	Set 2nd factor equal to 0.
$x - 3 = 0 \implies x = 3$	Set 3rd factor equal to 0.

The solutions are $x = -3$ and $x = 3$. Check these in the original equation.

NOTE   Do not be tricked by the quadratic equation in Example 2(a). Even though the left side is factored, the right side is not zero. So, you must first multiply the left side, then rewrite the equation in general form, and finally refactor.

Example 3 ■ Solving a Quadratic Equation by Factoring

Solve $3x^2 - 12x = 0$ by factoring.

Solution

The left side of this equation has a common factor of $3x$.

$3x^2 - 12x = 0$	Write original equation.
$3x(x - 4) = 0$	Factor.
$3x = 0 \implies x = 0$	Set 1st factor equal to 0.
$x - 4 = 0 \implies x = 4$	Set 2nd factor equal to 0.

The solutions are $x = 0$ and $x = 4$. Check these in the original equation.

You can also check the solutions in Example 3 graphically by observing that the graph of

$$y = 3x^2 - 12x$$

has x-intercepts at $x = 0$ and $x = 4$, as shown in Figure 6.1.

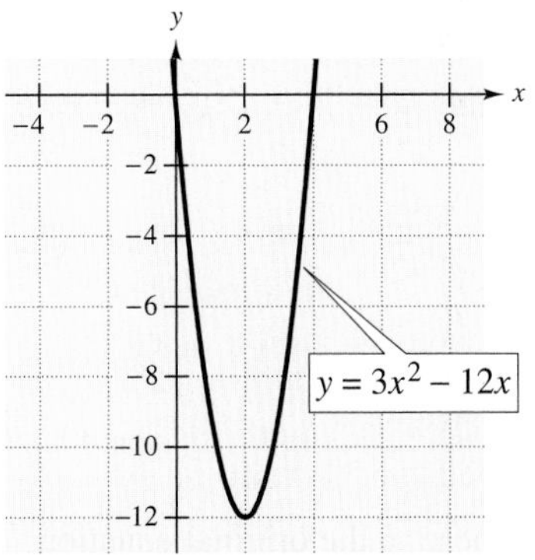

$y = 3x^2 - 12x$

Figure 6.1

Extracting Square Roots

Consider the equation below, where $d > 0$ and u is an algebraic expression.

$$u^2 = d$$ Write original equation.

$$u^2 - d = 0$$ Write in general form

$$\left(u + \sqrt{d}\right)\left(u - \sqrt{d}\right) = 0$$ Factor.

$$u + \sqrt{d} = 0 \quad\Longrightarrow\quad u = -\sqrt{d}$$ Set 1st factor equal to 0.

$$u - \sqrt{d} = 0 \quad\Longrightarrow\quad u = \sqrt{d}$$ Set 2nd factor equal to 0.

Because the solutions differ only in sign, they can be written together using a "plus or minus sign": $u = \pm\sqrt{d}$. This form of the solution is read as "u is equal to plus or minus the square root of d." Solving an equation of the form $u^2 = d$ *without* going through the steps of factoring is called **extracting square roots.**

Extracting Square Roots

The equation $u^2 = d$, where $d > 0$, has exactly two solutions:

$$u = \sqrt{d} \quad \text{and} \quad u = -\sqrt{d}.$$

These solutions can also be written as $u = \pm\sqrt{d}$.

Example 4 ■ Extracting Square Roots

Solve each quadratic equation by extracting square roots.

(a) $3x^2 = 15$ (b) $(x - 2)^2 = 10$ (c) $(2x - 1)^2 = 9$

Solution

(a) $3x^2 = 15$ Write original equation.

$\quad x^2 = 5$ Divide each side by 3.

$\quad x = \pm\sqrt{5}$ Extract square roots.

The solutions are $x = \sqrt{5}$ and $x = -\sqrt{5}$. Check these in the original equation.

(b) $(x - 2)^2 = 10$ Write original equation.

$\quad x - 2 = \pm\sqrt{10}$ Extract square roots.

$\quad x = 2 \pm \sqrt{10}$ Add 2 to each side.

The solutions are $x = 2 + \sqrt{10} \approx 5.16$ and $x = 2 - \sqrt{10} \approx -1.16$. Check these in the original equation.

(c) $(2x - 1)^2 = 9$ Write original equation.

$\quad 2x - 1 = \pm3$ Extract square roots.

$\quad 2x = 1 \pm 3$ Add 1 to each side.

$\quad x = \dfrac{1 \pm 3}{2}$ Divide each side by 2.

The solutions are $x = 2$ and $x = -1$. Check these in the original equation.

Example 5 ■ Solving a Quadratic Equation

Solve the quadratic equation

$$3(5x + 4)^2 - 81 = 0.$$

Algebraic Solution

In this case, some preliminary steps are needed before extracting the square roots.

$3(5x + 4)^2 - 81 = 0$	Write original equation.
$3(5x + 4)^2 = 81$	Add 81 to each side.
$(5x + 4)^2 = 27$	Divide each side by 3.
$5x + 4 = \pm\sqrt{27}$	Extract square roots.
$5x = -4 \pm 3\sqrt{3}$	Subtract 4 from each side.
$x = \dfrac{-4 \pm 3\sqrt{3}}{5}$	Divide each side by 5.

So, the solutions are

$$x = \frac{-4 + 3\sqrt{3}}{5} \approx 0.24 \text{ and } x = \frac{-4 - 3\sqrt{3}}{5} \approx -1.84.$$

Graphical Solution

Use a graphing utility to graph

$$y = 3(5x + 4)^2 - 81.$$

Use the *zoom* and *trace* features of the graphing utility to approximate the x-intercepts of the graph, as shown in Figure 6.2. You can approximate the solutions to be $x \approx -1.84$ and $x \approx 0.24$.

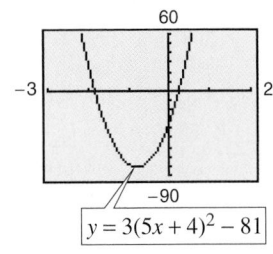

Figure 6.2

The built-in *zero* or *root* features of a graphing utility will approximate solutions of equations or approximate x-intercepts of graphs. If your graphing utility has such features, try using them to approximate the solutions in Example 5.

Equations with Imaginary Solutions

Prior to Section 5.6, the only solutions you had been finding had been real numbers. But now that you have studied complex numbers, it makes sense to look for other types of solutions. For instance, although the quadratic equation $x^2 + 1 = 0$ has no solutions that are real numbers, it does have two solutions that are imaginary numbers: i and $-i$. To check this, substitute i and $-i$ for x.

$(i)^2 + 1 = -1 + 1 = 0$	Solution checks. ✓
$(-i)^2 + 1 = -1 + 1 = 0$	Solution checks. ✓

One way to find imaginary solutions of a quadratic equation is to extend the *extraction of square roots* technique to cover the case where d is a negative number.

> **Extracting Imaginary Square Roots**
>
> The equation $u^2 = d$, where $d < 0$, has exactly two solutions:
>
> $$u = \sqrt{|d|}\, i \quad \text{and} \quad u = -\sqrt{|d|}\, i.$$
>
> These solutions can also be written as $u = \pm\sqrt{|d|}\, i$.

Example 6 ■ Extracting Imaginary Square Roots

Solve each quadratic equation.

(a) $x^2 + 8 = 0$ (b) $(x - 4)^2 = -3$ (c) $2(3x - 5)^2 + 32 = 0$

Solution

(a)
$$x^2 + 8 = 0 \qquad \text{Write original equation.}$$
$$x^2 = -8 \qquad \text{Subtract 8 from each side.}$$
$$x = \pm\sqrt{8}i \qquad \text{Extract imaginary square roots.}$$
$$x = \pm 2\sqrt{2}i \qquad \text{Simplify.}$$

The solutions are $x = 2\sqrt{2}i$ and $x = -2\sqrt{2}i$. Check these in the original equation.

(b)
$$(x - 4)^2 = -3 \qquad \text{Write original equation.}$$
$$x - 4 = \pm\sqrt{-3} \qquad \text{Extract imaginary square roots.}$$
$$x - 4 = \pm\sqrt{3}i \qquad \text{Write in } i\text{-form.}$$
$$x = 4 \pm \sqrt{3}i \qquad \text{Add 4 to each side.}$$

The solutions are $x = 4 + \sqrt{3}i$ and $x = 4 - \sqrt{3}i$. Check these in the original equation.

(c)
$$2(3x - 5)^2 + 32 = 0 \qquad \text{Write original equation.}$$
$$2(3x - 5)^2 = -32 \qquad \text{Subtract 32 from each side.}$$
$$(3x - 5)^2 = -16 \qquad \text{Divide each side by 2.}$$
$$3x - 5 = \pm 4i \qquad \text{Extract imaginary square roots.}$$
$$3x = 5 \pm 4i \qquad \text{Add 5 to each side.}$$
$$x = \frac{5 \pm 4i}{3} \qquad \text{Divide each side by 3.}$$

The solutions are

$$x = \frac{5}{3} + \frac{4}{3}i \text{ and } x = \frac{5}{3} - \frac{4}{3}i.$$

Check these in the original equation.

EXPLORATION

Solve each quadratic equation algebraically. Then use a graphing utility to check the solutions. Which equation(s) have real solutions? Which equation(s) have imaginary solutions? Which graph(s) have *x*-intercepts? Which graph(s) have no *x*-intercepts? Compare the type of solution(s) of a quadratic equation with the *x*-intercept(s) of the graph of the equation.

a. $y = 2x^2 + 3x - 5$

b. $y = 2x^2 + 4x + 2$

c. $y = x^2 + 4$

d. $y = x^2 + x + 2$

Equations of Quadratic Form

Both the factoring and extraction of square roots methods can be applied to nonquadratic equations that are of **quadratic form.** An equation is said to be of quadratic form if it has the form

$$au^2 + bu + c = 0$$

where u is an algebraic expression. Here are some examples.

Equation	Written in Quadratic Form
$x^4 + 5x^2 + 4 = 0$	$(x^2)^2 + 5(x^2) + 4 = 0$
$2x^{2/3} + 3x^{1/3} - 9 = 0$	$2(x^{1/3})^2 + 3(x^{1/3}) - 9 = 0$
$x - 5\sqrt{x} + 6 = 0$	$(\sqrt{x})^2 - 5(\sqrt{x}) + 6 = 0$

To solve an equation of quadratic form, it helps to make a substitution and rewrite the equation in terms of u, as demonstrated in Examples 7, 8, and 9.

Example 7 ■ Solving an Equation of Quadratic Form

Solve $x^4 - 13x^2 + 36 = 0$.

Algebraic Solution

This equation is of quadratic form with $u = x^2$.

$x^4 - 13x^2 + 36 = 0$	Write original equation.
$(x^2)^2 - 13(x^2) + 36 = 0$	Write in quadratic form.
$u^2 - 13u + 36 = 0$	Substitute u for x^2.
$(u - 4)(u - 9) = 0$	Factor.
$u - 4 = 0 \implies u = 4$	Set 1st factor equal to 0.
$u - 9 = 0 \implies u = 9$	Set 2nd factor equal to 0.

At this point you have found the "u-solutions." To find the "x-solutions," replace u with x^2, as follows.

$$u = 4 \implies x^2 = 4 \implies x = \pm 2$$

$$u = 9 \implies x^2 = 9 \implies x = \pm 3$$

The solutions are $x = 2$, $x = -2$, $x = 3$, and $x = -3$. Check these in the original equation.

Graphical Solution

Use a graphing utility to graph

$$y = x^4 - 13x^2 + 36.$$

From the graph in Figure 6.3, it appears that the graph has four x-intercepts. You can use the *zoom* and *trace* features of the graphing utility to approximate the corresponding solutions to be $x = \pm 2$ and $x = \pm 3$.

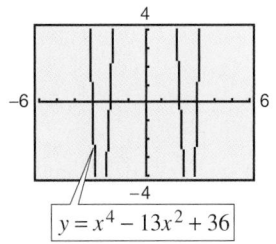

$$y = x^4 - 13x^2 + 36$$

Figure 6.3

NOTE Be sure you see in Example 7 that the u-solutions of 4 and 9 represent only a temporary step in the algebraic solution. They are not solutions of the original equation.

Instead of making a substitution in Example 7, you could also factor the expression on the left side of the original equation, as follows.

$$x^4 - 13x^2 + 36 = 0$$

$$(x^2 - 4)(x^2 - 9) = 0$$

$$x^2 - 4 = 0 \implies x^2 = 4 \implies x = \pm 2$$

$$x^2 - 9 = 0 \implies x^2 = 9 \implies x = \pm 3$$

Example 8 ■ Solving an Equation of Quadratic Form

Solve $x - 5\sqrt{x} + 6 = 0$.

Solution

This equation is of quadratic form with $u = \sqrt{x}$.

$$x - 5\sqrt{x} + 6 = 0 \qquad \text{Write original equation.}$$
$$\left(\sqrt{x}\right)^2 - 5\left(\sqrt{x}\right) + 6 = 0 \qquad \text{Write in quadratic form.}$$
$$u^2 - 5u + 6 = 0 \qquad \text{Substitute } u \text{ for } \sqrt{x}.$$
$$(u - 2)(u - 3) = 0 \qquad \text{Factor.}$$
$$u - 2 = 0 \implies u = 2 \qquad \text{Set 1st factor equal to 0.}$$
$$u - 3 = 0 \implies u = 3 \qquad \text{Set 2nd factor equal to 0.}$$

Now, using the u-solutions of 2 and 3, you obtain the x-solutions as follows.

$$u = 2 \implies \sqrt{x} = 2 \implies x = 4$$
$$u = 3 \implies \sqrt{x} = 3 \implies x = 9$$

The solutions are $x = 4$ and $x = 9$. Check these in the original equation.

The equation in Example 8 could also be solved using the techniques for solving radical equations as discussed in Section 5.5.

Example 9 ■ Solving an Equation of Quadratic Form

Solve $x^{2/3} - x^{1/3} - 6 = 0$.

Solution

This equation is of quadratic form with $u = x^{1/3}$.

$$x^{2/3} - x^{1/3} - 6 = 0 \qquad \text{Write original equation.}$$
$$\left(x^{1/3}\right)^2 - \left(x^{1/3}\right) - 6 = 0 \qquad \text{Write in quadratic form.}$$
$$u^2 - u - 6 = 0 \qquad \text{Substitute } u \text{ for } x^{1/3}.$$
$$(u + 2)(u - 3) = 0 \qquad \text{Factor.}$$
$$u + 2 = 0 \implies u = -2 \qquad \text{Set 1st factor equal to 0.}$$
$$u - 3 = 0 \implies u = 3 \qquad \text{Set 2nd factor equal to 0.}$$

Now, using the u-solutions of -2 and 3, you obtain the x-solutions as follows.

$$u = -2 \implies x^{1/3} = -2 \implies x = -8$$
$$u = 3 \implies x^{1/3} = 3 \implies x = 27$$

The solutions are $x = -8$ and $x = 27$. Check these in the original equation.

NOTE Remember that checking the solutions of a radical equation is especially important because the trial solutions often turn out to be extraneous.

Example 10 ■ An Application: The Dimensions of a Room

You are working on some house plans. You want the living room in the house to have 200 square feet of floor space. What dimensions should the living room have if you want it to be a rectangle whose width is two-thirds of its length?

Solution

If the room is rectangular (see Figure 6.4), let x represent the length. Because the width is two-thirds of the length, you can represent the width by $\frac{2}{3}x$.

$$x\left(\tfrac{2}{3}x\right) = 200 \qquad\qquad \text{Formula for area}$$

$$\tfrac{2}{3}x^2 = 200 \qquad\qquad \text{Simplify.}$$

$$x^2 = 300 \qquad\qquad \text{Multiply each side by } \tfrac{3}{2}.$$

$$x = \pm\sqrt{300} \qquad\qquad \text{Extract square roots.}$$

$$x \approx \pm 17.32 \qquad\qquad \text{Use a calculator.}$$

Choosing the positive value, you find that the room's length is $x \approx 17.32$ feet and its width is $\frac{2}{3}x \approx 11.55$ feet. Check this solution by multiplying 17.32 by 11.55 to see that you obtain approximately 200 square feet of floor space.

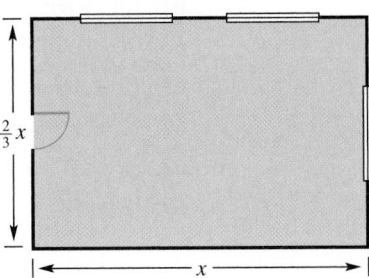

Figure 6.4

Example 11 ■ An Application: Surface Area of a Sphere

The surface area of a sphere of radius r is given by $S = 4\pi r^2$. The surface area of a softball is $144/\pi$ square inches. Find the diameter d of the softball.

Solution

$$S = 4\pi r^2 \qquad\qquad \text{Formula for surface area}$$

$$\frac{144}{\pi} = 4\pi r^2 \qquad\qquad \text{Substitute } 144/\pi \text{ for } S.$$

$$\frac{36}{\pi^2} = r^2 \quad\Longrightarrow\quad \pm\sqrt{\frac{36}{\pi^2}} = r \qquad\qquad \text{Divide each side by } 4\pi \text{ and extract square roots.}$$

Choosing the positive root, you obtain $r = 6/\pi$, and so the diameter of the softball is

$$d = 2r = 2\left(\frac{6}{\pi}\right) = \frac{12}{\pi} \approx 3.82 \text{ inches.}$$

6.1 ■ Exercises

Developing Skills

In Exercises 1–24, solve the quadratic equation by factoring.

1. $x^2 - 12x + 35 = 0$ **2.** $x^2 + 15x + 44 = 0$

3. $x^2 + x - 72 = 0$ **4.** $x^2 - 2x - 48 = 0$

5. $x^2 + 4x = 45$ **6.** $x^2 - 7x = 18$

7. $4x^2 - 12x = 0$ **8.** $25y^2 - 75y = 0$

9. $u(u - 9) - 12(u - 9) = 0$

10. $16x(x - 8) - 12(x - 8) = 0$

11. $4x^2 - 25 = 0$ **12.** $16y^2 - 121 = 0$

13. $8x^2 - 10x + 3 = 0$ **14.** $9x^2 + 24x + 16 = 0$

15. $x^2 + 60x + 900 = 0$ **16.** $x^2 - 12x + 36 = 0$

17. $(y - 4)(y - 3) = 6$

18. $(6 + u)(1 - u) = 10$

19. $2x(3x + 2) = 5 - 6x^2$

20. $8 - 3t^2 = 2t(6t - 7)$

21. $(3w + 2)(3w - 2) = -w^2 + 26w - 16$

22. $(2z + 1)(2z - 1) = -4z^2 - 5z + 2$

23. $3x(x - 6) - 5(x - 6) = 0$

24. $3(4 - x) - 2x(4 - x) = 0$

In Exercises 25–48, solve the quadratic equation by extracting square roots.

25. $x^2 = 64$ **26.** $z^2 = 169$

27. $6x^2 = 54$ **28.** $5t^2 = 125$

29. $\dfrac{y^2}{2} = 32$ **30.** $\dfrac{x^2}{6} = 24$

31. $25x^2 = 16$ **32.** $9z^2 = 121$

33. $2x^2 = 48$ **34.** $3x^2 = 144$

35. $4u^2 - 225 = 0$ **36.** $16x^2 - 1 = 0$

37. $(x + 4)^2 = 169$ **38.** $(y - 20)^2 = 625$

39. $(x - 3)^2 = 0.25$ **40.** $(x + 2)^2 = 0.81$

41. $(x - 2)^2 = 7$ **42.** $(x + 8)^2 = 28$

43. $(2x + 1)^2 = 50$ **44.** $(3x - 5)^2 = 48$

45. $(x - 5)^2 - 36 = 0$ **46.** $(x + 4)^2 - 121 = 0$

47. $(4x - 3)^2 - 98 = 0$

48. $(5x + 11)^2 - 300 = 0$

In Exercises 49–72, solve the quadratic equation by extracting imaginary square roots.

49. $z^2 = -36$ **50.** $x^2 = -9$

51. $x^2 + 4 = 0$ **52.** $y^2 + 16 = 0$

53. $(t - 3)^2 = -25$ **54.** $(y + 7)^2 = -225$

55. $(t - 1)^2 + 169 = 0$ **56.** $(x + 5)^2 + 81 = 0$

57. $(x + 4)^2 + 121 = 0$ **58.** $(z - 2)^2 + 1 = 0$

59. $(2x - 1)^2 + 4 = 0$ **60.** $(4t + 3)^2 + 49 = 0$

61. $(3z + 4)^2 + 144 = 0$ **62.** $(2y - 3)^2 + 25 = 0$

63. $9(x + 6)^2 = -121$ **64.** $4(x - 4)^2 = -169$

65. $(x - 1)^2 = -27$ **66.** $(2x + 3)^2 = -54$

67. $(x + 1)^2 + 0.04 = 0$

68. $(x - 3)^2 + 2.25 = 0$

69. $\left(c - \frac{2}{3}\right)^2 + \frac{1}{9} = 0$ **70.** $\left(u + \frac{5}{8}\right)^2 + \frac{49}{16} = 0$

71. $\left(x + \frac{7}{3}\right)^2 = -\frac{38}{9}$ **72.** $\left(y - \frac{5}{6}\right)^2 = -\frac{4}{5}$

In Exercises 73–82, use a graphing utility to graph the function. Set $y = 0$ and use the graphing utility to solve the resulting equation. Verify your results algebraically.

73. $y = x^2 - 9$ **74.** $y = 5x - x^2$

75. $y = x^2 - 2x - 15$ **76.** $y = 9 - 4(x - 3)^2$

77. $y = 4 - (x - 3)^2$ **78.** $y = 4(x + 1)^2 - 9$

79. $y = 2x^2 - x - 6$ **80.** $y = 4x^2 - x - 14$

81. $y = 3x^2 - 8x - 16$ **82.** $y = 5x^2 + 9x - 18$

In Exercises 83–88, use a graphing utility to graph the function and observe that the graph has no x-intercepts. Set $y = 0$ and solve the resulting equation. What type of roots does the equation have?

83. $y = (x - 1)^2 + 1$ **84.** $y = (x + 2)^2 + 3$

85. $y = (x + 3)^2 + 5$ **86.** $y = (x - 2)^2 + 3$

87. $y = x^2 + 7$ **88.** $y = x^2 + 5$

 Warm-up exercises, quizzes, and other study resources appropriate to this section are available at *college.hmco.com.*

The symbol ▦ indicates an exercise in which you are instructed to use a calculator or graphing utility.

In Exercises 89–104, find all real *and* imaginary solutions of the quadratic equation.

89. $3x^2 + 12 = 0$

90. $4y^2 + 20 = 0$

91. $3x^2 - 7x = 0$

92. $2x^2 - 5x = 0$

93. $3x^2 + 8x - 16 = 0$

94. $8y^2 + 42y - 11 = 0$

95. $t^2 - 225 = 0$

96. $x^2 - 100 = 0$

97. $(x + 1)^2 = 121$

98. $(z - 5)^2 = 25$

99. $(y + 12)^2 + 400 = 0$

100. $(x - 5)^2 + 100 = 0$

101. $14x = x^2 + 49$

102. $41x = 12x^2 + 35$

103. $(x - 9)^2 - 324 = 0$

104. $(y + 12)^2 - 400 = 0$

In Exercises 105–108, solve for *y* in terms of *x*. Let *f* and *g* be functions representing, respectively, the positive square root and the negative square root. Use a graphing utility to graph *f* and *g* in the same viewing window.

105. $x^2 + y^2 = 4$

106. $x^2 - y^2 = 4$

107. $x^2 + 4y^2 = 4$

108. $x - y^2 = 0$

In Exercises 109–130, solve the equation of quadratic form. (Find all real *and* imaginary solutions.)

109. $x^4 - 5x^2 + 4 = 0$

110. $4x^4 - 101x^2 + 25 = 0$

111. $x^4 - 5x^2 + 6 = 0$

112. $x^4 - 11x^2 + 30 = 0$

113. $x^4 - 3x^2 - 4 = 0$

114. $x^4 - x^2 - 6 = 0$

115. $x^4 - 9x^2 + 8 = 0$

116. $x^4 - 2x^2 - 15 = 0$

117. $(x^2 - 4)^2 + 2(x^2 - 4) - 3 = 0$

118. $(x^2 - 1)^2 + (x^2 - 1) - 6 = 0$

119. $x^{2/3} - x^{1/3} - 6 = 0$

120. $x^{2/3} + 3x^{1/3} - 10 = 0$

121. $x^{2/3} + 2x^{1/3} - 8 = 0$

122. $x^{2/3} - 7x^{1/3} - 18 = 0$

123. $2x^{2/3} - 7x^{1/3} + 5 = 0$

124. $3x^{2/3} + 8x^{1/3} + 5 = 0$

125. $x^{2/5} + 7x^{1/5} + 12 = 0$

126. $x^{2/5} - 6x^{1/5} + 5 = 0$

127. $x^{2/5} - 3x^{1/5} + 2 = 0$

128. $x^{2/5} + 5x^{1/5} + 6 = 0$

129. $2x^{2/5} - 7x^{1/5} + 3 = 0$

130. $2x^{2/5} + 3x^{1/5} + 1 = 0$

Solving Problems

Geometry   In Exercises 131–134, find the length of the side labeled *x*. (Round the result to two decimal places.)

131.

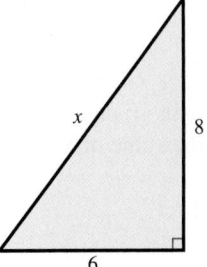

132.

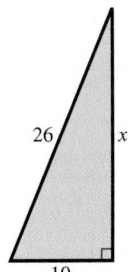

133.

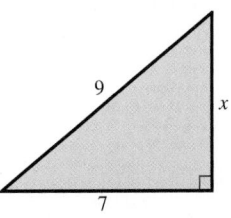

134.

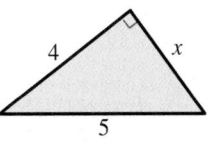

135. *Basketball*   A basketball court is 50 feet wide and 94 feet long. Find the length of the diagonal of the court.

136. *Plumbing*   A house has dimensions of 26 feet by 32 feet. The gas hot water heater and furnace are diagonally across the basement from where the natural gas line enters the house. Find the length of the gas line across the basement.

137. *Television Screens*   For a square 25-inch television screen, it is the diagonal measurement of the screen that is 25 inches. What are the dimensions of the screen?

138. *Geometry*   Determine the length and width of a rectangle with a perimeter of 68 inches and a diagonal of length 26 inches.

139. *Geometry*   The surface area of a baseball is $81/\pi$ square inches. Find the radius of the baseball.

140. *Geometry*   The surface area of a soccer ball is 784 square inches. Find the diameter of the soccer ball.

Free-Falling Object   **In Exercises 141 and 142, find the time required for an object to reach the ground if it is dropped from a height of s_0 feet. The height h (in feet) is given by**

$$h = -16t^2 + s_0$$

where t measures the time in seconds from the time the object is released.

141. $s_0 = 128$

142. $s_0 = 500$

143. *Free-Falling Object*   The height h (in feet) of an object propelled upward from a building 144 feet high is given by

$$h = 144 + 128t - 16t^2$$

where t measures the time in seconds from the time the object is released. Find the time it takes for the object to reach the ground.

144. *Revenue*   The revenue R (in dollars) from sales of x units of a product is given by

$$R = x\left(120 - \frac{1}{2}x\right).$$

Determine the number of units that must be sold to produce a revenue of $7000.

National Health Expenditures   **In Exercises 145 and 146, use the model, which gives the national expenditures for health care in the United States from 1993 through 1998.**

$$y = (0.77t + 27.7)^2, \qquad 3 \le t \le 8$$

In this model, y represents the expenditures (in billions of dollars) and t represents the year, with $t = 3$ corresponding to 1993 (see figure).   *(Source: U.S. Health Care Financing Administration)*

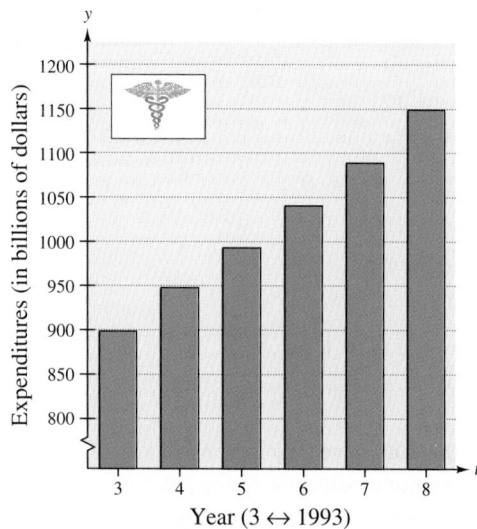

Year (3 ↔ 1993)

145. Algebraically determine the year when expenditures were approximately $900 billion. Graphically confirm the result.

146. Algebraically determine the year when expenditures were approximately $1095 billion. Graphically confirm the result.

147. *Geometry*   The area of the right triangle in the figure is 6 square units. Find the value of x.

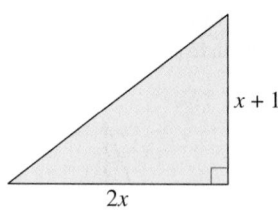

148. *Geometry*   The area of the right triangle in the figure is 84 square units. Find the value of x.

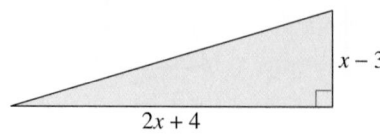

149. *Compound Interest* A principal of \$3500 is deposited in an account at an annual interest rate r compounded annually. If the amount after 2 years is \$5000, the annual interest rate is the solution of the equation

$$5000 = 3500(1 + r)^2.$$

Find r.

150. *Compound Interest* A principal of \$1500 is deposited in an account at an annual interest rate r compounded annually. If the amount after 2 years is \$1685.40, the annual interest rate is the solution of the equation

$$1685.40 = 1500(1 + r)^2.$$

Find r.

Explaining Concepts

151. Explain the Zero-Factor Property and how it can be used to solve a quadratic equation.

152. Is $x = 5$ the only solution of the equation $x^2 = 25$? Explain.

153. Is it possible for a quadratic equation to have only one solution? If so, give an example.

154. For a quadratic equation $ax^2 + bx + c = 0$, where a, b, and c are real numbers with $a \neq 0$, explain why b and c can equal zero but a cannot.

Ongoing Review

In Exercises 155–160, completely factor the expression.

155. $16r^4 - t^4$

156. $6x^2 + 25x + 4$

157. $57y^2 + y - 6$

158. $6x^2 + xy - 40y^2$

159. $5x(2x + 7) - (2x + 7)^2$

160. $x^2 + 10x + 25$

In Exercises 161–164, solve the equation.

161. $-3y + 20 = 2$

162. $5(x - 4) - 7 = -2(x - 3)$

163. $(x - 3)(x + 4) = 0$

164. $(3x - 7)(2x + 5) = 0$

165. *Speed* A jogger leaves a point on a fitness trail running at a rate of 6 miles per hour. Five minutes later a second jogger leaves from the same location running at 8 miles per hour. How long will it take the second runner to overtake the first, and how far will each have run at that point?

166. *Property Tax* The tax on a property with an assessed value of \$145,000 is \$2400. Find the tax on a property with an assessed value of \$90,000.

Looking Further

Suppose that you are tutoring a student in algebra and want to make up several practice equations for the student to solve by factoring. You might want to start by writing several simple equations with easy solutions such as 1 and 2. Then you might want to put in some tougher problems with solutions that include irrational or imaginary numbers.

(a) Find a quadratic equation that has $x = 1$ and $x = 2$ as solutions.

(b) Find a quadratic equation that has $x = 5$ and $x = -2$ as solutions.

(c) Find a quadratic equation that has $x = -3$ and $x = \frac{1}{2}$ as solutions.

(d) Find a quadratic equation that has $x = 1 + \sqrt{2}$ and $x = 1 - \sqrt{2}$ as solutions.

(e) Find a quadratic equation that has $x = 2 + 5i$ and $x = 2 - 5i$ as solutions.

(f) Write a short paragraph describing a general procedure for creating a quadratic equation that has two given numbers as solutions.

6.2 Completing the Square

Constructing Perfect Square Trinomials

Consider the quadratic equation

$$(x - 2)^2 = 10. \qquad \text{Completed square form}$$

You know from Example 4(b) in the preceding section that this equation has two solutions:

$$x = 2 + \sqrt{10} \text{ and } x = 2 - \sqrt{10}.$$

Suppose you were given the equation in its general form

$$x^2 - 4x - 6 = 0. \qquad \text{General form}$$

How would you solve this equation if you were given only the general form? You could try factoring, but after attempting to do so you would find that the left side of the equation is not factorable (using integer coefficients).

In this section, you will study a technique for rewriting an equation in a completed square form. This technique is called **completing the square.** To complete the square, you use the fact that all perfect square trinomials have a similar form. For instance, consider the perfect square trinomials listed below.

$$x^2 + 6x + 9 = (x + 3)^2$$
$$x^2 - 12x + 36 = (x - 6)^2$$
$$x^2 + 5x + \frac{25}{4} = \left(x + \frac{5}{2}\right)^2$$

In each case, note that the constant term of the perfect square trinomial is the square of half the coefficient of the x-term. That is, a perfect square trinomial has the following form.

Perfect Square Trinomial   Square of Binomial

$$x^2 + bx + \left(\frac{b}{2}\right)^2 = \left(x + \frac{b}{2}\right)^2$$

$$\underset{(\text{half})^2}{\underbrace{\qquad\qquad}}$$

So, to complete the square for an expression of the form $x^2 + bx$, you must add $(b/2)^2$ to the expression.

Study Tip

In Section 1.4, a *perfect square trinomial* was defined as the square of a binomial. For instance, the perfect square trinomial

$$x^2 + 10x + 25 = (x + 5)^2$$

is the square of the binomial $(x + 5)$.

Completing the Square

To **complete the square** for the expression $x^2 + bx$, add $(b/2)^2$, which is the square of half the coefficient of x. Consequently,

$$x^2 + bx + \left(\frac{b}{2}\right)^2 = \left(x + \frac{b}{2}\right)^2.$$

Example 1 ■ Creating a Perfect Square Trinomial

What term should be added to each expression so that it becomes a perfect square trinomial?

(a) $x^2 - 8x$ (b) $x^2 + 14x$ (c) $x^2 - 9x$ (d) $x^2 + \frac{7}{3}x$

Solution

(a) For this expression, the coefficient of the x-term is -8. By taking half of this coefficient and squaring the result, you can see that $(-4)^2 = 16$. This is the term that should be added to the expression to make it a perfect square trinomial.

$$x^2 - 8x + 16 = x^2 - 8x + (-4)^2 \qquad \text{Add 16 to the expression.}$$
$$= (x - 4)^2 \qquad \text{Completed square form}$$

(b) For this expression, the coefficient of the x-term is 14. By taking half of this coefficient and squaring the result, you can see that $7^2 = 49$ should be added to the expression to make it a perfect square trinomial.

$$x^2 + 14x + 49 = x^2 + 14x + 7^2 \qquad \text{Add 49 to the expression.}$$
$$= (x + 7)^2 \qquad \text{Completed square form}$$

(c) For this expression, the coefficient of the x-term is -9. By taking half of this coefficient and squaring the result, you can see that $\left(-\frac{9}{2}\right)^2 = \frac{81}{4}$ should be added to the expression to make it a perfect square trinomial.

$$x^2 - 9x + \tfrac{81}{4} = x^2 - 9x + \left(-\tfrac{9}{2}\right)^2 \qquad \text{Add } \tfrac{81}{4} \text{ to the expression.}$$
$$= \left(x - \tfrac{9}{2}\right)^2 \qquad \text{Completed square form}$$

(d) For this expression, the coefficient of the x-term is $\frac{7}{3}$. By taking half of this coefficient and squaring the result, you can see that $\left(\frac{1}{2} \cdot \frac{7}{3}\right)^2 = \left(\frac{7}{6}\right)^2 = \frac{49}{36}$ should be added to the expression to make it a perfect square trinomial.

$$x^2 + \tfrac{7}{3}x + \tfrac{49}{36} = x^2 + \tfrac{7}{3}x + \left(\tfrac{7}{6}\right)^2 \qquad \text{Add } \tfrac{49}{36} \text{ to the expression.}$$
$$= \left(x + \tfrac{7}{6}\right)^2 \qquad \text{Completed square form}$$

A geometric approach to completing the square is as follows.

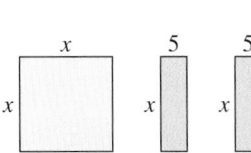

Figure 6.5

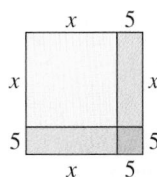

Figure 6.6

From Figure 6.5, you can see that the combined area of the three figures is $x^2 + 5x + 5x$ or $x^2 + 10x$. If you add a square whose area is $5^2 = 25$, the four figures form a square whose area is $x^2 + 10x + 25 = (x + 5)^2$ (see Figure 6.6).

Solving Equations by Completing the Square

When completing the square to solve an equation, remember that it is essential to *preserve the equality*. So, when you add a constant term to one side of the equation, you must be sure to add the same constant to the other side of the equation.

Example 2 ■ Completing the Square: Leading Coefficient Is 1

Solve $x^2 + 12x = 0$ by completing the square.

Solution

$$x^2 + 12x = 0 \qquad\qquad \text{Write original equation.}$$

$$x^2 + 12x + (6)^2 = 36 \qquad\qquad \text{Add } 6^2 = 36 \text{ to each side.}$$

$$\underset{\text{(half)}^2}{\underline{\hspace{1.2cm}}\uparrow}$$

$$(x + 6)^2 = 36 \qquad\qquad \text{Completed square form}$$

$$x + 6 = \pm\sqrt{36} \qquad\qquad \text{Extract square roots.}$$

$$x = -6 \pm 6 \qquad\qquad \text{Subtract 6 from each side.}$$

$$x = 0 \text{ or } x = -12 \qquad\qquad \text{Solutions}$$

The solutions are $x = 0$ and $x = -12$. Check these in the original equation.

NOTE In Example 2, completing the square is used for the sake of illustration. This particular equation would be easier to solve by factoring. Try reworking the problem by factoring to see that you obtain the same two solutions. In Example 3, the equation cannot be solved by factoring (using integer coefficients).

Example 3 ■ Completing the Square: Leading Coefficient Is 1

Solve $x^2 - 6x + 7 = 0$ by completing the square.

Solution

$$x^2 - 6x + 7 = 0 \qquad\qquad \text{Write original equation.}$$

$$x^2 - 6x = -7 \qquad\qquad \text{Subtract 7 from each side.}$$

$$x^2 - 6x + (-3)^2 = -7 + 9 \qquad\qquad \text{Add } (-3)^2 = 9 \text{ to each side.}$$

$$\underset{\text{(half)}^2}{\underline{\hspace{1.2cm}}\uparrow}$$

$$(x - 3)^2 = 2 \qquad\qquad \text{Completed square form}$$

$$x - 3 = \pm\sqrt{2} \qquad\qquad \text{Extract square roots.}$$

$$x = 3 \pm \sqrt{2} \qquad\qquad \text{Solutions}$$

Technology

Use a graphing utility to graph the equation in Example 3. Use the *zoom* and *trace* features of the graphing utility to approximate the x-intercepts of the equation to check your solutions.

The solutions are $x = 3 + \sqrt{2}$ and $x = 3 - \sqrt{2}$. Check these in the original equation.

Example 4 ■ Completing the Square: Leading Coefficient Is Not 1

Solve $3x^2 + 5x = 2$ by completing the square.

Solution

$$3x^2 + 5x = 2 \qquad \text{Write original equation.}$$

$$x^2 + \frac{5}{3}x = \frac{2}{3} \qquad \text{Divide each side by 3.}$$

$$x^2 + \frac{5}{3}x + \left(\frac{5}{6}\right)^2 = \frac{2}{3} + \frac{25}{36} \qquad \text{Add } \left(\frac{5}{6}\right)^2 = \frac{25}{36} \text{ to each side.}$$

$$\left(x + \frac{5}{6}\right)^2 = \frac{49}{36} \qquad \text{Completed square form}$$

$$x + \frac{5}{6} = \pm\frac{7}{6} \qquad \text{Extract square roots.}$$

$$x = -\frac{5}{6} \pm \frac{7}{6} \qquad \text{Subtract } \frac{5}{6} \text{ from both sides.}$$

$$x = \frac{1}{3} \text{ or } x = -2 \qquad \text{Solutions}$$

The solutions are $x = \frac{1}{3}$ and $x = -2$. Check these in the original equation.

Example 5 ■ Quadratic Equation: Leading Coefficient Is Not 1

Solve $2x^2 - x - 2 = 0$.

Algebraic Solution

$$2x^2 - x - 2 = 0 \qquad \text{Write original equation.}$$

$$2x^2 - x = 2 \qquad \text{Add 2 to each side.}$$

$$x^2 - \frac{1}{2}x = 1 \qquad \text{Divide each side by 2.}$$

$$x^2 - \frac{1}{2}x + \left(-\frac{1}{4}\right)^2 = 1 + \frac{1}{16} \qquad \text{Add } \left(-\frac{1}{4}\right)^2 = \frac{1}{16} \text{ to each side.}$$

$$\left(x - \frac{1}{4}\right)^2 = \frac{17}{16} \qquad \text{Completed square form}$$

$$x - \frac{1}{4} = \pm\frac{\sqrt{17}}{4} \qquad \text{Extract square roots.}$$

$$x = \frac{1}{4} \pm \frac{\sqrt{17}}{4} \qquad \text{Add } \frac{1}{4} \text{ to each side.}$$

The solutions are $x = \frac{1}{4}\left(1 + \sqrt{17}\right)$ and $x = \frac{1}{4}\left(1 - \sqrt{17}\right)$. Check these in the original equation. You can use a calculator to obtain the decimal approximations $x \approx 1.28$ and $x \approx -0.78$.

Graphical Solution

Use a graphing utility to graph

$$y = 2x^2 - x - 2.$$

From the graph in Figure 6.7, it appears that the graph has two x-intercepts. You can use the *zoom* and *trace* features of the graphing utility to approximate the corresponding solutions to be $x \approx 1.28$ and $x \approx -0.78$.

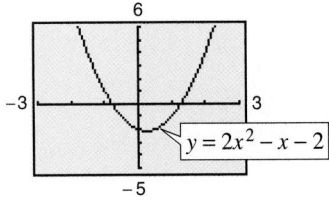

Figure 6.7

Example 6 ■ A Quadratic Equation with Imaginary Solutions

Solve $x^2 - 4x + 8 = 0$ by completing the square.

Solution

$$x^2 - 4x + 8 = 0 \qquad \text{Write original equation.}$$

$$x^2 - 4x = -8 \qquad \text{Subtract 8 from each side.}$$

$$x^2 - 4x + (-2)^2 = -8 + 4 \qquad \text{Add } (-2)^2 = 4 \text{ to each side.}$$

$$(x - 2)^2 = -4 \qquad \text{Completed square form}$$

$$x - 2 = \pm 2i \qquad \text{Extract imaginary square roots.}$$

$$x = 2 \pm 2i \qquad \text{Add 2 to each side.}$$

The solutions are $x = 2 + 2i$ and $x = 2 - 2i$. Check these in the original equation.

Example 7 ■ An Application: Dimensions of a Cereal Box

A cereal box has a volume of 441 cubic inches. Its height is 12 inches and its base has the dimensions x by $x + 7$ (see Figure 6.8). Find the dimensions of the base in inches.

Solution

$$lwh = V \qquad \text{Formula for volume of a rectangular box}$$

$$(x + 7)(x)(12) = 441 \qquad \begin{array}{l}\text{Substitute 441 for } V, x + 7 \text{ for length,}\\ x \text{ for width, and 12 for height.}\end{array}$$

$$12x^2 + 84x = 441 \qquad \text{Multiply factors.}$$

$$x^2 + 7x = \frac{441}{12} \qquad \text{Divide each side by 12.}$$

$$x^2 + 7x + \left(\frac{7}{2}\right)^2 = \frac{147}{4} + \frac{49}{4} \qquad \text{Add } \left(\frac{7}{2}\right)^2 = \frac{49}{4} \text{ to each side.}$$

$$\left(x + \frac{7}{2}\right)^2 = \frac{196}{4} \qquad \text{Completed square form}$$

$$\left(x + \frac{7}{2}\right)^2 = 49 \qquad \text{Simplify.}$$

$$x + \frac{7}{2} = \pm\sqrt{49} \qquad \text{Extract square roots.}$$

$$x = -\frac{7}{2} \pm 7 \qquad \text{Subtract } \tfrac{7}{2} \text{ from each side.}$$

Choosing the positive root, you obtain

$$x = -\frac{7}{2} + 7 = \frac{7}{2} = 3.5 \text{ inches} \qquad \text{Width of base}$$

and

$$x + 7 = 3.5 + 7 = 10.5 \text{ inches.} \qquad \text{Length of base}$$

Figure 6.8

12 in.

$x + 7$ x

LOW FAT OATIES NET WT 11.25 OZ (318g)

6.2 ■ Exercises

Developing Skills

In Exercises 1–16, add a term to the expression so that it becomes a perfect square trinomial.

1. $x^2 + 8x +$

2. $x^2 + 12x +$

3. $y^2 - 20y +$

4. $y^2 - 2y +$

5. $x^2 + 4x +$

6. $x^2 + 18x +$

7. $t^2 + 5t +$

8. $u^2 + 7u +$

9. $z^2 - 9z +$

10. $t^2 - 3t +$

11. $x^2 - \frac{6}{5}x +$

12. $y^2 + \frac{4}{3}y +$

13. $y^2 - \frac{3}{5}y +$

14. $a^2 - \frac{1}{3}a +$

15. $r^2 - 0.4r +$

16. $s^2 + 4.5s +$

In Exercises 17–26, solve the quadratic equation first by completing the square and then by factoring.

17. $x^2 + 6x = 0$

18. $t^2 - 9t = 0$

19. $t^2 - 8t + 7 = 0$

20. $y^2 - 8y + 12 = 0$

21. $x^2 + 2x - 24 = 0$

22. $x^2 + 12x + 27 = 0$

23. $x^2 + 7x + 12 = 0$

24. $z^2 + 3z - 10 = 0$

25. $5x^2 - 3x - 8 = 0$

26. $3x^2 - 3x - 6 = 0$

In Exercises 27–68, solve the quadratic equation by completing the square. Give the solutions in exact form and in decimal form rounded to two decimal places. (The solutions may be imaginary numbers.)

27. $x^2 - 4x - 3 = 0$

28. $x^2 - 6x + 7 = 0$

29. $x^2 + 6x + 7 = 0$

30. $x^2 + 4x - 3 = 0$

31. $u^2 - 4u + 1 = 0$

32. $a^2 - 10a - 15 = 0$

33. $x^2 + 2x + 3 = 0$

34. $x^2 - 6x + 12 = 0$

35. $x^2 - 10x - 2 = 0$

36. $x^2 + 8x - 4 = 0$

37. $y^2 + 20y + 10 = 0$

38. $y^2 + 6y - 24 = 0$

39. $t^2 + 5t + 3 = 0$

40. $u^2 - 9u - 1 = 0$

41. $v^2 + 3v - 2 = 0$

42. $z^2 - 7z + 9 = 0$

43. $x^2 - 11x + 3 = 0$

44. $u^2 + 15u + 2 = 0$

45. $-x^2 + x - 1 = 0$

46. $1 - x - x^2 = 0$

47. $x^2 - 5x + 5 = 0$

48. $x^2 + 3x - 1 = 0$

49. $x^2 - \frac{2}{3}x - 3 = 0$

50. $x^2 + \frac{4}{5}x - 1 = 0$

51. $v^2 + \frac{3}{4}v - 2 = 0$

52. $u^2 - \frac{2}{3}u + 5 = 0$

53. $2x^2 + 8x + 3 = 0$

54. $3x^2 - 24x - 5 = 0$

55. $3x^2 + 9x + 5 = 0$

56. $5x^2 - 15x + 7 = 0$

57. $4y^2 + 4y - 9 = 0$

58. $4z^2 - 3z + 2 = 0$

59. $2x^2 + 5x - 8 = 0$

60. $3t^2 - 7t - 1 = 0$

61. $5x^2 - 3x + 10 = 0$

62. $4x^2 + 4x - 7 = 0$

63. $x(x - 7) = 2$

64. $2x\left(x + \frac{4}{3}\right) = 5$

65. $\frac{1}{2}t^2 + t + 2 = 0$

66. $0.1x^2 + 0.5x = -0.2$

67. $0.1x^2 + 0.2x + 0.5 = 0$

68. $0.02x^2 + 0.10x - 0.05 = 0$

In Exercises 69–72, find the real solutions.

69. $\dfrac{x^2}{4} = \dfrac{x + 1}{2}$

70. $\dfrac{x^2 + 2}{24} = \dfrac{x - 1}{3}$

71. $\sqrt{2x + 1} = x - 3$

72. $\sqrt{3x - 2} = x - 2$

In Exercises 73–76, use a graphing utility to graph the function. Set $y = 0$ and use the graphing utility to solve the resulting equation. Then verify your results algebraically.

73. $y = x^2 + 6x - 4$

74. $y = x^2 - 2x - 5$

75. $y = \frac{1}{3}x^2 + 2x - 6$

76. $y = \frac{1}{2}x^2 - 3x + 1$

Solving Problems

77. *Geometric Modeling*

(a) Find the area of the large square and the two adjoining rectangles in the figure.

(b) Find the area of the small square in the lower right-hand corner of the figure and add it to the area found in part (a).

(c) Find the dimensions and the area of the entire figure after adjoining the small square in the lower right-hand corner. Note that you have shown completing the square geometrically.

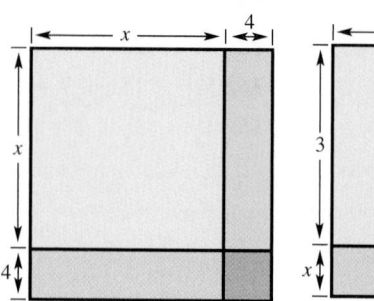

Figure for 77   Figure for 78

78. *Geometric Modeling* Repeat Exercise 77 for the model shown above.

79. *Geometry* Find the base and height of the triangle in the figure if its area is 12 square centimeters.

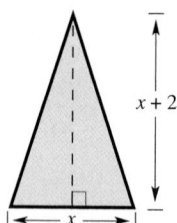

80. *Geometry* The area of the rectangle in the figure is 160 square feet. Find the rectangle's dimensions.

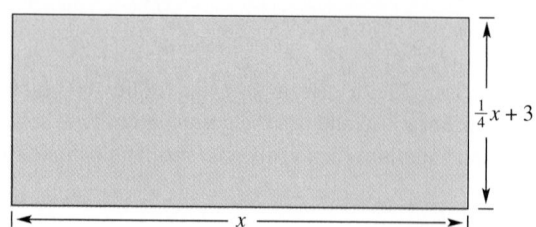

81. *Geometry* An open box with a rectangular base of x inches by $x + 4$ inches has a height of 6 inches (see figure). Find the dimensions of the box if its volume is 840 cubic inches.

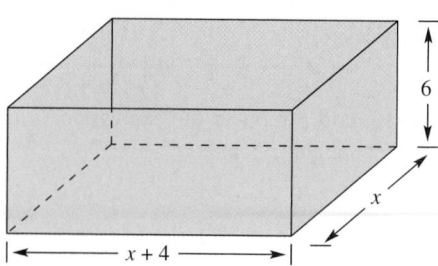

Figure for 81

82. *Geometry* An open box with a rectangular base of $x + 2$ inches by $x - 2$ inches has a height of 12 inches (see figure). Find the dimensions of the box if its volume is 540 cubic inches.

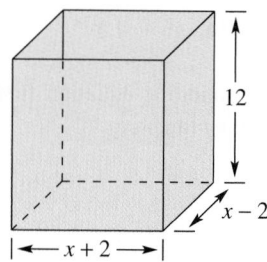

83. *Number Problem* Find two consecutive positive even integers whose product is 288.

84. *Number Problem* Find two consecutive positive odd integers whose product is 323.

85. *Revenue* The revenue R from selling x units of a product is

$$R = x\left(50 - \tfrac{1}{2}x\right).$$

Find the number of units that must be sold to produce a revenue of $1218.

86. *Revenue* The revenue R from selling x units of a product is

$$R = x\left(100 - \tfrac{1}{10}x\right).$$

Find the number of units that must be sold to produce a revenue of $12,040.

87. *ATM Terminals* The number y of ATM terminals (in thousands) in the United States from 1990 through 1999 can be modeled by

$$y = 1.904t^2 - 1.36t + 81.0, \ 0 \le t \le 9$$

where $t = 0$ represents 1990. From this model, in what year did the number of terminals reach 200,000? *(Source: Bank Network News)*

Explaining Concepts

88. Is it possible for a quadratic equation to have no real-number solution? If so, give an example and describe the graph.

89. When completing the square to solve a quadratic equation, what is the first step if the leading coefficient is not 1? Is the resulting equation equivalent to the original equation? Explain.

90. Explain the use of extracting square roots when solving a quadratic equation by the method of completing the square.

91. If you solve a quadratic equation by completing the square and obtain solutions that are rational numbers, then could you have solved the equation by factoring? Explain.

Ongoing Review

In Exercises 92–95, expand and simplify the expression.

92. $(t - 7)^2 + 5$

93. $(x + 6)^2 - 10$

94. $\left(y + \frac{1}{2}\right)^2 - 9$

95. $(2x + 1)^2 - 4$

In Exercises 96–101, solve the equation. (Find all real *and* imaginary solutions.)

96. $x^2 = -\frac{81}{121}$

97. $y^2 = \frac{4}{225}$

98. $(x + 7)^2 = 50$

99. $(t - 3)^2 = -64$

100. $(5x - 3)^2 = 7$

101. $5(3 - t) + 2t(3 - t) = 0$

102. *Geometry*   Find the length and the width of a rectangle whose length is 5 inches more than its width and whose diagonal is $5\sqrt{13}$ inches long.

103. *Demand*   The demand equation for a certain product is given by

$$p = 75 - \sqrt{1.2(x - 10)}$$

where x is the number of units demanded per day and p is the price per unit. Find the demand if the price is $59.90.

Looking Further

Completed Square Form   Consider the quadratic equation

$$(x - 1)^2 = d.$$

(a) What value(s) of d will produce a quadratic equation that has exactly one (repeated) solution?

(b) Describe the values of d that will produce two different solutions, both of which are *rational* numbers.

(c) Describe the values of d that will produce two different solutions, both of which are *irrational* numbers.

(d) Describe the values of d that will produce two different solutions, both of which are *imaginary* numbers.

6.3 The Quadratic Formula

What you should learn:

- How to derive the Quadratic Formula by completing the square for a general quadratic equation
- How to use the Quadratic Formula to solve quadratic equations
- How to determine the types of solutions of quadratic equations using the discriminant

Why you should learn it:

Quadratic equations can be used to model data for analysis of consumer behavior. For instance, in Exercise 92 on page 390 a quadratic equation can model the number of cellular phone subscribers in the United States.

The Quadratic Formula

You have now learned three techniques for solving quadratic equations: factoring, extracting square roots, and completing the square. A fourth technique involves the **Quadratic Formula.** This formula is obtained by completing the square for a general quadratic equation.

$$ax^2 + bx + c = 0 \qquad \text{General form, } a \neq 0$$

$$ax^2 + bx = -c \qquad \text{Subtract } c \text{ from each side.}$$

$$x^2 + \frac{b}{a}x = -\frac{c}{a} \qquad \text{Divide each side by } a.$$

$$x^2 + \frac{b}{a}x + \left(\frac{b}{2a}\right)^2 = -\frac{c}{a} + \left(\frac{b}{2a}\right)^2 \qquad \text{Add } \left(\frac{b}{2a}\right)^2 \text{ to each side.}$$

$$\left(x + \frac{b}{2a}\right)^2 = \frac{b^2 - 4ac}{4a^2} \qquad \text{Simplify.}$$

$$x + \frac{b}{2a} = \pm\sqrt{\frac{b^2 - 4ac}{4a^2}} \qquad \text{Extract square roots.}$$

$$x + \frac{b}{2a} = \pm\frac{\sqrt{b^2 - 4ac}}{2|a|} \qquad \text{Simplify radical.}$$

$$x = -\frac{b}{2a} \pm \frac{\sqrt{b^2 - 4ac}}{2|a|} \qquad \text{Subtract } \frac{b}{2a} \text{ from each side.}$$

$$x = \frac{-b \pm \sqrt{b^2 - 4ac}}{2a} \qquad \text{Simplify.}$$

NOTE Notice in the derivation of the Quadratic Formula that, because $\pm 2|a|$ represents the same numbers as $\pm 2a$, you can omit the absolute value bars.

The Quadratic Formula

The solutions of $ax^2 + bx + c = 0$, $a \neq 0$, are given by the **Quadratic Formula**

$$x = \frac{-b \pm \sqrt{b^2 - 4ac}}{2a}.$$

The Quadratic Formula is one of the most important formulas in algebra, and you should memorize it. It helps to try to memorize a verbal statement of the rule. For instance, you might try to remember the following verbal statement of the Quadratic Formula: "The opposite of b, plus or minus the square root of b squared minus $4ac$, all divided by $2a$."

Solving Equations by the Quadratic Formula

NOTE In Example 1, the solutions are rational numbers, which means that the equation could have been solved by factoring.

When using the Quadratic Formula, remember that *before* the formula can be applied, you must first write the quadratic equation in general form.

Example 1 ■ The Quadratic Formula: Two Distinct Solutions

Solve $x^2 + 6x = 16$ by using the Quadratic Formula.

Solution

$$x^2 + 6x = 16 \qquad \text{Write original equation.}$$
$$x^2 + 6x - 16 = 0 \qquad \text{Write in general form.}$$
$$x = \frac{-b \pm \sqrt{b^2 - 4ac}}{2a} \qquad \text{Quadratic Formula}$$
$$x = \frac{-6 \pm \sqrt{6^2 - 4(1)(-16)}}{2(1)} \qquad \text{Substitute 1 for } a, \text{ 6 for } b, \text{ and } -16 \text{ for } c.$$
$$x = \frac{-6 \pm \sqrt{100}}{2} \qquad \text{Simplify.}$$
$$x = \frac{-6 \pm 10}{2} \qquad \text{Simplify.}$$
$$x = 2 \ \text{ or } \ x = -8 \qquad \text{Solutions}$$

The solutions are $x = 2$ and $x = -8$. Check these in the original equation. Or, try using a graphic check, as shown in Figure 6.9.

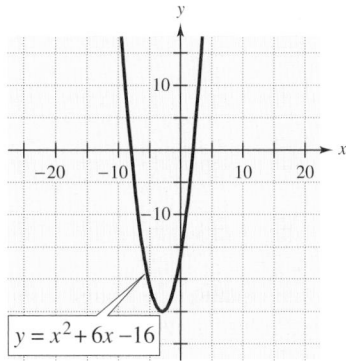

$y = x^2 + 6x - 16$

Figure 6.9

NOTE If the leading coefficient of a quadratic equation is negative, we suggest that you begin by multiplying each side of the equation by -1, as shown in Example 2. This will produce a positive leading coefficient, which is less cumbersome to work with.

Example 2 ■ The Quadratic Formula: Two Distinct Solutions

Solve $-x^2 - 4x + 8 = 0$ by using the Quadratic Formula.

Solution

$$-x^2 - 4x + 8 = 0 \qquad \text{Leading coefficient is negative.}$$
$$x^2 + 4x - 8 = 0 \qquad \text{Multiply each side by } -1.$$
$$x = \frac{-b \pm \sqrt{b^2 - 4ac}}{2a} \qquad \text{Quadratic Formula}$$
$$x = \frac{-4 \pm \sqrt{4^2 - 4(1)(-8)}}{2(1)} \qquad \text{Substitute 1 for } a, \text{ 4 for } b, \text{ and } -8 \text{ for } c.$$
$$x = \frac{-4 \pm \sqrt{48}}{2} \qquad \text{Simplify.}$$
$$x = \frac{-4 \pm 4\sqrt{3}}{2} \qquad \text{Simplify.}$$
$$x = \frac{2\left(-2 \pm 2\sqrt{3}\right)}{2} \qquad \text{Divide out common factor.}$$
$$x = -2 \pm 2\sqrt{3} \qquad \text{Solutions}$$

The solutions are $x = -2 + 2\sqrt{3}$ and $x = -2 - 2\sqrt{3}$. Check these in the original equation.

Example 3 ■ The Quadratic Formula: One Repeated Solution

Solve $18x^2 - 24x + 8 = 0$ by using the Quadratic Formula.

Solution

$$18x^2 - 24x + 8 = 0 \qquad \text{Write original equation.}$$

$$9x^2 - 12x + 4 = 0 \qquad \text{Divide each side by 2.}$$

$$x = \frac{-b \pm \sqrt{b^2 - 4ac}}{2a} \qquad \text{Quadratic Formula}$$

$$x = \frac{-(-12) \pm \sqrt{(-12)^2 - 4(9)(4)}}{2(9)} \qquad \begin{array}{l}\text{Substitute 9 for } a, -12 \\ \text{for } b, \text{ and 4 for } c.\end{array}$$

$$x = \frac{12 \pm \sqrt{144 - 144}}{18} \qquad \text{Simplify.}$$

$$x = \frac{12 \pm \sqrt{0}}{18} \qquad \text{Simplify.}$$

$$x = \frac{2}{3} \qquad \text{Solution}$$

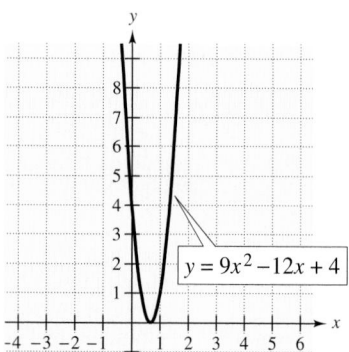

Figure 6.10

The only solution is $x = \frac{2}{3}$. Check this in the original equation. In Figure 6.10, note that when a quadratic equation has one repeated solution, its graph touches the x-axis at a single point.

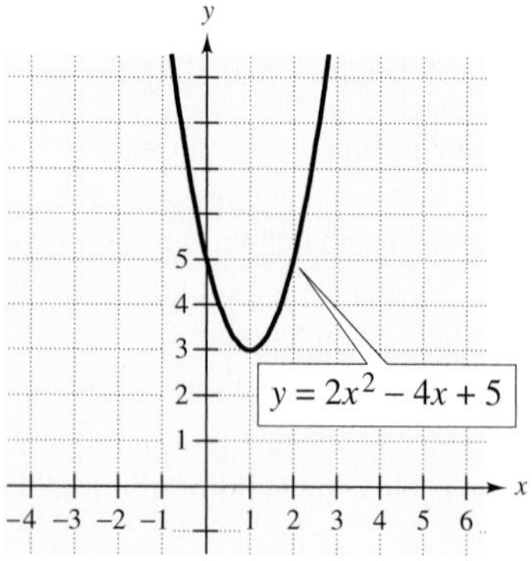

Example 4 ■ The Quadratic Formula: Imaginary Solutions

Solve $2x^2 - 4x + 5 = 0$ by using the Quadratic Formula.

Solution

$$2x^2 - 4x + 5 = 0 \qquad \text{Write original equation.}$$

$$x = \frac{-b \pm \sqrt{b^2 - 4ac}}{2a} \qquad \text{Quadratic Formula}$$

$$x = \frac{-(-4) \pm \sqrt{(-4)^2 - 4(2)(5)}}{2(2)} \qquad \begin{array}{l}\text{Substitute 2 for } a, -4 \\ \text{for } b, \text{ and 5 for } c.\end{array}$$

$$x = \frac{4 \pm \sqrt{-24}}{4} \qquad \text{Simplify.}$$

$$x = \frac{4 \pm 2\sqrt{6}i}{4} \qquad \text{Write in } i\text{-form.}$$

$$x = \frac{2(2 \pm \sqrt{6}i)}{2 \cdot 2} \qquad \begin{array}{l}\text{Divide out common} \\ \text{factor.}\end{array}$$

$$x = 1 \pm \frac{\sqrt{6}}{2}i \qquad \text{Solutions}$$

Figure 6.11

NOTE In Examples 1 through 4, notice the relationship between the real-number *solutions* of a quadratic equation $ax^2 + bx + c = 0$ and the *x-intercepts* of the graph of $y = ax^2 + bx + c$.

The solutions are $x = 1 \pm \frac{1}{2}\sqrt{6}i$. Check these in the original equation. In Figure 6.11, note that the graph has no x-intercepts. This verifies that the equation has no *real* solutions.

Example 5 ■ Solving a Quadratic Equation

Solve $3x^2 + 4x = 8$.

Algebraic Solution

$$3x^2 + 4x = 8 \qquad \text{Write original equation.}$$

$$3x^2 + 4x - 8 = 0 \qquad \text{Write in general form.}$$

$$x = \frac{-b \pm \sqrt{b^2 - 4ac}}{2a} \qquad \text{Quadratic Formula}$$

$$x = \frac{-4 \pm \sqrt{4^2 - 4(3)(-8)}}{2(3)} \qquad \text{Substitute 3 for } a, 4 \text{ for } b, \text{ and } -8 \text{ for } c.$$

$$x = \frac{-4 \pm \sqrt{112}}{6} \qquad \text{Simplify.}$$

$$x = \frac{-4 \pm 4\sqrt{7}}{6} \qquad \text{Simplify radical.}$$

$$x = \frac{2(-2 \pm 2\sqrt{7})}{2(3)} \qquad \text{Divide out common factor.}$$

$$x = \frac{-2 \pm 2\sqrt{7}}{3} \qquad \text{Solutions}$$

You can approximate the solutions to be $x \approx 1.10$ and $x \approx -2.43$. Check these in the original equation.

Graphical Solution

Use a graphing utility to graph

$$y_1 = 3x^2 + 4x \quad \text{and} \quad y_2 = 8$$

in the same viewing window. From Figure 6.12, it appears that the graphs intersect at two points. Use the *intersect* feature of the graphing utility to approximate the x-coordinates of these two points. The x-coordinates, $x \approx 1.10$ and $x \approx -2.43$, are the solutions of the equation $3x^2 + 4x = 8$.

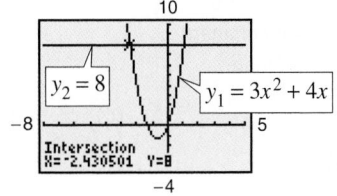

Figure 6.12

The Discriminant

The radicand in the Quadratic Formula, $b^2 - 4ac$, is called the **discriminant** because it allows you to "discriminate" among different types of solutions.

Study Tip

By reexamining Examples 1 through 5, you can see that the equations with rational or repeated solutions could have been solved by *factoring*. In general, quadratic equations (with integer coefficients) for which the discriminant is either zero or a perfect square are factorable using integer coefficients. Consequently, a quick test of the discriminant will help decide which solution method to use to solve a quadratic equation.

Using the Discriminant

Let a, b, and c be real numbers such that $a \neq 0$. The **discriminant** of the quadratic equation $ax^2 + bx + c = 0$ is given by $b^2 - 4ac$ and can be used to classify the solutions of the equation as follows.

Discriminant	Solution Type
1. Perfect square	Two distinct rational solutions (Example 1)
2. Positive nonperfect square	Two distinct irrational solutions (Examples 2 and 5)
3. Zero	One repeated rational solution (Example 3)
4. Negative number	Two distinct imaginary solutions (Example 4)

Example 6 ■ Using the Discriminant

Determine the type of solution(s) for each quadratic equation.

(a) $x^2 - x + 2 = 0$ (b) $2x^2 - 3x - 2 = 0$

(c) $x^2 - 2x + 1 = 0$ (d) $x^2 - 2x - 1 = 9$

Solution

Equation	Discriminant	Solution Type
(a) $x^2 - x + 2 = 0$	$\begin{aligned} b^2 - 4ac &= (-1)^2 - 4(1)(2) \\ &= 1 - 8 \\ &= -7 \end{aligned}$	Two distinct imaginary solutions
(b) $2x^2 - 3x - 2 = 0$	$\begin{aligned} b^2 - 4ac &= (-3)^2 - 4(2)(-2) \\ &= 9 + 16 \\ &= 25 \end{aligned}$	Two distinct rational solutions
(c) $x^2 - 2x + 1 = 0$	$\begin{aligned} b^2 - 4ac &= (-2)^2 - 4(1)(1) \\ &= 4 - 4 \\ &= 0 \end{aligned}$	One repeated rational solution
(d) $x^2 - 2x - 1 = 9$	$\begin{aligned} b^2 - 4ac &= (-2)^2 - 4(1)(-10) \\ &= 4 + 40 \\ &= 44 \end{aligned}$	Two distinct irrational solutions

---

When choosing one of these methods below, first check to see whether the equation is in a form in which you can extract square roots. Next, you can try factoring. If neither of these two methods works, use the Quadratic Formula (or completing the square), which will work for any quadratic equation. For *real* solutions, remember that you can use a graph to approximate the solutions.

Summary of Methods for Solving Quadratic Equations

Method	Example
1. Factoring	$3x^2 + x = 0$ $x(3x + 1) = 0 \implies x = 0$ and $3x + 1 = 0 \implies x = -\frac{1}{3}$
2. Extracing square roots	$(x + 2)^2 = 7$ $x + 2 = \pm\sqrt{7} \implies x = -2 + \sqrt{7}$ and $x = -2 - \sqrt{7}$
3. Completing the square	$x^2 + 6x = 2$ $x^2 + 6x + \left(\frac{1}{2} \cdot 6\right)^2 = 2 + \left(\frac{1}{2} \cdot 6\right)^2$ $(x + 3)^2 = 11 \implies x = -3 + \sqrt{11}$ and $x = -3 - \sqrt{11}$
4. Using the Quadratic Formula	$3x^2 - 2x + 2 = 0 \implies x = \dfrac{-(-2) \pm \sqrt{(-2)^2 - 4(3)(2)}}{2(3)} = \dfrac{1}{3} \pm \dfrac{\sqrt{5}}{3}i$

6.3 ■ Exercises

Developing Skills

In Exercises 1–4, write the quadratic equation in general form.

1. $2x^2 = 7 - 2x$　　　**2.** $7x^2 + 15x = 5$

3. $x(10 - x) = 5$　　　**4.** $x(3x + 8) = 15$

In Exercises 5–12, solve the quadratic equation first by using the Quadratic Formula and then by factoring.

5. $x^2 - 11x + 28 = 0$　　　**6.** $x^2 - 12x + 27 = 0$

7. $4x^2 + 4x + 1 = 0$　　　**8.** $9x^2 + 12x + 4 = 0$

9. $6x^2 - x - 2 = 0$　　　**10.** $10x^2 - 11x + 3 = 0$

11. $x^2 - 5x - 300 = 0$　　　**12.** $x^2 + 20x - 300 = 0$

In Exercises 13–42, solve the quadratic equation by using the Quadratic Formula. (Find all real *and* imaginary solutions.)

13. $x^2 - 2x - 4 = 0$　　　**14.** $x^2 - 2x - 6 = 0$

15. $t^2 + 4t + 1 = 0$　　　**16.** $y^2 + 6y + 4 = 0$

17. $x^2 + 6x - 3 = 0$　　　**18.** $x^2 + 8x - 4 = 0$

19. $x^2 - 10x + 23 = 0$　　　**20.** $u^2 - 12u + 29 = 0$

21. $x^2 + 3x + 3 = 0$　　　**22.** $2x^2 - x + 1 = 0$

23. $2v^2 - 2v - 1 = 0$　　　**24.** $4x^2 + 6x + 1 = 0$

25. $2x^2 + 4x - 3 = 0$　　　**26.** $2x^2 + 3x + 3 = 0$

27. $2x^2 + 4x + 1 = 0$　　　**28.** $2x^2 + 4x - 1 = 0$

29. $3x^2 - 5x + 3 = 0$

30. $3x^2 - 5x + 4 = 0$

31. $4x^2 - 4x + 5 = 0$

32. $3x^2 - 8x + 1 = 0$

33. $9z^2 + 6z - 4 = 0$

34. $8y^2 - 8y - 1 = 0$

35. $x^2 - 0.4x - 0.16 = 0$

36. $x^2 + 0.6x - 0.41 = 0$

37. $2.5x^2 + x - 0.9 = 0$

38. $0.09x^2 - 0.12x - 0.26 = 0$

39. $4x^2 - 6x + 3 = 0$

40. $-5x^2 - 15x + 10 = 0$

41. $9x^2 = 1 + 9x$

42. $x - x^2 = 1 - 6x^2$

In Exercises 43–52, use the discriminant to determine the type of solution(s) of the quadratic equation.

43. $x^2 + x + 1 = 0$　　　**44.** $x^2 + x - 1 = 0$

45. $2x^2 - 5x - 4 = 0$　　　**46.** $10x^2 + 5x + 1 = 0$

47. $x^2 + 7x + 15 = 0$　　　**48.** $3x^2 - 2x - 5 = 0$

49. $4x^2 - 12x + 9 = 0$　　　**50.** $2x^2 + 10x + 6 = 0$

51. $3x^2 - x + 2 = 0$　　　**52.** $9x^2 - 24x + 16 = 0$

In Exercises 53–72, solve the quadratic equation by the most convenient method. (Find all real *and* imaginary solutions.)

53. $z^2 - 169 = 0$　　　**54.** $t^2 = 150$

55. $y^2 + 15y = 0$　　　**56.** $x^2 + 22x = 0$

57. $9t^2 + 25 = 0$　　　**58.** $4u^2 + 49 = 0$

59. $25(x - 3)^2 - 36 = 0$

60. $4(y + 6)^2 + 9 = 0$

61. $3x(x + 10) - 4(x + 10) = 0$

62. $2y(y - 18) + 3(y - 18) = 0$

63. $(x + 4)^2 + 16 = 0$　　　**64.** $(y - 1)^2 - 25 = 0$

65. $t^2 + 7t + 12 = 0$　　　**66.** $x^2 - 3x - 4 = 0$

67. $18x^2 + 15x - 50 = 0$

68. $2x^2 - 15x + 225 = 0$

69. $x^2 - 24x + 128 = 0$

70. $x^2 + 8x + 25 = 0$

71. $1.2x^2 - 0.8x - 5.5 = 0$

72. $2x^2 + 8x + 4.5 = 0$

In Exercises 73–80, use a graphing utility to graph the function. Set $y = 0$ and use the graphing utility to solve the resulting equation. Verify your results algebraically.

73. $y = 3x^2 - 6x + 1$　　　**74.** $y = x^2 + x + 1$

75. $y = -(4x^2 - 20x + 25)$

76. $y = -(3x^2 + 5x - 6)$

77. $y = 5x^2 - 18x + 6$

78. $y = 15x^2 + 3x - 105$

79. $y = -0.04x^2 + 4x - 0.8$

80. $y = 3.7x^2 - 10.2x + 3.2$

In Exercises 81–84, use a graphing utility to determine the number of real solutions of the quadratic equation. Then verify your results algebraically.

81. $2x^2 - 5x + 5 = 0$

82. $2x^2 - x - 1 = 0$

83. $\frac{1}{5}x^2 + \frac{6}{5}x - 8 = 0$

84. $\frac{1}{3}x^2 - 5x + 25 = 0$

In Exercises 85–88, solve the equation.

85. $\dfrac{2x^2}{5} - \dfrac{x}{2} = 1$

86. $\dfrac{x^2 - 9x}{6} = \dfrac{x - 1}{2}$

87. $\sqrt{x + 3} = x - 1$

88. $\sqrt{2x - 3} = x - 2$

Solving Problems

89. *Geometry* A rectangle (see figure) has a width of x centimeters, a length of $x + 6.3$ centimeters, and an area of 58.14 square centimeters. Find its dimensions.

x

$x + 6.3$

90. *Geometry* A rectangle (see figure) has a width of x inches, a length of $x - 2.4$ inches, and an area of 133.12 square inches. Find its dimensions.

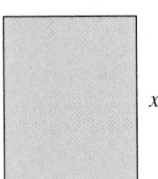

x

$x - 2.4$

91. *Analyzing Data* Use the model, which approximates the number of scientists employed in the electrical equipment industry in the United States in the years 1993 through 1998.

$$y = 81.8 - 1.64t + 1.648t^2, \quad 3 \le t \le 8$$

In this model, y represents the number of scientists employed in the electrical equipment industry (in thousands) and t represents the year, with $t = 3$ corresponding to 1993. *(Source: U.S. Department of Commerce)*

(a) Use a graphing utility to graph the model.

(b) Use the graph in part (a) to find the year in which there were approximately 100,000 scientists employed in the electrical equipment industry in the United States.

(c) Verify your answer from part (b) algebraically.

92. *Analyzing Data* The numbers of cellular phone subscribers s (in thousands) in the United States in the years 1994 through 1999 are shown in the table. *(Source: Cellular Telecommunications Industry Association)*

Year	1994	1995	1996
Subscribers	24,134	33,786	44,043

Year	1997	1998	1999
Subscribers	55,312	69,209	86,047

The data can be approximated by the model

$$s = 901.61t^2 + 482.1t + 8233, \quad 4 \le t \le 9$$

where $t = 4$ corresponds to 1994.

(a) Use a graphing utility to graph the model.

(b) Use the model to determine the year in which the cellular phone companies had approximately 70 million subscribers.

(c) Verify your answer from part (b) graphically.

Explaining Concepts

93. State the Quadratic Formula in words.

94. Explain how completing the square can be used to develop the Quadratic Formula.

In Exercises 95–98, describe the value of c such that the equation has (a) two real-number solutions, (b) one real-number solution, and (c) two complex-number solutions.

95. $x^2 - 6x + c = 0$ **96.** $x^2 - 12x + c = 0$

97. $x^2 + 8x + c = 0$ **98.** $x^2 + 2x + c = 0$

Ongoing Review

In Exercises 99–102, solve the quadratic equation by factoring.

99. $x^2 + 6x + 8 = 0$

100. $x^2 + 9x + 14 = 0$

101. $4x^2 + 12x + 9 = 0$

102. $9x^2 - 30x + 25$

In Exercises 103 and 104, solve the quadratic equation by completing the square. (Find all real *and* imaginary solutions.)

103. $x^2 + 7x - 2 = 0$

104. $2x^2 + 10x - 3 = 0$

In Exercises 105–108, simplify the radical.

105. $\sqrt{16 - 4(3)(1)}$

106. $\sqrt{4 - 4(3)(2)}$

107. $\sqrt{9 - 4(-1)(3)}$

108. $\sqrt{25 - 4(-5)(1)}$

109. *Mixture Problem*   A bag contains 42 coins, with a total weight of 246 grams. If the bag contains only gold coins that weigh 8 grams each and silver coins that weigh 5 grams each, how many gold and silver coins are in the bag?

110. *Number Problem*   One-fifth of a number plus one-fourth of the number is 5 less than one-half the number. What is the number?

Looking Further

Solutions of a Quadratic Equation   Determine the solutions x_1 and x_2 of each quadratic equation. Use the values of x_1 and x_2 to fill in the boxes.

Equation	x_1, x_2	$x_1 + x_2$	$x_1 x_2$
(a) $x^2 - x - 6 = 0$			
(b) $2x^2 + 5x - 3 = 0$			
(c) $4x^2 - 9 = 0$			
(d) $x^2 - 10x + 34 = 0$			

Consider a general quadratic equation

$$ax^2 + bx + c = 0$$

whose solutions are x_1 and x_2. Can you determine a relationship among the coefficients a, b, and c and the sum $(x_1 + x_2)$ and product $(x_1 x_2)$ of the solutions?

Mid-Chapter Quiz

Take this quiz as you would take a quiz in class. After you are done, check your work against the answers given in the back of the book.

In Exercises 1–8, solve the quadratic equation by the specified method.

1. Factor: $2x^2 - 72 = 0$

2. Factor: $2x^2 + 3x - 20 = 0$

3. Extract square roots: $t^2 = 12$

4. Extract square roots: $(u - 3)^2 - 16 = 0$

5. Complete the square: $s^2 + 10s + 1 = 0$

6. Complete the square: $2y^2 + 6y - 5 = 0$

7. Quadratic Formula: $x^2 + 4x - 6 = 0$

8. Quadratic Formula: $6v^2 - 3v - 4 = 0$

In Exercises 9–16, solve the equation by using the most convenient method. (Find all real *and* imaginary solutions.)

9. $x^2 + 5x + 7 = 0$

10. $36 - (t - 4)^2 = 0$

11. $x(x - 10) + 3(x - 10) = 0$

12. $x(x - 3) = 10$

13. $4b^2 - 12b + 9 = 0$

14. $3m^2 + 10m + 5 = 0$

15. $x^4 + 5x^2 - 14 = 0$

16. $x^{2/3} - 8x^{1/3} + 15 = 0$

In Exercises 17 and 18, use a graphing utility to graph the function. Set $y = 0$ and use the graphing utility to solve the resulting equation. Verify your results algebraically.

17. $y = \frac{1}{2}x^2 - 3x - 1$

18. $y = x^2 + 0.45x - 4$

19. The revenue R from selling x units of a certain product is given by

$$R = x(20 - 0.2x).$$

Find the number of units that must be sold to produce a revenue of $500.

In Exercises 20 and 21, find the length of the side labeled x. (Round the result to two decimal places.)

20.

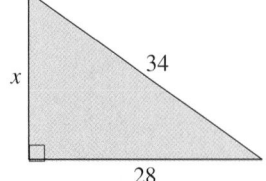

21.

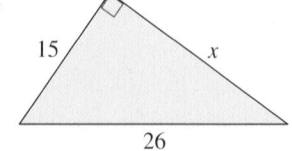

6.4 Applications of Quadratic Equations

Applications of Quadratic Equations

Example 1 ■ An Interest Problem

The formula

$$A = P(1 + r)^2$$

represents the amount of money A in an account in which P dollars is deposited for 2 years at an annual interest rate of r (in decimal form). Find the interest rate if a deposit of \$6000 increases to \$6933.75 over a two-year period.

Solution

$$A = P(1 + r)^2 \qquad \text{Write given formula.}$$
$$6933.75 = 6000(1 + r)^2 \qquad \text{Substitute for } A \text{ and } P.$$
$$1.155625 = (1 + r)^2 \qquad \text{Divide each side by 6000.}$$
$$\pm 1.075 = 1 + r \qquad \text{Extract square roots.}$$
$$0.075 = r \qquad \text{Choose positive solution.}$$

The annual interest rate is $r = 0.075 = 7.5\%$. Check this in the original statement of the problem.

Example 2 ■ Geometry

A picture is 4 inches taller than it is wide and has an area of 192 square inches. What are the dimensions of the picture?

Solution

Begin by drawing a diagram, as shown in Figure 6.13.

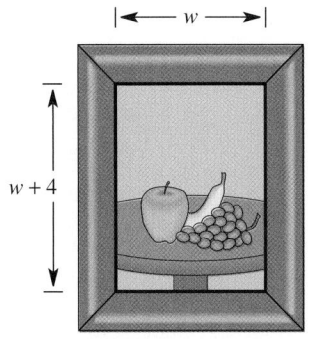

Figure 6.13

Verbal Model: Area of picture = Width · Height

Labels: Picture width $= w$ (inches)
Picture height $= w + 4$ (inches)
Area $= 192$ (square inches)

Equation:
$$192 = w(w + 4)$$
$$0 = w^2 + 4w - 192$$
$$0 = (w + 16)(w - 12)$$
$$w + 16 = 0 \quad \Longrightarrow \quad w = -16$$
$$w - 12 = 0 \quad \Longrightarrow \quad w = 12$$

Of the two possible solutions, choose the positive value of w and conclude that the picture is $w = 12$ inches wide and $w + 4 = 16$ inches tall. Check these dimensions in the original statement of the problem.

Example 3 ■ A Mathematical Puzzle

Use algebra to solve the puzzle below.

The product of two consecutive positive integers is 10 more than four times the smaller integer.

Solution

Verbal Model: $\boxed{\text{Smaller integer}} \cdot \boxed{\text{Larger integer}} = 10 + 4 \cdot \boxed{\text{Smaller integer}}$

Labels:
Smaller integer $= n$
Larger integer $= n + 1$

Equation:
$$n(n + 1) = 10 + 4n$$
$$n^2 + n = 10 + 4n$$
$$n^2 - 3n - 10 = 0$$
$$(n - 5)(n + 2) = 0$$
$$n - 5 = 0 \implies n = 5$$
$$n + 2 = 0 \implies n = -2$$

Of the two possible solutions, choose the positive value of n and conclude that the two consecutive integers are $n = 5$ and $n + 1 = 6$.

Example 4 ■ Dimensions of a Rectangle

The perimeter of a rectangle is 112 centimeters and the length of the diagonal is 40 centimeters (see Figure 6.14). Find the dimensions of the rectangle.

Solution

Verbal Model: $\left(\boxed{\text{Length}} \right)^2 + \left(\boxed{\text{Width}} \right)^2 = \left(\boxed{\text{Diagonal}} \right)^2$

Labels:
Length $= l$ (centimeters)
Width $= 56 - l$ (centimeters)
Diagonal $= 40$ (centimeters)

Equation:
$$l^2 + (56 - l)^2 = 40^2$$
$$l^2 + 3136 - 112l + l^2 = 1600$$
$$2l^2 - 112l + 1536 = 0$$
$$2(l^2 - 56l + 768) = 0$$
$$2(l - 32)(l - 24) = 0$$
$$l - 32 = 0 \implies l = 32$$
$$l - 24 = 0 \implies l = 24$$

Both solutions are positive, and it does not matter which one you choose. If you let the length $l = 32$ centimeters, then the width $= 56 - 32 = 24$ centimeters. Likewise, if you let the length $l = 24$ centimeters, then the width $= 56 - 24 = 32$ centimeters. So, the dimensions of the rectangle are 32 centimeters by 24 centimeters.

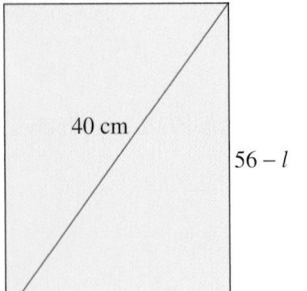

40 cm $56 - l$ l

Figure 6.14

NOTE In Example 4, the expression for the width of the rectangle was found by solving the equation for the perimeter of the rectangle for w as follows.

$$2l + 2w = 112$$
$$2w = 112 - 2l$$
$$w = 56 - l$$

Example 5 ■ An Application Involving the Pythagorean Theorem

An L-shaped sidewalk from building A to building B on a college campus is 200 meters long, as shown in Figure 6.15. By cutting diagonally across the grass, students shorten the walking distance to 150 meters. What are the lengths of the two legs of the sidewalk?

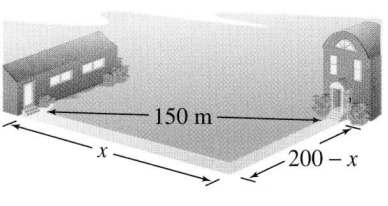

Figure 6.15

Solution

Verbal Model: $a^2 + b^2 = c^2$ Pythagorean Theorem

Labels: Length of one leg $= x$ (meters)
Length of other leg $= 200 - x$ (meters)
Length of diagonal $= 150$ (meters)

Equation:
$$x^2 + (200 - x)^2 = (150)^2$$
$$x^2 + 40{,}000 - 400x + x^2 = 22{,}500$$
$$2x^2 - 400x + 40{,}000 = 22{,}500$$
$$2x^2 - 400x + 17{,}500 = 0$$
$$x^2 - 200x + 8750 = 0$$

By the Quadratic Formula, you can find the solutions as follows.

$$x = \frac{-(-200) \pm \sqrt{(-200)^2 - 4(1)(8750)}}{2(1)} \qquad a = 1, b = -200, c = 8750$$

$$= \frac{200 \pm \sqrt{5000}}{2}$$

$$= \frac{200 \pm 50\sqrt{2}}{2}$$

$$= \frac{2\left(100 \pm 25\sqrt{2}\right)}{2}$$

$$= 100 \pm 25\sqrt{2}$$

Both solutions are positive, and it does not matter which one you choose. If you let

$$x = 100 + 25\sqrt{2} \approx 135.4 \text{ meters}$$

the length of the other leg is

$$200 - x \approx 200 - 135.4$$
$$\approx 64.6 \text{ meters.}$$

NOTE   In Example 5, notice that you obtain the same dimensions if you choose the other value of x. That is, if the length of one leg is

$$x = 100 - 25\sqrt{2} \approx 64.6 \text{ meters}$$

the length of the other leg is

$$200 - x \approx 200 - 64.6 \approx 135.4 \text{ meters.}$$

Example 6 ■ The Height of a Model Rocket

A model rocket is projected straight upward from ground level according to the height equation

$$h = -16t^2 + 192t, \ t \geq 0$$

where h is the height in feet and t is the time in seconds.

(a) After how many seconds is the height of the rocket 432 feet?

(b) After how many seconds does the rocket hit the ground?

Solution

(a)

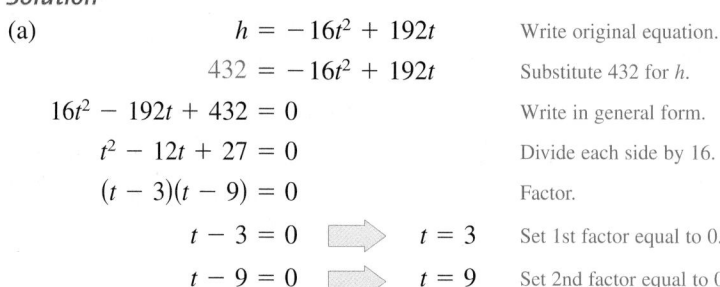

$h = -16t^2 + 192t$	Write original equation.
$432 = -16t^2 + 192t$	Substitute 432 for h.
$16t^2 - 192t + 432 = 0$	Write in general form.
$t^2 - 12t + 27 = 0$	Divide each side by 16.
$(t - 3)(t - 9) = 0$	Factor.
$t - 3 = 0 \implies t = 3$	Set 1st factor equal to 0.
$t - 9 = 0 \implies t = 9$	Set 2nd factor equal to 0.

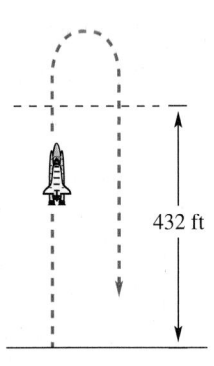

Figure 6.16

The rocket attains a height of 432 feet at two different times—once (going up) after 3 seconds and again (coming down) after 9 seconds. (See Figure 6.16.)

(b) To find the time it takes for the rocket to hit the ground, let the height be zero.

$0 = -16t^2 + 192t$	Substitute 0 for h in original equation.
$0 = t^2 - 12t$	Divide each side by -16.
$0 = t(t - 12)$	Factor.
$t = 0 \quad \text{or} \quad t = 12$	Solutions

The rocket hits the ground after 12 seconds. (Note that the time of $t = 0$ seconds corresponds to the time of lift-off.)

Technology

Use a graphing utility to estimate graphically the times when the rocket in Example 6 (a) is at a height of 432 feet, (b) reaches its maximum height, and (c) hits the ground.

Collaborate!

Exploring with Technology

Use a graphing utility to graph

$$y_1 = 3x^2 + 2x - 1 \text{ and } y_2 = -x^2 + 5x + 4.$$

For each function, use the *maximum* or *minimum* feature to find either the maximum or minimum function value, and the time at which it occurs. Discuss other methods that you could use to find these values.

6.4 ■ Exercises

Solving Problems

Number Problems   In Exercises 1–4, find two positive integers that satisfy the requirement.

1. The product of two consecutive integers is 8 less than 10 times the smaller integer.

2. The product of two consecutive integers is 80 more than 15 times the larger integer.

3. The product of two consecutive even integers is 50 more than three times the larger integer.

4. The product of two consecutive odd integers is 22 less than 15 times the smaller integer.

5. *Geometry*   A picture frame is two-and-one-half times as long as it is wide. If its area is 250 centimeters, find its perimeter.

6. *Geometry*   A rectangular garden has width w and length $1.5w$. The area of the garden is 216 square feet. Find the amount of fencing required to enclose the garden.

7. *Geometry*   A rectangular plot of land has an area of 720 square miles. If the width of the plot of land is 6 miles less than the length, what is the perimeter of the plot of land?

8. *Geometry*   The backyard of a neighbor's house is rectangular, and its length is 5 feet more than its width. If the area of the yard is 500 square feet, how much fencing would be required to enclose the yard?

9. *Fenced Area*   A farmer has 50 feet of fencing that he will use to make a rectangular pen for his sheep. He plans to use a stream on his property as one side of the pen. If he wishes to have 312 square feet enclosed by the fencing and the stream, what should the dimensions of his pen be?

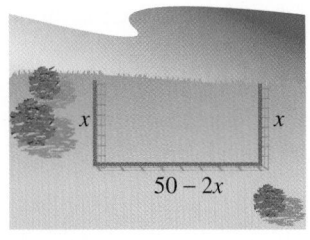

10. *Storage Area*   A retail lumberyard plans to store lumber in a rectangular region adjoining the sales office (see figure). The region will be fenced on three sides, and the fourth side will be bounded by the wall of the office building. Find the dimensions of the region if 350 feet of fencing is available and the area of the region is to be 12,500 square feet.

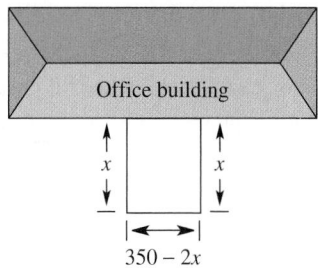

11. *Fenced Area*   You have 100 feet of fencing. Do you have enough to enclose a rectangular region whose area is 630 square feet? Is there enough to enclose a circular region of area 630 square feet? Explain.

12. *Fenced Area*   A family has built a fence around three sides of their property (see figure). In total, they used 550 feet of fencing. By their calculations, the lot is one acre (43,560 square feet). Is this correct? Explain your reasoning.

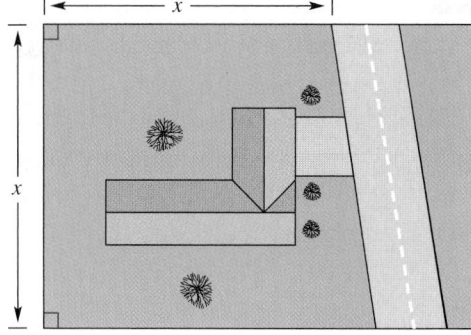

13. *Geometry*   Two circles are tangent to each other (intersect at one point) (see figure). The distance between the centers is 7 yards. Find the radius of each circle if their combined area is 29π square yards.

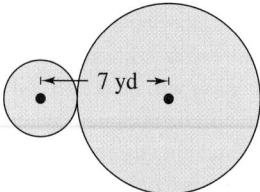

14. *Geometry*   Two circles are tangent to each other (intersect at one point) (see figure). The distance between the centers is 10 feet.

(a) Find the radius of each circle if their combined area is 52π square feet.

(b) Suppose the distance between the centers remained the same but the radii were made equal. Would the combined area of the circles increase or decrease?

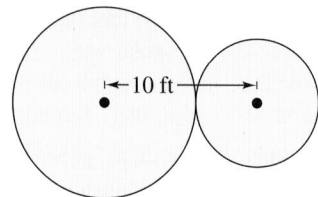

15. *Solving Graphically and Algebraically*   An adjustable rectangular form has minimum dimensions of 3 meters by 4 meters. The length and width can be expanded by equal amounts x (see figure).

(a) Write the length d of the diagonal as a function of x.

(b) Use a graphing utility to graph the function.

(c) Use the graph to approximate the value of x when $d = 10$ meters.

(d) Find x algebraically when $d = 10$.

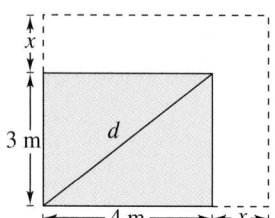

16. *Solving Graphically and Numerically*   A meteorologist is positioned 100 feet from the point where a weather balloon is launched (see figure). The instrument package lifted vertically by the balloon transmits data to the meteorologist.

(a) Write the distance d between the balloon and the meteorologist as a function of the height h of the balloon.

(b) Use a graphing utility to graph the function.

(c) Use the graph to approximate the value of h when $d = 200$ feet.

(d) Complete the table.

h	0	100	200	300
d				

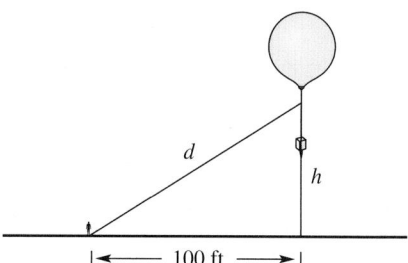

17. *Distance*   You are delivering pizza to offices B and C in your city (see figure), and you are required to keep a log of the mileage between stops. You forget to look at the odometer at stop B, but after getting to stop C you record the total distance traveled from the pizza shop as 18 miles. The return distance from C to A is 16 miles. If the route approximates a right triangle, estimate the distance from A to B.

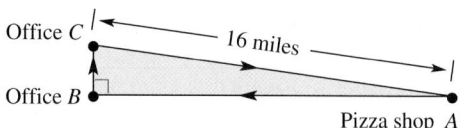

18. *Geometry*   The perimeter of a rectangle is 102 inches, and the length of the diagonal is 39 inches. Find the dimensions of the rectangle.

19. *Geometry*   The height of a triangle is twice its base, and the area of the triangle is 625 square inches. Find the dimensions of the triangle.

20. *Geometry*   The base of a triangle is four times its height. The area of the triangle is 98 square centimeters. Find the dimensions of the triangle.

21. *Fencing in an Area*   You have 200 meters of fencing to enclose two adjacent rectangular corrals (see figure). The total area of the enclosed region is 1400 square meters. What are the dimensions of each corral? (The corrals are the same size.)

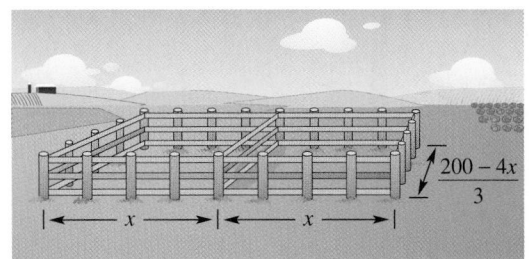

22. *Fencing in an Area*   A gardener wishes to use 600 feet of fencing to enclose a rectangular region and subdivide the region into two smaller rectangles. The total enclosed area is 15,000 square feet. Find the dimensions of the enclosed region.

23. *Distance*   A windlass is used to pull a boat to a dock (see figure). The rope is attached to the boat at a point 15 feet below the level of the windlass. Find the distance from the boat to the dock when the length of the rope is 75 feet.

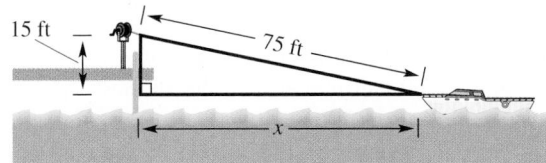

24. *Geometry*   On the sidewalk, the distance from the dormitory to the cafeteria is 400 meters (see figure). By cutting across the lawn, the walking distance is shortened to 300 meters. How long is each part of the L-shaped sidewalk?

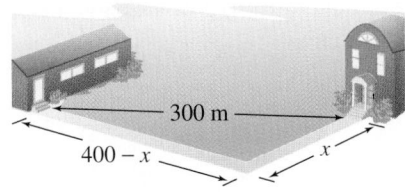

Compound Interest   In Exercises 25–30, find the interest rate r. Use the formula $A = P(1 + r)^2$, where A is the amount after 2 years in an account earning r percent (in decimal form) compounded annually, and P is the original investment.

25. $P = \$3000$
   $A = \$3499.20$

26. $P = \$10,000$
   $A = \$11,990.25$

27. $P = \$250$
   $A = \$280.90$

28. $P = \$500$
   $A = \$572.45$

29. $P = \$8000$
   $A = \$8420.20$

30. $P = \$6500$
   $A = \$7370.46$

Free-Falling Object   In Exercises 31–34, find the time necessary for an object to fall to ground level from an initial height of h_0 feet, if its height h at any time t (in seconds) is given by $h = h_0 - 16t^2$.

31. $h_0 = 144$

32. $h_0 = 625$

33. $h_0 = 1454$ (height of the Sears Tower)

34. $h_0 = 984$ (height of the Eiffel Tower)

35. *Free-Falling Object*   A ball is thrown upward at a velocity of 40 feet per second from a point that is 50 feet above the level of the water. The height h (in feet) of the ball at time t (in seconds) after it is thrown is

$$h = -16t^2 + 40t + 50.$$

(a) Find the time when the ball is again 50 feet above the water.

(b) Find the time when the ball strikes the water.

36. *Hitting Baseballs*   You are hitting baseballs. When you toss the ball into the air, your hand is 5 feet above the ground (see figure). You hit the ball when it falls back to a height of 4 feet. If you toss the ball with an initial velocity of 25 feet per second, the height h of the ball t seconds after it leaves your hand is given by

$$h = 5 + 25t - 16t^2.$$

How much time passes before you hit the ball?

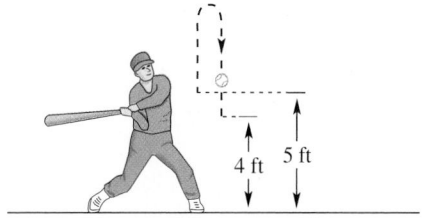

Cost, Revenue, and Profit In Exercises 37 and 38, you are given the cost C of producing x units, the revenue R from selling x units, and the profit P. Find the value of x that will produce the profit P.

37. $C = 100 + 30x$
$R = x(90 - x)$
$P = \$800$

38. $C = 4000 - 40x + 0.02x^2$
$R = x(50 - 0.01x)$
$P = \$63{,}500$

39. *Circular Area* A television station claims that it covers a circular region of approximately 25,000 square miles. Assume that the station is located at the center of the circular region. How far is the station from its farthest listener?

40. *Circular Area* Enough paint is purchased to cover an area of about 200 square feet. You use all of the paint to paint a circle. What is the radius of the circle?

Explaining Concepts

41. In your own words, describe strategies for solving word problems.

42. Describe four methods that can be used to solve a quadratic equation.

43. Describe the units of the product.
$$\frac{9 \text{ dollars}}{\text{hour}} \cdot 20 \text{ hours}$$

44. Describe the units of the product.
$$\frac{20 \text{ feet}}{\text{minute}} \cdot \frac{1 \text{ minute}}{60 \text{ seconds}} \cdot 45 \text{ seconds}$$

Ongoing Review

In Exercises 45–54, solve the equation. (Find all real *and* imaginary solutions.)

45. $2(x - 3) - 5 = 4(x - 5)$

46. $6x(4x - 12) = 0$

47. $3t(t - 7) + 5(t - 7) = 0$

48. $5u(u - 3) = 20$

49. $2(x + 4)^2 = 242$

50. $x - 2\sqrt{x} - 15 = 0$

51. $v^2 - 5v - 24 = 0$

52. $y^2 + y + 1 = 0$

53. $2w^2 + 4 = 9$

54. $x = \sqrt{12x - 35}$

55. *Ticket Sales* Admission prices at a football game were $6 for adults and $2 for children. The total value of the tickets sold was $2528, and 454 tickets were sold. How many adult and children's tickets were sold?

56. *Geometry* A lot is in the shape of a triangle. One side is 100 feet longer than the shortest side, while the third side is 200 feet longer than the shortest side. The perimeter of the lot is 1200 feet. Find the lengths of the sides of the lot.

Looking Further

The area A of an ellipse is given by the equation $A = \pi ab$ (see figure). For a certain ellipse it is required that $a + b = 20$.

(a) Show that $A = \pi a(20 - a)$.

(b) Complete the table.

a	4	7	10	13	16
A					

(c) Find two values of a such that $A = 300$.

(d) Use a graphing utility to graph the area function. Then use the graph to verify the results in part (c).

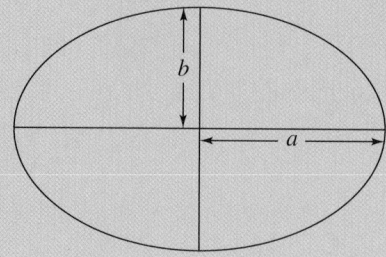

6.5 Graphs of Quadratic Functions

What you should learn:

- How to determine the vertices of parabolas by completing the square
- How to sketch parabolas
- How to write the equation of a parabola given the vertex and a point on the graph
- How to use parabolas to solve application problems

Why you should learn it:

Graphs of quadratic functions can be used to model real-life situations. For instance, In Exercise 96 on page 410, you will graph a quadratic model for the number of cable television systems in the United States.

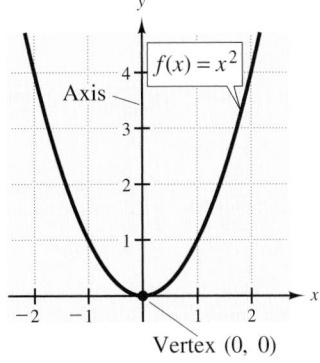

Graphs of Quadratic Functions

In this section, you will study the graphs of quadratic functions of the form

$$f(x) = ax^2 + bx + c.$$ Quadratic function

Figure 6.17 shows the graph of a simple quadratic function, $f(x) = x^2$. This graph is an example of a **parabola.** The lowest point on this graph, $(0, 0)$, is the **vertex** of the parabola, and the vertical line that passes through the vertex (the y-axis, in this case) is the **axis** of the parabola. Every parabola is *symmetric* about its axis, which means that if it were folded along its axis, the two parts would match.

EXPLORATION

Use a graphing utility to graph each parabola.

a. $y = (x - 1)^2 + 5$

b. $y = (x - 3)^2 - 3$

c. $y = (x + 4)^2 - 7$

d. $y = (x + 2)^2 + 1$

What do you notice about the equation of the parabola and the vertex of the parabola?

For each equation below, find the vertex of the parabola without graphing the parabola. Then use a graphing utility to verify each vertex.

a. $y = (x - 4)^2 + 6$

b. $y = (x - 8)^2 - 1$

c. $y = (x + 5)^2 - 3$

d. $y = (x + 3)^2 + 2$

Figure 6.17

Graphs of Quadratic Functions

The graph of

$$f(x) = ax^2 + bx + c, a \neq 0$$

is a **parabola.** The completed square form

$$f(x) = a(x - h)^2 + k$$ Standard form

is called the **standard form** of the function. The **vertex** of the parabola occurs at the point (h, k), and the vertical line passing through the vertex is the **axis** of the parabola.

Study Tip

When a number is added to a function and then that same number is subtracted from the function, the value of the function remains unchanged. Notice in Example 1 that $(-3)^2$ is added to the function to complete the square and then $(-3)^2$ is subtracted from the function so that the value of the function remains the same.

If a is positive, the graph of $f(x) = ax^2 + bx + c$ opens upward, and if a is negative, the graph opens downward, as shown in Figure 6.18. Observe in Figure 6.18 that the y-coordinate of the vertex identifies the minimum function value if $a > 0$ and the maximum function value if $a < 0$.

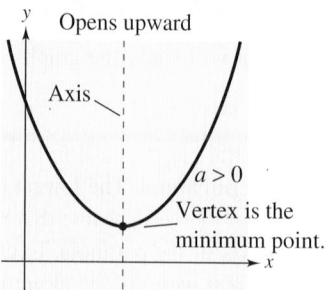 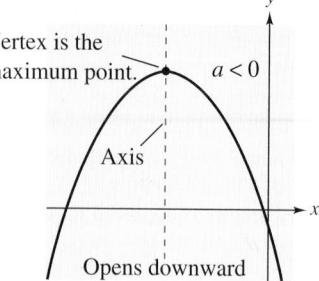

Figure 6.18

Example 1 ■ Finding the Vertex by Completing the Square

Find the vertex of the parabola given by $y = x^2 - 6x + 5$.

Solution

Begin by writing the equation in standard form.

$$y = x^2 - 6x + 5 \qquad \text{Write original equation.}$$
$$y = x^2 - 6x + (-3)^2 - (-3)^2 + 5 \qquad \text{Complete the square.}$$
$$y = (x^2 - 6x + 9) - 9 + 5 \qquad \text{Regroup terms.}$$
$$y = (x - 3)^2 - 4 \qquad \text{Standard form}$$

Now, from the standard form, you can see that the vertex of the parabola occurs at the point $(3, -4)$, as shown in Figure 6.19.

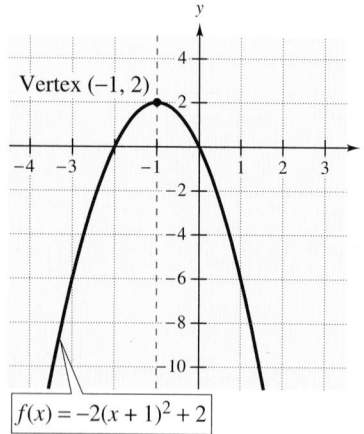

$y = (x-3)^2 - 4$

Vertex $(3, -4)$

Figure 6.19

Example 2 ■ Finding the Vertex by Completing the Square

Find the vertex of the parabola given by $f(x) = -2x^2 - 4x$.

Solution

Begin by writing the function in standard form.

$$f(x) = -2x^2 - 4x \qquad \text{Write original equation.}$$
$$f(x) = -2(x^2 + 2x) \qquad \text{Factor out leading coefficient.}$$
$$f(x) = -2(x^2 + 2x + 1^2 - 1^2) \qquad \text{Complete the square.}$$
$$f(x) = -2(x^2 + 2x + 1) - (-2)(1) \qquad \text{Regroup terms.}$$
$$f(x) = -2(x + 1)^2 + 2 \qquad \text{Write in standard form.}$$

Now, from the standard form, you can see that the vertex of the parabola occurs at the point $(-1, 2)$, as shown in Figure 6.20. Notice that the parabola opens downward because the leading coefficient is negative.

Vertex $(-1, 2)$

$f(x) = -2(x + 1)^2 + 2$

Figure 6.20

In Examples 1 and 2, you found the vertex of the given parabola by completing the square *for each* function. Another technique you could use for this purpose is to complete the square once for the general equation, as follows.

$$y = ax^2 + bx + c$$

$$y = ax^2 + bx + \frac{b^2}{4a} - \frac{b^2}{4a} + c$$

$$y = a\left(x^2 + \frac{bx}{2} + \frac{b^2}{4a^2}\right) + c - \frac{b^2}{4a}$$

$$y = a\left(x + \frac{b}{2a}\right)^2 + c - \frac{b^2}{4a}$$

From this form, you can see that the vertex occurs at $x = -b/2a$. You can use this method to verify the vertices in Examples 1 and 2.

Example 1

$$y = x^2 - 6x + 5$$
$$x = -b/2a = -(-6)/2(1) = 3$$
$$y = (3)^2 - 6(3) + 5 = -4$$
Vertex: $(3, -4)$

Example 2

$$f(x) = -2x^2 - 4x$$
$$x = -b/2a = -(-4)/2(-2) = -1$$
$$f(x) = -2(-1)^2 - 4(-1) = 2$$
Vertex: $(-1, 2)$

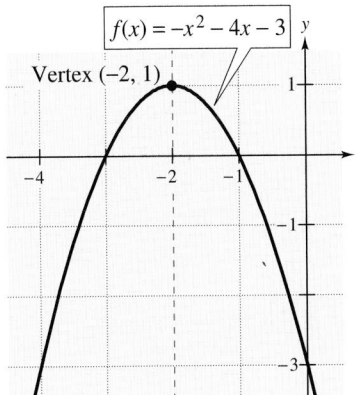

Figure 6.21

Example 3 ■ Finding the Vertex with a Formula

Find the vertex of the parabola given by the equation $f(x) = -x^2 - 4x - 3$.

Solution

From the original function, it follows that $a = -1$ and $b = -4$. So, the x-coordinate of the vertex is

$$x = \frac{-b}{2a} = \frac{-(-4)}{2(-1)} = -2$$

and the y-coordinate is $f\left(\dfrac{-b}{2a}\right) = f(-2) = -(-2)^2 - 4(-2) - 3 = 1$.

So, the vertex of the parabola is $(-2, 1)$, as shown in Figure 6.21.

Example 4 ■ Finding the Vertex with a Formula

Find the vertex of the parabola given by $f(x) = 3x^2 - 9x$.

Solution

From the original function, it follows that $a = 3$ and $b = -9$. So, the x-coordinate of the vertex is

$$x = \frac{-b}{2a} = \frac{-(-9)}{2(3)} = \frac{3}{2}$$

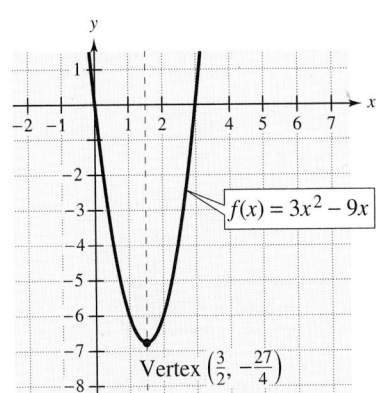

Figure 6.22

and the y-coordinate is $f\left(\dfrac{-b}{2a}\right) = f\left(\dfrac{3}{2}\right) = 3\left(\dfrac{3}{2}\right)^2 - 9\left(\dfrac{3}{2}\right) = -\dfrac{27}{4}$.

So, the vertex of the parabola is $\left(\dfrac{3}{2}, -\dfrac{27}{4}\right)$ as shown in Figure 6.22.

Sketching a Parabola

To obtain an accurate sketch of a parabola, the following guidelines are useful.

> **Guidelines for Sketching a Parabola**
>
> 1. Determine the vertex and axis of the parabola by completing the square or by using the formula $x = -b/2a$.
>
> 2. Plot the vertex, axis, x- and y-intercepts, and a few additional points of the parabola. (Using the symmetry about the axis can reduce the number of points you need to plot.)
>
> 3. Use the fact that the parabola opens *upward* if $a > 0$ and opens *downward* if $a < 0$ to complete the sketch.

NOTE The x-coordinate of the vertex of a parabola lies halfway between the x-intercepts.

Example 5 ■ Sketching a Parabola

Sketch the parabola given by $y = x^2 + 6x + 8$.

Solution

Begin by writing the equation in standard form.

$$y = x^2 + 6x + 8 \qquad \text{Write original equation.}$$

$$y = (x^2 + 6x + 3^2 - 3^2) + 8 \qquad \text{Complete the square.}$$

half of 6

$$y = (x^2 + 6x + 9) - 9 + 8 \qquad \text{Regroup terms.}$$

$$y = (x + 3)^2 - 1 \qquad \text{Write in standard form.}$$

So, the vertex occurs at the point $(-3, -1)$ and the axis is the line $x = -3$. After plotting this information, calculate a few additional points of the parabola, as shown in the table. Note that the y-intercept is $(0, 8)$ and the x-intercepts are solutions of the equation

$$x^2 + 6x + 8 = (x + 4)(x + 2) = 0.$$

x	-5	-4	-3	-2	-1
$y = (x + 3)^2 - 1$	3	0	-1	0	3
Solution point	$(-5, 3)$	$(-4, 0)$	$(-3, -1)$	$(-2, 0)$	$(-1, 3)$

The graph of the parabola is shown in Figure 6.23. Note that the parabola opens upward because the leading coefficient (in the standard form) is positive.

NOTE The graph of the parabola in Example 5 can also be obtained by shifting the graph of $y = x^2$ to the left three units and downward one unit, as discussed in Section 2.6.

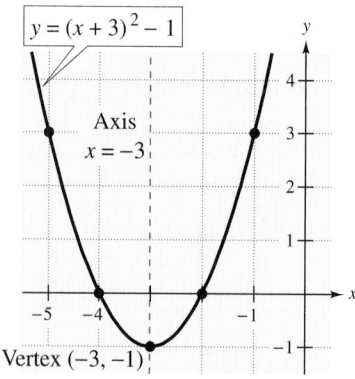

$y = (x + 3)^2 - 1$

Axis
$x = -3$

Vertex $(-3, -1)$

Figure 6.23

NOTE Remember that you can find any x-intercepts by setting y equal to zero and solving for x. See Example 5 in Section 2.2.

Example 6 ■ Sketching a Parabola

Sketch the parabola given by $x^2 + 2y - 4x + 8 = 0$.

Solution

Begin by writing the equation in standard form.

$x^2 + 2y - 4x + 8 = 0$	Write original equation.
$2y = -x^2 + 4x - 8$	Subtract $x^2 - 4x + 8$ from each side.
$y = -\dfrac{1}{2}x^2 + 2x - 4$	Divide each side by 2.
$y = -\dfrac{1}{2}(x^2 - 4x) - 4$	Factor out leading coefficient from first two terms.
$y = -\dfrac{1}{2}[x^2 - 4x + (-2)^2 - (-2)^2] - 4$	Complete the square.

$$\underbrace{\text{(half of } -4)^2}$$

$y = -\dfrac{1}{2}(x^2 - 4x + 4) + 2 - 4$	Regroup terms.
$y = -\dfrac{1}{2}(x - 2)^2 - 2$	Write in standard form.

So, the vertex occurs at the point $(2, -2)$ and the axis is the line $x = 2$. After plotting this information, calculate a few additional points of the parabola, as shown in the table. Note that the y-intercept is $(0, -4)$ and that there are no x-intercepts because the equation $-\frac{1}{2}x^2 + 2x - 4 = 0$ has complex solutions.

x	0	1	2	3	4
$y = -\frac{1}{2}(x - 2)^2 - 2$	-4	$-\frac{5}{2}$	-2	$-\frac{5}{2}$	-4
Solution point	$(0, -4)$	$\left(1, -\frac{5}{2}\right)$	$(2, -2)$	$\left(3, -\frac{5}{2}\right)$	$(4, -4)$

NOTE In Example 6, from the fifth line of the solution to the sixth line, the terms have been regrouped. In the fifth line. the last term in the bracket is $-(-2)^2$. To write this term outside the brackets, as in the sixth line, you must multiply the term by the coefficient of $-\frac{1}{2}$. Notice the following.

$$-\tfrac{1}{2}[-(-2)^2] = -\tfrac{1}{2}(-4) = 2$$

The graph of the parabola is shown in Figure 6.24. Note that the parabola opens downward because the leading coefficient (in the standard form) is negative.

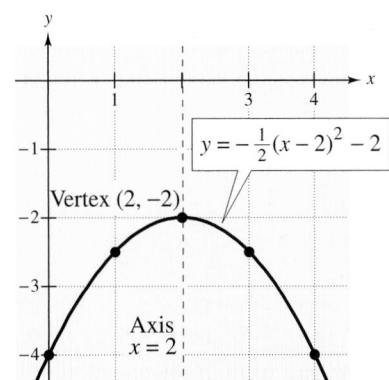

Figure 6.24

Finding the Equation of a Parabola

To write the equation of a parabola with a vertical axis, use the fact that its standard equation has the form $y = a(x - h)^2 + k$, where (h, k) is the vertex.

Example 7 ■ Finding the Equation of a Parabola

Find the equation of the parabola that has a vertex of $(-2, 1)$ and a y-intercept of $(0, -3)$, as shown in Figure 6.25.

Solution

Because the vertex occurs at $(h, k) = (-2, 1)$, the equation has the form

$$y = a(x - h)^2 + k \qquad \text{Standard form}$$
$$y = a[x - (-2)]^2 + 1 \qquad \text{Substitute } -2 \text{ for } h \text{ and } 1 \text{ for } k.$$
$$y = a(x + 2)^2 + 1. \qquad \text{Simplify.}$$

Now, to find the value of a, use the fact that the y-intercept is $(0, -3)$.

$$y = a(x + 2)^2 + 1 \qquad \text{Write standard form.}$$
$$-3 = a(0 + 2)^2 + 1 \qquad \text{Substitute } 0 \text{ for } x \text{ and } -3 \text{ for } y.$$
$$-3 = 4a + 1 \qquad \text{Simplify.}$$
$$-4 = 4a \qquad \text{Subtract 1 from each side.}$$
$$-1 = a \qquad \text{Divide each side by 4.}$$

So, the standard form of the equation of the parabola is $y = -(x + 2)^2 + 1$.

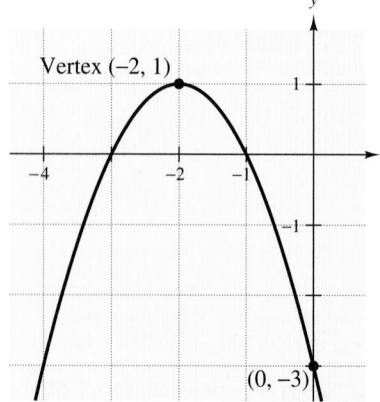

Vertex $(-2, 1)$

$(0, -3)$

Figure 6.25

Example 8 ■ Finding the Equation of a Parabola

Find the equation of the parabola that has a vertex of $(0, -4)$ and passes through the point $(3, 1)$, as shown in Figure 6.26.

Solution

Because the vertex occurs at $(h, k) = (0, -4)$, the equation has the form

$$y = a(x - h)^2 + k \qquad \text{Standard form}$$
$$y = a(x - 0)^2 - 4 \qquad \text{Substitute } 0 \text{ for } h \text{ and } -4 \text{ for } k.$$
$$y = ax^2 - 4. \qquad \text{Simplify.}$$

To find the value of a, use the fact that the parabola passes through the point $(3, 1)$.

$$y = ax^2 - 4 \qquad \text{Write standard form.}$$
$$1 = a(3)^2 - 4 \qquad \text{Substitute 3 for } x \text{ and 1 for } y.$$
$$1 = 9a - 4 \qquad \text{Simplify.}$$
$$5 = 9a \qquad \text{Add 4 to each side.}$$
$$\tfrac{5}{9} = a \qquad \text{Divide each side by 9.}$$

So, the standard form of the equation of the parabola is $y = \tfrac{5}{9}x^2 - 4$.

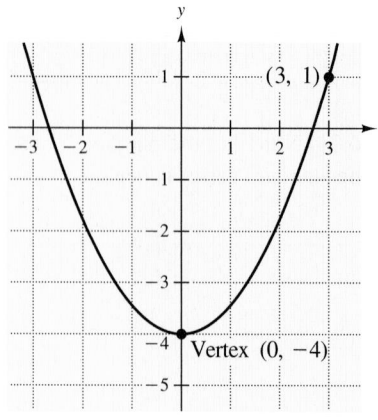

$(3, 1)$

Vertex $(0, -4)$

Figure 6.26

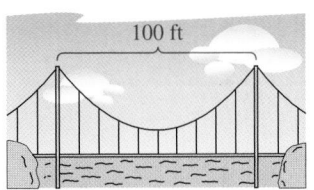

Figure 6.27

Application

Example 9 ■ An Application Involving a Parabola

A suspension bridge is 100 feet long, as shown in Figure 6.27. The bridge is supported by cables attached to the tops of the towers at each end of the bridge. Each cable hangs in the shape of a parabola given by

$$y = 0.01x^2 - x + 35$$

where x and y are both measured in feet. (a) Find the distance between the lowest point of the cable and the roadbed of the bridge. (b) How tall are the towers?

Solution

(a) By writing the equation of the parabola in standard form

$$y = 0.01x^2 - x + 35$$
$$y = 0.01(x^2 - 100x) + 35$$
$$y = 0.01[x^2 - 100x + (-50)^2 - (-50)^2] + 35$$
$$y = 0.01(x^2 - 100x + 2500) - 25 + 35$$
$$y = 0.01(x - 50)^2 + 10$$

you can see that the vertex occurs at the point $(50, 10)$. The minimum distance between the cable and the roadbed is 10 feet (see Figure 6.28).

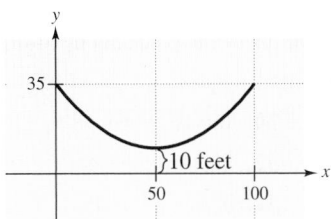

Figure 6.28

(b) Because the vertex of the parabola occurs at the midpoint of the bridge, the two towers are located at the points where $x = 0$ and $x = 100$. Substituting $x = 0$, you can determine that the corresponding y-value is

$$y = 0.01(0)^2 - 0 + 35 = 35 \text{ feet.}$$

So, the towers are each 35 feet tall. (Try substituting $x = 100$ into the equation to see that you obtain the same y-value.)

Collaborate!

Problem Solving

The data in the table represents the average monthly temperature y in degrees Fahrenheit in Savannah, Georgia for the month x, with $x = 1$ corresponding to November. *(Source: National Climate Data Center)*

Plot the data. Find a quadratic model for the data and use it to find the average temperatures for December and February. The actual average temperature for both December and February is 52°F. How well do you think the model fits the data? Use the model to predict the average temperature for June. How useful do you think the model would be for the whole year?

x	1	3	5
y	59	49	59

6.5 ■ Exercises

Developing Skills

In Exercises 1–6, match the equation with its graph. [The graphs are labeled (a), (b), (c), (d), (e), and (f).]

(a)
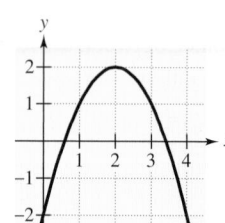

(b)
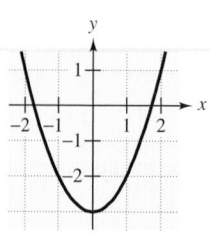

(c)
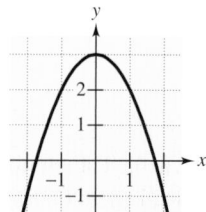

(d)
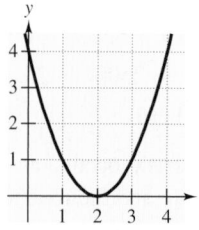

(e)
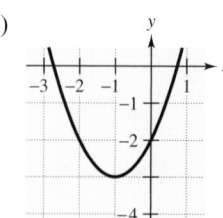

(f)
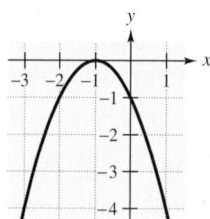

1. $y = x^2 - 3$

2. $y = -x^2 + 3$

3. $y = (x + 1)^2 - 3$

4. $y = -(x + 1)^2$

5. $y = (x - 2)^2$

6. $y = 2 - (x - 2)^2$

In Exercises 7–16, state whether the graph of the parabola opens upward or downward and find the vertex.

7. $y = 2(x - 0)^2 + 2$

8. $y = -3(x + 5)^2 - 3$

9. $f(x) = -4(x + 2)^2 - 3$

10. $g(x) = 7(x - 6)^2 + 1$

11. $y = 4 - (x - 10)^2$

12. $y = 2(x - 12)^2 + 3$

13. $h(x) = x^2 - 6$

14. $f(x) = -(x + 1)^2$

15. $y = -(x - 3)^2$

16. $y = (x - 3)^2 - 9$

In Exercises 17–28, write the equation of the parabola in standard form and find the vertex of its graph.

17. $f(x) = x^2 + 2$

18. $g(x) = x^2 + 2x$

19. $y = x^2 - 4x + 7$

20. $y = x^2 + 6x - 5$

21. $h(x) = x^2 + 6x + 5$

22. $g(x) = x^2 - 4x + 5$

23. $y = -x^2 + 6x - 10$

24. $y = 4 - 8x - x^2$

25. $f(x) = -x^2 + 2x - 7$

26. $f(x) = -x^2 - 10x + 10$

27. $y = 2x^2 + 6x + 2$

28. $y = 3x^2 - 3x - 9$

In Exercises 29–40, find the vertex of the parabola by using the formula $x = -b/2a$.

29. $y = x^2 - 10x$

30. $y = x^2 - 6x$

31. $g(x) = x^2 - 10$

32. $h(x) = x^2 - 4$

33. $y = -x^2 + 6x + 1$

34. $y = -x^2 + 4x + 1$

35. $f(x) = 2x^2 - 3x + 7$

36. $f(x) = 3x^2 - 10x + 2$

37. $y = -4x^2 + x + 1$

38. $y = -5x^2 - 6x + 3$

39. $f(x) = -3x^2 + 3x + 7$

40. $g(x) = -2x^2 - 4x + 5$

In Exercises 41–64, sketch the parabola. Identify the vertex and any intercepts. Use a graphing utility to verify your results.

41. $f(x) = x^2 - 4$

42. $h(x) = x^2 - 9$

43. $y = -x^2 + 4$

44. $y = -x^2 + 9$

45. $f(x) = x^2 - 3x$

46. $g(x) = x^2 - 4x$

47. $y = -x^2 + 3x$

48. $y = -x^2 + 4x$

49. $g(x) = (x - 4)^2$

50. $f(x) = -(x + 4)^2$

51. $x^2 - y - 8x + 15 = 0$

52. $x^2 - y + 4x + 2 = 0$

53. $h(x) = -(x^2 + 6x + 5)$

54. $f(x) = -x^2 + 2x + 8$

55. $x^2 + y - 6x + 7 = 0$

56. $x^2 - y + 4x + 7 = 0$

57. $f(x) = 2(x^2 + 6x + 8)$

58. $g(x) = 3x^2 - 6x + 4$

59. $y = \frac{1}{2}(x^2 - 2x - 3)$ **60.** $y = -\frac{1}{2}(x^2 - 6x + 7)$

61. $h(x) = 5 - \dfrac{x^2}{3}$ **62.** $g(x) = \dfrac{x^2}{3} - 2$

63. $y = \frac{1}{5}(3x^2 - 24x + 38)$

64. $y = \frac{1}{5}(2x^2 - 4x + 7)$

In Exercises 65–68, use a graphing utility to approximate the vertex of the graph. Verify the result algebraically.

65. $y = \frac{1}{6}(2x^2 - 8x + 11)$

66. $y = -\frac{1}{4}(4x^2 - 20x + 13)$

67. $y = -0.7x^2 - 2.7x + 2.3$

68. $y = 0.75x^2 - 7.50x + 23.00$

In Exercises 69–72, use a graphing utility to graph the two functions in the same viewing window. Do the graphs intersect? If so, approximate the point(s) of intersection.

69. $y_1 = -x^2 + 6$
   $y_2 = 2$

70. $y_1 = x^2 - 6x + 8$
   $y_2 = 3$

71. $y_1 = \frac{1}{2}x^2 - 3x + \frac{13}{2}$
   $y_2 = 3$

72. $y_1 = -2x^2 - 4x$
   $y_2 = 1$

In Exercises 73–78, write an equation of the parabola.

73.
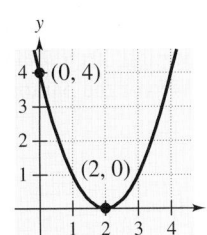

74.
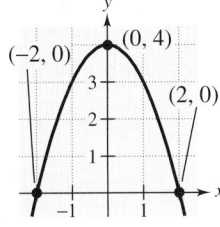

75.
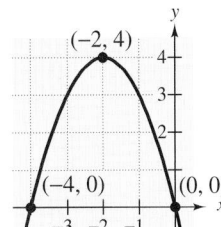

76.
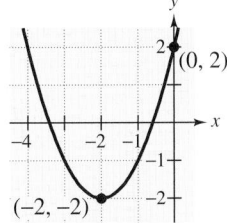

77.
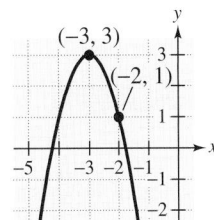

78.
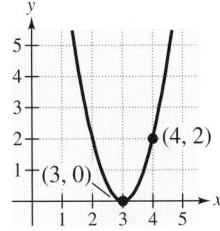

In Exercises 79–90, write an equation of the parabola

$$y = ax^2 + bx + c$$

that satisfies the conditions.

79. Vertex: $(2, 1)$; $a = 1$

80. Vertex: $(-3, -3)$; $a = 1$

81. Vertex: $(-3, 4)$; $a = -1$

82. Vertex: $(3, -2)$; $a = -1$

83. Vertex: $(2, -4)$;   Point on graph: $(0, 0)$

84. Vertex: $(-2, -4)$;   Point on graph: $(0, 0)$

85. Vertex: $(3, 2)$;   Point on graph: $(1, 4)$

86. Vertex: $(-1, -1)$;   Point on graph: $(0, 4)$

87. Vertex: $(-1, 5)$;   Point on graph: $(0, 1)$

88. Vertex: $(5, 10)$;   Point on graph: $(6, 6)$

89. Vertex: $(5, 2)$;   Point on graph: $(10, 3)$

90. Vertex: $(0, 20)$;   Point on graph: $(10, 15)$

Solving Problems

91. *Path of a Ball*   The height y (in feet) of a ball thrown by a child is given by

$$y = -\frac{1}{12}x^2 + 2x + 4$$

where x is the horizontal distance (in feet) from where the ball is thrown.

(a) Use a graphing utility to graph the path of the ball.

(b) From the graph in part (a), how high is the ball when it leaves the child's hand?

(c) From the graph in part (a), how high is the ball when it reaches its maximum height?

(d) From the graph in part (a), how far from the child does the ball strike the ground?

92. *Path of a Diver*   The path of a diver is given by

$$y = -\frac{4}{9}x^2 + \frac{24}{9}x + 10$$

where y is the height in feet and x is the horizontal distance from the end of the diving board in feet (see figure on page 410). What is the maximum height of the diver?

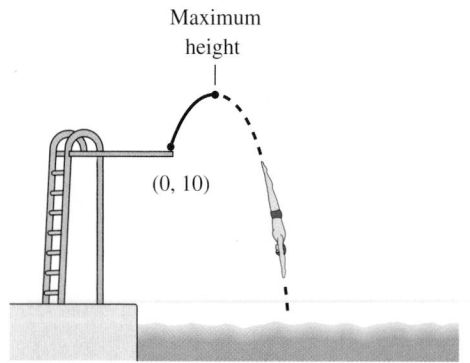

Maximum height

(0, 10)

Figure for 92

93. *Profit* The profit P (in thousands of dollars) for a company is given by

$$P = 230 + 20s - \frac{1}{2}s^2$$

where s is the amount (in hundreds of dollars) spent on advertising. Use a graphing utility to graph the profit function and approximate the amount of advertising that yields a maximum profit. Verify the maximum algebraically.

94. *Cost* The cost C of producing x units of a product is given by

$$C = 800 - 10x + \frac{1}{4}x^2, \quad 0 < x < 40.$$

Use a graphing utility to graph the cost function and approximate the value of x when C is minimum.

95. *Graphical Interpretation* The amounts of money that were kept in personal savings accounts in the United States in the years 1996 through 1999 are approximated by the model

$$S = -471.6 + 232.62t - 18.100t^2, \quad 6 \le t \le 9.$$

In this model, S is the amount in personal savings accounts in billions of dollars and t is the time in years, with $t = 6$ corresponding to 1996. *(Source: U.S. Bureau of Economic Analysis)*

(a) Use a graphing utility to graph the model.

(b) Determine the year in which personal savings were greatest. Approximate the amount for that year.

96. *Graphical Interpretation* The numbers of cable television systems in the United States in the years 1992 through 2000 are approximated by the model

$$C = 10.8 + 0.18t - 0.022t^2.$$

In this model, C is the number of cable television

systems in millions and t is the time in years, with $t = 2$ corresponding to 1992. *(Source: Warren Communication News, Inc.)*

(a) Use a graphing utility to graph the model.

(b) Determine the year in which the number of cable television systems was greatest. Approximate the number of systems for that year.

97. *Highway Design* A highway department engineer must design a parabolic arc to create a turn in a freeway around a city. The vertex of the parabola is placed at the origin, and the parabola must connect with roads represented by the equations

$$y = -0.4x - 100, \quad x < -500$$
$$y = 0.4x - 100, \quad x > 500$$

(see figure). Find an equation for the parabolic arc.

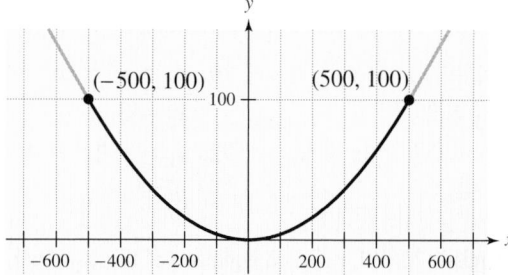

(−500, 100)   (500, 100)

98. *Bridge Design* A bridge is to be constructed over a gorge with the main supporting arch being a parabola (see figure). The equation of the parabola is

$$y = 4\left(100 - \frac{x^2}{2500}\right)$$

where x and y are measured in feet.

(a) Find the length of the road across the gorge.

(b) Find the height of the parabolic arch at the center of the span.

(c) Find the lengths of the vertical girders at intervals of 100 feet from the center of the bridge.

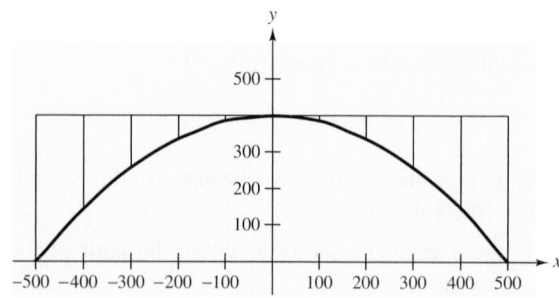

Explaining Concepts

99. Explain how to determine whether the graph of a quadratic function opens upward or downward.

100. Is it possible for the graph of a quadratic function to have two y-intercepts? Explain.

Ongoing Review

In Exercises 101–104, sketch the graph of the function.

101. $f(x) = 2x - 3$

102. $f(x) = -\dfrac{1}{3}x + 5$

103. $y = -x + 4$

104. $y = \dfrac{3}{4}x - 2$

In Exercises 105–108, solve the equation.

105. $6(x + 3) = 10$

106. $12 - 4x = 28 - 3x$

107. $5x(x - 4) = 0$

108. $7x(2x - 3) = 9$

In Exercises 109 and 110, expand and simplify the expression.

109. $-6(x + 5)^2 + 12$

110. $4(x - 3)^2 - 18$

111. *Geometry*   The length of a rectangle is 1 centimeter more than twice the width and its area is 36 square centimeters. Find the length and width of the rectangle.

112. *Number Problem*   The square of a positive real number is 18 more than 7 times the number. What is the number?

Looking Further

Maximum Profit   A company manufactures radios that cost (the company) $60 each. For buyers who purchase 100 or fewer radios, the purchase price is $90 per radio. To encourage large orders, the company will reduce the price *per radio* for orders over 100, as follows. If 101 radios are purchased, the price is $89.85 per radio. If 102 radios are purchased, the price is $89.70 per radio. If $(100 + x)$ radios are purchased, the price per unit is $p = 90 - x(0.15)$, where x is the amount over 100 in the order.

(a) Show that the profit P is given by

$$P = (100 + x)[90 - x(0.15)] - (100 + x)60$$

$$= 3000 + 15x - \frac{3}{20}x^2.$$

(b) Use a graphing utility to graph the profit function.

(c) Using the graph found in part (b), find the vertex of the profit curve and determine the order size for maximum profit.

(d) Would you recommend this pricing scheme? Explain your reasoning.

6.6 Quadratic Inequalities in One Variable

Finding Test Intervals

When you are working with polynomial inequalities, it is important to realize that the value of a polynomial can change signs only at its **zeros.** That is, a polynomial can change signs only at the x-values for which the value of the polynomial is zero. For instance, the first-degree polynomial $x + 2$ has a zero at $x = -2$, and it changes sign at that zero. You can picture this result on the real number line, as shown in Figure 6.29.

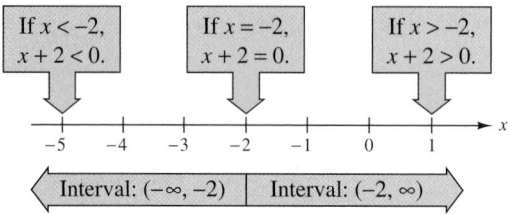

Figure 6.29

Note in Figure 6.29 that the zero of the polynomial partitions the real number line into two **test intervals.** The polynomial is negative for every x-value in the first test interval $(-\infty, -2)$ and is positive for every x-value in the second test interval $(-2, \infty)$. You can use the same basic approach to determine the test intervals for any polynomial.

Finding Test Intervals for a Polynomial

1. Find all real zeros of the polynomial, and arrange the zeros in increasing order. The zeros of a polynomial are called its **critical numbers.**

2. Use the critical numbers of the polynomial to determine its **test intervals.**

3. Choose a representative x-value in each test interval and evaluate the polynomial at that value. If the value of the polynomial is negative, the polynomial will have negative values for *every* x-value in the interval. If the value of the polynomial is positive, the polynomial will have positive values for *every* x-value in the interval.

EXPLORATION

Use a graphing utility to determine the intervals on the x-axis where

$$y = x^2 - x - 6$$

is negative and where it is positive.

Example 1 ■ Finding Test Intervals for a Quadratic Polynomial

Determine the intervals on which the polynomial $x^2 - x - 6$ is entirely negative or entirely positive.

Algebraic Solution

By factoring the given polynomial as

$$x^2 - x - 6 = 0$$
$$(x + 2)(x - 3) = 0$$

you can see that the critical numbers occur at $x = -2$ and $x = 3$. So, the test intervals for the polynomial are

$$(-\infty, -2), \quad (-2, 3), \quad \text{and} \quad (3, \infty). \qquad \text{Test intervals}$$

In each test interval, choose a representative x-value and evaluate the polynomial, as shown in the table.

Test interval	Representative x-value	Value of polynomial	Conclusion
$(-\infty, -2)$	$x = -3$	$(-3)^2 - (-3) - 6 = 6$	Polynomial is positive.
$(-2, 3)$	$x = 0$	$(0)^2 - (0) - 6 = -6$	Polynomial is negative.
$(3, \infty)$	$x = 4$	$(4)^2 - (4) - 6 = 6$	Polynomial is positive.

So, the polynomial has positive values for every x in the intervals $(-\infty, -2)$ and $(3, \infty)$ and negative values for every x in the interval $(-2, 3)$. This result is shown graphically in Figure 6.30.

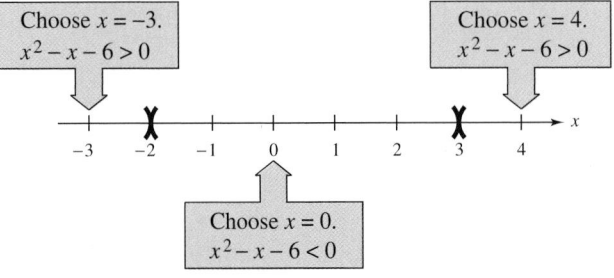

Figure 6.30

Graphical Solution

Begin by sketching the graph of

$$y = x^2 - x - 6$$

as shown in Figure 6.31. The portions of the graph that lie above the x-axis correspond to

$$x^2 - x - 6 > 0$$

and the portion that lies below the x-axis corresponds to

$$x^2 - x - 6 < 0.$$

From the graph, you can see that the value of the polynomial changes signs at $x = -2$ and $x = 3$. So, the polynomial has negative values for every x in the interval $(-2, 3)$ and positive values for every x in the intervals $(-\infty, -2)$ and $(3, \infty)$.

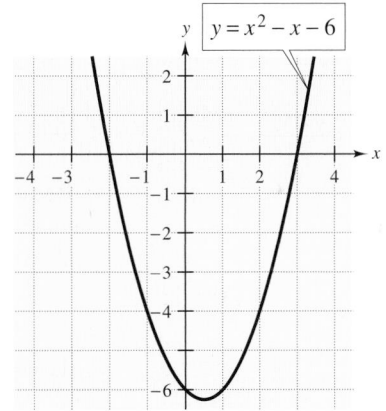

Figure 6.31

Quadratic Inequalities

The concepts of critical numbers and test intervals can be used to solve nonlinear inequalities, as demonstrated in Examples 2, 3, and 4.

Example 2 ■ Solving a Quadratic Inequality

Solve the inequality $x^2 - 5x < 0$.

Solution

First find the critical numbers of $x^2 - 5x < 0$ by finding the solutions of the equation $x^2 - 5x = 0$

$x^2 - 5x = 0$	Write corresponding equation.
$x(x - 5) = 0$	Factor.
$x = 0, x = 5$	Critical numbers

This implies that the test intervals are

$$(-\infty, 0), \quad (0, 5), \quad \text{and} \quad (5, \infty). \qquad \text{Test intervals}$$

To test an interval, choose a convenient number in the interval and determine if the number satisfies the inequality.

Test interval	Representative x-value	Is inequality satisfied?
$(-\infty, 0)$	$x = -1$	$(-1)^2 - 5(-1) \overset{?}{<} 0$ $6 \not< 0$
$(0, 5)$	$x = 1$	$1^2 - 5(1) \overset{?}{<} 0$ $-4 < 0$
$(5, \infty)$	$x = 6$	$6^2 - 5(6) \overset{?}{<} 0$ $6 \not< 0$

Because the inequality $x^2 - 5x < 0$ is satisfied only by the middle test interval, you can conclude that the solution set of the inequality is the interval $(0, 5)$, as shown in Figure 6.32.

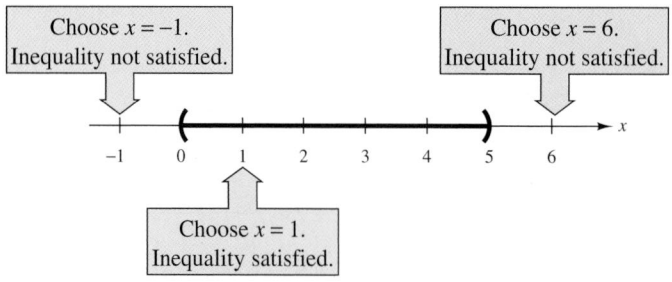

Figure 6.32

In Example 2, note that you would have used the same basic procedure if the inequality symbol had been $\leq$, $>$, or $\geq$. For instance, in Figure 6.32, you can see that the solution set of the inequality $x^2 - 5x \geq 0$ consists of the union of the intervals $(-\infty, 0]$ and $[5, \infty)$, which is written as $(-\infty, 0] \cup [5, \infty)$.

Just as in solving quadratic *equations*, the first step in solving a quadratic *inequality* is to write the inequality in **general form,** with the polynomial on the left and zero on the right, as demonstrated in Example 3.

Example 3 ■ Solving a Quadratic Inequality

Solve the inequality $2x^2 + 5x > 12$.

Algebraic Solution

Begin by writing the inequality in general form.

$$2x^2 + 5x > 12 \qquad \text{Write original inequality.}$$

$$2x^2 + 5x - 12 > 0 \qquad \text{Write in general form.}$$

Next, find the critical numbers for $2x^2 + 5x - 12 > 0$ by finding the solutions to the equation $2x^2 + 5x - 12 = 0$.

$$2x^2 + 5x - 12 = 0 \qquad \text{Write corresponding equation.}$$

$$(x + 4)(2x - 3) = 0 \qquad \text{Factor.}$$

$$x = -4, \ x = \tfrac{3}{2} \qquad \text{Critical numbers}$$

This implies the test intervals are $(-\infty, -4)$, $\left(-4, \tfrac{3}{2}\right)$, and $\left(\tfrac{3}{2}, \infty\right)$. To test an interval, choose a convenient number in the interval and determine if the number satisfies the inequality.

Test interval	Representative x-value	Is inequality satisfied?
$(-\infty, -4)$	$x = -5$	$2(-5)^2 + 5(-5) \overset{?}{>} 12$ $25 > 12$
$\left(-4, \tfrac{3}{2}\right)$	$x = 0$	$2(0)^2 + 5(0) \overset{?}{>} 12$ $0 \not> 12$
$\left(\tfrac{3}{2}, \infty\right)$	$x = 2$	$2(2)^2 + 5(2) \overset{?}{>} 12$ $18 > 12$

From this you can see that the inequality is satisfied for the intervals $(-\infty, -4)$ and $\left(\tfrac{3}{2}, \infty\right)$. So, the solution set is $(-\infty, -4) \cup \left(\tfrac{3}{2}, \infty\right),$ as

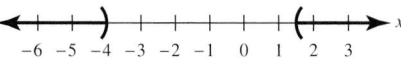

Figure 6.33

Graphical Solution

First, write the inequality

$$2x^2 + 5x > 12$$

in general form as

$$2x^2 + 5x - 12 > 0.$$

Then use a graphing utility to graph

$$y = 2x^2 + 5x - 12.$$

In Figure 6.34, you can see that the graph is above the x-axis when $x < -4$ or when $x > \tfrac{3}{2}$. So, you can graphically approximate the solution to be $(-\infty, -4) \cup \left(\tfrac{3}{2}, \infty\right)$.

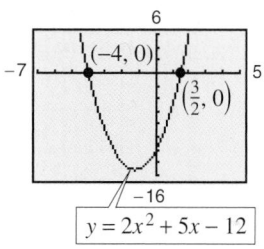

Figure 6.34

In Examples 2 and 3, the critical numbers were found by factoring. With quadratic polynomials that do not factor, you can use the Quadratic Formula to find the critical numbers. For instance, to solve the inequality $x^2 + 4x + 1 \le 0$ you can use the Quadratic Formula to determine that the critical numbers are $x = -2 - \sqrt{3} \approx -3.732$ and $x = -2 + \sqrt{3} \approx -0.268$.

Example 4 shows how the Quadratic Formula can be used to solve a quadratic inequality.

Example 4 ■ Using the Quadratic Formula

Solve the inequality $x^2 - 2x - 1 \le 0$.

Solution

Because the quadratic equation $x^2 - 2x - 1 = 0$ does not factor (using integer coefficients), you can use the Quadratic Formula to find its zeros. That is, the solutions of $x^2 - 2x - 1 = 0$ are given by

$$x = \frac{-(-2) \pm \sqrt{(-2)^2 - 4(1)(-1)}}{2(1)} = \frac{2 \pm \sqrt{8}}{2} = \frac{2 \pm 2\sqrt{2}}{2} = 1 \pm \sqrt{2}.$$

So, the critical numbers for $x^2 - 2x - 1 \le 0$ are $x = 1 - \sqrt{2}$ and $x = 1 + \sqrt{2}$. This implies that the test intervals are

$$\left(-\infty, 1 - \sqrt{2}\right), \quad \left(1 - \sqrt{2}, 1 + \sqrt{2}\right), \quad \text{and} \quad \left(1 + \sqrt{2}, \infty\right).$$

To test an interval, choose a convenient number in the interval and decide if the number satisfies the inequality.

Test interval	Representative x-value	Is inequality satisfied?
$\left(-\infty, 1 - \sqrt{2}\right)$	$x = -2$	$(-2)^2 - 2(-2) - 1 \overset{?}{\le} 0$ $7 \not\le 0$
$\left(1 - \sqrt{2}, 1 + \sqrt{2}\right)$	$x = 0$	$(0)^2 - 2(0) - 1 \overset{?}{\le} 0$ $-1 \le 0$
$\left(1 + \sqrt{2}, \infty\right)$	$x = 3$	$3^2 - 2(3) - 1 \overset{?}{\le} 0$ $2 \not\le 0$

After testing these intervals, you can see that the polynomial $x^2 - 2x - 1$ is less than or equal to zero in the *closed* interval $\left[1 - \sqrt{2}, 1 + \sqrt{2}\right]$. So, the solution set of the inequality is $\left[1 - \sqrt{2}, 1 + \sqrt{2}\right]$, as shown in Figure 6.35. Note that $1 - \sqrt{2} \approx -0.414$ and $1 + \sqrt{2} \approx 2.414$.

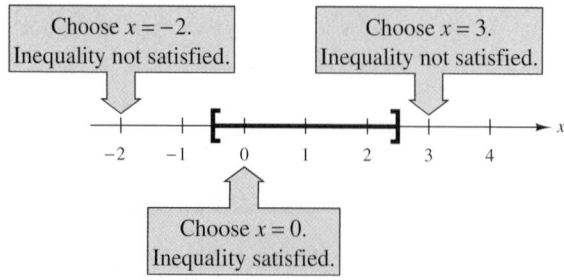

Choose $x = -2$. Inequality not satisfied.

Choose $x = 3$. Inequality not satisfied.

Choose $x = 0$. Inequality satisfied.

Figure 6.35

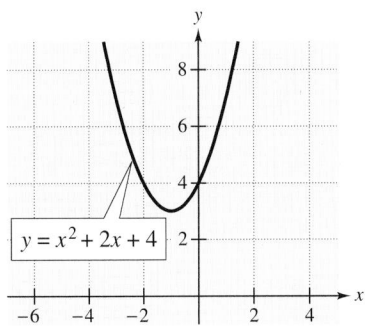

(a)

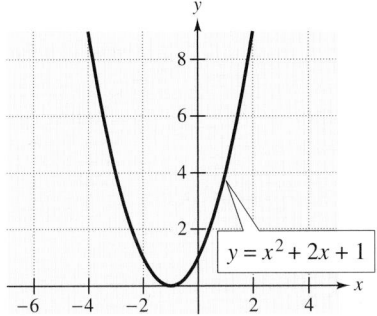

(b)

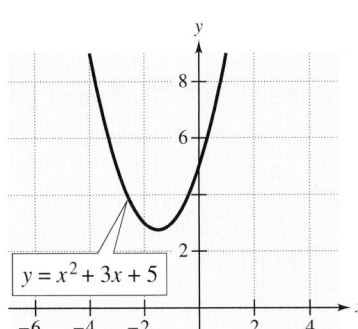

(c)

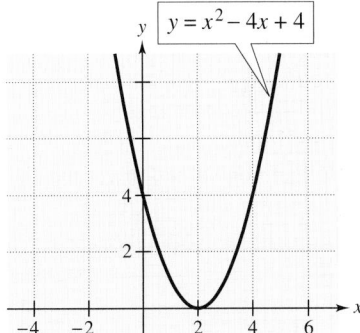

(d)
Figure 6.36

NOTE When solving a quadratic inequality, be sure you have accounted for the particular type of inequality symbol given in the inequality. For instance, in Example 4, note that the solution is a closed interval because the original inequality contained a "*less than or equal to*" symbol. If the original inequality had been $x^2 - 2x - 1 < 0$, the solution would have been the *open* interval $\left(1 - \sqrt{2}, 1 + \sqrt{2}\right)$.

The solutions of the quadratic inequalities in Examples 2, 3, and 4 consist of a single interval or the union of two intervals. When solving the exercises for this section, you should watch for some unusual solution sets, as illustrated in Example 5.

Example 5 ■ Unusual Solution Sets

Solve each quadratic inequality.

(a) The solution set of the quadratic inequality

$$x^2 + 2x + 4 > 0$$

consists of the entire set of real numbers, $(-\infty, \infty)$. This is true because the value of the quadratic $x^2 + 2x + 4$ is positive for every real value of x, as shown in Figure 6.36(a).

(b) The solution set of the quadratic inequality

$$x^2 + 2x + 1 \le 0$$

consists of the single number $\{-1\}$. This is true because $x^2 + 2x + 1 = (x + 1)^2$ has just one critical number, $x = -1$, and it is the only value that satisfies the inequality. [See Figure 6.36(b)].

(c) The solution set of the quadratic inequality

$$x^2 + 3x + 5 < 0$$

is empty. This is true because the value of the quadratic $x^2 + 3x + 5$ is not less than zero for any value of x, as shown in Figure 6.36(c).

(d) The solution set of the quadratic inequality

$$x^2 - 4x + 4 > 0$$

consists of all real numbers except the number 2. In interval notation, this solution set can be written as $(-\infty, 2) \cup (2, \infty)$. [See Figure 6.36(d).]

Remember that checking the solution set of an inequality is not as straightforward as checking the solutions of an equation, because inequalities tend to have infinitely many solutions. Even so, you should check several x-values in your solution set to confirm that they satisfy the inequality. Also try checking x-values that are not in the solution set to confirm that they do not satisfy the inequality.

For instance, the solution of $x^2 - 5x < 0$ is $(0, 5)$. Try checking some numbers in this interval to verify that they satisfy the inequality. Then check some numbers outside the interval to verify that they do not satisfy the inequality.

Application

Example 6 ■ The Height of a Projectile

A projectile is fired straight upward from ground level with an initial velocity of 256 feet per second, as shown in Figure 6.37, so that its height h at any time t is given by

$$h = -16t^2 + 256t$$

where h is measured in feet and t is measured in seconds. During what interval of time will the height of the projectile exceed 960 feet?

Solution

To solve this problem, begin by writing the inequality in general form.

$$-16t^2 + 256t > 960 \qquad \text{Write original inequality.}$$
$$-16t^2 + 256t - 960 > 0 \qquad \text{Write in general form.}$$

Next, find the critical numbers for $-16t^2 + 256t - 960 > 0$ by finding the solutions to the equation $-16t^2 + 256t - 960 = 0$.

$$-16t^2 + 256t - 960 = 0 \qquad \text{Write corresponding equation.}$$
$$t^2 - 16t + 60 = 0 \qquad \text{Divide each side by } -16.$$
$$(t - 6)(t - 10) = 0 \qquad \text{Factor.}$$
$$t = 6, \ t = 10 \qquad \text{Critical numbers}$$

This implies that the test intervals are

$$(-\infty, 6), \ (6, 10), \text{ and } (10, \infty).$$

To test an interval, choose a convenient number in the interval and decide if the number satisfies the inequality.

Test interval	Representative x-value	Is inequality satisfied?
$(-\infty, 6)$	$t = 0$	$-16(0)^2 + 256(0) \overset{?}{>} 960$ $0 \not> 960$
$(6, 10)$	$t = 7$	$-16(7)^2 + 256(7) \overset{?}{>} 960$ $1008 > 960$
$(10, \infty)$	$t = 11$	$-16(11)^2 + 256(11) \overset{?}{>} 960$ $880 \not> 960$

So, the height of the projectile will exceed 960 feet for values of t such that $6 < t < 10$.

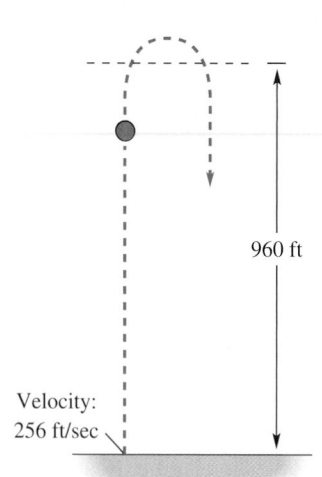

960 ft

Velocity: 256 ft/sec

Figure 6.37

6.6 ■ Exercises

Developing Skills

In Exercises 1–8, find the critical numbers.

1. $4x^2 - 81$

2. $y(y - 4) - 3(y - 4)$

3. $x(2x - 5)$

4. $5x(x - 3)$

5. $x^2 - 4x + 3$

6. $3x^2 - 2x - 8$

7. $4x^2 - 20x + 25$

8. $4x^2 - 4x - 3$

In Exercises 9–16, determine the intervals where the polynomial is entirely negative and where it is entirely positive.

9. $x - 4$

10. $3 - x$

11. $2x(x - 4)$

12. $7x(3 - x)$

13. $x^2 - 9$

14. $16 - x^2$

15. $x^2 - 4x - 5$

16. $2x^2 - 4x - 3$

In Exercises 17–42, solve the inequality and graph the solution on the real number line. (Some of the inequalities have no solution.)

17. $2x + 6 \geq 0$

18. $5x - 20 < 0$

19. $-\frac{3}{4}x + 6 < 0$

20. $3x - 2 \geq 0$

21. $3x(x - 2) < 0$

22. $2x(x - 6) > 0$

23. $3x(2 - x) < 0$

24. $2x(6 - x) > 0$

25. $x^2 - 4 \geq 0$

26. $z^2 \leq 9$

27. $x^2 + 3x \leq 10$

28. $t^2 - 15t + 50 < 0$

29. $-2u^2 + 7u + 4 < 0$

30. $-3x^2 - 4x + 4 \leq 0$

31. $x^2 + 4x + 5 < 0$

32. $x^2 + 6x + 10 > 0$

33. $x^2 + 2x + 1 \geq 0$

34. $25 \geq (x - 3)^2$

35. $x^2 - 4x + 2 > 0$

36. $-x^2 + 8x - 11 \leq 0$

37. $(x - 5)^2 < 0$

38. $(y + 3)^2 \geq 0$

39. $6 - (x - 5)^2 < 0$

40. $(y + 3)^2 - 6 \geq 0$

41. $x^2 - 6x + 9 \geq 0$

42. $x^2 + 8x + 16 < 0$

In Exercises 43–48, use a graphing utility to graph the function and solve the inequality. Verify your result algebraically.

43. $y = 0.5x^2 + 1.25x - 3, \quad y > 0$

44. $y = \frac{1}{3}x^2 - 3x, \quad y < 0$

45. $y = x^2 + 4x + 4, \quad y \geq 9$

46. $y = x^2 - 6x + 9, \quad y < 16$

47. $y = 9 - 0.2(x - 2)^2, \quad y < 4$

48. $y = 8x - x^2, \quad y > 12$

Solving Problems

49. *Height of a Projectile* A projectile is fired straight upward from ground level with an initial velocity of 128 feet per second, so that its height h at any time t is given by

$$h = -16t^2 + 128t$$

where h is measured in feet and t is measured in seconds. During what interval of time will the height of the projectile exceed 240 feet?

50. *Height of a Projectile* A projectile is fired straight upward from ground level with an initial velocity of 88 feet per second, so that its height h at any time t is given by

$$h = -16t^2 + 88t$$

where h is measured in feet and t is measured in seconds. During what interval of time will the height of the projectile exceed 50 feet?

51. *Annual Interest Rate* You are investing $1000 in a certificate of deposit for 2 years and you want the interest to exceed $150. The interest is compounded annually. What interest rate should you have? [*Hint:* Solve the inequality $1000(1 + r)^2 > 1150$.]

52. *Geometry* You have 64 feet of fencing to enclose a rectangular region. Determine the interval for the length such that the area will exceed 240 square feet.

53. *Geometry* A rectangular playing field with a perimeter of 100 meters must have an area of at least 500 square meters. Within what bounds must the length of the field lie?

54. *Profit* The revenue and cost equations for a product are given by

$$R = x(50 - 0.0002x)$$
$$C = 12x + 150,000$$

where R and C are measured in dollars and x represents the number of units sold (see figure). How many units must be sold to obtain a profit of at least $1,650,000?

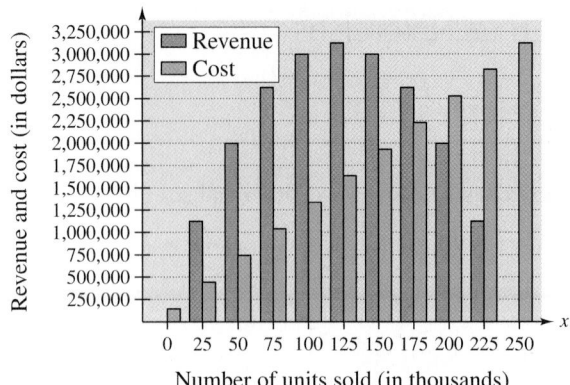

Figure for 54

Explaining Concepts

55. Give an example of a quadratic inequality that has no real solution.

56. Define the term *critical number* and explain how critical numbers are used in solving quadratic inequalities.

Ongoing Review

In Exercises 57–62, solve the inequality and sketch the graph of the solution on the real number line.

57. $4x + 20 \le 0$

58. $5x - 6 \ge 0$

59. $8 - 4x < 9 - x$

60. $5(x - 3) + 12 > 2$

61. $|x + 8| > 9$

62. $|x - 4| < 7$

In Exercises 63–66, perform the multiplication.

63. $(2x - 1)(2x + 1)$

64. $(3x - 4)(x + 6)$

65. $(x + 5)(x^2 + 10x + 15)$ **66.** $(x + 3)(x^2 - 2x + 4)$

67. *Geometry* The length of the sides of a square have been measured accurately to within 0.01 foot. This measured length is 4.25 feet.

(a) Write an absolute value inequality that describes the relationship between the actual length of each side of the square s and its measured length.

(b) Solve for s in the absolute value inequality you found in part (a).

Looking Further

Two circles are tangent to each other (intersect at one point) (see figure). The distance between their centers is 12 centimeters.

(a) If x is the radius of one of the circles, express the combined area A of the circles as a function of x. What is the domain of the function?

(b) Use a graphing utility to graph the area function in part (a).

(c) Determine the radii of the circles if the combined area of the circles must be at least 300 square centimeters but no more than 400 square centimeters.

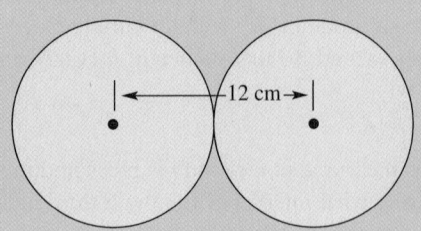

Chapter Summary: Key Concepts

6.1 ■ Extracting square roots

The equation $u^2 = d$, where $d > 0$, has exactly two solutions: $u = \sqrt{d}$ and $u = -\sqrt{d}$.

6.1 ■ Extracting imaginary square roots

The equation $u^2 = d$, where $d < 0$, has exactly two solutions, $u = \sqrt{|d|}\, i$ and $u = -\sqrt{|d|}\, i$.

6.2 ■ Completing the square

To complete the square for the expression $x^2 + bx$, add $(b/2)^2$, which is the square of half the coefficient of x. Consequently,

$$x^2 + bx + \left(\frac{b}{2}\right)^2 = \left(x + \frac{b}{2}\right)^2.$$

6.3 ■ The Quadratic Formula

The solutions of $ax^2 + bx + c = 0$, $a \neq 0$, are given by the Quadratic Formula

$$x = \frac{-b \pm \sqrt{b^2 - 4ac}}{2a}.$$

6.3 ■ Using the discriminant

The discriminant of the quadratic equation

$$ax^2 + bc + c = 0, \quad a \neq 0$$

can be used to classify the solutions of the equation as follows.

Discriminant	*Solution Type*
1. Perfect square	Two distinct rational solutions
2. Positive nonperfect square	Two distinct irrational solutions
3. Zero	One repeated rational solution
4. Negative number	Two distinct imaginary solutions

6.3 ■ Summary of methods for solving quadratic equations

1. Factoring

2. Extracting square roots

3. Completing the square

4. Using the Quadratic Formula

6.5 ■ Guidelines for sketching a parabola

1. Determine the vertex and axis of the parabola by completing the square or by using the formula $x = -b/2a$.

2. Plot the vertex, axis, x- and y-intercepts, and a few additional points of the parabola.

3. Use the fact that the parabola opens upward if $a > 0$ and opens downward if $a < 0$ to complete the sketch.

6.6 ■ Finding test intervals for a polynomial

1. Find all real zeros of the polynomial, and arrange the zeros in increasing order. The zeros of a polynomial are called its critical numbers.

2. Use the critical numbers of the polynomial to determine its test intervals.

3. Choose a representative x-value in each test interval and evaluate the polynomial at that value. If the value of the polynomial is negative, the polynomial will have negative values for every x-value in the interval. If the value of the polynomial is positive, the polynomial will have positive values for every x-value in the interval.

Review Exercises

Reviewing Skills

6.1 ■ *How to solve quadratic equations by factoring*

In Exercises 1–12, solve the equation by factoring.

1. $x^2 + 12x = 0$ **2.** $u^2 - 18u = 0$

3. $3z(z + 10) - 8(z + 10) = 0$

4. $7x(2x - 9) + 4(2x - 9) = 0$

5. $4y^2 - 1 = 0$ **6.** $2z^2 - 72 = 0$

7. $4y^2 + 20y + 25 = 0$ **8.** $x^2 + \frac{8}{3}x + \frac{16}{9} = 0$

9. $t^2 - t - 20 = 0$ **10.** $z^2 + \frac{2}{3}z - \frac{8}{9} = 0$

11. $2x^2 - 2x - 180 = 0$ **12.** $15x^2 - 30x - 45 = 0$

■ *How to solve quadratic equations by extracting square roots*
■ *How to solve quadratic equations with complex solutions*

In Exercises 13–24, solve the equation by extracting square roots.

13. $x^2 = 10{,}000$ **14.** $x^2 = 98$

15. $y^2 - 8 = 0$ **16.** $y^2 - 2.25 = 0$

17. $(x - 16)^2 = 400$ **18.** $(x + 3)^2 = 0.04$

19. $t^2 + 28 = 0$ **20.** $x^2 + 50 = 0$

21. $(2x - 3)^2 = -75$ **22.** $(3y - 5)^2 = -100$

23. $(x + 6)^2 + 20 = 0$ **24.** $(u - 7)^2 + 18 = 0$

■ *How to use substitution to solve equations of quadratic form*

In Exercises 25–32, solve the equation of quadratic form.

25. $x^4 - 4x^2 - 5 = 0$

26. $x^4 - 9x^2 + 14 = 0$

27. $x + 2\sqrt{x} - 3 = 0$

28. $x - 10\sqrt{x} + 9 = 0$

29. $x^{2/5} + 4x^{1/5} + 3 = 0$

30. $x^{2/5} + x^{1/5} - 6 = 0$

31. $6\left(\frac{1}{x}\right)^2 + 7\left(\frac{1}{x}\right) - 3 = 0$

32. $6\left(\frac{1}{x}\right)^2 - 7\left(\frac{1}{x}\right) + 2 = 0$

6.2 ■ *How to rewrite quadratic expressions in completed square form*

In Exercises 33–38, add a term to the expression so that it becomes a perfect square trinomial.

33. $x^2 + 6x + \quad$ **34.** $x^2 - 10x + \quad$

35. $x^2 + 7x + \quad$ **36.** $y^2 - 9y + \quad$

37. $x^2 - \frac{3}{4}x + \quad$ **38.** $t^2 + \frac{4}{5}t + \quad$

■ *How to solve quadratic equations by completing the square*

In Exercises 39–46, solve the equation by completing the square. (Find all real *and* imaginary solutions.)

39. $x^2 - 6x - 3 = 0$ **40.** $x^2 + 12x + 6 = 0$

41. $x^2 - 3x + 3 = 0$ **42.** $t^2 + t - 1 = 0$

43. $2y^2 + 10y + 3 = 0$ **44.** $3x^2 - 2x + 2 = 0$

45. $2y^2 + 3y + 1 = 0$ **46.** $4x^2 - 5x + 2 = 0$

6.3 ■ *How to use the Quadratic Formula to solve quadratic equations*

In Exercises 47–54, use the Quadratic Formula to solve the equation. (Find all real *and* imaginary solutions.)

47. $y^2 + y - 30 = 0$

48. $x^2 - x - 72 = 0$

49. $2y^2 + y - 21 = 0$

50. $2x^2 - 3x - 20 = 0$

51. $-x^2 + 5x + 84 = 0$

52. $-x^2 - 7x + 3 = 0$

53. $0.3t^2 - 2t + 5 = 0$

54. $-u^2 + 2.5u + 3 = 0$

■ *How to determine the types of solutions of quadratic equations using the discriminant*

In Exercises 55–60, use the discriminant to determine the type of solutions of the quadratic equation.

55. $2x^2 - 5x - 7 = 0$ **56.** $x^2 + 3x - 11 = 0$

57. $3x^2 - 2x + 10 = 0$ **58.** $4x^2 + 12x + 9 = 0$

59. $-x^2 - 20x + 100 = 0$ **60.** $-x^2 + 3x + 3 = 0$

6.5 ■ *How to determine the vertices of parabolas by completing the square*

In Exercises 61–64, state whether the graph opens upward or downward and find the vertex.

61. $y = 4(x - 3)^2 + 6$ **62.** $y = -(x - 7)^2 - 1$

63. $y = -3(x + 4)^2$ **64.** $y = 5(x + 2)^2 + 9$

In Exercises 65–68, find the vertex of the parabola given by the equation.

65. $y = 3x^2 + 6x$ **66.** $y = 2x^2 - 8x$

67. $y = -x^2 + 7x + 4$ **68.** $y = -x^2 - 5x + 6$

■ *How to sketch parabolas*

In Exercises 69–74, sketch the parabola. Identify the vertex and any *x*-intercepts. Use a graphing utility to verify your results.

69. $y = x^2 + 4$

70. $y = -(x - 2)^2$

71. $y = x^2 + 8x$

72. $y = -x^2 + 3x$

73. $y = x^2 - 6x + 5$

74. $y = x^2 + 3x - 10$

■ *How to write the equation of a parabola given the vertex and a point on the graph*

In Exercises 75–78, write an equation of the parabola

$$y = ax^2 + bx + c$$

that satisfies the conditions.

75. Vertex: $(3, 5)$; Point on graph: $(1, -3)$

76. Vertex: $(-2, 3)$; Point on graph: $(-3, 6)$

77. Vertex: $(5, 0)$; Point on graph: $(1, 1)$

78. Vertex: $(-2, 5)$; Point on graph: $(0, 1)$

6.6 ■ *How to determine the test intervals for polynomials*

In Exercises 79–84, determine the intervals on which the polynomial is entirely negative or entirely positive.

79. $12 - x$ **80.** $x + 6$

81. $4x(x + 1)$ **82.** $-3x(x - 9)$

83. $x^2 - 10x + 21$ **84.** $x^2 + 5x - 24$

■ *How to use test intervals to solve quadratic inequalities*

In Exercises 85–92, solve the inequality and sketch the graph of the solution on the real number line.

85. $4x - 12 < 0$ **86.** $3(x + 2) > 0$

87. $5x(7 - x) > 0$ **88.** $-2x(x - 10) \le 0$

89. $(x - 5)^2 - 36 > 0$ **90.** $16 - (x - 2)^2 \le 0$

91. $2x^2 + 3x - 20 < 0$

92. $3x^2 - 2x - 8 > 0$

In Exercises 93–96, use a graphing utility to graph the function and solve the inequality. Verify your result algebraically.

93. $y = x^2 - 6x, y < 0$

94. $y = 2x^2 + 5x, y > 0$

95. $f(x) = 2x^2 + 3x - 10, y \le 10$

96. $g(x) = 3x^2 - 2x - 5, y \ge 3$

Solving Problems

97. *Vertical Motion*   The height h in feet of an object above the ground is

$$h = 200 - 16t^2, \qquad t \ge 0$$

where t is the time in seconds.

(a) How high is the object when $t = 0$?

(b) Was the object thrown upward, dropped, or thrown downward? Explain.

(c) Find the time when the object strikes the ground.

98. *Path of a Projectile*   The path y of a projectile is

$$y = -\frac{1}{16}x^2 + 5x$$

where y is the height (in feet) and x is the horizontal distance (in feet) from where the projectile is launched.

(a) Sketch the path of the projectile.

(b) How high is the projectile when it is at its maximum height?

(c) How far from the launch point does the projectile strike the ground?

99. *Geometry* Find the dimensions of a triangle if its height is $1\frac{2}{3}$ times its base and its area is 3000 square centimeters.

100. *Geometry* The length of a flower garden is 12 feet greater than its width (see figure). The area of the garden is 108 square feet. Find the dimensions of the garden.

$$w$$

$$w + 12$$

101. *Solving Graphically, Numerically, and Algebraically* A launch control team is located 3 miles from the point where a rocket is launched.

(a) Write the distance d between the rocket and the team as a function of the height h of the rocket. Use a graphing utility to graph the function. Use the graph to approximate the value of h when $d = 4$ miles.

(b) Complete the table. Use the table to approximate the value of h when $d = 4$ miles.

h	1	2	3	4	5	6	7
d							

(c) Find h algebraically when $d = 4$ miles.

102. *Median Home Price* The median sale price S (in thousands of dollars) of privately owned one-family homes in the United States in selected years from 1980 to 2000 is shown in the table. *(Sources: U.S. Census Bureau and U.S. Department of Housing and Urban Development)*

t	1980	1985	1990	1995	2000
S	64.6	84.3	122.9	133.9	169.0

The data can be represented by the model

$$S = 0.009t^2 + 4.99t + 63.7, \ 0 \le t \le 20$$

where $t = 0$ corresponds to 1980.

(a) Use a graphing utility to graph the model.

(b) Use the model to determine the year in which the median price of a home reached $150,000.

(c) Verify your answer from part (b) graphically.

103. *College Completion* The percent P of the American population that graduated from college between 1960 and 2000 is approximated by the model

$$P = 7.3 + 0.40t + 0.002t^2, \ 0 \le t \le 40$$

where $t = 0$ represents 1960. *(Source: U.S. Census Bureau)*

(a) Use a graphing utility to graph the model.

(b) Use the model to determine the year in which the percent of the population that graduated from college reached 20%.

(c) Verify you answer from part (b) graphically.

104. *Cross Sectional Area* An open-topped rectangular conduit for carrying water in a manufacturing process is made by folding up the edges of a sheet of aluminum 48 inches wide (see figure). A cross section of the conduit must have an area of 288 square inches. Find the width and height of the conduit.

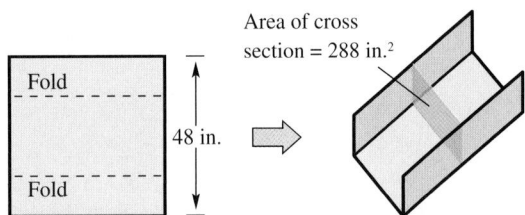

Area of cross section = 288 in.²

Fold

48 in.

Fold

105. *Property Area* Your home is on a square lot. To add more space to your yard, you purchase an additional 20 feet along the side of your property (see figure). The area of the lot is now 25,500 square feet. What are the dimensions of the new lot?

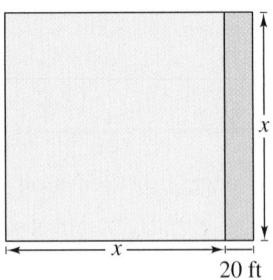

$$x$$

$$x$$

20 ft

106. *Number Problem* Find two consecutive positive integers whose product is 20 less than 13 times the smaller integer.

6 Chapter Test

Take this test as you would take a test in class. After you are done, check your work against the answers given in the back of the book.

In Exercises 1 and 2, solve the quadratic equation by factoring.

1. $x(x + 5) - 10(x + 5) = 0$ **2.** $8x^2 - 21x - 9 = 0$

In Exercises 3 and 4, solve the quadratic equation by extracting square roots.

3. $(x - 2)^2 = 50$ **4.** $(x + 3)^2 + 81 = 0$

5. What term should be added to $x^2 - 3x$ so that it becomes a perfect square trinomial?

6. Solve by completing the square: $2x^2 - 6x + 3 = 0$.

7. Find the discriminant of the quadratic equation $5x^2 - 12x + 10 = 0$, and explain how discriminants are used to determine the types of solutions of quadratic equations.

In Exercises 8 and 9, solve the equations by using the Quadratic Formula.

8. $3x^2 - 8x + 3 = 0$ **9.** $2y(y - 2) = 7$

10. Solve: $x^{2/3} - 6x^{1/3} + 8 = 0$.

In Exercises 11 and 12, sketch the graph of the equation. Identify the vertex and any x-intercepts. Use a graphing utility to verify your results.

11. $y = -x^2 + 16$ **12.** $y = x^2 - 2x - 15$

In Exercises 13 and 14, solve the inequality and sketch the solution on the real number line.

13. $2x(x - 3) < 0$ **14.** $x^2 - 4x \geq 12$

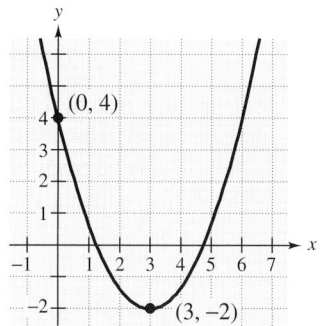

Figure for 15

15. Find the quadratic function, $y = ax^2 + bx + c$, whose graph is shown in the figure.

16. The width of a rectangle is 8 feet less than its length. The area of the rectangle is 240 square feet. Find the dimensions of the rectangle.

17. An object is dropped from a height of 75 feet. Its height h (in feet) at any time t is given by $h = -16t^2 + 75$, where t is measured in seconds. Find the time at which the object has fallen to a height of 35 feet.

18. The area of a rectangle is given by the equation

$$A = \frac{2}{\pi}(100x - x^2), \quad 0 < x < 100$$

where x is the length of the rectangle in feet. Use a graphing utility to graph the area function and approximate the value of x when A is maximum.

Cumulative Test: Chapters 1–6

Take this test as you would take a test in class. After you are done, check your work against the answers given in the back of the book.

In Exercises 1–4, factor the expression.

1. $2x^2 + 5x - 7$

2. $11x^2 + 6x - 5$

3. $12x^3 - 27x$

4. $8x^3 + 125$

In Exercises 5–10, solve the equation.

5. $125 - 50x = 0$

6. $t^2 - 8t = 0$

7. $x^2(x + 2) - (x + 2) = 0$

8. $x(10 - x) = 25$

9. $\dfrac{x + 3}{7} = \dfrac{x - 1}{4}$

10. $\dfrac{x}{4} - \dfrac{x + 2}{6} = \dfrac{3}{2}$

In Exercises 11 and 12, write the slope-intercept form of the equation of the line.

11. Passes through $(4, -2)$; $m = \frac{5}{2}$

12. Passes through $(-5, 8)$ and $(1, 2)$

13. Does the graph in the figure represent y as a function of x? Explain.

14. The base of a triangle is $5x$ and its height is $2x + 9$. Write the area A of the triangle as a function of x.

15. A sales representative indicates that if a customer waits another month for a new car that currently costs \$23,500, the price will increase by 4%. However, the customer will pay an interest penalty of \$725 for the early withdrawal of a certificate of deposit if the car is purchased now. Determine whether the customer should buy now or wait another month.

16. Solve the system of equations: $\begin{cases} 2x + 0.5y = 8 \\ 3x + 2y = 22 \end{cases}$.

17. How many liters of water should be evaporated from 160 liters of a 12% saline solution so that the solution that remains is a 20% solution?

In Exercises 18 and 19, evaluate the determinant of the matrix.

18. $\begin{bmatrix} 3 & 7 \\ -2 & 6 \end{bmatrix}$

19. $\begin{bmatrix} 3 & -2 & 1 \\ 0 & 5 & 3 \\ 6 & 1 & 1 \end{bmatrix}$

20. Simplify the expressions.

(a) $\sqrt{75x^3y}$

(b) $\dfrac{40}{3 - \sqrt{5}}$

21. Simplify and write the complex number in standard form.

(a) $(5 - 3i)(-2 - 4i)$

(b) $(-3 - 5i)(-3 + 5i)$

22. Find an equation of the parabola with vertex at $(2, -1)$ that passes through the origin.

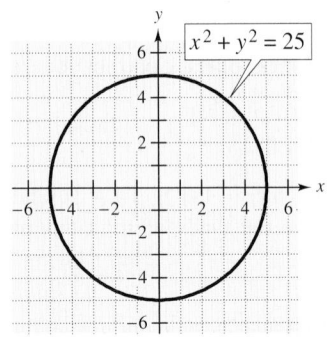

Figure for 13

$x^2 + y^2 = 25$

Rational Expressions and Rational Functions

7

THE BIG PICTURE

In this chapter you will learn how to:

■ determine the domain of a rational function and simplify rational expressions.

■ add, subtract, multiply, and divide rational expressions.

■ divide a polynomial by a monomial or a polynomial.

■ solve rational equations.

■ sketch graphs of rational functions.

■ solve rational inequalities.

■ model and solve real-life situations using rational equations and inequalities.

Key Terms

As you encounter each new vocabulary term in this chapter, add the term and its definition to your notebook glossary.

rational expression (p. 428)

domain of a rational expression (p. 428)

rational function (p. 428)

domain of a rational function (p. 428)

simplified form (p. 431)

complex fractions (p. 442)

least common multiple (LCM) (p. 447)

least common denominator (LCD) (p. 448)

dividend (p. 457)

divisor (p. 457)

quotient (p. 457)

remainder (p. 457)

synthetic division (p. 461)

branches (p. 477)

asymptote (p. 478)

vertical asymptote (p. 478)

horizontal asymptote (p. 478)

critical numbers (p. 487)

test intervals (p. 487)

Additional text-specific resources are available to help you do well in this course. See page xvi for details.

7.1 Simplifying Rational Expressions

What you should learn:

- How to find the domain of a rational function
- How to simplify rational expressions
- How to use rational expressions in real-life applications

Why you should learn it:

Rational expressions can be used to solve real-life problems. For instance, in Exercise 77 on page 437 you will find a rational expression that models the average cost of Medicare per person aged 65 and older.

The Domain of a Rational Function

A fraction whose numerator and denominator are polynomials is called a **rational expression.** Some examples are

$$\frac{3}{x+4}, \quad \frac{2x}{x^2-4x+4}, \quad \text{and} \quad \frac{x^2-5x}{x^2+2x-3}.$$

> ### EXPLORATION
>
> Evaluate each of the three rational expressions above for all integer values of x from -4 to 4. Organize your results in a table. Then, discuss your results. What do you observe?

In Section 2.4, you learned that because division by zero is undefined, the denominator of a rational expression cannot be zero. So, in your work with rational expressions, you must assume that all real number values of the variable that make the denominator zero are excluded. For the three fractions above, $x = -4$ is excluded from the first fraction, $x = 2$ from the second, and both $x = 1$ and $x = -3$ from the third. The set of *usable* values of the variable is called the **domain** of the rational expression.

> **Definition of a Rational Expression**
>
> Let u and v be polynomials. The algebraic expression
>
> $$\frac{u}{v}$$
>
> is a **rational expression.** The **domain** of this rational expression is the set of all real numbers for which $v \neq 0$.

Like polynomials, rational expressions can be used to describe functions. Such functions are called **rational functions.** Later in this chapter you will sketch graphs of rational functions.

> **Definition of a Rational Function**
>
> Let $u(x)$ and $v(x)$ be polynomial functions. The function
>
> $$f(x) = \frac{u(x)}{v(x)}$$
>
> is a **rational function.** The **domain** of f is the set of all real numbers for which $v(x) \neq 0$.

Example 1 ■ Finding the Domain of a Rational Function

Find the domain of each rational function.

(a) $f(x) = \dfrac{4}{x - 2}$

(b) $g(x) = \dfrac{2x + 5}{8}$

(c) $h(x) = 3x^2 + 2x - 5$

Solution

(a) The denominator is zero when $x = 2$. So, the domain is all real values of x such that $x \neq 2$. In interval notation, you can write the domain as

$$\text{Domain} = (-\infty, 2) \cup (2, \infty).$$

(b) The denominator, 8, is never zero, and so the domain is the set of *all* real numbers. In interval notation, you can write the domain as

$$\text{Domain} = (-\infty, \infty).$$

(c) Note that any polynomial is also a rational expression, because you can consider its denominator to be 1. The domain of this function is the set of all real numbers. In interval notation, you can write the domain as

$$\text{Domain} = (-\infty, \infty).$$

Example 2 ■ Finding the Domain of a Rational Function

Find the domain of each rational function.

(a) $f(x) = \dfrac{3x}{x^2 - 25}$

(b) $g(x) = \dfrac{x^2 + 3x}{x^2 + 5x - 6}$

Solution

(a) The denominator of this rational function is

$$x^2 - 25 = (x + 5)(x - 5).$$

Because this denominator is zero when $x = -5$ or when $x = 5$, the domain is all real values of x such that $x \neq -5$ and $x \neq 5$. In interval notation, you can write the domain as

$$\text{Domain} = (-\infty, -5) \cup (-5, 5) \cup (5, \infty).$$

(b) The denominator of this rational function is

$$x^2 + 5x - 6 = (x + 6)(x - 1).$$

Because this denominator is zero when $x = -6$ or when $x = 1$, the domain is all real values of x such that $x \neq -6$ and $x \neq 1$. In interval notation, you can write the domain as

$$\text{Domain} = (-\infty, -6) \cup (-6, 1) \cup (1, \infty).$$

Technology

Use a graphing utility to graph the equation

$$y = \dfrac{4}{x - 2}.$$

Then use the *trace* or *table* feature of the utility to determine the behavior of the graph near $x = 2$. Graph the equations that correspond to parts (b) and (c) of Example 1. How does each of these graphs differ from the graph of $y = 4 / (x - 2)$?

Study Tip

When a rational function is written, the domain is usually not listed with the function. It is *implied* that the real numbers that make the denominator zero are excluded from the function. For instance, you know to exclude $x = 2$ and $x = -2$ from the function

$$f(x) = \frac{3x + 2}{x^2 - 4}$$

without having to list this information with the function.

In applications involving rational functions, it is often necessary to further restrict the domain. To indicate such a restriction, write the domain to the right of the fraction. For instance, the domain of the rational function

$$f(x) = \frac{x^2 + 20}{x + 4}, \quad x > 0$$

is the set of positive real numbers $(0, \infty)$, as indicated by the inequality $x > 0$. Note that the normal domain of this function would be all real values of x such that $x \neq -4$. However, because "$x > 0$" is listed to the right of the function, the domain is restricted by this inequality.

Example 3 ■ An Application Involving a Restricted Domain

You have started a small manufacturing business. The initial investment for the business is $120,000. The cost of each unit that you manufacture is $15. So, your total cost of producing x units is

$$C = 15x + 120{,}000. \qquad \text{Cost function}$$

Your average cost per unit depends on the number of units produced. For instance, the average cost per unit $\overline{C}$ for producing 100 units is

$$\overline{C} = \frac{15(100) + 120{,}000}{100} \qquad \text{Substitute 100 for } x.$$

$$= \$1215. \qquad \text{Average cost per unit for 100 units}$$

The average cost per unit decreases as the number of units increases. For instance, the average cost per unit $\overline{C}$ for producing 1000 units is

$$\overline{C} = \frac{15(1000) + 120{,}000}{1000} \qquad \text{Substitute 1000 for } x.$$

$$= \$135. \qquad \text{Average cost per unit for 1000 units}$$

In general, the average cost of producing x units is

$$\overline{C} = \frac{15x + 120{,}000}{x}. \qquad \text{Average cost per unit for } x \text{ units}$$

What is the domain of this rational function?

Solution

If you were considering this function from only a mathematical point of view, you would say that the domain is all real values of x such that $x \neq 0$. However, because this fraction is a mathematical model representing a real-life situation, you must consider which values of x make sense in real life. For this model, the variable x represents the number of units that you produce. Assuming that you cannot produce a fractional number of units, you conclude that the domain is the set of positive integers. That is,

$$\text{Domain} = \{1, 2, 3, 4, \ldots\}.$$

Simplifying Rational Expressions

As with numerical fractions, a rational expression is said to be in **simplified** (or **reduced**) **form** if its numerator and denominator have no factors in common (other than ± 1). To simplify rational expressions, you can apply the rule below.

NOTE You can verify the rule shown at the right as follows.

$$\frac{uw}{vw} = \frac{u}{v} \cdot \frac{w}{w}$$

$$= \frac{u}{v} \cdot 1$$

$$= \frac{u}{v}$$

> ### Simplifying Rational Expressions
>
> Let u, v, and w represent numbers, variables, or algebraic expressions such that $v \neq 0$ and $w \neq 0$. Then the following is valid.
>
> $$\frac{u\cancel{w}}{v\cancel{w}} = \frac{u}{v}$$

Be sure you divide out only factors, not terms. For instance, consider the expressions below.

$$\frac{\cancel{2} \cdot 2}{\cancel{2}(x + 5)} \qquad \text{You can divide out the common factor 2.}$$

$$\frac{3 + x}{3 + 2x} \qquad \text{You cannot divide out the common term 3.}$$

Simplifying a rational expression requires two steps: (1) completely factor the numerator and denominator and (2) divide out any *factors* that are common to both the numerator and the denominator. So, your success in simplifying rational expressions actually lies in your ability to *factor completely* the polynomials in both the numerator and denominator. You may want to review the factoring techniques discussed in Sections 1.3 and 1.4.

NOTE As you study the examples and work the exercises in this and the next three sections, keep in mind that you are *rewriting expressions in simpler forms;* you are not solving equations. Equal signs are used in the steps of the simplification process only to indicate that the new form of the expression is *equivalent* to the previous form.

Example 4 ■ Simplifying a Rational Expression

Simplify

$$\frac{2x^3 - 6x}{6x^2}.$$

Solution

First note that the domain of the rational expression is all real values of x such that $x \neq 0$. Then, completely factor both the numerator and the denominator.

$$\frac{2x^3 - 6x}{6x^2} = \frac{2x(x^2 - 3)}{2x(3x)} \qquad \text{Factor numerator and denominator.}$$

$$= \frac{2\cancel{x}(x^2 - 3)}{2\cancel{x}(3x)} \qquad \text{Divide out common factor } 2x.$$

$$= \frac{x^2 - 3}{3x} \qquad \text{Simplified form}$$

In simplified form, the domain of the rational expression is the same as that of the original expression—all real values of x such that $x \neq 0$.

Example 5 ■ Adjusting the Domain After Simplifying

Simplify

$$\frac{x^2 + 2x - 15}{4x - 12}.$$

Solution

The domain of the rational expression is all real values of x such that $x \neq 3$.

$$\frac{x^2 + 2x - 15}{4x - 12} = \frac{(x + 5)(x - 3)}{4(x - 3)} \qquad \text{Factor numerator and denominator.}$$

$$= \frac{(x + 5)\cancel{(x - 3)}}{4\cancel{(x - 3)}} \qquad \text{Divide out common factor } (x - 3).$$

$$= \frac{x + 5}{4}, \quad x \neq 3 \qquad \text{Simplified form}$$

Dividing out common factors from the numerator and denominator of a rational expression can change its domain. In Example 5, for instance, the domain of the original expression is all real values of x such that $x \neq 3$. So, the original expression is equal to the simplified expression for all real numbers *except* 3.

Example 6 ■ Simplifying a Rational Expression

Simplify

$$\frac{x^3 - 16x}{x^2 - 2x - 8}.$$

Solution

The domain of the rational expression is all real values of x such that $x \neq -2$ and $x \neq 4$.

$$\frac{x^3 - 16x}{x^2 - 2x - 8} = \frac{x(x^2 - 16)}{(x + 2)(x - 4)} \qquad \text{Partially factor.}$$

$$= \frac{x(x + 4)(x - 4)}{(x + 2)(x - 4)} \qquad \text{Factor completely.}$$

$$= \frac{x(x + 4)\cancel{(x - 4)}}{(x + 2)\cancel{(x - 4)}} \qquad \text{Divide out common factor } (x - 4).$$

$$= \frac{x(x + 4)}{x + 2}, \quad x \neq 4 \qquad \text{Simplified form}$$

In this text, when simplifying a rational expression, the convention of listing *by the simplified expression* all values of x that must be specifically excluded from the domain in order to make the domains of the simplified and original expressions agree is followed. In Example 6, for instance, the restriction $x \neq 4$ is listed with the simplified expression in order to make the two domains agree. Note that the value of $x = -2$ is excluded from both domains, so it is not necessary to list this value.

Study Tip

Be sure to factor *completely* the numerator and denominator of a rational expression before concluding that there is no common factor. This may involve making a change in sign to see if further reduction is possible. Note that the Distributive Property allows you to write $(b - a)$ as $-(a - b)$. Watch for this in Example 7.

NOTE Here is another way to factor and simplify the expression in Example 7.

$$\frac{2x^2 - 9x + 4}{12 + x - x^2}$$

$$= \frac{2x^2 - 9x + 4}{-x^2 + x + 12}$$

$$= \frac{2x^2 - 9x + 4}{-1(x^2 - x - 12)}$$

$$= \frac{(2x - 1)(x - 4)}{-1(x - 4)(x + 3)}$$

$$= -\frac{2x - 1}{x + 3}, \quad x \neq 4$$

Example 7 ■ Simplification Involving a Change in Sign

Simplify $\dfrac{2x^2 - 9x + 4}{12 + x - x^2}$.

Solution

The domain of the rational expression is all real values of x such that $x \neq -3$ and $x \neq 4$.

$$\frac{2x^2 - 9x + 4}{12 + x - x^2} = \frac{(2x - 1)(x - 4)}{(4 - x)(3 + x)} \qquad \text{Factor numerator and denominator.}$$

$$= \frac{(2x - 1)(x - 4)}{-(x - 4)(3 + x)} \qquad (4 - x) = -(x - 4)$$

$$= \frac{(2x - 1)(x - 4)}{-(x - 4)(3 + x)} \qquad \text{Divide out common factor } (x - 4).$$

$$= -\frac{2x - 1}{3 + x}, \quad x \neq 4 \qquad \text{Simplified form}$$

The simplified form is equivalent to the original expression for all values of x such that $x \neq 4$. Note that $x = -3$ is excluded from the domains of both the original and simplified expressions.

In Example 7, be sure you see that when dividing the numerator and denominator by the common factor of $(x - 4)$, you keep the minus sign. In the simplified form of the fraction, this text uses the convention of moving the minus sign out in front of the fraction. However, this is a personal preference. All of the forms below are legitimate.

$$-\frac{2x - 1}{3 + x} = \frac{-(2x - 1)}{3 + x} = \frac{2x - 1}{-3 - x} = \frac{2x - 1}{-(3 + x)}$$

In the next two examples, rational expressions that involve more than one variable are simplified.

Example 8 ■ A Rational Expression Involving Two Variables

Simplify

$$\frac{3xy + y^2}{2y}.$$

Solution

The domain of the rational expression is all real values of y such that $y \neq 0$.

$$\frac{3xy + y^2}{2y} = \frac{y(3x + y)}{2y} \qquad \text{Factor numerator and denominator.}$$

$$= \frac{y(3x + y)}{2y} \qquad \text{Divide out common factor } y.$$

$$= \frac{3x + y}{2}, \quad y \neq 0 \qquad \text{Simplified form}$$

Example 9 ■ A Rational Expression Involving Two Variables

The domain of the rational expression is all real numbers such that $x \neq 0$ and $x \neq \pm y$.

$$\frac{2x^2 + 2xy - 4y^2}{5x^3 - 5xy^2} = \frac{2(x^2 + xy - 2y^2)}{5x(x^2 - y^2)} \qquad \text{Partially factor.}$$

$$= \frac{2(x - y)(x + 2y)}{5x(x - y)(x + y)} \qquad \text{Factor numerator and denominator.}$$

$$= \frac{2(x - y)(x + 2y)}{5x(x - y)(x + y)} \qquad \text{Divide out common factor } (x - y).$$

$$= \frac{2(x + 2y)}{5x(x + y)}, \quad x \neq y \qquad \text{Simplified form}$$

Application

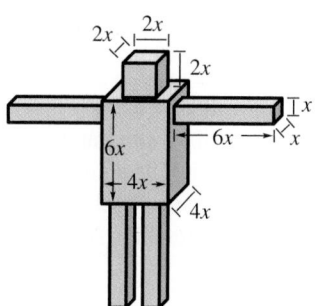

Figure 7.1

The geometric model shown in Figure 7.1 can be used to find an algebraic model for the ratio of surface area to volume for a person or animal. The model is created from six rectangular boxes. The surface area S and volume V of each box are as follows.

	Surface Area	*Volume*
Each Arm and Leg:	$S = x^2 + 4(6x^2) = 25x^2$	$V = 6x^3$
Head:	$S = 5(4x^2) = 20x^2$	$V = 8x^3$
Trunk:	$S = 2(16x^2) + 4(24x^2) - 4x^2 - 4x^2$	$V = 96x^3$
	$\quad = 120x^2$	

Example 10 ■ Comparing Surface Area to Volume

Find a rational expression that represents the ratio of surface area to volume for the model shown in Figure 7.1. Simplify the expression. Then evaluate the simplified expression for several values of x, where x is measured in feet.

Solution

$$\frac{\text{Surface area}}{\text{Volume}} = \frac{4(25x^2) + 20x^2 + 120x^2}{4(6x^3) + 8x^3 + 96x^3} = \frac{240x^2}{128x^3} = \frac{15}{8x}$$

In the table shown below, notice that small animals have large ratios and large animals have small ratios.

Animal	Mouse	Squirrel	Cat	Human	Elephant
x	$\frac{1}{24}$	$\frac{1}{12}$	$\frac{1}{6}$	$\frac{2}{5}$	2
Ratio	45	22.5	11.25	4.7	0.9

NOTE The air resistance of an animal that is falling through the air depends on the ratio of the animal's surface area to its volume. The greater the ratio, the greater the air resistance. For instance, if a mouse and a human each falls 30 feet, the mouse will encounter a greater air resistance, which means that it will hit the ground at a lower speed.

7.1 ■ Exercises

Developing Skills

In Exercises 1–14, find the domain of the rational function.

1. $f(x) = \dfrac{5}{x - 8}$

2. $f(x) = \dfrac{9}{x - 13}$

3. $f(x) = \dfrac{7x}{x + 4}$

4. $f(x) = x^4 + 2x^2 - 5$

5. $g(x) = \dfrac{3x + 6}{4}$

6. $f(x) = \dfrac{x^2 - 3}{7}$

7. $h(t) = \dfrac{5t}{t^2 - 16}$

8. $g(z) = \dfrac{z + 2}{z(z - 4)}$

9. $f(x) = \dfrac{x^2}{x^2 - 4x - 5}$

10. $h(y) = \dfrac{y + 5}{4y^2 - 5y - 6}$

11. $g(x) = \dfrac{4}{x^2 + 9}$

12. $g(x) = \dfrac{x}{x^2 - 4}$

13. $f(x) = \dfrac{x + 5}{x^2 - 3x}$

14. $f(t) = \dfrac{3t}{t^2 - 2t - 3}$

In Exercises 15–54, simplify the rational expression.

15. $\dfrac{32y^2}{12y}$

16. $\dfrac{15z^3}{25z^3}$

17. $\dfrac{18x^2y}{15xy^4}$

18. $\dfrac{16y^2z^2}{60y^5z}$

19. $\dfrac{x^2(x - 8)}{x(x - 8)}$

20. $\dfrac{a^2b(b - 3)}{b^3(b - 3)^2}$

21. $\dfrac{6y^2 + 3y^3}{3y^2}$

22. $\dfrac{15x^4 - 10x^2}{5x^2}$

23. $\dfrac{2x - 3}{4x - 6}$

24. $\dfrac{x - 5}{3x - 15}$

25. $\dfrac{81 - y^2}{2y - 18}$

26. $\dfrac{x^2 - 36}{6 - x}$

27. $\dfrac{x^2 - 25z^2}{x + 5z}$

28. $\dfrac{y^2 - 64}{5(3y + 24)}$

29. $\dfrac{u^2 - 12u + 36}{u - 6}$

30. $\dfrac{a + 3}{a^2 + 6a + 9}$

31. $\dfrac{x^2 - x - 20}{3x - 15}$

32. $\dfrac{x^2 + 7x + 12}{3x + 12}$

33. $\dfrac{z^2 + 22z + 121}{3z + 33}$

34. $\dfrac{x^2 - 7x}{x^2 - 14x + 49}$

35. $\dfrac{y^3 - 4y}{y^2 + 4y - 12}$

36. $\dfrac{x^2 - 7x}{x^2 - 4x - 21}$

37. $\dfrac{3 - x}{2x^2 - 3x - 9}$

38. $\dfrac{2y^2 + 13y + 20}{2y^2 + 17y + 30}$

39. $\dfrac{x^2 + 3x - 40}{10 + 3x - x^2}$

40. $\dfrac{7 + 20x - 3x^2}{2x^2 - 11x - 21}$

41. $\dfrac{56z^2 - 3z - 20}{49z^2 - 16}$

42. $\dfrac{15x^2 + 7x - 4}{15x^2 + x - 2}$

43. $\dfrac{4x^2y + x}{3x}$

44. $\dfrac{8yz + 2z^2}{4z}$

45. $\dfrac{5xy + 3x^2y^2}{xy^3}$

46. $\dfrac{4u^2v - 12uv^2}{18uv}$

47. $\dfrac{3m^2 - 12n^2}{m^2 + 4mn + 4n^2}$

48. $\dfrac{x^2 + xy - 2y^2}{x^2 + 3xy + 2y^2}$

49. $\dfrac{x^3 + 27z^3}{x^2 + xz - 6z^2}$

50. $\dfrac{x^3 - 8y^3}{x^2 - 4y^2}$

51. $\dfrac{x^2 + 4xy}{x^2 - 16y^2}$

52. $\dfrac{u^2 - 4v^2}{u^2 + uv - 2v^2}$

53. $\dfrac{mn + 3m - n^2 - 3n}{m^2 - n^2}$

54. $\dfrac{uv - 3u - 4v + 12}{v^2 - 5v + 6}$

In Exercises 55–58, evaluate the rational function as indicated. If not possible, state the reason.

55. $f(x) = \dfrac{4x}{x + 3}$ (a) $f(1)$ (b) $f(-2)$ (c) $f(-3)$ (d) $f(0)$

56. $g(t) = \dfrac{t - 2}{2t - 5}$ (a) $g(2)$ (b) $g\left(\tfrac{5}{2}\right)$ (c) $g(-2)$ (d) $g(0)$

57. $h(s) = \dfrac{s^2}{s^2 - s - 2}$ (a) $h(10)$ (b) $h(0)$ (c) $h(-1)$ (d) $h(2)$

58. $f(x) = \dfrac{x^3 + 1}{x^2 - 6x + 9}$ (a) $f(-1)$ (b) $f(3)$ (c) $f(-2)$ (d) $f(2)$

In Exercises 59–62, determine whether the reduction shown is valid or invalid. If it is invalid, state what is wrong.

59. $\dfrac{5 + 12}{5} = \dfrac{\cancel{5} + 12}{\cancel{5}} = 12$

60. $\dfrac{6 - 3}{6 + 11} = \dfrac{\cancel{6} - 3}{\cancel{6} + 11} = \dfrac{-3}{11}$

61. $\dfrac{9 \cdot 12}{19 \cdot 9} = \dfrac{\cancel{9} \cdot 12}{19 \cdot \cancel{9}} = \dfrac{12}{19}$

62. $\dfrac{28}{83} = \dfrac{2\cancel{8}}{8\cancel{3}} = \dfrac{2}{3}$

In Exercises 63–66, write the rational expression in reduced form. (Assume n is a positive integer.)

63. $\dfrac{x^{n+1} - 3x}{x}$

64. $\dfrac{x^{2n}}{x^{2n+1} + x^{2n}}$

65. $\dfrac{x^{2n} - 4}{x^n + 2}$

66. $\dfrac{x^{2n} + x^n - 12}{x^{n+1} + 4x}$

Solving Problems

In Exercises 67–70, describe the domain.

67. *Cost* The inventory cost I when x units of a product are ordered from a supplier is

$$I = \frac{0.25x + 2000}{x}.$$

68. *Pollution* The cost C in dollars of removing $p\%$ of the air pollutants from the stack emissions of a utility company is

$$C = \frac{80{,}000p}{100 - p}.$$

69. *Geometry* A rectangle of length x inches has an area of 500 square inches. The perimeter P of the rectangle is

$$P = 2\left(x + \frac{500}{x}\right).$$

70. *Cost* The cost C in millions of dollars for the government to seize $p\%$ of a certain illegal drug as it enters the country is

$$C = \frac{528p}{100 - p}.$$

Geometry In Exercises 71–74, find the ratio of the area of the shaded portion of the figure to the total area of the figure.

71.

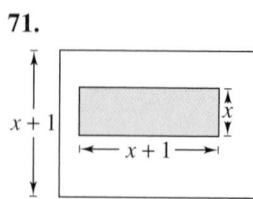

72.

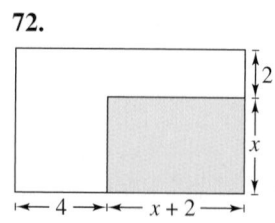

73.

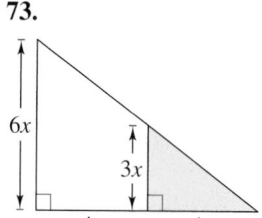

74.

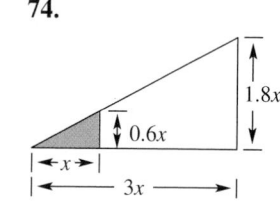

75. *Geometry* One swimming pool is circular and another is rectangular. The rectangular pool's width is three times its depth, and its length is 6 feet more than its width. The circular pool has a diameter that is twice the width of the rectangular pool, and it is 2 feet deeper. Find the ratio of the volume of the circular pool to the volume of the rectangular pool.

76. *Average Cost* A machine shop has a setup cost of $2500 for the production of a new product. The cost of the labor and material needed to produce each unit is $9.25.

(a) Write a rational expression that gives the average cost per unit when x units are produced.

(b) Determine the domain of the expression in part (a).

(c) Find the average cost per unit when $x = 100$ units.

Cost of Medicare   In Exercises 77 and 78, use the models below, which give the cost of Medicare and the U.S. population aged 65 and older from 1993 through 1998.

$$C = 111.5 + 13.91t \quad 3 \leq t \leq 8 \qquad \text{Cost of Medicare}$$

$$P = 31.9 + 0.33t \quad 3 \leq t \leq 8 \qquad \text{Population}$$

In these models, C represents the total annual cost of Medicare (in billions of dollars), P represents the U.S. population aged 65 and older (in millions), and t represents the year, with $t = 3$ corresponding to 1993. *(Sources: U.S. Healthcare Financing Administration and U.S. Census Bureau)*

77. Find a rational model that represents the average cost of Medicare per person aged 65 and older during the years 1993 through 1998.

78. Use the model found in Exercise 77 to find the average cost of Medicare per person aged 65 and older for the years 1993 through 1998. Organize your results in a table.

Explaining Concepts

79. How do you determine whether a rational expression is in reduced form?

80. Can you divide out common terms of the numerator and denominator of a rational expression? Explain.

81. Describe the error.

$$\frac{2x^2}{x^2 + 4} = \frac{2x^2}{x^2 + 4} = \frac{2}{1 + 4} = \frac{2}{5}$$

82. Give an example of a rational function whose domain is the set of all real numbers.

Ongoing Review

In Exercises 83–86, write the fraction in simplified form.

83. $\frac{54}{62}$

84. $\frac{48}{81}$

85. $\frac{112}{200}$

86. $\frac{174}{297}$

In Exercises 87 and 88, factor the algebraic expression.

87. $8x^2y - 24xy - 80y$

88. $3ax^2 + 2ax - 5a$

89. *Chemical Mixture*   A chemist mixes an 11% hydrochloric acid solution with a 6% hydrochloric acid solution. How many milliliters of each solution should the chemist use to make a 600-milliliter solution that is 8% hydrochloric acid?

90. *Work Rate*   Pump A can fill a pool in 6 hours and pump B can fill the same pool in 3 hours. How long will it take to fill the pool if both pumps are used?

Looking Further

(a) Complete the table.

x	-2	-1	0	1	2	3	4
$\dfrac{x^2 - x - 2}{x - 2}$							
$x + 1$							

(b) Write a paragraph describing the equivalence (or nonequivalence) of the two expressions in the table. Support your argument with appropriate algebra from this section *and* a discussion of the domains of the two expressions.

(c) Repeat parts (a) and (b) for the table below.

x	-2	-1	0	1	2	3	4
$\dfrac{x^2 + 5x}{x}$							
$x + 5$							

7.2 Multiplying and Dividing Rational Expressions

What you should learn:

- How to multiply rational expressions
- How to divide rational expressions
- How to simplify complex fractions

Why you should learn it:

Multiplication and division of rational expressions can be used to solve real-life applications. For instance, in Exercise 65 on page 444, you will use a rational expression to find the average amount Americans spend on books and maps.

Multiplying Rational Expressions

The rule for multiplying rational expressions is the same as the rule for multiplying numerical fractions.

$$\frac{3}{4} \cdot \frac{7}{6} = \frac{21}{24} = \frac{3 \cdot 7}{3 \cdot 8} = \frac{7}{8}$$

That is, you *multiply numerators, multiply denominators, and write the new fraction in simplified form.*

Multiplying Rational Expressions

Let u, v, w, and z be real numbers, variables, or algebraic expressions such that $v \neq 0$ and $z \neq 0$. Then the product of u/v and w/z is

$$\frac{u}{v} \cdot \frac{w}{z} = \frac{uw}{vz}.$$

In order to recognize common factors, write the numerators and denominators in factored form, as demonstrated in Example 1.

Example 1 ■ Multiplying Rational Expressions

Multiply the rational expressions.

$$\frac{4x^3y}{3xy^4} \cdot \frac{-6x^2y^2}{10x^4}$$

Solution

$$\frac{4x^3y}{3xy^4} \cdot \frac{-6x^2y^2}{10x^4} = \frac{(4x^3y) \cdot (-6x^2y^2)}{(3xy^4) \cdot (10x^4)} \qquad \text{Multiply numerators and denominators.}$$

$$= \frac{-24x^5y^3}{30x^5y^4} \qquad \text{Simplify.}$$

$$= \frac{-4(6)(x^5)(y^3)}{5(6)(x^5)(y^3)(y)} \qquad \text{Factor and divide out common factors.}$$

$$= -\frac{4}{5y}, \qquad x \neq 0 \qquad \text{Simplified form}$$

NOTE The result in Example 1 can be read as

$$\frac{4x^3y}{3xy^4} \cdot \frac{-6x^2y^2}{10x^4}$$

is equal to $-4/5y$ for all values of x except 0.

Technology

You can use a graphing utility to check your results when multiplying rational expressions. In Example 2, for instance, try graphing the equations

$$y_1 = \frac{4x^2 - 4x}{x^2 + 2x - 3} \cdot \frac{x^2 + x - 6}{4x}$$

and

$$y_2 = x - 2$$

in the same viewing window and use the *table* feature to create a table of values for the two equations. If the two graphs coincide, and the values of y_1 and y_2 are the same in the table except where a common factor has been divided out, as shown below, you can conclude that the solution checks.

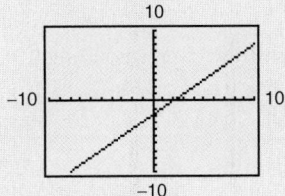

Example 2 ■ Multiplying Rational Expressions

Multiply the rational expressions.

$$\frac{4x^2 - 4x}{x^2 + 2x - 3} \cdot \frac{x^2 + x - 6}{4x}$$

Solution

$$\frac{4x^2 - 4x}{x^2 + 2x - 3} \cdot \frac{x^2 + x - 6}{4x}$$

$$= \frac{4x(x - 1)(x + 3)(x - 2)}{(x - 1)(x + 3)(4x)} \qquad \text{Multiply and factor.}$$

$$= \frac{4x(x - 1)(x + 3)(x - 2)}{(x - 1)(x + 3)(4x)} \qquad \text{Divide out common factors.}$$

$$= x - 2, \ x \neq 0, \ x \neq 1, \ x \neq -3 \qquad \text{Simplified form}$$

The rule for multiplying rational expressions can be extended to cover products involving expressions that are not in fractional form. To do this, rewrite the nonfractional expression as a fraction whose denominator is 1. Here is a simple example.

$$\frac{x + 3}{x - 2} \cdot (5x) = \frac{x + 3}{x - 2} \cdot \frac{5x}{1} = \frac{(x + 3)(5x)}{x - 2} = \frac{5x(x + 3)}{x - 2}$$

In the next example, note how to divide out factors that differ only in sign. The Distributive Property is used in the step in which $(y - x)$ is rewritten as $(-1)(x - y)$.

Example 3 ■ Multiplying Rational Expressions

Multiply the rational expressions.

$$\frac{x - y}{y^2 - x^2} \cdot \frac{x^2 - xy - 2y^2}{3x - 6y}$$

Solution

$$\frac{x - y}{y^2 - x^2} \cdot \frac{x^2 - xy - 2y^2}{3x - 6y}$$

$$= \frac{(x - y)(x - 2y)(x + y)}{(y + x)(y - x)(3)(x - 2y)} \qquad \text{Multiply and factor.}$$

$$= \frac{(x - y)(x - 2y)(x + y)}{(y + x)(-1)(x - y)(3)(x - 2y)} \qquad y - x = -1(x - y)$$

$$= \frac{(x - y)(x - 2y)(x + y)}{(x + y)(-1)(x - y)(3)(x - 2y)} \qquad \text{Divide out common factors.}$$

$$= -\frac{1}{3}, \ x \neq y, \ x \neq -y, \ x \neq 2y \qquad \text{Simplified form}$$

The rule for multiplying rational expressions can be extended to cover the product of three or more fractions, as shown in Example 4.

Example 4 ■ Multiplying Three Rational Expressions

Multiply the rational expressions.

$$\frac{x^2 - 3x + 2}{x + 2} \cdot \frac{3x}{x - 2} \cdot \frac{2x + 4}{x^2 - 5x}$$

Solution

$$\frac{x^2 - 3x + 2}{x + 2} \cdot \frac{3x}{x - 2} \cdot \frac{2x + 4}{x^2 - 5x}$$

$$= \frac{(x - 1)(x - 2)(3)(x)(2)(x + 2)}{(x + 2)(x - 2)(x)(x - 5)} \qquad \text{Multiply and factor.}$$

$$= \frac{(x - 1)(x - 2)(3)(x)(2)(x + 2)}{(x + 2)(x - 2)(x)(x - 5)} \qquad \text{Divide out common factors.}$$

$$= \frac{6(x - 1)}{x - 5}, \; x \neq 0, \; x \neq 2, \; x \neq -2 \qquad \text{Simplified form}$$

Dividing Rational Expressions

To divide two rational expressions, multiply the first fraction by the *reciprocal* of the second. That is, simply *invert the divisor and multiply.*

Dividing Rational Expressions

Let u, v, w, and z be real numbers, variables, or algebraic expressions such that $v \neq 0$, $w \neq 0$, and $z \neq 0$. Then the quotient of u/v and w/z is

$$\frac{u}{v} \div \frac{w}{z} = \frac{u}{v} \cdot \frac{z}{w} = \frac{uz}{vw}.$$

Example 5 ■ Dividing Rational Expressions

Divide the rational expressions.

$$\frac{x}{x + 3} \div \frac{4}{x - 1}$$

Solution

$$\frac{x}{x + 3} \div \frac{4}{x - 1}$$

$$= \frac{x}{x + 3} \cdot \frac{x - 1}{4} \qquad \text{Invert divisor and multiply.}$$

$$= \frac{x(x - 1)}{(x + 3)(4)} \qquad \text{Multiply numerators and denominators.}$$

$$= \frac{x(x - 1)}{4(x + 3)}, \; x \neq 1 \qquad \text{Simplify.}$$

Example 6 ■ Dividing Rational Expressions

$$\frac{2x}{3x - 12} \div \frac{x^2 - 2x}{x^2 - 6x + 8}$$

NOTE Remember that the original expression in Example 6 is equivalent to $\frac{2}{3}$ except for $x = 0$, $x = 2$, and $x = 4$.

$$= \frac{2x}{3x - 12} \cdot \frac{x^2 - 6x + 8}{x^2 - 2x}$$ Invert divisor and multiply.

$$= \frac{(2)(x)(x - 2)(x - 4)}{(3)(x - 4)(x)(x - 2)}$$ Factor.

$$= \frac{(2)(x)(x - 2)(x - 4)}{(3)(x - 4)(x)(x - 2)}$$ Divide out common factors.

$$= \frac{2}{3}, \ x \neq 0, \ x \neq 2, \ x \neq 4$$ Simplified form

Example 7 ■ Dividing Rational Expressions

Divide the rational expressions.

(a) $\dfrac{x^2 - y^2}{2x + 2y} \div \dfrac{2x^2 - 3xy + y^2}{6x + 2y}$ (b) $\dfrac{x^2 - 14x + 49}{x^2 - 49} \div \dfrac{3x - 21}{x^2 + 2x - 35}$

Solution

(a) $\dfrac{x^2 - y^2}{2x + 2y} \div \dfrac{2x^2 - 3xy + y^2}{6x + 2y}$

$$= \frac{x^2 - y^2}{2x + 2y} \cdot \frac{6x + 2y}{2x^2 - 3xy + y^2}$$ Invert divisor and multiply.

$$= \frac{(x + y)(x - y)(2)(3x + y)}{(2)(x + y)(2x - y)(x - y)}$$ Factor.

$$= \frac{(x + y)(x - y)(2)(3x + y)}{(2)(x + y)(2x - y)(x - y)}$$ Divide out common factors.

$$= \frac{3x + y}{2x - y}, \ x \neq y, \ x \neq -y$$ Simplified form

(b) $\dfrac{x^2 - 14x + 49}{x^2 - 49} \div \dfrac{3x - 21}{x^2 + 2x - 35}$

$$= \frac{x^2 - 14x + 49}{x^2 - 49} \cdot \frac{x^2 + 2x - 35}{3x - 21}$$ Invert divisor and multiply.

$$= \frac{(x - 7)(x - 7)(x + 7)(x - 5)}{(x + 7)(x - 7)3(x - 7)}$$ Factor.

$$= \frac{(x - 7)(x - 7)(x + 7)(x - 5)}{(x + 7)(x - 7)3(x - 7)}$$ Divide out common factors.

$$= \frac{x - 5}{3}, \ x \neq 7, \ x \neq -7$$ Simplified form

Study Tip

Note that for complex fractions you make the main fraction line slightly longer than the fraction lines in the numerator and denominator.

Complex Fractions

Problems involving the division of two rational expressions are sometimes written as **complex fractions.** A complex fraction is one that has a fraction in its numerator or denominator, or both. The rules for dividing fractions still apply in such cases. An example of a complex fraction is shown below.

$$\dfrac{\left(\dfrac{x-1}{5}\right)}{\left(\dfrac{x+3}{x}\right)} \quad \begin{array}{l} \text{Numerator fraction} \\[4pt] \rightarrow \text{Main fraction line} \\[4pt] \text{Denominator fraction} \end{array}$$

To perform the division implied by this complex fraction, invert the denominator and multiply:

$$\dfrac{\left(\dfrac{x-1}{5}\right)}{\left(\dfrac{x+3}{x}\right)} = \dfrac{x-1}{5} \cdot \dfrac{x}{x+3}$$

$$= \dfrac{x(x-1)}{5(x+3)}, \ x \neq 0.$$

Example 8 ■ Simplifying a Complex Fraction

Simplify the complex fraction $\dfrac{\left(\dfrac{x^2+2x-3}{x-3}\right)}{4x+12}$.

Solution

Begin by converting the denominator to fractional form.

$$\dfrac{\left(\dfrac{x^2+2x-3}{x-3}\right)}{4x+12} = \dfrac{\left(\dfrac{x^2+2x-3}{x-3}\right)}{\left(\dfrac{4x+12}{1}\right)} \qquad \text{Rewrite denominator.}$$

$$= \dfrac{x^2+2x-3}{x-3} \cdot \dfrac{1}{4x+12} \qquad \begin{array}{l}\text{Invert divisor and} \\ \text{multiply.}\end{array}$$

$$= \dfrac{(x-1)(x+3)}{(x-3)(4)(x+3)} \qquad \text{Factor.}$$

$$= \dfrac{(x-1)\cancel{(x+3)}}{(x-3)(4)\cancel{(x+3)}} \qquad \begin{array}{l}\text{Divide out common} \\ \text{factors.}\end{array}$$

$$= \dfrac{x-1}{4(x-3)}, \qquad x \neq -3 \qquad \text{Simplified form}$$

NOTE In Example 8, the domain of the complex fraction is restricted by the two denominators in the expression, $x-3$ and $4x+12$. So, the domain of the original expression is all real values of x such that $x \neq 3$ and $x \neq -3$.

7.2 ■ Exercises

Developing Skills

In Exercises 1–34, multiply and simplify.

1. $16u^4 \cdot \dfrac{12}{8u^2}$

2. $\dfrac{6}{5a} \cdot (25a)$

3. $\dfrac{8s^3}{9s} \cdot \dfrac{6s^2}{32s}$

4. $\dfrac{7x^2}{3} \cdot \dfrac{9}{14x}$

5. $\dfrac{-3x^4y}{7xy^4} \cdot \dfrac{8x^2y^2}{9y^3}$

6. $\dfrac{25xy}{8xy^4} \cdot \dfrac{-8xy^2}{35x^3}$

7. $\dfrac{1 - 3x}{4} \cdot \dfrac{46}{15 - 45x}$

8. $\dfrac{8}{3 + 4x} \cdot (9 + 12x)$

9. $\dfrac{x + 25}{8} \cdot \dfrac{8}{x + 25}$

10. $\dfrac{8u^2v}{3u + v} \cdot \dfrac{u + v}{12u}$

11. $\dfrac{12 - r}{3} \cdot \dfrac{3}{r - 12}$

12. $\dfrac{8 - z}{8 + z} \cdot \dfrac{z + 8}{z - 8}$

13. $\dfrac{6r}{r - 2} \cdot \dfrac{r^2 - 4}{33r^2}$

14. $\dfrac{5y - 20}{5y + 15} \cdot \dfrac{2y + 6}{y - 4}$

15. $\dfrac{(2x - 3)(x + 8)}{x^3} \cdot \dfrac{x}{3 - 2x}$

16. $\dfrac{x + 14}{x^3(10 - x)} \cdot \dfrac{x(x - 10)}{5}$

17. $\dfrac{3x - 15}{2x^2 - 50} \cdot \dfrac{2x^2 + 16x + 30}{6x + 9}$

18. $\dfrac{3x^2 + 21x + 36}{x - 2} \cdot \dfrac{5x^2 - 10x}{3x}$

19. $\dfrac{x^2 + 5x - 6}{x^2 + 4x} \cdot \dfrac{2x^2 + 4x - 16}{3x - 3}$

20. $\dfrac{5x + 10}{x^2 - 8x + 15} \cdot \dfrac{x^2 - 6x + 9}{10x^2 - 20x}$

21. $\dfrac{2t^2 - t - 15}{t + 2} \cdot \dfrac{t^2 - t - 6}{t^2 - 6t + 9}$

22. $\dfrac{y^2 - 16}{2y^3} \cdot \dfrac{4y}{y^2 - 6y + 8}$

23. $\dfrac{x^2 + x}{2x + 3} \cdot \dfrac{3x^2 + 19x + 28}{x^2 + 5x + 4}$

24. $\dfrac{x^2 - 16}{x^2 + 7x + 12} \cdot \dfrac{x^2 - 4x - 21}{x^2 - 4x}$

25. $\dfrac{2x^2 - 2x - 12}{x^2 - 5x - 6} \cdot \dfrac{x^2 - 7x + 6}{3x^2 - 12x + 9}$

26. $\dfrac{x^2 + 6x + 5}{4x^2 - 8x - 32} \cdot \dfrac{2x^2 - 8x - 24}{x^2 - 4x - 5}$

27. $(x^2 - 4y^2) \cdot \dfrac{xy}{(x - 2y)^2}$

28. $\dfrac{x^2 + 2xy - 3y^2}{(x + y)^2} \cdot \dfrac{x^2 - y^2}{x + 3y}$

29. $(u - 2v)^2 \cdot \dfrac{u + 2v}{u - 2v}$

30. $\dfrac{x - 2y}{x + 2y} \cdot \dfrac{x^2 + 4y^2}{x^2 - 4y^2}$

31. $\dfrac{x + 5}{x - 5} \cdot \dfrac{2x^2 - 9x - 5}{3x^2 + x - 2} \cdot \dfrac{x^2 - 1}{x^2 + 7x + 10}$

32. $\dfrac{t^2 + 4t + 3}{2t^2 - t - 10} \cdot \dfrac{t}{t^2 + 3t + 2} \cdot \dfrac{2t^2 + 4t^3}{t^2 + 3t}$

33. $\dfrac{x^3 + 3x^2 - 4x - 12}{x^3 - 3x^2 - 4x + 12} \cdot \dfrac{x^2 - 9}{x}$

34. $\dfrac{xu - yu + xv - yv}{xu + yu - xv - yv} \cdot \dfrac{xu + yu + xv + yv}{xu - yu - xv + yv}$

In Exercises 35–54, divide and simplify.

35. $\dfrac{7xy^2}{10u^2v} \div \dfrac{21x^3}{45uv}$

36. $\dfrac{25x^2y}{60x^3y^2} \div \dfrac{5x^4y^3}{16x^2y}$

37. $\dfrac{3(a + b)}{4} \div \dfrac{(a + b)^2}{2}$

38. $\dfrac{x^2 + 9}{5x + 10} \div \dfrac{x + 3}{5x^2 - 20}$

39. $\dfrac{(x^3y)^2}{(x + 2y)^2} \div \dfrac{x^2y}{(x + 2y)^3}$

40. $\dfrac{x^2 - y^2}{2x^2 - 8x} \div \dfrac{(x - y)^2}{2xy}$

41. $\dfrac{\left(\dfrac{x^2}{12}\right)}{\left(\dfrac{5x}{18}\right)}$

42. $\dfrac{\left[\dfrac{3(u^2v)^2}{6v^3}\right]}{\left[\dfrac{(uv^3)^2}{3uv}\right]}$

43. $\dfrac{\left(\dfrac{25x^2}{x - 5}\right)}{\left(\dfrac{10x}{5 - x}\right)}$

44. $\dfrac{\left(\dfrac{5x}{x + 7}\right)}{\left(\dfrac{10}{x^2 + 8x + 7}\right)}$

45. $\dfrac{16x^2 + 8x + 1}{3x^2 + 8x - 3} \div \dfrac{4x^2 - 3x - 1}{x^2 + 6x + 9}$

46. $\dfrac{x^2 - 25}{x} \div \dfrac{x^3 - 5x^2}{x^2 + x}$

47. $\dfrac{y^3 - 8}{y^2 + y - 6} \div \dfrac{y^3 + 2y^2 + 4y}{y^2 + 3y}$

48. $\dfrac{z^2 - 81}{z^2 - 16} \div \dfrac{z^2 - z - 20}{z^2 + 5z - 36}$

49. $\dfrac{x(x + 3) - 2(x + 3)}{x^2 - 4} \div \dfrac{x}{x^2 + 4x + 4}$

50. $\dfrac{t^3 + t^2 - 9t - 9}{t^2 - 5t + 6} \div \dfrac{t^2 + 6t + 9}{t - 2}$

51. $\dfrac{\left(\dfrac{x^2 - 4}{x - 7}\right)}{\left(\dfrac{3x + 6}{x^2 - 6x - 7}\right)}$

52. $\dfrac{\left(\dfrac{2x^2 + 18x + 40}{6x - 3}\right)}{\left(\dfrac{x^2 - x - 20}{2x - 1}\right)}$

53. $\dfrac{\left(\dfrac{x^2 + 3x - 10}{x^2 + x - 6}\right)}{\left(\dfrac{x^2 - x - 30}{2x^2 - 15x + 18}\right)}$

54. $\dfrac{\left(\dfrac{2y^2 + 11y + 15}{y^2 - 4y - 21}\right)}{\left(\dfrac{6y^2 + 11y - 10}{3y^2 - 23y + 14}\right)}$

In Exercises 55–62, perform the operations and simplify. (In Exercises 61 and 62, n is a positive integer.)

55. $\left[\dfrac{x^2}{9} \cdot \dfrac{3(x + 4)}{x^2 + 2x}\right] \div \dfrac{x}{x + 2}$

56. $\left(\dfrac{x^2 + 6x + 9}{x^2} \cdot \dfrac{2x + 1}{x^2 - 9}\right) \div \dfrac{4x^2 + 4x + 1}{x^2 - 3x}$

57. $\left[\dfrac{xy + y}{4x} \div (3x + 3)\right] \div \dfrac{y}{3x}$

58. $\left(\dfrac{3u^2 - u - 4}{u^2}\right)^2 \div \dfrac{3u^2 + 12u + 4}{u^4 - 3u^3}$

59. $\dfrac{2x^2 + 5x - 25}{3x^2 + 5x + 2} \cdot \dfrac{3x^2 + 2x}{x + 5} \div \left(\dfrac{x}{x + 1}\right)^2$

60. $\dfrac{t^2 - 100}{4t^2} \cdot \dfrac{t^3 - 5t^2 - 50t}{t^4 + 10t^3} \div \dfrac{(t - 10)^2}{5t}$

61. $x^3 \cdot \dfrac{x^{2n} - 9}{x^{2n} + 4x^n + 3} \div \dfrac{x^{2n} - 2x^n - 3}{x}$

62. $\dfrac{x^{n+1} - 8x}{x^{2n} + 2x^n + 1} \cdot \dfrac{x^{2n} - 4x^n - 5}{x} \div x^n$

In Exercises 63 and 64, use a graphing utility to graph the two equations in the same viewing window. Use the graphs and a table of values to verify that the expressions are equivalent. Verify the results algebraically.

63. $y_1 = \dfrac{3x + 2}{x} \cdot \dfrac{x^2}{9x^2 - 4}$

$y_2 = \dfrac{x}{3x - 2}, x \neq 0, x \neq -\dfrac{2}{3}$

64. $y_1 = \dfrac{x^2 - 10x + 25}{x^2 - 25} \cdot \dfrac{x + 5}{2}$

$y_2 = \dfrac{x - 5}{2}, x \neq \pm 5$

Solving Problems

65. *Analyzing Data* The amount A Americans spent on books and maps (in billions of dollars) and the population P of the United States (in millions) for the period 1993 through 1998 can be modeled by

$A = -0.114t^2 + 3.09t + 10.5, \quad 3 \leq t \leq 8$

and

$P = 2.47t + 250.7, \quad 3 \leq t \leq 8$

where t is time in years, with $t = 3$ corresponding to 1993. *(Sources: U.S. Bureau of Economic Analysis and U.S. Census Bureau)*

(a) Use a graphing utility to graph the two models in the same viewing window.

(b) Find a model for the average amount Americans spent per person on books and maps.

(c) Use the model in part (b) to complete the table.

Year, t	3	4	5	6	7	8
Amount per person						

(d) Why do you think the amount spent per person is so low?

The symbol ▦ indicates an exercise in which you are instructed to use a calculator or graphing utility.

Explaining Concepts

66. In your own words, explain how to divide rational expressions.

67. Explain how to divide a rational expression by a polynomial.

68. Define the term *complex fraction*. Give an example and show how to simplify the fraction.

69. Describe the error.

$$\frac{x^2 - 4}{5x} \div \frac{x + 2}{x - 2} = \frac{5x}{x^2 - 4} \cdot \frac{x + 2}{x - 2}$$
$$= \frac{5x}{(x + 2)(x - 2)} \cdot \frac{x + 2}{x - 2}$$
$$= \frac{5x}{(x - 2)^2}$$

Ongoing Review

In Exercises 70–75, evaluate the quantity and write your answer in simplified form.

70. $\frac{4}{15} \cdot \frac{5}{8}$

71. $\frac{4}{9} \cdot \frac{15}{16}$

72. $-\frac{3}{5} \cdot \frac{10}{21}$

73. $-\frac{8}{15} \cdot -\frac{4}{25}$

74. $\frac{8}{9} \div -\frac{16}{3}$

75. $-\frac{4}{225} \div -\frac{2}{45}$

In Exercises 76–79, factor the algebraic expression.

76. $x^2 + 7x - 18$

77. $3x^2 + 5x - 8$

78. $6x^2 + 7x + 2$

79. $9x^2 - 24x + 16$

80. *Number Problem*   The numerator of a fraction is 4 less than the denominator. If the numerator is increased by 14 and the denominator is decreased by 10, the resulting number is 5. What is the original fraction?

81. *Distance*   A motorboat left a harbor and traveled to an island at an average rate of 15 miles per hour. The average speed on the return trip was 10 miles per hour. If the total trip took 7.5 hours, how far is the island from the harbor?

Looking Further

Probability   Consider an experiment in which a marble is tossed into a rectangular box with dimensions x centimeters by $2x + 1$ centimeters. The probability that the marble will come to rest in the *unshaded* portion of the box is equal to the ratio of the unshaded area to the total area of the figure. Find the probability.

(a)

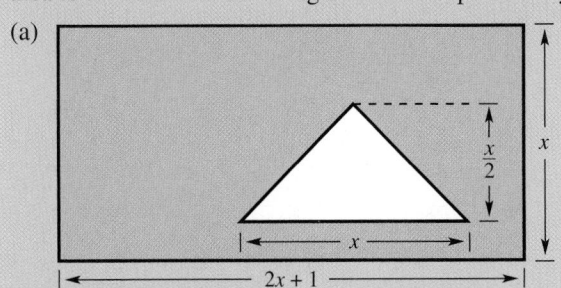

(b)

(c)

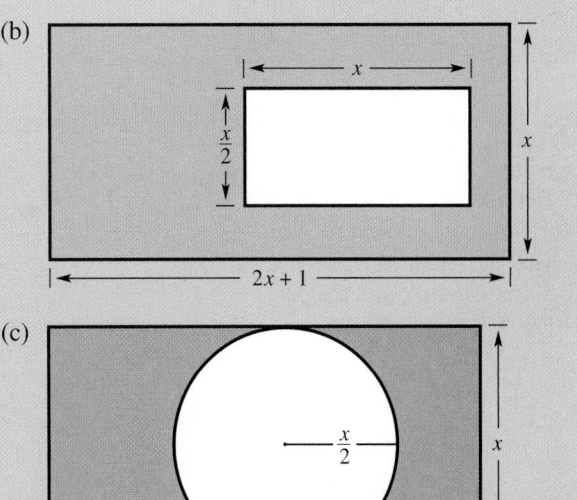

7.3 Adding and Subtracting Rational Expressions

Adding or Subtracting with Like Denominators

As with numerical fractions, the procedure used to add or subtract two rational expressions depends on whether the expressions have *like* or *unlike* denominators. To add or subtract two rational expressions with *like* denominators, simply combine their numerators and place the result over the common denominator.

Adding or Subtracting with Like Denominators

If u, v, and w are real numbers, variables, or algebraic expressions, and $w \neq 0$, the rules below are valid.

1. $\dfrac{u}{w} + \dfrac{v}{w} = \dfrac{u + v}{w}$ Add fractions with like denominators.

2. $\dfrac{u}{w} - \dfrac{v}{w} = \dfrac{u - v}{w}$ Subtract fractions with like denominators.

Example 1 ■ Adding and Subtracting with Like Denominators

Perform the indicated operation and simplify.

(a) $\dfrac{x}{4} + \dfrac{5 - x}{4}$ (b) $\dfrac{7}{2x - 3} - \dfrac{3x}{2x - 3}$

Solution

(a) $\dfrac{x}{4} + \dfrac{5 - x}{4} = \dfrac{x + (5 - x)}{4} = \dfrac{5}{4}$

(b) $\dfrac{7}{2x - 3} - \dfrac{3x}{2x - 3} = \dfrac{7 - 3x}{2x - 3}$

Example 2 ■ Subtracting Rational Expressions and Simplifying

Subtract: $\dfrac{x}{x^2 - 2xy - 3y^2} - \dfrac{3y}{x^2 - 2xy - 3y^2}$.

Solution

$$\dfrac{x}{x^2 - 2xy - 3y^2} - \dfrac{3y}{x^2 - 2xy - 3y^2} = \dfrac{x - 3y}{x^2 - 2xy - 3y^2} \quad \text{Subtract.}$$

$$= \dfrac{x - 3y}{(x - 3y)(x + y)} \quad \text{Factor.}$$

$$= \dfrac{(x - 3y)(1)}{(x - 3y)(x + y)} \quad \text{Divide out common factor.}$$

$$= \dfrac{1}{x + y}, \quad x \neq 3y \quad \text{Simplified form}$$

The rule for adding or subtracting rational expressions with like denominators can be extended to cover sums and differences involving three or more rational expressions, as illustrated in Example 3.

Example 3 ■ Combining Three Rational Expressions

$$\frac{x^2 - 26}{x - 5} - \frac{2x + 4}{x - 5} + \frac{10 + x}{x - 5} = \frac{(x^2 - 26) - (2x + 4) + (10 + x)}{x - 5}$$

$$= \frac{x^2 - 26 - 2x - 4 + 10 + x}{x - 5}$$

$$= \frac{x^2 - x - 20}{x - 5}$$

$$= \frac{(x - 5)(x + 4)}{x - 5}$$

$$= x + 4, \quad x \neq 5$$

Adding or Subtracting with Unlike Denominators

To add or subtract rational expressions with *unlike* denominators, you must first rewrite each expression using the **least common multiple (LCM)** of the denominators of the individual expressions. The least common multiple of two (or more) polynomials is the simplest polynomial that is a multiple of each of the original polynomials.

Example 4 ■ Finding Least Common Multiples

Find the least common multiple of each pair of expressions.

(a) $6x$ and $2x^2$ (b) $x^2 - x$ and $2x - 2$ (c) $3x^2 + 6x$ and $x^2 + 4x + 4$

Solution

(a) The least common multiple of

$$6x = 2 \cdot 3 \cdot x \quad \text{and} \quad 2x^2 = 2 \cdot x^2$$

is $2 \cdot 3 \cdot x^2 = 6x^2$.

(b) The least common multiple of

$$x^2 - x = x(x - 1) \quad \text{and} \quad 2x - 2 = 2(x - 1)$$

is $2x(x - 1)$.

(c) The least common multiple of

$$3x^2 + 6x = 3x(x + 2) \quad \text{and} \quad x^2 + 4x + 4 = (x + 2)^2$$

is $3x(x + 2)^2$.

To add or subtract rational expressions with *unlike* denominators, you must first rewrite the rational expressions so that they have *like* denominators. The like denominator that you use is the least common multiple of the original denominators and is called the **least common denominator (LCD)** of the original rational expressions. Once the rational expressions have been written with like denominators, you can simply add or subtract the rational expressions using the rule given at the beginning of this section.

Technology

You can use a graphing utility to check your results when adding or subtracting rational expressions. In Example 5, for instance, try graphing the equations

$$y_1 = \frac{7}{6x} + \frac{5}{8x}$$

and

$$y_2 = \frac{43}{24x}$$

in the same viewing window. If the two graphs coincide, as shown below, you can conclude that the solution checks.

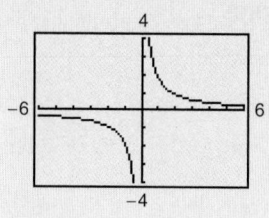

Example 5 ■ Adding with Unlike Denominators

Add the rational expressions.

$$\frac{7}{6x} + \frac{5}{8x}$$

Solution

By factoring the denominators, $6x = 2 \cdot 3 \cdot x$ and $8x = 2^3 \cdot x$, you can conclude that the least common denominator is $2^3 \cdot 3 \cdot x = 24x$.

$$\frac{7}{6x} + \frac{5}{8x} = \frac{7(4)}{6x(4)} + \frac{5(3)}{8x(3)}$$ Rewrite fractions using LCD of $24x$.

$$= \frac{28}{24x} + \frac{15}{24x}$$ Like denominators

$$= \frac{28 + 15}{24x}$$ Add fractions.

$$= \frac{43}{24x}$$ Simplified form

Example 6 ■ Subtracting with Unlike Denominators

Subtract the rational expressions.

$$\frac{3}{x - 3} - \frac{5}{x + 2}$$

Solution

The only factors of the denominators are $(x - 3)$ and $(x + 2)$. So, the least common denominator is $(x - 3)(x + 2)$.

$$\frac{3}{x - 3} - \frac{5}{x + 2} = \frac{3(x + 2)}{(x - 3)(x + 2)} - \frac{5(x - 3)}{(x - 3)(x + 2)}$$ Rewrite fractions using LCD of $(x - 3)(x + 2)$.

$$= \frac{3x + 6}{(x - 3)(x + 2)} - \frac{5x - 15}{(x - 3)(x + 2)}$$ Distributive Property

$$= \frac{(3x + 6) - (5x - 15)}{(x - 3)(x + 2)}$$ Subtract fractions.

$$= \frac{3x + 6 - 5x + 15}{(x - 3)(x + 2)}$$ Distributive Property

$$= \frac{-2x + 21}{(x - 3)(x + 2)}$$ Simplified form

Example 7 ■ Adding with Unlike Denominators

Add the rational expressions.

$$\frac{6x}{x^2 - 4} + \frac{3}{2 - x}$$

NOTE In Example 7, note that the factors in the denominator are $x^2 - 4 = (x + 2)(x - 2)$ and $2 - x$. Using the fact that

$$(2 - x) = (-1)(x - 2),$$

you can rewrite the second fraction by multiplying its numerator and denominator by -1.

Solution

$$\frac{6x}{x^2 - 4} + \frac{3}{2 - x} = \frac{6x}{x^2 - 4} + \frac{(-1)(3)}{(-1)(2 - x)} \qquad \text{Multiply numerator and denominator of second fraction by } -1.$$

$$= \frac{6x}{(x + 2)(x - 2)} + \frac{-3}{x - 2} \qquad \text{Simplify.}$$

$$= \frac{6x}{(x + 2)(x - 2)} + \frac{-3(x + 2)}{(x + 2)(x - 2)} \qquad \text{Rewrite fractions using LCD of } (x + 2)(x - 2).$$

$$= \frac{6x}{(x + 2)(x - 2)} + \frac{-3x - 6}{(x + 2)(x - 2)} \qquad \text{Distributive Property}$$

$$= \frac{6x - 3x - 6}{(x + 2)(x - 2)} \qquad \text{Add fractions.}$$

$$= \frac{3x - 6}{(x + 2)(x - 2)} \qquad \text{Combine like terms.}$$

$$= \frac{3(x - 2)}{(x + 2)(x - 2)} \qquad \text{Factor and divide out common factor } (x - 2).$$

$$= \frac{3}{x + 2}, \quad x \neq 2 \qquad \text{Simplified form}$$

Example 8 ■ Subtracting with Unlike Denominators

Subtract the rational expressions.

$$\frac{x}{x^2 - 5x + 6} - \frac{1}{x^2 - x - 2}$$

Solution

$$\frac{x}{x^2 - 5x + 6} - \frac{1}{x^2 - x - 2}$$

$$= \frac{x}{(x - 3)(x - 2)} - \frac{1}{(x - 2)(x + 1)} \qquad \text{Factor denominators.}$$

$$= \frac{x(x + 1)}{(x - 3)(x - 2)(x + 1)} - \frac{1(x - 3)}{(x - 3)(x - 2)(x + 1)} \qquad \text{Rewrite fractions using LCD of } (x - 3)(x - 2)(x + 1).$$

$$= \frac{x^2 + x}{(x - 3)(x - 2)(x + 1)} - \frac{x - 3}{(x - 3)(x - 2)(x + 1)} \qquad \text{Distributive Property}$$

$$= \frac{(x^2 + x) - (x - 3)}{(x - 3)(x - 2)(x + 1)} \qquad \text{Subtract fractions.}$$

$$= \frac{x^2 + x - x + 3}{(x - 3)(x - 2)(x + 1)} \qquad \text{Distributive Property}$$

$$= \frac{x^2 + 3}{(x - 3)(x - 2)(x + 1)} \qquad \text{Simplified form}$$

Example 9 ■ Combining Three Rational Expressions

Perform the indicated operations and simplify.

$$\frac{2x-5}{6x+9} - \frac{4}{2x^2+3x} + \frac{1}{x}$$

Solution

$$\frac{2x-5}{6x+9} - \frac{4}{2x^2+3x} + \frac{1}{x} = \frac{(2x-5)(x)}{3(2x+3)(x)} - \frac{(4)(3)}{x(2x+3)(3)} + \frac{3(2x+3)}{(x)(3)(2x+3)}$$

$$= \frac{2x^2-5x}{3x(2x+3)} - \frac{12}{3x(2x+3)} + \frac{6x+9}{3x(2x+3)}$$

$$= \frac{2x^2-5x-12+6x+9}{3x(2x+3)}$$

$$= \frac{2x^2+x-3}{3x(2x+3)}$$

$$= \frac{(x-1)(2x+3)}{3x(2x+3)}$$

$$= \frac{(x-1)\cancel{(2x+3)}}{3x\cancel{(2x+3)}}$$

$$= \frac{x-1}{3x}, \quad x \neq -\frac{3}{2}$$

Complex Fractions

Complex fractions can have numerators or denominators that are the sums or differences of fractions. To simplify a complex fraction, first combine its numerator and its denominator into single fractions. Then divide by inverting the denominator and multiplying.

Example 10 ■ Simplifying a Complex Fraction

$$\frac{\left(\dfrac{x}{4}+\dfrac{3}{2}\right)}{\left(2-\dfrac{3}{x}\right)} = \frac{\left(\dfrac{x}{4}+\dfrac{6}{4}\right)}{\left(\dfrac{2x}{x}-\dfrac{3}{x}\right)} \qquad \text{Find least common denominators.}$$

$$= \frac{\left(\dfrac{x+6}{4}\right)}{\left(\dfrac{2x-3}{x}\right)} \qquad \text{Add fractions in numerator and denominator.}$$

$$= \frac{x+6}{4} \cdot \frac{x}{2x-3} \qquad \text{Invert divisor and multiply.}$$

$$= \frac{x(x+6)}{4(2x-3)}, \quad x \neq 0 \qquad \text{Simplified form}$$

NOTE Another way to simplify the complex fraction given in Example 10 is to multiply the numerator and denominator by the least common denominator of *every* fraction in the numerator and denominator. For this fraction, when you multiply the numerator and denominator by $4x$, you obtain the same result.

$$\frac{\left(\dfrac{x}{4}+\dfrac{3}{2}\right)}{\left(2-\dfrac{3}{x}\right)} = \frac{\left(\dfrac{x}{4}+\dfrac{3}{2}\right) \cdot 4x}{\left(2-\dfrac{3}{x}\right) \cdot 4x}$$

$$= \frac{\dfrac{x}{4}(4x)+\dfrac{3}{2}(4x)}{2(4x)-\dfrac{3}{x}(4x)}$$

$$= \frac{x^2+6x}{8x-12}$$

$$= \frac{x(x+6)}{4(2x-3)}, \quad x \neq 0$$

Example 11 ■ Simplifying a Complex Fraction

$$\frac{\left(\dfrac{2}{x+2}\right)}{\left(\dfrac{1}{x+2}+\dfrac{2}{x}\right)} = \frac{\left(\dfrac{2}{x+2}\right)(x)(x+2)}{\dfrac{1}{x+2}(x)(x+2)+\dfrac{2}{x}(x)(x+2)} \qquad \text{Multiply numerator and denominator by LCD of } x(x+2).$$

$$= \frac{2x}{x+2(x+2)} \qquad \text{Simplify.}$$

$$= \frac{2x}{3x+4}, \ x \neq -2, \ x \neq 0 \qquad \text{Simplified form}$$

Example 12 ■ An Application: Marriage and Divorce Rates

For the years 1980 through 1998, the marriage rate M (per 1000 people) and the divorce rate D (per 1000 people) in the United States can be modeled by

$$M = 3.4 + \frac{235t + 990}{t^2 + 31t + 137} \quad \text{and} \quad D = 0.6 + \frac{337t + 2}{t^2 + 74t + 0.4}, \quad 0 \le t \le 18$$

where $t = 0$ represents 1980. Find a model for the difference R between marriages and divorces (per 1000 people per year). (You do not need to simplify the model.)   (*Source: U.S. Department of Health and Human Services*)

Solution

You can find the model for R by subtracting D from M.

$$R = M - D$$

$$= \left(3.4 + \frac{235t + 990}{t^2 + 31t + 137}\right) - \left(0.6 + \frac{337t + 2}{t^2 + 74t + 0.4}\right)$$

$$= 2.8 + \frac{235t + 990}{t^2 + 31t + 137} - \frac{337t + 2}{t^2 + 74t + 0.4}$$

The graphs of the two models are shown in Figure 7.2.

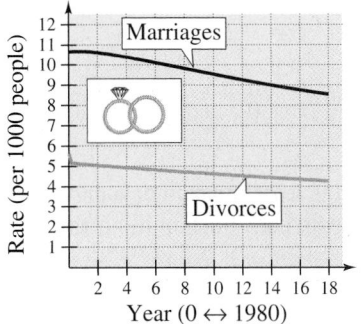

Figure 7.2 *Marriage (and divorce) rates are often given in terms of the number of marriages (or divorces) per 1000 people. Since 1900, the marriage rate has remained relatively constant: the year with the lowest rate was 1958 (a rate of 8.4), and the year with the highest rate was 1945 (a rate of 12.2). The divorce rate, however, has increased greatly since 1900—from a low of 0.7 in 1900 to a high of 5.3 around 1980.*
(*Source: U.S. Census Bureau*)

Collaborate!

Extending the Concept

Each person in your group should evaluate each of the expressions for the indicated variable in two different ways: (1) simplify the rational expressions first and then evaluate the expression at the indicated value, and (2) substitute the indicated value for the variable first and then simplify the resulting expression. Do you obtain the same result with each method? Discuss which method you prefer and why.

a. $\dfrac{1}{m-4} - \dfrac{1}{m+4} + \dfrac{3m}{m^2-16},$ $m = 2$ b. $\dfrac{x-2}{x^2-9} + \dfrac{3x+2}{x^2-5x+6},$ $x = 4$

c. $\dfrac{3y^2+16y-8}{y^2+2y-8} - \dfrac{y-1}{y-2} + \dfrac{y}{y+4},$ $y = 3$

7.3 ■ Exercises

Developing Skills

In Exercises 1–16, combine and simplify.

1. $\dfrac{5x}{8} + \dfrac{7x}{8}$

2. $\dfrac{7y}{12} - \dfrac{5y}{12}$

3. $\dfrac{x}{9} - \dfrac{x+2}{9}$

4. $\dfrac{4-y}{4} + \dfrac{3y}{4}$

5. $\dfrac{16+z}{5z} - \dfrac{11-z}{5z}$

6. $\dfrac{2x+5}{3} + \dfrac{1-x}{3}$

7. $\dfrac{3y}{3} - \dfrac{3y-3}{3} - \dfrac{7}{3}$

8. $\dfrac{-16u}{9} - \dfrac{27-16u}{9} + \dfrac{2}{9}$

9. $\dfrac{2x-1}{x(x-3)} + \dfrac{1-x}{x(x-3)}$

10. $\dfrac{5x-1}{x+4} + \dfrac{5-4x}{x+4}$

11. $\dfrac{3y-22}{y-6} - \dfrac{2y-16}{y-6}$

12. $\dfrac{7s-5}{2s+5} + \dfrac{3(s+10)}{2s+5}$

13. $\dfrac{3x^2-4x+6}{x^2-2x-15} - \dfrac{2x^2+2x+1}{x^2-2x-15}$

14. $\dfrac{5x^2-x+1}{2x^2-3x+1} + \dfrac{x^2-2}{2x^2-3x+1}$

15. $\dfrac{x^2-3}{x+3} + \dfrac{10x+9}{x+3} - \dfrac{2x-9}{x+3}$

16. $\dfrac{7x-1}{x-2} - \dfrac{4x+3}{x-2} + \dfrac{x^2+5}{x-2}$

In Exercises 17–26, find the least common multiple of the expressions.

17. $5x^2,\ 20x^3$

18. $14t^2,\ 42t^5$

19. $15x^2,\ 3(x+5)$

20. $6x^2,\ 15x(x-1)$

21. $9y^3(y+1)^2,\ 12y(y+1)$

22. $18y^3,\ 27y(y-3)^2$

23. $6(x^2-4),\ 2x(x+2)$

24. $t^3+3t^2+9t,\ 2t^2(t^2-9)$

25. $8t^2+16t,\ 14t^2-56t$

26. $2y^2+y-1,\ 4y^2-2y$

In Exercises 27–32, find the missing algebraic expression that makes the two fractions equivalent.

27. $\dfrac{7}{3y} = \dfrac{7x^2}{3y()},\quad x \neq 0$

28. $\dfrac{2x}{x-3} = \dfrac{14x(x-3)^2}{(x-3)()}$

29. $\dfrac{3u}{7v} = \dfrac{3u()}{7v(u+1)},\quad u \neq -1$

30. $\dfrac{3t+5}{t} = \dfrac{(3t+5)()}{5t^2(3t-5)},\quad t \neq \dfrac{5}{3}$

31. $\dfrac{13x}{x-2} = \dfrac{13x()}{4-x^2},\quad x \neq -2$

32. $\dfrac{x^2}{10-x} = \dfrac{x^2()}{x^2-10x},\quad x \neq 0$

In Exercises 33–40, find the least common denominator of the two fractions and rewrite each fraction using the least common denominator.

33. $\dfrac{n+8}{3n-12},\ \dfrac{10}{6n^2}$

34. $\dfrac{v}{2v^2+2v},\ \dfrac{4}{3v^2}$

35. $\dfrac{2}{x^2(x-3)},\ \dfrac{5}{x(x+3)}$

36. $\dfrac{5t}{2t(t-3)^2},\ \dfrac{4}{t(t-3)}$

37. $\dfrac{8s}{(s+2)^2},\ \dfrac{3}{s^3+s^2-2s}$

38. $\dfrac{4x}{(x+5)^2},\ \dfrac{x-2}{x^2-25}$

39. $\dfrac{x-8}{x^2-25},\ \dfrac{9x}{x^2-10x+25}$

40. $\dfrac{3y}{y^2-y-12},\ \dfrac{y-4}{y^2+3y}$

In Exercises 41–74, combine and simplify.

41. $\dfrac{7}{a} + \dfrac{14}{a^2}$

42. $\dfrac{10}{b} + \dfrac{1}{10b^2}$

43. $\dfrac{3x}{x-8} - \dfrac{6}{8-x}$

44. $\dfrac{1}{y-6} + \dfrac{y}{6-y}$

45. $\dfrac{3x}{3x-2} + \dfrac{2}{2-3x}$

46. $\dfrac{y}{5y-3} - \dfrac{3}{3-5y}$

47. $25 + \dfrac{10}{x+4}$

48. $\dfrac{100}{x-10} - 8$

49. $-\dfrac{1}{6x} + \dfrac{1}{6(x-3)}$

50. $\dfrac{1}{x} - \dfrac{1}{x+2}$

51. $\dfrac{x}{x+3} - \dfrac{5}{x-2}$

52. $\dfrac{3}{x+1} - \dfrac{2}{x}$

53. $\dfrac{3}{x-5y} + \dfrac{2}{x+5y}$

54. $\dfrac{7}{2x-3y} + \dfrac{3}{2x+3y}$

55. $\dfrac{4}{x^2} - \dfrac{4}{x^2+1}$

56. $\dfrac{3}{2t^2} - \dfrac{4}{2t^2-1}$

57. $\dfrac{x}{x^2-9} + \dfrac{3}{x(x-3)}$

58. $\dfrac{x}{x^2-x-30} - \dfrac{1}{x+5}$

59. $\dfrac{4}{x-4} + \dfrac{16}{(x-4)^2}$

60. $\dfrac{3}{x-2} - \dfrac{1}{(x-2)^2}$

61. $\dfrac{y}{x^2+xy} - \dfrac{x}{xy+y^2}$

62. $\dfrac{5}{x+y} + \dfrac{5}{x-y}$

63. $\dfrac{3u}{u^2-2uv+v^2} + \dfrac{2}{u-v}$

64. $\dfrac{5}{x+y} + \dfrac{2x}{x^2+3xy+2y^2}$

65. $\dfrac{x}{x^2-9} - \dfrac{3x-1}{x^2+7x+12}$

66. $\dfrac{2x-1}{x^2-3x-10} + \dfrac{x}{x^2-4}$

67. $\dfrac{4}{x} - \dfrac{2}{x^2} + \dfrac{4}{x+3}$

68. $\dfrac{5}{2(x+1)} - \dfrac{1}{2x} - \dfrac{3}{2(x+1)^2}$

69. $\dfrac{1}{x} - \dfrac{3}{y} + \dfrac{3x-y}{xy}$

70. $\dfrac{x+2}{x-1} - \dfrac{2}{x+6} - \dfrac{14}{x^2+5x-6}$

71. $\dfrac{x}{x^2+15x+50} + \dfrac{7}{2(x+10)} - \dfrac{3}{2(x+5)}$

72. $\dfrac{5}{5(x-3)} - \dfrac{6}{5(x+4)} + \dfrac{2x}{x^2+x-12}$

73. $\dfrac{1}{x^2+7x+12} + \dfrac{1}{x^2-9} + \dfrac{1}{x^2-16}$

74. $\dfrac{2}{y^2-3y+2} + \dfrac{3}{y^2-1} - \dfrac{5}{y^2+3y-10}$

In Exercises 75–78, simplify the algebraic expression.

75. $\dfrac{4x^4}{x^3} - 2x$

76. $\dfrac{x^2+2x-3}{x-1} - (3x-4)$

77. $\dfrac{15x^3y}{10x^2} + \dfrac{3xy^2}{2y}$

78. $\dfrac{8u^2v}{2u} + \dfrac{3(uv)^2}{uv}$

In Exercises 79–94, simplify the complex fraction.

79. $\dfrac{\dfrac{1}{2}}{\left(3+\dfrac{1}{x}\right)}$

80. $\dfrac{\dfrac{2}{3}}{\left(4-\dfrac{1}{x}\right)}$

81. $\dfrac{\left(\dfrac{4}{x}+3\right)}{\left(\dfrac{4}{x}-3\right)}$

82. $\dfrac{\left(\dfrac{1}{t}-1\right)}{\left(\dfrac{1}{t}+1\right)}$

83. $\dfrac{\left(16x-\dfrac{1}{x}\right)}{\left(\dfrac{1}{x}-4\right)}$

84. $\dfrac{\left(\dfrac{36}{y}-y\right)}{6+y}$

85. $\dfrac{\left(3+\dfrac{9}{x-3}\right)}{\left(4+\dfrac{12}{x-3}\right)}$

86. $\dfrac{\left(x+\dfrac{2}{x-3}\right)}{\left(x+\dfrac{6}{x-3}\right)}$

87. $\dfrac{\left(1-\dfrac{1}{y^2}\right)}{\left(1-\dfrac{4}{y}+\dfrac{3}{y^2}\right)}$

88. $\dfrac{\left(\dfrac{x+1}{x+2}-\dfrac{1}{x}\right)}{\left(\dfrac{2}{x+2}\right)}$

89. $\dfrac{\left(\dfrac{y}{x}-\dfrac{x}{y}\right)}{\left(\dfrac{x+y}{xy}\right)}$

90. $\dfrac{\left(x-\dfrac{2y^2}{x-y}\right)}{x-2y}$

91. $\dfrac{\left(\dfrac{3}{x^2}+\dfrac{1}{x}\right)}{\left(2-\dfrac{4}{5x}\right)}$

92. $\dfrac{\left(16-\dfrac{1}{x^2}\right)}{\left(\dfrac{1}{4x^2}-4\right)}$

93. $\dfrac{\left(\dfrac{x}{x-3}-\dfrac{2}{3}\right)}{\left(\dfrac{10}{3x}+\dfrac{x^2}{x-3}\right)}$

94. $\dfrac{\left(\dfrac{1}{2x}-\dfrac{6}{x+5}\right)}{\left(\dfrac{x}{x-5}+\dfrac{1}{x}\right)}$

In Exercises 95–98, simplify the expression.

95. $(u+v^{-2})^{-1}$

96. $x^{-2}(x^2+y^2)$

97. $\dfrac{a+b}{ba^{-1}-ab^{-1}}$

98. $\dfrac{u^{-1}-v^{-1}}{u^{-1}+v^{-1}}$

In Exercises 99 and 100, use the function to find and simplify the expression for

$$\dfrac{f(2+h)-f(2)}{h}.$$

99. $f(x) = \dfrac{1}{x}$

100. $f(x) = \dfrac{x}{x-1}$

Solving Problems

101. *Average* Determine the average of the two real numbers $x/4$ and $x/6$.

102. *Average* Determine the average of the three real numbers x, $x/2$, and $x/3$.

103. *Number Problem* Find three real numbers that divide the real number line between $x/6$ and $x/2$ into four equal parts.

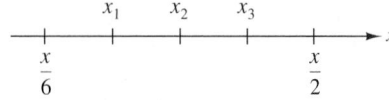

104. *Number Problem* Find two real numbers that divide the real number line between $x/5$ and $x/3$ into three equal parts.

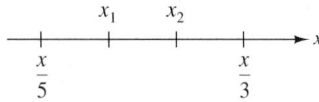

105. *Work Rate* After two workers have worked together for t hours on a common task, the fractional parts of the task done by the two workers are $t/4$ and $t/6$. What fractional part of the task has been completed?

106. *Work Rate* After two workers have worked together for t hours on a common task, the fractional parts of the task done by the two workers are $t/3$ and $t/5$. What fractional part of the task has been completed?

107. *Electronics* When two resistors are connected in parallel (see figure), the total resistance is

$$\frac{1}{\left(\dfrac{1}{R_1} + \dfrac{1}{R_2}\right)}.$$

Simplify this complex fraction.

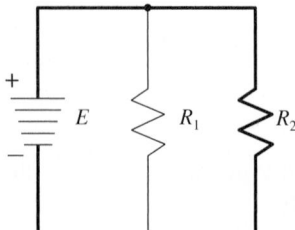

108. *Monthly Payment* The approximate annual percent rate r of a monthly installment loan is

$$r = \frac{\left[\dfrac{24(MN - P)}{N}\right]}{\left(P + \dfrac{MN}{12}\right)}$$

where N is the total number of payments, M is the monthly payment, and P is the amount financed.

(a) Approximate the annual percent rate for a four-year car loan of $10,000 that has monthly payments of $300.

(b) Simplify the expression for the annual percent rate r, and then rework part (a).

109. *Rewriting a Fraction* The fraction $4/(x^3 - x)$ can be rewritten as a sum of three fractions, as follows.

$$\frac{4}{x^3 - x} = \frac{A}{x} + \frac{B}{x + 1} + \frac{C}{x - 1}$$

The numbers A, B, and C are the solutions of the system

$$\begin{cases} A + B + C = 0 \\ \quad\;\; -B + C = 0 \\ -A \qquad\quad = 4. \end{cases}$$

Solve the system and verify that the sum of the three resulting fractions is the original fraction.

110. *Rewriting a Fraction* The fraction

$$\frac{x + 1}{x^3 - x^2}$$

can be rewritten as a sum of three fractions, as follows.

$$\frac{x + 1}{x^3 - x^2} = \frac{A}{x} + \frac{B}{x^2} + \frac{C}{x - 1}$$

The numbers A, B, and C are the solutions of the system

$$\begin{cases} A \qquad\;\; + C = 0 \\ -A + B \qquad = 1. \\ \quad\;\; -B \qquad = 1 \end{cases}$$

Solve the system and verify that the sum of the three resulting fractions is the original fraction.

Explaining Concepts

111. Is it possible for the least common denominator of two fractions to be the same as one of the fraction's denominators? If so, give an example.

112. In your own words, describe how to add or subtract rational expressions with unlike denominators.

Ongoing Review

In Exercises 113–120, perform the multiplication and simplify.

113. $-6x(10 - 7x)$

114. $(2 - y)(3 + 2y)$

115. $(11 - x)(11 + x)$

116. $(4 - 5z)(4 + 5z)$

117. $(x + 1)^2$

118. $t(t^2 + 1) - t(t^2 - 1)$

119. $(x - 2)(x^2 + 2x + 4)$

120. $t(t - 4)(2t + 3)$

Geometry In Exercises 121 and 122, find expressions for the perimeter and area of the region. Simplify the expressions.

121.

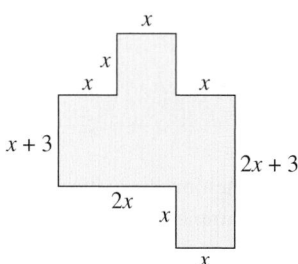

122.

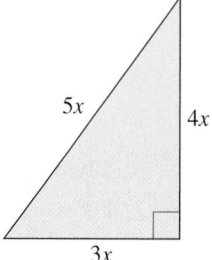

123. *Test Scores* A student has test scores of 80, 82, 94, and 71. What score does the student need on the next test to produce an average score of 85?

124. *Profit* It costs a manufacturer of sunglasses $8.95 to produce sunglasses that sell for $29.99. How many pairs of sunglasses must the manufacturer sell to make a profit of $17,884?

Looking Further

(a) Simplify the complex fraction $\dfrac{\left(1 - \dfrac{1}{x}\right)}{\left(1 - \dfrac{1}{x^2}\right)}$.

(b) Use a graphing utility to complete the table.

x	-3	-2	-1	0	1	2	3
$\dfrac{\left(1 - \dfrac{1}{x}\right)}{\left(1 - \dfrac{1}{x^2}\right)}$							
$\dfrac{x}{x + 1}$							

(c) Discuss the domains and the equivalence of the two expressions from the table in part (b).

(d) Simplify the complex fraction $\dfrac{\left(1 + \dfrac{4}{x} + \dfrac{4}{x^2}\right)}{\left(1 - \dfrac{4}{x^2}\right)}$.

(e) Make a table showing the values of the complex fraction in part (d) for integer values of x from -3 through 3. Also show the values of the simplified form of the fraction for these same values of x.

(f) Discuss the domains and the equivalence of the two expressions from the table in part (e).

7.4 Dividing Polynomials

What you should learn:

- How to divide polynomials by monomials
- How to use long division to divide polynomials by polynomials
- How to use synthetic division to divide polynomials by polynomials of the form $x - k$
- How to use synthetic division to factor polynomials

Why you should learn it:

Long division of polynomials is useful in higher-level mathematics when factoring, and finding zeros of, polynomials.

Dividing a Polynomial by a Monomial

To divide a polynomial by a monomial, reverse the procedure used to add or subtract two rational expressions. Here is an example.

$$2 + \frac{1}{x} = \frac{2x}{x} + \frac{1}{x} = \frac{2x + 1}{x} \qquad \text{Add fractions.}$$

$$\frac{2x + 1}{x} = \frac{2x}{x} + \frac{1}{x} = 2 + \frac{1}{x} \qquad \text{Divide by monomial.}$$

Dividing a Polynomial by a Monomial

Let u, v, and w be real numbers, variables, or algebraic expressions such that $w \neq 0$.

1. $\dfrac{u + v}{w} = \dfrac{u}{w} + \dfrac{v}{w}$

2. $\dfrac{u - v}{w} = \dfrac{u}{w} - \dfrac{v}{w}$

When dividing a polynomial by a monomial, remember to reduce the resulting expressions to simplest form, as illustrated in Example 1.

Example 1 ■ Dividing a Polynomial by a Monomial

Perform the division and simplify.

(a) $\dfrac{12x^2 - 20x + 8}{4x}$ (b) $\dfrac{5y^3 + 8y^2 - 10y}{5y^2}$

Solution

(a) $\dfrac{12x^2 - 20x + 8}{4x} = \dfrac{12x^2}{4x} - \dfrac{20x}{4x} + \dfrac{8}{4x}$ \qquad Divide each term in the numerator by $4x$.

$\qquad = \dfrac{3(4x)(x)}{4x} - \dfrac{5(4x)}{4x} + \dfrac{2(4)}{4x}$ \qquad Divide out common factors.

$\qquad = 3x - 5 + \dfrac{2}{x}$ \qquad Simplified form

(b) $\dfrac{5y^3 + 8y^2 - 10y}{5y^2} = \dfrac{5y^3}{5y^2} + \dfrac{8y^2}{5y^2} - \dfrac{10y}{5y^2}$ \qquad Divide each term in the numerator by $5y^2$.

$\qquad = \dfrac{5y^2(y)}{5y^2} + \dfrac{8y^2}{5y^2} - \dfrac{2(5y)}{5y(y)}$ \qquad Divide out common factors.

$\qquad = y + \dfrac{8}{5} - \dfrac{2}{y}$ \qquad Simplified form

Example 2 ■ Dividing a Polynomial by a Monomial

$$\frac{-12x^3y^2 + 6x^2y + 9y^3}{3xy^2} = -\frac{12x^3y^2}{3xy^2} + \frac{6x^2y}{3xy^2} + \frac{9y^3}{3xy^2}$$

$$= -\frac{4(3x)(x^2)(y^2)}{(3x)(y^2)} + \frac{2(3x)(x)(y)}{3x(y)(y)} + \frac{3y(3y^2)}{x(3y^2)}$$

$$= -4x^2 + \frac{2x}{y} + \frac{3y}{x}$$

Study Tip

There are different ways to check division problems such as those shown in Examples 1 and 2. One way is to combine terms using the denominator of the original expression. For instance, you can check the result of Example 1(b) as follows.

$$y + \frac{8}{5} - \frac{2}{y} = \frac{y \cdot 5y^2}{5y^2} + \frac{8 \cdot y^2}{5 \cdot y^2}$$

$$- \frac{2 \cdot 5y}{y \cdot 5y}$$

$$= \frac{5y^3}{5y^2} + \frac{8y^2}{5y^2} - \frac{10y}{5y^2}$$

$$= \frac{5y^3 + 8y^2 - 10y}{5y^2}$$

Long Division

In Section 7.1, you learned how to divide one polynomial by another by factoring and dividing out common factors. For instance, you can divide $x^2 - 2x - 3$ by $x - 3$ as follows.

$$(x^2 - 2x - 3) \div (x - 3) = \frac{x^2 - 2x - 3}{x - 3} \quad \text{Write as fraction.}$$

$$= \frac{(x + 1)(x - 3)}{x - 3} \quad \text{Factor numerator.}$$

$$= \frac{(x + 1)(x \!\!\!\!\diagdown 3)}{x \diagdown 3} \quad \text{Divide out common factor.}$$

$$= x + 1, \quad x \neq 3 \quad \text{Simplified form}$$

This procedure works well for polynomials that factor easily. For those that do not, you can use a more general procedure that follows a "long division algorithm" similar to the algorithm used for dividing positive integers, which is reviewed in Example 3.

Example 3 ■ Long Division Algorithm for Positive Integers

Use the long division algorithm to divide 6584 by 28.

Solution

Think $\frac{65}{28} \approx 2$.
Think $\frac{98}{28} \approx 3$.
Think $\frac{144}{28} \approx 5$.

$$
\begin{array}{r}
235 \\
28\,\overline{)6584} \\
\underline{56} \\
98 \\
\underline{84} \\
144 \\
\underline{140} \\
4
\end{array}
$$

Multiply 2 by 28.
Subtract and bring down 8.
Multiply 3 by 28.
Subtract and bring down 4.
Multiply 5 by 28.
Remainder

So, you have $6584 \div 28 = 235 + \frac{4}{28} = 235 + \frac{1}{7}$.

NOTE   In Example 3, the number 6584 is the **dividend,** 28 is the **divisor,** 235 is the **quotient,** and 4 is the **remainder.**

In the next several examples, you will see how the long division algorithm can be extended to cover division of one polynomial by another.

Along with the long division algorithm, follow these steps when performing long division of polynomials.

1. Write the dividend and divisor in descending powers of the variable.

2. Insert placeholders with zero coefficients for missing powers of the variable. (See Example 6.)

3. Perform the long division of the polynomials as you would with integers.

4. Continue the process until the degree of the remainder is less than that of the divisor.

Example 4 ■ Long Division Algorithm for Polynomials

Use the long division algorithm to divide $x^2 + 2x + 4$ by $x - 1$.

Solution

$$
\begin{array}{r}
\text{Think } \dfrac{x^2}{x} = x. \\
\text{Think } \dfrac{3x}{x} = 3.
\end{array}
$$

$$
\begin{array}{r}
x + 3 \\
x - 1 \overline{)\, x^2 + 2x + 4} \\
\underline{x^2 - x} \qquad \text{Multiply } x \text{ by } (x-1). \\
3x + 4 \qquad \text{Subtract and bring down 4.} \\
\underline{3x - 3} \qquad \text{Multiply 3 by } (x-1). \\
7 \qquad \text{Remainder}
\end{array}
$$

Considering the remainder as a fractional part of the divisor, the result is

$$
\underbrace{\frac{x^2 + 2x + 4}{\underbrace{x - 1}_{\text{Divisor}}}}_{\text{Dividend}} = \underbrace{x + 3}_{\text{Quotient}} + \underbrace{\frac{7}{\underbrace{x - 1}_{\text{Divisor}}}}_{\text{Remainder}}.
$$

You can check a long division problem by multiplying. For instance, you can check the result of Example 4 as follows.

$$
\frac{x^2 + 2x + 4}{x - 1} \stackrel{?}{=} x + 3 + \frac{7}{x - 1}
$$

$$
(x - 1)\left(\frac{x^2 + 2x + 4}{x - 1}\right) \stackrel{?}{=} (x - 1)\left(x + 3 + \frac{7}{x - 1}\right)
$$

$$
x^2 + 2x + 4 \stackrel{?}{=} (x + 3)(x - 1) + 7
$$

$$
x^2 + 2x + 4 \stackrel{?}{=} (x^2 + 2x - 3) + 7
$$

$$
x^2 + 2x + 4 = x^2 + 2x + 4 \;\checkmark
$$

When using the long division algorithm for polynomials, be sure that both the divisor and the dividend are written in standard form before beginning the division process, as shown in Example 5.

Example 5 ■ Writing in Standard Form Before Dividing

Divide $-13x^3 + 10x^4 + 8x - 7x^2 + 4$ by $3 - 2x$.

Solution

First write the divisor and dividend in standard polynomial form.

$$
\begin{array}{r}
-5x^3 - x^2 + 2x - 1 \\
-2x + 3 \overline{\smash{)}10x^4 - 13x^3 - 7x^2 + 8x + 4}
\end{array}
$$

$10x^4 - 15x^3$	Multiply $-5x^3$ by $(-2x + 3)$.
$2x^3 - 7x^2$	Subtract and bring down $-7x^2$.
$2x^3 - 3x^2$	Multiply $-x^2$ by $(-2x + 3)$.
$-4x^2 + 8x$	Subtract and bring down $8x$.
$-4x^2 + 6x$	Multiply $2x$ by $(-2x + 3)$.
$2x + 4$	Subtract and bring down 4.
$2x - 3$	Multiply -1 by $(-2x + 3)$.
7	Remainder

This shows that

$$
\underbrace{\overbrace{10x^4 - 13x^3 - 7x^2 + 8x + 4}^{\text{Dividend}}}_{\text{Divisor}} \Big/ \underbrace{(-2x + 3)}_{} = \overbrace{-5x^3 - x^2 + 2x - 1}^{\text{Quotient}} + \overbrace{\underbrace{\frac{7}{-2x + 3}}_{\text{Divisor}}}^{\text{Remainder}}.
$$

Check this result by multiplying.

Technology

You can check the result of a division problem *graphically* with a graphing utility by comparing the graphs of the original quotient and the simplified form. For example, the figure at the left shows the graphs of

$$
y_1 = \frac{10x^4 - 13x^3 - 7x^2 + 8x + 4}{-2x + 3}
$$

and

$$
y_2 = -5x^3 - x^2 + 2x - 1 + \frac{7}{-2x + 3}
$$

from Example 5. Because the graphs coincide, you can conclude that the solution checks.

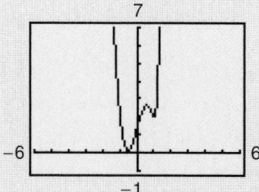

When the dividend is missing one or more powers of x, the long division algorithm requires that you account for the missing powers, as shown in Example 6.

Example 6 ■ Accounting for Missing Powers of x

Divide $x^3 - 2$ by $x - 1$.

Solution

Note how the missing x^2- and x-terms are accounted for.

$$
\begin{array}{r}
x^2 + x + 1 \\
x - 1 \,\overline{\smash{)}\, x^3 + 0x^2 + 0x - 2} \\
\end{array}
$$

$x^3 - x^2$	Insert $0x^2$ and $0x$.
$x^2 + 0x$	Multiply x^2 by $(x - 1)$.
$x^2 - x$	Subtract and bring down $0x$.
$x - 2$	Multiply x by $(x - 1)$.
$x - 1$	Subtract and bring down -2.
-1	Multiply 1 by $(x - 1)$.
	Remainder

So, you have

$$\frac{x^3 - 2}{x - 1} = x^2 + x + 1 - \frac{1}{x - 1}.$$

In each of the long division examples presented so far, the divisor has been a first-degree polynomial. The long division algorithm works just as well with polynomial divisors of degree 2 or more, as shown in Example 7.

Example 7 ■ A Second-Degree Divisor

Divide $x^4 + 6x^3 + 6x^2 - 10x - 3$ by $x^2 + 2x - 3$.

Solution

$$
\begin{array}{r}
x^2 + 4x + 1 \\
x^2 + 2x - 3 \,\overline{\smash{)}\, x^4 + 6x^3 + 6x^2 - 10x - 3} \\
\end{array}
$$

$x^4 + 2x^3 - 3x^2$	Multiply x^2 by $(x^2 + 2x - 3)$.
$4x^3 + 9x^2 - 10x$	Subtract and bring down $-10x$.
$4x^3 + 8x^2 - 12x$	Multiply $4x$ by $(x^2 + 2x - 3)$.
$x^2 + 2x - 3$	Subtract and bring down -3.
$x^2 + 2x - 3$	Multiply 1 by $(x^2 + 2x - 3)$.
0	

NOTE   If the remainder is zero, the divisor is said to **divide evenly** into the dividend.

So, $x^2 + 2x - 3$ divides evenly into $x^4 + 6x^3 + 6x^2 - 10x - 3$. That is,

$$\frac{x^4 + 6x^3 + 6x^2 - 10x - 3}{x^2 + 2x - 3} = x^2 + 4x + 1, \; x \neq -3, \; x \neq 1.$$

Synthetic Division

There is a nice shortcut for division by polynomials of the form $x - k$. It is called **synthetic division** and is outlined for a third-degree polynomial as follows.

Synthetic Division for a Third-Degree Polynomial

Use synthetic division to divide $ax^3 + bx^2 + cx + d$ by $x - k$ as follows.

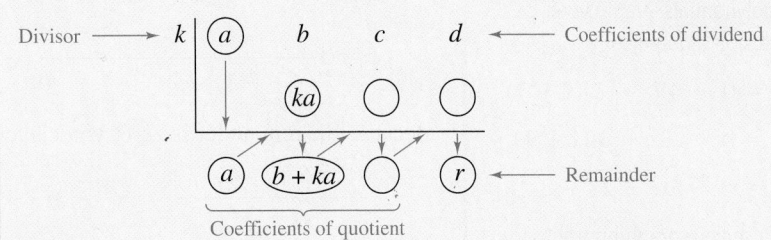

Vertical Pattern: Add terms.
Diagonal Pattern: Multiply by k.

NOTE   Be sure you see that synthetic division works *only* for divisors of the form $x - k$. Remember that $x + k = x - (-k)$. Moreover, the degree of the quotient is always 1 less than the degree of the dividend.

Example 8 ■ Using Synthetic Division

Use synthetic division to divide $x^3 + 3x^2 - 4x - 10$ by $x - 2$.

Solution

The coefficients of the dividend form the top row of the synthetic division array. Because you are dividing by $x - 2$, write 2 at the top left of the array. To begin the algorithm, bring down the first coefficient. Then multiply this coefficient by 2, write the result in the second row, and add the two numbers in the second column. By continuing this pattern, you obtain the following.

$$
\begin{array}{c|cccc}
2 & 1 & 3 & -4 & -10 \\
  &   & 2 & 10 & 12 \\
\hline
  & 1 & 5 & 6 & 2
\end{array}
$$

The bottom row shows that the quotient is

$$(1)x^2 + (5)x + (6)$$

and the remainder is 2. So, the result of the division problem is

$$\frac{x^3 + 3x^2 - 4x - 10}{x - 2} = x^2 + 5x + 6 + \frac{2}{x - 2}.$$

EXPLORATION

Factor the polynomial $x^3 - 64$. Use synthetic division to divide $x^3 - 64$ by $x - 4$. What can you conclude?

Example 9 ■ Using Synthetic Division

Use synthetic division to divide $x^4 - 10x^2 - 2x + 3$ by $x + 3$.

Solution

The coefficients of the dividend form the top row of the synthetic division array. Because you are dividing by $x + 3$, write -3 at the top left of the array. The completed synthetic division array is as follows. (Note that a zero is included in place of the missing x^3-term in the dividend.)

$$
\begin{array}{c|ccccc}
-3 & 1 & 0 & -10 & -2 & 3 \\
   &   & -3 & 9 & 3 & -3 \\
\hline
   & 1 & -3 & -1 & 1 & \underset{\text{Remainder}}{\langle 0 \rangle}
\end{array}
$$

Because the remainder is zero, you can conclude that

$$\frac{x^4 - 10x^2 - 2x + 3}{x + 3} = x^3 - 3x^2 - x + 1, \quad x \neq -3.$$

Factoring and Division

One of the uses of synthetic division (or long division) is in factoring of polynomials. If the remainder is zero in a synthetic division problem, you know that the divisor divides evenly into the dividend. So, the original polynomial can be factored as the product of two polynomials of lesser degrees. This is demonstrated in Example 10.

Example 10 ■ Factoring a Polynomial

Factor $x^3 - 7x + 6$ completely. Use the fact that $x - 1$ is one of the factors.

Solution

Because you are given one of the factors, you should divide this factor into the given polynomial. (In this case we use synthetic division, but long division could have been used also.)

$$
\begin{array}{c|cccc}
1 & 1 & 0 & -7 & 6 \\
  &   & 1 & 1 & -6 \\
\hline
  & 1 & 1 & -6 & \underset{\text{Remainder}}{\langle 0 \rangle}
\end{array}
$$

Because the remainder is zero, the divisor divides evenly into the dividend and you obtain

$$\frac{x^3 - 7x + 6}{x - 1} = x^2 + x - 6, \quad x \neq 1$$

which implies that $x^3 - 7x + 6 = (x - 1)(x^2 + x - 6)$. So, you have used division to factor the polynomial $x^3 - 7x + 6$ into the product of a first-degree polynomial and a second-degree polynomial. To complete the factorization, factor the second-degree polynomial as follows.

$$x^3 - 7x + 6 = (x - 1)(x^2 + x - 6) = (x - 1)(x + 3)(x - 2)$$

7.4 ■ Exercises

Developing Skills

In Exercises 1–54, perform the division.

1. $\dfrac{10z^2 + 4z - 12}{4}$

2. $\dfrac{4u^2 + 8u - 24}{16}$

3. $(7x^3 - 2x^2) \div x$

4. $(6a^2 + 7a) \div a$

5. $\dfrac{50z^3 + 30z}{-5z}$

6. $\dfrac{18c^4 - 24c^2}{-6c}$

7. $\dfrac{8z^3 + 3z^2 - 2z}{2z}$

8. $\dfrac{6x^4 + 8x^3 - 18x^2}{3x^2}$

9. $\dfrac{m^4 + 2m^2 - 7}{-m}$

10. $\dfrac{l^2 - 8}{-l}$

11. $(5x^2y - 8xy + 7xy^2) \div 2xy$

12. $(-14s^4t^2 + 7s^2t^2 - 18t) \div 2s^2t$

13. $\dfrac{x^2 - 8x + 15}{x - 3}$

14. $\dfrac{t^2 - 18t + 72}{t - 6}$

15. $(x^2 + 15x + 50) \div (x + 5)$

16. $(y^2 - 6y - 16) \div (y + 2)$

17. $(21 - 4x - x^2) \div (3 - x)$

18. $(5 + 4x - x^2) \div (1 + x)$

19. $(2y^2 + 7y + 3) \div (2y + 1)$

20. $(10t^2 - 7t - 12) \div (2t - 3)$

21. $\dfrac{16x^2 - 1}{4x + 1}$

22. $\dfrac{81y^2 - 25}{9y - 5}$

23. $\dfrac{x^3 + 125}{x + 5}$

24. $\dfrac{x^3 - 27}{x - 3}$

25. $\dfrac{x^3 - 2x^2 + 4x - 8}{x - 2}$

26. $\dfrac{x^3 - 28x - 48}{x + 4}$

27. $\dfrac{x^3 - 5x^2 - 2x + 10}{x - 5}$

28. $\dfrac{2x^3 + 12x^2 + 5x + 30}{x + 6}$

29. $(2x + 9) \div (x + 2)$

30. $(12x - 5) \div (2x + 3)$

31. $\dfrac{x^2 + 16}{x + 4}$

32. $\dfrac{y^2 + 8}{y + 2}$

33. $\dfrac{5x^2 + 2x + 3}{x + 2}$

34. $\dfrac{2x^2 + 5x + 2}{x + 4}$

35. $\dfrac{12x^2 + 17x - 5}{3x + 2}$

36. $\dfrac{8x^2 + 2x + 3}{4x - 1}$

37. $\dfrac{5x^3 + 6x^2 - 17x + 20}{x + 3}$

38. $\dfrac{6x^3 + 15x^2 - 8x + 2}{x + 4}$

39. $\dfrac{2x^3 - 5x^2 + x - 6}{x - 3}$

40. $\dfrac{5x^3 + 3x^2 + 12x + 20}{x + 1}$

41. $(x^6 - 1) \div (x - 1)$

42. $(x^5 - 1) \div (x - 1)$

43. $x^3 \div (x - 1)$

44. $x^4 \div (x - 2)$

45. $x^5 \div (x^2 + 1)$

46. $(x^6 + 4x^2) \div (x^2 + 3)$

47. $(x^3 + 4x^2 + 7x + 6) \div (x^2 + 2x + 3)$

48. $(2x^3 + 2x^2 - 2x - 15) \div (2x^2 + 4x + 5)$

49. $(3x^4 + x^3 - 5x^2 - 30) \div (3x^2 + x + 1)$

50. $(2x^4 - x^3 - 23x^2 + 9x + 45) \div (2x^2 - x - 5)$

51. $(x^3 + 5x^2 + 6x - 19) \div (x^2 + x - 4)$

52. $(2x^4 + 3x^3 - 7x - 10) \div (x^2 - 2x - 5)$

53. $(x^4 + 3x^2 - 6x - 10) \div (x^2 + 3x - 5)$

54. $(3x^4 - 8x^3 + 7x + 2) \div (3x^2 - 2x + 2)$

In Exercises 55 and 56, perform the division. (Assume *n* is a positive integer.)

55. $\dfrac{x^{3n} + 3x^{2n} + 6x^n + 8}{x^n + 2}$

56. $\dfrac{x^{3n} - x^{2n} + 5x^n - 5}{x^n - 1}$

In Exercises 57–70, use synthetic division to divide.

57. $\dfrac{x^3 + 3x^2 - 1}{x + 4}$

58. $\dfrac{x^4}{x + 2}$

59. $\dfrac{x^4 - 4x^3 + x + 10}{x - 2}$

60. $\dfrac{2x^5 - 3x^3 + x}{x - 3}$

61. $\dfrac{5x^3 + 12}{x + 5}$

62. $\dfrac{x^4 - 25x^2 + 144}{x + 3}$

63. $\dfrac{12x - 25}{x + 2}$

64. $\dfrac{8x + 35}{x - 10}$

65. $\dfrac{5x^3 - 6x^2 + 8}{x - 4}$

66. $\dfrac{5x^3 + 6x + 8}{x + 2}$

67. $\dfrac{10x^4 - 50x^3 - 800}{x - 6}$

68. $\dfrac{x^5 - 13x^4 - 120x + 80}{x + 3}$

69. $\dfrac{0.1x^2 + 0.8x + 1}{x - 0.2}$

70. $\dfrac{x^3 - 0.8x + 2.4}{x + 1}$

In Exercises 71–80, use synthetic division to factor the polynomial completely, given one of its factors.

Polynomial	Factor
71. $x^3 - 8x^2 + x + 42$	$x - 3$
72. $x^3 - 5x^2 - 2x + 24$	$x + 2$
73. $3z^3 - 20z^2 + 36z - 16$	$z - 4$
74. $2t^3 + 15t^2 + 19t - 30$	$t + 5$
75. $5t^3 - 27t^2 - 14t - 24$	$t - 6$
76. $y^3 + y^2 - 4y - 4$	$y - 2$
77. $x^4 - 16$	$x - 2$
78. $x^4 + 3x^3 - 8x - 24$	$x + 3$

Polynomial	Factor
79. $15x^2 - 2x - 8$	$x - \frac{4}{5}$
80. $18x^2 - 9x - 20$	$x + \frac{5}{6}$

In Exercises 81 and 82, use a graphing utility to graph the two equations in the same viewing window. Use the graphs to verify that the expressions are equivalent. Verify the results algebraically.

81. $y_1 = \dfrac{x^3}{x^2 + 1}$

$y_2 = x - \dfrac{x}{x^2 + 1}$

82. $y_1 = \dfrac{x^2 + 2}{x + 1}$

$y_2 = x - 1 + \dfrac{3}{x + 1}$

83. When a polynomial is divided by $x - 6$, the quotient is $x^2 + x + 1$ and the remainder is -4. Find the dividend.

84. When a polynomial is divided by $x + 3$, the quotient is $x^3 + x^2 - 4$ and the remainder is 8. Find the dividend.

In Exercises 85 and 86, find the constant c such that the denominator divides evenly into the numerator.

85. $\dfrac{x^3 + 2x^2 - 4x + c}{x - 2}$

86. $\dfrac{x^4 - 3x^2 + c}{x + 6}$

Solving Problems

87. *Geometry*   A rectangle's area is
$$2x^3 + 3x^2 - 6x - 9.$$
Find its width if its length is $2x + 3$.

88. *Geometry*   A rectangular house has a volume of
$$x^3 + 55x^2 + 650x + 2000$$
cubic feet (the space in the attic is not included). The height of the house is $x + 5$ (see figure). Find the number of square feet of floor space *on the first floor* of the house.

Geometry   In Exercises 89 and 90, you are given the expression for the volume of the solid shown. Find the expression for the missing dimension.

89. $V = x^3 + 18x^2 + 80x + 96$

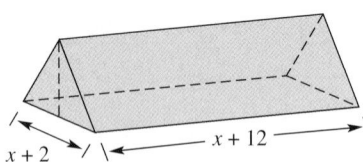

90. $V = h^4 + 3h^3 + 2h^2$

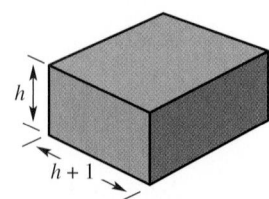

Explaining Concepts

91. *Error Analysis*   Describe the error.

$$\frac{6x + 5y}{x} = \frac{6x + 5y}{x} = 6 + 5y$$

92. Explain how you can check polynomial division.

93. If the divisor divides evenly into the dividend, are the divisor and quotient factors of the dividend? Explain.

94. Explain what it means for a divisor to divide evenly into a dividend.

Ongoing Review

In Exercises 95–98, rewrite the fraction in simplified form.

95. $\dfrac{20t^2}{15t}$

96. $\dfrac{-3x^6}{21x^3}$

97. $\dfrac{8x^2y^3}{20xy^2}$

98. $\dfrac{22u^5v^3}{-55uv^3}$

In Exercises 99–104, solve the equation.

99. $3(2 - x) = 5x$

100. $125 - 50x = 0$

101. $8y^2 - 50 = 0$

102. $t^2 - 8t = 0$

103. $x^2 + x - 42 = 0$

104. $x(10 - x) = 25$

105. *Distance*   A plane leaves Chicago headed for Los Angeles at 540 miles per hour. One hour later, a second plane leaves Los Angeles headed for Chicago at 660 miles per hour. If the air route from Chicago to Los Angeles is 1800 miles, how long will it take for the planes to pass each other? How far from Chicago will they be at that time?

106. *Simple Interest*   An investment advisor invested $14,000 in two accounts. One investment earned 8% annual simple interest and the other investment earned 6.5% annual simple interest. The amount of interest earned for 1 year was $1024. How much was invested in each account?

Looking Further

Consider the function

$$f(x) = 2x^4 - x^3 - 9x^2 + 4x + 4.$$

(a) Use synthetic division to divide $f(x)$ by each of the divisors in the table. Then complete the table.

k	-3	-2	-1	$-\frac{1}{2}$
Divisor $(x - k)$	$x + 3$	$x + 2$	$x + 1$	$x + \frac{1}{2}$
Remainder				

k	0	$\frac{1}{2}$	1	2	3
Divisor $(x - k)$	x	$x - \frac{1}{2}$	$x - 1$	$x - 2$	$x - 3$
Remainder					

(b) Evaluate $f(x)$ at each of the x-values in the table.

x	-3	-2	-1	$-\frac{1}{2}$	0	$\frac{1}{2}$	1	2	3
$f(x)$									

(c) Compare the function values from the table in part (b) with the remainders from the table in part (a). What can you conclude?

(d) Use the complete factorization

$$f(x) = (x - 2)(x + 2)(2x + 1)(x - 1)$$

to find all the solutions of the polynomial equation $2x^4 - x^3 - 9x^2 + 4x + 4 = 0$.

(e) Compare the solutions of the equation $f(x) = 0$ in part (d) with your observations in parts (a) through (c).

(f) Use a graphing utility to graph $f(x)$. What observations can you make about the x-intercepts of the graph?

Mid-Chapter Quiz

Take this quiz as you would take a quiz in class. After you are done, check your work against the answers given in the back of the book.

1. Determine the domain of the rational function $f(x) = \dfrac{x + 2}{x(x - 4)}$.

2. Evaluate the rational function $h(x) = (x^2 - 9)/(x^2 - x - 2)$ as indicated. If not possible, state the reason.

    (a) $h(-3)$ (b) $h(0)$ (c) $h(-1)$ (d) $h(5)$

In Exercises 3–8, write the expression in simplified form.

3. $\dfrac{9y^2}{6y}$ **4.** $\dfrac{8u^3v^2}{36uv^3}$ **5.** $\dfrac{4x^2 - 1}{x - 2x^2}$

6. $\dfrac{(z + 3)^2}{2z^2 + 5z - 3}$ **7.** $\dfrac{7ab + 3a^2b^2}{a^2b}$ **8.** $\dfrac{2mn^2 - n^3}{2m^2 + mn - n^2}$

In Exercises 9–16, perform the operations and simplify.

9. $\dfrac{11t^2}{6} \cdot \dfrac{9}{33t}$ **10.** $\dfrac{4}{3(x - 1)} \cdot \dfrac{2x}{(x^2 + 2x - 3)}$

11. $\dfrac{x - 1}{x^2 + 2x} \div \dfrac{1 - x^2}{x^2 + 5x + 6}$ **12.** $\dfrac{5u}{3(u + v)} \cdot \dfrac{2(u^2 - v^2)}{3v} \div \dfrac{25u^2}{18(u - v)}$

13. $\dfrac{\left(\dfrac{10}{x^2 + 2x}\right)}{\left(\dfrac{15}{x^2 + 3x + 2}\right)}$ **14.** $\dfrac{4x}{x + 5} - \dfrac{3x}{4}$

15. $\dfrac{x - y}{x^2 - xy - 6y^2} + \dfrac{3x - 5y}{x^2 + xy - 2y^2}$ **16.** $\dfrac{\left(1 - \dfrac{2}{x}\right)}{\left(\dfrac{3}{x} - \dfrac{4}{5}\right)}$

17. Divide $6x^3 - 16x^2 + 17x - 6$ by $3x - 2$.

18. Use synthetic division to divide $3x^3 + 4x^2 - 27x + 20$ by $x + 4$.

19. You start a business with a setup cost of $6000. The cost of material for producing each unit of your product is $10.50.

    (a) Write a rational function that gives the average cost per unit when x units are produced. Explain your reasoning.

    (b) Find the average cost per unit when $x = 500$ units are produced.

20. Find the ratio of the area of the shaded portion of the figure to the total area of the figure.

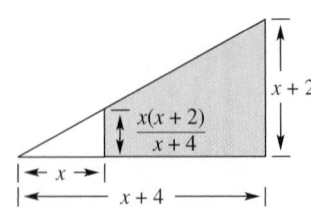

Figure for 20

7.5 Solving Rational Equations

What you should learn:

- How to solve rational equations containing constant denominators
- How to solve rational equations containing variable denominators
- How to solve application problems involving rational equations

Why you should learn it:

Rational equations can be used to model and solve real-life problems relating to the environment. For instance, in Exercise 76 on page 476 a rational equation can be used to model the cost of removing air pollutants from stack emissions.

Equations Containing Constant Denominators

In Section 1.5, you studied a strategy for solving equations that contain fractions with *constant* denominators. That procedure is reviewed here because it is the basis for solving more general equations involving fractions. Recall from Section 1.5 that you can "clear an equation of fractions" by multiplying each side of the equation by the least common denominator (LCD) of the fractions in the equation. Note how this is done in Example 1.

Example 1   ■   An Equation Containing Constant Denominators

Solve $\dfrac{3}{5} = \dfrac{x}{2}$.

Solution

The least common denominator of the two fractions is 10, so begin by multiplying each side of the equation by 10.

$$\frac{3}{5} = \frac{x}{2} \qquad \text{Write original equation.}$$

$$10\left(\frac{3}{5}\right) = 10\left(\frac{x}{2}\right) \qquad \text{Multiply each side by LCD of 10.}$$

$$6 = 5x \qquad \text{Simplify.}$$

$$\frac{6}{5} = x \qquad \text{Divide each side by 5.}$$

The solution is $x = \frac{6}{5}$. Check this solution in the original equation.

In Example 1, the original equation involves only two fractions, one on each side of the equation. To solve this type of equation, you can use an alternative technique called *cross-multiplication*.

$$\frac{3}{5} = \frac{x}{2}$$

$$3(2) = 5x$$

$$6 = 5x$$

$$\frac{6}{5} = x$$

Although this technique can be a little quicker than multiplying by the least common denominator, remember that *it can be used only with equations that have a single fraction on each side* of the equation.

Example 2 ■ An Equation Containing Constant Denominators

Solve $\dfrac{x}{6} = 7 - \dfrac{x}{12}$.

Solution

The least common denominator of the fractions is 12, so begin by multiplying each side of the equation by 12.

$$\frac{x}{6} = 7 - \frac{x}{12}$$ Write original equation.

$$12\left(\frac{x}{6}\right) = 12\left(7 - \frac{x}{12}\right)$$ Multiply each side by LCD of 12.

$$2x = 84 - x$$ Simplify.

$$3x = 84$$ Add x to each side.

$$x = 28$$ Divide each side by 3.

The solution is $x = 28$. Check this solution in the original equation.

Historical Note:

The Rhind Papyrus is a major source of knowledge about ancient Egyptian mathematicians. In one of its many documents, we find the *rule of false position*. Problem 24 states "A quantity and its $\frac{1}{7}$ added together become 19. What is the quantity?" It is solved as follows:

$$x + \frac{x}{7} = 19$$

Guess that x is, for example, 14.

$$14 + \frac{14}{7} = 14 + 2 = 16$$

As $19 = 16\left(\frac{19}{16}\right)$, the solution is $14\left(\frac{19}{16}\right)$ or $\frac{133}{8}$. Try the problem below from the Rhind Papyrus (using either the rule of false position or an equation involving rational expressions).

Problem 25: A quantity and its $\frac{1}{2}$ added together become 16. What is the quantity? $\left(\text{Solution: } \frac{32}{3}\right)$

Example 3 ■ An Equation Containing Constant Denominators

Solve $\dfrac{x+2}{6} - \dfrac{x-4}{8} = \dfrac{2}{3}$.

Solution

The least common denominator of the fractions is 24, so begin by multiplying each side of the equation by 24.

$$\frac{x+2}{6} - \frac{x-4}{8} = \frac{2}{3}$$ Write original equation.

$$24\left(\frac{x+2}{6} - \frac{x-4}{8}\right) = 24\left(\frac{2}{3}\right)$$ Multiply each side by LCD of 24.

$$4(x+2) - 3(x-4) = 8(2)$$ Simplify.

$$4x + 8 - 3x + 12 = 16$$ Distributive Property

$$x + 20 = 16$$ Combine like terms.

$$x = -4$$ Subtract 20 from each side.

The solution is $x = -4$. You can check this solution as follows.

Check

$$\frac{x+2}{6} - \frac{x-4}{8} = \frac{2}{3}$$ Write original equation.

$$\frac{-4+2}{6} - \frac{-4-4}{8} \stackrel{?}{=} \frac{2}{3}$$ Substitute -4 for x.

$$-\frac{1}{3} + 1 \stackrel{?}{=} \frac{2}{3}$$ Simplify.

$$\frac{2}{3} = \frac{2}{3}$$ Solution checks. ✓

Example 4 ■ An Equation That Has Two Solutions

Solve the equation.

$$\frac{x^2}{3} + \frac{x}{2} = \frac{5}{6}$$

Solution

The least common denominator is 6, so begin by multiplying each side of the equation by 6.

$$\frac{x^2}{3} + \frac{x}{2} = \frac{5}{6}$$ Write original equation.

$$6\left(\frac{x^2}{3} + \frac{x}{2}\right) = 6\left(\frac{5}{6}\right)$$ Multiply each side by LCD of 6.

$$\frac{6x^2}{3} + \frac{6x}{2} = \frac{30}{6}$$ Distributive Property

$$2x^2 + 3x = 5$$ Simplify.

$$2x^2 + 3x - 5 = 0$$ Write in general form.

$$(2x + 5)(x - 1) = 0$$ Factor.

$$2x + 5 = 0 \qquad \Longrightarrow \qquad x = -\frac{5}{2}$$ Set 1st factor equal to 0.

$$x - 1 = 0 \qquad \Longrightarrow \qquad x = 1$$ Set 2nd factor equal to 0.

So, the equation has two solutions: $x = -\frac{5}{2}$ and $x = 1$. You can check these solutions in the original equation as follows.

Check First Solution:

$$\frac{x^2}{3} + \frac{x}{2} = \frac{5}{6}$$ Write original equation.

$$\frac{(-5/2)^2}{3} + \frac{(-5/2)}{2} \stackrel{?}{=} \frac{5}{6}$$ Substitute $-\frac{5}{2}$ for x.

$$\left(\frac{25}{4}\right)\left(\frac{1}{3}\right) + \left(-\frac{5}{2}\right)\left(\frac{1}{2}\right) \stackrel{?}{=} \frac{5}{6}$$ Simplify.

$$\frac{25}{12} - \frac{5}{4} \stackrel{?}{=} \frac{5}{6}$$ Multiply.

$$\frac{5}{6} = \frac{5}{6}$$ Solution checks. ✓

Check Second Solution:

$$\frac{x^2}{3} + \frac{x}{2} = \frac{5}{6}$$ Write original equation.

$$\frac{(1)^2}{3} + \frac{(1)}{2} \stackrel{?}{=} \frac{5}{6}$$ Substitute 1 for x.

$$\frac{1}{3} + \frac{1}{2} \stackrel{?}{=} \frac{5}{6}$$ Simplify.

$$\frac{5}{6} = \frac{5}{6}$$ Solution checks. ✓

Equations Containing Variable Denominators

Remember that you always *exclude* those values of a variable that make the denominator of a rational expression zero. This is especially critical for solving equations that contain variable denominators. You will see why in the examples that follow.

Example 5 ■ An Equation Containing Variable Denominators

Solve $\dfrac{7}{x} - \dfrac{1}{3x} = \dfrac{8}{3}$.

Algebraic Solution

The least common denominator for this equation is $3x$. So, begin by multiplying each side of the equation by $3x$.

$$\frac{7}{x} - \frac{1}{3x} = \frac{8}{3} \qquad \text{Write original equation.}$$

$$3x\left(\frac{7}{x} - \frac{1}{3x}\right) = 3x\left(\frac{8}{3}\right) \qquad \begin{array}{l}\text{Multiply each side by LCD}\\ \text{of } 3x.\end{array}$$

$$\frac{21x}{x} - \frac{3x}{3x} = \frac{24x}{3} \qquad \text{Distributive Property}$$

$$21 - 1 = 8x \qquad \text{Simplify.}$$

$$\frac{20}{8} = x \qquad \text{Divide each side by 8.}$$

$$\frac{5}{2} = x \qquad \text{Simplify.}$$

The solution is $x = \frac{5}{2}$. Check this solution in the original equation.

Graphical Solution

To begin, use a graphing utility to graph each side of the equation in the same viewing window.

$$y_1 = \frac{7}{x} - \frac{1}{3x} \quad \text{and} \quad y_2 = \frac{8}{3}$$

In Figure 7.3, the graphs appear to have one point of intersection. Use the *intersect* feature of the graphing utility to approximate the point of intersection to be $\left(\frac{5}{2}, \frac{8}{3}\right)$. The x-coordinate of the intersection point is the solution of the equation. So, the solution is $x = \frac{5}{2}$. Check this solution in the original equation.

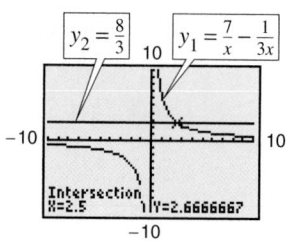

Figure 7.3

Example 6 ■ An Equation Containing Variable Denominators

$$\frac{3}{x + 1} + \frac{5}{2} = 1 \qquad \text{Write original equation.}$$

$$[2(x + 1)]\left(\frac{3}{x + 1} + \frac{5}{2}\right) = [2(x + 1)](1) \qquad \begin{array}{l}\text{Multiply each side by LCD of}\\ 2(x + 1).\end{array}$$

$$6 + 5(x + 1) = 2(x + 1) \qquad \text{Distributive Property}$$

$$6 + 5x + 5 = 2x + 2 \qquad \text{Distributive Property}$$

$$5x + 11 = 2x + 2 \qquad \text{Combine like terms.}$$

$$3x = -9 \qquad \text{Subtract } 2x \text{ and 11 from each side.}$$

$$x = -3 \qquad \text{Divide each side by 3.}$$

So, the solution is $x = -3$. Check this solution in the original equation.

Throughout the text, the importance of checking solutions is emphasized. One reason for checking is to make sure that you did not make errors in the solution process. In the next example you will see that there is another reason for checking solutions in the *original* equation. That is, even with no mistakes in the solution process, it can happen that a "trial solution" does not satisfy the original equation. As discussed in Section 5.5, this type of "solution" is called *extraneous*. Recall that an extraneous solution of an equation does not, by definition, satisfy the original equation, and so *must not* be listed as an actual solution.

Example 7 ■ An Equation with No Solution

Solve $\dfrac{5x}{x-2} = 7 + \dfrac{10}{x-2}$.

Solution

The least common denominator for this equation is $x - 2$. So, begin by multiplying each side of the equation by $x - 2$.

$$\frac{5x}{x-2} = 7 + \frac{10}{x-2} \qquad \text{Write original equation.}$$

$$(x-2)\left(\frac{5x}{x-2}\right) = (x-2)\left(7 + \frac{10}{x-2}\right) \qquad \text{Multiply each side by LCD of } x-2.$$

$$5x = 7(x-2) + 10, \ x \neq 2 \qquad \text{Simplify.}$$

$$5x = 7x - 14 + 10 \qquad \text{Distributive Property}$$

$$5x = 7x - 4 \qquad \text{Simplify.}$$

$$-2x = -4 \qquad \text{Subtract } 7x \text{ from each side.}$$

$$x = 2 \qquad \text{Divide each side by } -2.$$

At this point, the solution appears to be $x = 2$. However, by performing a check, you will see that this "trial solution" is extraneous.

Check

$$\frac{5x}{x-2} = 7 + \frac{10}{x-2} \qquad \text{Write original equation.}$$

$$\frac{5(2)}{2-2} \stackrel{?}{=} 7 + \frac{10}{2-2} \qquad \text{Substitute 2 for } x.$$

$$\frac{10}{0} \stackrel{?}{=} 7 + \frac{10}{0} \qquad \text{Solution does not check. } \bcancel{}$$

Because the check results in *division by zero*, 2 is extraneous. So, the original equation has no solution. Figure 7.4 shows that the graphs of "each side" of the equation have no point of intersection.

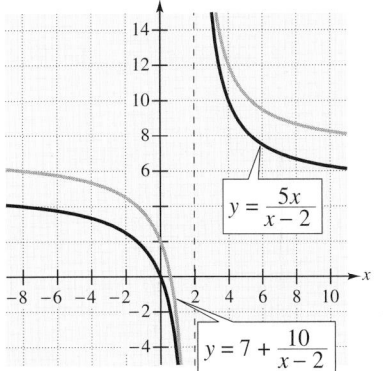

Figure 7.4

$y = \dfrac{5x}{x-2}$

$y = 7 + \dfrac{10}{x-2}$

NOTE   In Example 7, can you see why $x = 2$ is extraneous? By looking back at the original equation, you can see that $x = 2$ is excluded from the domain of two of the fractions that occur in the equation. When solving rational equations, you may find it helpful to list the domain restrictions before beginning the solution process.

Example 8 ■ An Equation Containing Variable Denominators

Solve $\dfrac{4}{x-2} + \dfrac{3x}{x+1} = 3$.

Solution

The domain is all real values of x such that $x \neq 2$ and $x \neq -1$. The least common denominator is $(x-2)(x+1)$.

$$\frac{4}{x-2} + \frac{3x}{x+1} = 3$$

$$(x-2)(x+1)\left(\frac{4}{x-2} + \frac{3x}{x+1}\right) = (x-2)(x+1)(3)$$

$$4(x+1) + 3x(x-2) = 3(x^2 - x - 2), \quad x \neq 2,\ x \neq -1$$

$$4x + 4 + 3x^2 - 6x = 3x^2 - 3x - 6$$

$$3x^2 - 2x + 4 = 3x^2 - 3x - 6$$

$$-2x + 4 = -3x - 6$$

$$x + 4 = -6$$

$$x = -10$$

The solution is $x = -10$. Check this solution in the original equation.

Example 9 ■ An Equation That Has Two Solutions

Solve $\dfrac{3x}{x+1} = \dfrac{12}{x^2 - 1} + 2$.

Solution

The domain is all real values of x such that $x \neq 1$ and $x \neq -1$. The least common denominator is $(x+1)(x-1) = x^2 - 1$.

$$\frac{3x}{x+1} = \frac{12}{x^2 - 1} + 2 \qquad \text{Write original equation.}$$

$$(x^2 - 1)\left(\frac{3x}{x+1}\right) = (x^2 - 1)\left(\frac{12}{x^2 - 1} + 2\right) \qquad \text{Multiply each side by LCD of } x^2 - 1.$$

$$(x-1)(3x) = 12 + 2(x^2 - 1), \quad x \neq \pm 1 \qquad \text{Simplify.}$$

$$3x^2 - 3x = 12 + 2x^2 - 2 \qquad \text{Distributive Property}$$

$$x^2 - 3x = 10 \qquad \text{Subtract } 2x^2 \text{ from each side.}$$

$$x^2 - 3x - 10 = 0 \qquad \text{Subtract 10 from each side.}$$

$$(x+2)(x-5) = 0 \qquad \text{Factor.}$$

$$x + 2 = 0 \implies x = -2 \qquad \text{Set 1st factor equal to 0.}$$

$$x - 5 = 0 \implies x = 5 \qquad \text{Set 2nd factor equal to 0.}$$

The solutions are $x = -2$ and $x = 5$. Check these solutions in the original equation.

EXPLORATION

Use a graphing utility to graph the equation

$$y = \frac{3x}{x+1} - \frac{12}{x^2 - 1} - 2.$$

Then use the *zero* or *root* feature of the graphing utility to determine the x-intercepts. How do the x-intercepts compare with the solutions to Example 9? What can you conclude?

Applications

Example 10 ■ A Work-Rate Problem

With only the cold water valve open, it takes 8 minutes to fill the tub of a washing machine. With both the hot and cold water valves open, it takes only 5 minutes. How long will it take the tub to fill with only the hot water valve open?

Solution

Verbal Model:	Rate for cold water	$+$	Rate for hot water	$=$	Rate for warm water

Labels: Rate for cold water $= \frac{1}{8}$ (tub per minute)

         Rate for hot water $= 1/t$ (tub per minute)

         Rate for warm water $= \frac{1}{5}$ (tub per minute)

Equation:
$$\frac{1}{8} + \frac{1}{t} = \frac{1}{5}$$
$$40t\left(\frac{1}{8} + \frac{1}{t}\right) = 40t\left(\frac{1}{5}\right)$$
$$5t + 40 = 8t$$
$$40 = 3t$$
$$\frac{40}{3} = t$$

So, it will take about $13\frac{1}{3}$ minutes to fill the tub with hot water alone. Check this in the original statement of the problem.

Example 11 ■ Average Cost

A manufacturing plant can produce x units of a certain item for $26 per unit *plus* an initial investment of $80,000. How many units must be produced to have an average cost of $30 per unit?

Solution

Verbal Model:	Average cost per unit	$=$	Total cost	$\div$	Number of units

Labels: Number of units $= x$ (units)

         Average cost per unit $= 30$ (dollars per unit)

         Total cost $= 26x + 80,000$ (dollars)

Equation:
$$30 = \frac{26x + 80,000}{x}$$
$$30x = 26x + 80,000$$
$$4x = 80,000$$
$$x = 20,000$$

The plant should produce 20,000 units (see Figure 7.5). Check this in the original statement of the problem.

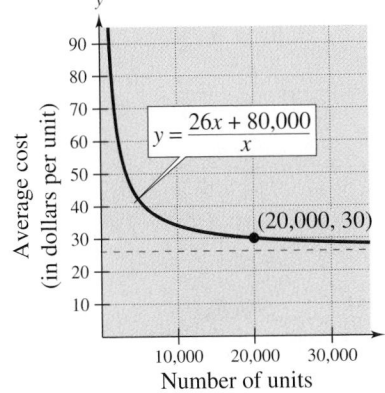

Figure 7.5

NOTE In Figure 7.5, notice that the average cost per unit drops as the number of units increases.

7.5 ■ Exercises

Developing Skills

In Exercises 1–4, determine whether the values of x are solutions of the equation.

    Equation   *Values*

1. $\dfrac{x}{3} - \dfrac{x}{5} = \dfrac{4}{3}$ (a) $x = 0$ (b) $x = -1$

                       (c) $x = \frac{1}{8}$ (d) $x = 10$

2. $x = 4 + \dfrac{21}{x}$ (a) $x = 0$ (b) $x = -3$

                       (c) $x = 7$ (d) $x = -1$

3. $\dfrac{x}{4} + \dfrac{3}{4x} = 1$ (a) $x = -1$ (b) $x = 1$

                       (c) $x = 3$ (d) $x = \frac{1}{2}$

4. $5 - \dfrac{1}{x - 3} = 2$ (a) $x = \frac{10}{3}$ (b) $x = -\frac{1}{3}$

                       (c) $x = 0$ (d) $x = 1$

In Exercises 5–12, solve the rational equation.

5. $\dfrac{x}{4} = \dfrac{3}{8}$ **6.** $\dfrac{x}{10} = \dfrac{12}{5}$

7. $6 - \dfrac{x}{4} = x - 9$ **8.** $5 + \dfrac{y}{3} = y + 2$

9. $\dfrac{4t}{3} = 15 - \dfrac{t}{6}$ **10.** $\dfrac{5x}{6} = 7 - \dfrac{x}{3}$

11. $\dfrac{h + 2}{5} - \dfrac{h - 1}{9} = \dfrac{2}{3}$ **12.** $\dfrac{u - 2}{6} + \dfrac{2u + 5}{15} = 3$

In Exercises 13–58, solve the rational equation. (Check for extraneous solutions.)

13. $\dfrac{9}{25 - y} = -\dfrac{1}{4}$ **14.** $\dfrac{2}{u + 4} = \dfrac{5}{8}$

15. $5 - \dfrac{12}{a} = \dfrac{5}{3}$ **16.** $\dfrac{6}{b} + 22 = 24$

17. $\dfrac{12}{y + 5} + \dfrac{1}{2} = 2$ **18.** $\dfrac{7}{8} - \dfrac{16}{t - 2} = \dfrac{3}{4}$

19. $\dfrac{5}{x} = \dfrac{25}{3(x + 2)}$ **20.** $\dfrac{10}{x + 4} = \dfrac{15}{4(x + 1)}$

21. $\dfrac{8}{3x + 5} = \dfrac{1}{x + 2}$ **22.** $\dfrac{500}{3x + 5} = \dfrac{50}{x - 3}$

23. $\dfrac{3}{x + 2} - \dfrac{1}{x} = \dfrac{1}{5x}$ **24.** $\dfrac{12}{x + 5} + \dfrac{5}{x} = \dfrac{20}{x}$

25. $\dfrac{1}{2} = \dfrac{18}{x^2}$ **26.** $\dfrac{1}{6} = \dfrac{150}{z^2}$

27. $\dfrac{32}{t} = 2t$ **28.** $\dfrac{45}{u} = \dfrac{u}{5}$

29. $x + 1 = \dfrac{72}{x}$ **30.** $\dfrac{x + 42}{x} = x$

31. $x - \dfrac{24}{x} = 5$ **32.** $\dfrac{t}{4} + \dfrac{2t + 5}{t + 4} = 0$

33. $\dfrac{10}{x(x - 2)} + \dfrac{4}{x} = \dfrac{5}{x - 2}$

34. $\dfrac{x}{x + 4} + \dfrac{4}{x + 4} + 2 = 0$

35. $\dfrac{10}{x + 3} + \dfrac{10}{3} = 6$ **36.** $\dfrac{7}{x - 4} + \dfrac{3}{4} = 4$

37. $\dfrac{2}{(x - 4)(x - 2)} = \dfrac{1}{x - 4} + \dfrac{2}{x - 2}$

38. $\dfrac{4}{(x - 3)(x + 5)} = \dfrac{1}{x - 3} - \dfrac{2}{x + 5}$

39. $\dfrac{1}{x - 5} + \dfrac{1}{x + 5} = \dfrac{x + 3}{x^2 - 25}$

40. $\dfrac{x}{x - 2} + \dfrac{1}{x - 4} = \dfrac{2}{x^2 - 6x + 8}$

41. $\dfrac{2}{x - 10} - \dfrac{3}{x - 2} = \dfrac{6}{x^2 - 12x + 20}$

42. $\dfrac{5}{x + 2} + \dfrac{2}{x^2 - 6x - 16} = \dfrac{-4}{x - 8}$

43. $\dfrac{2(x + 1)}{x^2 - 4x + 3} + \dfrac{6x}{x - 3} = \dfrac{3x}{x - 1}$

44. $\dfrac{2x - 10}{x - 8} + \dfrac{3x}{x^2 - 7x - 8} = \dfrac{x}{x + 1}$

45. $\dfrac{4}{2x + 3} + \dfrac{17}{5(2x + 3)} = 3$

46. $\dfrac{2}{6q + 5} - \dfrac{3}{4(6q + 5)} = \dfrac{1}{28}$

47. $1 = \dfrac{16}{y} - \dfrac{39}{y^2}$ **48.** $\dfrac{3x}{2} + \dfrac{4}{x} = 5$

49. $\dfrac{1}{x - 1} + \dfrac{3}{x + 1} = 2$ **50.** $\dfrac{2x}{3x + 10} - \dfrac{5}{x} = 0$

51. $\dfrac{x^2 - x}{x^4 + 4x^2 + 3} = \dfrac{1}{x^2 + 1} - \dfrac{1}{x^2 + 3}$

52. $\dfrac{x^2 + 5x}{x^4 + 3x^2 - 10} = \dfrac{2}{x^2 - 2} - \dfrac{2}{x^2 + 5}$

53. $\dfrac{2x}{5} = \dfrac{x^2 - 5x}{5x}$

54. $\dfrac{8(x - 1)}{x^2 - 4} = \dfrac{4}{x - 2}$

55. $\dfrac{x}{2} = \dfrac{2 - \dfrac{3}{x}}{1 - \dfrac{1}{x}}$

56. $\dfrac{2x}{3} = \dfrac{1 + \dfrac{2}{x}}{1 + \dfrac{1}{x}}$

57. $\dfrac{2(x + 7)}{x + 4} - 2 = \dfrac{2x + 20}{2x + 8}$

58. $\dfrac{2x^2 - 5}{x^2 - 4} + \dfrac{6}{x + 2} = \dfrac{4x - 7}{x - 2}$

Solving Problems

59. *Number Problem*   Find a number such that the sum of the number and its reciprocal is $\frac{65}{8}$.

60. *Number Problem*   Find a number such that the sum of the number and its reciprocal is $\frac{145}{12}$.

61. *Number Problem*   Find a number such that the sum of two times the number and three times its reciprocal is $\frac{97}{4}$.

62. *Number Problem*   Find a number such that the sum of three times the number and two times its reciprocal is $\frac{77}{5}$.

63. *Wind Speed*   A plane has a speed of 300 miles per hour in still air. Find the speed of the wind if the plane traveled a distance of 680 miles with a tailwind in the same time it took to travel 520 miles into a headwind.

64. *Current Speed*   A boat has a speed of 60 miles per hour in still water. Find the speed of the current in a stream if the boat traveled a distance of 11 miles upstream in the same time it took to travel 13 miles downstream.

65. *Average Speed*   During the first part of a six-hour trip, you travel 240 miles at an average speed of r miles per hour. For the next 72 miles of the trip, you increase your speed by 10 miles per hour. What were your two average speeds?

66. *Average Speed*   During the first part of a half-hour jog, you run 2 miles at an average speed of r miles per hour. For the next 3 miles of your jog, you decrease your speed by 3 miles per hour. What were your two average speeds?

67. *Comparing Two Speeds*   One person runs 2 miles per hour faster than another person. The first person runs 5 miles in the same time the second person runs 4 miles. Find the speed of each person.

68. *Comparing Two Speeds*   The speed of a commuter plane is 150 miles per hour slower than that of a passenger jet. The commuter plane travels 450 miles in the same time the passenger jet travels 1150 miles. Find the speed of each aircraft.

69. *Partnership Costs*   A group plans to start a new business that will require $240,000 for start-up capital. The individuals in the group will share the cost equally. If two additional people were to join the group, the cost per person would decrease by $4000. How many people are presently in the group?

70. *Population Growth*   A biologist introduces 100 insects into a culture. The population P of the culture is approximated by the model

$$P = \dfrac{500(1 + 3t)}{5 + t}$$

where t is the time in hours. Find the time required for the population to increase to 1000 insects.

71. *Work Rate*   One person can complete a task in 6 hours. A second person can complete the same task in 4 hours. How long will it take to complete the task if the two people work together?

72. *Work Rate*   One person can complete a task in 3 minutes. A second person can complete the same task in 5 minutes. How long will it take to complete the task if the two people work together?

73. *Work Rate*   An electrician can install the electric wires in a house in 14 hours. A second electrician requires 18 hours. How long would it take both electricians working together to install the wires?

74. *Work Rate*   Printer A can print a report in 3 hours. Printer B can print the same report in 4 hours. How long would it take both printers working together to print the report?

75. *Flow Rate*   The flow rate of one pipe is $1\frac{1}{4}$ times that of another pipe. A swimming pool can be filled in 5 hours using both pipes. Find the time required to fill the pool using only the pipe with the lower flow rate.

76. *Pollution Removal* The cost C in dollars of removing $p\%$ of the air pollutants from the stack emissions of a utility company is modeled by

$$C = \frac{120{,}000p}{100 - p}.$$

Determine the percent of the air pollutants in the stack emissions that can be removed for $680{,}000.

77. *Average Cost* The average cost of producing x units of a product is given by

$$\overline{C} = 1.50 + \frac{4200}{x}.$$

Determine the number of units that must be produced to have an average cost of $2.90.

78. *Learning Rate* For a particular person enrolled in a typing class, the number of words the person is able to type can be modeled by

$$y = \frac{50(t + 1)}{(t + 5)}, \quad t \geq 0.$$

In this model, t is the number of weeks the person takes lessons and y is the number of words per minute the person is able to type (see figure).

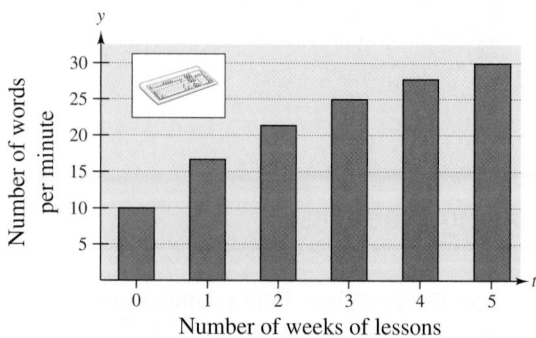

(a) Use the graph to determine graphically the number of weeks of lessons the person needs to be able to type 25 words per minute.

(b) Use the model to confirm the result in part (a) algebraically.

Explaining Concepts

79. Define the term *extraneous solution*. How do you identify an extraneous solution?

80. Explain when you can use cross-multiplication to solve a rational equation.

Ongoing Review

In Exercises 81 and 82, solve the equation.

81. $(4x - 7)(3x + 2) = 0$ **82.** $(2x + 5)(x - 9) = 0$

In Exercises 83–86, solve the inequality.

83. $7 - 3x > 4 - x$ **84.** $2(x + 6) - 20 < 2$

85. $|x - 3| < 2$ **86.** $|x - 5| > 3$

87. *Simple Interest* An investment of $2500 is made at an annual simple interest rate of 5.5%. How much additional money must be invested at an annual simple interest rate of 8% so that the total interest earned is 7% of the total investment?

Looking Further

Equations and Expressions It is important to distinguish between equations and expressions. In parts (a) through (d), if the exercise is an equation, solve it; if the exercise is an expression, simplify it.

(a) $\dfrac{16}{x^2 - 16} + \dfrac{x}{2x - 8} = \dfrac{1}{2}$

(b) $\dfrac{16}{x^2 - 16} + \dfrac{x}{2x - 8} + \dfrac{1}{2}$

(c) $\dfrac{5}{3} - \dfrac{5}{x + 3} + 3$

(d) $\dfrac{5}{3} - \dfrac{5}{x + 3} = 3$

(e) Explain the difference between an equation and an expression.

(f) Compare the use of a common denominator in solving rational equations with its use in adding or subtracting rational expressions.

7.6 Graphing Rational Functions

What you should learn

- How to use a table of values to sketch graphs of rational functions
- How to determine horizontal and vertical asymptotes of rational functions
- How to use asymptotes and intercepts to sketch graphs of rational functions
- How to use graphs of rational functions to solve application problems

Why you should learn it

You can use the graph of a rational function to model the concentration of a chemical in the bloodstream. For example, see Exercise 77 on page 485.

Introduction

Recall that the domain of a rational function consists of all values of x for which the denominator is not zero. For instance, the domain of

$$f(x) = \frac{x + 2}{x - 1}$$

is all real numbers except $x = 1$. When graphing a rational function, pay special attention to the shape of the graph near x-values that are not in the domain.

Example 1 ■ Point Plotting the Graph of a Rational Function

Use point plotting to sketch the graph of

$$f(x) = \frac{x + 2}{x - 1}.$$

Solution

Notice that the domain is all real numbers except $x = 1$. Next, construct a table of values, including x-values that are close to 1 on the left *and* the right.

x-Values to the Left of 1

x	-3	-2	-1	0	0.5	0.9	0.99	0.999
$f(x)$	0.25	0	-0.5	-2	-5	-29	-299	-2999

x-Values to the Right of 1

x	1.001	1.01	1.1	1.5	2	3	4	5
$f(x)$	3001	301	31	7	4	2.5	2	1.75

Plot several points to the left of 1 and connect them with a smooth curve, as shown in Figure 7.6. Do the same for several points to the right of 1. *Do not connect the two portions of the graph, which are called its* **branches.**

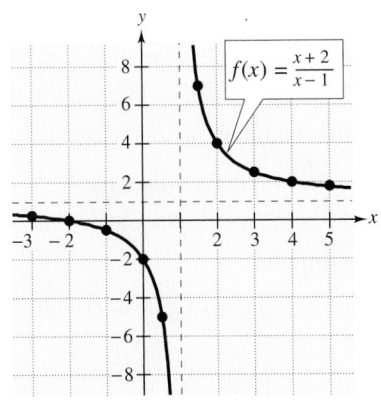

Figure 7.6

From the tables above, note that as x approaches 1 from the left, $f(x)$ decreases without bound, and as x approaches 1 from the right, $f(x)$ increases without bound. In other words, the values of $f(x)$ approach negative infinity as x approaches 1 from the left, and the values of $f(x)$ approach positive infinity as x approaches 1 from the right.

Horizontal and Vertical Asymptotes

An **asymptote** of a graph is a line to which the graph becomes arbitrarily close as $|x|$ or $|y|$ increases without bound. In other words, if a graph has an asymptote, it is possible to move far enough out on the graph so that there is almost no difference between the graph and the asymptote.

The graph in Example 1 has two asymptotes: the line $x = 1$ is a **vertical asymptote,** and the line $y = 1$ is a **horizontal asymptote.** Other examples of asymptotes are shown in Figure 7.7.

EXPLORATION

Use a graphing utility to graph $f(x)$ and its transformation $g(x)$ shown below.

$$f(x) = \frac{1}{x}$$

$$g(x) = \frac{1}{x - 4}$$

What can you conclude about the asymptotes of these functions? Could you have arrived at this conclusion algebraically by simply looking at the functions and not by graphing them?

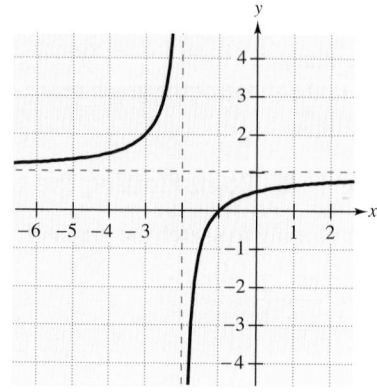

Graph of $y = \dfrac{x + 1}{x + 2}$

Horizontal asymptote: $y = 1$
Vertical asymptote: $x = -2$

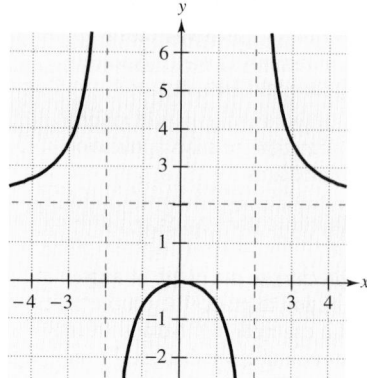

Graph of $y = \dfrac{2x^2}{x^2 - 4}$

Horizontal asymptote: $y = 2$
Vertical asymptotes: $x = \pm 2$

Figure 7.7

The graph of a rational function may have no horizontal or vertical asymptotes, or it may have several.

Guidelines for Finding Asymptotes

Let $f(x) = p(x)/q(x)$, where $p(x)$ and $q(x)$ have no common factors.

1. The graph of f has a vertical asymptote at each x-value for which the denominator is zero.

2. The graph of f has at most one horizontal asymptote.

    a. If the degree of $p(x)$ is less than the degree of $q(x)$, the line $y = 0$ is a horizontal asymptote.

    b. If the degree of $p(x)$ is equal to the degree of $q(x)$, the line $y = a/b$ is a horizontal asymptote, where a is the leading coefficient of $p(x)$ and b is the leading coefficient of $q(x)$.

    c. If the degree of $p(x)$ is greater than the degree of $q(x)$, the graph has no horizontal asymptote.

Example 2 ■ Finding Horizontal and Vertical Asymptotes

Find all horizontal and vertical asymptotes of the graph of

$$f(x) = \frac{2x}{3x^2 + 1}.$$

Solution

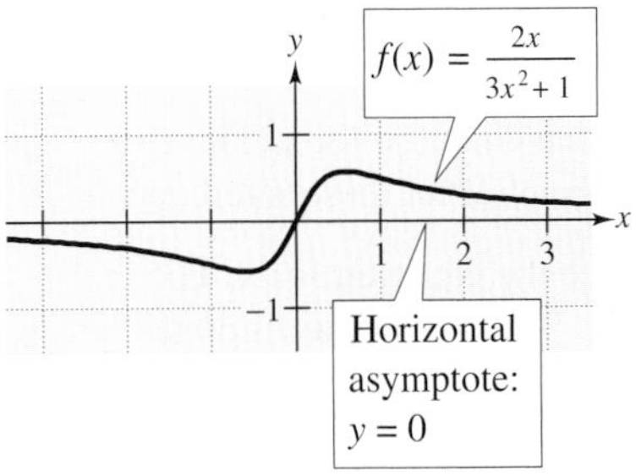

For this rational function, the degree of the numerator is less than the degree of the denominator. This implies that the graph has the line

$$y = 0 \qquad\qquad \text{Horizontal asymptote}$$

as a horizontal asymptote, as shown in Figure 7.8. To find any vertical asymptotes, set the denominator equal to zero and solve the resulting equation for x.

$$3x^2 + 1 = 0 \qquad\qquad \text{Set denominator equal to zero.}$$

Because this equation has no real solution, you can conclude that the graph has no vertical asymptote.

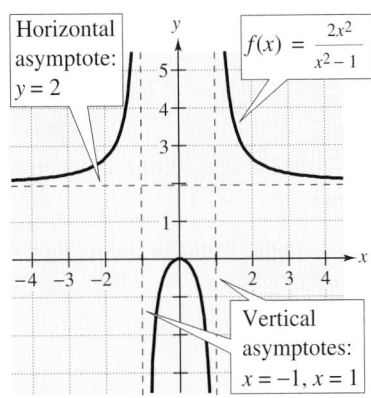

$f(x) = \dfrac{2x}{3x^2 + 1}$

Horizontal asymptote: $y = 0$

Figure 7.8

Remember that the graph of a rational function can have at most one horizontal asymptote, but it can have several vertical asymptotes. For instance, the graph in Example 3 has two vertical asymptotes.

Example 3 ■ Finding Horizontal and Vertical Asymptotes

Find all horizontal and vertical asymptotes of the graph of

$$f(x) = \frac{2x^2}{x^2 - 1}.$$

Solution

For this rational function, the degree of the numerator is equal to the degree of the denominator. The leading coefficient of the numerator is 2, and the leading coefficient of the denominator is 1. So, the graph has the line

$$y = \frac{2}{1} = 2 \qquad\qquad \text{Horizontal asymptote}$$

as a horizontal asymptote, as shown in Figure 7.9. Then, to find any vertical asymptotes, set the denominator equal to zero and solve the resulting equation for x.

$$x^2 - 1 = 0 \qquad\qquad \text{Set denominator equal to zero.}$$
$$(x + 1)(x - 1) = 0 \qquad\qquad \text{Factor.}$$
$$x + 1 = 0 \implies x = -1 \qquad \text{Set 1st factor equal to 0.}$$
$$x - 1 = 0 \implies x = 1 \qquad \text{Set 2nd factor equal to 0.}$$

This equation has two real solutions: $x = -1$ and $x = 1$. So, the graph has two vertical asymptotes: the lines $x = -1$ and $x = 1$.

Horizontal asymptote: $y = 2$

$f(x) = \dfrac{2x^2}{x^2 - 1}$

Vertical asymptotes: $x = -1, x = 1$

Figure 7.9

Graphing Rational Functions

To sketch the graph of a rational function, use the guidelines below.

> **Guidelines for Graphing Rational Functions**
>
> Let $f(x) = p(x)/q(x)$, where $p(x)$ and $q(x)$ have no common factors.
>
> 1. Find and plot the y-intercept (if any) by evaluating $f(0)$.
>
> 2. Set the numerator equal to zero and solve the equation for x. The real solutions represent the x-intercepts of the graph. Plot these intercepts.
>
> 3. Set the denominator equal to zero and solve the equation for x. The real solutions represent the vertical asymptotes. Sketch these asymptotes.
>
> 4. Find and sketch the horizontal asymptote of the graph.
>
> 5. Plot at least one point between and one point beyond each x-intercept and vertical asymptote.
>
> 6. Use smooth curves to complete the graph between and beyond the vertical asymptotes.

Example 4 ■ Sketching the Graph of a Rational Function

Sketch the graph of

$$f(x) = \frac{2}{x - 3}.$$

Solution

Begin by noting that the numerator and denominator have no common factors. Following the guidelines above produces the following.

- Because $f(0) = \dfrac{2}{0 - 3} = -\dfrac{2}{3}$, the y-intercept is $\left(0, -\dfrac{2}{3}\right)$.

- Because the numerator is never zero, there are no x-intercepts.

- Because the denominator is zero when $x - 3 = 0$ or $x = 3$, the line $x = 3$ is a vertical asymptote.

- Because the degree of the numerator is less than the degree of the denominator, the line $y = 0$ is a horizontal asymptote.

Plot the intercept, asymptotes, and the additional points from the table shown below. Then complete the graph by drawing two branches, as shown in Figure 7.10. Note that the two branches are not connected.

x	-2	1	2	4	5
$f(x)$	$-\frac{2}{5}$	-1	-2	2	1

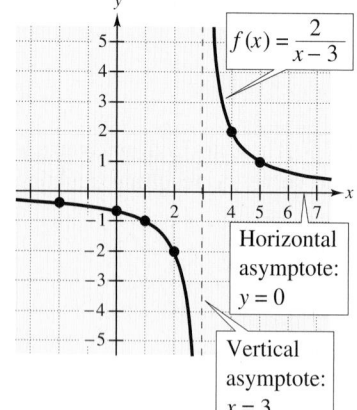

Figure 7.10

Example 5 ■ Sketching the Graph of a Rational Function

Sketch the graph of

$$f(x) = \frac{2x - 1}{x}.$$

Solution

Begin by noting that the numerator and denominator have no common factors.

- Because $x = 0$ is not in the domain of the function, $f(0)$ is undefined and there is no y-intercept.
- Because the numerator is zero when $2x - 1 = 0$ or $x = \frac{1}{2}$, the x-intercept is $\left(\frac{1}{2}, 0\right)$.
- Because the denominator is zero when $x = 0$, the line $x = 0$ is a vertical asymptote.
- Because the degree of the numerator is equal to the degree of the denominator and the ratio of the leading coefficients is 2, the line $y = 2$ is a horizontal asymptote.

By plotting the intercept, asymptotes, and the additional points from the table shown below, you can obtain the graph shown in Figure 7.11.

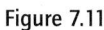

Figure 7.11

x	-4	-2	-1	$\frac{1}{4}$	4
$f(x)$	$\frac{9}{4}$	$\frac{5}{2}$	3	-2	$\frac{7}{4}$

Example 6 ■ Sketching the Graph of a Rational Function

Sketch the graph of $f(x) = \dfrac{2}{x^2 + 1}$.

Solution

Begin by noting that the numerator and denominator have no common factors.

- Because $f(0) = \dfrac{2}{0^2 + 1} = 2$, the y-intercept is $(0, 2)$.
- Because the numerator is never zero, there are no x-intercepts.
- Because the equation $x^2 + 1 = 0$ has no real solution, there are no vertical asymptotes.
- Because the degree of the numerator is less than the degree of the denominator, the line $y = 0$ is a horizontal asymptote.

By plotting the intercept, asymptote, and the additional points from the table shown below, you can obtain the graph shown in Figure 7.12.

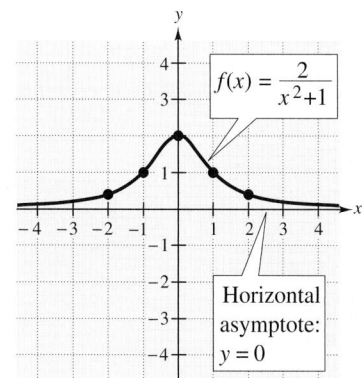

Figure 7.12

x	-2	-1	1	2
$f(x)$	$\frac{2}{5}$	1	1	$\frac{2}{5}$

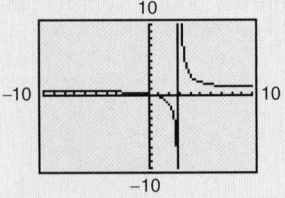
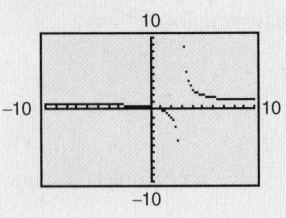
NOTE From the graph shown in Figure 7.13, notice that the average cost decreases as the number of calendars increases.

Application

Example 7 ■ Finding the Average Cost

As a fundraising project, a club is publishing a calendar. The costs of photography and typesetting total $850. In addition to these "one-time" costs, the unit cost of printing each calendar is $3.25. Let x represent the number of calendars printed. Write a model that represents the average cost per calendar.

Solution

To begin, you need to find the total cost of printing the calendars.

Verbal Model:	Total cost	=	Unit cost	·	Number of calendars	+	Cost of photography and typesetting

Labels: Total cost = C (dollars)

Unit cost = 3.25 (dollars)

Number of calendars = x

Cost of photography and typesetting = 850 (dollars)

The total cost C of printing x calendars is

Equation: $C = 3.25x + 850.$ Total cost function

Next, find the average cost per calendar.

Verbal Model:	Average cost	=	$\dfrac{\text{Total cost}}{\text{Number of calendars}}$

Labels: Average cost = $\overline{C}$ (dollars)

Total cost = C (dollars)

Number of calendars = x

Substitute $3.25x + 850$ for C to obtain the equation for the average cost per calendar $\overline{C}$.

Equation: $\overline{C} = \dfrac{C}{x} = \dfrac{3.25x + 850}{x}$ Average cost function

The graph of $\overline{C}$ is shown in Figure 7.13.

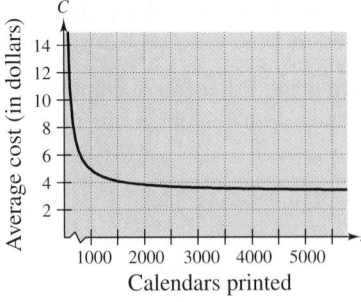

Figure 7.13

7.6 ■ Exercises

Developing Skills

Numerical Analysis   In Exercises 1 and 2, (a) complete each table, (b) determine the vertical and horizontal asymptotes of the graph, and (c) find the domain of the function.

x	0	0.5	0.9	0.99	0.999
y					

x	2	1.5	1.1	1.01	1.001
y					

x	2	5	10	100	1000
y					

1. $f(x) = \dfrac{4}{x - 1}$

2. $f(x) = \dfrac{2x}{x - 1}$

In Exercises 3–6, identify the horizontal and vertical asymptotes of the function. Use the asymptotes to match the rational function with its graph. [The graphs are labeled (a), (b), (c), and (d).]

(a)
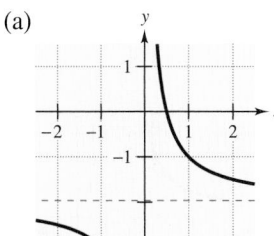

(b)
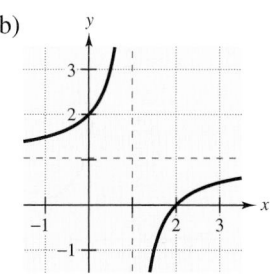

(c)
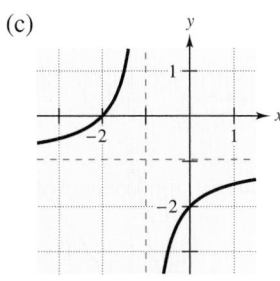

(d)
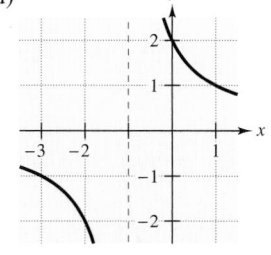

3. $f(x) = \dfrac{2}{x + 1}$

4. $f(x) = \dfrac{1 - 2x}{x}$

5. $f(x) = \dfrac{x - 2}{x - 1}$

6. $f(x) = -\dfrac{x + 2}{x + 1}$

In Exercises 7–20, find the domain of the function and identify any horizontal and vertical asymptotes.

7. $f(x) = \dfrac{5}{x^2}$

8. $g(x) = \dfrac{3}{x - 5}$

9. $f(x) = \dfrac{x}{x + 8}$

10. $f(u) = \dfrac{u^2}{u - 10}$

11. $g(t) = \dfrac{3}{t(t - 1)}$

12. $h(x) = \dfrac{4x - 3}{x}$

13. $h(s) = \dfrac{2s^2}{s + 3}$

14. $g(t) = \dfrac{3}{t^2 + 1}$

15. $y = \dfrac{3 - 5x}{1 - 3x}$

16. $y = \dfrac{3x + 2}{2x - 1}$

17. $y = \dfrac{5x^2}{x^2 - 1}$

18. $y = \dfrac{2x^2}{x^2 + 1}$

19. $f(x) = \dfrac{1}{x^2 + 2x - 3}$

20. $g(x) = \dfrac{1}{x^2 - 2x - 8}$

In Exercises 21–24, match the function with its graph. [The graphs are labeled (a), (b), (c), and (d).]

(a)
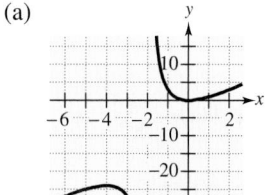

(b)
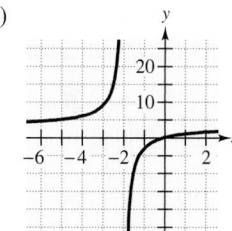

(c)
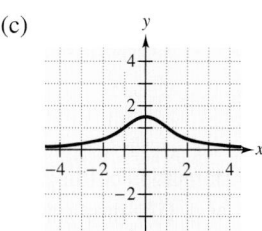

(d)
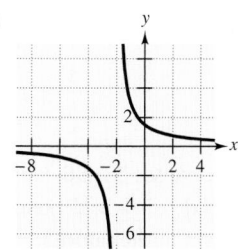

21. $f(x) = \dfrac{3}{x + 2}$

22. $f(x) = \dfrac{3x}{x + 2}$

23. $f(x) = \dfrac{3x^2}{x + 2}$

24. $f(x) = \dfrac{3}{x^2 + 2}$

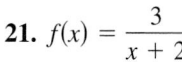

In Exercises 25–44, sketch the graph of the function. As sketching aids, check for intercepts, vertical asymptotes, and horizontal asymptotes.

25. $g(x) = \dfrac{5}{x}$

26. $g(x) = \dfrac{5}{x - 4}$

27. $f(x) = \dfrac{5}{x^2}$

28. $f(x) = \dfrac{5}{(x - 4)^2}$

29. $f(x) = \dfrac{1}{x - 2}$

30. $f(x) = \dfrac{3}{x + 1}$

31. $g(x) = \dfrac{1}{2 - x}$

32. $g(x) = \dfrac{-3}{x + 1}$

33. $y = \dfrac{2x + 4}{x}$

34. $y = \dfrac{x - 2}{x}$

35. $y = \dfrac{3x}{x + 4}$

36. $y = \dfrac{2x}{x + 4}$

37. $y = \dfrac{2x^2}{x^2 + 1}$

38. $y = \dfrac{10}{(x - 2)^2}$

39. $g(t) = 3 - \dfrac{2}{t}$

40. $g(v) = \dfrac{2v}{v + 1}$

41. $y = \dfrac{4}{x^2 + 1}$

42. $y = \dfrac{4x^2}{x^2 + 1}$

43. $y = -\dfrac{x}{x^2 - 4}$

44. $y = \dfrac{3x^2}{x^2 - x - 2}$

In Exercises 45–52, use a graphing utility to graph the function. Give the domain of the function and identify any horizontal or vertical asymptotes.

45. $h(x) = \dfrac{x - 3}{x - 1}$

46. $h(x) = \dfrac{x^2}{x - 2}$

47. $f(t) = \dfrac{6}{t^2 + 1}$

48. $g(t) = 2 + \dfrac{3}{t + 1}$

49. $y = \dfrac{2(x^2 + 1)}{x^2}$

50. $y = \dfrac{2(x^2 - 1)}{x^2}$

51. $y = \dfrac{3}{x} + \dfrac{1}{x - 2}$

52. $y = \dfrac{x}{2} - \dfrac{2}{x}$

In Exercises 53–56, identify the transformation of the graph of $f(x) = 1/x$ and sketch the graph of g. (The graph of f is shown in the figure.)

53. $g(x) = -\dfrac{1}{x}$

54. $g(x) = \dfrac{1}{x} + 2$

55. $g(x) = \dfrac{1}{x - 2}$

56. $g(x) = \dfrac{1}{x + 3}$

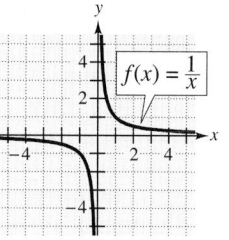
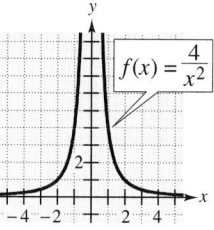

Figure for 53–56   Figure for 57–60

In Exercises 57–60, identify the transformation of the graph of $f(x) = 4/x^2$ and sketch the graph of g. (The graph of f is shown in the figure.)

57. $g(x) = 2 + \dfrac{4}{x^2}$

58. $g(x) = -\dfrac{4}{x^2}$

59. $g(x) = -\dfrac{4}{(x - 2)^2}$

60. $g(x) = 5 - \dfrac{4}{x^2}$

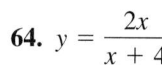 In Exercises 61 and 62, use a graphing utility to graph the rational function. Give the domain of the function and identify any asymptotes. Then zoom out far enough so that the graph appears as a line. Identify the line.

61. $y = \dfrac{2x^2 + x}{x + 1}$

62. $y = \dfrac{x^2 + 5x + 8}{x + 3}$

In Exercises 63–66, (a) use the graph to determine any x-intercepts of the function, and (b) set $y = 0$ and solve the resulting equation to confirm the result.

63. $y = \dfrac{x + 2}{x - 2}$

64. $y = \dfrac{2x}{x + 4}$

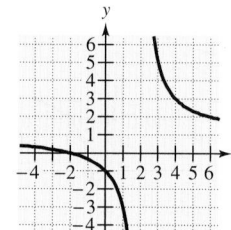
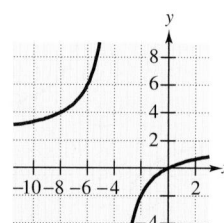

65. $y = x - \dfrac{1}{x}$

66. $y = x - \dfrac{2}{x} - 1$

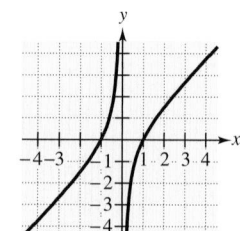
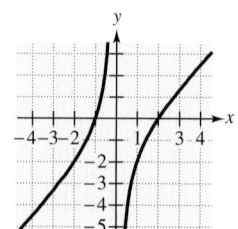

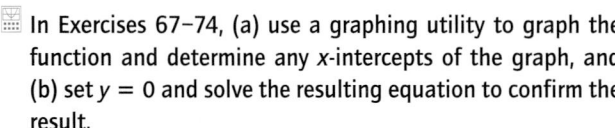

In Exercises 67–74, (a) use a graphing utility to graph the function and determine any *x*-intercepts of the graph, and (b) set $y = 0$ and solve the resulting equation to confirm the result.

67. $y = \dfrac{1}{x} + \dfrac{4}{x - 5}$

68. $y = \dfrac{x}{2} - \dfrac{4}{x} - 1$

69. $y = \dfrac{x - 4}{x + 5}$

70. $y = \dfrac{1}{x} - \dfrac{3}{x + 4}$

71. $y = (x + 1) - \dfrac{6}{x}$

72. $y = \dfrac{x^2 - 4}{x}$

73. $y = (x - 1) - \dfrac{12}{x}$

74. $y = 20\left(\dfrac{2}{x} - \dfrac{3}{x - 1}\right)$

Solving Problems

75. *Average Cost* A college club raises funds by selling candy bars. The club pays $0.60 for each candy bar and an annual cost of $250. Let *x* be the number of candy bars produced.

(a) Write the total cost *C* as a function of *x*.

(b) Write the average cost $\overline{C}$ as a function of *x*.

(c) Find the average costs of producing $x = 1000$ and $x = 10,000$ candy bars.

(d) Determine the horizontal asymptote of the average cost function algebraically and explain the meaning of the horizontal asymptote in the context of the problem.

(e) Use a graphing utility to graph the average cost function.

76. *Average Cost* A company spends $98,000 for equipment to produce a product. Each unit of the product costs $12.30. Let *x* be the number of units produced.

(a) Write the total cost *C* as a function of *x*.

(b) Write the average cost $\overline{C}$ as a function of *x*.

(c) Find the average costs of producing $x = 10,000$ and $x = 100,000$ units.

(d) Determine the horizontal asymptote of the average cost function algebraically and explain the meaning of the horizontal asymptote in the context of the problem.

(e) Use a graphing utility to graph the average cost function.

77. *Medicine* The concentration of a certain chemical in the bloodstream *t* hours after injection into muscle tissue is given by

$$C = \dfrac{2t}{4t^2 + 25}, \qquad t \geq 0.$$

(a) Determine the horizontal asymptote of the function and interpret its meaning in the context of the problem.

(b) Use a graphing utility to graph the function.

(c) Approximate the time when the concentration is the greatest.

78. *Concentration of a Mixture* A 25-liter container contains 5 liters of a 25% brine solution. You add *x* liters of a 75% brine solution to the container. The concentration *C* of the resulting mixture is

$$C = \dfrac{3x + 5}{4(x + 5)}.$$

(a) Determine the domain of the function based on the physical constraints of the problem.

(b) Use a graphing utility to graph the function.

(c) As the container is filled, what percent does the concentration of the brine appear to approach?

79. *Geometry* A rectangular region of length *x* and width *y* has an area of 400 square meters.

(a) Verify that the perimeter *P* is given by

$$P = 2\left(x + \dfrac{400}{x}\right).$$

(b) Determine the domain of the function based on the physical constraints of the problem.

(c) Use a graphing utility to graph the function.

(d) Approximate the dimensions of the rectangle that has a minimum perimeter.

80. *Sales* The cumulative number *N* (in thousands) of units of a product sold over a period of *t* years on the market is modeled by

$$N = \dfrac{150t(1 + 4t)}{1 + 0.15t^2}, \qquad t \geq 0.$$

(a) Estimate *N* when $t = 1$, $t = 2$, and $t = 4$.

(b) Use a graphing utility to graph the function.

(c) Determine the horizontal asymptote. Explain the meaning of the horizontal asymptote in the context of the problem.

Explaining Concepts

81. In your own words, describe what is meant by an asymptote of a graph.

82. In your own words, describe how to determine the domain of a rational function. Give an example of a rational function whose domain is all real numbers except 2.

83. If the graph of a rational function f has a vertical asymptote at $x = 3$, is it possible to sketch the graph without lifting your pencil from the paper? Explain.

84. Does every rational function have a vertical asymptote? Explain.

Think About It In Exercises 85 and 86, use a graphing utility to graph the function. Explain why there is no vertical asymptote when a superficial examination of the function may indicate that there should be one.

85. $g(x) = \dfrac{4 - 2x}{x - 2}$

86. $h(x) = \dfrac{x^2 - 9}{x + 3}$

Ongoing Review

In Exercises 87–90, find the product.

87. $(2x - 15)^2$

88. $(3x + 2)(7x - 10)$

89. $[(x + 1) - y][(x + 1) + y]$

90. $(x + 3)(x^2 - 3x + 9)$

In Exercises 91–94, sketch the graph of the equation.

91. $y = -4$

92. $x = 3$

93. $y = 2x + 1$

94. $y = -x - 4$

In Exercises 95 and 96, use long division to write the rational expression as the sum of a constant and a rational expression.

95. $\dfrac{4x - 3}{x + 2}$

96. $\dfrac{-6x + 1}{x - 3}$

97. *Geometry* The height of a triangle is 12 meters less than its base. Find the base and height of the triangle if its area is 80 square meters.

98. *Geometry* An open box with a square base is to be constructed from 825 square inches of material. What should be the dimensions of the base if the height of the box is to be 10 inches?

Looking Further

Write a rational function satisfying each set of criteria. Use a graphing utility to verify the results.

(a) Vertical asymptote: $x = 3$

Horizontal asymptote: $y = 2$

x-intercept of the function: $x = -1$

(b) Vertical asymptote: $x = -2$

Horizontal asymptote: $y = 0$

x-intercept of the function: $x = 3$

(c) Vertical asymptotes: $x = 1$ and $x = -1$

Horizontal asymptote: $y = 1$

x-intercept of the function: $x = 0$

(d) Vertical asymptotes: $x = 4$ and $x = -2$

Horizontal asymptote: $y = 0$

x-intercept of the function: $x = 6$

(e) Vertical asymptote: none

Horizontal asymptote: $y = -1$

x-intercept of the function: none

7.7 Rational Inequalities in One Variable

What you should learn:

- How to use test intervals to solve rational inequalities
- How to solve rational inequalities

Why you should learn it:

Rational inequalities can be used to model and solve real-life problems. For instance, in Exercise 38 on page 491 you will use a rational inequality to determine the number of units that should be produced to keep the average production cost below a specific value.

Test Intervals for Rational Inequalities

In Section 6.6, you studied strategies for solving polynomial inequalities. For instance, to solve the quadratic inequality

$$2x^2 + 5x > 12 \qquad \text{Original inequality}$$

you wrote the inequality in general form

$$2x^2 + 5x - 12 > 0 \qquad \text{General form}$$

and found the solutions of the corresponding equation

$$2x^2 + 5x - 12 = 0. \qquad \text{Corresponding equation}$$

These solutions, $x = -4$ and $x = \frac{3}{2}$, are called *critical numbers* and are used to form test intervals for the inequality.

$$(-\infty, -4), \quad \left(-4, \tfrac{3}{2}\right), \quad \left(\tfrac{3}{2}, \infty\right) \qquad \text{Test intervals}$$

For a review of how to use these test intervals to solve the inequality, see Example 3 on page 415.

The strategy for solving a rational inequality is similar to that for solving a polynomial inequality, and is summarized as follows.

Test Intervals for a Rational Inequality

To find the test intervals for a rational inequality, proceed as follows.

1. Write the inequality in general form, with a single rational expression on the left and zero on the right.

2. Find all real zeros of the numerator of the rational expression *and* all real zeros of the denominator of the rational expression. These zeros are the **critical numbers** of the rational expression.

3. Arrange the critical numbers in increasing order. Then use the critical numbers to determine the **test intervals** for the rational expression.

4. Choose a representative *x*-value in each test interval and evaluate the rational expression at that value. If the value of the rational expression is negative, it will have negative values for *every* *x*-value in the interval. If the value of the rational expression is positive, it will have positive values for *every* *x*-value in the interval.

NOTE   The value of a rational expression can change sign only at its *real zeros* (the real *x*-values for which its numerator is zero) and its *undefined values* (the real *x*-values for which its denominator is zero). These two types of numbers make up the critical numbers of a rational inequality.

Example 1 ■ Finding Test Intervals for a Rational Expression

Determine the intervals on which the rational expression

$$\frac{x + 2}{x - 3}$$

is entirely negative or entirely positive.

Algebraic Solution

Begin by finding the critical numbers of the rational expression. To do this, set the numerator equal to zero and the denominator equal to zero and solve the resulting equations.

$$x + 2 = 0 \implies x = -2 \qquad \text{Set numerator equal to 0.}$$
$$x - 3 = 0 \implies x = 3 \qquad \text{Set denominator equal to 0.}$$

The rational expression has two critical numbers: $x = -2$ and $x = 3$. These critical numbers determine the test intervals, as follows.

$$(-\infty, -2), \quad (-2, 3), \quad (3, \infty) \qquad \text{Test intervals}$$

In each test interval, choose a representative x-value and evaluate the rational expression, as shown in the table.

Test interval	Representative x-value	Value of rational expression	Conclusion
$(-\infty, -2)$	$x = -3$	$\dfrac{-3 + 2}{-3 - 3} = \dfrac{1}{6}$	Expression is positive.
$(-2, 3)$	$x = 0$	$\dfrac{0 + 2}{0 - 3} = -\dfrac{2}{3}$	Expression is negative.
$(3, \infty)$	$x = 4$	$\dfrac{4 + 2}{4 - 3} = 6$	Expression is positive.

So, the rational expression has positive values for every x in the intervals $(-\infty, -2)$ and $(3, \infty)$, and negative values for every x in the interval $(-2, 3)$. This result is shown graphically in Figure 7.14.

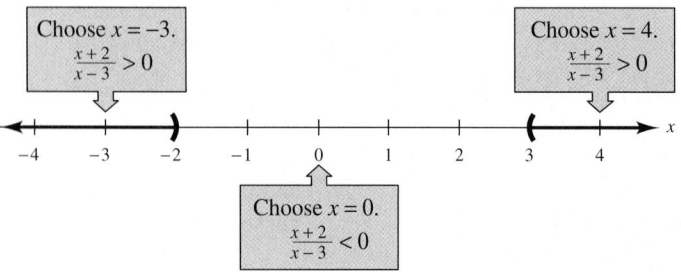

Figure 7.14

Graphical Solution

Sketch the graph of

$$y = \frac{x + 2}{x - 3}$$

as shown in Figure 7.15. The portions of the graph that lie above the x-axis correspond to

$$\frac{x + 2}{x - 3} > 0$$

and the portion that lies below the x-axis corresponds to

$$\frac{x + 2}{x - 3} < 0.$$

From the graph you can see that the value of the rational expression is positive for every x in the interval $(-\infty, -2)$, negative for every x in the interval $(-2, 3)$, and positive for every x in the interval $(3, \infty)$. Note that the value of the rational expression changes signs at $x = -2$ and $x = 3$, the critical numbers of the expression.

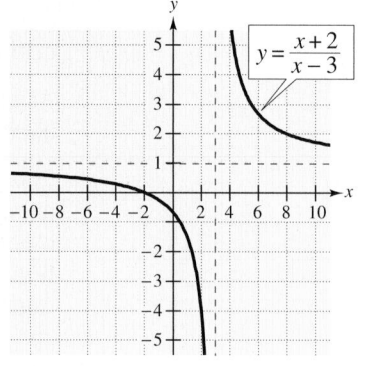

Figure 7.15

Solving Rational Inequalities

The concepts of critical numbers and test intervals can be used to solve rational inequalities, as demonstrated in Examples 2 and 3.

Example 2 ■ Solving a Rational Inequality

Solve the rational inequality $\dfrac{x}{x-2} > 0$.

Solution

The numerator is zero when $x = 0$ and the denominator is zero when $x = 2$. So, the two critical numbers are 0 and 2, which implies that the test intervals are

$$(-\infty, 0), \quad (0, 2), \quad \text{and} \quad (2, \infty). \qquad \text{Test intervals}$$

To test an interval, choose a convenient number in the interval and determine if the number satisfies the inequality, as shown in the table.

Test interval	Representative x-value	Is inequality satisfied?
$(-\infty, 0)$	$x = -1$	$\dfrac{-1}{-1-2} \overset{?}{>} 0$ $\frac{1}{3} > 0$
$(0, 2)$	$x = 1$	$\dfrac{1}{1-2} \overset{?}{>} 0$ $-1 \not> 0$
$(2, \infty)$	$x = 3$	$\dfrac{3}{3-2} \overset{?}{>} 0$ $3 > 0$

From this, you can see that the inequality is satisfied for the intervals $(-\infty, 0)$ and $(2, \infty)$. So, the solution set of the inequality is $(-\infty, 0) \cup (2, \infty)$, as shown in Figure 7.16.

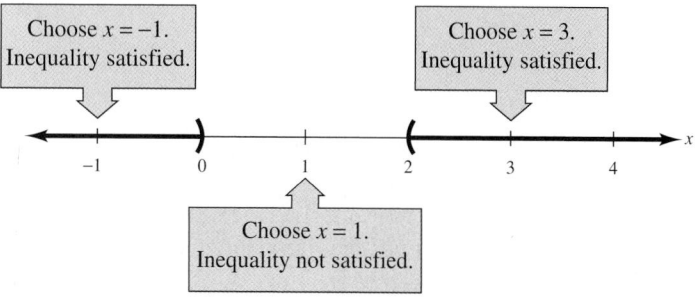

Choose $x = -1$. Inequality satisfied.

Choose $x = 3$. Inequality satisfied.

Choose $x = 1$. Inequality not satisfied.

Figure 7.16

When solving a rational inequality, you should begin by writing the inequality in general form, with the rational expression (as a single fraction) on the left and zero on the right. This is demonstrated in Example 3.

Example 3 ■ Solving a Rational Inequality

Solve the rational inequality $\dfrac{2x}{x+3} \leq 4$.

Algebraic Solution

Begin by writing the inequality in general form.

$$\frac{2x}{x+3} \leq 4 \qquad \text{Write original inequality.}$$

$$\frac{2x}{x+3} - 4 \leq 0 \qquad \text{Subtract 4 from each side.}$$

$$\frac{2x - 4(x+3)}{x+3} \leq 0 \qquad \text{Subtract fractions.}$$

$$\frac{-2x - 12}{x+3} \leq 0 \qquad \text{Simplify.}$$

$$\frac{-2(x+6)}{x+3} \leq 0 \qquad \text{Factor.}$$

From the factored general form, you can see that the critical numbers are $x = -6$ and $x = -3$. This implies that the test intervals are $(-\infty, -6)$, $(-6, -3)$, and $(-3, \infty)$. To test an interval, choose a convenient number in the interval and determine if the number satisfies the inequality.

Test interval	Representative x-value	Is inequality satisfied?
$(-\infty, -6)$	$x = -7$	$\dfrac{-2(-7+6)}{-7+3} \overset{?}{\leq} 0$ $-\frac{1}{2} \leq 0$
$(-6, -3)$	$x = -4$	$\dfrac{-2(-4+6)}{-4+3} \overset{?}{\leq} 0$ $4 \not\leq 0$
$(-3, \infty)$	$x = 0$	$\dfrac{-2(0+6)}{0+3} \overset{?}{\leq} 0$ $-4 \leq 0$

You can see that the inequality is satisfied for the intervals $(-\infty, -6]$ and $(-3, \infty)$. Because the rational expression is undefined when $x = -3$, that value is not included in the solution set. So, the solution set of the inequality is $(-\infty, -6] \cup (-3, \infty)$, as shown in Figure 7.17.

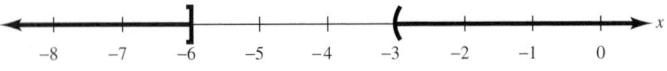

Figure 7.17

Graphical Solution

Use a graphing utility to graph

$$y_1 = \frac{2x}{x+3} \quad \text{and} \quad y_2 = 4$$

in the same viewing window. In Figure 7.18, you can see that the graphs appear to intersect at the point $(-6, 4)$. Use the *intersect* feature of the graphing utility to confirm this. The graph of y_1 lies below the graph of y_2 in the intervals $(-\infty, -6]$ and $(-3, \infty)$. So, you can graphically approximate the solution set to be all real numbers less than or equal to -6 *or* all real numbers greater than -3. That is,

$$(-\infty, -6] \cup (-3, \infty). \qquad \text{Solution set}$$

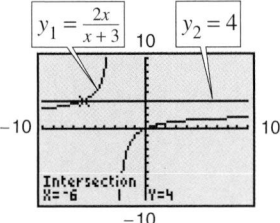

Figure 7.18

7.7 ■ Exercises

Developing Skills

In Exercises 1 and 2, determine whether the x-values are solutions of the inequality.

1. $\dfrac{2}{3-x} \le 0$ (a) $x = 3$ (b) $x = 4$

                (c) $x = -4$ (d) $x = -\frac{1}{3}$

2. $\dfrac{x-2}{x+1} > 0$ (a) $x = 0$ (b) $x = -1$

                (c) $x = -4$ (d) $x = -\frac{3}{2}$

In Exercises 3–6, find the critical numbers of the rational expression and plot them on the real number line.

3. $\dfrac{5}{x-3}$ **4.** $\dfrac{-6}{x+2}$

5. $\dfrac{2x}{x+5}$ **6.** $\dfrac{x-2}{x-10}$

In Exercises 7–24, solve the inequality and sketch the graph of the solution on the real number line.

7. $\dfrac{5}{x-3} > 0$ **8.** $\dfrac{3}{4-x} > 0$

9. $\dfrac{-5}{x-3} > 0$ **10.** $\dfrac{-3}{4-x} > 0$

11. $\dfrac{4}{x-3} < 0$ **12.** $\dfrac{5}{2-x} < 0$

13. $\dfrac{x}{x-4} \le 0$ **14.** $\dfrac{z-1}{z+3} < 0$

15. $\dfrac{y-3}{2y-11} \ge 0$ **16.** $\dfrac{x+5}{3x+2} \ge 0$

17. $\dfrac{3(u-3)}{u+1} < 0$ **18.** $\dfrac{2(4-t)}{4+t} > 0$

19. $\dfrac{6}{x-4} > 2$ **20.** $\dfrac{1}{x+2} > -3$

21. $\dfrac{x}{x-3} \le 2$ **22.** $\dfrac{x+4}{x-5} \ge 10$

23. $\dfrac{x-5}{x^2-6x+9} \le 0$ **24.** $\dfrac{x+1}{x^2+10x+25} \ge 0$

In Exercises 25–32, use a graphing utility to solve the rational inequality. Verify your results algebraically.

25. $\dfrac{1}{x} - x > 0$ **26.** $\dfrac{1}{x} - 4 < 0$

27. $\dfrac{x+6}{x+1} - 2 < 0$ **28.** $\dfrac{x+12}{x+2} - 3 \ge 0$

29. $\dfrac{6x}{x+5} < 2$ **30.** $x + \dfrac{1}{x} > 3$

31. $\dfrac{3x-4}{x-4} < -5$ **32.** $4 - \dfrac{1}{x^2} > 1$

Graphical Analysis In Exercises 33–36, use a graphing utility to graph the function. Use the graph to approximate the values of x that satisfy the specified inequalities.

Function	Inequalities	
33. $f(x) = \dfrac{3x}{x-2}$	(a) $f(x) \le 0$	(b) $f(x) \ge 6$
34. $f(x) = \dfrac{2(x-2)}{x+1}$	(a) $f(x) \le 0$	(b) $f(x) \ge 8$
35. $f(x) = \dfrac{2x^2}{x^2+4}$	(a) $f(x) \ge 1$	(b) $f(x) \le 2$
36. $f(x) = \dfrac{5x}{x^2+4}$	(a) $f(x) \ge 1$	(b) $f(x) \ge 0$

Solving Problems

37. *Average Speed* During the first 144 miles of a 260-mile trip, you travel at an average speed of r miles per hour. During the last part of the trip, you increase your average speed by 10 miles per hour. Find the interval for the two average speeds if you want to make the trip in less than 5 hours. (Assume that the speed limit is 65 miles per hour.)

38. *Average Cost* The cost C of producing x units of a product is $C = 3000 + 0.75x$, $x > 0$.

   (a) Write the average cost $\overline{C}$ as a function of x.

   (b) Use a graphing utility to graph the average cost function in part (a).

   (c) How many units must be produced if the average cost per unit is to be less than $2?

39. *Data Analysis*   The temperature T (in degrees Fahrenheit) of a metal in a laboratory experiment was recorded every 2 minutes for a period of 16 minutes. The table gives the experimental data, where t is the time in minutes.

t	0	2	4	6	8	10	12	14	16
T	250	290	338	410	498	560	530	370	160

A model for this data is

$$T = \frac{244.20 - 13.23t}{1 - 0.13t + 0.005t^2}.$$

(a) Use a graphing utility to plot the data and graph the model in the same viewing window.

(b) Use the graph to approximate the times when the temperature was at least 400°F.

Explaining Concepts

40. Explain the change in an inequality when each side is multiplied by a negative real number.

41. Define the term *critical number* and explain the use of critical numbers in solving rational inequalities.

42. Give a verbal description of the intervals $(-\infty, 5] \cup (10, \infty)$.

Ongoing Review

In Exercises 43–52, solve the inequality.

43. $3x + 15 \geq 0$

44. $5 + 2x < 3 - x$

45. $4(2 - 7x) > 10x - 1$

46. $2(4 + 3x) \leq 7x - 4$

47. $|x - 6| < 3$

48. $|x - 8| > 5$

49. $|x + 9| < 7$

50. $|x + 5| > 10$

51. $x^2 + 7x \geq 0$

52. $4x + x^2 > 0$

53. *Period of a Pendulum*   The period T in seconds of a pendulum is given by

$$T = 2\pi \sqrt{\frac{L}{32}}$$

where L is the length of the pendulum in feet. How long is the pendulum of a mantle clock with a period of 0.8 second?

54. *Cost*   Three people decide to share the cost of a yacht. By bringing in an additional partner, they can reduce the cost for each by $4000. What is the total cost of the yacht?

Looking Further

The cost of producing x units of a product is

$$C = 8000 + 0.6x, \quad x > 0.$$

(a) Write the average cost $\overline{C}$ as a function of x.

(b) Find the number of units that must be produced if $\overline{C} < 3$.

(c) Find the number of units that must be produced if $\overline{C} < 2$.

(d) Find the number of units that must be produced if $\overline{C} < 1$.

(e) Find the number of units that must be produced if $\overline{C} < 0.50$.

(f) Use a graphing utility to graph the average cost function.

(g) Determine the horizontal asymptote of the graph in part (f).

(h) Interpret what the horizontal asymptote in part (g) represents in the context of the problem. Describe how this result is related to the result in part (e).

Chapter Summary: Key Concepts

7.1 ■ Simplifying rational expressions

Let u, v, and w represent numbers, variables, or algebraic expressions such that $v \neq 0$ and $w \neq 0$. Then the following is valid.

$$\frac{u\cancel{w}}{v\cancel{w}} = \frac{u}{v}$$

7.2 ■ Multiplying rational expressions

Let u, v, w, and z represent numbers, variables, or algebraic expressions such that $v \neq 0$ and $z \neq 0$. Then the product of u/v and w/z is

$$\frac{u}{v} \cdot \frac{w}{z} = \frac{uw}{vz}.$$

7.2 ■ Dividing rational expressions

Let u, v, w, and z represent numbers, variables, or algebraic expressions such that $v \neq 0$, $w \neq 0$, and $z \neq 0$. Then the quotient of u/v and w/z is

$$\frac{u}{v} \div \frac{w}{z} = \frac{u}{v} \cdot \frac{z}{w} = \frac{uz}{vw}.$$

7.3 ■ Adding or subtracting with like denominators

If u, v, and w are real numbers, variables, or algebraic expressions, and $w \neq 0$, the rules below are valid.

1. $\dfrac{u}{w} + \dfrac{v}{w} = \dfrac{u+v}{w}$ 2. $\dfrac{u}{w} - \dfrac{v}{w} = \dfrac{u-v}{w}$

7.3 ■ Adding or subtracting with unlike denominators

Rewrite the rational expressions with like denominators by finding the least common denominator. Then add or subtract as with like denominators.

7.4 ■ Dividing a polynomial by a monomial

Let u, v, and w be real numbers, variables, or algebraic expressions such that $w \neq 0$.

1. $\dfrac{u+v}{w} = \dfrac{u}{w} + \dfrac{v}{w}$

2. $\dfrac{u-v}{w} = \dfrac{u}{w} - \dfrac{v}{w}$

7.5 ■ Solving rational equations

1. Determine the domain of each of the fractions in the equation.

2. Obtain an equivalent equation by multiplying each side of the equation by the least common denominator of all the fractions in the equation.

3. Solve the resulting equation.

4. Check your solution(s) in the original equation.

7.6 ■ Guidelines for graphing rational functions

Let $f(x) = p(x)/q(x)$, where $p(x)$ and $q(x)$ have no common factors.

1. Find and plot the y-intercept (if any) by evaluating $f(0)$.

2. Set the numerator equal to zero and solve the equation for x. The real solutions represent the x-intercepts of the graph. Plot these intercepts.

3. Set the denominator equal to zero and solve the equation for x. The real solutions represent the vertical asymptotes. Sketch these asymptotes.

4. Find and sketch the horizontal asymptote of the graph.

5. Plot at least one point between and one point beyond each x-intercept and vertical asymptote.

6. Use smooth curves to complete the graph between and beyond the vertical asymptotes.

7.7 ■ Test intervals for a rational inequality

1. Write the inequality in general form, with a single rational expression on the left and zero on the right.

2. Find all real zeros of the numerator of the rational expression and all real zeros of the denominator of the rational expression. These zeros are the critical numbers of the rational expression.

3. Then use the critical numbers, in increasing order, to determine the test intervals for the rational expression.

4. Choose a representative x-value in each test interval and evaluate the rational expression at that value. If the value of the rational expression is negative, it will have negative values for every x-value in the interval. If the value of the rational expression is positive, it will have positive values for every x-value in the interval.

Review Exercises

Reviewing Skills

7.1 ■ *How to find the domain of a rational function*

In Exercises 1–4, find the domain of the rational function.

1. $f(x) = \dfrac{3x}{x - 8}$

2. $f(t) = \dfrac{t + 4}{t + 12}$

3. $f(x) = \dfrac{x}{x^2 - 7x + 6}$

4. $f(x) = \dfrac{x - 12}{x(x^2 - 16)}$

■ *How to simplify rational expressions*

In Exercises 5–12, simplify the rational expression.

5. $\dfrac{6x^4y^2}{15xy^2}$

6. $\dfrac{2(y^3z)^2}{28(yz^2)^2}$

7. $\dfrac{5b - 15}{30b - 120}$

8. $\dfrac{4a}{10a^2 + 26a}$

9. $\dfrac{9x - 9y}{y - x}$

10. $\dfrac{x + 3}{x^2 - x - 12}$

11. $\dfrac{x^2 - 5x}{2x^2 - 50}$

12. $\dfrac{x^2 + 3x + 9}{x^3 - 27}$

7.2 ■ *How to multiply rational expressions*
    ■ *How to divide rational expressions*

In Exercises 13–20, multiply or divide the rational expressions and simplify.

13. $\dfrac{7}{8} \cdot \dfrac{2x}{y} \cdot \dfrac{y^2}{14x^2}$

14. $\dfrac{15(x^2y)^3}{3y^3} \cdot \dfrac{12y}{x}$

15. $\dfrac{60z}{z + 6} \cdot \dfrac{z^2 - 36}{5}$

16. $\dfrac{u}{u - 3} \cdot \dfrac{3u - u^2}{4u^2}$

17. $25y^2 \div \dfrac{xy}{5}$

18. $\dfrac{6}{z^2} \div 4z^2$

19. $\dfrac{x^2 - 7x}{x + 1} \div \dfrac{x^2 - 14x + 49}{x^2 - 1}$

20. $\dfrac{x^2 - x}{x + 1} \div \dfrac{5x - 5}{x^2 + 6x + 5}$

■ *How to simplify complex fractions*

In Exercises 21–24, simplify the complex fraction.

21. $\dfrac{\left(\dfrac{6}{x}\right)}{\left(\dfrac{2}{x^3}\right)}$

22. $\dfrac{xy}{\left(\dfrac{5x^2}{2y}\right)}$

23. $\dfrac{\left(\dfrac{6x^2}{x^2 + 2x - 35}\right)}{\left(\dfrac{x^3}{x^2 - 25}\right)}$

24. $\dfrac{\left[\dfrac{24 - 18x}{(2 - x)^2}\right]}{\left(\dfrac{60 - 45x}{x^2 - 4x + 4}\right)}$

7.3 ■ *How to add or subtract rational expressions with like denominators*
    ■ *How to add or subtract rational expressions with unlike denominators*

In Exercises 25–38, add or subtract the rational expressions and simplify.

25. $\dfrac{4a}{9} - \dfrac{11a}{9}$

26. $\dfrac{12x}{5} + \dfrac{2x}{5}$

27. $\dfrac{x + 7}{3x - 1} - \dfrac{2x + 3}{3x - 1}$

28. $\dfrac{2(3y + 4)}{2y + 1} + \dfrac{3 - y}{2y + 1}$

29. $\dfrac{15}{16x} - \dfrac{5}{24x} - 1$

30. $-\dfrac{3y}{8x} + \dfrac{7y}{6x} - \dfrac{1y}{12x}$

31. $\dfrac{1}{x + 5} + \dfrac{3}{x - 12}$

32. $\dfrac{2}{x - 10} + \dfrac{3}{4 - x}$

33. $5x + \dfrac{2}{x - 3} - \dfrac{3}{x + 2}$

34. $4 - \dfrac{4x}{x + 6} + \dfrac{7}{x - 5}$

35. $\dfrac{6}{x} - \dfrac{6x - 1}{x^2 + 4}$

36. $\dfrac{5}{x + 2} + \dfrac{25 - x}{x^2 - 3x - 10}$

37. $\dfrac{5}{x + 3} - \dfrac{4x}{(x + 3)^2} - \dfrac{1}{x - 3}$

38. $\dfrac{8}{y} - \dfrac{3}{y + 5} + \dfrac{4}{y - 2}$

■ *How to simplify complex fractions*

In Exercises 39–42, simplify the complex fraction.

39. $\dfrac{3t}{\left(5 - \dfrac{2}{t}\right)}$

40. $\dfrac{\left(x - 3 + \dfrac{2}{x}\right)}{\left(1 - \dfrac{2}{x}\right)}$

41. $\dfrac{\left(\dfrac{1}{a^2 - 16} - \dfrac{1}{a}\right)}{\left(\dfrac{1}{a^2 + 4a} + 4\right)}$

42. $\dfrac{\left(\dfrac{1}{x^2} - \dfrac{1}{y^2}\right)}{\left(\dfrac{1}{x} + \dfrac{1}{y}\right)}$

7.4 ■ *How to divide polynomials by monomials*

In Exercises 43–46, perform the division and simplify.

43. $\dfrac{4x^3 - x}{2x}$

44. $\dfrac{12y^2 + 7y}{2y}$

45. $\dfrac{5a^6b^4 - 2a^4b^3 + a^3b^2}{a^3b^2}$

46. $\dfrac{10x^7y^8 - 40x^6y^5 - 25x^3y^3}{5xy^2}$

■ *How to use long division to divide polynomials by polynomials*

In Exercises 47–50, use long division to divide.

47. $\dfrac{6x^3 + x^2 - 4x + 2}{3x - 1}$

48. $\dfrac{4x^4 - x^3 - 7x^2 + 18x}{x - 2}$

49. $\dfrac{x^4 - 3x^2 + 2}{x^2 - 1}$

50. $\dfrac{3x^6}{x^2 - 1}$

■ *How to use synthetic division to divide polynomials by polynomials of the form x − k*

In Exercises 51 and 52, use synthetic division to divide.

51. $(x^4 - 3x^2 - 25) \div (x - 3)$

52. $(2x^3 + 5x - 2) \div \left(x + \frac{1}{2}\right)$

■ *How to use synthetic division to factor polynomials*

In Exercises 53 and 54, use synthetic division to factor the polynomial completely given one of its factors.

Polynomial	*Factor*
53. $x^3 + 2x^2 - 5x - 6$	$x - 2$
54. $2x^3 + x^2 - 2x - 1$	$x + 1$

7.5 ■ *How to solve rational equations containing constant denominators*

■ *How to solve rational equations containing variable denominators*

In Exercises 55–66, solve the rational equation.

55. $\dfrac{3x}{8} = -15$

56. $\dfrac{t + 1}{8} = \dfrac{1}{2}$

57. $\dfrac{1}{3y - 4} = \dfrac{6}{4(y + 1)}$

58. $\dfrac{2}{y} - \dfrac{1}{3y} = \dfrac{1}{3}$

59. $r = 2 + \dfrac{24}{r}$

60. $\dfrac{2}{x} - \dfrac{x}{6} = \dfrac{2}{3}$

61. $\dfrac{3}{y + 1} - \dfrac{8}{y} = 1$

62. $\dfrac{2x}{x + 3} - \dfrac{3}{x} = 0$

63. $\dfrac{12}{x^2 + x - 12} - \dfrac{1}{x - 3} = -1$

64. $\dfrac{3}{x - 1} + \dfrac{6}{x^2 - 3x + 2} = 2$

65. $\dfrac{5}{x^2 - 4} - \dfrac{6}{x - 2} = -5$

66. $\dfrac{3}{x^2 - 9} + \dfrac{4}{x + 3} = 1$

7.6 ■ *How to use a table of values to sketch graphs of rational functions*

In Exercises 67 and 68, use point plotting to sketch the graph of the rational function.

67. $f(x) = \dfrac{x - 3}{x + 1}$

68. $f(x) = \dfrac{x + 6}{x - 4}$

■ *How to determine horizontal and vertical asymptotes of rational functions*

In Exercises 69–72, identify the horizontal and vertical asymptotes of the function.

69. $f(x) = \dfrac{5}{x - 6}$

70. $f(x) = \dfrac{6}{x + 5}$

71. $f(x) = \dfrac{6x}{x - 5}$

72. $f(x) = \dfrac{2x}{x + 6}$

In Exercises 73–76, find the domain of the function and identify any horizontal and vertical asymptotes.

73. $f(x) = \dfrac{x}{x + 4}$

74. $h(x) = \dfrac{x}{x - 2}$

75. $g(x) = \dfrac{2x^2 - 2}{x^2 - 9}$

76. $f(x) = \dfrac{x}{x^2 - 16}$

■ *How to use asymptotes and intercepts to sketch graphs of rational functions*

In Exercises 77–88, sketch the graph of the function. Use a graphing utility to confirm the graph.

77. $g(x) = \dfrac{2 + x}{1 - x}$

78. $h(x) = \dfrac{x - 3}{x - 2}$

79. $f(x) = \dfrac{x}{x^2 + 1}$

80. $f(x) = \dfrac{2x}{x^2 + 4}$

81. $f(x) = \dfrac{3x + 6}{x - 2}$

82. $s(x) = \dfrac{2x - 6}{x + 4}$

83. $h(x) = \dfrac{4}{(x - 1)^2}$

84. $g(x) = \dfrac{-2}{(x + 3)^2}$

85. $f(x) = -\dfrac{5}{x^2}$

86. $f(x) = \dfrac{4}{x}$

87. $y = \dfrac{x}{x^2 - 1}$

88. $y = \dfrac{2x}{x^2 - 4}$

7.7 ■ *How to use test intervals to solve rational inequalities*

In Exercises 89–92, determine the intervals on which the rational expression is entirely negative or entirely positive.

89. $\dfrac{x + 4}{x - 1}$

90. $\dfrac{x - 2}{x + 3}$

91. $\dfrac{x}{2x + 7}$

92. $\dfrac{x}{3x - 5}$

■ *How to solve rational inequalities*

In Exercises 93–96, solve the inequality and graph the solution on the real number line.

93. $\dfrac{x}{2x - 7} \geq 0$

94. $\dfrac{2x - 9}{x - 1} \leq 0$

95. $\dfrac{x}{x + 6} + 2 < 0$

96. $\dfrac{3x + 1}{x - 2} > 4$

Solving Problems

97. *Batting Average* In this year's playing season, a baseball player has been at bat 150 times and has hit the ball safely 45 times. So, the "batting average" for the player is $45/150 = 0.300$. How many consecutive times must the player hit safely to obtain a batting average of 0.400?

98. *Average Speed* You drive 56 miles on a service call for your company. On the return trip, which takes 10 minutes less than the original trip, your average speed is 8 miles per hour faster. What is your average speed on the return trip?

99. *Work Rate* Suppose that in 12 minutes your supervisor can complete a task that you require 15 minutes to complete. Determine the time required to complete the task if you work together.

100. *Partnership Costs* A group of people agree to share equally in the cost of a $60,000 piece of machinery. If they find two more people to join the group, each person's share of the cost will decrease by $5000. How many people are presently in the group?

101. *Average Cost* A business produces x units of a product at a cost of $C = 0.5x + 500$.

(a) Write the average cost $\overline{C}$ as a function of x.

(b) Find the horizontal asymptote and explain the meaning in the context of the problem.

102. *Population of Fish* The Parks and Wildlife Commission introduces 80,000 fish into a large lake. The population of the fish (in thousands) is

$$N = \dfrac{20(4 + 3t)}{1 + 0.05t}, \quad t \geq 0$$

where t is the time in years.

(a) Find the populations when $t = 5$, $t = 10$, and $t = 25$.

(b) Find the horizontal asymptote and explain its meaning in the context of the problem.

103. *Cost* The cost (in millions of dollars) for a government agency to seize $p\%$ of a certain illegal drug as it enters the country is

$$C = \dfrac{528p}{100 - p}, \quad 0 \leq p < 100.$$

(a) Find the cost of seizing 25% and 75%.

(b) According to this model, would it be possible to seize 100% of the drug?

(c) Find the percent of drugs seized if $C < 1500$.

104. *Average Cost* The cost of producing x units of a product is $C = 100,000 + 0.9x$.

(a) Write the average cost per unit $\overline{C}$ as a function of x.

(b) Find the number of units that must be produced if $\overline{C} < 2$.

7 Chapter Test

Take this test as you would take a test in class. After you are done, check your work against the answers given in the back of the book.

1. Find the domain of the rational function $f(x) = \dfrac{3x}{x^2 - 25}$.

2. Simplify the rational expression $\dfrac{2 - x}{3x - 6}$.

In Exercises 3–10, perform the operation(s) and simplify.

3. $\dfrac{y^2 + 8y + 16}{2(y - 2)} \cdot \dfrac{8y - 16}{(y + 4)^3}$

4. $(4x^2 - 9) \cdot \dfrac{2x + 3}{2x^2 - x - 3}$

5. $\dfrac{x^2 + 4x}{2x^2 - 7x + 3} \div \dfrac{x^2 - 16}{x - 3}$

6. $\dfrac{\left(\dfrac{3x}{x + 2}\right)}{\left(\dfrac{12}{x^3 + 2x^2}\right)}$

7. $\dfrac{\left(9x - \dfrac{1}{x}\right)}{\left(\dfrac{1}{x} - 3\right)}$

8. $2x + \dfrac{1 - 4x^2}{x + 1}$

9. $\dfrac{5x}{x + 2} - \dfrac{2}{x^2 - x - 6}$

10. $\dfrac{3}{x} - \dfrac{5}{x^2} + \dfrac{2x}{x^2 + 2x + 1}$

In Exercises 11 and 12, perform the division.

11. $\dfrac{24a^7 + 42a^4 - 6a^3}{6a^2}$

12. $\dfrac{t^4 + t^2 - 6t}{t^2 - 2}$

13. Use synthetic division to divide $2x^4 - 15x^2 - 7$ by $x - 3$.

14. Sketch the graph of each function.

(a) $f(x) = \dfrac{3}{x - 3}$ (b) $g(x) = \dfrac{3x}{x - 3}$

In Exercises 15 and 16, solve the rational equation.

15. $\dfrac{2}{x + 5} - \dfrac{3}{x + 3} = \dfrac{1}{x}$

16. $\dfrac{1}{x + 1} + \dfrac{1}{x - 1} = \dfrac{2}{x^2 - 1}$

In Exercises 17 and 18, solve the inequality and sketch the solution.

17. $\dfrac{x - 5}{x + 8} \geq 0$

18. $\dfrac{3u + 2}{u - 3} \leq 2$

19. One painter works $1\frac{1}{2}$ times as fast as another. Find their individual rates for painting a room if together it takes them 4 hours.

Cumulative Test: Chapters 1–7

Take this test as you would take a test in class. After you are done, check your work against the answers given in the back of the book.

In Exercises 1 and 2, simplify the expression.

1. $5(x + 2) - 4(2x - 3)$

2. $0.12x + 0.05(2000 - 2x)$

In Exercises 3 and 4, use the function to find and simplify the expression for $f(a + 2)$.

3. $f(x) = x^2 - 3$

4. $f(x) = \dfrac{3}{x + 5}$

In Exercises 5 and 6, simplify the rational expression.

5. $\dfrac{-16x^2}{12x}$

6. $\dfrac{6u^5 v^{-3}}{7uv^3}$

In Exercises 7–9, perform the operation and simplify. (Assume that all variables are positive.)

7. $\left(\sqrt{x} + 3\right)\left(\sqrt{x} - 3\right)$

8. $\sqrt{u}\left(\sqrt{20} - \sqrt{5}\right)$

9. $\left(2\sqrt{t} + 3\right)^2$

In Exercises 10 and 11, solve the system of equations.

10. $\begin{cases} 5x - 2y = -25 \\ -3x + 7y = 44 \end{cases}$

11. $\begin{cases} 3x - 2y + z = 1 \\ x + 5y - 6z = 4 \\ 4x - 3y + 2z = 2 \end{cases}$

In Exercises 12–14, solve the equation.

12. $x + \dfrac{4}{x} = 4$

13. $\sqrt{x + 10} = x - 2$

14. $(x - 5)^2 + 50 = 0$

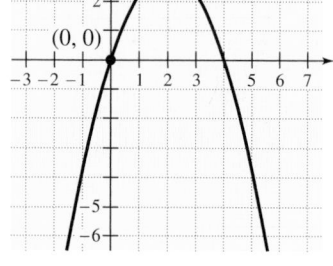

Figure for 16

15. Use a graphing utility to graph the equation $y = x^2 - 6x - 8$. Use the graph to approximate any x-intercepts. Verify algebraically.

16. Find an equation of the parabola shown in the figure.

17. Sketch the graph of the rational function $f(x) = \dfrac{4}{x - 2}$.

18. You can mow a lawn in 4 hours, and your friend can mow it in 5 hours. What fractional part of the lawn can each of you mow in 1 hour? How long will it take both of you to mow the lawn?

19. The volume of the box in the figure is given by $h^4 - 10h^3 + 21h^2$. The height of the box is h and the width is $h - 7$. Find the length of the box.

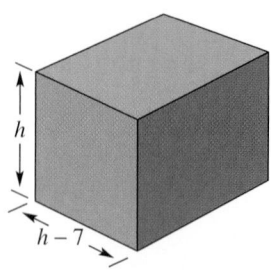

Figure for 19

More About Functions and Relations

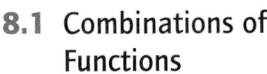

8.1 Combinations of Functions

8.2 Inverse Functions

8.3 Variation and Mathematical Models

8.4 Polynomial Functions and Their Graphs

8.5 Circles

8.6 Ellipses and Hyperbolas

8.7 Parabolas

8.8 Nonlinear Systems of Equations

THE BIG PICTURE

In this chapter you will learn how to:

- find combinations and compositions of functions.

- find inverse functions.

- solve real-life problems involving variation.

- sketch and identify the characteristics of graphs of polynomial functions.

- graph and find equations of circles.

- graph and find equations of ellipses and hyperbolas.

- graph and find equations of parabolas.

- solve nonlinear systems of equations.

Key Terms

As you encounter each new vocabulary term in this chapter, add the term and its definition to your notebook glossary.

inverse function (p. 510)
one-to-one function (p. 515)
direct variation (p. 520)
constant of proportionality (p. 520)
inverse variation (p. 523)
combined variation (p. 525)
polynomial function (p. 531)
zeros (of a function) (p. 534)
circle (p. 540)
center (of a circle) (p. 540)
radius (p. 540)
standard form of the equation of a circle (pp. 541, 542)
ellipse (p. 546)
focus (of an ellipse) (p. 546)
vertices (of an ellipse) (p. 546)
major axis (p. 546)
center (of an ellipse) (p. 546)

minor axis (p. 546)
co-vertices (p. 546)
standard form of the equation of an ellipse (pp. 547, 549)
hyperbola (p. 551)
foci (of a hyperbola) (p. 551)
transverse axis (of a hyperbola) (p. 551)
standard form of the equation of a hyperbola (pp. 551, 555)
branch (of a hyperbola) (p. 551)
parabola (p. 560)
directrix (p. 560)
focus (of a parabola) (p. 560)
vertex (of a parabola) (p. 560)
axis (of a parabola) (p. 560)
standard form of the equation of a parabola (p. 560)
nonlinear system of equations (p. 566)

Additional text-specific resources are available to help you do well in this course. See page xvi for details.

499

8.1 Combinations of Functions

Why you should learn it:

Combinations of functions can be used to model and solve real-life problems. For instance, in Exercise 80 on page 508 you will use combinations of functions to analyze the cost of operating an automobile.

Arithmetic Combinations of Functions

In Section 2.4, you were introduced to the concept of a function. In this section and in the next section, you will learn more about functions.

Just as two real numbers can be combined by the operations of addition, subtraction, multiplication, and division to form other real numbers, two functions can be combined to form new functions. For example, if

$$f(x) = 3x - 1 \quad \text{and} \quad g(x) = x^2 - 4$$

you can form the sum, difference, product, and quotient of f and g as follows.

$$f(x) + g(x) = (3x - 1) + (x^2 - 4) = x^2 + 3x - 5 \qquad \text{Sum}$$
$$f(x) - g(x) = (3x - 1) - (x^2 - 4) = -x^2 + 3x + 3 \qquad \text{Difference}$$
$$f(x)g(x) = (3x - 1)(x^2 - 4) = 3x^3 - x^2 - 12x + 4 \qquad \text{Product}$$
$$\frac{f(x)}{g(x)} = \frac{3x - 1}{x^2 - 4}, \quad x \neq \pm 2 \qquad \text{Quotient}$$

The domain of an arithmetic combination of functions f and g consists of all real numbers that are common to the domains of f and g. In the case of the quotient $f(x)/g(x)$, there is the further restriction that $g(x) \neq 0$.

EXPLORATION

Enter the functions

$$f(x) = 3x - 1 \quad \text{and} \quad g(x) = x^2 - 4$$

as y_1 and y_2 in your graphing utility, and let $y_3 = y_1 + y_2$. Algebraically find $f(x) + g(x)$ and enter the result as y_4 in your graphing utility. Compare the graphs of y_3 and y_4. Are y_3 and y_4 equivalent? Do you think they should be? Repeat this process using the difference, product, and quotient of the two functions.

Definitions of Sum, Difference, Product, and Quotient of Functions

Let f and g be two functions with overlapping domains. Then, for all x common to both domains, the **sum, difference, product,** and **quotient** of f and g are defined as follows.

1. *Sum:* $(f + g)(x) = f(x) + g(x)$

2. *Difference:* $(f - g)(x) = f(x) - g(x)$

3. *Product:* $(fg)(x) = f(x) \cdot g(x)$

4. *Quotient:* $\left(\dfrac{f}{g}\right)(x) = \dfrac{f(x)}{g(x)}, \quad g(x) \neq 0$

Example 1 ■ Finding the Sum and Difference of Two Functions

Given the functions $f(x) = 4x - 3$ and $g(x) = 2x^2 + x$, find each combination.

(a) $(f + g)(x)$ (b) $(f - g)(x)$

Then evaluate each of these combinations when $x = 3$.

Solution

(a) The sum of the functions f and g is given by

$$(f + g)(x) = f(x) + g(x) = (4x - 3) + (2x^2 + x) = 2x^2 + 5x - 3.$$

When $x = 3$, the value of this sum is

$$(f + g)(3) = 2(3)^2 + 5(3) - 3 = 30.$$

(b) The difference of the functions f and g is given by

$$(f - g)(x) = f(x) - g(x) = (4x - 3) - (2x^2 + x) = -2x^2 + 3x - 3.$$

When $x = 3$, the value of this difference is

$$(f - g)(3) = -2(3)^2 + 3(3) - 3 = -12.$$

NOTE   In Example 1, $(f + g)(3)$ also could have been evaluated as

$$(f + g)(3) = f(3) + g(3) = [4(3) - 3] + [2(3)^2 + 3] = 9 + 21 = 30.$$

Similarly, $(f - g)(3)$ also could have been evaluated as

$$(f - g)(3) = f(3) - g(3) = [4(3) - 3] - [2(3)^2 + 3] = 9 - 21 = -12.$$

In Example 1, both f and g have domains that consist of all real numbers. So, the domains of both $(f + g)$ and $(f - g)$ are also the set of all real numbers. In the next example, the domain of the sum of f and g is smaller than the domains of f and g.

Example 2 ■ Finding the Domain of the Sum of Two Functions

Given the functions $f(x) = \sqrt{x + 1}$ and $g(x) = \sqrt{1 - x}$, find the sum of f and g. Then find the domain of $(f + g)$.

Solution

The sum of the functions f and g is given by

$$(f + g)(x) = f(x) + g(x) = \sqrt{x + 1} + \sqrt{1 - x}.$$

Now, because f is defined for $x + 1 \geq 0$ and g is defined for $1 - x \geq 0$, it follows that the domains of f and g are as follows.

Domain of $f = \{x: x \geq -1\}$

Domain of $g = \{x: x \leq 1\}$

The set that is common to both of these domains is the closed interval $[-1, 1]$, as shown in Figure 8.1. So, the domain of $(f + g)$ is

Domain of $(f + g) = \{x: -1 \leq x \leq 1\}.$

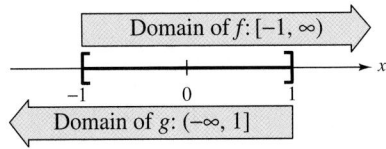

Figure 8.1

Example 3 ■ Finding the Product of Two Functions

Given the functions $f(x) = x^2$ and $g(x) = x - 3$, find the product of f and g. Then evaluate the product when $x = 4$ and when $x = -2$.

Solution

The product of the functions f and g is given by

$$(fg)(x) = f(x)g(x)$$
$$= (x^2)(x - 3)$$
$$= x^3 - 3x^2.$$

When $x = 4$, the value of this product is

$$(fg)(4) = 4^3 - 3(4)^2 = 16.$$

When $x = -2$, the value of this product is

$$(fg)(-2) = (-2)^3 - 3(-2)^2 = -20.$$

In Example 3, the domains of both f and g are the set of all real numbers. So, the domain of (fg) is also the set of all real numbers.

Example 4 ■ Finding the Quotient of Two Functions

Given the functions $f(x) = x^2$ and $g(x) = x - 3$, find the quotient of f and g. Then evaluate the quotient when $x = 5$ and when $x = -2$.

Solution

The quotient of the functions f and g is given by

$$\left(\frac{f}{g}\right)(x) = \frac{f(x)}{g(x)}$$
$$= \frac{x^2}{x - 3}, \quad x \neq 3.$$

(Note that the domain of the quotient has to be restricted to exclude the number 3.)

When $x = 5$, the value of this quotient is

$$\left(\frac{f}{g}\right)(5) = \frac{5^2}{5 - 3}$$
$$= \frac{25}{2}.$$

When $x = -2$, the value of this quotient is

$$\left(\frac{f}{g}\right)(-2) = \frac{(-2)^2}{-2 - 3}$$
$$= -\frac{4}{5}.$$

Composition of Functions

Another way of combining two functions is to form the **composition** of one with the other. For instance, if $f(x) = 2x^2$ and $g(x) = x - 1$, the composition of f with g is given by

$$f(g(x)) = f(x - 1)$$
$$= 2(x - 1)^2.$$

This composition is denoted as $(f \circ g)$.

Definition of Composition of Two Functions

The **composition** of the two functions f and g is given by

$$(f \circ g)(x) = f(g(x)).$$

The domain of $(f \circ g)$ is the set of all x in the domain of g such that $g(x)$ is in the domain of f (see Figure 8.2).

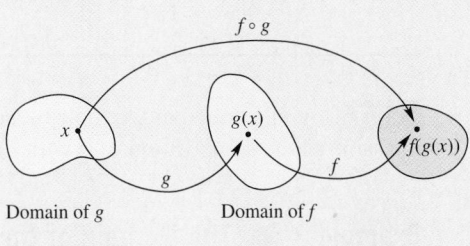

Domain of g Domain of f

Figure 8.2

Example 5 ■ Forming the Composition of Two Functions

Given the functions $f(x) = 2x + 4$ and $g(x) = 3x - 1$, find the composition of f with g. Then find the value of the composition when $x = 1$ in two ways.

Solution

$$
\begin{aligned}
(f \circ g)(x) &= f(g(x)) &&\text{Definition of } f \circ g \\
&= f(3x - 1) &&\text{Definition of } g(x) \\
&= 2(3x - 1) + 4 &&\text{Definition of } f(x) \\
&= 6x - 2 + 4 &&\text{Distributive Property} \\
&= 6x + 2 &&\text{Simplify.}
\end{aligned}
$$

To evaluate this composition when $x = 1$, you can use either of the two methods described below. Note that you obtain the same value with either method.

(a) You can use the result above, $(f \circ g)(x) = 6x + 2$, and write

$$(f \circ g)(1) = 6(1) + 2 = 8.$$

(b) You can use the fact that $(f \circ g)(x) = f(g(x))$, and write

$$(f \circ g)(1) = f(3(1) - 1) = f(2) = 2(2) + 4 = 8.$$

Example 6 ■ Comparing the Compositions of Functions

Given the functions $f(x) = 2x - 3$ and $g(x) = x^2 + 1$, find each composition.

(a) $(f \circ g)(x)$ (b) $(g \circ f)(x)$

Solution

(a) $(f \circ g)(x) = f(g(x))$ Definition of $f \circ g$

$\qquad\qquad = f(x^2 + 1)$ Definition of $g(x)$

$\qquad\qquad = 2(x^2 + 1) - 3$ Definition of $f(x)$

$\qquad\qquad = 2x^2 + 2 - 3$ Distributive Property

$\qquad\qquad = 2x^2 - 1$ Simplify.

(b) $(g \circ f)(x) = g(f(x))$ Definition of $g \circ f$

$\qquad\qquad = g(2x - 3)$ Definition of $f(x)$

$\qquad\qquad = (2x - 3)^2 + 1$ Definition of $g(x)$

$\qquad\qquad = 4x^2 - 12x + 9 + 1$ Expand.

$\qquad\qquad = 4x^2 - 12x + 10$ Simplify.

Technology

You can use a graphing utility to graph the composition of two functions and determine the domain of the composition. For the functions

$$f(x) = x^2 - 4$$

and

$$g(x) = \sqrt{4 - x^2}$$

the composition of f with g is given by

$$(f \circ g)(x) = \left(\sqrt{4 - x^2}\right)^2 - 4.$$

Enter the functions into your graphing utility as follows.

$$y_1 = \sqrt{(4 - x^2)}$$
$$y_2 = y_1{}^2 - 4$$

Graph y_2 as shown below. Use the *trace* feature to determine that the x-coordinates of points on the graph extend from -2 to 2. So, the domain of

$$(f \circ g)(x) \text{ is } -2 \leq x \leq 2.$$

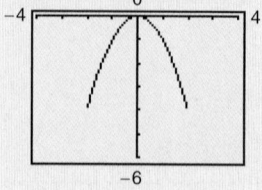

NOTE As illustrated in Example 6, the composition of f with g is generally *not* the same as the composition of g with f.

Example 7 ■ The Domain of the Composition of Two Functions

Given the functions

$$f(x) = \frac{1}{x - 2}$$

and

$$g(x) = \sqrt{x}$$

find the composition of f with g. Then find the domain of the composition.

Solution

The composition of f with g is given by

$$(f \circ g)(x) = f(g(x))$$
$$= f\left(\sqrt{x}\right)$$
$$= \frac{1}{\sqrt{x} - 2}.$$

Because $g(x) = \sqrt{x}$, the domain of g is the set of all nonnegative real numbers. You can see that if $x = 4$, the denominator of the composite function would be $\sqrt{4} - 2 = 0$. So, the domain of $(f \circ g)$ is the set of all nonnegative real numbers, except $x = 4$. In interval notation, you can write the domain as

$$\text{Domain of } (f \circ g) = \{x: x \geq 0 \text{ and } x \neq 4\}.$$

Application

In real-life applications, the order in which two procedures are performed will often affect the outcome. Example 8 illustrates this possibility.

Example 8 ■ Forming Composite Functions in Different Orders

The regular price of a certain new car is $15,800. The dealership advertises a factory rebate of $1500 *and* a 12% discount. Compare the sale price obtained by subtracting the rebate first and then taking the discount with the sale price obtained by taking the discount first and then subtracting the rebate.

Solution

Using function notation, the price after rebate $f(x)$ and the price after discount $g(x)$ can be represented by

$$f(x) = x - 1500 \qquad \text{Rebate of \$1500}$$
$$g(x) = 0.88x \qquad \text{Discount of 12\%}$$

where x is the price of the car.

Rebate First:

If you subtract the rebate first, the sale price is given by the composition of g with f.

$$
\begin{aligned}
g(f(15{,}800)) &= g(15{,}800 - 1500) & \text{Subtract rebate first.} \\
&= g(14{,}300) & \text{Simplify.} \\
&= 0.88 \cdot 14{,}300 = \$12{,}584 & \text{Take discount.}
\end{aligned}
$$

Discount First:

If you take the discount first, the sale price is given by the composition of f with g.

$$
\begin{aligned}
f(g(15{,}800)) &= f(0.88 \cdot 15{,}800) & \text{Take discount first.} \\
&= f(13{,}904) & \text{Simplify.} \\
&= 13{,}904 - 1500 = \$12{,}404 & \text{Subtract rebate.}
\end{aligned}
$$

So, given an option, you should take the discount first.

Collaborate!

Comparing the Compositions of Functions

You have seen that the composition of f with g is generally not the same as the composition of g with f. There are, however, special cases in which these two compositions are the same. For instance, let

$$f(x) = 2x \text{ and } g(x) = \tfrac{1}{2}x.$$

Show that for these two functions, the compositions $(f \circ g)$ and $(g \circ f)$ are the same. Can you and others in your group find other examples of functions f and g such that the compositions $(f \circ g)$ and $(g \circ f)$ are the same?

8.1 ■ Exercises

Developing Skills

In Exercises 1–10, use the functions f and g to find each combination.

(a) $(f + g)(x)$ (b) $(f - g)(x)$

(c) $(fg)(x)$ (d) $(f/g)(x)$

Then find the domain of each combination.

1. $f(x) = 2x$
$g(x) = x^2$

2. $f(x) = x + 1$
$g(x) = \sqrt{x}$

3. $f(x) = 4x - 3$
$g(x) = x^2 - 9$

4. $f(x) = x^3$
$g(x) = x^2 + 1$

5. $f(x) = \dfrac{1}{x}$

$g(x) = 5x$

6. $f(x) = \dfrac{1}{x + 2}$

$g(x) = \dfrac{1}{x - 2}$

7. $f(x) = \sqrt{x - 5}$

$g(x) = -\dfrac{3}{x}$

8. $f(x) = \dfrac{1}{x^2}$

$g(x) = \sqrt{x - 1}$

9. $f(x) = \sqrt{x + 4}$
$g(x) = \sqrt{4 - x}$

10. $f(x) = \sqrt{x}$
$g(x) = \sqrt{x^2 - 16}$

In Exercises 11–20, evaluate the combinations of the functions f and g at the indicated x-values. (If not possible, state the reason.)

11. $f(x) = x^2$, $g(x) = 2x + 3$
(a) $(f + g)(2)$ (b) $(f - g)(3)$

12. $f(x) = x^3$, $g(x) = x - 5$
(a) $(f + g)(-2)$ (b) $(f - g)(5)$

13. $f(x) = x^2 - 3x + 2$, $g(x) = 2x - 4$
(a) $(f + g)(-3)$ (b) $(f - g)(-1)$

14. $f(x) = 3x^2 - 1$, $g(x) = 9 - x^2$
(a) $(f + g)(3)$ (b) $(f - g)(5)$

15. $f(x) = \sqrt{x}$, $g(x) = x^2 + 1$
(a) $(f + g)(4)$ (b) $(f - g)(9)$

16. $f(x) = \sqrt{x - 3}$, $g(x) = x^2 - 4$
(a) $(f + g)(7)$ (b) $(f - g)(1)$

17. $f(x) = |x|$, $g(x) = 5$
(a) $(fg)(-2)$ (b) $\left(\dfrac{f}{g}\right)(-2)$

18. $f(x) = |x - 3|$, $g(x) = (x - 2)^3$
(a) $(fg)(3)$ (b) $\left(\dfrac{f}{g}\right)\left(\dfrac{5}{2}\right)$

19. $f(x) = \dfrac{1}{x - 4}$, $g(x) = x$

(a) $(fg)(-2)$ (b) $\left(\dfrac{f}{g}\right)(4)$

20. $f(x) = \dfrac{1}{x^2}$, $g(x) = \dfrac{1}{x + 1}$

(a) $(fg)\left(\dfrac{1}{2}\right)$ (b) $\left(\dfrac{f}{g}\right)(0)$

In Exercises 21–28, find the compositions.

21. $f(x) = x - 3$, $g(x) = x^2$
(a) $(f \circ g)(x)$ (b) $(g \circ f)(x)$
(c) $(f \circ g)(4)$ (d) $(g \circ f)(7)$

22. $f(x) = x + 5$, $g(x) = x^3$
(a) $(f \circ g)(x)$ (b) $(g \circ f)(x)$
(c) $(f \circ g)(2)$ (d) $(g \circ f)(-3)$

23. $f(x) = |x - 3|$, $g(x) = 3x$
(a) $(f \circ g)(x)$ (b) $(g \circ f)(x)$
(c) $(f \circ g)(1)$ (d) $(g \circ f)(2)$

24. $f(x) = |x|$, $g(x) = 2x + 5$
(a) $(f \circ g)(x)$ (b) $(g \circ f)(x)$
(c) $(f \circ g)(-2)$ (d) $(g \circ f)(-4)$

25. $f(x) = \sqrt{x}$, $g(x) = x + 5$
(a) $(f \circ g)(x)$ (b) $(g \circ f)(x)$
(c) $(f \circ g)(4)$ (d) $(g \circ f)(9)$

26. $f(x) = \sqrt{x + 6}$, $g(x) = 2x - 3$
(a) $(f \circ g)(x)$ (b) $(g \circ f)(x)$
(c) $(f \circ g)(3)$ (d) $(g \circ f)(-2)$

27. $f(x) = \dfrac{1}{x - 3}$, $g(x) = \sqrt{x}$

(a) $(f \circ g)(x)$ (b) $(g \circ f)(x)$

(c) $(f \circ g)(49)$ (d) $(g \circ f)(12)$

28. $f(x) = \dfrac{4}{x^2 - 4}$, $g(x) = \dfrac{1}{x}$

(a) $(f \circ g)(x)$ (b) $(g \circ f)(x)$

(c) $(f \circ g)(-2)$ (d) $(g \circ f)(1)$

In Exercises 29–34, find the compositions (a) $f \circ g$ and (b) $g \circ f$. Then find the domain of each composition.

29. $f(x) = x^2 + 1$ **30.** $f(x) = 2 - 3x$

$g(x) = 2x$ $g(x) = 5x + 3$

31. $f(x) = \sqrt{x}$ **32.** $f(x) = \sqrt{x - 5}$

$g(x) = x - 2$ $g(x) = x^2$

33. $f(x) = \dfrac{9}{x + 9}$ **34.** $f(x) = \dfrac{x}{x - 4}$

$g(x) = x^2$ $g(x) = \sqrt{x}$

In Exercises 35–44, find the indicated combination of the functions $f(x) = x^2 - 3x$ and $g(x) = 5x + 3$.

35. $(f - g)(t)$ **36.** $(f + g)(t - 2)$

37. $(f/g)(2)$ **38.** $(fg)(z)$

39. $\dfrac{g(x + h) - g(x)}{h}$ **40.** $\dfrac{f(x + h) - f(x)}{h}$

41. $(f \circ g)(-1)$ **42.** $(g \circ f)(3)$

43. $(g \circ f)(y)$ **44.** $(f \circ g)(z)$

In Exercises 45–54, find the indicated combination of the functions $f(x) = x + 3$ and $g(x) = x^2 - 2x - 15$.

45. $(f + g)(t)$ **46.** $(f - g)(h + 1)$

47. $(f/g)(t)$ **48.** $(fg)(-2)$

49. $\dfrac{f(x + h) - f(x)}{h}$ **50.** $\dfrac{g(x + h) - g(x)}{h}$

51. $(f \circ g)(5)$ **52.** $(g \circ f)(-3)$

53. $(g \circ f)(t)$ **54.** $(f \circ g)(y)$

In Exercises 55–60, use the graphs to sketch the graph of $h(x) = (f + g)(x)$ and $k(x) = (f - g)(x)$.

55.
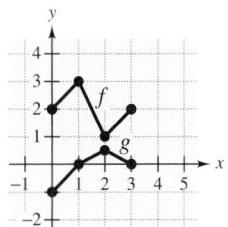

56.
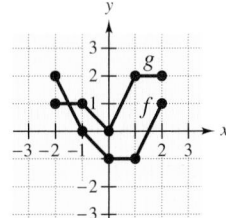

57.
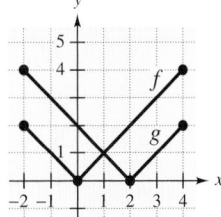

58.
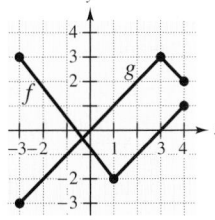

59.
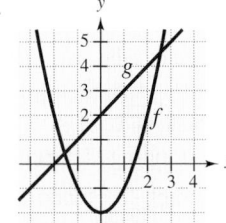

60.
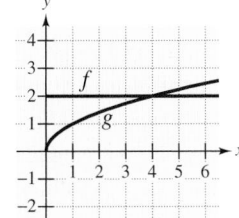

In Exercises 61 and 62, use a graphing utility to graph f, g, and $f + g$ in the same viewing window. Observe from the graphs which of the two functions contributes more to the magnitude of the sum for different values of x.

61. $f(x) = x^3$

$g(x) = 3(1 - x^2)$

62. $f(x) = x^3$

$g(x) = -3x$

In Exercises 63–66, use the functions f and g to find the indicated values.

$$f = \{(-2, 3), (-1, 1), (0, 0), (1, -1), (2, -3)\}$$
$$g = \{(-3, 1), (-1, -2), (0, 2), (2, 2), (3, 1)\}$$

63. (a) $f(1)$ (b) $g(-1)$ (c) $(g \circ f)(1)$

64. (a) $g(0)$ (b) $f(2)$ (c) $(f \circ g)(0)$

65. (a) $(f \circ g)(-3)$ (b) $(g \circ f)(-2)$

66. (a) $(f \circ g)(2)$ (b) $(g \circ f)(2)$

The symbol ▦ indicates an exercise in which you are instructed to use a calculator or graphing utility.

In Exercises 67–70, use the functions f and g to find the indicated values.

$f = \{(0, 1), (1, 2), (2, 5), (3, 10), (4, 17)\}$

$g = \{(5, 4), (10, 1), (2, 3), (17, 0), (1, 2)\}$

67. (a) $f(3)$ (b) $g(10)$ (c) $(g \circ f)(3)$

68. (a) $g(2)$ (b) $f(0)$ (c) $(f \circ g)(10)$

69. (a) $(g \circ f)(4)$ (b) $(f \circ g)(2)$

70. (a) $(f \circ g)(1)$ (b) $(g \circ f)(0)$

In Exercises 71–74, use the functions f and g to find the indicated values.

$f = \{(3, 6), (4, -2), (5, 0), (6, -2), (7, 10)\}$

$g = \{(-2, 5), (0, 7), (6, 4), (10, 7), (15, 1)\}$

71. (a) $f(6)$ (b) $g(6)$ (c) $(f \circ g)(6)$

72. (a) $g(15)$ (b) $f(7)$ (c) $(g \circ f)(3)$

73. (a) $(f \circ g)(0)$ (b) $(g \circ f)(5)$

74. (a) $(g \circ f)(6)$ (b) $(f \circ g)(-2)$

In Exercises 75–78, use a graphing utility to graph $f \circ g$. Use the graph to find the domain of $f \circ g$.

75. $f(x) = x^3 + 2x, \quad g(x) = -5x$

76. $f(x) = -x^3 - 7, \quad g(x) = x + 1$

77. $f(x) = \dfrac{2}{x - 1}, \quad g(x) = \sqrt{x}$

78. $f(x) = \sqrt{x}, \quad g(x) = 2x + 5$

Solving Problems

79. *Stopping Distance*   A car traveling at x miles per hour stops quickly. The distance the car travels during the driver's reaction time is $R(x) = \frac{3}{4}x$. The distance the car travels while braking is $B(x) = \frac{1}{15}x^2$.

(a) Find a model for the total stopping distance T.

(b) Use a graphing utility to graph the functions R, B, and T in the interval $0 \le x \le 60$.

(c) Which function contributes more to the magnitude of the sum at higher speeds?

80. *Data Analysis*   The table shows the variable costs of operating an automobile in the United States for the years 1992 through 1997. The functions $y_1, y_2,$ and y_3 represent the costs in cents per mile for gas and oil, maintenance, and tires, respectively, and t represents the year, with $t = 2$ corresponding to 1992. *(Source: American Automobile Manufacturers Association Inc.)*

t	2	3	4	5	6	7
y_1	6.0	6.0	5.6	6.0	5.9	6.6
y_2	2.2	2.4	2.5	2.6	2.8	2.8
y_3	0.9	0.9	1.1	1.4	1.4	1.4

Mathematical models that approximate the data are given by

$y_1 = 0.0194t^3 - 0.179t^2 + 0.42t + 5.7, \ 2 \le t \le 7$

$y_2 = 1.8 + 0.22t - 0.011t^2, \ 2 \le t \le 7$

$y_3 = 0.4 + 0.25t - 0.014t^2, \ 2 \le t \le 7$

Use a graphing utility to graph $y_1, y_2, y_3,$ and $y_1 + y_2 + y_3$ in the same viewing window. Use the model to estimate the total variable cost per mile in 2000.

81. *Sales Bonus*   You are a sales representative for a clothing manufacturer. You are paid an annual salary plus a bonus of 2% of your sales over $200,000. Consider the two functions

$$f(x) = x - 200,000 \quad \text{and} \quad g(x) = 0.02x.$$

If x is greater than $200,000, find each composition and determine which represents your bonus. Explain.

(a) $f(g(x))$ (b) $g(f(x))$

82. *Production Cost*   The daily cost of producing x units in a manufacturing process is $C(x) = 8.5x + 300$. The number of units produced in t hours during a day is given by $x(t) = 12t, \ 0 \le t \le 8$. Find, simplify, and interpret $(C \circ x)(t)$.

83. *Geometry*   You are standing on a bridge over a calm pond and drop a pebble, causing ripples of concentric circles in the water (see figure). The radius (in feet) of the outer ripple is given by $r(t) = 0.6t$, where t is the time in seconds after the pebble hits the water. The area of the circle is given by $A(r) = \pi r^2$. Find an equation for the composition $A(r(t))$.

Explaining Concepts

84. Describe the error.

If $f(x) = 3x^2 + x$ and $g(x) = x - 2$, find $(f \circ g)(3)$.

$(f \circ g)(3) = f(3) - 2$

$\qquad = [3(3)^2 + 3] - 2$

$\qquad = (27 + 3) - 2$

$\qquad = 28$

85. Describe the error.

If $f(x) = 2x - 1$ and $g(x) = x^3 + 1$, find $(f \circ g)(2)$.

$(f \circ g)(2) = (2 \cdot 2 - 1)(2^3 + 1)$

$\qquad = (4 - 1)(8 + 1)$

$\qquad = (3)(9)$

$\qquad = 27$

86. Give an example showing that the composite functions $(f \circ g)(x)$ and $(g \circ f)(x)$ are not necessarily the same.

Ongoing Review

In Exercises 87–90, find the domain of the function.

87. $f(x) = -5x^2 + 3x - 6$

88. $g(x) = \sqrt{4x - 12}$

89. $h(x) = \dfrac{4}{3 + x}$

90. $f(x) = \dfrac{x + 4}{x^2 - 16}$

In Exercises 91–96, sketch the graph of the function.

91. $f(x) = 6x + 1$

92. $f(x) = -x - 3$

93. $f(x) = 2x^2 - 3$

94. $f(x) = 4 - x^2$

95. $f(x) = x^3 - 2$

96. $f(x) = |x + 2|$

97. *Geometry*   A triangle has a perimeter of 84 centimeters. Each of the two longer sides of the triangle is three times as long as the shortest side. Find the length of each side of the triangle.

98. *Work-Rate Problem*   A worker can build a fence in 8 hours. With the help of an assistant, the fence can be built in 5 hours. How long would it take the assistant to build the fence alone?

Looking Further

Rebate and Discount   The suggested retail price of a new car is p dollars. The dealership advertises a factory rebate of \$2000 and a 5% discount.

(a) Write a function R in terms of p, giving the cost of the car after receiving the factory rebate.

(b) Write a function S in terms of p, giving the cost of the car after receiving the dealership discount.

(c) Form the compositions $(R \circ S)(p)$ and $(S \circ R)(p)$ and interpret each.

(d) Find $(R \circ S)(26,000)$ and $(S \circ R)(26,000)$. Which yields the lower cost for the car? Explain.

8.2 Inverse Functions

What you should learn:

- How to find inverse functions informally and verify that two functions are inverse functions of each other
- How to find inverse functions algebraically
- How to use graphs of functions to decide whether functions have inverse functions

Why you should learn it:

Inverse functions can be used to model and solve real-life problems. For instance, in Exercise 89 on page 519 you will use an inverse function to determine the number of units produced for a certain hourly wage.

Inverse Functions

When functions were introduced in Section 2.4, you saw that one way to represent a function is by a set of ordered pairs. For instance, the function $f(x) = x + 2$ from the set $A = \{1, 2, 3, 4\}$ to the set $B = \{3, 4, 5, 6\}$ can be written as follows.

$$f(x) = x + 2: \quad \{(1, 3), (2, 4), (3, 5), (4, 6)\}$$

By interchanging the first and second coordinates in each of these ordered pairs, you can form another function that is called the **inverse function** of f. This function is denoted by f^{-1}. It is a function from the set B to the set A, and can be written as follows.

$$f^{-1}(x) = x - 2: \quad \{(3, 1), (4, 2), (5, 3), (6, 4)\}$$

Note that the domain of f is equal to the range of f^{-1}, and vice versa, as shown in Figure 8.3. Also note that the functions f and f^{-1} have the effect of "undoing" each other. In other words, when you form the composition of f with f^{-1}, or the composition of f^{-1} with f, you obtain the identity function, as follows.

$$f(f^{-1}(x)) = f(x - 2) = (x - 2) + 2 = x$$
$$f^{-1}(f(x)) = f^{-1}(x + 2) = (x + 2) - 2 = x$$

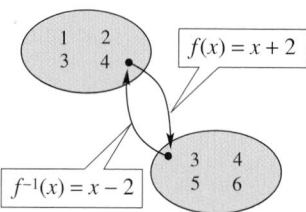

Figure 8.3

Example 1 ■ Determining Inverse Functions Informally

Find the inverse function of $f(x) = 3x$.

Solution

The given function is a function that *multiplies* each input by 3. To "undo" this function, you need to *divide* each input by 3. So, the inverse function of $f(x) = 3x$ is

$$f^{-1}(x) = \frac{x}{3}.$$

NOTE   In Example 1, you can verify that both $f(f^{-1}(x))$ and $f^{-1}(f(x))$ are equal to the identity function, as follows.

$$f(f^{-1}(x)) = f\left(\frac{x}{3}\right) = 3\left(\frac{x}{3}\right) = x$$

$$f^{-1}(f(x)) = f^{-1}(3x) = \frac{3x}{3} = x$$

Example 2 ■ Determining Inverse Functions Informally

Find the inverse function of $f(x) = x - 4$ and verify that both $f(f^{-1}(x))$ and $f^{-1}(f(x))$ are equal to the identity function.

Solution

The given function is a function that *subtracts* 4 from each input. To "undo" this function, you need to *add* 4 to each input. So, the inverse function of $f(x) = x - 4$ is

$$f^{-1}(x) = x + 4.$$

You can verify that both $f(f^{-1}(x))$ and $f^{-1}(f(x))$ are equal to the identity function, as follows.

$$f(f^{-1}(x)) = f(x + 4)$$
$$= (x + 4) - 4 = x$$
$$f^{-1}(f(x)) = f^{-1}(x - 4)$$
$$= (x - 4) + 4 = x$$

Do not be confused by the use of -1 to denote the inverse function f^{-1}. Whenever f^{-1} is written, it *always* refers to the inverse function of f and *not* to the reciprocal of $f(x)$.

The formal definition of the inverse function is given as follows.

Definition of Inverse Function

Let f and g be two functions such that

$$f(g(x)) = x \quad \text{for every } x \text{ in the domain of } g$$

and

$$g(f(x)) = x \quad \text{for every } x \text{ in the domain of } f.$$

Then the function g is called the **inverse function** of the function f, and is denoted by f^{-1} (read "f-inverse"). So, $f(f^{-1}(x)) = x$ and $f^{-1}(f(x)) = x$. The domain of f must be equal to the range of f^{-1}, and vice versa.

Note from this definition that if the function g is the inverse function of the function f, it must also be true that the function f is the inverse function of the function g. For this reason, the functions f and g are sometimes called *inverse functions of each other*.

You can use a graphing utility to get a visual image of two functions being inverse functions of each other. For example, enter the functions from Example 3 as follows.

$$y_1 = x^3 + 1$$
$$y_2 = (x - 1)^{1/3}$$
$$y_3 = y_1(y_2)$$
$$y_4 = y_2(y_1)$$

y_3 and y_4 both yield the line $y = x$, which means that y_1 and y_2 are inverse functions of each other.

Example 3 ■ Verifying Inverse Functions

Show that $f(x) = x^3 + 1$ and $g(x) = \sqrt[3]{x - 1}$ are inverse functions of each other.

Solution

Begin by noting that the domains and ranges of both functions are the entire set of real numbers. To show that f and g are inverse functions of each other, you need to show that $f(g(x)) = x$ and $g(f(x)) = x$, as follows.

$$\begin{aligned} f(g(x)) &= f\left(\sqrt[3]{x - 1}\right) \\ &= \left(\sqrt[3]{x - 1}\right)^3 + 1 \\ &= (x - 1) + 1 \\ &= x \end{aligned}$$

$$\begin{aligned} g(f(x)) &= g(x^3 + 1) \\ &= \sqrt[3]{(x^3 + 1) - 1} \\ &= \sqrt[3]{x^3} \\ &= x \end{aligned}$$

Note that the two functions f and g "undo" each other in the verbal sense as follows. The function f first cubes the input x and then adds 1, whereas the function g first subtracts 1, and then takes the cube root of the result.

NOTE Some functions are their own inverse functions. For instance, the function $f(x) = 1/(5x)$ in Example 4 is its own inverse function.

Example 4 ■ Verifying Inverse Functions

Which of the functions is the inverse function of $f(x) = 1/(5x)$?

$$g(x) = \frac{5}{x} \quad \text{or} \quad h(x) = \frac{1}{5x}$$

Solution

By forming the composition of f with g, you find that

$$\begin{aligned} f(g(x)) &= f\left(\frac{5}{x}\right) \\ &= \frac{1}{5\left(\dfrac{5}{x}\right)} = \frac{1}{\left(\dfrac{25}{x}\right)} = \frac{x}{25}. \end{aligned}$$

Because this composition does not yield x, you can conclude that g *is not* the inverse function of f. By forming the composition of f with h, you find that

$$\begin{aligned} f(h(x)) &= f\left(\frac{1}{5x}\right) \\ &= \frac{1}{5\left(\dfrac{1}{5x}\right)} = \frac{1}{\left(\dfrac{1}{x}\right)} = x. \end{aligned}$$

So, it appears that h *is* the inverse function of f. You can confirm this by showing that the composition of h with f is also equal to the identity function. (Try doing this.)

Finding an Inverse Function

For simple functions (such as the ones in Examples 1 and 2), you can find inverse functions by inspection. For instance, the inverse function of $f(x) = 10x$ is $f^{-1}(x) = x/10$. For more complicated functions, however, it is best to use the steps below for finding an inverse function. The key step in these guidelines is switching the roles of x and y. This step corresponds to the fact that inverse functions have ordered pairs with the coordinates reversed.

Finding an Inverse Function

To find the inverse function of a function f, use the steps below.

1. In the equation for $f(x)$, replace $f(x)$ by y.

2. Interchange the roles of x and y.

3. If the new equation does not represent y as a function of x, the function f does not have an inverse function. If the new equation does represent y as a function of x, solve the new equation for y.

4. Replace y by $f^{-1}(x)$.

5. Verify that f and f^{-1} are inverse functions of each other by showing that $f(f^{-1}(x)) = x = f^{-1}(f(x))$.

NOTE   Note in Step 3 of the guidelines for finding an inverse function that it is possible that a function has no inverse function. This possibility is illustrated in Example 6.

Historical Note

Sonya Kovalevsky was a brilliant Russian mathematician and mathematical physicist who lived from 1850 until 1891. She worked extensively with inverse functions in higher mathematics.

Example 5 ■ Finding an Inverse Function

Determine whether $f(x) = 2x + 3$ has an inverse function. If it does, find its inverse function.

Solution

$$f(x) = 2x + 3 \qquad \text{Write original function.}$$
$$y = 2x + 3 \qquad \text{Replace } f(x) \text{ by } y.$$
$$x = 2y + 3 \qquad \text{Interchange } x \text{ and } y.$$
$$y = \frac{x - 3}{2} \qquad \text{Solve for } y.$$
$$f^{-1}(x) = \frac{x - 3}{2} \qquad \text{Replace } y \text{ by } f^{-1}(x).$$

So, the inverse function of $f(x) = 2x + 3$ is

$$f^{-1}(x) = \frac{x - 3}{2}.$$

Verify that $f(f^{-1}(x)) = x = f^{-1}(f(x))$.

Example 6 ■ A Function That Has No Inverse Function

Determine whether

$$f(x) = x^2$$

has an inverse function. If it does, find its inverse function.

Solution

$$f(x) = x^2 \qquad \text{Write original function.}$$
$$y = x^2 \qquad \text{Replace } f(x) \text{ by } y.$$
$$x = y^2 \qquad \text{Interchange } x \text{ and } y.$$

Because the equation $x = y^2$ does not represent y as a function of x, you can conclude that the original function f does not have an inverse function.

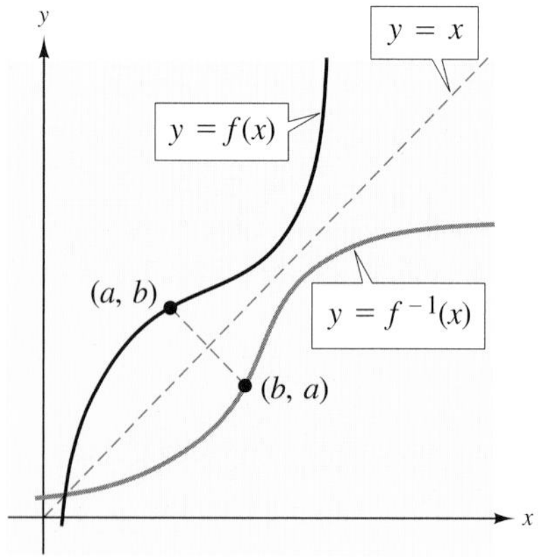

NOTE Recall from Section 2.5 that the equation $x = y^2$ does not represent y as a function of x because you can find two different y-values that correspond to the same x-value. For example, when $x = 9$, y can be 3 or -3.

The Graph of an Inverse Function

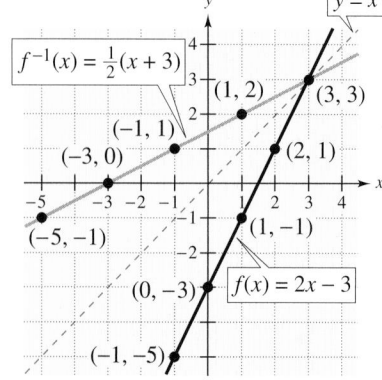

Figure 8.4 *The graph of f^{-1} is a reflection of the graph of f in the line $y = x$.*

The graphs of f and f^{-1} are related to each other in the following way. If the point (a, b) lies on the graph of f, then the point (b, a) must lie on the graph of f^{-1}, and vice versa. This means that the graph of f^{-1} is a reflection of the graph of f in the line $y = x$, as shown in Figure 8.4. This "reflective property" of the graphs of f and f^{-1} is illustrated in Examples 7 and 8.

Example 7 ■ The Graphs of f and f^{-1}

Sketch the graphs of the inverse functions

$$f(x) = 2x - 3 \text{ and } f^{-1}(x) = \tfrac{1}{2}(x + 3)$$

on the same rectangular coordinate system, and show that the graphs are reflections of each other in the line $y = x$.

Solution

The graphs of f and f^{-1} are shown in Figure 8.5. Visually, it appears that the graphs are reflections of each other in the line $y = x$. You can further verify this reflective property by testing a few points on each graph. Note in the list below that if the point (a, b) is on the graph of f, then the point (b, a) is on the graph of f^{-1}.

$f(x) = 2x - 3$	$f^{-1}(x) = \tfrac{1}{2}(x + 3)$
$(-1, -5)$	$(-5, -1)$
$(0, -3)$	$(-3, 0)$
$(1, -1)$	$(-1, 1)$
$(2, 1)$	$(1, 2)$
$(3, 3)$	$(3, 3)$

Figure 8.5

In Example 6, you saw that the function

$$f(x) = x^2$$

has no inverse function. A more complete way of saying this is "*assuming that the domain of f is the entire real line*, the function $f(x) = x^2$ has no inverse function." If, however, you restrict the domain of f to the nonnegative real numbers, then f does have an inverse function, as demonstrated in Example 8.

Example 8 ■ Verifying Inverse Functions Graphically and Numerically

Verify that f and g are inverse functions of each other graphically and numerically.

$$f(x) = x^2, \quad x \geq 0 \quad \text{and} \quad g(x) = \sqrt{x}$$

Graphical Solution

You can graphically verify that f and g are inverse functions of each other by graphing the functions on the same rectangular coordinate system. From the graph in Figure 8.6, you can verify that the graph of g is the reflection of the graph of f in the line $y = x$.

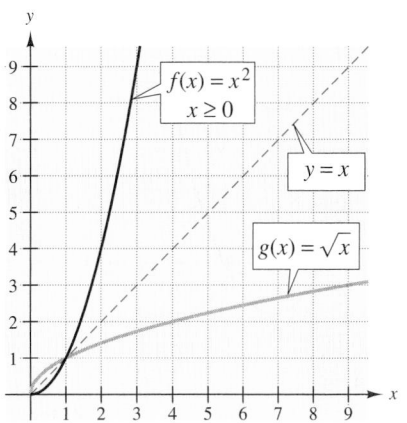

Figure 8.6

Numerical Solution

You can numerically verify that f and g are inverse functions of each other by creating two tables, as shown below.

x	0	1	2	3
$f(x)$	0	1	4	9

x	0	1	4	9
$g(x)$	0	1	2	3

Note that the entries in the tables are the same except that their rows are interchanged. From the tables, you can verify that f and g are inverse functions of each other.

In the guidelines for finding an inverse function, an *algebraic* test for determining whether a function has an inverse function was included. The reflective property of the graphs of inverse functions gives you a nice *geometric* test for determining whether a function has an inverse function. This is called the **Horizontal Line Test** for inverse functions.

> **Horizontal Line Test for Inverse Functions**
>
> A function f has an inverse function if and only if no *horizontal* line intersects the graph of f at more than one point. Such a function is called **one-to-one**.

The Horizontal Line Test for inverse functions is demonstrated in Example 9.

Example 9　■　Applying the Horizontal Line Test

Use the Horizontal Line Test to determine if the function is one-to-one and so has an inverse function.

(a) $f(x) = 2x - 1$　　(b) $f(x) = x^2 - 1$

Solution

(a) The graph of the function $f(x) = 2x - 1$ is shown in Figure 8.7. Because no horizontal line intersects the graph of f at more than one point, you can conclude that f *is* a one-to-one function and *does* have an inverse function.

(b) The graph of the function $f(x) = x^2 - 1$ is shown in Figure 8.8. Because it is possible to find a horizontal line that intersects the graph of f at more than one point, you can conclude that f *is not* a one-to-one function and *does not* have an inverse function.

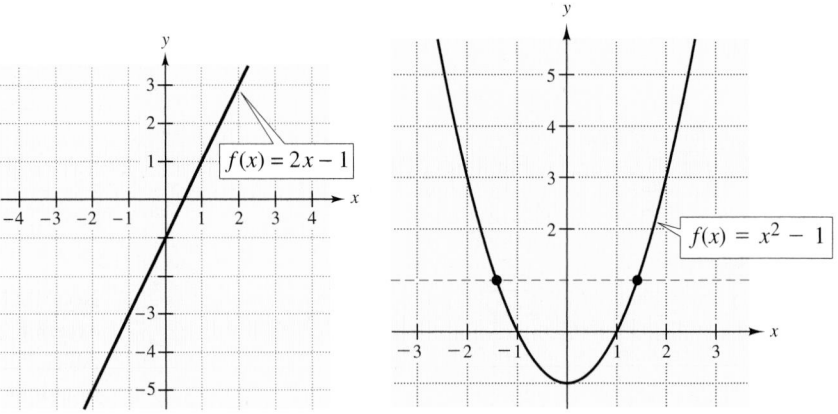

Figure 8.7　　　　　　　　　　Figure 8.8

Collaborate!

The Existence of an Inverse Function

In your group, describe why the functions below do or do not have inverse functions.

a. Let x represent the retail price of an item (in dollars), and let $f(x)$ represent the sales tax on the item. Assume that the sales tax is 6% of the retail price *and* that the sales tax is rounded to the nearest cent. Does this function have an inverse function? (*Hint:* Can you undo this function? For instance, if you know that the sales tax is $0.12, can you determine *exactly* what the retail price is?)

b. Let x represent the temperature in degrees Celsius, and let $f(x)$ represent the temperature in degrees Fahrenheit. Does this function have an inverse function? (*Hint:* The formula for converting from degrees Celsius to degrees Fahrenheit is $F = \frac{9}{5}C + 32$.)

8.2 ■ Exercises

Developing Skills

In Exercises 1–10, find the inverse function of f informally. Verify that $f(f^{-1}(x))$ and $f^{-1}(f(x))$ are equal to the identity function.

1. $f(x) = 5x$

2. $f(x) = \frac{1}{3}x$

3. $f(x) = x + 10$

4. $f(x) = x - 5$

5. $f(x) = \frac{1}{2}x$

6. $f(x) = 2x$

7. $f(x) = x^7$

8. $f(x) = x^5$

9. $f(x) = \sqrt[3]{x}$

10. $f(x) = x^{1/5}$

In Exercises 11–22, verify algebraically that the functions f and g are inverse functions of each other.

11. $f(x) = 10x$

$g(x) = \frac{1}{10}x$

12. $f(x) = \frac{2}{3}x$

$g(x) = \frac{3}{2}x$

13. $f(x) = x + 15$

$g(x) = x - 15$

14. $f(x) = 3 - x$

$g(x) = 3 - x$

15. $f(x) = 1 - 2x$

$g(x) = \dfrac{1 - x}{2}$

16. $f(x) = 2x - 1$

$g(x) = \frac{1}{2}(x + 1)$

17. $f(x) = 2 - 3x$

$g(x) = \frac{1}{3}(2 - x)$

18. $f(x) = -\frac{1}{4}x + 3$

$g(x) = -4(x - 3)$

19. $f(x) = \sqrt[3]{x + 1}$

$g(x) = x^3 - 1$

20. $f(x) = x^5$

$g(x) = \sqrt[5]{x}$

21. $f(x) = \dfrac{1}{x}$

$g(x) = \dfrac{1}{x}$

22. $f(x) = \dfrac{1}{x - 3}$

$g(x) = 3 + \dfrac{1}{x}$

In Exercises 23–50, find the inverse function.

23. $f(x) = 8x$

24. $f(x) = \dfrac{x}{10}$

25. $g(x) = x + 25$

26. $f(x) = 7 - x$

27. $g(x) = 3 - 4x$

28. $g(t) = 6t + 1$

29. $f(x) = 4x + 1$

30. $g(x) = 3x + 7$

31. $f(x) = -6x + 5$

32. $h(t) = -7t + 2$

33. $g(s) = 2s - 9$

34. $f(t) = -3t + 8$

35. $g(t) = \frac{1}{4}t + 2$

36. $h(s) = 5 - \frac{3}{2}s$

37. $h(x) = \sqrt{x}$

38. $h(x) = \sqrt{x + 5}$

39. $f(x) = 2\sqrt{x + 1}$

40. $f(x) = 3\sqrt{2x - 1}$

41. $h(t) = 1 + t^3$

42. $u(t) = 2t^3 + 5$

43. $f(t) = t^3 - 1$

44. $h(t) = t^5$

45. $g(s) = \dfrac{5}{s}$

46. $f(s) = \dfrac{2}{3 - s}$

47. $g(s) = \dfrac{-3s}{s + 4}$

48. $f(s) = \dfrac{3s}{s - 5}$

49. $h(x) = \dfrac{x - 5}{x}$

50. $f(x) = \dfrac{2x}{x + 1}$

In Exercises 51–54, match the graph with the graph of its inverse function. [The graphs of the inverse functions are labeled (a), (b), (c), and (d).]

(a)
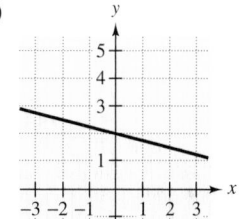

(b)
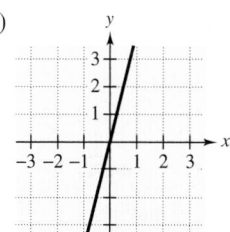

(c)
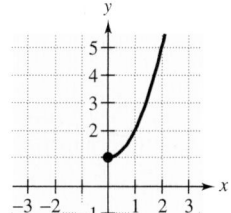

(d)
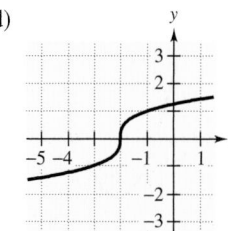

51.
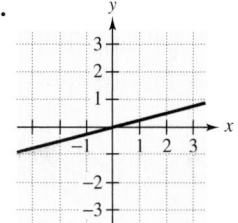

52.
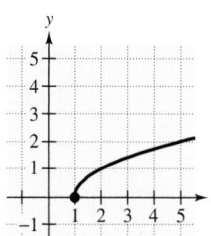

53.
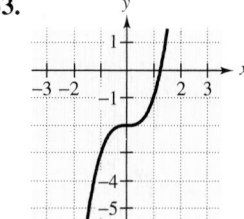

54.
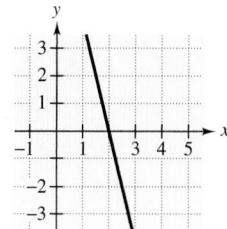

In Exercises 55–62, use a graphing utility to verify that f and g are inverse functions of each other.

55. $f(x) = \frac{1}{3}x$

$g(x) = 3x$

56. $f(x) = \frac{1}{5}x - 1$

$g(x) = 5x + 5$

57. $f(x) = \sqrt{x + 1}$

$g(x) = x^2 - 1, \ x \geq 0$

58. $f(x) = \sqrt{4 - x}$

$g(x) = 4 - x^2, \ x \geq 0$

59. $f(x) = \frac{1}{8}x^3$

$g(x) = 2\sqrt[3]{x}$

60. $f(x) = \sqrt[3]{x + 2}$

$g(x) = x^3 - 2$

61. $f(x) = |x + 4|, \ x \geq -4$

$g(x) = x - 4, \ x \geq 0$

62. $f(x) = |x - 2|, \ x \geq 2$

$g(x) = x + 2, \ x \geq 0$

In Exercises 63–66, find the inverse function of f. Use a graphing utility to graph f and f^{-1} to confirm the result.

63. $f(x) = x^3 + 1$

64. $f(x) = 5 - x^3$

65. $f(x) = \sqrt{x^2 - 4}, \ x \geq 2$

66. $f(x) = \sqrt{x^2 - 25}, \ x \geq 5$

In Exercises 67–70, determine whether the function has an inverse function.

67. $f(x) = x^2 - 2$

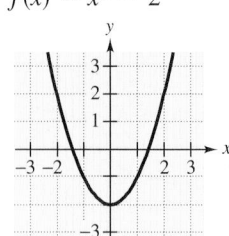

68. $f(x) = \frac{1}{5}x$

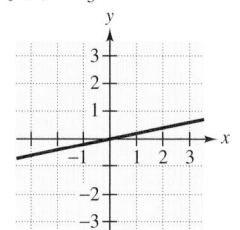

69. $g(x) = \sqrt{25 - x^2}$

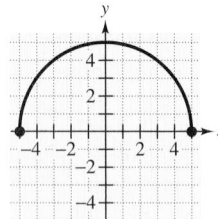

70. $g(x) = |x - 4|$

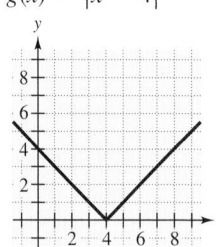

In Exercises 71–80, use a graphing utility to graph the function and determine whether the function is one-to-one.

71. $f(x) = 3 + \frac{1}{4}x^3$

72. $f(x) = 9 - x^2$

73. $f(t) = \sqrt[3]{5 - t}$

74. $h(t) = 4 - \sqrt[3]{t}$

75. $f(x) = (x - 1)^4$

76. $f(x) = (x + 2)^5$

77. $h(t) = \dfrac{5}{t}$

78. $g(t) = \dfrac{5}{t^2}$

79. $f(s) = \dfrac{4}{s^2 + 1}$

80. $f(x) = \dfrac{1}{x - 2}$

In Exercises 81–84, delete part of the graph of the function so that the remaining part is one-to-one. Find the inverse function of the remaining part and give the domain of the inverse function. (*Note*: There is more than one correct answer.)

81. $f(x) = x^4$

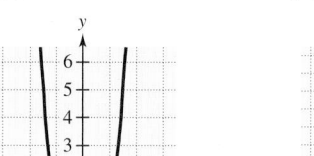

82. $f(x) = 9 - x^2$

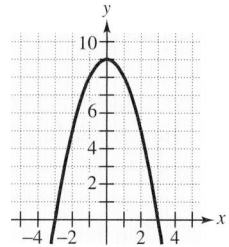

83. $f(x) = (x - 2)^2$

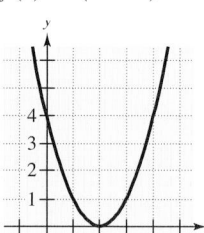

84. $f(x) = |x - 2|$

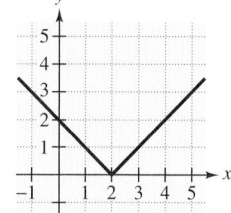

In Exercises 85–88, use the graph of f to sketch the graph of f^{-1}.

85.

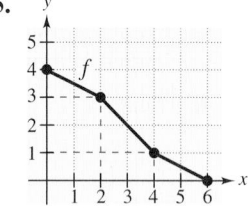

86.

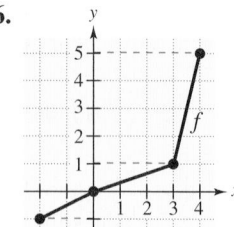

87.

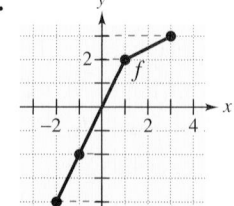

88.

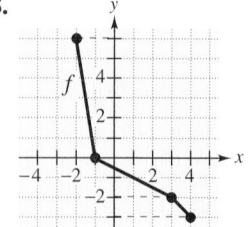

Solving Problems

89. *Hourly Wage*   Your wage is $9.00 per hour plus $0.65 for each unit produced per hour. So, your hourly wage y in terms of the number of units produced x is $y = 9 + 0.65x$.

(a) Find the inverse function.

(b) What does each variable represent in the inverse function?

(c) Determine the number of units produced when your hourly wage averages $14.20.

90. *Cost*   You need 100 pounds of two commodities that cost $0.50 and $0.75 per pound.

(a) Verify that your total cost is $y = 0.50x + 0.75(100 - x)$, where x is the number of pounds of the less expensive commodity.

(b) Find the inverse function. What does each variable represent in the inverse function?

(c) Use the context of the problem to determine the domain of the inverse function.

(d) Determine the number of pounds of the less expensive commodity purchased if the total cost is $60.

Explaining Concepts

91. If the inverse function of f exists, is the y-intercept of f an x-intercept of f^{-1}? Explain.

92. If the inverse function of f exists, are the domains of f and f^{-1} the same? Explain.

93. If the inverse function of f exists and its graph passes through the point $(2, 2)$, does the graph of f^{-1} also pass through the point $(2, 2)$?

94. Describe the relationship between the graph of a function and the graph of its inverse function.

Ongoing Review

In Exercises 95–98, find the domain of the function.

95. $f(x) = 3x - x^3$ **96.** $h(x) = \sqrt[3]{2x + 1}$

97. $g(x) = \sqrt{4 - x^2}$ **98.** $f(x) = \dfrac{-1}{x^2 + 3x + 2}$

99. *Telephone Charges*   The cost of a long distance telephone call is $0.95 for the first minute and $0.35 for each additional minute. The total cost of a call is $5.15. Find the length of the call.

100. *Free-Falling Object*   The velocity of a free-falling object is given by

$$v = \sqrt{2gh}$$

where v is the velocity measured in feet per second, $g = 32$ feet per second squared, and h is the distance (in feet) the object has fallen. Find the distance an object has fallen if its velocity is 80 feet per second.

Looking Further

Consider the functions $f(x) = 4x$ and $g(x) = x + 6$.

(a) Find $(f \circ g)(x)$.

(b) Find $(f \circ g)^{-1}(x)$.

(c) Find $f^{-1}(x)$ and $g^{-1}(x)$.

(d) Find $(g^{-1} \circ f^{-1})(x)$. Compare the result with that of part (b).

(e) Repeat parts (a) through (d) for $f(x) = x^3 + 1$ and $g(x) = 2x$.

(f) Write two one-to-one functions f and g, and repeat parts (a) through (d) for these functions.

(g) Make a conjecture about $(f \circ g)^{-1}(x)$ and $(g^{-1} \circ f^{-1})(x)$.

8.3 Variation and Mathematical Models

Direct Variation

In this section, you will continue your study of methods for creating mathematical models by looking at models that are related to the concept of **variation.**

The mathematical model for **direct variation** is a *linear* function of x. That is,

$$y = kx.$$

To use this mathematical model in applications involving direct variation, you need to use the given values of x and y to find the value of the constant k. This procedure is demonstrated in Examples 1 and 2.

Direct Variation

The statements below are equivalent.

1. y **varies directly** as x.

2. y is **directly proportional** to x.

3. $y = kx$ for some constant k.

The number k is called the **constant of proportionality.**

Example 1 ■ Direct Variation

The total revenue R (in dollars) obtained from selling x units of a given product is directly proportional to the number of units sold x. When 10,000 units are sold, the total revenue is $142,500. (a) Find a mathematical model that relates R to x. (b) Find the total revenue obtained from selling 12,000 units.

Solution

(a) Because the total revenue is directly proportional to the number of units sold, you have the model

$$R = kx.$$

To find the value of k, use the fact that $R = 142,500$ when $x = 10,000$. Substituting these values into the model produces

$$142,500 = k(10,000)$$

which implies that $k = 142,500/10,000 = 14.25$. So, the equation relating the total revenue to the total number of units sold is

$$R = 14.25x. \qquad \text{Direct variation model}$$

The graph of this equation is shown in Figure 8.9.

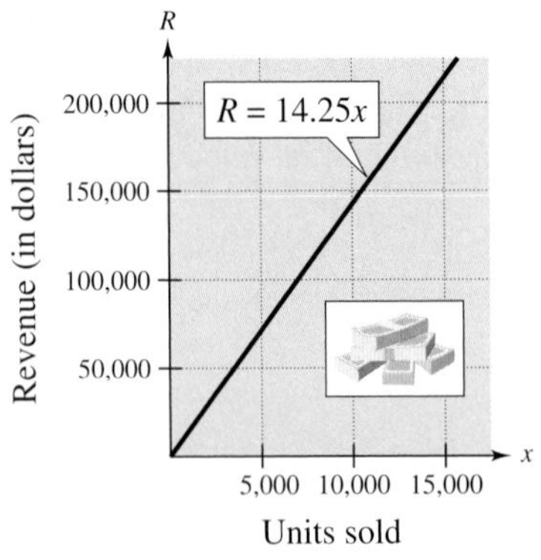

(b) When $x = 12,000$, the total revenue is $R = 14.25(12,000) = \$171,000$.

Figure 8.9

Revenue (in dollars)

$R = 14.25x$

200,000
150,000
100,000
50,000

5,000 10,000 15,000

Units sold

Example 2 ■ Direct Variation

Hooke's Law for springs states that the distance a spring is stretched (or compressed) is directly proportional to the force on the spring. A force of 20 pounds stretches a certain spring 5 inches.

(a) Find a mathematical model that relates the distance the spring is stretched to the force applied to the spring.

(b) How far will a force of 30 pounds stretch the spring?

Solution

(a) For this problem, let d represent the distance (in inches) that the spring is stretched and let F represent the force (in pounds) that is applied to the spring. Because the distance d is directly proportional to the force F, the model is

$$d = kF.$$

To find the value of k, use the fact that $d = 5$ when $F = 20$. Substituting these values into the model produces

$$5 = k(20) \qquad \text{Substitute 5 for } d \text{ and 20 for } F.$$

$$\frac{5}{20} = k \qquad \text{Divide each side by 20.}$$

$$\frac{1}{4} = k. \qquad \text{Simplify.}$$

So, the equation relating distance and force is

$$d = \frac{1}{4}F. \qquad \text{Direct variation model}$$

(b) When $F = 30$, the distance is

$$d = \frac{1}{4}(30) = 7.5 \text{ inches.}$$

(See Figure 8.10.)

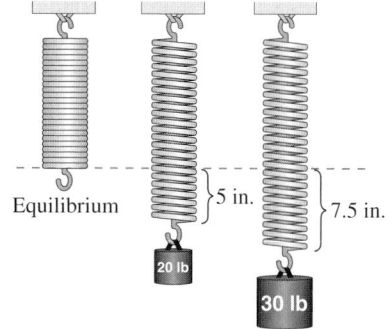

Equilibrium $\}$ 5 in.   7.5 in.   20 lb   30 lb

Figure 8.10

In Example 2, you can get a clearer understanding of Hooke's Law by using the model $d = \frac{1}{4}F$ to create a table or a graph (see Figure 8.11). From the table or from the graph, you can see what it means for the distance to be "directly proportional to the force."

Force, F	10 lb	20 lb	30 lb	40 lb	50 lb	60 lb
Distance, d	2.5 in.	5.0 in.	7.5 in.	10.0 in.	12.5 in.	15.0 in.

In Examples 1 and 2, the direct variations were such that an *increase* in one variable corresponded to an *increase* in the other variable. There are, however, other applications of direct variation in which an *increase* in one variable corresponds to a *decrease* in the other variable. For instance, in the model $y = -2x$, an increase in x will yield a decrease in y.

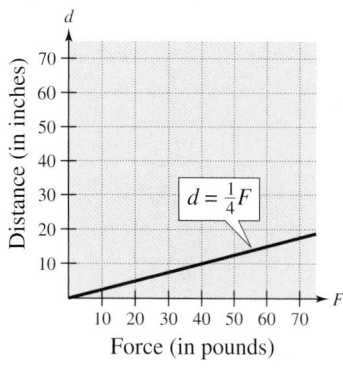

$d = \frac{1}{4}F$

Distance (in inches)

Force (in pounds)

Figure 8.11

Another type of direct variation relates one variable to a *power* of another variable.

Direct Variation as *n*th Power

The statements below are equivalent.

1. y **varies directly as the *n*th power** of x.

2. y is **directly proportional to the *n*th power** of x.

3. $y = kx^n$ for some constant k.

Example 3 ■ Direct Variation as a Power

The distance a ball rolls down an inclined plane is directly proportional to the square of the time it rolls. During the first second, the ball rolls 6 feet.

(a) Find a mathematical model that relates the distance traveled to the time.

(b) How far will the ball roll during the first 2 seconds?

Solution

(a) Letting d be the distance (in feet) that the ball rolls and letting t be the time (in seconds), you obtain the model

$$d = kt^2.$$

Because $d = 6$ when $t = 1$, you obtain

$d = kt^2$	Write original equation.
$6 = k(1)^2$	Substitute 6 for d and 1 for t.
$6 = k.$	Simplify.

So, the equation relating distance to time is

$$d = 6t^2. \qquad \text{Direct variation as 2nd power model}$$

The graph of this equation is shown in Figure 8.12.

(b) When $t = 2$, the distance traveled is

$$d = 6(2)^2$$
$$= 6(4)$$
$$= 24 \text{ feet.}$$

(See Figure 8.13.)

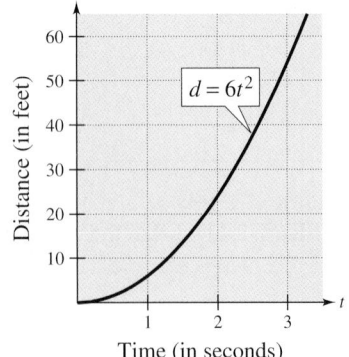

Figure 8.12

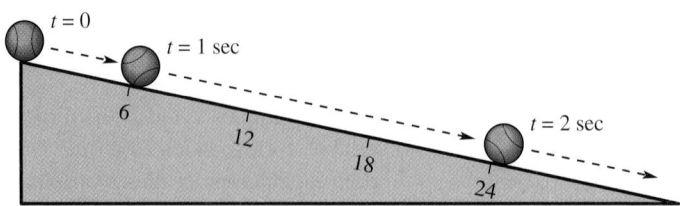

Figure 8.13

Inverse Variation

A second type of variation is called **inverse variation.** With this type of variation, one of the variables is said to be inversely proportional to the other variable.

> **Inverse Variation**
>
> The statements below are equivalent.
>
> 1. y **varies inversely** as x.
>
> 2. y is **inversely proportional** to x.
>
> 3. $y = \dfrac{k}{x}$ for some constant k.

If x and y are related by an equation of the form $y = k/x^n$, it is said that y varies inversely as the nth power of x (or that y is inversely proportional to the nth power of x).

Example 4 ■ Inverse Variation

The marketing department of a large company has found that the demand for one of its products varies inversely with the price of the product. (When the price is low, more people are willing to buy the product than when the price is high.) When the price of the product is \$7.50, the monthly demand is 50,000 units. Approximate the monthly demand if the price is reduced to \$6.00.

Solution

Let x represent the number of units sold each month (the demand), and let p represent the price per unit (in dollars). Because the demand is inversely proportional to the price, you obtain the model $x = k/p$. By substituting $x = 50{,}000$ when $p = 7.50$, you obtain

$$50{,}000 = \frac{k}{7.50} \qquad \text{Substitute 50,000 for } x \text{ and 7.50 for } p.$$

$$375{,}000 = k. \qquad \text{Multiply each side by 7.50.}$$

So, the model is

$$x = \frac{375{,}000}{p}. \qquad \text{Inverse variation model}$$

The graph of this equation is shown in Figure 8.14. To find the demand that corresponds to a price of \$6.00, substitute 6 for p in the equation and obtain a demand of

$$x = \frac{375{,}000}{6} = 62{,}500 \text{ units.}$$

So, if the price is lowered from \$7.50 per unit to \$6.00 per unit, you can expect the monthly demand to increase from 50,000 units to 62,500 units.

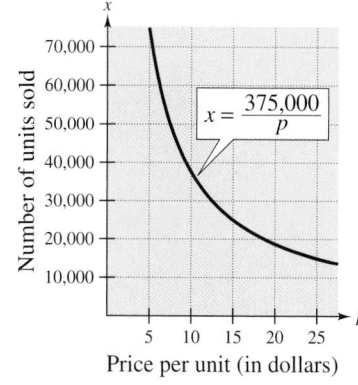

Figure 8.14

The graph in Figure 8.15 shows the "banking angles" for a bicycle at a speed of 15 miles per hour. The banking angle and the radius of the turn vary inversely. As the radius of the turn gets smaller, the bicyclist must lean at greater angles to avoid falling over.

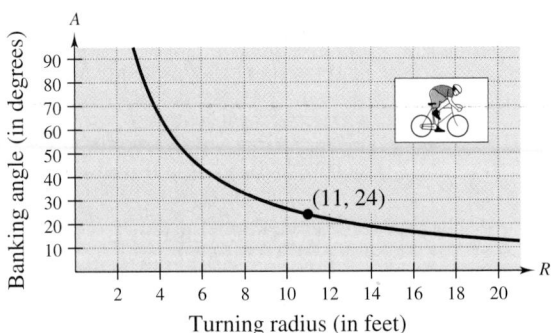

Turning radius (in feet)

Figure 8.15

Example 5 ■ Finding a Model for "Banking Angles"

The banking angle for a bicycle at a speed of 15 miles per hour varies inversely with the turning radius.

(a) Use the coordinates of the point shown in Figure 8.15 to find a model that relates the banking angle A to the turning radius R.

(b) Use the model to find the banking angle for a turning radius of 8 feet.

Solution

From Figure 8.15, you can see that A is $24°$ when the turning radius R is 11 feet.

(a) $A = \dfrac{k}{R}$ Model for inverse variation

$24 = \dfrac{k}{11}$ Substitute 24 for A and 11 for R.

$264 = k$ Solve for k.

The model is given by

$A = \dfrac{264}{R}$ Inverse variation model

where A is measured in degrees and R is measured in feet.

(b) When the turning radius is 8 feet, the banking angle is

$A = \dfrac{264}{R}$

$A = \dfrac{264}{8}$ Substitute 8 for R.

$= 33°$.   Simplify.

Some applications of variation involve problems with *both* direct and inverse variation in the same model. These types of models are said to have **combined variation.** For instance, in the model

$$z = k\left(\frac{x}{y}\right)$$

z is directly proportional to x *and* inversely proportional to y.

Example 6 ■ Direct and Inverse Variation

A company determines that the demand for one of its products is directly proportional to the amount spent on advertising and inversely proportional to the price of the product. When \$40,000 is spent on advertising and the price per unit is \$20, the monthly demand is 10,000 units.

(a) If the amount spent on advertising were increased to \$50,000, how much could the price be increased to maintain a monthly demand of 10,000 units?

(b) If you were in charge of the advertising department, would you recommend this increased expenditure for advertising?

Solution

(a) Let x represent the number of units sold each month (the demand), let a represent the amount spent on advertising (in dollars), and let p represent the price per unit (in dollars). Because the demand is directly proportional to the advertising expenditure and inversely proportional to the price, you obtain the model

$$x = \frac{ka}{p}.$$ Combined variation model

By substituting 10,000 for x when $a = 40,000$ and $p = 20$, you obtain

$$x = \frac{ka}{p}$$ Write combined variation model.

$$10,000 = \frac{k(40,000)}{20}$$ Substitute 10,000 for x, 40,000 for a, and 20 for p.

$$200,000 = 40,000k$$ Multiply each side by 20.

$$5 = k.$$ Divide each side by 40,000.

So, the model is $x = 5a/p$. To find the price that corresponds to a demand of 10,000 and an advertising expenditure of \$50,000, substitute 10,000 for x and 50,000 for a in the model and solve for p.

$$10,000 = \frac{5(50,000)}{p} \quad \Longrightarrow \quad p = \frac{5(50,000)}{10,000} = \$25$$

(b) The total revenue for selling 10,000 units at \$20 is \$200,000, and the total revenue for selling 10,000 units at \$25 is \$250,000. So, by increasing the advertising expenditure from \$40,000 to \$50,000, the company can increase its revenue by \$50,000. This implies that you should recommend the increased expenditure for advertising.

Joint Variation

The model used in Example 6 involved both direct and inverse variation, and the word "and" was used to couple the two types of variation together. To describe two different *direct* variations in the same statement, the word "jointly" is used. For instance, the model $z = kxy$ can be described by saying that z is *jointly* proportional to x and y.

Joint Variation

The statements below are equivalent.

1. z **varies jointly** as x and y.

2. z is **jointly proportional** to x and y.

3. $z = kxy$ for some constant k.

NOTE If x, y, and z are related by an equation of the form $z = kx^n y^m$, it is said that z varies jointly as the nth power of x and the mth power of y.

Example 7 ■ Joint Variation

The *simple interest* for a certain savings account is jointly proportional to the time and the principal. After one quarter (3 months), the interest for a principal of $6000 is $120. How much interest would a principal of $7500 earn in 5 months?

Solution

To begin, let I represent the interest earned (in dollars), let P represent the principal (in dollars), and let t represent the time (in years). Then, because the interest is jointly proportional to the time and the principal, you obtain the model $I = ktP$. Because $I = 120$ when $P = 6000$ and $t = \frac{1}{4}$, you have

$$120 = k\left(\tfrac{1}{4}\right)(6000) \qquad \text{Substitute 120 for } I, \tfrac{1}{4} \text{ for } t, \text{ and 6000 for } P.$$
$$120 = 1500k \qquad \text{Simplify.}$$
$$0.08 = k. \qquad \text{Divide each side by 1500.}$$

So, the model is

$$I = 0.08tP. \qquad \text{Joint variation model}$$

To find the interest earned for a principal of $7500 over a five-month period of time, substitute 7500 for P and $\frac{5}{12}$ for t in the model to obtain an interest of

$$I = 0.08\left(\frac{5}{12}\right)(7500) = \$250.$$

Collaborate!

You Be the Instructor

Suppose you are teaching an algebra class and are writing a test that covers the material in this section. In your group, write two test questions that give a fair representation of this material. (Assume that your students can spend 10 minutes on each question.)

8.3 ■ Exercises

Developing Skills

In Exercises 1–12, write a model for the statement.

1. I varies directly as V.

2. C varies directly as r.

3. u is directly proportional to the square of v.

4. s varies directly as the cube of t.

5. p varies inversely as d.

6. S is inversely proportional to the square of v.

7. P is inversely proportional to the square root of $1 + r$.

8. A varies inversely as the fourth power of t.

9. A varies jointly as l and w.

10. V varies jointly as h and the square of r.

11. *Boyle's Law*   If the temperature of a gas is not allowed to change, its absolute pressure P is inversely proportional to its volume V.

12. *Newton's Law of Universal Gravitation*   The gravitational attraction F between two particles of masses m_1 and m_2 is directly proportional to the product of the masses and inversely proportional to the square of the distance r between the particles.

In Exercises 13–20, write a verbal sentence using variation terminology to describe the formula.

13. *Area of a Triangle* $A = \frac{1}{2}bh$

14. *Area of a Circle* $A = \pi r^2$

15. *Area of a Rectangle* $A = lw$

16. *Surface Area of a Sphere* $A = 4\pi r^2$

17. *Volume of a Right Circular Cylinder* $V = \pi r^2 h$

18. *Volume of a Sphere* $V = \frac{4}{3}\pi r^3$

19. *Average Speed* $r = \dfrac{d}{t}$

20. *Height of a Cylinder* $h = \dfrac{V}{\pi r^2}$

In Exercises 21–40, find the constant of proportionality and write an equation that relates the variables.

21. s varies directly as t, and $s = 20$ when $t = 4$.

22. h is directly proportional to r, and $h = 28$ when $r = 12$.

23. F is directly proportional to the square of x, and $F = 500$ when $x = 40$.

24. v varies directly as the square root of s, and $v = 24$ when $s = 16$.

25. H is directly proportional to u, and $H = 100$ when $u = 40$.

26. M varies directly as the cube of n, and $M = 0.012$ when $n = 0.2$.

27. n varies inversely as m, and $n = 32$ when $m = 1.5$.

28. q is inversely proportional to p, and $q = \frac{3}{2}$ when $p = 50$.

29. g varies inversely as the square root of z, and $g = \frac{4}{5}$ when $z = 25$.

30. u varies inversely as the square of v, and $u = 40$ when $v = \frac{1}{2}$.

31. L varies inversely as the cube of x, and $L = 18$ when $x = 3$.

32. T is inversely proportional to the cube root of d, and $T = 41$ when $d = 8$.

33. P varies directly as the square of x, and $P = 10$ when $x = 5$.

34. q is directly proportional to the square root of s, and $q = 112$ when $s = 64$.

35. F varies jointly as x and y, and $F = 500$ when $x = 15$ and $y = 8$.

36. V varies jointly as h and the square of b, and $V = 288$ when $h = 6$ and $b = 12$.

37. c varies jointly as a and the square root of b, and $c = 24$ when $a = 9$ and $b = 4$.

38. L varies jointly as R and the square of M, and $L = 45$ when $R = 10$ and $M = 3$.

39. d varies directly as the square of x and inversely as r, and $d = 3000$ when $x = 10$ and $r = 4$.

40. z is directly proportional to x and inversely proportional to the square root of y, and $z = 720$ when $x = 48$ and $y = 81$.

Solving Problems

41. *Revenue* The total revenue R is directly proportional to the number of units sold x. When 500 units are sold, the revenue is $3875.

(a) Find the revenue when 635 units are sold.

(b) Interpret the constant of proportionality.

42. *Revenue* The total revenue R is directly proportional to the number of units sold x. When 25 units are sold, the revenue is $300.

(a) Find the revenue when 42 units are sold.

(b) Interpret the constant of proportionality.

43. *Hooke's Law* A force of 50 pounds stretches a spring 3 inches.

(a) How far will a force of 20 pounds stretch the spring?

(b) What force will stretch the spring 1.5 inches?

44. *Hooke's Law* A force of 50 pounds stretches a spring 5 inches.

(a) How far will a force of 20 pounds stretch the spring?

(b) What force will stretch the spring 1.5 inches?

45. *Hooke's Law* A baby weighing $10\frac{1}{2}$ pounds compresses the spring of a baby scale 7 millimeters. Determine the weight of a baby that compresses the spring 12 millimeters.

46. *Hooke's Law* A force of 50 pounds stretches a spring 1.5 inches.

(a) Write the force F as a function of the distance x the spring is stretched.

(b) Graph the function in part (a) where $0 \le x \le 5$. Identify the graph.

47. *Stopping Distance* The stopping distance d of an automobile is directly proportional to the square of its speed s. On a certain road surface, a car requires 75 feet to stop when its speed is 30 miles per hour. Estimate the stopping distance if the brakes are applied when the car is traveling at 50 miles per hour under similar road conditions.

48. *Free-Falling Object* Neglecting air resistance, the distance d that an object falls varies directly as the square of the time t it has been falling. If an object falls 64 feet in 2 seconds, determine the distance it will fall in 6 seconds.

49. *Free-Falling Object* The velocity v of a free-falling object is directly proportional to the time that it has fallen. The constant of proportionality is the acceleration due to gravity. Find the acceleration due to gravity if the velocity of a falling object is 96 feet per second after the object has fallen for 3 seconds.

50. *Velocity of a Stream* The diameter d of the largest particle that can be moved by a stream is directly proportional to the square of the velocity v of the stream. A stream with a velocity of $\frac{1}{4}$ mile per hour can move coarse sand particles about 0.02 inch in diameter. What must the velocity be to carry particles with a diameter of 0.12 inch?

51. *Power Generation* The power P generated by a wind turbine varies directly as the cube of the wind speed w. The turbine generates 750 watts of power in a 25-mile-per-hour wind. Find the power it generates in a 40-mile-per-hour wind.

52. *Travel Time* The travel time between two cities is inversely proportional to the average speed. A train travels between two cities in 3 hours at an average speed of 65 miles per hour. How long would it take at an average speed of 80 miles per hour? What does the constant of proportionality measure in this problem?

53. *Demand* A company has found that the daily demand x for its product is inversely proportional to the price p. When the price is $5, the demand is 800 units. Approximate the demand if the price is increased to $6.

54. *Predator-Prey* The number N of prey t months after a natural predator is introduced into a test area is inversely proportional to $t + 1$. If $N = 500$ when $t = 0$, find N when $t = 4$.

55. *Weight* A person's weight on the moon varies directly with his or her weight on Earth. Neil Armstrong, the first man on the moon, weighed 360 pounds on Earth, including his heavy equipment. On the moon, he weighed only 60 pounds with equipment. If the first woman in space, Valentina V. Tereshkova, had landed on the moon and weighed 54 pounds with equipment, how much would she have weighed on Earth with her equipment?

56. *Weight*　The gravitational force F with which an object is attracted to Earth is inversely proportional to the square of its distance r from the center of Earth. An astronaut weighs 190 pounds on the surface of Earth ($r \approx 4000$ miles). What will the astronaut weigh 1000 miles above Earth's surface?

57. *Best Buy*　The prices of 9-inch, 12-inch, and 15-inch diameter pizzas at a certain pizza shop are $6.78, $9.78, and $12.18. One would expect that the price of a certain size pizza would be directly proportional to its surface area. Is this the case for this pizza shop? If not, which size pizza is the best buy?

58. *Pressure*　When a person walks, the pressure P on each sole varies inversely as the area A of the sole. Denise is trudging through deep snow, wearing boots that have a sole area of 29 square inches each. The sole pressure is 4 pounds per square inch. Denise is wearing snowshoes, each with an area 11 times that of her boot soles. What is the pressure on each snowshoe? The constant of proportionality in this problem is Denise's weight. How much does she weigh?

59. *Projection Area*　The area of a projected picture on a movie screen varies directly as the square of the distance from the projector from the screen. A distance of 20 feet produces a picture with an area of 64 square feet. What distance produces an area of 100 square feet?

60. *Sound*　The loudness, measured in decibels, of a stereo speaker is inversely proportional to the square of the distance of the listener from the speaker. The loudness is 28 decibels at a distance of 8 feet. What is the loudness when the listener is 4 feet from the speaker?

61. *Airline Passengers*　A commuter airline has found that the average number of passengers per month between any two cities on its service routes is jointly proportional to each city's population and inversely proportional to the square of the distance between them. Alameda has a population of 50,000 and is 80 miles from Baltic, which has a population of 64,000. Airline records indicate that an average of 3840 passengers per month travel between Alameda and Baltic. Estimate the average number of passengers per month that the airline could expect between Alameda and Crystal Lake, which has a population of 144,000 and is 160 miles from Alameda.

62. *Music*　The frequency of vibration of a piano string varies directly as the square root of the tension on the string and inversely as the length of the string. The middle A string has a frequency of 440 vibrations per second. Find the frequency of a string that has 1.25 times as much tension and is 1.2 times as long.

63. *Oil Spill*　The graph shows the percent p of oil that remained in Chedabucto Bay, Nova Scotia after an oil spill. Clean-up was left primarily to natural actions such as wave motion, evaporation, photochemical decomposition, and bacterial decomposition. After about a year, the percent that remained varied inversely with time. Find a model that relates p and t, where t is the number of years since the spill. Then use the model to find the amount of oil that remained $6\frac{1}{2}$ years after the spill, and compare the result with the graph.

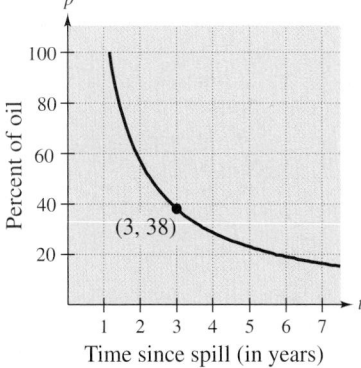

Time since spill (in years)

64. *Ocean Temperatures*　The graph shows the temperature of the water in the north central Pacific Ocean. At depths greater than 900 meters, the water temperature varies inversely with the depth. Find a model that relates the temperature T to the depth d. Then use the model to find the water temperature at a depth of 4385 meters, and compare the result with the graph.

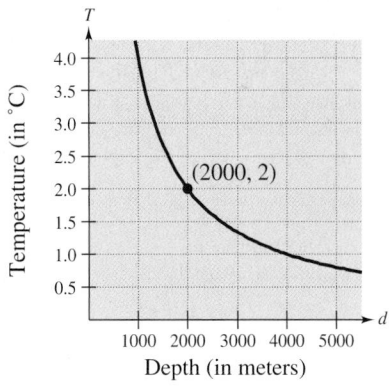

Depth (in meters)

Explaining Concepts

65. Suppose the constant of proportionality is positive and y varies directly as x. If one of the variables increases, how will the other change? Explain.

66. Suppose the constant of proportionality is positive and y varies inversely as x. If one of the variables increases, how will the other change? Explain.

67. If y varies inversely as the square of x and x is doubled, how will y change? Use the properties of exponents to explain your answer.

68. If y varies directly as the square of x and x is doubled, how will y change? Use the properties of exponents to explain your answer.

Ongoing Review

In Exercises 69–72, use the functions f and g to find the combinations (a) $(f + g)(x)$, (b) $(f - g)(x)$, (c) $(fg)(x)$, (d) $(f/g)(x)$, (e) $(f \circ g)(x)$, and (f) $(g \circ f)(x)$.

69. $f(x) = 3x^2 + x$

$g(x) = -5x$

70. $f(x) = 7x - 1$

$g(x) = 2x^2$

71. $f(x) = \dfrac{1}{x + 4}$

$g(x) = \dfrac{1}{6x}$

72. $f(x) = \dfrac{1}{x}$

$g(x) = 3x + 1$

73. *Cost* The inventor of a new game believes that the variable cost of producing the game is \$5.75 per unit and the fixed costs are \$12,000. Find the number of games produced if the total cost is \$17,060.

74. *Geometry* The length of a rectangle is 5 inches more than 1.5 times its width. Find the dimensions of the rectangle if its perimeter is 47.5 inches.

Looking Further

Engineering The load P that can be safely supported by a horizontal beam varies jointly as the product of the width W of the beam and the square of its depth D and inversely as its length L (see figure).

(a) Write a model for the statement.

(b) How does P change when the width and length of the beam are doubled?

(c) How does P change when the width and depth of the beam are doubled?

(d) How does P change when all three of the dimensions of the beam are doubled?

(e) How does P change when the depth of the beam is cut in half?

(f) A beam of width 3 inches, depth 8 inches, and length 10 feet can safely support 2000 pounds. Determine the safe load for a beam made from the same material if its depth is increased to 10 inches.

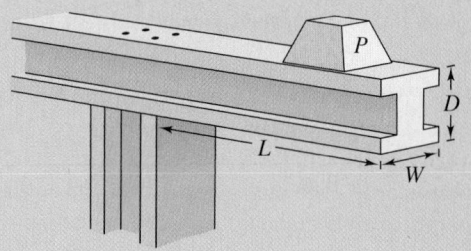

8.4 Polynomial Functions and Their Graphs

Graphs of Polynomial Functions

A function of the form $f(x) = a_n x^n + a_{n-1} x^{n-1} + \cdots + a_1 x + a_0,\ a_n \neq 0$, is a **polynomial function** of degree n. You already know how to sketch the graph of a polynomial function of degree 0, 1, or 2.

Function	Degree	Graph
$f(x) = a$	0	Horizontal line
$f(x) = ax + b$	1	Line of slope a
$f(x) = ax^2 + bx + c$	2	Parabola

Example 1 ■ Polynomial Functions of Degrees 0, 1, and 2

Identify and sketch each polynomial function.

(a) $f(x) = 3$ (b) $f(x) = -\frac{1}{2}x + 3$ (c) $f(x) = x^2 + 2x - 3$

Solution

(a) The graph of $f(x) = 3$ is a horizontal line, as shown in Figure 8.16(a). This polynomial function is of degree 0 and is called a *constant* function.

(b) The graph of $f(x) = -\frac{1}{2}x + 3$ is a line with a slope of $-\frac{1}{2}$ and a y-intercept of 3, as shown in Figure 8.16(b). This polynomial function is of degree 1 and is called a *linear* function.

(c) The graph of $f(x) = x^2 + 2x - 3$ is a parabola with a vertex of $(-1, -4)$, as shown in Figure 8.16(c). This polynomial function is of degree 2 and is called a *quadratic* function.

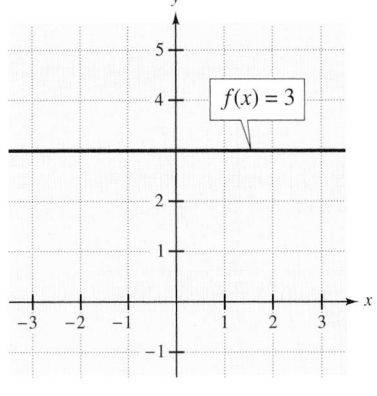

(a)

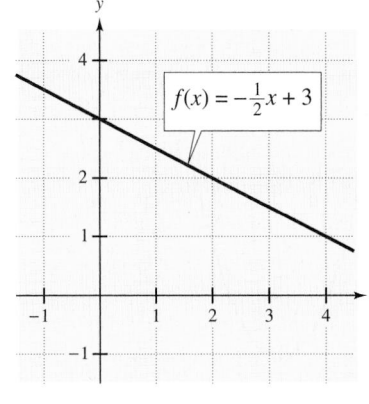

(b)

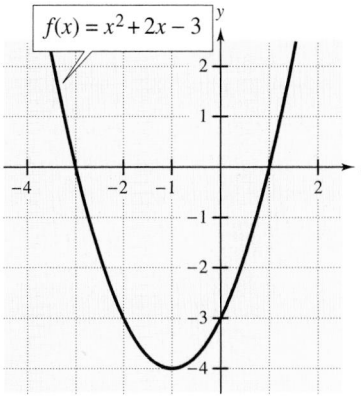

(c)

Figure 8.16

The graphs of polynomial functions of degree greater than 2 are more difficult to sketch, as illustrated in Figure 8.17. In this section, however, you will learn to recognize some of their basic features. With these features and point plotting, you will be able to make rough sketches *by hand*.

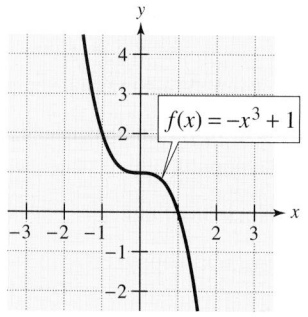

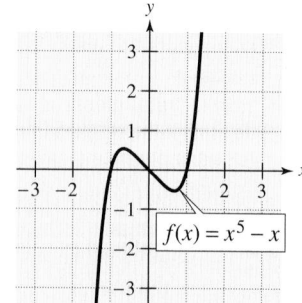

Figure 8.17

NOTE If you go on to take a course in calculus, you will learn to sketch more accurate graphs of polynomial functions of degree greater than 2.

NOTE Examine the graphs in Figure 8.17 and note that the features described at the right hold true.

Features of Graphs of Polynomial Functions

1. The graph of a polynomial function is **continuous.** This means that the graph has no breaks—you could sketch the graph without lifting your pencil from the paper.

2. The graph of a polynomial function has only smooth turns. The graph of a function of degree n has *at most* $n - 1$ turns.

3. If the leading coefficient of the polynomial function is positive, the graph rises to the right. If the leading coefficient is negative, the graph falls to the right.

The polynomial functions (of degree 2 or greater) that have the simplest graphs are monomial functions of the form $f(x) = a_n x^n$. When n is *even*, the graph is similar to the graph of $f(x) = x^2$, as shown in Figure 8.18(a). When n is *odd*, the graph is similar to the graph of $f(x) = x^3$, as shown in Figure 8.18(b). Moreover, the greater the value of n, the flatter the graph of a monomial is on the interval $-1 \le x \le 1$.

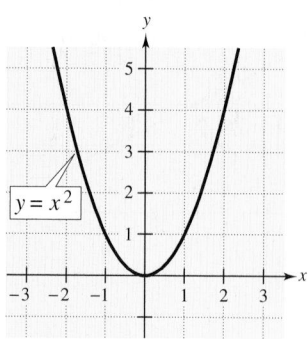

(a) For n even, the graph of $y = x^n$ is similar to the graph of $y = x^2$.

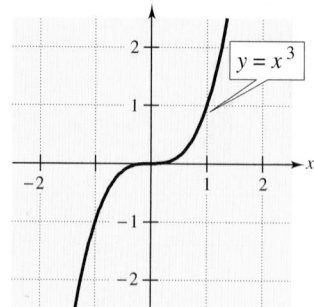

(b) For n odd, the graph of $y = x^n$ is similar to the graph of $y = x^3$.

Figure 8.18

Example 2 ■ Sketching Transformations of Monomial Functions

Describe the transformation of f that produces g, then sketch the transformation.

(a) $f(x) = x^5$ (b) $f(x) = x^4$ (c) $f(x) = x^3$

     $g(x) = -x^5$ $g(x) = x^4 + 1$ $g(x) = (x + 1)^3$

Solution

(a) *Reflection:* To sketch the graph of

$$g(x) = -x^5$$

reflect the graph of $f(x) = x^5$ in the x-axis, as shown in Figure 8.19(a).

(b) *Vertical Shift:* To sketch the graph of

$$g(x) = x^4 + 1$$

shift the graph of $f(x) = x^4$ upward one unit, as shown in Figure 8.19(b).

(c) *Horizontal Shift:* To sketch the graph of

$$g(x) = (x + 1)^3$$

shift the graph of $f(x) = x^3$ left one unit, as shown in Figure 8.19(c).

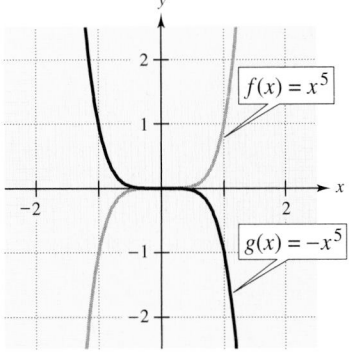

(a)

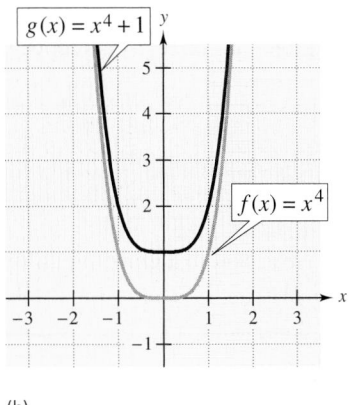

(b)

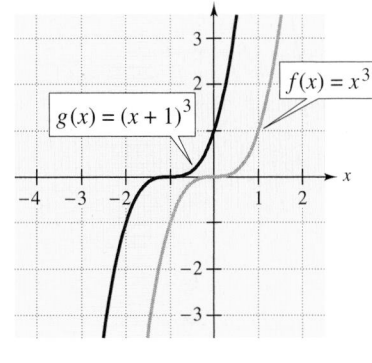

(c)

Figure 8.19

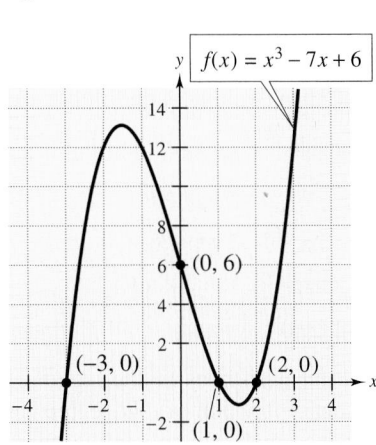

Figure 8.20

Sketching Graphs of Polynomial Functions

In sketching the graph of a polynomial function, it helps to label the intercepts. The y-intercept of the graph of a polynomial function is easy to find—it is the value of the function when $x = 0$. For instance, the y-intercept of the graph of $f(x) = x^3 - 7x + 6$ is

$$f(0) = 0^3 - 7(0) + 6$$

$$= 6 \qquad \text{\textit{y}-intercept}$$

as shown in Figure 8.20. The graph of a polynomial function always has exactly one y-intercept.

The x-intercepts of the graph of a polynomial function are not as easy to find. Moreover, the graph can have no x-intercepts, one x-intercept, or several x-intercepts. The x-intercepts of the graph of a polynomial function are the x-values for which $f(x) = 0$. You can find these values by solving the equation

$$f(x) = 0. \qquad \text{\textit{x}-intercept(s)}$$

These values are also called the **zeros** of the function. If you go on to take a course in college algebra, you will study techniques for solving polynomial equations of degree 3 or greater. You will also learn that the graph of a polynomial function of degree n has *at most* n x-intercepts. For instance, the graph of $f(x) = x^3 - 7x + 6$ has three x-intercepts, as shown in Figure 8.20. Try showing that $f(x) = 0$ when $x = -3$, $x = 1$, and $x = 2$.

Example 3 ■ Graphing a Cubic Polynomial Function

Sketch the graph of $f(x) = x^3 - 9x$.

Solution

Because the leading coefficient is positive, you know that the graph rises to the right. Also, because the degree of the polynomial function is 3, you know that the graph can have at most two turns. Using these general observations with the values shown in the table, you can sketch the graph, as shown in Figure 8.21. Notice that the graph has three x-intercepts: $(-3, 0)$, $(0, 0)$, and $(3, 0)$.

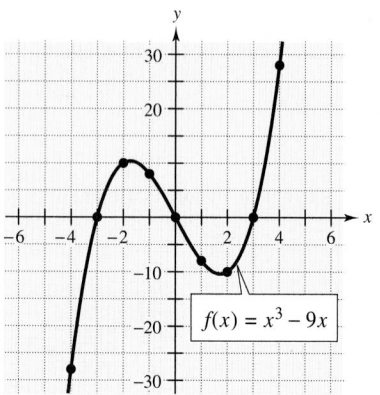

Figure 8.21

x	-4	-3	-2	-1	0	1	2	3	4
$f(x)$	-28	0	10	8	0	-8	-10	0	28

If you have access to a graphing utility, try using it to confirm the graph shown in Figure 8.21.

The x-intercepts of the graphs of some polynomial functions can be found by setting $f(x) = 0$ and factoring. For the equation in Example 3, this is done as follows.

$$
\begin{array}{ll}
x^3 - 9x = 0 & \text{Set } f(x) \text{ equal to 0.} \\
x(x^2 - 9) = 0 & \text{Factor out common monomial factor.} \\
x(x + 3)(x - 3) = 0 & \text{Factor difference of two squares.} \\
x = 0 & \text{Set 1st factor equal to 0.} \\
x + 3 = 0 \implies x = -3 & \text{Set 2nd factor equal to 0.} \\
x - 3 = 0 \implies x = 3 & \text{Set 3rd factor equal to 0.}
\end{array}
$$

So, the three x-intercepts are $(0, 0)$, $(-3, 0)$, and $(3, 0)$.

In Figure 8.21, note that the graph of the function rises to the right and falls to the left. In general, the graph of a polynomial function of *odd* degree has opposite behaviors to the right and left, whereas the graph of a polynomial function of *even* degree has the same behavior to the right and left.

Example 4 ■ Identifying the Right-Hand and Left-Hand Behavior of a Graph

Describe the right-hand and left-hand behavior of the graph of each function.

(a) $f(x) = -x^3 + 4x$ (b) $f(x) = x^4 - 5x^2 + 4$ (c) $f(x) = x^5 - x$

Solution

(a) Because the leading coefficient of

$$f(x) = -x^3 + 4x$$

is negative, the graph falls to the right. Because the degree is odd, the graph rises to the left, as shown in Figure 8.22(a).

(b) Because the leading coefficient of

$$f(x) = x^4 - 5x^2 + 4$$

is positive, the graph rises to the right. Because the degree is even, the graph also rises to the left, as shown in Figure 8.22(b).

(c) Because the leading coefficient of

$$f(x) = x^5 - x$$

is positive, the graph rises to the right. Because the degree is odd, the graph falls to the left, as shown in Figure 8.22(c).

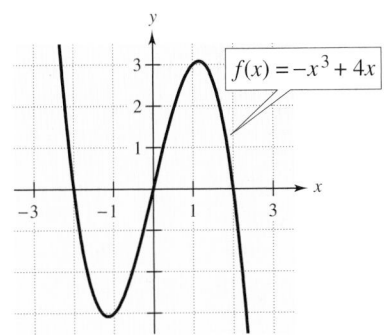

(a)

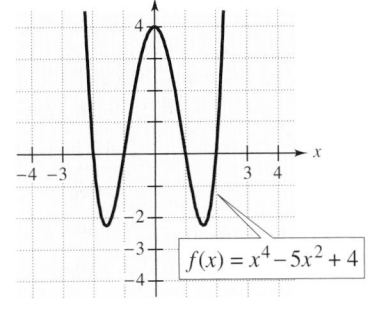

(b)

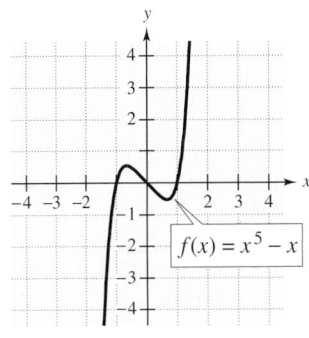

(c)

Figure 8.22

Collaborate!

Finding *x*-Intercepts of Polynomial Functions

Choose four integers between -10 and 10. Then, in your group, discuss how to find a fourth–degree polynomial function that has these integers as its zeros. Find the polynomial function and use a graphing utility to confirm your result.

　　Is there only one fourth-degree polynomial that has these four real numbers as its zeros? If so, explain why. If not, find another and graph the polynomial in the same viewing window.

8.4 ■ Exercises

Developing Skills

In Exercises 1–8, identify and sketch the graph of the polynomial function.

1. $g(x) = -2$ **2.** $f(x) = 8$

3. $f(x) = 7 - 3x$ **4.** $h(x) = \frac{1}{2}x + 1$

5. $f(x) = -(x - 3)^2 + 4$ **6.** $g(x) = (x + 1)^2 - 3$

7. $f(x) = 2x + 1$ **8.** $g(x) = -6x + 9$

In Exercises 9–14, state the maximum possible number of turns in the graph of the polynomial function.

9. $f(x) = 2x^5 - x + 6$ **10.** $f(x) = \frac{1}{2}x^3 + 1$

11. $g(s) = -3s^2 + 2s + 4$

12. $h(x) = x^4 + 9x - 1$

13. $f(x) = -2x^6 + 3x + 7$

14. $f(x) = 4x^{10} + 3x^8 - 2x^7 + 1$

In Exercises 15–18, describe the transformation of g that produces f.

15. $f(x) = (x - 5)^3$ **16.** $f(x) = x^4 + 3$

     $g(x) = x^3$ $g(x) = x^4$

17. $f(x) = x^5 - 2$ **18.** $f(x) = -x^6$

     $g(x) = x^5$ $g(x) = x^6$

In Exercises 19 and 20, use the graph to sketch the specified transformations.

19. $y = x^3$ **20.** $y = x^4$

   (a) $f(x) = (x - 2)^3$ (a) $f(x) = (x + 3)^4$

   (b) $f(x) = x^3 - 2$ (b) $f(x) = x^4 - 3$

   (c) $f(x) = -(x - 2)^3 - 2$ (c) $f(x) = 4 - x^4$

   (d) $f(x) = -x^3$ (d) $f(x) = -(x - 1)^4$

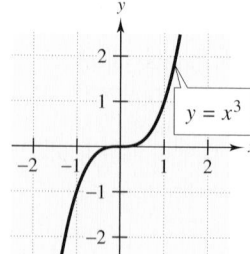

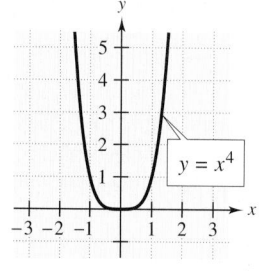

In Exercises 21–30, determine the right-hand and left-hand behavior of the graph of the polynomial function.

21. $f(x) = 2x^2 - 3x + 1$

22. $g(x) = 5 - \frac{7}{2}x - 3x^2$

23. $f(x) = \frac{1}{3}x^3 + 5x$

24. $f(x) = -2.1x^5 + 4x^3 - 2$

25. $f(x) = 6 - 2x + 4x^2 - 5x^3$

26. $h(x) = 1 - x^6$

27. $g(x) = -x^4 + 3x^3 + x^2 - 7x + 5$

28. $h(x) = x^7 + 5x^4 - x^2 + 9x$

29. $f(x) = 5 + 6x^4$

30. $g(x) = 1 - x + 5x^2 + 6x^3 + 10x^5$

In Exercises 31–42, find the x- and y-intercepts of the graph of the polynomial function.

31. $f(x) = x^2 - 25$ **32.** $f(x) = 49 - x^2$

33. $h(x) = x^2 - 6x + 9$ **34.** $f(x) = x^2 + 10x + 25$

35. $f(x) = x^2 + x - 2$

36. $f(x) = 3x^2 - 12x + 3$

37. $f(x) = x^3 - 4x^2 + 4x$

38. $f(x) = x^4 - x^3 - 20x^2$

39. $g(x) = \frac{1}{2}x^4 - \frac{1}{2}$

40. $f(x) = 2x^4 - 2x^2 - 40$

41. $f(x) = 5x^4 + 15x^2 + 10$

42. $f(x) = x^3 - 4x^2 - 25x + 100$

Graphical Analysis   In Exercises 43–50, use a graphing utility to graph the function and approximate any x-intercepts of the graph.

43. $g(t) = \frac{1}{2}t^4 - \frac{1}{2}$ **44.** $f(x) = x^5 + x^3 - 6x$

45. $f(x) = 2x^4 - 2x^2 - 40$

46. $g(t) = t^5 - 6t^3 + 9t$

47. $f(x) = 5x^4 + 15x^2 + 10$

48. $f(x) = x^3 - 4x^2 - 25x + 100$

49. $y = 4x^3 + 4x^2 - 7x + 2$

50. $y = x^5 - 5x^3 + 4x$

In Exercises 51–58, match the polynomial function with the corresponding graph. [The graphs are labeled (a), (b), (c), (d), (e), (f), (g), and (h).]

(a)
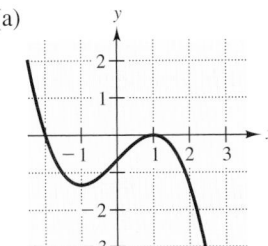

(b)
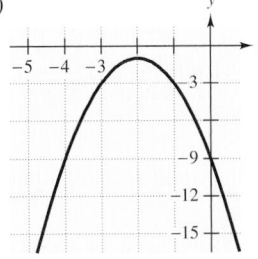

(g)
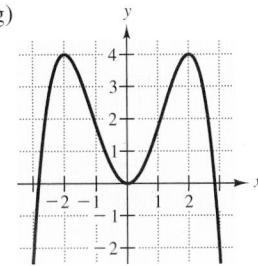

(h)
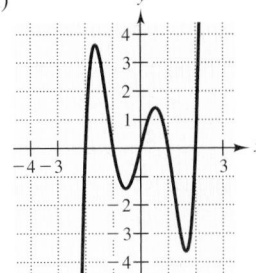

(c)
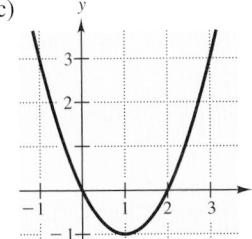

(d)
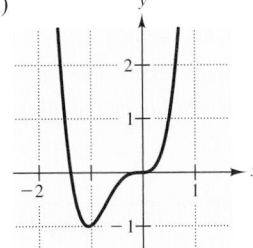

51. $f(x) = -3x + 5$ **52.** $f(x) = x^2 - 2x$

53. $f(x) = -2x^2 - 8x - 9$ **54.** $f(x) = 3x^3 - 9x + 1$

55. $f(x) = -\frac{1}{3}x^3 + x - \frac{2}{3}$ **56.** $f(x) = -\frac{1}{4}x^4 + 2x^2$

57. $f(x) = 3x^4 + 4x^3$ **58.** $f(x) = x^5 - 5x^3 + 4x$

In Exercises 59–70, sketch the graph of the polynomial function.

59. $f(x) = -\frac{3}{2}$ **60.** $h(x) = \frac{1}{3}x - 3$

61. $f(x) = -x^3$ **62.** $g(x) = -x^4$

63. $f(x) = x^5 + 1$ **64.** $f(x) = x^4 - 2$

65. $h(x) = (x + 2)^4$ **66.** $g(x) = (x - 4)^3$

67. $f(x) = 1 - x^6$ **68.** $g(x) = 1 - (x + 1)^6$

69. $f(x) = x^3 - 3x^2$ **70.** $f(x) = x^3 - 4x$

(e)
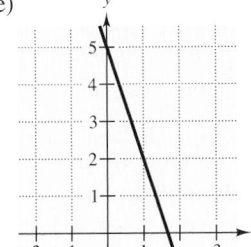

(f)
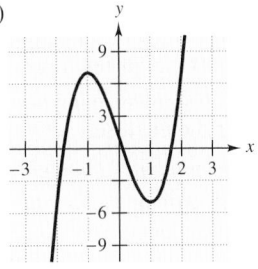

In Exercises 71–76, use a graphing utility to graph the function. Does the graph have the maximum number of turns possible for the degree of the polynomial function?

71. $f(x) = -x^3 + 4x^2$ **72.** $f(x) = \frac{1}{4}x^4 - 2x^2$

73. $g(t) = -\frac{1}{4}(t - 2)^2(t + 2)^2$

74. $f(x) = x^2(x - 4)$

75. $h(s) = s^5 + s$ **76.** $g(t) = t^5 + 3t^3 - t$

Solving Problems

77. *Geometry* An open box is made from a 12-inch-square piece of material by cutting equal squares from the corners and turning up the sides (see figure).

(a) Verify that the volume of the box is

$$V(x) = 4x(6 - x)^2.$$

(b) Determine the domain of the function V.

(c) Use a graphing utility to graph the function and use the graph to estimate the value of x for which $V(x)$ is a maximum.

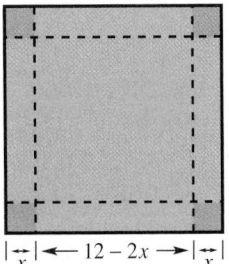

Figure for 77

78. *Environment* The growth of a red oak tree is approximated by the function

$$G = -0.003t^3 + 0.137t^2 + 0.458t - 0.839,$$

$$2 \leq t \leq 34$$

where G is the height of the tree (in feet) and t is its age (in years).

(a) Use a graphing utility to graph the function.

(b) Use the graph to estimate the age of the tree when it is growing most rapidly. This point is called the *point of diminishing returns* because the increase in size will be less with each additional year. (*Hint:* Use a viewing window in which $2 \leq x \leq 34$ and $0 \leq y \leq 60$.)

Explaining Concepts

79. Explain how to find the x- and y-intercepts of a polynomial function.

80. Explain how to determine the right-hand and left-hand behavior of a polynomial function without looking at its graph.

81. Does a polynomial function of degree n have $n - 1$ x-intercepts? Explain.

82. Can a fifth-degree polynomial have five turning points in its graph? Explain.

Ongoing Review

In Exercises 83–88, factor the expression completely.

83. $6x^2 + 23x - 13$

84. $5x^3 - 22x^2 + 24x$

85. $10x^4 + 22x^3 + 4x^2$

86. $8t^3 + 27$

87. $x^3 + 6x^2 - 2x - 12$

88. $t^3 - 4t^2 + 2t - 8$

In Exercises 89–92, solve the equation. Give all real and imaginary solutions of the equation.

89. $3x^2 + 5 = 0$

90. $x^2 - 8x + 6 = 0$

91. $6x^2 + 3x - 5 = 0$

92. $x^4 + 14x^2 + 49 = 0$

93. *Parking Lot Construction* A community college wants to construct a rectangular parking lot on land bordered on one side by a highway. The school has 320 feet of fencing with which to fence off the other three sides. What should the dimensions of the lot be if the enclosed area is to be maximized?

94. *Electrical Current* The current in a simple electrical circuit is inversely proportional to the resistance. The current is 20 amps when the resistance is 5 ohms. Find the current when the resistance is 8 ohms.

Looking Further

(a) Find the zeros of the quadratic function $g(x)$.

(i) $g(x) = x^2 - 4x - 12$

(ii) $g(x) = x^2 + 5x$

(iii) $g(x) = x^2 + 3x - 10$

(iv) $g(x) = x^2 - 4x + 4$

(v) $g(x) = x^2 - 2x - 6$

(vi) $g(x) = x^2 + 3x + 4$

(b) For each function in part (a), use a graphing utility to graph $f(x) = (x - 2) \cdot g(x)$. Verify that $(2, 0)$ is an x-intercept of the graph of $f(x)$. Describe the behavior of each function at this x-intercept.

(c) For each function in part (b), use the graph of $f(x)$ to approximate the other x-intercepts of the graph.

(d) Describe the connections that you find among the results of parts (a), (b), and (c).

Mid-Chapter Quiz

Take this quiz as you would take a quiz in class. After you are done, check your work against the answers given in the back of the book.

In Exercises 1–6, let $f(x) = 2x - 8$ and $g(x) = x^2$. Find the specified combination of the functions and find its domain.

1. $(f + g)(x)$ **2.** $(fg)(x)$ **3.** $(g - f)(x)$

4. $(f/g)(x)$ **5.** $(f \circ g)(x)$ **6.** $(g \circ f)(x)$

In Exercises 7–12, let $f(x) = \sqrt{x}$ and $g(x) = x^2 + 2x$. Evaluate the combination of the functions f and g at the indicated x-value.

7. $(f + g)(9)$ **8.** $(f - g)(4)$ **9.** $(fg)(1)$

10. $(f/g)(16)$ **11.** $(f \circ g)(-5)$ **12.** $(g \circ f)(7)$

13. The weekly cost of producing x units in a manufacturing process is given by the function $C(x) = 60x + 750$. The number of units produced in t hours during a week is given by $x(t) = 50t$, $0 \le t \le 40$. Find, simplify, and interpret $(C \circ x)(t)$.

14. Verify algebraically that the functions $f(x) = 100 - 15x$ and $g(x) = \frac{1}{15}(100 - x)$ are inverse functions of each other.

15. Find the inverse function of $f(x) = x^3 - 8$. Verify your result algebraically, graphically, and numerically.

In Exercises 16–18, find the constant of proportionality and give the equation relating the variables.

16. z varies directly as t, and $z = 12$ when $t = 4$.

17. S varies jointly as h and the square of r, and $S = 120$ when $h = 6$ and $r = 2$.

18. N varies directly as the square of t and inversely as s, and $N = 300$ when $t = 10$ and $s = 5$.

19. One of the dangers of coal mining is the methane gas that can leak out of seams in the rock. Methane forms an explosive mixture with air at a concentration of 5% or greater. A steady leak of methane begins in a coal mine such that the concentration of methane gas varies directly as time. Twelve minutes after the leak begins, the concentration of methane in the air is 2%. If the leak continues at the same rate, when could an explosion occur?

20. Does the graph of $h(x) = 3x^5 - 2x^2 + 5x - 7$ rise to the right or fall to the right? Explain your reasoning.

21. Use the graph of $y = x^3 - 2x^2$ shown in the figure to sketch the graphs of the specified transformations.

(a) $f(x) = -x^3 + 2x^2$ (b) $f(x) = (x - 2)^3 - 2(x - 2)^2$

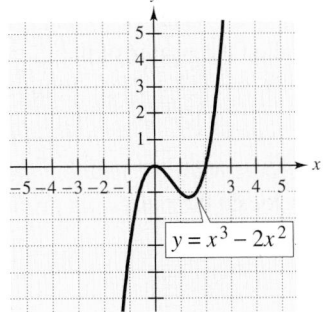

$y = x^3 - 2x^2$

Figure for 21

8.5 Circles

What you should learn:

- How to recognize the four basic conics: circles, parabolas, ellipses, and hyperbolas
- How to graph and find equations of circles centered at the origin
- How to graph and find equations of circles centered at (h, k)

Why you should learn it:

Circles can be used to model and solve scientific problems. For instance, in Exercise 43 on page 545 you will find the equation of a circular orbit of a satellite.

The Conics

In Section 6.5, you saw that the graph of a second-degree equation

$$y = ax^2 + bx + c$$

is a parabola. A parabola is one of four types of curves that are called **conics** or **conic sections.** The other three types are circles, ellipses, and hyperbolas. All four types have equations that are of second degree. As indicated in Figure 8.23, the name "conic" relates to the fact that each of these figures can be obtained by intersecting a plane with a double-napped cone.

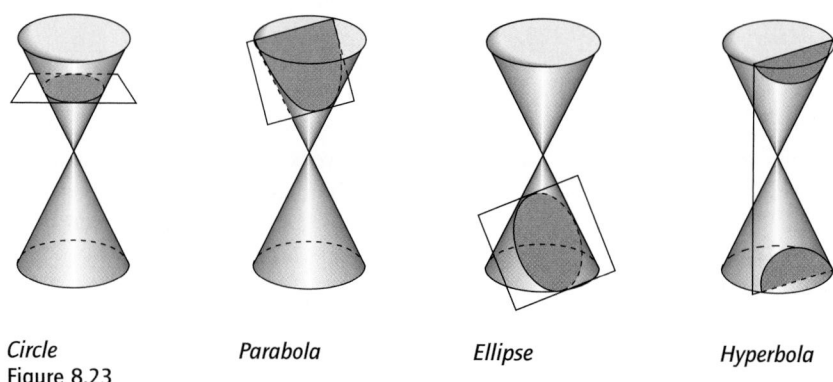

| *Circle* | *Parabola* | *Ellipse* | *Hyperbola* |

Figure 8.23

Conic sections occur in many practical applications. Reflective surfaces in satellite dishes, flashlights, and telescopes often have a parabolic shape. The orbits of planets are elliptical, and the orbits of comets are usually either elliptical or hyperbolic. Ellipses and parabolas are also used in the construction of archways and bridges.

Circles Centered at the Origin

> **Definition of a Circle**
>
> A **circle** in the rectangular coordinate plane consists of all points (x, y) that are a given positive distance r from a fixed point, called the **center** of the circle. The distance r is called the **radius** of the circle.

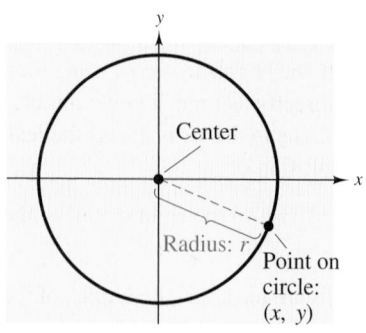

Figure 8.24

If the center of the circle is the origin, as shown in Figure 8.24, the relationship between the coordinates of any point (x, y) on the circle and the radius r is given by

$$\text{Radius} = r = \sqrt{(x - 0)^2 + (y - 0)^2} \qquad \text{Distance Formula (See Section 2.1.)}$$
$$= \sqrt{x^2 + y^2}.$$

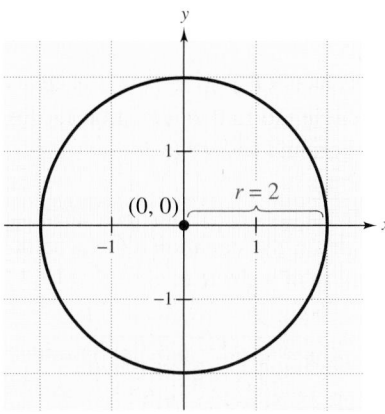

Figure 8.25

By squaring each side of this equation, you obtain the equation shown below, which is called the **standard form of the equation of a circle centered at the origin.**

> **Standard Equation of a Circle (Center at Origin)**
>
> The **standard form of the equation of a circle centered at the origin** is
>
> $$x^2 + y^2 = r^2.$$
>
> The positive number r is called the **radius** of the circle.

Example 1 ■ Finding an Equation of a Circle

Find an equation of the circle that is centered at the origin and has a radius of 2 (see Figure 8.25).

Solution

Using the standard form of the equation of a circle (with center at the origin) and $r = 2$, you obtain

$x^2 + y^2 = r^2$	Standard form with center at $(0, 0)$
$x^2 + y^2 = 2^2$	Substitute 2 for r.
$x^2 + y^2 = 4.$	Equation of circle

To sketch the circle for a given equation, first write the equation in standard form. Then, from the standard form, you can identify the center and radius, and sketch the circle.

Example 2 ■ Sketching a Circle

Identify the center and radius of the circle given by the equation, and sketch the circle.

$$4x^2 + 4y^2 - 25 = 0$$

Solution

Begin by writing the equation in standard form.

$4x^2 + 4y^2 - 25 = 0$	Write original equation.
$4x^2 + 4y^2 = 25$	Add 25 to each side.
$x^2 + y^2 = \dfrac{25}{4}$	Divide each side by 4.
$x^2 + y^2 = \left(\dfrac{5}{2}\right)^2$	Standard form

Now, from this standard form, you can see that the graph of the equation is a circle that is centered at the origin and has a radius of $\frac{5}{2}$. The graph of the equation of the circle is shown in Figure 8.26.

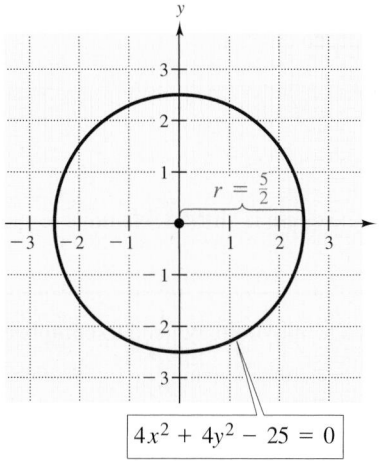

Figure 8.26

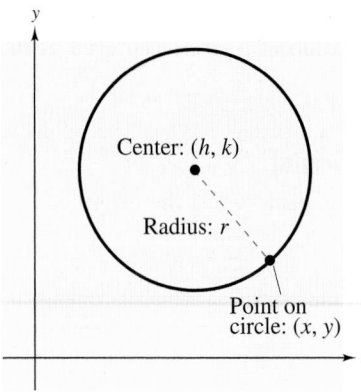

Figure 8.27

Circles Centered at (h, k)

Consider a circle whose radius is r and whose center is the point (h, k), as shown in Figure 8.27. Let (x, y) be any point on the circle. To find an equation for this circle, you can use a variation of the Distance Formula and write

$$\text{Radius} = r = \sqrt{(x - h)^2 + (y - k)^2}. \qquad \text{Distance Formula (See Section 2.1.)}$$

By squaring each side of this equation, you obtain the equation shown below, which is called the **standard form of the equation of a circle centered at (h, k).**

Standard Equation of a Circle [Center at (h, k)]

The **standard form of the equation of a circle centered at (h, k)** is

$$(x - h)^2 + (y - k)^2 = r^2.$$

> **NOTE** When $h = 0$ and $k = 0$, the circle is centered at the origin. Otherwise, you can use the rules on horizontal and vertical shifts from Section 2.6 to shift the center of the circle h units horizontally and k units vertically from the origin.

Example 3 ■ Finding an Equation of a Circle

The point $(2, 5)$ lies on a circle whose center is at $(5, 1)$, as shown in Figure 8.28. Find the standard form of the equation of this circle.

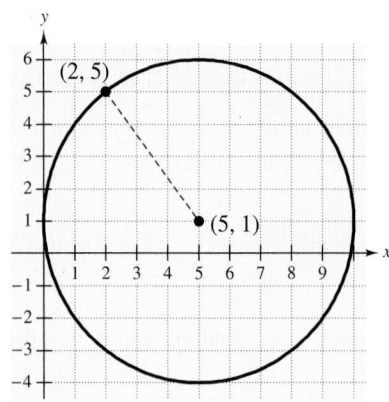

Figure 8.28

Solution

The radius r of the circle is the distance between $(2, 5)$ and $(5, 1)$.

$$\begin{aligned}
r &= \sqrt{(2 - 5)^2 + (5 - 1)^2} && \text{Distance Formula} \\
&= \sqrt{(-3)^2 + 4^2} && \text{Simplify.} \\
&= \sqrt{9 + 16} && \text{Simplify.} \\
&= \sqrt{25} && \text{Simplify.} \\
&= 5 && \text{Radius}
\end{aligned}$$

Using $(h, k) = (5, 1)$ and $r = 5$, the equation of the circle is

$$\begin{aligned}
(x - h)^2 + (y - k)^2 &= r^2 && \text{Standard form} \\
(x - 5)^2 + (y - 1)^2 &= 5^2 && \text{Substitute for } h, k, \text{ and } r. \\
(x - 5)^2 + (y - 1)^2 &= 25. && \text{Equation of circle}
\end{aligned}$$

From the graph, you can see that the center of the circle is shifted five units to the right and one unit upward from the origin.

To write the equation of a circle in standard form, you may need to complete the square, as demonstrated in Example 4.

Example 4 ■ Writing an Equation in Standard Form

Write the equation

$$x^2 + y^2 - 2x + 4y - 4 = 0$$

in standard form, and sketch the circle represented by the equation.

Solution

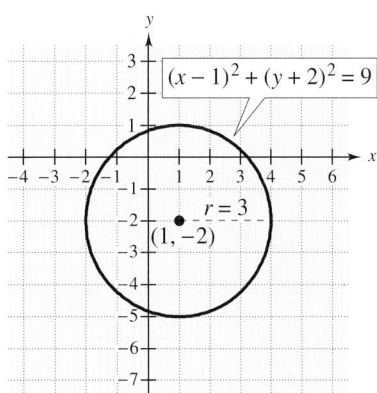

$$x^2 + y^2 - 2x + 4y - 4 = 0 \qquad \text{Write original equation.}$$
$$\left(x^2 - 2x + \boxed{}\right) + \left(y^2 + 4y + \boxed{}\right) = 4 \qquad \text{Group terms.}$$
$$\left[x^2 - 2x + (-1)^2\right] + (y^2 + 4y + 2^2) = 4 + 1 + 4 \qquad \text{Complete the square.}$$
$$\underbrace{\qquad}_{(\text{half})^2} \qquad \underbrace{\qquad}_{(\text{half})^2}$$
$$(x - 1)^2 + (y + 2)^2 = 9 \qquad \text{Standard form}$$

From this standard form, you can see that the circle has a radius of 3 and that the center of the circle is $(1, -2)$. The graph of the equation of the circle is shown in Figure 8.29. From the graph you can see that the center of the circle is shifted one unit to the right and two units downward from the origin.

Figure 8.29

Example 5 ■ An Application: Mechanical Drawing

You are in a mechanical drawing class and are asked to help program a computer to model the metal piece shown in Figure 8.30. Part of your assignment is to find an equation for the semicircular upper portion of the hole in the metal piece. What is the equation?

Solution

From the drawing, you can see that the center of the circle is $(h, k) = (5, 2)$ and that the radius of the circle is $r = 1.5$. This implies that the equation of the entire circle is

$$(x - h)^2 + (y - k)^2 = r^2 \qquad \text{Standard form}$$
$$(x - 5)^2 + (y - 2)^2 = 1.5^2 \qquad \text{Substitute for } h, k, \text{ and } r.$$
$$(x - 5)^2 + (y - 2)^2 = 2.25 \qquad \text{Equation of circle}$$

To find the equation of the upper portion of the circle, solve this standard equation for y.

$$(x - 5)^2 + (y - 2)^2 = 2.25$$
$$(y - 2)^2 = 2.25 - (x - 5)^2$$
$$y - 2 = \pm\sqrt{2.25 - (x - 5)^2}$$
$$y = 2 \pm \sqrt{2.25 - (x - 5)^2}$$

Finally, take the positive square root to obtain the upper portion of the circle

$$y = 2 + \sqrt{2.25 - (x - 5)^2}.$$

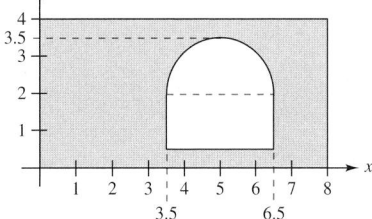

Figure 8.30

NOTE   In Example 5, if you had wanted the lower portion of the circle, you would have taken the negative square root $y = 2 - \sqrt{2.25 - (x - 5)^2}$.

8.5 ■ Exercises

Developing Skills

In Exercises 1–6, match the equation with its graph. [The graphs are labeled (a), (b), (c), (d), (e), and (f).]

(a)
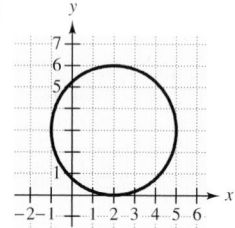

(b)
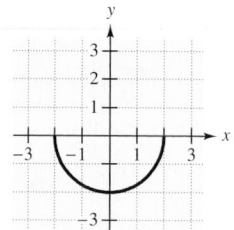

(c)
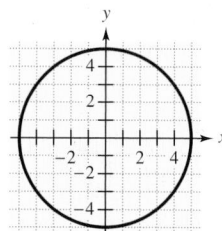

(d)
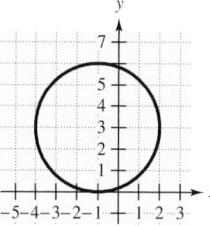

(e)
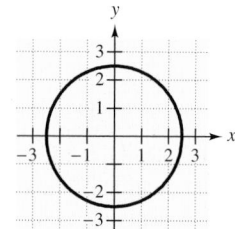

(f)
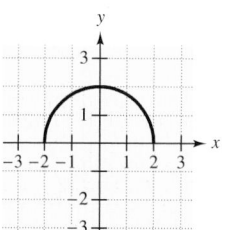

1. $x^2 + y^2 = 25$
2. $4x^2 + 4y^2 = 25$
3. $(x - 2)^2 + (y - 3)^2 = 9$
4. $(x + 1)^2 + (y - 3)^2 = 9$
5. $y = -\sqrt{4 - x^2}$
6. $y = \sqrt{4 - x^2}$

In Exercises 7–14, find an equation of the circle with center at $(0, 0)$ that satisfies the criterion.

7. Radius: 5
8. Radius: 7
9. Radius: $\frac{2}{3}$
10. Radius: $\frac{5}{2}$
11. Passes through the point $(0, 8)$
12. Passes through the point $(-2, 0)$
13. Passes through the point $(5, 2)$
14. Passes through the point $(-1, -4)$

In Exercises 15–22, find an equation of the circle with center at (h, k) that satisfies the criteria.

15. Center: $(4, 3)$; Radius: 10
16. Center: $(-2, 5)$; Radius: 6
17. Center: $(5, -3)$; Radius: 9
18. Center: $(-5, -2)$; Radius: $\frac{5}{2}$
19. Center: $(-2, 1)$; Passes through the point $(0, 1)$
20. Center: $(8, 2)$; Passes through the point $(8, 0)$
21. Center: $(3, 2)$; Passes through the point $(4, 6)$
22. Center: $(-3, -5)$; Passes through the point $(0, 0)$

In Exercises 23–36, identify the center and radius of the circle and sketch its graph.

23. $x^2 + y^2 = 16$
24. $x^2 + y^2 = 25$
25. $x^2 + y^2 = 36$
26. $x^2 + y^2 = 10$
27. $4x^2 + 4y^2 = 1$
28. $9x^2 + 9y^2 = 64$
29. $25x^2 + 25y^2 - 144 = 0$
30. $\dfrac{x^2}{4} + \dfrac{y^2}{4} - 1 = 0$
31. $(x + 1)^2 + (y - 5)^2 = 64$
32. $(x + 10)^2 + (y + 1)^2 = 100$
33. $(x - 2)^2 + (y - 3)^2 = 4$
34. $(x + 4)^2 + (y - 3)^2 = 25$
35. $\left(x + \frac{5}{2}\right)^2 + (y + 3)^2 = 9$
36. $(x - 5)^2 + \left(y + \frac{3}{4}\right)^2 = 1$

In Exercises 37–42, write the equation in standard form. Then sketch the circle.

37. $x^2 + y^2 - 4x - 2y + 1 = 0$
38. $x^2 + y^2 + 6x - 4y - 3 = 0$
39. $x^2 + y^2 + 2x + 6y + 6 = 0$
40. $x^2 + y^2 - 2x + 6y - 15 = 0$
41. $x^2 + y^2 + 8x + 4y - 5 = 0$
42. $x^2 + y^2 - 14x + 8y + 56 = 0$

Solving Problems

43. *Satellite Orbit*   Find an equation of the circular orbit of a satellite 500 miles above the surface of Earth. Place the origin of the rectangular coordinate system at the center of Earth and assume that the radius of Earth is 4000 miles.

44. *Height of an Arch*   A semicircular arch for a tunnel under a river has a diameter of 100 feet (see figure). Determine the height of the arch 5 feet from the edge of the tunnel.

45. *Architecture*   The top portion of a stained-glass window is in the form of a pointed Gothic arch (see figure). Each side of the arch is an arc of a circle of radius 12 feet with its center at the base of the

opposite arch. Find an equation of one of the circles and use it to determine the height of the point of the arch above the rectangular portion of the window.

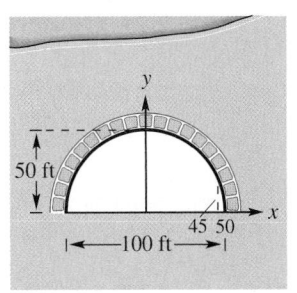

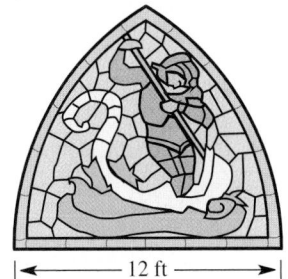

Figure for 44   Figure for 45

Explaining Concepts

46. Name the four types of conics.

47. Define a circle and give the standard form of the equation of a circle centered at the origin.

48. Explain how to use the method of completing the square to write an equation of a circle in standard form.

Ongoing Review

In Exercises 49 and 50, solve the quadratic equation by completing the square.

49. $x^2 + 6x - 4 = 0$

50. $2x^2 - 16x + 5 = 0$

In Exercises 51–54, sketch the graph of the equation.

51. $y = \frac{2}{5}x + 3$

52. $y = -2x - 1$

53. $y = x^2 - 12x + 36$

54. $y = 25 - x^2$

55. *Recreation*   A service organization paid $288 for a block of tickets to a ball game. The block contained three more tickets than the organization needed for its members. By inviting three more people to attend (and share in the cost), the organization lowered the price per ticket by $8. How many people are going to the game?

Looking Further

Graphical Analysis   A rectangle centered at the origin with sides parallel to the coordinate axes is placed in a circle of radius 25 inches centered at the origin (see figure). The length of the rectangle is $2x$ inches. Verify that the width of the rectangle is $2\sqrt{625 - x^2}$. Find a model for the area of the rectangle. Use a graphing utility to graph the model. Approximate the value of x for which the area is a maximum. Repeat this process for a circle of radius 4 inches, and compare the results.

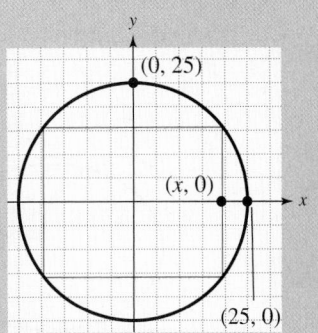

Ellipses and Hyperbolas

Ellipses Centered at the Origin

The second type of conic is called an **ellipse** and is defined as follows.

Definition of an Ellipse

An **ellipse** in the rectangular coordinate system consists of all points (x, y) such that the *sum* of the distances between (x, y) and two distinct fixed points is a constant, as shown in Figure 8.31(a). Each of the two fixed points is called a **focus** of the ellipse. (The plural of focus is **foci.**)

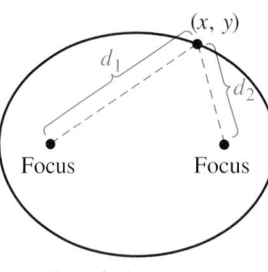

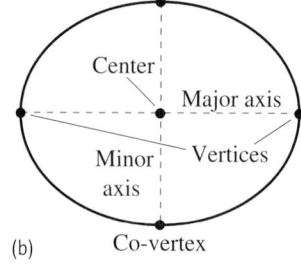

(a) $d_1 + d_2$ is constant.

(b)

Figure 8.31

The line through the foci intersects the ellipse at two points, called the **vertices,** as shown in Figure 8.31(b). The line segment joining the vertices is called the **major axis,** and its midpoint is called the **center** of the ellipse. The line segment perpendicular to the major axis at the center is called the **minor axis** of the ellipse, and the points at which the minor axis intersects the ellipse are called **co-vertices.**

EXPLORATION

Drawing an Ellipse   You can visualize the definition of an ellipse by imagining two thumbtacks placed at the foci, as shown below. If the ends of a fixed length of string are fastened to the thumbtacks and the string is drawn taut with a pencil, the path traced by the pencil will be an ellipse. Try doing this. Vary the length of the string and the distance between the thumbtacks. Explain how to obtain ellipses that are almost circular. Explain how to obtain ellipses that are long and narrow.

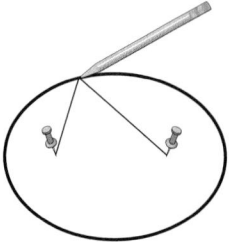

The standard form of the equation of an ellipse takes one of two forms, depending on whether the major axis is horizontal or vertical.

Standard Equation of an Ellipse (Center at Origin)

The **standard form of the equation of an ellipse centered at the origin** with major and minor axes of lengths $2a$ and $2b$ is

$$\frac{x^2}{a^2} + \frac{y^2}{b^2} = 1 \qquad \text{or} \qquad \frac{x^2}{b^2} + \frac{y^2}{a^2} = 1, \qquad 0 < b < a.$$

The vertices lie on the major axis, a units from the center, and the co-vertices lie on the minor axis, b units from the center, as shown in Figure 8.32.

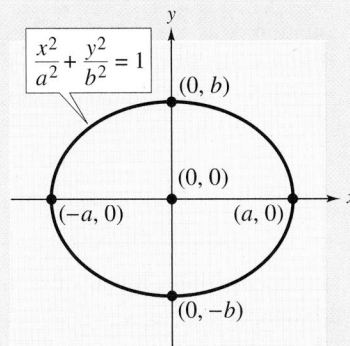

Major axis is horizontal.
Minor axis is vertical.

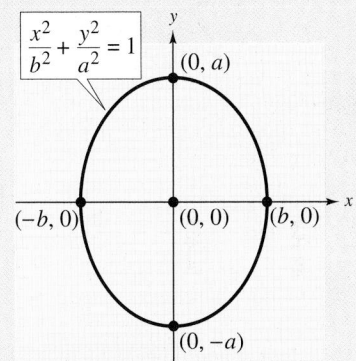

Major axis is vertical.
Minor axis is horizontal.

Figure 8.32

Technology

You can use a graphing utility to graph an ellipse by graphing the upper and lower portions in the same viewing window. For instance, to graph the ellipse $x^2 + 4y^2 = 4$, first solve for y to obtain

$$y_1 = \tfrac{1}{2}\sqrt{4 - x^2}$$

and

$$y_2 = -\tfrac{1}{2}\sqrt{4 - x^2}.$$

Use a viewing window in which $-3 \le x \le 3$ and $-2 \le y \le 2$. You should obtain the graph shown below.

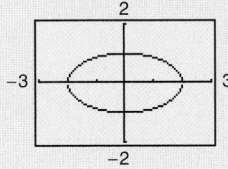

Example 1 ■ Finding the Standard Equation of an Ellipse

Find the standard form of the equation of the ellipse with vertices $(-3, 0)$ and $(3, 0)$ and co-vertices $(0, -2)$ and $(0, 2)$.

Solution

Begin by plotting the two vertices and the two co-vertices, as shown in Figure 8.33. The center of the ellipse is $(0, 0)$, because it is the point that lies halfway between the vertices (or halfway between the co-vertices). From the figure, you can see that the major axis is horizontal. So, the equation of the ellipse has the form

$$\frac{x^2}{a^2} + \frac{y^2}{b^2} = 1.$$

In this equation, a is the distance between the center and either vertex, which implies that $a = 3$. Similarly, b is the distance between the center and either co-vertex, which implies that $b = 2$. So, the standard form of the equation of the ellipse is

$$\frac{x^2}{3^2} + \frac{y^2}{2^2} = 1. \qquad \text{Standard form}$$

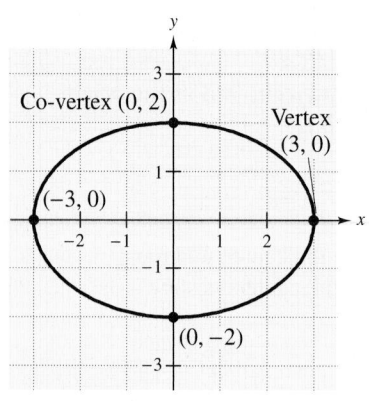

Figure 8.33

To sketch an ellipse, it helps first to write its equation in standard form, as shown in Example 2.

Example 2 ■ Sketching an Ellipse

Sketch the ellipse given by

$$4x^2 + y^2 = 36$$

and identify the vertices and co-vertices.

Solution

Begin by writing the equation in standard form.

$$4x^2 + y^2 = 36 \qquad \text{Write original equation.}$$

$$\frac{4x^2}{36} + \frac{y^2}{36} = \frac{36}{36} \qquad \text{Divide each side by 36.}$$

$$\frac{x^2}{9} + \frac{y^2}{36} = 1 \qquad \text{Simplify.}$$

$$\frac{x^2}{3^2} + \frac{y^2}{6^2} = 1 \qquad \text{Standard form}$$

Because the denominator of the y^2-term is larger than the denominator of the x^2-term, you can conclude that the major axis is vertical. Moreover, because $a = 6$, the vertices are $(0, -6)$ and $(0, 6)$. Finally, because $b = 3$, the co-vertices are $(-3, 0)$ and $(3, 0)$, as shown in Figure 8.34.

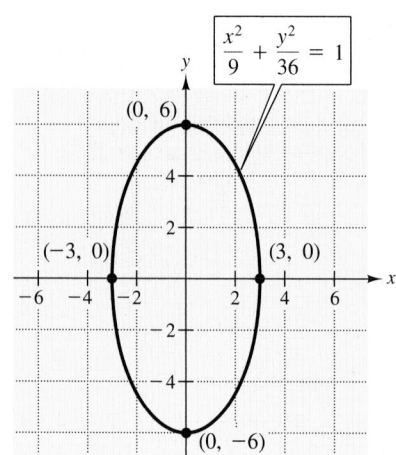

Figure 8.34

Ellipses Centered at (h, k)

> **Standard Equation of an Ellipse [Center at (h, k)]**
> The **standard form of the equation of an ellipse centered at (h, k)** with major and minor axes of lengths $2a$ and $2b$, where $0 < b < a$, is
>
> $$\frac{(x - h)^2}{a^2} + \frac{(y - k)^2}{b^2} = 1 \qquad \text{Major axis is horizontal.}$$
>
> or
>
> $$\frac{(x - h)^2}{b^2} + \frac{(y - k)^2}{a^2} = 1. \qquad \text{Major axis is vertical.}$$
>
> The foci lie on the major axis, c units from the center, with $c^2 = a^2 - b^2$.

Figure 8.35 shows both the horizontal and vertical orientations for an ellipse.

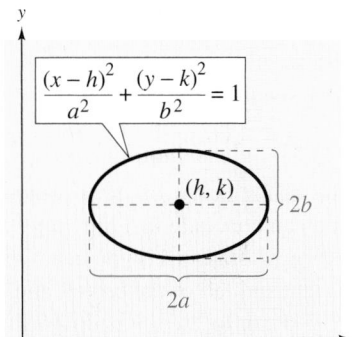 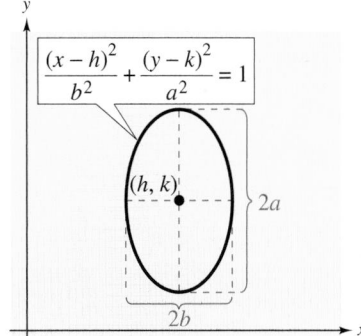

Figure 8.35

NOTE   When $h = 0$ and $k = 0$, the ellipse is centered at the origin. Otherwise, you can use the rules on horizontal and vertical shifts from Section 2.6 to shift the center of the ellipse h units horizontally and k units vertically from the origin.

Example 3 ■ Finding the Standard Equation of an Ellipse

Find the standard form of the equation of the ellipse with vertices $(-2, 2)$ and $(4, 2)$ and co-vertices $(1, 3)$ and $(1, 1)$, as shown in Figure 8.36.

Solution
Because the vertices are $(-2, 2)$ and $(4, 2)$, the center of the ellipse is $(h, k) = (1, 2)$. The distance from the center to either vertex is $a = 3$, and the distance to either co-vertex is $b = 1$. Because the major axis is horizontal, the standard form of the equation is

$$\frac{(x - 1)^2}{9} + \frac{(y - 2)^2}{1} = 1. \qquad \text{Standard form}$$

From the graph, you can see that the center of the ellipse is shifted one unit to the right and two units upward from the origin.

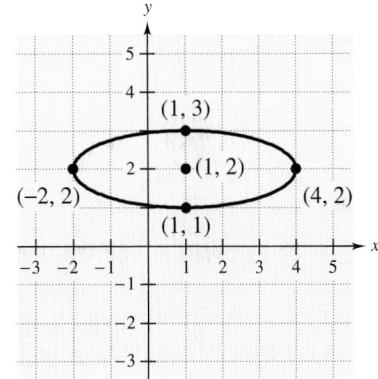

Figure 8.36

To write an equation of an ellipse in standard form, you must group the x-terms and the y-terms and then complete each square, as shown in Example 4.

Example 4 ■ Sketching an Ellipse

Sketch the ellipse given by $4x^2 + y^2 - 8x + 6y + 9 = 0$.

Solution

Begin by writing the equation in standard form. In the fourth step, note that 9 and 4 are added to *each* side of the equation.

$$4x^2 + y^2 - 8x + 6y + 9 = 0 \qquad \text{Write original equation.}$$

$$(4x^2 - 8x + \quad) + (y^2 + 6y + \quad) = -9 \qquad \text{Group terms.}$$

$$4(x^2 - 2x + \quad) + (y^2 + 6y + \quad) = -9 \qquad \text{Factor 4 out of } x\text{-terms.}$$

$$4(x^2 - 2x + 1) + (y^2 + 6y + 9) = -9 + 4(1) + 9 \qquad \text{Complete each square.}$$

$$4(x - 1)^2 + (y + 3)^2 = 4 \qquad \text{Simplify.}$$

$$\frac{(x-1)^2}{1} + \frac{(y+3)^2}{4} = 1 \qquad \text{Divide each side by 4.}$$

Now you can see that the center of the ellipse is at

$$(h, k) = (1, -3). \qquad \text{Center of ellipse}$$

Because the denominator of the y^2-term is larger than the denominator of the x^2-term, you can conclude that the major axis is vertical. Because the denominator of the x^2-term is $b^2 = 1^2$, you can locate the endpoints of the minor axis one unit to the right and left of the center, and because the denominator of the y^2-term is $a^2 = 2^2$, you can locate the endpoints of the major axis two units upward and downward from the center, as shown in Figure 8.37(a). To complete the graph, sketch an oval shape that is determined by the vertices and co-vertices, as shown in Figure 8.37(b).

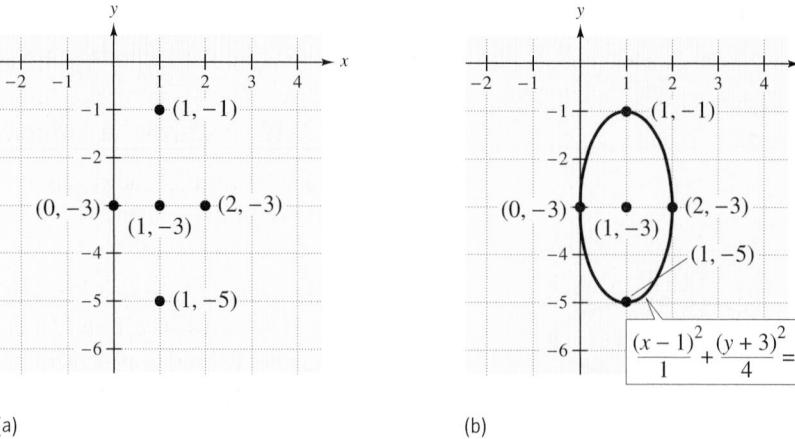

(a) (b)

Figure 8.37

From Figure 8.37(b), you can see that the center of the ellipse is shifted one unit to the right and three units downward from the origin.

Hyperbolas Centered at the Origin

The third basic type of conic is called a **hyperbola** and is defined as follows.

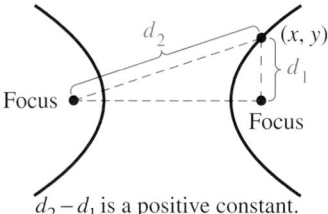

Figure 8.38

> ### Definition of a Hyperbola
>
> A **hyperbola** on the rectangular coordinate system consists of all points (x, y) such that the *difference* of the distances between (x, y) and two fixed points is a positive constant, as shown in Figure 8.38. The two fixed points are called the **foci** of the hyperbola. The line on which the foci lie is called the **transverse axis** of the hyperbola.

> ### Standard Equation of a Hyperbola (Center at Origin)
>
> The **standard form of the equation of a hyperbola centered at the origin** is
>
> $$\frac{x^2}{a^2} - \frac{y^2}{b^2} = 1 \qquad \text{Transverse axis is horizontal.}$$
>
> or
>
> $$\frac{y^2}{a^2} - \frac{x^2}{b^2} = 1 \qquad \text{Transverse axis is vertical.}$$
>
> where a and b are positive real numbers. The **vertices** of the hyperbola lie on the transverse axis, a units from the center, as shown in Figure 8.39.

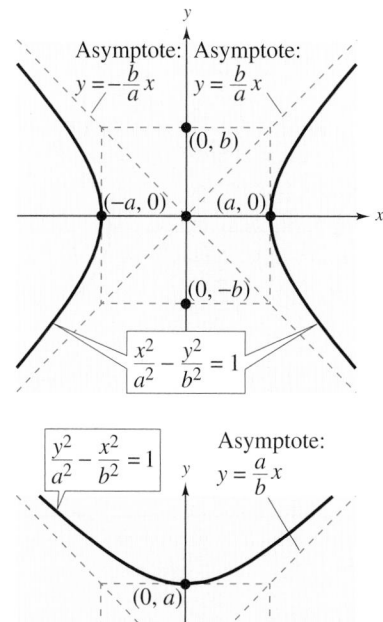

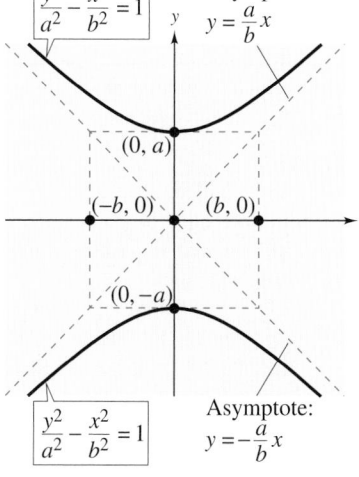

Figure 8.40

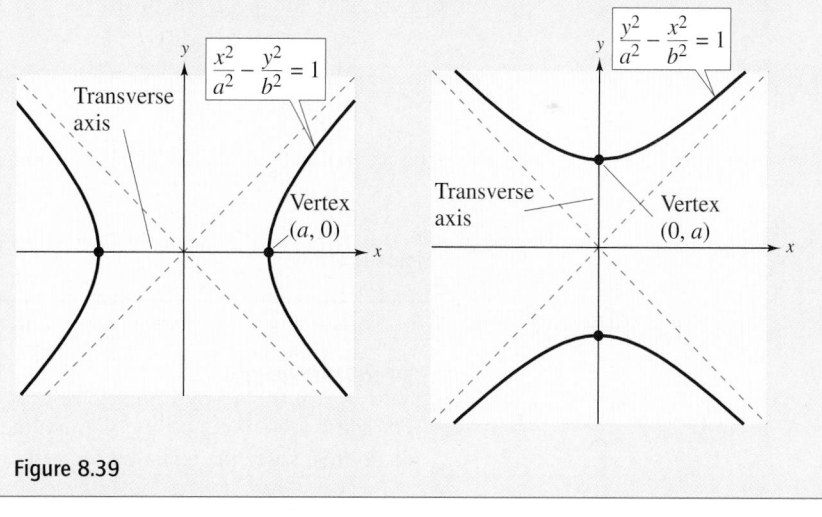

Figure 8.39

A hyperbola has two disconnected parts, each of which is called a **branch** of the hyperbola. The two branches approach a pair of intersecting lines called the **asymptotes** of the hyperbola. The two asymptotes intersect at the center of the hyperbola. To sketch a hyperbola, form a **central rectangle** that is centered at the origin and has side lengths of $2a$ and $2b$. Note in Figure 8.40 that the asymptotes pass through the corners of the central rectangle and that the vertices of the hyperbola lie at the centers of opposite sides of the central rectangle.

Example 5 ■ Sketching a Hyperbola

Identify the vertices of the hyperbola given by the equation, and sketch the hyperbola.

$$\frac{x^2}{36} - \frac{y^2}{16} = 1$$

Solution

From the standard form of the equation

$$\frac{x^2}{6^2} - \frac{y^2}{4^2} = 1$$

you can see that the center of the hyperbola is the origin and the transverse axis is horizontal. So, the vertices lie six units to the left and right of the center at the points $(-6, 0)$ and $(6, 0)$. Because $a = 6$ and $b = 4$, you can sketch the hyperbola by first drawing a central rectangle with a width of $2a = 12$ and a height of $2b = 8$. Next, draw the asymptotes of the hyperbola through the corners of the central rectangle and plot the vertices, as shown in Figure 8.41(a). Finally, sketch the hyperbola, as shown in Figure 8.41(b).

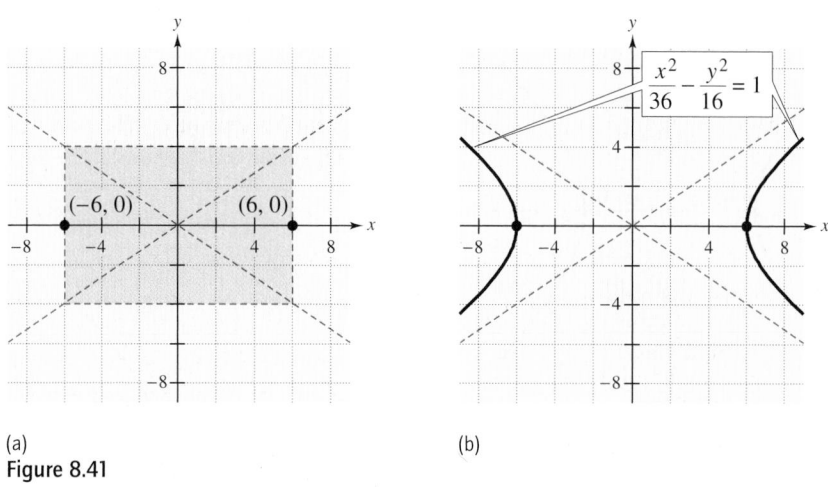

(a)   (b)

Figure 8.41

Technology

Try using a graphing utility to graph the hyperbola shown in Example 5. To do this, solve the equation for y to obtain two functions of y.

$$y_1 = \sqrt{\frac{4x^2}{9} - 16}$$

and

$$y_2 = -\sqrt{\frac{4x^2}{9} - 16}$$

Then graph both functions in the same viewing window.

In Example 5, the transverse axis of the hyperbola is horizontal and the two branches of the hyperbola open to the right and to the left. A hyperbola of the form

$$\frac{y^2}{a^2} - \frac{x^2}{b^2} = 1$$

has a transverse axis that is vertical and its two branches open upward and downward, as shown in Example 6.

Example 6 ■ Sketching a Hyperbola

Identify the vertices of the hyperbola given by the equation, and sketch the hyperbola.

$$4y^2 - 9x^2 = 36$$

Solution

To begin, write the equation in standard form.

$$4y^2 - 9x^2 = 36 \qquad \text{Write original equation.}$$

$$\frac{4y^2}{36} - \frac{9x^2}{36} = \frac{36}{36} \qquad \text{Divide each side by 36.}$$

$$\frac{y^2}{9} - \frac{x^2}{4} = 1 \qquad \text{Simplify.}$$

$$\frac{y^2}{3^2} - \frac{x^2}{2^2} = 1 \qquad \text{Standard form}$$

Now, from the standard form, you can see that the center of the hyperbola is the origin and the transverse axis is vertical. So, the vertices lie three units above and below the center at $(0, 3)$ and $(0, -3)$. Because $a = 3$ and $b = 2$, you can sketch the hyperbola by first drawing a central rectangle with a width of $2b = 4$ and a height of $2a = 6$. Next, draw the asymptotes of the hyperbola through the corners of the central rectangle and plot the vertices, as shown in Figure 8.42(a). Finally, sketch the hyperbola, as shown in Figure 8.42(b).

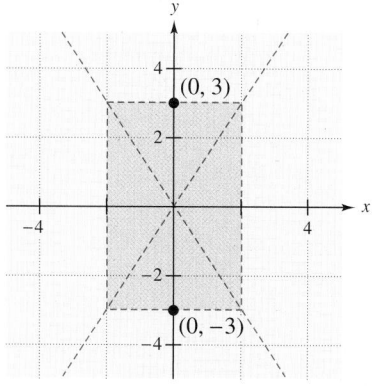

(a)

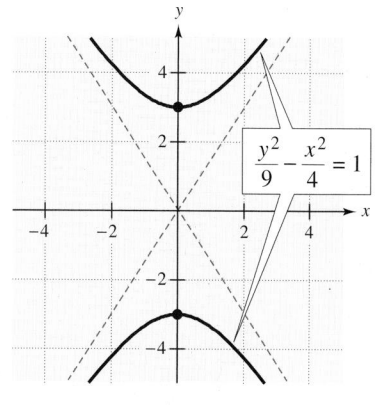

(b)

Figure 8.42

Finding the equation of a hyperbola is a little more difficult than finding the equation of one of the other types of conics. However, if you know the vertices and the asymptotes, you can find the values of a and b, which enable you to write the equation. Notice in Example 7 that the key to this procedure is knowing that the central rectangle has a width of $2b$ and a height of $2a$.

Example 7 ■ Finding the Standard Equation of a Hyperbola

Find the standard form of the equation of the hyperbola with a vertical transverse axis and vertices $(0, 3)$ and $(0, -3)$. The equations of the asymptotes of the hyperbola are $y = \frac{3}{5}x$ and $y = -\frac{3}{5}x$.

Solution

To begin, sketch the lines that represent the asymptotes, as shown in Figure 8.43(a). Note that these two lines intersect at the origin, which implies that the center of the hyperbola is $(0, 0)$. Next, plot the two vertices at the points $(0, 3)$ and $(0, -3)$. Because you know where the vertices are located, you can sketch the central rectangle of the hyperbola, as shown in Figure 8.43(a). Note that the corners of the central rectangle occur at the points

$$(-5, 3), (5, 3), (-5, -3), \text{ and } (5, -3).$$

Because the width of the central rectangle is $2b = 10$, it follows that $b = 5$. Similarly, because the height of the central rectangle is $2a = 6$, it follows that $a = 3$. Now that you know the values of a and b, you can use the standard form of the equation of the hyperbola to write the equation.

$$\frac{y^2}{a^2} - \frac{x^2}{b^2} = 1 \qquad \text{Transverse axis is vertical.}$$

$$\frac{y^2}{3^2} - \frac{x^2}{5^2} = 1 \qquad \text{Substitute 3 for } a \text{ and 5 for } b.$$

$$\frac{y^2}{9} - \frac{x^2}{25} = 1 \qquad \text{Equation of the hyperbola}$$

The graph is shown in Figure 8.43(b).

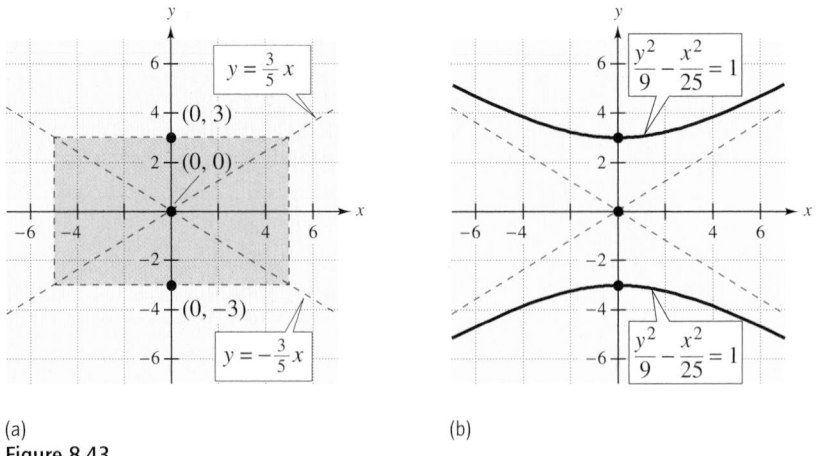

(a)   (b)

Figure 8.43

Hyperbolas Centered at (h, k)

Standard Equation of a Hyperbola [Center at (h, k)]

The **standard form of the equation of a hyperbola centered at (h, k)** is

$$\frac{(x - h)^2}{a^2} - \frac{(y - k)^2}{b^2} = 1 \qquad \text{Transverse axis is horizontal.}$$

or

$$\frac{(y - k)^2}{a^2} - \frac{(x - h)^2}{b^2} = 1 \qquad \text{Transverse axis is vertical.}$$

where a and b are positive real numbers. The vertices lie on the transverse axis, a units from the center, as shown in Figure 8.44.

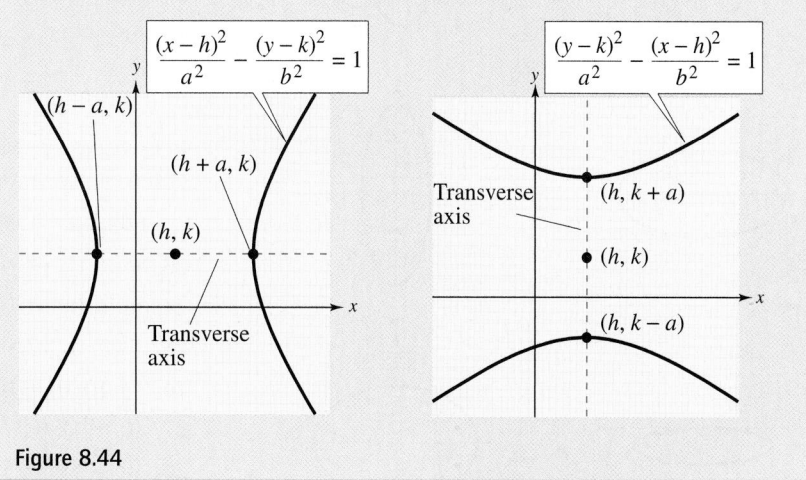

Figure 8.44

NOTE   When $h = 0$ and $k = 0$, the hyperbola is centered at the origin. Otherwise, you can use the rules on horizontal and vertical shifts from Section 2.6 to shift the center of the hyperbola h units horizontally and k units vertically from the origin.

Example 8   ■   Sketching a Hyperbola

Sketch the graph of the hyperbola given by $\dfrac{(y - 1)^2}{9} - \dfrac{(x + 2)^2}{4} = 1.$

Solution

From the form of the equation, you can see that the transverse axis is vertical. The center of the hyperbola is $(h, k) = (-2, 1)$. Because $a = 3$ and $b = 2$, you can begin by sketching a central rectangle that is six units high and four units wide, centered at $(-2, 1)$. Then, sketch the asymptotes by drawing lines through the corners of the central rectangle. Sketch the hyperbola, as shown in Figure 8.45. From the graph, you can see that the center of the hyperbola is shifted two units to the left and one unit upward from the origin.

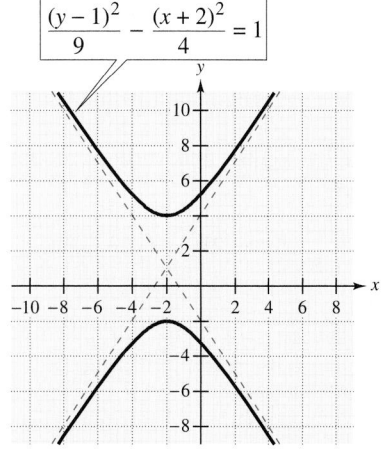

Figure 8.45

8.6 ■ Exercises

Developing Skills

In Exercises 1–6, match the equation with its graph. [The graphs are labeled (a), (b), (c), (d), (e), and (f).]

(a)

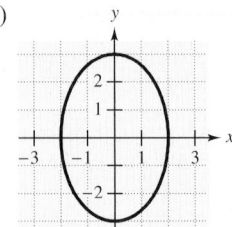

(b)

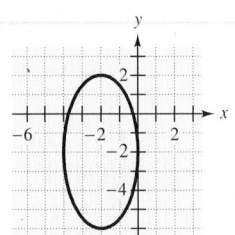

(c)

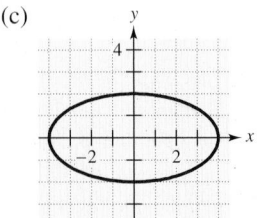

(d)

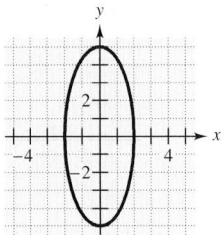

(e)

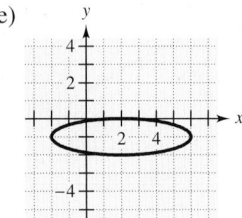

(f)

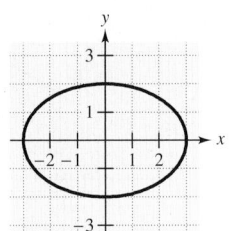

1. $\dfrac{x^2}{4} + \dfrac{y^2}{9} = 1$ **2.** $\dfrac{x^2}{9} + \dfrac{y^2}{4} = 1$

3. $\dfrac{x^2}{4} + \dfrac{y^2}{25} = 1$ **4.** $\dfrac{y^2}{4} + \dfrac{x^2}{16} = 1$

5. $\dfrac{(x - 2)^2}{16} + \dfrac{(y + 1)^2}{1} = 1$

6. $\dfrac{(x + 2)^2}{4} + \dfrac{(y + 2)^2}{16} = 1$

In Exercises 7–18, sketch the ellipse. Identify its vertices and co-vertices.

7. $\dfrac{x^2}{16} + \dfrac{y^2}{4} = 1$ **8.** $\dfrac{x^2}{25} + \dfrac{y^2}{9} = 1$

9. $\dfrac{x^2}{4} + \dfrac{y^2}{16} = 1$ **10.** $\dfrac{x^2}{9} + \dfrac{y^2}{25} = 1$

11. $\dfrac{x^2}{25/9} + \dfrac{y^2}{16/9} = 1$ **12.** $\dfrac{x^2}{1} + \dfrac{y^2}{1/4} = 1$

13. $\dfrac{9x^2}{4} + \dfrac{25y^2}{16} = 1$ **14.** $\dfrac{36x^2}{49} + \dfrac{16y^2}{9} = 1$

15. $16x^2 + 25y^2 - 9 = 0$

16. $64x^2 + 36y^2 - 49 = 0$

17. $4x^2 + y^2 - 4 = 0$ **18.** $4x^2 + 9y^2 - 36 = 0$

In Exercises 19–22, use a graphing utility to graph the equation. Identify the vertices. (*Note:* Solve for *y*.)

19. $x^2 + 2y^2 = 4$ **20.** $9x^2 + y^2 = 64$

21. $3x^2 + y^2 - 12 = 0$ **22.** $5x^2 + 2y^2 - 10 = 0$

In Exercises 23–34, write the standard form of the equation of the ellipse centered at the origin.

	Vertices	*Co-vertices*
23.	$(-4, 0), (4, 0)$	$(0, -3), (0, 3)$
24.	$(-4, 0), (4, 0)$	$(0, -1), (0, 1)$
25.	$(-2, 0), (2, 0)$	$(0, -1), (0, 1)$
26.	$(-10, 0), (10, 0)$	$(0, -4), (0, 4)$
27.	$(0, -4), (0, 4)$	$(-3, 0), (3, 0)$
28.	$(0, -5), (0, 5)$	$(-1, 0), (1, 0)$
29.	$(0, -2), (0, 2)$	$(-1, 0), (1, 0)$
30.	$(0, -8), (0, 8)$	$(-4, 0), (4, 0)$

31. Major axis (vertical) 10 units, minor axis 6 units

32. Major axis (horizontal) 24 units, minor axis 10 units

33. Major axis (horizontal) 20 units, minor axis 12 units

34. Major axis (horizontal) 50 units, minor axis 30 units

In Exercises 35–40, find the center and vertices of the ellipse and sketch its graph.

35. $\dfrac{(x + 5)^2}{16} + y^2 = 1$ **36.** $\dfrac{(x - 2)^2}{4} + \dfrac{(y - 3)^2}{9} = 1$

37. $\dfrac{(x - 1)^2}{9} + \dfrac{(y - 5)^2}{25} = 1$

38. $\dfrac{(x + 2)^2}{1/4} + \dfrac{(y + 4)^2}{1} = 1$

39. $9x^2 + 4y^2 + 36x - 24y + 36 = 0$

40. $9x^2 + 4y^2 - 36x + 8y + 31 = 0$

In Exercises 41–44, find an equation of the ellipse.

41.

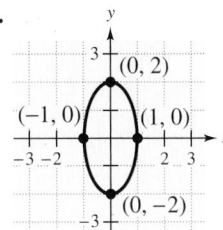

42.

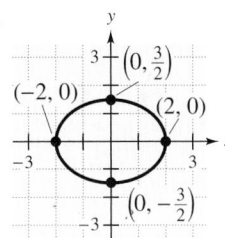

43.

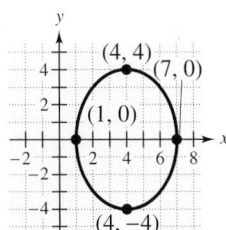

44.

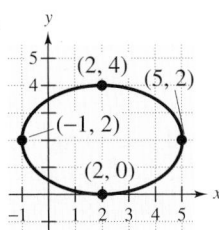

In Exercises 45–50, match the equation with its graph. [The graphs are labeled (a), (b), (c), (d), (e), and (f).]

(a)

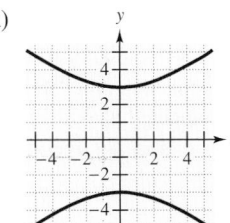

(b)

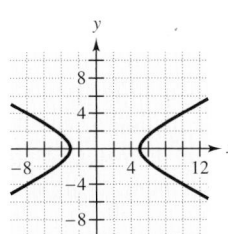

(c)

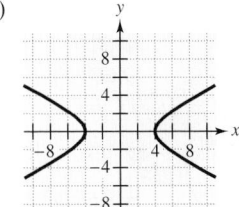

(d)

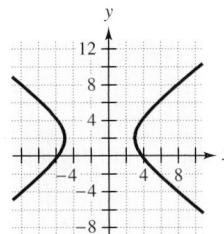

(e)

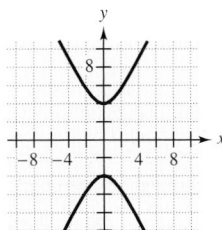

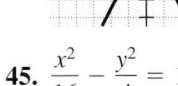

(f)

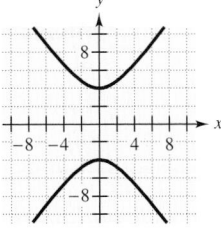

45. $\dfrac{x^2}{16} - \dfrac{y^2}{4} = 1$

46. $\dfrac{y^2}{16} - \dfrac{x^2}{4} = 1$

47. $\dfrac{y^2}{9} - \dfrac{x^2}{16} = 1$

48. $\dfrac{y^2}{16} - \dfrac{x^2}{9} = 1$

49. $\dfrac{(x - 1)^2}{16} - \dfrac{y^2}{4} = 1$

50. $\dfrac{(x + 1)^2}{16} - \dfrac{(y - 2)^2}{9} = 1$

In Exercises 51–62, sketch the hyperbola. Identify its vertices and asymptotes.

51. $x^2 - y^2 = 9$

52. $x^2 - y^2 = 1$

53. $y^2 - x^2 = 9$

54. $y^2 - x^2 = 1$

55. $\dfrac{x^2}{9} - \dfrac{y^2}{25} = 1$

56. $\dfrac{x^2}{4} - \dfrac{y^2}{9} = 1$

57. $\dfrac{y^2}{9} - \dfrac{x^2}{25} = 1$

58. $\dfrac{y^2}{4} - \dfrac{x^2}{9} = 1$

59. $\dfrac{x^2}{1} - \dfrac{y^2}{9/4} = 1$

60. $\dfrac{y^2}{1/4} - \dfrac{x^2}{25/4} = 1$

61. $4y^2 - x^2 + 16 = 0$

62. $4y^2 - 9x^2 - 36 = 0$

In Exercises 63–70, write the standard form of the equation of the hyperbola centered at the origin.

	Vertices	Asymptotes	
63.	$(-4, 0), (4, 0)$	$y = 2x$	$y = -2x$
64.	$(-2, 0), (2, 0)$	$y = \frac{1}{3}x$	$y = -\frac{1}{3}x$
65.	$(0, -4), (0, 4)$	$y = \frac{1}{2}x$	$y = -\frac{1}{2}x$
66.	$(0, -2), (0, 2)$	$y = 3x$	$y = -3x$
67.	$(-9, 0), (9, 0)$	$y = \frac{2}{3}x$	$y = -\frac{2}{3}x$
68.	$(-1, 0), (1, 0)$	$y = \frac{1}{2}x$	$y = -\frac{1}{2}x$
69.	$(0, -1), (0, 1)$	$y = 2x$	$y = -2x$
70.	$(0, -5), (0, 5)$	$y = x$	$y = -x$

In Exercises 71–76, find the center and vertices of the hyperbola and sketch its graph. Use a graphing utility to verify the graph.

71. $(y + 4)^2 - (x - 3)^2 = 25$

72. $(y + 6)^2 - (x - 2)^2 = 1$

73. $\dfrac{(x - 1)^2}{4} - \dfrac{(y + 2)^2}{1} = 1$

74. $\dfrac{(x - 2)^2}{4} - \dfrac{(y - 3)^2}{9} = 1$

75. $9x^2 - y^2 - 36x - 6y + 18 = 0$

76. $x^2 - 9y^2 + 36y - 72 = 0$

In Exercises 77–80, find an equation of the hyperbola.

77.

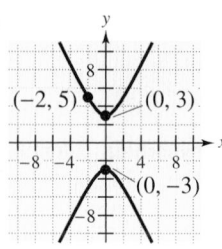

78.

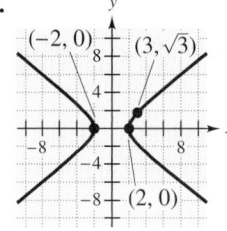

79.

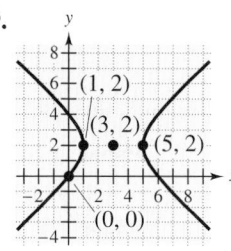

80.

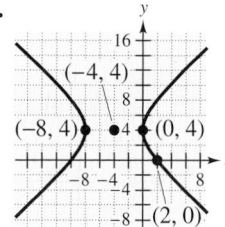

Solving Problems

81. *Wading Pool* You are building a wading pool that is in the shape of an ellipse. Your plans give an equation for the elliptical shape of the pool measured in feet as

$$\frac{x^2}{324} + \frac{y^2}{196} = 1.$$

Find the longest distance and shortest distance across the pool.

82. *Height of an Arch* A semielliptical arch for a tunnel under a river has a width of 100 feet and a height of 40 feet (see figure). Determine the height of the arch 5 feet from the edge of the tunnel.

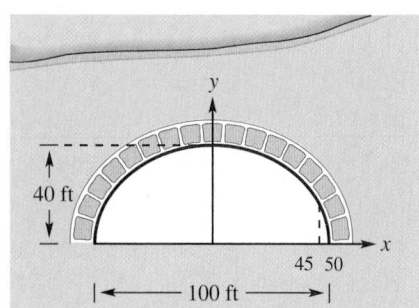

83. *Athletics* In Australia, football by *Australian Rules* (or rugby) is played on elliptical fields. The field can be a maximum of 170 yards wide and a maximum of 200 yards long. Let the center of a field of maximum size be represented by the point $(0, 85)$. Write an equation of the ellipse that represents this field. *(Source: Oxford Companion to World Sports and Games)*

84. *Bicycle Chainwheel* The pedals of a bicycle drive a chainwheel, which drives a smaller sprocket wheel on the rear axle (see figure). Many chainwheels are circular. Some, however, are slightly elliptical, which tends to make pedaling easier. Find an equation of an elliptical chainwheel that is 8 inches in diameter at its widest point and $7\frac{1}{2}$ inches in diameter at its narrowest point.

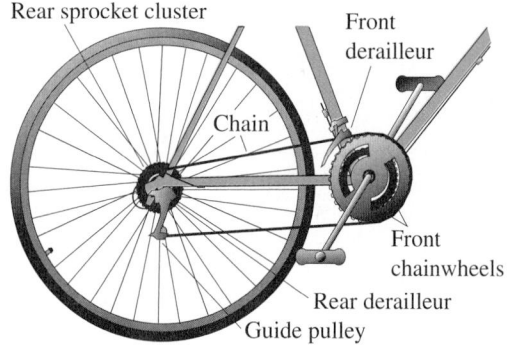

Explaining Concepts

85. Describe the relationship between circles and ellipses. How are they similar? How do they differ?

86. Explain the significance of the foci in an ellipse.

87. Explain the significance of the foci in a hyperbola.

88. Explain how to write an equation of an ellipse if you know the coordinates of the vertices and co-vertices.

In Exercises 89 and 90, describe the part of the hyperbola

$$\frac{(x-3)^2}{4} - \frac{(y-1)^2}{9} = 1$$

given by the equation.

89. $x = 3 - \frac{2}{3}\sqrt{9 + (y-1)^2}$

90. $y = 1 + \frac{3}{2}\sqrt{(x-3)^2 - 4}$

Ongoing Review

In Exercises 91 and 92, find the distance between the points.

91. $(5, 2), (-1, 4)$ **92.** $(-4, -3), (6, 10)$

In Exercises 93–96, graph the lines on the same set of coordinate axes.

93. $y = \pm 4x$

94. $y = 6 \pm \dfrac{1}{3}x$

95. $y = 5 \pm \dfrac{1}{2}(x - 2)$

96. $y = \pm \dfrac{1}{3}(x - 6)$

In Exercises 97–100, Find the unknown in the equation $c^2 = a^2 - b^2$. (Assume that a, b, and c are positive.)

97. $a = 25, b = 7$ **98.** $a = \sqrt{41}, c = 4$

99. $b = 5, c = 12$ **100.** $a = 6, b = 3$

101. *Average Speed*   From a point on a straight road, two people ride bicycles in opposite directions. One person rides at 10 miles per hour and the other rides at 12 miles per hour. In how many hours will they be 55 miles apart?

102. *Mixture Problem*   You have a collection of 30 gold coins. Some of the coins are worth $10 each, and the rest are worth $20 each. The value of the entire collection is $540. How many of each type of coin do you have?

Looking Further

(a) Sketch a graph of the ellipse that consists of all points (x, y) such that the sum of the distances between (x, y) and two fixed points is 15 units and the foci are located at the centers of the two sets of concentric circles in the figure.

(b) Sketch a graph of the hyperbola that consists of all points (x, y) such that the difference of the distances between (x, y) and two fixed points is eight units and the foci are located at the centers of the two sets of concentric circles in the figure.

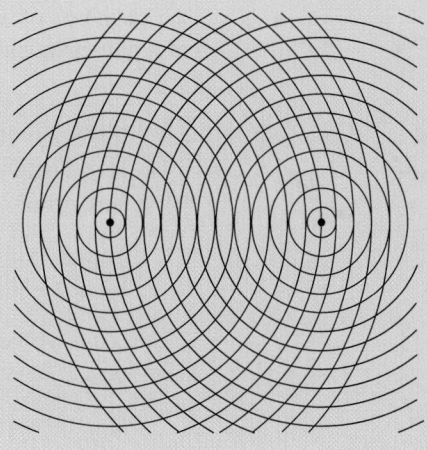

8.7 Parabolas

What you should learn:

- How to graph and find equations of parabolas

Why you should learn it:

Parabolas can be used to model and solve real-life problems. For instance, in Exercise 54 on page 565, a parabola is used to model the path of a softball.

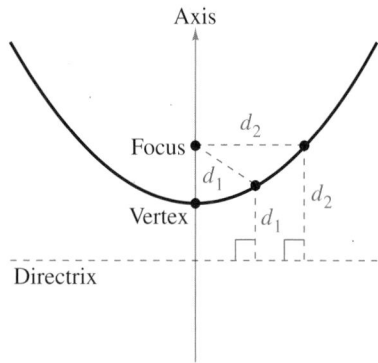

Figure 8.46

NOTE If the focus is above or to the right of the vertex, p is positive. If the focus is below or to the left of the vertex, p is negative.

Equations of Parabolas

The fourth basic type of conic is a **parabola.** In Section 6.5, you studied some of the properties of parabolas. There, you saw that the graph of a quadratic function of the form $y = ax^2 + bx + c$ is a parabola, which opens upward if a is positive and downward if a is negative. You also learned that each parabola has a vertex and that the vertex of the graph of $y = ax^2 + bx + c$ occurs when $x = -b/2a$.

In this section, you will study the technical definition of a parabola, and you will study the equations of parabolas that open to the right and to the left.

> **Definition of a Parabola**
>
> A **parabola** is the set of all points (x, y) that are equidistant from a fixed line (**directrix**) and a fixed point (**focus**) not on the line.

The midpoint between the focus and the directrix is called the **vertex,** and the line passing through the focus and the vertex is called the **axis** of the parabola. Note in Figure 8.46 that a parabola is symmetric with respect to its axis. Using the definition of a parabola, you can derive the **standard form of the equation of a parabola** whose directrix is parallel to the x-axis or to the y-axis.

> **Standard Equation of a Parabola**
>
> The **standard form of the equation of a parabola** with vertex at the origin $(0, 0)$ is
>
> $$x^2 = 4py, \quad p \neq 0 \qquad \text{Vertical axis}$$
> $$y^2 = 4px, \quad p \neq 0. \qquad \text{Horizontal axis}$$
>
> The focus lies on the axis p units (*directed distance*) from the vertex. If the vertex is at (h, k), then the standard form of the equation is
>
> $$(x - h)^2 = 4p(y - k), \quad p \neq 0 \qquad \text{Vertical axis; directrix: } y = k - p$$
> $$(y - k)^2 = 4p(x - h), \quad p \neq 0. \qquad \text{Horizontal axis; directrix: } x = h - p$$
>
> (See Figure 8.47.)

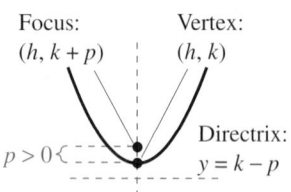

(a) Parabola with vertical axis
Figure 8.47

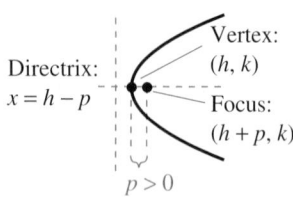

(b) Parabola with horizontal axis

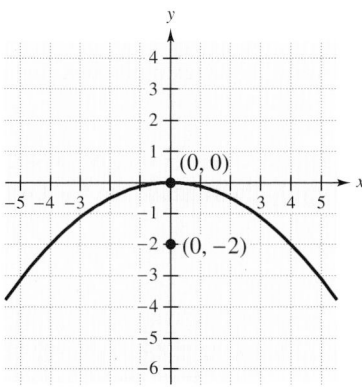

Figure 8.48

Example 1 ■ Finding the Standard Equation of a Parabola

Find the standard form of the equation of the parabola with vertex $(0, 0)$ and focus $(0, -2)$, as shown in Figure 8.48.

Solution

Because the vertex is at the origin and the axis of the parabola is vertical, consider the equation

$$x^2 = 4py$$

where p is the directed distance from the vertex to the focus. Because the focus is two units *below* the vertex, you have $p = -2$. So, the equation of the parabola is

$$x^2 = 4py$$
$$x^2 = 4(-2)y$$
$$x^2 = -8y.$$

Example 2 ■ Finding the Standard Equation of a Parabola

Find the standard form of the equation of the parabola with vertex $(3, -2)$ and focus $(4, -2)$, as shown in Figure 8.49.

Solution

Because the vertex is at $(h, k) = (3, -2)$ and the axis of the parabola is horizontal, consider the equation

$$(y - k)^2 = 4p(x - h)$$

where $h = 3$, $k = -2$, and $p = 1$. So, the equation of the parabola is

$$(y - k)^2 = 4p(x - h)$$
$$[y - (-2)]^2 = 4(1)(x - 3)$$
$$(y + 2)^2 = 4(x - 3).$$

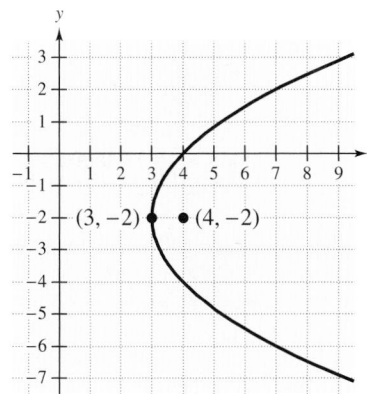

Figure 8.49

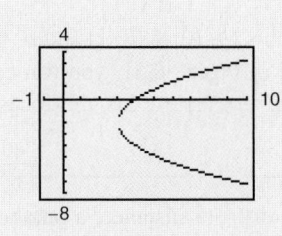

Technology

You can use a graphing utility to graph the parabola in Example 2 by graphing the upper and lower portions in the same viewing window. First solve for y to obtain

$$y_1 = -2 + 2\sqrt{x - 3}$$

and

$$y_2 = -2 - 2\sqrt{x - 3}.$$

Use a viewing window in which $-1 \le x \le 10$ and $-8 \le y \le 4$. You should obtain the graph shown at the left. Notice that the graphing utility does not connect the two portions of the parabola.

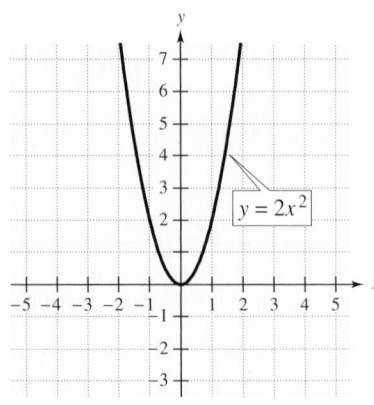

Figure 8.50

Example 3 ■ Analyzing a Parabola

Sketch the graph of the parabola $y = 2x^2$ and identify its vertex and focus.

Solution

Because the equation can be written in the standard form $x^2 = 4py$, it is a parabola whose vertex is at the origin. You can identify the focus of the parabola by writing its equation in standard form.

$y = 2x^2$	Write original equation.
$2x^2 = y$	Interchange sides of the equation.
$x^2 = \frac{1}{2}y$	Divide each side by 2.
$x^2 = 4\left(\frac{1}{8}\right)y$	Rewrite $\frac{1}{2}$ in the form $4p$.

From this standard form, you can see that $p = \frac{1}{8}$. Because the parabola opens upward, as shown in Figure 8.50, you can conclude that the focus lies $p = \frac{1}{8}$ unit above the vertex. So, the focus is $\left(0, \frac{1}{8}\right)$.

Example 4 ■ Analyzing a Parabola

Sketch the graph of the parabola $x = 2y^2 - 4y + 10$ and identify its vertex and focus.

Solution

This equation can be written in the standard form $(y - k)^2 = 4p(x - h)$. To do this, you can complete the square, as follows.

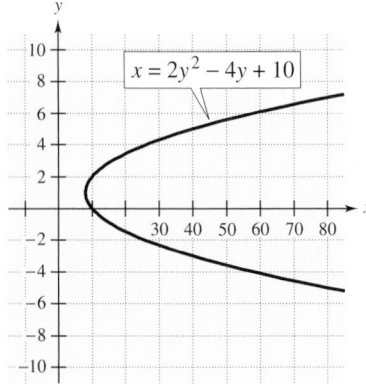

Figure 8.51

$x = 2y^2 - 4y + 10$	Write original equation.
$2y^2 - 4y + 10 = x$	Interchange sides of the equation.
$y^2 - 2y + 5 = \frac{1}{2}x$	Divide each side by 2.
$y^2 - 2y = \frac{1}{2}x - 5$	Subtract 5 from each side.
$y^2 - 2y + 1 = \frac{1}{2}x - 5 + 1$	Complete the square on left side.
$(y - 1)^2 = \frac{1}{2}x - 4$	Simplify.
$(y - 1)^2 = \frac{1}{2}(x - 8)$	Factor.
$(y - 1)^2 = 4\left(\frac{1}{8}\right)(x - 8)$	Rewrite $\frac{1}{2}$ in the form $4p$.

From this standard form, you can see that the vertex is $(h, k) = (8, 1)$ and $p = \frac{1}{8}$. Because the parabola opens to the right, as shown in Figure 8.51, you can conclude that the focus lies $p = \frac{1}{8}$ unit to the right of the vertex. So, the focus is $\left(8\frac{1}{8}, 1\right)$.

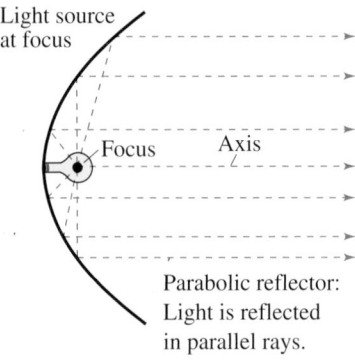

Figure 8.52

Parabolas occur in a wide variety of applications. For instance, a parabolic reflector can be formed by revolving a parabola around its axis. The resulting surface has the property that all incoming rays parallel to the axis are reflected through the focus of the parabola. This is the principle behind the construction of the parabolic mirrors used in reflecting telescopes. Conversely, the light rays emanating from the focus of a parabolic reflector used in a flashlight are all parallel to one another, as shown in Figure 8.52.

8.7 ■ Exercises

Developing Skills

In Exercises 1–6, match the equation with its graph. [The graphs are labeled (a), (b), (c), (d), (e), and (f).]

(a)

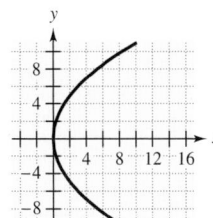

(b)

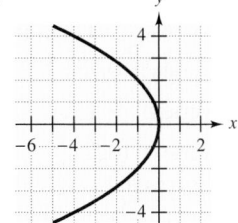

(c)

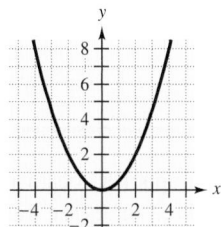

(d)

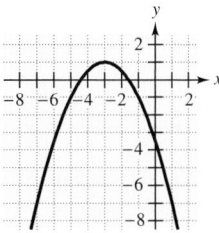

(e)

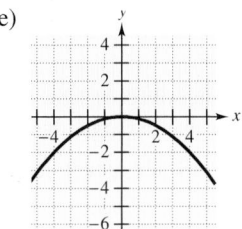

(f)

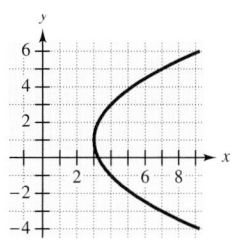

1. $y^2 = -4x$ **2.** $x^2 = 2y$

3. $x^2 = -8y$ **4.** $y^2 = 12x$

5. $(y - 1)^2 = 4(x - 3)$ **6.** $(x + 3)^2 = -2(y - 1)$

In Exercises 7–20, find the vertex and focus of the parabola and sketch its graph.

7. $y = \frac{1}{2}x^2$ **8.** $y = 2x^2$

9. $y^2 = -6x$ **10.** $y^2 = 3x$

11. $x^2 + 8y = 0$ **12.** $x + y^2 = 0$

13. $(x - 1)^2 + 8(y + 2) = 0$

14. $(x + 3) + (y - 2)^2 = 0$

15. $\left(y + \frac{1}{2}\right)^2 = 2(x - 5)$

16. $\left(x + \frac{1}{2}\right)^2 = 4(y - 3)$

17. $y = \frac{1}{4}(x^2 - 2x + 5)$

18. $4x - y^2 - 2y - 33 = 0$

19. $y^2 + 6y + 8x + 25 = 0$

20. $y^2 - 4y - 4x = 0$

In Exercises 21–24, use a graphing utility to graph the parabola. Identify the vertex and focus.

21. $y = -\frac{1}{6}(x^2 + 4x - 2)$

22. $x^2 - 2x + 8y + 9 = 0$

23. $y^2 + x + y = 0$

24. $y^2 - 4x - 4 = 0$

In Exercises 25–28, change the equation so that its graph matches the given graph.

25. $y^2 = -4x$

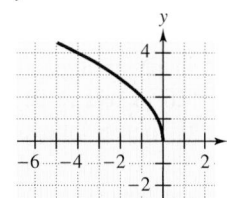

26. $y^2 = 9x$

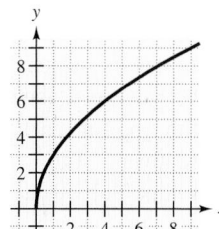

27. $y^2 = 16x$

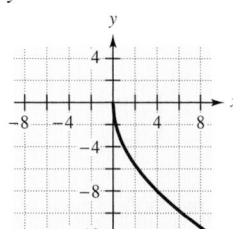

28. $y^2 = -25x$

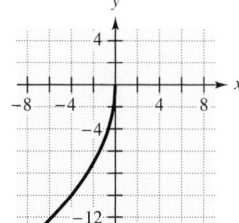

In Exercises 29–40, find an equation of the parabola with its vertex at the origin.

29.

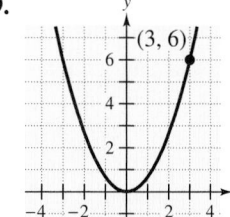

(3, 6)

30.

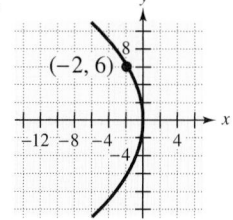

(−2, 6)

31. Focus: $\left(0, -\frac{3}{2}\right)$ **32.** Focus: $(2, 0)$

33. Focus: $(-2, 0)$ **34.** Focus: $(0, -2)$

35. Focus: $(0, 1)$

36. Focus: $(-3, 0)$

37. Focus: $(4, 0)$

38. Focus: $(0, 2)$

39. Horizontal axis and passes through the point $(4, 6)$

40. Vertical axis and passes through the point $(-2, -2)$

In Exercises 41–50, find an equation of the parabola with its vertex at (h, k).

41.

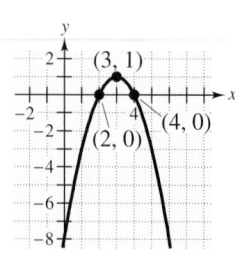

42.

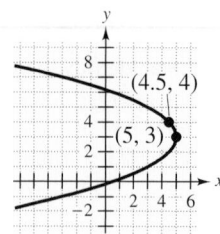

43.

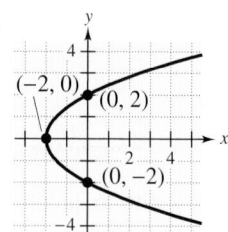

44.

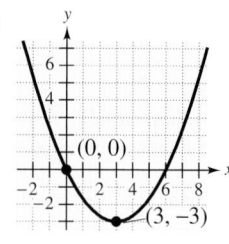

45. Vertex: $(3, 2)$; Focus: $(1, 2)$

46. Vertex: $(-1, 2)$; Focus: $(-1, 0)$

47. Vertex: $(0, 4)$; Focus: $(0, 6)$

48. Vertex: $(-2, 1)$; Focus: $(-5, 1)$

49. Vertex: $(0, 2)$;
Horizontal axis and passes through $(1, 3)$

50. Vertex: $(0, 2)$; Vertical axis and passes through $(6, 0)$

Solving Problems

51. *Suspension Bridge* Each cable of a suspension bridge is suspended (in the shape of a parabola) between two towers that are 120 meters apart, and the top of each tower is 20 meters above the roadway. The cables touch the roadway at the midpoint between the towers (see figure).

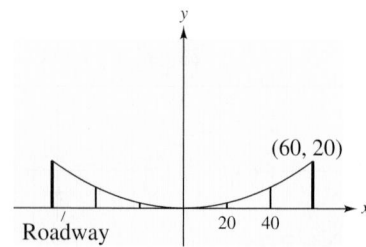

(a) Find an equation for the parabolic shape of each cable.

(b) Complete the table by finding the height of the suspension cables y over the roadway at a distance of x meters from the center of the bridge.

x	0	20	40	60
y				

52. *Beam Deflection* A simply supported beam is 16 meters long and has a load at the center (see figure). The deflection of the beam at its center is 3 centimeters. Assume that the shape of the deflected beam is parabolic.

(a) Find an equation of the parabola. (Assume that the origin is at the center of the deflected beam.)

(b) How far from the center of the beam is the deflection 1 centimeter?

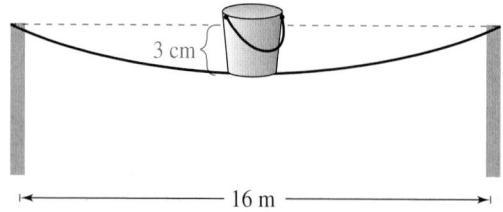

Not drawn to scale

53. *Revenue* The revenue R generated by the sale of x units is given by

$$R = 375x - \tfrac{3}{2}x^2.$$

(a) Use a graphing utility to graph the function.

(b) Use the graph to approximate the number of sales that will maximize revenue.

54. *Path of a Projectile*   The path of a softball is given by

$$y = -0.08x^2 + x + 4.$$

The coordinates x and y are measured in feet, with $x = 0$ corresponding to the position from which the ball was thrown.

(a) Use a graphing utility to graph the trajectory of the softball.

(b) Move the cursor along the path to approximate the highest point and the range of the trajectory.

Explaining Concepts

55. Explain the significance of a parabola's directrix and focus.

56. Is y a function of x in the equation $y^2 = 6x$? Explain.

57. Is it possible for a parabola to intersect its directrix? Explain.

58. If the vertex and focus of a parabola are on a horizontal line, is the directrix of the parabola vertical? Explain.

Ongoing Review

In Exercises 59–62, Expand and simplify the expression.

59. $(x + 6)^2 - 5$

60. $(x - 7)^2 - 2$

61. $12 - (x - 8)^2$

62. $16 - (x + 1)^2$

In Exercises 63–66, complete the square for the quadratic expression.

63. $x^2 - 4x + 1$

64. $x^2 + 12x - 3$

65. $-x^2 + 6x + 5$

66. $-2x^2 + 10x - 14$

In Exercises 67 and 68, find an equation of the line passing through the point with the specified slope.

67. $(-2, 5), m = \dfrac{5}{8}$

68. $(1, 7), m = -\dfrac{2}{5}$

69. *Test Scores*   A student has test scores of 90, 74, 82, and 90. The next examination is the final examination, which counts as two tests. What score does the student need on the final examination to produce an average score of 85?

70. *Simple Interest*   An investment of $2500 is made at an annual simple interest rate of 5.5%. How much additional money must be invested at an annual simple interest rate of 8% so that the total interest earned is 7% of the total investment?

Looking Further

Consider the parabola $x^2 = 4py$.

(a) Use a graphing utility to graph the parabola for $p = 1$, $p = 2$, $p = 3$, and $p = 4$. Describe the effect on the graph when p increases.

(b) Locate the focus for each parabola in part (a).

(c) For each parabola in part (a), find the length of the chord passing through the focus perpendicular to the axis of the parabola. How can the length of this chord be determined directly from the standard form of the equation of the parabola?

(d) Explain how the result in part (c) can be used as a sketching aid when graphing parabolas.

8.8 Nonlinear Systems of Equations

What you should learn:

- How to solve nonlinear systems of equations graphically
- How to solve nonlinear systems of equations by substitution
- How to solve nonlinear systems of equations by elimination
- How to use nonlinear systems of equations to model and solve real-life problems

Why you should learn it:

Nonlinear systems of equations can be used to analyze real-life data sets. For instance, in Exercise 48 on page 574 models represent the populations of two states in the United States.

Solving Nonlinear Systems of Equations by Graphing

In Chapter 4, you studied several methods for solving systems of linear equations. For instance, the linear system

$$\begin{cases} 2x - 3y = 7 \\ x + 4y = -2 \end{cases}$$

has one solution: $(2, -1)$. Graphically, this means that $(2, -1)$ is a point of intersection of the two lines represented by the system.

In Chapter 4, you also learned that a linear system can have no solution, exactly one solution, or infinitely many solutions. A **nonlinear system of equations** is a system that contains at least one nonlinear equation. Nonlinear systems of equations can have no solution, one solution, or two or more solutions. For instance, the hyperbola and line in Figure 8.53(a) have no point of intersection, the circle and line in Figure 8.53(b) have one point of intersection, and the parabola and line in Figure 8.53(c) have two points of intersection.

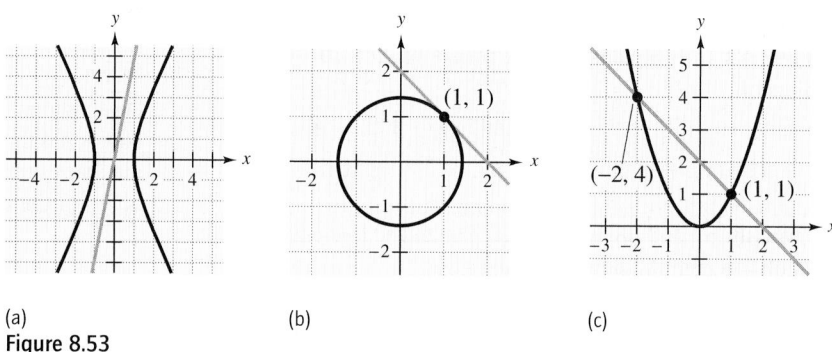

(a) (b) (c)

Figure 8.53

To solve a nonlinear system of equations graphically, you can use the following steps.

Solving a Nonlinear System Graphically

1. Sketch the graph of each equation in the system.

2. Locate the point(s) of intersection of the graphs (if any) and graphically approximate the coordinates of the points.

3. Check the coordinate values by substituting them into each equation in the original system. If the coordinate values do not check, you may have to use an algebraic approach, as discussed later in this section.

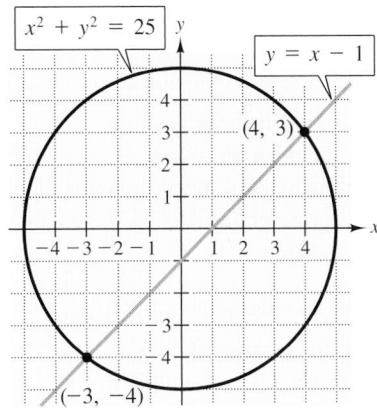

Figure 8.54

Technology

Try using a graphing utility to solve the system described in Example 1. When you do this, remember that the circle needs to be entered as two separate equations.

$y_1 = \sqrt{25 - x^2}$	Top half of circle
$y_2 = -\sqrt{25 - x^2}$	Bottom half of circle
$y_3 = x - 1$	Line

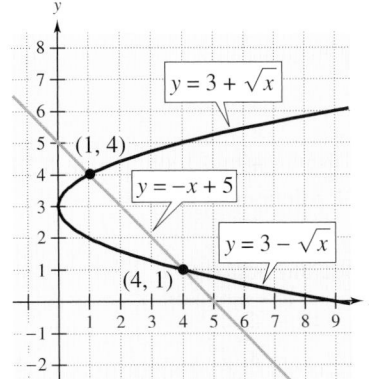

Figure 8.55

Example 1 ■ Solving a Nonlinear System Graphically

Find all solutions of the nonlinear system of equations.

$$\begin{cases} x^2 + y^2 = 25 & \text{Equation 1} \\ x - y = 1 & \text{Equation 2} \end{cases}$$

Solution

Begin by sketching the graph of each equation. The first equation graphs as a circle centered at the origin and having a radius of 5. The second equation, which can be written as $y = x - 1$, graphs as a line with a slope of 1 and a y-intercept of $(0, -1)$. From the graphs shown in Figure 8.54, you can see that the system appears to have two solutions: $(-3, -4)$ and $(4, 3)$. You can check these solutions by substituting for x and y in the original system, as follows.

Check

To check $(-3, -4)$, substitute -3 for x and -4 for y in each equation.

$$(-3)^2 + (-4)^2 \overset{?}{=} 25 \qquad \text{Substitute } -3 \text{ for } x \text{ and } -4 \text{ for } y \text{ in Equation 1.}$$
$$9 + 16 = 25 \qquad \text{Solution checks in Equation 1. } \checkmark$$
$$(-3) - (-4) \overset{?}{=} 1 \qquad \text{Substitute } -3 \text{ for } x \text{ and } -4 \text{ for } y \text{ in Equation 2.}$$
$$-3 + 4 = 1 \qquad \text{Solution checks in Equation 2. } \checkmark$$

To check $(4, 3)$, substitute 4 for x and 3 for y in each equation.

$$4^2 + 3^2 \overset{?}{=} 25 \qquad \text{Substitute 4 for } x \text{ and 3 for } y \text{ in Equation 1.}$$
$$16 + 9 = 25 \qquad \text{Solution checks in Equation 1. } \checkmark$$
$$4 - 3 \overset{?}{=} 1 \qquad \text{Substitute 4 for } x \text{ and 3 for } y \text{ in Equation 2.}$$
$$1 = 1 \qquad \text{Solution checks in Equation 2. } \checkmark$$

So, you can conclude that both points are solutions of the system of equations.

Example 2 ■ Solving a Nonlinear System Graphically

Find all solutions of the nonlinear system of equations.

$$\begin{cases} x = (y - 3)^2 & \text{Equation 1} \\ x + y = 5 & \text{Equation 2} \end{cases}$$

Solution

Begin by sketching the graph of each equation. Solve the first equation for y as follows.

$$x = (y - 3)^2 \qquad \text{Write original equation.}$$
$$\pm\sqrt{x} = y - 3 \qquad \text{Take the square root of each side.}$$
$$3 \pm \sqrt{x} = y \qquad \text{Add 3 to each side.}$$

The graph of $y = 3 \pm \sqrt{x}$ is a parabola with its vertex at $(0, 3)$. The second equation, which can be written as $y = -x + 5$, graphs as a line with a slope of -1 and a y-intercept of $(0, 5)$. From the graphs shown in Figure 8.55, you can see that the system appears to have two solutions: $(4, 1)$ and $(1, 4)$. Check these solutions in the original system.

EXPLORATION

Graphical Reasoning A circle and an ellipse can have no points of intersection, or they can have one, two, three, or four points of intersection. Draw examples of each of these five possibilities.

Now, discuss the number of possible points of intersection of each pair of graphs described below. Sketch examples of each possibility.

a. A parabola and a circle

b. A hyperbola and a circle

c. A line and a hyperbola

Solving Nonlinear Systems of Equations by Substitution

The graphical approach to solving any type of system (linear or nonlinear) in two variables is very useful for helping you see the number of solutions and their approximate coordinates. For systems with solutions having "messy" coordinates, however, a graphical approach is usually not accurate enough to produce exact solutions. In such cases, you should use an algebraic approach. (With an algebraic approach, you should still sketch the graph of each equation in the system.)

As with systems of *linear* equations, there are two basic algebraic approaches: substitution and elimination. Substitution usually works well for systems in which one of the equations is linear, as shown in Example 3.

Example 3 ■ Using Substitution to Solve a Nonlinear System

Solve the nonlinear system of equations.

$$\begin{cases} 4x^2 + y^2 = 4 & \text{Equation 1} \\ -2x + y = 2 & \text{Equation 2} \end{cases}$$

Solution

Begin by solving for y in Equation 2.

$$y = 2x + 2 \qquad \text{Revised Equation 2}$$

Next, substitute this expression for y into Equation 1.

$$4x^2 + y^2 = 4 \qquad \text{Write Equation 1.}$$

$$4x^2 + (2x + 2)^2 = 4 \qquad \text{Substitute } 2x + 2 \text{ for } y.$$

$$4x^2 + 4x^2 + 8x + 4 = 4 \qquad \text{Multiply.}$$

$$8x^2 + 8x = 0 \qquad \text{Simplify.}$$

$$8x(x + 1) = 0 \qquad \text{Factor.}$$

$$8x = 0 \Rightarrow x = 0 \qquad \text{Set 1st factor equal to 0.}$$

$$x + 1 = 0 \Rightarrow x = -1 \qquad \text{Set 2nd factor equal to 0.}$$

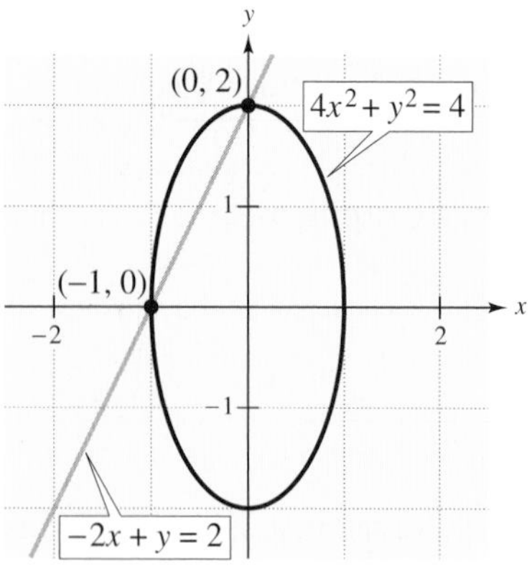

Figure 8.56

Finally, back-substitute these values of x into the revised Equation 2 to solve for y.

For $x = 0$: $\qquad y = 2(0) + 2 = 2$

For $x = -1$: $\qquad y = 2(-1) + 2 = 0$

So, the system of equations has two solutions: $(0, 2)$ and $(-1, 0)$. Figure 8.56 shows the graph of the system. You can check the solutions as follows.

Check First Solution

$$4(0)^2 + 2^2 \stackrel{?}{=} 4$$
$$0 + 4 = 4 \checkmark$$
$$-2(0) + 2 \stackrel{?}{=} 2$$
$$2 = 2 \checkmark$$

Check Second Solution

$$4(-1)^2 + 0^2 \stackrel{?}{=} 4$$
$$4 + 0 = 4 \checkmark$$
$$-2(-1) + 0 \stackrel{?}{=} 2$$
$$2 = 2 \checkmark$$

The steps for using the method of substitution to solve a system of two equations involving two variables are summarized as follows.

Method of Substitution

To solve a system of two equations in two variables, use the steps below.

1. Solve one of the equations for one variable in terms of the other.

2. Substitute the expression found in Step 1 into the other equation to obtain an equation of one variable.

3. Solve the equation obtained in Step 2.

4. Back-substitute the solution from Step 3 into the expression obtained in Step 1 to find the value of the other variable.

5. Check the solution to see that it satisfies *each* of the original equations.

Example 4 shows how the method of substitution and graphing can be used to determine that a nonlinear system of equations has no solution.

Example 4 ■ Solving a Nonlinear System: No-Solution Case

Solve the nonlinear system of equations.

$$\begin{cases} x^2 - y = 0 & \text{Equation 1} \\ x - y = 1 & \text{Equation 2} \end{cases}$$

Algebraic Solution

Begin by solving for y in Equation 2.

$$y = x - 1 \qquad \text{Revised Equation 2}$$

Next, substitute this expression for y into Equation 1.

$$x^2 - y = 0 \qquad \text{Equation 1}$$
$$x^2 - (x - 1) = 0 \qquad \text{Substitute } x - 1 \text{ for } y.$$
$$x^2 - x + 1 = 0 \qquad \text{Simplify.}$$
$$x = \frac{-(-1) \pm \sqrt{(-1)^2 - 4(1)(1)}}{2(1)} \qquad \text{Use Quadratic Formula.}$$
$$x = \frac{1 \pm \sqrt{1 - 4}}{2} \qquad \text{Simplify.}$$
$$x = \frac{1 \pm \sqrt{-3}}{2} \qquad \text{Simplify.}$$

Now, because the Quadratic Formula yields a negative number inside the radical, you can conclude that the equation $x^2 - x + 1 = 0$ has no (real) solution. So, the system has no (real) solution.

Graphical Solution

Begin by solving each equation for y. Then use a graphing utility to graph $y_1 = x^2$ and $y_2 = x - 1$ in the same viewing window, as shown in Figure 8.57. From the graph, it appears that the parabola and the line have no point of intersection. Try using the *intersect* feature to find a point of intersection. Because the graphing utility cannot find a point of intersection, you will get an error. Use the *zoom* feature and then try using the *intersect* feature again. Because repeated zooming yields the same error message, you can conclude that the parabola and the line have no point of intersection.

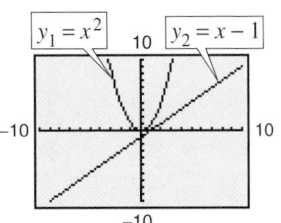

Figure 8.57

Solving Nonlinear Systems of Equations by Elimination

In Section 4.1, you learned how to use the method of elimination to solve a linear system. This method can also be used with special types of nonlinear systems, as demonstrated in Example 5.

Example 5 ■ Using Elimination to Solve a Nonlinear System

Solve the nonlinear system of equations.

$$\begin{cases} 4x^2 + y^2 = 64 & \text{Equation 1} \\ x^2 + y^2 = 52 & \text{Equation 2} \end{cases}$$

Solution

Because both equations have y^2 as a term (and no other terms containing y), you can eliminate y by subtracting Equation 2 from Equation 1.

$$
\begin{array}{rl}
4x^2 + y^2 = & 64 \\
-x^2 - y^2 = & -52 \qquad \text{Subtract Equation 2 from Equation 1.} \\
\hline
3x^2 \quad\;\; = & 12
\end{array}
$$

After eliminating y, solve the remaining equation for x.

$$3x^2 = 12 \qquad \text{Write resulting equation.}$$
$$x^2 = 4 \qquad \text{Divide each side by 3.}$$
$$x = \pm 2 \qquad \text{Take square root of each side.}$$

To find the corresponding values of y, substitute these values of x into either of the original equations. By substituting $x = 2$, you obtain

$$x^2 + y^2 = 52 \qquad \text{Write Equation 2.}$$
$$(2)^2 + y^2 = 52 \qquad \text{Substitute 2 for } x.$$
$$y^2 = 48 \qquad \text{Subtract 4 from each side.}$$
$$y = \pm\sqrt{48} \qquad \text{Take square root of each side.}$$
$$y = \pm 4\sqrt{3}. \qquad \text{Simplify.}$$

By substituting $x = -2$, you obtain the same values of y, as follows.

$$x^2 + y^2 = 52 \qquad \text{Write Equation 2.}$$
$$(-2)^2 + y^2 = 52 \qquad \text{Substitute } -2 \text{ for } x.$$
$$y^2 = 48 \qquad \text{Subtract 4 from each side.}$$
$$y = \pm\sqrt{48} \qquad \text{Take square root of each side.}$$
$$y = \pm 4\sqrt{3} \qquad \text{Simplify.}$$

This implies that the system has four solutions:

$$\left(2, 4\sqrt{3}\right), \quad \left(2, -4\sqrt{3}\right), \quad \left(-2, 4\sqrt{3}\right), \quad \left(-2, -4\sqrt{3}\right).$$

Check these in the original system. Figure 8.58 shows the graph of the system. Notice that the graph of Equation 1 is an ellipse and the graph of Equation 2 is a circle.

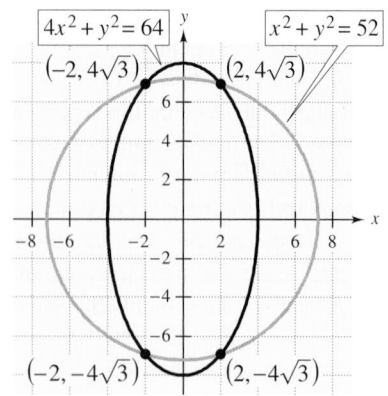

Figure 8.58

Example 6 ■ Solving a Nonlinear System

Solve the nonlinear system of equations.

$$\begin{cases} x^2 - 2y = 4 & \text{Equation 1} \\ x^2 - y^2 = 1 & \text{Equation 2} \end{cases}$$

Algebraic Solution

Because both equations have x^2 as a term (and no other terms containing x), you can eliminate x by subtracting Equation 2 from Equation 1.

$$\begin{array}{rcl} x^2 - 2y &=& 4 \\ -x^2 + y^2 &=& -1 \\ \hline y^2 - 2y &=& 3 \end{array}$$

Subtract Equation 2 from Equation 1.

After eliminating x, solve the remaining equation for y.

$$y^2 - 2y = 3 \qquad \text{Write resulting equation.}$$
$$y^2 - 2y - 3 = 0 \qquad \text{Write in general form.}$$
$$(y - 3)(y + 1) = 0 \qquad \text{Factor.}$$
$$y - 3 = 0 \Rightarrow y = 3 \qquad \text{Set 1st factor equal to 0.}$$
$$y + 1 = 0 \Rightarrow y = -1 \qquad \text{Set 2nd factor equal to 0.}$$

When $y = -1$, you obtain

$$x^2 - y^2 = 1 \qquad \text{Write Equation 2.}$$
$$x^2 - (-1)^2 = 1 \qquad \text{Substitute } -1 \text{ for } y.$$
$$x^2 = 2 \qquad \text{Add 1 to each side.}$$
$$x = \pm\sqrt{2} \qquad \text{Take square root of each side.}$$

When $y = 3$, you obtain

$$x^2 - y^2 = 1 \qquad \text{Write Equation 2.}$$
$$x^2 - (3)^2 = 1 \qquad \text{Substitute 3 for } y.$$
$$x^2 = 10 \qquad \text{Add 9 to each side.}$$
$$x = \pm\sqrt{10} \qquad \text{Take square root of each side.}$$

This implies that the system has four solutions:

$$\left(\sqrt{2}, -1\right), \quad \left(-\sqrt{2}, -1\right), \quad \left(\sqrt{10}, 3\right), \quad \left(-\sqrt{10}, 3\right).$$

Check these in the original system.

Graphical Solution

Begin by solving each equation for y, as follows.

$$y = \tfrac{1}{2}x^2 - 2 \qquad y = \pm\sqrt{x^2 - 1}$$

Use a graphing utility to graph all three equations

$$y_1 = \tfrac{1}{2}x^2 - 2$$
$$y_2 = \sqrt{x^2 - 1}$$

and

$$y_3 = -\sqrt{x^2 - 1}$$

in the same viewing window. In Figure 8.59, you can see that the graphs appear to intersect at four points. Use the *intersect* feature of the graphing utility to approximate the points of intersection to be $(1.41, -1)$, $(-1.41, -1)$, $(3.16, 3)$, and $(-3.16, 3)$.

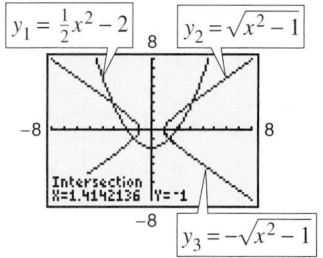

Figure 8.59

In Example 6, the method of elimination yields the four exact solutions to the system of equations

$$\left(\sqrt{2}, -1\right), \left(-\sqrt{2}, -1\right), \left(\sqrt{10}, 3\right), \text{ and } \left(-\sqrt{10}, 3\right)$$

whereas the graphical approach yields decimal approximations of the four solutions

$$(1.41, -1), (-1.41, -1), (3.16, 3), \text{ and } (-3.16, 3).$$

If you use the decimal approximations to check your solutions in the original system, be aware that they may not check.

Application

There are many examples of the use of nonlinear systems of equations in business and science. For instance, in Example 7 a nonlinear system of equations is used to compare the revenues of two companies.

Example 7 ■ Comparing the Revenues of Two Companies

From 1990 through 2003, the revenues R (in millions of dollars) of Company A and Company B can be modeled by

$$\begin{cases} R = 0.1t + 2.5 & \text{Company A} \\ R = 0.02t^2 - 0.2t + 3.2 & \text{Company B} \end{cases}$$

where t represents the year, with $t = 0$ corresponding to 1990. Sketch the graphs of these two models. During which two years did the companies have approximately equal revenues?

Solution

The graphs of the two models are shown in Figure 8.60. From the graph, you can see that Company A's revenue followed a linear pattern. It had a revenue of $2.5 million in 1990 and had an increase of $0.1 million each year. Company B's revenue followed a quadratic pattern. Between 1990 and 1995, the company's revenue was decreasing. Then, from 1995 through 2003, the revenue was increasing. From the graph, you can see that the two companies had approximately equal revenues in 1993 (Company A had $2.8 million and Company B had $2.78 million) and again in 2002 (Company A had $3.7 million and Company B had $3.68 million).

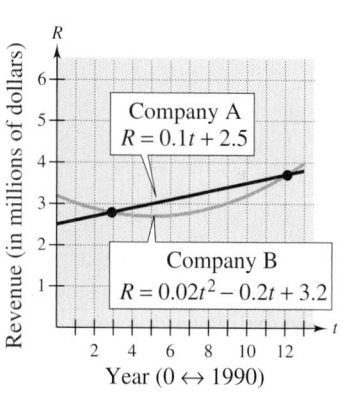

Figure 8.60

Collaborate!

Creating Examples

A circle and a parabola could have 0, 1, 2, 3, or 4 points of intersection. Sketch the circle given by

$$x^2 + y^2 = 4. \qquad \text{Circle}$$

Discuss how this circle could intersect a parabola with an equation of the form

$$y = x^2 + C. \qquad \text{Parabola}$$

Then find values of C for each of the five cases described below.

a. No points of intersection

b. One point of intersection

c. Two points of intersection

d. Three points of intersection

e. Four points of intersection

Use a graphing utility to confirm your results.

8.8 ■ Exercises

Developing Skills

In Exercises 1–6, graph the equations to determine whether the system has any solutions. Find any solutions that exist.

1. $\begin{cases} x + y = 2 \\ x^2 - y = 0 \end{cases}$
2. $\begin{cases} 2x + y = 10 \\ x^2 + y^2 = 25 \end{cases}$

3. $\begin{cases} y = \sqrt{x - 2} \\ x - 2y = 1 \end{cases}$
4. $\begin{cases} x - 2y = 4 \\ x^2 - y = 0 \end{cases}$

5. $\begin{cases} 9x^2 - 4y^2 = 36 \\ 5x - 2y = 0 \end{cases}$
6. $\begin{cases} 9x^2 + 4y^2 = 36 \\ 3x - 2y + 6 = 0 \end{cases}$

In Exercises 7–16, use a graphing utility to graph the equations and find any solutions of the system.

7. $\begin{cases} y = x \\ y = x^3 \end{cases}$
8. $\begin{cases} y = x^2 \\ y = x + 2 \end{cases}$

9. $\begin{cases} y = x^2 \\ y = -x^2 + 4x \end{cases}$
10. $\begin{cases} y = 8 - x^2 \\ y = 6 - x \end{cases}$

11. $\begin{cases} y = x^2 + 2 \\ y = -x^2 + 4 \end{cases}$
12. $\begin{cases} \sqrt{x + 1} = y \\ 2x + y = 4 \end{cases}$

13. $\begin{cases} x^2 - y^2 = 12 \\ x - 2y = 0 \end{cases}$
14. $\begin{cases} x^2 + y = 4 \\ x + y = 6 \end{cases}$

15. $\begin{cases} y = x^3 \\ y = x^3 - 3x^2 + 3x \end{cases}$
16. $\begin{cases} y = -2(x^2 - 1) \\ y = 2(x^4 - 2x^2 + 1) \end{cases}$

In Exercises 17–34, solve the system by the method of substitution.

17. $\begin{cases} y = 2x^2 \\ y = -2x + 12 \end{cases}$
18. $\begin{cases} y = 5x^2 \\ y = -15x - 10 \end{cases}$

19. $\begin{cases} x^2 + y = 9 \\ x - y = -3 \end{cases}$
20. $\begin{cases} x - y^2 = 0 \\ x - y = 2 \end{cases}$

21. $\begin{cases} x^2 + 2y = 6 \\ x - y = -4 \end{cases}$
22. $\begin{cases} x^2 + y^2 = 100 \\ x = 12 \end{cases}$

23. $\begin{cases} x^2 + y^2 = 25 \\ 2x - y = -5 \end{cases}$
24. $\begin{cases} x^2 + y^2 = 169 \\ x + y = 7 \end{cases}$

25. $\begin{cases} x^2 + y^2 = 64 \\ -3x + y = 8 \end{cases}$
26. $\begin{cases} x^2 + y^2 = 81 \\ x + 3y = 27 \end{cases}$

27. $\begin{cases} 4x + y^2 = 2 \\ 2x - y = -11 \end{cases}$
28. $\begin{cases} x^2 + y^2 = 10 \\ 2x - y = 5 \end{cases}$

29. $\begin{cases} y = \sqrt{4 - x} \\ x + 3y = 6 \end{cases}$
30. $\begin{cases} y = \sqrt[3]{x} \\ y = x \end{cases}$

31. $\begin{cases} 16x^2 + 9y^2 = 144 \\ 4x + 3y = 12 \end{cases}$
32. $\begin{cases} y = 2x^2 \\ y = x^4 - 2x^2 \end{cases}$

33. $\begin{cases} x^2 - y^2 = 9 \\ x^2 + y^2 = 1 \end{cases}$
34. $\begin{cases} x^2 - y^2 = 16 \\ 3x - y = 12 \end{cases}$

In Exercises 35–44, solve the system by the method of elimination.

35. $\begin{cases} x^2 + 2y = 1 \\ x^2 + y^2 = 4 \end{cases}$
36. $\begin{cases} x + y^2 = 5 \\ 2x^2 + y^2 = 6 \end{cases}$

37. $\begin{cases} -x + y^2 = 10 \\ x^2 - y^2 = -8 \end{cases}$
38. $\begin{cases} x^2 + y = 9 \\ x^2 - y^2 = 7 \end{cases}$

39. $\begin{cases} x^2 + y^2 = 7 \\ x^2 - y^2 = 1 \end{cases}$
40. $\begin{cases} x^2 + y^2 = 25 \\ y^2 - x^2 = 7 \end{cases}$

41. $\begin{cases} \dfrac{x^2}{4} + y^2 = 1 \\ x^2 + \dfrac{y^2}{4} = 1 \end{cases}$
42. $\begin{cases} x^2 - y^2 = 1 \\ \dfrac{x^2}{2} + y^2 = 1 \end{cases}$

43. $\begin{cases} y^2 - x^2 = 10 \\ x^2 + y^2 = 16 \end{cases}$
44. $\begin{cases} x^2 + y^2 = 25 \\ x^2 + 2y^2 = 36 \end{cases}$

Solving Problems

45. *Hyperbolic Mirror*　In a hyperbolic mirror, light rays directed to one focus are reflected to the other focus. The mirror in the figure has the equation

$$\frac{x^2}{9} - \frac{y^2}{16} = 1.$$

At which point on the mirror will light from the point $(0, 10)$ reflect to the focus?

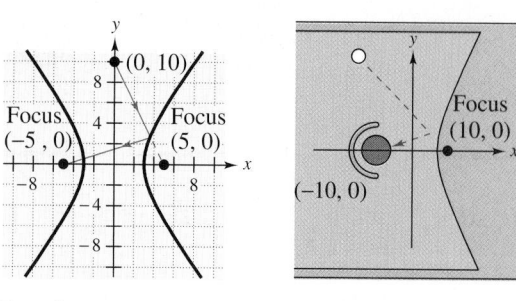

Figure for 45   Figure for 46

46. *Miniature Golf*   You are playing miniature golf and your golf ball is at $(-15, 25)$ (see figure). A wall at the end of the enclosed area is part of a hyperbola whose equation is

$$\frac{x^2}{19} - \frac{y^2}{81} = 1.$$

Using the reflective property of hyperbolas given in Exercise 45, at which point on the wall must your ball hit for it to go into the hole? (The ball bounces off the wall only once.)

47. *Busing Boundary*   To be eligible to ride the school bus to East High School, a student must live at least 1 mile from the school (see figure). Describe the

portion of Clarke Street for which the residents are *not* eligible to ride the school bus. Use a coordinate system in which the school is at $(0, 0)$ and each unit represents 1 mile.

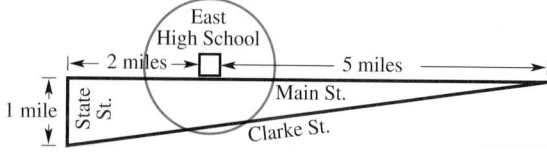

Figure for 47

48. *Data Analysis*   From 1990 through 1999, the population of the state of New York grew at a lower rate than the population of Texas. Two models that represent the populations of the two states are

$$P = 17,999 + 46.1t - 3.05t^2 \quad \text{New York}$$
$$P = 16,992 + 332.8t + 0.75t^2 \quad \text{Texas}$$

where P is the population in thousands and t is the calendar year, with $t = 0$ corresponding to 1990. Use a graphing utility to determine the year in which the population of Texas overtook the population of New York.   (*Source: U.S. Census Bureau*)

Explaining Concepts

49. Explain how to solve a nonlinear system of equations using the method of substitution.

50. Explain how to solve a nonlinear system of equations using the elimination method.

Ongoing Review

In Exercises 51 and 52, solve the system of linear equations.

51. $\begin{cases} 3x + 5y = 9 \\ 2x - 3y = -13 \end{cases}$ **52.** $\begin{cases} x - 2y + z = 7 \\ 2x + y - z = 0 \\ 3x + 2y - 2z = -2 \end{cases}$

53. *Simple Interest*   An investment of $4600 is made at an annual simple interest rate of 6.8%. How much additional money must be invested at an annual simple interest rate of 9% so that the total interest earned is 8% of the total investment?

Looking Further

Geometry   A theorem from geometry states that if a triangle is inscribed in a circle such that one side of the triangle is a diameter of the circle, then the triangle is a right triangle. Show that this theorem is true for the circle $x^2 + y^2 = 100$ and the triangle formed by the lines

$$y = 0, \quad y = \tfrac{1}{2}x + 5, \quad \text{and} \quad y = -2x + 20.$$

(Find the vertices of the triangle and verify that it is a right triangle.)

Chapter Summary: Key Concepts

8.1 ■ Definitions of sum, difference, product, and quotient of functions

Let f and g be two functions with overlapping domains.

1. Sum or difference: $(f \pm g)(x) = f(x) \pm g(x)$

2. Product: $(fg)(x) = f(x) \cdot g(x)$

3. Quotient: $\left(\dfrac{f}{g}\right)(x) = \dfrac{f(x)}{g(x)}, \; g(x) \neq 0$

8.1 ■ Composition of two functions

The composition of two functions f and g is given by $(f \circ g)(x) = f(g(x))$. The domain of the composite function $(f \circ g)$ is the set of all x in the domain of g such that $g(x)$ is in the domain of f.

8.2 ■ Finding an inverse function

To find the inverse function of a function f, use the steps below.

1. In the equation for $f(x)$, replace $f(x)$ by y.

2. Interchange the roles of x and y.

3. If the new equation represents y as a function of x, solve the new equation for y.

4. Replace y by $f^{-1}(x)$.

5. Verify that f and f^{-1} are inverse functions of each other by showing that $f(f^{-1}(x)) = x = f^{-1}(f(x))$.

8.2 ■ Horizontal Line Test for inverse functions

A function f has an inverse function if and only if no horizontal line intersects the graph of f at more than one point. Such a function is called one-to-one.

8.3 ■ Variation models

1. *Direct variation*: $y = kx$, k is a constant.

2. *Inverse variation*: $y = k/x$, k is a constant.

3. *Joint variation*: $z = kxy$, k is a constant.

8.4 ■ Features of graphs of polynomial functions

1. The graph of a polynomial function is continuous.

2. The graph of a polynomial function has only smooth turns. The graph of a function of degree n has at most $n - 1$ turns.

3. If the leading coefficient of the polynomial function is positive, the graph rises to the right. If the leading coefficient is negative, the graph falls to the right.

8.5 ■ Standard forms of the equations of circles

1. Center at the origin and radius r: $x^2 + y^2 = r^2$

2. Center at (h, k) and radius r:
$(x - h)^2 + (y - k)^2 = r^2$

8.6 ■ Standard forms of the equations of ellipses and hyperbolas

1. Ellipse with center at the origin $(0 < b < a)$:

$$x^2/a^2 + y^2/b^2 = 1 \text{ or } x^2/b^2 + y^2/a^2 = 1$$

2. Ellipse with center at (h, k) $(0 < b < a)$:

$$(x - h)^2/a^2 + (y - k)^2/b^2 = 1 \text{ or}$$
$$(x - h)^2/b^2 + (y - k)^2/a^2 = 1$$

3. Hyperbola with center at the origin $(a > 0, b > 0)$:

$$x^2/a^2 - y^2/b^2 = 1 \text{ or } y^2/a^2 - x^2/b^2 = 1$$

4. Hyperbola with center at (h, k) $(a > 0, b > 0)$:

$$(x - h)^2/a^2 - (y - k)^2/b^2 = 1 \text{ or}$$
$$(y - k)^2/a^2 - (x - h)^2/b^2 = 1$$

8.7 ■ Standard forms of the equations of parabolas

1. Vertex at the origin:

$$x^2 = 4py, \; p \neq 0 \quad \text{Vertical axis}$$
$$y^2 = 4px, \; p \neq 0 \quad \text{Horizontal axis}$$

2. Vertex at (h, k):

$$(x - h)^2 = 4p(y - k), \; p \neq 0 \quad \text{Vertical axis}$$
$$(y - k)^2 = 4p(x - h), \; p \neq 0 \quad \text{Horizontal axis}$$

8.8 ■ Solving a nonlinear system graphically

1. Sketch the graph of each equation in the system.

2. Approximate the point(s) of intersection of the graphs (if any).

3. Check the coordinate values in the original system.

Review Exercises

Reviewing Skills

8.1 ■ *How to add, subtract, multiply, and divide functions*

In Exercises 1–4, use the functions f and g to find each combination.

(a) $(f + g)(x)$ (b) $(f - g)(x)$

(c) $(fg)(x)$ (d) $(f/g)(x)$

Find the domain of each combination.

1. $f(x) = x + 2$, $g(x) = x - 2$
2. $f(x) = x^2 + 6$, $g(x) = \sqrt{1 - x}$
3. $f(x) = \dfrac{2}{x - 1}$, $g(x) = x$
4. $f(x) = \dfrac{1}{x}$, $g(x) = \dfrac{1}{x - 4}$

In Exercises 5–8, evaluate the combinations of the functions f and g at the indicated x-values. (If not possible, state the reason.)

5. $f(x) = x^2$, $g(x) = 4x - 5$
   (a) $(f + g)(-5)$ (b) $(f - g)(0)$
   (c) $(fg)(2)$ (d) $\left(\dfrac{f}{g}\right)(1)$

6. $f(x) = \frac{3}{4}x^3$, $g(x) = x + 1$
   (a) $(f + g)(-1)$ (b) $(f - g)(2)$
   (c) $(fg)\left(\dfrac{1}{3}\right)$ (d) $\left(\dfrac{f}{g}\right)(2)$

7. $f(x) = \frac{2}{3}\sqrt{x}$, $g(x) = -x^2$
   (a) $(f + g)(1)$ (b) $(f - g)(9)$
   (c) $(fg)\left(\dfrac{1}{4}\right)$ (d) $\left(\dfrac{f}{g}\right)(2)$

8. $f(x) = |x|$, $g(x) = 3$
   (a) $(f + g)(-2)$ (b) $(f - g)(3)$
   (c) $(fg)(-10)$ (d) $\left(\dfrac{f}{g}\right)(-3)$

■ *How to find compositions of one function with another function*

In Exercises 9–12, find the compositions.

9. $f(x) = x + 2$, $g(x) = x^2$
   (a) $(f \circ g)(x)$ (b) $(g \circ f)(x)$
   (c) $(f \circ g)(2)$ (d) $(g \circ f)(-1)$

10. $f(x) = \sqrt[3]{x}$, $g(x) = x + 2$
   (a) $(f \circ g)(x)$ (b) $(g \circ f)(x)$
   (c) $(f \circ g)(6)$ (d) $(g \circ f)(64)$

11. $f(x) = \sqrt{x + 1}$, $g(x) = x^2 - 1$
   (a) $(f \circ g)(x)$ (b) $(g \circ f)(x)$
   (c) $(f \circ g)(5)$ (d) $(g \circ f)(-1)$

12. $f(x) = \dfrac{1}{x - 5}$, $g(x) = \dfrac{5x + 1}{x}$
   (a) $(f \circ g)(x)$ (b) $(g \circ f)(x)$
   (c) $(f \circ g)(1)$ (d) $(g \circ f)\left(\dfrac{1}{5}\right)$

In Exercises 13 and 14, find the domains of the compositions (a) $f \circ g$ and (b) $g \circ f$.

13. $f(x) = \sqrt{x - 4}$, $g(x) = 2x$

14. $f(x) = \dfrac{2}{x - 4}$, $g(x) = x^2$

8.2 ■ *How to find inverse functions informally and verify that two functions are inverse functions of each other*

In Exercises 15–18, find the inverse function of f informally. Verify that $f(f^{-1}(x))$ and $f^{-1}(f(x))$ are equal to the identity function.

15. $f(x) = \frac{1}{6}x$ 16. $f(x) = 7 + x$

17. $f(x) = x^3$ 18. $f(x) = \sqrt[5]{x}$

■ *How to find inverse functions algebraically*

In Exercises 19–24, find the inverse function.

19. $f(x) = \frac{1}{4}x$ 20. $f(x) = 2x - 3$

21. $h(x) = \sqrt{x}$ 22. $g(x) = x^2 + 2$, $x \geq 0$

23. $f(t) = |t + 3|$ 24. $h(t) = t$

In Exercises 25 and 26, restrict the domain of the function f to an interval over which the function is increasing, and determine f^{-1} over that interval. Use a graphing utility to graph f and f^{-1} in the same viewing window.

25. $f(x) = 2(x - 4)^2$ **26.** $f(x) = |x - 2|$

■ *How to use graphs of functions to decide whether functions have inverse functions*

In Exercises 27–30, use a graphing utility to decide if the function has an inverse function. Explain your reasoning.

27. $f(x) = x^2 - 25$ **28.** $f(x) = \frac{1}{4}x^3$

29. $h(x) = 4\sqrt[3]{x}$ **30.** $g(x) = \sqrt{9 - x^2}$

8.3 ■ *How to write mathematical models for direct variation*
   ■ *How to write mathematical models for inverse variation*
   ■ *How to write mathematical models for joint variation*

In Exercises 31–34, find the constant of proportionality and write an equation that relates the variables.

31. y varies directly as the cube root of x, and $y = 12$ when $x = 8$.

32. r varies inversely as s, and $r = 45$ when $s = \frac{3}{5}$.

33. T varies jointly as r and the square of s, and $T = 5000$ when $r = 0.09$ and $s = 1000$.

34. D is directly proportional to the cube of x and inversely proportional to y, and $D = 810$ when $x = 3$ and $y = 25$.

8.4 ■ *How to use transformations to sketch graphs of polynomial functions*

In Exercises 35 and 36, use the graph to sketch the graphs of the specified transformations.

35. $y = x^5$ **36.** $y = x^6$
   (a) $f(x) = (x - 3)^5$ (a) $f(x) = (x + 4)^6$
   (b) $f(x) = x^5 - 3$ (b) $f(x) = -x^6$
   (c) $f(x) = (x - 3)^5 - 3$ (c) $f(x) = x^6 + 4$
   (d) $f(x) = 3 - x^5$ (d) $f(x) = -(x + 4)^6 - 4$

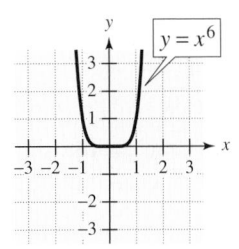

■ *How to sketch graphs of and determine the right-hand and left-hand behavior of polynomial functions*

In Exercises 37–42, sketch the graph of the polynomial function and identify any intercepts. Use a graphing utility to verify your result.

37. $f(x) = -(x - 2)^3$ **38.** $f(x) = (x + 1)^3$
39. $g(x) = x^4 - x^3 - 2x^2$ **40.** $f(x) = x^3 - 4x$
41. $f(x) = x(x + 3)^2$ **42.** $f(x) = x^4 - 4x^2$

In Exercises 43–46, determine the right-hand and left-hand behavior of the graph of the polynomial function.

43. $f(x) = -x^2 + 6x + 9$
44. $f(x) = \frac{1}{2}x^3 + 2x$
45. $g(x) = \frac{3}{4}(x^4 + 3x^2 + 2)$
46. $h(x) = -x^5 - 7x^2 + 10x$

8.5 ■ *How to recognize the four basic conics: circles, parabolas, ellipses, and hyperbolas*

In Exercises 47–54, identify the conic.

47.   **48.**

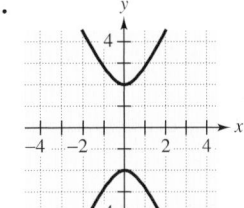   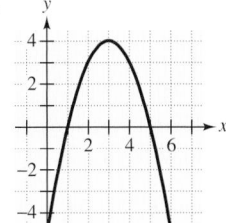

49.   **50.**

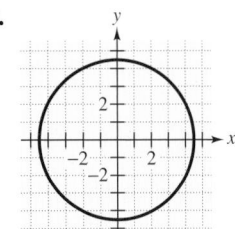   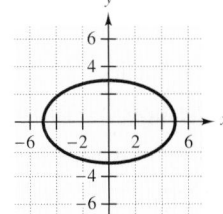

51.   **52.**

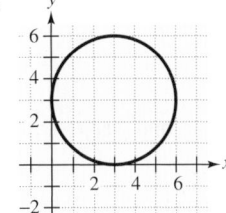   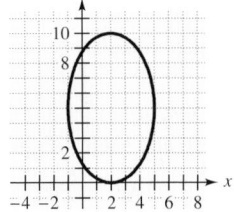

53. 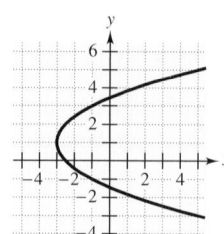   **54.**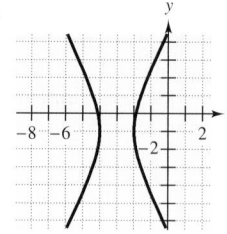

■ *How to graph and find equations of circles centered at the origin*

■ *How to graph and find equations of circles centered at (h, k)*

In Exercises 55–58, find an equation of the circle. Then sketch the circle.

55. Center: $(0, 0)$;   Radius: 12

56. Center: $(3, 5)$;   Radius: 5

57. Center: $(0, 0)$;   Passes through the point $(-1, 3)$

58. Center: $(-2, 3)$;   Passes through the point $(1, 1)$

In Exercises 59 and 60, write the equation of the circle in standard form. Then sketch the circle.

59. $x^2 + y^2 = 64$ **60.** $4x^2 + 4y^2 - 9 = 0$

8.6 ■ *How to graph and find equations of ellipses centered at the origin*

■ *How to graph and find equations of ellipses centered at (h, k)*

In Exercises 61–64, write the standard form of the equation of the ellipse. Then sketch the ellipse.

61. Vertices: $(0, -5), (0, 5)$;

Co-vertices: $(-2, 0), (2, 0)$

62. Vertices: $(-10, 0), (10, 0)$;

Co-vertices: $(0, -6), (0, 6)$

63. Vertices: $(0, 3), (10, 3)$;

Co-vertices: $(5, 0), (5, 6)$

64. Vertices: $(0, 0), (0, 8)$;

Co-vertices: $(-3, 4), (3, 4)$

In Exercises 65–68, find the center and vertices of the ellipse and sketch its graph.

65. $x^2 + 4y^2 = 64$ **66.** $x^2 + 9y^2 - 9 = 0$

67. $9x^2 + 4y^2 - 18x + 16y - 299 = 0$

68. $x^2 + 25y^2 - 4x - 21 = 0$

■ *How to graph and find equations of hyperbolas centered at the origin*

■ *How to graph and find equations of hyperbolas centered at (h, k)*

In Exercises 69–72, write the standard form of the equation of the hyperbola. Then sketch the hyperbola.

69. Vertices: $(\pm 6, 0)$;   Asymptotes: $y = \pm\frac{1}{3}x$

70. Vertices: $(0, \pm 4)$;   Asymptotes: $y = \pm 2x$

71. Vertices: $(\pm 1, 3)$;   Asymptotes: $y = \pm 2x + 3$

72. Vertices: $(0, 0), (0, -4)$;

Asymptotes: $y = \pm 2x - 2$

In Exercises 73 and 74, find the center and vertices of the hyperbola and sketch its graph.

73. $\dfrac{x^2}{25} - \dfrac{y^2}{4} = 1$

74. $\dfrac{(y + 1)^2}{4} - \dfrac{(x - 3)^2}{9} = -1$

8.7 ■ *How to graph and find equations of parabolas*

In Exercises 75 and 76, identify the vertex and focus of the parabola.

75. $y = x^2 - 4x + 2$

76. $x = y^2 + 10y - 4$

In Exercises 77–80, find an equation of the parabola. Then sketch the parabola.

77. Vertex: $(0, 0)$;   Focus: $(-2, 0)$

78. Vertex: $(0, 0)$;   Focus: $(0, 4)$

79. Vertex: $(-6, 4)$;   Focus: $(-6, -1)$

80. Vertex: $(0, 5)$;   Focus: $(2, 5)$

8.8 ■ *How to solve nonlinear systems of equations graphically*

In Exercises 81–84, use a graphing utility to graph the equations and find any solutions of the system.

81. $\begin{cases} y = x^2 \\ y = 3x \end{cases}$ **82.** $\begin{cases} y = 2 + x^2 \\ y = 8 - x \end{cases}$

83. $\begin{cases} x^2 + y^2 = 16 \\ -x + y = 4 \end{cases}$ **84.** $\begin{cases} 2x^2 - y^2 = -8 \\ y = x + 6 \end{cases}$

■ How to solve nonlinear systems of equations by substitution

In Exercises 85–88, solve the nonlinear system of equations by the substitution method.

85. $\begin{cases} y = 5x^2 \\ y = -15x - 10 \end{cases}$ **86.** $\begin{cases} y^2 = 16x \\ 4x - y = -24 \end{cases}$

87. $\begin{cases} x^2 + y^2 = 1 \\ x + y = -1 \end{cases}$ **88.** $\begin{cases} x^2 + y^2 = 100 \\ x + y = 0 \end{cases}$

■ How to solve nonlinear systems of equations by elimination

In Exercises 89–92, solve the nonlinear system of equations by the method of elimination.

89. $\begin{cases} \dfrac{x^2}{16} + \dfrac{y^2}{4} = 1 \\ y = x + 2 \end{cases}$ **90.** $\begin{cases} \dfrac{x^2}{100} + \dfrac{y^2}{25} = 1 \\ y = -x - 5 \end{cases}$

91. $\begin{cases} \dfrac{x^2}{25} + \dfrac{y^2}{9} = 1 \\ \dfrac{x^2}{25} - \dfrac{y^2}{9} = 1 \end{cases}$ **92.** $\begin{cases} x^2 + y^2 = 16 \\ -x^2 + \dfrac{y^2}{16} = 1 \end{cases}$

Solving Problems

93. *Power Generation*   The power generated by a wind turbine is given by the function $P = kw^3$, where P is the number of kilowatts produced at a wind speed of w miles per hour and k is the constant of proportionality.

(a) Find k if $P = 1000$ when $w = 20$.

(b) Find the output for a wind speed of 25 miles per hour.

94. *Hooke's Law*   A force of 100 pounds stretches a spring 4 inches.

(a) How far will a force of 200 pounds stretch the spring?

(b) What force will stretch the spring 2.5 inches?

95. *Demand*   A company has found that the daily demand x for its product varies inversely as the square root of the price p. When the price is \$25, the demand is approximately 1000 units. Approximate the demand if the price is increased to \$28.

96. *Weight*   The gravitational force F with which an object is attracted to Earth is inversely proportional to the square of its distance r from the center of Earth. An astronaut weighs 200 pounds on Earth's surface ($r \approx 4000$ miles). What does the astronaut weigh 500 miles above Earth's surface?

97. *Satellite Orbit*   Find an equation of the circular orbit of a satellite 1000 miles above the surface of Earth. Place the origin of the rectangular coordinate system at the center of Earth and assume the radius of Earth to be 4000 miles.

98. *Satellite Orbit*   Find an equation of the elliptical orbit of a satellite that varies in altitude from 500 miles to 1000 miles above the surface of Earth. Place the origin of the rectangular coordinate system at the center of Earth and assume the radius of Earth to be 4000 miles.

99. *Path of a Ball*   The height y (in feet) of a ball thrown by a child is

$$y = -\frac{1}{10}x^2 + 3x + 6$$

where x is the horizontal distance (in feet) from where the ball was thrown.

(a) Use a graphing utility to graph the path of the ball.

(b) How high is the ball when it leaves the child's hand?

(c) How high is the ball when it is at its maximum height?

(d) How far from the child does the ball strike the ground?

100. *Dimensions of a Corral*   Suppose that you have 250 feet of fencing to enclose two corrals of equal size (see figure). The combined area of the corrals is 2400 square feet. Find the dimensions of each corral.

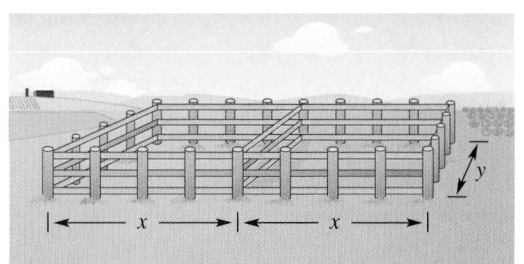

8 Chapter Test

Take this test as you would take a test in class. After you are done, check your work against the answers given in the back of the book.

In Exercises 1–4, use the functions $f(x) = \frac{1}{2}x$ and $g(x) = x^2 - 1$ to find each combination.

1. $(f - g)(x)$ **2.** $(fg)(x)$ **3.** $\left(\dfrac{f}{g}\right)(x)$ **4.** $(f \circ g)(x)$

5. Find the domain of the composite function $(f \circ g)(x)$ if $f(x) = \sqrt{25 - x}$ and $g(x) = x^2$.

6. Find the inverse function of $f(x) = \frac{1}{2}x - 1$. Verify that $f(f^{-1}(x)) = x = f^{-1}(f(x))$.

7. Write a mathematical model for the statement "S varies directly as the square of x and inversely as y."

8. Find a mathematical model that relates u and v if v varies directly as the square root of u, and $v = \frac{3}{2}$ when $u = 36$.

9. Determine the right-hand and left-hand behavior of the polynomial function $f(x) = -2x^3 + 3x^2 - 4$.

10. Sketch the graph of the function $f(x) = 1 - (x - 2)^3$. Identify any intercepts of the graph.

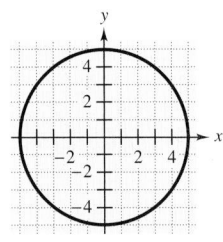

Figure for 11

11. Write an equation of the circle shown in the figure.

12. Find the standard form of the equation of the ellipse with vertices $(0, -10)$ and $(0, 10)$ and co-vertices $(-3, 0)$ and $(3, 0)$.

13. Find the standard form of the equation of the hyperbola with vertices $(-3, 0)$ and $(3, 0)$ and asymptotes $y = \frac{1}{2}x$ and $y = -\frac{1}{2}x$.

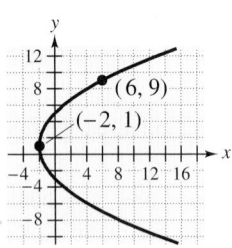

Figure for 14

14. Write an equation of the parabola shown in the figure.

15. Sketch the graph of each equation.

(a) $(x + 5)^2 + (y - 1)^2 = 9$ (b) $\dfrac{x^2}{9} + \dfrac{y^2}{16} = 1$ (c) $\dfrac{x^2}{9} - \dfrac{y^2}{16} = 1$

16. Find any solutions of each nonlinear system of equations.

(a) $\begin{cases} y = \frac{1}{2}x^2 \\ y = -4x + 6 \end{cases}$ (b) $\begin{cases} x^2 + y^2 = 100 \\ x + y = 14 \end{cases}$

17. If the temperature of a gas is not allowed to change, its absolute pressure P is inversely proportional to its volume V, according to Boyle's Law. A large balloon is filled with 180 cubic meters of helium at atmospheric pressure (1 atm) at sea level. What is the volume of the helium if the balloon rises to an altitude at which the atmospheric pressure is 0.75 atm? (Assume that the temperature does not change.)

Cumulative Test: Chapters 1–8

Take this test as you would take a test in class. After you are done, check your work against the answers given in the back of the book.

In Exercises 1–6, simplify the expression.

1. $-(-3x^2)^3(2x^4)$

2. $-\dfrac{(2u^2v)^2}{-3uv^2}$

3. $\dfrac{5}{\sqrt{12}-2}$

4. $\sqrt{\dfrac{14xz^5}{6xz}}$

5. $\dfrac{6x-5}{x-3}-\dfrac{3x-8}{x-3}$

6. $\dfrac{x}{x-5}+\dfrac{7x}{x+3}$

In Exercises 7 and 8, write the number in scientific notation.

7. 3,770,000,000

8. 0.00000000026

In Exercises 9 and 10, find the distance between the two points.

9. $(8,1),(3,6)$

10. $(-6,7),(6,-2)$

In Exercises 11 and 12, write an equation of the line passing through the two points.

11. $(-1,-2),(3,6)$

12. $(1,5),(6,0)$

13. Write an equation of the parabola $y = ax^2 + bx + c$ whose vertex is $(1,3)$ and that passes through the point $(0,5)$.

14. Determine which sets of ordered pairs represent functions from A to B.

$$A = \{1,2,3,4\} \qquad B = \{-2,-1,0,1,2\}$$

(a) $\{(1,1),(2,-1),(3,-2),(4,2)\}$

(b) $\{(1,-2),(2,0),(3,-2),(4,0)\}$

(c) $\{(1,-1),(3,0),(1,1),(3,2)\}$

(d) $\{(1,2),(2,2),(3,2),(4,2)\}$

In Exercises 15–18, factor the expression completely.

15. $3x^2 - 21x$

16. $y^3 + 6y^2 + 9y$

17. $16x^4 - 1$

18. $5x(2x-5) - (2x-5)^2$

In Exercises 19 and 20, simplify the complex fraction.

19. $\dfrac{\left(\dfrac{9}{x}\right)}{\left(\dfrac{6}{x}+2\right)}$

20. $\dfrac{\left(1+\dfrac{2}{x}\right)}{\left(x-\dfrac{4}{x}\right)}$

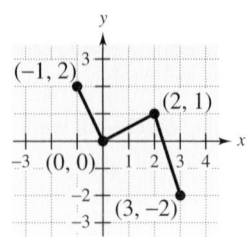

Figure for 25

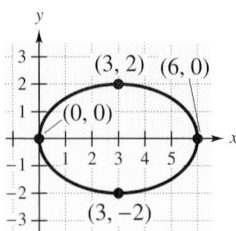

Figure for 26

21. Evaluate the function $f(x) = x^2 - 11$ for each value of x.

(a) $f(7)$ (b) $f(-5)$ (c) $f(3h)$ (d) $f(w + 3)$

In Exercises 22–24, sketch a graph of the equation.

22. $y = 3 - \dfrac{1}{2}x$ **23.** $(x - 5)^2 + (y + 2)^2 = 25$ **24.** $\dfrac{x^2}{9} - \dfrac{y^2}{25} = 1$

25. Use the graph of f shown in the figure to sketch each graph.

(a) $y = f(x) - 3$ (b) $y = -f(x)$ (c) $y = f(x + 1)$

(d) $y = f(x - 2)$ (e) $y = f(-x)$ (f) $y = 3 + f(x)$

26. Find an equation of the ellipse shown in the figure.

In Exercises 27–30, solve the equation.

27. $x^2 + 3x - 2 = 0$ **28.** $|2x + 14| = 60$

29. $x = \sqrt{12x - 35}$ **30.** $\dfrac{5x}{x - 3} - 7 = \dfrac{15}{x - 3}$

In Exercises 31 and 32, solve the inequality.

31. $6x^2 - 4 \le 5x$ **32.** $|2x - 9| < 7$

In Exercises 33 and 34, solve the system of equations.

33. $\begin{cases} 6x + 7y = -4 \\ 2x + 5y = 4 \end{cases}$ **34.** $\begin{cases} y = x^2 - x - 1 \\ 3x - y = 4 \end{cases}$

35. Find the determinant of the matrix.

$$\begin{bmatrix} 5 & -3 & -1 \\ -2 & 1 & -1 \\ 1 & 0 & 2 \end{bmatrix}$$

36. The cost of a long distance telephone call is \$1.10 for the first minute and \$0.45 for each additional minute. The total cost of the call cannot exceed \$11. Find the interval of time that is available for the call.

37. A company produces a product for which the variable cost is \$5.35 per unit and the fixed costs are \$30,000. The product is sold for \$11.60. How many units must be sold before the company breaks even?

38. Find two positive consecutive odd integers such that their sum is 47 less than their product.

39. The maximum load a cylindrical column of circular cross section can support varies directly as the fourth power of the diameter and inversely as the square of the height. A column 2 feet in diameter and 10 feet high supports up to 6 tons. How much of a load does a column 3 feet in diameter and 14 feet high support?

Exponential and Logarithmic Functions and Equations

9

THE BIG PICTURE

In this chapter you will learn how to:

- evaluate and sketch the graphs of exponential functions.

- evaluate and sketch the graphs of logarithmic functions.

- evaluate and rewrite logarithmic expressions using properties of logarithms.

- solve exponential and logarithmic equations.

- solve real-life problems using exponential and logarithmic functions.

Key Terms

As you encounter each new vocabulary term in this chapter, add the term and its definition to your notebook glossary.

exponential function (p. 584)
exponential function with base a (p. 584)
natural base (p. 588)
natural exponential function (p. 588)
radioactive decay (p. 589)
compound interest (p. 589)
continuous compounding (p. 590)
logarithmic function with base a (p. 595)
common logarithmic function (p. 598)

natural logarithmic function (p. 601)
change-of-base formula (p. 602)
human memory model (p. 603)
exponentiate (p. 618)
exponential growth model (p. 626)
exponential decay (p. 626)
intensity model (p. 629)
Richter scale (p. 629)

Additional text-specific resources are available to help you do well in this course.
See page xvi for details.

9.1 Exponential Functions and Their Graphs

Exponential Functions

In this section, you will study a new type of function called an **exponential function.** Whereas polynomial and rational functions have terms with variable bases and constant exponents, exponential functions have terms with *constant bases* and *variable exponents.* Here are some examples.

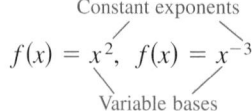

Polynomial or Rational Function

Constant exponents

$$f(x) = x^2, \quad f(x) = x^{-3}$$

Variable bases

Exponential Function

Variable exponents

$$f(x) = 2^x, \quad f(x) = 3^{-x}$$

Constant bases

Definition of Exponential Function

The **exponential function f with base a** is denoted by

$$f(x) = a^x$$

where $a > 0$, $a \neq 1$, and x is any real number.

NOTE The base $a = 1$ is excluded because $f(x) = 1^x = 1$ is a constant function, *not* an exponential function.

In Chapter 5, you learned to evaluate a^x for integer and rational values of x. For example, you know that

$$a^3 = a \cdot a \cdot a \quad \text{and} \quad a^{2/3} = \left(\sqrt[3]{a}\right)^2.$$

However, to evaluate a^x for any real number x, you need to interpret forms with *irrational* exponents. For the purpose of this text, it is sufficient to think of a number such as

$$a^{\sqrt{2}}$$

where $\sqrt{2} \approx 1.414214$, as the number that has the successively closer approximations

$$a^{1.4}, a^{1.41}, a^{1.414}, a^{1.4142}, a^{1.41421}, a^{1.414214}, \ldots .$$

The properties of exponents that were discussed in Section 5.1 can be extended to cover exponential functions, as shown below.

Properties of Exponential Functions

Let a be a positive real number, and let x and y be real numbers, variables, or algebraic expressions.

1. $a^x \cdot a^y = a^{x+y}$ 2. $(a^x)^y = a^{xy}$ 3. $\dfrac{a^x}{a^y} = a^{x-y}$ 4. $a^{-x} = \dfrac{1}{a^x} = \left(\dfrac{1}{a}\right)^x$

Example 1 ■ Evaluating Exponential Functions

Evaluate each function at the indicated values of x. Use a calculator only if it is necessary or more efficient.

Function	Values
(a) $f(x) = 2^x$	$x = 3, x = -4, x = \pi$
(b) $g(x) = 12^x$	$x = 3, x = -0.1, x = \frac{6}{7}$
(c) $h(x) = (1.085)^x$	$x = 0, x = -3, x = \sqrt{2}$

Solution

Evaluation	Comment
(a) $f(3) = 2^3 = 8$	Calculator is not necessary.
$f(-4) = 2^{-4} = \dfrac{1}{2^4} = \dfrac{1}{16}$	Calculator is not necessary.
$f(\pi) = 2^\pi \approx 8.8250$	Calculator is necessary.
(b) $g(3) = 12^3 = 1728$	Calculator is more efficient.
$g(-0.1) = 12^{-0.1} \approx 0.7800$	Calculator is necessary.
$g\left(\dfrac{6}{7}\right) = 12^{6/7} \approx 8.4142$	Calculator is necessary.
(c) $h(0) = (1.085)^0 = 1$	Calculator is not necessary.
$h(-3) = (1.085)^{-3} \approx 0.7829$	Calculator is more efficient.
$h(\sqrt{2}) = (1.085)^{\sqrt{2}} \approx 1.1223$	Calculator is necessary.

Graphs of Exponential Functions

The basic nature of the graph of an exponential function can be determined by the point-plotting method or by using a graphing utility.

Example 2 ■ The Graphs of Exponential Functions

In the same coordinate plane, sketch the graph of each function.

(a) $f(x) = 2^x$ (b) $g(x) = 4^x$

Solution

The table lists some values of each function, and Figure 9.1 shows their graphs.

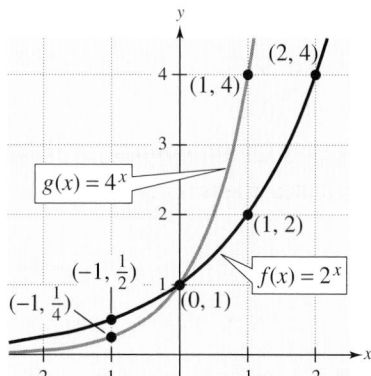

Figure 9.1

x	-2	-1	0	1	2	3
2^x	$\frac{1}{4}$	$\frac{1}{2}$	1	2	4	8
4^x	$\frac{1}{16}$	$\frac{1}{4}$	1	4	16	64

You know from your study of functions in Section 2.6 that the graph of $h(x) = f(-x) = 2^{-x}$ is a reflection of the graph of $f(x) = 2^x$ in the y-axis. This is reinforced in the next example.

Example 3 ■ The Graphs of Exponential Functions

In the same coordinate plane, sketch the graph of each function. Determine the domains and ranges.

(a) $f(x) = 2^{-x}$ (b) $g(x) = 4^{-x}$

Solution

The table lists some values of each function, and Figure 9.2 shows their graphs. From the graphs, you can see that the domain of each function is the set of all real numbers and that the range of each function is the set of all positive real numbers.

x	-3	-2	-1	0	1	2
2^{-x}	8	4	2	1	$\frac{1}{2}$	$\frac{1}{4}$
4^{-x}	64	16	4	1	$\frac{1}{4}$	$\frac{1}{16}$

Graph showing curves with points $(-1, 4)$, $(-2, 4)$, $(-1, 2)$, $(0, 1)$, $(1, \frac{1}{2})$, $(1, \frac{1}{4})$ labeled. $f(x) = 2^{-x}$ and $g(x) = 4^{-x}$.

Figure 9.2

EXPLORATION

Use a graphing utility to investigate the function $f(x) = k^x$ for different values of k. Discuss the effect that k has on the shape of the graph.

Examples 2 and 3 suggest that, for $a > 1$, the values of the function $y = a^x$ increase as x increases and the values of the function $y = a^{-x}$ decrease as x increases. The graphs shown in Figure 9.3 are typical of the graphs of exponential functions. Note that each has a y-intercept at $(0, 1)$ and a horizontal asymptote (the x-axis).

Graph of $y = a^x$

- Domain: $(-\infty, \infty)$
- Range: $(0, \infty)$
- Intercept: $(0, 1)$
- Increasing (moves up to the right)
- Asymptote: x-axis

Graph of $y = a^{-x} = \left(\dfrac{1}{a}\right)^x$

- Domain: $(-\infty, \infty)$
- Range: $(0, \infty)$
- Intercept: $(0, 1)$
- Decreasing (moves down to the right)
- Asymptote: x-axis

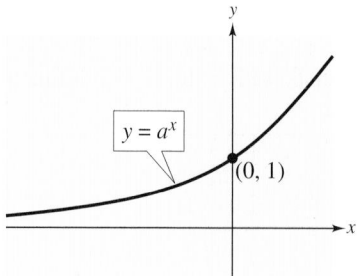
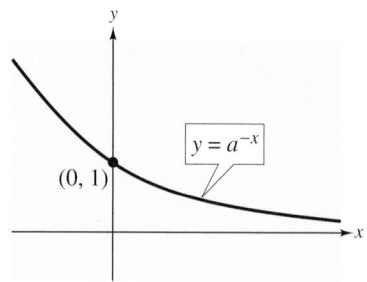

Figure 9.3 *Characteristics of the exponential functions $y = a^x$ and $y = a^{-x}$ ($a > 1$)*

In the next two examples, notice how the graph of $y = a^x$ can be used to sketch the graphs of functions of the form $f(x) = b \pm a^{x+c}$. Also note that the transformation in Example 4(a) keeps the x-axis as a horizontal asymptote, but the transformation in Example 4(b) yields a new horizontal asymptote of $y = -3$. Also, be sure to note how the y-intercept is affected by each transformation.

Example 4 ■ Transformations of Graphs of Exponential Functions

Use transformations to analyze and sketch the graph of each function.

(a) $g(x) = 2^{x-3}$

(b) $h(x) = 2^x - 3$

Solution

Consider the function $f(x) = 2^x$.

(a) The function g is related to f by $g(x) = f(x - 3)$. To sketch the graph of g, shift the graph of f three units to the right, as shown in Figure 9.4(a).

(b) The function h is related to f by $h(x) = f(x) - 3$. To sketch the graph of h, shift the graph of f three units downward, as shown in Figure 9.4(b).

Example 5 ■ Transformations of Graphs of Exponential Functions

Use transformations to analyze and sketch the graph of each function.

(a) $g(x) = -2^x$

(b) $h(x) = 2 + 2^{-x}$

Solution

(a) The function g is related to $f(x) = 2^x$ by $g(x) = -f(x)$. To sketch the graph of g, reflect the graph of f about the x-axis, as shown in Figure 9.5(a).

(b) The function h is related to $f(x) = 2^x$ by $h(x) = 2 + f(-x)$. To sketch the graph of h, shift the graph of f two units upward and reflect it about the y-axis, as shown in Figure 9.5(b).

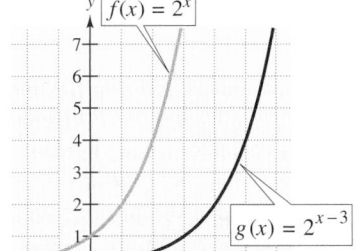

(a)

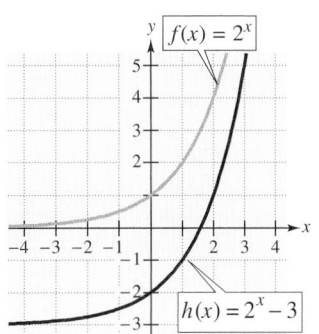

(b)

Figure 9.4

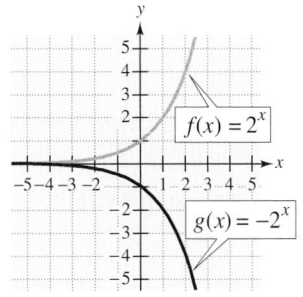

(a)

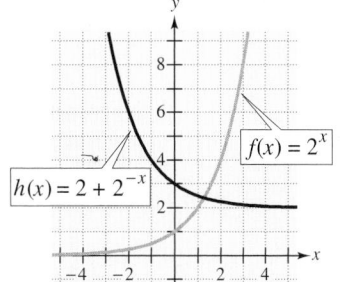

(b)

Figure 9.5

The Natural Exponential Function

So far, we have used integers or rational numbers as bases of exponential functions. In many applications of exponential functions, the convenient choice for a base is the irrational number below, denoted by the letter "e."

$$e \approx 2.71828. \ . \ . \ . \qquad \text{Natural base}$$

This number is called the **natural base.** The function

$$f(x) = e^x \qquad \text{Natural exponential function}$$

is called the **natural exponential function.** Be sure you understand that for this function, e is the constant number 2.71828. . . . , and x is a variable. To evaluate the natural exponential function, you need a calculator, preferably one having a natural exponential key $\boxed{e^x}$.

After evaluating the natural exponential function at several values, as shown in the table, you can sketch its graph, as shown in Figure 9.6.

x	-1.5	-1.0	-0.5	0.0	0.5	1.0	1.5
$f(x) = e^x$	0.223	0.368	0.607	1.000	1.649	2.718	4.482

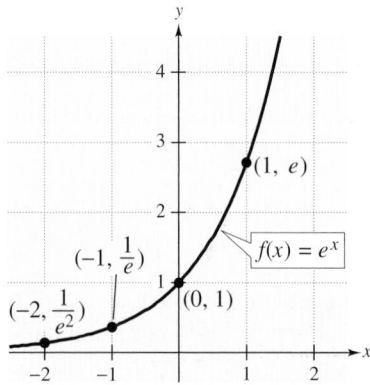

Figure 9.6 *Natural exponential function*

From the graph, notice the characteristics of the natural exponential function listed below.

- The domain is the set of all real numbers.
- The range is the set of positive real numbers.
- The y-intercept is $(0, 1)$.
- Increasing (moves up to the right)
- Asymptote: x-axis

Notice that these characteristics are consistent with those listed for the exponential function $y = a^x$ on page 586.

NOTE The number e can be approximated by the expression $\left(1 + \dfrac{1}{x}\right)^x$ for large values of x.

Applications

A common scientific application of exponential functions is that of **radioactive decay.**

Example 6 ■ Radioactive Decay

After t years, the remaining mass y (in grams) of 10 grams of a radioactive element whose half-life is 25 years is given by

$$y = 10\left(\tfrac{1}{2}\right)^{t/25}, \quad t \geq 0.$$

How much of the initial mass remains after 120 years?

Algebraic Solution

When $t = 120$, the mass is given by

$y = 10\left(\tfrac{1}{2}\right)^{t/25}$	Write original equation.
$= 10\left(\tfrac{1}{2}\right)^{120/25}$	Substitute 120 for t.
$= 10\left(\tfrac{1}{2}\right)^{4.8}$	Simplify.
$\approx 0.359.$	Use a calculator.

So, after 120 years, the mass has decayed from an initial amount of 10 grams to only 0.359 gram.

Graphical Solution

Use a graphing utility to graph $y = 10\left(\tfrac{1}{2}\right)^{x/25}$. Note that the graph of the function shows the 25-year half-life. That is, after 25 years the mass is 5 grams (half of the original), after another 25 years the mass is 2.5 grams, and so on. Use the *value* feature or the *zoom* and *trace* features of the graphing utility to determine that the value of y when $x = 120$ is about 0.359, as shown in Figure 9.7. So, about 0.359 gram is present after 120 years.

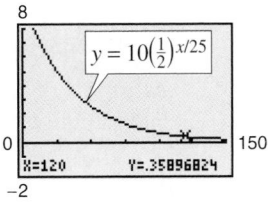

Figure 9.7

One of the most familiar applications of exponential functions involves **compound interest.** Suppose a principal P is invested at an annual interest rate r (in decimal form), compounded once a year. If the interest is added to the principal at the end of the year, the balance is

$$A = P + Pr = P(1 + r).$$

This pattern of multiplying the previous principal by $(1 + r)$ is then repeated each successive year, as shown below.

Time in Years	*Balance at Given Time*
0	$A = P$
1	$A = P(1 + r)$
2	$A = P(1 + r)(1 + r) = P(1 + r)^2$
3	$A = P(1 + r)^2(1 + r) = P(1 + r)^3$
$\vdots$	$\vdots$
t	$A = P(1 + r)^t$

To account for more frequent compounding of interest (such as quarterly or monthly compounding), let n be the number of compoundings per year and let t be the number of years. Then the rate per compounding is r/n and the account balance after t years is

$$A = P\left(1 + \frac{r}{n}\right)^{nt}.$$

Example 7 ■ Finding the Balance for Compound Interest

A sum of $10,000 is invested at an annual interest rate of 7.5%, compounded monthly. Find the balance in the account after 10 years.

Solution

Using the formula for compound interest, with $P = 10{,}000$, $r = 0.075$, $n = 12$ (for monthly compounding), and $t = 10$, you obtain the balance as follows.

$$A = 10{,}000\left(1 + \frac{0.075}{12}\right)^{12(10)} \approx \$21{,}120.65.$$

A second method that banks use to compute interest is called **continuous compounding.** The formula for the balance for this type of compounding is

$$A = Pe^{rt}.$$

Formulas for Compound Interest

After t years, the balance A in an account with principal P and annual interest rate r (in decimal form) is given by the two formulas below.

1. For n compoundings per year: $A = P\left(1 + \dfrac{r}{n}\right)^{nt}$

2. For continuous compounding: $A = Pe^{rt}$

Example 8 ■ Three Types of Compounding

A total of $15,000 is invested at an annual interest rate of 8%. Find the balance after 6 years if the interest is compounded (a) quarterly, (b) monthly, and (c) continuously.

Solution

(a) Letting $P = 15{,}000$, $r = 0.08$, $n = 4$, and $t = 6$, the balance after 6 years is

$$A = 15{,}000\left(1 + \frac{0.08}{4}\right)^{4(6)} = \$24{,}126.56.$$

(b) Letting $P = 15{,}000$, $r = 0.08$, $n = 12$, and $t = 6$, the balance after 6 years is

$$A = 15{,}000\left(1 + \frac{0.08}{12}\right)^{12(6)} = \$24{,}202.53.$$

(c) Letting $P = 15{,}000$, $r = 0.08$, and $t = 6$, the balance after 6 years is

$$A = 15{,}000e^{0.08(6)} = \$24{,}241.12.$$

NOTE   Example 8 illustrates a general rule for compound interest. For a given principal, interest rate, and time, the more often the interest is compounded per year, the greater the balance will be. Moreover, the balance obtained by continuous compounding is larger than the balance obtained by compounding n times per year.

9.1 ■ Exercises

Developing Skills

In Exercises 1–8, evaluate the expression. (Round to three decimal places.)

1. $4^{\sqrt{3}}$

2. $6^{-\pi}$

3. $e^{1/3}$

4. $e^{-1/3}$

5. $\dfrac{4e^3}{12e^2}$

6. $\dfrac{20e}{15e^4}$

7. $(-8e^3)^{1/3}$

8. $(9e^2)^{3/2}$

In Exercises 9–24, evaluate the function as indicated. Use a calculator only if it is necessary. (Round to three decimal places.)

9. $f(x) = 3^x$
  (a) $x = -2$
  (b) $x = 0$
  (c) $x = 1$

10. $F(x) = 3^{-x}$
  (a) $x = -2$
  (b) $x = 0$
  (c) $x = 1$

11. $g(x) = 5^x$
  (a) $x = -1$
  (b) $x = 1$
  (c) $x = 3$

12. $G(x) = 5^{-x}$
  (a) $x = -1$
  (b) $x = 1$
  (c) $x = \sqrt{3}$

13. $f(t) = 500\left(\tfrac{1}{2}\right)^t$
  (a) $t = 0$
  (b) $t = 1$
  (c) $t = \pi$

14. $g(s) = 1200\left(\tfrac{2}{3}\right)^s$
  (a) $s = 0$
  (b) $s = 2$
  (c) $s = 4$

15. $f(x) = 1000(1.05)^{2x}$
  (a) $x = 0$
  (b) $x = 5$
  (c) $x = 10$

16. $h(x) = 750(1.035)^{1.5x}$
  (a) $x = 0$
  (b) $x = 4$
  (c) $x = 20$

17. $g(x) = \dfrac{500}{(1.02)^{5x}}$
  (a) $x = 1$
  (b) $x = 8$
  (c) $x = 25$

18. $P(t) = \dfrac{10,000}{(1.01)^{12t}}$
  (a) $t = 2$
  (b) $t = 10$
  (c) $t = 20$

19. $f(t) = 6e^{3t}$
  (a) $t = -1$
  (b) $t = 0$

20. $A(t) = 200e^{0.1t}$
  (a) $t = 10$
  (b) $t = 20$

21. $g(x) = 10e^{-0.5x}$
  (a) $x = -4$
  (b) $x = 4$

22. $h(t) = 14e^{-3t}$
  (a) $t = -7$
  (b) $t = 7$

23. $A(x) = \dfrac{42}{1 + e^{-1.2x}}$
  (a) $x = 0$
  (b) $x = 20$

24. $f(z) = \dfrac{100}{1 + e^{-0.05z}}$
  (a) $z = 0$
  (b) $z = 10$

In Exercises 25–32, match the function with its graph. [The graphs are labeled (a), (b), (c), (d), (e), (f), (g), and (h).]

(a)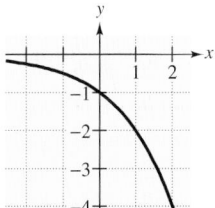

(b)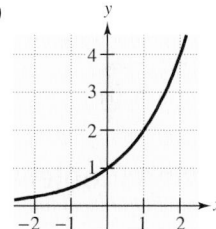

(c)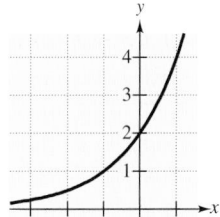

(d)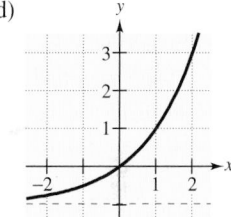

(e)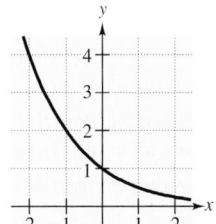

(f)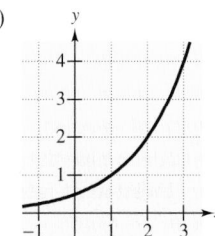

(g)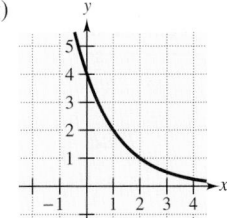

(h)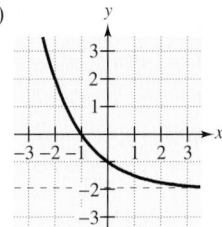

25. $f(x) = 2^x$

26. $f(x) = -2^x$

27. $f(x) = 2^{-x}$

28. $f(x) = 2^x - 1$

29. $f(x) = 2^{x-1}$

30. $f(x) = 2^{x+1}$

31. $f(x) = \left(\frac{1}{2}\right)^x - 2$

32. $f(x) = \left(\frac{1}{2}\right)^{x-2}$

In Exercises 33–44, identify the transformation of the graph of $f(x) = 3^x$ and sketch the graph of h.

33. $h(x) = 3^x - 2$

34. $h(x) = 3^x + 1$

35. $h(x) = 3^{x+5}$

36. $h(x) = 3^{x-3}$

37. $h(x) = 3^{-x}$

38. $h(x) = -3^x$

39. $h(x) = 3^{-(x+4)}$

40. $h(x) = 3^{6-x}$

41. $h(x) = -3^x + 10$

42. $h(x) = -3^x - 1$

43. $h(x) = 3^{2+x} - 4$

44. $h(x) = 3^{x-1} + 3$

In Exercises 45–54, sketch the graph of the function and identify the horizontal asymptote.

45. $f(x) = 4^{x-5}$

46. $g(x) = 4^{x+1}$

47. $h(x) = 2^x - 3$

48. $f(x) = 5 + 2^x$

49. $f(x) = -5^x$

50. $g(x) = 5^{-x}$

51. $g(x) = 7 + 10^{x-4}$

52. $f(x) = 10^{x+1} - 3$

53. $f(x) = \left(\frac{1}{2}\right)^{x+4} + 1$

54. $h(x) = 4 + \left(\frac{1}{2}\right)^{x-5}$

In Exercises 55–72, use a graphing utility to graph the function.

55. $h(x) = \frac{1}{2}(3^x)$

56. $h(x) = \frac{1}{2}(3^{-x})$

57. $f(t) = 2^{-t^2}$

58. $f(t) = 2^{t^2}$

59. $f(x) = -2^{0.5x}$

60. $h(t) = -2^{-0.5t}$

61. $f(x) = 5^{x/3}$

62. $f(x) = 5^{-x/3} + 2$

63. $g(t) = 200\left(\frac{1}{2}\right)^t$

64. $h(y) = 27\left(\frac{2}{3}\right)^y$

65. $h(t) = 500(1.06)^t$

66. $f(t) = 100(1.06)^{-t}$

67. $g(x) = 3e^{0.2x}$

68. $P(t) = 100e^{-0.1t}$

69. $f(x) = 6e^{-x^2/3}$

70. $h(t) = 11e^{t^3/5}$

71. $g(x) = \dfrac{15}{1 + e^{-2.4x}}$

72. $g(t) = \dfrac{10}{1 + e^{-0.5t}}$

Solving Problems

Compound Interest In Exercises 73 and 74, complete the table to determine the balance A for P dollars invested at rate r for t years, compounded n times per year.

n	1	4	12	365	Continuous compounding
A					

	Principal	Rate	Time
73.	$P = \$100$	$r = 8\%$	$t = 20$ years
74.	$P = \$2000$	$r = 9\%$	$t = 10$ years

Compound Interest In Exercises 75 and 76, complete the table to determine the principal P that yields a balance of A dollars invested at rate r for t years, compounded n times per year.

n	1	4	12	365	Continuous compounding
P					

	Balance	Rate	Time
75.	$A = \$5000$	$r = 7\%$	$t = 10$ years
76.	$A = \$100,000$	$r = 9\%$	$t = 20$ years

77. *Compound Interest* Determine how much principal would need to be deposited in an account for 20 years in order to yield $1,000,000 if the account earns 10.5% interest compounded (a) quarterly, (b) monthly, (c) daily, and (d) continuously.

78. *Compound Interest* Determine how much principal would need to be deposited in an account for 2 years in order to yield $2500 if the account earns 7.5% interest compounded (a) quarterly, (b) monthly, (c) daily, and (d) continuously.

79. *Property Value* Suppose that the value of a piece of property doubles every 15 years. If you buy the property for $64,000, its value t years after the date of purchase should be $V(t) = 64,000(2)^{t/15}$. Use the model to approximate the value of the property (a) 5 years and (b) 20 years after it is purchased.

80. *Price and Demand* The daily demand x and the price p for a certain product are related by $p = 25 - 0.4e^{0.02x}$. Find the prices for demands of (a) $x = 100$ units and (b) $x = 125$ units.

The symbol ⊞ indicates an exercise in which you are instructed to use a calculator or graphing utility.

81. *Depreciation*   After t years, the value of a car that originally cost \$16,000 depreciates so that each year it is worth $\frac{3}{4}$ of its value for the previous year. Find a model for $V(t)$, the value of the car for year t. Sketch a graph of the model and determine the value of the car 2 years after it was purchased.

82. *Inflation Rate*   Suppose the annual rate of inflation averages 5% over the next 10 years. With this rate of inflation, the approximate cost C of goods or services during any year in that decade will be given by

$$C(t) = P(1.05)^t, \quad 0 \le t \le 10$$

where t is the time in years and P is the present cost. The price of an oil change for your car is presently \$24.95. Estimate the price 10 years from now.

83. *Radioactive Decay*   After t years, the remaining mass y (in grams) of 16 grams of a radioactive element whose half-life is 30 years is given by

$$y = 16\left(\tfrac{1}{2}\right)^{t/30}, \quad t \ge 0.$$

How much of the initial mass remains after 80 years?

84. *Radioactive Decay*   After t years, the remaining mass y (in grams) of 23 grams of a radioactive element whose half-life is 45 years is given by

$$y = 23\left(\tfrac{1}{2}\right)^{t/45}, \quad t \ge 0.$$

How much of the initial mass remains after 150 years?

85. *Parachute Drop*   A parachutist jumps from a plane and opens the parachute at a height of 2000 feet. The height of the parachutist is

$$h = 1950 + 50e^{-1.6t} - 20t$$

where h is the height in feet and t is the time in seconds. (The time $t = 0$ corresponds to the time when the parachute is opened.)

(a) Use a graphing utility to graph the function.

(b) Find the height of the parachutist when $t = 0$, 25, 50, and 75.

(c) Approximate the time when the parachutist reaches the ground.

86. *Graphical Interpretation*   Investments of \$500 in two different accounts with interest rates of 6% and 8% are compounded continuously. The balances in the accounts after t years are modeled by

$$A_1 = 500e^{0.06t}$$

and

$$A_2 = 500e^{0.08t}.$$

(a) Use a graphing utility to graph each of the models in the same viewing window.

(b) Use a graphing utility to graph the function $A_2 - A_1$ in the same viewing window with the graphs in part (a).

(c) Use the graphs to discuss the rates of increase of the balances in the two accounts.

87. *Data Analysis*   A meteorologist measures the atmospheric pressure P (in kilograms per square meter) at altitude h (in kilometers). The data is shown in the table.

h	0	5	10	15	20
P	10,332	5583	2376	1240	517

(a) Use a graphing utility to plot the data points.

(b) A model for the data is given by

$$P = 10{,}958e^{-0.15h}.$$

Use a graphing utility to graph the model in the same viewing window as in part (a). How well does the model fit the data?

(c) Use a graphing utility to create a table comparing the model with the data points.

(d) Estimate the atmospheric pressure at a height of 8 kilometers.

(e) Use the graph to estimate the altitude at which the atmospheric pressure is 2000 kilograms per square meter.

88. *Median Home Prices*   For the years 1994 through 1999, the median prices of a home in the United States are shown in the table. *(Sources: U.S. Census Bureau and U.S. Dept. of Urban Development)*

Year	1994	1995	1996
Price	\$130,000	\$133,900	\$140,000

Year	1997	1998	1999
Price	\$146,000	\$152,500	\$159,800

A model for this data is given by $y = 109,205e^{0.0424t}$, where t is the time in years, with $t = 4$ representing 1994.

(a) Use the model to complete the table, and compare the results with the actual data.

Year	1994	1995	1996	1997	1998	1999
Price						

(b) Use a graphing utility to graph the model.

(c) If the model were used to predict home prices in the years ahead, would the predictions be increasing at a faster rate or a slower rate with increasing t? Do you think the model would be reliable for predicting the future prices of homes? Explain.

Explaining Concepts

89. Is $e = \dfrac{271,801}{99,990}$? Explain.

90. Without using a calculator, explain why $2^{\sqrt{2}}$ is greater than 2, but less than 4.

91. Explain why $f(x) = 1^x$ is not an exponential function.

92. Compare the graphs of $f(x) = 3^x$ and $g(x) = \left(\frac{1}{3}\right)^x$.

93. Which of the functions are exponential? Explain your reasoning.

(a) $f(x) = 2x$ (b) $f(x) = 2x^2$

(c) $f(x) = 2^x$ (d) $f(x) = 2^{-x}$

Ongoing Review

In Exercises 94–97, use the properties of exponents to simplify the expression.

94. $(10t)^0$

95. $5x^4 \cdot x^3$

96. $\dfrac{9t^5}{3t^2}$

97. $\left(-\dfrac{4}{x^2}\right)^2$

In Exercises 98–103, graph the inequality.

98. $y < \frac{2}{3}x - 1$

99. $x > 6 - y$

100. $y \le 2$

101. $x > 7$

102. $2x + 3y \ge 6$

103. $5x - 2y < 5$

104. *Work Rate* Working together, two people can complete a task in 10 hours. Working alone, one person takes 3 hours longer than the other. How long would it take each person to do the task alone?

105. *Geometry* A family is setting up the boundaries for a backyard volleyball court. The court is to be 60 feet long and 30 feet wide. To be assured that the court is rectangular, someone suggests that they measure the diagonals of the court. What should be the length of each diagonal?

Looking Further

(a) Use a calculator to complete the table.

x	1	10	100	1000	10,000
$\left(1 + \dfrac{1}{x}\right)^x$					

(b) Use the table to sketch the graph of the function
$$f(x) = \left(1 + \frac{1}{x}\right)^x.$$

Does this graph appear to be approaching a horizontal asymptote?

(c) From parts (a) and (b), what conjecture can you make about the value of
$$\left(1 + \frac{1}{x}\right)^x$$
as x gets larger and larger?

9.2 Logarithmic Functions and Their Graphs

Logarithmic Functions

In Section 8.2, you were introduced to the concept of an inverse function. Moreover, you saw that if a function has the property that no horizontal line intersects the graph of the function more than once, the function must have an inverse function. By looking back at the graphs of the exponential functions introduced in Section 9.1, you will see that every function of the form

$$f(x) = a^x$$

passes the Horizontal Line Test and so must have an inverse function. To describe the inverse function of $f(x) = a^x$, follow the steps used in Section 8.2.

$$y = a^x \qquad \text{Replace } f(x) \text{ by } y.$$
$$x = a^y \qquad \text{Interchange } x \text{ and } y.$$

At this point, there is no way to solve for y. A verbal description of y in the equation $x = a^y$ is "y equals the exponent needed on base a to get x." This inverse function of $f(x) = a^x$ is denoted by

$$f^{-1}(x) = \log_a x.$$

This inverse function is called the **logarithmic function with base a.**

Definition of Logarithmic Function

Let a and x be positive real numbers such that $a \neq 1$. The **logarithm of x with base a** is denoted by $\log_a x$ and is defined as follows.

$$y = \log_a x \qquad \text{if and only if} \qquad x = a^y$$

The function $f(x) = \log_a x$ is the **logarithmic function with base a.**

From the definition of a logarithmic function, you can see that the equations $y = \log_a x$ and $x = a^y$ are equivalent.

Logarithmic Equation	*Exponential Equation*
$y = \log_a x$	$x = a^y$

The first equation is in **logarithmic** form and the second equation is in *exponential* form. From these equivalent equations it should also be clear that *a logarithm is an exponent*. For instance, because the exponent in the expression

$$2^3 = 8 \qquad \text{The exponent of } 2^3 \text{ is 3.}$$

is 3, the value of the logarithm $\log_2 8$ is 3. That is

$$\log_2 8 = 3. \qquad \text{A logarithm is an exponent.}$$

Example 1 ■ Rewriting Logarithmic Equations in Exponential Form

Rewrite each logarithmic equation in exponential form.

(a) $\log_2 16 = 4$ (b) $\log_{49} 7 = \frac{1}{2}$ (c) $\log_{10} \frac{1}{100} = -2$

Solution

(a) For this equation, the base is 2 and the logarithm is 4. Using the fact that a logarithm is an exponent, you can write

$$2^4 = 16.$$

(b) For this equation, the base is 49 and the logarithm is $\frac{1}{2}$. Using the fact that a logarithm is an exponent, you can write

$$49^{1/2} = 7.$$

(c) For this equation, the base is 10 and the logarithm is -2. Using the fact that a logarithm is an exponent, you can write

$$10^{-2} = \frac{1}{100}.$$

In many applications involving exponential equations, it is helpful to rewrite the equation in logarithmic form.

Historical Note

John Napier (1550–1617), a Scottish mathematician, developed logarithms as a way to simplify some of the tedious calculations of his day. Beginning in 1594, Napier worked about 20 years on logarithms, but he was only partially successful in his quest. Nonetheless, the development of logarithms was a step forward and received immediate recognition.

Study Tip

Remember that a logarithm is an exponent. So, to evaluate the logarithmic expression $\log_a x$, you need to ask the question, "To what power must a be raised to obtain x?"

Example 2 ■ Rewriting Exponential Equations in Logarithmic Form

Rewrite each exponential equation in logarithmic form.

(a) $4^3 = 64$ (b) $2^0 = 1$ (c) $3^{-1} = \frac{1}{3}$ (d) $9^{1/2} = 3$

Solution

(a) For this equation, the base is 4 and the exponent is 3. Using the fact that a logarithm is an exponent, you can write

$$3 = \log_4 64.$$

(b) For this equation, the base is 2 and the exponent is 0. Using the fact that a logarithm is an exponent, you can write

$$0 = \log_2 1.$$

(c) For this equation, the base is 3 and the exponent is -1. Using the fact that a logarithm is an exponent, you can write

$$-1 = \log_3 \frac{1}{3}.$$

(d) For this equation, the base is 9 and the exponent is $\frac{1}{2}$. Using the fact that a logarithm is an exponent, you can write

$$\frac{1}{2} = \log_9 3.$$

Example 3 ■ Evaluating Logarithms

Evaluate each logarithm.

(a) $\log_2 32$ (b) $\log_3 9$ (c) $\log_{25} 5$

Solution

In each case you should answer the question, "To what power must the base be raised to obtain the given number?"

(a) The power to which 2 must be raised to obtain 32 is 5. That is,

$$2^5 = 32 \quad \Longrightarrow \quad \log_2 32 = 5.$$

(b) The power to which 3 must be raised to obtain 9 is 2. That is,

$$3^2 = 9 \quad \Longrightarrow \quad \log_3 9 = 2.$$

(c) The power to which 25 must be raised to obtain 5 is $\frac{1}{2}$. That is,

$$25^{1/2} = 5 \quad \Longrightarrow \quad \log_{25} 5 = \frac{1}{2}.$$

Each of the logarithms in Example 4 involves a special important case.

Example 4 ■ Evaluating Logarithms

Evaluate each logarithm.

(a) $\log_5 1$ (b) $\log_{10} \dfrac{1}{10}$

(c) $\log_3(-1)$ (d) $\log_4 0$

Solution

(a) The power to which 5 must be raised to obtain 1 is 0. That is,

$$5^0 = 1 \quad \Longrightarrow \quad \log_5 1 = 0.$$

(b) The power to which 10 must be raised to obtain $\frac{1}{10}$ is -1. That is,

$$10^{-1} = \frac{1}{10} \quad \Longrightarrow \quad \log_{10} \frac{1}{10} = -1.$$

(c) There is no power to which 3 can be raised to obtain -1. The reason for this is that for any value of x, 3^x is a positive number. So, $\log_3(-1)$ is undefined.

(d) There is no power to which 4 can be raised to obtain 0. So, $\log_4 0$ is undefined.

NOTE   Be sure you see that a logarithm can be zero or negative, but that you cannot take the logarithm of zero or a negative number.

The properties of logarithms below follow directly from the definition of the logarithmic function with base a.

Properties of Logarithms

Let a and x be positive real numbers such that $a \neq 1$. Then the properties below are true.

1. $\log_a 1 = 0$ because $a^0 = 1$.

2. $\log_a a = 1$ because $a^1 = a$.

3. $\log_a a^x = x$ because $a^x = a^x$.

The logarithmic function with base 10 is called the **common logarithmic function.**

Example 5 ■ Evaluating Common Logarithms

Evaluate each logarithm. Use a calculator only if necessary.

(a) $\log_{10} 100$

(b) $\log_{10} 0.01$

(c) $\log_{10} 10^{16}$

(d) $\log_{10} 5$

(e) $\log_{10} 2.5$

Solution

(a) The power to which 10 must be raised to obtain 100 is 2. That is,

$$10^2 = 100 \quad \Longrightarrow \quad \log_{10} 100 = 2.$$

(b) The power to which 10 must be raised to obtain 0.01 or $\frac{1}{100}$ is -2. That is,

$$10^{-2} = \tfrac{1}{100} \quad \Longrightarrow \quad \log_{10} 0.01 = -2.$$

(c) The power to which 10 must be raised to obtain 10^{16} is 16. That is,

$$10^{16} = 10^{16} \quad \Longrightarrow \quad \log_{10} 10^{16} = 16.$$

(d) There is no simple power to which 10 can be raised to obtain 5, so you should use a calculator to evaluate $\log_{10} 5$. So, rounded to three decimal places,

$$\log_{10} 5 \approx 0.699.$$

(e) There is no simple power to which 10 can be raised to obtain 2.5, so you should use a calculator to evaluate $\log_{10} 2.5$. So, rounded to three decimal places,

$$\log_{10} 2.5 \approx 0.398.$$

Graphs of Logarithmic Functions

To sketch the graph of $y = \log_a x$, you can use the fact that the graphs of inverse functions are reflections of each other in the line $y = x$.

Example 6 ■ Graphs of Exponential and Logarithmic Functions

On the same rectangular coordinate system, sketch the graphs of each function.

(a) $f(x) = 2^x$

(b) $g(x) = \log_2 x$

Solution

(a) Begin by making a table of values for $f(x) = 2^x$.

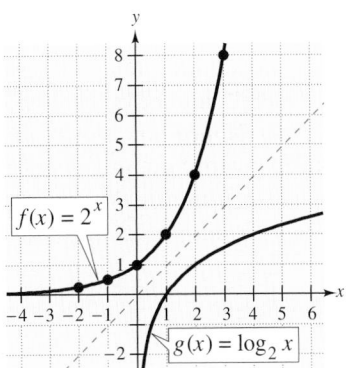

Figure 9.8

x	-2	-1	0	1	2	3
$f(x) = 2^x$	$\frac{1}{4}$	$\frac{1}{2}$	1	2	4	8

By plotting these points and connecting them with a smooth curve, you obtain the graph shown in Figure 9.8.

(b) Because $g(x) = \log_2 x$ is the inverse function of $f(x) = 2^x$, the graph of g is obtained by reflecting the graph of f in the line $y = x$, as shown in Figure 9.8.

Study Tip

In Example 6, the inverse property of logarithmic functions was used to sketch the graph of $g(x) = \log_2 x$. You could also use a standard point-plotting approach or a graphing utility.

Notice from the graph of

$$g(x) = \log_2 x$$

shown in Figure 9.8 that the domain of the function is the set of positive numbers and the range is the set of all real numbers. The basic characteristics of the graph of a logarithmic function are summarized in Figure 9.9. In this figure, note that the graph has one x-intercept, at $(1, 0)$. Also note that the y-axis is a vertical asymptote of the graph.

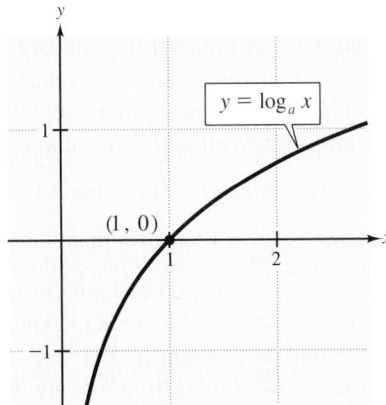

Graph of $y = \log_a x, \ a > 1$

- Domain: $(0, \infty)$
- Range: $(-\infty, \infty)$
- Intercept: $(1, 0)$
- Increasing
- Asymptote: y-axis

Figure 9.9 *Characteristics of the logarithmic function $y = \log_a x$*

Example 7 uses a standard point-plotting approach to sketch the graph of a logarithmic function.

Example 7 ■ Sketching the Graph of a Logarithmic Function

Sketch the graph of the common logarithmic function $f(x) = \log_{10} x$.

Solution

Begin by making a table of values. Note that some of the values can be obtained without a calculator, whereas others require a calculator. Using the points listed in the table, sketch the graph as shown in Figure 9.10. Notice how slowly the graph rises for $x > 1$. In Figure 9.10, you would need to move out to $x = 1000$ before the graph would rise to $y = 3$.

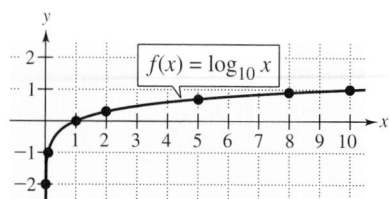

Figure 9.10

	Without calculator				With calculator		
x	$\frac{1}{100}$	$\frac{1}{10}$	1	10	2	5	8
$\log_{10} x$	-2	-1	0	1	0.301	0.699	0.903

NOTE In Example 8, the graph of $y = \log_a x$ is used to sketch the graphs of functions of the form $y = b \pm \log_a(x + c)$. Notice how each transformation affects the vertical asymptote.

Example 8 ■ Sketching the Graphs of Logarithmic Functions

Sketch the graphs of (a) $g(x) = 2 + \log_{10} x$, (b) $h(x) = \log_{10}(x - 1)$, and (c) $k(x) = \log_{10}(-x)$.

Solution

(a) You can sketch the graph of g by shifting the graph of $f(x) = \log_{10} x$ two units upward, as shown in Figure 9.11. The vertical asymptote of the graph of g is $x = 0$, and the domain of g is $(0, \infty)$. Once you have sketched the graph of g, you should check a few points to make sure that you have shifted the graph properly. For instance, when $x = 1$, you have

$$g(1) = 2 + \log_{10} 1 = 2 + 0 = 2.$$

This implies that $(1, 2)$ is a point on the graph of g, as shown in Figure 9.11.

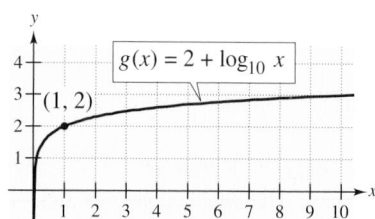

Figure 9.11

(b) You can sketch the graph of h by shifting the graph of $f(x) = \log_{10} x$ one unit to the right, as shown in Figure 9.12. The vertical asymptote of the graph of h is $x = 1$, and the domain of h is $(1, \infty)$. After sketching the graph of h, try checking some points to make sure that you have shifted the graph correctly. For instance, when $x = 2$, you have

$$h(2) = \log_{10}(2 - 1) = \log_{10} 1 = 0.$$

This implies that $(2, 0)$ is a point on the graph of h, as shown in Figure 9.12.

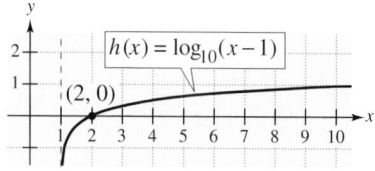

Figure 9.12

(c) You can sketch the graph of k by reflecting the graph of $f(x) = \log_{10} x$ in the y-axis, as shown in Figure 9.13. The vertical asymptote of the graph of k is $x = 0$, and the domain of k is $(-\infty, 0)$. After sketching the graph of k, try checking some points to make sure that you reflected the graph correctly.

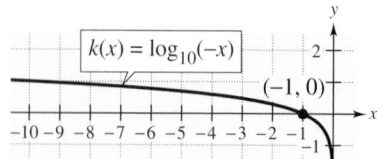

Figure 9.13

The Natural Logarithmic Function

As with exponential functions, the most widely used base for logarithmic functions is the number e. The logarithmic function with base e is the **natural logarithmic function** and is denoted by the special symbol $\ln x$, which is read as "el en of x."

Graph of $g(x) = \ln x$
- Domain: $(0, \infty)$
- Range: $(-\infty, \infty)$
- Intercept: $(1, 0)$
- Increasing (moves up to the right)
- Asymptote: y-axis

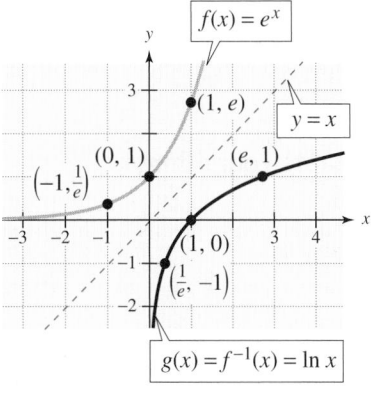

Figure 9.14   *Characteristics of the natural logarithm function $g(x) = \ln x$*

> **The Natural Logarithmic Function**
>
> The function defined by
>
> $$f(x) = \log_e x = \ln x$$
>
> where $x > 0$, is called the **natural logarithmic function.**

The definition above implies that the natural logarithmic function and the natural exponential function are inverse functions of each other. So, every logarithmic equation can be written in an equivalent exponential form and every exponential equation can be written in logarithmic form.

Because the functions $f(x) = e^x$ and $g(x) = \ln x$ are inverse functions of each other, their graphs are reflections of each other in the line $y = x$. This reflective property is illustrated in Figure 9.14. The figure also contains a summary of several characteristics of the graph of the natural logarithmic function.

Notice that the domain of the natural logarithmic function is the set of *positive real numbers*—be sure you see that $\ln x$ is not defined for zero or for negative numbers.

The three properties of logarithms listed earlier in this section are also valid for natural logarithms.

> **Properties of Natural Logarithms**
>
> Let x be a positive real number. Then the properties below are true.
>
> 1. $\ln 1 = 0$ because $e^0 = 1$.
>
> 2. $\ln e = 1$ because $e^1 = e$.
>
> 3. $\ln e^x = x$ because $e^x = e^x$.

Example 9 ■ Evaluating Natural Logarithmic Functions

Evaluate each expression.

(a) $\ln e^2$ (b) $\ln \dfrac{1}{e}$

Solution

Using the property that $\ln e^x = x$, you obtain the values as follows.

(a) $\ln e^2 = 2$ Note: $e^2 \approx 7.39$

(b) $\ln \dfrac{1}{e} = \ln e^{-1} = -1$ Note: $\dfrac{1}{e} \approx 0.37$

On most calculators, the natural logarithm key is denoted by $\boxed{\text{LN}}$.

Change of Base

Although 10 and e are the most frequently used bases, you occasionally need to evaluate logarithms with other bases. In such cases, the **change-of-base formula** shown below is useful.

Change-of-Base Formula

Let a, b, and x be positive real numbers such that $a \neq 1$ and $b \neq 1$. Then $\log_a x$ is given as follows.

$$\log_a x = \frac{\log_b x}{\log_b a} \quad \text{or} \quad \log_a x = \frac{\ln x}{\ln a}$$

The usefulness of this change-of-base formula is that you can use a calculator that has only the common logarithm key $\boxed{\text{LOG}}$ and the natural logarithm key $\boxed{\text{LN}}$ to evaluate logarithms to any base.

Example 10 ■ Changing Bases to Evaluate Logarithms

(a) Use *common* logarithms to evaluate $\log_3 5$.

(b) Use *natural* logarithms to evaluate $\log_6 2$.

Solution

Using the change-of-base formula, you can convert to common and natural logarithms as follows.

(a) $\log_3 5 = \dfrac{\log_{10} 5}{\log_{10} 3} \approx 1.465$

(b) $\log_6 2 = \dfrac{\ln 2}{\ln 6} \approx 0.387$

Historical Note

The English mathematician Henry Briggs (1561–1630) was also a founder of calculations by logarithms. He worked with John Napier, and they both published papers on the advantage of having tables of logarithms using the base 10.

Technology

You can use a graphing utility to graph logarithmic functions that do not have a base of 10 by using the change-of-base formula. Use the change-of-base formula to rewrite $g(x) = \log_2 x$ in Example 6(b) on page 599 (with $b = 10$) and graph the function. You should obtain a graph similar to the graph shown below.

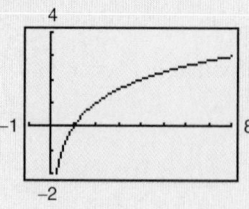

Example 11 ■ An Application: Human Memory Model

In an experiment, students attended several lectures on a subject and then were tested every month for a year to see how much of the material they remembered. The average scores for the group are given by the **human memory model**

$$s(t) = 80 - 9 \ln(t + 1), \ 0 \le t \le 12$$

where t is the time in months. Find the average scores for the group after (a) 2 months and (b) 8 months.

Algebraic Solution

You can find the average scores as follows.

(a) $s(2) = 80 - 9 \ln 3$

≈ 70.1

So, the average score after 2 months was about 70.1.

(b) $s(8) = 80 - 9 \ln 9$

≈ 60.2

So, the average score after 8 months was about 60.2.

Graphical Solution

Use a graphing utility to graph the function $y = 80 - 9 \ln(x + 1)$, as shown in Figure 9.15. Then use the *value* feature or the *zoom* and *trace* features to approximate y when $x = 2$ and $x = 8$.

(a) When $x = 2$, $y \approx 70.1$. So, the average score after 2 months was about 70.1.

(b) When $x = 8$, $y \approx 60.2$. So, the average score after 8 months was about 60.2.

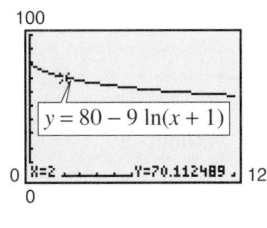

(a)

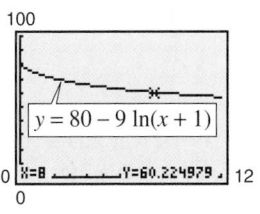

(b)

Figure 9.15

At this point, you have been introduced to all the basic types of functions that are covered in this course: polynomial functions, radical functions, rational functions, exponential functions, and logarithmic functions.

Collaborate!

Reviewing the Major Concepts

Suppose you work for a research and development firm that deals with a wide variety of disciplines. Your supervisor has asked your group to give a presentation to your department on four basic kinds of mathematical models. Identify each of the models shown below. Develop a presentation describing the types of data sets that each model would best represent. Include distinctions in domain, range, and intercepts, and a discussion of the types of applications to which each model is suited.

a.
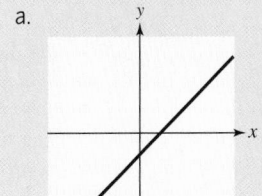

b.
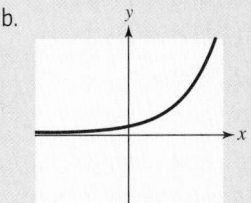

c.
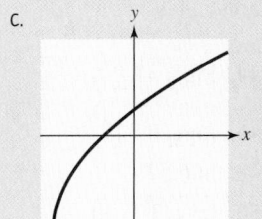

d.
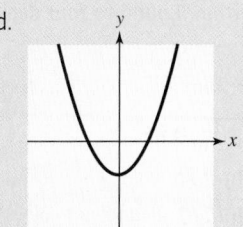

9.2 ■ Exercises

Developing Skills

In Exercises 1–8, write the logarithmic equation in exponential form.

1. $\log_2 8 = 3$

2. $\log_3 81 = 4$

3. $\log_4 256 = 4$

4. $\log_7 49 = 2$

5. $\log_4 \frac{1}{16} = -2$

6. $\log_8 \frac{1}{8} = -1$

7. $\log_{36} 6 = \frac{1}{2}$

8. $\log_{32} 4 = \frac{2}{5}$

In Exercises 9–18, write the exponential equation in logarithmic form.

9. $7^2 = 49$

10. $6^4 = 1296$

11. $3^{-2} = \frac{1}{9}$

12. $6^{-3} = \frac{1}{216}$

13. $5^4 = 625$

14. $2^5 = 32$

15. $8^{2/3} = 4$

16. $81^{3/4} = 27$

17. $25^{-1/2} = \frac{1}{5}$

18. $27^{-1/3} = \frac{1}{3}$

In Exercises 19–42, evaluate the logarithm without a calculator. (If not possible, state the reason.)

19. $\log_2 8$

20. $\log_3 27$

21. $\log_{10} 10$

22. $\log_8 8$

23. $\log_2 \frac{1}{4}$

24. $\log_3 \frac{1}{9}$

25. $\log_3 \frac{1}{81}$

26. $\log_5 \frac{1}{25}$

27. $\log_4 \frac{1}{64}$

28. $\log_2 \frac{1}{2}$

29. $\log_2(-3)$

30. $\log_3 1$

31. $\log_{10} 1000$

32. $\log_{10} \frac{1}{100}$

33. $\log_4 1$

34. $\log_5(-6)$

35. $\log_9 3$

36. $\log_{25} 125$

37. $\log_4(-4)$

38. $\log_2 0$

39. $\log_{16} 4$

40. $\log_{64} 8$

41. $\log_9 9^4$

42. $\log_5 5^3$

In Exercises 43–50, use a calculator to evaluate the logarithm. Round to four decimal places.

43. $\log_{10} \dfrac{\sqrt{3}}{2}$

44. $\log_{10} 31$

45. $\log_{10} 0.85$

46. $\log_{10} 0.345$

47. $\ln 0.75$

48. $\ln 6.57$

49. $\ln\left(\sqrt{2} + 4\right)$

50. $\ln\left(\sqrt{3} - 1\right)$

In Exercises 51–56, match the function with its graph. [The graphs are labeled (a), (b), (c), (d), (e), and (f).]

(a)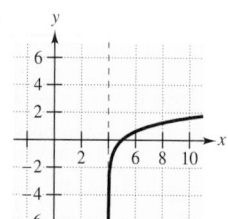

(b)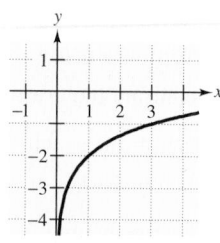

(c)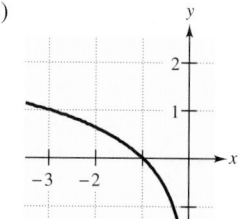

(d)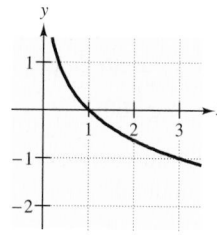

(e)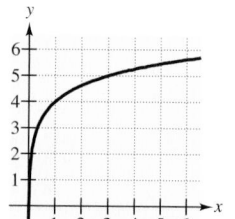

(f)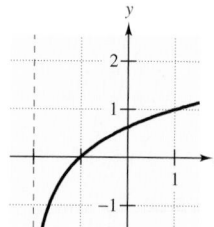

51. $f(x) = 4 + \log_3 x$

52. $f(x) = -2 + \log_3 x$

53. $f(x) = -\log_3 x$

54. $f(x) = \log_3(-x)$

55. $f(x) = \log_3(x - 4)$

56. $f(x) = \log_3(x + 2)$

In Exercises 57–66, identify the transformation of the graph of $g(x) = \log_2 x$ from Example 6(b) and sketch the graph of h.

57. $h(x) = 5 + \log_2 x$

58. $h(x) = \log_2 x - 3$

59. $h(x) = \log_2(x - 7)$

60. $h(x) = \log_2(x + 1)$

61. $h(x) = \log_2(-x)$

62. $h(x) = -\log_2 x$

63. $h(x) = \log_2(x - 2) + 5$

64. $h(x) = \log_2(x + 3) - 6$

65. $h(x) = \log_2(x + 4) - 8$

66. $h(x) = 1 + \log_2(x - 9)$

In Exercises 67–76, sketch the graph of the function. Determine the domain and the vertical asymptote.

67. $f(x) = \log_5(x - 3)$

68. $f(x) = \log_5(3 + x)$

69. $g(x) = \log_{10} x - 4$

70. $h(t) = \log_4 t + 1$

71. $f(x) = -\log_4 x$

72. $g(x) = \log_3(-x)$

73. $h(x) = \log_5(x - 1) + 4$

74. $f(x) = \log_3(x + 5) - 2$

75. $f(x) = -\log_{10}(x + 2)$

76. $f(x) = -\log_3 x + 2$

In Exercises 77–82, match the function with its graph. [The graphs are labeled (a), (b), (c), (d), (e), and (f).]

77. $f(x) = \ln(x + 1)$

78. $f(x) = 4 - \ln(x + 4)$

79. $f(x) = \ln\left(x - \frac{3}{2}\right)$

80. $f(x) = 6 + \ln(-x)$

81. $f(x) = -\ln x - 2$

82. $f(x) = \ln(-x)$

In Exercises 83–92, identify the transformation of the graph of $f(x) = \ln x$ and sketch the graph of h. Determine the domain and the vertical asymptote of h.

83. $h(x) = \ln x + 7$

84. $h(x) = \ln x - 5$

85. $h(x) = \ln(x - 2)$

86. $h(x) = \ln(x + 4)$

87. $h(x) = -\ln x$

88. $h(x) = \ln(-x)$

89. $h(x) = 3 + \ln(-x)$

90. $h(x) = -\ln(x - 2)$

91. $h(x) = \ln(x + 1) - 5$

92. $h(x) = 6 + \ln(x - 3)$

(a)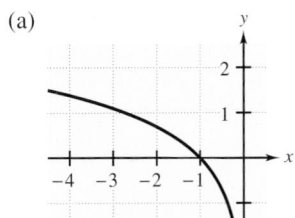

(b)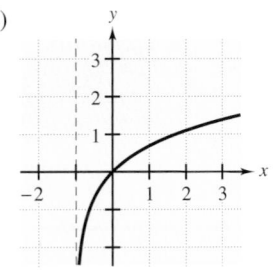

In Exercises 93–100, use a graphing utility to graph the function. Determine the domain and the vertical asymptote.

93. $f(x) = 5 \log_{10}(x - 3)$

94. $f(x) = -3 + 5 \log_{10} x$

95. $f(x) = \log_{10}(-5x)$

96. $h(t) = 5 \log_{10}(3t)$

97. $g(t) = 2 \ln(t - 4)$

98. $g(x) = -3 \ln(x + 3)$

99. $f(t) = 3 + 2 \ln t$

100. $g(t) = \ln(3 - t)$

(c)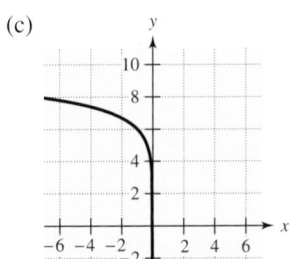

(d)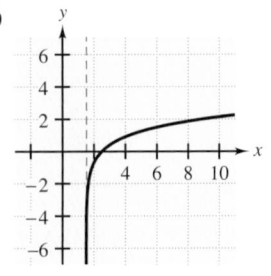

In Exercises 101–112, use a calculator to evaluate the logarithm by means of the change-of-base formula. Use the common logarithm key and the natural logarithm key. Round to four decimal places.

101. $\log_8 132$

102. $\log_5 510$

103. $\log_3 7$

104. $\log_7 4$

105. $\log_2 0.72$

106. $\log_{12} 0.6$

107. $\log_{15} 1250$

108. $\log_{20} 125$

109. $\log_{1/2} 4$

110. $\log_{1/3}(0.015)$

111. $\log_4 \sqrt{42}$

112. $\log_3(1 + e^2)$

(e)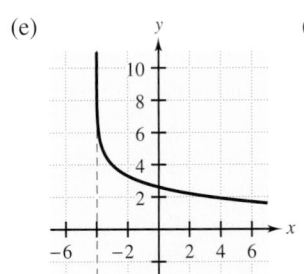

(f)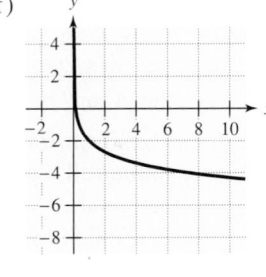

Solving Problems

113. *American Elk* The antler spread a (in inches) and the shoulder height h (in inches) of an adult male elk are related by the model

$$h = 116 \log_{10}(a + 40) - 176.$$

Approximate the shoulder height of an adult male elk with an antler spread of 55 inches.

114. *Intensity of Sound* The relationship between the number of decibels B and the intensity of a sound I in watts per meter squared is given by

$$B = 10 \log_{10}\left(\frac{I}{10^{-16}}\right).$$

Determine the number of decibels of a sound with an intensity of 10^{-4} watts per meter squared.

115. *Doubling Time* The time t in years for an investment to double in value when compounded continuously at annual rate r is given by

$$t = \frac{\ln 2}{r}.$$

Complete the table, which shows the "doubling times" for several annual percentage rates.

r	0.07	0.08	0.09	0.10	0.11	0.12
t						

116. *Tractrix* A person walking along a dock (the y-axis) drags a boat by a 10-foot rope (see figure). The boat travels along a path known as a *tractrix*. The equation of this path is

$$y = 10 \ln\left(\frac{10 + \sqrt{100 - x^2}}{x}\right) - \sqrt{100 - x^2}.$$

(a) Use a graphing utility to graph the function. What is the domain of the function?

(b) Identify any asymptotes of the graph.

(c) Determine the position of the person when the x-coordinate of the position of the boat is $x = 2$.

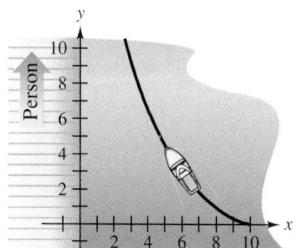

Explaining Concepts

117. Explain the relationship between

$$f(x) = 2^x \quad \text{and} \quad g(x) = \log_2 x.$$

118. Explain why $\log_a a^x = x$.

119. Explain why $\log_a a = 1$.

120. What are common logarithms and natural logarithms?

Ongoing Review

In Exercises 121–124, rewrite the expression using only positive exponents.

121. $(a^{-3}b^4)^{-2}$

122. $(5xy^{-3})(4x^{-6}y^{-7})$

123. $\left(\dfrac{a^2}{3b^{-1}}\right)^{-3}$

124. $\left(\dfrac{a^{-2}}{2b^{-4}}\right)\left(\dfrac{b^2}{6a^{-1}}\right)^3$

In Exercises 125 and 126, graph the equation.

125. $y = -(x - 2)^2 + 1$

126. $y = x^2 - 6x + 5$

127. *Endowment* A group of people agrees to share equally in the cost of a $250,000 endowment to a college. If they could find two more people to join the group, each person's share of the cost would decrease by $6250. How many people are presently in the group?

128. *Speed* A boat travels at a speed of 18 miles per hour in the still water. It travels 35 miles upstream and then returns to the starting point in a total of 4 hours. Find the speed of the current.

Looking Further

Answer the questions for the function $f(x) = \log_{10} x$. (Do not use a calculator.)

(a) Describe the values of $f(x)$ for $1000 \le x \le 10,000$.

(b) Describe the values of x, given that $f(x)$ is negative.

(c) By what amount will x increase, given that $f(x)$ is increased by one unit?

(d) Find the ratio of a to b, given that $f(a) = 3f(b)$.

9.3 Properties of Logarithms

What you should learn:

- How to use the properties of logarithms to evaluate logarithms
- How to use the properties of logarithms to rewrite, expand, or condense logarithmic expressions
- How to use the properties of logarithms to solve application problems

Why you should learn it:

Logarithmic functions are often used to model scientific observations. For instance, in Exercise 126 on page 613 you will use a logarithmic function to model the relationship between the number of decibels and the intensity of a sound.

Properties of Logarithms

You know from the preceding section that the logarithmic function with base a is the *inverse function* of the exponential function with base a. So, it makes sense that each property of exponents should have a corresponding property of logarithms. For instance, the exponential property

$$a^0 = 1 \qquad \text{Exponential property}$$

has the corresponding logarithmic property

$$\log_a 1 = 0. \qquad \text{Corresponding logarithmic property}$$

In this section, you will study the logarithmic properties that correspond to the three exponential properties below.

Base a	*Natural Base*
1. $a^m a^n = a^{m+n}$	$e^m e^n = e^{m+n}$
2. $\dfrac{a^m}{a^n} = a^{m-n}$	$\dfrac{e^m}{e^n} = e^{m-n}$
3. $(a^m)^n = a^{mn}$	$(e^m)^n = e^{mn}$

Properties of Logarithms

Let a be a positive real number such that $a \neq 1$, and let n be a real number. If u and v are real numbers, variables, or algebraic expressions such that $u > 0$ and $v > 0$, the properties below are true.

Logarithm with Base a

1. $\log_a(uv) = \log_a u + \log_a v$

2. $\log_a \dfrac{u}{v} = \log_a u - \log_a v$

3. $\log_a u^n = n \log_a u$

Natural Logarithm

1. $\ln(uv) = \ln u + \ln v$

2. $\ln \dfrac{u}{v} = \ln u - \ln v$

3. $\ln u^n = n \ln u$

NOTE There is no general property of logarithms that can be used to simplify $\log_a(u + v)$. Specifically,

$$\log_a(u + v) \text{ does not equal } \log_a u + \log_a v.$$

Example 1 ■ Using Properties of Logarithms

Use $\ln 2 \approx 0.693$, $\ln 3 \approx 1.099$, and $\ln 5 \approx 1.609$ to approximate each expression.

(a) $\ln \dfrac{2}{3}$ (b) $\ln 10$ (c) $\ln 30$

Solution

(a) $\ln \dfrac{2}{3} = \ln 2 - \ln 3$ Property 2

$\quad\quad \approx 0.693 - 1.099$ Substitute for $\ln 2$ and $\ln 3$.

$\quad\quad = -0.406$ Simplify.

(b) $\ln 10 = \ln(2 \cdot 5)$ Factor.

$\quad\quad = \ln 2 + \ln 5$ Property 1

$\quad\quad \approx 0.693 + 1.609$ Substitute for $\ln 2$ and $\ln 5$.

$\quad\quad = 2.302$ Simplify.

(c) $\ln 30 = \ln(2 \cdot 3 \cdot 5)$ Factor.

$\quad\quad = \ln 2 + \ln 3 + \ln 5$ Property 1

$\quad\quad \approx 0.693 + 1.099 + 1.609$ Substitute for $\ln 2$, $\ln 3$, and $\ln 5$.

$\quad\quad = 3.401$ Simplify.

NOTE   When using the properties of logarithms, it helps to state the properties *verbally*. For instance, the verbal form of the property

$$\ln(uv) = \ln u + \ln v$$

is: *The log of a product is the sum of the logs of the factors.* Similarly, the verbal form of the property

$$\ln\frac{u}{v} = \ln u - \ln v$$

is: *The log of a quotient is the difference of the logs of the numerator and denominator.*

Study Tip

Remember that you can verify results such as those given in Examples 1 and 2 with a calculator.

Example 2 ■ Using Properties of Logarithms

Use the properties of logarithms to verify that $-\ln 2 = \ln \frac{1}{2}$.

Solution

Using Property 3, you can write the expression $-\ln 2$ as follows.

$$-\ln 2 = (-1)\ln 2 \qquad \text{Rewrite coefficient as } -1.$$

$$= \ln 2^{-1} \qquad \text{Property 3}$$

$$= \ln \frac{1}{2} \qquad \text{Rewrite } 2^{-1} \text{ as } \tfrac{1}{2}.$$

Rewriting Logarithmic Expressions

In Examples 1 and 2, the properties of logarithms were used to rewrite logarithmic expressions involving the logs of *constants*. A more common use of the properties of logarithms is to rewrite the logs of *variable expressions*.

Example 3 ■ Rewriting a Logarithmic Expression

Use the properties of logarithms to rewrite each expression.

(a) $\log_{10} 7x^3$ (b) $\ln \dfrac{8x^3}{y}$

Solution

NOTE   When you rewrite a logarithmic expression as in Example 3, you are **expanding** the expression. When you rewrite a logarithmic expression as in Example 4, you are **condensing** the expression.

(a) $\log_{10} 7x^3 = \log_{10} 7 + \log_{10} x^3$ Property 1

$\phantom{(a) \log_{10} 7x^3} = \log_{10} 7 + 3 \log_{10} x$ Property 3

(b) $\ln \dfrac{8x^3}{y} = \ln 8x^3 - \ln y$ Property 2

$\phantom{(b) \ln \dfrac{8x^3}{y}} = \ln 8 + \ln x^3 - \ln y$ Property 1

$\phantom{(b) \ln \dfrac{8x^3}{y}} = \ln 8 + 3 \ln x - \ln y$ Property 3

Example 4 ■ Condensing a Logarithmic Expression

Use the properties of logarithms to condense each expression.

(a) $\ln x - \ln 3$ (b) $2 \log_3 x + \log_3 5$

Solution

(a) Using Property 2, you can write

$$\ln x - \ln 3 = \ln \frac{x}{3}.$$ Property 2

Technology

When you are rewriting a logarithmic expression, remember that you can use a graphing utility to check your result graphically. For instance, in Example 4(a), try graphing the functions

$$y_1 = \ln x - \ln 3$$

and

$$y_2 = \ln \frac{x}{3}$$

in the same viewing window. You should obtain the same graph for each function.

(b) $2 \log_3 x + \log_3 5 = \log_3 x^2 + \log_3 5$ Property 3

$ = \log_3 5x^2$ Property 1

Example 5 ■ Expanding a Logarithmic Expression

Use the properties of logarithms to expand $\log_6 \dfrac{\sqrt{3x - 5}}{7}$.

Solution

$$\log_6 \frac{\sqrt{3x - 5}}{7} = \log_6 \left[\frac{(3x - 5)^{1/2}}{7} \right]$$ Rewrite using rational exponent.

$$= \log_6 (3x - 5)^{1/2} - \log_6 7$$ Property 2

$$= \tfrac{1}{2} \log_6 (3x - 5) - \log_6 7$$ Property 3

Sometimes expanding or condensing logarithmic expressions involves several steps. In the next example, be sure that you can justify each step in the solution. Also, notice how different the expanded expressions are from the original expressions.

Example 6 ■ Expanding a Logarithmic Expression

Use the properties of logarithms to expand each expression.

(a) $\log_2 3xy^2, x > 0, y > 0$ (b) $\ln \sqrt{x^2 - 1}, x > 1$

Solution

(a) $\log_2 3xy^2 = \log_2 3 + \log_2 x + \log_2 y^2$ Property 1

$\qquad\qquad\quad = \log_2 3 + \log_2 x + 2\log_2 y$ Property 3

(b) $\ln \sqrt{x^2 - 1} = \ln(x^2 - 1)^{1/2}$ Rewrite using rational exponent.

$\qquad\qquad\quad = \frac{1}{2}\ln(x^2 - 1)$ Property 3

$\qquad\qquad\quad = \frac{1}{2}\ln[(x - 1)(x + 1)]$ Factor.

$\qquad\qquad\quad = \frac{1}{2}[\ln(x - 1) + \ln(x + 1)]$ Property 1

$\qquad\qquad\quad = \frac{1}{2}\ln(x - 1) + \frac{1}{2}\ln(x + 1)$ Distributive Property

Example 7 ■ Condensing a Logarithmic Expression

Use the properties of logarithms to condense each expression.

(a) $\ln 2 - 2\ln x$ (b) $3(\ln 4 + \ln x)$

Solution

(a) $\ln 2 - 2\ln x = \ln 2 - \ln x^2, \quad x > 0$ Property 3

$\qquad\qquad\quad = \ln \dfrac{2}{x^2}, \quad x > 0$ Property 2

(b) $3(\ln 4 + \ln x) = 3(\ln 4x)$ Property 1

$\qquad\qquad\quad = \ln(4x)^3$ Property 3

$\qquad\qquad\quad = \ln 64x^3$ Simplify.

When you expand or condense a logarithmic expression, it is possible to change the domain of the expression. For instance, the domain of the function

$\qquad f(x) = 2\ln x$ Domain is the set of positive real numbers.

is the set of positive real numbers, whereas the domain of

$\qquad g(x) = \ln x^2$ Domain is the set of nonzero real numbers.

is the set of nonzero real numbers. So, when you expand or condense a logarithmic expression, you should check to see whether the rewriting has changed the domain of the expression. In such cases, you should restrict the domain appropriately. For instance, you can write

$\qquad f(x) = 2\ln x = \ln x^2, \ x > 0.$

EXPLORATION

Use the properties of logarithms to rewrite the logarithmic expression $\ln x^4$. Use a graphing utility to graph the two expressions. Should both have the same graph? What can you conclude?

Application

Example 8 ■ Estimating the Time of Death

At 8:30 A.M., a coroner was called to the home of a person who had died during the night. To estimate the time of death, the coroner took the person's body temperature twice. At 9:00 A.M. the temperature was 85.7°F, and at 9:30 A.M. the temperature was 82.8°F. The room temperature was 70°F. When did the person die?

Solution

The coroner used Newton's Law of Cooling

$$kt = \ln \frac{T - S}{T_0 - S}$$

where k is a constant, S is the temperature of the surrounding air, and t is the time it takes for the body temperature to cool from $T_0 = 98.6$ (normal body temperature) to T. From the first temperature reading, the coroner could obtain

$$kt = \ln \frac{T - S}{T_0 - S} = \ln \frac{85.7 - 70}{98.6 - 70} \approx -0.6.$$

From the second temperature reading, the coroner could obtain

$$k\left(t + \frac{1}{2}\right) = \ln \frac{82.8 - 70}{98.6 - 70} \approx -0.8.$$

Solving for k in the equation $kt = -0.6$ produces $k = -0.6/t$. Substitute this value into the equation $k(t + 1/2) = -0.8$ to find t, as follows.

$$k\left(t + \frac{1}{2}\right) = -0.8 \qquad \text{Write original equation.}$$

$$\frac{-0.6}{t}\left(t + \frac{1}{2}\right) = -0.8 \qquad \text{Substitute } -0.6/t \text{ for } k.$$

$$-0.6 - \frac{0.3}{t} = -0.8 \qquad \text{Distributive Property}$$

$$\frac{-0.3}{t} = -0.2 \qquad \text{Add 0.6 to each side.}$$

$$-0.2t = -0.3 \qquad \text{Cross multiply.}$$

$$t = 1.5 \qquad \text{Divide each side by } -0.2.$$

To find k, substitute $t = 1.5$ into the equation $kt = -0.6$ to obtain $k = -0.4$. So, the coroner could approximate that the first temperature reading took place 1.5 hours after death, which implies that the death occurred at about 7:30 A.M.

9.3 ■ Exercises

Developing Skills

In Exercises 1–26, use properties of logarithms to evaluate the expression without a calculator.

1. $\log_5 5^2$

2. $\log_8 8^5$

3. $\log_{18} 9 + \log_{18} 2$

4. $\log_6 2 + \log_6 3$

5. $\log_4 8 + \log_4 2$

6. $\log_{10} 5 + \log_{10} 20$

7. $\log_4 8 - \log_4 2$

8. $\log_5 50 - \log_5 2$

9. $\log_6 72 - \log_6 2$

10. $\log_3 324 - \log_3 4$

11. $\log_3 \frac{2}{3} + \log_3 \frac{1}{2}$

12. $\log_7 \frac{1}{5} + \log_7 \frac{5}{49}$

13. $\ln 1$

14. $\log_{10} 1$

15. $\log_3 9$

16. $\log_6 \sqrt{6}$

17. $\log_2 \frac{1}{8}$

18. $\log_5 \frac{1}{625}$

19. $\ln e^5 - \ln e^2$

20. $\ln e^8 + \ln e^4$

21. $\ln e^4$

22. $\ln e^{-3}$

23. $\ln \frac{e^3}{e^2}$

24. $\ln \frac{e^5}{e^2}$

25. $\ln(e^2 \cdot e^4)$

26. $\ln(e^3 \cdot e^5)$

In Exercises 27–64, use the properties of logarithms to expand the expression.

27. $\log_3 11x$

28. $\ln 5x$

29. $\ln y^3$

30. $\log_7 x^2$

31. $\log_2 \frac{z}{17}$

32. $\log_{10} \frac{7}{y}$

33. $\log_5 x^{-2}$

34. $\log_3 x^{-4}$

35. $\log_5 \sqrt[3]{x}$

36. $\log_2 \sqrt{s}$

37. $\log_3 \sqrt[3]{x+1}$

38. $\log_6 \sqrt{4-y}$

39. $\log_5 \frac{1}{\sqrt[3]{x}}$

40. $\log_4 \frac{1}{\sqrt{t}}$

41. $\ln 3x^2 y$

42. $\ln \left[y(y-1)^2 \right]$

43. $\ln \sqrt{x(x+2)}$

44. $\ln \sqrt[3]{x^2(x-1)}$

45. $\ln \frac{t+4}{15}$

46. $\ln \frac{5}{x-2}$

47. $\log_2 \frac{x^2}{x-3}$

48. $\log_8 \frac{t+5}{t^3}$

49. $\log_3 \sqrt{\frac{4x}{y^2}}$

50. $\log_5 \sqrt{\frac{x}{y}}$

51. $\ln \sqrt[3]{x(x+5)}$

52. $\ln \sqrt[3]{\frac{x^2}{x+1}}$

53. $\ln \left(\frac{x+1}{x-1} \right)^2$

54. $\ln[3x(x-5)]^2$

55. $\log_4[x^6(x-7)^2]$

56. $\log_3 \frac{x^2 y}{z^7}$

57. $\ln \frac{a^3(b-4)}{c^2}$

58. $\log_8[(x-y)^4 z^6]$

59. $\log_{10} \frac{(4x)^3}{x-7}$

60. $\log_4 \frac{\sqrt[3]{a+1}}{(ab)^4}$

61. $\ln \frac{x\sqrt[3]{y}}{(wz)^4}$

62. $\ln \left[(x+y) \frac{\sqrt[5]{w+2}}{3t} \right]$

63. $\log_5[(xy)^2(x+3)^4]$

64. $\log_6[a\sqrt{b}(c-d)^3]$

In Exercises 65–92, use the properties of logarithms to condense the expression.

65. $\log_2 3 + \log_2 x$

66. $\log_5 2x + \log_5 3y$

67. $\log_{10} 4 - \log_{10} x$

68. $\ln 10x - \ln z$

69. $4 \ln b$

70. $10 \log_4 z$

71. $-2 \log_5 2x$

72. $-\frac{1}{2} \log_3 5y$

73. $\frac{1}{3} \ln(2x+1)$

74. $-5 \ln(x+3)$

75. $\log_3 2 + \frac{1}{2} \log_3 y$

76. $\ln 6 - 3 \ln z$

77. $2 \ln x + 3 \ln y - \ln z$

78. $4 \ln 3 - 2 \ln x - \ln y$

79. $4(\ln x + \ln y)$

80. $2[\ln x - \ln(x+1)]$

81. $\frac{1}{2}(\ln 8 + \ln 2x)$

82. $5[\ln x - \frac{1}{2} \ln(x+4)]$

83. $\log_4(x+8) - 3 \log_4 x$

84. $5 \log_3 x + \log_3(x-6)$

85. $\frac{1}{2} \log_5 x + \log_5(x-3)$

86. $\frac{1}{4} \log_6(x+1) - \log_6 x$

87. $5 \log_6(c+d) - \log_6(m-n)$

88. $2 \log_5(x+y) + 3 \log_5 w$

89. $\frac{1}{5}(3 \log_2 x - \log_2 y)$

90. $\frac{1}{3}[\ln(x - 6) - \ln y - 2 \ln z]$

91. $\frac{1}{5} \log_6(x - 3) - 2 \log_6 x - 3 \log_6(x + 1)$

92. $3[\frac{1}{2} \log_9(a + 6) - \log_9(a - 1)]$

In Exercises 93–106, use $\log_4 2 = 0.5000$ and $\log_4 3 \approx 0.7925$ to approximate the logarithm. Do not use a calculator.

93. $\log_4 4$

94. $\log_4 8$

95. $\log_4 6$

96. $\log_4 24$

97. $\log_4 \frac{3}{2}$

98. $\log_4 \frac{9}{2}$

99. $\log_4 \sqrt{2}$

100. $\log_4 \sqrt[3]{9}$

101. $\log_4 (3 \cdot 2^4)$

102. $\log_4 \sqrt{3 \cdot 2^5}$

103. $\log_4 3^0$

104. $\log_4 4^3$

105. $\log_4 \sqrt[3]{\frac{1}{9}}$

106. $\log_4 \sqrt{\frac{1}{6}}$

In Exercises 107–112, use $\log_{10} 3 \approx 0.477$ and $\log_{10} 12 \approx 1.079$ to approximate the logarithm. Use a calculator to verify your result.

107. $\log_{10} 9$

108. $\log_{10} 144$

109. $\log_{10} 36$

110. $\log_{10} \frac{1}{4}$

111. $\log_{10} \sqrt[3]{36}$

112. $\log_{10} 27$

In Exercises 113–124, simplify the expression.

113. $\log_4 \frac{4}{x}$

114. $\log_7 \frac{49}{x}$

115. $\log_2(5 \cdot 2^3)$

116. $\log_3(3^2 \cdot 4)$

117. $\log_5 \sqrt{50}$

118. $\log_2 \sqrt{22}$

119. $\ln 3e^2$

120. $\ln 10\, e^3$

121. $\ln \frac{12}{e^3}$

122. $\ln \frac{6}{e^5}$

123. $\log_{10} 50 + \log_{10} 4$

124. $\log_{12} 9 + \log_{12} 4$

Solving Problems

125. *Human Memory Model*   Students participating in a psychology experiment attended several lectures on a subject. Every month for a year after that, the students were tested to see how much of the material they remembered. The average scores for the group are given by the human memory model

$$s(t) = 80 - \log_{10}(t + 1)^{12}, \quad 0 \le t \le 12$$

where t is the time in months.

   (a) Find the average scores when $t = 2$ and $t = 8$.

   (b) Use a graphing utility to graph the function.

126. *Intensity of Sound*   The relationship between the number of decibels B and the intensity of a sound I in watts per centimeter squared is given by

$$B = 10 \log_{10} \frac{I}{10^{-16}}.$$

   (a) Use the properties of logarithms to write the formula in simpler form.

   (b) Determine the number of decibels of a sound with an intensity of 10^{-10} watts per centimeter squared.

Molecular Transport   In Exercises 127 and 128, use the following information. The energy E (in kilocalories per gram molecule) required to transport a substance from the outside to the inside of a living cell is given by

$$E = 1.4(\log_{10} C_2 - \log_{10} C_1)$$

where C_1 and C_2 are the concentrations of the substance outside and inside the cell, respectively.

127. Condense the expression.

128. The concentration of a particular substance inside a cell is twice the concentration outside the cell. How much energy is required to transport the substance from outside to inside the cell?

129. *Newton's Law of Cooling*   At 7:00 A.M., a coroner was called to the home of a person who had died during the night. To estimate the time of death, the coroner took the person's body temperature twice. At 8:30 A.M. the temperature was 84.8°F, and at 9:30 A.M. the temperature was 81.8°F. The room temperature was 68°F. Use Newton's Law of Cooling

$$kt = \ln \frac{T - S}{T_0 - S}$$

to determine when the person died. (See Example 8.)

Explaining Concepts

In Exercises 130–138, determine whether the equation is true or false. If it is false, state why or give an example to show that it is false.

130. $\ln e^{2-x} = 2 - x$

131. $\log_2 8x = 3 + \log_2 x$

132. $\log_8 4 + \log_8 16 = 2$

133. $\log_3(u + v) = \log_3 u + \log_3 v$

134. $\log_3(u + v) = \log_3 u \cdot \log_3 v$

135. $\log_5 x^2 = \log_5 x + \log_5 x$

136. $\log_3 t^3 = \log_3 t^5 - \log_3 t^2$

137. $\ln(xy) = \ln x \cdot \ln y$

138. $\ln(xy) = \ln x + \ln y$

In Exercises 139–142, determine whether the statement is true or false given that $f(x) = \ln x$. If it is false, state why or give an example to show that it is false.

139. $f(0) = 0$

140. $\sqrt{f(x)} = \frac{1}{2} \ln x$

141. $f(2x) = \ln 2 + \ln x, \quad x > 0$

142. $f(x - 3) = \ln x - \ln 3, \quad x > 3$

143. If $f(x) = \log_a x$, does $f(ax) = 1 + f(x)$? Explain.

144. Explain how you can show that
$$\frac{\ln x}{\ln y} \neq \ln \frac{x}{y}.$$

Ongoing Review

In Exercises 145–148, rewrite the equation in exponential form.

145. $\log_7 49 = 2$

146. $\log_5 \frac{1}{5} = -1$

147. $\ln e^{4x} = 4x$

148. $\log_8 1 = 0$

In Exercises 149–154, perform the operation and simplify. (Assume all variables are positive.)

149. $25\sqrt{3x} - 3\sqrt{12x}$

150. $(\sqrt{x} + 3)(\sqrt{x} - 3)$

151. $\sqrt{u}(\sqrt{20} - \sqrt{5})$

152. $(2\sqrt{t} + 3)^2$

153. $\dfrac{50x}{\sqrt{2}}$

154. $\dfrac{12}{\sqrt{t + 2} + \sqrt{t}}$

155. *Demand* The demand equation for a product is given by
$$p = 30 - \sqrt{0.5(x - 1)}$$
where x is the number of units demanded per day and p is the price per unit. Find the demand if the price is $26.76.

156. *Discount* The sale price of a computer is $1955. The discount is 15% of the list price. Find the list price.

Looking Further

Without using a calculator, approximate the natural logarithms of as many integers as possible between 1 and 30 using $\ln 2 \approx 0.6931$, $\ln 3 \approx 1.0986$, $\ln 5 \approx 1.6094$, and $\ln 7 \approx 1.9459$. Explain the method you used. Then verify your results with a calculator and explain any differences in the results.

Mid-Chapter Quiz

Take this quiz as you would take a quiz in class. After you are done, check your work against the answers given in the back of the book.

1. Evaluate the function $f(x) = \left(\frac{4}{3}\right)^x$ at the indicated values of the independent variable. Use a calculator only if it is more efficient.

(a) $f(2)$ (b) $f(0)$ (c) $f(-1)$ (d) $f(1.5)$

2. Determine the domain and range of the function $g(x) = 2^{-0.5x}$.

In Exercises 3–6, sketch the graph of the function. Determine the domain and the horizontal asymptote. Use a graphing utility for Exercises 5 and 6.

3. $y = 4^{x+3}$

4. $y = 2^{-x} + 1$

5. $f(t) = 12e^{-0.4t}$

6. $g(x) = 100(1.08)^x$

7. You deposit \$750 in an account at an annual interest rate of $7\frac{1}{2}\%$. Complete the table giving the account balance A after 20 years if the interest is compounded n times per year.

n	1	4	12	365	Continuous compounding
A					

8. After t years, the remaining mass y (in grams) of 14 grams of a radioactive element whose half-life is 40 years is given by

$$y = 14\left(\frac{1}{2}\right)^{t/40}, \quad t \geq 0.$$

How much of the initial mass remains after 125 years?

9. Write the logarithmic equation $\log_4\left(\frac{1}{16}\right) = -2$ in exponential form.

10. Write the exponential equation $3^4 = 81$ in logarithmic form.

11. Evaluate $\log_5 125$ without a calculator.

In Exercises 12 and 13, sketch the graph of the function. Determine the domain and the vertical asymptote.

12. $f(t) = \ln(t + 4)$

13. $h(x) = 3 - \ln x$

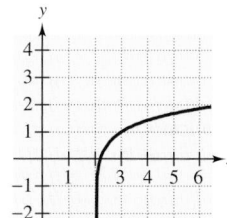

Figure for 14

14. Use the graph of f shown in the figure at the left to determine h and k if $f(x) = \log_5(x - h) + k$.

15. Use a calculator and the change-of-base formula to evaluate $\log_6 450$.

16. Use the properties of logarithms to expand $\ln\left(\dfrac{6x^2}{\sqrt{x^2 + 1}}\right)$.

17. Use the properties of logarithms to condense $2(\ln x - 3 \ln y)$.

9.4 Solving Exponential and Logarithmic Equations

What you should learn:

- How to solve basic exponential and logarithmic equations
- How to use inverse properties to solve exponential equations
- How to use inverse properties to solve logarithmic equations

Why you should learn it:

Applications of exponential and logarithmic equations are found in many branches of science. For instance, in Exercise 119 on page 622 you will use an exponential equation to model the force required by a person to restrain an untrained horse.

Exponential and Logarithmic Equations

So far in this chapter, you have studied the definitions, graphs, and properties of exponential and logarithmic functions. In this section, you will study procedures for *solving equations* that involve exponential or logarithmic expressions. As a simple example, consider the exponential equation $2^x = 16$. By rewriting this equation in the form $2^x = 2^4$, you can see that the solution is $x = 4$. To solve this equation, you can use one of the properties below.

Properties of Exponential and Logarithmic Equations

Let a be a positive real number such that $a \neq 1$, and let x and y be real numbers. Then the properties below are true.

1. $a^x = a^y$ if and only if $x = y$.

2. $\log_a x = \log_a y$ if and only if $x = y$ $(x > 0,\ y > 0)$.

Example 1 ■ Solving Exponential and Logarithmic Equations

Solve each equation.

(a) $4^{x+2} = 4^5$ (b) $\ln(2x - 3) = \ln 11$

Solution

(a) $4^{x+2} = 4^5$ Write original equation.

     $x + 2 = 5$ Property 1

         $x = 3$ Subtract 2 from each side.

The solution is $x = 3$.

(b) $\ln(2x - 3) = \ln 11$ Write original equation.

     $2x - 3 = 11$ Property 2

        $2x = 14$ Add 3 to each side.

           $x = 7$ Divide each side by 2.

The solution is $x = 7$. You can check these solutions as follows.

Check

(a) $4^{x+2} = 4^5$ Write original equation.

     $4^{3+2} \overset{?}{=} 4^5$ Substitute 3 for x.

       $4^5 = 4^5$ Solution checks. ✓

(b) $\ln(2x - 3) = \ln 11$ Write original equation.

   $\ln[2(7) - 3] \overset{?}{=} \ln 11$ Substitute 7 for x.

    $\ln(14 - 3) \overset{?}{=} \ln 11$ Simplify.

        $\ln 11 = \ln 11$ Solution checks. ✓

Solving Exponential Equations

In Example 1(a), you were able to solve the original equation because each side of the equation was written in exponential form (with the same base). However, if only one side of the equation is written in exponential form, it is more difficult to solve the equation. For example, how would you solve the equation $2^x = 7$? You must find the power to which 2 can be raised to obtain 7. To do this, you can take the logarithm of each side and use one of the following inverse properties of exponents and logarithms.

Solving Exponential Equations

To solve an exponential equation, first isolate the exponential expression, then **take the logarithm of each side of the equation** (or write the equation in logarithmic form) and solve for the variable.

Inverse Properties of Exponents and Logarithms

Base a	*Natural Base e*
1. $\log_a(a^x) = x$	$\ln(e^x) = x$
2. $a^{(\log_a x)} = x$	$e^{(\ln x)} = x$

Recall that $x = a^y$ if and only if $\log_a x = y$.

Example 2 ■ Solving Exponential Equations

Solve each exponential equation.

(a) $2^x = 7$ (b) $2e^x = 10$ (c) $4^{3x} = 20$

Solution

(a) To isolate the x, take the $\log_2$ of each side of the equation or write the equation in logarithmic form, as follows.

$$2^x = 7 \qquad \text{Write original equation.}$$
$$x = \log_2 7 \qquad \text{Inverse property}$$

The solution is $x = \log_2 7 \approx 2.807$. Check this in the original equation.

(b)
$$2e^x = 10 \qquad \text{Write original equation.}$$
$$e^x = 5 \qquad \text{Divide each side by 2.}$$
$$x = \ln 5 \qquad \text{Inverse property}$$

The solution is $x = \ln 5 \approx 1.609$. Check this in the original equation.

(c)
$$4^{3x} = 20 \qquad \text{Write original equation.}$$
$$3x = \log_4 20 \qquad \text{Inverse property}$$
$$x = \frac{\log_4 20}{3} \qquad \text{Divide each side by 3.}$$

The solution is $x = (\log_4 20)/3 \approx 0.720$. Check this in the original equation.

Example 3 ■ Solving an Exponential Equation

$$4^{x-3} = 9 \qquad \text{Write original equation.}$$
$$x - 3 = \log_4 9 \qquad \text{Inverse property}$$
$$x = \log_4 9 + 3 \qquad \text{Add 3 to each side.}$$

The solution is $x = \log_4 9 + 3 \approx 4.585$. Check this in the original equation.

Example 4 ■ Solving an Exponential Equation

Solve $5 + e^{x+1} = 20$.

Algebraic Solution

$$5 + e^{x+1} = 20 \qquad \text{Write original equation.}$$
$$e^{x+1} = 15 \qquad \text{Subtract 5 from each side.}$$
$$x + 1 = \ln 15 \qquad \text{Inverse property}$$
$$x = -1 + \ln 15 \qquad \text{Subtract 1 from each side.}$$

The solution is $x = -1 + \ln 15 \approx 1.708$. Check this in the original equation, as follows.

Check

$$5 + e^{x+1} = 20 \qquad \text{Write original equation.}$$
$$5 + e^{-1+\ln 15} \overset{?}{=} 20 \qquad \text{Substitute } -1 + \ln 15 \text{ for } x.$$
$$5 + e^{\ln 15} \overset{?}{=} 20 \qquad \text{Simplify.}$$
$$5 + 15 = 20 \qquad \text{Solution checks. } ✓$$

Graphical Solution

Use a graphing utility to graph the left- and right-hand sides of the equation as $y_1 = 5 + e^{x+1}$ and $y_2 = 20$ in the same viewing window. Use the *intersect* feature or the *zoom* and *trace* features of the graphing utility to approximate the intersection point, as shown in Figure 9.16. So, the approximate solution is $x \approx 1.708$.

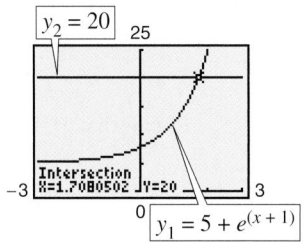

Figure 9.16

Solving Logarithmic Equations

To solve a logarithmic equation, you can **exponentiate** each side. For instance, to solve a logarithmic equation such as $\ln x = 2$, you can exponentiate each side of the equation as follows.

$$e^{\ln x} = e^2 \qquad \text{Exponentiate each side.}$$
$$x = e^2 \qquad \text{Inverse property}$$

Notice that you can obtain the same result by writing the equation in exponential form. This procedure is demonstrated in the next three examples. The following guideline can be used for solving logarithmic equations.

> **Solving Logarithmic Equations**
>
> To solve a logarithmic equation, first isolate the logarithmic expression, then **exponentiate each side of the equation** (or write the equation in exponential form) and solve for the variable.

Example 5 ■ Solving a Logarithmic Equation

Solve $2 \ln x = 5$.

Solution

$$2 \ln x = 5 \qquad \text{Write original equation.}$$

$$\ln x = \frac{5}{2} \qquad \text{Divide each side by 2.}$$

$$x = e^{5/2} \qquad \text{Inverse property}$$

The solution is $x = e^{5/2} \approx 12.182$. Check this in the original equation.

Example 6 ■ Solving a Logarithmic Equation

Solve $3 \log_{10} x = 6$.

Algebraic Solution

$$3 \log_{10} x = 6 \qquad \text{Write original equation.}$$

$$\log_{10} x = 2 \qquad \text{Divide each side by 3.}$$

$$x = 10^2 \qquad \text{Inverse property}$$

$$x = 100 \qquad \text{Simplfiy.}$$

The solution is $x = 100$. Check this in the original equation, as follows.

Check

$$3 \log_{10} x = 6 \qquad \text{Write original equation.}$$

$$3 \log_{10} 100 \overset{?}{=} 6 \qquad \text{Substitute 100 for } x.$$

$$3(2) \overset{?}{=} 6 \qquad \text{Evaluate logarithm.}$$

$$6 = 6 \qquad \text{Solution checks. } \checkmark$$

Graphical Solution

First rewrite the equation as $3 \log_{10} x - 6 = 0$. Then use a graphing utility to graph $y = 3 \log_{10} x - 6$, as shown in Figure 9.17. Use the *zero* or *root* feature or the *zoom* and *trace* features of the graphing utility to determine that 100 is an approximate solution. You can verify that 100 is an exact solution by substituting $x = 100$ in the original equation.

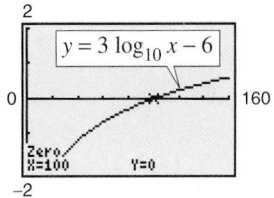

Figure 9.17

NOTE   When checking approximate solutions to exponential and logarithmic equations, be aware of the fact that because the solution is approximate, the check will not be exact.

Example 7 ■ Solving a Logarithmic Equation

Solve $20 \ln 0.2x = 30$.

Solution

$$20 \ln 0.2x = 30 \qquad \text{Write original equation.}$$

$$\ln 0.2x = 1.5 \qquad \text{Divide each side by 20.}$$

$$0.2x = e^{1.5} \qquad \text{Inverse property}$$

$$x = 5e^{1.5} \qquad \text{Divide each side by 0.2.}$$

The solution is $x = 5e^{1.5} \approx 22.408$. Check this in the original equation.

Example 8 ■ Solving a Logarithmic Equation

Solve $\log_{10} 2x - \log_{10}(x - 3) = 1$.

Solution

$$\log_{10} 2x - \log_{10}(x - 3) = 1 \qquad \text{Write original equation.}$$

$$\log_{10} \frac{2x}{x - 3} = 1 \qquad \text{Condense the left side.}$$

$$\frac{2x}{x - 3} = 10^1 \qquad \text{Inverse property}$$

$$2x = 10x - 30 \qquad \text{Multiply each side by } x - 3.$$

$$-8x = -30 \qquad \text{Subtract } 10x \text{ from each side.}$$

$$x = \frac{15}{4} \qquad \text{Divide each side by } -8.$$

The solution is $x = \frac{15}{4}$. Check this in the original equation.

Example 9 ■ Checking for Extraneous Solutions

Solve $\log_6 x + \log_6(x - 5) = 2$.

Algebraic Solution

$$\log_6 x + \log_6(x - 5) = 2 \qquad \text{Write original equation.}$$

$$\log_6[x(x - 5)] = 2 \qquad \text{Condense the left side.}$$

$$x(x - 5) = 6^2 \qquad \text{Inverse property}$$

$$x^2 - 5x - 36 = 0 \qquad \text{Write in general form.}$$

$$(x - 9)(x + 4) = 0 \qquad \text{Factor.}$$

$$x - 9 = 0 \implies x = 9 \qquad \text{Set 1st factor equal to 0.}$$

$$x + 4 = 0 \implies x = -4 \qquad \text{Set 2nd factor equal to 0.}$$

From this it appears that the solutions are $x = 9$ and $x = -4$. To be sure, you need to check each in the original equation.

Check First Solution:

$$\log_6 x + \log_6(x - 5) = 2 \qquad \text{Write original equation.}$$

$$\log_6(9) + \log_6(9 - 5) \overset{?}{=} 2 \qquad \text{Substitute 9 for } x.$$

$$\log_6(9 \cdot 4) \overset{?}{=} 2 \qquad \text{Condense the left side.}$$

$$\log_6 36 = 2 \qquad \text{Solution checks. } \checkmark$$

Check Second Solution:

$$\log_6 x + \log_6(x - 5) = 2 \qquad \text{Write original equation.}$$

$$\log_6(-4) + \log_6(-4 - 5) \overset{?}{=} 2 \qquad \text{Substitute } -4 \text{ for } x.$$

$$\log_6(-4) + \log_6(-9) \neq 2 \qquad \text{Left side is not defined. } \times$$

Of the two possible solutions, only $x = 9$ checks. The other possible solution is extraneous.

Graphical Solution

Use a graphing utility to graph each side of the equation, $y_1 = \log_6 x + \log_6(x - 5)$ and $y_2 = 2$, in the same viewing window. Remember that in order to graph $y_1 = \log_6 x + \log_6(x - 5)$ using a graphing utility, you use the change-of-base formula to enter

$$y_1 = \left(\frac{\log_{10} x}{\log_{10} 6}\right) + \left[\frac{\log_{10}(x - 5)}{\log_{10} 6}\right].$$

From the graph in Figure 9.18, it appears that the graphs have only one point of intersection. Use the *intersect* feature or the *zoom* and *trace* features to approximate the point of intersection. So, the approximate solution is 9. You can verify that 9 is an exact solution by substituting $x = 9$ into the original equation.

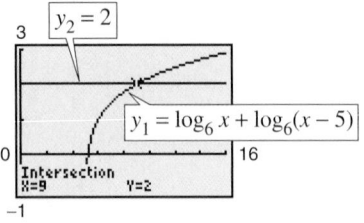

Figure 9.18

9.4 ■ Exercises

Developing Skills

In Exercises 1–6, determine whether the x-values are solutions of the equation.

1. $3^{2x-5} = 27$

  (a) $x = 1$

  (b) $x = 4$

2. $4^{x+3} = 16$

  (a) $x = -1$

  (b) $x = 0$

3. $e^{x+5} = 45$

  (a) $x = -5 + \ln 45$

  (b) $x = -5 + e^{45}$

4. $2^{3x-1} = 324$

  (a) $x \approx 3.1133$

  (b) $x \approx 2.4327$

5. $\log_9(6x) = \frac{3}{2}$

  (a) $x = 27$

  (b) $x = \frac{9}{2}$

6. $\ln(x + 3) = 2.5$

  (a) $x = -3 + e^{2.5}$

  (b) $x \approx 9.1825$

In Exercises 7–30, solve the equation. (Do not use a calculator.)

7. $2^x = 2^5$

8. $5^x = 5^3$

9. $3^{x+4} = 3^{12}$

10. $10^{1-x} = 10^4$

11. $3^{x-1} = 3^7$

12. $4^{2x} = 64$

13. $4^{x-1} = 16$

14. $3^{2x} = 81$

15. $2^{x+2} = \frac{1}{16}$

16. $3^{2-x} = 9$

17. $5^x = \frac{1}{125}$

18. $4^{x+1} = \frac{1}{64}$

19. $\log_2(x + 3) = \log_2 7$

20. $\log_5 2x = \log_5 36$

21. $\log_4(x - 4) = \log_4 12$

22. $\log_3(2 - x) = \log_3 2$

23. $\ln 5x = \ln 22$

24. $\ln(2x - 3) = \ln 17$

25. $\ln(2x - 3) = \ln 15$

26. $\ln 3x = \ln 24$

27. $\log_3 x = 4$

28. $\log_5 x = 3$

29. $\log_{10} 2x = 6$

30. $\log_2(3x - 1) = 5$

In Exercises 31–34, simplify the expression.

31. $\ln e^{2x-1}$

32. $\log_3 3^{x^2}$

33. $10^{\log_{10} 2x}$

34. $e^{\ln(x+1)}$

In Exercises 35–74, solve the exponential equation. (Round the result to two decimal places.)

35. $2^x = 45$

36. $5^x = 212$

37. $10^{2y} = 52$

38. $12^{x-1} = 1500$

39. $4^x = 8$

40. $2^x = 1.5$

41. $3^{x+4} = 6$

42. $5^{3-x} = 15$

43. $2 \cdot 10^{x+6} = 500$

44. $3 \cdot 5^{x+6} = 48$

45. $4 \cdot 7^{x-2} = 428$

46. $5 \cdot 6^{x+1} = 300$

47. $2^{x-1} - 6 = 1$

48. $5^{x+6} - 4 = 12$

49. $\frac{1}{5}(4^{x+2}) = 300$

50. $3(2^{t+4}) = 350$

51. $5(2)^{3x} - 4 = 13$

52. $-16 + 0.2(10)^x = 35$

53. $3e^x = 42$

54. $6e^{-x} = 3$

55. $\frac{1}{4}e^x = 5$

56. $\frac{2}{3}e^x = 1$

57. $2e^{3x} = 80$

58. $4e^{-3x} = 6$

59. $4 + e^{2x} = 150$

60. $500 - e^{x/2} = 35$

61. $8 - 12e^{-x} = 7$

62. $4 - 2e^x = -23$

63. $23 - 5e^{x+1} = 3$

64. $2e^x + 5 = 115$

65. $300e^{x/2} = 9000$

66. $1000^{0.12x} = 25{,}000$

67. $6000e^{-2t} = 1200$

68. $10{,}000e^{-0.1t} = 4000$

69. $250(1.04)^x = 1000$

70. $32(1.5)^x = 640$

71. $\dfrac{1600}{(1.1)^x} = 200$

72. $\dfrac{5000}{(1.05)^x} = 250$

73. $4(1 + e^{x/3}) = 84$

74. $50(3 - e^{2x}) = 125$

In Exercises 75–110, solve the logarithmic equation. (Round the result to two decimal places.)

75. $\log_{10} x = 0$

76. $\ln x = 1$

77. $\log_{10} 4x = \frac{3}{2}$

78. $\log_{10}(x + 3) = \frac{5}{3}$

79. $\log_2 x = 4.5$

80. $\log_4(25x) = 7$

81. $4 \log_3 x = 28$

82. $5 \log_{10}(x + 2) = 18$

83. $2 \log_{10}(x + 5) = 15$

84. $-1 + 3 \log_{10} \dfrac{x}{2} = 8$

85. $16 \ln x = 30$

86. $\ln x = 2.1$

87. $\ln 2x = 3$

88. $\ln\!\left(\frac{1}{2}t\right) = \frac{1}{4}$

89. $1 - 2 \ln x = -4$

90. $-5 + 2 \ln 3x = 5$

91. $\frac{2}{3}\ln(x + 1) = -1$

92. $8 \ln(3x - 2) = 1.5$

93. $\ln x^2 = 6$

94. $\ln \sqrt{x} = 6.5$

95. $\log_4 x + \log_4 5 = 2$

96. $\log_5 x - \log_5 4 = 2$

97. $\log_6(x + 8) + \log_6 3 = 2$

98. $\log_7(x - 1) - \log_7 4 = 1$

99. $\log_{10} x + \log_{10}(x - 3) = 1$

100. $\log_{10}(25x) - \log_{10}(x - 1) = 2$

101. $\log_5(x + 3) - \log_5 x = 1$

102. $\log_3(x - 2) + \log_3 5 = 3$

103. $\log_6(x - 5) + \log_6 x = 2$

104. $\log_{10} x + \log_{10}(x + 1) = 0$

105. $\log_2(x - 1) + \log_2(x + 3) = 3$

106. $\log_5(3 - x) + \log_5(x + 9) = 1$

107. $\log_4 3x + \log_4(x - 2) = \frac{1}{2}$

108. $\log_{16} 2x + \log_{16}(x + 4) = \frac{3}{4}$

109. $\log_3 2x + \log_3(x - 1) - \log_3 4 = 1$

110. $\log_2 x + \log_2(x + 2) - \log_2 3 = 4$

In Exercises 111–114, use a graphing utility to solve the equation. (Round the result to two decimal places.)

111. $e^x = 2$

112. $\ln x = 2$

113. $2 \ln(x + 3) = 3$

114. $1000e^{-x/2} = 200$

Solving Problems

115. *Doubling Time* Solve the exponential equation $5000 = 2500e^{0.09t}$ for t to determine the number of years for an investment of \$2500 to double in value when compounded continuously at a rate of 9%.

116. *Doubling Time* Solve the exponential equation $10,000 = 5000e^{10r}$ for r to determine the interest rate required for an investment of \$5000 to double in value when compounded continuously for 10 years.

117. *Intensity of Sound* The relationship between the number of decibels B and the intensity of a sound I in watts per centimeter squared is given by

$$B = 10 \log_{10}\left(\frac{I}{10^{-16}}\right).$$

Determine the intensity of a sound I if it registers 75 decibels on an intensity meter.

118. *Human Memory Model* The average score A for a group of students who took a test t months after the completion of a course is given by the human memory model $A = 80 - \log_{10}(t + 1)^{12}$. How long after completing the course will the average score fall to $A = 72$?

(a) Answer the question algebraically by letting $A = 72$ and solving the resulting equation.

(b) Answer the question graphically by using a graphing utility to graph the equations $y_1 = 80 - \log_{10}(t + 1)^{12}$ and $y_2 = 72$ and finding their point(s) of intersection.

(c) Which strategy works better for this problem? Explain your reasoning.

119. *Friction* In order to restrain an untrained horse, a person partially wraps a rope around a cylindrical post in a corral (see figure). If the horse is pulling on the rope with a force of 200 pounds, the force F in pounds required by the person is $F = 200e^{-0.2\pi\theta/180}$, where θ is the angle of the wrap in degrees. Find the smallest value of θ if F cannot exceed 80 pounds.

120. *Oceanography* Oceanographers use the density d (in grams per cubic centimeter) of seawater to obtain information about the circulation of water masses and the rates at which waters of different densities mix. For water with a salinity of 30%, the water temperature T (in degrees Celsius) is related to the density by

$$T = 7.9 \ln(1.0245 - d) + 61.84.$$

Find the densities of the subantarctic water and the antarctic bottom water shown in the figure.

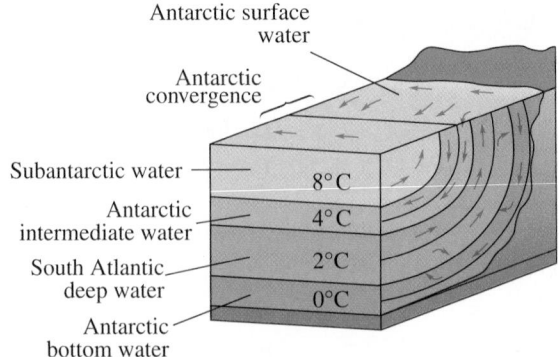

The cross section shows complex currents at various depths in the South Atlantic Ocean off Antarctica.

121. *Muon Decay*   A muon is an elementary particle that is similar to an electron, but much heavier. Muons are unstable—they quickly decay to form electrons and other particles. In an experiment conducted in 1943, the number of muon decays m (of an original 5000 muons) was related to the time T (in microseconds) by the model $T = 15.7 - 2.48 \ln m$. How many decays were recorded when $T = 2.5$?

122. *Military Personnel*   The numbers N (in thousands) of United States military personnel on active duty in the years 1995 through 1998 are modeled by the equation

$$N = 1716e^{-0.0250t}, \quad 5 \le t \le 8$$

where t is the time in years, with $t = 5$ corresponding to 1995. *(Source: U.S. Department of Defense)*

(a) Use a graphing utility to graph the equation over the specified domain.

(b) Use the graph in part (a) to estimate the value of t when $N = 1450$.

Explaining Concepts

123. Which of the equations requires logarithms for its solution? Explain.

$$2^{x-1} = 32 \text{ or } 2^{x-1} = 30$$

124. Explain how to solve $10^{2x-1} = 5316$.

125. State the three basic properties of logarithms.

126. In your own words, state the guidelines for solving exponential and logarithmic equations.

Ongoing Review

In Exercises 127–132, solve the equation.

127. $\frac{2}{3}x + \frac{2}{3} = 4x - 6$

128. $x^2 - 10x + 17 = 0$

129. $\frac{5}{2x} - \frac{4}{x} = 3$

130. $\frac{1}{x} + \frac{2}{x-5} = 0$

131. $|x - 4| = 3$

132. $\sqrt{x + 2} = 7$

133. *Mixture Problem*   A pharmacist needs 100 liters of a 50% alcohol solution. She has on hand a 30% alcohol solution and an 80% alcohol solution, which she can mix. How many liters of each will be required to make the 100 liters of a 50% alcohol solution?

Looking Further

Making Ice Cubes   You place a tray of 60°F water in a freezer that is set at 0°F. The water cools according to Newton's Law of Cooling

$$kt = \ln \frac{T - S}{T_0 - S}$$

where T is the temperature (in °F), t is the number of hours the tray is in the freezer, S is the temperature of the surrounding air, and T_0 is the original temperature of the water.

(a) The water freezes in 4 hours. What is the constant k? (*Hint:* Water freezes at 32°F.)

(b) You lower the temperature in the freezer to −10°F. At this temperature, how long will it take for the ice cubes to form?

(c) The initial temperature of the water is 50°F. The freezer temperature is 0°F. How long will it take for the ice cubes to form?

9.5 Exponential and Logarithmic Applications

Compound Interest

In Section 9.1, you were introduced to two formulas for compound interest. Recall that in these formulas, A is the balance, P is the principal, r is the annual interest rate (in decimal form), and t is the time in years.

n Compoundings per Year

$$A = P\left(1 + \frac{r}{n}\right)^{nt}$$

Continuous Compounding

$$A = Pe^{rt}$$

Example 1 ■ Finding the Annual Interest Rate

An investment of $50,000 is made in an account that compounds interest quarterly. After 4 years, the balance in the account is $71,381.07. What is the annual interest rate for this account?

Solution

Formula: $A = P\left(1 + \frac{r}{n}\right)^{nt}$

Labels:

Principal $= P = 50{,}000$	(dollars)
Balance $= A = 71{,}381.07$	(dollars)
Time $= t = 4$	(years)
Number of compoundings per year $= n = 4$	
Annual interest rate $= r$	(percent in decimal form)

Equation:

$71{,}381.07 = 50{,}000\left(1 + \frac{r}{4}\right)^{(4)(4)}$ Substitute for A, P, n, and t.

$1.42762 \approx \left(1 + \frac{r}{4}\right)^{16}$ Divide each side by 50,000.

$(1.42762)^{1/16} \approx 1 + \frac{r}{4}$ Raise each side to $\frac{1}{16}$ power.

$1.0225 \approx 1 + \frac{r}{4}$ Simplify.

$0.0225 \approx \frac{r}{4}$ Subtract 1 from each side.

$0.09 \approx r$ Multiply each side by 4.

The annual interest rate is approximately 9%. Check this in the original problem.

Solving an exponential equation often requires "getting rid of" the exponent in the variable expression. This can be accomplished by raising each side of the equation to the reciprocal power. For instance, in Example 1 the variable expression had power 16, so each side was raised to the reciprocal power $\frac{1}{16}$.

Example 2 ■ Doubling Time for Continuous Compounding

An investment is made in a trust fund at an annual interest rate of 8.75%, compounded continuously. How long will it take for the investment to double?

NOTE In Example 2, note that you do not need to know the principal to find the doubling time. In other words, the doubling time is the same for *any* principal.

Solution

$$A = Pe^{rt}$$ Formula for continuous compounding

$$2P = Pe^{0.0875t}$$ Substitute known values.

$$2 = e^{0.0875t}$$ Divide each side by P.

$$\ln 2 = 0.0875t$$ Inverse property

$$\frac{\ln 2}{0.0875} = t$$ Divide each side by 0.0875.

$$7.92 \approx t$$

It will take approximately 7.92 years for the investment to double. Check this in the original problem.

Example 3 ■ Finding the Type of Compounding

You deposit $1000 in an account. At the end of 1 year, your balance is $1077.63. The bank tells you that the annual interest rate for the account is 7.5%. How was the interest compounded?

Numerical Solution

If the interest had been compounded continuously at 7.5%, the balance would have been $A = 1000e^{(0.075)(1)} = \$1077.88$. Because the actual balance is slightly less than this, you should use the formula for interest that is compounded n times per year.

$$A = 1000\left(1 + \frac{0.075}{n}\right)^n = 1077.63$$

At this point, it is not clear what you should do to solve the equation for n. You can use the *table* feature of a graphing utility to evaluate

$$y = 1000\left(1 + \frac{0.075}{x}\right)^x$$

for the most common types of compounding. From the table in Figure 9.19, you can see that $x = 12$. So, the interest was compounded monthly.

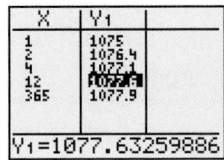

Figure 9.19

Graphical Solution

Use a graphing utility to graph

$$y_1 = 1000\left(1 + \frac{0.075}{x}\right)^x$$

and

$$y_2 = 1077.63$$

in the same viewing window, as shown in Figure 9.20. Use the *intersect* feature or the *zoom* and *trace* features of the graphing utility to approximate the point of intersection of the two graphs. The approximate solution is 12. So, the interest was compounded monthly.

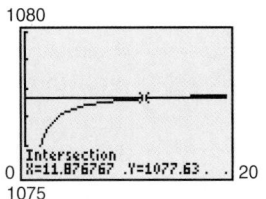

Figure 9.20

In Example 3, notice that an investment of $1000 compounded monthly produced a balance of $1077.63 at the end of 1 year. Because $77.63 of this amount is interest, the **effective yield** for the investment is

$$\text{Effective yield} = \frac{\text{year's interest}}{\text{amount invested}} = \frac{77.63}{1000} = 0.07763 = 7.763\%.$$

In other words, the effective yield for an investment collecting compound interest is the *simple interest rate* that would yield the same balance at the end of 1 year.

Example 4 ■ Finding the Effective Yield

An investment is made in an account that pays 6.75% interest, compounded continuously. What is the effective yield for this investment?

Solution

Notice that you do not have to know the principal or the time that the money will be left in the account. Instead, you can choose an arbitrary principal, such as $1000. Then, because effective yield is based on the balance at the end of 1 year, you can use the formula.

$$A = Pe^{rt} = 1000e^{0.0675(1)} = 1069.83.$$

Now, because the account would earn $69.83 in interest after 1 year for a principal of $1000, you can conclude that the effective yield is

$$\text{Effective yield} = \frac{69.83}{1000} = 0.06983 = 6.983\%.$$

NOTE   In Example 4, you can find the effective yield *without* substituting a value for P.

$$A = Pe^{rt}$$
$$A = Pe^{(0.0675)(1)}$$
$$A \approx P(1.06983)$$
$$A \approx P + P(0.06983)$$

From the last equation, you can see that the principal increased by 6.983%.

Growth and Decay

The balance in an account earning *continuously* compounded interest is one example of a quantity that increases over time according to the **exponential growth model** $y = Ce^{kt}$.

NOTE   In the model at the right, if $k > 0$ the model represents **exponential growth,** and if $k < 0$ it represents **exponential decay.**

Exponential Growth and Decay

The mathematical model for exponential growth or decay is given by

$$y = Ce^{kt}.$$

For this model, t is the time, C is the original amount of the quantity, and y is the amount after the time t. The number k is a constant that is determined by the rate of growth.

One common application of exponential growth is in modeling the growth of a population. Example 5 illustrates the use of the growth model

$$y = Ce^{kt}, \ k > 0.$$

Example 5 ■ Population Growth

The population of California was 30 million in 1990 and 34 million in 2000. What would you predict the population of California to be in the year 2010? *(Source: U.S. Census Bureau)*

Solution

If you assumed a *linear growth model*, you would simply predict the population in the year 2010 to be 38 million. However, social scientists and demographers have discovered that *exponential growth models* are better than linear growth models for representing population growth. So, you can use the exponential growth model

$$y = Ce^{kt}.$$

In this model, let $t = 0$ represent 1990. The given information about the population can be described by the table shown below.

t (year)	0	10	20
Ce^{kt} (million)	$Ce^{k(0)} = 30$	$Ce^{k(10)} = 34$	$Ce^{k(20)} = ?$

To find the population when $t = 20$, you must first find the values of C and k. From the table, you can use the fact that $Ce^{k(0)} = Ce^0 = 30$ to conclude that $C = 30$. Then, using this value of C, you can solve for k as follows.

$$
\begin{aligned}
Ce^{k(10)} &= 34 & &\text{From table}\\
30e^{10k} &= 34 & &\text{Substitute value of } C.\\
e^{10k} &= \frac{17}{15} & &\text{Divide each side by 30.}\\
10k &= \ln\frac{17}{15} & &\text{Inverse property}\\
k &= \frac{1}{10}\ln\frac{17}{15} & &\text{Divide each side by 10.}\\
k &\approx 0.0125 & &\text{Simplify.}
\end{aligned}
$$

Finally, you can use this value of k in the model from the table for 2010 (for $t = 20$) to conclude that the population in the year 2010 is given by

$$
\begin{aligned}
30e^{0.0125(20)} &\approx 30(1.28)\\
&= 38.4 \text{ million.}
\end{aligned}
$$

The graph of the population model is shown in Figure 9.21.

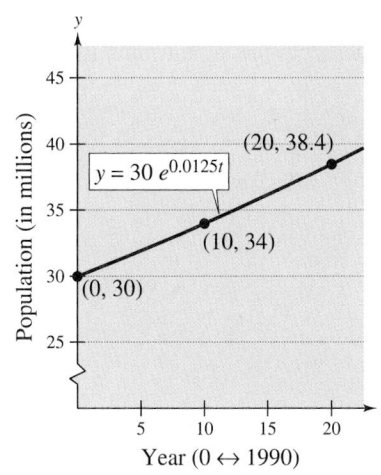

Figure 9.21

Example 6 ■ Radioactive Decay

Radioactive iodine is a by-product of some types of nuclear reactors. Its **half-life** is 60 days. That is, after 60 days, an initial amount of radioactive iodine will have decayed to half the original amount. Suppose a nuclear accident occurs and releases 20 grams of radioactive iodine. How long will it take for the radioactive iodine to decay to a level of 1 gram?

Solution

To solve this problem, use the model for exponential decay.

$$y = Ce^{kt}$$

Next, use the information given in the problem to set up the table shown below.

t (days)	0	60	?
Ce^{kt} (grams)	$Ce^{k(0)} = 20$	$Ce^{k(60)} = 10$	$Ce^{k(t)} = 1$

Because $Ce^{k(0)} = Ce^0 = 20$, you can conclude that $C = 20$. Then, using this value of C, you can solve for k as follows.

$Ce^{k(60)} = 10$	From table
$20e^{60k} = 10$	Substitute value of C.
$e^{60k} = \dfrac{1}{2}$	Divide each side by 20.
$60k = \ln \dfrac{1}{2}$	Inverse property
$k = \dfrac{1}{60} \ln \dfrac{1}{2}$	Divide each side by 60.
≈ -0.01155	Simplify.

Finally, you can use this value of k in the model from the table to find the time when the amount is 1 gram, as follows.

$Ce^{k(t)} = 1$	From table
$20e^{-0.01155t} = 1$	Substitute values of C and k.
$e^{-0.01155t} = \dfrac{1}{20}$	Divide each side by 20.
$-0.01155t = \ln \dfrac{1}{20}$	Inverse property
$t = \dfrac{1}{-0.01155} \ln \dfrac{1}{20}$	Divide each side by -0.01155.
≈ 259.4 days	Simplify.

So, 20 grams of radioactive iodine will have decayed to 1 gram after about 259.4 days. This solution is shown graphically in Figure 9.22.

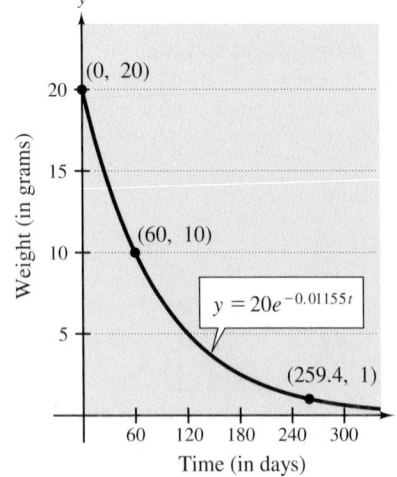

Figure 9.22 *Radioactive decay*

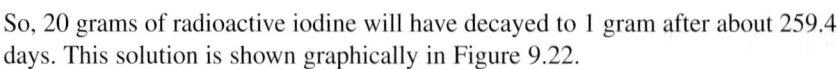

Intensity Models

On the **Richter scale,** the magnitude R of an earthquake can be measured by the **intensity model**

$$R = \log_{10} I$$

where I is the intensity of the shock wave.

Example 7 ■ Earthquake Intensity

In 1993, Guam experienced an earthquake that measured 8.1 on the Richter scale. In 1999, northwest Turkey experienced an earthquake that measured 7.4 on the Richter scale. Compare the intensities of these two earthquakes.

Solution

The intensity of the 1993 earthquake is given as follows.

$$
\begin{aligned}
8.1 &= \log_{10} I && \text{Given} \\
10^{8.1} &= I && \text{Inverse property}
\end{aligned}
$$

The intensity of the 1999 earthquake can be found in a similar way.

$$
\begin{aligned}
7.4 &= \log_{10} I && \text{Given} \\
10^{7.4} &= I && \text{Inverse property}
\end{aligned}
$$

The ratio of these two intensities is

$$
\begin{aligned}
\frac{I \text{ for } 1993}{I \text{ for } 1999} &= \frac{10^{8.1}}{10^{7.4}} \\
&= 10^{8.1 - 7.4} \\
&= 10^{0.7} \\
&\approx 5.01.
\end{aligned}
$$

So, the 1993 earthquake had an intensity that was about five times as great as that of the 1999 earthquake.

Collaborate!

You Be the Instructor

Write a problem that could be answered by investigating the exponential growth model

$$y = 10e^{0.08t}$$

or the exponential decay model

$$y = 5e^{-0.25t}.$$

Exchange your problem with that of another group member, and solve one another's problems.

9.5 ■ Exercises

Solving Problems

Annual Interest Rate In Exercises 1–10, find the annual interest rate.

	Principal	Balance	Time	Compounding
1.	$500	$1004.83	10 years	Monthly
2.	$3000	$21,628.70	20 years	Quarterly
3.	$1000	$36,581.00	40 years	Daily
4.	$200	$314.85	5 years	Yearly
5.	$750	$8267.38	30 years	Continuous
6.	$2000	$2718.28	10 years	Continuous
7.	$5000	$22,405.68	25 years	Daily
8.	$10,000	$110,202.78	30 years	Daily
9.	$1500	$24,666.97	40 years	Continuous
10.	$7500	$15,877.50	15 years	Continuous

Doubling Time In Exercises 11–20, find the time for the investment to double. Use a graphing utility to verify the result graphically.

	Principal	Rate	Compounding
11.	$6000	8%	Quarterly
12.	$500	$5\frac{1}{4}$%	Monthly
13.	$2000	10.5%	Daily
14.	$10,000	9.5%	Yearly
15.	$1500	7.5%	Continuous
16.	$100	6%	Continuous
17.	$300	5%	Yearly
18.	$12,000	4%	Continuous
19.	$6000	7%	Quarterly
20.	$500	9%	Daily

Type of Compounding In Exercises 21–24, determine the type of compounding. Solve the problem by trying the more common types of compounding.

	Principal	Balance	Time	Rate
21.	$750	$1587.75	10 years	7.5%
22.	$10,000	$73,890.56	20 years	10%
23.	$100	$141.48	5 years	7%
24.	$4000	$4788.76	2 years	9%

Effective Yield In Exercises 25–34, find the effective yield.

	Rate	Compounding
25.	8%	Continuous
26.	9.5%	Daily
27.	7%	Monthly
28.	8%	Yearly
29.	6%	Quarterly
30.	9%	Quarterly
31.	8%	Monthly
32.	$5\frac{1}{4}$%	Daily
33.	7.5%	Continuous
34.	4%	Continuous

35. *Doubling Time* Is it necessary to know the principal P to find the doubling time in Exercises 11–20? Explain.

36. *Effective Yield*

(a) Is it necessary to know the principal P to find the effective yield in Exercises 25–34? Explain.

(b) When the interest is compounded more frequently, what inference can you make about the difference between the effective yield and the stated annual percent rate?

Compound Interest In Exercises 37–46, find the principal that must be deposited in an account to obtain the given balance.

	Balance	Rate	Time	Compounding
37.	$10,000	9%	20 years	Continuous
38.	$5000	8%	5 years	Continuous
39.	$750	6%	3 years	Daily
40.	$3000	7%	10 years	Monthly
41.	$25,000	7%	30 years	Monthly
42.	$8000	6%	2 years	Monthly
43.	$1000	5%	1 year	Daily
44.	$100,000	9%	40 years	Daily
45.	$500,000	8%	25 years	Continuous
46.	$1,000,000	10%	50 years	Continuous

Balance After Monthly Deposits   In Exercises 47–50, you make monthly deposits of P dollars in a savings account at an annual interest rate r, compounded continuously. Find the balance A after t years given that

$$A = \frac{P(e^{rt} - 1)}{e^{r/12} - 1}.$$

	Principal	Rate	Time
47.	$P = \$30$	$r = 8\%$	$t = 10$ years
48.	$P = \$100$	$r = 9\%$	$t = 30$ years
49.	$P = \$50$	$r = 10\%$	$t = 40$ years
50.	$P = \$20$	$r = 7\%$	$t = 20$ years

Exponential Growth and Decay   In Exercises 51–54, find the constant k such that the graph of $y = Ce^{kt}$ passes through the points.

51.
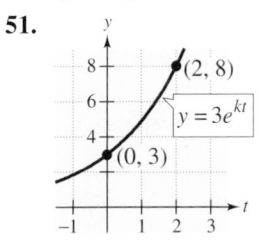

52.
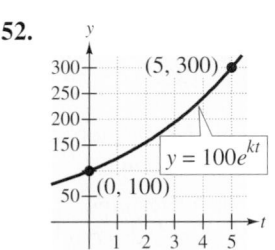

53.
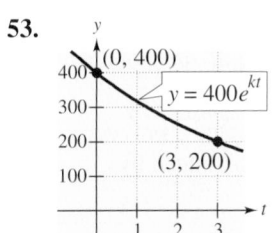

54.
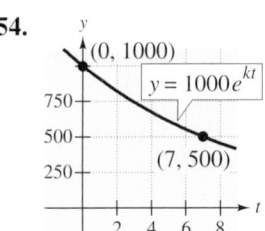

Population of a Region   In Exercises 55–62, the population (in millions) of an urban region in the year 2000 and the predicted population (in millions) for the year 2015 are given. Find the constants C and k to obtain the exponential model $y = Ce^{kt}$ for the population. (Let $t = 0$ correspond to 2000.) Use the model to predict the population of the region in the year 2020.   *(Source: United Nations)*

	Region	2000	2015
55.	Los Angeles	13.1	14.1
56.	New York	16.6	17.4
57.	Shanghai, China	12.9	14.6
58.	Jakarta, Indonesia	11.0	17.3
59.	Bombay, India	18.1	26.1
60.	Cairo, Egypt	10.6	13.8
61.	Mexico City, Mexico	18.1	19.2
62.	São Paulo, Brazil	17.8	20.4

63. *Rate of Growth*   Compare the values of k in Exercises 57 and 59. Which is larger? Explain.

64. *Rate of Growth*   What variable in the continuous compound interest formula is equivalent to k in the model for population growth? Use your answer to give an interpretation of k.

65. *Radioactive Decay*   Radioactive radium (^{226}Ra) has a half-life of 1620 years. If you start with 5 grams of the isotope, how much will remain after 1000 years?

66. *Radioactive Decay*   The half-life of plutonium 239 (^{239}Pu) is about 24,360 years. If you start with 10 grams of the isotope, how much will remain after 10,000 years?

67. *Radioactive Decay*   Carbon 14 (^{14}C) has a half-life of 5730 years. If you start with 5 grams of the isotope, how much will remain after 1000 years?

68. *Radioactive Decay*   The half-life of phosphorus 32 (^{32}P) is about 14 days. If you start with 6.6 grams of the isotope, how much will remain after 38 days?

69. *Radioactive Decay*   The half-life of plutonium 241 (^{241}Pu) is about 13 years. If you start with 100 grams of the isotope, how much will remain after 10 years?

70. *Radioactive Decay*   The half-life of iodine 131 (^{131}I) is about 8 years. If you start with 2000 grams of the isotope, how much will remain after 30 years?

71. *Carbon 14 Dating*   Carbon 14 dating assumes that the carbon dioxide on Earth today has the same radioactive content as it did centuries ago. If this is true, the amount of ^{14}C absorbed by a tree that grew several centuries ago should be the same as the amount of ^{14}C absorbed by a tree growing today. A piece of ancient charcoal contains only 15% as much of the radioactive carbon as a piece of modern charcoal. How long ago did the tree burn to make the ancient charcoal if the half-life of ^{14}C is 5730 years? (Round the result to the nearest 100 years.)

72. *Depreciation*   A car that cost \$22,000 new has a depreciated value of \$16,500 after 1 year. Find the value of the car when it is 3 years old by using the exponential model $y = Ce^{kt}$.

73. *Depreciation*   After x years, the value y of a truck that cost \$32,000 new is given by $y = 32,000(0.8)^x$.

(a) Use a graphing utility to graph the equation.

(b) Use the graph from part (a) to approximate the value of the truck after 1 year.

(c) Use the graph from part (a) to approximate the time when the truck's value will be \$16,000.

74. *World Population*   The figure shows the population P (in billions) of the world as projected by the Population Reference Bureau. The bureau's projection can be modeled by

$$P = \frac{11.14}{1 + 1.101e^{-0.051t}}$$

where $t = 0$ represents 1990. Use the model to estimate the population in 2020.

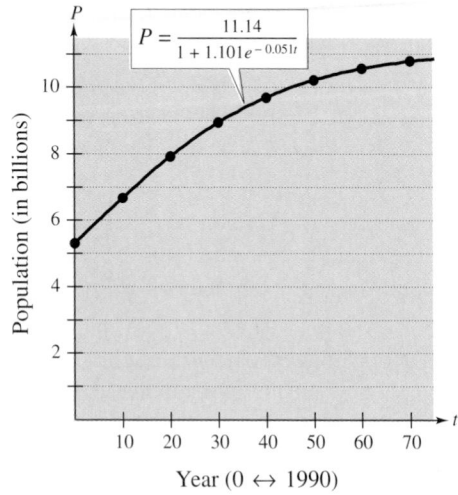

Balance After Monthly Deposits   In Exercises 75 and 76, you make monthly deposits of $30 in a savings account at an annual interest rate of 8%, compounded continuously (see figure).

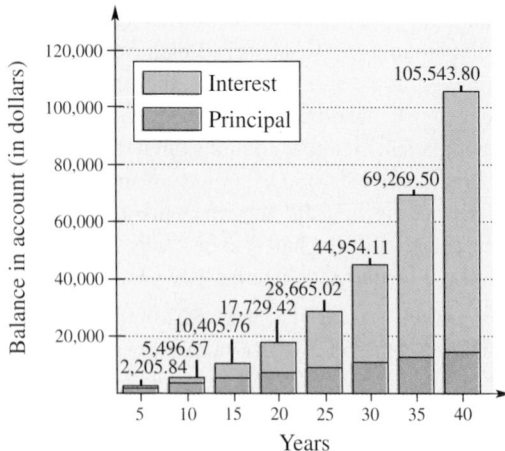

75. Find the total amount that has been deposited in the account in 20 years and the total interest earned.

76. Find the total amount that has been deposited in the account in 40 years and the total interest earned.

Earthquake Intensity   In Exercises 77–80, compare the intensities of the two earthquakes.

Location	Date	Magnitude
77. Alaska	3/27/64	8.4
San Fernando Valley	1/17/94	6.6
78. El Salvador	1/13/2001	7.7
Armenia, Colombia	1/25/99	6.0
79. Mexico City, Mexico	9/19/85	8.1
Nepal	8/20/88	6.5
80. Chile	8/16/06	8.6
Armenia, USSR	12/7/88	6.8

81. *Population Growth*   The population P of a certain species t years after it is introduced into a new habitat is given by

$$P(t) = \frac{5000}{1 + 4e^{-t/6}}.$$

(a) Use a graphing utility to graph the population function.

(b) Determine the population size that was introduced into the habitat.

(c) Determine the population size after 9 years.

(d) After how many years will the population be 2000?

82. *Sales Growth*   Annual sales y of a product x years after it is introduced are approximated by

$$y = \frac{2000}{1 + 4e^{-x/2}}.$$

(a) Use a graphing utility to graph the equation.

(b) Use the graph from part (a) to approximate annual sales when $x = 4$.

(c) Use the graph from part (a) to approximate the time when annual sales are $y = 1100$ units.

(d) Use the graph from part (a) to estimate the maximum level that annual sales will approach.

83. *Advertising Effect*   The sales S (in thousands of units) of a product after spending x hundred dollars on advertising are given by $S = 10(1 - e^{kx})$.

(a) Find S as a function of x if 2500 units are sold when $500 is spent on advertising.

(b) How many units will be sold if advertising expenditures are raised to $700?

Explaining Concepts

84. If the equation $y = Ce^{kt}$ models exponential growth, what must be true about k?

85. If the equation $y = Ce^{kt}$ models exponential decay, what must be true about k?

86. In your own words, explain what is meant by the half-life of a radioactive isotope.

87. If the reading on the Richter scale is increased by 1, the intensity of the earthquake is increased by what factor?

88. What is meant by the effective yield of an investment? Explain how it is computed.

Ongoing Review

In Exercises 89–92, evaluate the expression without using a calculator.

89. $\log_{10} 10,000$

90. $\log_4 64$

91. $\log_3 \frac{1}{81}$

92. $\ln e^3$

In Exercises 93–98, solve the system of equations.

93. $\begin{cases} x - y = 0 \\ x + 2y = 9 \end{cases}$

94. $\begin{cases} 2x + 5y = 15 \\ 3x + 6y = 20 \end{cases}$

95. $\begin{cases} y = x^2 \\ -3x + 2y = 2 \end{cases}$

96. $\begin{cases} x - y^3 = 0 \\ x - 2y^2 = 0 \end{cases}$

97. $\begin{cases} x - y = -1 \\ x + 2y - 2z = 3 \\ 3x - y + 2z = 3 \end{cases}$

98. $\begin{cases} 2x - y - 2z = 1 \\ x - z = 1 \\ 3x + 3y + z = 12 \end{cases}$

99. *Average Speed*   Two cars start out from the same location and travel in opposite directions. At the end of 6 hours, they are 690 kilometers apart. One car travels 15 kilometers per hour faster than the other. What are their speeds?

100. *Geometry*   If the radius of a certain circle is tripled and then increased by 8.2 inches, the result is the circumference of the circle. Find the radius of the circle.

Looking Further

Acidity Model   Use the acidity model

$$pH = -\log_{10} [H^+]$$

where acidity (pH) is a measure of the hydrogen ion concentration $[H^+]$ (measured in moles of hydrogen per liter) of a solution.

(a) Find the pH of a solution that has a hydrogen ion concentration of 9.2×10^{-8}.

(b) Compute the hydrogen ion concentration if the pH of a solution is 4.7.

(c) A certain fruit has a pH of 2.5 and an antacid tablet has a pH of 9.5. The hydrogen ion concentration of the fruit is how many times the hydrogen ion concentration of the tablet?

(d) If the pH of a solution is decreased by one unit, the hydrogen ion concentration is increased by what factor?

Chapter Summary: Key Concepts

9.1 ■ Properties of exponential functions

Let a be a positive real number, and let x and y be real numbers, variables, or algebraic expressions.

1. $a^x \cdot a^y = a^{x+y}$

2. $(a^x)^y = a^{xy}$

3. $\dfrac{a^x}{a^y} = a^{x-y}$

4. $a^{-x} = \dfrac{1}{a^x} = \left(\dfrac{1}{a}\right)^x$

9.1 ■ Formulas for compound interest

After t years, the balance A in an account with principal P and annual interest rate r (in decimal form) is given by the two formulas below.

1. For n compoundings per year: $A = P\left(1 + \dfrac{r}{n}\right)^{nt}$

2. For continuous compounding: $A = Pe^{rt}$

9.2 ■ Properties of logarithms and natural logarithms

Let a and x be positive real numbers such that $a \neq 1$. Then the properties below are true.

1. $\log_a 1 = 0$ because $a^0 = 1$.
   $\ln 1 = 0$ because $e^0 = 1$.

2. $\log_a a = 1$ because $a^1 = a$.
   $\ln e = 1$ because $e^1 = e$.

3. $\log_a a^x = x$ because $a^x = a^x$.
   $\ln e^x = x$ because $e^x = e^x$.

9.2 ■ Change-of-base formula

Let a, b, and x be positive real numbers such that $a \neq 1$ and $b \neq 1$. Then

$$\log_a x = \frac{\log_b x}{\log_b a} \quad \text{or} \quad \log_a x = \frac{\ln x}{\ln a}.$$

9.3 ■ Properties of logarithms

Let a be a positive real number such that $a \neq 1$, and let n be a real number. If u and v are real numbers, variables, or algebraic expressions such that $u > 0$ and $v > 0$, the following properties are true.

Logarithm with base a	*Natural logarithm*
1. $\log_a(uv) = \log_a u + \log_a v$	$\ln(uv) = \ln u + \ln v$
2. $\log_a \dfrac{u}{v} = \log_a u - \log_a v$	$\ln \dfrac{u}{v} = \ln u - \ln v$
3. $\log_a u^n = n \log_a u$	$\ln u^n = n \ln u$

9.4 ■ Properties of exponential and logarithmic equations

Let a be a positive real number such that $a \neq 1$, and let x and y be real numbers. Then the properties below are true.

1. $a^x = a^y$ if and only if $x = y$.

2. $\log_a x = \log_a y$ if and only if $x = y$ $(x > 0, y > 0)$.

9.4 ■ Solving exponential and logarithmic equations

1. To solve an exponential equation, first isolate the exponential expression, then take the logarithm of each side of the equation (or write the equation in logarithmic form) and solve for the variable.

2. To solve a logarithmic equation, first isolate the logarithmic expression, then exponentiate each side of the equation (or write the equation in exponential form) and solve for the variable.

9.4 ■ Inverse properties of exponents and logarithms

Base a	*Natural base e*
1. $\log_a(a^x) = x$	$\ln(e^x) = x$
2. $a^{(\log_a x)} = x$	$e^{(\ln x)} = x$

9.5 ■ Exponential growth and decay

The mathematical model for exponential growth or decay is given by

$$y = Ce^{kt}.$$

For this model, t is the time, C is the original amount of the quantity, and y is the amount after the time t. The number k is a constant that is determined by the rate of growth. If $k > 0$ the model represents exponential growth, and if $k < 0$ it represents exponential decay.

Review Exercises

Reviewing Skills

9.1 ■ *How to evaluate exponential functions*

In Exercises 1–4, evaluate the function as indicated.

1. $f(x) = 2^x$
    (a) $x = -3$ (b) $x = 1$ (c) $x = 2$

2. $g(x) = 2^{-x}$
    (a) $x = -2$ (b) $x = 0$ (c) $x = 2$

3. $f(x) = 4 + 3^x$
    (a) $x = 1$ (b) $x = -2$ (c) $x = 10$

4. $g(x) = 10 - 3^x$
    (a) $x = 0$ (b) $x = 5$ (c) $x = -1$

■ *How to graph exponential functions*

In Exercises 5–10, sketch the graph of the function and identify the horizontal asymptote.

5. $f(x) = 3^x + 1$

6. $f(x) = 3^x - 5$

7. $g(x) = -3^{-x}$

8. $h(x) = -3^{x-2}$

9. $f(x) = 3^{x+4}$

10. $g(x) = 3^{x-3} + 4$

In Exercises 11 and 12, use a graphing utility to graph the function.

11. $f(x) = 3^{-x^2}$ **12.** $g(t) = 3^{|t|}$

■ *How to evaluate the natural base e and graph natural exponential functions*

In Exercises 13–16, evaluate the function as indicated. (Round the result to three decimal places.)

13. $g(t) = e^{-t/3}$
    (a) $t = -3$ (b) $t = \pi$ (c) $t = 6$

14. $f(x) = e^{5x}$
    (a) $x = 0$ (b) $x = 2$ (c) $x = -1$

15. $f(x) = 1 - e^{0.2x}$
    (a) $x = 0$ (b) $x = 2$ (c) $x = \sqrt{10}$

16. $g(x) = 4 + e^{1.1x}$
    (a) $x = 1$ (b) $x = 4.5$ (c) $x = \sqrt{15}$

In Exercises 17–20, use a graphing utility to graph the function.

17. $f(x) = 5e^{-x/4}$

18. $g(t) = 6 - e^{t/2}$

19. $h(t) = \dfrac{5}{1 - e^{-t}}$

20. $f(x) = \dfrac{8}{1 + e^{-x/5}}$

9.2 ■ *How to rewrite logarithmic and exponential equations and evaluate logarithms*

In Exercises 21–24, rewrite the logarithmic equation in exponential form.

21. $\ln e = 1$

22. $\log_3 9 = 2$

23. $\log_5 \frac{1}{25} = -2$

24. $\log_9 \frac{1}{9} = -1$

In Exercises 25–28, rewrite the exponential equation in logarithmic form.

25. $4^3 = 64$

26. $7^4 = 2401$

27. $25^{3/2} = 125$

28. $8^{2/3} = 4$

In Exercises 29–44, evaluate the logarithm.

29. $\log_{10} 1000$ **30.** $\log_9 3$

31. $\log_3 \frac{1}{9}$ **32.** $\log_4 \frac{1}{16}$

33. $\log_2 64$ **34.** $\log_a \frac{1}{a}$

35. $\log_{10} \frac{1}{1000}$ **36.** $\log_5 125$

37. $\log_3 27$ **38.** $\log_2 (0.5)$

39. $\log_{10} 0.01$ **40.** $\log_{10} \sqrt{10}$

41. $\log_4 \frac{1}{64}$ **42.** $\log_4 16$

43. $\log_2 \sqrt{2}$ **44.** $\log_2 \sqrt[5]{2}$

■ *How to graph logarithmic functions*

In Exercises 45 and 46, sketch the graphs of *f* and *g* on the same rectangular coordinate system.

45. $f(x) = \log_5 x$
$\quad g(x) = 5^x$

46. $f(x) = 3^x$
$\quad g(x) = \log_3 x$

In Exercises 47–50, sketch the graph of the function. Determine the domain and the vertical asymptote.

47. $f(x) = -2 + \log_3 x$

48. $f(x) = 2 + \log_3 x$

49. $f(x) = \log_2(x - 4)$

50. $f(x) = -\log_4(x + 1)$

■ *How to evaluate natural logarithms and graph natural logarithmic functions*

In Exercises 51–54, evaluate the expression.

51. $\ln e^3$

52. $\ln \dfrac{1}{e^2}$

53. $\ln \dfrac{5}{4}$

54. $\ln e^{-6}$

In Exercises 55 and 56, use a graphing utility to graph the function.

55. $y = 3 \ln x + 2$

56. $y = -\ln(x - 2)$

■ *How to use the change-of-base formula to evaluate logarithms*

In Exercises 57–60, evaluate the logarithm using the change-of-base formula. (Round the result to three decimal places.)

57. $\log_4 9$

58. $\log_{1/2} 5$

59. $\log_{12} 200$

60. $\log_3 0.28$

9.3 ■ *How to use the properties of logarithms to evaluate logarithms*

In Exercises 61–66, approximate the logarithm given that $\log_5 2 \approx 0.4307$ and $\log_5 3 \approx 0.6826$.

61. $\log_5 18$

62. $\log_5 \sqrt{6}$

63. $\log_5 \dfrac{1}{2}$

64. $\log_5 \dfrac{2}{3}$

65. $\log_5(12)^{2/3}$

66. $\log_5(5^2 \cdot 6)$

■ *How to use the properties of logarithms to rewrite, expand, or condense logarithmic expressions*

In Exercises 67–74, use the properties of logarithms to expand the expression.

67. $\log_4 6x^4$

68. $\log_{10} 2x^{-3}$

69. $\log_5 \sqrt{x + 2}$

70. $\ln \sqrt[3]{\dfrac{x}{5}}$

71. $\ln \dfrac{x + 2}{x - 2}$

72. $\ln[x(x - 3)^2]$

73. $\ln\left[\sqrt{2x}(x + 3)^5\right]$

74. $\log_3 \dfrac{a^2 \sqrt{b}}{cd^5}$

In Exercises 75–84, use the properties of logarithms to condense the expression.

75. $\log_4 x - \log_4 10$

76. $5 \log_2 y$

77. $\log_8 16x + \log_8 2x^2$

78. $4(1 + \ln x + \ln x)$

79. $-2(\ln 2x - \ln 3)$

80. $-\dfrac{2}{3} \ln 3y$

81. $3 \ln x + 4 \ln y + \ln z$

82. $\dfrac{1}{3}(\log_8 a + 2 \log_8 b)$

83. $4[\log_2 k - \log_2(k - t)]$

84. $\ln(x + 4) - 3 \ln x - \ln y$

In Exercises 85–90, determine whether the equation is true or false. If it is false, state why or give an example to show that it is false.

85. $\log_2 4x = 2 \log_2 x$

86. $\dfrac{\ln 5x}{\ln 10x} = \ln \dfrac{1}{2}$

87. $\log_{10} 10^{2x} = 2x$

88. $e^{\ln t} = t$

89. $\log_4 \dfrac{16}{x} = 2 - \log_4 x$

90. $e^{2x} - 1 = (e^x + 1)(e^x - 1)$

9.4 ■ *How to solve basic exponential and logarithmic equations*

In Exercises 91–98, solve the equation.

91. $2^{x+1} = 64$

92. $3^{x-2} = 81$

93. $4^{x-3} = \dfrac{1}{16}$

94. $5^{x+2} = \dfrac{1}{125}$

95. $\log_3 5x = \log_3 85$

96. $\log_2 2x = \log_2 100$

97. $\ln(x + 4) = \ln(17)$

98. $\log_{10}(x - 3) = \log_{10} 6$

■ *How to use inverse properties to solve exponential equations*

In Exercises 99–104, solve the equation. (Round the result to two decimal places.)

99. $3^x = 500$

100. $8^x + 6 = 10$

101. $2e^{x/2} = 45$

102. $100e^{-0.6x} = 20$

103. $\dfrac{500}{(1.05)^x} = 100$

104. $25(1 - e^t) = 12$

■ *How to use inverse properties to solve logarithmic equations*

In Exercises 105–118, solve the equation. (Round the result to two decimal places.)

105. $\log_3 x = 5$

106. $\log_8 x = 3$

107. $\log_5(x - 10) = 2$

108. $\log_2(x + 7) = 6$

109. $\log_{10} 2x = 1.5$

110. $\frac{1}{3} \log_2 x + 5 = 7$

111. $\ln x = 7.25$

112. $\ln x = -0.5$

113. $\log_2 2x = -0.65$

114. $\log_5(x + 1) = 4.8$

115. $\log_2 x + \log_2 3 = 3$

116. $\log_5 12 + \log_5 x = 4$

117. $\log_3(x + 2) - \log_3 x = 3$

118. $2 \log_4 x - \log_4(x - 1) = 1$

Solving Problems

Compound Interest In Exercises 119 and 120, complete the table to determine the balance A for P dollars invested at rate r for t years, compounded n times per year.

n	1	4	12	365	Continuous compounding
A					

	Principal	Rate	Time
119.	$P = \$500$	$r = 7\%$	$t = 30$ years
120.	$P = \$2500$	$r = 8\%$	$t = 1$ year

Compound Interest In Exercises 121 and 122, complete the table to determine the principal P that yields a balance of A dollars invested at rate r for t years, compounded n times per year.

n	1	4	12	365	Continuous compounding
P					

	Balance	Rate	Time
121.	$A = \$50,000$	$r = 8\%$	$t = 40$ years
122.	$A = \$1000$	$r = 6\%$	$t = 1$ year

123. *Effective Yield* An investment is made in an account that pays 8.5% interest, compounded monthly. What is the effective yield for this investment?

124. *Doubling Time* Find the time for an investment of $1000 to double in value when invested at 8% compounded monthly.

125. *Radioactive Decay* The half-life of uranium 234 (^{234}U) is about 233,000 years. If you start with 5 grams of the isotope, how much will remain after 100,000 years?

126. *Radioactive Decay* The half-life of bismuth 210 (^{210}Bi) is about 5 days. If you start with 60 grams of the isotope, how much will remain after 12 days?

127. *Product Demand* The daily demand x and price p of a product are related by $p = 25 - 0.4e^{0.02x}$. Approximate the demand when the price is $16.97.

128. *Sound Intensity* The relationship between the number of decibels B and the intensity of a sound I in watts per centimeter squared is given by

$$B = 10 \log_{10}\left(\frac{I}{10^{-16}}\right).$$

Determine the intensity of a sound in watts per centimeter squared if the decibel level is 125.

129. *Deer Herd* The state Parks and Wildlife Department releases 100 deer into a wilderness area. The population P of the herd can be modeled by

$$P = \frac{500}{1 + 4e^{-0.36t}}$$

where t is measured in years.

(a) Find the population after 5 years.

(b) After how many years will the population be 250 deer?

(c) Use a graphing utility to graph the model.

(d) Determine the asymptote of the graph of the model and interpret its meaning in the context of the problem.

9 Chapter Test

Take this test as you would take a test in class. After you are done, check your work against the answers given in the back of the book.

1. Evaluate $f(t) = 54\left(\frac{2}{3}\right)^t$ when $t = -1, 0, \frac{1}{2}$, and 2.

2. Sketch a graph of the function $f(x) = 2^{x+3}$ and identify the horizontal asymptote.

3. Sketch a graph of the function $g(x) = -\log_2(x - 1)$ and determine the domain and the vertical asymptote.

4. Write the logarithmic equation $\log_5 125 = 3$ in exponential form.

5. Write the exponential equation $4^{-2} = \frac{1}{16}$ in logarithmic form.

6. Evaluate $\log_8 2$ without a calculator.

7. Describe the relationship between the graphs of $f(x) = \log_5 x$ and $g(x) = 5^x$.

8. Expand the expression: $\log_4\left(5x^2/\sqrt{y}\right)$.

9. Condense the expression: $8 \ln a + \ln b - 3 \ln c$.

In Exercises 10–17, solve the equation. Round your answer to three decimal places, if necessary.

10. $\log_4 x = 3$

11. $10^{3y} = 832$

12. $400e^{0.08t} = 1200$

13. $\dfrac{1.06^x - 1}{0.06} = 2.56$

14. $\dfrac{1 - 1.0045^{-x}}{0.0045} = 92.57$

15. $3 \ln(2x - 3) = 10$

16. $\log_2 x + \log_2(x + 4) = 5$

17. $2 \log_{10} x - \log_{10} 9 = 2$

18. Determine the balance after 20 years if $2000 is invested at 7% compounded (a) quarterly and (b) continuously.

19. Determine the principal that will yield $100,000 if it is invested at 9% compounded quarterly for 25 years.

20. A principal of $500 yields a balance of $1006.88 in 10 years when the interest is compounded continuously. What is the annual interest rate?

21. A car that cost $18,000 new has a depreciated value of $14,000 after 1 year. Use the model $y = Ce^{kt}$ to find the value of the car when it is 3 years old.

22. The population P of a certain species t years after it is introduced into a new habitat is given by

$$P(t) = \frac{2400}{1 + 3e^{-t/4}}.$$

(a) Determine the population after 4 years.

(b) After how many years will the population be 1200?

Cumulative Test: Chapters 1–9

Take this test as you would take a test in class. After you are done, check your work against the answers given in the back of the book.

In Exercises 1–3, solve the equation.

1. $4 - \frac{1}{2}x = 6$ **2.** $12(3 - x) = 5 - 7(2x + 1)$ **3.** $x^2 - 10 = 0$

In Exercises 4–6, solve the inequality.

4. $-2 \le 1 - 2x \le 2$ **5.** $|x + 7| \ge 2$ **6.** $x^2 + 3x - 10 < 0$

In Exercises 7–9, simplify the expression.

7. $(x^2 \cdot x^3)^4$ **8.** $\left(\dfrac{3x^2}{2y}\right)^{-2}$ **9.** $(9x^8)^{3/2}$

In Exercises 10 and 11, simplify the complex fraction.

10. $\dfrac{\left(\dfrac{4}{x^2 - 9} + \dfrac{2}{x - 2}\right)}{\left(\dfrac{1}{x + 3} + \dfrac{1}{x - 3}\right)}$ **11.** $\dfrac{\left(\dfrac{1}{x + 1} + \dfrac{1}{2}\right)}{\left(\dfrac{3}{2x^2 + 4x + 2}\right)}$

In Exercises 12–15, graph the function.

12. $y = \frac{1}{2}x - 3$ **13.** $y = x^2 + 3x - 8$

14. $f(x) = \dfrac{4}{x - 2}$ **15.** $f(x) = \dfrac{4x^2}{x^2 + 1}$

In Exercises 16–19, solve the equation.

16. $\sqrt{x} - x + 12 = 0$ **17.** $\sqrt{5 - x} + 10 = 11$

18. $\dfrac{1}{x} + \dfrac{4}{10 - x} = 1$ **19.** $\dfrac{x - 3}{x} + 1 = \dfrac{x - 4}{x - 6}$

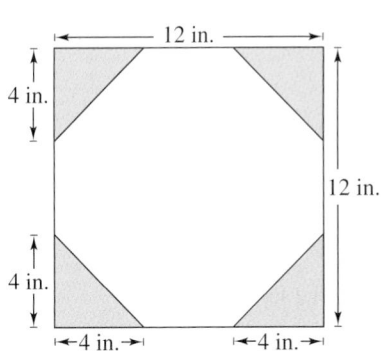

Figure for 20

20. The four corners are cut from a 12-inch-by-12-inch piece of glass, as shown in the figure. Find the perimeter of the remaining piece of glass.

In Exercises 21 and 22, identify the transformation of the graph of $f(x) = \sqrt{x}$. Use a graphing utility to verify the transformation.

21. $h(x) = \sqrt{4 + x}$ **22.** $h(x) = -\sqrt{x}$

In Exercises 23–26, perform the operation and write the result in standard form.

23. $\sqrt{-36}\sqrt{-3}$ **24.** $(7 + 4i) - (6 - 2i)$

25. $(2 + 9i)(5 - i)$ **26.** $\dfrac{1 + 2i}{2 + 3i}$

In Exercises 27 and 28, solve the system of equations using any method.

27. $\begin{cases} 4x - 3y + 2z = -2 \\ -2x + y + z = 1 \\ x - 2y - 6z = -12 \end{cases}$

28. $\begin{cases} 2x - y = 4 \\ 3x + y = -5 \end{cases}$

In Exercises 29 and 30, evaluate the determinant of the matrix.

29. $\begin{bmatrix} 10 & 25 \\ 6 & -5 \end{bmatrix}$

30. $\begin{bmatrix} 4 & 3 & 5 \\ 3 & 2 & -2 \\ 5 & -2 & 0 \end{bmatrix}$

31. Use a determinant to find the equation of the line passing through the points $(3, 5)$ and $(1, 2)$.

In Exercises 32 and 33, find (a) $(f + g)(x)$, (b) $(f - g)(x)$, (c) $(fg)(x)$, and (d) $(f/g)(x)$. What is the domain of f/g?

32. $f(x) = x - 3, \ g(x) = 4x + 1$

33. $f(x) = \sqrt{x - 1}, \ g(x) = x^2 + 1$

In Exercises 34 and 35, find (a) $f \circ g$ and (b) $g \circ f$.

34. $f(x) = 2x^2, \ g(x) = \sqrt{x + 6}$

35. $f(x) = x - 2, \ g(x) = |x|$

36. Determine whether $h(x) = 5x - 2$ has an inverse function. If so, find it.

37. The stopping distance d of a car is directly proportional to the square of its speed s. On a certain type of pavement, a car requires 50 feet to stop when its speed is 25 miles per hour. Estimate the stopping distance when the speed of the car is 40 miles per hour. Explain your reasoning.

In Exercises 38–41, graph the equation.

38. $x^2 + y^2 = 8$

39. $x^2 + 2y = 0$

40. $\dfrac{(x - 1)^2}{1} + \dfrac{(y + 2)^2}{4} = 1$

41. $\dfrac{(x + 3)^2}{4} - \dfrac{(y - 5)^2}{1} = 1$

In Exercises 42–44, rewrite the equation in exponential form.

42. $\log_4 64 = 3$

43. $\log_3 \frac{1}{81} = -4$

44. $\ln 1 = 0$

45. Use the properties of logarithms to condense $3(\log_2 x + \log_2 y) - \log_2 z$.

46. Use the properties of logarithms to expand $\ln \dfrac{5x}{(x + 1)^2}$.

47. Solve each equation.

(a) $\log_3 x = -2$ (b) $4 \ln x = 10$ (c) $3(1 + e^{2x}) = 20$

48. Determine the numbers of gallons of a 30% solution and a 60% solution that must be mixed together to obtain 20 gallons of a 40% solution.

49. After t years, the value of a car that cost $22,000 new is given by $V(t) = 22,000(0.8)^t$. Sketch a graph of the function and determine when the value of the car is $15,000.

Sequences, Series, and the Binomial Theorem

10

THE BIG PICTURE

In this chapter you will learn how to:

■ write and evaluate sequences and series using sequence and sigma notation.

■ identify, write, and evaluate sums of arithmetic sequences.

■ identify, write, and evaluate sums of geometric sequences.

■ evaluate binomial coefficients and expand binomial expression using the Binomial Theorem.

Key Terms

As you encounter each new vocabulary term in this chapter, add the term and its definition to your notebook glossary.

sequence (p. 642)
term (of a sequence) (p. 642)
infinite sequence (p. 642)
finite sequence (p. 642)
recursively (p. 644)
n factorial (p. 645)
series (p. 647)
partial sum (p. 647)
infinite series (p. 647)
sigma notation (p. 648)
index of summation (p. 648)
upper limit of summation (p. 648)

lower limit of summation (p. 648)
arithmetic sequence (p. 654)
common difference (p. 654)
recursion formula (p. 656)
geometric sequence (p. 664)
common ratio (p. 664)
infinite geometric series (p. 667)
binomial (p. 674)
binomial coefficients (p. 674)
Binomial Theorem (p. 674)
Pascal's Triangle (p. 676)

Additional text-specific resources are available to help you do well in this course.
See page xvi for details.

10.1 Sequences and Series

What you should learn:

- How to use sequence notation to write the terms of sequences
- How to write the terms of sequences involving factorials
- How to find the apparent nth term of a sequence
- How to sum the terms of sequences to obtain series and use sigma notation to represent partial sums

Why you should learn it:

Sequences and series are useful in modeling sets of values in order to identify patterns. For instance, in Exercise 122 on page 652, a sequence can be used to model the depreciation of a car.

Sequences

Suppose you were given two choices of a contract offer for the next 5 years of employment.

Contract A $20,000 the first year and a $2200 raise each year

Contract B $20,000 the first year and a 10% raise each year

Which contract offers the largest salary over the five-year period? The salaries for each contract are shown below.

Year	Contract A	Contract B
1	$20,000	$20,000
2	$22,200	$22,000
3	$24,400	$24,200
4	$26,600	$26,620
5	$28,800	$29,282
Total	$122,000	$122,102

Notice that after 5 years contract B is a better contract offer than contract A. The salaries for each contract option represent a sequence.

A mathematical **sequence** is simply an ordered list of numbers. Each number in the list is a **term** of the sequence. A sequence can have a finite number of terms or an infinite number of terms. For instance, the sequence of positive odd integers that are less than 15 is a *finite* sequence

$$1, 3, 5, 7, 9, 11, 13 \qquad \text{Finite sequence}$$

whereas the sequence of positive odd integers is an *infinite* sequence.

$$1, 3, 5, 7, 9, 11, 13, \ldots \qquad \text{Infinite sequence}$$

Note that the three dots indicate that the sequence continues and has an infinite number of terms.

Because each term of a sequence is matched with its location, a sequence can be defined as a function whose domain is a subset of positive integers.

Sequences

An **infinite sequence** $a_1, a_2, a_3, \ldots, a_n, \ldots$ is a function whose domain is the set of positive integers.

A **finite sequence** $a_1, a_2, a_3, \ldots, a_n$ is a function whose domain is the finite set $\{1, 2, 3, \ldots, n\}$.

Example 1 ■ Finding the Terms of a Sequence

Write the first six terms of the sequence whose nth term is

$$a_n = n^2 - 1.$$ Begin sequence with $n = 1$.

Solution

$$a_1 = (1)^2 - 1 = 0 \qquad a_2 = (2)^2 - 1 = 3 \qquad a_3 = (3)^2 - 1 = 8$$
$$a_4 = (4)^2 - 1 = 15 \qquad a_5 = (5)^2 - 1 = 24 \qquad a_6 = (6)^2 - 1 = 35$$

The entire sequence can be written as follows.

$$0,\ 3,\ 8,\ 15,\ 24,\ 35, \ldots, n^2 - 1, \ldots$$

NOTE   Sometimes it is convenient to begin subscripting an infinite sequence with 0 instead of 1. In such cases, the terms of the sequence are denoted by

$$a_0, a_1, a_2, a_3, a_4, a_5, \ldots, a_n, \ldots$$

Example 2 ■ Finding the Terms of a Sequence

Write the first six terms of the sequence whose nth term is

$$a_n = 3(2^n).$$ Begin sequence with $n = 0$.

Solution

$$a_0 = 3(2^0) = 3 \cdot 1 = 3 \qquad a_1 = 3(2^1) = 3 \cdot 2 = 6$$
$$a_2 = 3(2^2) = 3 \cdot 4 = 12 \qquad a_3 = 3(2^3) = 3 \cdot 8 = 24$$
$$a_4 = 3(2^4) = 3 \cdot 16 = 48 \qquad a_5 = 3(2^5) = 3 \cdot 32 = 96$$

The entire sequence can be written as follows.

$$3,\ 6,\ 12,\ 24,\ 48,\ 96, \ldots, 3(2^n), \ldots$$

Example 3 ■ A Sequence Whose Terms Alternate in Sign

Write the first six terms of the sequence whose nth term is

$$a_n = \frac{(-1)^n}{2n - 1}.$$ Begin sequence with $n = 1$.

Solution

$$a_1 = \frac{(-1)^1}{2(1) - 1} = -\frac{1}{1} \qquad a_2 = \frac{(-1)^2}{2(2) - 1} = \frac{1}{3} \qquad a_3 = \frac{(-1)^3}{2(3) - 1} = -\frac{1}{5}$$
$$a_4 = \frac{(-1)^4}{2(4) - 1} = \frac{1}{7} \qquad a_5 = \frac{(-1)^5}{2(5) - 1} = -\frac{1}{9} \qquad a_6 = \frac{(-1)^6}{2(6) - 1} = \frac{1}{11}$$

The entire sequence can be written as follows.

$$-1,\ \frac{1}{3},\ -\frac{1}{5},\ \frac{1}{7},\ -\frac{1}{9},\ \frac{1}{11}, \ldots, \frac{(-1)^n}{2n - 1}, \ldots$$

Technology

Most graphing utilities have a "sequence graphing mode" that allows you to plot the terms of a sequence as points on a rectangular coordinate system. For instance, the graph of the first six terms of the sequence given by

$$a_n = n^2 - 1$$

is shown below.

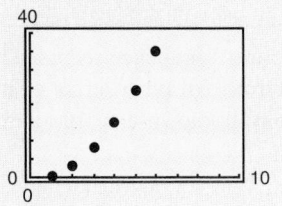

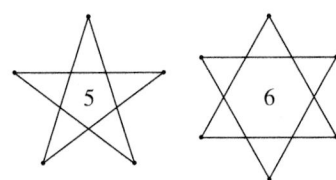

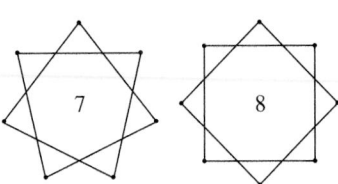

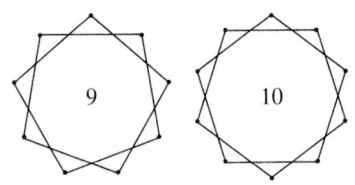

Figure 10.1

Example 4 ■ Finding the Terms of a Sequence

The number of degrees d_n in the angle at each point of each of the six n-pointed stars shown in Figure 10.1 is

$$d_n = \frac{180(n - 4)}{n}, \qquad n \geq 5.$$

Write the first six terms of the sequence.

Solution

$$d_5 = \frac{180(5 - 4)}{5} = 36° \qquad d_6 = \frac{180(6 - 4)}{6} = 60°$$

$$d_7 = \frac{180(7 - 4)}{7} \approx 77.14° \qquad d_8 = \frac{180(8 - 4)}{8} = 90°$$

$$d_9 = \frac{180(9 - 4)}{9} = 100° \qquad d_{10} = \frac{180(10 - 4)}{10} = 108°$$

Some sequences are defined **recursively.** To define a sequence recursively, you need to be given one or more of the first few terms. All other terms of the sequence are then defined using previous terms. A well-known example is the Fibonacci Sequence shown in Example 5.

Example 5 ■ A Sequence That Is Defined Recursively

The Fibonacci Sequence is defined recursively, as follows.

$$a_0 = 1, \ a_1 = 1, \ a_k = a_{k-2} + a_{k-1}, \text{ where } k \geq 2$$

Write the first five terms of the sequence.

Solution

$a_0 = 1$	0th term is given.
$a_1 = 1$	1st term is given.
$a_2 = a_0 + a_1 = 1 + 1 = 2$	Use recursion formula.
$a_3 = a_1 + a_2 = 1 + 2 = 3$	Use recursion formula.
$a_4 = a_2 + a_3 = 2 + 3 = 5$	Use recursion formula.

Example 6 ■ A Sequence That Is Defined Recursively

Write the first five terms of the sequence that is defined recursively, as follows.

$$a_1 = 6, \ a_{k+1} = 2a_k - 5$$

Solution

$a_1 = 6$	1st term is given.
$a_2 = 2a_1 - 5 = 2(6) - 5 = 7$	Use recursion formula.
$a_3 = 2a_2 - 5 = 2(7) - 5 = 9$	Use recursion formula.
$a_4 = 2a_3 - 5 = 2(9) - 5 = 13$	Use recursion formula.
$a_5 = 2a_4 - 5 = 2(13) - 5 = 21$	Use recursion formula.

Factorial Notation

Technology

Most graphing utilities have a built-in function that will display the terms of a sequence. Consult the user's guide of your graphing utility for specific keystrokes.

Some very important sequences in mathematics involve terms that are defined with special types of products called **factorials.**

Definition of Factorial

If n is a positive integer, **n factorial** is defined as

$$n! = 1 \cdot 2 \cdot 3 \cdot 4 \cdot \cdots \cdot (n - 1) \cdot n.$$

As a special case, zero factorial is defined as $0! = 1$.

The first several factorial values are as follows.

$0! = 1$	$1! = 1$
$2! = 1 \cdot 2 = 2$	$3! = 1 \cdot 2 \cdot 3 = 6$
$4! = 1 \cdot 2 \cdot 3 \cdot 4 = 24$	$5! = 1 \cdot 2 \cdot 3 \cdot 4 \cdot 5 = 120$

Many calculators have a factorial key, denoted by $\boxed{N!}$. If your calculator has such a key, try using it to evaluate $n!$ for several values of n. You will see that the value of n does not have to be very large before the value of $n!$ becomes huge. For instance

$$10! = 3,628,800.$$

Example 7 ■ A Sequence Involving Factorials

Write the first five terms of the sequence whose nth term is

$$a_n = \frac{1}{n!}. \qquad \text{Begin sequence with } n = 1.$$

Solution

$$a_1 = \frac{1}{1!} = \frac{1}{1} = 1$$

$$a_2 = \frac{1}{2!} = \frac{1}{2}$$

$$a_3 = \frac{1}{3!} = \frac{1}{1 \cdot 2 \cdot 3} = \frac{1}{6}$$

$$a_4 = \frac{1}{4!} = \frac{1}{1 \cdot 2 \cdot 3 \cdot 4} = \frac{1}{24}$$

$$a_5 = \frac{1}{5!} = \frac{1}{1 \cdot 2 \cdot 3 \cdot 4 \cdot 5} = \frac{1}{120}$$

The entire sequence can be written as follows.

$$1, \frac{1}{2}, \frac{1}{6}, \frac{1}{24}, \frac{1}{120}, \cdots, \frac{1}{n!}, \cdots$$

Example 8 ■ A Sequence Involving Factorials

Write the first six terms of the sequence with the given nth term.

$$a_n = \frac{2^n}{n!} \qquad \text{Begin sequence with } n = 0.$$

Solution

$$a_0 = \frac{2^0}{0!} = \frac{1}{1} = 1 \qquad\qquad a_1 = \frac{2^1}{1!} = \frac{2}{1} = 2$$

$$a_2 = \frac{2^2}{2!} = \frac{2 \cdot 2}{1 \cdot 2} = 2 \qquad\qquad a_3 = \frac{2^3}{3!} = \frac{2 \cdot 2 \cdot 2}{1 \cdot 2 \cdot 3} = \frac{4}{3}$$

$$a_4 = \frac{2^4}{4!} = \frac{2 \cdot 2 \cdot 2 \cdot 2}{1 \cdot 2 \cdot 3 \cdot 4} = \frac{2}{3} \qquad a_5 = \frac{2^5}{5!} = \frac{2 \cdot 2 \cdot 2 \cdot 2 \cdot 2}{1 \cdot 2 \cdot 3 \cdot 4 \cdot 5} = \frac{4}{15}$$

The entire sequence can be written as follows.

$$1, 2, 2, \frac{4}{3}, \frac{2}{3}, \frac{4}{15}, \cdot \cdot \cdot \frac{2^n}{n!}, \cdot \cdot \cdot$$

Factorials follow the same conventions for order of operations as do exponents. For instance, $2n!$ means $2(n!)$, not $(2n)!$.

Example 9 ■ A Sequence Involving Factorials

Write the first six terms of the sequence whose nth term is

$$a_n = \frac{2n!}{(2n)!}. \qquad \text{Begin sequence with } n = 0.$$

Solution

$$a_0 = \frac{2(0!)}{(2 \cdot 0)!} = \frac{2(0!)}{0!} = \frac{2}{1} = 2 \qquad a_1 = \frac{2(1!)}{(2 \cdot 1)!} = \frac{2(1!)}{2!} = \frac{2}{2} = 1$$

$$a_2 = \frac{2(2!)}{(2 \cdot 2)!} = \frac{2(2!)}{4!} = \frac{4}{24} = \frac{1}{6} \qquad a_3 = \frac{2(3!)}{(2 \cdot 3)!} = \frac{2(3!)}{6!} = \frac{12}{720} = \frac{1}{60}$$

$$a_4 = \frac{2(4!)}{(2 \cdot 4)!} = \frac{2(4!)}{8!} = \frac{48}{40{,}320} = \frac{1}{840}$$

$$a_5 = \frac{2(5!)}{(2 \cdot 5)!} = \frac{2(5!)}{10!} = \frac{240}{3{,}628{,}800} = \frac{1}{15{,}120}$$

The entire sequence can be written as follows.

$$2, 1, \frac{1}{6}, \frac{1}{60}, \frac{1}{840}, \frac{1}{15{,}120}, \cdot \cdot \cdot, \frac{2n!}{(2n)!}, \cdot \cdot \cdot$$

In Example 9, the numerators and denominators were multiplied before the fractions were reduced. When you are finding the terms of a sequence, reducing is often easier if you leave the numerator and denominator in factored form. For instance, notice how the like factors are divided out in the fraction below.

$$a_4 = \frac{2(4!)}{8!} = \frac{2 \cdot 1 \cdot 2 \cdot 3 \cdot 4}{1 \cdot 2 \cdot 3 \cdot 4 \cdot 5 \cdot 6 \cdot 7 \cdot 8} = \frac{1}{5 \cdot 3 \cdot 7 \cdot 8} = \frac{1}{840}$$

Finding the *n*th Term of a Sequence

Sometimes you will have the first several terms of a sequence and need to find a formula (the *n*th term) that will generate those terms. *Pattern recognition* is crucial in finding a form for the *n*th term.

Study Tip

Simply listing the first few terms is not sufficient to define a unique sequence—the *n*th term *must be given.* Consider the sequence

$$\frac{1}{2}, \frac{1}{4}, \frac{1}{8}, \frac{1}{15}, \cdots$$

The first three terms are identical to the first three terms of the sequence in Example 10(a). However, the *n*th term of the sequence is defined as

$$a_n = \frac{6}{(n+1)(n^2 - n + 6)}.$$

Example 10 ■ Finding the *n*th Term of a Sequence

Write an expression for the *n*th term of each sequence.

(a) $\dfrac{1}{2}, \dfrac{1}{4}, \dfrac{1}{8}, \dfrac{1}{16}, \dfrac{1}{32}, \cdots$ (b) $1, -4, 9, -16, 25, \ldots$

Solution

(a)

n:	1	2	3	4	5	. . .	*n*
Terms:	$\frac{1}{2}$	$\frac{1}{4}$	$\frac{1}{8}$	$\frac{1}{16}$	$\frac{1}{32}$	$\cdots$	a_n

Pattern: The numerator is 1 and each denominator is an increasing power of 2.

$$a_n = \frac{1}{2^n}$$

(b)

n:	1	2	3	4	5	. . .	*n*
Terms:	1	-4	9	-16	25	$\cdots$	a_n

Pattern: The terms have alternating signs, with those in the even positions being negative. Each term is the square of *n*.

$$a_n = (-1)^{n+1} n^2$$

Series

In the example involving contract offers at the beginning of this section, the terms of the finite sequence were *added*. If you add all the terms of an infinite sequence, you obtain a **series.**

Definition of a Series

For an infinite sequence

$$a_1, a_2, a_3, \ldots, a_n, \ldots$$

1. the sum of the first *n* terms

$$S_n = a_1 + a_2 + a_3 + \cdots + a_n$$

is called a **partial sum,** and

2. the sum of all the terms

$$a_1 + a_2 + a_3 + \cdots + a_n + \cdots$$

is called an **infinite series,** or simply a **series.**

Example 11 ■ Finding Partial Sums

Find the indicated partial sums for each sequence.

(a) Find S_1, S_2, and S_5 for $a_n = 3n - 1$.

(b) Find S_2, S_3, and S_4 for $a_n = \dfrac{(-1)^n}{n + 1}$.

Solution

(a) The first five terms of the sequence $a_n = 3n - 1$ are

$$a_1 = 2, a_2 = 5, a_3 = 8, a_4 = 11, \text{ and } a_5 = 14.$$

So, the partial sums are

$$S_1 = 2, S_2 = 2 + 5 = 7, \text{ and } S_5 = 2 + 5 + 8 + 11 + 14 = 40.$$

(b) The first four terms of the sequence $a_n = \dfrac{(-1)^n}{n + 1}$ are

$$a_1 = -\tfrac{1}{2}, a_2 = \tfrac{1}{3}, a_3 = -\tfrac{1}{4}, \text{ and } a_4 = \tfrac{1}{5}.$$

So, the partial sums are

$$S_2 = -\tfrac{1}{2} + \tfrac{1}{3} = -\tfrac{1}{6}, S_3 = -\tfrac{1}{2} + \tfrac{1}{3} - \tfrac{1}{4} = -\tfrac{5}{12}, \text{ and }$$
$$S_4 = -\tfrac{1}{2} + \tfrac{1}{3} - \tfrac{1}{4} + \tfrac{1}{5} = -\tfrac{13}{60}.$$

A convenient shorthand notation for denoting a partial sum is called **sigma notation.** This name comes from the use of the uppercase Greek letter sigma, which is written as Σ.

Definition of Sigma Notation

The sum of the first n terms of the sequence whose nth term is a_n is

$$\sum_{i=1}^{n} a_i = a_1 + a_2 + a_3 + a_4 + \cdots + a_n$$

where i is the **index of summation,** n is the **upper limit of summation,** and 1 is the **lower limit of summation.**

Example 12 ■ Sigma Notation for Sums

Find the sum $\displaystyle\sum_{i=1}^{6} 2i$

Solution

$$\sum_{i=1}^{6} 2i = 2(1) + 2(2) + 2(3) + 2(4) + 2(5) + 2(6)$$
$$= 2 + 4 + 6 + 8 + 10 + 12$$
$$= 42$$

NOTE In Example 12, the index of summation is i and the summation begins with $i = 1$. Any letter can be used as the index of summation, and the summation can begin with any integer. For instance, in Example 13, the index of summation is k and the summation begins with $k = 0$.

Example 13 ■ Sigma Notation for Sums

Find the sum $\displaystyle\sum_{k=0}^{8} \frac{1}{k!}$.

Solution

$$\sum_{k=0}^{8} \frac{1}{k!} = \frac{1}{0!} + \frac{1}{1!} + \frac{1}{2!} + \frac{1}{3!} + \frac{1}{4!} + \frac{1}{5!} + \frac{1}{6!} + \frac{1}{7!} + \frac{1}{8!}$$

$$= 1 + 1 + \frac{1}{2} + \frac{1}{6} + \frac{1}{24} + \frac{1}{120} + \frac{1}{720} + \frac{1}{5040} + \frac{1}{40{,}320}$$

$$\approx 2.71828$$

Note that this sum is approximately $e = 2.71828\ldots$.

Example 14 ■ Writing a Sum in Sigma Notation

Write each sum in sigma notation.

(a) $\dfrac{2}{2} + \dfrac{2}{3} + \dfrac{2}{4} + \dfrac{2}{5} + \dfrac{2}{6}$ (b) $1 - \dfrac{1}{3} + \dfrac{1}{9} - \dfrac{1}{27} + \dfrac{1}{81}$

Solution

(a) To write this sum in sigma notation, you must find a pattern for the terms. After examining the terms, you can see that they have numerators of 2 and denominators that range over the integers from 2 to 6. So, one possible sigma notation is

$$\sum_{i=1}^{5} \frac{2}{i+1} = \frac{2}{2} + \frac{2}{3} + \frac{2}{4} + \frac{2}{5} + \frac{2}{6}.$$

(b) To write this sum in sigma notation, you must find a pattern for the terms. After examining the terms, you can see that the numerators alternate in sign and the denominators are integer powers of 3, starting with 3^0 and ending with 3^4. So, one possible sigma notation is

$$\sum_{i=0}^{4} \frac{(-1)^i}{3^i} = \frac{1}{3^0} + \frac{-1}{3^1} + \frac{1}{3^2} + \frac{-1}{3^3} + \frac{1}{3^4}.$$

Collaborate!

Communicating Mathematically

You learned in this section that a sequence is an ordered list of numbers. Study the following sequence and see if you can guess what its next term should be.

    Z, O, T, T, F, F, S, S, E, N, T, E, T, . . .

Construct another sequence with letters. Can the other members of your group guess the next term?

10.1 ■ Exercises

Developing Skills

In Exercises 1–26, write the first five terms of the sequence. (Assume that n begins with 1.)

1. $a_n = 2n$ **2.** $a_n = 3n$

3. $a_n = (-1)^n 2n$ **4.** $a_n = (-1)^{n+1} 3n$

5. $a_n = \left(\frac{1}{2}\right)^n$ **6.** $a_n = \left(\frac{1}{3}\right)^n$

7. $a_n = \left(-\frac{1}{2}\right)^{n+1}$ **8.** $a_n = \left(\frac{2}{3}\right)^{n-1}$

9. $a_n = (-0.2)^{n-1}$ **10.** $a_n = \left(-\frac{2}{3}\right)^n$

11. $a_n = \dfrac{1}{n+1}$ **12.** $a_n = \dfrac{3}{2n+1}$

13. $a_n = \dfrac{2n}{3n+2}$ **14.** $a_n = \dfrac{5n}{4n+3}$

15. $a_n = \dfrac{(-1)^n}{n^2}$ **16.** $a_n = \dfrac{(-1)^{n+1}}{2n^2}$

17. $a_n = \dfrac{2 + (-1)^{n+1}}{n}$ **18.** $a_n = \dfrac{5 + (-1)^n}{4n}$

19. $a_n = 5 - \dfrac{1}{2^n}$ **20.** $a_n = 7 + \dfrac{1}{3^n}$

21. $a_n = \dfrac{2^n}{n!}$ **22.** $a_n = \dfrac{n!}{(n-1)!}$

23. $a_n = 2 + (-2)^n$ **24.** $a_n = 6 + \left(-\dfrac{1}{2}\right)^n$

25. $a_n = \dfrac{5 + (-1)^{n+1}}{2n^2}$ **26.** $a_n = \dfrac{1 + (-1)^n}{n^2}$

In Exercises 27–36, write the first six terms of the sequence defined recursively. (Assume that n begins with 1.)

27. $a_1 = 1, \ a_{n+1} = a_n + 4$

28. $a_1 = 6, \ a_{n+1} = a_n - 3$

29. $a_1 = 2, \ a_{n+1} = 2a_n + 1$

30. $a_1 = 1, \ a_{n+1} = 3 - a_n$

31. $a_1 = 4, \ a_{n+1} = \dfrac{1}{a_n} + 1$

32. $a_1 = 1, \ a_{n+1} = \dfrac{1}{a_n + 1}$

33. $a_1 = 1, \ a_{n+1} = \dfrac{1}{2}a_n + 5$

34. $a_1 = 12, \ a_{n+1} = \dfrac{4 - a_n}{2}$

35. $a_1 = -2, \ a_2 = 0, \ a_{n+2} = a_{n+1} - 2a_n$

36. $a_1 = 2, \ a_2 = -1, \ a_{n+2} = \dfrac{a_{n+1}}{a_n}$

In Exercises 37–42, find the indicated term of the sequence.

37. $a_n = (-1)^n(5n - 3)$ **38.** $a_n = (-1)^{n+1}(9 - n)$

    $a_{15} = \ \blacksquare$ $a_{10} = \ \blacksquare$

39. $a_n = \dfrac{n}{5 + \sqrt{n}}$ **40.** $a_n = \dfrac{4 - n}{3\sqrt{n}}$

    $a_{25} = \ \blacksquare$ $a_{36} = \ \blacksquare$

41. $a_n = \dfrac{n!}{(n-2)!}$ **42.** $a_n = \dfrac{n^2}{n!}$

    $a_8 = \ \blacksquare$ $a_{12} = \ \blacksquare$

In Exercises 43–54, simplify the expression.

43. $\dfrac{5!}{4!}$ **44.** $\dfrac{18!}{17!}$

45. $\dfrac{10!}{12!}$ **46.** $\dfrac{5!}{8!}$

47. $\dfrac{25!}{27!}$ **48.** $\dfrac{20!}{15! \cdot 5!}$

49. $\dfrac{n!}{(n+1)!}$ **50.** $\dfrac{(n+2)!}{n!}$

51. $\dfrac{(n+1)!}{(n-1)!}$ **52.** $\dfrac{(3n)!}{(3n+2)!}$

53. $\dfrac{(2n)!}{(2n-1)!}$ **54.** $\dfrac{(2n+2)!}{(2n)!}$

In Exercises 55–62, write an expression for the nth term of the sequence. (Assume that n begins with 1.) There are many correct answers.

55. $1, -1, 1, -1, 1, \ldots$

56. $-2, 2, -2, 2, -2, \ldots$

57. $1, 4, 9, 16, 25, \ldots$ **58.** $-1, 2, 7, 14, 23, \ldots$

59. $4, 2, \dfrac{4}{3}, 1, \dfrac{4}{5}, \ldots$

60. $1, -\dfrac{1}{2}, \dfrac{1}{3}, -\dfrac{1}{4}, \dfrac{1}{5}, \ldots$

61. $1, 3, 5, 7, 9, \ldots$

62. $-2, 0, 2, 4, 6, \ldots$

In Exercises 63–82, find the partial sum.

63. $\displaystyle\sum_{k=1}^{6} 3k$

64. $\displaystyle\sum_{k=1}^{4} 5k$

65. $\displaystyle\sum_{i=0}^{6} (2i + 5)$

66. $\displaystyle\sum_{j=3}^{7} (6j - 10)$

67. $\displaystyle\sum_{i=0}^{4} (3i + 2)$

68. $\displaystyle\sum_{i=2}^{7} (4i - 1)$

69. $\displaystyle\sum_{i=0}^{5} (i^2 + 3)$

70. $\displaystyle\sum_{i=1}^{5} (25 - i^2)$

71. $\displaystyle\sum_{i=0}^{4} 2^i$

72. $\displaystyle\sum_{i=0}^{4} (1 + 3^i)$

73. $\displaystyle\sum_{j=1}^{5} \dfrac{(-1)^{j+1}}{j}$

74. $\displaystyle\sum_{j=0}^{3} \dfrac{1}{j^2 + 1}$

75. $\displaystyle\sum_{m=2}^{6} \dfrac{2m}{2(m - 1)}$

76. $\displaystyle\sum_{k=1}^{5} \dfrac{10k}{k + 2}$

77. $\displaystyle\sum_{k=1}^{6} (-8)$

78. $\displaystyle\sum_{n=3}^{12} 10$

79. $\displaystyle\sum_{i=1}^{8} \left(\dfrac{1}{i} - \dfrac{1}{i + 1} \right)$

80. $\displaystyle\sum_{k=1}^{5} \left(\dfrac{2}{k} - \dfrac{2}{k + 2} \right)$

81. $\displaystyle\sum_{n=0}^{5} \left(-\dfrac{1}{3} \right)^n$

82. $\displaystyle\sum_{n=0}^{6} \left(\dfrac{3}{2} \right)^n$

In Exercises 83–88, use a graphing utility to find the partial sum.

83. $\displaystyle\sum_{n=1}^{6} n(n + 1)$

84. $\displaystyle\sum_{n=0}^{5} 2n^2$

85. $\displaystyle\sum_{j=2}^{6} (j! - j)$

86. $\displaystyle\sum_{j=0}^{4} \dfrac{6}{j!}$

87. $\displaystyle\sum_{k=1}^{6} \ln k$

88. $\displaystyle\sum_{k=2}^{4} \dfrac{k}{\ln k}$

In Exercises 89–106, write the sum using sigma notation. (Begin with $k = 0$ or $k = 1$.) There are many correct answers.

89. $1 + 2 + 3 + 4 + 5$

90. $8 + 9 + 10 + 11 + 12 + 13 + 14$

91. $2 + 4 + 6 + 8 + 10$

92. $24 + 30 + 36 + 42$

93. $\dfrac{1}{2(1)} + \dfrac{1}{2(2)} + \dfrac{1}{2(3)} + \dfrac{1}{2(4)} + \cdots + \dfrac{1}{2(10)}$

94. $\dfrac{3}{1 + 1} + \dfrac{3}{1 + 2} + \dfrac{3}{1 + 3} + \dfrac{3}{1 + 4} + \cdots + \dfrac{3}{1 + 50}$

95. $\dfrac{1}{1^2} + \dfrac{1}{2^2} + \dfrac{1}{3^2} + \dfrac{1}{4^2} + \cdots + \dfrac{1}{20^2}$

96. $\dfrac{1}{2^0} + \dfrac{1}{2^1} + \dfrac{1}{2^2} + \dfrac{1}{2^3} + \cdots + \dfrac{1}{2^{12}}$

97. $\dfrac{1}{3^0} - \dfrac{1}{3^1} + \dfrac{1}{3^2} - \dfrac{1}{3^3} + \cdots - \dfrac{1}{3^9}$

98. $\left(-\dfrac{2}{3} \right)^0 + \left(-\dfrac{2}{3} \right)^1 + \left(-\dfrac{2}{3} \right)^2 + \cdots + \left(-\dfrac{2}{3} \right)^{20}$

99. $\dfrac{4}{1 + 3} + \dfrac{4}{2 + 3} + \dfrac{4}{3 + 3} + \cdots + \dfrac{4}{20 + 3}$

100. $\dfrac{1}{2^3} - \dfrac{1}{4^3} + \dfrac{1}{6^3} - \dfrac{1}{8^3} + \cdots + \dfrac{1}{14^3}$

101. $\dfrac{1}{2} + \dfrac{2}{3} + \dfrac{3}{4} + \dfrac{4}{5} + \dfrac{5}{6} + \cdots + \dfrac{11}{12}$

102. $\dfrac{2}{4} + \dfrac{4}{7} + \dfrac{6}{10} + \dfrac{8}{13} + \dfrac{10}{16} + \cdots + \dfrac{20}{31}$

103. $\dfrac{2}{4} + \dfrac{4}{5} + \dfrac{6}{6} + \dfrac{8}{7} + \cdots + \dfrac{40}{23}$

104. $\left(2 + \dfrac{1}{1} \right) + \left(2 + \dfrac{1}{2} \right) + \left(2 + \dfrac{1}{3} \right) + \cdots + \left(2 + \dfrac{1}{25} \right)$

105. $1 + 1 + 2 + 6 + 24 + 120 + 720$

106. $1 + 1 + \dfrac{1}{2} + \dfrac{1}{6} + \dfrac{1}{24} + \dfrac{1}{120} + \dfrac{1}{720}$

In Exercises 107–110, match the sequence with the graph of its first 10 terms. [The graphs are labeled (a), (b), (c), and (d).]

(a)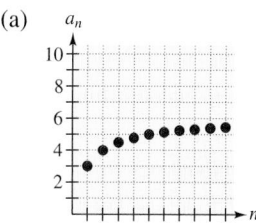

(b)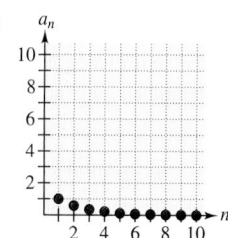

(c)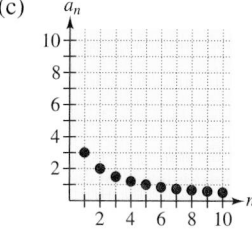

(d)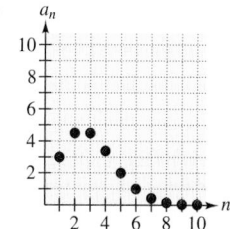

107. $a_n = \dfrac{6}{n + 1}$

108. $a_n = \dfrac{6n}{n + 1}$

109. $a_n = (0.6)^{n - 1}$

110. $a_n = \dfrac{3^n}{n!}$

The symbol ▦ indicates an exercise in which you are instructed to use a calculator or graphing utility.

In Exercises 111–116, use a graphing utility to graph the first 10 terms of the sequence.

111. $a_n = (-0.8)^{n-1}$

112. $a_n = \dfrac{2n^2}{n^2 + 1}$

113. $a_n = \dfrac{1}{2}n$

114. $a_n = \dfrac{n + 2}{n}$

115. $a_n = 3 - \dfrac{4}{n}$

116. $a_n = 10\left(\dfrac{3}{4}\right)^{n-1}$

Arithmetic Mean In Exercises 117–120, find the arithmetic mean of the set. The *arithmetic mean* of a set of n measurements $x_1, x_2, x_3, \ldots, x_n$ is

$$\bar{x} = \frac{1}{n}\sum_{i=1}^{n} x_i.$$

117. 3, 7, 2, 1, 5

118. 84, 69, 66, 96

119. 0.5, 0.8, 1.1, 0.8, 0.7, 0.7, 1.0

120. -1.0, 4.2, 5.4, -3.2, 3.6

Solving Problems

121. *Compound Interest* A deposit of \$500 is made in an account that earns 7% interest compounded yearly. The balance in the account after N years is given by

$$A_N = 500(1 + 0.07)^N, \quad N = 1, 2, 3, \ldots.$$

(a) Compute the first eight terms of the sequence.

(b) Find the balance in the account after 40 years by computing A_{40}.

(c) Use a graphing utility to graph the first 40 terms of the sequence.

(d) The terms of the sequence are increasing. Is the rate of growth of the terms increasing? Explain.

122. *Depreciation* At the end of each year, the value of a car with an initial cost of \$16,000 is three-fourths what it was at the beginning of the year. So, after n years, its value is given by

$$a_n = 16{,}000\left(\tfrac{3}{4}\right)^n, \quad n = 1, 2, 3, \ldots.$$

(a) Find the value of the car 3 years after it was purchased by computing a_3.

(b) Find the value of the car 6 years after it was purchased by computing a_6. Is this value half of what it was after 3 years?

123. *Soccer Ball* The number of degrees a_n in each angle of a regular n-sided polygon is

$$a_n = \frac{180(n - 2)}{n}, \quad n \geq 3.$$

The surface of a soccer ball is made of regular hexagons and pentagons. If the ball is taken apart and flattened, as shown in the figure, the sides don't meet each other. Use the terms a_5 and a_6 to explain why there are gaps between adjacent hexagons.

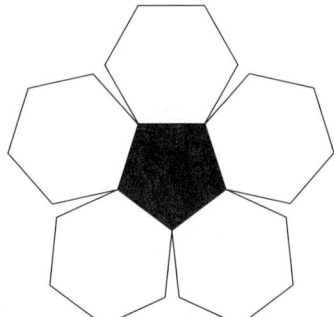

Explaining Concepts

124. Give an example of an infinite sequence.

125. The nth term of a sequence is $a_n = (-1)^n n$. Which terms of the sequence are negative? Explain your answer.

In Exercises 126–128, decide whether the statement is true. Explain your reasoning.

126. $\displaystyle\sum_{i=1}^{4} (i^2 + 2i) = \sum_{i=1}^{4} i^2 + \sum_{i=1}^{4} 2i$

127. $\displaystyle\sum_{k=1}^{4} 3k = 3\sum_{k=1}^{4} k$

128. $\displaystyle\sum_{j=1}^{4} 2^j = \sum_{j=3}^{6} 2^{j-2}$

Ongoing Review

In Exercises 129–132, perform the indicated operations and simplify.

129. $\dfrac{1}{2} \cdot \dfrac{2}{3} \cdot \dfrac{3}{4}$

130. $3 - \dfrac{3}{2} + 1 - \dfrac{3}{4} + \dfrac{3}{5}$

131. $\dfrac{1}{n} + \dfrac{1}{n+1}$

132. $\dfrac{n+1}{n^2} + \dfrac{n+2}{2n^2}$

In Exercises 133–138, simplify the expression.

133. $(x + 10)^{-2}$

134. $\dfrac{18(x-3)^5}{(x-3)^2}$

135. $(a^2)^{-4}$

136. $(8x^3)^{1/3}$

137. $\sqrt{128x^3}$

138. $\dfrac{5}{\sqrt{x} - 2}$

139. *Depreciation*   After t years, the value of a car that originally cost \$22,000 is given by

$$V(t) = 22,000(0.8)^t.$$

Sketch a graph of the function and determine when the value of the car is \$15,000.

140. *Radioactive Decay*   Carbon 14 has a half-life of 5730 years. If you start with 10 grams of this isotope, how much remains after 3000 years?

Looking Further

Stars   The stars in Example 4 were formed by placing n equally spaced points on a circle and connecting each point with the second point from it on the circle. The stars in the figure below are formed in a similar way except that each point is connected with the third point from it. For these stars, the number of degrees in each point is given by

$$d_n = \frac{180(n-6)}{n}, \qquad n \geq 7.$$

(a) Write the first five terms of the sequence.

(b) If you form the stars by connecting each point with the fourth point from it, you obtain stars with the following numbers of points and degrees: 9 points, $20°$; 10 points, $36°$; 11 points, $49\frac{1}{11}°$; 12 points, $60°$. Find a formula for the number of degrees in each point of an n-pointed star.

10.2 Arithmetic Sequences

What you should learn:

- How to recognize, write, and find the nth terms of arithmetic sequences
- How to find the nth partial sum of an arithmetic sequence
- How to use arithmetic sequences to solve application problems

Why you should learn it:

An arithmetic sequence can reduce the amount of time it takes to find the sum of a sequence of numbers with a common difference. For example, in Exercise 115 on page 661 you will use an arithmetic sequence to find the number of times a clock chimes in a 12-hour period.

Arithmetic Sequences

A sequence whose consecutive terms have a common difference is called an **arithmetic sequence.**

Definition of an Arithmetic Sequence

A sequence is called **arithmetic** if the differences between consecutive terms are the same. So, the sequence

$$a_1, a_2, a_3, a_4, \ldots, a_n, \ldots$$

is arithmetic if there is a number d such that

$$a_2 - a_1 = d, \quad a_3 - a_2 = d, \quad a_4 - a_3 = d$$

and so on. The number d is the **common difference** of the sequence.

Example 1 ■ Examples of Arithmetic Sequences

(a) The sequence whose nth term is $3n + 2$ is arithmetic. For this sequence, the common difference between consecutive terms is 3.

$$5, 8, 11, 14, \ldots, 3n + 2, \ldots \qquad \text{Begin with } n = 1.$$
$$8 - 5 = 3$$

(b) The sequence whose nth term is $7 - 5n$ is arithmetic. For this sequence, the common difference between consecutive terms is -5.

$$2, -3, -8, -13, \ldots, 7 - 5n, \ldots \qquad \text{Begin with } n = 1.$$
$$-3 - 2 = -5$$

(c) The sequence whose nth term is $\frac{1}{4}(n + 3)$ is arithmetic. For this sequence, the common difference between consecutive terms is $\frac{1}{4}$.

$$1, \frac{5}{4}, \frac{3}{2}, \frac{7}{4}, \ldots, \frac{1}{4}(n + 3), \ldots \qquad \text{Begin with } n = 1.$$
$$\tfrac{5}{4} - 1 = \tfrac{1}{4}$$

The sequence $1, 4, 9, 16, \ldots$, whose nth term is n^2, is *not* arithmetic. The difference between the first two terms is

$$a_2 - a_1 = 4 - 1 = 3$$

but the difference between the second and third terms is

$$a_3 - a_2 = 9 - 4 = 5.$$

> **The nth Term of an Arithmetic Sequence**
>
> The nth term of an arithmetic sequence has the form
>
> $$a_n = a_1 + (n - 1)d$$
>
> where d is the common difference between the terms of the sequence, and a_1 is the first term.

Example 2 ■ Finding the nth Term of an Arithmetic Sequence

Find a formula for the nth term of the arithmetic sequence whose common difference is 2 and whose first term is 5.

Solution

You know that the formula for the nth term is of the form $a_n = a_1 + (n - 1)d$. Moreover, because the common difference is $d = 2$, and the first term is $a_1 = 5$, the formula must have the form

$$a_n = 5 + 2(n - 1).$$

So, the formula for the nth term is

$$a_n = 2n + 3.$$

The sequence therefore has the following form.

$$5, 7, 9, 11, 13, \ldots, 2n + 3, \ldots$$

Example 3 ■ Finding the nth Term of an Arithmetic Sequence

Find a formula for the nth term of the arithmetic sequence whose common difference is 6 and whose *second* term is 11.

Solution

You know that the formula for the nth term is of the form $a_n = a_1 + (n - 1)d$. Moreover, because the common difference is $d = 6$, the formula must have the form

$$a_n = a_1 + 6(n - 1).$$

Because the second term is

$$a_2 = 11 = a_1 + 6(2 - 1)$$
$$= a_1 + 6$$

it follows that $a_1 = 5$. So, the formula for the nth term is

$$a_n = 5 + 6(n - 1)$$
$$= 6n - 1.$$

The sequence therefore has the following form.

$$5, 11, 17, 23, 29, \ldots, 6n - 1, \ldots$$

If you know the nth term and the common difference of an arithmetic sequence, you can find the $(n + 1)$th term by using the **recursion formula**

$$a_{n+1} = a_n + d.$$

Example 4 ■ Using a Recursion Formula

The 12th term of an arithmetic sequence is 52 and the common difference is 3.

(a) What is the 13th term of the sequence?

(b) What is the first term?

Solution

(a) You know that $a_{12} = 52$ and $d = 3$. So, using the recursion formula $a_{13} = a_{12} + d$, you can determine that the 13th term of the sequence is

$$a_{13} = 52 + 3$$
$$= 55.$$

(b) Using $n = 12$, $d = 3$, and $a_{12} = 52$ in the formula $a_n = a_1 + (n - 1)d$ yields

$$52 = a_1 + (12 - 1)(3)$$
$$52 = a_1 + 33$$
$$19 = a_1.$$

Historical Note

Carl Friedrich Gauss (1777–1855) was asked by a teacher to add all the integers from 1 to 100. When Gauss returned with the correct answer after only a few moments, the teacher could only look at him in astounded silence. This is what Gauss did:

$$1 + 2 + 3 + \cdots + 100$$
$$100 + 99 + 98 + \cdots + 1$$
$$\overline{101 + 101 + 101 + \cdots + 101}$$
$$\frac{100 \times 101}{2} = 5050$$

The Partial Sum of an Arithmetic Sequence

The sum of the first n terms of an arithmetic sequence is called the **nth partial sum** of the sequence. For instance, the fifth partial sum of the arithmetic sequence whose nth term is $3n + 4$ is

$$\sum_{i=1}^{5} (3i + 4) = 7 + 10 + 13 + 16 + 19$$
$$= 65.$$

A formula for the nth partial sum of an arithmetic sequence is given below.

Study Tip

You can use the formula for the nth partial sum of an arithmetic sequence to find the sum of consecutive numbers. For instance, the sum of the integers from 1 to 100 is

$$\sum_{i=1}^{100} i = \frac{100}{2}(1 + 100)$$
$$= 50(101)$$
$$= 5050.$$

> **The nth Partial Sum of an Arithmetic Sequence**
>
> The nth partial sum of the arithmetic sequence whose nth term is a_n is
>
> $$\sum_{i=1}^{n} a_i = a_1 + a_2 + a_3 + a_4 + \cdots + a_n$$
> $$= \frac{n}{2}(a_1 + a_n).$$
>
> In other words, to find the sum of the first n terms of an arithmetic sequence, find the average of the first and nth terms, and multiply by n.

Example 5 ■ Finding the *n*th Partial Sum

Find the sum of the first 20 terms of the arithmetic sequence whose *n*th term is $4n + 1$.

Solution

The first term of this sequence is $a_1 = 4(1) + 1 = 5$ and the 20th term is $a_{20} = 4(20) + 1 = 81$. So, the sum of the first 20 terms is given by

$$\sum_{i=1}^{n} a_i = \frac{n}{2}(a_1 + a_n)$$ *n*th partial sum formula

$$\sum_{i=1}^{20} (4i + 1) = \frac{20}{2}(a_1 + a_{20})$$ Substitute 20 for *n*.

$$= 10(5 + 81)$$ Substitute 5 for a_1 and 81 for a_{20}.

$$= 10(86)$$ Simplify.

$$= 860.$$ *n*th partial sum

Example 6 ■ Finding the *n*th Partial Sum

Find the following sum.

$$7 + 10 + 13 + 16 + 19 + 22 + 25 + 28 + 31 + 34 + 37 + 40 + 43$$

Solution

One way to find this sum is simply to add all of the numbers. However, by recognizing that the numbers form an arithmetic sequence that has 13 terms, you can find the sum using the formula for the *n*th partial sum of an arithmetic sequence.

$$\text{Sum} = \frac{13}{2}(7 + 43) = \frac{13}{2}(50) = 13(25) = 325$$

Check this result on your calculator by actually adding the 13 terms.

Example 7 ■ Finding the *n*th Partial Sum

Find the sum of the even integers from 2 to 100.

Solution

Because the integers

$$2, 4, 6, 8, \ldots, 100$$

form an arithmetic sequence, you can find the sum as follows.

$$\sum_{i=1}^{n} a_i = \frac{n}{2}(a_1 + a_n)$$ *n*th partial sum formula

$$\sum_{i=1}^{50} 2i = \frac{50}{2}(a_1 + a_{50})$$ Substitute 50 for *n*.

$$= 25(2 + 100)$$ Substitute 2 for a_1 and 100 for a_{50}.

$$= 25(102)$$ Simplify.

$$= 2550$$ *n*th partial sum

Applications

Example 8 ■ Total Sales

Your business sells $100,000 worth of products during its first year. You have a goal of increasing annual sales by $25,000 each year for 9 years. If you meet this goal, how much will you sell during your first 10 years of business?

Solution

The annual sales during the first 10 years form an arithmetic sequence, as follows.

$100,000, $125,000, $150,000, $175,000, $200,000,

$225,000, $250,000, $275,000, $300,000, $325,000

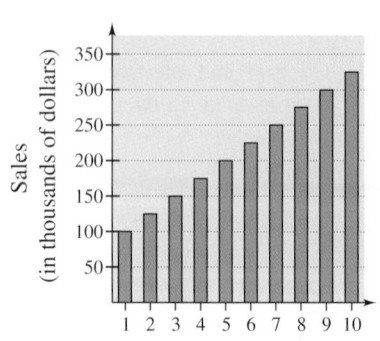

Figure 10.2

Using the formula for the nth partial sum of an arithmetic sequence, you can find the total sales during the first 10 years as follows.

$$\text{Total sales} = \frac{n}{2}(a_1 + a_n) \qquad \text{\textit{n}th partial sum formula}$$

$$= \frac{10}{2}(100{,}000 + 325{,}000) \qquad \text{Substitute for } n, a_1, \text{ and } a_n.$$

$$= 5(425{,}000) \qquad \text{Simplify.}$$

$$= \$2{,}125{,}000 \qquad \text{Simplify.}$$

From the bar graph shown in Figure 10.2, notice that the annual sales for this company follow a *linear growth* pattern. In other words, saying that a quantity increases arithmetically is the same as saying that it increases linearly.

Example 9 ■ Seating Capacity

You are organizing a concert at Red Rocks Park. How much should you charge per ticket in order to receive $50,000 in ticket sales for a sold-out performance?

Solution

From the information given in the caption of the photo, you know that there are an estimated 3320 seats in the last 24 rows. To approximate the number of seats in the first 45 rows, you can use the formula for the nth partial sum of an arithmetic sequence.

Red Rocks Park is an open-air amphitheater carved out of rock near Denver, Colorado. The amphitheater has 69 rows of seats. Rows 46 through 69 have seats for an estimated 3320 people. The number of seats in the first 45 rows can be modeled by the arithmetic sequence whose nth term is $87\frac{1}{2} + \frac{3}{2}n$.

$$\sum_{n=1}^{45}\left(87\frac{1}{2} + \frac{3}{2}n\right) = 89 + 90\frac{1}{2} + 92 + 93\frac{1}{2} + \cdots + 155$$

$$= \frac{45}{2}(89 + 155)$$

$$= 5490$$

So, the total number of seats is about $3320 + 5490 = 8810$. To bring in $50,000, you should charge about

$$\frac{1}{8810}(50{,}000) \approx \$5.68 \text{ per ticket.}$$

Collaborate!

Extending the Concept

6	1	8
7	5	3
2	9	4

A *magic square* is a square table of positive integers in which each row, column, and diagonal adds up to the same number. One example is shown at the left. In addition, the values in the middle row, in the middle column, and along both diagonals form arithmetic sequences. See if you can complete each magic square.

a.
	11	14
	10	
		15

b.
8		
	9	
	13	

c.
		20
	13	
6		

10.2 ■ Exercises

Developing Skills

In Exercises 1–12, find the common difference of the arithmetic sequence.

1. $2, 5, 8, 11, \ldots$

2. $-8, 0, 8, 16, \ldots$

3. $100, 94, 88, 82, \ldots$

4. $3200, 2800, 2400, 2000, \ldots$

5. $-5, -2, 1, 4, \ldots$

6. $-12, -7, -2, 3, \ldots$

7. $5, 1, -3, -7, \ldots$

8. $8, -1, -10, -19, \ldots$

9. $1, \frac{5}{3}, \frac{7}{3}, 3, \ldots$

10. $\frac{1}{2}, \frac{5}{4}, 2, \frac{11}{4}, \ldots$

11. $\frac{7}{2}, \frac{9}{4}, 1, -\frac{1}{4}, -\frac{3}{2}, \ldots$

12. $\frac{5}{2}, \frac{11}{6}, \frac{7}{6}, \frac{1}{2}, -\frac{1}{6}, \ldots$

In Exercises 13–28, determine whether the sequence is arithmetic. If so, find the common difference.

13. $2, 4, 6, 8, \ldots$

14. $1, 2, 4, 8, 16, \ldots$

15. $32, 16, 0, -16, \ldots$

16. $32, 16, 8, 4, \ldots$

17. $1789, 1732, 1675, 1618, \ldots$

18. $-184, 181, 546, 911, \ldots$

19. $-504, -213, 76, 363, \ldots$

20. $407, 526, 645, 664, \ldots$

21. $3.2, 4, 4.8, 5.6, \ldots$

22. $8, 4, 2, 1, 0.5, 0.25, \ldots$

23. $2, \frac{7}{2}, 5, \frac{13}{2}, \ldots$

24. $3, \frac{5}{2}, 2, \frac{3}{2}, 1, \ldots$

25. $\frac{1}{3}, \frac{2}{3}, \frac{4}{3}, \frac{8}{3}, \frac{16}{3}, \ldots$

26. $\frac{9}{4}, 2, \frac{7}{4}, \frac{3}{2}, \frac{5}{4}, \ldots$

27. $\ln 4, \ln 8, \ln 12, \ln 16, \ldots$

28. $e, e^2, e^3, e^4, \ldots$

In Exercises 29–40, write the first five terms of the arithmetic sequence. (Assume that n begins with 1.)

29. $a_n = 3n + 4$

30. $a_n = 5n - 4$

31. $a_n = -2n + 8$

32. $a_n = -10n + 100$

33. $a_n = \frac{5}{2}n - 1$

34. $a_n = \frac{2}{3}n + 2$

35. $a_n = \frac{3}{5}n + 1$

36. $a_n = \frac{3}{4}n - 2$

37. $a_n = 3(n + 3) - 1$

38. $a_n = -\frac{1}{2}(n - 2) + 10$

39. $a_n = -\frac{1}{4}(n - 1) + 4$

40. $a_n = 4(n + 2) + 24$

In Exercises 41–48, write the first five terms of the arithmetic sequence defined recursively.

41. $a_1 = 25$
$a_{k+1} = a_k + 3$

42. $a_1 = 12$
$a_{k+1} = a_k - 6$

43. $a_1 = 9$
$a_{k+1} = a_k - 3$

44. $a_1 = 8$
$a_{k+1} = a_k + 7$

45. $a_1 = -10$
$a_{k+1} = a_k + 6$

46. $a_1 = -20$
$a_{k+1} = a_k - 4$

47. $a_1 = 100$
$a_{k+1} = a_k - 20$

48. $a_1 = 4.2$
$a_{k+1} = a_k + 0.4$

In Exercises 49–54, match the sequence with its graph. [The graphs are labeled (a), (b), (c), (d), (e), and (f).]

(a)
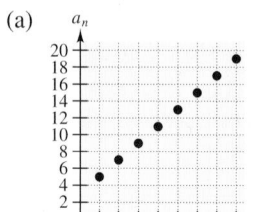

(b)
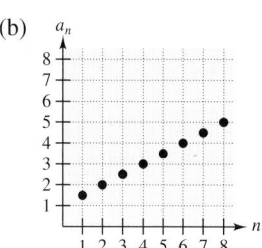

(c)
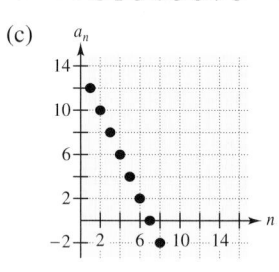

(d)
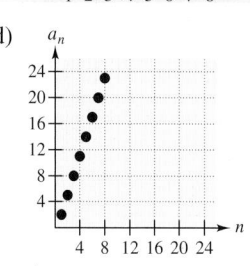

(e)
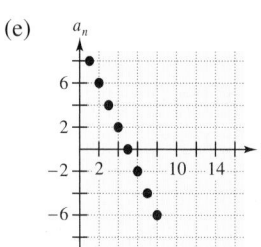

(f)
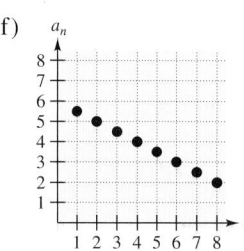

49. $a_n = \frac{1}{2}n + 1$ **50.** $a_n = -\frac{1}{2}n + 6$

51. $a_n = -2n + 10$ **52.** $a_n = 2n + 3$

53. $a_1 = 12$ **54.** $a_1 = 2$
     $a_{n+1} = a_n - 2$ $a_{n+1} = a_n + 3$

In Exercises 55–60, use a graphing utility to graph the first 10 terms of the sequence.

55. $a_n = -2n + 21$ **56.** $a_n = \frac{3}{2}n + 1$

57. $a_n = \frac{3}{5}n + \frac{3}{2}$ **58.** $a_n = -25n + 500$

59. $a_n = 2.5n - 8$ **60.** $a_n = 6.2n + 3$

In Exercises 61–78, find a formula for the nth term of the arithmetic sequence.

61. $a_1 = 3$, $d = \frac{1}{2}$

62. $a_1 = -1$, $d = 1.2$

63. $a_1 = 1000$, $d = -25$

64. $a_1 = 64$, $d = -8$

65. $a_1 = 3$, $d = \frac{3}{2}$ **66.** $a_1 = 12$, $d = -3$

67. $a_3 = 20$, $d = -4$ **68.** $a_6 = 5$, $d = \frac{3}{2}$

69. $a_1 = 5$, $a_5 = 15$ **70.** $a_2 = 93$, $a_6 = 65$

71. $a_3 = 16$, $a_4 = 20$ **72.** $a_5 = 30$, $a_4 = 25$

73. $a_1 = 50$, $a_3 = 30$ **74.** $a_{10} = 32$, $a_{12} = 48$

75. $a_2 = 10$, $a_6 = 8$ **76.** $a_7 = 8$, $a_{13} = 6$

77. $a_1 = 0.35$, $a_2 = 0.30$

78. $a_1 = 0.08$, $a_2 = 0.082$

In Exercises 79–88, find the partial sum.

79. $\displaystyle\sum_{k=1}^{20} k$ **80.** $\displaystyle\sum_{k=1}^{30} 4k$

81. $\displaystyle\sum_{k=1}^{10} 5k$ **82.** $\displaystyle\sum_{n=1}^{30} \left(\tfrac{1}{2}n + 2\right)$

83. $\displaystyle\sum_{k=1}^{50} (k + 3)$ **84.** $\displaystyle\sum_{k=1}^{100} (4k - 1)$

85. $\displaystyle\sum_{n=1}^{500} \frac{n}{2}$ **86.** $\displaystyle\sum_{n=1}^{600} \frac{2n}{3}$

87. $\displaystyle\sum_{n=1}^{30} \left(\tfrac{1}{3}n - 4\right)$ **88.** $\displaystyle\sum_{n=1}^{75} (0.3n + 5)$

In Exercises 89–94, use a graphing utility to find the partial sum.

89. $\displaystyle\sum_{j=1}^{20} \left(750 - \tfrac{1}{2}j\right)$ **90.** $\displaystyle\sum_{i=1}^{60} \left(300 - \tfrac{8}{3}i\right)$

91. $\displaystyle\sum_{n=1}^{40} (1000 - 25n)$ **92.** $\displaystyle\sum_{n=1}^{20} (500 - 10n)$

93. $\displaystyle\sum_{n=1}^{50} (2.15n + 5.4)$ **94.** $\displaystyle\sum_{n=1}^{60} (200 - 3.4n)$

In Exercises 95–106, find the nth partial sum of the arithmetic sequence.

95. $5, 12, 19, 26, 33, \ldots, n = 12$

96. $2, 12, 22, 32, 42, \ldots, n = 20$

97. $2, 8, 14, 20, \ldots, n = 25$

98. $500, 480, 460, 440, \ldots, n = 20$

99. $200, 175, 150, 125, 100, \ldots, n = 8$

100. $800, 785, 770, 755, 740, \ldots, n = 25$

101. $-50, -38, -26, -14, -2, \ldots, n = 50$

102. $-16, -8, 0, 8, 16, \ldots, n = 30$

103. $1, 4.5, 8, 11.5, 15, \ldots, n = 12$

104. $2.2, 2.8, 3.4, 4.0, 4.6, \ldots, n = 12$

105. $0.5, 0.9, 1.3, 1.7, \ldots, n = 10$

106. $1.5, 2.6, 3.7, 4.8, \ldots, n = 10$

Solving Problems

107. Find the sum of the first 75 positive integers.

108. Find the sum of the integers from 35 to 100.

109. Find the sum of the first 50 positive even integers.

110. Find the sum of the first 100 positive odd integers.

111. *Salary*   In your new job you are told that your starting salary will be $36,000 with an increase of $2000 at the end of each of the first 5 years. How much will you be paid through the end of your first 6 years of employment with the company?

112. *Salary*   Suppose that you receive 25 cents the first day of the month, 50 cents the second day, 75 cents the third day, and so on. Determine the total amount that you will receive during a 30-day month.

113. *Ticket Prices*   There are 20 rows of seats on the main floor of a concert hall: 20 seats in the first row, 21 seats in the second row, 22 seats in the third row, and so on (see figure). How much should you charge per ticket in order to obtain $15,000 for the sale of all of the seats on the main floor?

— 22 seats
— 21 seats
— 20 seats

114. *Pile of Logs*   Logs are stacked in a pile as shown in the figure. The top row has 15 logs and the bottom row has 21 logs. How many logs are in the pile?

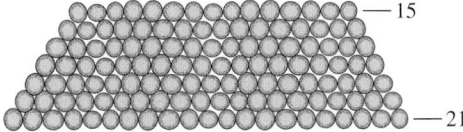

— 15
— 21

115. *Clock Chimes*   A clock chimes once at 1:00, twice at 2:00, three times at 3:00, and so on. The clock also chimes once at 15-minute intervals that are not on the hour. How many times does the clock chime in a 12-hour period?

116. *Clock Chimes*   A clock chimes once at 1:00, twice at 2:00, three times at 3:00, and so on. The clock

also chimes once on the half-hour. How many times does the clock chime in a 12-hour period?

117. *Baling Hay*   In the first two trips baling hay around a large field (see figure), a farmer obtains 93 bales and 89 bales, respectively. The farmer estimates that the same pattern will continue. Estimate the total number of bales made if there are another six trips around the field.

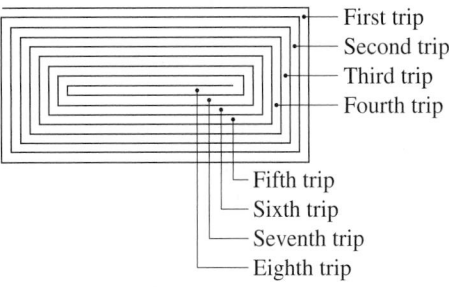

— First trip
— Second trip
— Third trip
— Fourth trip
— Fifth trip
— Sixth trip
— Seventh trip
— Eighth trip

118. *Baling Hay*   In the first two trips baling hay around a field (see figure), a farmer obtains 64 bales and 60 bales, respectively. The farmer estimates that the same pattern will continue. Estimate the total number of bales made if there are another four trips around the field.

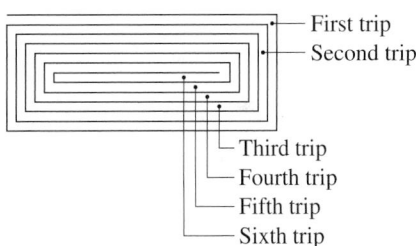

— First trip
— Second trip
— Third trip
— Fourth trip
— Fifth trip
— Sixth trip

119. *Free-Falling Object*   A free-falling object will fall 16 feet during the first second, 48 more feet during the second second, 80 more feet during the third second, and so on. What is the total distance the object will fall in 8 seconds if this pattern continues?

120. *Free-Falling Object*   A free-falling object will fall 4.9 meters during the first second, 14.7 more meters during the second second, 24.5 more meters during the third second, and so on. What is the total distance the object will fall in 5 seconds if this pattern continues?

121. *Pattern Recognition*

(a) Compute the sums of the positive odd integers.

$$1 + 3 = \boxed{}$$

$$1 + 3 + 5 = \boxed{}$$

$$1 + 3 + 5 + 7 = \boxed{}$$

$$1 + 3 + 5 + 7 + 9 = \boxed{}$$

$$1 + 3 + 5 + 7 + 9 + 11 = \boxed{}$$

(b) Use the sums in part (a) to make a conjecture about the sums of positive odd integers. Check your conjecture for the sum

$$1 + 3 + 5 + 7 + 9 + 11 + 13 = \boxed{}\ .$$

(c) Verify your conjecture in part (b) analytically.

Explaining Concepts

122. In your own words, explain what makes a sequence arithmetic.

123. Explain what is meant by a recursion formula.

124. Explain how the first two terms of an arithmetic sequence can be used to find the nth term.

125. The second and third terms of an arithmetic sequence are 12 and 15, respectively. What is the first term?

126. Explain how to find the sum of the integers from 100 to 200.

Ongoing Review

In Exercises 127–130, evaluate the expression.

127. $10\left(\dfrac{7 + 15}{2}\right)$

128. $15\left(\dfrac{9 + 111}{2}\right)$

129. $\dfrac{3}{2}[4(-3) + 5(4)]$

130. $\dfrac{5}{2}[2(8) + 6(5)]$

In Exercises 131–136, find the domain of the function.

131. $f(x) = x^3 - 2x$

132. $g(x) = \sqrt[3]{x}$

133. $h(x) = \sqrt{16 - x^2}$

134. $A(x) = \dfrac{3}{36 - x^2}$

135. $g(t) = \ln(t - 2)$

136. $f(s) = 630e^{-0.2s}$

137. *Compound Interest*   Determine the balance when \$10,000 is invested at $7\frac{1}{2}\%$ compounded daily for 15 years.

138. *Compound Interest*   Determine the amount in an account after 5 years if \$4000 is invested earning 6% compounded monthly.

Looking Further

The following sequence of perfect squares is *not* arithmetic.

$$1,\ 4,\ 9,\ 16,\ 25,\ 36,\ 49,\ 64,\ 81,\dots$$

However, you can form a related sequence that is arithmetic by finding the differences of consecutive terms as follows.

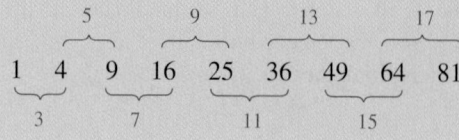

So, the related arithmetic sequence is

$$3,\ 5,\ 7,\ 9,\ 11,\ 13,\ 15,\ 17,\dots$$

(a) Can you find an arithmetic sequence that is related to the following sequence of perfect cubes?

$$1,\ 8,\ 27,\ 64,\ 125,\ 216,\ 343,\ 512,\ 729,\dots$$

(b) Can you find an arithmetic sequence that is related to the following sequence of perfect fourth powers?

$$1,\ 16,\ 81,\ 256,\ 625,\ 1296,\ 2401,\ 4096,\dots$$

Mid-Chapter Quiz

Take this quiz as you would take a quiz in class. After you are done, check your work against the answers given in the back of the book.

In Exercises 1–4, write the first five terms of the sequence. (Assume that n begins with 1.)

1. $a_n = 32\left(\frac{1}{4}\right)^{n-1}$

2. $a_n = \frac{(-3)^n n}{n + 4}$

3. $a_n = \frac{n!}{(n + 1)!}$

4. $a_n = \frac{n}{(n + 1)^2}$

In Exercises 5 and 6, write the first six terms of the sequence defined recursively.

5. $a_1 = 4,\ a_{n+1} = 3a_n - 2$

6. $a_1 = 1,\ a_2 = 3,\ a_{n+2} = \dfrac{3a_n}{a_{n+1}}$

In Exercises 7–10, write an expression for the nth term of the sequence. (Assume that n begins with 1.) There are many correct answers.

7. $1, 4, 7, 10, 13, \ldots$

8. $0, 3, 8, 15, 24, \ldots$

9. $\frac{2}{3}, \frac{3}{4}, \frac{4}{5}, \frac{5}{6}, \frac{6}{7}, \ldots$

10. $\frac{1}{2}, -\frac{1}{4}, \frac{1}{8}, -\frac{1}{16}, \ldots$

In Exercises 11–13, find the partial sum.

11. $\displaystyle\sum_{k=1}^{4} 10k^3$

12. $\displaystyle\sum_{i=1}^{10} (i^2 + 4)$

13. $\displaystyle\sum_{j=1}^{5} \frac{60}{j + 1}$

In Exercises 14 and 15, write the sum using sigma notation.

14. $\dfrac{2}{3(1)} + \dfrac{2}{3(2)} + \dfrac{2}{3(3)} + \cdots + \dfrac{2}{3(20)}$

15. $\dfrac{1}{1^3} - \dfrac{1}{2^3} + \dfrac{1}{3^3} - \cdots + \dfrac{1}{25^3}$

In Exercises 16 and 17, find a formula for the nth term of the arithmetic sequence.

16. $a_1 = 5, d = 7$

17. $a_1 = 20, a_4 = 11$

In Exercises 18–21, find the partial sum.

18. $\displaystyle\sum_{i=1}^{50} (3i + 5)$

19. $\displaystyle\sum_{j=1}^{300} \frac{j}{5}$

20. $\displaystyle\sum_{i=1}^{60} (0.2i + 3)$

21. $\displaystyle\sum_{i=1}^{50} \left(\tfrac{1}{3}i - 4\right)$

22. The temperature of a coolant in an experiment decreased by 25.75°F the first hour. For each subsequent hour, the temperature decreased by 2.25°F less than the previous hour. Determine how much the temperature decreased during the 10th hour.

10.3 Geometric Sequences and Series

Geometric Sequences

In Section 10.2, you studied sequences whose consecutive terms have a common *difference*. In this section, you will study sequences whose consecutive terms have a common *ratio*.

Definition of a Geometric Sequence

A sequence is called **geometric** if the ratios of consecutive terms are the same. So, the sequence $a_1, a_2, a_3, a_4, \ldots, a_n, \ldots$ is geometric if there is a number r, $r \neq 0$, such that

$$\frac{a_2}{a_1} = r, \quad \frac{a_3}{a_2} = r, \quad \frac{a_4}{a_3} = r$$

and so on. The number r is the **common ratio** of the sequence.

Example 1 ■ Examples of Geometric Sequences

(a) The sequence whose nth term is 2^n is geometric. For this sequence, the common ratio between consecutive terms is 2.

$$\underbrace{2, 4, 8, 16, \ldots, 2^n, \ldots}_{} \qquad \text{Begin with } n = 1.$$
$$\tfrac{4}{2} = 2$$

(b) The sequence whose nth term is $4(3^n)$ is geometric. For this sequence, the common ratio between consecutive terms is 3.

$$\underbrace{12, 36, 108, 324, \ldots, 4(3^n), \ldots}_{} \qquad \text{Begin with } n = 1.$$
$$\tfrac{36}{12} = 3$$

(c) The sequence whose nth term is $\left(-\tfrac{1}{3}\right)^n$ is geometric. For this sequence, the common ratio between consecutive terms is $-\tfrac{1}{3}$.

$$\underbrace{-\frac{1}{3}, \frac{1}{9}, -\frac{1}{27}, \frac{1}{81}, \ldots, \left(-\frac{1}{3}\right)^n, \ldots}_{} \qquad \text{Begin with } n = 1.$$
$$\frac{1/9}{-1/3} = -\frac{1}{3}$$

The sequence $1, 4, 9, 16, \ldots$, whose nth term is n^2, is *not* geometric. The ratio of the second term to the first term is

$$\frac{a_2}{a_1} = \frac{4}{1} = 4$$

but the ratio of the third term to the second term is

$$\frac{a_3}{a_2} = \frac{9}{4}.$$

The nth Term of a Geometric Sequence

The nth term of a geometric sequence has the form

$$a_n = a_1 r^{n-1}$$

where r is the common ratio of consecutive terms of the sequence. So, every geometric sequence can be written in the following form.

$$a_1, a_1 r, a_1 r^2, a_1 r^3, a_1 r^4, \ldots, a_1 r^{n-1}, \ldots$$

Example 2 ■ Finding the nth Term of a Geometric Sequence

(a) Find a formula for the nth term of the geometric sequence whose common ratio is 3 and whose first term is 1.

(b) What is the eighth term of the sequence?

Solution

(a) You know that the formula for the nth term is of the form $a_n = a_1 r^{n-1}$. Moreover, because the common ratio is $r = 3$, and the first term is $a_1 = 1$, the formula must have the form

$$\begin{aligned} a_n &= a_1 r^{n-1} & \text{Formula for geometric sequence} \\ &= (1)(3)^{n-1} & \text{Substitute 1 for } a_1 \text{ and 3 for } r. \\ &= 3^{n-1}. & \text{Simplify.} \end{aligned}$$

The sequence therefore has the following form.

$$1, 3, 9, 27, 81, \ldots, 3^{n-1}, \ldots$$

(b) The eighth term of the sequence is $a_8 = 3^{8-1} = 3^7 = 2187$.

Example 3 ■ Finding the nth Term of a Geometric Sequence

Find a formula for the nth term of the geometric sequence whose first two terms are 4 and 2.

Solution

Because the common ratio is

$$r = \frac{a_2}{a_1} = \frac{2}{4} = \frac{1}{2}$$

the formula for the nth term must be

$$\begin{aligned} a_n &= a_1 r^{n-1} & \text{Formula for geometric sequence} \\ &= 4\left(\frac{1}{2}\right)^{n-1}. & \text{Substitute 4 for } a_1 \text{ and } \tfrac{1}{2} \text{ for } r. \end{aligned}$$

The sequence therefore has the following form.

$$4, 2, 1, \frac{1}{2}, \frac{1}{4}, \ldots, 4\left(\frac{1}{2}\right)^{n-1}, \ldots$$

The Partial Sum of a Geometric Sequence

In Section 10.2, you saw that there is a simple formula for finding the sum of the first n terms of an arithmetic sequence. There is also a formula for finding the sum of the first n terms of a geometric sequence.

The nth Partial Sum of a Geometric Sequence

The nth partial sum of the geometric sequence whose nth term is $a_n = a_1 r^{n-1}$ is given by

$$\sum_{i=1}^{n} a_1 r^{i-1} = a_1 + a_1 r + a_1 r^2 + a_1 r^3 + \cdots + a_1 r^{n-1} = a_1 \left(\frac{r^n - 1}{r - 1} \right).$$

Example 4 ■ Finding the nth Partial Sum

Find the sum.

$$1 + 2 + 4 + 8 + 16 + 32 + 64 + 128$$

Solution

This is a geometric sequence whose common ratio is $r = 2$. Because the first term of the sequence is $a_1 = 1$, it follows that the sum is

$$\sum_{i=1}^{8} 2^{i-1} = (1) \left(\frac{2^8 - 1}{2 - 1} \right) \qquad \text{Substitute 1 for } a_1 \text{ and 2 for } r.$$

$$= \frac{256 - 1}{2 - 1} \qquad \text{Simplify.}$$

$$= 255. \qquad \text{Simplify.}$$

Example 5 ■ Finding the nth Partial Sum

Find the sum of the first five terms of the geometric sequence whose nth term is

$$a_n = \left(\frac{2}{3} \right)^n.$$

Solution

$$\sum_{i=1}^{5} \left(\frac{2}{3} \right)^i = \frac{2}{3} \left[\frac{(2/3)^5 - 1}{(2/3) - 1} \right] \qquad \text{Substitute } \tfrac{2}{3} \text{ for } a_1 \text{ and } \tfrac{2}{3} \text{ for } r.$$

$$= \frac{2}{3} \left[\frac{(32/243) - 1}{-1/3} \right] \qquad \text{Simplify.}$$

$$= \frac{2}{3} \left(-\frac{211}{243} \right)(-3) \qquad \text{Simplify.}$$

$$= \frac{422}{243} \qquad \text{Simplify.}$$

$$\approx 1.737 \qquad \text{Use a calculator.}$$

Example 6 ■ Finding the *n*th Partial Sum

Find the sum.

$$\sum_{i=1}^{7} \frac{3^{i-1}}{2}$$

Solution

By writing out a few terms of the sum

$$\sum_{i=1}^{7} \frac{3^{i-1}}{2} = \sum_{i=1}^{7} \frac{1}{2}(3^{i-1}) = \frac{1}{2} + \frac{1}{2}(3) + \frac{1}{2}(3^2) + \cdots + \frac{1}{2}(3^6)$$

you can see that $a_1 = \frac{1}{2}$ and $r = 3$. So, the sum is as follows.

$$\sum_{i=1}^{n} a_1 r^{i-1} = a_1 \left(\frac{r^n - 1}{r - 1} \right) \qquad \text{Formula for } n\text{th partial sum}$$

$$\sum_{i=1}^{7} \frac{1}{2}(3^{i-1}) = \frac{1}{2} \left(\frac{3^7 - 1}{3 - 1} \right) \qquad \text{Substitute } \tfrac{1}{2} \text{ for } a_1 \text{ and 3 for } r.$$

$$= \frac{1}{2} \left(\frac{2187 - 1}{2} \right) \qquad \text{Simplify.}$$

$$= \frac{2186}{4} \qquad \text{Simplify.}$$

$$= 546.5 \qquad \text{Simplify.}$$

Geometric Series

Suppose that in Example 5 you were to find the sum of all the terms of the *infinite* geometric sequence

$$\frac{2}{3}, \frac{4}{9}, \frac{8}{27}, \frac{16}{81}, \cdots, \left(\frac{2}{3} \right)^n, \cdots$$

A summation of all the terms of an infinite geometric sequence is called an **infinite geometric series,** or simply a **geometric series.**

In your mind, would this sum be infinitely large or would it be a finite number? Consider the formula for the *n*th partial sum of a geometric sequence.

$$S_n = a_1 \left(\frac{r^n - 1}{r - 1} \right) = a_1 \left(\frac{1 - r^n}{1 - r} \right)$$

Suppose that $|r| < 1$ and you let n become larger and larger. It follows that r^n gets closer and closer to zero, so that the term r^n drops out of the formula above. You then get the sum

$$S = a_1 \left(\frac{1}{1 - r} \right) = \frac{a_1}{1 - r}.$$

Notice that this sum is not dependent on the *n*th term of the sequence. In the case of Example 5, $r = \left(\frac{2}{3} \right) < 1$, and so the sum of the infinite geometric sequence is

$$S = \sum_{i=1}^{\infty} \left(\frac{2}{3} \right)^i = \frac{a_1}{1 - r} = \frac{2/3}{1 - (2/3)} = \frac{2/3}{1/3} = 2.$$

> **Sum of an Infinite Geometric Series**
>
> If $a_1, a_1 r, a_1 r^2, \ldots, a_1 r^n, \ldots$ is an infinite geometric sequence, then for $|r| < 1$, the sum of the terms of the corresponding infinite geometric series is
>
> $$S = \sum_{i=0}^{\infty} a_1 r^i = \frac{a_1}{1 - r}.$$

Example 7 ■ Finding the Sum of an Infinite Geometric Series

Find the value of each sum.

(a) $\displaystyle\sum_{i=1}^{\infty} 5\left(\frac{3}{4}\right)^{i-1}$ (b) $\displaystyle\sum_{n=0}^{\infty} 4\left(\frac{3}{10}\right)^n$ (c) $\displaystyle\sum_{i=0}^{\infty} \left(-\frac{3}{5}\right)^i$

Solution

(a) This series is geometric, with $a_1 = 5\left(\frac{3}{4}\right)^{1-1} = 5$ and $r = \frac{3}{4}$. So,

$$\sum_{i=1}^{\infty} 5\left(\frac{3}{4}\right)^{i-1} = \frac{5}{1 - (3/4)}$$

$$= \frac{5}{1/4} = 20.$$

(b) This series is geometric, with $a_1 = 4\left(\frac{3}{10}\right)^0 = 4$ and $r = \frac{3}{10}$. So,

$$\sum_{n=0}^{\infty} 4\left(\frac{3}{10}\right)^n = \frac{4}{1 - (3/10)}$$

$$= \frac{4}{7/10} = \frac{40}{7}.$$

(c) This series is geometric, with $a_1 = \left(-\frac{3}{5}\right)^0 = 1$ and $r = -\frac{3}{5}$. So,

$$\sum_{i=0}^{\infty} \left(-\frac{3}{5}\right)^i = \frac{1}{1 - (-3/5)}$$

$$= \frac{1}{1 + (3/5)} = \frac{5}{8}.$$

EXPLORATION

Any repeating decimal can be written in rational form (as the ratio of two integers). Here is an example. Consider the repeating decimal

$x = 0.414141\ldots.$

To write this number in rational form, multiply by 100 to obtain an expression for $100x$. Then subtract x from $100x$ to obtain an expression for $99x$. Finally, divide by 99 to obtain a rational expression for x. Try doing this. A repeating decimal can also be written as the sum of an *infinite* geometric series. For example,

$x = 0.414141\ldots = 41\left(\frac{1}{100}\right) + 41\left(\frac{1}{100}\right)^2 + 41\left(\frac{1}{100}\right)^3 + \cdots.$

Try using the formula for the sum of an infinite geometric series with $a_1 = 41\left(\frac{1}{100}\right)$ and $r = \frac{1}{100}$ to rewrite x as a rational number.

Applications

Example 8 ■ A Lifetime Salary

You have accepted a job that pays a salary of \$28,000 the first year. During the next 39 years, suppose you receive a 6% raise each year. What will your total salary be over the 40-year period?

Solution

Using a geometric sequence, your salary during the first year will be $a_1 = 28{,}000$. Then, with a 6% raise each year, your salary for the next 2 years will be as follows.

$$a_2 = 28{,}000 + 28{,}000(0.06)$$
$$= 28{,}000(1.06)^1$$

$$a_3 = 28{,}000(1.06) + 28{,}000(1.06)(0.06)$$
$$= 28{,}000(1.06)^2$$

From this pattern, you can see that the common ratio of the geometric sequence is $r = 1.06$. Using the formula for the *n*th partial sum of a geometric sequence, you will find that the total salary over the 40-year period is given by

$$\text{Total salary} = a_1\left(\frac{r^n - 1}{r - 1}\right) = 28{,}000\left[\frac{(1.06)^{40} - 1}{1.06 - 1}\right]$$

$$= 28{,}000\left[\frac{(1.06)^{40} - 1}{0.06}\right]$$

$$\approx \$4{,}333{,}335.$$

Example 9 ■ Increasing Annuity

You deposit \$100 in an account each month for 2 years. The account pays an annual interest rate of 9%, compounded monthly. What is your balance at the end of 2 years? (This type of savings plan is called an **increasing annuity.**)

Solution

The first deposit would earn interest for the full 24 months, the second deposit would earn interest for 23 months, the third deposit would earn interest for 22 months, and so on. Using the formula for compound interest, you can see that the total of the 24 deposits would be

$$\text{Total} = a_1 + a_2 + \cdots + a_{24}$$

$$= 100\left(1 + \frac{0.09}{12}\right)^1 + 100\left(1 + \frac{0.09}{12}\right)^2 + \cdots + 100\left(1 + \frac{0.09}{12}\right)^{24}$$

$$= 100(1.0075)^1 + 100(1.0075)^2 + \cdots + 100(1.0075)^{24}$$

$$= 100(1.0075)\left(\frac{1.0075^{24} - 1}{1.0075 - 1}\right) \qquad a_1\left(\frac{r^n - 1}{r - 1}\right)$$

$$= \$2638.49.$$

10.3 ■ Exercises

Developing Skills

In Exercises 1–14, find the common ratio of the geometric sequence.

1. $2, 6, 18, 54, \ldots$ **2.** $5, -10, 20, -40, \ldots$

3. $1, -3, 9, -27, \ldots$ **4.** $10, 20, 40, 80, \ldots$

5. $192, -48, 12, -3, \ldots$

6. $54, 18, 6, 2, \ldots$

7. $1, -\frac{3}{2}, \frac{9}{4}, -\frac{27}{8}, \ldots$ **8.** $12, -6, 3, -\frac{3}{2}, \ldots$

9. $9, 6, 4, \frac{8}{3}, \ldots$ **10.** $5, -\frac{5}{2}, \frac{5}{4}, -\frac{5}{8}, \ldots$

11. $1, \pi, \pi^2, \pi^3, \ldots$ **12.** $e, e^2, e^3, e^4, \ldots$

13. $1.1, (1.1)^2, (1.1)^3, (1.1)^4, \ldots$

14. $500(1.06), 500(1.06)^2, 500(1.06)^3, 500(1.06)^4, \ldots$

In Exercises 15–28, determine whether the sequence is geometric. If so, find the common ratio.

15. $64, 32, 16, 8, \ldots$ **16.** $64, 32, 0, -32, \ldots$

17. $10, 15, 20, 25, \ldots$ **18.** $10, 20, 40, 80, \ldots$

19. $5, 10, 20, 40, \ldots$

20. $54, -18, 6, -2, \ldots$

21. $1, 8, 27, 64, 125, \ldots$

22. $12, 7, 2, -3, -8, \ldots$

23. $1, -\frac{2}{3}, \frac{4}{9}, -\frac{8}{27}, \ldots$

24. $1, \frac{3}{5}, \frac{9}{25}, \frac{27}{125}, \ldots$

25. $\frac{1}{4}, \frac{2}{5}, \frac{3}{6}, \frac{4}{7}, \ldots$

26. $\frac{1}{3}, -\frac{2}{3}, \frac{4}{3}, -\frac{8}{3}, \ldots$

27. $10(1 + 0.02), 10(1 + 0.02)^2, 10(1 + 0.02)^3, \ldots$

28. $1, 0.2, 0.04, 0.008, \ldots$

In Exercises 29–44, write the first five terms of the geometric sequence. If necessary, round answers to two decimal places.

29. $a_1 = 4, \ r = 2$ **30.** $a_1 = 3, \ r = 4$

31. $a_1 = 5, \ r = -2$ **32.** $a_1 = -2, r = -4$

33. $a_1 = 6, \ r = \frac{1}{3}$ **34.** $a_1 = 4, \ r = \frac{1}{2}$

35. $a_1 = 1, \ r = -\frac{1}{2}$ **36.** $a_1 = 32, \ r = -\frac{3}{4}$

37. $a_1 = 4, \ r = -\frac{1}{2}$ **38.** $a_1 = 4, \ r = \frac{3}{2}$

39. $a_1 = 1000, \ r = 1.01$

40. $a_1 = 200, \ r = 1.07$

41. $a_1 = 4000, \ r = \frac{1}{1.01}$

42. $a_1 = 1000, \ r = \frac{1}{1.05}$

43. $a_1 = 10, \ r = \frac{3}{5}$

44. $a_1 = 36, \ r = \frac{2}{3}$

In Exercises 45–60, find the nth term of the geometric sequence.

45. $a_1 = 3, \ r = -3, \ a_8 =$

46. $a_1 = -2, \ r = -5, \ a_9 =$

47. $a_1 = 120, \ r = -\frac{1}{3}, \ a_{10} =$

48. $a_1 = 240, \ r = -\frac{1}{4}, \ a_{13} =$

49. $a_1 = 6, \ r = \frac{1}{2}, \ a_{10} =$

50. $a_1 = 8, \ r = \frac{3}{4}, \ a_8 =$

51. $a_1 = 3, \ r = \sqrt{2}, \ a_{10} =$

52. $a_1 = 5, \ r = \sqrt{3}, \ a_9 =$

53. $a_1 = 200, \ r = 1.2, \ a_{12} =$

54. $a_1 = 500, \ r = 1.06, \ a_{40} =$

55. $a_1 = 4, \ a_2 = 3, \ a_5 =$

56. $a_1 = 1, \ a_2 = 9, \ a_7 =$

57. $a_1 = 1, \ a_3 = \frac{9}{4}, \ a_6 =$

58. $a_2 = 12, \ a_3 = 16, \ a_4 =$

59. $a_3 = 6, \ a_5 = \frac{8}{3}, \ a_6 =$

60. $a_4 = 100, \ a_5 = -25, \ a_7 =$

In Exercises 61–74, find the formula for the nth term of the geometric sequence. (Assume that n begins with 1.)

61. $a_1 = 2, \ r = 3$ **62.** $a_1 = 5, \ r = 4$

63. $a_1 = 1, \ r = 2$ **64.** $a_1 = 1, \ r = -5$

65. $a_1 = 4, \ r = -\frac{1}{2}$ **66.** $a_1 = 9, \ r = \frac{2}{3}$

67. $a_1 = 8, \ a_2 = 2$ **68.** $a_1 = 18, \ a_2 = 8$

69. $a_1 = 12, \ a_2 = 8$ **70.** $a_1 = 15, \ a_2 = 3$

71. $4, 12, 36, 108, \ldots$

72. $-2, 10, -50, 250, \ldots$

73. $4, -6, 9, -\frac{27}{2}, \ldots$

74. $1, \frac{3}{2}, \frac{9}{4}, \frac{27}{8}, \ldots$

In Exercises 75–78, match the sequence with its graph. [The graphs are labeled (a), (b), (c), and (d).]

(a)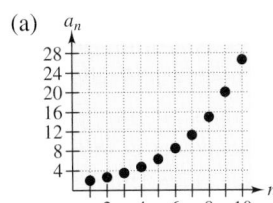

(b)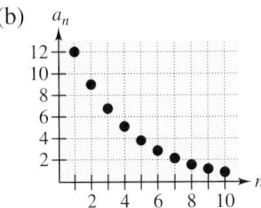

(c)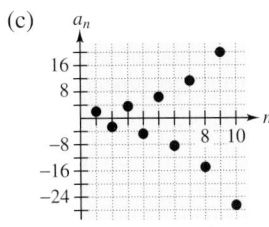

(d)

75. $a_n = 12\left(\frac{3}{4}\right)^{n-1}$

76. $a_n = 12\left(-\frac{3}{4}\right)^{n-1}$

77. $a_n = 2\left(\frac{4}{3}\right)^{n-1}$

78. $a_n = 2\left(-\frac{4}{3}\right)^{n-1}$

In Exercises 79–82, use a graphing utility to graph the first 10 terms of the sequence.

79. $a_n = 20(-0.6)^{n-1}$

80. $a_n = 4(1.4)^{n-1}$

81. $a_n = 15(0.6)^{n-1}$

82. $a_n = 8(-0.6)^{n-1}$

In Exercises 83–92, find the partial sum.

83. $\displaystyle\sum_{i=1}^{10} 2^{i-1}$

84. $\displaystyle\sum_{i=1}^{6} 3^{i-1}$

85. $\displaystyle\sum_{i=1}^{12} 3\left(\frac{3}{2}\right)^{i-1}$

86. $\displaystyle\sum_{i=1}^{20} 12\left(\frac{2}{3}\right)^{i-1}$

87. $\displaystyle\sum_{i=1}^{15} 3\left(-\frac{1}{3}\right)^{i-1}$

88. $\displaystyle\sum_{i=1}^{8} 8\left(-\frac{1}{4}\right)^{i-1}$

89. $\displaystyle\sum_{i=1}^{8} 6(0.1)^{i-1}$

90. $\displaystyle\sum_{i=1}^{24} 1000(1.06)^{i-1}$

91. $\displaystyle\sum_{i=1}^{10} 15(0.3)^{i-1}$

92. $\displaystyle\sum_{i=1}^{20} 5(1.25)^{i-1}$

In Exercises 93–96, use a graphing utility to find the partial sum.

93. $\displaystyle\sum_{i=1}^{30} 100(0.75)^{i-1}$

94. $\displaystyle\sum_{i=1}^{24} 5000(1.08)^{-(i-1)}$

95. $\displaystyle\sum_{i=1}^{20} 100(1.1)^{i}$

96. $\displaystyle\sum_{i=1}^{40} 50(1.07)^{i}$

In Exercises 97–110, find the nth partial sum of the geometric sequence.

97. $1, -3, 9, -27, 81, \ldots, \quad n = 10$

98. $3, -6, 12, -24, 48, \ldots, \quad n = 12$

99. $5, 10, 20, 40, 80, \ldots, \quad n = 9$

100. $1, 5, 25, 125, 625, \ldots, \quad n = 6$

101. $8, 4, 2, 1, \frac{1}{2}, \ldots, \quad n = 15$

102. $9, 6, 4, \frac{8}{3}, \frac{16}{9}, \ldots, \quad n = 10$

103. $4, 12, 36, 108, \ldots, \quad n = 8$

104. $\frac{1}{36}, -\frac{1}{12}, \frac{1}{4}, -\frac{3}{4}, \ldots, \quad n = 20$

105. $60, -15, \frac{15}{4}, -\frac{15}{16}, \ldots, \quad n = 12$

106. $40, -10, \frac{5}{2}, -\frac{5}{8}, \frac{5}{32}, \ldots, \quad n = 10$

107. $30, 30(1.06), 30(1.06)^2, \ldots, \quad n = 20$

108. $100, 100(1.08), 100(1.08)^2, \ldots, \quad n = 40$

109. $500, 500(1.04), 500(1.04)^2, \ldots, \quad n = 18$

110. $1, \sqrt{2}, 2, 2\sqrt{2}, 4, \ldots, \quad n = 12$

In Exercises 111–118, find the sum.

111. $\displaystyle\sum_{n=0}^{\infty} \left(\frac{1}{2}\right)^{n}$

112. $\displaystyle\sum_{n=0}^{\infty} 2\left(\frac{2}{3}\right)^{n}$

113. $\displaystyle\sum_{n=0}^{\infty} \left(-\frac{1}{2}\right)^{n}$

114. $\displaystyle\sum_{n=0}^{\infty} \left(\frac{1}{10}\right)^{n}$

115. $\displaystyle\sum_{i=0}^{\infty} 2\left(-\frac{2}{3}\right)^{i}$

116. $\displaystyle\sum_{i=0}^{\infty} 4\left(\frac{1}{4}\right)^{i}$

117. $8 + 6 + \frac{9}{2} + \frac{27}{8} + \cdots$

118. $3 - 1 + \frac{1}{3} - \frac{1}{9} + \cdots$

Solving Problems

119. *Depreciation* A company buys a machine for $250,000. During the next 5 years, the machine depreciates at a rate of 25% per year. (That is, at the end of each year, the depreciated value is 75% of what it was at the beginning of the year.)

(a) Find a formula for the nth term of the geometric sequence that gives the value of the machine n full years after it was purchased.

(b) Find the depreciated value of the machine at the end of 5 full years.

(c) During which year did the machine depreciate most?

120. *Population Growth* A city of 500,000 people is growing at a rate of 1% per year. (That is, at the end of each year, the population is 1.01 times what it was at the beginning of the year.)

(a) Find a formula for the nth term of the geometric sequence that gives the population t years from now.

(b) Estimate the population 20 years from now.

121. *Salary Increases* You accept a job that pays a salary of $30,000 the first year. During the next 39 years, you receive a 5% raise each year. What is your total salary over the 40-year period?

122. *Salary Increases* You accept a job that pays a salary of $30,000 the first year. During the next 39 years, you receive a 5.5% raise each year.

(a) What is your total salary over the 40-year period?

(b) How much more income did the extra 0.5% provide than the result in Exercise 121?

Increasing Annuity In Exercises 123–128, find the balance in an increasing annuity in which a principal of P dollars is invested each month for t years, compounded monthly at rate r.

123. $P = \$100$ $t = 10$ years $r = 9\%$
124. $P = \$50$ $t = 5$ years $r = 7\%$
125. $P = \$30$ $t = 40$ years $r = 8\%$
126. $P = \$200$ $t = 30$ years $r = 10\%$
127. $P = \$75$ $t = 30$ years $r = 6\%$
128. $P = \$100$ $t = 25$ years $r = 8\%$

129. *Wages* You start work at a company that pays $0.01 the first day, $0.02 the second day, $0.04 the third day, and so on. If the daily wage keeps doubling, what would your total income be for working (a) 29 days and (b) 30 days?

130. *Wages* You start work at a company that pays $0.01 the first day, $0.03 the second day, $0.09 the third day, and so on. If the daily wage keeps tripling, what would your total income be for working (a) 25 days and (b) 26 days?

131. *Power Supply* The electrical power for an implanted medical device decreases by 0.1% each day.

(a) Find a formula for the nth term of the geometric sequence that gives the power n days after the device is implanted.

(b) What percent of the initial power is still available 1 year after the device is implanted?

(c) The power supply needs to be changed when half the power is depleted. Use a graphing utility to graph the first 750 terms of the sequence and estimate when the power source should be changed.

132. *Cooling* The temperature of water in an ice-cube tray is 70°F when it is placed in a freezer. Its temperature n hours after being placed in the freezer is 20% less than 1 hour earlier.

(a) Find a formula for the nth term of the geometric sequence that gives the temperature of the water n hours after it is placed in the freezer.

(b) Find the temperature of the water 6 hours after it is placed in the freezer.

(c) Use a graphing utility to estimate the time when the water freezes. Explain your reasoning.

133. *Bungee Jumping* A bungee jumper jumps from a bridge and stretches a cord 100 feet. Successive bounces stretch the cord 75% of each previous length (see figure). Find the total distance traveled by the bungee jumper during 10 bounces.

$$100 + 2(100)(0.75) + \cdots + 2(100)(0.75)^{10}$$

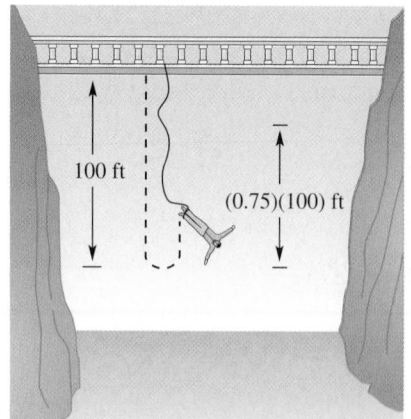

Explaining Concepts

134. In your own words, explain what makes a sequence geometric.

135. Give an example of a geometric sequence whose terms alternate in sign.

136. Explain why the terms of a geometric sequence decrease when $a_1 > 0$ and $0 < r < 1$.

137. Explain what is meant by the nth partial sum of a sequence.

Ongoing Review

In Exercises 138 and 139, simplify the expression.

138. $\dfrac{(n + 1)!}{(n - 1)!}$

139. $\dfrac{n!}{n \cdot (n + 1)!}$

In Exercises 140–143, solve the inequality.

140. $3x - 5 > 0$

141. $\frac{3}{2}y + 11 < 20$

142. $2x^2 - 7x + 5 > 0$

143. $2x - \dfrac{5}{x} > 3$

144. *Geometry*   A television set is advertised as having a 19-inch screen. Determine the dimensions of the square screen if its diagonal is 19 inches.

145. *Construction*   A construction worker is building the forms for the rectangular foundation of a home that is 25 feet wide and 40 feet long. To make sure that the corners are square, the worker measures the diagonal of the foundation. What should that measurement be?

Looking Further

Number of Ancestors   The number of direct ancestors a person has had is as follows.

$$2 + 2^2 + 2^3 + 2^4 + \cdots + 2^n + \cdots$$

Parents   Grand-parents   Great-grand-parents   Great-great-grand-parents

This formula is valid provided the person has no common ancestors. (A common ancestor is one to whom you are related in more than one way.) During the past 2000 years, suppose your ancestry can be traced through 66 generations. During that time, your total number of ancestors would be

$$2 + 2^2 + 2^3 + 2^4 + \cdots + 2^{66}.$$

Considering the total, do you think that you have had no common ancestors in the past 2000 years? Explain.

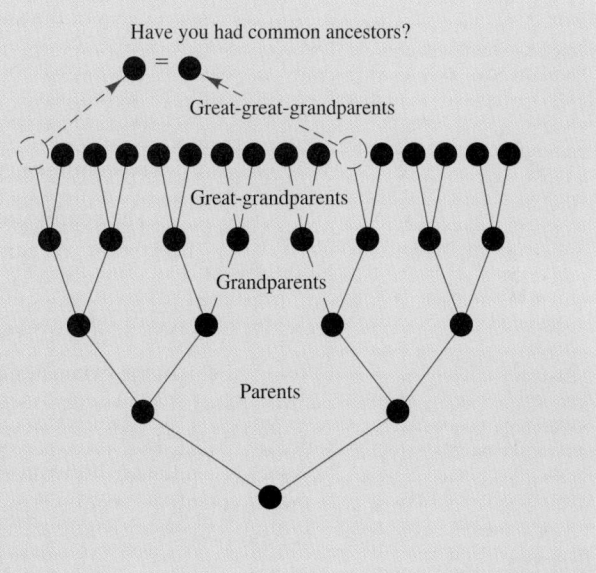

Have you had common ancestors?

= Great-great-grandparents

Great-grandparents

Grandparents

Parents

10.4 The Binomial Theorem

What you should learn:

- How to use the Binomial Theorem to calculate binomial coefficients
- How to use Pascal's Triangle to calculate binomial coefficients
- How to expand binomial expressions

Why you should learn it:

You can use binomial coefficients to approximate quantities. See Exercises 69–72 on page 679.

Binomial Coefficients

Recall that a **binomial** is a polynomial that has two terms. In this section, you will study a formula that provides a quick method of raising a binomial to a power. To begin, let's look at the expansion of $(x + y)^n$ for several values of n.

$$(x + y)^0 = 1$$
$$(x + y)^1 = x + y$$
$$(x + y)^2 = x^2 + 2xy + y^2$$
$$(x + y)^3 = x^3 + 3x^2y + 3xy^2 + y^3$$
$$(x + y)^4 = x^4 + 4x^3y + 6x^2y^2 + 4xy^3 + y^4$$
$$(x + y)^5 = x^5 + 5x^4y + 10x^3y^2 + 10x^2y^3 + 5xy^4 + y^5$$

There are several observations you can make about these expansions.

1. In each expansion, there are $n + 1$ terms.

2. In each expansion, x and y have symmetrical roles. The powers of x decrease by 1 in successive terms, whereas the powers of y increase by 1.

3. The sum of the powers of each term is n. For instance, in the expansion of $(x + y)^5$, the sum of the powers of each term is 5.

$$4 + 1 = 5 \qquad 3 + 2 = 5$$
$$(x + y)^5 = x^5 + 5x^4y^1 + 10x^3y^2 + 10x^2y^3 + 5xy^4 + y^5$$

4. The coefficients increase and then decrease in a symmetrical pattern.

The coefficients of a binomial expansion are called **binomial coefficients.** To find them, you can use the **Binomial Theorem.**

The Binomial Theorem

In the expansion of $(x + y)^n$

$$(x + y)^n = x^n + nx^{n-1}y + \cdots + {}_nC_r x^{n-r}y^r + \cdots + nxy^{n-1} + y^n$$

the coefficient of $x^{n-r}y^r$ is given by

$${}_nC_r = \frac{n!}{(n - r)!r!}.$$

NOTE Other notations that are commonly used for ${}_nC_r$ are $\binom{n}{r}$ and $C(n, r)$.

Technology

The formula for the binomial coefficient is the same as the formula for combinations in the study of probability. Most graphing utilities have the capability to evaluate a binomial coefficient. Consult the user's guide for your graphing utility.

Example 1 ■ Finding Binomial Coefficients

Find each binomial coefficient.

(a) $_8C_2$ (b) $_{10}C_3$ (c) $_7C_0$ (d) $_8C_8$ (e) $_9C_5$

Solution

(a) $_8C_2 = \dfrac{8!}{6! \cdot 2!} = \dfrac{(8 \cdot 7) \cdot 6!}{6! \cdot 2!} = \dfrac{8 \cdot 7}{2 \cdot 1} = 28$

(b) $_{10}C_3 = \dfrac{10!}{7! \cdot 3!} = \dfrac{(10 \cdot 9 \cdot 8) \cdot 7!}{7! \cdot 3!} = \dfrac{10 \cdot 9 \cdot 8}{3 \cdot 2 \cdot 1} = 120$

(c) $_7C_0 = \dfrac{7!}{7! \cdot 0!} = 1$

(d) $_8C_8 = \dfrac{8!}{0! \cdot 8!} = 1$

(e) $_9C_5 = \dfrac{9!}{4! \cdot 5!} = \dfrac{(9 \cdot 8 \cdot 7 \cdot 6) \cdot 5!}{4! \cdot 5!} = \dfrac{9 \cdot 8 \cdot 7 \cdot 6}{4 \cdot 3 \cdot 2 \cdot 1} = 126$

When $r \neq 0$ and $r \neq n$, there is a simple pattern for evaluating binomial coefficients. Note how this is used in parts (a) and (b) of Example 2.

Example 2 ■ Finding Binomial Coefficients

Find each binomial coefficient.

(a) $_7C_3$ (b) $_7C_4$

(c) $_{12}C_1$ (d) $_{12}C_{11}$

Solution

(a) $_7C_3 = \dfrac{7 \cdot 6 \cdot 5}{3 \cdot 2 \cdot 1} = 35$

(b) $_7C_4 = \dfrac{7 \cdot 6 \cdot 5 \cdot 4}{4 \cdot 3 \cdot 2 \cdot 1} = 35$ 　　　　　 $_7C_4 = {}_7C_3$

(c) $_{12}C_1 = \dfrac{12!}{11! \cdot 1!} = \dfrac{(12) \cdot 11!}{11! \cdot 1!} = \dfrac{12}{1} = 12$

(d) $_{12}C_{11} = \dfrac{12!}{1! \cdot 11!} = \dfrac{(12) \cdot 11!}{1! \cdot 11!} = \dfrac{12}{1} = 12$ 　　 $_{12}C_{11} = {}_{12}C_1$

NOTE　In Example 2, it is not a coincidence that the results in parts (a) and (b) are the same *and* that the results in parts (c) and (d) are the same. In general, it is true that

$$_nC_r = {}_nC_{n-r}.$$

This shows the symmetric property of binomial coefficients.

Pascal's Triangle

There is a convenient way to remember a pattern for binomial coefficients. By arranging the coefficients in a triangular pattern, you obtain the array shown below, which is called **Pascal's Triangle.** This triangle is named after the famous French mathematician Blaise Pascal (1623–1662).

NOTE The top row in Pascal's Triangle is called the *zero row* because it corresponds to the binomial expansion

$$(x + y)^0 = 1.$$

Similarly, the next row is called the *first row* because it corresponds to the binomial expansion

$$(x + y)^1 = 1(x) + 1(y).$$

In general, the *nth row* in Pascal's Triangle gives the coefficients of $(x + y)^n$.

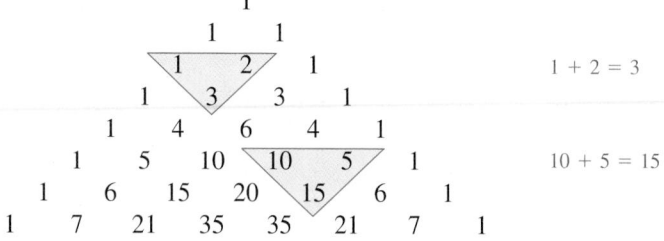

$$1 + 2 = 3$$

$$10 + 5 = 15$$

The first and last numbers in each row of Pascal's Triangle are 1. Every other number in each row is formed by adding the two numbers immediately above the number. Pascal noticed that numbers in this triangle are precisely the same numbers that are the coefficients of binomial expansions, as follows.

$$(x + y)^0 = 1$$
$$(x + y)^1 = 1x + 1y$$
$$(x + y)^2 = 1x^2 + 2xy + 1y^2$$
$$(x + y)^3 = 1x^3 + 3x^2y + 3xy^2 + 1y^3$$
$$(x + y)^4 = 1x^4 + 4x^3y + 6x^2y^2 + 4xy^3 + 1y^4$$
$$(x + y)^5 = 1x^5 + 5x^4y + 10x^3y^2 + 10x^2y^3 + 5xy^4 + 1y^5$$
$$(x + y)^6 = 1x^6 + 6x^5y + 15x^4y^2 + 20x^3y^3 + 15x^2y^4 + 6xy^5 + 1y^6$$
$$(x + y)^7 = 1x^7 + 7x^6y + 21x^5y^2 + 35x^4y^3 + 35x^3y^4 + 21x^2y^5 + 7xy^6 + 1y^7$$

You can use the seventh row of Pascal's Triangle to find the binomial coefficients of the eighth row, as follows.

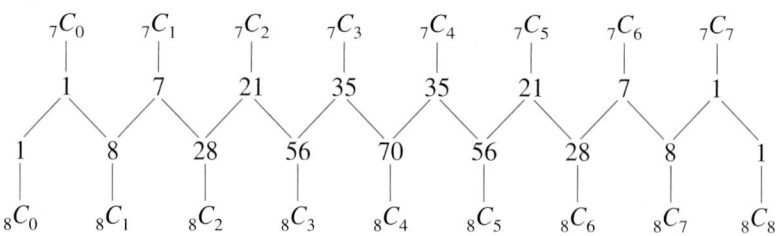

Example 3 ■ Using Pascal's Triangle

Use the fifth row of Pascal's Triangle to evaluate $_5C_2$.

Solution

$$
\begin{array}{cccccc}
1 & 5 & 10 & 10 & 5 & 1 \\
_5C_0 & _5C_1 & _5C_2 & _5C_3 & _5C_4 & _5C_5
\end{array}
$$

So, $_5C_2 = 10$.

Binomial Expansions

As mentioned at the beginning of this section, when you write out the coefficients for a binomial that is raised to a power, you are **expanding a binomial.** The formulas for binomial coefficients give you an easy way to expand binomials, as demonstrated in the next three examples.

Example 4 ■ Expanding a Binomial

Write the expansion for the expression.

$$(x + 1)^3$$

Solution

The binomial coefficients from the third row of Pascal's Triangle are

1, 3, 3, 1.

So, the expansion is as follows.

$$(x + 1)^3 = (1)x^3 + (3)x^2(1) + (3)x(1^2) + (1)(1^3)$$
$$= x^3 + 3x^2 + 3x + 1$$

To expand binomials representing *differences*, rather than sums, you alternate signs. Here are two examples.

$$(x - 1)^3 = x^3 - 3x^2 + 3x - 1$$
$$(x - 1)^4 = x^4 - 4x^3 + 6x^2 - 4x + 1$$

Example 5 ■ Expanding a Binomial

Write the expansion for each expression.

(a) $(x - 3)^4$ (b) $(2x + 1)^3$

Solution

(a) The binomial coefficients from the fourth row of Pascal's Triangle are

1, 4, 6, 4, 1.

So, the expansion is as follows.

$$(x - 3)^4 = (1)x^4 - (4)x^3(3) + (6)x^2(3^2) - (4)x(3^3) + (1)(3^4)$$
$$= x^4 - 12x^3 + 54x^2 - 108x + 81$$

(b) The binomial coefficients from the third row of Pascal's Triangle are

1, 3, 3, 1.

So, the expansion is as follows.

$$(2x + 1)^3 = (1)(2x)^3 + (3)(2x)^2(1) + (3)(2x)(1^2) + (1)(1^3)$$
$$= 8x^3 + 12x^2 + 6x + 1$$

Example 6 ■ Expanding a Binomial

Write the expansion for $(3x - y)^4$.

Solution

Use the fourth row of Pascal's Triangle, as follows.

$$(3x - y)^4 = (1)(3x)^4 - (4)(3x)^3y + (6)(3x)^2y^2 - (4)3xy^3 + (1)y^4$$
$$= 81x^4 - 108x^3y + 54x^2y^2 - 12xy^3 + y^4$$

Example 7 ■ Finding a Term in the Binomial Expansion

(a) Find the sixth term of $(a + 2b)^8$.

(b) Find the coefficient of the term a^6b^5 in the expansion of $(3a - 2b)^{11}$.

Solution

(a) From the Binomial Theorem, you can see that the $(r + 1)$th term is $_nC_rx^{n-r}y^r$. So in this case, $6 = r + 1$ means that $r = 5$. Because $n = 8$, $x = a$, and $y = 2b$, the sixth term in the binomial expansion is

$$_8C_5a^{8-5}(2b)^5 = 56 \cdot a^3 \cdot (2b)^5$$
$$= 56(2^5)a^3b^5$$
$$= 1792a^3b^5$$

(b) From the Binomial Theorem, you can see that the $(r + 1)$th term is $_nC_rx^{n-r}y^r$. So in this case, $n = 11$, $r = 5$, $x = 3a$, and $y = -2b$. Substitute these values to obtain

$$_nC_rx^{n-r}y^r = {}_{11}C_5(3a)^6(-2b)^5$$
$$= (462)(729a^6)(-32b^5)$$
$$= -10,777,536a^6b^5.$$

So, the coefficient is $-10,777,536$.

Collaborate!

Extending the Concept

By adding the numbers in each of the rows of Pascal's Triangle, you obtain the following.

Row 0: $1 = 1$

Row 1: $1 + 1 = 2$

Row 2: $1 + 2 + 1 = 4$

Row 3: $1 + 3 + 3 + 1 = 8$

Row 4: $1 + 4 + 6 + 4 + 1 = 16$

Find a pattern for the sequence. Then use the pattern to find the sum of the numbers in the 10th row of Pascal's Triangle. Finally, check your answer by actually adding the numbers in the 10th row.

10.4 ■ Exercises

Developing Skills

In Exercises 1–12, evaluate the binomial coefficient $_nC_r$.

1. $_6C_4$ **2.** $_7C_3$

3. $_{10}C_5$ **4.** $_{12}C_9$

5. $_{20}C_{20}$ **6.** $_{15}C_0$

7. $_{18}C_{18}$ **8.** $_{200}C_1$

9. $_{50}C_{48}$ **10.** $_{75}C_1$

11. $_{25}C_4$ **12.** $_{18}C_5$

In Exercises 13–16, use a graphing utility to evaluate $_nC_r$.

13. $_{52}C_5$ **14.** $_{100}C_6$

15. $_{200}C_{195}$ **16.** $_{500}C_4$

In Exercises 17–26, evaluate the binomial coefficient $_nC_r$. Also, evaluate its symmetric coefficient $_nC_{n-r}$.

17. $_{15}C_3$ **18.** $_9C_4$

19. $_{25}C_5$ **20.** $_{30}C_3$

21. $_5C_2$ **22.** $_8C_6$

23. $_{12}C_5$ **24.** $_{14}C_8$

25. $_{10}C_0$ **26.** $_{25}C_{25}$

In Exercises 27–32, use Pascal's Triangle to evaluate $_nC_r$.

27. $_6C_2$ **28.** $_9C_3$

29. $_7C_3$ **30.** $_9C_5$

31. $_8C_4$ **32.** $_{10}C_6$

In Exercises 33–48, expand the expression.

33. $(x + 3)^6$ **34.** $(x - 5)^4$

35. $(x + 1)^5$ **36.** $(x + 2)^5$

37. $(x - 4)^6$ **38.** $(x - 8)^4$

39. $(x + y)^8$ **40.** $(u - v)^6$

41. $(u - 2v)^3$ **42.** $(2x + y)^5$

43. $(2x - 1)^4$ **44.** $(4t - 1)^4$

45. $(2y + z)^6$ **46.** $(2t - s)^5$

47. $(3a + 2b)^4$ **48.** $(4u - 3v)^3$

In Exercises 49–60, find the specified nth term in the expansion of the binomial.

49. $(x + y)^{10}$, $n = 4$ **50.** $(x - y)^6$, $n = 7$

51. $(a + 4b)^9$, $n = 6$ **52.** $(a + 5b)^{12}$, $n = 8$

53. $(x - 6y)^5$, $n = 3$ **54.** $(x - 10z)^7$, $n = 4$

55. $(3a - b)^{12}$, $n = 10$ **56.** $(8x - y)^4$, $n = 3$

57. $(4x + 3y)^9$, $n = 8$ **58.** $(5a + 6b)^5$, $n = 4$

59. $(10x - 3y)^{12}$, $n = 9$ **60.** $(3x - 2y)^{15}$, $n = 7$

In Exercises 61–68, find the coefficient of the term in the expansion of the binomial.

Binomial	*Term*
61. $(x + 1)^{10}$	x^7
62. $(x + 3)^{12}$	x^9
63. $(x - y)^{15}$	x^4y^{11}
64. $(x + y)^{10}$	x^7y^3
65. $(2x + y)^{12}$	x^3y^9
66. $(x - 3y)^{14}$	x^3y^{11}
67. $(x^2 - 3)^4$	x^4
68. $(3 - y^3)^5$	y^9

Solving Problems

In Exercises 69–72, use the Binomial Theorem to approximate the quantity accurate to three decimal places. For example:

$$(1.02)^{10} = (1 + 0.02)^{10}$$
$$\approx 1 + 10(0.02) + 45(0.02)^2 + 120(0.02)^3.$$

69. $(1.02)^8$ **70.** $(2.005)^{10}$

71. $(2.99)^{12}$ **72.** $(1.98)^9$

Graphical Reasoning   In Exercises 73 and 74, use a graphing utility to graph f and g in the same viewing window. What is the relationship between the two graphs? Use the Binomial Theorem to write the polynomial function g in standard form.

73. $f(x) = -x^2 + 3x + 2$, $g(x) = f(x - 2)$

74. $f(x) = x^3 - 4x$, $g(x) = f(x + 4)$

Probability In Exercises 75–80, use the Binomial Theorem to expand the expression. In the study of probability, it is sometimes necessary to use the expansion $(p + q)^n$, where $p + q = 1$.

75. $\left(\frac{1}{2} + \frac{1}{2}\right)^5$

76. $\left(\frac{2}{3} + \frac{1}{3}\right)^4$

77. $\left(\frac{1}{4} + \frac{3}{4}\right)^4$

78. $\left(\frac{2}{5} + \frac{3}{5}\right)^3$

79. $(0.7 + 0.3)^3$

80. $(0.4 + 0.6)^6$

81. *Pascal's Triangle* Describe the pattern.

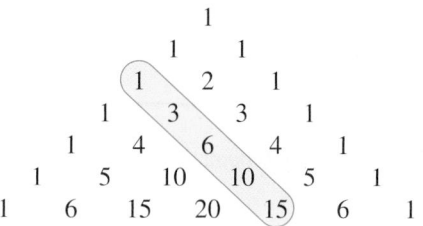

Explaining Concepts

82. How many terms are in the expansion of $(x + y)^n$?

83. How do the expansions of $(x + y)^n$ and $(x - y)^n$ differ?

84. What is the relationship between $_nC_r$ and $_nC_{n-r}$?

85. In your own words, explain how to form the rows in Pascal's Triangle.

Ongoing Review

In Exercises 86–89, evaluate the determinant of the matrix.

86. $\begin{bmatrix} 10 & 25 \\ 6 & -5 \end{bmatrix}$

87. $\begin{bmatrix} 3 & 7 \\ -2 & 6 \end{bmatrix}$

88. $\begin{bmatrix} 3 & -2 & 1 \\ 0 & 5 & 3 \\ 6 & 1 & 1 \end{bmatrix}$

89. $\begin{bmatrix} 4 & 3 & 5 \\ 3 & 2 & -2 \\ 5 & -2 & 0 \end{bmatrix}$

In Exercises 90–95, solve the equation. (If necessary, round the result to two decimal places.)

90. $\log_4 64 = x$

91. $\log_3 \frac{1}{81} = x$

92. $3^x = 50$

93. $e^{x/2} = 8$

94. $\log_2(x - 5) = 6$

95. $\ln(x + 3) = 10$

96. *Area of a Triangle* The vertices of a triangle are located at coordinates $(-5, 8)$, $(10, 0)$, and $(3, -4)$. Use a determinant to find the area of the triangle.

97. *Path of a Ball* The path of a ball passes through the points $(0, 2)$, $(10, 8)$, and $(20, 0)$. Use Cramer's Rule to find the quadratic equation that models the path of the ball.

Looking Further

Patterns in Pascal's Triangle Use each encircled group of numbers to form a 2×2 matrix. Find the determinant of each matrix. Describe the pattern.

Chapter Summary: Key Concepts

10.1 ■ Sequences

An infinite sequence $a_1, a_2, a_3, \ldots, a_n, \ldots$ is a function whose domain is the set of positive integers.

A finite sequence $a_1, a_2, a_3, \ldots, a_n$ is a function whose domain is the finite set $\{1, 2, 3, \ldots, n\}$.

10.1 ■ Definition of factorial

If n is a positive integer, n factorial is defined as

$$n! = 1 \cdot 2 \cdot 3 \cdot 4 \cdot \cdots \cdot (n - 1) \cdot n.$$

As a special case, zero factorial is defined as $0! = 1$.

10.1 ■ Definition of a series

For an infinite sequence $a_1, a_2, a_3, \ldots, a_n, \ldots$
1. the sum of the first n terms

$$S_n = a_1 + a_2 + a_3 + \cdots + a_n$$

is called a partial sum, and

2. the sum of all the terms

$$a_1 + a_2 + a_3 + \cdots + a_n + \cdots$$

is called an infinite series, or simply a series.

10.1 ■ Definition of sigma notation

The sum of the first n terms of the sequence whose nth term is a_n is

$$\sum_{i=1}^{n} a_i = a_1 + a_2 + a_3 + a_4 + \cdots + a_n$$

where i is the index of summation, n is the upper limit of summation, and 1 is the lower limit of summation.

10.2 ■ The nth term of an arithmetic sequence

The nth term of an arithmetic sequence has the form

$$a_n = a_1 + (n - 1)d$$

where d is the common difference between the terms of the sequence, and a_1 is the first term.

10.2 ■ The nth partial sum of an arithmetic sequence

The nth partial sum of the arithmetic sequence whose nth term is a_n is

$$\sum_{i=1}^{n} a_i = a_1 + a_2 + a_3 + a_4 + \cdots + a_n = \frac{n}{2}(a_1 + a_n).$$

10.3 ■ The nth term of a geometric sequence

The nth term of a geometric sequence has the form

$$a_n = a_1 r^{n-1}$$

where r is the common ratio of consecutive terms of the sequence. So, every geometric sequence can be written in the following form.

$$a_1, a_1 r, a_1 r^2, a_1 r^3, a_1 r^4, \ldots, a_1 r^{n-1}, \ldots$$

10.3 ■ The nth partial sum of a geometric sequence

The nth partial sum of the geometric sequence whose nth term is $a_n = a_1 r^{n-1}$ is given by

$$\sum_{i=1}^{n} a_1 r^{i-1} = a_1 + a_1 r + a_1 r^2 + a_1 r^3 + \cdots + a_1 r^{n-1}$$
$$= a_1 \left(\frac{r^n - 1}{r - 1} \right).$$

10.3 ■ Sum of an infinite geometric series

If $a_1, a_1 r, a_1 r^2, \ldots, a_1 r^n, \ldots$ is an infinite geometric sequence, then for $|r| < 1$, the sum of the terms of the corresponding infinite geometric series is

$$S = \sum_{i=0}^{\infty} a_1 r^i = \frac{a_1}{1 - r}.$$

10.4 ■ The Binomial Theorem

In the expansion of $(x + y)^n$

$$(x + y)^n = x^n + nx^{n-1}y + \cdots +$$
$$\quad _nC_r x^{n-r}y^r + \cdots + nxy^{n-1} + y^n$$

the coefficient of $x^{n-r}y^r$ is given by

$$_nC_r = \frac{n!}{(n - r)!\,r!}.$$

Review Exercises

Reviewing Skills

10.1 ■ *How to use sequence notation to write the terms of sequences*

In Exercises 1–4, write the first five terms of the sequence. (Assume that n begins with 1.)

1. $a_n = 3n + 5$

2. $a_n = \frac{1}{2}n - 4$

3. $a_n = \frac{1}{2^n} + \frac{1}{2}$

4. $a_n = n + (-1)^n$

In Exercises 5–8, write the first six terms of the sequence defined recursively. (Assume that n begins with 1.)

5. $a_1 = 1$, $a_{n+1} = 5a_n - 2$

6. $a_1 = 2$, $a_{n+1} = 1 - a_n$

7. $a_1 = 2$, $a_2 = 3$, $a_{n+2} = a_{n+1} - a_n$

8. $a_1 = 1$, $a_2 = 4$, $a_{n+2} = 2a_n + a_{n+1}$

In Exercises 9 and 10, find the indicated term of the sequence.

9. $a_n = (-1)^n(n + 4)$

$a_{10} = $ ▨

10. $a_n = \dfrac{(-1)^{n+1}}{n^2}$

$a_{10} = $ ▨

▨ In Exercises 11 and 12, use a graphing utility to graph the first 10 terms of the sequence.

11. $a_n = \dfrac{3n}{n + 1}$

12. $a_n = \dfrac{3}{n + 1}$

■ *How to write the terms of sequences involving factorials*

In Exercises 13–16, write the first five terms of the sequence. (Assume that n begins with 1.)

13. $a_n = (n + 1)!$

14. $a_n = 2(n!)$

15. $a_n = \dfrac{n}{n!}$

16. $a_n = \dfrac{(n - 1)!}{n^2}$

■ *How to find the apparent nth term of a sequence*

In Exercises 17–20, write an expression for the nth term of the sequence. There are many correct answers.

17. $1, 3, 5, 7, 9, \ldots$

18. $3, -6, 9, -12, 15, \ldots$

19. $\frac{1}{4}, \frac{2}{9}, \frac{3}{16}, \frac{4}{25}, \frac{5}{36}, \ldots$

20. $\frac{0}{2}, \frac{1}{3}, \frac{2}{4}, \frac{3}{5}, \frac{4}{6}, \ldots$

■ *How to sum the terms of sequences to obtain series and use sigma notation to represent partial sums*

In Exercises 21–24, find the partial sum.

21. $\displaystyle\sum_{k=1}^{4} 7$

22. $\displaystyle\sum_{k=1}^{4} \frac{(-1)^k}{k}$

23. $\displaystyle\sum_{n=1}^{4} \left(\frac{1}{n} - \frac{1}{n + 1} \right)$

24. $\displaystyle\sum_{n=1}^{4} \left(\frac{1}{n} - \frac{1}{n + 2} \right)$

In Exercises 25–28, use sigma notation to write the sum. There are many correct answers.

25. $[5(1) - 3] + [5(2) - 3] + [5(3) - 3] + [5(4) - 3]$

26. $[9 - 2(1)] + [9 - 2(2)] + [9 - 2(3)] + [9 - 2(4)]$

27. $\dfrac{1}{3(1)} + \dfrac{1}{3(2)} + \dfrac{1}{3(3)} + \dfrac{1}{3(4)} + \dfrac{1}{3(5)} + \dfrac{1}{3(6)}$

28. $\left(-\frac{1}{3}\right)^0 + \left(-\frac{1}{3}\right)^1 + \left(-\frac{1}{3}\right)^2 + \left(-\frac{1}{3}\right)^3 + \left(-\frac{1}{3}\right)^4$

10.2 ■ *How to recognize, write, and find the nth terms of arithmetic sequences*

In Exercises 29–36, write the first five terms of the arithmetic sequence. (Assume that n begins with 1.)

29. $a_n = 132 - 5n$

30. $a_n = 2n + 3$

31. $a_n = \frac{3}{4}n + \frac{1}{2}$

32. $a_n = -\frac{3}{5}n + 1$

33. $a_1 = 5$
$a_{k+1} = a_k + 3$

34. $a_1 = 12$
$a_{k+1} = a_k + 1.5$

35. $a_1 = 80$
$a_{k+1} = a_k - \frac{5}{2}$

36. $a_1 = 25$
$a_{k+1} = a_k - 6$

In Exercises 37–40, find a formula for the nth term of the arithmetic sequence.

37. $a_1 = 10$, $d = 4$ **38.** $a_1 = 32$, $d = -2$

39. $a_1 = 1000$, $a_2 = 950$ **40.** $a_1 = 12$, $a_2 = 20$

■ *How to find the nth partial sum of an arithmetic sequence*

In Exercises 41–44, find the nth partial sum of the arithmetic sequence.

41. $\displaystyle\sum_{k=1}^{12} (7k - 5)$ **42.** $\displaystyle\sum_{k=1}^{10} (100 - 10k)$

43. $\displaystyle\sum_{j=1}^{100} \frac{j}{4}$ **44.** $\displaystyle\sum_{j=1}^{50} \frac{3j}{2}$

In Exercises 45 and 46, use a graphing utility to evaluate the partial sum.

45. $\displaystyle\sum_{i=1}^{60} (1.25i + 4)$ **46.** $\displaystyle\sum_{i=1}^{100} (5000 - 3.5i)$

10.3 ■ *How to recognize, write, and find the nth terms of geometric sequences*

In Exercises 47–52, write the first five terms of the geometric sequence.

47. $a_1 = 10$, $r = 3$ **48.** $a_1 = 2$, $r = -5$

49. $a_1 = 100$, $r = -\frac{1}{2}$ **50.** $a_1 = 12$, $r = \frac{1}{6}$

51. $a_1 = 3$, $a_{k+1} = 2a_k$ **52.** $a_1 = 36$, $a_{k+1} = \frac{1}{2}a_k$

In Exercises 53–58, find a formula for the nth term of the geometric sequence.

53. $a_1 = 1$, $r = -\frac{2}{3}$

54. $a_1 = 100$, $r = 1.07$

55. $a_1 = 24$, $a_2 = 48$

56. $a_1 = 16$, $a_2 = -4$

57. $a_1 = 12$, $a_4 = -\frac{3}{2}$

58. $a_2 = 1$, $a_3 = \frac{1}{3}$

■ *How to find the nth partial sums of geometric sequences*

In Exercises 59–66, find the nth partial sum of the geometric sequence. If necessary, round answers to three decimal places.

59. $\displaystyle\sum_{n=1}^{12} 2^n$ **60.** $\displaystyle\sum_{n=1}^{12} (-2)^n$

61. $\displaystyle\sum_{k=1}^{8} 5\left(-\frac{3}{4}\right)^k$ **62.** $\displaystyle\sum_{k=1}^{10} 4\left(\frac{3}{2}\right)^k$

63. $\displaystyle\sum_{i=1}^{8} (1.25)^{i-1}$ **64.** $\displaystyle\sum_{i=1}^{8} (-1.25)^{i-1}$

65. $\displaystyle\sum_{n=1}^{120} 500(1.01)^n$ **66.** $\displaystyle\sum_{n=1}^{40} 1000(1.1)^n$

In Exercises 67 and 68, use a graphing utility to evaluate the partial sum. If necessary, round answers to three decimal places.

67. $\displaystyle\sum_{k=1}^{50} 50(1.2)^{k-1}$ **68.** $\displaystyle\sum_{j=1}^{60} 25(0.9)^{j-1}$

■ *How to find the sums of infinite geometric series*

In Exercises 69–72, find the sum of the infinite geometric series.

69. $\displaystyle\sum_{i=1}^{\infty} \left(\frac{7}{8}\right)^{i-1}$ **70.** $\displaystyle\sum_{i=1}^{\infty} \left(\frac{1}{3}\right)^{i-1}$

71. $\displaystyle\sum_{k=1}^{\infty} 4\left(\frac{2}{3}\right)^{k-1}$ **72.** $\displaystyle\sum_{k=1}^{\infty} 1.3\left(\frac{1}{10}\right)^{k-1}$

10.4 ■ *How to use the Binomial Theorem to calculate binomial coefficients*

In Exercises 73–76, evaluate the binomial coefficient $_nC_r$.

73. $_8C_3$ **74.** $_{12}C_2$

75. $_{12}C_0$ **76.** $_{100}C_1$

In Exercises 77–80, use a graphing utility to evaluate $_nC_r$.

77. $_{40}C_4$ **78.** $_{15}C_9$

79. $_{25}C_6$ **80.** $_{32}C_2$

■ *How to use Pascal's Triangle to calculate binomial coefficients*

In Exercises 81–84, use Pascal's Triangle to expand the expression.

81. $(x + 9)^4$ **82.** $(b - 7)^5$

83. $(5 - 2x)^3$ **84.** $(3 + 4y)^4$

■ *How to expand binomial expressions*

In Exercises 85–94, use the Binomial Theorem to expand the expression. Simplify the result.

85. $(x + 1)^{10}$ **86.** $(u - v)^9$

87. $(y - 2)^6$ **88.** $(x + 3)^5$

89. $\left(\frac{1}{2} - x\right)^8$ **90.** $\left(\frac{1}{3} + a\right)^5$

91. $(10x + 3y)^4$
92. $(3x - 2y)^4$
93. $(u^2 + v^3)^9$
94. $(x^4 - y^5)^8$

In Exercises 95–98, find the coefficient of the term in the expansion of the binomial.

Expression	Term
95. $(x - 3)^{10}$	x^5
96. $(x + 4)^9$	x^6
97. $(a + b)^6$	ab^5
98. $(x - y)^5$	x^3y^2

In Exercises 99–102, find the specified nth term in the expansion of the binomial.

99. $(2x - 3y)^5$, $n = 4$
100. $(4a + 3b)^4$, $n = 3$
101. $(10x - 3y)^9$, $n = 7$
102. $(5a + 10b)^{10}$, $n = 10$

Solving Problems

103. *Compound Interest* A deposit of $5000 is made in an account that earns 8% interest compounded quarterly. The balance in the account after n quarters is

$$A_n = 5000\left(1 + \frac{0.08}{4}\right)^n, \quad n = 1, 2, 3, \ldots$$

(a) Compute the first eight terms of the sequence.

(b) Find the balance in this account after 10 years by computing A_{40}.

(c) Use a graphing utility to graph the first 40 terms of the sequence.

104. *Number Problem* Find the sum of the first 50 positive integers that are multiples of 4.

105. *Number Problem* Find the sum of the integers from 225 to 300.

106. *Auditorium Seating* Each row in a small auditorium has three more seats than the preceding row. Find the seating capacity of the auditorium if the front row seats 22 people and there are 12 rows of seats.

107. *Free-Falling Object* A free-falling object will fall 4.9 meters during the first second, 14.7 more meters during the second second, 24.5 more meters during the third second, and so on. What is the total distance the object will fall in 10 seconds if this pattern continues?

108. *Depreciation* A company pays $120,000 for a machine. During the next 5 years, the machine depreciates at a rate of 30% per year. (That is, at the end of each year, the depreciated value is 70% of what it was at the beginning of the year.)

(a) Find a formula for the nth term of the geometric sequence that gives the value of the machine n full years after it was purchased.

(b) Find the depreciated value of the machine at the end of 5 full years.

109. *Population Growth* A city of 85,000 people is growing at a rate of 1.2% per year. (That is, at the end of each year, the population is 1.012 times what it was at the beginning of the year.)

(a) Find a formula for the nth term of the geometric sequence that gives the population n years from now.

(b) Estimate the population 50 years from now.

110. *Salary Increase* You accept a job that pays a salary of $32,000 the first year. During the next 39 years, you receive a 5.5% raise each year. What is your total salary over the 40-year period?

Increasing Annuity In Exercises 111–114, find the balance in an increasing annuity in which a principal of P dollars is invested each month for t years, compounded monthly at rate r.

111. $P = \$50$, $t = 20$ years, $r = 7\%$
112. $P = \$75$, $t = 25$ years, $r = 9\%$
113. $P = \$100$, $t = 40$ years, $r = 10\%$
114. $P = \$20$, $t = 50$ years, $r = 6\%$

10 Chapter Test

Take this test as you would take a test in class. After you are done, check your work against the answers given in the back of the book.

1. Write the first five terms of the sequence $a_n = \left(-\frac{2}{3}\right)^{n-1}$. (Assume that n begins with 1.)

2. Write the first five terms of the sequence defined recursively as

$$a_1 = 76, \ a_{k+1} = \frac{a_k}{2} + 6.$$

3. Write an expression for the nth term of the sequence

$$\frac{1}{2 \cdot 3}, \frac{1}{3 \cdot 4}, \frac{1}{4 \cdot 5}, \frac{1}{5 \cdot 6}, \ldots$$

4. Evaluate: $\displaystyle\sum_{n=1}^{5} \frac{n!}{7}$.

5. Use sigma notation to write $\dfrac{2}{3(1) + 1} + \dfrac{2}{3(2) + 1} + \cdots + \dfrac{2}{3(12) + 1}$.

6. Write the first five terms of the arithmetic sequence whose first term is $a_1 = 12$ and whose common difference is $d = 4$.

7. Find a formula for the nth term of the arithmetic sequence whose first term is $a_1 = 5000$ and whose common difference is $d = -100$.

8. Find the common ratio of the geometric sequence $2, -3, \frac{9}{2}, -\frac{27}{4}, \ldots$.

9. Find a formula for the nth term of the geometric sequence whose first term is $a_1 = 4$ and whose common ratio is $r = \frac{1}{2}$.

In Exercises 10 and 11, evaluate the sum.

10. $\displaystyle\sum_{j=0}^{4} (3j + 1)$

11. $\displaystyle\sum_{n=1}^{10} 3\left(\frac{1}{2}\right)^{n}$

12. Evaluate: $\displaystyle\sum_{i=1}^{\infty} 4\left(\frac{2}{3}\right)^{i-1}$.

13. Find the sum of the first 50 positive integers that are multiples of 3.

14. Fifty dollars is deposited each month in an increasing annuity that pays 8%, compounded monthly. What is the balance after 25 years?

15. Evaluate $_{20}C_3$.

16. Use Pascal's Triangle to expand $(x - 2)^5$.

17. Find the coefficient of the term x^3y^5 in the expansion of $(x + y)^8$.

18. You accept a job that pays a salary of $35,000 the first year. During the next 14 years, you receive a 5% raise each year. What is your total salary over the 15-year period?

Cumulative Test: Chapters 1–10

Take this test as you would take a test in class. After you are done, check your work against the answers given in the back of the book.

In Exercises 1–6, simplify the expression.

1. $(-3x^2y^3)^2 \cdot (4xy^2)$

2. $\frac{3}{8}x - \frac{1}{12}x + 8$

3. $\frac{64r^2s^4}{16rs^2}$, $r \neq 0, s \neq 0$

4. $\left(\frac{3x}{4y^3}\right)^2$, $y \neq 0$

5. $\frac{8}{\sqrt{10}}$

6. $\log_4 64$

In Exercises 7–10, factor the expression completely.

7. $5x - 20x^2$

8. $64 - (x - 6)^2$

9. $15x^2 - 16x - 15$

10. $8x^3 + 1$

In Exercises 11–14, sketch the graph of the equation.

11. $y = 2x - 3$

12. $y = -\frac{3}{4}x + 2$

13. $9x^2 + 4y^2 = 36$

14. $(x - 2)y = 5$

15. Write the equation of the line through the points $\left(\frac{3}{2}, 8\right)$ and $\left(\frac{11}{2}, \frac{5}{2}\right)$.

16. Find the domain of the function $h(x) = \sqrt{16 - x^2}$.

17. Find $g(c - 6)$ if $g(x) = \frac{x}{x + 10}$.

18. The number N of prey t months after a predator is introduced into an area is inversely proportional to $t + 1$. If $N = 300$ when $t = 0$, find N when $t = 5$.

19. Find an equation of the parabola shown in the figure.

20. A game manufacturer determines that the profit P for a board game that is on the market for t years is given by the equation

$$P = 6000 + 20{,}000(3)^{-0.2t}.$$

(a) Determine the time when the profit is $13,000.

(b) Use a graphing utility to graph the profit function. Use the graph to verify your answer to part (a).

21. At age 5, a girl's height is approximately 62% of her full adult height. At age 15, she has reached about 98% of her adult height. The logarithmic function below gives an approximate percentage H of adult height a girl has reached at any age x from 5 to 15 years.

$$H = 62 + 35 \log_{10}(x - 4)$$

A girl is $4'6''$ at age 10. How tall can she expect to be as an adult?

22. Use the Binomial Theorem to expand and simplify $(z - 3)^4$.

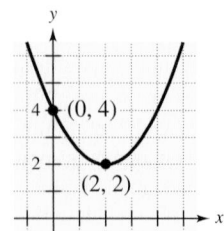

Figure for 19

Appendix

The remaining appendices are located on the website that accompanies this text at *college.hmco.com*.

Appendix A Real Numbers

A.1 Operations with Real Numbers

What you should learn:

- How to understand the set of real numbers and the subsets of real numbers
- How to use the real number line to order real numbers and find distances between real numbers
- How to determine the absolute value of a real number
- How to add, subtract, multiply, and divide real numbers
- How to write repeated multiplication in exponential form and evaluate exponential expressions
- How to find the principal nth root of a number
- How to use order of operations to evaluate expressions

Why you should learn it:

Real numbers are used to represent many real-life quantities, such as a company's profits (see Exercise 171 on page A15).

Real Numbers

The formal term that is used in mathematics to talk about a collection of objects is the word **set.** For instance, the set

$$\{1, 2, 3\}$$

contains the three numbers 1, 2, and 3. In this text, a *pair* of braces { } always indicates the members of a set. Parentheses ( ) and brackets [ ] are used to represent other ideas.

The set of numbers that is used in arithmetic is the set of **real numbers.** The term *real* distinguishes real numbers from *imaginary* numbers—a type of number that is discussed in Chapter 5. One of the most commonly used **subsets** of real numbers is the set of **natural numbers** or **positive integers**

$$\{1, 2, 3, 4, \ldots\}. \qquad \text{Set of natural numbers}$$

Note that the three dots indicate that the pattern continues. For instance, the set also contains the numbers 5, 6, 7, and so on.

Positive integers can be used to describe many quantities that you encounter in everyday life—you might be taking four classes this term, or you might be paying \$420 per month for rent. But even in everyday life, positive integers cannot describe some concepts accurately. For instance, you could have a zero balance in your checking account, or the temperature could be $-10°$ (10 degrees below zero). To describe such quantities, you need to expand the set of positive integers to include **zero** and the **negative integers.** The positive integers and zero make up the set of **whole numbers.** The expanded set containing the whole numbers and the negative integers is called the set of **integers,** which is written as follows.

$$\underbrace{\{\ldots, -3, -2, -1,}_{\text{Negative integers}} \overbrace{0, 1, 2, 3, \ldots\}}^{\text{Whole Numbers}}_{\text{Positive integers}} \qquad \text{The set of integers}$$

The set of integers is a subset of the set of real numbers, which is another way of saying that every integer is a real number. Even when the entire set of integers is used, there are still many quantities in everyday life that cannot be described accurately. The costs of many items are not in whole-dollar amounts, but in parts of dollars, such as \$1.19 or \$39.98. You might work $8\frac{1}{2}$ hours, or you might miss the first *half* of a movie. To describe quantities, you can expand the set of integers to include **fractions.** The expanded set is called the set of **rational numbers.** Formally, a real number is called **rational** if it can be written as the ratio p/q of two integers, where $q \neq 0$ (the symbol $\neq$ means **does not equal**). Here are some examples of rational numbers.

$$2 = \frac{2}{1}, \quad \frac{1}{3} = 0.333\ldots, \quad \frac{1}{8} = 0.125, \quad \text{and} \quad \frac{125}{111} = 1.126126\ldots$$

Real numbers that cannot be written as ratios of two integers are called **irrational.** For instance, the numbers

$$\sqrt{2} = 1.4142135 \ldots \quad \text{and} \quad \pi = 3.1415926 \ldots$$

are irrational. The decimal representation of a rational number is either **terminating** or **repeating.** For instance, the decimal representation of $\frac{1}{4} = 0.25$ is terminating, and the decimal representation of

$$\frac{4}{11} = 0.363636 \cdots$$
$$= 0.\overline{36}$$

is repeating. (The line over 36 indicates which digits repeat.)

The decimal representation of an irrational number neither terminates nor repeats. When performing operations with such numbers, you usually use a decimal approximation that has been **rounded** to a certain number of decimal places. The rounding rule used in this text is to round *up* if the succeeding digit is 5 or more and round *down* if the succeeding digit is 4 or less. For example, if you wanted to round 7.35 to one decimal place, you would round up to 7.4. Similarly, if you wanted to round 2.364 to two decimal places, you would round down to 2.36. Rounded to four decimal places, the decimal approximations of $\frac{2}{3}$ and π are

$$\frac{2}{3} \approx 0.6667 \quad \text{and} \quad \pi \approx 3.1416.$$

NOTE   The symbol $\approx$ means **is approximately equal to.**

Figure A.1 shows several commonly used subsets of real numbers and their relationships to each other.

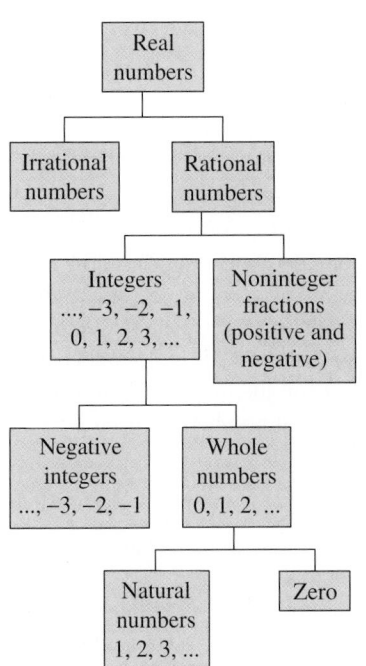

Figure A.1   *Subsets of real numbers*

Example 1 ■ Identifying Real Numbers

Which of the numbers in the set

$$\left\{ -7, -\sqrt{3}, -1, -\frac{1}{5}, 0, \frac{3}{4}, \sqrt{2}, \pi, 5 \right\}$$

are (a) natural numbers, (b) integers, (c) rational numbers, and (d) irrational numbers?

Solution

(a) Natural numbers: $\{5\}$

(b) Integers: $\{-7, -1, 0, 5\}$

(c) Rational numbers: $\left\{ -7, -1, -\frac{1}{5}, 0, \frac{3}{4}, 5 \right\}$

(d) Irrational numbers: $\left\{ -\sqrt{3}, \sqrt{2}, \pi \right\}$

Order and Distance

A **real number line** can be used to picture the real numbers. It consists of a horizontal line with a point (the **origin**) labeled as 0. Numbers to the left of 0 are **negative** and numbers to the right of 0 are **positive,** as shown in Figure A.2.

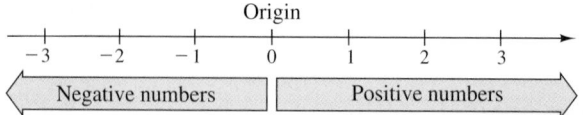

Figure A.2

The real number zero is neither positive nor negative. So, when you want to talk about real numbers that might be positive or zero, you should use the term **nonnegative real number.**

Each point on the real number line corresponds to exactly one real number, and each real number corresponds to exactly one point on the real number line, as shown in Figure A.3. When you draw the point (on the real number line) that corresponds to a real number, you are **plotting** the real number.

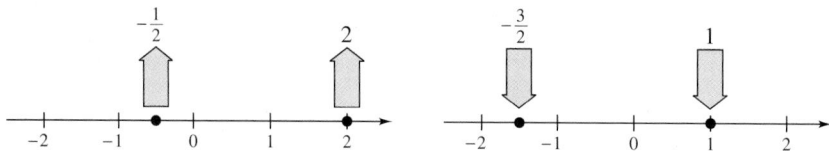

Figure A.3

The real number line provides you with a way of comparing any two real numbers. For instance, if you choose any two (different) numbers on the real number line, one of the numbers must be to the left of the other number. The number to the left is **less than** the number to the right. Similarly, the number to the right is **greater than** the number to the left, as shown in Figure A.4.

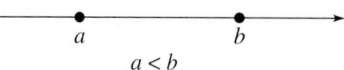

Figure A.4   *a is to the left of b.*

Definition of Order on the Real Number Line

If a and b are real numbers, a is **less than** b if $b - a$ is positive. The **order** of a and b is denoted by the inequality

$$a < b.$$

This relationship can also be described by saying that b is **greater than** a and by writing $b > a$. The inequality $a \leq b$ means that a is **less than or equal to** b, and the inequality $b \geq a$ means that b is **greater than or equal to** a. The symbols $<$, $>$, $\leq$, and $\geq$ are called **inequality symbols.**

When you are asked to order two numbers, you are simply being asked to say which of the two numbers is greater.

Example 2 ■ Ordering Real Numbers

Place the correct inequality symbol ($<$ or $>$) between each pair.

(a) -3 -5 (b) $\frac{1}{5}$ $\frac{1}{3}$ (c) -4 0

Solution

(a) Because -3 lies to the right of -5 on the real number line, it follows that -3 is *greater than* -5.

$$-3 > -5 \qquad \text{See Figure A.5(a).}$$

(b) Because $\frac{1}{5}$ lies to the left of $\frac{1}{3}$ on the real number line, it follows that $\frac{1}{5}$ is *less than* $\frac{1}{3}$.

$$\frac{1}{5} < \frac{1}{3} \qquad \text{See Figure A.5(b).}$$

(c) Because -4 lies to the left of 0 on the real number line, it follows that -4 is *less than* 0.

$$-4 < 0 \qquad \text{See Figure A.5(c).}$$

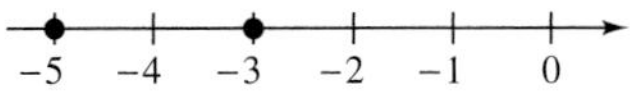

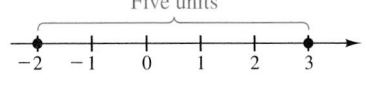

(a)

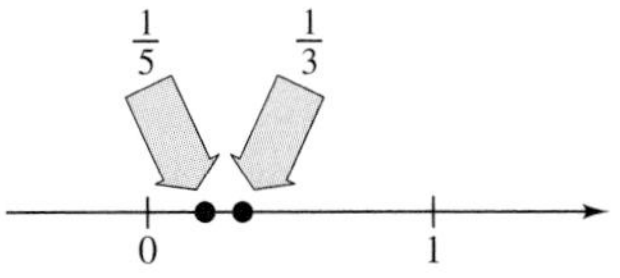

(b)

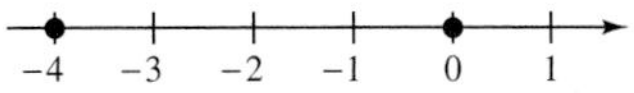

(c)

Figure A.5

Once you know how to represent real numbers as points on the real number line, it is natural to talk about the **distance between two real numbers.** Specifically, if a and b are two real numbers such that $a \leq b$, the distance between a and b is defined as $b - a$.

> **NOTE** Note from this definition that if $a = b$, the distance between a and b is zero. If $a < b$, the distance between a and b is positive.

> **Definition of Distance Between Two Real Numbers**
>
> If a and b are two real numbers such that $a \leq b$, then the **distance between a and b** is
>
> (Distance between a and b) $= b - a$.

Example 3 ■ Finding the Distance Between Two Real Numbers

Find the distance between the real numbers in each pair.

(a) -2 and 3 (b) 0 and 4 (c) 1 and $-\frac{1}{2}$

Solution

(a) Because $-2 \leq 3$, the distance between -2 and 3 is

$$3 - (-2) = 3 + 2 = 5. \qquad \text{See Figure A.6(a).}$$

(b) Because $0 \leq 4$, the distance between 0 and 4 is

$$4 - 0 = 4. \qquad \text{See Figure A.6(b).}$$

(c) Because $-\frac{1}{2} \leq 1$, the distance between 1 and $-\frac{1}{2}$ is

$$1 - \left(-\frac{1}{2}\right) = 1 + \frac{1}{2} = 1\frac{1}{2}. \qquad \text{See Figure A.6(c).}$$

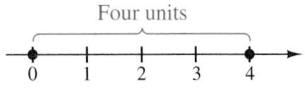

(a)

(b)

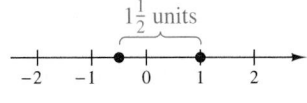

(c)

Figure A.6

Absolute Value

The distance between a real number a and 0 (the origin) is called the **absolute value** of a. Absolute value is denoted by double vertical bars, $|\ \ |$. For example

$$|5| = \text{"distance between 5 and 0"} = 5$$

and

$$|-8| = \text{"distance between } -8 \text{ and 0"} = 8.$$

Study Tip

Be sure you see from this definition that the absolute value of a real number is never negative. For instance, if $a = -3$, then $|-3| = -(-3) = 3$. Moreover, the only real number whose absolute value is zero is 0. That is, $|0| = 0$.

Absolute Value of a Real Number

The **absolute value** of a real number a is the distance between a and 0 on the real number line.

1. If $a > 0$, then $|a| = a - 0 = a$.

2. If $a = 0$, then $|a| = 0 - 0 = 0$.

3. If $a < 0$, then $|a| = 0 - a = -a$.

Two real numbers are called **opposites** of each other if they lie the same distance from, but on opposite sides of, zero on the real number line. For instance, -2 is the opposite of 2. (See Figure A.7.) Because *opposite* numbers lie the same distance from zero on the real number line, they have the same absolute value. So, $|5| = 5$ and $|-5| = 5$. The use of a negative sign to denote the *opposite* of a number gives meaning to expressions such as $-(-3)$ and $-|-3|$, as follows.

$$-(-3) = (\text{opposite of } -3) = 3$$

$$-|-3| = \left(\text{opposite of } |-3|\right) = -3$$

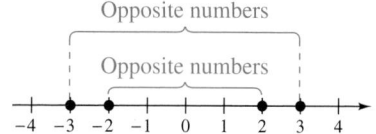

Opposite numbers

Opposite numbers

Figure A.7

Opposite numbers are also referred to as **additive inverses** because their sum is zero. For instance, $3 + (-3) = 0$, or, in general, $b + (-b) = 0$.

Example 4 ■ Finding Absolute Values

Evaluate each expression.

(a) $|-10|$ (b) $\left|\dfrac{3}{4}\right|$ (c) $|-3.2|$ (d) $-|-6|$

Solution

(a) $|-10| = 10$ The absolute value of -10 is 10.

(b) $\left|\dfrac{3}{4}\right| = \dfrac{3}{4}$ The absolute value of $\frac{3}{4}$ is $\frac{3}{4}$.

(c) $|-3.2| = 3.2$ The absolute value of -3.2 is 3.2.

(d) $-|-6| = -(6) = -6$ The opposite of $|-6|$ is -6.

Note that part (d) does not contradict the fact that the absolute value of a number cannot be negative. The expression $-|-6|$ calls for the *opposite* of an absolute value and so it must be negative.

For any two real numbers a and b, exactly one of the following orders must be true.

$a < b$, $a = b$, or $a > b$.

This property of real numbers is called the **Law of Trichotomy.** In words, this property tells you that if a and b are any two real numbers, then a is less than b, a is equal to b, or a is greater than b.

Example 5 ■ Comparing Real Numbers

Place the correct symbol ($<$, $>$, or $=$) between each pair.

(a) $|-9|$ $|9|$ (b) $|-2|$ 1

(c) -4 $-|-4|$ (d) $|12|$ $|-15|$

(e) 2 $-|-2|$ (f) $|-3|$ 5

Solution

(a) $|-9| = |9|$, because $|-9| = 9$ and $|9| = 9$.

(b) $|-2| > 1$, because $|-2| = 2$ and 2 is greater than 1.

(c) $-4 = -|-4|$, because $-|-4| = -4$ and -4 is equal to -4.

(d) $|12| < |-15|$, because $|12| = 12$ and $|-15| = 15$, and 12 is less than 15.

(e) $2 > -|-2|$, because $-|-2| = -2$ and 2 is greater than -2.

(f) $|-3| < 5$, because $|-3| = 3$ and 3 is less than 5.

When the distance between two real numbers a and b was defined as $b - a$, the definition included the restriction $a \leq b$. Using absolute value, you can generalize this definition as follows. If a and b are *any* two real numbers, the distance between a and b is

Distance between a and $b = |b - a| = |a - b|$.

For instance, the distance between -2 and 1 is

$|-2 - 1| = |-3| = 3$.

NOTE You could also find the distance between -2 and 1 as follows.

$|1 - (-2)| = |3| = 3$

Operations with Real Numbers

There are four basic operations of arithmetic: addition, subtraction, multiplication, and division.

The result of **adding** two real numbers is called the **sum** of the two numbers, and the two real numbers are called **terms** of the sum. **Subtraction** of one real number from another can be described as *adding the opposite* of the second number to the first number. For instance

$7 - 5 = 7 + (-5) = 2$ and $10 - (-13) = 10 + 13 = 23$.

The result of subtracting one real number from another is called the **difference** of the two numbers.

Example 6 ■ Adding and Subtracting Real Numbers

(a) $-84 + 14 = -70$

(b) $6 + (-13) + 10 = 3$

(c) $-13.8 - 7.02 = -13.8 + (-7.02) = -20.82$

(d) $\dfrac{1}{5} - \dfrac{2}{5} = \dfrac{1-2}{5} = -\dfrac{1}{5}$

NOTE In the fraction

$$\frac{a}{b}$$

a is the **numerator** and b is the **denominator.**

(e) To add two fractions with *unlike denominators*, you must first rewrite one (or both) of the fractions so that they have a common denominator.

$$\frac{3}{8} + \frac{5}{12} = \frac{3(3)}{8(3)} + \frac{5(2)}{12(2)} \qquad \text{Common denominator is 24.}$$

$$= \frac{9}{24} + \frac{10}{24} \qquad \text{Fractions have like denominators.}$$

$$= \frac{9 + 10}{24} \qquad \text{Add numerators.}$$

$$= \frac{19}{24}$$

The result of **multiplying** two real numbers is called their **product,** and each of the two numbers is called a **factor** of the product. The product of zero and any other number is zero. For instance, if you multiply 0 and 4, you obtain $(0)(4) = 0$.

Multiplication is denoted in a variety of ways. For instance,

$$7 \times 3, \quad 7 \cdot 3, \quad 7(3), \quad (7)3, \quad \text{and} \quad (7)(3)$$

all denote the product of "7 times 3," which you know is 21.

Example 7 ■ Multiplying Real Numbers

(a) $(-5)(-7) = 35$

(b) $(-1.2)(0.4) = -0.48$

(c) To find the product of more than two numbers, find the product of their absolute values. If there is an *even* number of negative factors, the product is positive. If there is an *odd* number of negative factors, the product is negative. For instance, in the product below there are two negative factors, so the product must be positive, and you can write

$$5(-3)(-4)(7) = 420.$$

(d) To multiply two fractions, multiply their numerators and their denominators. For instance, the product of $\frac{2}{3}$ and $\frac{4}{5}$ is

$$\left(\frac{2}{3}\right)\left(\frac{4}{5}\right) = \frac{(2)(4)}{(3)(5)}$$

$$= \frac{8}{15}.$$

The **reciprocal** of a nonzero real number a is defined as the number by which a must be multiplied to obtain 1. For instance, the reciprocal of 3 is $\frac{1}{3}$ because

$$3\left(\frac{1}{3}\right) = 1.$$

Similarly, the reciprocal of $-\frac{4}{5}$ is $-\frac{5}{4}$ because

$$-\frac{4}{5}\left(-\frac{5}{4}\right) = 1.$$

In general, the reciprocal of a/b is b/a. Note that the reciprocal of a positive number is positive, and the reciprocal of a negative number is negative. Also, be sure you see that zero does not have a reciprocal because there is no number that can be multiplied by zero to obtain 1.

To divide one real number by a second (nonzero) real number, multiply the first number by the reciprocal of the second number. The result of dividing two real numbers is called the **quotient** of the two numbers. The quotient of a and b can be written as

$$a \div b, \quad a/b, \quad \frac{a}{b}, \quad \text{or} \quad b\overline{)a}.$$

The numerator a is called the **dividend**, and the denominator b is called the **divisor**.

Example 8   ■   Dividing Real Numbers

(a) $-30 \div 5 = -30\left(\frac{1}{5}\right) = -\frac{30}{5} = -6$

(b) $-\frac{9}{14} \div -\frac{1}{3} = -\frac{9}{14}\left(-\frac{3}{1}\right) = \frac{27}{14}$

(c) $\frac{5}{16} \div 2\frac{3}{4} = \frac{5}{16} \div \frac{11}{4} = \frac{5}{16}\left(\frac{4}{11}\right) = \frac{5(4)}{4(4)(11)} = \frac{5}{44}$

(d) $\dfrac{-\frac{2}{3}}{\frac{3}{5}} = -\frac{2}{3} \div \frac{3}{5} = -\frac{2}{3}\left(\frac{5}{3}\right) = -\frac{10}{9}$

The answers to parts (b) and (d) of Example 8 are written as *improper fractions*. They could also have been written as *mixed numbers*. For instance,

$$\frac{27}{14} = 1\frac{13}{14}$$

and

$$-\frac{10}{9} = -1\frac{1}{9}.$$

NOTE   Division by zero is *not* defined. For instance, the expression $\frac{3}{0}$ is undefined.

Positive Integer Exponents

Let n be a positive integer and let a be a real number. Then the product of n factors of a is given by

$$a^n = \underbrace{a \cdot a \cdot a \cdots a.}_{n \text{ factors}}$$

In the **exponential form** a^n, a is called the **base** and n is the **exponent.** When you write the exponential form a^n, you are **raising a to the nth power.**

When a number is raised to the *first* power, you usually do not write the exponent 1. For instance, you would usually write 5 rather than 5^1. Raising a number to the *second* power is called **squaring** the number. Raising a number to the *third* power is called **cubing** the number.

Example 9 ■ Evaluating Exponential Expressions

Evaluate each expression.

(a) $(-3)^4$　　　(b) -3^4　　　(c) $(-2)^5$　　　(d) $\left(\frac{2}{5}\right)^3$

Solution

(a) $(-3)^4 = (-3)(-3)(-3)(-3) = 81$

(b) $-3^4 = -(3)(3)(3)(3) = -81$

(c) $(-2)^5 = (-2)(-2)(-2)(-2)(-2) = -32$

(d) $\left(\frac{2}{5}\right)^3 = \left(\frac{2}{5}\right)\left(\frac{2}{5}\right)\left(\frac{2}{5}\right) = \frac{8}{125}$

Roots

When you multiply a number by itself, you are squaring the number. For instance, $5^2 = 5 \cdot 5 = 25$. To undo this "squaring operation," you can take the **square root** of 25. Similarly, you can undo a cubing operation by taking a **cube root.**

Number	Equal Factors	Root
$25 = 5^2$	$5 \cdot 5$	5 (square root)
$25 = (-5)^2$	$(-5)(-5)$	-5 (square root)
$-27 = (-3)^3$	$(-3)(-3)(-3)$	-3 (cube root)
$16 = 2^4$	$2 \cdot 2 \cdot 2 \cdot 2$	2 (fourth root)

In general, the ***n*th root** of a number is defined as follows.

Definition of *n*th Root of a Number

Let a and b be real numbers and let n be a positive integer such that $n \geq 2$. If

$$a = b^n$$

then b is an ***n*th root of a.** If $n = 2$, the root is a **square root,** and if $n = 3$, the root is a **cube root.**

By applying this definition, you can see that some numbers have more than one nth root. For example, both 5 and -5 are square roots of 25 because $25 = 5^2$ and $25 = (-5)^2$. Similarly, both 2 and -2 are fourth roots of 16 because $16 = 2^4$ and $16 = (-2)^4$. To avoid ambiguity, if a is a real number, the **principal nth root of a** is defined as the nth root that has the same sign as a, and it is denoted by the radical symbol

$$\sqrt[n]{a}. \qquad \text{Principal } n\text{th root}$$

The positive integer n is called the **index** of the radical, and the number a is called the **radicand.** If $n = 2$, the index is omitted and written as $\sqrt{a}$ rather than $\sqrt[2]{a}$. "Having the same sign" means that the principal nth root of a is positive if a is positive and negative if a is negative. For example, $\sqrt{4} = 2$ and $\sqrt[3]{-8} = -2$.

NOTE The number -4 has no real square root because there is no real number that can be squared to produce -4. So, $\sqrt{-4}$ is not a real number. This same comment can be made about any negative number. That is, if a is negative, $\sqrt{a}$ is not a real number.

Example 10 ■ Finding the Principal nth Root of a Number

Find the principal nth root of each number.

(a) $\sqrt{36}$ (b) $\sqrt{0}$ (c) $\sqrt[3]{8}$

(d) $\sqrt[3]{-27}$ (e) $\sqrt[4]{256}$ (f) $\sqrt[4]{625}$

Solution

(a) $\sqrt{36} = 6$ (b) $\sqrt{0} = 0$

(c) $\sqrt[3]{8} = 2$ (d) $\sqrt[3]{-27} = -3$

(e) $\sqrt[4]{256} = 4$ (f) $\sqrt[4]{625} = 5$

Order of Operations

One way to help avoid confusion when communicating algebraic ideas is to establish an **order of operations.** This is done by giving priorities to different operations. First priority is given to exponents, second priority is given to multiplication and division, and third priority is given to addition and subtraction. To distinguish between operations with the same priority, use the *Left-to-Right Rule.*

Order of Operations

To evaluate an expression involving more than one operation, use the steps below.

1. First do operations that occur within symbols of grouping. The most common grouping symbols are parentheses ( ), brackets [ ], and fraction bars.

2. Then evaluate powers.

3. Then do multiplications and divisions from left to right.

4. Finally, do additions and subtractions from left to right.

Study Tip

The order of operations for multiplication applies when multiplication is written with the symbols $\times$ or $\cdot$. When multiplication is implied by parentheses, it has a higher priority than the Left-to-Right Rule. For instance

$$8 \div 4(2) = 8 \div 8 = 1$$

but

$$8 \div 4 \cdot 2 = 2 \cdot 2 = 4.$$

Example 11 ■ Order of Operations Without Symbols of Grouping

Evaluate each expression.

(a) $20 - 2 \cdot 3^2$ (b) $5 - 6 - 2$ (c) $8 \div 2 \cdot 2$

Solution

(a) $20 - 2 \cdot 3^2 = 20 - 2 \cdot 9 = 20 - 18 = 2$

(b) $5 - 6 - 2 = (5 - 6) - 2 = -1 - 2 = -3$

(c) $8 \div 2 \cdot 2 = (8 \div 2) \cdot 2 = 4 \cdot 2 = 8$

When you want to change the established order of operations, you must use parentheses or other symbols of grouping. Part (e) in the next example shows that a fraction bar acts as a symbol of grouping.

Example 12 ■ Order of Operations with Symbols of Grouping

(a) $-4 + 2(-2 + 5)^2 = -4 + 2(3)^2$ Add within parentheses.

$\qquad\qquad\qquad\quad = -4 + 2(9)$ Evaluate the power.

$\qquad\qquad\qquad\quad = -4 + 18$ Multiply.

$\qquad\qquad\qquad\quad = 14$ Add.

(b) $(-4 + 2)(-2 + 5)^2 = (-2)(3)^2$ Add within parentheses.

$\qquad\qquad\qquad\quad = (-2)(9)$ Evaluate the power.

$\qquad\qquad\qquad\quad = -18$ Multiply.

(c) $(-3)(-2)^2 = (-3)(4)$ Evaluate the power.

$\qquad\qquad\quad = -12$ Multiply.

(d) $(-3)(-2^2) = (-3)(-4)$ Evaluate the power.

$\qquad\qquad\quad = 12$ Multiply.

(e) $\dfrac{2 \cdot 5^2 - 10}{3^2 - 4} = (2 \cdot 5^2 - 10) \div (3^2 - 4)$ Rewrite using parentheses.

$\qquad\qquad\quad = (50 - 10) \div (9 - 4)$ Evaluate powers and multiply within symbols of grouping.

$\qquad\qquad\quad = 40 \div 5$ Subtract within parentheses.

$\qquad\qquad\quad = 8$ Divide.

The symbols for absolute value and radicals are given the same order of operation as that of grouping symbols. Here are two examples.

$$5 - |3 - 4| = 5 - |-1| = 5 - 1 = 4$$

$$\sqrt{4^2 + 9} + 1 = \sqrt{16 + 9} + 1 = \sqrt{25} + 1 = 5 + 1 = 6$$

A.1 ■ Exercises

In Exercises 1 and 2, determine which of the real numbers are (a) natural numbers, (b) integers, (c) rational numbers, and (d) irrational numbers.

1. $\left\{-10, -\sqrt{5}, -\frac{2}{3}, -\frac{1}{4}, 0, \frac{5}{8}, 1, \sqrt{3}, 4, 6, 2\pi\right\}$

2. $\left\{-\frac{7}{2}, -\sqrt{6}, -\frac{\pi}{2}, -\frac{3}{8}, 0, \frac{10}{3}, \sqrt{15}, 8, 245\right\}$

In Exercises 3–6, list the numbers satisfying the specified requirements.

3. The integers between -5.8 and 3.2

4. The even integers between -2.1 and 10.5

5. The odd integers between 0 and 9.1

6. All prime numbers between 0 and 25

In Exercises 7 and 8, plot the real numbers on the real number line.

7. (a) 3 (b) $\frac{5}{2}$ (c) $-\frac{7}{2}$ (d) -5.2

8. (a) 8 (b) $\frac{4}{3}$ (c) -6.75 (d) $-\frac{9}{2}$

In Exercises 9–12, approximate the two numbers and order them.

9.

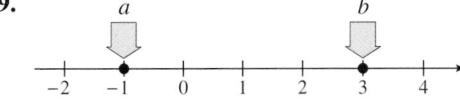

10.

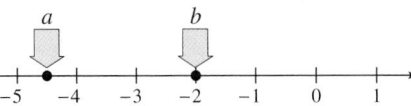

11.

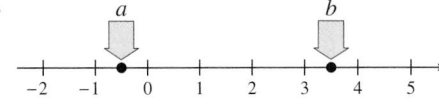

12.

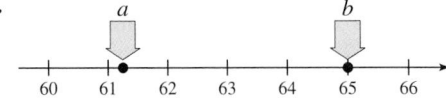

In Exercises 13–24, plot the two real numbers on the real number line and place the correct inequality symbol ($<$ or $>$) between the two numbers.

13. $2 \quad\quad 5$

14. $8 \quad\quad 3$

15. $-7 \quad\quad -2$

16. $-2 \quad\quad -5$

17. $-5 \quad\quad -2$

18. $-8 \quad\quad 3$

19. $\frac{1}{3} \quad\quad \frac{1}{4}$

20. $\frac{4}{5} \quad\quad 1$

21. $-\frac{2}{3} \quad\quad -\frac{10}{3}$

22. $-\frac{11}{4} \quad\quad \frac{7}{6}$

23. $2.75 \quad\quad \pi$

24. $-\pi \quad\quad -3.1$

In Exercises 25–32, find the distance between the two real numbers.

25. 4 and 10

26. 75 and 20

27. 18 and -32

28. -54 and 32

29. -35 and 0

30. 0 and 35

31. $-\frac{7}{2}$ and 0

32. $\frac{3}{4}$ and $\frac{9}{4}$

In Exercises 33–42, evaluate the expression.

33. $|10|$

34. $|62|$

35. $|-225|$

36. $|-6|$

37. $-|-25|$

38. $-|16|$

39. $-\left|\frac{3}{8}\right|$

40. $-\left|-\frac{3}{4}\right|$

41. $-|3.5|$

42. $|-1.4|$

In Exercises 43–50, place the correct symbol ($<$, $>$, or $=$) between the two real numbers.

43. $|-6| \quad\quad |2|$

44. $|150| \quad\quad |-310|$

45. $|47| \quad\quad |-27|$

46. $|-2| \quad\quad |2|$

47. $\left|-\frac{3}{4}\right| \quad\quad -\left|\frac{4}{5}\right|$

48. $-\left|-\frac{7}{3}\right| \quad\quad -\left|\frac{1}{3}\right|$

49. $-|-16.8| \quad\quad -|16.8|$

50. $|12.5| \quad\quad -|-25.1|$

In Exercises 51–58, find the opposite and the absolute value of the number.

51. 14

52. 34

53. -160

54. -52

55. $-\frac{5}{4}$

56. $\frac{4}{3}$

57. -0.4

58. π

In Exercises 59–66, plot the number and its opposite on the real number line.

59. -3

60. 7

61. 6

62. -4

63. $\frac{5}{3}$

64. $-\frac{3}{4}$

65. -4.25

66. 3.5

In Exercises 67–72, write the statement using inequality notation.

67. x is negative.

68. y is more than 25.

69. u is at least 16.

70. z is greater than 2 and no more than 10.

71. The price p of a coat will be less than \$225 during the sale.

72. The tire pressure p must be at least 30 pounds per square inch and no more than 35 pounds per square inch.

In Exercises 73–94, perform the indicated operation(s).

73. $13 + 32$

74. $16 + 84$

75. $-13 + 32$

76. $16 + (-84)$

77. $-7 - 15$

78. $-5 + (-52)$

79. $4 - 16 + (-8)$

80. $-15 + (-6) + 32$

81. $5.8 - 6.2 + 1.1$

82. $46.08 - 35.1 - 16.25$

83. $\frac{3}{4} - \frac{1}{4}$

84. $\frac{5}{6} + \frac{7}{6}$

85. $\frac{5}{8} + \frac{1}{4} - \frac{5}{6}$

86. $\frac{3}{11} + \frac{-5}{2}$

87. $5\frac{3}{4} + 7\frac{3}{8}$

88. $8\frac{1}{2} - 24\frac{2}{3}$

89. $85 - |-25|$

90. $-36 + |-8|$

91. $-(-11.325) + |34.625|$

92. $|-16.25| - 54.78$

93. $-|-15.667| - 12.333$

94. $-\left|-15\frac{2}{3}\right| - 12\frac{1}{3}$

In Exercises 95–106, find the product.

95. $5(-6)$

96. $-7(3)$

97. $(-8)(-6)$

98. $(-4)(-7)$

99. $6(12)$

100. $7(10)$

101. $\left(-\frac{5}{8}\right)\left(-\frac{4}{5}\right)$

102. $-\frac{3}{2}\left(\frac{8}{5}\right)$

103. $-\frac{9}{8}\left(\frac{16}{27}\right)\left(\frac{1}{2}\right)$

104. $\frac{2}{3}\left(-\frac{18}{5}\right)\left(-\frac{5}{6}\right)$

105. $6.3(5.1)$

106. $(-4.4)(-3.2)$

In Exercises 107–116, find the quotient.

107. $\dfrac{-18}{-3}$

108. $-\dfrac{30}{-15}$

109. $-48 \div 16$

110. $-27 \div (-9)$

111. $-\frac{4}{5} \div \frac{8}{25}$

112. $-\frac{11}{12} \div \frac{5}{24}$

113. $\left(-\frac{1}{3}\right) \div \left(-\frac{5}{6}\right)$

114. $\frac{8}{15} \div \frac{32}{5}$

115. $5\frac{3}{4} \div 2\frac{1}{8}$

116. $-3\frac{5}{6} \div -2\frac{2}{3}$

In Exercises 117 and 118, find the quotient and round the result to two decimal places.

117. $\dfrac{25.5}{6.325}$

118. $\dfrac{265.45}{25.6}$

In Exercises 119–124, write the expression as a repeated multiplication problem.

119. $(-3)^4$

120. 6^5

121. $\left(-\frac{3}{4}\right)^4$

122. $\left(\frac{2}{3}\right)^3$

123. $(0.67)^6$

124. $(-0.8)^4$

In Exercises 125–128, write the repeated multiplication in exponential form.

125. $(-5)(-5)(-5)(-5)$

126. $(-4)(-4)(-4)(-4)(-4)(-4)$

127. $-(5 \times 5 \times 5 \times 5 \times 5 \times 5)$

128. $-(7 \cdot 7 \cdot 7)$

In Exercises 129–138, evaluate the exponential expression.

129. $(-4)^3$

130. $(-3)^4$

131. -5^2

132. -3^5

133. $\left(-\frac{7}{8}\right)^2$

134. $-\left(\frac{2}{3}\right)^4$

135. $\left(\frac{4}{5}\right)^3$

136. $-\left(-\frac{1}{2}\right)^5$

137. $(0.3)^3$

138. $(0.2)^4$

In Exercises 139–148, find the root.

139. $\sqrt{81}$

140. $\sqrt{100}$

141. $-\sqrt{49}$

142. $-\sqrt{144}$

143. $\sqrt[3]{125}$

144. $\sqrt[3]{64}$

145. $\sqrt[3]{-8}$

146. $-\sqrt[3]{-125}$

147. $-\sqrt[4]{1}$

148. $\sqrt[5]{32}$

In Exercises 149–162, evaluate the expression.

149. $16 - 6 - 10$

150. $18 - 12 + 4$

151. $24 - 5 \cdot 2^2$

152. $18 + 3^2 - 12$

153. $14 - 2(8 - 4)$

154. $21 - 7(7 - 5)$

155. $2 + [8 - (14 \div 2)]$

156. $18 - [4 + (17 - 12)]$

157. $0.2(6 - 10)^3 + 85$

158. $-0.3(5 - 3)^3 + 7.1$

159. $5^3 + |-14 + 4|$

160. $|(-2)^5| - (25 + 7)$

161. $\dfrac{4^2 - 5}{11} - 7$

162. $\dfrac{5^3 - 50}{-15} + 27$

In Exercises 163–170, evaluate the expression using a calculator. Round the result to two decimal places.

163. $5.6[13 - 2.5(-6.3)]$ **164.** $35(1032 - 4650)$

165. $5^6 - 3(400)$ **166.** $5(100 - 3.6^4) \div 4.1$

167. $\dfrac{500}{(1.055)^{20}}$ **168.** $\dfrac{265.45}{25.6}$

169. $\sqrt{9^2 + 7.5^2}$ **170.** $\sqrt{9.4^2 - 4(2)(6)}$

171. *Company Profits* The annual profits for a company (in millions of dollars) for the years 1998 through 2002 are shown in the bar graph. Complete the table showing the increase or decrease in profit from the preceding year.

Year	Yearly gain or loss
1999	
2000	
2001	
2002	

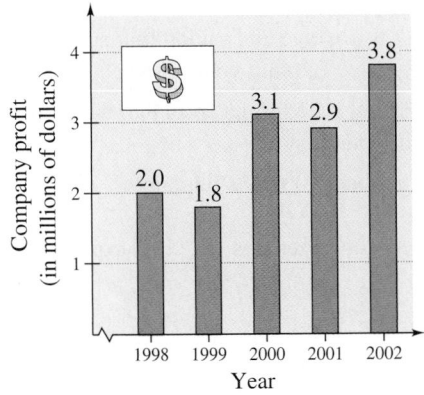

172. *Balance in an Account* During one month, you made the following transactions in your non-interest-bearing checking account.

Initial Balance:	$2618.68
Deposit:	$1236.45
Withdrawal:	$25.62
Withdrawal:	$455.00
Withdrawal:	$125.00
Withdrawal:	$715.95

Find the balance at the end of the month.

In Exercises 173 and 174, determine the unknown fractional part of the circle graph.

173. **174.**

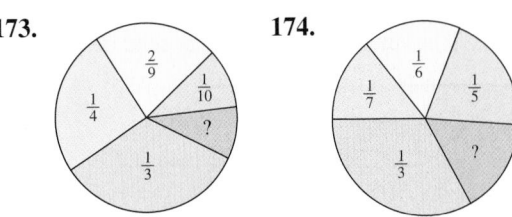

Area In Exercises 175–178, find the area of the figure. (The area of a rectangle is $A = lw$, and the area of a triangle is $A = \frac{1}{2}bh$.)

175.

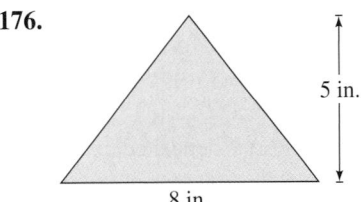

176.

177.

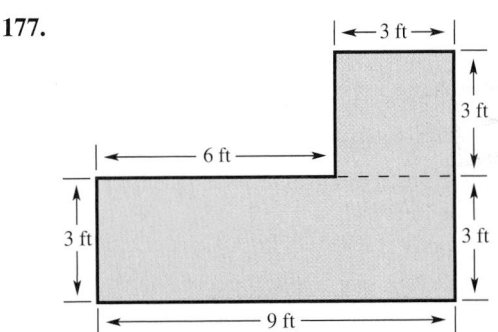

178.

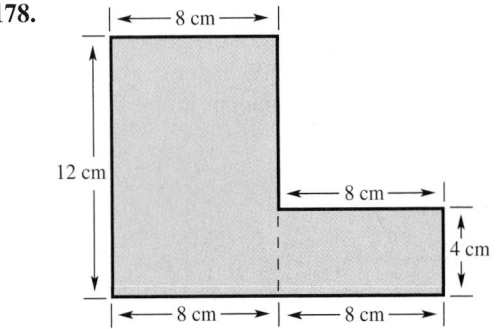

The symbol ⬛ indicates an exercise in which you are instructed to use a calculator or graphing utility.

Volume In Exercises 179 and 180, use the following information. A bale of hay is a rectangular solid having the dimensions shown in the figure and weighing approximately 50 pounds. (The volume of a rectangular solid is $V = lwh$.)

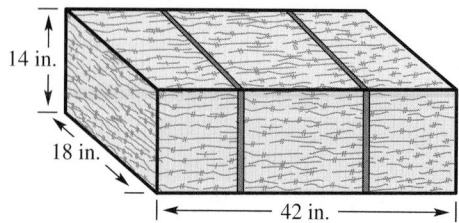

14 in.

18 in.

42 in.

179. Find the volume of a bale of hay in cubic feet (1728 cubic inches equals 1 cubic foot).

180. Approximate the number of bales in a ton of hay. Then approximate the volume of a stack of baled hay that weighs 12 tons.

181. *Savings Plan*

(a) You save \$50 per month for 18 years. How much money has been set aside during the 18 years?

(b) If the money in part (a) is deposited in a savings account earning 9% interest compounded monthly, the total amount in the fund after 18 years will be

$$50\left[\left(1 + \frac{0.09}{12}\right)^{216} - 1\right]\left(1 + \frac{12}{0.09}\right).$$

Use a calculator to determine this amount.

(c) How much of the amount in part (b) is earnings from interest?

182. *Savings Plan*

(a) You save \$75 per month for 30 years. How much money has been set aside during the 30 years?

(b) If the money in part (a) is deposited in a savings account earning 8% interest compounded monthly, the total amount in the fund after 30 years will be

$$75\left[\left(1 + \frac{0.08}{12}\right)^{360} - 1\right]\left(1 + \frac{12}{0.08}\right).$$

Use a calculator to determine this amount.

(c) How much of the amount in part (b) is earnings from interest?

True or False? In Exercises 183–188, determine whether the statement is true or false. If the statement is false, give an example of a real number that makes the statement false.

183. Every integer is a rational number.

184. Every rational number is an integer.

185. The reciprocal of every nonzero integer is an integer.

186. The reciprocal of every nonzero rational number is a rational number.

187. If x and y are real numbers, $|x + y| = |x| + |y|$.

188. If a negative real number is raised to the 11th power, the result will be positive.

189. Is there a difference between saying that a real number is positive and saying that a real number is nonnegative? Explain your answer.

190. Does -8 or 6 lie farther from the real number -4? Explain your answer.

191. Is it true that $\frac{1}{6} = 0.17$? Explain.

Error Analysis In Exercises 192–195, explain the error.

192. $\dfrac{2}{3} + \dfrac{3}{2} = \dfrac{2 + 3}{3 + 2} = 1$

193. $\dfrac{5 + 12}{5} = \dfrac{5 + 12}{5} = 12$

194. $\dfrac{28}{83} = \dfrac{28}{83} = \dfrac{2}{3}$

195. $3 \cdot 4^2 = 12^2$

A.2 Properties of Real Numbers

Mathematical Systems

In this section, you will review the properties of real numbers. These properties make up the third component of what is called a **mathematical system.** These three components are a set of numbers, operations with the set of numbers, and properties of the numbers (and operations).

Figure A.8 is a diagram that represents different mathematical systems. Note that the set of numbers for the system can vary. The set can consist of whole numbers, integers, rational numbers, real numbers, or algebraic expressions.

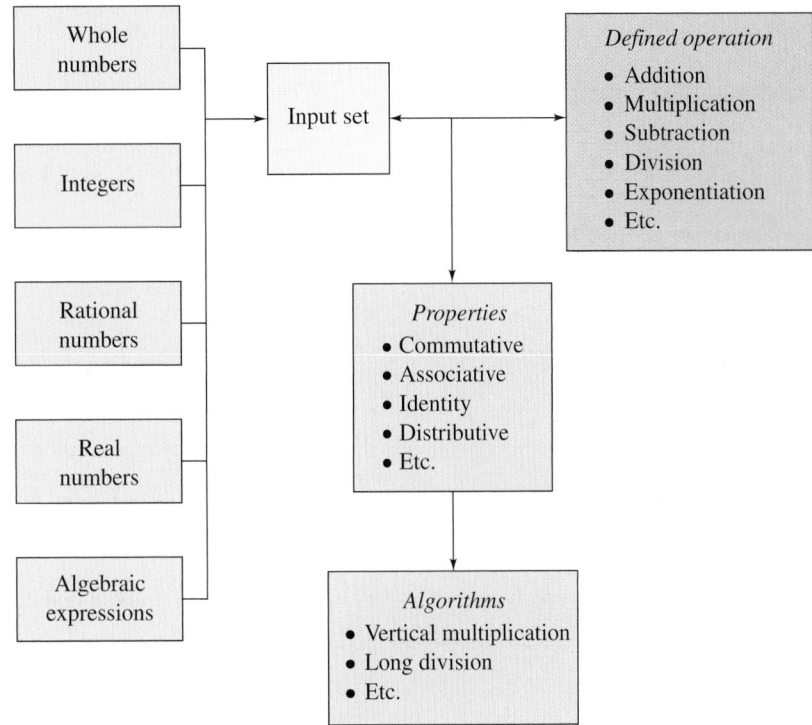

Figure A.8

Basic Properties of Real Numbers

For the mathematical system that consists of the set of real numbers together with the operations of addition, subtraction, multiplication, and division, the resulting properties are called the **properties of real numbers.** In the list on page A18, a verbal description of each property is given, as well as one or two examples.

Properties of Real Numbers

Let a, b, and c represent real numbers.

Property	*Verbal Description*
Closure Property of Addition	The sum of two real numbers is a real number.
$a + b$ is a real number.	Example: $1 + 5 = 6$, and 6 is a real number.
Closure Property of Multiplication	The product of two real numbers is a real number.
ab is a real number.	Example: $7 \cdot 3 = 21$, and 21 is a real number.
Commutative Property of Addition	Two real numbers can be added in either order.
$a + b = b + a$	Example: $2 + 6 = 6 + 2$
Commutative Property of Multiplication	Two real numbers can be multiplied in either order.
$a \cdot b = b \cdot a$	Example: $3 \cdot (-5) = -5 \cdot 3$
Associative Property of Addition	When three real numbers are added, it makes no difference which two are added first.
$(a + b) + c = a + (b + c)$	Example: $(1 + 7) + 4 = 1 + (7 + 4)$
Associative Property of Multiplication	When three real numbers are multiplied, it makes no difference which two are multiplied first.
$(ab)c = a(bc)$	Example: $(4 \cdot 3) \cdot 9 = 4 \cdot (3 \cdot 9)$
Distributive Property	Multiplication distributes over addition.
$a(b + c) = ab + ac$ $(b + c)a = ba + ca$	Examples: $2(3 + 4) = 2 \cdot 3 + 2 \cdot 4$ $(3 + 4)2 = 3 \cdot 2 + 4 \cdot 2$
Additive Identity Property	The sum of zero and a real number equals the number itself.
$a + 0 = 0 + a = a$	Example: $4 + 0 = 0 + 4 = 4$
Multiplicative Identity Property	The product of 1 and a real number equals the number itself.
$a \cdot 1 = 1 \cdot a = a$	Example: $5 \cdot 1 = 1 \cdot 5 = 5$
Additive Inverse Property	The sum of a real number and its opposite is zero.
$a + (-a) = 0$	Example: $5 + (-5) = 0$
Multiplicative Inverse Property	The product of a nonzero real number and its reciprocal is 1.
$a \cdot \dfrac{1}{a} = 1, \quad a \neq 0$	Example: $7 \cdot \dfrac{1}{7} = 1$

NOTE   The operations of subtraction and division are not listed above because they fail to possess many of the properties described in the list. For instance, subtraction and division are not commutative. To see this, consider $4 - 3 \neq 3 - 4$ and $15 \div 5 \neq 5 \div 15$. Similarly, the examples $8 - (6 - 2) \neq (8 - 6) - 2$ and $20 \div (4 \div 2) \neq (20 \div 4) \div 2$ illustrate the fact that subtraction and division are not associative.

Example 1 ■ Identifying Properties of Real Numbers

Name the property of real numbers that justifies each statement. (*Note: a* and *b* are real numbers.)

(a) $9 \cdot 5 = 5 \cdot 9$

(b) $4(a + 3) = 4 \cdot a + 4 \cdot 3$

(c) $6 \cdot \frac{1}{6} = 1$

(d) $-3 + (2 + b) = (-3 + 2) + b$

(e) $(b + 8) + 0 = b + 8$

Solution

(a) This statement is justified by the Commutative Property of Multiplication.

(b) This statement is justified by the Distributive Property.

(c) This statement is justified by the Multiplicative Inverse Property.

(d) This statement is justified by the Associative Property of Addition.

(e) This statement is justified by the Additive Identity Property.

Example 2 ■ Identifying Properties of Real Numbers

The area of the rectangle in Figure A.9 can be represented in two ways: as the area of a single rectangle, or as the sum of the areas of the two rectangles.

(a) Find this area in both ways.

(b) What property of real numbers does this demonstrate?

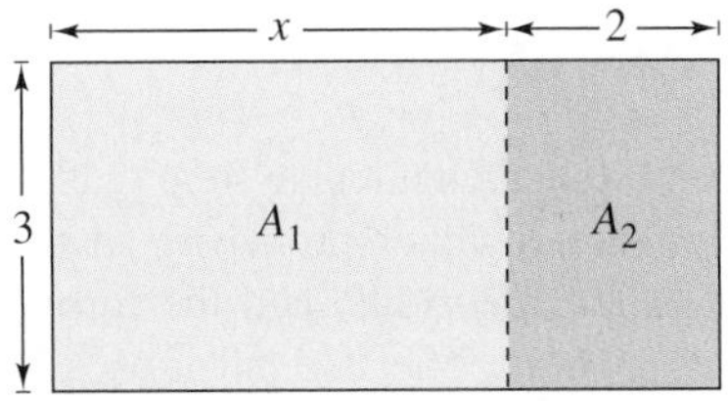

Solution

(a) The area of the single rectangle with width 3 and length $x + 2$ is

$$A = 3(x + 2).$$

The areas of the two rectangles are

$$A_1 = 3(x) \text{ and } A_2 = 3(2).$$

The sum of these two areas represents the area of the single rectangle. That is,

$$A = A_1 + A_2$$
$$3(x + 2) = 3(x) + 3(2)$$
$$= 3x + 6.$$

(b) Because the area of the single rectangle and the sum of the areas of the two rectangles are equal, you can write $3(x + 2) = 3x + 6$. This demonstrates the Distributive Property.

Figure A.9

To help you understand each property of real numbers, try stating the properties in your own words.

Example 3 ■ Using the Properties of Real Numbers

Complete each statement using the specified property of real numbers.

(a) Multiplicative Identity Property

$$(4a)1 = \boxed{}$$

(b) Associative Property of Addition

$$(a + 9) + 1 = \boxed{}$$

(c) Additive Inverse Property

$$0 = 5c + \boxed{}$$

(d) Distributive Property

$$4 \cdot b + 4 \cdot 5 = \boxed{}$$

Solution

(a) By the Multiplicative Identity Property, you can write

$$(4a)1 = 4a.$$

(b) By the Associative Property of Addition, you can write

$$(a + 9) + 1 = a + (9 + 1).$$

(c) By the Additive Inverse Property, you can write

$$0 = 5c + (-5c).$$

(d) By the Distributive Property, you can write

$$4 \cdot b + 4 \cdot 5 = 4(b + 5).$$

Additional Properties of Real Numbers

Once you have determined the basic properties of a mathematical system (called the **axioms** of the system), you can go on to develop other properties of the system. These additional properties are often called **theorems,** and the formal arguments that justify the theorems are called **proofs.** The list on page A21 summarizes several additional properties of real numbers.

Example 4 ■ Proof of a Property of Equality

Prove that if $a + c = b + c$, then $a = b$. (Use the Addition Property of Equality.)

Solution

$a + c = b + c$	Write original equation.
$(a + c) + (-c) = (b + c) + (-c)$	Addition Property of Equality
$a + [c + (-c)] = b + [c + (-c)]$	Associative Property of Addition
$a + 0 = b + 0$	Additive Inverse Property
$a = b$	Additive Identity Property

NOTE Notice in Example 4 how each step is justified from the previous step by means of a property of real numbers.

Additional Properties of Real Numbers

Let a, b, and c be real numbers.

Properties of Equality

Addition Property of Equality

If $a = b$, then $a + c = b + c$.

Verbal Description

Adding a real number to each side of a true equation produces another true equation.

Multiplication Property of Equality

If $a = b$, then $ac = bc$, $c \neq 0$.

Multiplying each side of a true equation by a nonzero real number produces another true equation.

Cancellation Property of Addition

If $a + c = b + c$, then $a = b$.

Subtracting a real number from each side of a true equation produces another true equation.

Cancellation Property of Multiplication

If $ac = bc$ and $c \neq 0$, then $a = b$.

Dividing each side of a true equation by a nonzero real number produces another true equation.

Reflexive Property of Equality

$a = a$

A real number always equals itself.

Symmetric Property of Equality

If $a = b$, then $b = a$.

If a real number equals a second real number, then the second real number equals the first.

Transitive Property of Equality

If $a = b$ and $b = c$, then $a = c$.

If a real number equals a second real number and the second real number equals a third real number, then the first real number equals the third real number.

Properties of Zero

Multiplication Property of Zero

$0 \cdot a = 0$

Verbal Description

The product of zero and any real number is zero.

Division Property of Zero

$\dfrac{0}{a} = 0$, $a \neq 0$

If zero is divided by any *nonzero* real number, the result is zero.

Division by Zero Is Undefined

$\dfrac{a}{0}$ is undefined.

We do not define division by zero.

Properties of Negation

Multiplication by -1

$(-1)a = -a$

$(-1)(-a) = a$

Verbal Description

The opposite of a real number a can be obtained by multiplying the real number by -1.

Placement of Negative Signs

$-(ab) = (-a)(b) = (a)(-b)$

The opposite of the product of two numbers is equal to the product of one of the numbers and the opposite of the other.

Product of Two Opposites

$(-a)(-b) = ab$

The product of the opposites of two real numbers is equal to the product of the two real numbers.

Example 5 ■ Proof of a Property of Negation

Prove that

$$(-1)a = -a.$$

(You may use any of the properties of equality and properties of zero.)

Solution

At first glance, it is a little difficult to see what you are being asked to prove. However, a good way to start is to consider carefully the definitions of each of the three numbers in the equation.

$$a = \text{given real number}$$
$$-1 = \text{the additive inverse of 1}$$
$$-a = \text{the additive inverse of } a$$

By showing that $(-1)a$ has the same properties as the additive inverse of a, you will be showing that $(-1)a$ must be the additive inverse of a.

$(-1)a + a = (-1)a + (1)(a)$	Multiplicative Identity Property
$= (-1 + 1)a$	Distributive Property
$= (0)a$	Additive Inverse Property
$= 0$	Multiplication Property of Zero

Because you have shown that $(-1)a + a = 0$, you can now use the fact that $-a + a = 0$ to conclude that $(-1)a + a = -a + a$. From this, you can complete the proof as follows.

$(-1)a + a = -a + a$	Shown in first part of proof
$(-1)a = -a$	Cancellation Property of Addition

The list of additional properties of real numbers forms a very important part of algebra. Knowing the names of the properties is not especially important, but knowing how to use each property is extremely important. The next two examples show how several of the properties are used to solve common problems in algebra.

Example 6 ■ Applying Properties of Real Numbers

In the solution of the equation $b + 2 = 6$, identify the property of real numbers that justifies each step.

Solution

$$b + 2 = 6 \qquad \text{Original equation}$$

Solution Step	*Property*
$(b + 2) + (-2) = 6 + (-2)$	Addition Property of Equality
$b + [2 + (-2)] = 4$	Associative Property of Addition
$b + 0 = 4$	Additive Inverse Property
$b = 4$	Additive Identity Property

Example 7 ■ Applying the Properties of Real Numbers

In the solution of the equation $3a = 9$, identify the property of real numbers that justifies each step.

$$3a = 9 \qquad \text{Original equation}$$

Solution Step	*Property*
$\left(\dfrac{1}{3}\right)(3a) = \left(\dfrac{1}{3}\right)(9)$	Multiplication Property of Equality
$\left(\dfrac{1}{3} \cdot 3\right)(a) = 3$	Associative Property of Multiplication
$(1)(a) = 3$	Multiplicative Inverse Property
$a = 3$	Multiplicative Identity Property

A.2 ■ Exercises

In Exercises 1–28, name the property of real numbers that justifies the statement.

1. $3 + (-5) = -5 + 3$

2. $-5(7) = 7(-5)$

3. $25 - 25 = 0$

4. $5 + 0 = 5$

5. $6(-10) = -10(6)$

6. $2(6 \cdot 3) = (2 \cdot 6)3$

7. $7 \cdot 1 = 7$

8. $4 \cdot \dfrac{1}{4} = 1$

9. $25 + 35 = 35 + 25$

10. $(-4 \cdot 10) \cdot 8 = -4(10 \cdot 8)$

11. $3 + (12 - 9) = (3 + 12) - 9$

12. $(16 + 8) - 5 = 16 + (8 - 5)$

13. $(8 - 5)(10) = 8 \cdot 10 - 5 \cdot 10$

14. $7(9 + 15) = 7 \cdot 9 + 7 \cdot 15$

15. $(10 + 8) + 3 = 10 + (8 + 3)$

16. $(5 + 10)(8) = 8(5 + 10)$

17. $5(2a) = (5 \cdot 2)a$

18. $10(2x) = (10 \cdot 2)x$

19. $1 \cdot (5t) = 5t$

20. $8y \cdot 1 = 8y$

21. $3x + 0 = 3x$

22. $0 + 8w = 8w$

23. $\dfrac{1}{y} \cdot y = 1$

24. $10x \cdot \dfrac{1}{10x} = 1$

25. $3(6 + b) = 3 \cdot 6 + 3 \cdot b$

26. $(x + 1) - (x + 1) = 0$

27. $3(2 + x) = 3 \cdot 2 + 3x$

28. $(6 + x) - m = 6 + (x - m)$

In Exercises 29–38, use the property of real numbers to fill in the missing part of the statement.

29. Associative Property of Multiplication

$3(6y) = $ ▮

30. Commutative Property of Addition

$10 + (-6) = $ ▮

31. Commutative Property of Multiplication

$15(-3) = $ ▮

32. Associative Property of Addition

$6 + (5 - y) = $ ▮

33. Distributive Property

$(6 + z)5 = $ ▮

34. Distributive Property

$-3(4 + x) = $ ▮

35. Commutative Property of Addition

$25 + (-x) = $ ▮

36. Additive Inverse Property

$13x - 13x = $ ▮

37. Multiplicative Identity Property

$(x + 8) \cdot 1 = $ ▮

38. Additive Identity Property

$(8x) + 0 = $ ▮

In Exercises 39–46, give (a) the additive inverse and (b) the multiplicative inverse of the quantity.

39. 10

40. 18

41. -16

42. -52

43. $6z, \ z \neq 0$

44. $2y, \ y \neq 0$

45. $x + 1, \ x \neq -1$

46. $y - 4, \ y \neq 4$

In Exercises 47–54, rewrite the expression using the Associative Property of Addition or the Associative Property of Multiplication.

47. $(x + 5) - 3$

48. $(z - 6) + 10$

49. $32 + (-4 + y)$

50. $15 + (3 + x)$

51. $3(4 \cdot 5)$

52. $(10 \cdot 8) \cdot 5$

53. $6(2y)$

54. $8(3x)$

In Exercises 55–62, rewrite the expression using the Distributive Property.

55. $20(2 + 5)$

56. $-3(4 - 8)$

57. $5(3x - 4)$

58. $6(2x + 5)$

59. $(x + 6)(-2)$

60. $(z - 10)(12)$

61. $-6(2y - 5)$

62. $-4(10 - b)$

In Exercises 63–68, the right side of the equation is *not* equal to the left side. Change the right side so that it *is* equal to the left side.

63. $3(x + 5) \neq 3x + 5$

64. $4(x + 2) \neq 4x + 2$

65. $-2(x + 8) \neq -2x + 16$

66. $-9(x + 4) \neq -9x + 36$

67. $3\left(\frac{0}{3}\right) \neq 1$

68. $6\left(\frac{1}{6}\right) \neq 0$

In Exercises 69 and 70, use the properties of real numbers to prove the statement.

69. If $ac = bc$ and $c \neq 0$, then $a = b$.

70. $(-1)(-a) = a$

In Exercises 71–74, identify the property of real numbers that justifies each step.

71.
$$x + 5 = 3 \qquad \text{Original equation}$$
$$(x + 5) + (-5) = 3 + (-5)$$
$$x + [5 + (-5)] = -2$$
$$x + 0 = -2$$
$$x = -2$$

72.
$$x - 8 = 20 \qquad \text{Original equation}$$
$$(x - 8) + 8 = 20 + 8$$
$$x + (-8 + 8) = 28$$
$$x + 0 = 28$$
$$x = 28$$

73.
$$2x - 5 = 6 \qquad \text{Original equation}$$
$$(2x - 5) + 5 = 6 + 5$$
$$2x + (-5 + 5) = 11$$
$$2x + 0 = 11$$
$$2x = 11$$
$$\tfrac{1}{2}(2x) = \tfrac{1}{2}(11)$$
$$\left(\tfrac{1}{2} \cdot 2\right)x = \tfrac{11}{2}$$
$$1 \cdot x = \tfrac{11}{2}$$
$$x = \tfrac{11}{2}$$

74.
$$3x + 4 = 10 \qquad \text{Original equation}$$
$$(3x + 4) + (-4) = 10 + (-4)$$
$$3x + [4 + (-4)] = 6$$
$$3x + 0 = 6$$
$$3x = 6$$
$$\tfrac{1}{3}(3x) = \tfrac{1}{3}(6)$$
$$\left(\tfrac{1}{3} \cdot 3\right)x = 2$$
$$1 \cdot x = 2$$
$$x = 2$$

In Exercises 75–80, use the Distributive Property to perform the arithmetic mentally. For example, suppose you work in an industry where the wage is \$14 per hour with "time and a half" for overtime. So, your hourly wage for overtime is

$$14(1.5) = 14\left(1 + \tfrac{1}{2}\right)$$
$$= 14 + 7$$
$$= \$21.$$

75. $16(1.75) = 16\left(2 - \tfrac{1}{4}\right)$

76. $15\left(1\tfrac{2}{3}\right) = 15\left(2 - \tfrac{1}{3}\right)$

77. $7(62) = 7(60 + 2)$

78. $5(49) = 5(50 - 1)$

79. $9(6.98) = 9(7 - 0.02)$

80. $12(19.95) = 12(20 - 0.05)$

Dividends   In Exercises 81–84, the dividends paid per share of common stock by the Proctor & Gamble Company for the years 1994 through 2001 are approximated by the expression

$0.113t + 0.13$.

In this expression, the dividend per share is measured in dollars and t represents the year, with $t = 4$ corresponding to 1994 (see figure). *(Source: Proctor & Gamble Company)*

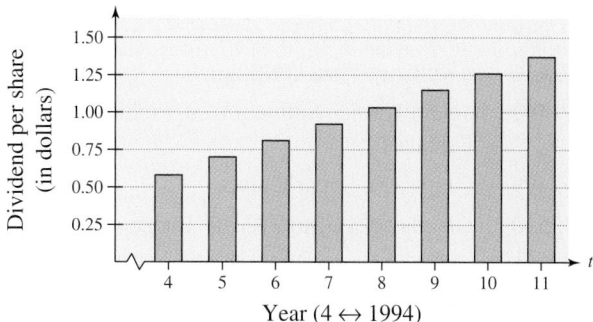

Year (4 ↔ 1994)

81. Use the graph to approximate the dividend paid in 1999.

82. Use the expression to approximate the annual increase in the dividend paid per share.

83. Use the expression to forecast the dividend per share in 2004.

84. In 2000, the actual dividend paid per share of common stock was $1.28. Compare this with the approximation given by the expression.

85. *Geometry*   The figure shows two adjoining rectangles. Find the total area of the rectangles in two ways.

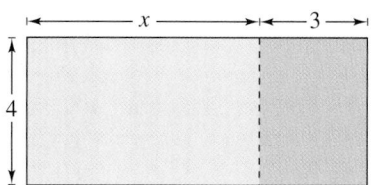

86. *Geometry*   The figure shows two adjoining rectangles. Find the total area of the two rectangles in two ways.

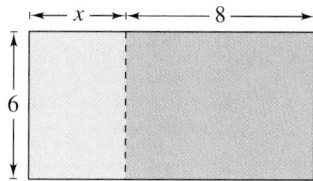

True or False?   In Exercises 87–90, decide whether the statement is true or false. Explain your reasoning.

87. $-6x + 6x = 0$

88. $-9 + 5 = -5 + 9$

89. $6(7 + 2) = 6(7) + 2$

90. $-4(8 + 1) = -4(8) - 4(1)$

91. Does every real number have a multiplicative inverse? Explain.

92. What is the additive inverse of a real number? Give an example of the Additive Inverse Property.

93. What is the multiplicative inverse of a real number? Give an example of the Multiplicative Inverse Property.

94. State the Multiplication Property of Zero.

95. Explain how the Addition Property of Equality can be used to allow you to subtract the same number from each side of an equation.

96. *Investigation*   Suppose you define a new mathematical operation using the symbol $\odot$. This operation is defined as $a \odot b = 2 \cdot a + b$.

   (a) Is this operation commutative? Explain.

   (b) Is this operation associative? Explain.

Photo Credits

P. 1	Composition I, 1920 (1872–1944) Private Collection/Lauros-Giraudon-Bridgeman Art Library. ©2002 Mondrian/Holtzman Trust, c/o Beeldrect/Artists Rights Society (ARS), New York
P. 2	Frank Grant/International Stock
P. 15	Richard B. Levine
P. 51	Paul Seheult/Corbis
P. 63	Vince Streano/Corbis
P. 79	Hideo Kurihara/Stone
P. 80	Walter Hodges/Stone
P. 96	Mary Kate Denny/PhotoEdit
P. 109	Stewart Cohen/Stone
P. 123	Spencer Grant/PhotoEdit
P. 143	Richard T. Nowitz/Corbis
P. 159	Dennis Degnan/Corbis
P. 160	Roger Ressmeyer/Corbis
P. 171	Michelle D. Bridewell/PhotoEdit
P. 186	Jacksonville Courier Journal/The Image Works
P. 200	Spencer Grant/PhotoEdit
P. 214	BSIP/Laurent/Yakou/Photo Researchers, Inc.
P. 231	James P. Blair/Corbis
P. 232	David Young Wolff/PhotoEdit
P. 248	Richard Hamilton Smith/Corbis
P. 262	Michael S. Yamashita/Corbis
P. 286	Michael Newman/PhotoEdit
P. 305	Darrell Gulin/The Image Bank
P. 306	NASA
P. 313	Scott Barrow/International Stock
P. 324	Robert Mathena/Fundamental Photographs
P. 333	Michael Crockett/Corbis
P. 340	Paul Merideth/Stone
P. 363	Richard Hamilton Smith/Corbis
P. 364	Chronis Jons/Stone
P. 376	Jeff Greenberg/PhotoEdit
P. 384	Spencer Grant/PhotoEdit
P. 401	Rob Crandall/The Image Works
P. 427	Johnny Johnson/Stone
P. 428	Don Smetzer/Stone
P. 438	Spencer Grant/PhotoEdit
P. 446	Kaluzny/Thatcher/Stone
P. 467	Sam Kittner/National Geographic/Getty Images
P. 499	James A. Sugar/Corbis
P. 510	Andy Sacks/Stone/Getty Images
P. 531	Christi Carter/Grant Hellman Photography
P. 546	Nathan Bilow/Allsport/Getty Images
P. 560	Donald Miralle/Allsport
P. 583	Peter/Stef Lamberti/Stone/Getty Images
P. 584	©SuperStock
P. 595	Gordon & Cathy Illg/Animals Animals
P. 616	Michael S. Lewis/Corbis
P. 624	Reuters/New Media Inc./Corbis
P. 641	Paul A. Souders/Corbis
P. 642	Bill Bachman/Photo Researchers, Inc.
P. 658	Carl & Ann Purcell/Corbis
P. 664	Buddy Mays/Corbis

Answers to Odd-Numbered Exercises, Quizzes, and Tests

Chapter 1

Section 1.1 (page 11)

1. $10x, 5$ **3.** $-3y^2, 2y, -8$ **5.** $4x^2, -3y^2, -5x, 2y$

7. $x^2, -2.5x, -\dfrac{1}{x}$ **9.** 5 **11.** $-\dfrac{3}{4}$

13. Commutative Property of Addition

15. Associative Property of Multiplication

17. Multiplicative Inverse Property

19. Distributive Property

21. Multiplicative Identity Property

23. (a) $5x + 5 \cdot 6$ or $5x + 30$ (b) $(x + 6)5$

25. (a) $(xy)6$ (b) $(6x)y$

27. (a) 0 (b) $(-4t^2) + 4t^2$

29. $(x \cdot x \cdot x) \cdot (x \cdot x \cdot x \cdot x)$ **31.** $(2y)(2y)(2y)$

33. $(-2x)(-2x)(-2x)$ **35.** $\left(\dfrac{y}{5}\right)\left(\dfrac{y}{5}\right)\left(\dfrac{y}{5}\right)\left(\dfrac{y}{5}\right)$

37. $\left(\dfrac{6}{x}\right)\left(\dfrac{6}{x}\right)\left(\dfrac{6}{x}\right)\left(\dfrac{6}{x}\right)$ **39.** $(5x)^4$ **41.** y^3y^4

43. $(-z)^7$ **45.** x^7 **47.** a^8 **49.** x^4 **51.** $\dfrac{a^4}{b^2}$

53. $27y^6$ **55.** $16x^2$ **57.** $-125z^6$ **59.** $a^{16}b^{16}$

61. $-x^4$ **63.** $6x^3y^4$ **65.** $81x^2$ **67.** $\dfrac{16}{3}x^2y^2$

69. $-25y^6$ **71.** $-3125z^8$ **73.** $16a^4$ **75.** $8xy$

77. $\dfrac{2}{3}a^7b$ **79.** $-y^1$ **81.** $-\dfrac{4x^8}{25y^2}$ **83.** x

85. x^{4n} **87.** x^{n+3} **89.** $r^{n-1}s^{m-3}$ **91.** $7x$

93. $8y$ **95.** $8x + 18y$ **97.** $\dfrac{11}{2}z^2 + \dfrac{3}{2}z + 10$

99. $4u^2v^2 + uv$ **101.** $5a^2b^2 - 2ab$ **103.** $12x - 35$

105. $2x - 1$ **107.** $-2z^2 + z - 6$

109. $-3y^2 - 7y - 7$ **111.** $-12x^2 - 100$

113. $-2b^2 + 4b - 36$ **115.** $2y^3 + y^2 + y$

117. $-x^2y^2 - xy$ **119.** $-51a^7$

121. (a) -10 (b) 3 **123.** (a) 7 (b) 7

125. (a) 11 (b) 7 **127.** (a) 0 (b) $\dfrac{3}{10}$

129. (a) 13 (b) -36 **131.** (a) -9 (b) 81

133. (a) 3 (b) 0 **135.** (a) 210 (b) 140

137. (a) $4x + 8$ (b) $6x$

139. (a) $\dfrac{5}{2}b + 2$ (b) $\dfrac{1}{4}b^2 + \dfrac{1}{2}b$ **141.** $\dfrac{1}{2}b^2 - \dfrac{3}{2}b;\ 90$

143. $\dfrac{5}{4}h^2 + 10h;\ 300$

145. Graph: 6500 million; Model: 6498 million

147. $\begin{aligned} b_1 h + \tfrac{1}{2}(b_2 - b_1)h &= b_1 h + \left(\tfrac{1}{2}b_2 - \tfrac{1}{2}b_1\right)h \\ &= b_1 h + \tfrac{1}{2}b_2 h - \tfrac{1}{2}b_1 h \\ &= \left(b_1 h - \tfrac{1}{2}b_1 h\right) + \tfrac{1}{2}b_2 h \\ &= \tfrac{1}{2}b_1 h + \tfrac{1}{2}b_2 h \\ &= \tfrac{1}{2}h(b_1 + b_2) \\ &= \tfrac{1}{2}(b_1 + b_2)h \end{aligned}$

149. $a(b + c) = ab + ac$

151. Constants are real numbers and variables are letters used to represent numbers.

153. The base of the exponent 3 in the expression $(2x)^3$ is $2x$. The base of the exponent 3 in the expression $2x^3$ is x.

155. True, Distributive Property

Section 1.2 (page 24)

1. $10x - 4;\ 1;\ 10$ **3.** $-3y^4 + 5;\ 4;\ -3$

5. $-16z^2 + 8z;\ 2;\ -16$ **7.** $4t^5 - t^2 + 6t + 3;\ 5;\ 4$

9. $x^3 - 5x^2 + x - 5;\ 3;\ 1$ **11.** $x;\ 1;\ 1$

13. Binomial   **15.** Trinomial   **17.** Monomial

19. Answers will vary. Sample answer: $5x^3$

21. Answers will vary. Sample answer: $-2x^4 + x^2 + 1$

23. Answers will vary. Sample answer: $7x$

25. Answers will vary. Sample answer: 14

27. (a) 16 (b) 0 **29.** (a) -27 (b) $\dfrac{9}{16}$

31. $7x^2 + 3$ **33.** $6x^2 - 7x + 8$ **35.** $2x^2 - 3x$

37. 4 **39.** $x^2 - 3x + 2$ **41.** $4x^3 + x + 4$

43. $x^2 - 2x + 5$ **45.** $-4x^3 - 2x + 13$

47. $-2x^2 + 15$ **49.** $2x^2 + 9x - 11$

51. $7x^3 + 22x^2 + 4$ **53.** $29s + 8$ **55.** $7x^3 + 2x$

57. $7x^3 + 15x^2 + 2x$ **59.** $3t^2 + 29$

61. $z^2 + 23z - 16$ **63.** $11x^3 - 3x^2 - 41x + 15$

65. $4.29x^2$ **67.** $-7.148a^2 + 15.691a$ **69.** $16a^3$

71. $10y - 2y^2$ **73.** $8x^5 - 12x^4 + 20x^3$

75. $-10x^2 - 6x^4 + 14x^5$ **77.** $x^2 + 3x - 28$

79. $-x^2 + 2x + 15$ **81.** $2t^2 + 15t - 8$

83. $6a^8 - 11a^4 - 35$ **85.** $6x^2 + 7xy + 2y^2$

87. $5x^4 + 2x^2y - 3y^2$ **89.** $48y^2 + 32y - 3$

91. $75x^3 + 30x^2$ **93.** $-a^2 + 19a$

95. $2t^2 + 7t - 16$ **97.** $x^4 - 2x^3 - 3x^2 + 8x - 4$

99. $2u^3 + 13u^2 + 11u - 20$

101. $7x^3 + 7x^2 - 33x + 27$ **103.** $-2x^3 + 3x^2 - 1$

105. $x^2 - 16$ **107.** $a^2 - 36c^2$ **109.** $4t^2 - 81$

111. $4x^2 - \dfrac{1}{16}$ **113.** $0.04t^2 - 0.25$ **115.** $x^6 - 16$

117. $x^2 + 10x + 25$ **119.** $25x^2 - 20x + 4$

121. $4a^2 + 12ab + 9b^2$ **123.** $4x^8 - 12x^4 + 9$

125. $x^4 - 2x^3 - 8x - 16$

127. $2t^4 + 10t^3 - 7t^2 - 25t + 5$

129. $a^3 + 15a^2 + 75a + 125$

131. $8x^3 - 36x^2 + 54x - 27$

133. $a^4 + 8a^3 - 11a^2 + 32a - 15$

135. $x^2 + 4x + 4 - 2yx - 4y + y^2$

137. $4z^2 + 4yz + 4z + y^2 + 2y + 1$ **139.** $-8x - 7$

141. $12t$ **143.** $2x^2 + 2x$ **145.** $10y - 1$

147. $6x^2 - 13x - 5$ **149.** $8x^2 + 26x$

151. $6t^2 + 18t$

153. (a) $x^2 + 3x$ (b)

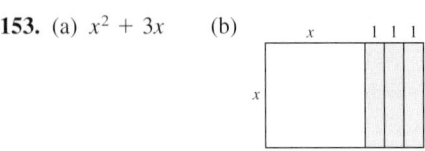

155. (a) $t^2 + 5t + 6$ (b)

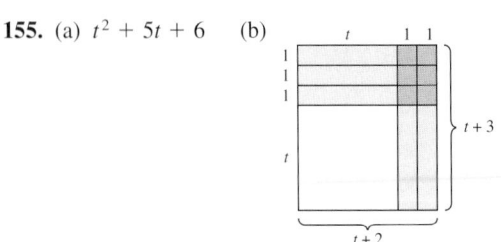

157. $(x + a)(x + b)$, $x^2 + ax + bx + ab$;
$(x + a)(x + b) = x^2 + bx + ax + ab$
This statement illustrates the FOIL Method.

159. (a) $5w$ (b) $\frac{3}{2}w^2$ **161.** $1000 + 2000r + 1000r^2$

163. Dropped, 100 feet   **165.** Thrown downward, 50 feet

167. 224 feet, 216 feet, 176 feet

169. $\sqrt{x}$ is not an integer power of x.

171. No. For example $x^3 + 2x + 3$ is a third degree trinomial.

173. (a) $x^2 - 1$ (b) $x^3 - 1$ (c) $x^4 - 1$; $x^5 - 1$

Section 1.3   (page 35)

1. 6   **3.** $3x$ **5.** $6z^2$ **7.** $14ab^2$ **9.** $21(x + 8)^2$

11. $x(1 - z)^2$ **13.** $8(z - 1)$ **15.** $6(4x^2 - 3)$

17. $x(2x + 1)$ **19.** $7u(3u - 2)$

21. No common factor other than 1 or -1

23. $3y(x^2y - 5)$ **25.** $4(7x^2 + 4x - 2)$

27. $x^2(14x^2 + 21x + 9)$ **29.** $y^2(17x^5y - x + 34)$

31. $3xy(3x^2 + 2y)$ **33.** $xy(3x^2y^2 - 2xy + 5)$

35. $-5(x - 2)$ **37.** $-7(2x - 1)$

39. $-2(x^2 - 2x - 4)$ **41.** $-1(4t^2 - 2t + 15)$

43. $6x + 5$ **45.** $3x + 2y$

47. $(y - 3)(2y + 5)$ **49.** $(t^2 + 1)(5t - 4)$

51. $(a + 6)(3a - 5)$ **53.** $(y - 6)(y^2 + 2)$

55. $(2z + 3)(7z^2 - 3)$ **57.** $(x + 2)(x^2 + 1)$

59. $(a - 4)(a^2 + 2)$ **61.** $(z + 3)(z^3 - 2)$

63. $(c - 3)(d + 3)$ **65.** $(x + 8)(x - 8)$

67. $(10 - 3y)(10 + 3y)$ **69.** $(11 + y)(11 - y)$

71. $(4y + 3z)(4y - 3z)$ **73.** $(x + 2y)(x - 2y)$

75. $(a^4 + 6)(a^4 - 6)$ **77.** $(ab - 4)(ab + 4)$

79. $(a + 11)(a - 3)$ **81.** $(4 - z)(14 + z)$

83. $(x - 2)(x^2 + 2x + 4)$ **85.** $(y + 5)(y^2 - 5y + 25)$

87. $(2t - 3)(4t^2 + 6t + 9)$

89. $(3s + 4)(9s^2 - 12s + 16)$

91. $(2x - y)(4x^2 + 2xy + y^2)$

93. $(y + 4z)(y^2 - 4yz + 16z^2)$ **95.** $2(5x + 2)(5x - 2)$

97. $x(x + 12)(x - 12)$ **99.** $(b + 2)(b - 2)(b^2 + 4)$

101. $(y + 3x)(y - 3x)(y^2 + 9x^2)$

103. $2(x - 3)(x^2 + 3x + 9)$

105. $3(a + 4)(a^2 - 4a + 16)$ **107.** $(2x^n - 5)(2x^n + 5)$

109. $x^2(3x + 4) - (3x + 4) = (x - 1)(x + 1)(3x + 4)$
$3x(x^2 - 1) + 4(x^2 - 1) = (x - 1)(x + 1)(3x + 4)$

111. $P(1 + rt)$ **113.** $l = 32 - w$

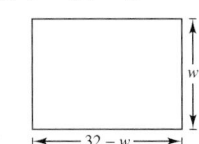

115. $l = w + 8$ **117.** $l = 5x + 2$

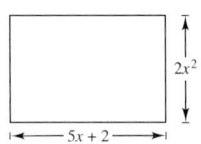

119. $S = 2\pi r(r + h)$ **121.** $kx(Q - x)$

123. It is written as a product of polynomials.

125. $(80 + 1)(80 - 1) = 80^2 - 1^2 = 6399$

Mid-Chapter Quiz   (page 38)

1. Distributive Property

2. Multiplicative Identity Property

3. Additive Inverse Property   **4.** $(5x)^4$ **5.** $4x^5$

6. $16x^4$ **7.** $\dfrac{y^6}{27}$ **8.** $\dfrac{3}{2}xy^2$ **9.** $-x^2 + 2xy$

10. $3x^2 + 2x + 8$ **11.** $7x - 15$ **12.** $-6x^3 - 3x$

13. $36r^2 - 25s^2$ **14.** $2x^3 + 6x^2 - 4x - 12$

15. $4x^2 - 12xy + 9y^2$ **16.** $x^3 + 1$

17. $2z^2 + 3z - 35$ **18.** $-12v$

19. (a) 225   (b) 5   **20.** $6x + 16$

21. $-16x^4 + 7x^2 + 4$
Degree: 4
Leading coefficient: -16

22. $2x + 2$ **23.** $4x(6x^2 + 7)$ **24.** $(y - 5)(3y + 1)$

25. $(x - 9)(x^2 + 5)$ **26.** $(4a + b)(16a^2 - 4ab + b^2)$

27. $-2(2x - 3)(4x^2 + 6x + 9)$ **28.** $(9x + 4y)(9x - 4y)$

29. $x(x + 3y)(x - 3y)$ **30.** $(t + 2)(t + 1)(t - 1)$

31. When $t = 1$, height $= 379$ feet.
When $t = 2$, height $= 376$ feet.
When $t = 3$, height $= 341$ feet.

Section 1.4   (page 48)

1. $x + 1$ **3.** $y - 5$ **5.** $z - 2$

7. $(x + 3)(x + 1)$ **9.** $(y + 10)(y - 3)$

11. $(t - 7)(t + 3)$ **13.** $(x - 2)(x - 10)$

15. $(t - 12)(t - 5)$ **17.** $(x - 8)(x - 12)$

19. $(u + 2v)(u + 3v)$ **21.** $(x - 7y)(x + 5y)$

23. $5x + 3$ **25.** $5a - 3$ **27.** $2y - 9$
29. $(3x + 1)(x + 1)$ **31.** $(2t + 1)(4t - 5)$
33. $(3b - 1)(2b + 7)$ **35.** $(2a - 5)(a - 4)$
37. $(2x + 3y)(x + y)$ **39.** $(11y + z)(y - 4z)$
41. $(3a + 2b)(2a - b)$ **43.** $(5x + 4)(4x - 3)$
45. $(2u - 5v)(u + 7v)$ **47.** Not factorable
49. $(-1)(2x - 3)(x + 2)$ **51.** $(-1)(4x + 1)(15x - 1)$
53. $(3x + 4)(x + 2)$ **55.** $(2x - 1)(3x + 2)$
57. $(3x - 1)(5x - 2)$ **59.** $(x + 2)^2$ **61.** $(a - 6)^2$
63. $(5y - 1)^2$ **65.** $(3b + 2)^2$ **67.** $(2x - y)^2$
69. $(u + 4v)^2$ **71.** $3x^3(x - 2)(x + 2)$
73. $2t(5t - 9)(t + 2)$ **75.** $5(a - 2)(a - 3)$
77. $3(2u + 7)(u - 3)$ **79.** $(3y - 11)^2$
81. $(3x - 2)(7x - 2)$ **83.** $x(x + 3)(x + 5)$
85. $(3 - z)(9 + z)$ **87.** $4(y + 1)(y - 2)$
89. $2(3x - 1)(9x^2 + 3x + 1)$ **91.** $v(v^2 + 3v + 5)$
93. $2xy(x - 7y)(x + 6y)$ **95.** $5y(x - 2y)(x + 2y)$
97. $(x + 2)(x + 4)(x - 4)$ **99.** $(x + 3)(x - 3)(x - 6)$
101. $(x - 5 + y)(x - 5 - y)$
103. $(a - b + 4)(a - b - 4)$
105. $(x^4 + 1)(x^2 + 1)(x + 1)(x - 1)$
107. $b(b - 6)(b^2 + 6b + 36)$ **109.** ± 18 **111.** ± 6
113. ± 12 **115.** 16   **117.** 9   **119.** 25   **121.** 14
123. $\pm 9, \pm 11, \pm 19$ **125.** $\pm 4, \pm 20$
127. $\pm 13, \pm 14, \pm 22, \pm 41$ **129.** $8, -16$ **131.** $2, -40$
133. $-5, -32$ **135.** $(6 + x)(6 - x)$
137. $(x + 3)(x - 3)$ **139.** $x + 3$ **141.** c   **142.** b
143. a   **144.** d   **145.** $(x + 3)(x + 1)$

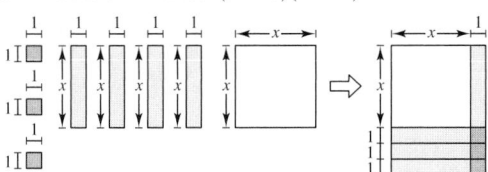

147. $9x^2 - 9x - 54 = 9(x^2 - x - 6)$
$\qquad\qquad = 9(x - 3)(x + 2)$
149. $52^2 = (53 - 1)^2$
$\qquad = 2809 - (2)(53)(1) + 1^2$
$\qquad = 2704$

Section 1.5   (page 59)

1. (a) No   (b) Yes   **3.** (a) No   (b) Yes
5. (a) Yes   (b) No   **7.** No solution
9. Identity   **11.** Linear
13. Not linear because the variable is in the denominator
15. Original equation
   Subtract 15 from each side.
   Combine like terms.
   Divide each side by 3.
   Simplify.

17. Original equation.
   Subtract 5 from each side.
   Combine like terms.
   Divide each side by -2.
   Simplify.
19. -7 **21.** 4   **23.** 2   **25.** 2   **27.** $\frac{1}{3}$ **29.** 0
31. Not possible because $2 \neq 0$ **33.** 0   **35.** 6
37. -2 **39.** 11   **41.** -3 **43.** $\frac{6}{5}$ **45.** -3
47. -2 **49.** -2 **51.** -3 **53.** 50   **55.** $\frac{19}{10}$
57. 0   **59.** $\frac{24}{91}$ **61.** 2   **63.** $-\frac{10}{3}$ **65.** $-\frac{20}{9}$
67. 72   **69.** $-\frac{24}{11}$ **71.** $\frac{1}{5}$ **73.** 23   **75.** 12
77. (a) $x = \dfrac{6 + 3y}{2}$ (b) $y = \dfrac{2x - 6}{3}$
79. (a) $x = \dfrac{10y - 11}{7}$ (b) $y = \dfrac{7x + 11}{10}$
81. (a) $x = \dfrac{42 - 7y}{12}$ (b) $y = \dfrac{42 - 12x}{7}$
83. (a) $x = \dfrac{10 - 2y}{5}$ (b) $y = \dfrac{10 - 5x}{2}$
85–89. Answers will vary.   **91.** $R = \dfrac{E}{I}$ **93.** $h = \dfrac{V}{lw}$
95. $r = \dfrac{C}{2\pi}$ **97.** $a = P - b - c$ **99.** $w = \dfrac{A}{l}$
101. $h = \dfrac{2A}{b}$ **103.** $n = \dfrac{2S}{a_1 + a_n}$ **105.** $C = \dfrac{S}{1 + r}$
107. $P = \dfrac{A}{1 + rt}$ **109.** $n = \dfrac{L - a}{d} + 1$
111. $m_2 = \dfrac{Fr^2}{km_1}$
113. (a)

t	1	1.5	2
Width	300	240	200
Length	300	360	400
Area	90,000	86,400	80,000

t	3	4	5
Width	150	120	100
Length	450	480	500
Area	67,500	57,600	50,000

(b) In a rectangle of fixed perimeter with length l equal to t times width w and $t \geq 1$, as t increases, w decreases, l increases, and the area A decreases. The maximum area occurs when the length and width are equal (when $t = 1$).

115. 1.5 seconds   **117.** 6 hours   **119.** 1991

121. (a) An expression is a collection of variables and constants using the operations of addition, subtraction, multiplication, and division. An equation is a statement of equality of two expressions.

(b) An equation whose solution set is not the entire set of real numbers is called a conditional equation. The solution set of an identity is all real numbers.

Section 1.6   (page 71)

1. $0, 8$ **3.** $-10, 3$ **5.** $-4, 2$ **7.** $-\frac{25}{2}, 0, \frac{3}{2}$

9. $-4, -\frac{1}{2}, 3$ **11.** $-2, 5$ **13.** $4, 5$ **15.** $-3, 0$

17. ± 5 **19.** ± 10 **21.** 4 **23.** $\frac{3}{2}$ **25.** $-\frac{1}{2}, 7$

27. $-2, 5$ **29.** $-12, 6$ **31.** $-2, 10$ **33.** $-2, 6$

35. $-\frac{1}{3}, 0, \frac{1}{2}$ **37.** $0, 7, 12$ **39.** $\pm 4, 25$

41. ± 2 **43.** ± 3 **45.** $-2, \pm 3$ **47.** $0, 5, \pm 3$

49. (a) and (b) $-5, -\frac{10}{3}$ (c) Answers will vary.

51. $-\frac{5}{4}, -\frac{1}{2}$ **53.** $0, 1$ **55.** $x^2 - 7x + 6 = 0$

57. $x^2 + 12x + 20 = 0$ **59.** 8 **61.** $\frac{39}{4}$ seconds

63. $\frac{5}{2}$ cm   **65.** $b = 14, h = 10$

67. 20 inches $\times$ 20 inches   **69.** 50 units

71. False. It is not an application of the Zero-Factor Property because there is an unlimited number of factors whose product is 1.

73. Maximum number: n. The third-degree equation $(x + 1)^3 = 0$ has one solution: $x = -1$.

Review Exercises   (page 75)

1. Terms: $14y^2, -9$ **3.** Terms: $15t^3, -2t^2, 19t$
Coefficients: $14, -9$ Coefficients: $15, -2, 19$

5. Associative Property of Addition

7. Additive Identity Property   **9.** x^6 **11.** $-2y^5$

13. $-64a^7$ **15.** $-3x^3y^4$ **17.** $4u^7v^3$ **19.** $8u^2v^2$

21. $3x - 20$ **23.** $4y^2 - 10y$ **25.** $20x - 80$

27. $5x - y$ **29.** $-8x + 23y$ **31.** $18b - 15a$

33. (a) 22 (b) -5 **35.** (a) -1 (b) 0

37. $12x^6 - 4x^2 + x + 15; 6; 12$ **39.** $6x - x^2$

41. $5x^2 - 5x + 17$ **43.** $-9x^3 + 9x - 4$

45. $6y^2 - 2y + 15$ **47.** $-8x^4 - 32x^3$

49. $6z^2 - z - 15$ **51.** $15x^2 - 11x - 12$

53. $4x^3 - 5x + 6$ **55.** $16x^2 - 56x + 49$

57. $25u^2 - 64$ **59.** $3x^2(2 + 5x)$

61. $-14(x + 5)(5x + 23)$ **63.** $(y + 5)(y^2 + 7)$

65. $(x + 4)(x - 4)$ **67.** $(3a + 10)(3a - 10)$

69. $(y - 7)(y + 1)$ **71.** $(u - 1)(u^2 + u + 1)$

73. $(3x + 4)(9x^2 - 12x + 16)$

75. $x(x + 3)(x - 3)(x + 7)$

77. $[x + 9 + 2y][x + 9 - 2y]$

79. $(x - 8)(x - 3)$ **81.** $(3x - 1)(x + 8)$

83. $(2x + 5)(x - 3)$ **85.** $(a + 2b)^2$

87. $2y(3y + 4)(3y - 4)$ **89.** $h(2h + 1)(3h - 13)$

91. $x(x - 4)(x^2 + 1)$ **93.** Equation with no solution

95. (a) Yes   (b) No   **97.** 3 **99.** -7

101. -2 **103.** 14 **105.** 24 **107.** $\frac{16}{3}$ **109.** 2

111. $\frac{213}{164}$ **113.** $h = \dfrac{V}{\pi r^2}$ **115.** $y = \dfrac{-3x + 19}{5}$

117. $0, 3$ **119.** $-5, 4$ **121.** $2, 3$ **123.** $-\frac{1}{2}, 7$

125. $-4, 9$ **127.** ± 10 **129.** ± 4 **131.** $-3, 0, 4$

133. $\pm 1, 6$ **135.** $P = 8x + 8; A = 15x$

137. (a) $12w$ (b) $5w^2$

139. $16 - l$ **141.** 24 inches $\times$ 18 inches

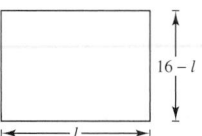

Chapter 1 Test   (page 78)

1. $3x^4y^3$ **2.** $-200x^7y^9$ **3.** $3xy^4$

4. $-2x^2 + 5x - 1$ **5.** $6x^2 - 11x + 3$

6. $10x^2 - 6x + 5$ **7.** $-2y^2 - 2y$

8. $8x^2 - 4x + 10$ **9.** $11t + 7$ **10.** $\dfrac{y^3}{8}$

11. $2x^2 + 7xy - 15y^2$ **12.** $6s^3 - 17s^2 + 26s - 21$

13. $16x^2 - 24x + 9$ **14.** $36 - 16y^2$

15. $6y(3y - 2)$ **16.** $(x - 2)(5x^2 - 6)$ **17.** $(3u - 1)^2$

18. $2(3x + 2)(x - 5)$ **19.** $(b - 8)(b + 6)$

20. $(7x - 6)(7x + 6)$ **21.** $(a + 2 - 3b)(a + 2 + 3b)$

22. $(2y + 3)(y + 6)$ **23.** $2x(x + 12)(x + 1)$

24. $2(x - 4y)(x^2 + 4xy + 16y^2)$ **25.** 4 **26.** 4

27. 24 **28.** $-\frac{4}{3}, 3$ **29.** $7, -10$ **30.** $-4, -1, 0$

31. $x^2 + 26x$ **32.** $b = \frac{1}{2}(3a + 17)$

Chapter 2

Section 2.1   (page 91)

1.   **3.**

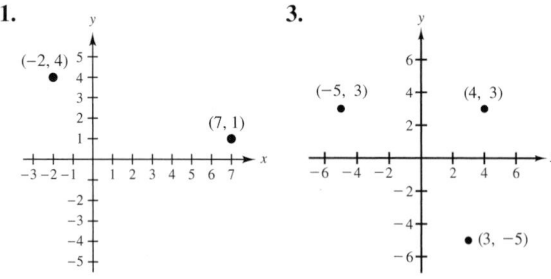

5.

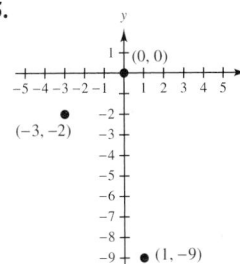

7.

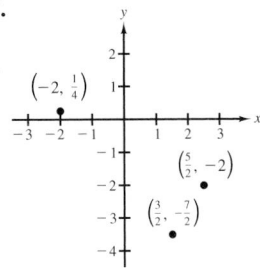

9. $A: (3, 0)$ $B: (-2, -2)$ $C: (-1, 5)$
11. $A: (4, -2)$ $B: \left(-3, -\frac{5}{2}\right)$ $C: \left(3, \frac{1}{2}\right)$

13.

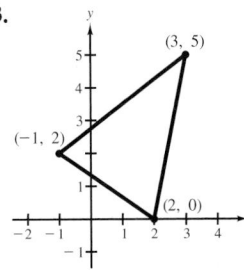

15.

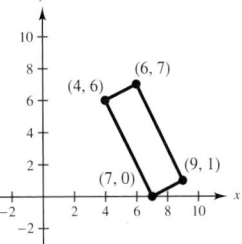

17.

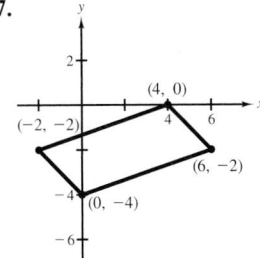

19. Quadrant III   **21.** Quadrant I   **23.** Quadrant IV
25. Quadrants I and II   **27.** Quadrants II and IV
29. Quadrant II   **31.** $(-5, 2)$ **33.** $(3, 6)$
35. $(10, 0)$ **37.** $(0, 12)$

39.

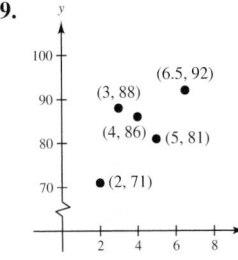

41.

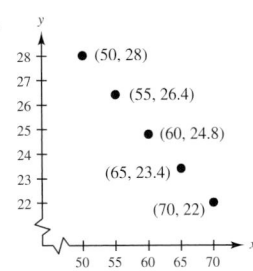

43. (a) Yes (b) No   (c) No   (d) Yes
45. (a) Yes (b) Yes (c) Yes (d) No
47. (a) No   (b) Yes (c) Yes (d) No

49.

x	-2	0	2	4	6
$y = 5x - 1$	-11	-1	9	19	29

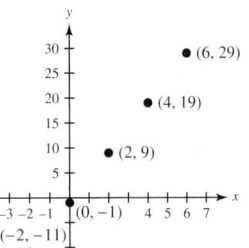

51.

x	-4	$\frac{2}{5}$	4	8	12
$y = -\frac{5}{2}x + 4$	14	3	-6	-16	-26

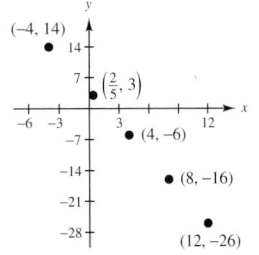

53.

x	-2	0	2	4	6
$y = 4x^2 + x - 2$	12	-2	16	66	148

55.

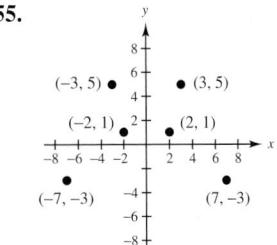

Reflection about the y-axis
57. $(-1, 1), (3, 2), (0, 4)$
59. $(1, -1), (3, -4), (5, -1), (3, 2)$

61. 7 **63.** 7

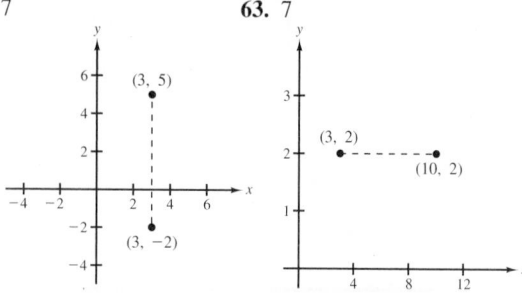

Vertical line   Horizontal line

65. 5 **67.** 15 **69.** 13 **71.** 7.81 **73.** 5.39
75. 5 **77.** 1.41
79. (a) Length of horizontal side: 8
  Length of vertical side: 6
  Length of hypotenuse: 10
  (b) $8^2 + 6^2 = 10^2$
     $100 = 100$ ✓
81. Right triangle **83.** Not a right triangle
85. $3 + \sqrt{26} + \sqrt{29} \approx 13.48$
87. Midpoint: $(1, 4)$ **89.** Midpoint: $(3, 3)$

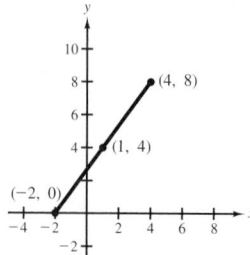 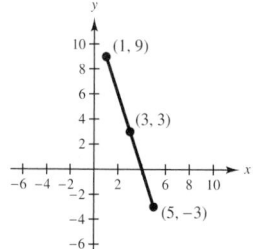

91. Midpoint: $\left(\frac{7}{2}, \frac{9}{2}\right)$ **93.** Midpoint: $\left(\frac{3}{2}, 4\right)$

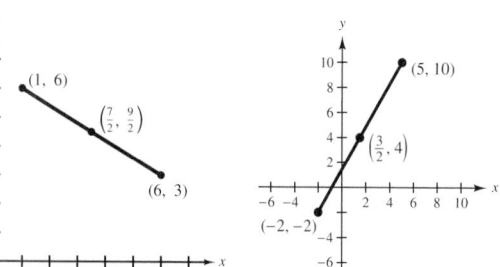

95. There appears to have been no change between the fourth and fifth hours.
97. Yes; About 500 cubic feet
99. Order is significant because each number in the pair has a particular interpretation. The first measures horizontal distance and the second measures vertical distance.
101. $4x^{10}y^{10}$ **103.** $x^2 + 19x - 15$
105. $2(x + 4)(x - 4)$ **107.** $(4x - 5)(x + 1)$
109. $-2, 3$ **111.** $2354.45

Section 2.2 (page 105)

1. e **2.** b **3.** f **4.** a **5.** d **6.** c
7.

x	-4	-2	0	2	4
y	11	7	3	-1	-5

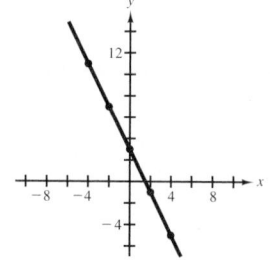

9.

x	± 2	-1	0	2	± 3
y	0	3	4	0	-5

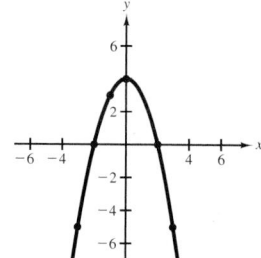

11. **13.**

15. **17.**

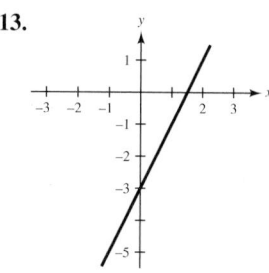

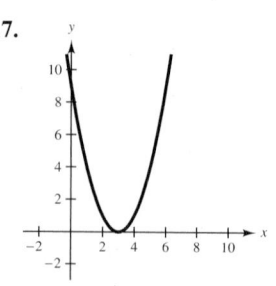

19.

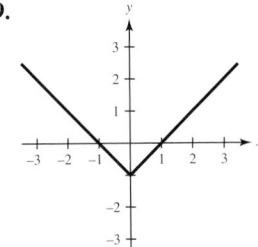

21. $(0, 3)$ **23.** $(2, 0), (0, 2)$ **25.** $(10, 0), (0, 5)$
27. $\left(-\frac{4}{3}, 0\right), \left(0, \frac{10}{9}\right)$ **29.** $(-20, 0), (0, 15)$
31. $\left(-\frac{7}{2}, 0\right), (4, 0), (0, -28)$ **33.** $(-7, 0), (6, 0), (0, -42)$
35. $(-4, 0), (4, 0), (0, 0)$

37.

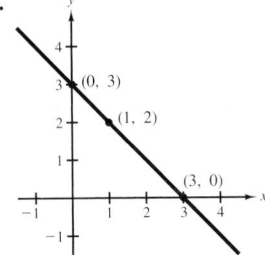

39.

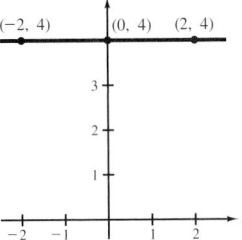

41.

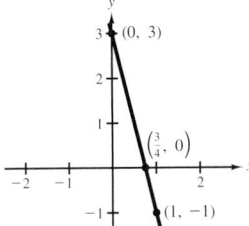

43.

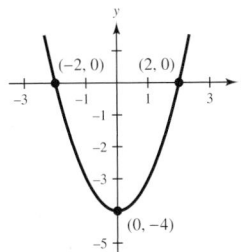

45.

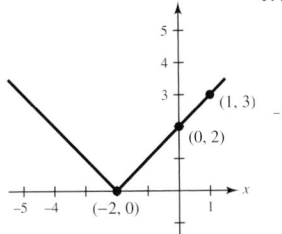

47.

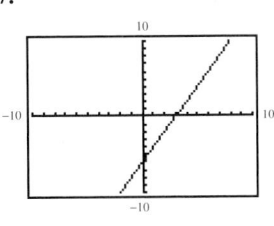

y-intercept: $(0, -6)$

49.

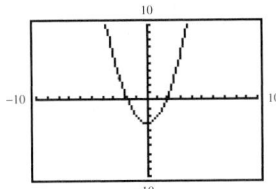

y-intercept: $(0, -3)$

51.

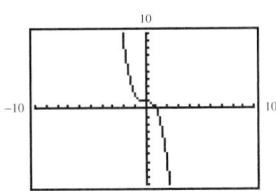

y-intercept: $(0, 1)$

53.

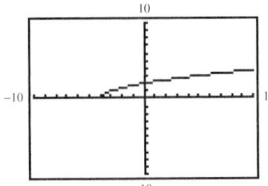

y-intercept: $(0, 2)$

55.

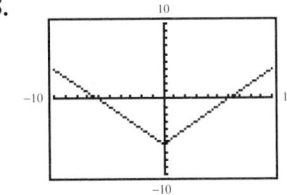

y-intercept: $(0, -6)$

57.

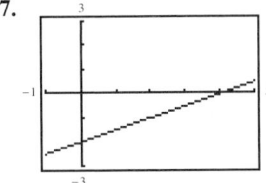

x-intercept: $(4, 0)$

59.

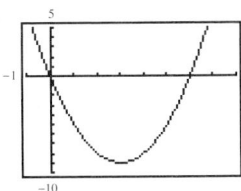

x-intercepts: $(0, 0), (6, 0)$

61. $\frac{9}{2}$ **63.** ± 2 **65.** 1 **67.** $-4, \frac{3}{2}$ **69.** $\pm 2, 0$
71. ± 5 **73.** $x \approx -4.1, x \approx 6.1$
75. (a)

x	0	3	6	9	12
F	0	4	8	12	16

(b)

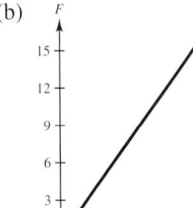

(c) The length is also doubled.

77. (a)

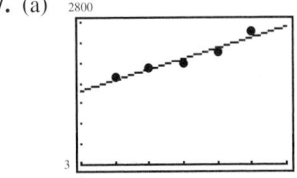

(b) The model fits the data well.

(c) $3,970,000

(d) The model increases to infinity for values of S.

79.

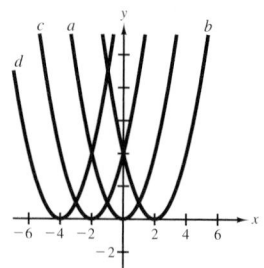

81. (a) Answers will vary.

(b)

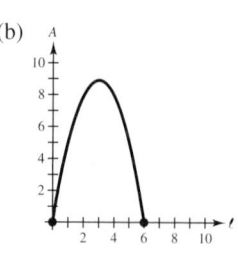

(c) 8 square meters

83. Identical graphs
Distributive Property

85. Identical graphs
Associative Property of Addition

87. x-intercepts: $(\pm 3, 0)$
Solutions: $x = \pm 3$

89. x-intercepts: $(-1, 0), (3, 0)$
Solutions: $x = -1, x = 3$

91.

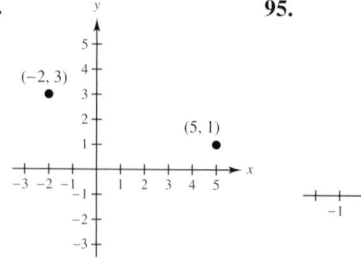

The graph of $y = (x + c)^2$, $c > 0$, is obtained by shifting the graph of $y = x^2$ to the *left* c units.
The graph of $y = (x - c)^2$, $c > 0$, is obtained by shifting the graph of $y = x^2$ to the *right* c units.

93.

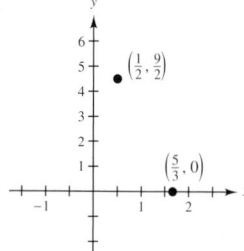

95.

97. $-\frac{32}{3}$　**99.** 17　**101.** $\frac{17}{3}$　**103.** 18.55 feet

Section 2.3 (page 118)

1. $\frac{2}{3}$　**3.** -2　**5.** Undefined

7. $-\frac{3}{2}$, falling

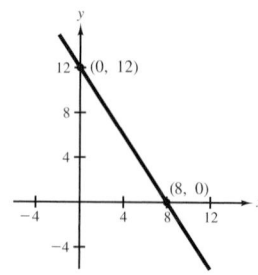

9. $\frac{1}{2}$, rising

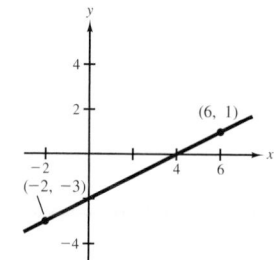

11. $-\frac{3}{4}$, falling

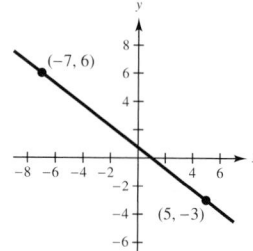

13. $\frac{5}{2}$, rising

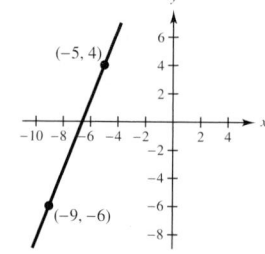

15. 0, horizontal

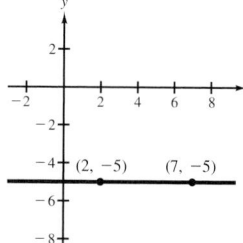

17. $\frac{5}{3}$, rising

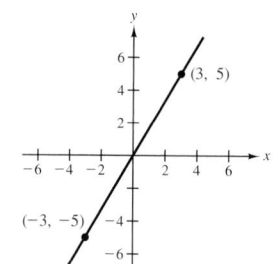

19. $\frac{6}{35}$, rising

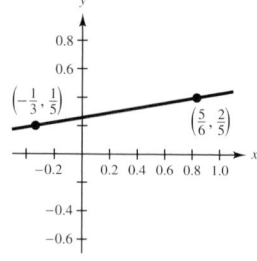

21. $-\frac{18}{17}$, falling

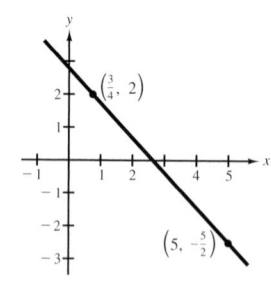

23. $-\frac{5}{6}$, falling

25. 0, horizontal

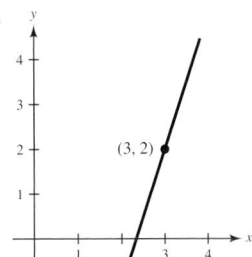

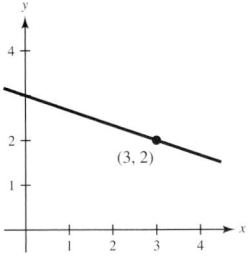

27. $x = 1$ **29.** $y = -15$

31.

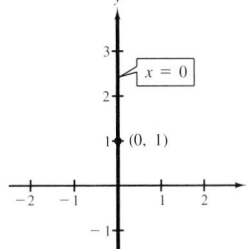

33.

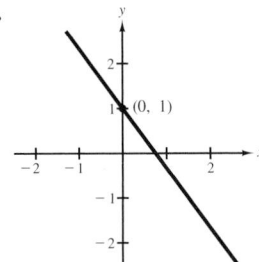

35.

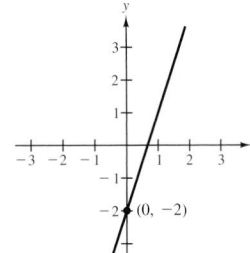

37.

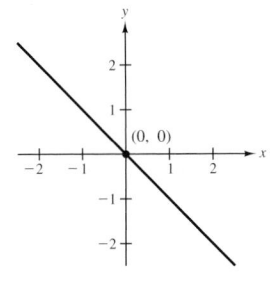

39. $(6, 2), (10, 2)$ **41.** $(4, -1), (5, 2)$
43. $(1, 2), (2, 1)$ **45.** $(-2, 4), (1, 8)$
47. $y = 2x - 3$
    Slope: 2
    y-intercept: $(0, -3)$

49. $y = \frac{1}{4}x - 1$
    Slope: $\frac{1}{4}$
    y-intercept: $(0, -1)$

51. $y = -\frac{2}{5}x + \frac{3}{5}$
    Slope: $-\frac{2}{5}$
    y-intercept: $\left(0, \frac{3}{5}\right)$

53. $y = \frac{1}{2}x + 2$
    Slope: $\frac{1}{2}$
    y-intercept: $(0, 2)$

55. $y = 3x - 2$

57. $y = -x$

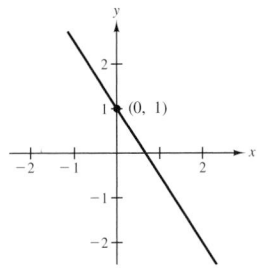 *(Note: positions 55/57 graphs at bottom left)*

59. $y = -\frac{3}{2}x + 1$

61. $y = \frac{1}{4}x + \frac{1}{2}$

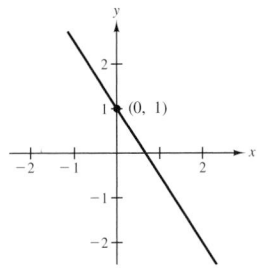

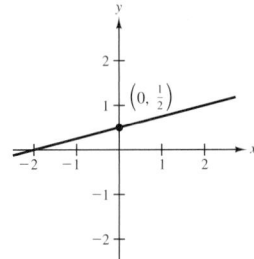

63. $y = 2$

65. $y = 5x - 5$

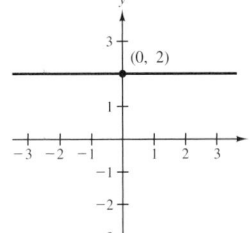

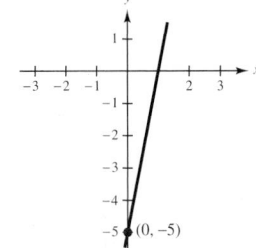

67. Parallel   **69.** Perpendicular   **71.** Neither
73. Parallel   **75.** Perpendicular   **77.** Parallel
79. $y = -x + 9$
    (a) -1 (b) 1

81. $y = 2x - \frac{11}{7}$
    (a) 2   (b) $-\frac{1}{2}$

83. $y = \frac{5}{2}x + 13$
    (a) $\frac{5}{2}$ (b) $-\frac{2}{5}$

85. $y = -\frac{3}{2}x - \frac{5}{6}$
    (a) $-\frac{3}{2}$ (b) $\frac{2}{3}$

87. b; slope: -10
    For every week that goes by, the amount owed on the loan decreases by \$10.

88. c; slope: 1.50
    For every unit produced in an hour, the hourly pay increases by \$1.50.

89. a; slope: 0.25
    For every mile traveled within a day, the daily pay increases by \$0.25.

90. d; slope: -100
    For every year that goes by, the worth of the word processor decreases by \$100.

91. $\frac{9}{5}$
93. (a) $P = 8x + 140$

    (b)

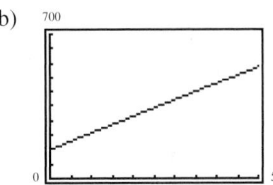

    (c) $m = 8$; 8 feet

95. 16,667 feet   **97.** 11.25   **99.** 11.25 feet

101. Negative slope: line falls to the right.

Zero slope: line is horizontal.

Positive slope: line rises to the right.

103. No. Their slopes must be negative reciprocals of each other.

105. $\frac{3}{4}$ **107.** $y = 4x + 46$ **109.** $P = 10x + 8$

Mid-Chapter Quiz   (page 122)

1. (a)

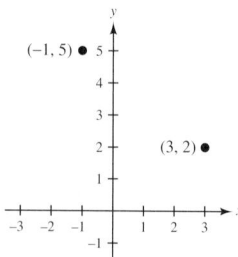

2. (a)

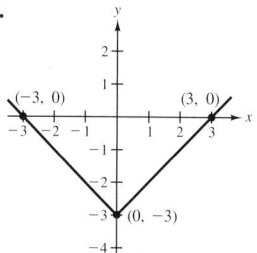

(b) 5

(c) $\left(1, \frac{7}{2}\right)$

(d) $-\frac{3}{4}$

(b) 13

(c) $\left(-\frac{1}{2}, 4\right)$

(d) $\frac{12}{5}$

3. $(10, -3)$ **4.** (a) No (b) Yes (c) Yes (d) Yes

5.

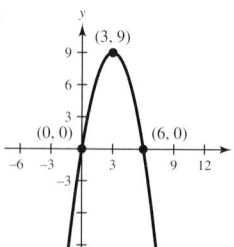

6.

7.

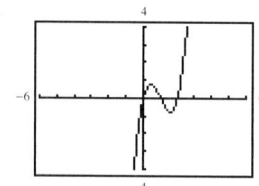

x-intercepts: $(0, 0), (1, 0), (2, 0)$

y-intercept: $(0, 0)$

8.

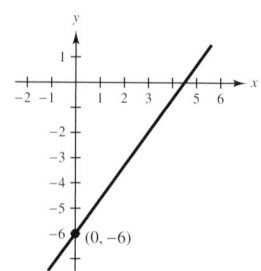

9. $y = -3x + 6$

Slope: -3

y-intercept: $(0, 6)$

10. $y = \frac{4}{3}x - 5$

Slope: $\frac{4}{3}$

y-intercept: $(0, -5)$

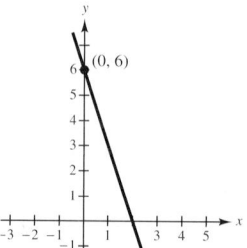

11. Perpendicular   **12.** Neither   **13.** Parallel

14. $y = -\frac{5}{3}x + 3$ (a) $-\frac{5}{3}$ (b) $\frac{3}{5}$

15. $-\$8100$ per year   **16.** $68°$F; 12:00 P.M.; $58°$F; 8:00 A.M.

Section 2.4   (page 130)

1. Domain $= \{-2, 0, 1\}$

Range $= \{-1, 0, 1, 4\}$

3. Domain $= \{0, 2, 4, 5, 6\}$

Range $= \{-3, 0, 5, 8\}$

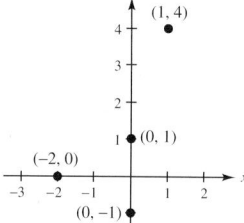

5. $(3, 150), (2, 100), (8, 400), (6, 300), \left(\frac{1}{2}, 25\right)$

7. $(1, 1), (2, 8), (3, 27), (4, 64), (5, 125), (6, 216), (7, 343)$

9. (1997, Florida Marlins), (1998, New York Yankees),

(1999, New York Yankees), (2000, New York Yankees),

(2001, Arizona Diamondbacks)

11. Not a function   **13.** Function   **15.** Not a function

17. Not a function

19. (a) Function from A to B

(b) Not a function from A to B

(c) Function from A to B

(d) Not a function from A to B

21. There are two values of y associated with one value of x.

23. There are two values of y associated with one value of x.

25. Both high school enrollment and college enrollment are functions of the year. For each there is exactly one enrollment for each year.

27. (a) $f(2) = 3(2) + 5 = 11$

(b) $f(-2) = 3(-2) + 5 = -1$

(c) $f(k) = 3(k) + 5 = 3k + 5$

(d) $f(k + 1) = 3(k + 1) + 5 = 3k + 8$

29. (a) 29   (b) 11   (c) $12a - 2$ (d) $12a + 5$

31. (a) 2   (b) 3   (c) $\sqrt{\frac{31}{3}}$ (d) $\sqrt{5z + 5}$

33. (a) 0   (b) $-\frac{3}{2}$ (c) $-\frac{5}{2}$ (d) $\dfrac{3x + 12}{x - 1}$

35. (a) 2   (b) -2 (c) 10   (d) -8

37. (a) -14 (b) 20   (c) 31   (d) 24

39. (a) $2x^2 + 8x$ (b) $2x - 12$

41. Domain $= \{0, 2, 4, 6\}$
Range $= \{0, 1, 8, 27\}$

43. Domain $= \{-5, -2, 0, 3, 6\}$
Range $= \{-2, 3, 9\}$

45. Domain $= \{r: r > 0\}$
Range $= \{C: C > 0\}$

47. Domain $= \{x: x$ is a real number$\}$

49. Domain $= \{x: x$ is all real numbers, except $x = 3\}$

51. Domain $= \{t: t$ is all real numbers, except $t = 0, -2\}$

53. Domain $= \{x: x$ is all real numbers, except $x = 2, 1\}$

55. Domain $= \{x: x$ is all real numbers, except $x = 1, -1\}$

57. Domain $= \{x: x \geq 2\}$ **59.** Domain $= \{x: x \leq 3\}$

61. Domain $= \{t: t$ is a real number$\}$

63. Domain $= \{t: t$ is a real number$\}$

65. $P = 4x$

67. $V = x(24 - 2x)^2$ **69.** $A = (32 - 2x)^2$
$\quad\;\; = 4x(12 - x)^2$ $\quad\;\;\;\; = 4(16 - x)^2$

71. $d(t) = 230t$ **73.** (a) \$700   (b) \$750

75. A relation is any set of ordered pairs. A function is a relation in which no two ordered pairs have the same first component and different second components.

77. (a) Correct   (b) Not correct

79. $2x + 5t + 7$ **81.** $7x^2 + 56x + 104$

83. (a) 3   (b) -12

85. (a) 0   (b) $x = 12$ is not in the domain of the function.

87. 22 feet $\times$ 14 feet

Section 2.5   (page 140)

1. $m = 2$ **3.** $m = -1$

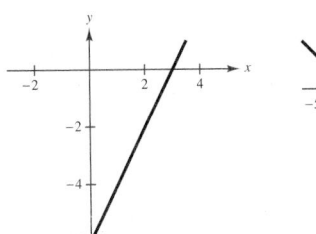
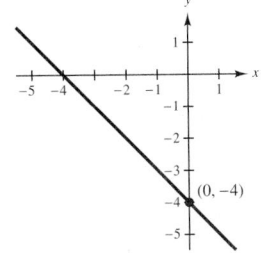

5. $m = -\frac{3}{4}$

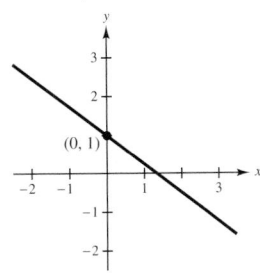

7. $f(x) = 3x + 10$ **9.** $f(x) = -\frac{1}{4}x + \frac{45}{8}$

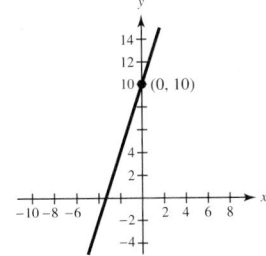
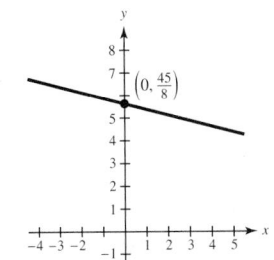

11. Function   **13.** Not a function

15. Not a function   **17.** Function

19. Domain $= \{x: x$ is a real number$\}$
Range $= \{y: y$ is a real number$\}$

21. Domain $= \{x: x \leq -2$ or $x \geq 2\}$
Range $= \{y: y \geq 0\}$

23. Domain $= \{x: -3 \leq x \leq 3\}$
Range $= \{y: 0 \leq y \leq 3\}$

25.

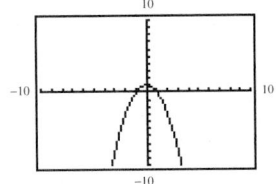

Domain $= \{x: x$ is a real number$\}$
Range $= \{y: y \leq 1\}$

27.

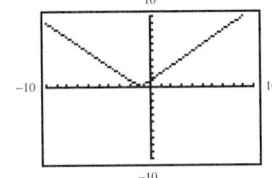

Domain $= \{x: x$ is a real number$\}$
Range $= \{y: y \geq 0\}$

29.

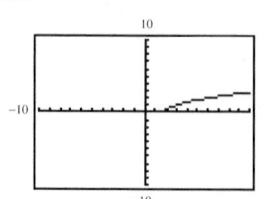

31.

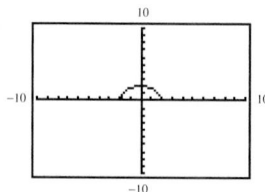

43.

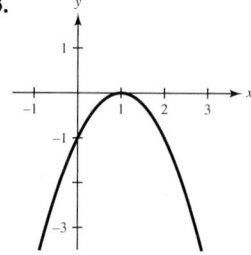

Domain $= \{x: x \geq 2\}$
Range $= \{y: y \geq 0\}$

Domain $= \{t: -2 \leq t \leq 2\}$
Range $= \{y: 0 \leq y \leq 2\}$

Domain $= \{x: x$ is a real number$\}$
Range $= \{y: y \leq 0\}$

33. b   **34.** c   **35.** d   **36.** a

37.

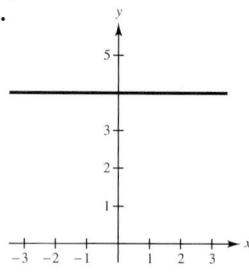

Domain $= \{x: x$ is a real number$\}$
Range $= \{4\}$

45.

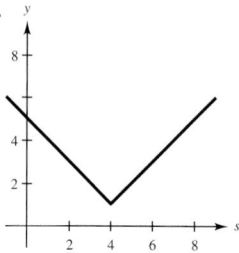

Domain $= \{s: s$ is a real number$\}$
Range $= \{y: y \geq 1\}$

39.

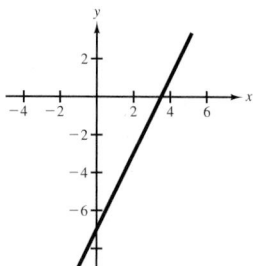

Domain $= \{x: x$ is a real number$\}$
Range $= \{y: y$ is a real number$\}$

47.

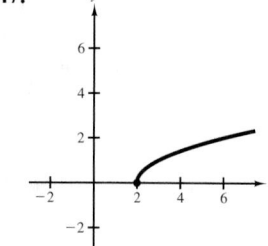

Domain $= \{t: t \geq 2\}$
Range $= \{y: y \geq 0\}$

41.

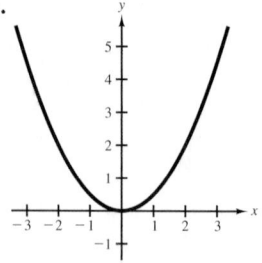

Domain $= \{x: x$ is a real number$\}$
Range $= \{y: y \geq 0\}$

49.

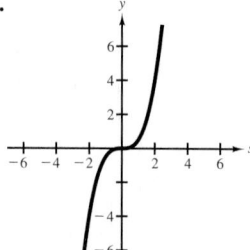

Domain $= \{s: s$ is a real number$\}$
Range $= \{y: y$ is a real number$\}$

51.

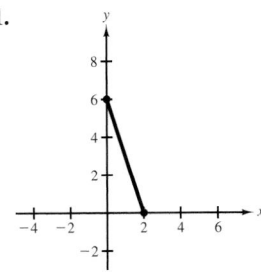

Domain = $\{x: 0 \le x \le 2\}$
Range = $\{y: 0 \le y \le 6\}$

53.

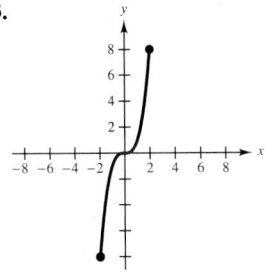

Domain = $\{x: -2 \le x \le 2\}$
Range = $\{y: -8 \le y \le 8\}$

55.

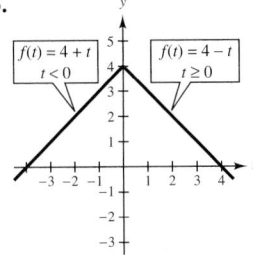

Domain = $\{t: t$ is a real number$\}$
Range = $\{y: y \le 4\}$

57.

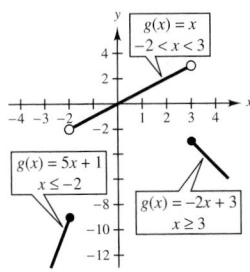

Domain = $\{x: x$ is a real number$\}$
Range = $\{y: y \le -3$ or $-2 < y < 3\}$

59. (a) Answers will vary.

(b)

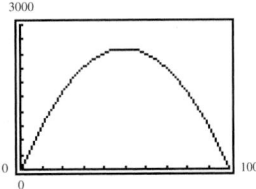

(c) $x = 50$. The figure is a square.

61. (a)

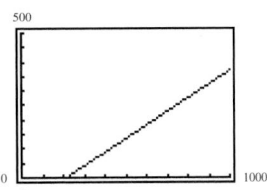

(b) 213   (c) 851

63. Function. For each value of y there corresponds one value of x.

65. A graph shows y as a function of x if and only if no vertical line passes through the graph more than once.

67. $x = 5, 8$ **69.** $a = -2, 0, 2$ **71.** (a) 36   (b) 0

73. (a) $\dfrac{10}{3}$ (b) $\dfrac{t+3}{t-4}$ **75.** $C = 2.05x + 9500$

Section 2.6   (page 149)

1.

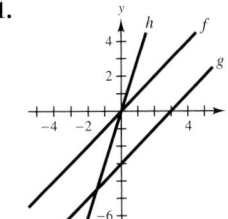

3.

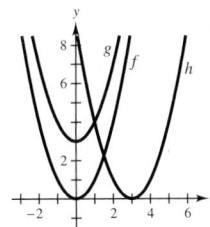

5.

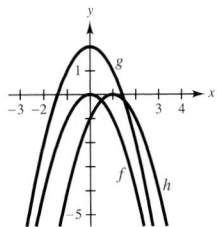

7.

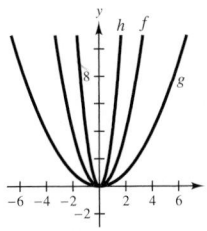

9.

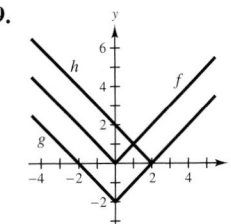

11.

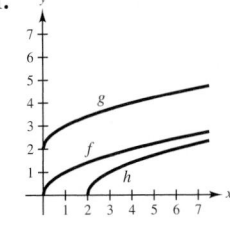

13. Horizontal shift of $y = x^3$
$y = (x - 2)^3$

15. Reflection in the x-axis of $y = x^2$
$y = -x^2$

17. Reflection in the x-axis and a vertical shift of $y = \sqrt{x}$
$y = 1 - \sqrt{x}$

19. Vertical shift of $y = x^2$
$y = x^2 - 1$

21. Reflection in the x- or y-axis and a vertical shift of $y = x^3$
$y = -x^3 + 1$

23. Vertical or horizontal shift of $y = x$
$y = x + 3$

25. Horizontal and vertical shifts of $y = |x|$
$y = |x - 2| - 2$

27. $y = -\sqrt{x}$ **29.** $y = \sqrt{x + 2}$ **31.** $y = \sqrt{-x}$

33. Vertical shift
three units upward

35. Horizontal shift
three units to the right

37. Reflection in
the y-axis

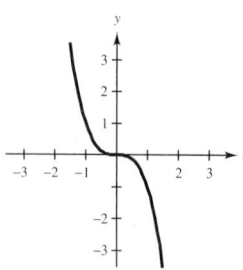

39. Reflection in the x-axis
followed by a horizontal
shift one unit to the right
followed by a vertical
shift two units upward

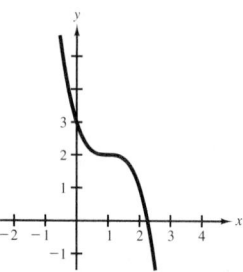

41. Horizontal shift
five units to the right

43. Vertical shift
five units downward

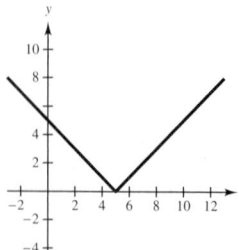

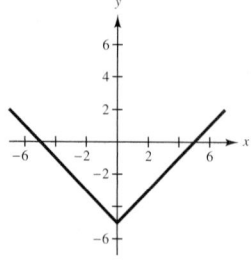

45. Reflection in the x-axis

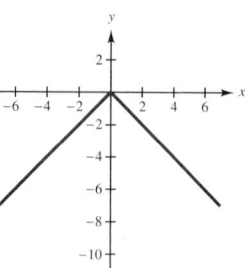

47.

49.

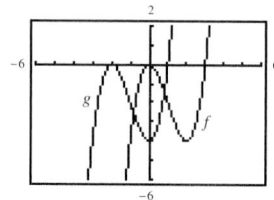

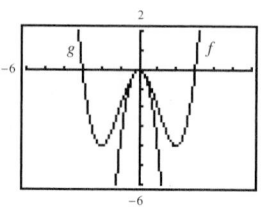

g is a horizontal shift
two units to the left.

g is a reflection in the
y-axis.

51. $g(x) = -x^3 + 3x^2 + 1$

53. (a)   (b)

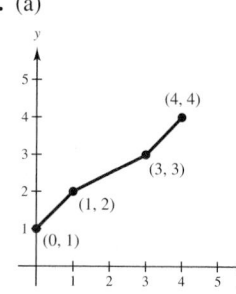

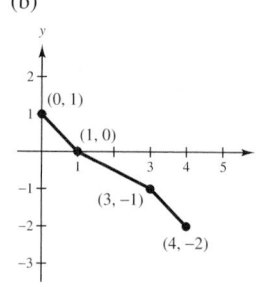

(c)   (d)

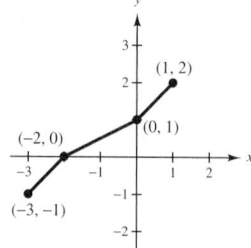

(e)   (f)

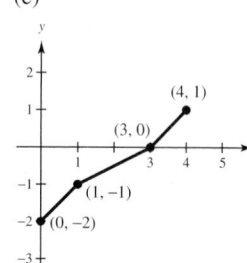

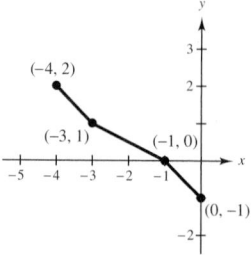

55. (a)

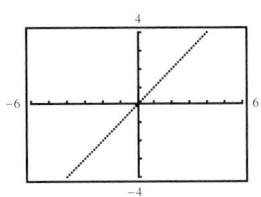

(b)

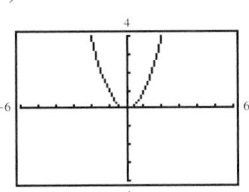

(c)

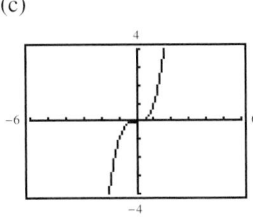

(d)

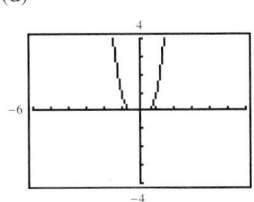

(e)

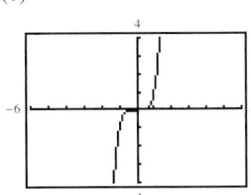

(f)

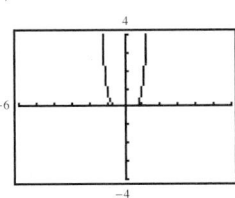

All the graphs pass through the origin. The graphs of the odd powers of x resemble the graphs of the cubing function and the graphs of the even powers resemble the graphs of the squaring function. As the powers increase, the graphs become flatter in the interval $-1 < x < 1$.

57.

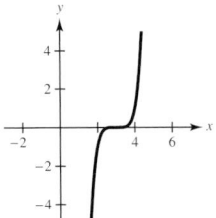

59. (a)

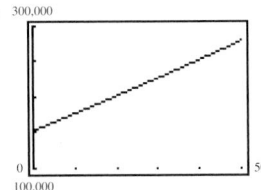

(b) 1980; P_2 is a horizontal shift 30 units to the left of P_1.

(c)

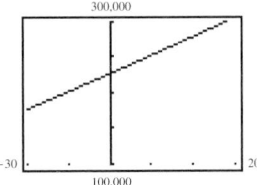

61.

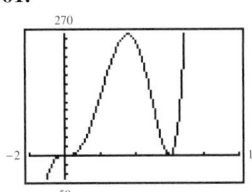

63.

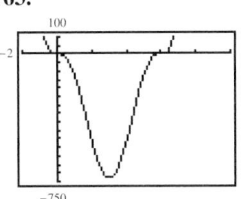

65. $4x + 34$ **67.** $-3x - 18$ **69.** $x^2 + 3x - 28$
71. $z^2 + z - 6$ **73.** $-x^2 + 16$ **75.** $A = 5x + 72$

Review Exercises   (page 153)

1.

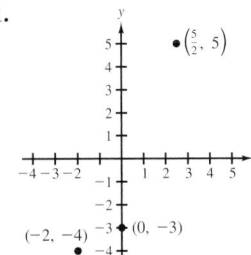

3.

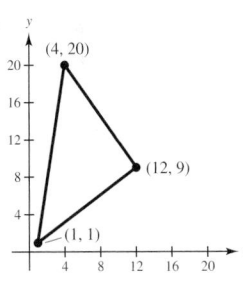

5. Quadrant IV   **7.** Quadrants I and IV
9. (a) Yes   (b) No   (c) No   (d) Yes
11. (a)

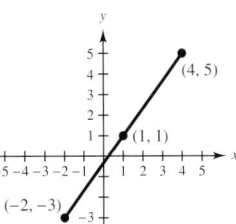

13. (a)

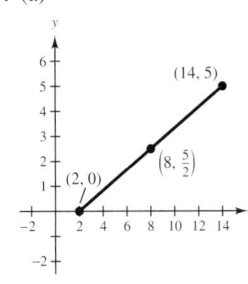

(b) 10   (c) (1, 1)   (b) 13   (c) $\left(8, \frac{5}{2}\right)$

15.

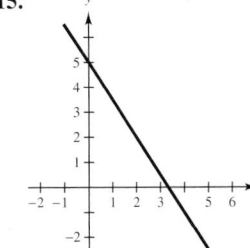

17.

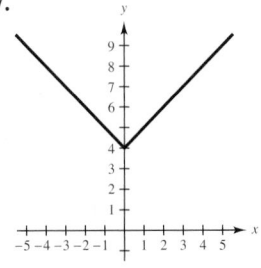

19.
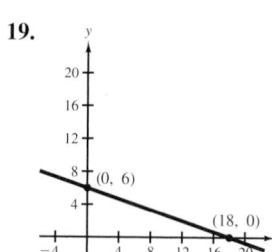

21.
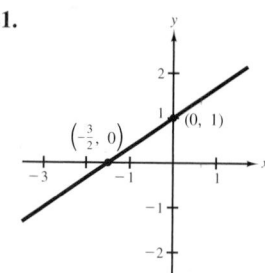

69. (a) 1   (b) 1   (c) $\dfrac{5}{4}$ (d) $\dfrac{|x+2|}{4}$

71. (a) -3 (b) 2   (c) 0   (d) -7

73. Domain $= \{1, 3, 5, 7\}$ **75.** Domain $= \{s: s > 0\}$
Range $= \{2, 7, 8\}$ Range $= \{P: P > 0\}$

77. Domain $= \{x: x \text{ is a real number}\}$

79. Domain $= \{x: x \geq 2\}$

81.
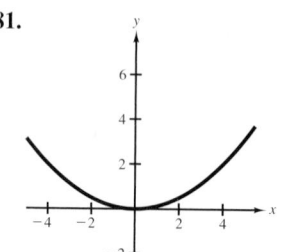

83.
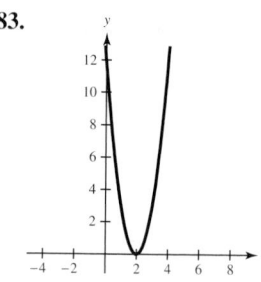

23.
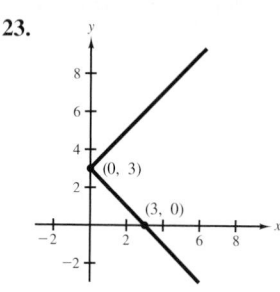

25.
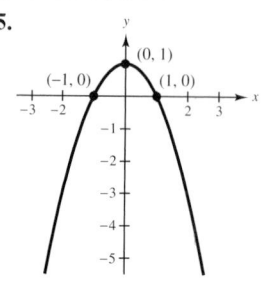

27. 4   **29.** $-3, 6$ **31.** $\frac{2}{7}$ **33.** 0   **35.** $-\frac{3}{4}$

37. Answers will vary. Sample answer: $(3, -7), (4, -10)$

39. Answers will vary. Sample answer: $(7, 6), (11, 11)$

41. Answers will vary. Sample answer: $(3, 0), (3, 5)$

43. $y = \frac{5}{2}x - 2$ **45.** $y = -\frac{1}{2}x + 1$

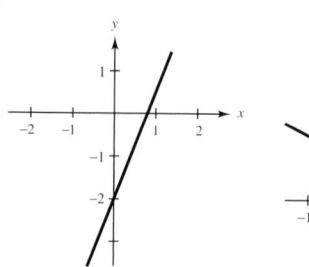

85.
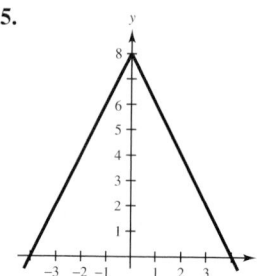

87.
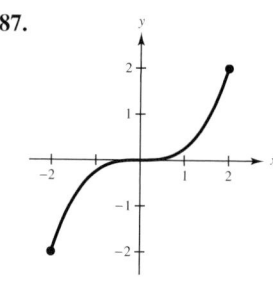

89.
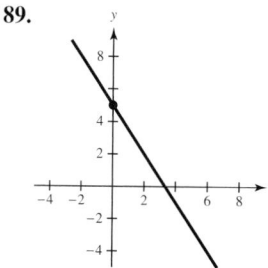

91.
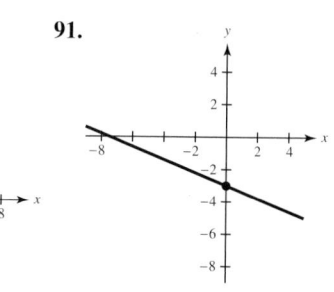

47. Neither   **49.** Perpendicular   **51.** Perpendicular

53. Domain: $\{1, 2, 3\}$
Range: $\{6, 7, 8, 10\}$

93. Not a function   **95.** Function   **97.** Function

99.
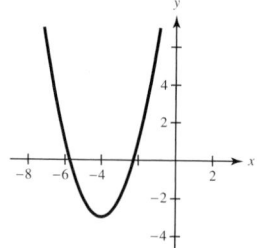

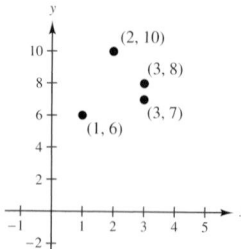

Domain $= \{x: x \text{ is a real number}\}$
Range $= \{y: y \geq 3\}$

55. Not a function   **57.** Function   **59.** Function

61. Not a function   **63.** Function

65. (a) 29   (b) 3   (c) $18 - \frac{5}{2}t$ (d) $4 - \frac{5}{2}x - \frac{5}{2}h$

67. (a) 3   (b) 0   (c) $\sqrt{2}$ (d) $\sqrt{5 - 5z}$

101.

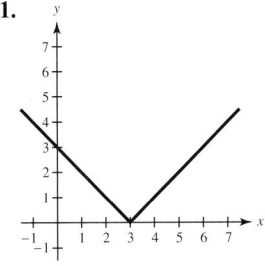

Domain = $\{x: x$ is a real number$\}$
Range = $\{y: y \geq 0\}$

103.

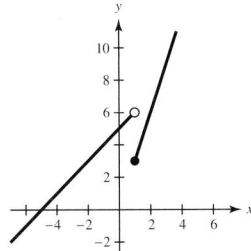

105. $f(x) = x$; Reflection in the x-axis followed by a vertical shift two units upward; $h(x) = -x + 2$

107. $f(x) = x^2$; Reflection in the x-axis and a horizontal shift four units to the right; $h(x) = -(x - 4)^2$

109. Constant function $f(x) = c$; $h(x) = -6$

111. Reflection in the x-axis   **113.** Horizontal shift one unit to the right

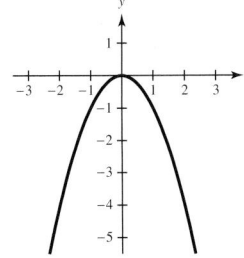   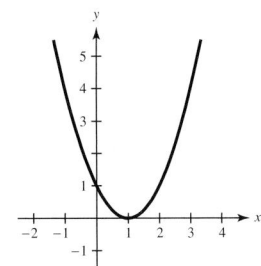

115. (a)   (b) 80 feet

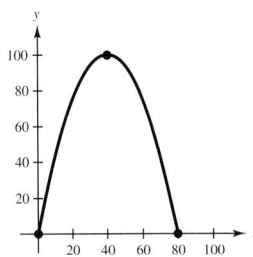

117. (a)

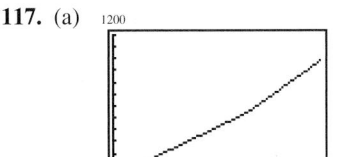

(b) \$490; \$560; \$665; \$770

(c) No, values of $h < 0$ are not included in the domain of the function.

119. \$1800 per year

Chapter 2 Test   (page 157)

1. Quadrant IV   **2.** Yes

3. (a) $\sqrt{73} \approx 8.54$ (b) $\left(\frac{3}{2}, 5\right)$ **4.** $(-1, 0), (0, -3)$

5.   **6.** (a) $-\frac{2}{3}$
                                              (b) Undefined

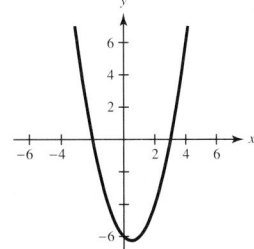

7. $y = \frac{1}{2}x - 3$

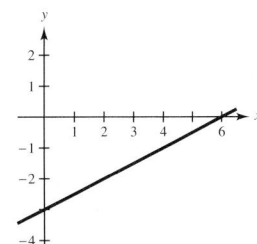

8. (a) $\frac{7}{5}$ (b) $-\frac{5}{7}$ **9.** Not a function

10. (a) -2 (b) 7   (c) Undefined   (d) $\dfrac{x + 2}{x - 1}$

11. (a) 104   (b) -25 (c) 29   (d) -1

12. (a) Domain = $\{t: t \geq -9\}$

(b) Domain = $\{x: x$ is all real numbers, except $x = 4\}$

(c) Domain = $\{r: r$ is a real number$\}$

13.

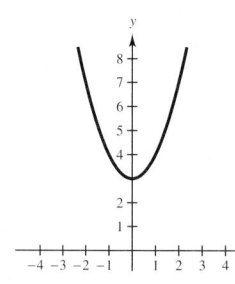

14. (a) $y = |x - 2|$
(b) $y = |x| - 2$
(c) $y = -|x| + 2$

Chapter 3

Section 3.1 (page 168)

1. Slope: $\frac{2}{3}$; y-intercept: $(0, -2)$
3. Slope: $\frac{3}{2}$; y-intercept: $(0, 0)$
5. Slope: $\frac{5}{2}$; y-intercept: $(0, 12)$ **7.** $y = -\frac{1}{2}x$
9. $y = 3x - 4$ **11.** $x = 10$ **13.** $y = 3$
15. b   **16.** d   **17.** a   **18.** c   **19.** f   **20.** e
21. $y = 2x - 4$ **23.** $y = \frac{3}{4}x + 7$

Cumulative Test: Chapters 1 and 2 (page 158)

1. 24   **2.** (a) $8a^8b^7$ (b) $-24x^7y^{11}$
3. (a) $t^2 - 9t$ (b) $2x^3 - 11x$
4. (a) $3x^3 + x^2 + x + 35$
(b) $x^2 - 2xy + y^2 + 4x - 4y + 4$
5. (a) $\frac{3}{2}$ (b) $-\frac{3}{2}$ **6.** (a) ± 8 (b) $-\frac{1}{2}, 3$
7. $y = \frac{2}{3}x + 3$ **8.** (a) -4 (b) -4 (c) $\frac{1}{4}$
9. (a) Multiplicative Inverse Property
(b) Associative Property of Addition
10. $8x + 12$ **11.** $(y + 3)(y - 3)^2$
12. $(x - 5)(3x + 7)$

13. (a)

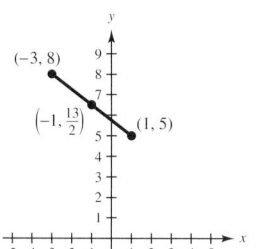

(b) 5
(c) $\left(-1, \frac{13}{2}\right)$
(d) $-\frac{3}{4}$

14. Function   **15.** Domain $= \{x : x \geq 2\}$
16. (a) 4   (b) $c^2 + 3c$

17.

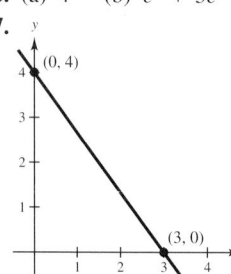

18.

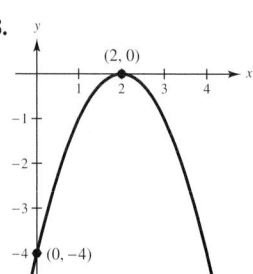

19.

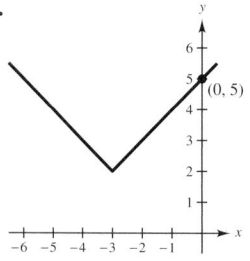

20.

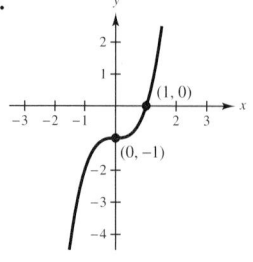

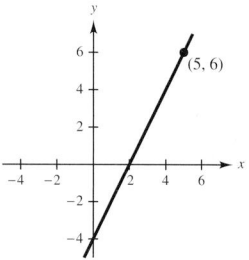

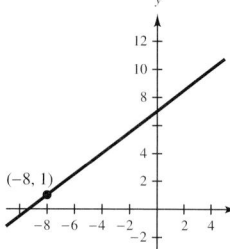

25. $y = \frac{2}{3}x - \frac{19}{3}$ **27.** $y = 5$

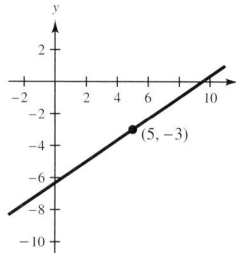

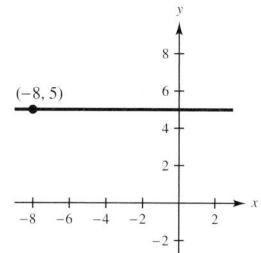

29. $x = 2$ **31.** $y = \frac{4}{3}x + \frac{3}{2}$

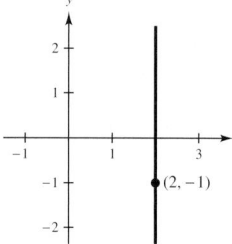

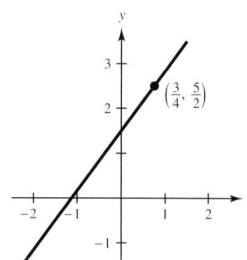

33. $y = \frac{3}{2}x$ **35.** $y = 4x - 25$ **37.** $y = -\frac{2}{5}x$
39. $y = 12$ **41.** $y = -\frac{3}{7}x + \frac{15}{7}$ **43.** $x = 1$
45. $y = -\frac{1}{2}x - \frac{19}{10}$ **47.** $y = -\frac{7}{3}x + \frac{13}{2}$
49. $y = \frac{85}{46}x + \frac{363}{46}$

51. $f(x) = \frac{1}{2}x + 3$ **53.** $f(x) = 3$

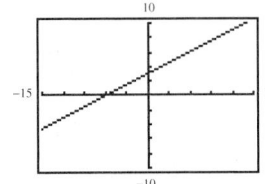 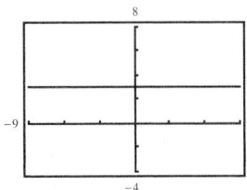

55. (a) $y = 3x - 5$ (b) $y = -\frac{1}{3}x + \frac{5}{3}$
57. (a) $y = -\frac{3}{2}x - 7$ (b) $y = \frac{2}{3}x + 6$
59. (a) $y = \frac{4}{3}x - \frac{25}{3}$ (b) $y = -\frac{3}{4}x - \frac{25}{4}$
61. (a) $y = 2$ (b) $x = -1$ **63.** $y = \frac{1}{5}x + 3$
65. $y = -2x + 10$ **67.** $y = \frac{3}{7}x + \frac{27}{7}$
69. $y = \frac{1}{4}x - \frac{29}{4}$ **71.** $y = 1$ **73.** $x = 9$
75. (a) $V = -2300t + 12,500$ (b) \$5600
77. (a) $x = -\frac{1}{15}p + 80$

(b) 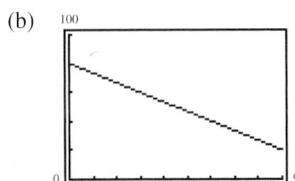   (c) 42
  (d) 48

79. (a) $N = 60t + 1500$

(b) 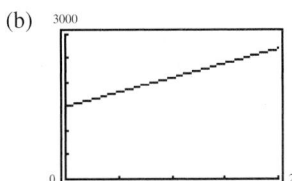   (c) 2700
  (d) 1800

81. Yes. When different pairs of points are selected, the change in y and the change in x are the lengths of the sides of similar triangles. Corresponding sides of similar triangles are proportional.

83. 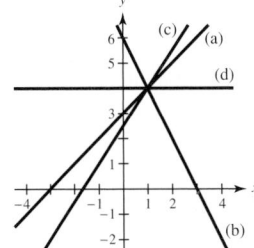   **85.** $C = 5.75x + 12,000$

Section 3.2 (page 181)

1. $8 + n$ **3.** $15 - 3n$ **5.** $\frac{1}{3}n$ **7.** $0.30L$
9. $\frac{x}{6}$ **11.** $\frac{3 + 4x}{8}$ **13.** $|n - 5|$

15. The sum of three times a number and 2
17. Eight times the difference of a number and 5
19. The ratio of a number to 8
21. The sum of a number and 10, all divided by 3
23. $0.25n$ **25.** $55t$ **27.** $\frac{100}{r}$ **29.** $0.45y$
31. $0.0125I$ **33.** $0.80L$ **35.** $8.25 + 0.60q$
37. 4 **39.** $\frac{15}{2}$ **41.** 16 **43.** $\frac{20}{3}$ **45.** $-\frac{34}{3}$
47. $\frac{11}{10}$ **49.** $6w$ **51.** $4x + 12$ **53.** $4x^2 - 12x + 9$
55. $2w^2 + 24w$ **57.** $6w^2$ square meters
59. Perimeter: $8w + 4$ feet; Area: $3w^2 + 2w$ square feet
61. 19, 20, 21 **63.** 44, 46, 48 **65.** 10 **67.** 42
69. 52 **71.** 62 **73.** 15, 17 **75.** 43 centimeters
77. 3 **79.** 15.625% **81.** 200 **83.** 52°, 128°
85. 72°, 108° **87.** 57°, 87°, 36° **89.** 5
91. (a) 1990: 377,045; 2000: 395,897
  (b) 8.5%. Different bases are used for the two calculations.
93. $\frac{9}{1}$ **95.** $\frac{3}{2}$ **97.** 46,400 votes **99.** $10\frac{1}{2}$ cups
101. No. $\frac{1}{2}\% = 0.5\% \neq 50\%$
103. If a and b have the same units, then a/b is the ratio of a to b. Price earnings ratio, gear ratio, . . .
105. $-4x + 16$ **107.** $19y + 22$ **109.** 2 **111.** 12
113. 8 **115.** 0.7 mile

Section 3.3 (page 194)

1. 2275 miles **3.** $\frac{125}{16}$ seconds **5.** Department store
7. 2.5 hours **9.** 9 minutes, \$2.06 **11.** 18.3%
13. 9% **15.** \$580
17. Surcharge $=$ \$172.40; Total premium $=$ \$1034.40
19. 16°C **21.** ≈ 461.8 cubic centimeters
23. 375 meters per minute **25.** 8.32 minutes
27. 1440 miles **29.** 3 hours **31.** 3:00 P.M.
33. 3 hours at 58 miles per hour, $2\frac{3}{4}$ hours at 52 miles per hour
35. $\frac{1}{5}, \frac{1}{8}; \frac{40}{13}$ hours **37.** $\frac{12}{7}$ hours ≈ 1 hour, 43 minutes
39. 12 hours **41.** 12 \$0.20 stamps; 58 \$0.32 stamps
43. 100 **45.** $4\frac{1}{2}\%$: \$2500; $5\frac{1}{2}\%$: \$3500
47. (a) 8%: \$9600; 10%: \$12,000 (b) \$15,000
49. 50 milliliters
51. 50 gallons at 20%; 50 gallons at 60%
53. 8 quarts at 15%; 16 quarts at 60% **55.** $\frac{5}{6}$ gallons
57. 220 **59.** ≈ 0.926 feet
61. (a) 1992
  (b) \$0.371 per year. This is the slope of the linear model.
63. Construction workers. The slope is greater in the model for construction workers.

65. (a) $A = \frac{1}{3}x^2 + \frac{1}{2}x$

(b) 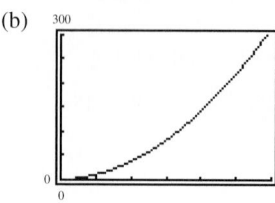   (c) 23.76 units

67. Yes. If the length of each side of a square is s, then the perimeter is $4s$. If the length of the sides are doubled, each side is $2s$ and the perimeter is $8s$.

69. $\frac{8}{5}$ **71.** $-\frac{27}{4}$ **73.** -40 **75.** -72

77. $\frac{120}{7}$ gallons

Mid-Chapter Quiz   (page 199)

1. $y = 2x - \frac{3}{2}$ **2.** $y = \frac{1}{2}x + 5$ **3.** $y = -\frac{3}{4}x + \frac{63}{8}$

4. $y = 3x + 8.7$ **5.** $y = 2x - 3$ **6.** $y = -\frac{31}{30}x + \frac{4}{5}$

7. $y = -1$ **8.** $x = 4$ **9.** $y = \frac{2}{3}x + 3$

10. $y = \frac{4}{3}x + \frac{16}{3}$ **11.** $P = 3.6x - 24{,}000$ **12.** $5n - 8$

13. Perimeter: $3.2l - 2$; Area: $0.6l^2 - l$

14. $3n + 6$ **15.** 2000

16. 40 gallons at 25%; 10 gallons at 50%

17. $\frac{12}{7}$ hours   **18.** $\frac{1}{3}$ hour

Section 3.4   (page 210)

1. (a) Yes   (b) No   (c) Yes   (d) No

3. (a) No   (b) Yes   (c) Yes   (d) No

5. d   **6.** a   **7.** c   **8.** b   **9.** c   **10.** a

11. d   **12.** b   **13.** d   **14.** b   **15.** c   **16.** a

17. $(-\infty, 2]$ **19.** $(3.5, \infty)$

21. $(-5, 3]$ **23.** $\left(0, \frac{3}{2}\right]$

25. $\left(-\frac{15}{4}, -\frac{5}{2}\right)$ **27.** $(-\infty, -5) \cup (-1, \infty)$

29. $(-\infty, 3] \cup (7, \infty)$ **31.** $x \leq 2$

33. $x < \frac{11}{2}$

35. $x \leq -4$

37. $x > 8$

39. $x > 7$

41. $x \geq 7$

43. $x > -\frac{2}{3}$

45. $x > \frac{9}{2}$

47. $x > -\frac{4}{3}$

49. $y > 2$

51. $y \leq -10$

53. $x \geq -12$

55. $\frac{5}{2} < x < 7$

57. $-\frac{3}{2} < x < \frac{9}{2}$

59. $1 < x < 10$

61. $-\frac{3}{5} < x < -\frac{1}{5}$

63. $-1 < x \leq 4$

65. $x \leq -6$

67. All real numbers

69. $x < -\frac{8}{3}$ or $x \geq \frac{5}{2}$

71. $x \geq 0$ **73.** $z \geq 2$ **75.** $n \leq 16$

77. x is at least $\frac{5}{2}$.

79. z is greater than 0 and no more than π.

81. $\left[3, \frac{15}{2}\right]$ **83.** \$2600

85. The average temperature in Miami is greater than the average temperature in New York.

87. 25,357 miles

89. $x \geq 31$

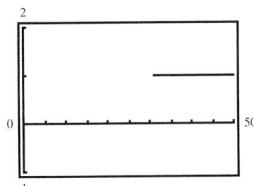

91. The call must be less than 6.38 minutes. If a portion of a minute is billed as a full minute, the call must be less than or equal to 6 minutes.

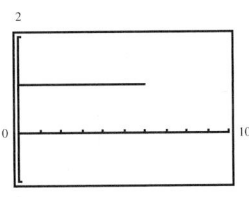

93. $[2, 8]$ **95.** $[4, 18]$

97. $8 + 0.75n > 12.5;\ n > 6$

99. (a) 1994, 1995, 1996, 1997, 1998   (b) 1990, 1991

101. The multiplication and division properties differ. The inequality symbol is reversed if both sides of the inequality are multiplied or divided by a negative real number.

103. $(-8, \infty)$ **105.** $<$ **107.** $<$ **109.** -4

111. $\frac{4}{5}$ **113.** 19.8 square meters

Section 3.5   (page 220)

1. Not a solution   **3.** Solution

5. $x - 10 = 17;\ x - 10 = -17$

7. $4x + 1 = \frac{1}{2};\ 4x + 1 = -\frac{1}{2}$ **9.** $45, -45$ **11.** 0

13. 21,11   **15.** $11, -14$ **17.** $\frac{16}{3}, 16$ **19.** No solution

21. $-\frac{17}{2}, -\frac{9}{2}$ **23.** $\frac{4}{3}$ **25.** $-\frac{39}{2}, \frac{15}{2}$ **27.** $18.75, -6.25$

29. $\frac{17}{5}, -\frac{11}{5}$ **31.** $\frac{12}{7}, -\frac{30}{7}$ **33.** $-\frac{5}{2}, -\frac{15}{2}$

35. $-\frac{11}{4}, -\frac{29}{4}$ **37.** $-\frac{5}{2}, -\frac{15}{2}$ **39.** No solution

41. $\frac{1}{3}, -\frac{11}{6}$ **43.** $-\frac{8}{5}, -2$ **45.** $-3, 7$

47. 11, 13   **49.** $\frac{3}{2}, -\frac{1}{4}$ **51.** $|2x + 3| = 5$

53. (a) Yes   (b) No   **55.** (a) No   (b) Yes

57. (a) Yes   (b) No   **59.** (a) No   (b) Yes

61. $-3 < y + 5 < 3$ **63.** $7 - 2h \geq 9;\ 7 - 2h \leq -9$

65. 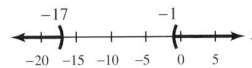   **67.**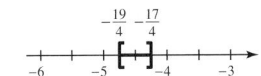

69. $-4 < y < 4$

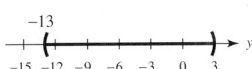

71. $-7 < x < 7$

73. $-9 \leq y \leq 9$

75. $-2 \leq y \leq 6$

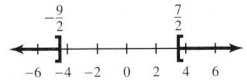

77. $-13 < y < 3$

79. $y < -2$ or $y > 10$

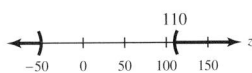

81. $y \leq -\frac{9}{2}$ or $y \geq \frac{7}{2}$

83. $t \leq -\frac{15}{2}$ or $t \geq \frac{5}{2}$

   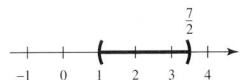

85. $z < -50$ or $z > 110$

87. $-5 < x < 35$

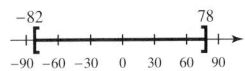   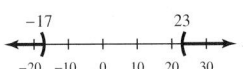

89. $x < -8$ or $x > 0$

91. $1 < x < \frac{7}{2}$

   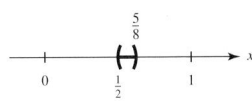

93. $-82 \leq x \leq 78$

95. $s > 23$ or $s < -17$

97. $x \leq -\frac{7}{6}$ or $x \geq \frac{1}{3}$

99. $\frac{1}{2} < x < \frac{5}{8}$

101. $x < -17$ or $x > -1$

103. $-\frac{19}{4} \leq x \leq -\frac{17}{4}$

105. $-2 < x < \frac{2}{3}$ **107.** $3 \leq x \leq 7$

109. $x > 3$ or $x < -6$

111. d   **112.** c   **113.** b   **114.** a

115. $|x| \leq 2$ **117.** $|x - 10| < 3$ **119.** $|x - 19| < 3$

121.   **123.** $|t - 98.6| < 1$

125. The absolute value of a real number n can be described graphically as the distance between n and zero on the real number line.

127. Answers will vary.   **129.** 8   **131.** -26

133. 8   **135.** $-\frac{12}{5} \le x \le -\frac{2}{5}$ **137.** $x < 0$ or $x > 2$

139. \$1002.6 million

(b) 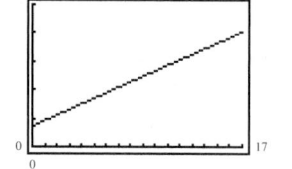   **(c)** \$1091.67

(d) \$741.67

Review Exercises   (page 224)

1. $y = -\frac{1}{3}x + 3$ **3.** $y = \frac{7}{9}x + 3$

   Slope: $-\frac{1}{3}$ Slope: $\frac{7}{9}$

   y-intercept: $(0, 3)$ y-intercept: $(0, 3)$

5. $y = \frac{3}{2}x - 7$ **7.** (a) $y = -4$ (b) $x = 3$

9. $y = 2x - 6$ **11.** $y = -4x$ **13.** $y = -\frac{2}{3}x + \frac{17}{3}$

15. $x = 7$ **17.** $y = -\frac{1}{2}x - 3$ **19.** $y = -\frac{9}{14}x - \frac{31}{7}$

21. $x = \frac{5}{2}$

23. (a) $y = -\frac{1}{2}x + \frac{9}{2}$ (b) $y = 2x + 7$

25. (a) $y = -3x + 1$ (b) $y = \frac{1}{3}x - 1$

27. $200 - 3n$ **29.** $n^2 + 49$ **31.** $|n + 10|$

33. The sum of twice a number and 7

35. The difference of a number and 5, all divided by 4

37. $0.18I$ **39.** $l(l - 5) = l^2 - 5l$ **41.** $30p$

43. $\dfrac{x + 9}{2} = 15$ **45.** $350 + 32x = 590$

47. $z \le 10$ **49.** $7 \le y < 14$

51. $V < 27$ **53.** $x \le 3$ or $x > 10$

55. $x > 3$ **57.** $y > -\frac{70}{3}$

59. $-20 < x \le 20$ **61.** $-16 < x < -1$

63. $x > 2$ **65.** $x \le -2$ or $x > 8$

67. $-4, 8$ **69.** $-4, 1$ **71.** 10, 14   **73.** $\frac{1}{2}, 3$

75. $-4 < x < 11$ **77.** $x < 1$ or $x > 7$

79. $b \le -9$ or $b \ge 5$

81. (a) $R = \frac{175}{3}t + \frac{650}{3}$

83. \$1856.25   **85.** 487.5 miles   **87.** 3   **89.** 80 feet

91. $\approx 175\%$ **93.** \$27,166.25   **95.** $142°; 38°$

97. 30% solution: $3\frac{1}{3}$ liters; 60% solution: $6\frac{2}{3}$ liters

99. ≈ 6.35 hours   **101.** 2800 miles

103. $\frac{20}{9} \approx 2.22$ hours   **105.** \$340   **107.** \$210,526.32

109. \$30,000   **111.** 8 inches $\times$ 6 inches

113. $5 \le d \le 11$

Chapter 3 Test   (page 228)

1. $y = \frac{1}{2}x - \frac{55}{2}$ **2.** $y = \frac{3}{5}x - 4$

3. $V = -4000t + 26,000$; 2.5 years

4. $T = 3t + 72$ **5.** 38 inches $\times$ 19 inches   **6.** 7

7. 26, 28   **8.** \$1466.67   **9.** $2\frac{1}{2}$ hours

10. 10% solution: $33\frac{1}{3}$ liters; 40% solution: $66\frac{2}{3}$ liters

11. 40 minutes   **12.** \$2000   **13.** $37°, 66°, 77°$

14. $-17, 5$ **15.** $-1, 4$

16. $x > 2$ **17.** $-7 < x \le 1$

18. $x < -\frac{39}{5}$ or $x \ge \frac{10}{3}$ **19.** $1 \le x \le 5$

20. $x > -3$ or $x < -\frac{17}{3}$

Cumulative Test: Chapters 1–3   (page 229)

1. $125x^7$ **2.** $4802z^{15}$ **3.** $-\frac{4}{5}ab^2$ **4.** $-\dfrac{27x^{18}}{8y^6}$

5. 10   **6.** $-8x - 10$ **7.** $6x^2 + 5x - 4$

8. $25x^2 - 1$ **9.** $(x + 5)(x^2 - 8)$ **10.** $(2x - 7y)^2$

11. $(2x + 3)(4x^2 - 6x + 9)$ **12.** $3(2x - 5)(2x + 5)$

13. 9   **14.** $\frac{22}{19}$ **15.** -1 **16.** $-\frac{13}{2}$ **17.** $-3, \frac{8}{3}$

18. $0, \pm 3$ **19.** $5, -11$ **20.** $\frac{2}{3}, -\frac{4}{3}$

21. $(0, 5), (4, 0)$ **22.** $(-3, 0), (7, 0), (0, -21)$

23.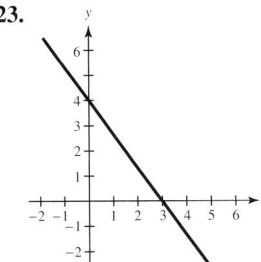

24.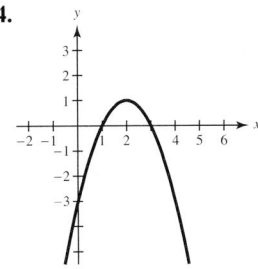

25. (a) -7 (b) $-2x$ **26.** No

27. Domain $= \{x: x \geq 4\}$ Range $= \{y: y \geq 2\}$

28. $g(x) = -\sqrt{x} + 2$ **29.** $\frac{7}{3}$ **30.** $y = -\frac{1}{4}x - \frac{21}{4}$

31. $y = -\frac{2}{7}x + \frac{45}{7}$

32. $x < \frac{3}{2}$ **33.** $-3 < x < 7$

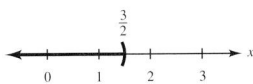 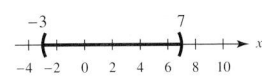

34. $x \leq -1$ or $x \geq 5$

35. 58 units **36.** $C = 5.75x + 12{,}000$ **37.** $\frac{18}{7}$ hours

38. 6.5 **39.** 7.5%: \$15,000; 9%: \$9000

40. Minimum: 20; Maximum: 80

Chapter 4

Section 4.1 (page 244)

1. (a) Yes (b) No **3.** (a) No (b) Yes

5. No solution **7.** $\left(1, \frac{1}{3}\right)$

9. Infinitely many solutions

11.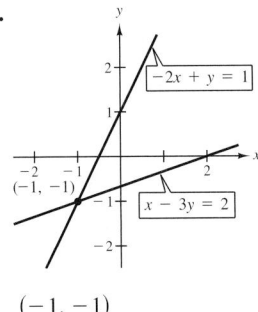

$(-1, -1)$

13.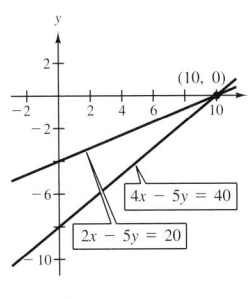

$(10, 0)$

15.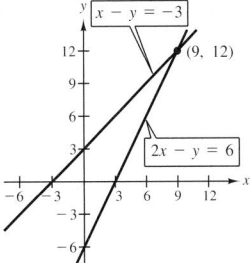

$(9, 12)$

17.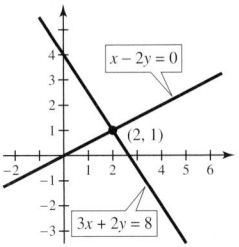

$(2, 1)$

19.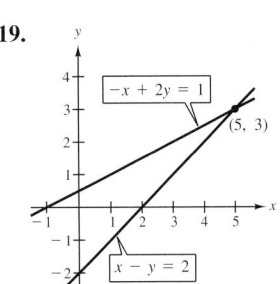

$(5, 3)$

21.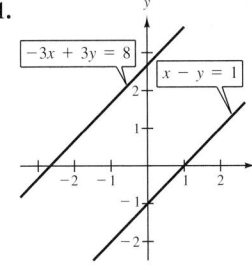

No solution

23.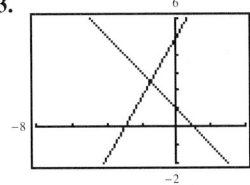

$\left(-\frac{3}{2}, \frac{5}{2}\right)$

25.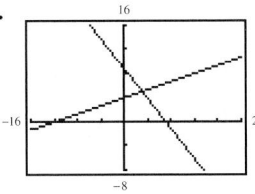

$(3, 5)$

27.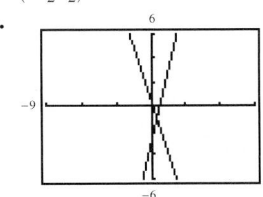

$\left(\frac{1}{2}, -1\right)$

29. Consistent **31.** Consistent

33.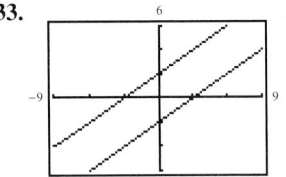

Inconsistent

35.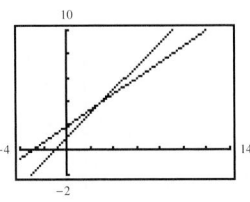

One solution

37. 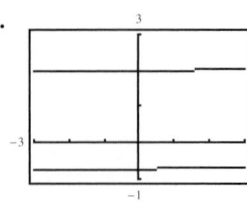 **39.**

$\left(2992, \frac{798}{25}\right)$ (50, 4)

41. (4, 3) **43.** (1, 2) **45.** (4, −2) **47.** (7, 2)

49. $\left(\frac{1}{3}, \frac{17}{3}\right)$ **51.** $\left(\frac{3}{2}, \frac{3}{2}\right)$ **53.** $\left(4, -\frac{1}{2}\right)$ **55.** (3, 2)

57. (−2, 5) **59.** (−1, −1) **61.** (7, −2)

63. No solution **65.** $\left(\frac{3}{2}, 1\right)$ **67.** (−2, −1)

69. Infinitely many solutions

71. (12.5, 4.948) **73.** No solution

75. 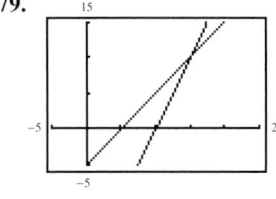 **77.**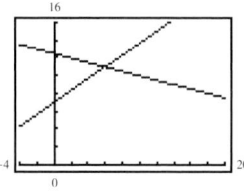

(2, −3) (2, 7)

79. 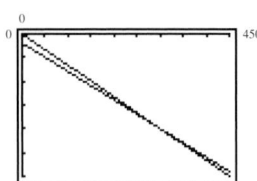 **81.**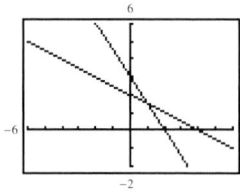

(15, 10) (6, 11)

83. **85.**

(3000, −2000) $\left(1, \frac{3}{2}\right)$

87. 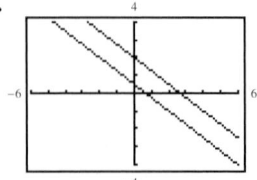 **89.** $k = 4$

No solution

91. Answers will vary. **93.** Answers will vary.

$$\begin{cases} 2x + 5y = 20 \\ 3x - 8y = -1 \end{cases}$$ $$\begin{cases} x + 2y = 0 \\ 4x + 2y = 9 \end{cases}$$

95. 50 meters × 60 meters **97.** 65°, 25°

99. Regular: $1.11; Premium: $1.22

101. 70 tons of the $75-per-ton hay;
30 tons of the $125-per-ton hay

103. 520 adult tickets; 280 student tickets

105. $15,000 at 8%; $5000 at 9.5% **107.** 10,000 items

109. 32 inches, 128 inches **111.** 75 miles, 225 miles

113. The system has no solution.

115. No. The lines represented by a consistent system of linear equations intersect at one point (one solution) or coincide (infinitely many solutions).

117. 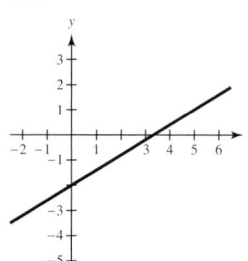 **119.**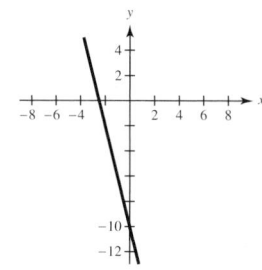

121. 2 **123.** $y = 3x - 2$ **125.** 19,555 miles or less

Section 4.2 (page 258)

1. (a) No (b) Yes (c) Yes (d) No

3. (22, −1, −5) **5.** (14, 3, −1)

7. (1, 2, 3) **9.** (1, 2, 3) **11.** (2, −3, −2)

13. No solution **15.** (−4, 8, 5) **17.** No solution

19. (−4, 2, 3) **21.** (1, 0, −2) **23.** (−1, 5, 5)

25. $\left(\frac{7}{5}a + 1, \frac{1}{5}a - 1, a\right)$ **27.** (2, 1, 3) **29.** (0, 3, 1)

31. $\left(-\frac{1}{2}a + \frac{1}{2}, \frac{3}{5}a + \frac{2}{5}, a\right)$ **33.** No solution

35. (1, 1, −1) **37.** (1, −1, 2)

39. $\begin{cases} x + 2y - z = -4 \\ y + 2z = 1 \\ 3x + y + 3z = 15 \end{cases}$ **41.** $y = 2x^2 + 3x - 4$

43. $y = x^2 - 4x + 3$ **45.** $y = -x^2 + 2x$

47. $s = -16t^2 + 144$ **49.** $s = -16t^2 + 48t$

51. $7000 at 5%, $4000 at 6%, $5000 at 7%

53. 20 gallons of spray X; 18 gallons of spray Y; 16 gallons of spray Z

55. Defense: 20; Offense: 30; Special teams: 10

57. 88°; 32°; 60°

59. $\dfrac{2x^2 - 9x}{(x - 2)^3} = \dfrac{2}{x - 2} - \dfrac{1}{(x - 2)^2} - \dfrac{10}{(x - 2)^3}$

61. $y = \frac{1}{2}x^2 - \frac{1}{2}x$; Yes

63. No. When the first equation was multiplied by −2 and added to the second equation, the constant term should have been −11.

65. $\begin{cases} x - 2y + 3z = 5 \\ \quad\quad y - 2z = 9 \\ 2x \quad\quad - 3z = 0 \end{cases}$

Eliminated the x-term in Equation 2

67. -56 **69.** $\frac{23}{35}$ **71.** $9x - y$ **73.** $-30y + 3z$

75. 4 hours

Section 4.3 (page 269)

1. 3×2 **3.** 2×3 **5.** 3×4

7. $\begin{bmatrix} 4 & -5 & \vdots & -2 \\ -1 & 8 & \vdots & 10 \end{bmatrix}$ **9.** $\begin{bmatrix} 1 & 10 & -3 & \vdots & 2 \\ 5 & -3 & 4 & \vdots & 0 \\ 2 & 4 & 0 & \vdots & 6 \end{bmatrix}$

11. $\begin{bmatrix} 4 & 13 & 0 & \vdots & -2 \\ 7 & -6 & 1 & \vdots & 0 \end{bmatrix}$

13. $\begin{cases} 4x + 3y = 8 \\ x - 2y = 3 \end{cases}$ **15.** $\begin{cases} x \quad\quad + 2z = -10 \\ \quad 3y - z = 5 \\ 4x + 2y \quad\quad = 3 \end{cases}$

17. $\begin{cases} 15x - 4y + 12z + 3w = 6 \\ -x - 3y \quad\quad + 10w = 14 \\ 9x + 2y - 7z \quad\quad = 5 \end{cases}$ **19.** $\begin{bmatrix} 1 & 4 & 3 \\ 0 & 2 & -1 \end{bmatrix}$

21. $\begin{bmatrix} 1 & 1 & 4 & -1 \\ 0 & 5 & -2 & 6 \\ 0 & 3 & 20 & 4 \end{bmatrix} \begin{bmatrix} 1 & 1 & 4 & -1 \\ 0 & 1 & -\frac{2}{5} & \frac{6}{5} \\ 0 & 3 & 20 & 4 \end{bmatrix}$

23. $\begin{bmatrix} 1 & 2 & 3 \\ 0 & 1 & 2 \end{bmatrix}$ **25.** $\begin{bmatrix} 1 & \frac{3}{2} & \frac{1}{4} \\ 0 & 1 & \frac{11}{10} \end{bmatrix}$

27. $\begin{bmatrix} 1 & 1 & 0 & 5 \\ 0 & 1 & 2 & 0 \\ 0 & 0 & 1 & -1 \end{bmatrix}$ **29.** $\begin{bmatrix} 1 & -1 & -1 & 1 \\ 0 & 1 & 6 & 3 \\ 0 & 0 & 1 & \frac{4}{5} \end{bmatrix}$

31. $\begin{bmatrix} 1 & 1 & -1 & 3 \\ 0 & 1 & -4 & 1 \\ 0 & 0 & 0 & 0 \end{bmatrix}$ **33.** $\begin{cases} x - 2y = 4 \\ \quad y = -3 \end{cases}$

$(-2, -3)$

35. $\begin{cases} x - y + 2z = 4 \\ \quad y - z = 2 \\ \quad\quad z = -2 \end{cases}$ **37.** $(3, 2)$ **39.** $(1, 1)$

$(8, 0, -2)$

41. $(1, -2)$ **43.** No solution **45.** $(2, -3, 2)$

47. $(1, 1, 2)$ **49.** No solution **51.** $(2a + 1, 3a + 2, a)$

53. $(1, 2, -1)$ **55.** $(1, -1, 2)$ **57.** $(34, -4, -4)$

59. No solution **61.** $(-12a - 1, 4a + 1, a)$

63. 8%: \$800,000; 9%: \$500,000; 12%: \$200,000

65. $y = x^2 + 2x + 4$

67. (a) $y = \frac{7}{20}t^2 - \frac{49}{20}t + \frac{297}{5}$

(b)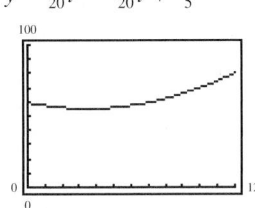

(c) 109.8 million short tons

69. (a) $y = -\frac{1}{250}x^2 + \frac{3}{5}x + 6$

(b)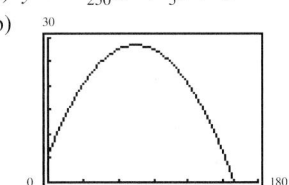

Maximum height: 28.5 feet

The ball struck the ground at approximately $(159.4, 0)$.

71. There will be a row in the matrix with all zero entries except in the last column.

73. The one matrix can be obtained from the other by using the elementary row operations.

75. (a) Interchange two rows

(b) Multiply a row by a non-zero constant.

(c) Add a multiple of a row to another row.

77. 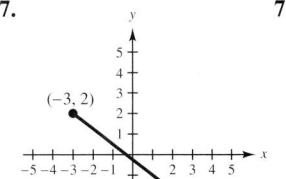 **79.**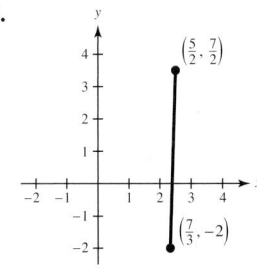

$-\frac{3}{4}$ 33

81.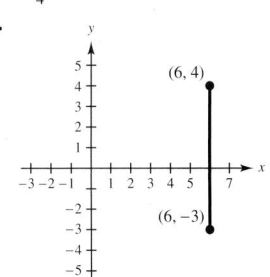

Slope is undefined.

83. 7 **85.** $-\frac{28}{5}$ **87.** 7650 members

Mid-Chapter Quiz (page 273)

1. $(2, 1)$ 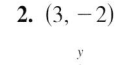 **2.** $(3, -2)$

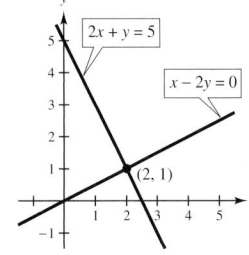 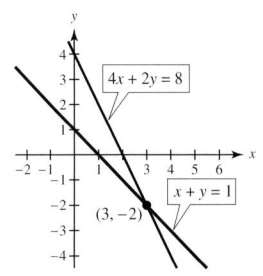

3. $(7, 3)$ **4.** $(5, 10)$ **5.** $(8, 1)$ **6.** $(-2, 4)$
7. $(5, -1, 3)$ **8.** $(2, 1, -2)$
9. Answers will vary.   **10.** Answers will vary.

$$\begin{cases} x + y = -2 \\ 2x + y = 8 \end{cases} \qquad \begin{cases} x + y + z = 0 \\ x - y + z = 6 \\ x - y - 3z = 2 \end{cases}$$

11. 75 miles, 225 miles
12. 20%: $13\frac{1}{3}$ gallons; 50%: $6\frac{2}{3}$ gallons   **13.** $75°, 105°$
14. $y = x^2 + 3x - 2$

Section 4.4   (page 282)

1. 5   **3.** 27   **5.** 0   **7.** 6   **9.** -24
11. -0.16 **13.** -24 **15.** -2 **17.** -30
19. 3   **21.** 0   **23.** -75 **25.** -58 **27.** -0.22
29. $x - 5y + 2$ **31.** 248   **33.** 19,185   **35.** 0
37. $-\frac{167}{8}$ **39.** 77   **41.** -6.37 **43.** $(1, 2)$
45. $(2, -2)$ **47.** $\left(\frac{13}{4}, \frac{3}{8}\right)$ **49.** $\left(\frac{3}{4}, -\frac{1}{2}\right)$ **51.** $\left(\frac{2}{3}, \frac{1}{2}\right)$
53. $(-1, 3, 2)$ **55.** $\left(1, \frac{1}{2}, \frac{3}{2}\right)$ **57.** $(1, -2, 1)$
59. $\left(\frac{22}{27}, \frac{22}{9}\right)$ **61.** $\left(\frac{1}{2}, 4\right)$ **63.** $\left(\frac{1}{3}, 1, -\frac{2}{3}\right)$
65. $\left(2, \frac{1}{2}, -\frac{1}{2}\right)$ **67.** $\left(\frac{51}{16}, -\frac{7}{16}, -\frac{13}{16}\right)$ **69.** 6, 1
71. 16   **73.** $\frac{31}{2}$ **75.** $\frac{41}{2}$ **77.** 3   **79.** $\frac{53}{2}$ **81.** 15
83. Collinear   **85.** Not collinear   **87.** Not collinear
89. $3x - 5y = 0$ **91.** $7x - 6y - 28 = 0$
93. $x + y - 9 = 0$ **95.** $2x - 6y - 3 = 0$
97. $y = 2x^2 - 6x + 1$ **99.** $y = -3x^2 + 2x$

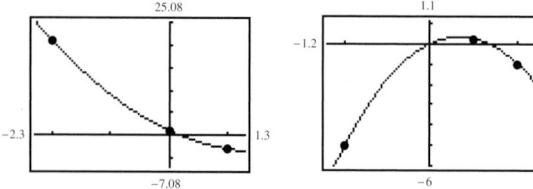

101. $I_1 = 1, I_2 = 2, I_3 = 1$
103. (a) $\left(\dfrac{4k - 3}{2k - 1}, \dfrac{4k - 1}{2k - 1}\right)$ (b) $\dfrac{1}{2}$
105. No. The matrix must be square.
107. The minor is the determinant of the matrix that remains after the deletion of the row and column in which the entry occurs.
109. (a) 12   (b) $\frac{3}{16}$
111. 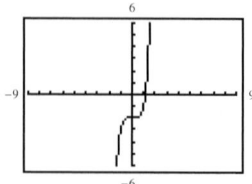   **113.**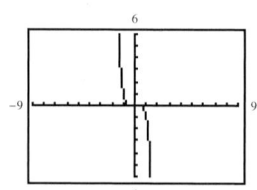

Vertical shift two units
downward

Reflection in the x-axis

115. $y = -\frac{5}{2}x + \frac{41}{2}$ **117.** 5, 8, 20

Section 4.5   (page 293)

1. b   **2.** a   **3.** d   **4.** e   **5.** f   **6.** c
7. (a) Yes   (b) No   (c) Yes   (d) Yes
9. (a) No   (b) No   (c) Yes   (d) Yes
11. (a) Yes   (b) No   (c) No   (d) Yes
13. (a) No   (b) Yes   (c) No   (d) Yes

15.   **17.**

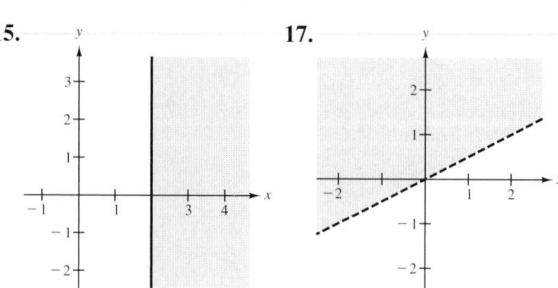

19.   **21.**

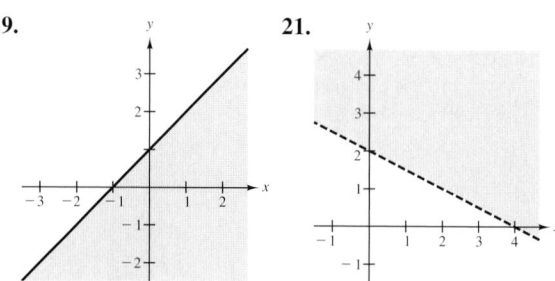

23.   **25.**

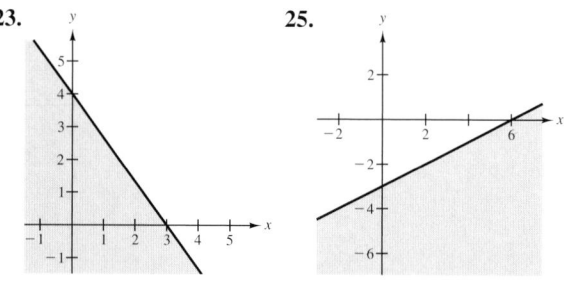

27.   **29.**

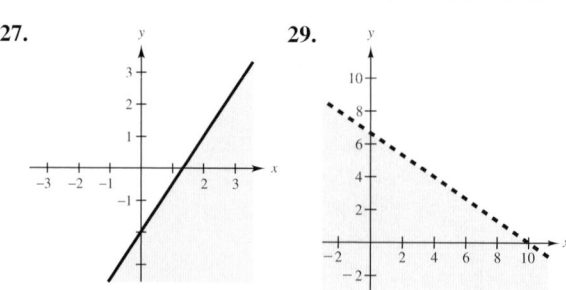

31.
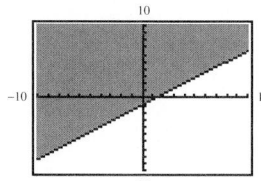

33.
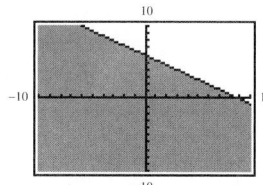

35.
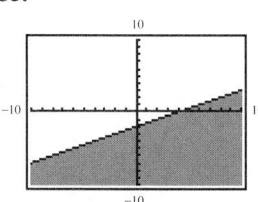

37.
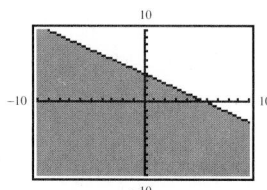

39. $3x + 4y > 17$ **41.** $y < 2$ **43.** $x - 2y < 0$
45. c   **46.** b   **47.** f   **48.** e   **49.** a   **50.** d

51.
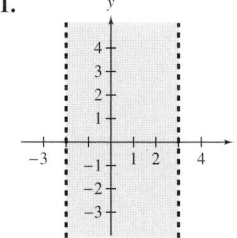

53.
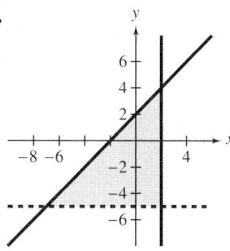

55.
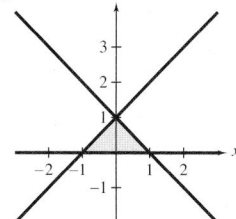

57.
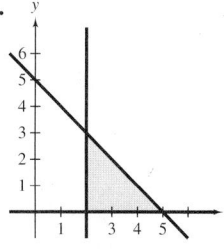

59.
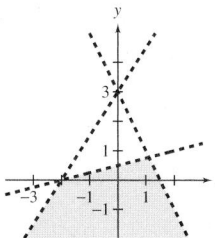

61.
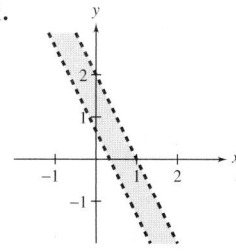

63.
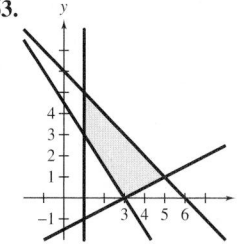

65.
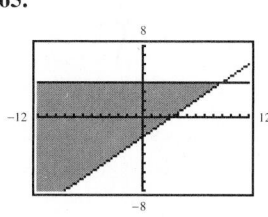

67.
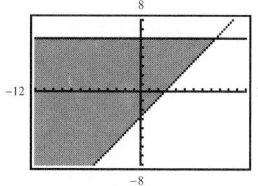

69.
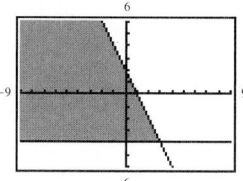

71. $\begin{cases} x \geq 1 \\ x \leq 8 \\ y \geq -5 \\ y \leq 3 \end{cases}$ **73.** $\begin{cases} y \leq \frac{9}{10}x + \frac{42}{5} \\ y \geq 3x \\ y \geq \frac{2}{3}x + 7 \end{cases}$

75. $2x + 2y \leq 500$ or
$y \leq -x + 250$
(*Note:* x and y cannot
be negative.)

77. $10x + 15y \leq 1000$
or $y \leq -\frac{2}{3}x + \frac{200}{3}$
(*Note:* x and y cannot
be negative.)

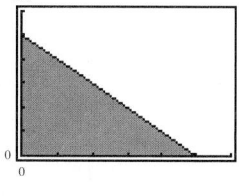

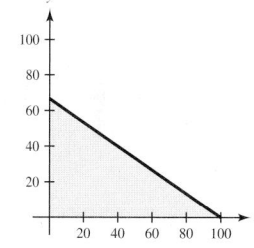

79. $\begin{cases} x + y \leq 20{,}000 \\ x \geq 5000 \\ y \geq 5000 \\ y \geq 2x \end{cases}$ **81.** $\begin{cases} x + y \geq 15{,}000 \\ 15x + 25y \geq 275{,}000 \\ x \geq 8000 \\ y \geq 4000 \end{cases}$

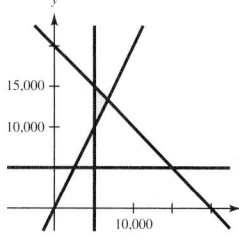

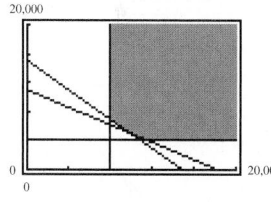

83. $\begin{cases} x < 90 \\ y \leq 0 \\ y \geq -10 \\ y \geq -\frac{1}{7}x \end{cases}$

85. The graph of a line divides the plane into two half-planes. In
graphing a linear inequality, graph the corresponding linear
equation. The solution to the inequality will be one of the
half-planes determined by the line. Example: $x - 2y > 1$.

87. Test a point in one of the half-planes.

89.
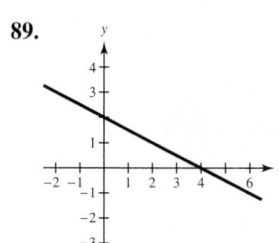

91.
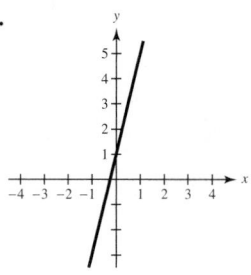

93.
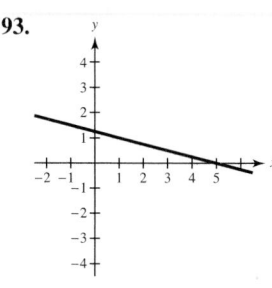

95.
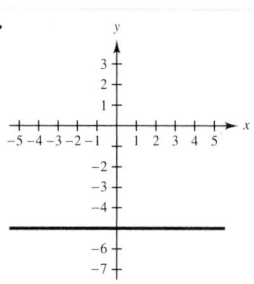

97. 10.4 hours

Review Exercises   (page 298)

1. (a) Yes   (b) No

3.
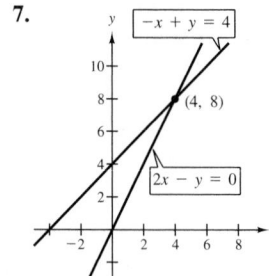

(1, 1)

5.
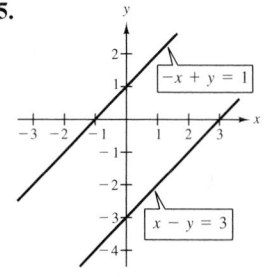

No solution

7.
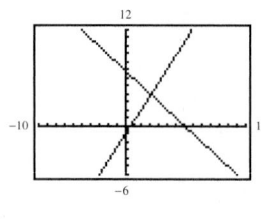

(4, 8)

9.
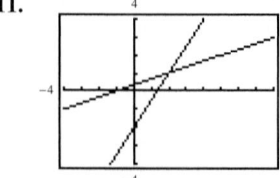

(3, 4)

11.

(2, 1)

13. $(2, -1)$ **15.** No solution   **17.** $(-10, -5)$
19. $(0, 0)$ **21.** $\left(\frac{5}{2}, 3\right)$ **23.** (a) No   (b) Yes
25. $(1, -1, 3)$ **27.** $(5, 2, -6)$ **29.** 1×2
31. 3×1

33. (a) $\begin{bmatrix} 12 & -2 \\ 3 & 8 \end{bmatrix}$ (b) $\begin{bmatrix} 12 & -2 & \vdots & 6 \\ 3 & 8 & \vdots & -5 \end{bmatrix}$

35. (a) $\begin{bmatrix} 3 & -3 & 1 \\ 10 & 7 & 0 \\ 6 & 14 & -9 \end{bmatrix}$ (b) $\begin{bmatrix} 3 & -3 & 1 & \vdots & 1 \\ 10 & 7 & 0 & \vdots & 4 \\ 6 & 14 & -9 & \vdots & 10 \end{bmatrix}$

37. $\begin{cases} 6x + 4y + z = 0 \\ 12x + 9y - 2z = 6 \\ x + 8y - 4z = -5 \end{cases}$ **39.** $(3, 1)$ **41.** $\left(-1, \frac{1}{3}, \frac{4}{3}\right)$

43. $(10, -12)$ **45.** $\left(\frac{24}{5}, \frac{22}{5}, -\frac{8}{5}\right)$ **47.** $\left(\frac{1}{2}, -\frac{1}{3}, 1\right)$
49. $(0.6, 0.5)$ **51.** $\left(\frac{5}{16}a + \frac{13}{16}, \frac{19}{16}a + \frac{11}{16}, a\right)$ **53.** 5
55. 1   **57.** 102   **59.** $(-3, 7)$
61. $D = 0$, cannot use Cramer's Rule   **63.** $(2, -3, 3)$
65. 16   **67.** 7   **69.** Collinear   **71.** Not collinear
73. $x - 2y + 4 = 0$ **75.** $2x + 6y - 13 = 0$
77. (a) Yes   (b) No   (c) Yes   (d) No

79.
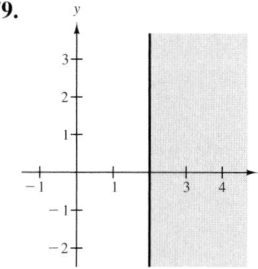

81.
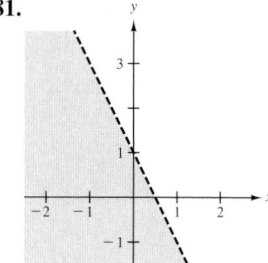

83.
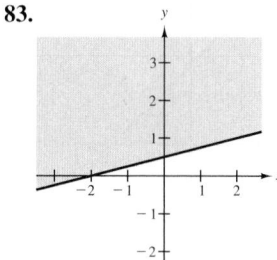

85.
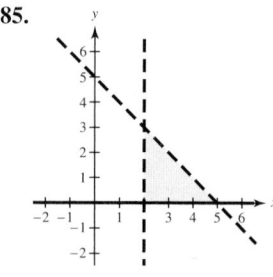

87.
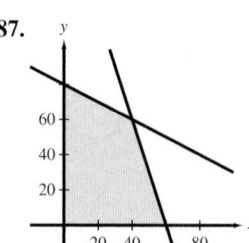

89. $\begin{cases} 2x + y \geq 7 \\ x - y \leq 2 \\ 2x + y \leq 22 \\ x - y \geq -4 \end{cases}$

91. 16,667 units   **93.** 96 meters $\times$ 144 meters
95. \$9.95 tapes: 400; \$14.95 tapes: 250
97. $70°; 50°; 60°$ **99.** $y = 2x^2 + x - 6$

101. (a) $y = -\frac{1}{45}x^2 + \frac{2}{3}x + 11$

(b)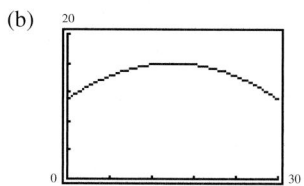

(c) $7\frac{1}{4}$ feet

103. $\begin{cases} x + y \le 1500 \\ x \ge 400 \\ y \ge 600 \end{cases}$

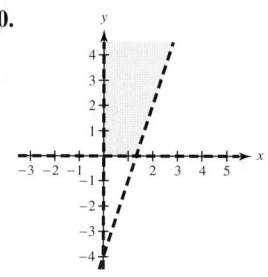

Chapter 4 Test   (page 302)

1. $(2, 4)$ **2.** $(2 - a, 2a - 1, a)$ **3.** $(2, -1, 1)$
4. $(1, -2, 3)$ **5.** $(-1, 2, 1)$ **6.** -62
7. $y = 2x^2 - 3x + 4$ **8.** 12

9.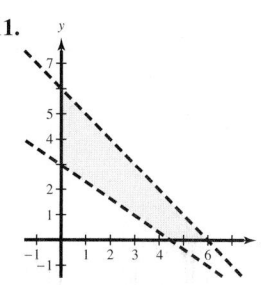

10.

11.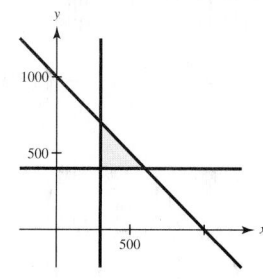

12. $\begin{cases} x + y = 200 \\ 4x - y = 0 \end{cases}$

40 miles, 160 miles

13. Regular: \$1.29 per gallon; Premium: \$1.46 per gallon

14. $\begin{cases} x + y \le 1000 \\ x \ge 300 \\ y \ge 400 \end{cases}$

Cumulative Test: Chapters 1–4   (page 303)

1. $-54a^{18}$ **2.** $\frac{8}{3}x^{10}y$ **3.** $-\frac{36x^4y^2}{25z^6}$ **4.** $2a^nb^m$
5. Perimeter: $b^2 + 17b$; Area: $4b^3 + 4b$;
Perimeter: 60; Area: 120
6. $x^2 + 10x - 8$ **7.** $5a(2 - 3a^3)$
8. $(2a - 3b)(a + 2b)$ **9.** $(x + 1)(x - 1)(x - 3)$
10. $(y - 4)(y^2 + 4y + 16)$ **11.** $\frac{8}{3}$ **12.** 1   **13.** 3
14. $-7, 6$ **15.** 5   **16.** $-18, 12$
17. Domain $= \{x: x$ is a real number$\}$
18. Domain $= \{x: x$ is a real number$\}$
19. Domain $= \{x: x \ne 8\}$
20. Domain $= \{t: t \ge 6\}$
21. Vertical shift two units downward
22. Horizontal shift two units to the right
23. Reflection in the x-axis
24. (a) $y = 250 - x$ (b) $A = x(250 - x)$
25. $y = 5x - 37$ **26.** $x = -2$
27. $V = -13{,}000t + 117{,}000$; when $t = 3$, $V = \$78{,}000$
28. \$650   **29.** 112 adult tickets; 28 student tickets
30. $-3 < x \le \frac{1}{2}$ **31.** $4 < x < 5$

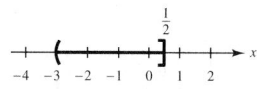   

32.   **33.**

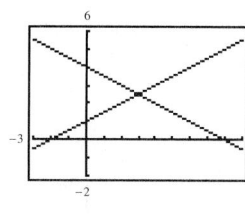   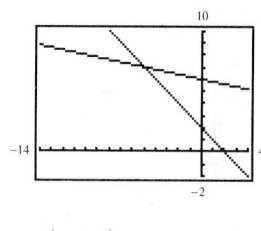

$\left(3, \frac{5}{2}\right)$ $(-5, 7)$

34. $(4, -1, 3)$ **35.** 18 feet $\times$ 16 feet
36. 8%: \$625,000; 9%: \$50,000; 10%: \$125,000
37. $x - 2y + 12 = 0$ **38.** Not collinear

39.

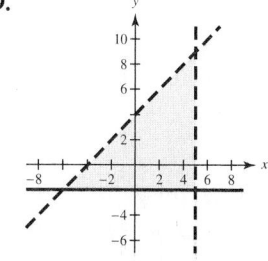

Chapter 5

Section 5.1 (page 311)

1. $\frac{1}{25}$ **3.** 64 **5.** 1 **7.** 1 **9.** 100,000

11. $\frac{1}{16}$ **13.** $\frac{7}{x^3}$ **15.** y^2 **17.** $\frac{1}{x^7}$ **19.** x^6

21. $\frac{y}{x^3}$ **23.** $\frac{10}{x}$ **25.** $\frac{4t}{3}$ **27.** t^2 **29.** $\frac{1}{4x^4}$

31. $-\frac{12}{xy^3}$ **33.** $-\frac{y^9}{64x^6}$ **35.** $\frac{y^4}{9x^4}$ **37.** $\frac{9y^8}{64x}$

39. $\frac{3x^5}{y^4}$ **41.** $\frac{81v^8}{u^6}$ **43.** $\frac{x^{12}}{16y^8}$ **45.** x^8y^{12}

47. $36m^5$ **49.** $12x^3y^7$ **51.** $125x^3y^{27}$ **53.** $\frac{2b^{10}}{25a^{12}}$

55. $\frac{8x}{y^7}$ **57.** $48xy^2$ **59.** 3.6×10^6

61. 3.81×10^{-3} **63.** 1.394×10^8 **65.** 60,000,000

67. 0.0000001359 **69.** 350,000,000 **71.** 6.80×10^5

73. 4.00×10^3 **75.** 9.00×10^{15} **77.** 4.70×10^{11}

79. 3.46×10^{10} **81.** 1.58×10^{-5} year ≈ 8.3 minutes

83. 3.33×10^5 or 333,000

85. $(-2x)^{-4} = (-2)^{-4}x^{-4} = \frac{1}{16}x^{-4} \neq -2x^{-4}$

87. All real numbers except $x = -1$

89. All real numbers such that $x \geq 4$

91. 7.5%: \$15,000; 9%: \$9000

Section 5.2 (page 321)

1. 7 **3.** 4.2 **5.** Square root **7.** 8

9. Not a real number **11.** $-\frac{2}{3}$ **13.** 0.3 **15.** 5

17. 10 **19.** $-\frac{1}{4}$ **21.** 3 **23.** -0.3

25. Irrational **27.** Irrational **29.** Rational

31. $16^{1/2} = 4$ **33.** $\sqrt[3]{125} = 5$ **35.** 5 **37.** $\frac{1}{4}$

39. $\frac{4}{9}$ **41.** $\frac{3}{11}$ **43.** 8.5440 **45.** 0.0038

47. 4.3004 **49.** 66.7213 **51.** $|t|$ **53.** y^3

55. t^3 **57.** x^6 **59.** $\frac{1}{x}$ **61.** x^3 **63.** $x^{3/4}y^{1/4}$

65. $y^{5/2}z^4$ **67.** 3 **69.** $\frac{1}{2}$ **71.** $\frac{4}{9}$ **73.** $6^{4/3}$

75. x^3 **77.** $\frac{9y^{3/2}}{x^{2/3}}$ **79.** $\frac{1}{a^{5/4}}$ **81.** $\frac{b^{3/5}}{2a^3}$ **83.** $x^{1/4}$

85. $\frac{y^3}{8x}$ **87.** $c^{1/2}$ **89.** $\frac{x^{2/5}}{y^{1/3}}$ **91.** $xy^{1/3}$ **93.** $y^{1/8}$

95. $x^{1/6}$ **97.** $(x + y)^{1/2}$ **99.** $\frac{1}{(3u - 2v)^{5/6}}$

101. (a) 0 (b) 6 (c) 12 (d) 30

103. (a) 2 (b) 1 (c) 4 (d) -3

105. Domain = $\{x: x \geq 0\}$

107. Domain = $\{x: x > 0\}$

109. Domain = $\{x: x \geq -10\}$

111. Domain = $\left\{x: x \geq -\frac{8}{3}\right\}$

113.

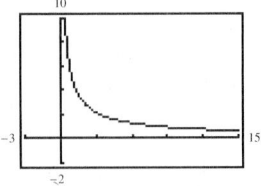

Domain: = $\{x: x > 0\}$

115.

Domain: = $\{x: x \text{ is a real number}\}$

117. $2x^{3/2} - 3x^{1/2}$ **119.** $1 + 5y$ **121.** 12.8%

123. 13 inches $\times$ 13 inches $\times$ 13 inches

125. ≈ 0.026 inch

127. If a and b are real numbers, n is an integer greater than or equal to 2, and $a = b^n$, then b is an nth root of a.

129. $x < 0$. If $x < 0$, then $\sqrt{x^2} = -x$.

131. $2x^2 - 7x - 4$ **133.** $x^2 - 4y^2$

135. $(3x - 4)(x - 1)$ **137.** $2(x^2 + 1)(x - 8)$

139. ≈ 49.09 miles/hour

Section 5.3 (page 329)

1. $2\sqrt{5}$ **3.** $3\sqrt{3}$ **5.** 0.2 **7.** $2\sqrt[3]{3}$ **9.** $10\sqrt[4]{3}$

11. $\frac{\sqrt{15}}{2}$ **13.** $\frac{\sqrt[3]{35}}{4}$ **15.** $\frac{\sqrt[5]{15}}{3}$ **17.** $\frac{\sqrt{13}}{5}$

19. $a^3\sqrt{a}$ **21.** $3x^2\sqrt{x}$ **23.** $4y^2\sqrt{3}$ **25.** $x^3\sqrt[3]{x}$

27. $2a^2\sqrt[3]{a}$ **29.** $3x\sqrt[3]{2x^2}$ **31.** $a^4b^2\sqrt{ab}$

33. $xy\sqrt[3]{x}$ **35.** $2|uv|\sqrt[4]{8v^3}$ **37.** $2xy\sqrt[5]{y}$ **39.** $\frac{2\sqrt[5]{x^2}}{y}$

41. $\frac{3a\sqrt[3]{2a}}{b^3}$ **43.** $\frac{4a^2\sqrt{2}}{|b|}$ **45.** $3x^2$

47.

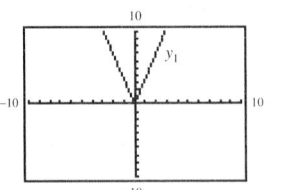 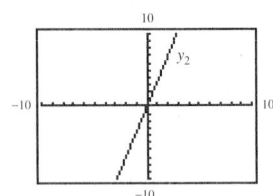

$y_1 \neq y_2$ because the range of y_1 is $[0, \infty)$ and the range of y_2 is $(-\infty, \infty)$.

49.

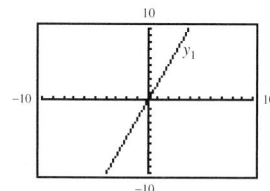 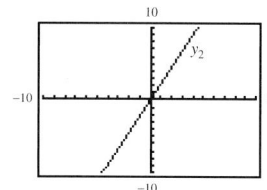

$y_1 = y_2$

51. $\dfrac{\sqrt{3}}{3}$ **53.** $4\sqrt{3}$ **55.** $\dfrac{\sqrt[4]{20}}{2}$ **57.** $\dfrac{3\sqrt[3]{2}}{2}$

59. $\dfrac{\sqrt{y}}{y}$ **61.** $\dfrac{2\sqrt{x}}{x}$ **63.** $\dfrac{2\sqrt{x}}{x^2}$ **65.** $\dfrac{\sqrt[3]{18xy^2}}{3y}$

67. $\dfrac{a^2\sqrt[3]{a^2b}}{b}$ **69.** $\dfrac{2\sqrt{3b}}{b^2}$ **71.** $\dfrac{|a|\sqrt{10ab}}{5b^4}$

73. $\dfrac{2u^2\sqrt[3]{9v^2}}{3v}$ **75.** $\dfrac{|x|y^2\sqrt{3xz}}{z^4}$ **77.** $\dfrac{x^4y\sqrt[3]{14y^2z^2}}{2z^4}$

79.

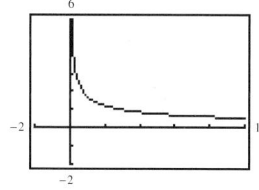

81. $2\sqrt{2}$ **83.** $18\sqrt{2}$ **85.** $7\sqrt{3}$ **87.** $21\sqrt{6}$
89. $30\sqrt[3]{2}$ **91.** $12\sqrt{x}$ **93.** $13\sqrt{y}$
95. $(3a + 6a^3)\sqrt{a}$ **97.** $(10 - z)\sqrt[3]{z}$ **99.** $7y^2\sqrt{y}$
101. $6x\sqrt[4]{2x^3}$ **103.** $15xy^2\sqrt{2xy}$ **105.** $27x\sqrt{5xy}$

107.

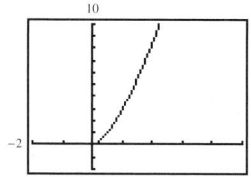

109. $\dfrac{2\sqrt{5}}{5}$

111. $\dfrac{10\sqrt{3x}}{3}$ **113.** $\sqrt{7} + \sqrt{18} > \sqrt{7 + 18}$
115. $5 > \sqrt{3^2 + 2^2}$ **117.** $3\sqrt{5}$
119. $16\pi\sqrt{5} \approx 112.40$ square feet
121. $8 + 8\sqrt{2} \approx 19.31$ feet **123.** 89.44 cycles per second
125. 1. All possible nth powers have been removed from each
radical.
2. No radical contains a fraction.
3. No denominator of a fraction contains a radical.
127. Two or more radical expressions are alike if they have the
same radicand and the same index.
129. Distance: $5\sqrt{5} \approx 11.18$; Midpoint: $\left(-\dfrac{3}{2}, 7\right)$
131. Distance: $2\sqrt{65} \approx 16.12$; Midpoint: $(-1, -1)$

133. $x \geq \dfrac{10}{3}$

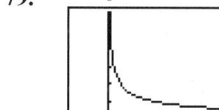

135. $-14 < x < 26$

137. 32 inches $\times$ 32 inches

Mid-Chapter Quiz   (page 332)

1. $-\dfrac{1}{144}$ **2.** $\dfrac{64}{27}$ **3.** $\dfrac{5}{3}$ **4.** 16 **5.** $3t^{3/2}$ **6.** $\dfrac{x^3}{8}$

7. $\dfrac{2}{3u^3}$ **8.** $\dfrac{3x^6}{16y^3}$

9. (a) 1.34×10^7 (b) 7.5×10^{-4}
10. (a) $5\sqrt{6}$ (b) $3\sqrt[3]{2}$
11. (a) $3x^2\sqrt{x}$ (b) $3x^2y\sqrt[4]{x^3y}$
12. (a) $\dfrac{\sqrt[4]{5}}{2}$ (b) $\dfrac{2\sqrt{6}}{7}$
13. (a) $\dfrac{2u\sqrt{10u}}{3z^5}$ (b) $\dfrac{2a\sqrt[3]{2a^2}}{b^4}$
14. (a) 0 (b) $2\sqrt{7}$ **15.** (a) 6 (b) $3\sqrt{2}$
16. (a) Domain $= \{x: x \leq 3\}$
(b) Domain $= \{x: x \text{ is a real number}\}$
17. (a) $\dfrac{\sqrt{6}}{3}$ (b) $4\sqrt{3}$ **18.** (a) $\dfrac{2\sqrt{5x}}{x}$ (b) $\dfrac{\sqrt[3]{12a}}{2a}$
19. $4\sqrt{2y}$ **20.** $10x\sqrt[3]{5x}$ **21.** $\dfrac{3\sqrt{10x}}{2}$
22. $\sqrt{5^2 + 12^2} = \sqrt{169} = 13$ **23.** $\left(23 + 8\sqrt{2}\right)$ inches

Section 5.4   (page 337)

1. 4 **3.** $3\sqrt{2}$ **5.** $2\sqrt{5} - \sqrt{15}$ **7.** $2\sqrt{10} + 8\sqrt{2}$
9. -1 **11.** 4 **13.** 4 **15.** $8\sqrt{5} + 24$
17. $2x + 20\sqrt{2x} + 100$ **19.** $y + 4\sqrt{y}$
21. $\sqrt{15} + 3\sqrt{3} - 5\sqrt{5} - 15$ **23.** $45x - 17\sqrt{x} - 6$
25. $x - y$ **27.** $2 - 7\sqrt[3]{4}$ **29.** $\sqrt[3]{4x^2} + 10\sqrt[3]{2x} + 25$
31. $2y - 10\sqrt[3]{2y} + 10\sqrt[3]{4y^2} - 100$ **33.** $(x + 3)$
35. $(4 - 3x)$ **37.** $\left(2u + \sqrt{2u}\right)$ **39.** $\dfrac{1 - 2\sqrt{x}}{3}$
41. $\dfrac{-1 + \sqrt{3y}}{4}$ **43.** $-2x + \sqrt{7x}$ **45.** $2 - \sqrt{5}; -1$
47. $\sqrt{6} - 10; -94$ **49.** $\sqrt{11} + \sqrt{3}; 8$
51. $\sqrt{x} + 3; x - 9$ **53.** $\sqrt{2u} + \sqrt{3}; 2u - 3$
55. $\sqrt{6x} - \sqrt{y}; 6x - y$ **57.** $\dfrac{6\sqrt{11} + 12}{7}$
59. $\dfrac{35 - 7\sqrt{3}}{22}$ **61.** $-4\sqrt{7} + 12$
63. $\dfrac{\sqrt{6} - \sqrt{2}}{2}$ **65.** $\dfrac{6 - \sqrt{2}}{17}$ **67.** $\dfrac{4\sqrt{7} + 11}{3}$

69. $\dfrac{x\sqrt{15} + x\sqrt{3}}{4}$ **71.** $\dfrac{t\sqrt{5t} + t\sqrt{t}}{2}$

73. $\dfrac{7\sqrt{5z} + 7\sqrt{z}}{4}$ **75.** $\dfrac{2x - 5\sqrt{x} + 2}{x - 4}$

77. $\dfrac{2x - 9\sqrt{x} - 5}{4x - 1}$ **79.** $-\dfrac{\sqrt{u+v}\left(\sqrt{u-v} + \sqrt{u}\right)}{v}$

81. (a) $2\sqrt{3} - 4$ (b) 0   **83.** (a) 0   (b) -1

85.

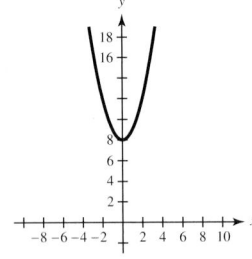

87.

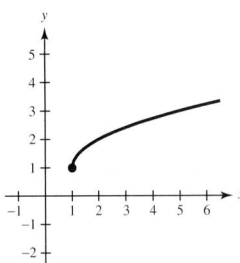

89. $16\sqrt{33}$ square inches   **91.** $192\sqrt{2}$ square inches

93. $\dfrac{\sqrt{3}}{2}$ **95.** $\dfrac{500k\sqrt{k^2 + 1}}{k^2 + 1}$

97. Answers will vary. The FOIL Method is the same for polynomial expressions and radical expressions. Some students have more difficulty using the method when radicals are involved.

99. $4x - 16$ **101.** $x^2 + x + \frac{1}{4}$ **103.** $9x^2 - 64$

105. Vertical shift eight units upward

107. Vertical shift one unit upward and horizontal shift one unit to the right

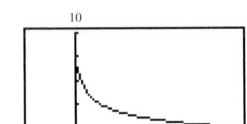
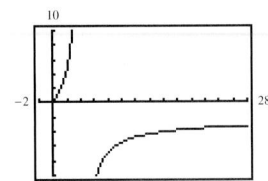

109. $8x^2y^3\sqrt{2y}$ **111.** $9x\sqrt{2x}$ **113.** 1.5 seconds

115. 30% solution: $13\frac{1}{3}$ gallons; 60% solution: $6\frac{2}{3}$ gallons

Section 5.5   (page 346)

1. (a) Not a solution
    (b) Not a solution
    (c) Not a solution
    (d) Solution

3. (a) Not a solution
    (b) Solution
    (c) Not a solution
    (d) Not a solution

5. 400   **7.** 49   **9.** No solution   **11.** 525

13. 90   **15.** $\frac{44}{3}$ **17.** $\frac{14}{25}$ **19.** $-\frac{2}{3}$ **21.** 8

23. 1   **25.** 5   **27.** No solution   **29.** 4   **31.** 2

33. No solution   **35.** 4   **37.** 7   **39.** -15

41. 1, 3   **43.** $\frac{4}{5}$ **45.** $\frac{1}{4}$ **47.** $\frac{1}{4}$ **49.** $\frac{1}{2}$ **51.** 6

53.

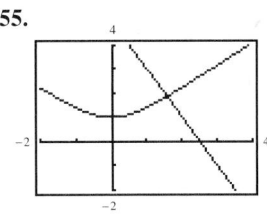

1.347

55.

1.569

57.

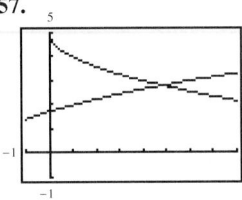

4.840

59.

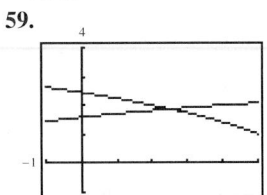

2.513

61. 1.82 feet   **63.** $h^2 = \dfrac{S^2 - \pi^2 r^4}{\pi^2 r^2}$

65. 56.57 feet per second   **67.** 56.25 feet

69. 500 units

71. No. Raising each side of an equation to the nth power sometimes introduces extraneous solutions, which indicates that the new equation is not equivalent to the original equation. This is why it is always necessary to check any solutions in the *original* equation.

73. 5   **75.** -6 **77.** $-7, 11$ **79.** \$12,720

Section 5.6   (page 354)

1. $2i$ **3.** $\frac{2}{5}i$ **5.** $2\sqrt{2}\,i$ **7.** $\sqrt{7}\,i$ **9.** $10i$

11. -4 **13.** $-3\sqrt{6}$ **15.** $-2\sqrt{3} - 3$

17. $5\sqrt{2} - 4\sqrt{5}$ **19.** -16 **21.** $a = 3, b = -4$

23. $a = 2, b = -3$ **25.** $a = -4, b = -2\sqrt{2}$

27. $a = 2, b = -2$ **29.** $10 + 4i$ **31.** $-14 - 40i$

33. $-14 + 20i$ **35.** $9 - 7i$ **37.** $-3 + 49i$

39. -36 **41.** $-36i$ **43.** $27i$ **45.** $-65 - 10i$

47. $20 - 12i$ **49.** $-40 - 5i$ **51.** $-14 + 42i$

53. $-7 - 24i$ **55.** $-21 + 20i$ **57.** 9

59. $2 + 11i$ **61.** 5   **63.** 68   **65.** 31   **67.** 100

69. 4   **71.** $2 + 2i$ **73.** $-\frac{24}{53} + \frac{84}{53}i$ **75.** $-\frac{72}{85} + \frac{84}{85}i$

77. $-10i$ **79.** $-\frac{1}{5} - \frac{9}{5}i$ **81.** $-\frac{6}{5} + \frac{2}{5}i$

83. $\frac{5}{13} + \frac{1}{13}i$ **85.** $\frac{8}{5} - \frac{1}{5}i$ **87.** $1 - \frac{6}{5}i$

89. $-\frac{53}{25} + \frac{29}{25}i$ **91.** (a) Solution   (b) Solution

93. (a) Solution   (b) Solution   **95.** $2a$ **97.** $2bi$

99. $i = \sqrt{-1}$

101. $\sqrt{-3}\sqrt{-3} = \left(\sqrt{3}\,i\right)\left(\sqrt{3}\,i\right)$
$= 3i^2 = -3$

103. -21 **105.** $-\dfrac{4}{3}, 4$ **107.** $t = \dfrac{360}{r}$

Review Exercises   (page 357)

1. $\dfrac{1}{72}$ **3.** $\dfrac{125}{8}$ **5.** $\dfrac{x^{12}}{y^{11}}$ **7.** $-\dfrac{12b^4}{a^6}$ **9.** $\dfrac{x^2}{2y^{10}}$

11. $\dfrac{5b^4c^5}{a}$ **13.** 1.46×10^9 **15.** 6.41×10^{-7}

17. 0.00000000000409 **19.** $95,800,000,000$

21. 3.6×10^7 **23.** 500 **25.** -4 **27.** -2

29. $\frac{2}{5}$ **31.** 0.1

33. Not possible, because you cannot take the 4th root of a negative number.

35. 81 **37.** 125 **39.** $\dfrac{1}{2}$ **41.** $x^{7/12}$ **43.** $\dfrac{4}{a^{2/3}}$

45. $\dfrac{3}{x^{1/4}y^{2/5}}$ **47.** $\sqrt[3]{3x+2}$ **49.** 0.04

51. (a) 1 (b) 2 (c) 5 (d) 4

53. Domain $= \{x : x \, 1 > 0\}$

55. Domain $= \{x : x \geq -3\}$ **57.** $6\sqrt{10}$ **59.** $5x^2\sqrt{2}$

61. $2ab\sqrt[3]{6b}$ **63.** $3x^3y^3\sqrt[4]{y^3}$ **65.** $2a^2b^3\sqrt[5]{ab}$

67. $\dfrac{\sqrt{30}}{6}$ **69.** $\dfrac{\sqrt{3x}}{2x}$ **71.** $\dfrac{\sqrt[3]{4x^2}}{x}$ **73.** $\dfrac{2xy\sqrt{15yz}}{5z}$

75. $\dfrac{x^2\sqrt[3]{2x^2y}}{2y}$ **77.** $-\dfrac{2\sqrt[4]{2xy^3}}{y}$ **79.** $-24\sqrt{10}$

81. $7\sqrt[4]{y}+3$ **83.** $\left(1-\dfrac{3}{5x}\right)\sqrt{5x}$

85. $5+12x\sqrt{5}+36x^2$ **87.** $2x-3\sqrt{x}-35$

89. $\sqrt{3}+\sqrt{x}; 3-x$ **91.** $\sqrt{5u}-\sqrt{v}; 5u-v$

93. $\dfrac{15\sqrt{x}-45}{x-9}$ **95.** $\dfrac{5\sqrt{x}-x\sqrt{6}}{25-6x}$

97. $\dfrac{x+20\sqrt{x}+100}{x-100}$ **99.** 225 **101.** 105

103. $-3, -5$ **105.** 5 **107.** $\frac{3}{32}$ **109.** $-\frac{16}{3} \approx -5.33$

111. $4i$ **113.** $4\sqrt{3}i$ **115.** $0.4i$ **117.** $7i$

119. $-5\sqrt{2}$ **121.** $-\sqrt{30}-3\sqrt{5}$ **123.** -7

125. $a=10, b=-4$ **127.** $a=4, b=7$

129. $8-3i$ **131.** -90 **133.** 25 **135.** $59+74i$

137. $-\frac{4}{5}i$ **139.** $\frac{9}{17}+\frac{2}{17}i$ **141.** 0.15 feet

143. $2\sqrt{61} \approx 15.62$ inches   **145.** ≈ 1.37 feet

Chapter Test   (page 360)

1. (a) $\frac{3}{8}$ (b) 3.0×10^{-5} **2.** (a) $\frac{1}{9}$ (b) $\frac{3}{2}$

3. (a) 3.2×10^{-5} (b) $30,400,000$

4. (a) $\dfrac{3s^7}{5t}$ (b) $-\dfrac{y^2}{2x^5}$

5. (a) $\dfrac{16x^{1/3}}{25y^{7/3}z^2}$ (b) $25x^2y^4$

6. (a) $\dfrac{4x\sqrt{2x}}{3y}$ (b) $2u^2v^4\sqrt[3]{3u^2v^2}$

7. (a) $\dfrac{5(\sqrt{6}+\sqrt{2})}{2}$ (b) $\dfrac{2\sqrt[3]{3y^2}}{3y}$ **8.** $-10\sqrt{3x}$

9. $5\sqrt{3x}+3\sqrt{5}$ **10.** $16-8\sqrt{2x}+2x$

11. No solution   **12.** 9 **13.** 29 **14.** $2-2i$

15. $-8+4i$ **16.** $7+9i$ **17.** $-5-12i$

18. $13+13i$ **19.** $\frac{13}{10}-\frac{11}{10}i$ **20.** 100 feet

Cumulative Test: Chapters 1–5   (page 361)

1. $y \leq 45$ **2.** $x \geq 15$ **3.** Additive Identity Property

4. Distributive Property   **5.** $9.35 + 0.75q$

6. $d = 48t$ **7.** $(4x+11)(4x-11)$ **8.** $(3t-4)^2$

9. $(x-10)(x-4)$ **10.** $4x(x^2-3x+4)$ **11.** $\frac{8}{3}$

12. 3 **13.** $-\frac{2}{3}, 5$ **14.** $-5, 8$ **15.** $-8, 3$ **16.** 41

17. x-intercept: $(6, 0)$ **18.** x-intercepts: $(-4, 0), (2, 0)$
       y-intercept: $(0, 6)$ y-intercept: $(0, -8)$

19. x-intercept: $(0, 0)$ **20.** $-t^2 - t$
       y-intercept: $(0, 0)$

21. (a) $g(x) = -\sqrt{x}$ (b) $g(x) = \sqrt{x}+2$
       (c) $g(x) = \sqrt{x-2}$

22.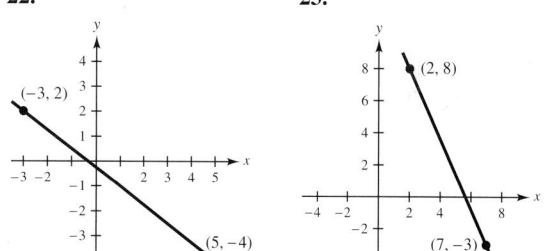

23.

$m = -\frac{3}{4}$ $m = -\frac{11}{5}$

24. $y = 10x + 38$ **25.** $y = \frac{1}{3}x + \frac{14}{3}$

26. $y = -\frac{30}{11}x + \frac{34}{11}$ **27.** 27 **28.** 8

29. $x > -8$ **30.** $x \leq -24$ or $x \geq 8$

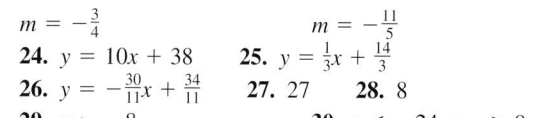

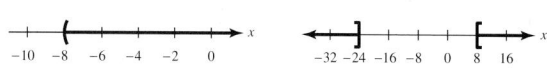

31. $\left(\frac{2}{5}, \frac{8}{5}\right)$ **32.** $(1, 2, 1)$

33. 75% solution: 40 gallons; 50% solution: 60 gallons

34. 335 **35.** 58

36.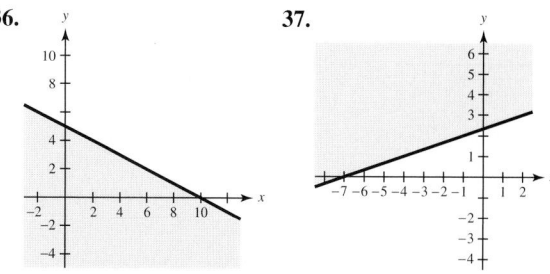

37.

38. $2|x|y\sqrt{6y}$ **39.** $2a^5b^2\sqrt[3]{10b^2}$ **40.** $\dfrac{2\sqrt{3}b^3}{a^2}$

41. 2 **42.** $-7 + 16i$ **43.** $-\frac{3}{5} + \frac{11}{5}i$

Chapter 6

Section 6.1   (page 372)

1. $5, 7$ **3.** $-9, 8$ **5.** $-9, 5$ **7.** $0, 3$ **9.** $9, 12$
11. $\pm\frac{5}{2}$ **13.** $\frac{1}{2}, \frac{3}{4}$ **15.** -30 **17.** $1, 6$ **19.** $-\frac{5}{6}, \frac{1}{2}$
21. $\frac{3}{5}, 2$ **23.** $\frac{5}{3}, 6$ **25.** ± 8 **27.** ± 3 **29.** ± 8
31. $\pm\frac{4}{5}$ **33.** $\pm 2\sqrt{6}$ **35.** $\pm\frac{15}{2}$ **37.** $9, -17$
39. $2.5, 3.5$ **41.** $2 \pm \sqrt{7}$ **43.** $\dfrac{-1 \pm 5\sqrt{2}}{2}$
45. $-1, 11$ **47.** $\dfrac{3 \pm 7\sqrt{2}}{4}$ **49.** $\pm 6i$ **51.** $\pm 2i$
53. $3 \pm 5i$ **55.** $1 \pm 13i$ **57.** $-4 \pm 11i$ **59.** $\frac{1}{2} \pm i$
61. $-\frac{4}{3} \pm 4i$ **63.** $-6 \pm \frac{11}{3}i$ **65.** $1 \pm 3\sqrt{3}i$
67. $-1 \pm 0.2i$ **69.** $\frac{2}{3} \pm \frac{1}{3}i$ **71.** $-\frac{7}{3} \pm \frac{\sqrt{38}}{3}i$

73.   **75.**

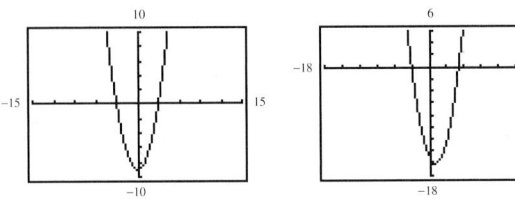

$(-3, 0), (3, 0)$ $(-3, 0), (5, 0)$

77.   **79.**

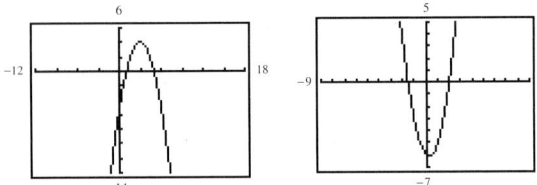

$(1, 0), (5, 0)$ $(2, 0), \left(-\frac{3}{2}, 0\right)$

81.   **83.**

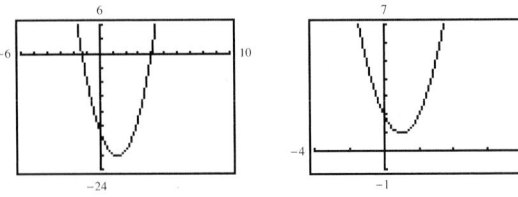

$\left(-\frac{4}{3}, 0\right), (4, 0)$ $1 \pm i$

85.   **87.**

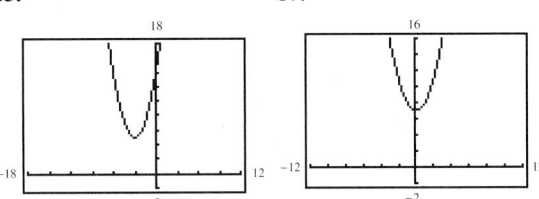

$-3 \pm \sqrt{5}i$ $\pm\sqrt{7}i$

89. $\pm 2i$ **91.** $0, \frac{7}{3}$ **93.** $-4, \frac{4}{3}$ **95.** ± 15
97. $10, -12$ **99.** $-12 \pm 20i$ **101.** 7 **103.** $-9, 27$
105. $f(x) = \sqrt{4 - x^2}$ **107.** $f(x) = \frac{1}{2}\sqrt{4 - x^2}$
     $g(x) = -\sqrt{4 - x^2}$ $g(x) = -\frac{1}{2}\sqrt{4 - x^2}$

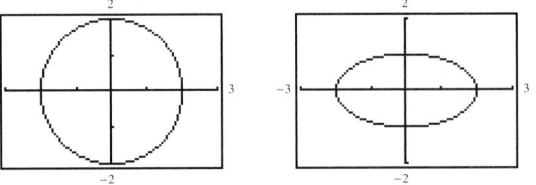

109. $\pm 1, \pm 2$ **111.** $\pm\sqrt{2}, \pm\sqrt{3}$ **113.** $\pm 2, \pm i$
115. $\pm 1, \pm 2\sqrt{2}$ **117.** $\pm 1, \pm\sqrt{5}$ **119.** $-8, 27$
121. $-64, 8$ **123.** $\frac{125}{8}, 1$ **125.** $-1024, -243$
127. $1, 32$ **129.** $\frac{1}{32}, 243$ **131.** 10
133. $4\sqrt{2} \approx 5.66$ **135.** $2\sqrt{2834} \approx 106.47$ feet
137. $\dfrac{25\sqrt{2}}{2}$ inches $\times \dfrac{25\sqrt{2}}{2}$ inches **139.** $\dfrac{9}{2\pi}$ inches
141. $2\sqrt{2} \approx 2.83$ seconds **143.** 9 seconds **145.** 1993
147. 2 **149.** $\approx 20\%$
151. Factoring and the Zero-Factor Property allow you to solve a quadratic equation by converting it into two linear equations that you already know how to solve.
153. Yes. If $(x - 1)^2 = 0$, the only solution is $x = 1$.
155. $(4r^2 + t^2)(2r + t)(2r - t)$ **157.** $(19y - 6)(3y + 1)$
159. $(2x + 7)(3x - 7)$ **161.** 6 **163.** $-4, 3$
165. 15 minutes; 2 miles

Section 6.2   (page 381)

1. 16 **3.** 100 **5.** 4 **7.** $\frac{25}{4}$ **9.** $\frac{81}{4}$ **11.** $\frac{9}{25}$
13. $\frac{9}{100}$ **15.** 0.04 **17.** $0, -6$ **19.** $1, 7$
21. $4, -6$ **23.** $-3, -4$ **25.** $-1, \frac{8}{5}$
27. $2 + \sqrt{7} \approx 4.65$ **29.** $-3 + \sqrt{2} \approx -1.59$
    $2 - \sqrt{7} \approx -0.65$ $-3 - \sqrt{2} \approx -4.41$
31. $2 + \sqrt{3} \approx 3.73$ **33.** $-1 + \sqrt{2}i \approx -1 + 1.41i$
    $2 - \sqrt{3} \approx 0.27$ $-1 - \sqrt{2}i \approx -1 - 1.41i$
35. $5 + 3\sqrt{3} \approx 10.20$ **37.** $-10 + 3\sqrt{10} \approx -0.51$
    $5 - 3\sqrt{3} \approx -0.20$ $-10 - 3\sqrt{10} \approx -19.49$

39. $\dfrac{-5 + \sqrt{13}}{2} \approx -0.70$ **41.** $\dfrac{-3 + \sqrt{17}}{2} \approx 0.56$

$\dfrac{-5 - \sqrt{13}}{2} \approx -4.30$ $\dfrac{-3 - \sqrt{17}}{2} \approx -3.56$

43. $\dfrac{11 + \sqrt{109}}{2} \approx 10.72$ **45.** $\dfrac{1}{2} + \dfrac{\sqrt{3}}{2}i \approx 0.5 + 0.87i$

$\dfrac{11 - \sqrt{109}}{2} \approx 0.28$ $\dfrac{1}{2} - \dfrac{\sqrt{3}}{2}i \approx 0.5 - 0.87i$

47. $\dfrac{5 + \sqrt{5}}{2} \approx 3.62$ **49.** $\dfrac{1 + 2\sqrt{7}}{3} \approx 2.10$

$\dfrac{5 - \sqrt{5}}{2} \approx 1.38$ $\dfrac{1 - 2\sqrt{7}}{3} \approx -1.43$

51. $\dfrac{-3 + \sqrt{137}}{8} \approx 1.09$ **53.** $\dfrac{-4 + \sqrt{10}}{2} \approx -0.42$

$\dfrac{-3 - \sqrt{137}}{8} \approx -1.84$ $\dfrac{-4 - \sqrt{10}}{2} \approx -3.58$

55. $\dfrac{-9 + \sqrt{21}}{6} \approx -0.74$ **57.** $\dfrac{-1 + \sqrt{10}}{2} \approx 1.08$

$\dfrac{-9 - \sqrt{21}}{6} \approx -2.26$ $\dfrac{-1 - \sqrt{10}}{2} \approx -2.08$

59. $\dfrac{-5 + \sqrt{89}}{4} \approx 1.11$ $\dfrac{-5 - \sqrt{89}}{4} \approx -3.61$

61. $\dfrac{3}{10} + \dfrac{\sqrt{191}}{10}i \approx 0.30 + 1.38i$

$\dfrac{3}{10} - \dfrac{\sqrt{191}}{10}i \approx 0.30 - 1.38i$

63. $\dfrac{7 + \sqrt{57}}{2} \approx 7.27$ **65.** $-1 + \sqrt{3}i \approx -1 + 1.73i$

$\dfrac{7 - \sqrt{57}}{2} \approx -0.27$ $-1 - \sqrt{3}i \approx -1 - 1.73i$

67. $-1 \pm 2i$ **69.** $1 \pm \sqrt{3}$ **71.** $4 + 2\sqrt{2}$

73.   **75.**

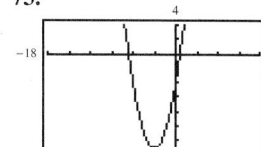 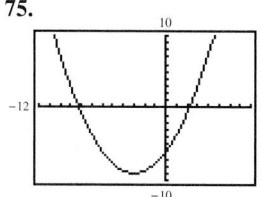

$\left(-3 \pm \sqrt{13},\, 0\right)$ $\left(-3 \pm 3\sqrt{3},\, 0\right)$

77. (a) $x^2 + 8x$ (b) $x^2 + 8x + 16$ (c) $(x + 4)^2$

79. Base: 4 centimeters; Height: 6 centimeters

81. 6 inches $\times$ 10 inches $\times$ 14 inches **83.** 16, 18

85. 42 units, 58 units **87.** 1998

89. Divide both sides of the equation by the leading coefficient. Dividing both sides of an equation by a nonzero constant yields an equivalent equation.

91. Yes, because a quadratic equation can be solved by factoring if the solutions are rational numbers.

93. $x^2 + 12x + 26$ **95.** $4x^2 + 4x - 3$ **97.** $\pm\dfrac{2}{15}$

99. $3 \pm 8i$ **101.** $-\dfrac{5}{2}, 3$ **103.** 200 units

Section 6.3 **(page 389)**

1. $2x^2 + 2x - 7 = 0$ **3.** $-x^2 + 10x - 5 = 0$

5. 4, 7 **7.** $-\dfrac{1}{2}$ **9.** $-\dfrac{1}{2}, \dfrac{2}{3}$ **11.** $-15, 20$

13. $1 \pm \sqrt{5}$ **15.** $-2 \pm \sqrt{3}$ **17.** $-3 \pm 2\sqrt{3}$

19. $5 \pm \sqrt{2}$ **21.** $-\dfrac{3}{2} \pm \dfrac{\sqrt{3}}{2}i$ **23.** $\dfrac{1 \pm \sqrt{3}}{2}$

25. $\dfrac{-2 \pm \sqrt{10}}{2}$ **27.** $\dfrac{-2 \pm \sqrt{2}}{2}$ **29.** $\dfrac{5}{6} \pm \dfrac{\sqrt{11}}{6}i$

31. $\dfrac{1}{2} \pm i$ **33.** $\dfrac{-1 \pm \sqrt{5}}{3}$ **35.** $\dfrac{1 \pm \sqrt{5}}{5}$

37. $\dfrac{-1 \pm \sqrt{10}}{5}$ **39.** $\dfrac{3}{4} \pm \dfrac{\sqrt{3}}{4}i$ **41.** $\dfrac{3 \pm \sqrt{13}}{6}$

43. Two distinct imaginary solutions

45. Two distinct irrational solutions

47. Two distinct imaginary solutions

49. One (repeated) rational solution

51. Two distinct imaginary solutions

53. ± 13 **55.** $0, -15$ **57.** $\pm\dfrac{5}{3}i$ **59.** $\dfrac{9}{5}, \dfrac{21}{5}$

61. $-10, \dfrac{4}{3}$ **63.** $-4 \pm 4i$ **65.** $-3, -4$

67. $\dfrac{-5 \pm 5\sqrt{17}}{12}$ **69.** 8, 16 **71.** $\dfrac{5}{2}, -\dfrac{11}{6}$

73.   **75.**

 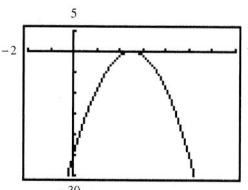

$(0.18, 0), (1.82, 0)$ $(2.50, 0)$

77.   **79.**

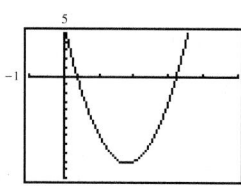 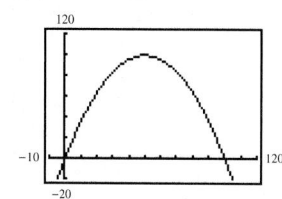

$(3.23, 0), (0.37, 0)$ $(99.80, 0), (0.20, 0)$

81. No real solutions **83.** Two real solutions

85. $\dfrac{5 \pm \sqrt{185}}{8}$ **87.** $\dfrac{3 + \sqrt{17}}{2}$

89. 5.1 centimeters $\times$ 11.4 centimeters

91. (a) 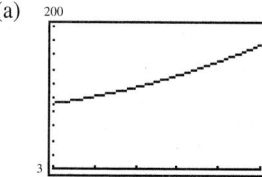 (b) 1994 (c) 1994

93. The opposite of b, plus or minus the square root of b squared minus $4ac$, all divided by $2a$.

95. (a) $c < 9$ (b) $c = 9$ (c) $c > 9$

97. (a) $c < 16$ (b) $c = 16$ (c) $c > 16$

99. $-2, -4$ **101.** $-\dfrac{3}{2}$ **103.** $\dfrac{-7 \pm \sqrt{57}}{2}$ **105.** 2

107. $\sqrt{21}$ **109.** Gold coins: 12; Silver coins: 30

Mid-Chapter Quiz   (page 392)

1. ± 6 **2.** $-4, \dfrac{5}{2}$ **3.** $\pm 2\sqrt{3}$ **4.** $-1, 7$

5. $-5 \pm 2\sqrt{6}$ **6.** $\dfrac{-3 \pm \sqrt{19}}{2}$ **7.** $-2 \pm \sqrt{10}$

8. $\dfrac{3 \pm \sqrt{105}}{12}$ **9.** $-\dfrac{5}{2} \pm \dfrac{\sqrt{3}}{2}i$ **10.** $-2, 10$

11. $-3, 10$ **12.** $-2, 5$ **13.** $\dfrac{3}{2}$ **14.** $\dfrac{-5 \pm \sqrt{10}}{3}$

15. $\pm\sqrt{2}, \pm\sqrt{7}i$ **16.** $27, 125$

17.   **18.**

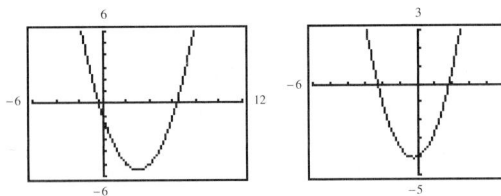

$(-0.32, 0), (6.32, 0)$ $(-2.24, 0), (1.79, 0)$

19. 50 units   **20.** $2\sqrt{93} \approx 19.29$ **21.** $\sqrt{451} \approx 21.24$

Section 6.4   (page 397)

1. 1, 2 or 8, 9   **3.** 8, 10

5. 70 centimeters   **7.** 108 miles

9. 24 feet $\times$ 13 feet or 26 feet $\times$ 12 feet

11. Rectangular region: no; Circular region: yes

13. 2 yards, 5 yards

15. (a) $d = \sqrt{(3 + x)^2 + (4 + x)^2}$

(b)

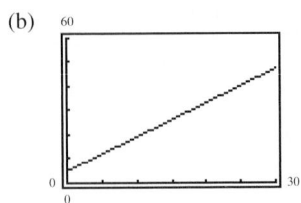

(c) 3.56 meters   (d) $\dfrac{-7 + \sqrt{199}}{2} \approx 3.55$ meters

17. 15.86 miles or 2.14 miles

19. Base: 25 inches; Height: 50 inches

21. 35 meters $\times$ 20 meters or 15 meters $\times$ 46.67 meters

23. 73.48 feet   **25.** 8%   **27.** 6%   **29.** 2.59%

31. 3 seconds   **33.** 9.5 seconds

35. (a) 2.5 seconds   (b) 3.42 seconds

37. 30 units   **39.** 89.2 miles   **41.** Answers will vary.

43. Dollars   **45.** $\dfrac{9}{2}$ **47.** $-\dfrac{5}{3}, 7$ **49.** $-15, 7$

51. $-3, 8$ **53.** $\pm\dfrac{\sqrt{10}}{2}$

55. Adult tickets: 405; Child tickets: 49

Section 6.5   (page 408)

1. b   **2.** c   **3.** e   **4.** f   **5.** d   **6.** a

7. Upward, $(0, 2)$ **9.** Downward, $(-2, -3)$

11. Downward, $(10, 4)$ **13.** Upward, $(0, -6)$

15. Downward, $(3, 0)$

17. $f(x) = (x - 0)^2 + 2, \ (0, 2)$

19. $y = (x - 2)^2 + 3, \ (2, 3)$

21. $h(x) = (x + 3)^2 - 4, \ (-3, -4)$

23. $y = -(x - 3)^2 - 1, \ (3, -1)$

25. $f(x) = -(x - 1)^2 - 6, \ (1, -6)$

27. $y = 2\left(x + \dfrac{3}{2}\right)^2 - \dfrac{5}{2}, \ \left(-\dfrac{3}{2}, -\dfrac{5}{2}\right)$

29. $(5, -25)$ **31.** $(0, -10)$ **33.** $(3, 10)$

35. $\left(\dfrac{3}{4}, \dfrac{47}{8}\right)$ **37.** $\left(\dfrac{1}{8}, \dfrac{17}{16}\right)$ **39.** $\left(\dfrac{1}{2}, \dfrac{31}{4}\right)$

41.   **43.**

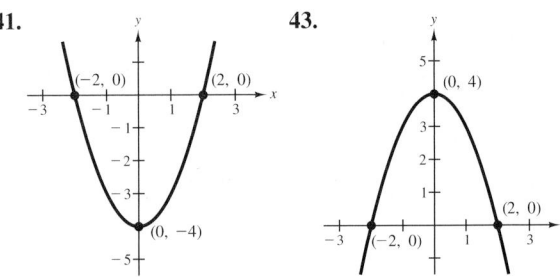

45.   **47.**

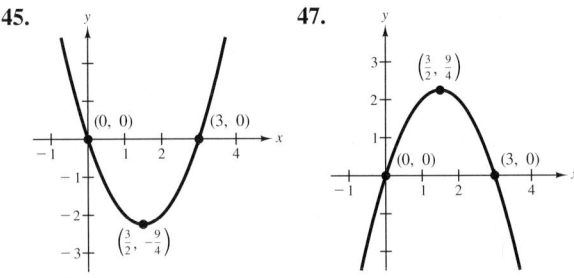

49.   **51.**

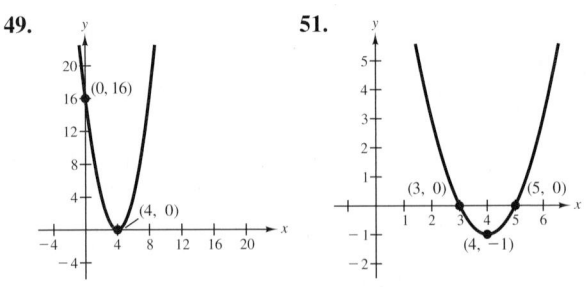

53.

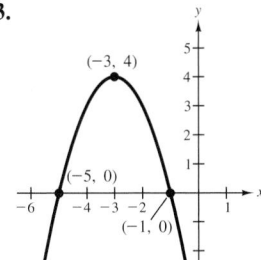

55.

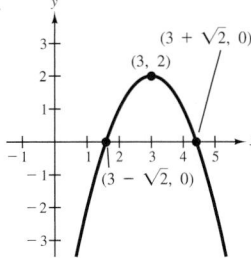

57.

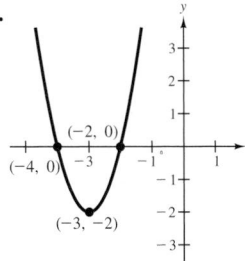

59.

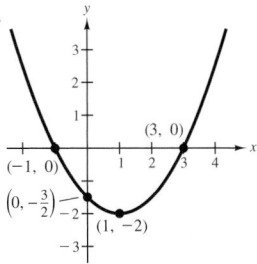

61.

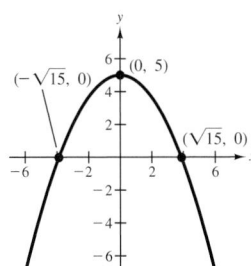

63.

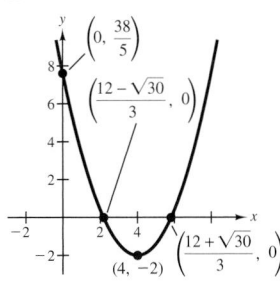

65.

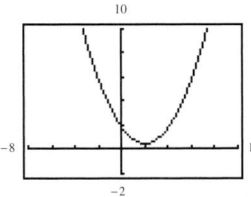

Vertex: $(2, 0.5)$

67.

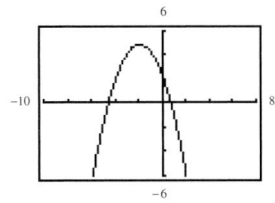

Vertex: $(-1.9, 4.9)$

69.

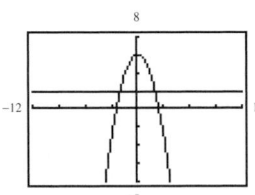

$(\pm 2, 2)$

71.

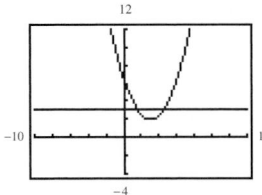

$(4.4, 3), (1.6, 3)$

73. $y = x^2 - 4x + 4$ **75.** $y = -x^2 - 4x$
77. $y = -2x^2 - 12x - 15$ **79.** $y = x^2 - 4x + 5$
81. $y = -x^2 - 6x - 5$ **83.** $y = x^2 - 4x$

85. $y = \frac{1}{2}x^2 - 3x + \frac{13}{2}$ **87.** $y = -4x^2 - 8x + 1$
89. $y = \frac{1}{25}x^2 - \frac{2}{5}x + 3$

91. (a)

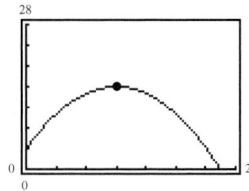

(b) 4 feet   (c) 16 feet   (d) ≈ 25.9 feet

93.

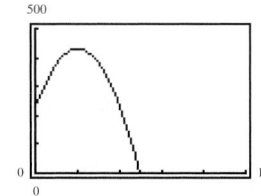

$2000

95. (a)

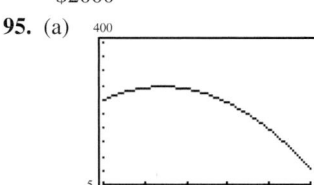

(b) 1996;
$275,804,000,000

97. $y = \frac{1}{2500}x^2$
99. If $a > 0$, the graph of $f(x) = ax^2 + bx + c$ opens upward, and if $a < 0$, it opens downward.

101.

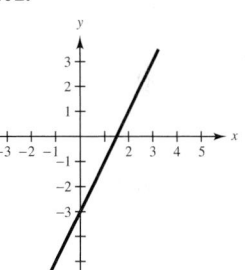

103.

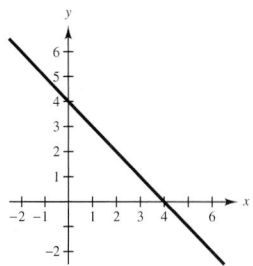

105. $-\frac{4}{3}$ **107.** 0, 4   **109.** $-6x^2 - 60x - 138$
111. 9 centimeters, 4 centimeters

Section 6.6   (page 419)

1. $\pm\frac{9}{2}$ **3.** $0, \frac{5}{2}$ **5.** 1, 3   **7.** $\frac{5}{2}$
9. Negative: $(-\infty, 4)$ **11.** Negative: $(0, 4)$
 Positive: $(4, \infty)$ Positive: $(-\infty, 0) \cup (4, \infty)$
13. Negative: $(-3, 3)$
 Positive: $(-\infty, -3) \cup (3, \infty)$
15. Negative: $(-1, 5)$
 Positive: $(-\infty, -1) \cup (5, \infty)$

17. $[-3, \infty)$ **19.** $(8, \infty)$

21. $(0, 2)$ **23.** $(-\infty, 0) \cup (2, \infty)$

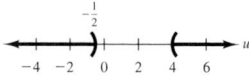

25. $(-\infty, -2] \cup [2, \infty)$ **27.** $[-5, 2]$

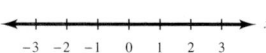

29. $\left(-\infty, -\frac{1}{2}\right) \cup (4, \infty)$ **31.** No solution

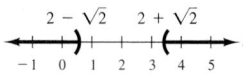

33. $(-\infty, \infty)$

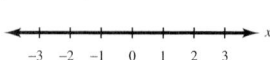

35. $\left(-\infty, 2 - \sqrt{2}\right) \cup \left(2 + \sqrt{2}, \infty\right)$ **37.** No solution

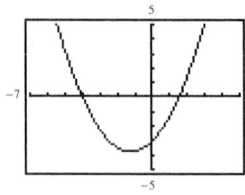

39. $\left(-\infty, 5 - \sqrt{6}\right) \cup \left(5 + \sqrt{6}, \infty\right)$

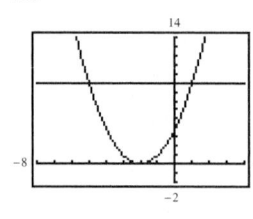

41. $(-\infty, \infty)$

43.   **45.**

$(-\infty, -4) \cup \left(\frac{3}{2}, \infty\right)$ $(-\infty, -5] \cup [1, \infty)$

47.

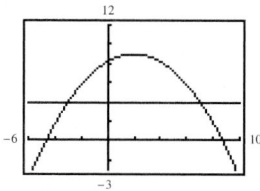

$(-\infty, -3) \cup (7, \infty)$

49. $(3, 5)$ **51.** $r > 7.24\%$

53. $\left[25 - 5\sqrt{5}, 25 + 5\sqrt{5}\right]$ **55.** $x^2 + 2x + 6 < 0$

57. $(-\infty, -5]$ **59.** $\left(-\frac{1}{3}, \infty\right)$

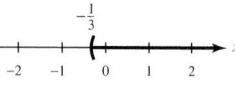

61. $(-\infty, -17) \cup (1, \infty)$ **63.** $4x^2 - 1$

65. $x^3 + 15x^2 + 65x + 75$

67. (a) $|s - 4.25| \le 0.01$ (b) $4.24 \le s \le 4.26$

Review Exercises   (page 422)

1. $0, -12$ **3.** $-10, \frac{8}{3}$ **5.** $\pm\frac{1}{2}$ **7.** $-\frac{5}{2}$

9. $-4, 5$ **11.** $-9, 10$ **13.** ±100 **15.** $\pm2\sqrt{2}$

17. $-4, 36$ **19.** $\pm2\sqrt{7}i$ **21.** $\frac{3}{2} \pm \frac{5\sqrt{3}i}{2}$

23. $-6 \pm 2\sqrt{5}i$ **25.** $\pm\sqrt{5}, \pm i$ **27.** 1

29. $-243, -1$ **31.** $-\frac{2}{3}, 3$ **33.** 9 **35.** $\frac{49}{4}$

37. $\frac{9}{64}$ **39.** $3 \pm 2\sqrt{3}$ **41.** $\frac{3}{2} \pm \frac{\sqrt{3}}{2}i$

43. $\frac{-5 \pm \sqrt{19}}{2}$ **45.** $-\frac{1}{2}, -1$ **47.** $-6, 5$

49. $-\frac{7}{2}, 3$ **51.** $-7, 12$ **53.** $\frac{10}{3} \pm \frac{5\sqrt{2}}{3}i$

55. Two distinct rational solutions

57. Two distinct imaginary solutions

59. Two distinct irrational solutions

61. Upward; $(3, 6)$ **63.** Downward; $(-4, 0)$

65. $(-1, -3)$ **67.** $\left(\frac{7}{2}, \frac{65}{4}\right)$

69.   **71.**

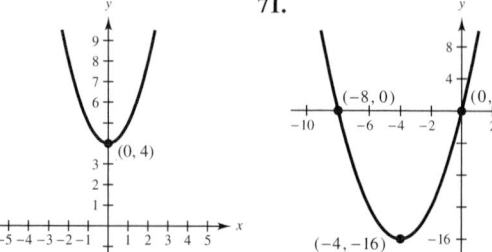

73.

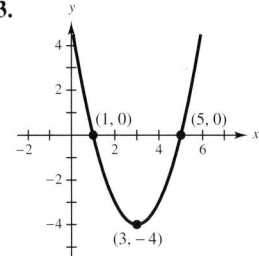

75. $y = -2x^2 + 12x - 13$

77. $y = \frac{1}{16}x^2 - \frac{5}{8}x + \frac{25}{16}$

79. $(-\infty, 12)$ Positive
  $(12, \infty)$ Negative

81. $(-\infty, -1)$ Positive
  $(-1, 0)$ Negative
  $(0, \infty)$ Positive

83. $(-\infty, 3)$ Positive
  $(3, 7)$ Negative
  $(7, \infty)$ Positive

85. $(-\infty, 3)$

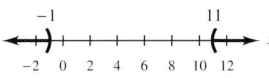

87. $(0, 7)$

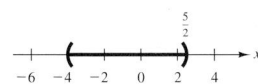

89. $(-\infty, -1) \cup (11, \infty)$

91. $\left(-4, \frac{5}{2}\right)$

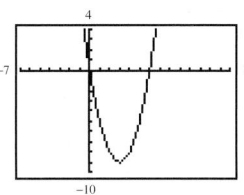

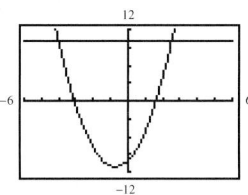

93.

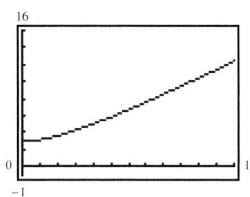

$(0, 6)$

95.

$\left[-4, \frac{5}{2}\right]$

97. (a) 200 feet
  (b) Dropped. The coefficient of the first-degree term is 0.
  (c) $\dfrac{5\sqrt{2}}{2} \approx 3.54$ seconds

99. Base: 60 centimeters; Height: 100 centimeters

101. (a) $d = \sqrt{9 + h^2}$

$h \approx 2.6$ miles

(b)

h	1	2	3	4	5	6	7
d	3.16	3.61	4.24	5	5.83	6.71	7.62

  $h \approx 2.5$ miles
  (c) $h = \sqrt{7} \approx 2.646$ miles

103. (a)

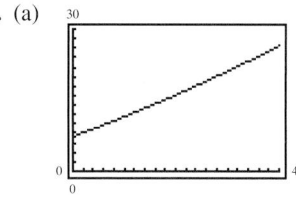

(b) and (c) 1987

105. 150 feet $\times$ 170 feet

Chapter 6 Test   (page 425)

1. $-5, 10$ **2.** $-\frac{3}{8}, 3$ **3.** $2 \pm 5\sqrt{2}$

4. $-3 \pm 9i$ **5.** $\dfrac{9}{4}$ **6.** $\dfrac{3 \pm \sqrt{3}}{2}$

7. -56; If the discriminant is a perfect square, there are two distinct rational solutions. If the discriminant is a positive nonperfect square, there are two distinct irrational solutions. If the discriminant is zero, there is one repeated rational solution. If the discriminant is a negative number, there are two distinct imaginary solutions.

8. $\dfrac{4 \pm \sqrt{7}}{3}$ **9.** $\dfrac{2 \pm 3\sqrt{2}}{2}$ **10.** $8, 64$

11.

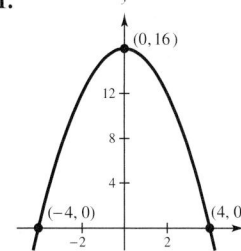

12.

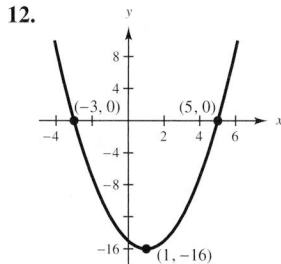

13. $(0, 3)$

14. $(-\infty, -2] \cup [6, \infty)$

15. $y = \frac{2}{3}x^2 - 4x + 4$ **16.** 12 feet $\times$ 20 feet

17. $\dfrac{\sqrt{10}}{2} \approx 1.58$ seconds

18. $x = 50$

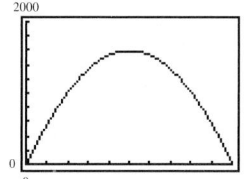

Cumulative Test: Chapters 1–6   (page 426)

1. $(2x + 7)(x - 1)$ **2.** $(11x - 5)(x + 1)$

3. $3x(2x + 3)(2x - 3)$ **4.** $(2x + 5)(4x^2 - 10x + 25)$

5. $\frac{5}{2}$ **6.** $0, 8$ **7.** $-2, \pm 1$ **8.** 5 **9.** $\frac{19}{3}$

10. 22 **11.** $y = \frac{5}{2}x - 12$ **12.** $y = -x + 3$

13. No. For some values of x there correspond two values of y.

14. $A = \frac{5}{2}x(2x + 9)$ **15.** Buy now.   **16.** $(2, 8)$

17. 64 liters   **18.** 32 **19.** -60

20. (a) $5x\sqrt{3xy}$ (b) $10(3 + \sqrt{5})$

21. (a) $-22 - 14i$ (b) 34 **22.** $y = \frac{1}{4}x^2 - x$

Chapter 7

Section 7.1   (page 435)

1. $(-\infty, 8) \cup (8, \infty)$ **3.** $(-\infty, -4) \cup (-4, \infty)$

5. $(-\infty, \infty)$ **7.** $(-\infty, -4) \cup (-4, 4) \cup (4, \infty)$

9. $(-\infty, -1) \cup (-1, 5) \cup (5, \infty)$ **11.** $(-\infty, \infty)$

13. $(-\infty, 0) \cup (0, 3) \cup (3, \infty)$ **15.** $\frac{8y}{3}$

17. $\frac{6x}{5y^3}, x \neq 0$ **19.** $x, x \neq 8, x \neq 0$

21. $2 + y, y \neq 0$ **23.** $\frac{1}{2}, x \neq \frac{3}{2}$

25. $-\frac{9 + y}{2}, y \neq 9$ **27.** $x - 5z, x \neq -5z$

29. $u - 6, u \neq 6$ **31.** $\frac{x + 4}{3}, x \neq 5$

33. $\frac{z + 11}{3}, z \neq -11$ **35.** $\frac{y(y + 2)}{y + 6}, y \neq 2$

37. $-\frac{1}{2x + 3}, x \neq 3$ **39.** $-\frac{x + 8}{x + 2}, x \neq 5$

41. $\frac{8z - 5}{7z - 4}, z \neq -\frac{4}{7}$ **43.** $\frac{1 + 4xy}{3}, x \neq 0$

45. $\frac{5 + 3xy}{y^2}, x \neq 0$ **47.** $\frac{3(m - 2n)}{m + 2n}$

49. $\frac{x^2 - 3xz + 9z^2}{x - 2z}, x \neq -3z$ **51.** $\frac{x}{x - 4y}, x \neq -4y$

53. $\frac{n + 3}{m + n}, m \neq n$

55. (a) 1 (b) -8 (c) Undefined   (d) 0

57. (a) $\frac{25}{22}$ (b) 0 (c) Undefined   (d) Undefined

59. Invalid; cannot divide out common terms, only common factors

61. Valid   **63.** $x^n - 3, x \neq 0$ **65.** $x^n - 2, x^n \neq -2$

67. $(0, \infty)$ **69.** $(0, \infty)$ **71.** $\frac{x}{x + 3}, x > 0$

73. $\frac{1}{4}, x > 0$ **75.** $\pi, x > 0$

77. $\frac{C}{P} = \frac{1000(111.5 + 13.91t)}{31.9 + 0.33t}$

79. The rational expression is in reduced form if the numerator and denominator have no common factors other than ± 1.

81. You can divide out only common factors.

83. $\frac{27}{31}$ **85.** $\frac{14}{25}$ **87.** $8y(x - 5)(x + 2)$

89. 11% solution: 240 milliliters; 6% solution: 360 milliliters

Section 7.2   (page 443)

1. $24u^2, u \neq 0$ **3.** $\frac{s^3}{6}, s \neq 0$ **5.** $-\frac{8x^5}{21y^4}, x \neq 0$

7. $\frac{23}{30}, x \neq \frac{1}{3}$ **9.** $1, x \neq -25$ **11.** $-1, r \neq 12$

13. $\frac{2(r + 2)}{11r}, r \neq 2$ **15.** $-\frac{x + 8}{x^2}, x \neq \frac{3}{2}$

17. $\frac{x + 3}{2x + 3}, x \neq 5, x \neq -5$

19. $\frac{2(x + 6)(x - 2)}{3x}, x \neq 1, x \neq -4$

21. $2t + 5, t \neq 3, t \neq -2$

23. $\frac{x(3x + 7)}{2x + 3}, x \neq -4, x \neq -1$

25. $\frac{2(x + 2)}{3(x + 1)}, x \neq 1, x \neq 3, x \neq 6$ **27.** $\frac{xy(x + 2y)}{(x - 2y)}$

29. $(u - 2v)(u + 2v), u \neq 2v$

31. $\frac{(x - 1)(2x + 1)}{(3x - 2)(x + 2)}, x \neq 5, x \neq -5, x \neq -1$

33. $\frac{(x + 3)^2}{x}, x \neq 3, x \neq -2, x \neq 2$ **35.** $\frac{3y^2}{2ux^2}, v \neq 0$

37. $\frac{3}{2(a + b)}$ **39.** $x^4y(x + 2y), x \neq 0, y \neq 0, x \neq -2y$

41. $\frac{3x}{10}, x \neq 0$ **43.** $-\frac{5x}{2}, x \neq 0, x \neq 5$

45. $\frac{(x + 3)(4x + 1)}{(3x - 1)(x - 1)}, x \neq -3, x \neq -\frac{1}{4}$

47. $1, y \neq -3, y \neq 0, y \neq 2$

49. $\frac{(x + 2)(x + 3)}{x}, x \neq -2, x \neq 2$

51. $\frac{(x - 2)(x + 1)}{3}, x \neq -2, x \neq -1, x \neq 7$

53. $\frac{2x - 3}{x + 3}, x \neq -5, x \neq \frac{3}{2}, x \neq 2, x \neq 6$

55. $\frac{x + 4}{3}, x \neq -2, x \neq 0$ **57.** $\frac{1}{4}, x \neq -1, x \neq 0, y \neq 0$

59. $\frac{(x + 1)(2x - 5)}{x}, x \neq -1, x \neq -5, x \neq -\frac{2}{3}$

61. $\frac{x^4}{(x^n + 1)^2}, x^n \neq -3, x^n \neq 3, x \neq 0$

63.

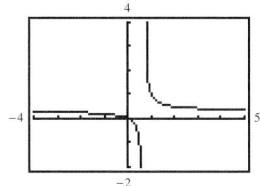

65. (a)

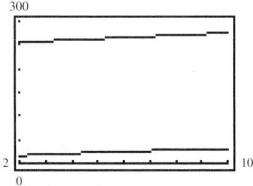

(b) Spending = $\dfrac{(-0.114t^2 + 3.09t + 10.5)(1000)}{2.47t + 250.7}$

(c)

Year, t	3	4	5
Amount per person	$72.62	$80.73	$87.82

Year, t	6	7	8
Amount per person	$93.91	$99.05	$103.25

(d) Answers will vary.

67. Multiply the rational expression by the reciprocal of the polynomial.

69. Invert the divisor, not the dividend.　　**71.** $\frac{5}{12}$

73. $\frac{32}{375}$　　**75.** $\frac{2}{5}$　　**77.** $(3x + 8)(x - 1)$

79. $(3x - 4)^2$　　**81.** 45 miles

Section 7.3　(page 452)

1. $\dfrac{3x}{2}$　　**3.** $-\dfrac{2}{9}$　　**5.** $\dfrac{5 + 2z}{5z}$　　**7.** $-\dfrac{4}{3}$

9. $\dfrac{1}{x - 3}$, $x \neq 0$　　**11.** 1, $y \neq 6$　　**13.** $\dfrac{x - 1}{x + 3}$, $x \neq 5$

15. $x + 5$, $x \neq -3$　　**17.** $20x^3$　　**19.** $15x^2(x + 5)$

21. $36y^3(y + 1)^2$　　**23.** $6x(x + 2)(x - 2)$

25. $56t(t + 2)(t - 4)$　　**27.** x^2　　**29.** $(u + 1)$

31. $-(x + 2)$　　**33.** $\dfrac{2n^2(n + 8)}{6n^2(n - 4)}$, $\dfrac{10(n - 4)}{6n^2(n - 4)}$

35. $\dfrac{2(x + 3)}{x^2(x + 3)(x - 3)}$, $\dfrac{5x(x - 3)}{x^2(x + 3)(x - 3)}$

37. $\dfrac{8s^2(s - 1)}{s(s - 1)(s + 2)^2}$, $\dfrac{3(s + 2)}{s(s - 1)(s + 2)^2}$

39. $\dfrac{(x - 8)(x - 5)}{(x + 5)(x - 5)^2}$, $\dfrac{9x(x + 5)}{(x + 5)(x - 5)^2}$

41. $\dfrac{7(a + 2)}{a^2}$　　**43.** $\dfrac{3(x + 2)}{x - 8}$　　**45.** 1, $x \neq \dfrac{2}{3}$

47. $\dfrac{5(5x + 22)}{x + 4}$　　**49.** $\dfrac{1}{2x(x - 3)}$　　**51.** $\dfrac{x^2 - 7x - 15}{(x + 3)(x - 2)}$

53. $\dfrac{5(x + y)}{(x + 5y)(x - 5y)}$　　**55.** $\dfrac{4}{x^2(x^2 + 1)}$

57. $\dfrac{x^2 + 3x + 9}{x(x - 3)(x + 3)}$　　**59.** $\dfrac{4x}{(x - 4)^2}$

61. $\dfrac{y - x}{xy}$, $x \neq -y$　　**63.** $\dfrac{5u - 2v}{(u - v)^2}$

65. $\dfrac{-2x^2 + 14x - 3}{(x - 3)(x + 3)(x + 4)}$　　**67.** $\dfrac{2(4x^2 + 5x - 3)}{x^2(x + 3)}$

69. 0, $x \neq 0$, $y \neq 0$　　**71.** $\dfrac{6x + 5}{2(x + 10)(x + 5)}$

73. $\dfrac{3x^2 - 7x - 13}{(x - 3)(x + 3)(x - 4)(x + 4)}$　　**75.** $2x$, $x \neq 0$

77. $3xy$, $x \neq 0$, $y \neq 0$　　**79.** $\dfrac{x}{2(3x + 1)}$, $x \neq 0$

81. $\dfrac{4 + 3x}{4 - 3x}$, $x \neq 0$　　**83.** $-4x - 1$, $x \neq 0$, $x \neq \frac{1}{4}$

85. $\dfrac{3}{4}$, $x \neq 0$, $x \neq 3$　　**87.** $\dfrac{y + 1}{y - 3}$, $y \neq 0$, $y \neq 1$

89. $y - x$, $x \neq 0$, $y \neq 0$, $x \neq -y$　　**91.** $\dfrac{5(x + 3)}{2x(5x - 2)}$

93. $\dfrac{x(x + 6)}{3x^3 + 10x - 30}$, $x \neq 0$, $x \neq 3$　　**95.** $\dfrac{v^2}{uv^2 + 1}$, $v \neq 0$

97. $\dfrac{ab}{b - a}$, $a \neq 0$, $b \neq 0$　　**99.** $-\dfrac{1}{2(2 + h)}$　　**101.** $\dfrac{5x}{24}$

103. $x_1 = \dfrac{x}{4}$, $x_2 = \dfrac{x}{3}$, $x_3 = \dfrac{5x}{12}$　　**105.** $\dfrac{5t}{12}$　　**107.** $\dfrac{R_1 R_2}{R_1 + R_2}$

109. $A = -4$, $B = 2$, $C = 2$;

$$\dfrac{4}{x^3 - x} = -\dfrac{4}{x} + \dfrac{2}{x + 1} + \dfrac{2}{x - 1}$$

111. Yes, $\dfrac{3}{2(x + 2)} + \dfrac{x}{x + 2}$　　**113.** $42x^2 - 60x$

115. $121 - x^2$　　**117.** $x^2 + 2x + 1$　　**119.** $x^3 - 8$

121. Perimeter: $12x + 6$; Area: $5x^2 + 9x$　　**123.** 98

Section 7.4　(page 463)

1. $\frac{5}{2}z^2 + z - 3$　　**3.** $7x^2 - 2x$, $x \neq 0$

5. $-10z^2 - 6$, $z \neq 0$　　**7.** $4z^2 + \frac{3}{2}z - 1$, $z \neq 0$

9. $-m^3 - 2m + \dfrac{7}{m}$　　**11.** $\dfrac{5}{2}x - 4 + \dfrac{7}{2}y$, $x \neq 0$, $y \neq 0$

13. $x - 5$, $x \neq 3$　　**15.** $x + 10$, $x \neq -5$

17. $x + 7$, $x \neq 3$　　**19.** $y + 3$, $y \neq -\frac{1}{2}$

21. $4x - 1$, $x \neq -\frac{1}{4}$　　**23.** $x^2 - 5x + 25$, $x \neq -5$

25. $x^2 + 4$, $x \neq 2$　　**27.** $x^2 - 2$, $x \neq 5$

29. $2 + \dfrac{5}{x + 2}$　　**31.** $x - 4 + \dfrac{32}{x + 4}$

33. $5x - 8 + \dfrac{19}{x + 2}$ **35.** $4x + 3 - \dfrac{11}{3x + 2}$

37. $5x^2 - 9x + 10 - \dfrac{10}{x + 3}$ **39.** $2x^2 + x + 4 + \dfrac{6}{x - 3}$

41. $x^5 + x^4 + x^3 + x^2 + x + 1, \ x \neq 1$

43. $x^2 + x + 1 + \dfrac{1}{x - 1}$ **45.** $x^3 - x + \dfrac{x}{x^2 + 1}$

47. $x + 2$ **49.** $x^2 - 2 + \dfrac{2x - 28}{3x^2 + x + 1}$

51. $x + 4 + \dfrac{6x - 3}{x^2 + x - 4}$

53. $x^2 - 3x + 17 + \dfrac{75 - 72x}{x^2 + 3x - 5}$

55. $x^{2n} + x^n + 4, \ x^n \neq -2$ **57.** $x^2 - x + 4 - \dfrac{17}{x + 4}$

59. $x^3 - 2x^2 - 4x - 7 - \dfrac{4}{x - 2}$

61. $5x^2 - 25x + 125 - \dfrac{613}{x + 5}$ **63.** $12 - \dfrac{49}{x + 2}$

65. $5x^2 + 14x + 56 + \dfrac{232}{x - 4}$

67. $10x^3 + 10x^2 + 60x + 360 + \dfrac{1360}{x - 6}$

69. $0.1x + 0.82 + \dfrac{1.164}{x - 0.2}$ **71.** $(x - 3)(x - 7)(x + 2)$

73. $(3z - 2)(z - 2)(z - 4)$ **75.** $(5t^2 + 3t + 4)(t - 6)$

77. $(x + 2)(x^2 + 4)(x - 2)$ **79.** $5(3x + 2)\left(x - \frac{4}{5}\right)$

81.

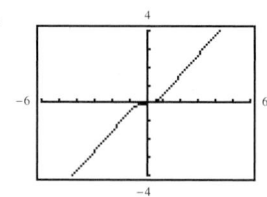

83. $x^3 - 5x^2 - 5x - 10$ **85.** -8 **87.** $x^2 - 3$

89. $2x + 8$

91. The x's cannot be divided out because they are terms, not factors.

93. Yes, if $\dfrac{n(x)}{d(x)} = q(x)$, then $n(x) = d(x) \cdot q(x)$.

95. $\dfrac{4t}{3}, \ t \neq 0$ **97.** $\dfrac{2xy}{5}, \ x \neq 0, y \neq 0$ **99.** $\dfrac{3}{4}$

101. $\pm\dfrac{5}{2}$ **103.** $-7, 6$

105. 2 hours 3 minutes; 1107 miles

Mid-Chapter Quiz   (page 466)

1. $(-\infty, 0) \cup (0, 4) \cup (4, \infty)$

2. (a) 0   (b) $\frac{9}{2}$ (c) Undefined   (d) $\frac{8}{9}$

3. $\dfrac{3}{2}y, \ y \neq 0$ **4.** $\dfrac{2u^2}{9v}, \ u \neq 0$ **5.** $-\dfrac{2x + 1}{x}, \ x \neq \dfrac{1}{2}$

6. $\dfrac{z + 3}{2z - 1}, \ z \neq -3$ **7.** $\dfrac{7 + 3ab}{a}, \ b \neq 0$

8. $\dfrac{n^2}{m + n}, \ n \neq 2m$ **9.** $\dfrac{t}{2}, \ t \neq 0$

10. $\dfrac{8x}{3(x - 1)^2(x + 3)}$

11. $-\dfrac{x + 3}{x(1 + x)}, \ x \neq -3, x \neq -2, x \neq 1$

12. $\dfrac{4(u - v)^2}{5uv}, \ u \neq v, u \neq -v$

13. $\dfrac{2(x + 1)}{3x}, \ x \neq -2, x \neq -1$ **14.** $\dfrac{-3x^2 + x}{4(x + 5)}$

15. $\dfrac{4(x - 2y)^2}{(x - 3y)(x + 2y)(x - y)}$ **16.** $\dfrac{5(2 - x)}{4x - 15}, \ x \neq 0$

17. $2x^2 - 4x + 3, \ x \neq \frac{2}{3}$ **18.** $3x^2 - 8x + 5, \ x \neq -4$

19. (a) $\dfrac{6000 + 10.50x}{x}$ (b) \$22.50   **20.** $\dfrac{8(x + 2)}{(x + 4)^2}$

Section 7.5   (page 474)

1. (a) Not a solution   (b) Not a solution
(c) Not a solution   (d) Solution

3. (a) Not a solution   (b) Solution
(c) Solution   (d) Not a solution

5. $\frac{3}{2}$ **7.** 12   **9.** 10   **11.** $\frac{7}{4}$ **13.** 61   **15.** $\frac{18}{5}$

17. 3   **19.** 3   **21.** $-\frac{11}{5}$ **23.** $\frac{4}{3}$ **25.** ± 6

27. ± 4 **29.** $8, -9$ **31.** $8, -3$ **33.** No solution

35. $\frac{3}{4}$ **37.** No solution   **39.** 3   **41.** 20

43. $-1, -\frac{2}{3}$ **45.** $-\frac{4}{15}$ **47.** 3, 13   **49.** 0, 2

51. $-1, 2$ **53.** -5 **55.** 2, 3   **57.** No solution

59. $8, \frac{1}{8}$ **61.** $12, \frac{1}{8}$ **63.** 40 miles per hour

65. 50 miles per hour, 60 miles per hour

67. 8 miles per hour, 10 miles per hour   **69.** 10 people

71. $2\frac{2}{5}$ hours   **73.** $7\frac{7}{8}$ hours   **75.** $11\frac{1}{4}$ hours

77. 3000 units

79. It is an extra solution found by multiplying both sides of the original equation by an expression containing the variable. It is identified by checking all solutions in the original equation.

81. $\frac{7}{4}, -\frac{2}{3}$ **83.** $x < \frac{3}{2}$ **85.** $1 < x < 5$ **87.** \$3750

Section 7.6   (page 483)

1. (a)

x	0	0.5	0.9	0.99	0.999
y	-4	-8	-40	-400	-4000

x	2	1.5	1.1	1.01	1.001
y	4	8	40	400	4000

x	2	5	10	100	1000
y	4	1	0.4444	0.0404	0.0040

(b) Vertical asymptote: $x = 1$;
Horizontal asymptote: $y = 0$
(c) $(-\infty, 1) \cup (1, \infty)$

3. d   **4.** a   **5.** b   **6.** c

7. Domain: $(-\infty, 0) \cup (0, \infty)$; Horizontal asymptote: $y = 0$;
Vertical asymptote: $x = 0$

9. Domain: $(-\infty, -8) \cup (-8, \infty)$; Horizontal asymptote:
$y = 1$; Vertical asymptote: $x = -8$

11. Domain: $(-\infty, 0) \cup (0, 1) \cup (1, \infty)$; Horizontal asymptote: $y = 0$; Vertical asymptotes: $t = 0, t = 1$

13. Domain: $(-\infty, -3) \cup (-3, \infty)$; Horizontal asymptotes: none; Vertical asymptote: $s = -3$

15. Domain: $\left(-\infty, \frac{1}{3}\right), \cup \left(\frac{1}{3}, \infty\right)$; Horizontal asymptote: $y = \frac{5}{3}$; Vertical asymptote: $x = \frac{1}{3}$

17. Domain: $(-\infty, -1) \cup (-1, 1) \cup (1, \infty)$; Horizontal asymptote: $y = 5$; Vertical asymptotes: $x = -1, x = 1$

19. Domain: $(-\infty, -3) \cup (-3, 1) \cup (1, \infty)$; Horizontal asymptote: $y = 0$; Vertical asymptotes: $x = -3, x = 1$

21. d   **22.** b   **23.** a   **24.** c

25.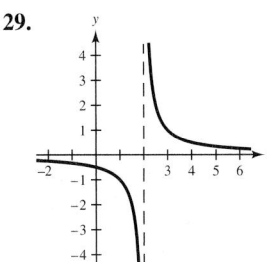

27.

29.

31.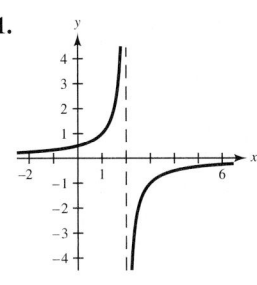

33.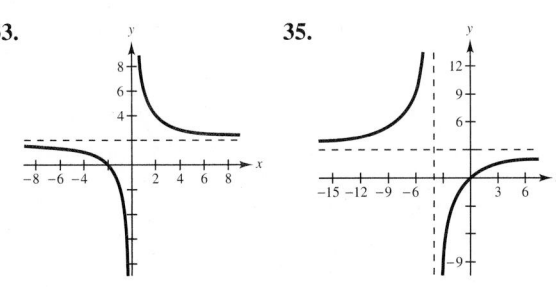

35.

37.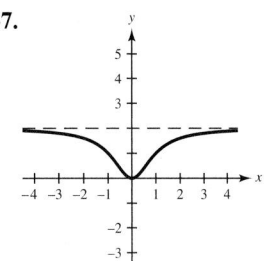

39.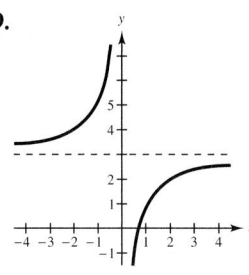

41.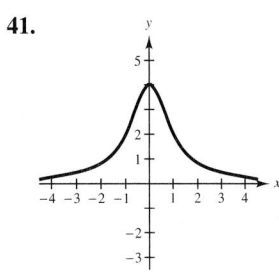

43.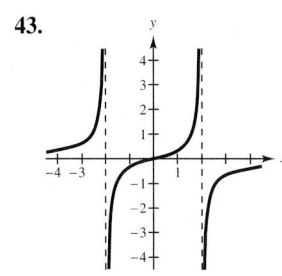

45.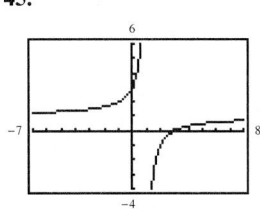

47.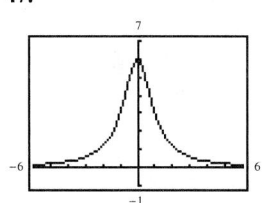

Domain: $(-\infty, 1) \cup (1, \infty)$ Domain: $(-\infty, \infty)$
Horizontal asymptote: $y = 1$ Horizontal asymptote: $y = 0$
Vertical asymptote: $x = 1$ Vertical asymptotes: none

49.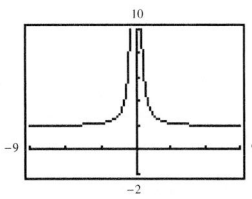

51.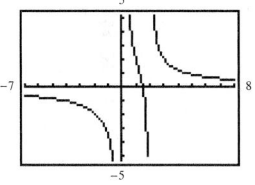

Domain: $(-\infty, 0) \cup (0, \infty)$ Domain:
Horizontal asymptote: $y = 2$ $(-\infty, 0) \cup (0, 2) \cup (2, \infty)$
Vertical asymptote: $x = 0$ Horizontal asymptote: $y = 0$
Vertical asymptotes:
$x = 0, x = 2$

53.

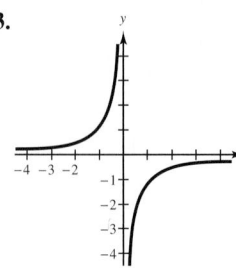

Reflection in the x-axis

55.

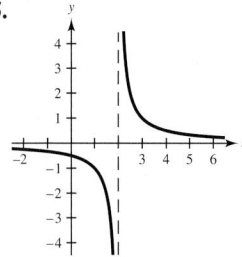

Horizontal shift two units to the right

57.

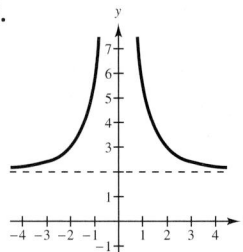

Vertical shift two units upward

59.

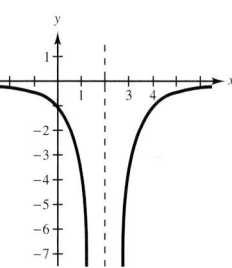

Reflection in the x-axis and a horizontal shift two units to the right

Domain: $(-\infty, -1), (-1, 0)$
Vertical asymptote: $x = -1$
$y = 2x - 1$

61.

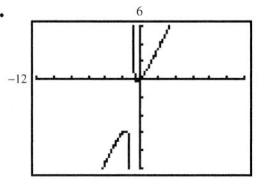

63. $(-2, 0)$ **65.** $(-1, 0), (1, 0)$

67. (a)

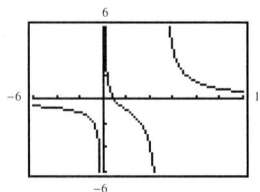

(b) $(1, 0)$

69. (a)

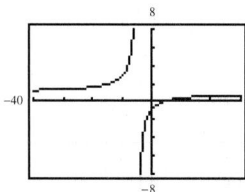

(b) $(4, 0)$

71. (a)

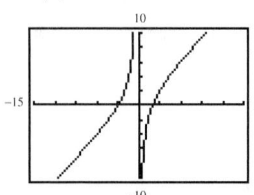

(b) $(-3, 0), (2, 0)$

73. (a)

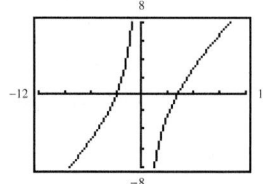

(b) $(-3, 0), (4, 0)$

75. (a) $C = 250 + 0.60x$ (b) $\overline{C} = \dfrac{250 + 0.60x}{x}$

(c) \$0.85; \$0.63

(d) $\overline{C} = \$0.60$; $\overline{C}$ will never be less than \$0.60.

(e)

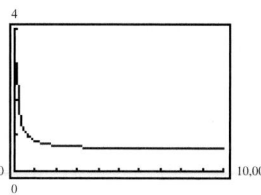

77. (a) $C = 0$. The chemical is eliminated from the body.

(b)

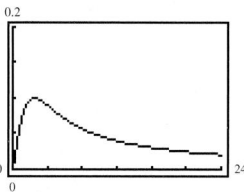

(c) $t \approx 2.5$ hours

79. (a) Answers will vary.   (b) $(0, \infty)$

(c)

(d) 20 units × 20 units

81. An asymptote of a graph is a line to which the graph becomes arbitrarily close as $|x|$ or $|y|$ increases without bound.

83. No. There are separate pieces of the graph on each side of the asymptote.

85.

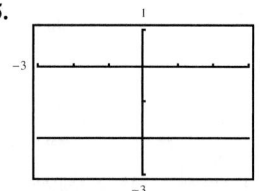

The fraction is not reduced to lowest terms.

87. $4x^2 - 60x + 225$ **89.** $x^2 - y^2 + 2x + 1$

91.

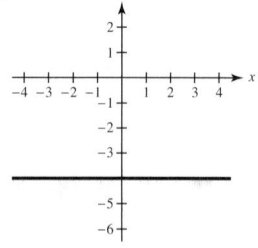

93.

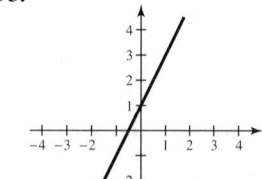

95. $4 - \dfrac{11}{x + 2}$ **97.** Base: 20 meters; Height: 8 meters

Section 7.7   (page 491)

1. (a) Not a solution   (b) Solution
(c) Not a solution   (d) Not a solution

3. 3   **5.** $0, -5$

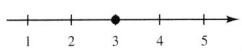

7. $(3, \infty)$ **9.** $(-\infty, 3)$

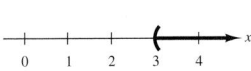

11. $(-\infty, 3)$ **13.** $[0, 4)$

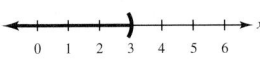

15. $(-\infty, 3] \cup \left(\frac{11}{2}, \infty\right)$ **17.** $(-1, 3)$

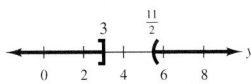

19. $(4, 7)$ **21.** $(-\infty, 3) \cup [6, \infty)$

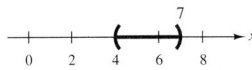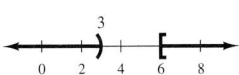

23. $(-\infty, 3) \cup (3, 5)$ **25.** $(-\infty, -1) \cup (0, 1)$

27. $(-\infty, -1) \cup (4, \infty)$ **29.** $\left(-5, \frac{5}{2}\right)$ **31.** $(3, 4)$

33.   **35.**

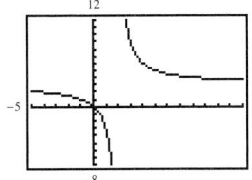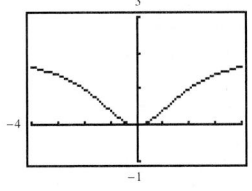

(a) $[0, 2)$ (b) $(2, 4]$ (a) $(-\infty, -2] \cup [2, \infty)$
(b) $(-\infty, \infty)$

37. $48 < r < 55;\ 58 < r + 10 < 65$

39. (a)

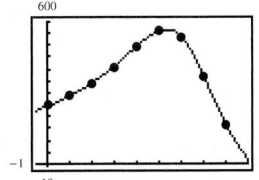

(b) $5.7 \le t \le 13.6$

41. Critical numbers are all real zeros of the numerator and all real zeros of the denominator of a rational expression. They are used to determine the test intervals of the rational expression, which are then used to solve the inequality.

43. $[-5, \infty)$ **45.** $\left(\frac{9}{38}, \infty\right)$ **47.** $(3, 9)$

49. $(-16, -2)$ **51.** $(-\infty, -7] \cup [0, \infty)$ **53.** 0.52 foot

Review Exercises   (page 494)

1. $(-\infty, 8) \cup (8, \infty)$ **3.** $(-\infty, 1) \cup (1, 6) \cup (6, \infty)$

5. $\dfrac{2x^3}{5},\ x \ne 0, y \ne 0$ **7.** $\dfrac{b - 3}{6(b - 4)}$ **9.** $-9,\ x \ne y$

11. $\dfrac{x}{2(x + 5)},\ x \ne 5$ **13.** $\dfrac{y}{8x},\ y \ne 0$

15. $12z(z - 6),\ z \ne -6$ **17.** $\dfrac{125y}{x},\ y \ne 0$

19. $\dfrac{x(x - 1)}{x - 7},\ x \ne -1, x \ne 1$ **21.** $3x^2,\ x \ne 0$

23. $\dfrac{6(x + 5)}{x(x + 7)},\ x \ne \pm 5$ **25.** $-\dfrac{7a}{9}$ **27.** $\dfrac{4 - x}{3x - 1}$

29. $\dfrac{-48x + 35}{48x}$ **31.** $\dfrac{4x + 3}{(x + 5)(x - 12)}$

33. $\dfrac{5x^3 - 5x^2 - 31x + 13}{(x + 2)(x - 3)}$ **35.** $\dfrac{x + 24}{x(x^2 + 4)}$

37. $\dfrac{6(x - 9)}{(x + 3)^2(x - 3)}$ **39.** $\dfrac{3t^2}{5t - 2},\ t \ne 0$

41. $\dfrac{-a^2 + a + 16}{(4a^2 + 16a + 1)(a - 4)},\ a \ne 0, a \ne -4$

43. $2x^2 - \dfrac{1}{2},\ x \ne 0$ **45.** $5a^3b^2 - 2ab + 1,\ a \ne 0, b \ne 0$

47. $2x^2 + x - 1 + \dfrac{1}{3x - 1}$ **49.** $x^2 - 2,\ x \ne 1, x \ne -1$

51. $x^3 + 3x^2 + 6x + 18 + \dfrac{29}{x - 3}$

53. $(x + 3)(x + 1)(x - 2)$ **55.** -40 **57.** 2

59. $-4, 6$ **61.** $-2, -4$ **63.** $-2, 2$ **65.** $-\frac{9}{5}, 3$

67.

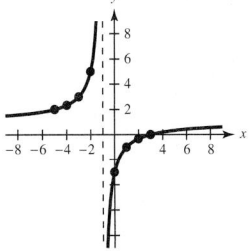

69. Horizontal asymptote: $y = 0$; Vertical asymptote: $x = 6$

71. Horizontal asymptote: $y = 6$; Vertical asymptote: $x = 5$

73. Domain: $(-\infty, -4) \cup (-4, \infty)$; Horizontal asymptote: $y = 1$; Vertical asymptote: $x = -4$

75. Domain: $(-\infty, -3) \cup (-3, 3) \cup (3, \infty)$; Horizontal asymptote: $y = 2$; Vertical asymptotes: $x = 3, x = -3$

77. 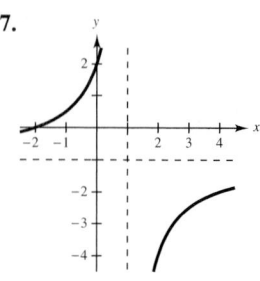 **79.**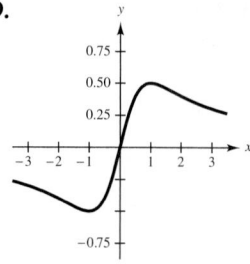

81. 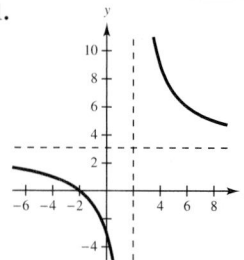 **83.**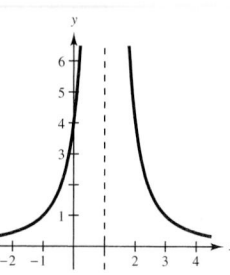

85. 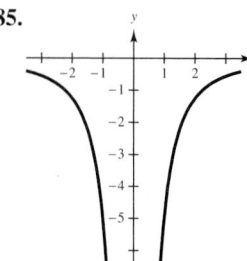 **87.**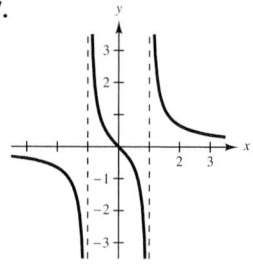

89. $(-4, 1)$ — Negative
$(-\infty, -4) \cup (1, \infty)$ — Positive

91. $\left(-\infty, -\frac{7}{2}\right) \cup (0, \infty)$ — Positive
$\left(-\frac{7}{2}, 0\right)$ — Negative

93. $(-\infty, 0] \cup \left(\frac{7}{2}, \infty\right)$ **95.** $(-6, -4)$

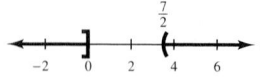

97. 25 **99.** $6\frac{2}{3}$ minutes

101. (a) $\overline{C} = \dfrac{C}{x} = \dfrac{0.5x + 500}{x}, \ x > 0$

(b) Horizontal asymptote: $\overline{C} = \frac{1}{2}$; As x gets larger, the average cost per unit approaches \$0.50.

103. (a) \$176 million; \$1584 million

(b) No. $p = 100$ is a vertical asymptote. (c) $[0, 74)$

Chapter 7 Test (page 497)

1. $(-\infty, -5) \cup (-5, 5) \cup (5, \infty)$ **2.** $-\frac{1}{3}, \ x \neq 2$

3. $\dfrac{4}{y + 4}, \ y \neq 2$ **4.** $\dfrac{(2x + 3)^2}{x + 1}, \ x \neq \dfrac{3}{2}$

5. $\dfrac{x}{(2x - 1)(x - 4)}, \ x \neq -4, x \neq 3$

6. $\dfrac{x^3}{4}, \ x \neq 0, x \neq -2$ **7.** $-(3x + 1), \ x \neq 0, \ x \neq \frac{1}{3}$

8. $\dfrac{-2x^2 + 2x + 1}{x + 1}$ **9.** $\dfrac{5x^2 - 15x - 2}{(x - 3)(x + 2)}$

10. $\dfrac{5x^3 + x^2 - 7x - 5}{x^2(x + 1)^2}$ **11.** $4a^5 + 7a^2 - a, \ a \neq 0$

12. $t^2 + 3 - \dfrac{6t - 6}{t^2 - 2}$ **13.** $2x^3 + 6x^2 + 3x + 9 + \dfrac{20}{x - 3}$

14. (a) (b)

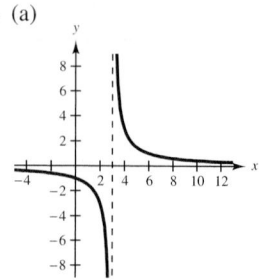 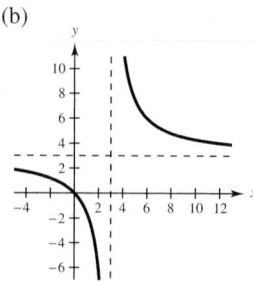

15. $-1, -\frac{15}{2}$ **16.** No solution

17. $(-\infty, -8) \cup [5, \infty)$ **18.** $[-8, 3)$

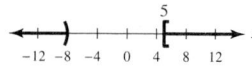 

19. $6\frac{2}{3}$ hours, 10 hours

Cumulative Test: Chapters 1–7 (page 498)

1. $-3x + 22$ **2.** $0.02x + 100$ **3.** $a^2 + 4a + 1$

4. $\dfrac{3}{a + 7}$ **5.** $-\dfrac{4x}{3}, \ x \neq 0$ **6.** $\dfrac{6u^4}{7v^6}, \ u \neq 0$

7. $x - 9$ **8.** $\sqrt{5u}$ **9.** $4t + 12\sqrt{t} + 9$

10. $(-3, 5)$ **11.** $(2, 4, 3)$ **12.** 2 **13.** 6

14. $5 \pm 5\sqrt{2}i$

15.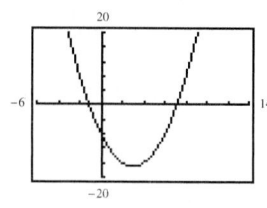

$x = 3 \pm \sqrt{17}$

16. $y = -\frac{3}{4}x^2 + 3x$

17.

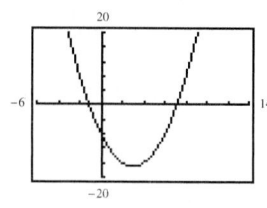

18. $\frac{1}{4}, \frac{1}{5}; \frac{20}{9}$ **19.** $h^2 - 3h$

Chapter 8

Section 8.1 (page 506)

1. (a) $(f + g)(x) = x^2 + 2x$;
  Domain = $\{x : x \text{ is a real number}\}$
(b) $(f - g)(x) = 2x - x^2$;
  Domain = $\{x : x \text{ is a real number}\}$
(c) $(fg)(x) = 2x^3$;
  Domain = $\{x : x \text{ is a real number}\}$
(d) $\left(\dfrac{f}{g}\right)(x) = \dfrac{2}{x}$; Domain = $\{x : x \neq 0\}$

3. (a) $(f + g)(x) = x^2 + 4x - 12$;
  Domain = $\{x : x \text{ is a real number}\}$
(b) $(f - g)(x) = -x^2 + 4x + 6$;
  Domain = $\{x : x \text{ is a real number}\}$
(c) $(fg)(x) = 4x^3 - 3x^2 - 36x + 27$;
  Domain = $\{x : x \text{ is a real number}\}$
(d) $\left(\dfrac{f}{g}\right)(x) = \dfrac{4x - 3}{x^2 - 9}$; Domain = $\{x : x \neq -3, 3\}$

5. (a) $(f + g)(x) = \dfrac{1}{x} + 5x$; Domain = $\{x : x \neq 0\}$
(b) $(f - g)(x) = \dfrac{1}{x} - 5x$; Domain = $\{x : x \neq 0\}$
(c) $(fg)(x) = 5$; Domain = $\{x : x \neq 0\}$
(d) $\left(\dfrac{f}{g}\right)(x) = \dfrac{1}{5x^2}$; Domain = $\{x : x \neq 0\}$

7. (a) $(f + g)(x) = \sqrt{x - 5} - \dfrac{3}{x}$; Domain = $\{x : x \geq 5\}$
(b) $(f - g)(x) = \sqrt{x - 5} + \dfrac{3}{x}$; Domain = $\{x : x \geq 5\}$
(c) $(fg)(x) = -\dfrac{3\sqrt{x - 5}}{x}$; Domain = $\{x : x \geq 5\}$
(d) $\left(\dfrac{f}{g}\right)(x) = -\dfrac{x\sqrt{x - 5}}{3}$; Domain = $\{x : x \geq 5\}$

9. (a) $(f + g)(x) = \sqrt{x + 4} + \sqrt{4 - x}$;
  Domain = $\{x : -4 \leq x \leq 4\}$
(b) $(f - g)(x) = \sqrt{x + 4} - \sqrt{4 - x}$;
  Domain = $\{x : -4 \leq x \leq 4\}$
(c) $(fg)(x) = \sqrt{16 - x^2}$;
  Domain = $\{x : -4 \leq x \leq 4\}$
(d) $\left(\dfrac{f}{g}\right)(x) = \dfrac{\sqrt{x + 4}}{\sqrt{4 - x}}$;
  Domain = $\{x : -4 \leq x < 4\}$

11. (a) 11 (b) 0 **13.** (a) 10 (b) 12
15. (a) 19 (b) -79 **17.** (a) 10 (b) $\frac{2}{5}$
19. (a) $\frac{1}{3}$ (b) Undefined (division by zero)
21. (a) $x^2 - 3$ (b) $(x - 3)^2$ (c) 13 (d) 16
23. (a) $3|x - 1|$ (b) $3|x - 3|$ (c) 0 (d) 3
25. (a) $\sqrt{x + 5}$ (b) $\sqrt{x} + 5$ (c) 3 (d) 8

27. (a) $\dfrac{1}{\sqrt{x - 3}}$ (b) $\dfrac{1}{\sqrt{x - 3}}$ (c) $\dfrac{1}{4}$ (d) $\dfrac{1}{3}$
29. (a) $(f \circ g)(x) = 4x^2 + 1$;
  Domain = $\{x : x \text{ is a real number}\}$
(b) $(g \circ f)(x) = 2(x^2 + 1)$;
  Domain = $\{x : x \text{ is a real number}\}$
31. (a) $(f \circ g)(x) = \sqrt{x - 2}$; Domain = $\{x : x \geq 2\}$
(b) $(g \circ f)(x) = \sqrt{x} - 2$; Domain = $\{x : x \geq 0\}$
33. (a) $(f \circ g)(x) = \dfrac{9}{x^2 + 9}$;
  Domain = $\{x : x \text{ is a real number}\}$

(b) $(g \circ f)(x) = \left(\dfrac{9}{x + 9}\right)^2$; Domain = $\{x : x \neq -9\}$
35. $t^2 - 8t - 3$ **37.** $-\frac{2}{13}$ **39.** 5 **41.** 10
43. $5y^2 - 15y + 3$ **45.** $t^2 - t - 12$
47. $\dfrac{1}{t - 5}, t \neq -3$ **49.** 1 **51.** 3 **53.** $t^2 + 4t - 12$

55. **57.**

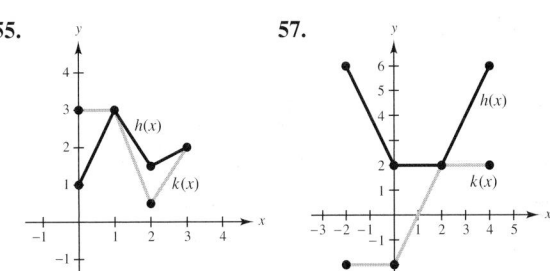

59. **61.**

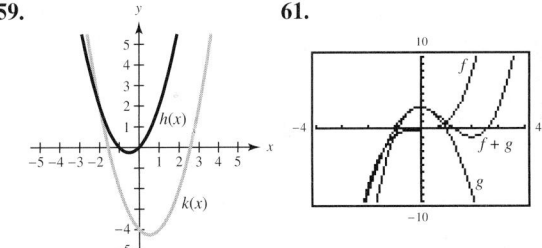

f is more significant for
large x.

63. (a) -1 (b) -2 (c) -2 **65.** (a) -1 (b) 1
67. (a) 10 (b) 1 (c) 1 **69.** (a) 0 (b) 10
71. (a) -2 (b) 4 (c) -2 **73.** (a) 10 (b) 7
75.

Domain = $\{x : x \text{ is a real number}\}$

77.

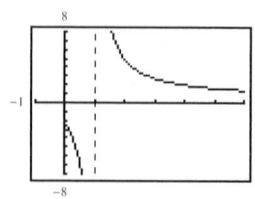

Domain $= \{x: x \neq 1\}$

79. (a) $T(x) = \frac{3}{4}x + \frac{1}{15}x^2$

(b)

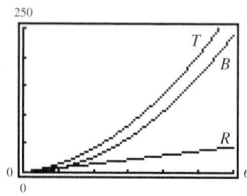

(c) B

81. (a) $f(g(x)) = 0.02x - 200{,}000$

(b) $g(f(x)) = 0.02(x - 200{,}000)$

The function $g(f(x))$ represents the bonus, because it gives 2% of sales over \$200,000.

83. $A(r(t)) = 0.36\pi t^2$

85. $(f \circ g)(2) = f(g(2))$
$= f(2^3 + 1)$
$= f(9)$
$= 2(9) - 1$
$= 17$

87. Domain $= \{x: x \text{ is a real number}\}$

89. Domain $= \{x: x \neq -3\}$

91.

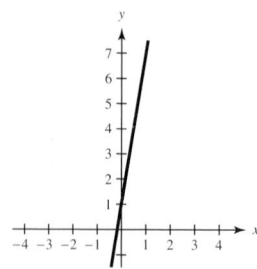

93.

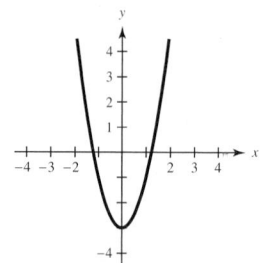

95.

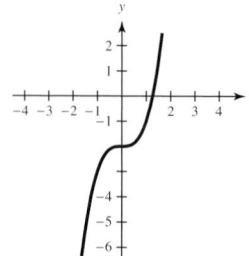

97. 12 centimeters, 36 centimeters, 36 centimeters

Section 8.2 (page 517)

1. $f^{-1}(x) = \dfrac{x}{5}$ **3.** $f^{-1}(x) = x - 10$

5. $f^{-1}(x) = 2x$ **7.** $f^{-1}(x) = \sqrt[3]{x}$ **9.** $f^{-1}(x) = x^3$

11. $f(g(x)) = f\left(\frac{1}{10}x\right) = 10\left(\frac{1}{10}x\right) = x$
$g(f(x)) = g(10x) = \frac{1}{10}(10x) = x$

13. $f(g(x)) = f(x - 15) = (x - 15) + 15 = x$
$g(f(x)) = g(x + 15) = (x + 15) - 15 = x$

15. $f(g(x)) = f\left(\dfrac{1 - x}{2}\right) = 1 - 2\left(\dfrac{1 - x}{2}\right)$
$= 1 - (1 - x) = x$
$g(f(x)) = g(1 - 2x) = \dfrac{1 - (1 - 2x)}{2} = \dfrac{2x}{2} = x$

17. $f(g(x)) = f\left[\frac{1}{3}(2 - x)\right] = 2 - 3\left[\frac{1}{3}(2 - x)\right]$
$= 2 - (2 - x) = x$
$g(f(x)) = g(2 - 3x) = \frac{1}{3}[2 - (2 - 3x)]$
$= \frac{1}{3}(3x) = x$

19. $f(g(x)) = f(x^3 - 1) = \sqrt[3]{(x^3 - 1) + 1} = \sqrt[3]{x^3} = x$
$g(f(x)) = g(\sqrt[3]{x + 1}) = (\sqrt[3]{x + 1})^3 - 1$
$= x + 1 - 1 = x$

21. $f(g(x)) = f\left(\dfrac{1}{x}\right) = \dfrac{1}{(1/x)} = x$
$g(f(x)) = g\left(\dfrac{1}{x}\right) = \dfrac{1}{(1/x)} = x$

23. $f^{-1}(x) = \dfrac{x}{8}$ **25.** $g^{-1}(x) = x - 25$

27. $g^{-1}(x) = \dfrac{3 - x}{4}$ **29.** $f^{-1}(x) = \dfrac{x - 1}{4}$

31. $f^{-1}(x) = -\dfrac{x - 5}{6}$ **33.** $g^{-1}(s) = \dfrac{s + 9}{2}$

35. $g^{-1}(t) = 4t - 8$ **37.** $h^{-1}(x) = x^2, \ x \geq 0$

39. $\dfrac{x^2 - 4}{4}, \ x \geq 0$ **41.** $h^{-1}(t) = \sqrt[3]{t - 1}$

43. $f^{-1}(t) = \sqrt[3]{t + 1}$ **45.** $g^{-1}(s) = \dfrac{5}{s}$

47. $g^{-1}(s) = \dfrac{-4s}{s + 3}$ **49.** $h^{-1}(x) = \dfrac{-5}{x - 1}$

51. b **52.** c **53.** d **54.** a

55.

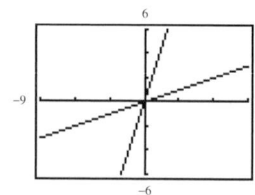

57.

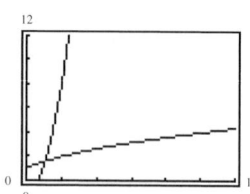

59.

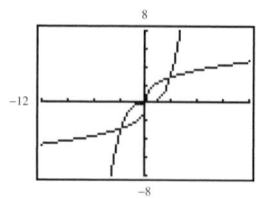

61.

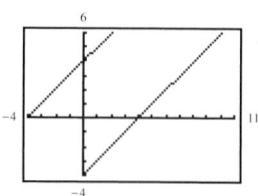

63. $f^{-1}(x) = \sqrt[3]{x - 1}$

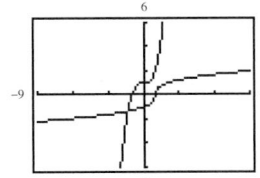

65. $f^{-1}(x) = \sqrt{x^2 + 4}, \; x \geq 0$

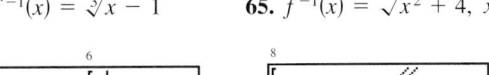

67. No **69.** No

71.

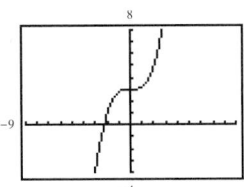

Yes

73.

Yes

75.

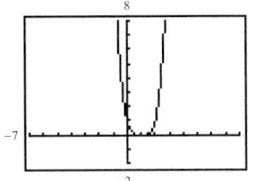

No

77.

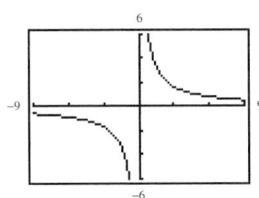

Yes

79.

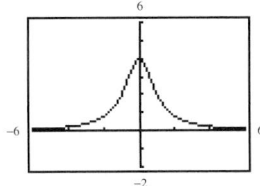

No

81. $x \geq 0; \; f^{-1}(x) = \sqrt[4]{x}$

83. $x \geq 2; \; f^{-1}(x) = \sqrt{x} + 2$

85.

x	0	1	3	4
f^{-1}	6	4	2	0

87.

x	-4	-2	2	3
f^{-1}	-2	-1	1	3

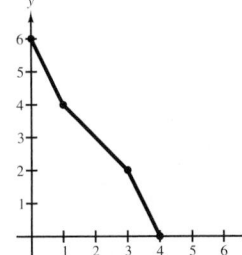

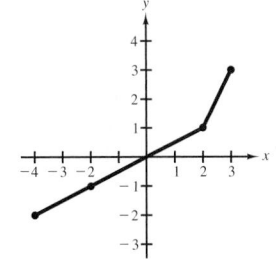

89. (a) $y = \frac{20}{13}(x - 9)$

(b) x: hourly wage; y: number of units produced

(c) 8 units

91. Yes; To form f^{-1}, interchange the first and second coordinates of the solution points of f. The y-intercept $(0, a)$ of f becomes the x-intercept $(a, 0)$ of f^{-1}.

93. Yes **95.** Domain $= \{x : x \text{ is a real number}\}$

97. Domain $= \{x : -2 \leq x \leq 2\}$ **99.** 13 minutes

Section 8.3 (page 527)

1. $I = kV$ **3.** $u = kv^2$ **5.** $p = \dfrac{k}{d}$

7. $P = \dfrac{k}{\sqrt{1 + r}}$ **9.** $A = klw$ **11.** $P = \dfrac{k}{V}$

13. A varies jointly as the base and height.

15. A varies jointly as the length and the width.

17. V varies jointly as the square of the radius and the height.

19. r varies directly as the distance and inversely as the time.

21. $s = 5t$ **23.** $F = \frac{5}{16}x^2$ **25.** $H = \frac{5}{2}u$

27. $n = \dfrac{48}{m}$ **29.** $g = \dfrac{4}{\sqrt{z}}$ **31.** $L = \dfrac{486}{x^3}$

33. $P = \frac{2}{5}x^2$ **35.** $F = \frac{25}{6}xy$ **37.** $c = \frac{4}{3}a\sqrt{b}$

39. $d = \dfrac{120x^2}{r}$ **41.** (a) \$4921.25 (b) Price per unit

43. (a) 1.2 inches (b) 25 pounds **45.** 18 pounds

47. $208\frac{1}{3}$ feet **49.** 32 feet per second squared

51. 3072 Watts **53.** 667 units **55.** 324 pounds

57. No, k is different for each pizza. The 15-inch pizza is the best buy.

59. 25 feet **61.** 2160 passengers **63.** $p = \dfrac{114}{t}; \; 17.5\%$

65. Increase. Because $y = kx$ and $k > 0$, the variables increase or decrease together.

67. y will be one-fourth as great. If $y = k/x^2$ and x is replaced by $2x$, you have $y = k/(2x)^2 = k/(4x^2)$.

69. (a) $(f + g)(x) = 3x^2 - 4x$

(b) $(f - g)(x) = 3x^2 + 6x$

(c) $(fg)(x) = -15x^3 - 5x^2$

(d) $\left(\dfrac{f}{g}\right)(x) = -\dfrac{(3x + 1)}{5}, \; x \neq 0$

(e) $(f \circ g)(x) = 75x^2 - 5x$

(f) $(g \circ f)(x) = -15x^2 - 5x$

71. (a) $(f + g)(x) = \dfrac{7x + 4}{6x(x + 4)}$

(b) $(f - g)(x) = \dfrac{5x - 4}{6x(x + 4)}$

(c) $(fg)(x) = \dfrac{1}{6x(x + 4)}$

(d) $\left(\dfrac{f}{g}\right)(x) = \dfrac{6x}{x + 4}$

(e) $(f \circ g)(x) = \dfrac{6x}{1 + 24x}$

(f) $(g \circ f)(x) = \dfrac{x + 4}{6}$

73. 880 games

Section 8.4 (page 536)

1. Horizontal line

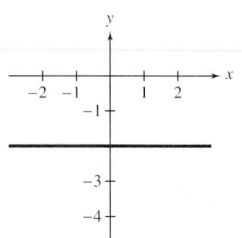

3. Line with slope -3

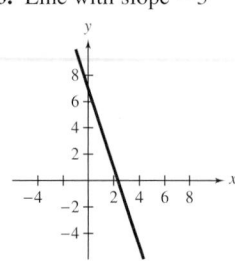

5. Parabola

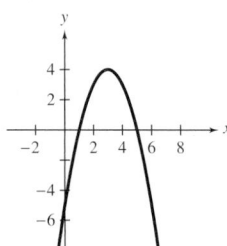

7. Line with slope 2

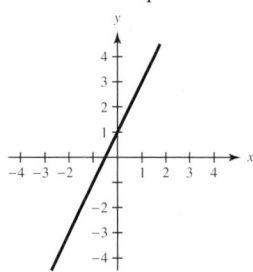

9. 4 **11.** 1 **13.** 5

15. Horizontal translation five units to the right

17. Vertical translation two units downward

19. (a)

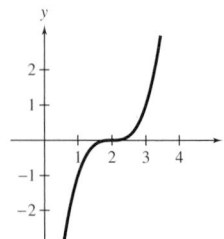

(b)

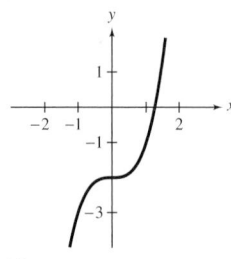

(c)

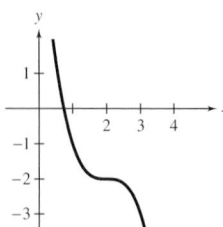

(d)

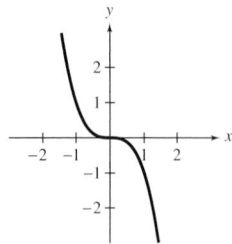

21. Rises to the left; Rises to the right

23. Falls to the left; Rises to the right

25. Rises to the left; Falls to the right

27. Falls to the left; Falls to the right

29. Rises to the left; Rises to the right

31. $(\pm 5, 0), (0, -25)$ **33.** $(3, 0), (0, 9)$

35. $(-2, 0), (1, 0), (0, -2)$ **37.** $(2, 0), (0, 0)$

39. $(\pm 1, 0), \left(0, -\frac{1}{2}\right)$ **41.** $(0, 10)$

43.

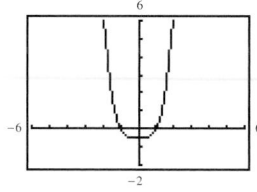

45.

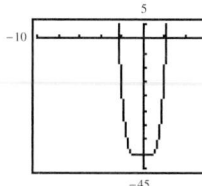

$(-1, 0), (1, 0)$ $(-2.236, 0), (2.236, 0)$

47.

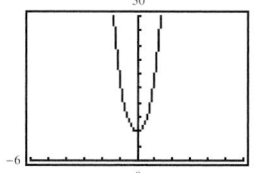

49.

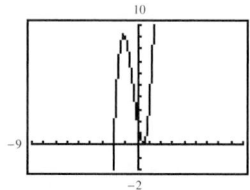

No x-intercepts  $(-2, 0), \left(\frac{1}{2}, 0\right)$

51. e **52.** c **53.** b **54.** f **55.** a

56. g **57.** d **58.** h

59.

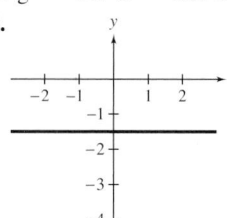

61.

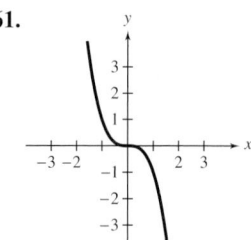

63.

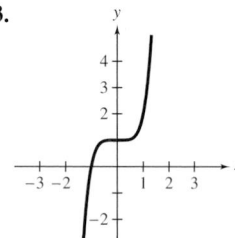

65.

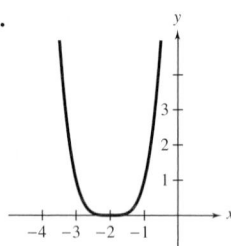

67.

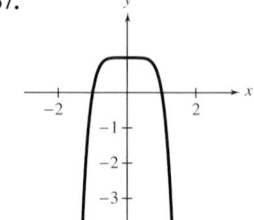

69.

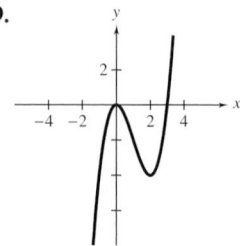

71.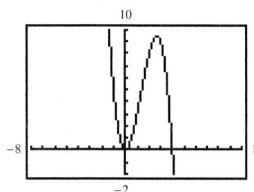

73.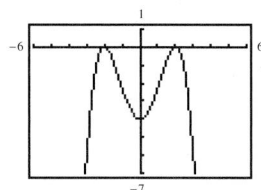

Yes   Yes

75.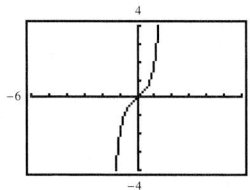

No

77. (a) Answers will vary.   (b) $(0, 6)$

(c) 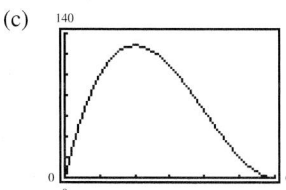 $x = 2$

79. The y-intercept is the value of the function when $x = 0$. The x-intercepts are the x-values for which $f(x) = 0$.

81. Not necessarily. A polynomial function of degree n can have any number of x-intercepts from zero to n.

83. $(2x - 1)(3x + 13)$ **85.** $2x^2(x + 2)(5x + 1)$

87. $(x - 6)(x^2 - 2)$ **89.** $\pm\dfrac{\sqrt{5}}{3}i$ **91.** $\dfrac{-3 \pm \sqrt{129}}{12}$

93. 80 feet $\times$ 160 feet

Mid-Chapter Quiz   (page 539)

1. $x^2 + 2x - 8;\ (-\infty, \infty)$ **2.** $2x^3 - 8x^2;\ (-\infty, \infty)$

3. $x^2 - 2x + 8;\ (-\infty, \infty)$

4. $\dfrac{2x - 8}{x^2};\ (-\infty, 0) \cup (0, \infty)$ **5.** $2x^2 - 8;\ (-\infty, \infty)$

6. $(2x - 8)^2;\ (-\infty, \infty)$ **7.** 102   **8.** -22

9. 3   **10.** $\frac{1}{72}$ **11.** $\sqrt{15}$ **12.** $7 + 2\sqrt{7}$

13. $(C \circ x)(t) = C(x(t)) = 3000t + 750,\ 0 \le t \le 40$

Cost of units produced in t hours

14. $f(g(x)) = 100 - 15\left[\frac{1}{15}(100 - x)\right]$

$\quad\quad\quad\ = 100 - (100 - x) = x$

$\quad g(f(x)) = \frac{1}{15}[100 - (100 - 15x)]$

$\quad\quad\quad\ = \frac{1}{15}(15x) = x$

15. $f^{-1}(x) = \sqrt[3]{x + 8}$ **16.** $z = 3t$

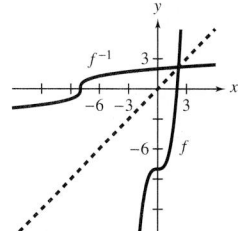

17. $S = 5hr^2$ **18.** $N = \dfrac{15t^2}{s}$ **19.** 30 minutes

20. Rises to the right because the leading coefficient is positive

21. (a) 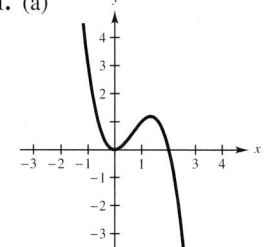   (b)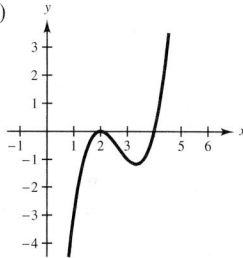

Section 8.5   (page 544)

1. c   **2.** e   **3.** a   **4.** d   **5.** b   **6.** f

7. $x^2 + y^2 = 25$ **9.** $x^2 + y^2 = \frac{4}{9}$

11. $x^2 + y^2 = 64$ **13.** $x^2 + y^2 = 29$

15. $(x - 4)^2 + (y - 3)^2 = 100$

17. $(x - 5)^2 + (y + 3)^2 = 81$

19. $(x + 2)^2 + (y - 1)^2 = 4$

21. $(x - 3)^2 + (y - 2)^2 = 17$

23. Center: $(0, 0)$ **25.** Center: $(0, 0)$

$\quad r = 4$ $r = 6$

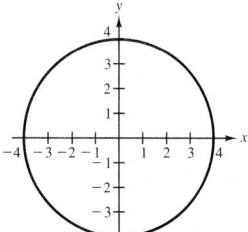   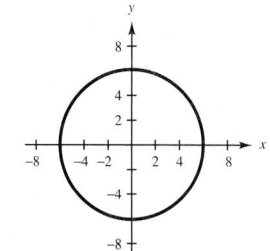

27. Center: $(0, 0)$
$r = \frac{1}{2}$

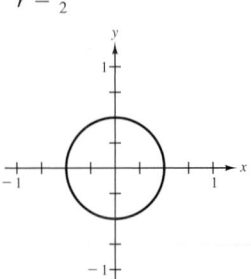

29. Center: $(0, 0)$
$r = \frac{12}{5}$

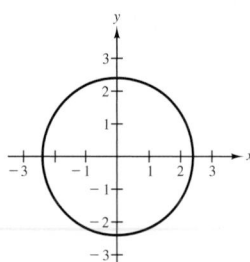

51.

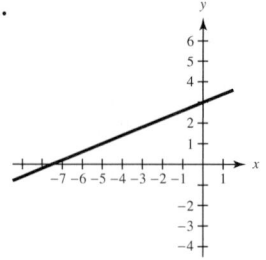

53.

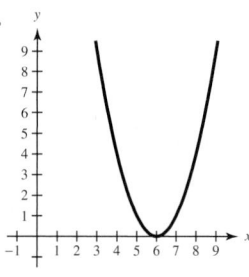

55. 12 people

Section 8.6   (page 556)

1. a   **2.** f   **3.** d   **4.** c   **5.** e   **6.** b
7. Vertices: $(\pm 4, 0)$ **9.** Vertices: $(0, \pm 4)$
Co-vertices: $(0, \pm 2)$ Co-vertices: $(\pm 2, 0)$

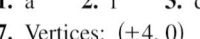

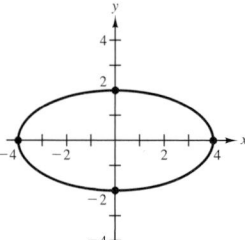

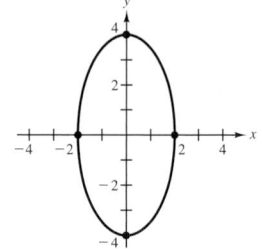

31. Center: $(-1, 5)$
$r = 8$

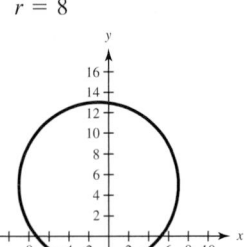

33. Center: $(2, 3)$
$r = 2$

 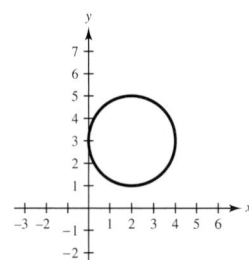

11. Vertices: $\left(\pm\frac{5}{3}, 0\right)$ **13.** Vertices: $\left(0, \pm\frac{4}{5}\right)$
Co-vertices: $\left(0, \pm\frac{4}{3}\right)$ Co-vertices: $\left(\pm\frac{2}{3}, 0\right)$

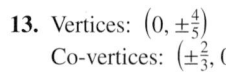

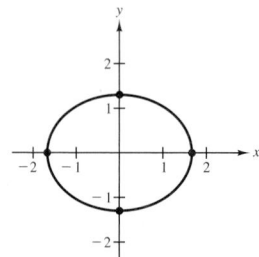

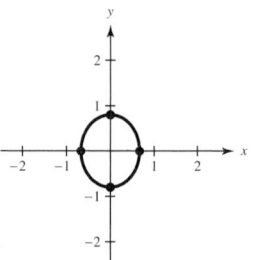

35. Center: $\left(-\frac{5}{2}, -3\right)$
$r = 3$

37. $(x - 2)^2 + (y - 1)^2 = 4$

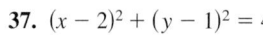 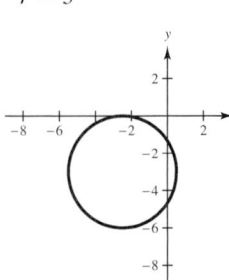

15. Vertices: $\left(\pm\frac{3}{4}, 0\right)$ **17.** Vertices: $(0, \pm 2)$
Co-vertices: $\left(0, \pm\frac{3}{5}\right)$ Co-vertices: $(\pm 1, 0)$

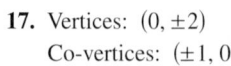

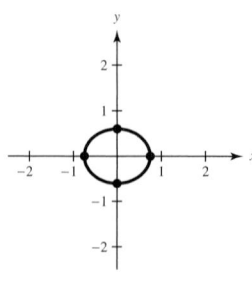

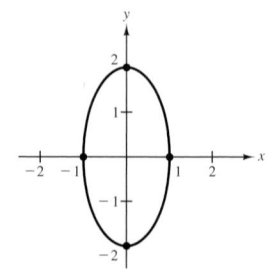

39. $(x + 1)^2 + (y + 3)^2 = 4$ **41.** $(x + 4)^2 + (y + 2)^2 = 25$

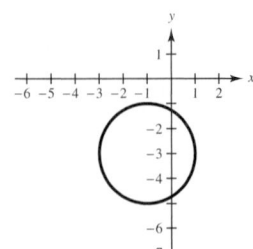

43. $x^2 + y^2 = 4500^2$ **45.** $x^2 + y^2 = 144;$ 10.4 feet
47. A circle consists of all points (x, y) in a plane that are a
given positive distance r from a fixed point (h, k) called the
center.
$x^2 + y^2 = r^2$
49. $-3 \pm \sqrt{13}$

19. Vertices: $(\pm 2, 0)$ **21.** Vertices: $\left(0, \pm 2\sqrt{3}\right)$

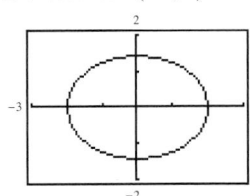
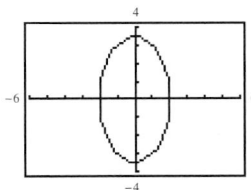

23. $\dfrac{x^2}{16} + \dfrac{y^2}{9} = 1$ **25.** $\dfrac{x^2}{4} + \dfrac{y^2}{1} = 1$

27. $\dfrac{x^2}{9} + \dfrac{y^2}{16} = 1$ **29.** $\dfrac{x^2}{1} + \dfrac{y^2}{4} = 1$

31. $\dfrac{x^2}{9} + \dfrac{y^2}{25} = 1$ **33.** $\dfrac{x^2}{100} + \dfrac{y^2}{36} = 1$

35. Center: $(-5, 0)$
Vertices: $(-9, 0), (-1, 0)$

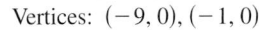

37. Center: $(1, 5)$
Vertices: $(1, 0), (1, 10)$

39. Center: $(-2, 3)$
Vertices:
$(-2, 6), (-2, 0)$

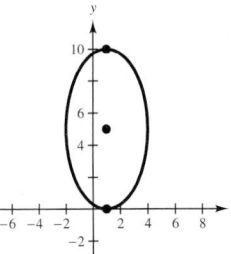
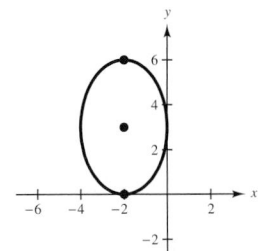

41. $\dfrac{x^2}{1} + \dfrac{y^2}{4} = 1$ **43.** $\dfrac{(x-4)^2}{9} + \dfrac{y^2}{16} = 1$

45. c   **46.** e   **47.** a   **48.** f   **49.** b   **50.** d

51. Vertices: $(\pm 3, 0)$
Asymptotes: $y = \pm x$

53. Vertices: $(0, \pm 3)$
Asymptotes: $y = \pm x$

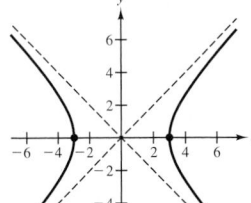
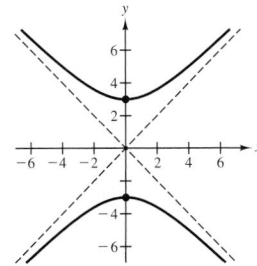

55. Vertices: $(\pm 3, 0)$
Asymptotes: $y = \pm \tfrac{5}{3}x$

57. Vertices: $(0, \pm 3)$
Asymptotes: $y = \pm \tfrac{3}{5}x$

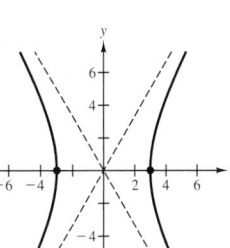
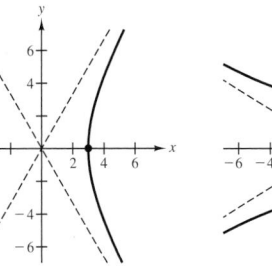

59. Vertices: $(\pm 1, 0)$
Asymptotes: $y = \pm \tfrac{3}{2}x$

61. Vertices: $(\pm 4, 0)$
Asymptotes: $y = \pm \tfrac{1}{2}x$

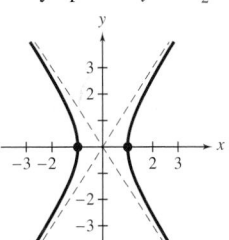

63. $\dfrac{x^2}{16} - \dfrac{y^2}{64} = 1$ **65.** $\dfrac{y^2}{16} - \dfrac{x^2}{64} = 1$

67. $\dfrac{x^2}{81} - \dfrac{y^2}{36} = 1$ **69.** $\dfrac{y^2}{1} - \dfrac{x^2}{1/4} = 1$

71. Center: $(3, -4)$
Vertices: $(3, 1), (3, -9)$

73. Center: $(1, -2)$
Vertices:
$(-1, -2), (3, -2)$

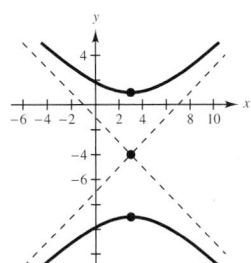
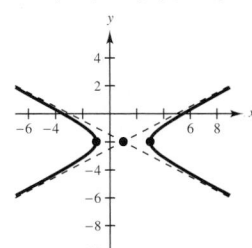

75. Center: $(2, -3)$
Vertices: $(3, -3), (1, -3)$

77. $\dfrac{y^2}{9} - \dfrac{x^2}{9/4} = 1$

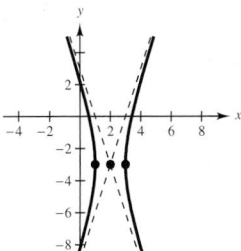

79. $\dfrac{(x-3)^2}{4} - \dfrac{(y-2)^2}{16/5} = 1$ **81.** 36 feet; 28 feet

83. $\dfrac{x^2}{7225} + \dfrac{(y-85)^2}{10,000} = 1$ or $\dfrac{x^2}{10,000} + \dfrac{(y-85)^2}{7225} = 1$

85. A circle is an ellipse in which the major axis and the minor axis have the same length. Both circles and ellipses have foci; however, in a circle the foci are both at the same point, whereas in an ellipse they are not.

87. The difference of the distances between each point on the hyperbola and the two foci is a positive constant.

89. Left half   **91.** $2\sqrt{10}$

93.

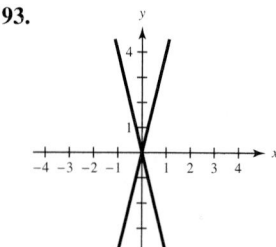

95.

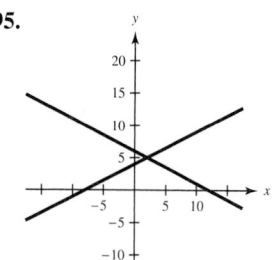

97. $c = 24$ **99.** $a = 13$ **101.** 2.5 hours

Section 8.7   (page 563)

1. b   **2.** c   **3.** e   **4.** a   **5.** f   **6.** d

7. Vertex: $(0, 0)$
   Focus: $\left(0, \frac{1}{2}\right)$

9. Vertex: $(0, 0)$
   Focus: $\left(-\frac{3}{2}, 0\right)$

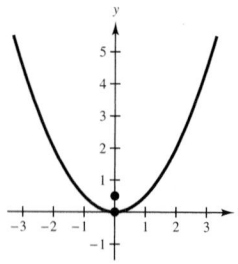
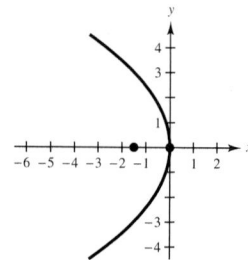

11. Vertex: $(0, 0)$
   Focus: $(0, -2)$

13. Vertex: $(1, -2)$
   Focus: $(1, -4)$

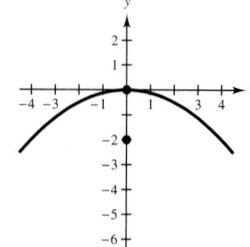
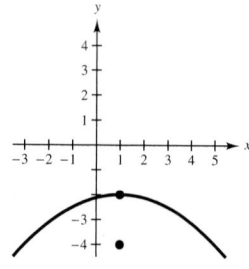

15. Vertex: $\left(5, -\frac{1}{2}\right)$
   Focus: $\left(\frac{11}{2}, -\frac{1}{2}\right)$

17. Vertex: $(1, 1)$
   Focus: $(1, 2)$

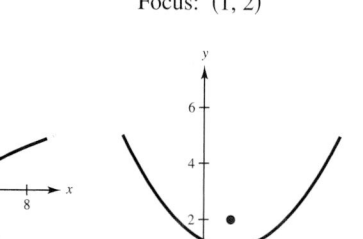

19. Vertex: $(-2, -3)$
   Focus: $(-4, -3)$

21. Vertex: $(-2, 1)$
   Focus: $\left(-2, -\frac{1}{2}\right)$

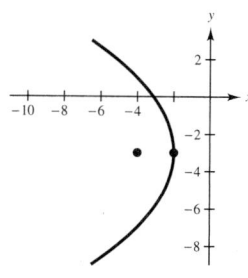
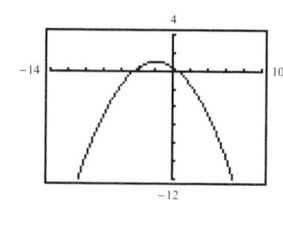

23. Vertex: $\left(\frac{1}{4}, -\frac{1}{2}\right)$
   Focus: $\left(0, -\frac{1}{2}\right)$

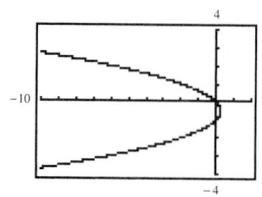

25. $y = 2\sqrt{-x}$ **27.** $y = -4\sqrt{x}$

29. $x^2 = \frac{3}{2}y$ **31.** $x^2 = -6y$ **33.** $y^2 = -8x$

35. $x^2 = 4y$ **37.** $y^2 = 16x$ **39.** $y^2 = 9x$

41. $(x-3)^2 = -(y-1)$ **43.** $y^2 = 2(x+2)$

45. $(y-2)^2 = -8(x-3)$ **47.** $x^2 = 8(y-4)$

49. $(y-2)^2 = x$

51. (a) $y = \dfrac{x^2}{180}$ (b)

x	0	20	40	60
y	0	$2\frac{2}{9}$	$8\frac{8}{9}$	20

53. (a)

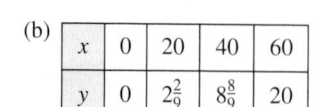

(b) $x = 125$

55. All points on the parabola are equidistant from the directrix and the focus.

57. No. If the graph intersected the directive, there would exist points nearer the directrix than the focus.

59. $x^2 + 12x + 31$ **61.** $-x^2 + 16x - 52$

63. $(x - 2)^2 - 3$ **65.** $-(x - 3)^2 + 14$

67. $y = \frac{5}{8}x + \frac{25}{4}$ **69.** 87

Section 8.8   (page 573)

1.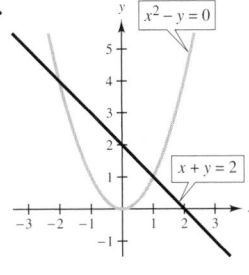

$(-2, 4), (1, 1)$

3.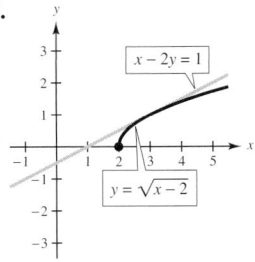

$(3, 1)$

5.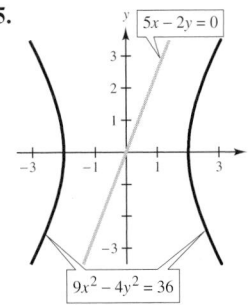

No real solution

7.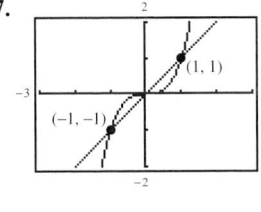

$(0, 0), (1, 1), (-1, -1)$

9.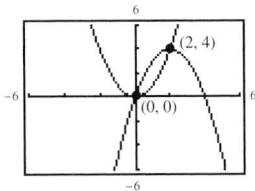

$(0, 0), (2, 4)$

11.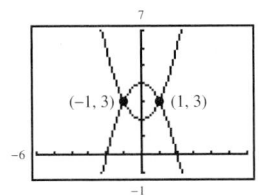

$(-1, 3), (1, 3)$

13.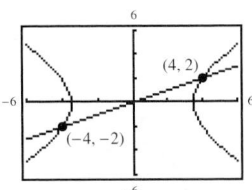

$(4, 2), (-4, -2)$

15.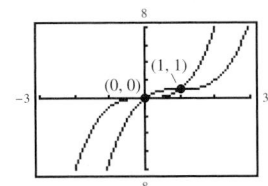

$(0, 0), (1, 1)$

17. $(-3, 18), (2, 8)$ **19.** $(2, 5), (-3, 0)$

21. No real solution   **23.** $(0, 5), (-4, -3)$

25. $(0, 8), \left(-\frac{24}{5}, -\frac{32}{5}\right)$ **27.** $\left(-\frac{17}{2}, -6\right), \left(-\frac{7}{2}, 4\right)$

29. $(0, 2), (3, 1)$ **31.** $(0, 4), (3, 0)$

33. No real solution   **35.** $\left(\pm\sqrt{3}, -1\right)$

37. $\left(2, \pm 2\sqrt{3}\right), (-1, \pm 3)$ **39.** $\left(\pm 2, \pm\sqrt{3}\right)$

41. $\left(\pm\dfrac{2\sqrt{5}}{5}, \pm\dfrac{2\sqrt{5}}{5}\right)$ **43.** $\left(\pm\sqrt{3}, \pm\sqrt{13}\right)$

45. $(3.633, 2.733)$

47. Between points $\left(-\frac{3}{5}, -\frac{4}{5}\right)$ and $\left(\frac{4}{5}, -\frac{3}{5}\right)$

49. Answers will vary.   **51.** $(-2, 3)$ **53.** $\$5520$

Review Exercises   (page 576)

1. (a) $(f + g)(x) = 2x; \; (-\infty, \infty)$
   (b) $(f - g)(x) = 4; \; (-\infty, \infty)$
   (c) $(fg)(x) = x^2 - 4; \; (-\infty, \infty)$
   (d) $\left(\dfrac{f}{g}\right)(x) = \dfrac{x + 2}{x - 2}; \; (-\infty, 2) \cup (2, \infty)$

3. (a) $(f + g)(x) = \dfrac{2}{x - 1} + x; \; (-\infty, 1) \cup (1, \infty)$
   (b) $(f - g)(x) = \dfrac{2}{x - 1} - x; \; (-\infty, 1) \cup (1, \infty)$
   (c) $(fg)(x) = \dfrac{2x}{x - 1}; \; (-\infty, 1) \cup (1, \infty)$
   (d) $\left(\dfrac{f}{g}\right)(x) = \dfrac{2}{x(x - 1)}; \; (-\infty, 0) \cup (0, 1) \cup (1, \infty)$

5. (a) 0 (b) 5 (c) 12 (d) -1

7. (a) $-\dfrac{1}{3}$ (b) 83 (c) $-\dfrac{1}{48}$ (d) $-\dfrac{\sqrt{2}}{6}$

9. (a) $x^2 + 2$ (b) $(x + 2)^2$ (c) 6 (d) 1

11. (a) $|x|$ (b) $x, x \geq -1$ (c) 5 (d) -1

13. (a) $[2, \infty)$ (b) $[4, \infty)$ **15.** $f^{-1}(x) = 6x$

17. $f^{-1}(x) = \sqrt[3]{x}$ **19.** $f^{-1}(x) = 4x$

21. $h^{-1}(x) = x^2, \; x \geq 0$ **23.** f is not one-to-one.

25. $f(x) = 2(x - 4)^2, \; x \geq 4$

$$f^{-1}(x) = \dfrac{\sqrt{2x}}{2} + 4$$

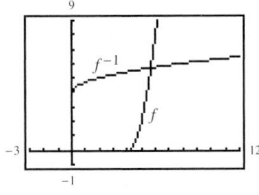

27. 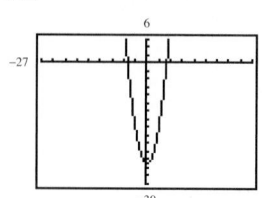 **29.**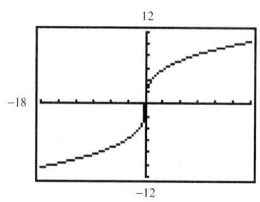

No; function is not one-to-one.   Yes; function is one-to-one.

31. $k = 6,\ y = 6\sqrt[3]{x}$ **33.** $k = \frac{1}{18},\ T = \frac{1}{18}rs^2$

35. (a) 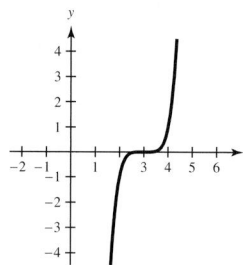   (b)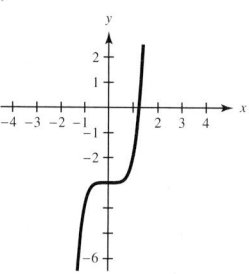

(c) 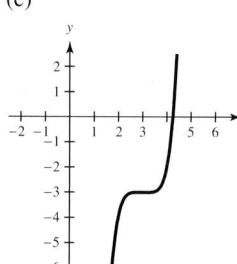   (d)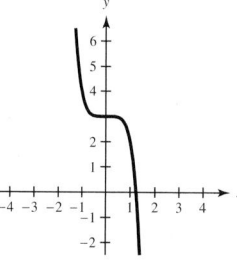

37. 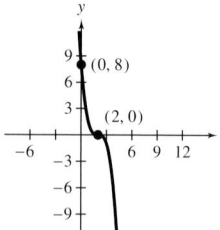   **39.**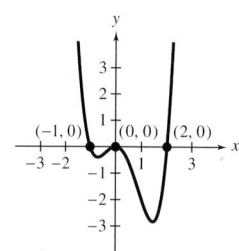

$(2, 0), (0, 8)$ $(0, 0), (-1, 0), (2, 0)$

41.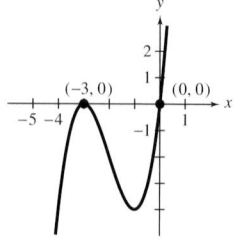

$(0, 0), (-3, 0)$

43. Falls to the left; Falls to the right

45. Rises to the left; Rises to the right   **47.** Hyperbola

49. Circle   **51.** Circle   **53.** Parabola

55. $x^2 + y^2 = 144$ **57.** $x^2 + y^2 = 10$

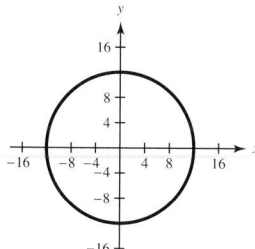   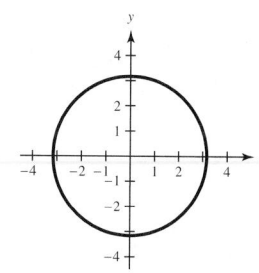

59. $x^2 + y^2 = 64$ **61.** $\dfrac{x^2}{4} + \dfrac{y^2}{25} = 1$

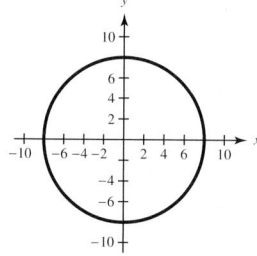   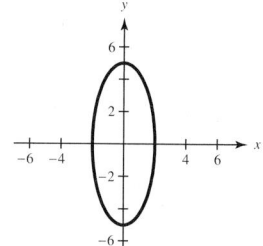

63. $\dfrac{(x-5)^2}{25} + \dfrac{(y-3)^2}{9} = 1$

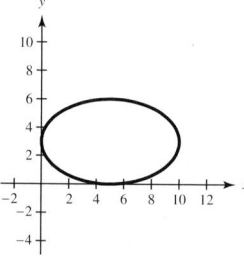

65. Center: $(0, 0)$ **67.** Center: $(1, -2)$
Vertices: $(-8, 0), (8, 0)$ Vertices: $(1, -11), (1, 7)$
Co-vertices: Co-vertices:
$(0, -4), (0, 4)$ $(-5, -2), (7, -2)$

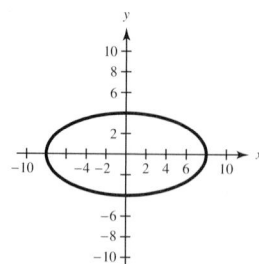   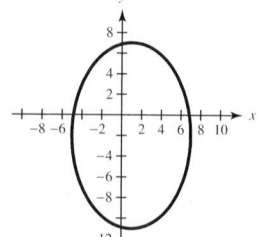

69. $\dfrac{x^2}{36} - \dfrac{y^2}{4} = 1$

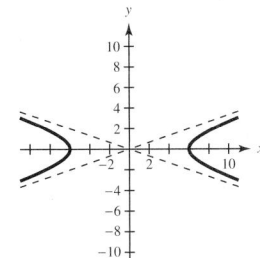

71. $\dfrac{x^2}{1} - \dfrac{(y-3)^2}{4} = 1$

73. Center: $(0, 0)$
Vertices: $(\pm 5, 0)$

75. Vertex: $(2, -2)$
Focus: $\left(2, -\dfrac{7}{4}\right)$

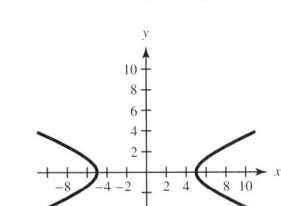

77. $y^2 = -8x$

79. $(x + 6)^2 = -20(y - 4)$

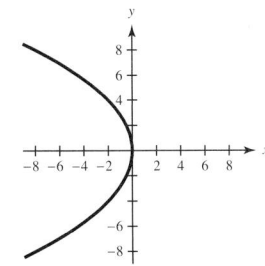
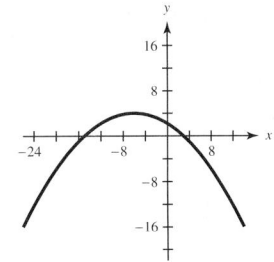

81.

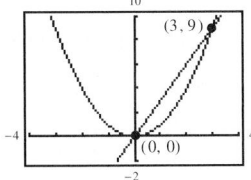

$(0, 0), (3, 9)$

83.

$(-4, 0), (0, 4)$

85. $(-1, 5), (-2, 20)$ **87.** $(-1, 0), (0, -1)$
89. $(0, 2), \left(-\dfrac{16}{5}, -\dfrac{6}{5}\right)$ **91.** $(\pm 5, 0)$
93. (a) $k = \dfrac{1}{8}$ (b) 1953.125 kilowatts   **95.** 945 units
97. $x^2 + y^2 = 25{,}000{,}000$

99. (a)

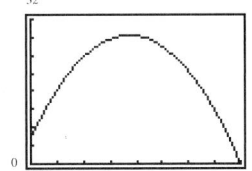

(b) 6 feet
(c) 28.5 feet
(d) 31.9 feet

Chapter 8 Test   (page 580)

1. $(f - g)(x) = -x^2 + \dfrac{1}{2}x + 1$ **2.** $(fg)(x) = \dfrac{1}{2}x^3 - \dfrac{1}{2}x$

3. $\left(\dfrac{f}{g}\right)(x) = \dfrac{x}{2(x^2 - 1)}$

4. $(f \circ g)(x) = \dfrac{1}{2}(x^2 - 1)$ **5.** $[-5, 5]$

6. $f^{-1}(x) = 2x + 2$ **7.** $S = \dfrac{kx^2}{y}$ **8.** $v = \dfrac{1}{4}\sqrt{u}$

9. Rises to the left; Falls to the right

10.

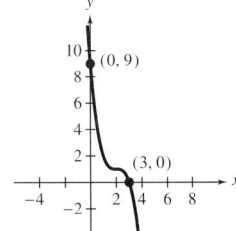

$(3, 0), (0, 9)$

11. $x^2 + y^2 = 25$

12. $\dfrac{x^2}{9} + \dfrac{y^2}{100} = 1$ **13.** $\dfrac{x^2}{9} - \dfrac{y^2}{\frac{9}{4}} = 1$
14. $(y - 1)^2 = 8(x + 2)$
15. (a)   (b)

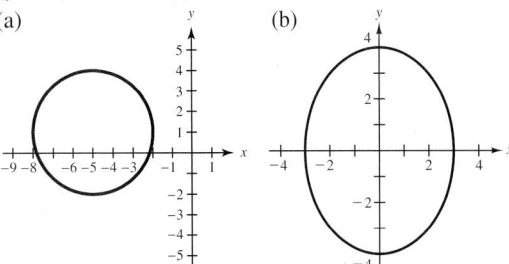

(c)

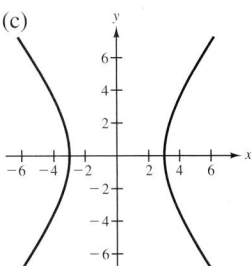

16. (a) $\left(-4 - 2\sqrt{7}, 22 + 8\sqrt{7}\right), \left(-4 + 2\sqrt{7}, 22 - 8\sqrt{7}\right)$
(b) $(6, 8), (8, 6)$

17. 240 cubic centimeters

Cumulative Test: Chapters 1–8   (page 581)

1. $54x^{10}$ **2.** $\dfrac{4u^3}{3}, v \neq 0, u \neq 0$ **3.** $\dfrac{5(1 + \sqrt{3})}{4}$

4. $\dfrac{z^2 \sqrt{21}}{3}, x \neq 0, z \neq 0$ **5.** $\dfrac{3(x + 1)}{x - 3}$

6. $\dfrac{8x(x - 4)}{(x - 5)(x + 3)}$ **7.** 3.77×10^9 **8.** 2.6×10^{-10}

9. $5\sqrt{2}$ **10.** 15   **11.** $y = 2x$ **12.** $y = -x + 6$

13. $y = 2x^2 - 4x + 5$

14. (a) Function   (b) Function
     (c) Not a function   (d) Function

15. $3x(x - 7)$ **16.** $y(y + 3)^2$

17. $(4x^2 + 1)(2x + 1)(2x - 1)$ **18.** $(2x - 5)(3x + 5)$

19. $\dfrac{9}{2(x + 3)}, x \neq 0$ **20.** $\dfrac{1}{x - 2}, x \neq 0, x \neq -2$

21. (a) 38   (b) 14   (c) $9h^2 - 11$ (d) $w^2 + 6w - 2$

22. 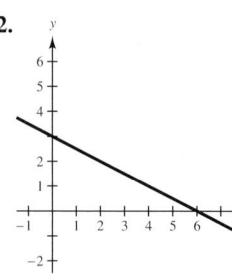   **23.**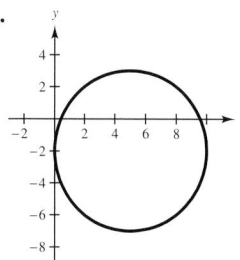

24.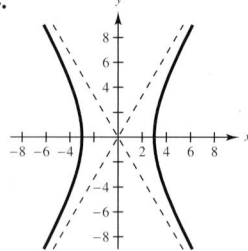

25. (a) 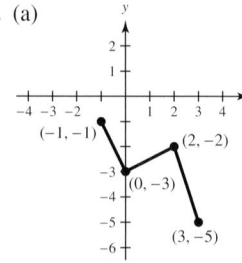   (b)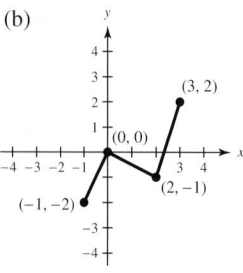

(c) 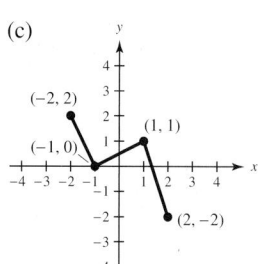   (d)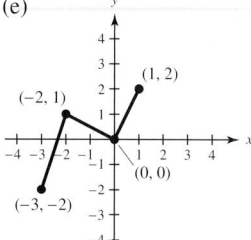

(e) 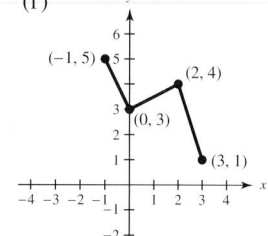   (f)

26. $\dfrac{(x - 3)^2}{9} + \dfrac{y^2}{4} = 1$ **27.** $\dfrac{-3 \pm \sqrt{17}}{2}$

28. $-37, 23$ **29.** 5, 7   **30.** No solution

31. $-\frac{1}{2} \leq x \leq \frac{4}{3}$ **32.** $1 < x < 8$ **33.** $(-3, 2)$

34. $(1, -1), (3, 5)$ **35.** 2   **36.** $0 < t \leq 23$ minutes

37. 4800 units   **38.** 7, 9   **39.** 15.5 tons

Chapter 9

Section 9.1   (page 591)

1. 11.036   **3.** 1.396   **5.** 0.906   **7.** -5.437

9. (a) $\frac{1}{9}$ (b) 1   (c) 3   **11.** (a) $\frac{1}{5}$ (b) 5   (c) 125

13. (a) 500   (b) 250   (c) 56.657

15. (a) 1000   (b) 1628.895   (c) 2653.298

17. (a) 452.865   (b) 226.445   (c) 42.068

19. (a) 0.299   (b) 6   **21.** (a) 73.891   (b) 1.353

23. (a) 21   (b) 42   **25.** b   **26.** a   **27.** e

28. d   **29.** f   **30.** c   **31.** h   **32.** g

33. Vertical shift two units   **35.** Horizontal shift five units
     downward   to the left

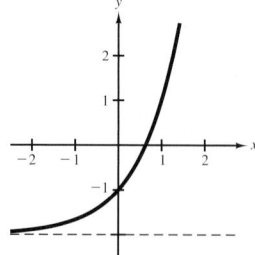   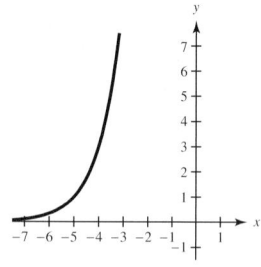

37. Reflection in the *y*-axis

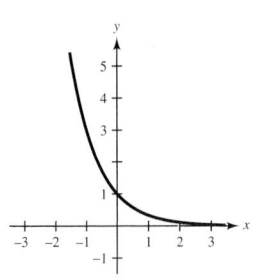

39. Horizontal shift four units to the left and reflection in the *y*-axis

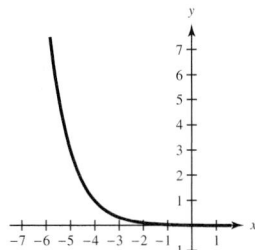

41. Vertical shift 10 units upward and reflection in the *x*-axis

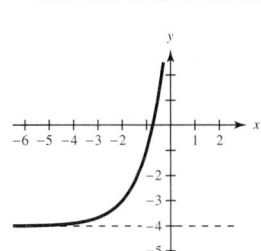

43. Vertical shift four units downward and horizontal shift two units to the left

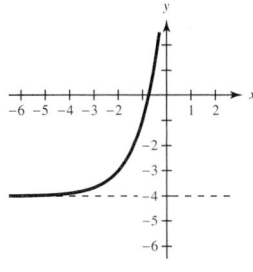

45.

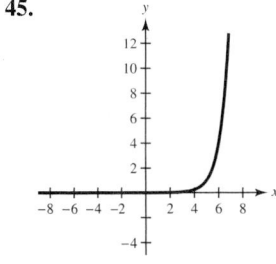

$y = 0$

47.

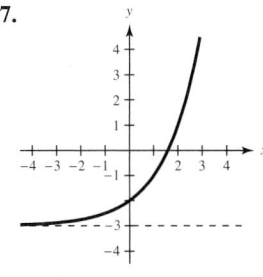

$y = -3$

49.

$y = 0$

51.

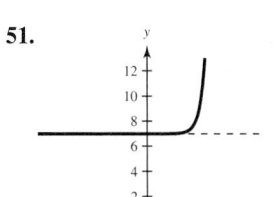

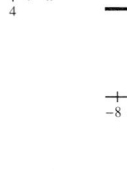

$y = 7$

53.

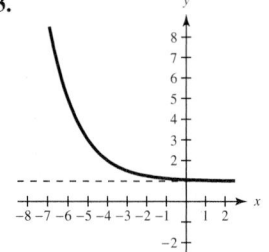

$y = 1$

55.

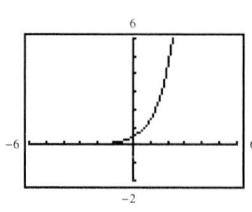

57.

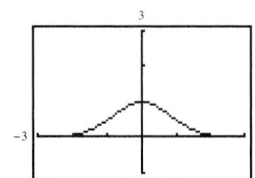

59.

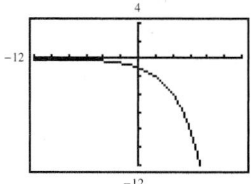

61.

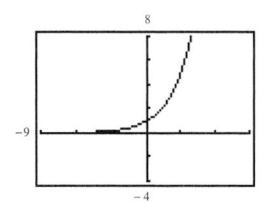

63.

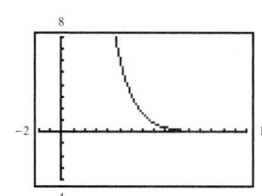

65.

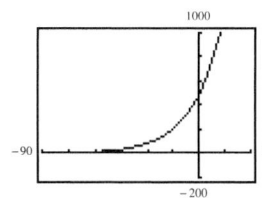

67.

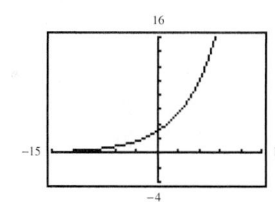

69.

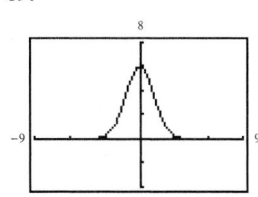

71.

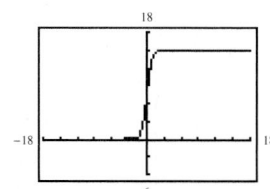

73.

n	1	4	12
A	\$466.10	\$487.54	\$492.68

n	365	Continuous compounding
A	\$495.22	\$495.30

75.

n	1	4	12
P	$2541.75	$2498.00	$2487.98

n	365	Continuous compounding
P	$2483.09	$2482.93

77. (a) $125,819.05 (b) $123,580.10
  (c) $122,493.42 (d) $122,456.43
79. (a) $80,634.95 (b) $161,269.89
81. $V(t) = 16,000\left(\frac{3}{4}\right)^t$

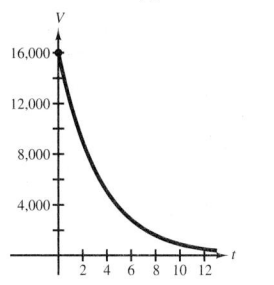

83. 2.52 grams

85. (a)

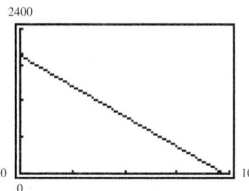

(b)

t	0 second	25 seconds	50 second	75 second
h	2000 feet	1450 feet	950 feet	450 feet

(c) Ground level: 97.5 seconds
87. (a) and (b)

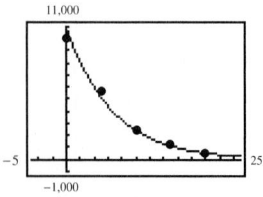

(c)

h	0	5	10	15	20
P	10,332	5583	2376	1240	517
Approx.	10,958	5176	2445	1155	546

(d) 3300 kilograms/square meters (e) 11.3 kilograms

89. No. e is an irrational number.
91. $f(x) = 1^x$ is a constant function.
93. Functions c and d are exponential. These two functions have terms with variable exponents.
95. $5x^7$ **97.** $\dfrac{16}{x^4}$

99.

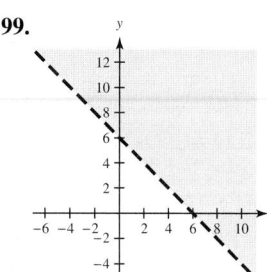

101.

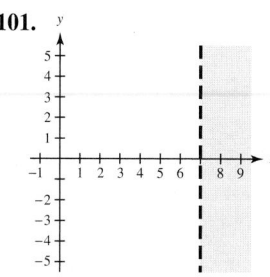

103.

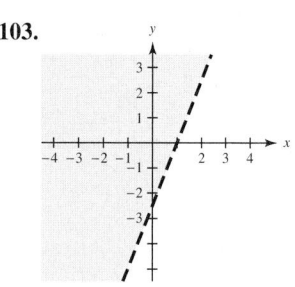

105. $30\sqrt{5} \approx 67.08$ feet

Section 9.2 (page 604)

1. $2^3 = 8$ **3.** $4^4 = 256$ **5.** $4^{-2} = \frac{1}{16}$
7. $36^{1/2} = 6$ **9.** $\log_7 49 = 2$ **11.** $\log_3 \frac{1}{9} = -2$
13. $\log_5 625 = 4$ **15.** $\log_8 4 = \frac{2}{3}$ **17.** $\log_{25} \frac{1}{5} = -\frac{1}{2}$
19. 3 **21.** 1 **23.** -2 **25.** -4 **27.** -3
29. There is no power to which 2 can be raised to obtain -3.
31. 3 **33.** 0 **35.** $\frac{1}{2}$
37. There is no power to which 4 can be raised to obtain -4.
39. $\frac{1}{2}$ **41.** 4 **43.** -0.0625 **45.** -0.0706
47. -0.2877 **49.** 1.6890
51. e **52.** b **53.** d **54.** c **55.** a **56.** f
57. Vertical shift five units upward **59.** Horizontal shift seven units to the right

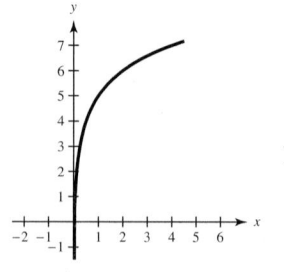

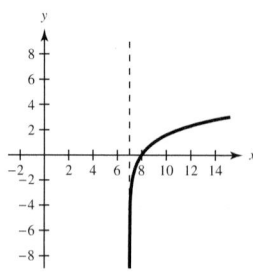

61. Reflection in the y-axis

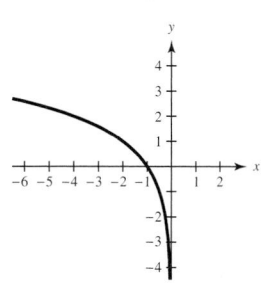

63. Horizontal shift two units to the right and vertical shift five units upward

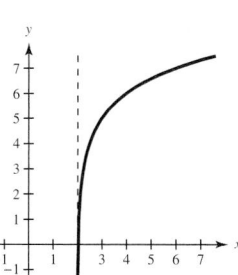

65. Horizontal shift four units to the left and vertical shift eight units downward

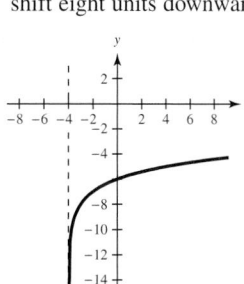

67.

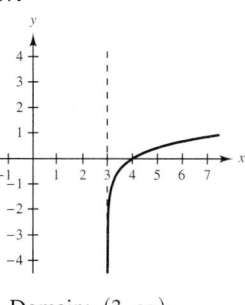

Domain: $(3, \infty)$
Vertical asymptote: $x = 3$

69. Domain: $(0, \infty)$
Vertical asymptote: $x = 0$

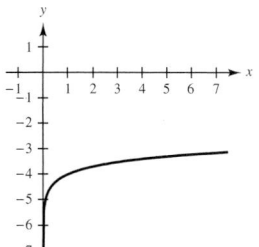

71. Domain: $(0, \infty)$
Vertical asymptote: $x = 0$

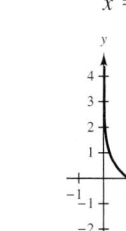

73. Domain: $(1, \infty)$
Vertical asymptote: $x = 1$

75. Domain: $(-2, \infty)$
Vertical asymptote: $x = -2$

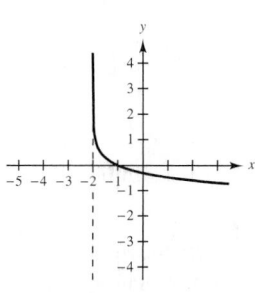

77. b   **78.** e   **79.** d   **80.** c   **81.** f   **82.** a

83. Vertical shift seven units upward

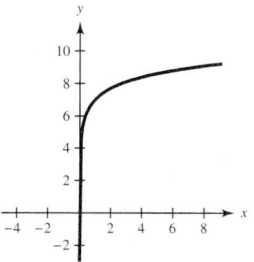

Domain: $(0, \infty)$
Vertical asymptote: $x = 0$

85. Horizontal shift two units to the right

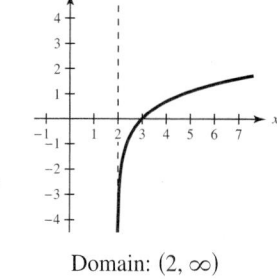

Domain: $(2, \infty)$
Vertical asymptote: $x = 2$

87. Reflection in the x-axis

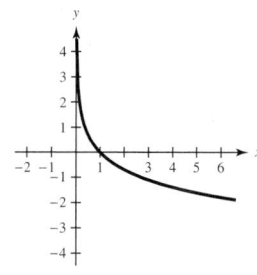

Domain: $(0, \infty)$
Vertical asymptote: $x = 0$

89. Reflection in the y-axis and vertical shift three units upward

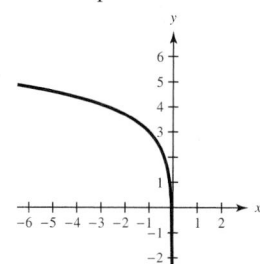

Domain: $(-\infty, 0)$
Vertical asymptote: $x = 0$

91. Horizontal shift one unit to the left and vertical shift five units downward

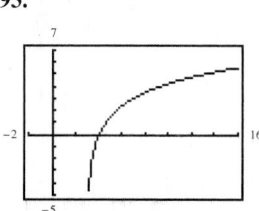

Domain: $(-1, \infty)$
Vertical asymptote: $x = -1$

93.

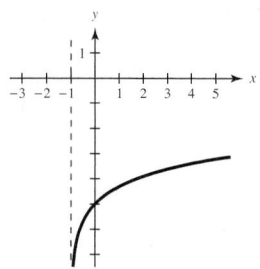

Domain: $(3, \infty)$
Vertical asymptote: $x = 3$

95.

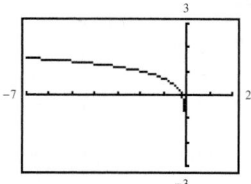

Domain: $(-\infty, 0)$
Vertical asymptote: $x = 0$

97.

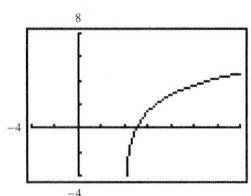

99.

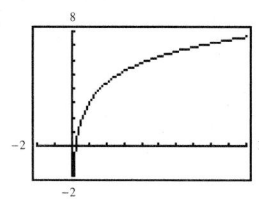

Domain: $(4, \infty)$ Domain: $(0, \infty)$

Vertical asymptote: $t = 4$ Vertical asymptote: $t = 0$

101. 2.3481   **103.** 1.7712   **105.** -0.4739

107. 2.6332   **109.** -2 **111.** 1.3481

113. 53.4 inches

115.

r	0.07	0.08	0.09	0.10	0.11	0.12
t	9.9	8.7	7.7	6.9	6.3	5.8

117. $g = f^{-1}$ **119.** $a = a^1$, so $\log_a a = 1$.

121. $\dfrac{a^6}{b^8}$ **123.** $\dfrac{27}{a^6 b^3}$

125.

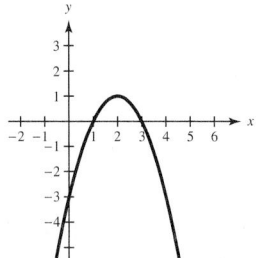

127. 8 people

Section 9.3   (page 612)

1. 2   **3.** 1   **5.** 2   **7.** 1   **9.** 2   **11.** -1

13. 0   **15.** 2   **17.** -3 **19.** 3   **21.** 4

23. 1   **25.** 6   **27.** $\log_3 11 + \log_3 x$ **29.** $3 \ln y$

31. $\log_2 z - \log_2 17$ **33.** $-2 \log_5 x$ **35.** $\frac{1}{3} \log_5 x$

37. $\frac{1}{3} \log_3(x + 1)$ **39.** $-\frac{1}{3} \log_5 x$

41. $\ln 3 + 2 \ln x + \ln y$ **43.** $\frac{1}{2}[\ln x + \ln(x + 2)]$

45. $\ln(t + 4) - \ln 15$ **47.** $2 \log_2 x - \log_2(x - 3)$

49. $\frac{1}{2}(\log_3 4 + \log_3 x - 2 \log_3 y)$ **51.** $\frac{1}{3}[\ln x + \ln(x + 5)]$

53. $2[\ln(x + 1) - \ln(x - 1)]$

55. $6 \log_4 x + 2 \log_4(x - 7)$

57. $3 \ln a + \ln(b - 4) - 2 \ln c$

59. $3(\log_{10} 4 + \log_{10} x) - \log_{10}(x - 7)$

61. $\ln x + \frac{1}{3} \ln y - 4(\ln w + \ln z)$

63. $2(\log_5 x + \log_5 y) + 4 \log_5(x + 3)$ **65.** $\log_2 3x$

67. $\log_{10} \dfrac{4}{x}$ **69.** $\ln b^4,\ b > 0$

71. $\log_5(2x)^{-2},\ x > 0$ **73.** $\ln \sqrt[3]{2x + 1}$

75. $\log_3 2\sqrt{y}$ **77.** $\ln \dfrac{x^2 y^3}{z},\ x > 0, y > 0, z > 0$

79. $\ln(xy)^4,\ x > 0, y > 0$ **81.** $\ln(4\sqrt{x})$

83. $\log_4 \dfrac{x + 8}{x^3},\ x > 0$ **85.** $\log_5\left[\sqrt{x}(x - 3)\right]$

87. $\log_6 \dfrac{(c + d)^5}{m - n},\ c > -d, m > n$

89. $\log_2 \sqrt[5]{\dfrac{x^3}{y}},\ x > 0, y > 0$

91. $\log_6 \dfrac{\sqrt[5]{x - 3}}{x^2(x + 1)^3},\ x > 3$ **93.** 1   **95.** 1.2925

97. 0.2925   **99.** 0.2500   **101.** 2.7925   **103.** 0

105. -0.5283 **107.** 0.954   **109.** 1.556

111. 0.5187   **113.** $1 - \log_4 x$ **115.** $\log_2 5 + 3$

117. $1 + \frac{1}{2} \log_5 2$ **119.** $2 + \ln 3$

121. $\ln 12 - 3$ **123.** $2 + \log_{10} 2$

125. (a) $s(2) = 74.27,\ s(8) = 68.55$

(b)

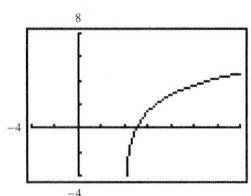

127. $E = 1.4 \log_{10} \dfrac{C_2}{C_1}$ **129.** 5:30 A.M.   **131.** True

133. False. There is no property of logarithms that can be used to simplify $\log_a(u + v)$.

135. True   **137.** False. $\ln(2 \cdot 3) \overset{?}{=} \ln 2 \cdot \ln 3$
$$1.7918 \neq 0.7615$$

139. False. 0 is not in the domain of f.   **141.** True

143. Yes. $f(ax) = \log_a(ax) = \log_a a + \log_a x = 1 + f(x)$

145. $7^2 = 49$ **147.** $e^{4x} = e^{4x}$ **149.** $19\sqrt{3x}$

151. $\sqrt{5u}$ **153.** $25\sqrt{2}x$ **155.** 22 units

Mid-Chapter Quiz   (page 615)

1. (a) $\frac{16}{9}$ (b) 1   (c) $\frac{3}{4}$ (d) 1.54

2. Domain: $(-\infty, \infty)$; Range: $(0, \infty)$

3.

4.

Domain: $(-\infty, \infty)$ Domain: $(-\infty, \infty)$

Horizontal asymptote: Horizontal asymptote:

$y = 0$ $y = 1$

5.

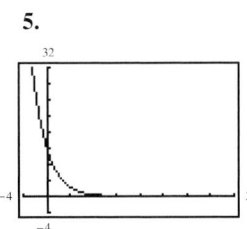

6.

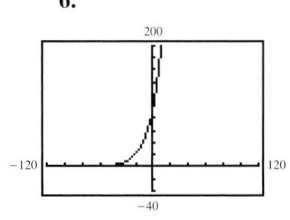

Domain: $(-\infty, \infty)$

Horizontal asymptote:
$y = 0$

Domain: $(-\infty, \infty)$

Horizontal asymptote:
$y = 0$

7.

n	1	4	12
A	\$3185.89	\$3314.90	\$3345.61

n	365	Continuous compounding
A	\$3360.75	\$3361.27

8. 1.6 grams **9.** $4^{-2} = \frac{1}{16}$ **10.** $\log_3 81 = 4$ **11.** 3

12.

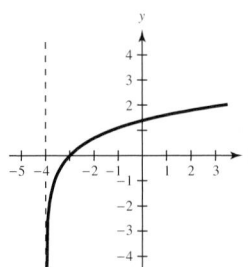

13.

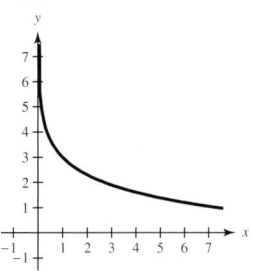

Domain: $(-4, \infty)$

Vertical asymptote: $t = -4$

Domain: $(0, \infty)$

Vertical asymptote: $x = 0$

14. $h = 2,\ k = 1$ **15.** 3.4096

16. $\ln 6 + 2 \ln x - \dfrac{1}{2} \ln(x^2 + 1)$ **17.** $\ln\left(\dfrac{x}{y^3}\right)^2$

Section 9.4 (page 621)

1. (a) Not a solution (b) Solution
3. (a) Solution (b) Not a solution
5. (a) Not a solution (b) Solution
7. 5 **9.** 8 **11.** 8 **13.** 3 **15.** -6
17. -3 **19.** 4 **21.** 16 **23.** $\frac{22}{5}$ **25.** 9
27. 81 **29.** 500,000 **31.** $2x - 1$ **33.** $2x,\ x > 0$
35. 5.49 **37.** 0.86 **39.** 1.5 **41.** -2.37
43. -3.60 **45.** 4.40 **47.** 3.81 **49.** 3.28
51. 0.59 **53.** 2.64 **55.** 3.00 **57.** 1.23
59. 2.49 **61.** 2.48 **63.** 0.39 **65.** 6.80
67. 0.80 **69.** 35.35 **71.** 21.82 **73.** 8.99
75. 1 **77.** 7.91 **79.** 22.63 **81.** 2187
83. 31,622,771.60 **85.** 6.52 **87.** 10.04

89. 12.18 **91.** -0.78 **93.** ± 20.09 **95.** 3.20
97. 4 **99.** 5 **101.** 0.75 **103.** 9 **105.** 2.46
107. 2.29 **109.** 3
111. **113.**

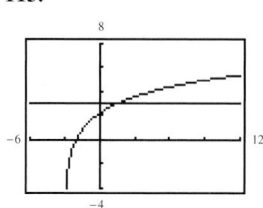

0.69 1.48
115. 7.70 years **117.** $10^{-8.5}$ watts per square centimeter
119. 262.5° **121.** 205 decays
123. $2^{x-1} = 30$ requires logarithms because x cannot be solved for easily.
125. 1. $\log_a(uv) = \log_a u + \log_a v$
  2. $\log_a\left(\dfrac{u}{v}\right) = \log_a u - \log_a v$
  3. $\log_a u^n = n \log_a u$
127. 2 **129.** $-\frac{1}{2}$ **131.** 1, 7
133. 60 liters of 30% solution; 40 liters of 80% solution

Section 9.5 (page 630)

1. 7% **3.** 9% **5.** 8% **7.** 6% **9.** 7%
11. 8.75 years **13.** 6.60 years **15.** 9.24 years
17. 14.21 years **19.** 9.99 years **21.** Continuous
23. Quarterly **25.** 8.33% **27.** 7.23%
29. 6.136% **31.** 8.30% **33.** 7.79% **35.** No
37. \$1652.99 **39.** \$626.46 **41.** \$3080.15
43. \$951.23 **45.** \$67,667.64 **47.** \$5496.57
49. \$320,250.81 **51.** $k = \frac{1}{2} \ln \frac{8}{3} \approx 0.4904$
53. $k = \frac{1}{3} \ln \frac{1}{2} \approx -0.2310$ **55.** $y = 13.1e^{0.0049t};\ 14.4$
57. $y = 12.9e^{0.0083t};\ 15.2$ **59.** $y = 18.1e^{0.0244t};\ 29.5$
61. $y = 18.1e^{0.0039t};\ 19.6$
63. k is larger in Exercise 59 because the population of Bombay is increasing faster than the population of Shanghai.
65. 3.3 grams **67.** 4.43 grams **69.** 58.7 grams
71. 15,700 years
73. (a)

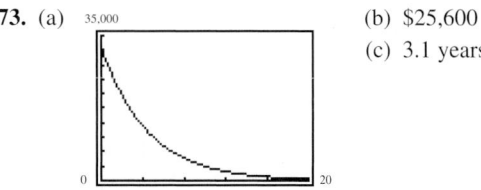

(b) \$25,600
(c) 3.1 years

75. Total deposits: \$7200.00; Total interest: \$10,529.42
77. The one in Alaska was 63 times as great.
79. The one in Mexico was 40 times as great.

81. 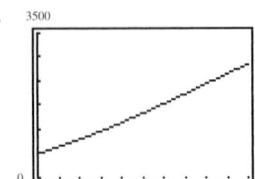 (b) 1000
(c) 2642
(d) 5.88 years

83. (a) $S = 10(1 - e^{-0.0575x})$ (b) 3300 units
85. $k < 0$ **87.** 10 **89.** 4 **91.** -4
93. $(3, 3)$ **95.** $\left(-\frac{1}{2}, \frac{1}{4}\right), (2, 4)$ **97.** $(1, 2, 1)$
99. 50 kilometers per hour and 65 kilometers per hour

Review Exercises (page 635)

1. (a) $\frac{1}{8}$ (b) 2 (c) 4
3. (a) 7 (b) $\frac{37}{9}$ (c) 59,053

5. 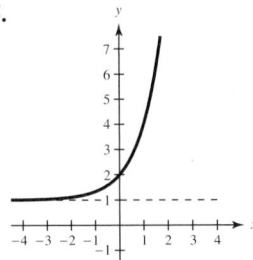 **7.**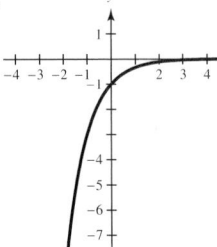

Horizontal asymptote: $y = 1$ Horizontal asymptote: $y = 0$

9. 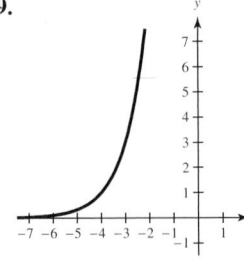 **11.**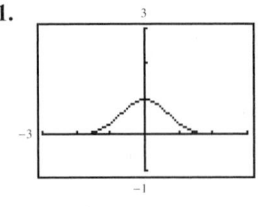

Horizontal asymptote: $y = 0$
13. (a) 2.718 (b) 0.351 (c) 0.135
15. (a) 0 (b) -0.492 (c) -0.882
17. 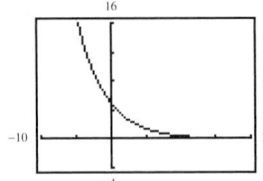 **19.**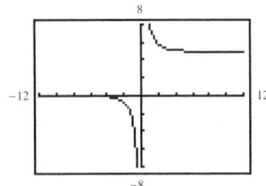

21. $e^1 = e$ **23.** $5^{-2} = \frac{1}{25}$ **25.** $\log_4 64 = 3$
27. $\log_{25} 125 = \frac{3}{2}$ **29.** 3 **31.** -2 **33.** 6
35. -3 **37.** 3 **39.** -2 **41.** -3 **43.** $\frac{1}{2}$

45. 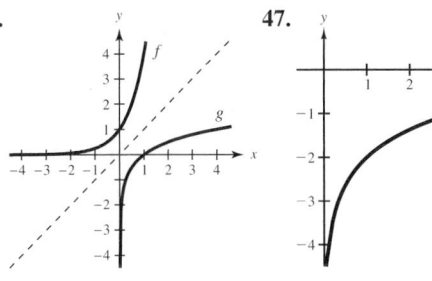 **47.**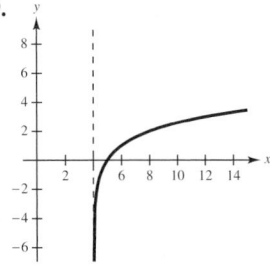

Domain: $(0, \infty)$
Vertical asymptote: $x = 0$

49.

Domain: $(4, \infty)$
Vertical asymptote: $x = 4$
51. 3 **53.** 0.223
55. 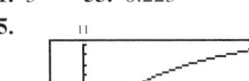 **57.** 1.585 **59.** 2.132

61. 1.7959 **63.** -0.4307 **65.** 1.0293
67. $\log_4 6 + 4 \log_4 x$ **69.** $\frac{1}{2} \log_5(x + 2)$
71. $\ln(x + 2) - \ln(x - 2)$
73. $\frac{1}{2}(\ln 2 + \ln x) + 5 \ln(x + 3)$ **75.** $\log_4 \dfrac{x}{10}$
77. $\log_8 32x^3$ **79.** $\ln \dfrac{9}{4x^2}, \; x > 0$ **81.** $\ln(x^3 y^4 z)$
83. $\log_2\left(\dfrac{k}{k - t}\right)^4$ **85.** False. $\log_2 4x = 2 + \log_2 x$
87. True **89.** True **91.** 5 **93.** 1 **95.** 17
97. 13 **99.** 5.66 **101.** 6.23 **103.** 32.99
105. 243 **107.** 35 **109.** 15.81 **111.** 1408.10
113. 0.32 **115.** 2.67 **117.** 0.08

119.

n	1	4	12
A	\$3806.13	\$4009.59	\$4058.25

n	365	Continuous compounding
A	\$4082.26	\$4083.08

121.

n	1	4	12
P	\$2301.55	\$2103.50	\$2059.87

n	365	Continuous compounding
P	\$2038.82	\$2038.11

123. 8.87%   **125.** 3.7 grams   **127.** 150 units
129. (a) 301 deer   (b) 3.85 years

(c)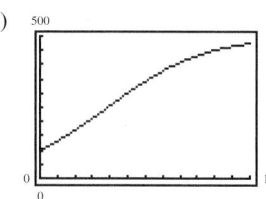

(d) There is a horizontal asymptote at $y = 0$ because the population cannot be less than zero. There is also a horizontal asymptote at $P = 500$ because the deer population will level off at 500.

Chapter 9 Test   (page 638)

1. $f(-1) = 81$
  $f(0) = 54$
  $f\left(\frac{1}{2}\right) = 18\sqrt{6} \approx 44.09$
  $f(2) = 24$

2. 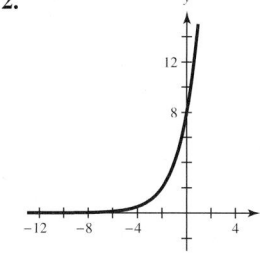   **3.**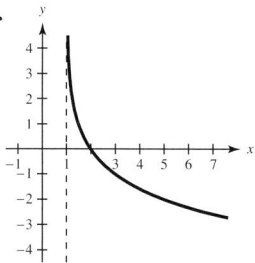

Horizontal asymptote: Domain: $(1, \infty)$
$y = 0$ Vertical asymptote: $x = 1$
4. $5^3 = 125$ **5.** $\log_4 \frac{1}{16} = -2$ **6.** $\frac{1}{3}$
7. f is the inverse   **8.** $\log_4 5 + 2\log_4 x - \frac{1}{2}\log_4 y$
function of g.

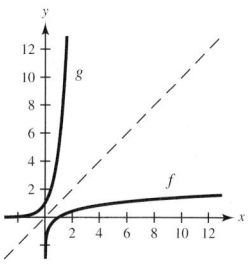

9. $\ln \frac{a^8 b}{c^3}$ **10.** 64   **11.** 0.973   **12.** 13.733
13. 2.452   **14.** 120.007   **15.** 15.516   **16.** 4
17. 30   **18.** (a) \$8012.78   (b) \$8110.40
19. \$10,806.08   **20.** 7%   **21.** \$8469.14
22. (a) 1141   (b) 4.4 years

Cumulative Test: Chapters 1–9   (page 639)

1. -4 **2.** -19 **3.** $\pm\sqrt{10}$ **4.** $-\frac{1}{2} \le x \le \frac{3}{2}$
5. $x \le -9$ or $x \ge -5$ **6.** $-5 < x < 2$ **7.** x^{20}
8. $\frac{4y^2}{9x^4}$ **9.** $27x^{12}$ **10.** $\frac{x^2 + 2x - 13}{x(x-2)}, x \ne -3, x \ne 3$
11. $\frac{(x+1)(x+3)}{3}, x \ne -1$

12. 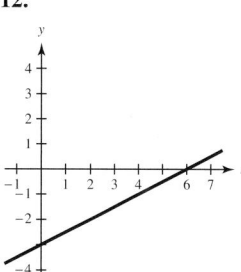   **13.**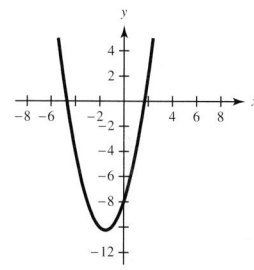

14. 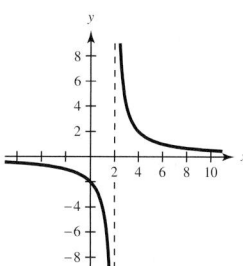   **15.**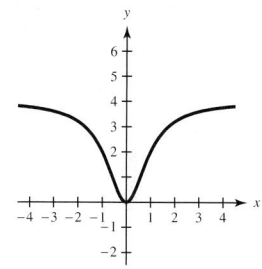

16. 16   **17.** 4   **18.** 2, 5   **19.** 2, 9
20. $16 + 16\sqrt{2} \approx 38.63$ inches
21. Horizontal shift four units to the left
22. Reflection in the x-axis   **23.** $-6\sqrt{3}$ **24.** $1 + 6i$
25. $19 + 43i$ **26.** $\frac{8}{13} + \frac{1}{13}i$ **27.** $(2, 4, 1)$
28. $\left(-\frac{1}{5}, -\frac{22}{5}\right)$ **29.** -200 **30.** -126
31. $3x - 2y + 1 = 0$
32. (a) $5x - 2$ (b) $-3x - 4$ (c) $4x^2 - 11x - 3$
   (d) $\frac{x-3}{4x+1}; \left(-\infty, -\frac{1}{4}\right) \cup \left(-\frac{1}{4}, \infty\right)$
33. (a) $\sqrt{x-1} + x^2 + 1$ (b) $\sqrt{x-1} - x^2 - 1$
   (c) $\sqrt{x-1}(x^2+1)$ (d) $\frac{\sqrt{x-1}}{x^2+1}; [1, \infty)$
34. (a) $2x + 12$ (b) $\sqrt{2x^2 + 6}$
35. (a) $|x| - 2$ (b) $|x - 2|$ **36.** $h^{-1}(x) = \frac{x+2}{5}$

37. 128 feet

38.

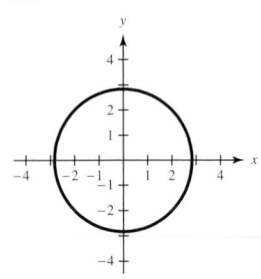

39.

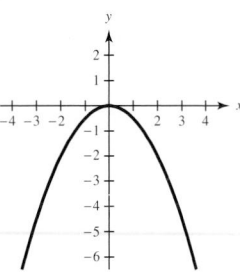

40.

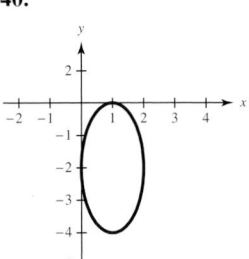

41.

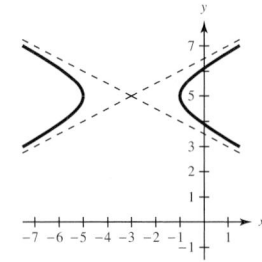

42. $4^3 = 64$ **43.** $3^{-4} = \frac{1}{81}$ **44.** $e^0 = 1$

45. $\log_2 \dfrac{x^3 y^3}{z}$ **46.** $\ln 5 + \ln x - 2\ln(x + 1)$

47. (a) $\frac{1}{9}$ (b) 12.182   (c) 0.867

48. 30% solution: $13\frac{1}{3}$ gallons; 60% solution: $6\frac{2}{3}$ gallons

49.

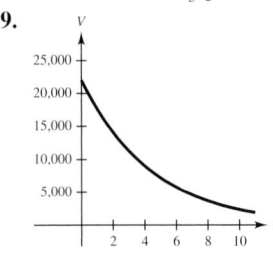

$t = 1.7$

Chapter 10

Section 10.1   (page 650)

1. 2, 4, 6, 8, 10   **3.** $-2, 4, -6, 8, -10$

5. $\frac{1}{2}, \frac{1}{4}, \frac{1}{8}, \frac{1}{16}, \frac{1}{32}$ **7.** $\frac{1}{4}, -\frac{1}{8}, \frac{1}{16}, -\frac{1}{32}, \frac{1}{64}$

9. $1, -0.2, 0.04, -0.008, 0.0016$ **11.** $\frac{1}{2}, \frac{1}{3}, \frac{1}{4}, \frac{1}{5}, \frac{1}{6}$

13. $\frac{2}{5}, \frac{1}{2}, \frac{6}{11}, \frac{4}{7}, \frac{10}{17}$ **15.** $-1, \frac{1}{4}, -\frac{1}{9}, \frac{1}{16}, -\frac{1}{25}$

17. $3, \frac{1}{2}, 1, \frac{1}{4}, \frac{3}{5}$ **19.** $\frac{9}{2}, \frac{19}{4}, \frac{39}{8}, \frac{79}{16}, \frac{159}{32}$ **21.** $2, 2, \frac{4}{3}, \frac{2}{3}, \frac{4}{15}$

23. $0, 6, -6, 18, -30$ **25.** $3, \frac{1}{2}, \frac{1}{3}, \frac{1}{8}, \frac{3}{25}$

27. 1, 5, 9, 13, 17, 21   **29.** 2, 5, 11, 23, 47, 95

31. $4, \frac{5}{4}, \frac{9}{5}, \frac{14}{9}, \frac{23}{14}, \frac{37}{23}$ **33.** $30, 20, 15, \frac{25}{2}, \frac{45}{4}, \frac{85}{8}$

35. $-2, 0, 4, 4, -4, -12$ **37.** -72 **39.** $\frac{5}{2}$

41. 56   **43.** 5   **45.** $\dfrac{1}{132}$ **47.** $\dfrac{1}{702}$ **49.** $\dfrac{1}{n + 1}$

51. $n(n + 1)$ **53.** $2n$ **55.** $a_n = (-1)^{n+1}$

57. $a_n = n^2$ **59.** $a_n = \dfrac{4}{n}$ **61.** $a_n = 2n - 1$

63. 63   **65.** 77   **67.** 40   **69.** 73   **71.** 31

73. $\frac{47}{60}$ **75.** $\frac{437}{60}$ **77.** -48 **79.** $\frac{8}{9}$ **81.** $\frac{182}{243}$

83. 112   **85.** 852   **87.** 6.5793   **89.** $\displaystyle\sum_{k=1}^{5} k$

91. $\displaystyle\sum_{k=1}^{5} 2k$ **93.** $\displaystyle\sum_{k=1}^{10} \dfrac{1}{2k}$ **95.** $\displaystyle\sum_{k=1}^{20} \dfrac{1}{k^2}$ **97.** $\displaystyle\sum_{k=0}^{9} \dfrac{1}{(-3)^k}$

99. $\displaystyle\sum_{k=1}^{20} \dfrac{4}{k + 3}$ **101.** $\displaystyle\sum_{k=1}^{11} \dfrac{k}{k + 1}$ **103.** $\displaystyle\sum_{k=1}^{20} \dfrac{2k}{k + 3}$

105. $\displaystyle\sum_{k=0}^{6} k!$ **107.** c   **108.** a   **109.** b   **110.** d

111.

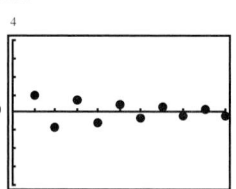

113.

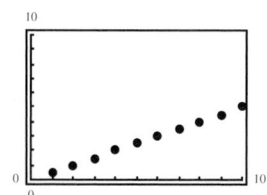

115.

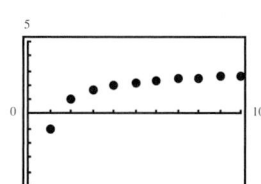

117. 3.6   **119.** 0.8

121. (a) \$535, \$572.45, \$612.52, \$655.40, \$701.28, \$750.37, \$802.89, \$859.09

(b) \$7487.23   (c)

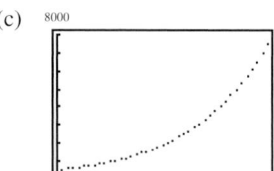

(d) Yes. An investment earning compound interest increases at an increasing rate.

123. $a_5 = 108°, a_6 = 120°$

At the point where any two hexagons and a pentagon meet, the sum of the three angles is $a_5 + 2a_6 = 348° < 360°$. Therefore, there is a gap of 12°.

125. The odd terms   **127.** True   **129.** $\frac{1}{4}$

131. $\dfrac{2n + 1}{n(n + 1)}$ **133.** $\dfrac{1}{(x + 10)^2}$ **135.** $\dfrac{1}{a^8}$

137. $8x\sqrt{2x}$

139.

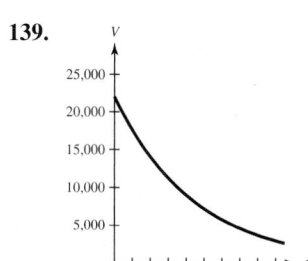

After 1.72 years

Section 10.2 (page 659)

1. 3 **3.** −6 **5.** 3 **7.** −4 **9.** $\frac{2}{3}$ **11.** $-\frac{5}{4}$
13. Arithmetic, 2 **15.** Arithmetic, −16
17. Arithmetic, −57 **19.** Not arithmetic
21. Arithmetic, 0.8 **23.** Arithmetic, $\frac{3}{2}$
25. Not arithmetic **27.** Not arithmetic
29. 7, 10, 13, 16, 19 **31.** 6, 4, 2, 0, −2
33. $\frac{3}{2}, 4, \frac{13}{2}, 9, \frac{23}{2}$ **35.** $\frac{8}{5}, \frac{11}{5}, \frac{14}{5}, \frac{17}{5}, 4$
37. 11, 14, 17, 20, 23 **39.** $4, \frac{15}{4}, \frac{7}{2}, \frac{13}{4}, 3$
41. 25, 28, 31, 34, 37 **43.** 9, 6, 3, 0, −3
45. −10, −4, 2, 8, 14 **47.** 100, 80, 60, 40, 20
49. b **50.** f **51.** e **52.** a **53.** c **54.** d
55. **57.**

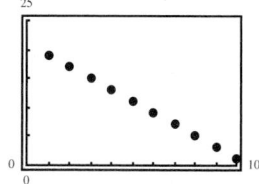

59. **61.** $a_n = \frac{1}{2}n + \frac{5}{2}$

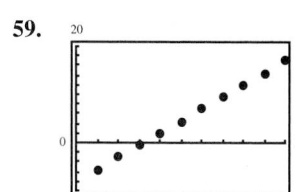

63. $a_n = -25n + 1025$ **65.** $a_n = \frac{3}{2}n + \frac{3}{2}$
67. $a_n = -4n + 32$ **69.** $a_n = \frac{5}{2}n + \frac{5}{2}$
71. $a_n = 4n + 4$ **73.** $a_n = -10n + 60$
75. $a_n = -\frac{1}{2}n + 11$ **77.** $a_n = -0.05n + 0.40$
79. 210 **81.** 275 **83.** 1425 **85.** 62,625
87. 35 **89.** 14,895 **91.** 19,500 **93.** 3011.25
95. 522 **97.** 1850 **99.** 900 **101.** 12,200
103. 243 **105.** 23 **107.** 2850 **109.** 2550
111. $246,000 **113.** $25.43 **115.** 114 times
117. 632 bales **119.** 1024 feet

121. (a) 4, 9, 16, 25, 36 (b) 49 (c) $\sum_{k=1}^{n} (2k-1) = n^2$

123. A recursion formula gives the relationship between the terms a_{n+1} and a_n.
125. 9 **127.** 110 **129.** 12 **131.** $(-\infty, \infty)$
133. $[-4, 4]$ **135.** $(2, \infty)$ **137.** $30,798.61

Mid-Chapter Quiz (page 663)

1. $32, 8, 2, \frac{1}{2}, \frac{1}{8}$ **2.** $-\frac{3}{5}, 3, -\frac{81}{7}, \frac{81}{2}, -135$
3. $\frac{1}{2}, \frac{1}{3}, \frac{1}{4}, \frac{1}{5}, \frac{1}{6}$ **4.** $\frac{1}{4}, \frac{2}{9}, \frac{3}{16}, \frac{4}{25}, \frac{5}{36}$
5. 4, 10, 28, 82, 244, 730 **6.** $1, 3, 1, 9, \frac{1}{3}, 81$
7. $a_n = 3n - 2$ **8.** $a_n = n^2 - 1$ **9.** $a_n = \frac{n+1}{n+2}$
10. $a_n = \frac{(-1)^{n+1}}{2^n}$ **11.** 1000 **12.** 425 **13.** 87
14. $\sum_{k=1}^{20} \frac{2}{3k}$ **15.** $\sum_{k=1}^{25} \frac{(-1)^{k-1}}{k^3}$ **16.** $a_n = 7n - 2$
17. $a_n = -3n + 23$ **18.** 4075 **19.** 9030
20. 546 **21.** 225 **22.** 5.5°

Section 10.3 (page 670)

1. 3 **3.** −3 **5.** $-\frac{1}{4}$ **7.** $-\frac{3}{2}$ **9.** $\frac{2}{3}$ **11.** π
13. 1.1 **15.** Geometric, $\frac{1}{2}$ **17.** Not geometric
19. Geometric, 2 **21.** Not geometric
23. Geometric, $-\frac{2}{3}$ **25.** Not geometric
27. Geometric, 1.02 **29.** 4, 8, 16, 32, 64
31. 5, −10, 20, −40, 80 **33.** $6, 2, \frac{2}{3}, \frac{2}{9}, \frac{2}{27}$
35. $1, -\frac{1}{2}, \frac{1}{4}, -\frac{1}{8}, \frac{1}{16}$ **37.** $4, -2, 1, -\frac{1}{2}, \frac{1}{4}$
39. 1000, 1010, 1020.1, 1030.30, 1040.60
41. 4000, 3960.40, 3921.18, 3882.36, 3843.92
43. $10, 6, \frac{18}{5}, \frac{54}{25}, \frac{162}{125}$ **45.** −6561 **47.** $-\frac{40}{6561}$
49. $\frac{3}{256}$ **51.** $48\sqrt{2}$ **53.** 1486.02 **55.** $\frac{81}{64}$
57. $\pm\frac{243}{32}$ **59.** $\pm\frac{16}{9}$ **61.** $a_n = 2(3)^{n-1}$
63. $a_n = 2^{n-1}$ **65.** $a_n = 4\left(-\frac{1}{2}\right)^{n-1}$
67. $a_n = 8\left(\frac{1}{4}\right)^{n-1}$ **69.** $a_n = 12\left(\frac{2}{3}\right)^{n-1}$
71. $a_n = 4(3)^{n-1}$ **73.** $a_n = 4\left(-\frac{3}{2}\right)^{n-1}$
75. b **76.** d **77.** a **78.** c
79. **81.**

83. 1023 **85.** 772.48 **87.** 2.25
89. 6.67 **91.** 21.43 **93.** 399.93 **95.** 6300.25
97. −14,762 **99.** 2555 **101.** 16.00 **103.** 13,120
105. 48.00 **107.** 1103.57 **109.** 12,822.71 **111.** 2
113. $\frac{2}{3}$ **115.** $\frac{6}{5}$ **117.** 32

119. (a) $250{,}000(0.75)^n$ (b) \$59,326.17   (c) The first year
121. \$3,623,993   **123.** \$19,496.56
125. \$105,428.44   **127.** \$75,715.32
129. (a) \$5,368,709.11   (b) \$10,737,418.23
131. (a) $P = (0.999)^n$ (b) 69.4%

(c) 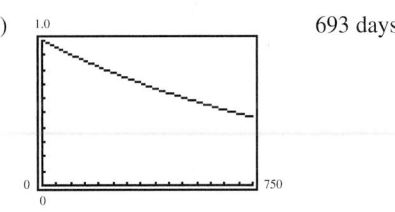   693 days

133. 666.21 feet   **135.** $a_n = \left(-\frac{2}{3}\right)^{n-1}$
137. The nth partial sum of a sequence is the sum of the first n terms of the sequence.

139. $\dfrac{1}{n(n+1)}$ **141.** $y < 6$

143. $-1 < x < 0$ or $x > \frac{5}{2}$ **145.** 47.17 feet

Section 10.4   (page 679)

1. 15   **3.** 252   **5.** 1   **7.** 1   **9.** 1225
11. 12,650   **13.** 2,598,960   **15.** 2,535,650,040
17. $_{15}C_3 = 455$; $_{15}C_{12} = 455$
19. $_{25}C_5 = 53{,}130$; $_{25}C_{20} = 53{,}130$
21. $_5C_2 = 10$; $_5C_3 = 10$ **23.** $_{12}C_5 = 792$; $_{12}C_7 = 792$
25. $_{10}C_0 = 1$; $_{10}C_{10} = 1$ **27.** 15   **29.** 35   **31.** 70
33. $x^6 + 18x^5 + 135x^4 + 540x^3 + 1215x^2 + 1458x + 729$
35. $x^5 + 5x^4 + 10x^3 + 10x^2 + 5x + 1$
37. $x^6 - 24x^5 + 240x^4 - 1280x^3 + 3840x^2 - 6144x + 4096$
39. $x^8 + 8x^7y + 28x^6y^2 + 56x^5y^3 + 70x^4y^4$
$\qquad + 56x^3y^5 + 28x^2y^6 + 8xy^7 + y^8$
41. $u^3 - 6u^2v + 12uv^2 - 8v^3$
43. $16x^4 - 32x^3 + 24x^2 - 8x + 1$
45. $64y^6 + 192y^5z + 240y^4z^2 + 160y^3z^3 + 60y^2z^4$
$\qquad + 12yz^5 + z^6$
47. $81a^4 + 216a^3b + 216a^2b^2 + 96ab^3 + 16b^4$
49. $120x^7y^3$ **51.** $129{,}024a^4b^5$ **53.** $360x^3y^2$
55. $-5940a^3b^9$ **57.** $1{,}259{,}712x^2y^7$
59. $32{,}476{,}950{,}000x^4y^8$ **61.** 120   **63.** -1365
65. 1760   **67.** 54   **69.** 1.172   **71.** 510,568.785
73.

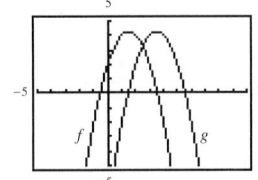

g is a shift of f two units to the right
$g(x) = -x^2 + 7x - 8$

75. $\frac{1}{32} + \frac{5}{32} + \frac{10}{32} + \frac{10}{32} + \frac{5}{32} + \frac{1}{32}$
77. $\frac{1}{256} + \frac{12}{256} + \frac{54}{256} + \frac{108}{256} + \frac{81}{256}$

79. $0.343 + 0.441 + 0.189 + 0.027$
81. In the shaded portion, the difference between consecutive numbers increases by 1.
83. The signs of the terms alternate in the expansion of $(x - y)^n$.
85. The first and last numbers in each row are 1. Every other number in the row is formed by adding the two numbers immediately above the number.
87. 32   **89.** -126
91. -4 **93.** 4.16   **95.** 22,023.47
97. $y = -0.07x^2 + 1.3x + 2$

Review Exercises   (page 682)

1. 8, 11, 14, 17, 20   **3.** $1, \frac{3}{4}, \frac{5}{8}, \frac{9}{16}, \frac{17}{32}$
5. 1, 3, 13, 63, 313, 1563   **7.** 2, 3, 1, -2, -3, -1
9. 14
11.

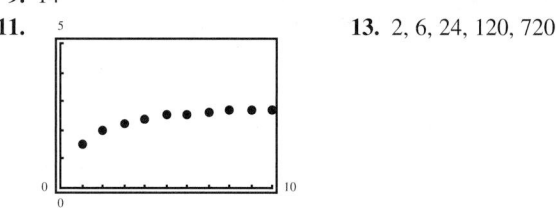

13. 2, 6, 24, 120, 720

15. $1, 1, \frac{1}{2}, \frac{1}{6}, \frac{1}{24}$ **17.** $a_n = 2n - 1$
19. $a_n = \dfrac{n}{(n+1)^2}$ **21.** 28   **23.** $\dfrac{4}{5}$
25. $\displaystyle\sum_{k=1}^{4} (5k - 3)$ **27.** $\displaystyle\sum_{k=1}^{6} \dfrac{1}{3k}$
29. 127, 122, 117, 112, 107   **31.** $\frac{5}{4}, 2, \frac{11}{4}, \frac{7}{2}, \frac{17}{4}$
33. 5, 8, 11, 14, 17   **35.** $80, \frac{155}{2}, 75, \frac{145}{2}, 70$
37. $a_n = 4n + 6$ **39.** $a_n = -50n + 1050$
41. 486   **43.** $\frac{2525}{2}$ **45.** 2527.5
47. 10, 30, 90, 270, 810   **49.** 100, -50, 25, -12.5, 6.25
51. 3, 6, 12, 24, 48   **53.** $a_n = \left(-\frac{2}{3}\right)^{n-1}$
55. $a_n = 24(2)^{n-1}$ **57.** $a_n = 12\left(-\frac{1}{2}\right)^{n-1}$
59. 8190   **61.** -1.928 **63.** 19.842
65. 116,169.538   **67.** 2,274,859.538   **69.** 8
71. 12   **73.** 56   **75.** 1   **77.** 91,390
79. 177,100
81. $x^4 + 36x^3 + 486x^2 + 2916x + 6561$
83. $25 - 150x + 60x^2 - 8x^3$
85. $x^{10} + 10x^9 + 45x^8 + 120x^7 + 210x^6 + 252x^5 + 210x^4$
$\qquad + 120x^3 + 45x^2 + 10x + 1$
87. $y^6 - 12y^5 + 60y^4 - 160y^3 + 240y^2 - 192y + 64$
89. $x^8 - 4x^7 + 7x^6 - 7x^5 + \frac{35}{8}x^4 - \frac{7}{4}x^3 + \frac{7}{16}x^2 - \frac{1}{16}x + \frac{1}{256}$
91. $10{,}000x^4 + 12{,}000x^3y + 5400x^2y^2 + 1080xy^3 + 81y^4$
93. $u^{18} + 9u^{16}v^3 + 36u^{14}v^6 + 84u^{12}v^9 + 126u^{10}v^{12}$
$\qquad + 126u^8v^{15} + 84u^6v^{18} + 36u^4v^{21} + 9u^2v^{24} + v^{27}$
95. $-61{,}236$ **97.** 6   **99.** $-1080x^2y^3$
101. $61{,}236{,}000x^3y^6$

103. (a) $5100.00, $5202.00, $5306.04, $5412.16, $5520.40,
$5630.81, $5743.43, $5858.30
   (b) $11,040.20
   (c)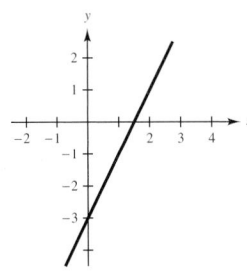

105. 19,950   **107.** 490 meters
109. (a) $a_n = 85,000(1.012)^n$ (b) 154,328
111. $26,198.27   **113.** $637,678.02

Chapter 10 Test   (page 685)

1. $1, -\frac{2}{3}, \frac{4}{9}, -\frac{8}{27}, \frac{16}{81}$ **2.** 76, 44, 28, 20, 16
3. $\dfrac{1}{(n+1)(n+2)}$ **4.** $\dfrac{153}{7}$ **5.** $\displaystyle\sum_{k=1}^{12} \dfrac{2}{3k+1}$
6. 12, 16, 20, 24, 28   **7.** $a_n = -100n + 5100$
8. $-\frac{3}{2}$ **9.** $a_n = 4\left(\frac{1}{2}\right)^{n-1}$ **10.** 35   **11.** $\frac{3069}{1024}$
12. 12   **13.** 3825   **14.** $47,868.33   **15.** 1140
16. $x^5 - 10x^4 + 40x^3 - 80x^2 + 80x - 32$ **17.** 56
18. $755,249.73

Cumulative Test: Chapters 1–10   (page 686)

1. $36x^5y^8$ **2.** $\dfrac{7}{24}x + 8$ **3.** $4rs^2$ **4.** $\dfrac{9x^2}{16y^6}$
5. $\dfrac{4\sqrt{10}}{5}$ **6.** 3   **7.** $5x(1-4x)$
8. $(2+x)(14-x)$ **9.** $(3x-5)(5x+3)$
10. $(2x+1)(4x^2-2x+1)$
11.   **12.**

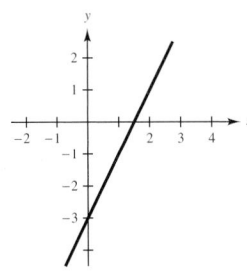

13.   **14.**

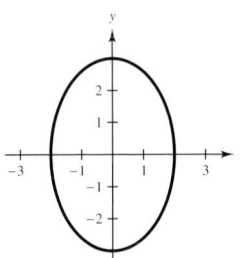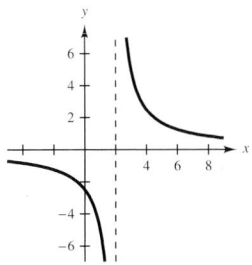

15. $y = -\dfrac{11}{8}x + \dfrac{161}{16}$ **16.** $[-4, 4]$ **17.** $\dfrac{c-6}{c+4}$
18. 50   **19.** $y = \frac{1}{2}(x-2)^2 + 2$
20. (a) During the fourth year

   (b)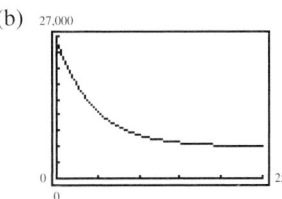

21. About 5 feet   **22.** $z^4 - 12z^3 + 54z^2 - 108z + 81$

Appendix A

Section A.1   (page A13)

1. (a) 1, 4, 6   (b) $-10, 0, 1, 4, 6$
   (c) $-10, -\frac{2}{3}, -\frac{1}{4}, 0, \frac{5}{8}, 1, 4, 6$ (d) $-\sqrt{5}, \sqrt{3}, 2\pi$
3. $-5, -4, -3, -2, -1, 0, 1, 2, 3$ **5.** 1, 3, 5, 7, 9
7. (a)   (b)

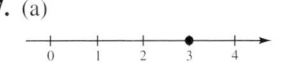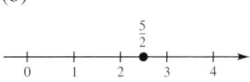

   (c)   (d)

9. $-1 < 3$ **11.** $-\frac{1}{2} < \frac{7}{2}$
13. $2 < 5$ **15.** $-7 < -2$

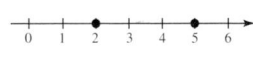

17. $-5 < -2$ **19.** $\frac{1}{3} > \frac{1}{4}$

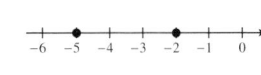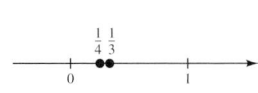

21. $-\frac{2}{3} > -\frac{10}{3}$ **23.** $2.75 < \pi$

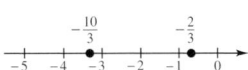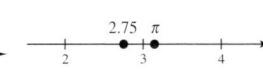

25. 6   **27.** 50   **29.** 35   **31.** $\frac{7}{2}$ **33.** 10
35. 225   **37.** -25 **39.** $-\frac{3}{8}$ **41.** -3.5
43. $|-6| > |2|$ **45.** $|47| > |-27|$
47. $\left|-\frac{3}{4}\right| > -\left|\frac{4}{5}\right|$ **49.** $-|-16.8| = -|16.8|$
51. $-14, 14$ **53.** 160, 160   **55.** $\frac{5}{4}, \frac{5}{4}$ **57.** 0.4, 0.4

59.

61.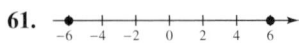

63.

$-\frac{5}{3}$ $\frac{5}{3}$

65.

-4.25 4.25

67. $x < 0$ **69.** $u \geq 16$ **71.** $p < 225$ **73.** 45

75. 19   **77.** -22 **79.** -20 **81.** 0.7   **83.** $\frac{1}{2}$

85. $\frac{1}{24}$ **87.** $\frac{105}{8}$ **89.** 60   **91.** 45.95   **93.** -28

95. -30 **97.** 48   **99.** 72   **101.** $\frac{1}{2}$ **103.** $-\frac{1}{3}$

105. 32.13   **107.** 6   **109.** -3 **111.** $-\frac{5}{2}$ **113.** $\frac{2}{5}$

115. $\frac{46}{17}$ **117.** 4.03   **119.** $(-3)(-3)(-3)(-3)$

121. $\left(-\frac{3}{4}\right)\left(-\frac{3}{4}\right)\left(-\frac{3}{4}\right)\left(-\frac{3}{4}\right)$

123. $(0.67)(0.67)(0.67)(0.67)(0.67)(0.67)$ **125.** $(-5)^4$

127. -5^6 **129.** -64 **131.** -25 **133.** $\frac{49}{64}$

135. $\frac{64}{125}$ **137.** 0.027   **139.** 9   **141.** -7 **143.** 5

145. -2 **147.** -1 **149.** 0   **151.** 4   **153.** 6

155. 3   **157.** 72.2   **159.** 135   **161.** -6

163. 161   **165.** 14,425   **167.** 171.36   **169.** 11.72

171.

Year	1999	2000	2001	2002
Yearly gain or loss (in millions)	-0.2	$+1.3$	-0.2	$+0.9$

173. $\frac{17}{180}$ **175.** 15 square meters   **177.** 36 square feet

179. 6.125 cubic feet

181. (a) \$10,800   (b) \$27,018.72   (c) \$16,218.72

183. True

185. False. The reciprocal of the integer 2 is $\frac{1}{2}$, which is not an integer.

187. False. $|5 + (-4)| \overset{?}{=} |5| + |-4|$
$$|1| \overset{?}{=} 5 + 4$$
$$1 \neq 9$$

189. Yes. The nonnegative real numbers include 0.

191. No. $\frac{1}{6}$ yields a repeating decimal, so $\frac{1}{6}$ is *approximately* 0.17.

193. $\dfrac{5 + 12}{5} = \dfrac{17}{5}$; Only common factors (not terms) of the numerator and denominator can be divided out.

195. $3 \cdot 4^2 = 3 \cdot 16 = 48$; Exponents must be evaluated before multiplication is done.

Section A.2   (page A23)

1. Commutative Property of Addition

3. Additive Inverse Property

5. Commutative Property of Multiplication

7. Multiplicative Identity Property

9. Commutative Property of Addition

11. Associative Property of Addition

13. Distributive Property

15. Associative Property of Addition

17. Associative Property of Multiplication

19. Multiplicative Identity Property

21. Additive Identity Property

23. Multiplicative Inverse Property

25. Distributive Property   **27.** Distributive Property

29. $(3 \cdot 6)y$ **31.** $-3(15)$

33. $5 \cdot 6 + 5 \cdot z$ or $30 + 5z$ **35.** $-x + 25$

37. $x + 8$ **39.** (a) -10 (b) $\frac{1}{10}$

41. (a) 16   (b) $-\dfrac{1}{16}$ **43.** (a) $-6z$ (b) $\dfrac{1}{6z}$

45. (a) $-x - 1$ or $-(x + 1)$ (b) $\dfrac{1}{x + 1}$

47. $x + (5 - 3)$ **49.** $(32 - 4) + y$ **51.** $(3 \cdot 4)5$

53. $(6 \cdot 2)y$ **55.** $20 \cdot 2 + 20 \cdot 5$ or $40 + 100$

57. $5 \cdot 3x - 5 \cdot 4$ or $15x - 20$

59. $x \cdot (-2) + 6 \cdot (-2)$ or $-2x - 12$

61. $-6 \cdot 2y - (-6) \cdot 5$ or $-12y + 30$ **63.** $3x + 15$

65. $-2x - 16$ **67.** 0   **69.** Answers will vary.

71. Original equation
Addition Property of Equality
Associative Property of Addition
Additive Inverse Property
Additive Identity Property

73. Original equation
Addition Property of Equality
Associative Property of Addition
Additive Inverse Property
Additive Identity Property
Multiplication Property of Equality
Associative Property of Multiplication
Multiplicative Inverse Property
Multiplicative Identity Property

75. 28   **77.** 434   **79.** 62.82   **81.** \$1.15

83. \$1.71   **85.** $4(x + 3) = 4 \cdot x + 4 \cdot 3 = 4x + 12$

87. True, Additive Inverse Property

89. False, $6(7 + 2) = 6(7) + 6(2)$

91. No. Zero does not have a multiplicative inverse.

93. The multiplicative inverse of a real number a is the number $1/a$. The product of a number and its multiplicative inverse is 1. For example, $4 \cdot (1/4) = 1$.

95. To subtract the number a from each side of an equation, use the Addition Property of Equality to add $(-a)$ to each side.

Index of Applications

Biology and Life Science Applications

10 second pulse count, 193
Air sacs in lungs, 311
American elk, 605
Carbon 14 dating, 631
Deer herd, 637
Environment, growth of a red oak tree, 538
Fertilizer mixture, 260
Fruit distribution, 301
Girl's height, 686
Growth of a bacterium, 94
Human memory model, 603, 613, 622
Molecular transport, 613
Nutrition, 295
Oil spill, 529
Pollution, 436
Pollution removal, 436, 476
Population
   of fish, 496
   of a species, 638
Population growth, 475
Predator-prey, 528, 686
Rattlesnake's pit-organ sensory system, 219
Recycling, 271
Width of human hair, 311

Business Applications

Advertising effect, 632
Aircraft sales, 142
Annual sales of recreational vehicles, 13
Average cost, 436, 473, 476, 485, 491, 496
   of a calendar, 482
   per candy bar, 485
   per unit, 466
Average and total costs of producing x units, 430
Break-even analysis, 73, 200, 243, 246, 301, 582
Company profits, A15
Comparing revenues, 572
Cost, 410, 436, 492, 496, 519, 539
   of delivery vans, 241
   of operating a truck, 247
Cost, revenue, profit, 400
Defective parts, 183, 199, 323
Demand, 347, 383, 523, 525, 528, 579, 614

for apartments, 169
for soft drinks, 169
Depreciation, 652, 653, 671, 684
   of business equipment, 156
   of a car, 228, 593, 631, 638, 640
   of a computer system, 169
   of a molding machine, 107, 120
   of a photocopier, 169
   of a printing press, 122, 322
   of a truck, 631
Depreciation rate, 322
Dividends, A25
Employment, company layoff, 183
Expenses of a broadcasting company, 226
Inflation, 193, 593
Inventory costs, 301
Manufacturing, 107
   area of a washer, 37
Markup rate, 227
Operating costs, 211, 212, 247
Partnership costs, 475, 496
Pizza delivery, 398
Price and demand, 592
Price increase, 226
Price index, 323, 359
Product demand, 347, 637
Production cost, 508
Profit, 133, 141, 199, 212, 410, 419, 455, 686
   for a company that manufactures radios, 411
Property value, 592
Quality control, 199, 323
Real estate agency commission, 176, 183, 194
Reimbursed expenses, 186
Revenue, 36, 226, 230, 374, 382, 392, 520, 528, 564, 565
Sales, 166, 170, 222, 301, 311, 485
Sales bonus, 508
Sales growth, 632
Straight-line depreciation, 304
Ticket prices, 661
Ticket sales, 196, 246, 261, 295, 304, 606
Total sales, 166, 658

Chemistry and Physics Applications

Acid mixture, 261
Acidity model, 633

Alcohol mixture, 192, 246, 623
Atmospheric pressure, 593
Average temperature in Duluth, 92
Banking angles, 524
Beam deflection, 564
Body temperature, 222
Boyle's Law, 527, 580
Brine solution, 273
Chemical mixture, 260, 362, 437
Chemical reaction, 37
Compression ratio, 184
Concentration of a mixture, 485
Concentration of a solution, 196, 197
Cooling, 672
Earthquake intensity, 629, 632
Electrical current, 538
Electrical networks, 284
Electronics, 454
Engineering, safe load on a beam, 133, 530
Flow rate, 475
Force, 106, 339
   required to stretch a spring, 106, 198
Free-falling object, 70, 72, 77, 339, 347, 374, 399, 425, 448, 519, 528, 661, 684, 685
Friction, 622
Fuel efficiency, 92, 94
Fuel usage, 198
Gasoline mixture, 246, 302
Gasoline-to-oil ratio, 184, 197, 226
Hearing frequencies, 213
Heating a home, 94, 151
Height
   of a free-falling object, 22, 27, 70
   of a model rocket, 396
   of an object, 38, 256, 259, 425
   of a projectile, 156, 418, 419
   of a rock, 360
   of a rocket, 424
Hitting baseballs, 399
Hooke's Law for springs, 521, 528, 579
Hyperbolic mirror, 573
Intensity of sound, 605, 613, 622, 637
Kepler's Third Law, 312
Kirchhoff's Laws, 284
Length of time an object has been falling, 345
Making ice cubes, 623
Masses of Earth and sun, 312
Maximum load, 582
Mechanics, gear ratio, 184
Medicine, in the bloodstream, 485

Index

Basic Rules of Algebra

Commutative Property of Addition

$$a + b = b + a$$

Commutative Property of Multiplication

$$ab = ba$$

Associative Property of Addition

$$(a + b) + c = a + (b + c)$$

Associative Property of Multiplication

$$(ab)c = a(bc)$$

Left Distributive Property

$$a(b + c) = ab + ac$$

Right Distributive Property

$$(a + b)c = ac + bc$$

Additive Identity Property

$$a + 0 = a$$

Multiplicative Identity Property

$$a \cdot 1 = 1 \cdot a = a$$

Additive Inverse Property

$$a + (-a) = 0$$

Multiplicative Inverse Property

$$a \cdot \frac{1}{a} = a, \qquad a \neq 0$$

Properties of Equality

Addition Property of Equality

If $a = b$, then $a + c = b + c$.

Multiplication Property of Equality

If $a = b$, then $ac = bc$.

Cancellation Property of Addition

If $a + c = b + c$, then $a = b$.

Cancellation Property of Multiplication

If $ac = bc$ and $c \neq 0$, then $a = b$.

Zero-Factor Property

If $ab = 0$, then $a = 0$ or $b = 0$.

Properties of Negation

Multiplication by -1

$$(-1)a = -a \qquad (-1)(-a) = a$$

Placement of Minus Signs

$$(-a)(b) = -(ab) = (a)(-b)$$

Product of Two Opposites

$$(-a)(-b) = ab$$

Operations with Fractions

$$\frac{a}{b} \cdot \frac{c}{d} = \frac{a \cdot c}{b \cdot d} \qquad\qquad \frac{\left(\dfrac{a}{b}\right)}{\left(\dfrac{c}{d}\right)} = \frac{a}{b} \cdot \frac{d}{c}$$

$$\frac{a}{b} + \frac{c}{d} = \frac{ad + bc}{bd} \qquad\qquad \frac{a}{b} - \frac{c}{d} = \frac{ad - bc}{bd}$$

Properties of Exponents

$$a^0 = 1 \qquad\qquad a^m \cdot a^n = a^{m+n}$$

$$(ab)^m = a^m \cdot b^m \qquad\qquad (a^m)^n = a^{mn}$$

$$\frac{a^m}{a^n} = a^{m-n}, \; a \neq 0 \qquad\qquad \left(\frac{a}{b}\right)^m = \frac{a^m}{b^m}, \; b \neq 0$$

$$a^{-n} = \frac{1}{a^n}, \; a \neq 0 \qquad\qquad \left(\frac{a}{b}\right)^{-n} = \frac{b^n}{a^n}, \; a \neq 0, \; b \neq 0$$

Special Products

Square of a Binomial

$$(u + v)^2 = u^2 + 2uv + v^2$$
$$(u - v)^2 = u^2 - 2uv + v^2$$

Difference of Two Squares

$$u^2 - v^2 = (u + v)(u - v)$$

Difference of Two Cubes

$$u^3 - v^3 = (u - v)(u^2 + uv + v^2)$$

Sum of Two Cubes

$$u^3 + v^3 = (u + v)(u^2 - uv + v^2)$$

The Quadratic Formula

Solutions of $ax^2 + bx + c = 0$

$$x = \frac{-b \pm \sqrt{b^2 - 4ac}}{2a}$$